Of Related Interest from
The Benjamin/Cummings Series in the Life Sciences

General Biology

N. A. Campbell
Biology (1987)

Biochemistry and Cell Biology

C. J. Avers
Molecular Cell Biology (1986)

W. M. Becker
The World of the Cell (1986)

W. B. Wood, J. H. Wilson, R. M. Benbow,
and L. E. Hood
Biochemistry: A Problems Approach, Second Edition
(1981)

Molecular Biology and Genetics

F. J. Ayala and J. A. Kiger, Jr.
Modern Genetics, Second Edition (1984)

L. E. Hood, I. L. Weissman, W. B. Wood,
and J. H. Wilson
Immunology, Second Edition (1984)

J. B. Jenkins
Human Genetics (1983)

R. Schleif
Genetics and Molecular Biology (1986)

J. D. Watson, N. H. Hopkins, J. W. Roberts,
J. A. Steitz, and A. M. Weiner
Molecular Biology of the Gene, Fourth Edition,
Volumes I and II (1987)

G. Zubay
Genetics (1987)

Microbiology

E. Alcamo
Fundamentals of Microbiology, Second Edition (1987)

R. M. Atlas and R. Bartha
Microbial Ecology: Fundamentals and Applications,
Second Edition (1987)

M. Dworkin
Developmental Biology of the Bacteria (1985)

G. J. Tortora, B. R. Funke, and C. L. Case
Microbiology: An Introduction, Second Edition (1986)

P. J. VanDemark and B. L. Batzing
The Microbes (1987)

Zoology

H. E. Evans
Insect Biology: A Textbook of Entomology (1984)

P. E. Lutz
Invertebrate Zoology (1986)

A. P. Spence
Basic Human Anatomy, Second Edition (1986)

A. P. Spence and E. B. Mason
Human Anatomy and Physiology, Third Edition (1987)

Evolution, Ecology, and Behavior

F. J. Ayala and J. W. Valentine
*Evolving: The Theory and Processes
of Organic Evolution* (1979)

D. D. Chiras
Environmental Science, Second Edition (1988)

R. J. Lederer
Ecology and Field Biology (1984)

M. Lerman
Marine Biology: Environment, Diversity, and Ecology
(1986)

D. McFarland
Animal Behavior (1985)

E. Minkoff
Evolutionary Biology (1983)

R. Trivers
Social Evolution (1985)

ZOOLOGY

Lawrence G. Mitchell

John A. Mutchmor

Warren D. Dolphin

Iowa State University

The Benjamin/Cummings Publishing Company, Inc.
Menlo Park, California • Reading, Massachusetts
Don Mills, Ontario • Wokingham, U.K. • Amsterdam • Sydney
Singapore • Tokyo • Madrid • Bogota • Santiago • San Juan

Editor-in-Chief: James W. Behnke
Sponsoring Editor: Robin J. Williams
Developmental Editor: Amy Satran
Production Supervisor: Betsy Dilernia
Copy Editor: Janet Greenblatt
Designer and Art Director: Hal Lockwood
Art Coordinator: Richard Mason
Photo Editor: Darcy Lanham
Photo Researchers: Carl May, Kevin Schafer,
   Larry Minden
Production Assistant: Bruce Lundquist
Artists: Martha Blake, Carl Brown, Cyndie Clark-
   Huegel, Cecile Duray-Bito, David Freifelder, Darwen
   and Vally Hennings, Linda McVay, Fran Milner,
   Elizabeth Morales, John Parsons, Carla Simmons,
   Peggy Skycraft (marbling), John and Judy Waller
Layout Artists: Wendy Goldberg, Judy Levinson
Composition, Camerawork, and Film Preparation:
   York Graphic Services
Printer and Binder: Von Hoffmann Press

The image on the cover is *Zebras* (1938) by Victor
Vasarely, reproduced with permission from the artist.

Credits for photographs and illustrations appear after
the Glossary.

Library of Congress Cataloging-in-Publication Data

Mitchell, Lawrence G.
   Zoology.

   Includes index.
   1. Zoology.   I. Mutchmor, John A.   II. Dolphin,
Warren D.   III. Title.
QL47.2.M58   1988   591   87-19355
ISBN 0-8053-2562-X

4 5 6 7 8 9 10 -VH- 95 94 93 92

The Benjamin/Cummings Publishing Company, Inc.
2727 Sand Hill Road
Menlo Park, California 94025

About the Authors

Lawrence G. Mitchell

Lawrence Mitchell is an Associate Professor of Zoology and Animal Ecology at Iowa State University. He received his Ph.D. in Zoology and Microbiology from the University of Montana in 1970. Dr. Mitchell has 18 years' experience teaching courses in biology, zoology, parasitology, aquatic ecology, and comparative vertebrate anatomy. Prior to coming to Iowa State, he was an Assistant Professor of Zoology at the University of Montana. In addition to his classroom experience, Dr. Mitchell has developed television courses in biology and has written and produced several programs on wildlife biology for public television. He received the Outstanding Teacher Award at Iowa State in 1982. Dr. Mitchell's primary research areas are host–parasite relationships of the Myxozoa and ecological relationships of freshwater mussels in the Upper Mississippi River. Dr. Mitchell conducts field research and teaches invertebrate zoology at the Iowa Lakeside Laboratory, the biological station of Iowa State University and the University of Iowa.

John A. Mutchmor

Professor of Zoology and Entomology, Dr. Mutchmor has been teaching at Iowa State University since 1962. He was a Research Officer with the Canada Department of Agriculture in Ontario for five years prior to receiving his Ph.D. in Entomology from the University of Minnesota in 1962. In addition to his teaching responsibilities in zoology, general biology, and insect physiology, Dr. Mutchmor served as an adviser of undergraduate students in the Iowa State College of Sciences and Humanities Advising Center from 1985 to 1987. He received the Outstanding Teacher Award at Iowa State in 1979. Dr. Mutchmor's research is in insect physiology. He concentrates on problems of low temperature acclimation and how low temperatures affect insect performance.

Warren D. Dolphin

Warren Dolphin is Professor of Zoology and Executive Officer for the Biology Program at Iowa State University. He received his Ph.D. in Zoology and Cell Biology from Ohio State University in 1968. Prior to his appointment at Iowa State, Dr. Dolphin was an Assistant Professor of Zoology at the University of Maine, Orono, for two years. His research interests include metabolism in amoebae and educational investigation in student achievement. Dr. Dolphin has experimented with innovative programs featuring videotaped lectures and self-paced testing for freshman biology. In addition to authoring laboratory manuals for general biology, Dr. Dolphin has 19 years' teaching experience in cell biology, general biology, and animal physiology.

This book is dedicated
to our students—
past, present, and future.

Preface

Zoology is an exceedingly broad, vibrant field of life science, and our chief motivation for writing this textbook has been to integrate some of the vibrancy with an introduction to fundamental concepts. We have sought to create a book that illustrates, verbally and visually, the excitement both of animal life and of zoologists' rapidly growing understanding of it. We have also attempted to include the flavor of experimentation and the vital role of questioning and controversy in zoology. Whenever possible we have posed questions, such as: What problems of existence do animals face in various environments? How do their evolutionary adaptations allow them to solve these problems? We hope that through this approach, our text will bring out some of the essence of zoology as a growing science in which much work remains to be done.

Another strong motivation for us has been that we believe an introduction to animal biology should be an integral part of an undergraduate curriculum. We feel this is especially true today, because life science is being driven so strongly toward molecular and cellular levels. In both teaching and research, our science concentrates more and more on questions about subcellular structure and function, and there is a growing tendency to postpone the quest for knowledge about whole organisms and their interrelationships. Most life scientists trained two decades ago received, at least in their early undergraduate programs, some introduction to plant and animal biology. Today, there is a tendency in some institutions to provide students with nothing more than a one- to two-week coverage of all of life's diversity in a large lecture course in general biology. Today's baccalaureate student may be introduced two or three times to the central concepts of molecular genetics but may not have an ample opportunity to discover organismal diversity. We believe this situation is unfair to students. It also seems unfair to the diversity of living species, many of which are threatened with extinction. Thus, we feel it is becoming increasingly important that an appreciation for animal diversity, form and function, and evolutionary relationships be gained early in an undergraduate program. Knowing something about the million and a half other species of animals with which we share space on this planet enriches our lives. It also helps us appreciate the need to conserve species.

As taught today, zoology and animal biology courses are quite varied. Although most courses seem to follow one or two introductory biology classes, some schools teach introductory zoology as the first life science course. This text can be used in a variety of course settings, although it is intended mainly for use by college freshmen and sophomores who are majors in the biological sciences. We envision it being used in a single- or two-term course that prepares the life science major for upper-level courses and at the same time provides a lasting impression of animal form and function. For all students, we hope our book and the course for which they use it will strengthen the inherent interest and curiosity that we believe all people have in animals.

AUTHORSHIP

Multiauthored texts vary greatly in the way they are written. We feel that our text, and its companion laboratory manual, have benefitted from having three conferring authors in close proximity. Although we all shared in the planning of the project and in preparation of the first draft, each author was responsible for different aspects of the total effort.

Many multiauthored books suffer from an uneven level and style of presentation from chapter to chapter. We felt that our book would be best integrated and most useful to its audience if the writing was accomplished by one author. Thus, the senior author transformed all early manuscript into final form.

As our book developed, it became clear that the art program was also best served by one author. We all recognized the vital role that instructive and clearly developed art can play in learning. Accordingly, the second author refined the initial art manuscript to fully complement the concepts presented in the text. He then worked closely with the artists and the photo researcher to make certain that the figures were accurately executed. The figure legends are an extensive and essential part of this book's total learning package. The legends have been carefully constructed in tandem with the text to fully explain each illustration. Thus, it is possible to preview or review each chapter of the text by studying the illustrations and figure legends. We feel that this scrutiny and consistent allocation of our resources has ensured a visual quality previously unseen in zoology textbooks.

As the laboratory is to the lecture, the accompanying laboratory manual is an important part of our learning package. The manual includes coverage of the major animal phyla and a considerable amount of problem-solving, experimental activities. We all participated in planning, but the third author wrote the manual and directed its illustrators.

CONTENT AND ORGANIZATION

Courses in a subject area as broad as zoology are bound to vary in both depth and breadth of coverage. The organization and content of this text reflect the collective bias of its three authors, but we feel that the text is adaptable for virtually any introductory course in zoology.

Following Chapter 1, which introduces zoology and provides a photographic introduction to the main groups of animals, the text is organized into four main units.

Unit I. Functional Systems of Animals

The overall goals of Unit I are to illustrate the major problems that animals must solve to survive in various environments, and to show some of the range of solutions that have evolved to cope with the problems of existence. We have placed this unit first in the book because we believe it is best to gain some understanding of how animals function before looking at the vast diversity of the animal kingdom. This is a matter of opinion, however, and as is done in some other textbooks, the instructor may prefer to start with diversity and work back to functional systems. For those who use our chronology, the short introduction to major animal groups in Chapter 1 provides the necessary background for reading Unit I.

The first two chapters of Unit I deal mainly with structure and function at the molecular and cellular levels. These topics are usually covered in introductory biology courses. We have included them for use in courses that do not have a biology course prerequisite, and for other students who wish to review this material, in the context of animal biology.

The bulk of Unit I covers the more strictly zoological topics, addressing such things as how animals obtain and use nourishment, dispose of wastes, detect and process stimuli in their environment, move about, and interact with other animals. Each chapter takes a comparative approach. We look first at the fundamentals of an organ system (e.g., circulatory or nervous system), then at the problems being solved (e.g., transport of materials or coordination within the body), and we compare and contrast the structure and function of the system in several different types of animals. We have tried to achieve a balance in these chapters between vertebrate and invertebrate material. We have also chosen to include the human species only where it seems appropriate as an example of mammalian form and function.

Unit II. Reproduction and Evolution

The four chapters in Unit II examine subjects relating to how species survive and change from generation to generation. In this context, the topics of reproduction, genetics, development, and evolution are interrelated and fundamentally different from the subjects of Unit I.

While Unit I deals with problems of individual survival, Unit II examines subjects vital to the survival of species. Likewise, Unit I is about the products of evolution, whereas Unit II describes the mechanisms by which evolution occurs. We hope that a deeper appreciation of the evolutionary process (Unit II) may be achieved by having a firm background in what the process accomplishes (Unit I). Placing evolution and closely related topics at this point in the text serves as a bridge between Units II and III.

Unit III. The Animal Kingdom

Unit III illustrates the vast diversity of animal life. Here we examine all of the phyla, or major groups, of animals, as well as several phyla of unicellular organisms, the protozoa. In this unit we have chosen not to follow the classical approach of presenting single species as representative types of each phylum. Rather, after a brief introduction to the phylum as a whole, each chapter includes a short survey of several species that represent diversity in the phylum. As in Unit I, we then compare some of the animals in terms of their environmental relationships, structure, function, and life history. Using this approach, we hope to give the student a clear overview of animal diversity and evolution.

The first chapter in Unit III examines animal classification and phylogeny, the evolutionary relationships among animal groups. This chapter begins where the previous chapter on evolutionary mechanisms leaves off, and leads into the study of animal phyla. Our organizing principle throughout Unit III is animal history, as it can be reconstructed from the fossil record and surmised by the comparative study of living animals. Ideally, students will come to Unit III armed with an appreciation of animals as problem-solving entities, and with a well-developed tendency to pose questions about such things as what body features allow a given animal to survive and reproduce in its particular environment. This approach will make Unit III more meaningful than if it is read simply for its facts about animals. At numerous points in the unit, to highlight the evolutionary theme, we have included special between-chapter commentaries called Trends and Strategies. At the end of each chapter, key features of each phylum are identified, and a classification section outlines the main groups within each phylum.

Unit IV. Ecology

We placed our unit on ecology last because we feel that it forms a logical endpiece for the book. Consisting of only two chapters, this is the shortest unit in the text. Yet much of zoology is about the roles of animals in their environments, and about adaptations that allow animals to survive in various environments. For this reason, discussions of environmental relationships form an integral part of virtually all other chapters in the book. Hopefully, with its discussion of the highest levels of biological organization—the population, community, and ecosystem—this brief unit will help tie the rest of the book together.

SPECIAL FEATURES FOR STUDENTS

In our early planning, we tried to imagine what special features a text like this should have to help students retain some of the key concepts and also impart some of the excitement of zoology. Several such features will, we hope, serve as useful teaching aids and interest stimulators. Within each chapter, key terms are boldfaced as they are introduced and defined. A brief summary and a reading list appear at the end of each chapter. The reading lists contain a mix of textbook references, pertinent review articles in such journals as *Scientific American* and *American Scientist,* and where appropriate, some popular literature, for example, articles from *National Geographic.* A study chart summarizing key features of the animal phyla and an extensive Glossary follow the last chapter. A world map inside the back cover of the text provides a reference to key sites and geographic features that are of importance in many sections of the text.

Chapter and Page Cross-References

At many places within the text it is appropriate to provide cross-references to material discussed in greater detail in earlier or later chapters. Cross-references are identified in each chapter with an open arrow in the margin of the text. The cross-references are intended to help the reader connect related material located in different parts of the book. They will also help students find information that adds depth and interest to the material at hand. Students may also find the cross-references useful in locating material needed for review while studying.

Trends and Strategies

At selected intervals throughout the book, we have included special commentaries that we call Trends and Strategies. These are positioned in the text to help students follow the main trends in animal evolution. We use the word "strategy" in this context, as it is often used in zoology, in reference to evolved mechanisms, or adaptations, that allow animals to survive and reproduce. Most Trends and Strategies sections appear as short passages between chapters in Unit III. Placing eight of them among the first ten chapters in this unit allows us to highlight the major events and milestones in animal history as they are reflected in the animal groups about to be discussed. A Trends and Strategies commentary on the evolution of multicellularity, for example, precedes our chapter on animals that exhibit multicellularity at the least complex levels. Likewise, a commentary on metamerism, or body segmentation, precedes chapters on annelids and arthropods, two strongly metameric groups. In many cases, our Trends and Strategies sections are keyed to, and supplemented by, within-chapter sections on the phylogeny of individual phyla.

The Trends and Strategies sections within Unit III fulfill our original intent of emphasizing major milestones in animal history. As the book began to take shape, it became clear that additional short commentaries placed at the beginnings of units and at certain other key points in the text would help emphasize or introduce other important concepts. As a result, we expanded our use of Trends and Strategies and wrote several more of these passages for the other units. We now view these commentaries as a key strength of our book. They have become places where we hold forth on some of the major "take-home" messages that tie the book's diverse subject matter together.

Essays

Twenty-two of our thirty-five chapters contain boxed essays focusing on specific topics that we think will be of special interest to students. An essay on filter feeding in whales is in our chapter on Feeding and Nutrition. One on diving adaptations in mammals augments our chapter on Respiratory Gas Exchange. Other essays explore such topics as ocean vent communities, how lampreys destroyed fish populations in the Great Lakes, how some animals survive drought and temperature extremes by means of suspended animation, and how some fishes maintain warm body temperature. We hope students will enjoy the essays and will share their reactions with their instructors and with us.

ACKNOWLEDGEMENTS

As must be true for any project this size, this text represents the collective efforts of many people. After nearly a year of early planning, the project began in earnest when we teamed up with the capable people at The Benjamin/Cummings Publishing Company. It has been a heartening and invigorating experience to be on the receiving end of the efforts of a company composed of energetic professionals truly committed to excellence in publishing.

Many people at Benjamin/Cummings have contributed more of their time and creativity to this book than we could have reasonably expected. Jim Behnke, General Manager of the company, made initial contact with us some five years ago, and has supported the project steadfastly ever since. Our developmental editor, Amy Satran, who was also involved from the beginning, provided tireless steerage through the entire review and rewriting process. Amy's ability to separate wheat from chaff in our early manuscript was uncanny. She was also sensitive to our individual strengths and weaknesses and made sure that the strengths worked for the good of the book. Working with Amy, Paul Elias made a major contribution in commissioning expert reviews for Unit III. Robin Williams, who took over as Sponsoring Editor from Amy and Paul as the final manuscript was being written and reviewed, brought a unique background in zoology and publishing to the project. Robin knows what it takes for a life science book to be a useful teaching tool, and we are most grateful to her for acting on her knowledge in giving her full support to our book. Assisting Robin and the authors in innumerable ways, Jim Donald has been a model of efficiency and a source of good spirits to us all.

We consider ourselves especially fortunate to have had Betsy Dilernia head the team that took our manuscript and made a book out of it. Working with Betsy has been a pleasure and an inspiration. She is a dedicated professional whose energy and skills in orchestrating all of the activities that go into timely book production are truly remarkable. Bruce Lundquist worked tirelessly with Betsy. Darcy Lanham did a superb job of finding just the right photographs to illustrate the text. We are also indebted to Janet Greenblatt for her sensitive and meticulous job of copy editing the final manuscript. Much of the "look of the book" is the result of the creative talents and tireless efforts of Hal Lockwood and Richard Mason of Bookman Productions. Hal designed the book and its cover, and Richard coordinated the art program, which is a special source of pride to us.

We regret that space prohibits our mentioning all the people at Benjamin/Cummings and Bookman who have made contributions to the project. We are especially grateful to Laura Argento, Linda Crockett, Donna Fitch, and Jane Gillen.

We are also deeply indebted to our colleagues who have provided consultation and critical reviews at each stage of manuscript preparation. Gary J. Atchison wrote drafts of the two chapters in Unit IV. Foremost among our reviewers, we would like to thank Gary J. Brusca of Humboldt State University, who read and critiqued the entire manuscript. We are grateful for his thousands of helpful comments and for his remarkable ability to remain constructive. Gary's seemingly boundless patience and refusal to become indignant, even with our weakest writing efforts, are truly exemplary.

In addition to reviews by colleagues, several students at Iowa State University have read and critiqued the manuscript at several stages. Special thanks go to Laura Molgaard, whose careful readings of every chapter made us rethink and clarify our presentation of numerous topics. James Altmaier made a similar contribution to about half the book, and Mark Sandheinrich provided key insights for the ecology unit.

In completing this text, we have gained a sense of accomplishment, but we have also acquired a humbling appreciation for the great breadth of modern zoology. We have discovered that we have been naive about many fields that are integral parts of our science. While to a certain degree the current growth rate of knowledge makes narrowness an axiom for most life scientists, our experience over the course of this project has fostered concern. Often we have found ourselves revisiting decisions that were made regarding organization, emphasis, and deletion. Most worrisome of all, we struggle with the textbook author's unavoidable concern for outright errors.

We have been fortunate in having the full support of a strong team of reviewers and a publisher dedicated to excellence, both of which have given us every opportunity to produce a first-rate book. We hope we have written a useful, accurate text. Recognizing our fallibility, however, we take full responsibility for any errors, and we hope that colleagues and students who find problems will help us improve future editions by telling us about them.

LAWRENCE G. MITCHELL
JOHN A. MUTCHMOR
WARREN D. DOLPHIN
Iowa State University
Ames, Iowa

Iowa State University Manuscript Consultants

Gary J. Atchison
Stephen H. Bishop
Richard J. Hoffmann
Edwin C. Powell
James R. Redmond
Kenneth C. Shaw
Carl L. Tipton

Manuscript Reviewers

Kenneth P. Abel
State University of New York, Albany

Steve Anderson
University of the Pacific

Peter August
University of Rhode Island

Joseph T. Bagnara
University of Arizona

Malcolm Braid
University of Montevallo

Ann Janice Brothers
University of California, Berkeley

Judy Brown
San Jacinto College North

Gary J. Brusca
Humboldt State University

Alan H. Brush
University of Connecticut, Storrs

Warren Burggren
University of Massachusetts, Amherst

William Bukovsan
State University of New York, Oneonta

Albert G. Canaris
University of Texas, El Paso

Gregory M. Capelli
College of William and Mary

Jonathan A. Coddington
Smithsonian Museum of Natural History

Richard J. Connett
University of Rochester

David J. Cotter
Georgia College

David L. Cox
Illinois Central College

John L. Crites
Ohio State University

Peter Dalby
Clarion University

Richard Daniels
University of Missouri

Bonnie J. Davis
San Francisco State University

Fred Diehl
University of Virginia

Robert Eaton
University of Colorado, Boulder

Mark Engstrom
Angelo State University

Norman A. Engstrom
Northern Illinois University

Steve Ervin
California State University, Fresno

Lincoln Fairchild
Ohio State University

Kristian Fauchald
Smithsonian Museum of Natural History

Howard Feder
University of Alaska, Fairbanks

David Fox
Biological Consultants ATAP Corporation

Ron Fritzsche
Humboldt State University

Kenneth L. Goodhue-McWilliams
California State University, Fullerton

Barbara Grimes
North Carolina State University

George L. Harp
Arkansas State University

George J. Hechtel
State University of New York, Stony Brook

Richard Heckmann
Brigham Young University

Ronald L. Hybertson
Mankato State University

Richard E. Johnsen
Colorado State University

Kathleen Karrer
Brandeis University

Donald R. Kirk
Shasta Junior College

John A. W. Kirsch
University of Wisconsin, Madison

David Klingerer
University of Massachusetts, Amherst

Astrid Kodric-Brown
University of Arizona

J. A. Lackey
State University of New York, Oswego

Peter Landres
University of Montana

Charles Z. Leavell
Fullerton College

Welton Lee
California Academy of Science

William D. Longest
University of Mississippi

Sheldon Lustick
Ohio State University

Richard N. Mariscal
Florida State University

Richard Marsh
Northeastern University

J. Forbes McClellan
Colorado State University, Fort Collins

Bayard H. McConnaughey
University of Oregon

V. Rick McDaniel
Arkansas State University

W. Brian O'Conner
University of Massachusetts, Amherst

David L. Pawson
Smithsonian Museum of Natural History

James F. Payne
Memphis State University

David W. Phillips
University of California, Davis

William P. Pielou
Furman University

Harold Rauch
University of Massachusetts, Amherst

William S. Romoser
Ohio University

Arthur Rourke
University of Idaho

Ronald L. Rutowski
Arizona State University

Robert E. Savage
Swarthmore College

William D. Schmid
University of Minnesota, Minneapolis

Richard Snyder
University of Washington, Seattle

Robert Spies
Lawrence Livermore National Laboratory

Samuel S. Sweet
University of California, Santa Barbara

John F. Tibbs
University of Montana

Stephen G. Tilley
Smith College

Judith Van Houten
University of Vermont

David R. Voth
Metropolitan State College

Archie M. Waterbury
California Polytechnic State University,
San Luis Obispo

James Waters
Humboldt State University

Keith H. Woodwick
California State University, Fresno

Harold L. Zimmack
Ball State University

Brief Contents

xvi

Detailed Contents

UNIT II
REPRODUCTION AND EVOLUTION   323

TRENDS AND STRATEGIES
Long-Term Survival and Evolution of Species   325

CHAPTER **14**
Reproductive Mechanisms and Strategies   327

CHAPTER **15**
Genetics   349

Introduction to Zoology and Animal Diversity

Fossils indicate the diversity of extinct animal life. Here, a fossil in a dinosaur quarry in Dinosaur National Monument, Utah, is being prepared for study.

Our opening chapter has four objectives. First, we want to provide a brief sketch of what zoology is like today. Our second objective is to present zoology as a scientific process—to describe briefly what zoologists do and how they do it. Next we introduce several broad concepts that provide an approach to the study of animals. Finally, we give a brief overview of the major groups of animals, thereby introducing our main cast of characters as well as highlighting some organizing concepts that will help keep animal diversity in perspective. Overall, we hope this chapter will give you an invigorating view of zoology's unique combination of scientific rigor and animal beauty.

WHAT IS ZOOLOGY?

Up until the late 1800s, **zoology**, the scientific study of animals, was essentially a descriptive science involving careful observations of animals made under both natural and experimental conditions. Knowledge about animals was limited enough that a dedicated zoologist could assimilate most of the information then available while at the same time keeping up with new developments. But with the publication of Charles Darwin's concepts of evolution in 1859 and the consolidation of genetic theory during the early twentieth century, knowledge about zoology and the other life sciences began to expand at an unprecedented rate. Moreover, modern research in zoology has become increasingly dependent on the physical and mathematical sciences, which have aided in transforming the field from a descriptive to a more predictive science. Particularly in the last four decades, zoologists have made many exciting discoveries in cell and molecular biology, as well as in animal **physiology**, the science that deals with the function of organs and organ systems. Important new concepts in ecology (the study of environmental relationships) have emerged during this time, many of which have global implications. Much of this progress has followed the development of new research techniques and technological advances. The life scientist of today can make precise measurements and predictions about phenomena that may have been unknown several decades ago.

The immensity of zoology as a field of knowledge and the rapid rate of new discoveries make it difficult for today's zoologists to be familiar with more than a limited research area. Some zoologists prefer to concentrate on a particular group: Entomologists study insects, ichthyologists study fishes, and protozoologists study protozoa. Other zoologists focus on phenomena that occur in many different kinds of animals. A neurobiologist might study nerve structure and function in several different animal species, or a physiologist might be interested in how various marine animals obtain oxygen. While these two approaches may seem very different, the distinction between them is increasingly less obvious. Today, most zoologists are specialists regardless of whether their starting point is a certain animal group or a particular phenomenon. Mammalogists, for example, do not try to learn everything about mammals; most concentrate on the biochemistry, physiology, ecology, or development of one or a few species. An entomolo-

gist may study the anatomy of insects, ways to control insect pests, or how insects transmit human diseases; but again, most work with only a small number of species.

Researchers studying molecules and cells approach life science at its most basic levels. A zoologist working in these areas might attempt to isolate the molecules that allow muscle cells to contract or might study the effect of an enzyme on cell metabolism. At the other end of the spectrum are zoologists who study how whole animals function in their environments (Figure 1.1). A behaviorist might study an animal's mating rituals. An ecologist might examine the impact of predators on a prey species or measure energy flow through food chains. A population geneticist might be concerned with how gene frequencies change as a population adapts to changing environmental conditions. Taxonomists and systematists attempt to name and classify animals in ways that reflect evolutionary relationships.

Figure 1.1 A zoological field study. Here a zoologist takes a census of terns as part of a study of this bird's behavior and ecology. Zoologists studying animal behavior often spend considerable time observing animals in their natural habitat.

Zoologists may have different philosophical approaches to their science. Many attempt to answer questions basic to a general understanding of animal life (**basic science**) and have chosen a research area for its own intrinsic value or interest. Others are concerned mainly with solving specific practical problems (**applied science**) and use animals to gain knowledge immediately applicable to medicine or human welfare. Both of these approaches have led to discoveries that have broadened our understanding of animal biology.

Although specialization is unavoidable, the modern zoologist must also keep abreast of major developments in life science as a whole. Scientists must incorporate new information from a broad range of subjects into their own research, for only then can they make significant progress.

ZOOLOGY AS A SCIENTIFIC PROCESS

As scientists, zoologists approach their work from a questioning standpoint and develop **hypotheses**, or tentative suppositions, to be tested by their research. Hypotheses may be based on a researcher's previous experience or on information gleaned from critical reading of the scientific literature. Questions often occur to scientists while they are reading an article in a research journal, performing experiments on a related topic, discussing research with colleagues, or simply observing animals. A zoologist interested in animal nutrition might become intrigued with how a parasitic animal with no digestive tract obtains and digests its food. An initial question might be, Is food taken up by cells on the animal's body surface? This question would then form the basis of a working hypothesis: Perhaps surface cells ingest food particles and digestion occurs within these cells. Because scientists approach their work by testing hypotheses, the next two steps would be to make observations and perform experiments that could disprove the working hypothesis. If microscopic observation reveals that the animal has a surface coat of dead cells, the original hypothesis is clearly invalid, and a new line of inquiry must be pursued. However, if the animal does have living cells at its surface, then the working hypothesis is supported rather than disproved. A logical next step would be to carry out radioactive tracer studies to see whether, and in what manner, these cells ingest certain food substances. Scientific research proceeds as observations and experimental results generate new questions and hypotheses.

A hypothesis that is not disproved by repeated tests is considered a tentatively acceptable description of a natural process. Eventually, hypotheses that have withstood exhaustive testing may be considered theories or parts of a theory. Whereas hypotheses are early guesses or tentative ideas to test, **theories** in science are generally accepted concepts or broad explanations of natural phenomena. Thus, Albert Einstein's "theory of relativity" and Darwin's "theory of natural selection" have both withstood repeated testing, have a high probability of being true, and have broad applicability. Natural **laws** or **principles** are descriptions of natural phenomena that do not vary, at least under certain conditions. In biology, few ideas are considered laws, and no theory or law is beyond challenge; generally accepted theories are often reexamined and retested in light of new discoveries and ideas.

Designing Research Projects

The foundation of any scientific investigation is its **experimental design**, a logical outline that guides the gathering and evaluation of information. The experimental design is based on a hypothesis, and it is the researcher's plan for testing the validity of that hypothesis. Much thought and hard work accompany the development of hypotheses and experiments that will yield unambiguous results. A scientist's ability to ask key questions and to formulate them into testable hypotheses may largely determine the success or failure of a given research project.

Three factors are central to good experimental design. First, proper *controls* must be incorporated into each experiment. A **control group** receives the same treatment as the experimental group except that the factor being tested is applied to the experimental group only, not to the control. Second, experiments must be *repeated* enough times to allow meaningful comparisons between experimental and control groups. It is through such repetition that data can be compared statistically and a high degree of accuracy obtained. Finally, experiments must be designed to avoid *bias*. A researcher must strive to prevent personal opinion about a hypothesis from influencing how tests are made and must also be aware of the bias that any technique or instrument may introduce in the outcome of an experiment.

Experimental design is also influenced by certain practicalities. Will animals be available, and

Figure 1.2 A modern marine research vessel. Deep-sea submersible vessels are designed to withstand the enormous pressures at great depths in the ocean. Here the pilot of the *Atlantis* steers the vessel over a reef.

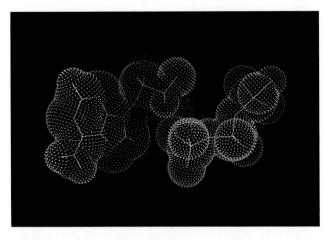

Figure 1.3 Computer graphic illustration. Computers are not only useful for "number crunching." They also allow scientists to model complex biological structures and systems. The high-energy compound adenosine triphosphate (ATP) is modeled in this photograph.

can they be maintained in the appropriate condition? Are instruments available to make and record the necessary measurements? Where will the research be done, and by whom? Often, the best place to collect animals or to study them in their natural habitats is an inland or marine field research station. Research on animals in the open ocean is often performed from specially equipped vessels that can stay at sea for several months. Submersible vessels designed for deep-sea research allow scientists to make significant discoveries about life at great depths (Figure 1.2).

Support for research is another major consideration. A zoologist may need to obtain funds to cover the costs of animals, chemicals, equipment, and personnel. Although several private foundations, such as the Ford and Rockefeller foundations, offer help in this respect, the bulk of support for zoological research in the United States comes from public institutions, such as the National Science Foundation (NSF) and the National Institutes of Health (NIH). Monetary grants are awarded on a selective basis.

Zoologists frequently use an array of modern instruments in their work, and obtaining the proper equipment is an important part of preparing for research. Computers have become invaluable research tools, allowing large amounts of data to be stored, analyzed, and retrieved rapidly (Figure 1.3). Electron microscopes are routinely used to study the detailed architecture of cells and cellular organelles; high-speed centrifuges and fast chromatographic techniques allow organelles and biomolecules to be separated; and other instruments and techniques can be used to make precise measurements of cellular and subcellular biochemical

processes. (See the Chapter 2 Essay on modern ◁ techniques for isolating parts of cells, p. 48.) The application of high technology to biological problems, and the development of new technologies applicable to biology are rapidly becoming a field of their own, known as biotechnology. This promises to be one of the major areas of employment in the life sciences in the future.

In field studies, some of the most exciting ecological and behavioral research is carried out with equipment as simple as a camera and binoculars, yet modern technology also plays a significant role. Radiotelemetry—the use of radio transmitters and directional receivers—makes it possible to monitor animals' movements and study their behavior (Figure 1.4). Recently, space satellites have been used to follow animals tagged with radio transmitters.

Figure 1.4 Radiotelemetry. By placing a small radio transmitter on an animal, researchers can study the animal's movements and behavior without interfering with its daily routine. Here a radio-collared Dall sheep has just been released.

Doing the Work

Researchers are meticulous notetakers. They make detailed notes in a record book that becomes a scientific diary of the research project in process. Data, or results, accumulate as tests outlined in the experimental design are completed. *Anecdotal data* relate in words what happens in an experiment, recording chance observations of how an animal, cell culture, or tissue extract reacts to changes in procedure, as well as describing mistakes and unexpected events. Serendipity, the making of important discoveries by chance or by accident, can be a significant factor in the advancement of knowledge. *Numerical data* consist of measurements, counts, or elapsed time determined by a person or instrument.

As results accumulate, the researcher tries to find patterns or relationships in the data, perhaps by plotting graphs and doing statistical tests. As patterns begin to emerge, the researcher must address critical questions: If two variables appear to change in a coordinated manner, are they truly related or is their apparent correlation merely the effect of chance? Does a change in one variable cause the observed change in another variable? Questions such as these are frequently anticipated in the experimental design, and appropriate experiments can sometimes be included to determine cause and effect. Often, however, problems arise that were not anticipated, and additional experiments must be carried out to clarify matters. No matter how foolproof an experimental design seems at the outset, pitfalls invariably confront the researcher. Instruments may break down; despite the best of care, animals may die in the middle of any experiment; or the results may show only that an experiment was inadequate. Collecting data is a time-consuming, tedious process, and patience is an essential ingredient in science. When the results of repeated tests are consistent and patterns become discernible, the next stage of the scientific process is reporting the results.

Reporting the Results

Science is a powerful group activity, a mode of inquiry providing many opportunities for correcting errors. The group process often starts when the experimental design is reviewed by other scientists prior to funding. It continues during the research phase as co-workers discuss experiments, but it is most highly developed in the reporting stages. Researchers formulate their ideas from data analysis, then describe these ideas at seminars and meetings pertaining to their particular fields. Thus, a researcher might give a short presentation at the annual meeting of the American Society of Zoologists, the Society of Protozoologists, or the Ecological Society of America. In fact, these are just a few of the many scientific societies whose meetings are attended by an international community of scientists. It is her that the researcher has an opportunity to interact with other experts in the field and to see how well his or her work stands up to peer scrutiny. The researcher then decides whether more experiments are needed or whether it is time to publish the results.

Publishing a scientific paper is the next step in the reporting process. Hundreds of professional societies throughout the world publish journals containing articles that describe original research. Work so published is then permanently available to the present and future scientific communities. The researcher prepares a manuscript and sends it to the editor of an appropriate journal. The editor then sends the manuscript to other scientists in the field, who review and evaluate the report in light of the soundness of the experimental design, the significance and originality of the results, and the appropriateness of the paper for the society's journal. Very few papers are accepted for publication as they stand; most are returned to the author for minor rewriting or major experimental reworking; some are rejected on scientific grounds or for lack of interest among the society's members. Scientific information has been made available in this way for hundreds of years.

The review process has served science well and no doubt will continue to do so, though the time may soon arrive when scientific papers are simply stored in a large central computer data bank to be transmitted by telephone lines to those who request them. Certainly, the technology to distribute information electronically already exists and could be of great benefit in dealing with the information explosion mentioned at the beginning of the chapter. During a recent year, as many as 90,000 articles of interest to zoologists appeared in some 9,000 different research journals. Obviously, keeping up with new information is a formidable task. Most zoologists subscribe to several journals and use abstracting services to find pertinent reports published elsewhere. Such services as *Biological Abstracts* and *Zoological Record*, for example, provide indexes and abstracts of thousands of published reports in all areas of zoology. Monitoring current developments and searching past and present literature for pertinent information are ongoing aspects of any scientist's work.

Approaching the Study of Animals

From our discussion so far, it is clear that the questioning approach is central to science. It allows zoologists to pose hypotheses and then devise experiments to test them. Developing a questioning attitude is also a logical way to begin studying zoology. About 1.5 million species of living animals are known to exist, and many more remain to be discovered. It is easy to become overwhelmed by this diversity, but by posing broad questions, it is possible to organize our knowledge about animals rather than memorize a huge assortment of isolated facts about them.

To a large extent, each animal can be viewed as an organized set of solutions to the problems of existence in a particular environment. Certain fundamental problems confront all animals, and each of these problems can be stated as a general question. To survive, an animal must maintain its internal systems in a dynamically balanced condition, different from that of its environment. Self-maintenance of an internal steady state is called **homeostasis**, a central concept in zoology. Many of the questions we can ask about how animals survive are ultimately questions about homeostasis. For example: How do animals exchange materials with their environment? How do they obtain nutrients and dispose of their wastes? Does a particular animal regulate its body temperature, and if so, how does it do this? These questions relate to animal physiology. Questions can also be posed about animal behavior, reproduction, ecology, and evolution: How do animals gather and use information about their environment? How do they communicate with other members of their species? How do they reproduce? What role does an animal species play in its environment, and how is its role related to that of other organisms? Which animals are most closely related, and what were their ancestors like?

Despite the diversity of animals, all species have an ultimate common ancestor and share certain basic traits. Cell structures and functions are fundamentally the same or very similar in all animals. All species reproduce, and the mechanism essential to all forms of reproduction is DNA replication. All animals are **heterotrophs**, meaning that they obtain organic nutrients synthesized by other organisms (called autotrophs). There are many different methods of obtaining food, but digestive mechanisms and enzymes are remarkably similar among different animals. Most animals are motile, and the kinds of molecules, tissues, and mechanisms underlying their different types of motility have many common features. These fundamental similarities unify the animal kingdom and also help make animal diversity seem less overwhelming.

Overview of the Major Phyla

The animal kingdom is one of five kingdoms of organisms (Table 1.1). All animals are *multicellular* and eukaryotic (their cells have a membrane-bounded nucleus); as a group they are called **metazoans**. Unicellular eukaryotic organisms, including the **protozoa** (which were once considered members of the animal kingdom), are placed in the Kingdom Protista. Like metazoans, most protozoa are motile and heterotrophic.

To begin our study of animals, it will help to have a working knowledge of the major subdivisions of the animal kingdom and of the way zoologists classify animals. The animal kingdom is classified in hierarchical groups called **taxonomic categories**. (Chapter 18 discusses methods and modern approaches in animal classification, p. 435.) Table 1.2 lists six major categories in the animal kingdom, from the largest, or most inclusive, the **phylum** (pl. phyla), down to the least inclusive, the **species**. Of these categories, only the species is defined by nature. Each species is defined according to its reproductive isolation; that is, individuals of a species can mate and produce fertile offspring only with members of their own species. (Hybrids of two species, if viable at all, are usually infertile.) Species names are Latinized **binominals**: The first (generic) name is that of the genus in which the species is classified, and the second (specific) name is a descriptive adjective. Thus, the human species is called *Homo sapiens* (literally, "knowing man").

Taxonomic categories above the species level are defined by human judgment. Species are placed in certain genera, genera are classified in families, and so forth, on the basis of careful consideration of anatomy, developmental patterns, biochemical makeup, and any other available data that provide clues about similarities or differences. Just as the animal kingdom consists of many phyla, each phylum may contain several classes, each class several orders, and so on. Phylum, class, order, family, genus, and species are the main categories, but each of these may have subgroupings. A phylum may be divided into subphyla, a class into

Table 1.1 The Five Kingdoms of Life

Kingdom	Organisms	Diagnostic Features
Monera	Bacteria, blue-green algae (cyanobacteria)	Prokaryotic (no cell nuclei or organelle membranes)
Protista	Protozoa, unicellular algae	Eukaryotic, unicellular
Fungi	Fungi	Eukaryotic, multicellular, heterotrophic,* cell walls made of chitin
Plantae	Plants, multicellular algae	Eukaryotic, multicellular, autotrophic,* cell walls made of cellulose
Animalia	Animals (metazoans)	Eukaryotic, multicellular, heterotrophic, motile, lacking cell walls

*Autotrophic organisms can produce their own food by using chemical or solar energy to synthesize organic compounds within their body cells. Heterotrophic organisms cannot make their own food and must ingest organic compounds in the form of plants or other animals. All animals are heterotrophic.

Table 1.2 How Animals Are Classified

Taxonomic Category	General Description
Phylum	The largest subcategory of the animal kingdom; animals included in a given phylum share a similar level of body organization and general pattern of development and represent a major line of evolutionary descent.
Class Order Family Genus	Classes, orders, families, and genera form the hierarchy of animal classification. Each phylum may contain several classes, each class several orders, and so forth, down to the species level. The features that define these taxonomic categories are different in each phylum. Definitions become narrower and more specific with each group approaching the species level.
Species	Each species is defined by its own unique genetic makeup and is reproductively isolated from other species.

subclasses, and even a species may have subspecies defined by a unique trait such as color or length of an appendage.

The 1.5 million living, named animal species are classified into about 34 phyla. (This number is approximate, because zoologists do not all agree on how many phyla are appropriate; some phyla probably remain to be discovered, and these groups are, after all, invented by humans.) All 34 phyla contain **invertebrates**, animals without a backbone. **Vertebrates**, animals with a backbone (ourselves included), are all classified in the single Phylum Chordata, which also includes some invertebrates. The word *invertebrate* is not a taxonomic category; it is simply a convenient term referring to all animals except vertebrates.

Of the 34 animal phyla, 9 contain by far the greatest number of species living today and have the greatest impact on world environments. (They also happen to contain the animals that we humans are most familiar with and that we usually consider the most important.) These nine dominant phyla are described briefly in this chapter.

In addition to the nine dominant animal phyla, three large protozoan phyla (in the Kingdom Protista) contain many species that are common and important in aquatic environments and others that act as parasites, infecting animals and plants. Because these organisms have certain features in common with animals, we also introduce them in this first chapter.

Table 1.3 shows approximately how many species each of the major animal and protistan phyla contains. The other phyla (25 animal and 3 protozoan) are not to be dismissed, however, for there is much to learn from them. Some of these groups were dominant in earlier geological times, and they are also interesting in their own right. All animal phyla are discussed in detail in later chapters on animal diversity.

Table 1.3   The Twelve Major Phyla of Protists and Animals

Phylum	Common Name	Approximate Number of Species
Kingdom Protista		
Sarcomastigophora	Amoebas, flagellates	20,000
Apicomplexa	Sporozoans	4,500
Ciliophora	Ciliates	7,500
	PROTOZOAN SUBTOTAL	**32,000**
Kingdom Animalia		
Porifera	Sponges	4,300
Cnidaria	Cnidarians (jellyfish, hydras, sea anemones, corals)	11,000
Platyhelminthes*	Flatworms (turbellarians, flukes, tapeworms)	15,000
Nematoda	Roundworms	90,000
Mollusca*	Molluscs	110,000
Annelida*	Segmented worms (earthworms, polychaetes, leeches)	15,000
Arthropoda*	Arthropods (insects, spiders, crabs)	1,000,000
Echinodermata	Echinoderms (sea stars, sea urchins)	6,000
Chordata*	Chordates	47,200
	Lancelets, tunicates	(2,100)
	Fishes	(21,100)
	Amphibians	(3,900)
	Reptiles	(7,000)
	Birds	(8,600)
	Mammals	(4,500)
	METAZOAN SUBTOTAL	**1,298,500**
	GRAND TOTAL	**1,330,500**
All other (25) animal phyla combined		8,500

*Indicates phyla with some terrestrial species. All protozoa are aquatic, and of the approximately 34 animal phyla, only the 5 indicated here and one minor phylum, the Onychophora (see Chapter 26, p. 643), contain truly terrestrial species. Other phyla include species that are often considered terrestrial but are actually aquatic because they live in pools or droplets of water on plants, or they live in moist soil or in the body fluids of terrestrial animals.

Three Protozoan Phyla

Protozoa have only one major feature in common: They are **unicellular**, so all their life processes occur within a single cell. In a sense, these organisms represent life organized at the subcellular or cytoplasmic level, because their organelles perform some of the functions handled by tissues and organs in animals. Until recently, protozoa were all classified together in the Phylum Protozoa because few of their truly distinctive features had been discerned. But in the past several decades, studies with the electron microscope have prompted a re-evaluation of the evolutionary relationships among protozoa. Zoologists now generally agree that several groups within Protozoa merit phylum status. Protozoa are usually referred to by their common names: the *amoebas*, which crawl around on pond and lake bottoms using cell extensions called pseudopodia; the *flagellates*, which move by means of long, whiplike cell processes called flagella; the *sporozoans*, spore-forming parasites that cause malaria, among other diseases; and the *ciliates*, which move by means of many short, whiplike processes called cilia. Amoebas and flagellates are classified together in the **Phylum Sarcomastigophora**, sporozoans in the **Phylum Apicomplexa**, and ciliates in the **Phylum Ciliophora** (Figure 1.5).

Multicellular Life

The nine dominant animal (metazoan, or multicellular) phyla are introduced in the following paragraphs in a sequence that highlights some of the major events in animal evolution. Animals vary

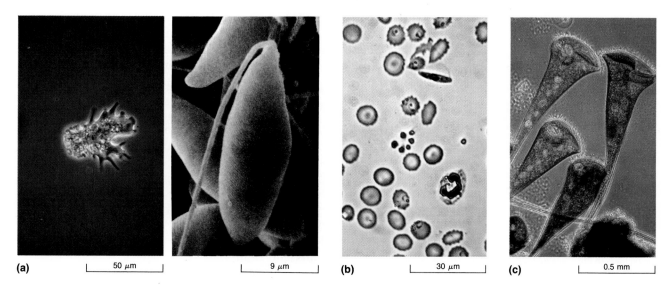

(a) ⊢ 50 μm ⊣ ⊢ 9 μm ⊣ **(b)** ⊢ 30 μm ⊣ **(c)** ⊢ 0.5 mm ⊣

Figure 1.5 Representatives of three dominant protozoan phyla. (a) Phylum Sarcomastigophora. A phase-contrast micrograph of a freshwater amoeba, *Mayorella* (about 30–50 μm in longest dimension) and a scanning electron micrograph of the flagellate *Euglena* (about 50 μm long). (b) Phylum Apicomplexa. In this photomicrograph of a mammal's blood, most of the cells are normal red blood cells (about 7.5 μm in diameter). Stages of *Plasmodium falciparum*, the causative agent of pernicious malaria, can be seen within the red blood cells. (c) Phylum Ciliophora. The freshwater ciliate, *Stentor* (about 1 mm long), as seen with the light microscope.

greatly in the degree of complexity of their body organization. Those with the least complex body plans are presented first. Their relative simplicity of organization—though not the living animals themselves—is thought to resemble that of ancestral metazoans. Animals whose body construction is more complex illustrate several architectural features that were central to animal evolution: bilateral symmetry, body cavities, and metamerism (repetition of body parts). These features are also discussed briefly.

Phylum Porifera

The *poriferans,* or *sponges* (Figure 1.6), differ from the protozoa in that they are multicellular; but they are not as highly organized as most other animals. Sponges lack highly specialized tissues and have no organs. As in protozoa, most of their life processes, such as digestion and excretion, occur within individual body cells. Adult sponges are **sessile**, meaning that they do not move from place to place. Most sit on the bottom of marine (or, in a few cases, freshwater) habitats and feed by filtering fine particles of organic material from the surrounding water. The body of many sponges grows irregularly over bottom substrates, but some are **radially symmetrical**; in geometric terms, this

Figure 1.6 Phylum Porifera. Marine sponges often live singly or in small clusters formed by budding. Here we see *Sycon lingua,* a vase-shaped sponge, measuring from 1 to 3 cm long. It bears a fringe of needle-sharp spicules around the opening at the upper end; these spicules help prevent small animals from entering the sponge's internal cavity, the spongocoel.

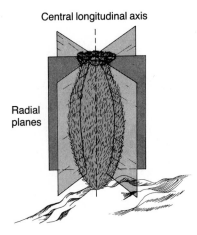

Figure 1.7 Radial symmetry. Animal bodies that are cylindrical, such as the body of this sponge, are radially symmetrical. Any plane passing through the center of the animal along its longitudinal axis divides the body into equal halves.

means that their body is arranged around a central longitudinal axis passing from the open end of the animal to the opposite, closed end (Figure 1.7).

Phylum Cnidaria

Cnidarians include the *jellyfishes, corals, sea anemones,* and *hydras* (Figure 1.8). Most are radially symmetrical, and all are more highly organized than the sponges, having well-developed tissues and some organs. And unlike the sponges, the cnidarian body plan includes an internal digestive (gut) cavity; there is no anus, however, and the gut is therefore said to be incomplete. These animals are called cnidarians because they have **cnidocytes**, specialized cells containing unique organelles (**nematocysts**) that function in prey capture, attachment, and protection. (See Figure 21.11 for a detailed drawing of nematocysts.) Cnidarians are abundant in marine environments, but only a few species, such as hydras, are found in freshwater habitats. The reef-forming corals are of great ecological importance, since their massive limestone reefs, including those off the South Atlantic coast of North America and the Australian Great Barrier Reef, provide habitats for many other tropical animals and plants.

Bilateral Symmetry

Unlike sponges and cnidarians, most animals, including those in the remaining seven phyla, are **bilaterally symmetrical**: They have a distinct head, or **anterior** end, and tail, or **posterior** end, as well as **dorsal** (back), **ventral** (belly), and **lateral** (side) surfaces (Figure 1.9). Whereas radial symmetry is adaptive for sessile animals, enabling them to interact with their surroundings equally in any direction, bilateral symmetry is adaptive for animals that move from place to place. A bilateral animal travels head first through its environment, so sensory elements concentrated on the head contact the

(a)   **(b)**

Figure 1.8 Phylum Cnidaria. (a) The bright orange hydroid, *Garveia annulata,* is an attached colonial species common on rocky shores of the U.S. Pacific Northwest. (b) *Cyanea capillata* is a marine jellyfish of northern oceans. It may be 50 cm in diameter with tentacles trailing down about 2 m; the sting of large individuals can be dangerous.

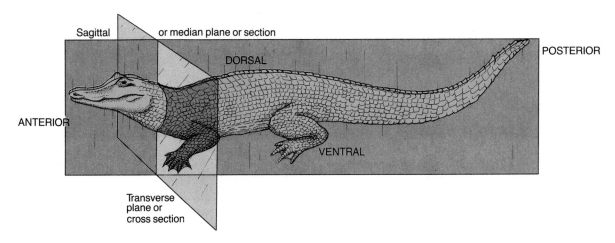

Figure 1.9 Bilateral symmetry. Only one plane passing through the center of the body from the anterior end to the posterior end divides this crocodile into symmetrical halves. This plane, called the sagittal plane, creates two equal and opposite, or mirror-image, halves.

environment first and allow the animal to respond appropriately. The next phylum contains the simplest bilateral animals.

Phylum Platyhelminthes

Platyhelminthes, or *flatworms,* have a more highly organized body than either the sponges or the cnidarians. Sensory organs are well developed, and there are reproductive and neuromuscular organ systems. Unlike most other bilateral animals, however, most flatworms have an incomplete digestive tract, as it lacks an anus. Also unlike other bilateral animals, flatworms are **acoelomate,** meaning they do not have a cavity between the gut and the body wall. Except for the space within the digestive tract, the flatworm body is a solid mass of cells, tissues, and organs.

There are three major groups of flatworms (Figure 1.10). *Turbellarians* are mostly free-living in freshwater and marine environments, whereas *flukes* and *tapeworms* are **parasitic** (live in or on, and at the expense of, other animals) and have complex life cycles, often involving two or more hosts. The larvae of these parasitic groups often inhabit invertebrate animals, while the adults of many species inhabit the gut of vertebrates and live on partially digested host nutrients, in effect stealing energy. Tapeworms are highly adapted to a parasitic existence. They do not even have a gut of their own, but instead absorb predigested nutrients across their body surface.

(a) |⎯⎯ 20 mm ⎯⎯|

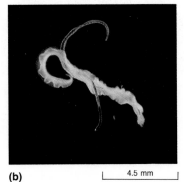

(b) |⎯⎯ 4.5 mm ⎯⎯|

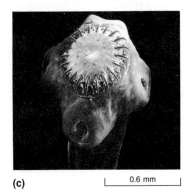

(c) |⎯⎯ 0.6 mm ⎯⎯|

Figure 1.10 Phylum Platyhelminthes. (a) The psychedelic flatworm, *Pseudoceros affinus,* is a marine turbellarian. (b) The human blood fluke *Schistosoma mansoni,* shown copulating in this scanning electron micrograph. The larger male holds the female in a groove of his body. (c) A scanning electron micrograph of the scolex ("head") of a tapeworm.

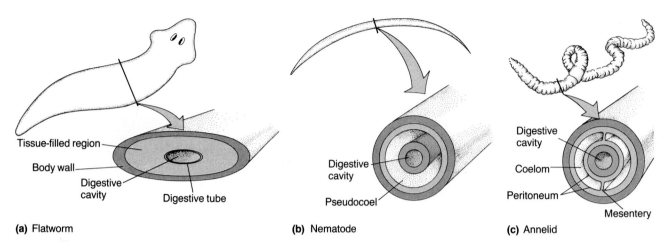

Tissue-filled region

Body wall

Digestive cavity

Digestive tube

(a) Flatworm

Digestive cavity

Pseudocoel

(b) Nematode

Digestive cavity

Coelom

Peritoneum

Mesentery

(c) Annelid

Figure 1.11 Body cavities. Cross sections of acoelomate, pseudocoelomate, and eucoelomate animals. (a) The acoelomate body is filled with a mass of cells. (b) In the pseudocoelomate, the internal organs lie in the body cavity (pseudocoel). (c) In the eucoelomate, the peritoneum lines the body cavity and envelops the body organs. Mesenteries (membranes) formed from the peritoneum suspend the organs from the body wall.

Body Cavities

All the animals in the preceding phyla lack a cavity between the gut and the body wall. Of the remaining six phyla discussed in this chapter, one—the Phylum Nematoda—contains animals with an internal cavity, called a **pseudocoel**, in which the organs lie. Nematodes and several other phyla (illustrated later in the book) are collectively called **pseudocoelomates**. All the animals in the other five dominant phyla (collectively called **eucoelomates**) have a **coelom**, which is a cavity lined with a sheet of tissue called **peritoneum**. The peritoneum forms membranes that cover and suspend the various organs, in effect separating the organs from the body cavity (Figure 1.11). As discussed in later chapters, body cavities have several advantages,

and their development is thought to reveal much about the evolutionary history of certain animal groups. (The importance of body cavities in the history of the animal kingdom is discussed in Chapter 18, p. 434, and in the Trends and Strategies section beginning on p. 531.)

Phylum Nematoda

Nematodes, or *roundworms,* are abundant in all types of moist environments, including mud at the bottom of lakes and oceans, water droplets in moist soil, and the bodies of other animals. Roundworms are elongate and cylindrical, with a mouth at one end and an anus near the other (Figure 1.12). Internally they have a "tube-within-a-tube" body plan:

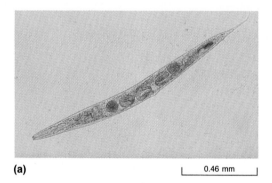

(a)

|———— 0.46 mm ————|

(b)

Figure 1.12 Phylum Nematoda. (a) The genus *Diplogaster* includes many free-living species common in organically rich, moist soil. (b) Shown here in human sputum are parasitic roundworms that spend most of their time in the intestine and may produce diarrhea and ulceration, a disease called strongyloidiasis.

(a)

(b)

(c)

(d)

Figure 1.13 Phylum Mollusca. (a) A terrestrial snail *(Helix)* on pumpkin; (b) a nudibranch *(Cyerce nigra)* feeding on algae; (c) an octopus *(Octopus rubescens);* (d) the file shell *(Lima scabra)* swims by jetting water out between its bivalve shell.

Their body wall forms an outside tube surrounding the fluid-filled pseudocoel in which organs are housed; the digestive tract, or gut, is the inside tube. Judging from the large number of nematode species (about 90,000) and their presence in a wide variety of environments, the roundworm body plan is a good illustration of the fact that evolutionary success may not depend on anatomical complexity.

Phylum Mollusca

Chitons, squids, and *octopuses* are marine members of this large phylum (Figure 1.13). *Clams* and other *bivalves* (molluscs with a two-part shell) live in lakes, streams, and oceans. *Snails* and *slugs* are mainly aquatic, but certain species are terrestrial. The limy shells of molluscs remain intact long after these animals die, and mollusc shells are readily preserved as fossils. Much of what is known about ancient environments has been learned by studying fossilized mollusc shells.

It may seem contradictory that while many molluscs have a hard shell, the phylum name Mollusca means soft, in reference to the soft body contained within the shell. In fact, not all molluscs have an outer shell, and three other less obvious features are more diagnostic of the phylum. These are the muscular **foot**, usually adapted for locomotion; the **mantle**, a sheet of tissue covering the dorsal or general body surface; and the **mantle cavity**, a space between the body wall and the mantle. The mantle cavity often contains gills in aquatic molluscs, whereas it is modified as a lung in terrestrial snails.

Metamerism

The following two phyla illustrate a significant development in animal architecture that molluscs do not display. **Metamerism** (segmentation) is the division of the body into repeated parts, called **metameres** or segments. It permits certain kinds of complex body movements and accounts to a large extent for the great success of the Annelida (segmented worms) and Arthropoda (arthropods, or joint-legged animals).

Phylum Annelida

Earthworms and other segmented worms making up the Phylum Annelida illustrate metamerism better than any other animal group. An annelid's entire body, except for the anterior and posterior ends, consists of more or less uniform segments. Earthworms and their close relatives, some of which inhabit fresh water, are collectively called *oligochaetes*. They are so named because they have relatively few bristles projecting from their body surface (*oligo* is from the Greek, meaning "few"; *chaeta* is from the Greek, meaning "long hair"). The bristles anchor the worms in their burrows and provide leverage for burrowing. The burrowing and feeding activities of the earthworm provide aeration and help to decompose organic matter in soil environments.

The largest group of annelids, the *polychaetes* (from the Greek *poly*, "many"), have many body bristles. They live in oceans and estuaries, where some species wriggle about on the bottom while others burrow in sand or mud or live sedentary lives in secreted tubes (Figure 1.14).

The *leeches*, a small but successful third group of annelids, are found in all types of freshwater, marine, and moist terrestrial habitats. They are notorious for their bloodsucking habits, although most species are carnivorous, feeding on small invertebrates such as snails and aquatic insects.

Phylum Arthropoda

The *arthropods* comprise the largest phylum in the animal kingdom, with about 75% of the known animal species in this group (Table 1.3). A jointed protective **exoskeleton** is the most distinctive feature of the arthropod body and in large measure accounts for the evolutionary success of the phylum. Common arthropods include the *insects, arachnids* (e.g., spiders and ticks), and *crustaceans* (aquatic forms such as crayfish, lobsters, shrimps, and barnacles) (Figure 1.15). Like annelids, with whom they probably share a common ancestor, these animals are metameric. However, arthropod body segments are much less uniform along the length of the body than are those of annelids. The typical arthropod body is formed of several distinct, often fused, groups of segments. Insects, for example, have three groups: a head; a middle region, or thorax, often bearing wings; and an abdomen.

(a)   20 mm

(b)   2 cm

Figure 1.14   Phylum Annelida. (a) Fan worms, or feather duster worms *(Sabella melanostigma)* are sedentary, tube-dwelling polychaetes; they extend a fan of anterior projections that trap fine organic food particles suspended in the water. (b) A leech *(Placobdella parasitica)* attached to an Eastern painted turtle *(Chrysemys picta).*

(a)

16 cm

(b)

5.6 cm

(c)

11 mm

(d)

Figure 1.15 Phylum Arthropoda. (a) The horseshoe crab *(Limulus)* is a marine scavenger; these individuals are mating. (b) Centipedes, such as this South American form, *Scolopendra,* are carnivorous. (c) The black widow spider, *Latrodectus mactans,* is distinguished by a red "hourglass" marking on the abdomen. (d) The scarlet lobster *(Enoplometopus occidentalis)* is found in Hawaiian waters. (e) A dragonfly *(Aeshna)* photographed in the Florida Everglades.

(e)

Phylum Echinodermata

The phylum name Echinodermata is derived from the Greek words meaning "like hedgehog or sea urchin skin," and most echinoderms, including *sea stars* and *sea urchins,* have a rough or spiny appearance (Figure 1.16). All echinoderms are marine animals and all possess a coelom, but they lack body

segments and most are radially symmetrical as adults. Many sea stars, for instance, have five arms radiating from a central mouth region. This may seem like an evolutionary step backward and appear to contradict our earlier statement that the last seven dominant phyla all consist of bilaterally symmetrical animals. There is no contradiction, however, because the larval stages of echinoderms are bilaterally symmetrical. Thus, the adults are **secondarily radial**. As discussed in Unit III, the larvae of many animals are a valuable source of information about both animal development and evolutionary history (see Chapter 18, p. 428).

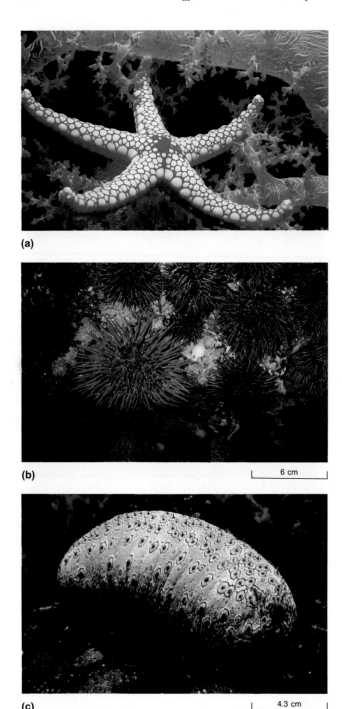

(a)

(b) 6 cm

(c) 4.3 cm

Figure 1.16 Phylum Echinodermata. (a) The candy cane sea star foraging on soft coral in the Red Sea. (b) The red sea urchin *(Strongylocentrotus franciscanus)* and the purple sea urchin *(S. purpuratus)* are common in tidal areas of the U.S. Pacific Northwest. (c) A sea cucumber *(Holothuria argus)* in waters off Borneo.

Phylum Chordata

Four distinctive features appear in the embryonic stages, and sometimes in the adults, of all members of the Phylum Chordata. All chordates have (1) a dorsal stiffening rod, the **notochord**; (2) a **dorsal**

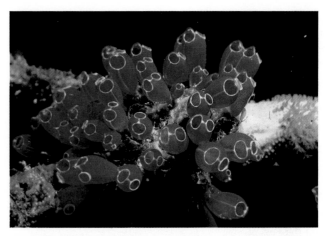

Figure 1.17 Phylum Chordata. The colonial tunicate, *Clavelina*, is an invertebrate chordate.

tubular nerve cord; (3) **gill structures** or pharyngeal pouches in the anterior part of the gut; and (4) a **postanal tail** (posterior to the anus). Together these features are diagnostic of the chordates.

Although the vertebrates are the most numerous and successful of the chordates, there are also two small groups of marine invertebrates, the *lancelets* and the *tunicates*, included in this phylum (Figure 1.17). Invertebrate chordates are filter feeders, using their pharyngeal apparatus to trap fine organic particles suspended in seawater. Their mode of life may reflect traits characteristic of ancestral chordates. (The significance of invertebrate chordates in chordate evolution is discussed in Chapter 28, p. 681.)

There are seven classes of vertebrates: one class of jawless fishlike forms called **agnathans** (see Chapter 28, p. 683), two classes of fishes, and one each of amphibians, reptiles, birds, and mammals. Among the fishes, one class consists of sharks and rays, all with a skeleton composed of cartilage. Bony fishes, a huge group, most with a hard, bony skeleton, constitute the other class. Amphibians, reptiles, birds, and mammals are collectively called **Tetrapods** because most of them have four limbs that provide support and movement on land or in the air (Figure 1.18). Reflecting our human bias, much more is known about certain birds and mammals (including *Homo sapiens*) than about any other group of animals. Viewing the animal kingdom with somewhat less bias, consider that there are only about 45,000 named vertebrate species, making up only about 4% of all known animals in the 34 phyla. The other 96% of the animal kingdom, well over a million species, are invertebrates.

Figure 1.18 The vertebrate chordates. (a) This sea horse *(Hippocampus)* is one of the bony fishes; (b) the horned tree frog *(Gastrotheca ceratophyrys),* seen here clinging to a tree branch in Costa Rica, is an amphibian; (c) the gila monster *(Heloderma suspectum)* is a colorful, but venomous, reptile of southwestern U.S. and Mexican deserts; (d) seagulls, like other birds, are members of the Class Aves; (e) the brown lemur *(Lemur macaco)* of Madagascar is a mammal; lemurs, monkeys, apes, and humans are members of the mammalian Order Primates.

☐ SUMMARY

1. Modern zoology is such a broad discipline that zoologists must specialize in their research studies. Some work with particular kinds of animals, while others focus on vital processes that occur in all animals.

2. As a science, zoology involves formulating and testing hypotheses within the framework of a carefully planned experimental design. Zoologists test hypotheses by measuring structures and processes in animals through repeated observation and controlled experiments. Their work is critiqued by other researchers, and results that withstand peer review become part of the accepted base of working knowledge. Today, zoological research relies heavily on the methods of chemistry, physics, and mathematics and has become increasingly quantitative and predictive.

3. The animal kingdom is organized in a hierarchical system composed of groups and subgroups called taxonomic categories. Animals within each category share certain common features. The largest category within a kingdom is the phylum, and there are about 34 animal phyla. Each phylum is divided into the increasingly smaller categories of class, order, family, genus, and species.

4. The species is the only taxonomic category defined by nature. Individuals belonging to the same species may interbreed with one another, but usually not with members of other species. Naturally occurring hybrids of two species are infertile.

5. Animals (metazoans) are multicellular and fundamentally different from the unicellular animal-like protozoa. Of roughly 34 animal and 6 protozoan phyla, 9 animal phyla and 3 protozoan phyla predominate because they contain the greatest number of species and have broad ecological impact.

6. Animals in the nine dominant phyla exhibit various degrees of body organization and other basic features reflecting major trends in animal evolution. Bilateral symmetry is a common denominator of most animal body plans. Most animals also have a body cavity, and many show metamerism, or body segmentation. Only one phylum, the Chordata, contains vertebrate animals (those with a backbone); all others contain only invertebrates.

☐ FURTHER READING

Davies, J. T. *The Scientific Approach.* 2d ed. New York: Academic Press, 1973. *Discussion of the theoretical approach of science and the interaction of science and society.*

Grzimek, B., ed. *Grzimek's Animal Life Encyclopedia.* New York: Van Nostrand Reinhold, 1972–1975. *General coverage with color drawings of the entire animal kingdom.*

Lederman, L. M. "The Value of Fundamental Science." *Scientific American* 251(1984): 40–47. *Makes the case that basic research enriches society culturally and promotes technological advances.*

Margulis, L., and K. V. Schwartz. *Five Kingdoms: An Illustrated Guide to the Phyla of Life on Earth.* San Francisco: W.H. Freeman and Company, 1982. *A modern, exceptionally well illustrated introduction to all forms of life; provides habitat sketches for each group.*

Marks, J. *Science and the Making of the Modern World.* London: Heinemann Educational Books, 1983. *Traces the development of science and the scientific method from origins in ancient Greece to modern times.*

Medawar, P. B., and J. S. Medawar. *Aristotle to Zoos: A Philosophical Dictionary of Biology.* Cambridge, Mass.: Harvard University Press, 1983. *Brief definitions and discussions of 175 key terms used in biology (from "adaptation" to "zoos").*

Wall, J. G., ed. *Encyclopedia of Marine Invertebrates.* Neptune City, N.J.: T.F.H. Publications, 1982. *Surveys all marine phyla; many color illustrations.*

Homeostatic Systems

In 1859, Claude Bernard, a French physiologist, suggested that complex animals live in two environments. One, the exterior environment, is beyond an animal's control, for it varies with season, geographical location, rainfall, and so on. The other, the internal environment, is composed of the body fluids that bathe an animal's cells. Bernard emphasized that it is the relative stability of the internal environment that allows an animal to live in different and changing external environments. With a stable internal environment, an animal's cells can function independently of the changing conditions outside the body.

In the 1930s, American physiologist Walter Cannon proposed the term **homeostasis** to describe the self-maintenance of internal stability. Cannon maintained that homeostasis is not a static condition. Instead, it is a dynamic interplay between those factors that tend to change the internal environment and the physiological processes that counteract such changes. Although Cannon originally coined the term *homeostasis* to describe processes regulating the composition of body fluids, the word is now used in a much broader sense to describe the steady-state condition fundamental to all aspects of normal body function. Thus, zoologists refer to the homeostatic condition of animals as it pertains to the gas and ion composition of body fluids, body temperature, blood pressure, and energy relationships.

This unit, Functional Systems of Animals, focuses on the challenges that animals must face in different environments. We use examples of protists and animals from the twelve dominant phyla described in Chapter 1 to compare different types of solutions to environmental problems. The next six chapters describe several aspects of homeostasis and some of the adaptations that have evolved that help maintain a steady internal environment. First we look at how animal bodies are organized and at life processes as they occur at the cellular level. Because animals maintain homeostasis by exchanging energy and matter with their environment, we discuss how animal cells use energy and how they obtain a steady supply of it and other resources from food. To obtain energy from food molecules, most animal cells need a constant supply of oxygen; gas exchange between the internal and external environments is therefore another vital aspect of homeostasis. In discussing gas exchange, we illustrate how the problems posed by aquatic environments differ from those posed by the dehydrating effects of air; and

21

we also see how the salt content of a marine environment poses challenges quite different from those of a freshwater environment.

As though dealing with external challenges were not enough, the internal environment is also subject to change. Cells produce wastes that must be removed to prevent the poisoning of body fluids. Moreover, fluids must be circulated throughout the body to maintain homeostasis in larger animals. Materials must be transported from areas of excess to regions where they are needed or to regions where they can be eliminated.

Generally speaking, any biological problem has a range of possible solutions that are dictated by the physical and chemical laws of nature and by the capabilities of a living organism. In considering the concept of homeostasis, we are focusing both on the major problems that animals face and on the major types of structural and functional adaptations that enable animals to solve the problems.

CHAPTER 2

Levels of Organization of Animals

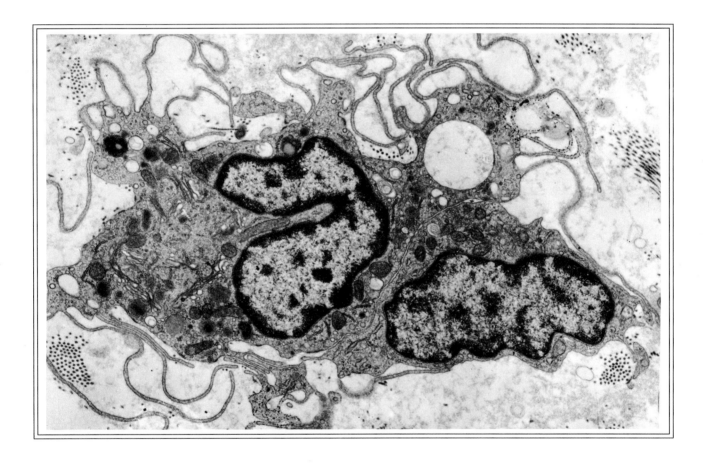

Questions that guide research can be formulated at several levels, corresponding to the levels of organization that characterize animal life. Atoms and molecules make up the most fundamental level of animal structure, and although these particles of matter are not themselves "alive," much has been

All life is composed of cells and all animals are multicellular. This photomicrograph shows a scavenger cell, called a macrophage, in the connective tissue of the mammalian testis; note the array of surface folds on the cell; some folds are engulfing droplets of fluid.

learned by studying animals at this level. Since the 1830s, when the German researchers Matthias Schleiden and Theodor Schwann clearly stated the cell theory, it has been recognized that the most basic unit of life is the cell. Thus, everything organisms do depends on the integrated activities of their cells and ultimately on the properties of the molecules produced by their cells. Except among the protists and bacteria, no single cell performs all the functions that a whole organism performs. The cells of multicellular organisms (including ani-

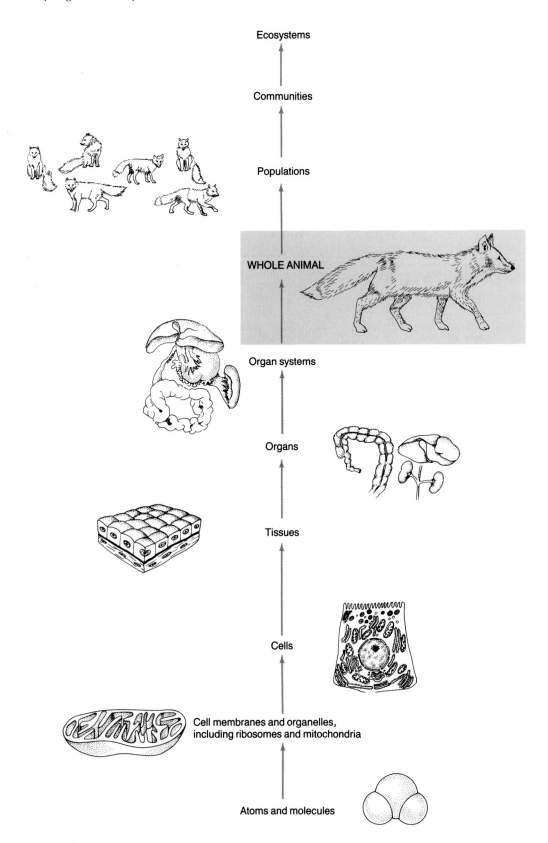

Ecosystems

Communities

Populations

WHOLE ANIMAL

Organ systems

Organs

Tissues

Cells

Cell membranes and organelles, including ribosomes and mitochondria

Atoms and molecules

Figure 2.1 Levels of organization. Animals can be studied at several levels. Biochemists study the structure and function of molecules in animal cells. At higher levels of organization, ecologists study the interactions of animals in populations, communities, and ecosystems.

mals) are structurally and functionally specialized, resulting in a division of labor in which each type of cell contributes uniquely to survival. Most animals have tissues (integrated aggregates of cells specialized for a specific function) organized as organs, which perform tasks that no one tissue could carry out alone. Organs, in turn, are often grouped into organ systems. Molecules, cells, tissues, organs, and organ systems thus represent five levels of organization within an animal (Figure 2.1). All of these levels play important roles in maintaining an animal's homeostatic internal environment, and it is these five levels that we will concentrate on in this chapter. Still higher levels of organization are the whole animal itself; populations, consisting of groups of individuals that interact with one another; and ecosystems, consisting of living organisms and their environment.

FUNDAMENTAL CHEMICAL CONCEPTS

Atomic Structure

All matter is composed of **elements**, substances that cannot be broken down into simpler materials by chemical reactions. An **atom** is the smallest quantity of an element that still retains the properties of that element, and all atoms of a given element have the same chemical properties. Table 2.1 lists the most common elements found in animals.

Table 2.1. Common Chemical Elements in Animals*

Element	Symbol	Approximate Atomic Weight	Atomic Number
Oxygen	O	16	8
Carbon	C	12	6
Hydrogen	H	1	1
Nitrogen	N	14	7
Calcium	Ca	40	20
Phosphorus	P	31	15
Chlorine	Cl	35	17
Sulfur	S	32	16
Potassium	K	39	19
Sodium	Na	23	11
Magnesium	Mg	24	12
Iodine	I	127	53
Iron	Fe	56	26

*Elements are listed from top to bottom in decreasing order of percent of human body weight. A similar trend holds for other animals, but actual percentages vary with the species.

All atoms consist of the same types of elementary particles. Every atom has a nucleus that contains positively charged particles called **protons**. The atomic nucleus of the simplest element, hydrogen, contains only one proton. The nuclei of all other elements contain protons and particles called **neutrons**, which have the same mass as protons but carry no electrical charge. The carbon atom, for example, contains six protons and six neutrons. Surrounding the nucleus of every atom are one or more **electrons**, negatively charged particles whose mass is only $1/1836$ that of a proton (or neutron).

It is possible to arrange the elements in a series based on their **atomic number**, which is the number of protons in their nucleus. The atomic number of an electrically neutral (uncharged) atom is also the number of electrons it contains, since by definition a neutral atom is one in which the number of protons and electrons is the same.

The **mass number** of an atom is the sum of the number of protons and neutrons in its nucleus. By subtracting the atomic number (number of protons) from the mass number (protons + neutrons), we can determine the number of neutrons in an element's atomic nucleus (Figure 2.2). Because the mass of electrons is virtually negligible, an atom's mass is determined by its protons and neutrons. Therefore, the atomic mass of an element (often referred to as its **atomic weight**) is essentially equal to its mass number. Atomic weights are arranged on a relative scale, where the mass ("weight") of carbon is assigned a value of 12 atomic mass units (amu). The weights of all other atoms are expressed in terms of how much heavier or lighter they are than carbon (see Table 2.1).

Some elements have variant forms called **isotopes**, which are atoms with the same number of protons and electrons but with a different number of neutrons. Diagrams of the three isotopes of carbon are shown in Figure 2.3. Isotopes with unstable nuclei break down radioactively to yield lighter forms of the same or some other element plus high-energy particles. Radioactive isotopes of hydrogen, carbon, sulfur, and phosphorus have been useful in tracing biological processes. It is possible, for example, to synthesize sugars containing radioactive carbon or hydrogen. If such sugars are fed to an animal, their physical and chemical fate can be determined by locating the radioactive atoms in various molecules and tissues.

Despite their small size, the number and arrangement of electrons in the atoms of any element determine the chemical properties of that element. The models of the atom proposed by Niels Bohr

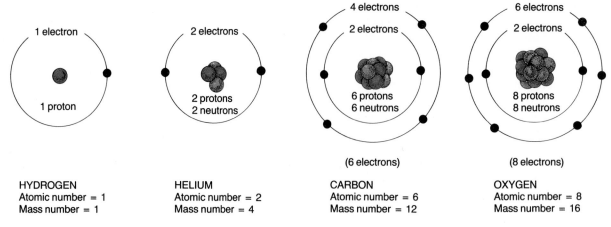

Figure 2.2 Structure of several atoms. The atomic number indicates the number of protons in the atomic nucleus and the total number of electrons (in a neutral atom) in the surrounding shells. The mass number is the sum of the protons and neutrons. Drawings are not to scale.

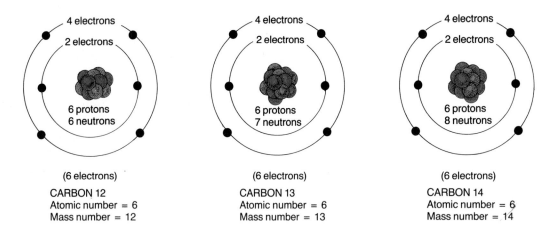

Figure 2.3 Isotopes of carbon. All three isotopes of carbon have an atomic number of 6, but the mass number may be 12, 13, or 14, depending on the number of neutrons.

and Ernest Rutherford at the turn of this century suggested that electrons orbit the atomic nucleus much as planets orbit the sun. Although modern quantum mechanics describes electron behavior more precisely, this planetary model of atomic structure is still useful today. It is convenient to depict electrons orbiting the atomic nucleus within concentric layers called **shells** (see Figure 2.2). Each shell can accommodate up to a certain number of electrons. The innermost shell is full with only two electrons, so in atoms containing more than this number, the remainder are found in shells farther from the nucleus. Oxygen, for instance, has eight electrons, two in the shell closest to the nucleus and six in a second shell. The maximum number of electrons that can be accommodated in the second shell is eight, so in a sodium atom, which has eleven electrons, there are two electrons in the

innermost shell, eight in the second shell, and one in the third shell. The maximum number of electrons in a shell n is $2n^2$. For example, shell 3 can hold 18 electrons (2×3^2).

Compounds, Molecules, and Chemical Bonds

An atom is most stable when its outer electron shell contains the maximum number of electrons. Such stability can be observed in the so-called **noble gases**, such as helium and neon, elements that have a filled outer shell. In nature, atoms with an unfilled outer electron shell form electrically neutral entities called compounds and molecules. A **compound** is any such entity that contains two or more kinds of elements; it is formed by chemical bonds involving the exchange or sharing of electrons between atoms. The term *molecule* is often used loosely, but strictly speaking, a **molecule** consists of two or more atoms of the same *or* different elements held together by chemical bonds that involve electron sharing. Such bonds are called **covalent bonds**. Atoms of carbon, nitrogen, sulfur, and phosphorus tend to share electrons (and thus form molecules by covalent bonding) when they react with each other or with other elements. The molecule O₂ (molecular oxygen), for example, is formed as two oxygen atoms share electrons in their outer orbits. Likewise, a molecule of the gas methane (which is also a compound) consists of four hydrogen atoms and one carbon atom sharing electrons (Figure 2.4).

In contrast to electron sharing, an atom may give up or accept one or more electrons. Thus, in forming certain compounds, one atom may donate its outer-shell electrons to another atom. In doing so, each atom becomes an **ion**, or charged atom, and **ionic bonding** occurs when oppositely charged ions attract one another. This forms **ionic compounds**, such as sodium chloride (table salt), a biologically important substance (Figure 2.5). Ionic compounds dissociate into separate ions when dissolved in solution, and in a strict sense, they are not molecules.

Hydrogen Bonds. Many molecules that contain two or more different types of atoms have local regions of electronegativity or electropositivity. Covalent bonds between hydrogen and either oxygen, nitrogen, or fluorine often result in the electrons being drawn away from hydrogen. For this reason, the oxygen, nitrogen, or fluorine in these molecules tends to carry a slight negative charge and hydrogen a positive charge. Bonds that yield such asymmetrical charge distributions are said to be **polar**. Since opposite charges attract, molecules with polar bonds will align themselves so that the positively charged areas of one molecule are electrostatically bound to the negative areas of another. This type of interaction produces a relatively weak bond called a **hydrogen bond** (Figure 2.6).

Molecular Weight and the Mole Concept

The **molecular weight** of a molecule equals the sum of the weights of the atoms composing it. Thus, methane (CH₄) has a molecular weight of 12 +

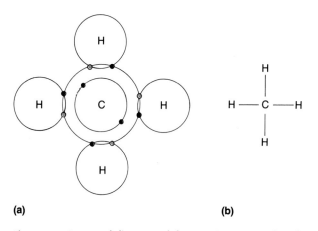

Figure 2.4 Structural diagrams of the organic compound methane. Carbon will covalently bond with four hydrogen atoms by sharing electrons to form methane. Thus, the hydrogen atoms will have two electrons and the carbon atoms eight in their outer shells, so that both elements achieve stable electron configurations. (a) A diagram of methane showing its electron shells. (b) The structural formula of methane showing only covalent bonds.

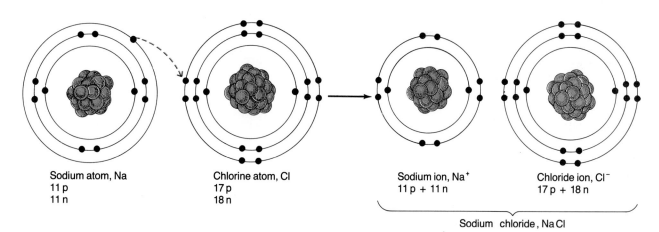

Figure 2.5 Ionic bond formation. Sodium reacts with chlorine by donating an electron, so that both atoms achieve a stable number of eight electrons in their outer shells. The oppositely charged sodium and chlorine ions attract each other, forming an ionic bond.

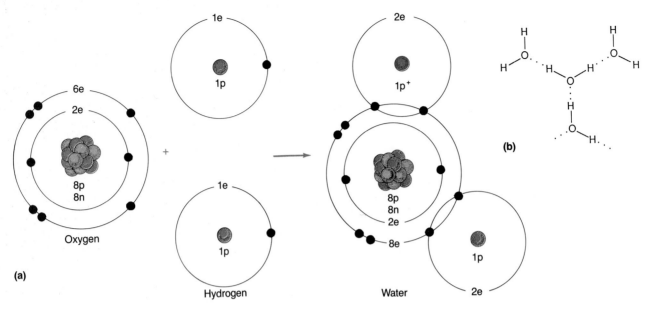

Figure 2.6   Hydrogen bonding in water. (a) When covalent bonds form between an oxygen atom and two hydrogen atoms, the distribution of electrons is not uniform throughout the molecule. (b) The oxygen nucleus tends to draw electrons from the small hydrogen atoms, giving oxygen a slight negative charge and exposing the positively charged hydrogen protons. The oppositely charged regions on different water molecules attract one another and form hydrogen bonds.

$4(1) = 16$ amu. Individual molecules cannot be weighed conveniently, and yet it is often useful to work with a known number of molecules. The mole concept addresses this problem. A **mole** is the amount of a substance that contains as many elementary units (atoms or molecules) as there are atoms in 12 grams (g) of carbon-12. It is also equivalent to the molecular weight of the substance in grams. Because the weights of all atoms are defined relative to the same standard—carbon—a mole of methane, weighing 16 g, contains the same number of molecules as a mole of the human hormone insulin, weighing 5700 g. The number of elementary units in a mole is known as **Avogadro's number**, after the originator of this concept, and it equals 6.023×10^{23}.

Most biochemical reactions take place in water, and their rates often depend on the concentration of molecules present. Concentrations are usually expressed on a scale that indicates **molarity**. For example, a 1-molar solution, written 1 M, consists of 1 mole (mol) of a substance dissolved in enough water to make up 1 liter (L) of solution. Most compounds in cells occur at less than millimolar concentrations (10^{-3} M = mM).

Acids and Bases

In any solution having water as its solvent, most of the water molecules will be hydrogen-bonded to each other and a few will dissociate into hydrogen ions (H^+) and hydroxyl ions (OH^-) according to the equation

$$H_2O \longleftrightarrow H^+ + OH^-$$

In 1 L of pure water, the hydrogen ion concentration is 10^{-7} M. Since the H^+ concentration of a chemical or biological system is often of interest, the **pH scale** has been devised to quantify it. The **pH** of a solution is defined as the negative logarithm of its hydrogen ion concentration. Thus, the pH of pure water is

$$pH = -\log[H^+] = -\log(10^{-7}) = -(-7) = 7$$

Acids are chemical compounds that donate hydrogen ions (H^+ can also be thought of as a proton, a hydrogen nucleus stripped of its electron). When added to pure water, acids increase the H^+ concentration, decreasing the pH value to less than 7. **Bases** are compounds that accept hydrogen ions (protons), thus removing them from solution and increasing the pH to a value greater than 7. The pH scale runs from 0 to 14, with values below 7 becoming increasingly acidic and values above 7 increasingly basic.

The balance of acid and base in cells and body fluids is critical to the proper functioning of many enzymes and other molecules. In most animals, the pH of body fluids is carefully regulated to preserve homeostasis. Body fluids also contain **buffers**, chemical compounds that resist changes in pH by acting either as acids, donating hydrogen ions when the local pH rises, or as bases, accepting hydrogen ions when the pH falls.

Organic Chemistry

How the preceding information applies to animals becomes clearer when we turn to **organic chemistry**, the study of chemical reactions involving carbon-containing compounds. A carbon atom has the special property of being able to form up to four covalent bonds with other carbon atoms or with atoms of other elements. These covalent bonds may be single, double, or triple, depending on whether one, two, or three pairs of electrons are shared. This bonding versatility enables carbon to form an enormous number of molecules, and animal cells contain many compounds with long chains of covalently bonded carbon atoms. Hemoglobin, the protein that carries oxygen in vertebrate blood, contains 3032 carbon atoms, as well as hydrogen, oxygen, nitrogen, and small amounts of sulfur and iron.

The chemical diversity of organic compounds is further enhanced by the occurrence of structural isomers. **Isomers** are molecules that contain the same atoms, but in different structural arrangements. Consider an organic molecule with the formula C_2H_6O. Here are two possible arrangements of its atoms:

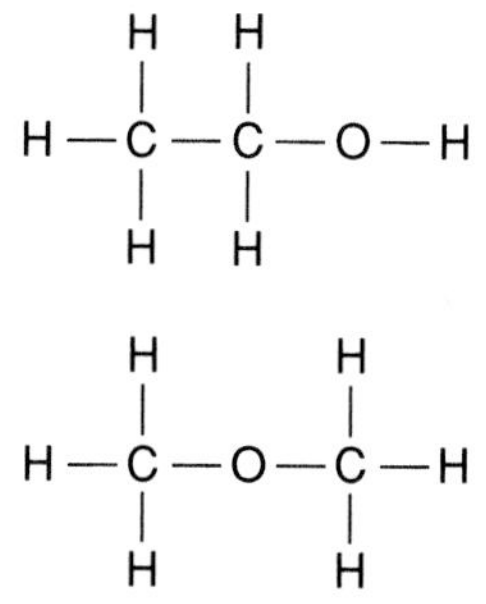

The molecule on the top is the liquid ethanol (ethyl alcohol); the molecule on the bottom is the gas dimethyl ether. Their structural differences seem minor, but they have completely different properties.

Large molecules with chains of carbon atoms forming their "backbone" participate in chemical reactions according to the types of atoms or groups of atoms attached to the backbone. Table 2.2 shows the structure of several side groups that are important in biological molecules.

Table 2.2 Important Functional Groups in Biological Molecules

Group	Molecular Example
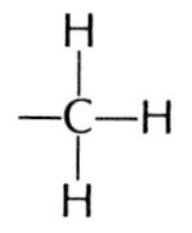 Methyl	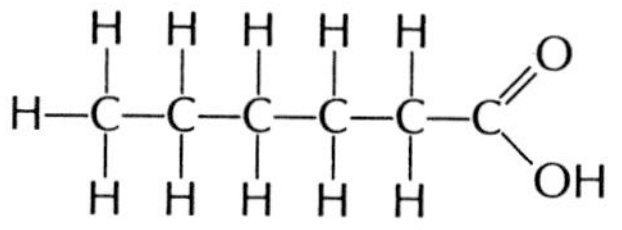 Fatty acid
—OH Hydroxyl	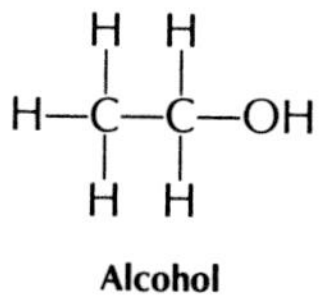 Alcohol
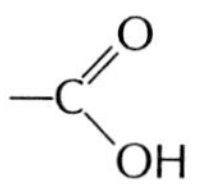 Carboxyl	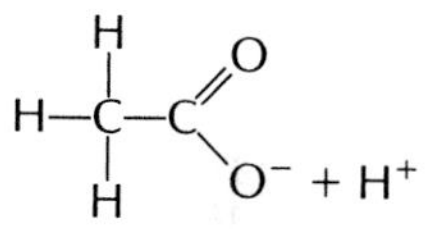 Organic acid
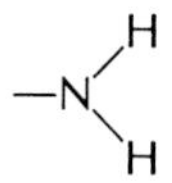 Amino	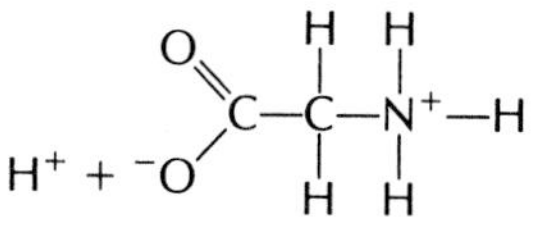 Amino acid
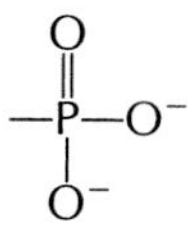 Phosphate	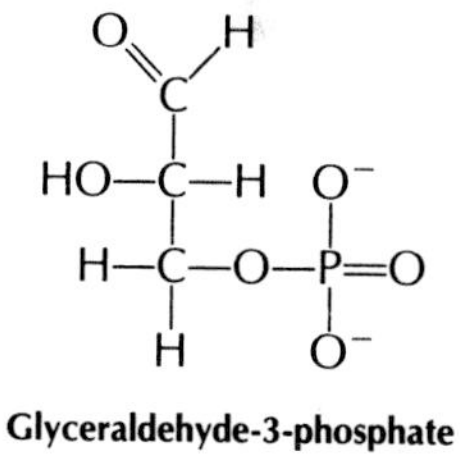 Glyceraldehyde-3-phosphate
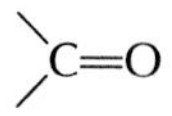 Carbonyl	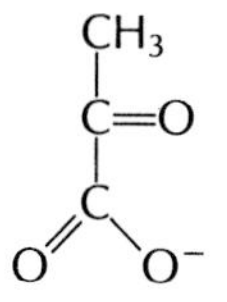 Pyruvic acid
—S—H Sulfhydryl	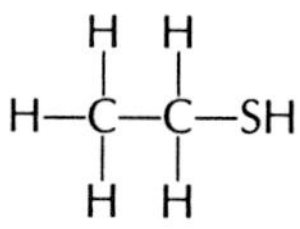 Thiol

CHEMICAL COMPOSITION OF ANIMAL CELLS

An animal's body contains vast numbers of different kinds of molecules, most of which occur within cells. Molecules also circulate in the bloodstream or other body fluids, and some molecules form extracellular (outside the cell) substances, such as the hard matrix of bone. The chemical composition of cells reveals much about the general chemical makeup of animals. Water, an inorganic substance, is generally the most prevalent molecule, followed by several organic compounds (proteins, lipids, carbohydrates, and nucleic acids) and salts (both organic and inorganic).

Water

All life as we know it requires water, and several special properties of water make it an ideal medium for life. Water is an excellent solvent for both polar and ionic compounds because it is polar and interacts with charged particles in solution. When a substance dissolves in water, each of its molecules or ions becomes surrounded by a **hydration shell** (Figure 2.7a). Molecules that attract water in this way are said to be **hydrophilic**, meaning "water-loving." Much of the water present in a cell is immobilized in hydration shells around dissolved substances. However, water is not a good solvent for molecules containing only carbon and hydrogen (Figure 2.7b). These molecules are electrically symmetrical, or **nonpolar**, because the carbon nucleus does not draw electrons away from the hydrogen atoms and so there are no charged regions that attract water molecules. Nonpolar substances are said to be **hydrophobic**, or "water-hating." Oils are hydrophobic, and you have probably noticed how they form a separate layer when added to water.

The hydrogen bonding of water has other biologically important effects as well. It gives water a property called **surface tension**, which causes water droplets to be spherically shaped and allows certain insects to "walk on water." Hydrogen bonding is also responsible for a property called **capillarity**, which causes water to rise in narrow tubes and climb up the threads of a piece of cloth.

Unlike most liquids, water does not contract uniformly with decreasing temperature. In fact, water has the highly unusual property of expanding as it freezes to form ice. As water expands, it becomes less dense, and this is why ice can float on

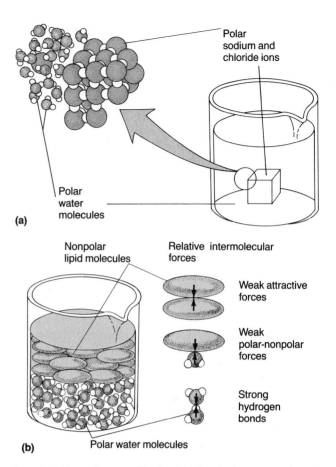

(a)

(b)

Figure 2.7 How substances dissolve. (a) When ionic compounds such as sodium chloride (NaCl) dissolve in water, the charged ions attract polar water molecules and are surrounded by hydration shells that keep the ions in solution. (b) Lipids do not dissolve in water because lipid molecules are nonpolar and are not strongly attracted to water molecules.

liquid water. The implications of this property for aquatic life are enormous; were it not for this property, ice would form on the bottom of bodies of water and might never melt completely in warmer seasons.

Biomolecules

Biomolecules include the proteins, lipids, carbohydrates, and nucleic acids synthesized by living cells. Many of these molecules consist of hundreds or thousands of atoms and collectively are called **macromolecules.** A class of relatively small biomolecules involved in energy passage in cells is introduced in Chapter 3 (p. 61).

Proteins. A typical animal cell contains about 10 billion protein molecules, many of which are identical copies, but all of which have some vital function. Some proteins located in membranes help recognize or transport other substances; thousands of proteins are enzymes that promote cellular chemical reactions (collectively called metabolism); structural proteins give cells their shape; and many other proteins regulate cellular processes. Although there are thousands of different types of proteins in animal cells, they all share certain structural features. First, regardless of their function, most are macromolecules that consist of **polymers** made of smaller molecules called **monomers**. Polymers are like chains whose links are the repeating units. In the case of proteins, the monomers are different kinds of **amino acids** (Figure 2.8).

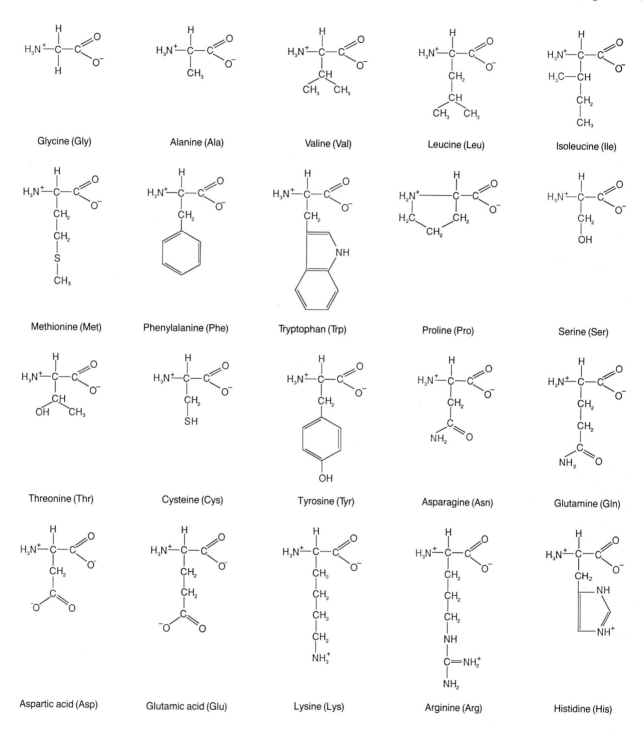

Figure 2.8 The chemical structures of 20 common amino acids. All amino acids have a common structural component and a distinctive side group, or R group. At the pH values common in cells, the carboxyl group (–COOH) ionizes, donating a hydrogen ion into solution; the H^+ binds to the electronegative nitrogen in the amino group (–NH_2).

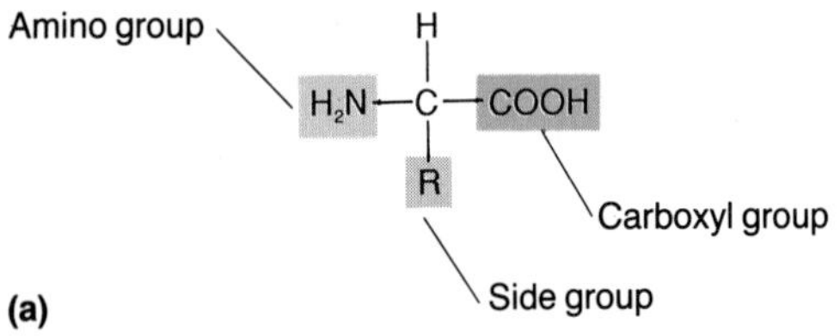

Figure 2.9 Amino acid structure and peptide bonds. (a) The amino and carboxyl groups, common to all amino acids, are covalently bonded to a central carbon atom. The R group, illustrated for each amino acid in Figure 2.8, is the side group that gives each amino acid its distinctive properties. (b) A peptide bond forms when the carboxyl group of one amino acid reacts with the amino group of another. Since the reaction product, a dipeptide, also has free amino and carboxyl groups, it can continue to react with other amino acids to yield long polypeptide chains.

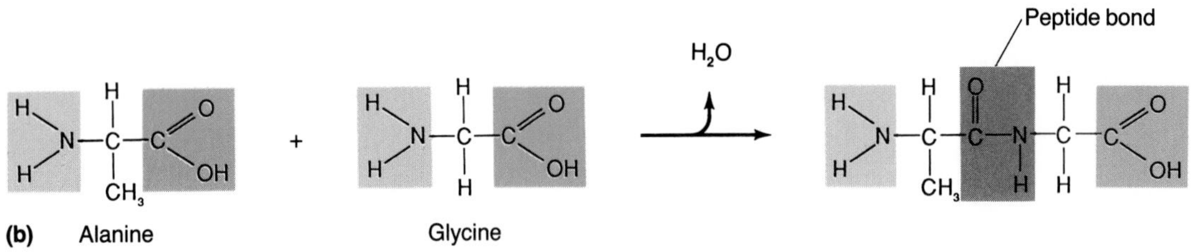

Although animal cells can synthesize a number of amino acids, they must obtain others—the so-called **essential amino acids**—from the animal's diet. Human adults require eight essential amino acids, while some amoebas require only five.

Each amino acid contains a central carbon atom to which are attached an amino group ($-NH_2$), a carboxyl group ($-COOH$), and an **R group** consisting of one or more atoms that make amino acids different from one another (Figure 2.8). Covalent **peptide bonds** can form between the amino group of one amino acid and the carboxyl group of another, resulting in molecules called **polypeptides** (Figure 2.9). Functional proteins consist of one or more polypeptide chains.

Because there are only about 20 different amino acids, the same amino acid often occurs in a given protein more than once. Two proteins with the same number of the same amino acids can be quite different structurally and functionally because their amino acid *sequences* are unique. The sequence of amino acids in a given protein is called that protein's **primary structure**, and it is genetically determined (Figure 2.10a). (Chapter 15 covers determination of protein structure by DNA, p. 362.) Proteins are not simply straight chains of amino acids. Most have a **secondary structure** taking the form of a pleated sheet or an α (right-handed) helix (Figure 2.10b). The α-helical structure is seen in certain **fibrous proteins**, including collagen (common in skin, ligaments, and bones in vertebrate animals) and certain keratins (those forming hooves, nails, and horns). Other keratins, such as those found in spider webs, bird feathers, beaks, and silk, have the pleated sheet arrangement.

In addition to a secondary structure, proteins may have a complexly folded, three-dimensional **tertiary structure** (Figure 2.10c). Tertiary folding may occur because certain amino acids in a polypeptide chain are hydrophobic and cluster tightly together to avoid water, while the hydrophilic amino acids extend away from the hydrophobic ones into the aqueous surroundings. Finally, when two or more polypeptide chains form a protein, the way in which the chains associate produces a **quaternary structure** (Figure 2.10d). Proteins with tertiary and quaternary structure form a large group called the **globular proteins**. Examples include enzymes, oxygen storage and transport molecules, antibodies, and certain hormones. The secondary, tertiary, and quaternary structures of proteins are maintained by chemical bonds that form between adjacent portions of the molecule: Hydrogen bonds form between adjacent polar amino acids, and covalent **disulfide bonds** form between two adjacent cysteine portions of proteins (see Figure 2.10). As discussed in Chapter 3, the shape of proteins largely determines their function.

Lipids. **Lipids** are a diverse class of compounds that includes fats, oils, and steroids. All are insoluble in water because the molecules consist mainly of nonpolar $-CH_2$ groups. Lipids form cell membranes, act as waterproofing agents on animal body surfaces, and serve as long-term energy storage compounds. Some steroids, such as estrogen in vertebrates and ecdysone in insects, act as hormones.

A typical energy storage lipid is shown in Figure 2.11. It consists of three long-chain fatty acids covalently bonded to a glycerol molecule, forming a **triglyceride**. All fatty acids consist of a hydrocarbon chain, typically 14 to 22 carbons long in animal cells, ending in a carboxyl group. Fatty acids

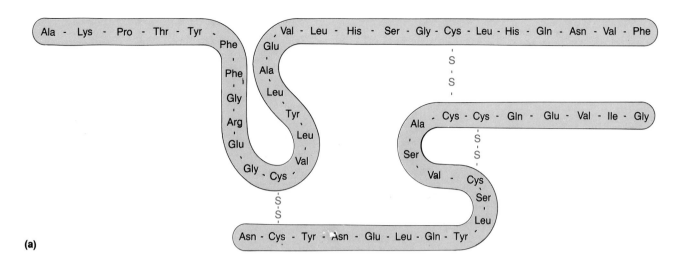

(a)

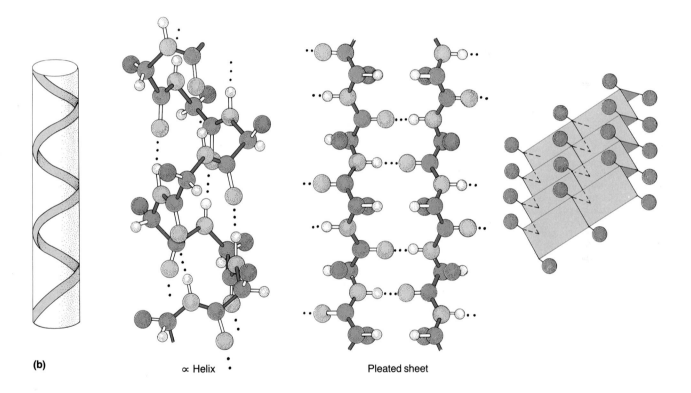

(b)   ∝ Helix ·   Pleated sheet

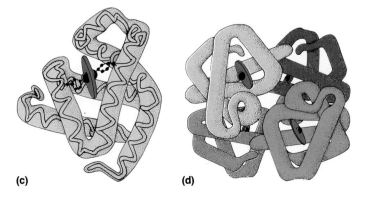

(c)   (d)

Figure 2.10 Levels of structure in proteins. (a) The primary structure of a protein is its sequence of amino acids. Insulin, a small protein, is shown here; note the disulfide bridges between cysteines. (b) Secondary structure is produced in many proteins when the amino acid chain forms a helix or a pleated sheet, with the amino acid R groups projecting away from the "backbone." In the structures shown, medium color represents C, light color represents N, medium gray represents the R group, and light gray and white are O and H, respectively. (c) Tertiary structure involves complex three-dimensional folding of the polypeptide chain. (d) Quaternary protein structure is the association of two or more polypeptide chains. Shown here are the four polypeptides found in vertebrate hemoglobin, the protein that transports oxygen in the blood.

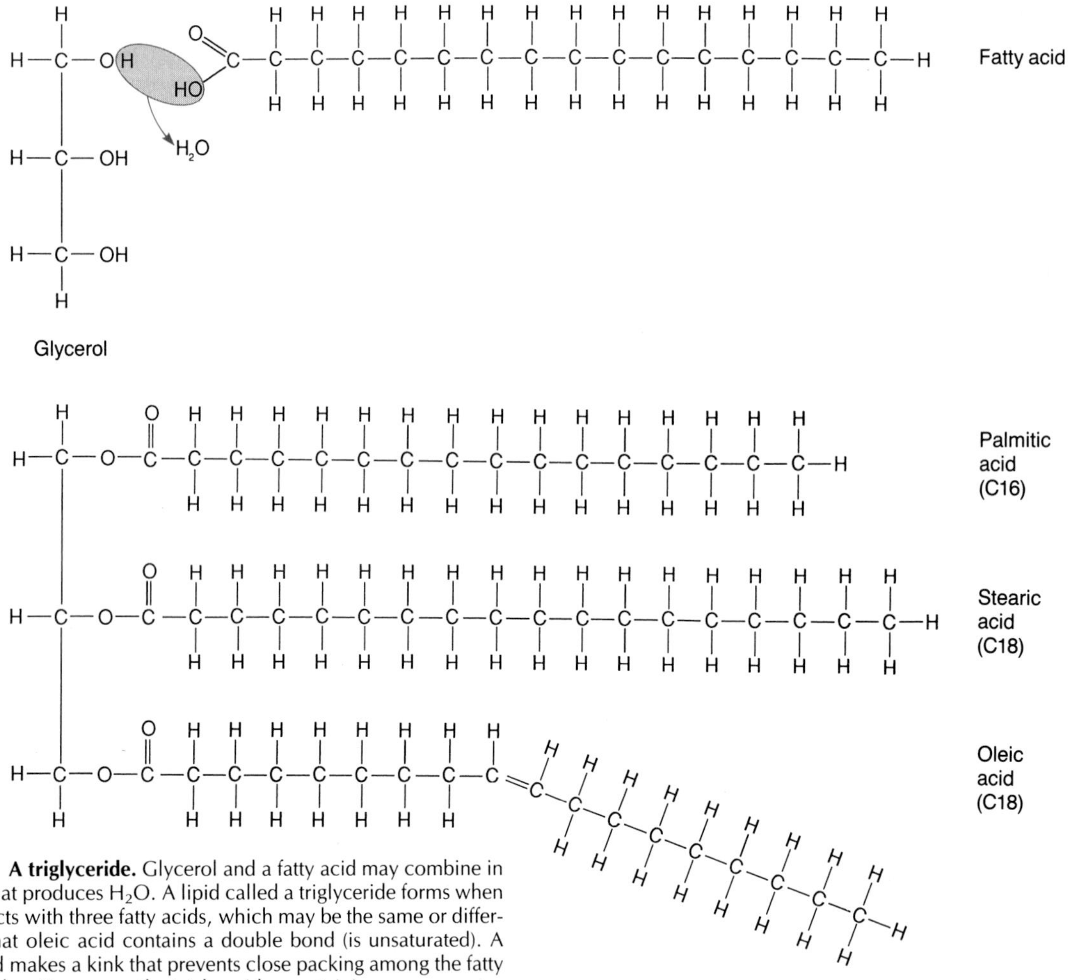

Figure 2.11 A triglyceride. Glycerol and a fatty acid may combine in a reaction that produces H_2O. A lipid called a triglyceride forms when glycerol reacts with three fatty acids, which may be the same or different. Note that oleic acid contains a double bond (is unsaturated). A double bond makes a kink that prevents close packing among the fatty acid molecules. Consequently, triglycerides containing many unsaturated fatty acids melt at lower temperatures than those containing saturated fatty acids.

can differ from one another in the length of the carbon chain and/or the number of single and double bonds between adjacent carbon atoms. **Saturated** fatty acids have only single covalent bonds between adjacent carbon atoms; when double bonds occur between carbons, the fatty acid is said to be **unsaturated**.

Phospholipids are similar to triglycerides except that a phosphate group replaces one of the fatty acids. Phospholipids are partially hydrophilic because the region near the phosphate group has a slight charge asymmetry. When placed in water, phospholipids form a bilayer film, with their hydrophobic fatty acid tails oriented toward the inside of the film and their hydrophilic phosphate heads toward the water. Phospholipid molecules are the main structural components of cell membranes (discussed in more detail later in this chapter, p. 37).

Phospholipids exist in a transitional form between the solid and liquid states. At the temperature extremes found in some habitats, they can melt or solidify, damaging the cell membrane structure. Zoologists interested in how animals adapt to their environments have found that many species exhibit differences in the lipid composition of their membranes in accordance with the climate. Animals living in hot springs or deserts, for example, have lipids containing long-chain, saturated fatty acids that melt only at very high temperatures, whereas animals inhabiting arctic environments have lipids containing short-chain, unsaturated fatty acids with low melting points. In temperate climates, animals such as crayfishes, amphibians, and fishes undergo seasonal changes in the lipid composition of their membranes. Research may help determine the molecular control mechanisms involved in these lipid changes.

Carbohydrates. **Carbohydrates**, including sugars, starches, and cellulose, often serve as energy sources and as starting materials for the synthesis of amino acids, fats, and nucleic acids. Simple sugars, or **monosaccharides**, contain from three to seven carbon atoms and usually have common names ending in *-ose*. Ribose and deoxyribose are common **pentoses** (five-carbon sugars) occurring in the nucleic acids of animal cells (Figure 2.12). Glucose and fructose are two nutritionally important **hexoses** (six-carbon sugars); they are structural isomers having the same atomic composition but different three-dimensional configurations. Since the carbon chains of pentoses and hexoses are not rigid, the carbon atoms constantly rotate about the axes of their covalent bonds with other carbons. When this happens, a relatively stable ring structure is formed, so most monosaccharides tend to exist as rings rather than as straight chains.

Simple sugars may combine to form larger molecules. Maltose, for example, is a **disaccharide** composed of two glucose monomers, while sucrose is a disaccharide composed of glucose and fructose (Figure 2.13a and b). **Polysaccharides** are large carbohydrates (polymers) consisting of many linked monosaccharides. **Glycogen** is a huge polymer of variable length (Figure 2.13c). It consists entirely of glucose units and is structurally similar to plant starches and cellulose. As the chief carbohydrate storage molecule in animals, glycogen can be synthesized rapidly or broken down to glucose as needed.

Glycoproteins. **Glycoproteins** are a large class of proteins that are covalently bonded to carbohydrates. Many of these substances are secreted by cells and have a wide variety of vital functions. For example, antibodies, albumins, and certain hormones are highly soluble glycoproteins that circulate in the blood, whereas membrane glycoproteins are associated with cell surfaces, where they are involved in cell-to-cell recognition.

Proteoglycans are glycoproteins with a linear structure. They consist of a protein core from which many sugars project like the bristles of a bottle brush. Proteoglycans are much less soluble than globular glycoproteins. One group of proteoglycans, referred to as "cell glues," help hold cells together to form tissues. Cell glues are common in cartilage and connective tissue (see Figure 2.27; Chapter 8, p. 188; and Figure 8.18). **Mucus** (also called mucoproteins) constitutes another group of proteoglycans that protect and lubricate cell sur-

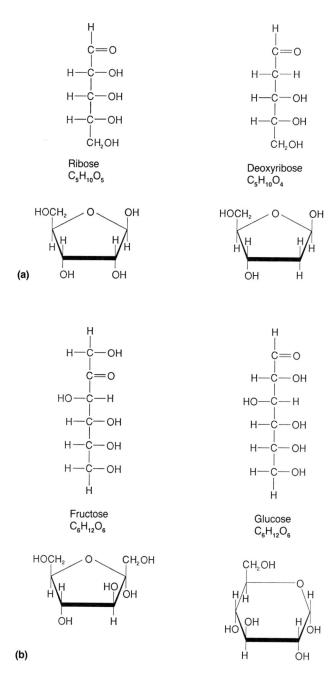

Figure 2.12 Monosaccharides. Simple sugars can be diagrammed as linear or ring-shaped molecules. (a) The pentoses ribose and deoxyribose. (b) The hexoses glucose and fructose are isomers having the same atomic composition but different structural arrangements of their atoms.

faces. The many hydroxyl groups on the sugars projecting from mucoprotein molecules form hydrogen bonds with water, creating hydration shells. Mucus is slippery because the hydrogen-bound water molecules slide past one another when a shearing force is applied.

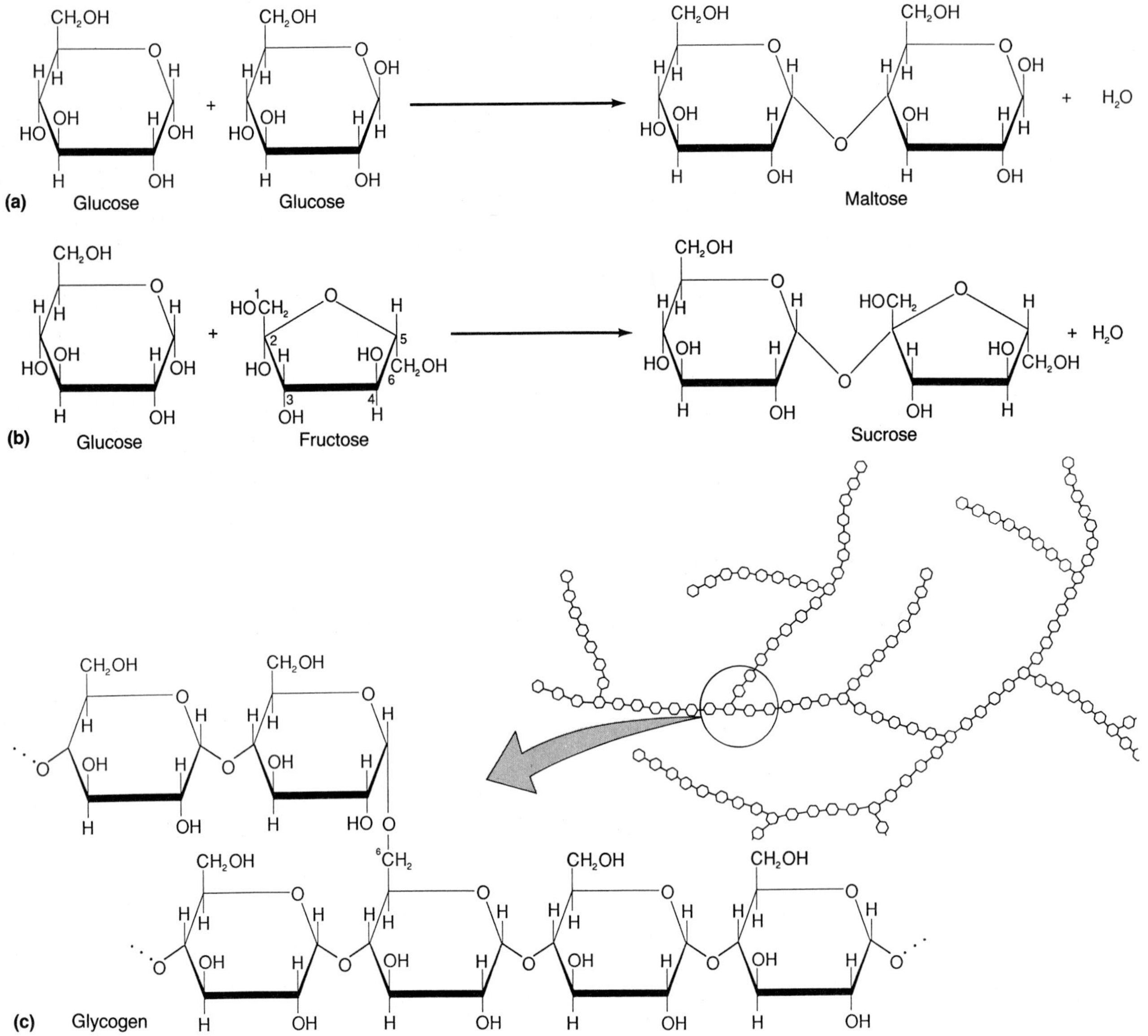

Figure 2.13 Disaccharides and polysaccharides. (a) Maltose and (b) sucrose are disaccharides. (c) Glycogen can be a huge, branched molecule; note that it is composed of glucose monomers. Sugars may have several kinds of reactive groups other than hydroxyl (–OH) groups attached to their carbons; carboxyl (–COOH), sulfate (–OSO$_3^-$), and amino (–NH$_2$) groups are also common.

Nucleic Acids. The structure and function of nucleic acids are discussed in detail in Chapter 15. **Deoxyribonucleic acid (DNA)** is the genetic material containing an animal's hereditary information (see Figure 15.9). DNA is present in all cell nuclei and mitochondria. **Ribonucleic acid (RNA)**, found in every animal cell, acts as a carrier of information between the genes (composed of DNA) and the cellular sites of protein synthesis.

CELL STRUCTURE

Cells, the basic organizational units of life, are remarkably similar throughout the animal kingdom. The same structures found in the liver cells of a lion also appear in the connective tissue cells of a clam and the muscle cells of a fly maggot. Most differences among animals result from differences in the number—not the kinds—of structures present in cells, and from differences in the number of cells making up the animals. The theoretical implication of this fact is that all animals are related through evolution from common ancestors. The practical implication is that facts about the structure and function of typical cells usually apply to all animals.

Cell Membranes

Membranes regulate the passage of materials into and out of cells, and also within cells. The **fluid-mosaic model**, developed in the early 1970s, has withstood considerable testing and is now accepted widely as a representation of biological membranes (Figure 2.14). According to this model, the fundamental membrane structure is a phospholipid bilayer, though the types of lipids forming the bilayer vary from cell to cell and from organism to organism. Because of their hydrophobic nature, lipids are barriers to the passage of many polar sugars and amino acids as well as ions. Proteins are associated with the bilayer, often in a manner that allows them to drift from one area to another within the plane of the membrane itself. Membrane functions are largely the result of activity by associated proteins. **Intrinsic**, or **integral**, **proteins** and

glycoproteins pass completely or partly through the membrane; some act as gates, admitting polar compounds into the cell, while others act as receptor sites for the binding of chemical messengers such as hormones. **Extrinsic**, or **peripheral**, **proteins** occur on only one surface of a membrane. Those on the outside include glycoproteins that function in cell recognition, whereas those on the inside provide anchoring sites for fibers that maintain cellular shape.

Plasmalemma. The membrane surrounding the cell is called the **plasmalemma**, and it is often structurally specialized to perform specific functions. For example, the surface membranes of absorptive cells in the kidneys or intestines of vertebrates are highly folded into fingerlike projections called microvilli that greatly increase the cell's surface area and therefore enhance its capacity to transport materials inward or outward. (Microvilli are discussed in Chapter 4, p. 94; see Figure 4.17.)

Cells that form the surfaces of organs or that occur in the skin are attached to each other by local cell membrane modifications called **junctional complexes** that prevent the cells from separating during stretching. One such complex is the **desmosome**, in effect an adhesive plate (Figure 2.15a). Protein fibers connect the desmosomes of adjacent

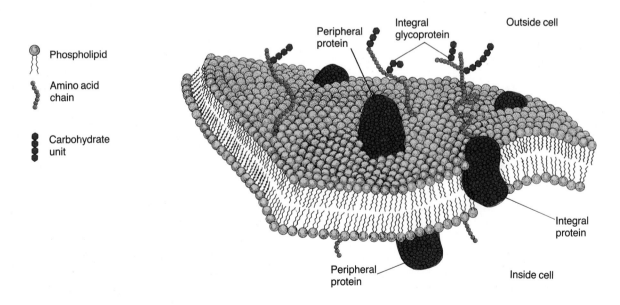

Figure 2.14 The fluid-mosaic model. Cell membranes consist of two layers of phospholipids oriented with their hydrophobic tails toward the membrane's center. Proteins can "float" within the plane of the membrane. Peripheral proteins are exposed on only one side of the membrane, and integral proteins and glycoproteins extend through the membrane.

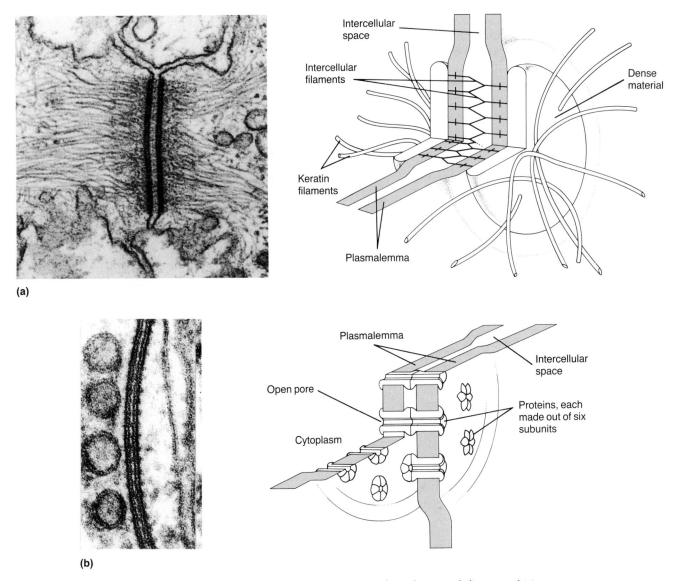

(a)

(b)

Figure 2.15 Intercellular junctions. Electron micrographs and structural diagrams of (a) a desmosome (64,700×) and (b) a gap junction (176,000×).

cells, giving them a cablelike strength. Cells may also be held together by **tight junctions**, where two cell membranes are joined by proteins on their outer surfaces. Other cells are linked by **gap junctions**, which not only hold the cells together but also allow small molecules, such as ions, simple sugars, and amino acids, to pass rapidly between adjacent cells (Figure 2.15b).

Transport Across Cell Membranes. Cells constantly exchange materials with their environment. Water passes in and out, ions are accumu-

lated or secreted, and nutrients are often acquired by various transport mechanisms.

Diffusion is the movement of molecules from a region of relatively high concentration to a region where their concentration is lower. Because of the phospholipid bilayer, fat-soluble materials diffuse through membranes readily, whereas many polar compounds do not. Ionic and most polar compounds (water is an exception) must enter cells by diffusion through channels formed by intrinsic proteins that selectively allow the passage of particular substances. Diffusion through such channels is called *facilitated* diffusion because it is aided by the proteins. The ability of a cell membrane to regulate this traffic is called **selective** or **differential permeability**.

Osmosis, the movement of water across differentially permeable membranes, is a special case of diffusion. In Figure 2.16, a cell membrane is shown having pure water on one side and water containing dissolved proteins on the other. (In a living animal, the concentration of these substances within the cell may or may not be the same as their concentration in the surrounding body fluids or environment.) The proteins cannot pass through the pores of the membrane, although the water molecules can. Because water is at a higher concentration outside the cell than inside, it will diffuse inward until the water concentration is equal on both sides of the membrane or until increasing pressure within the cell creates a backflow that is equal to the osmotic flow. Living cells cannot survive if their internal fluids become diluted. Consequently, a cell in a dilute environment must constantly remove excess water from its fluids. Freshwater protozoa and freshwater sponges constantly gain water from their environment, but their body cells have water expulsion vacuoles that remove excess water.

Neither diffusion nor osmosis requires an animal to expend energy, and because molecules are in constant, heat-dependent motion, they simply cross the cell membrane as they contact it. However, cells can also selectively accumulate materials *against* concentration gradients by a process called **active transport**. Active transport requires specific proteins to be embedded in cell membranes. Such proteins bind with a particular substance on one side of the membrane and then use energy produced by the cell to translocate the substance to the other side. An impressive example of active transport takes place in the cells that line the human stomach. These cells secrete hydrogen ions against a gradient whose relative concentration is one molecule inside the cell to 10 million outside. The hydrochloric acid concentration within the stomach approaches 1 M, but inside the cells it is only about 10^{-7} M.

The Cytoplasm

Membranes not only surround animal cells, but also divide them into internal compartments. The two major compartments are the nucleus, which houses the genetic material, and the **cytoplasm**, where most metabolic events occur. The cytoplasm is further divided into the **cytosol**, which is the nonstructured part of the cytoplasm containing many types of biological molecules in solution, and the **organelles**, which are the organized structures within the cell. Different metabolic activities occur in the cytosol and organelles. Figure 2.17 is a model of an animal cell constructed from studies with the electron microscope; you may want to refer to it as you read the following discussion about cell structure and function.

Mitochondria. **Mitochondria** are spherical to elongate structures found in the cytoplasm of virtually every animal cell. Cells active in synthesis, secretion, or movement tend to have the most mitochondria. Two membranes form the outer boundary of each mitochondrion. The inner membrane is folded to form projections called **cristae** that increase the surface area to which enzymes attach. The **matrix**, or innermost region, of this organelle contains additional enzymes in solution. All these enzymes participate in the process of oxidizing food materials and capturing their chemical energy in forms the cell can use. (Chapter 3 covers this process, especially the role of the energy currency molecule ATP, in more detail.) In animal cells, mitochondria are the only organelles besides the nucleus that contain DNA; when mitochondria

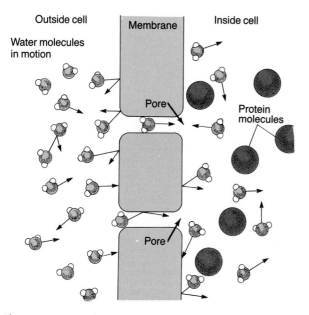

Figure 2.16 Osmosis. Osmosis occurs through differentially permeable membranes that allow water to pass while retaining proteins and many other dissolved substances. In this example, more water molecules diffuse into the cell than out because water is more highly concentrated outside the cell.

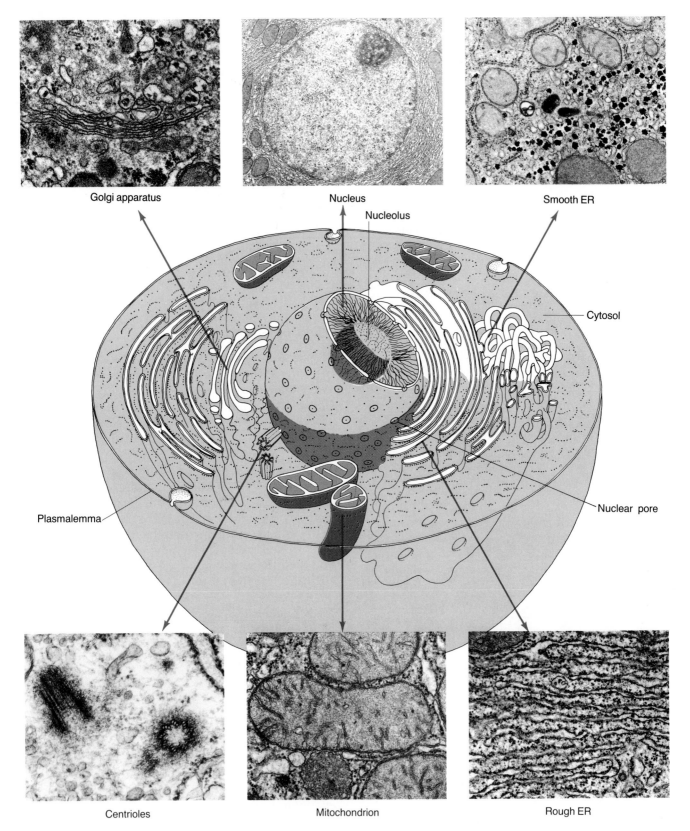

Golgi apparatus

Nucleus

Nucleolus

Smooth ER

Cytosol

Plasmalemma

Nuclear pore

Centrioles

Mitochondrion

Rough ER

Figure 2.17 Diagram of an animal cell. This interpretation of cell ultrastructure has arisen from studies with the transmission electron microscope. In addition to the structures shown, oil droplets, protein crystals, glycogen granules, and various microtubules and microfilaments also occur in the cytoplasm. (Golgi apparatus, 29,000×; nucleus, 6000×; centrioles, 46,000×; mitochondrion, 24,100×; rough ER, 21,300×.)

grow and divide, both nuclear and mitochondrial genetic information are used to make new mitochondrial components.

Endoplasmic Reticulum and Ribosomes. The **endoplasmic reticulum** (**ER**) is an interconnected system of membranes shaped like tubes and flattened sacs that form channels throughout the cytoplasm. There are two types of ER. **Smooth endoplasmic reticulum** (**smooth ER**) consists primarily of tubular membrane arrangements that perform a number of roles, including synthesizing steroid sex hormones in the cells of vertebrate ovaries and testes, detoxifying various drugs and breaking down glycogen during periods of starvation in vertebrate liver cells, and releasing calcium when nerve impulses arrive in the cells of skeletal muscles. The second type of ER is called **rough endoplasmic reticulum** (**rough ER**) because its outer surface is studded with nonmembranous organelles called **ribosomes**, the sites of protein synthesis. Only some of the 10 million or so ribosomes typically present in a cell are attached to membranes. Free ribosomes attach to certain RNA molecules in the cytoplasm, forming clusters called **polysomes**. Ribosomes on the rough ER surface synthesize proteins that will become membrane components or that will be secreted from the cell; free ribosomes make proteins that function within the cell.

Proteins synthesized on the rough ER surface pass through the rough ER membrane into enclosed membranous channels called **cisternae**, which are isolated from the rest of the cytoplasm. By passing into the cisternae, these proteins are separated from those made by free ribosomes. The sequestered proteins may then be chemically modified by enzymes that remove certain amino acids or add sugars. These modified proteins then pass from the rough ER to the Golgi apparatus in small vesicles.

Golgi Apparatus. The **golgi apparatus** is a stack of flattened membranous sacs often located near the nucleus. Protein vesicles from the rough ER pass

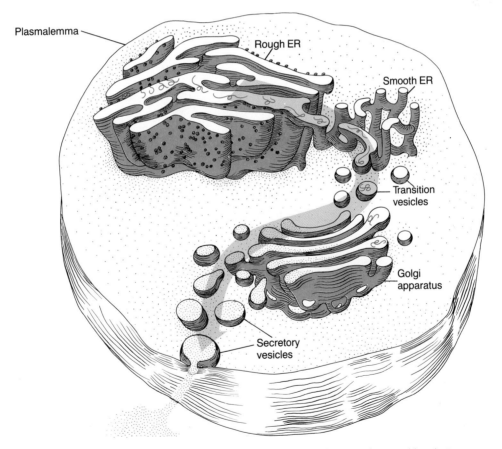

Figure 2.18 Synthesis and secretion of proteins. Protein synthesis can be traced by placing secretory cells, such as those found in the liver or pancreas, in solutions containing radioactively labeled amino acids. Radioactive proteins first appear in the vicinity of the ribosomes and then pass into the endoplasmic reticulum. From here, they travel to the Golgi apparatus; eventually, they appear in small vesicles, which fuse with the plasmalemma and are released (secreted).

into the Golgi, where the proteins are sorted, chemically modified, and packaged in specific types of vesicles (Figure 2.18). Some of the resulting vesicles pass into the cytoplasm. Others, called **secretory vesicles**, pass to the cell surface, fuse with the plasmalemma, and release their contents. Cells active in protein secretion have highly developed systems of rough ER and Golgi membranes.

Lysosomes. **Lysosomes**, produced by the Golgi apparatus, are membrane-bounded organelles containing enzymes that can digest macromolecules. About 40 or so enzymes occur in lysosomes, and all function best at an acidic pH. Lysosomes are common in cells active in digestion and in cells that have ceased to function and are decomposing.

Certain amoeboid cells, including white blood cells, engulf (ingest) materials from the medium that surrounds them by **endocytosis**. Endocytosis of solid materials is called **phagocytosis**, whereas **pinocytosis** is the engulfing of fluids and dissolved molecules. In both processes, receptors on the cell surface may first bind with and concentrate the material to be ingested. The plasmalemma then surrounds the material, forming a **endocytic (food) vacuole**. No digestion occurs in the vacuole until one or more lysosomes fuse with it. The products of digestion that diffuse from the vacuole into the cytosol are then used in metabolism (Figure 2.19).

Programmed cell death is often an essential feature of animal development. For example, when a tadpole changes (metamorphoses) into a frog, the tadpole's tail is resorbed as lysosomes release their enzymes and literally digest the cells from within. Moreover, a continual breakdown of organelles occurs in most cells; in vertebrate liver cells, for

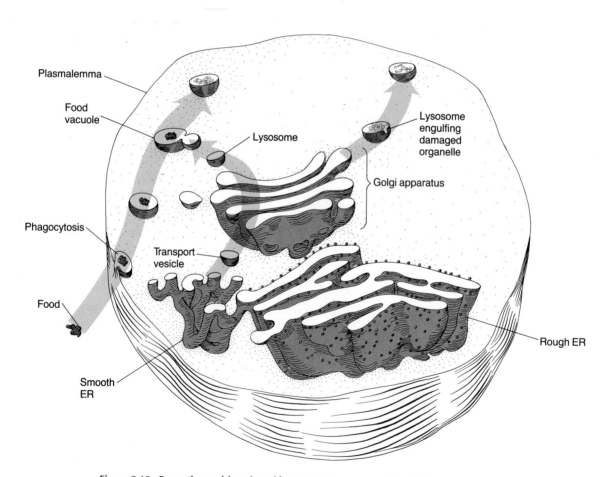

Figure 2.19 Formation and function of lysosomes. Enzymes contained in lysosomes are synthesized on the rough ER and pass to the Golgi apparatus. Lysosomes form as enzyme-filled vesicles that pinch off from the Golgi apparatus. When lysosomes fuse with endocytic (food) vacuoles, the lysosomal enzymes digest the contents of the vacuoles. Lysosomes may also engulf and recycle the molecular components of damaged organelles.

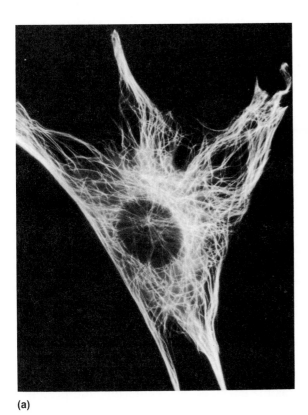

(a)

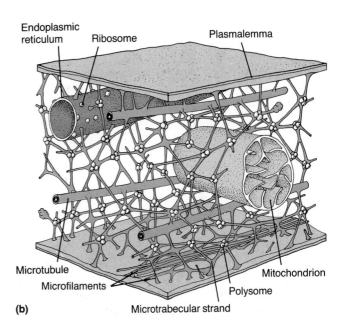

Endoplasmic reticulum Ribosome Plasmalemma

Microtubule Mitochondrion
Microfilaments Polysome
(b) Microtrabecular strand

Figure 2.20 The cytoskeleton. (a) This cell, called a fibroblast, has been stained to show the distribution of microfilaments, microtubules, and the complex network of the microtrabecular lattice in the cytoplasm. (b) Artist's interpretation of the spongelike cytoskeleton.

example, it is estimated that one mitochondrion breaks down about every 10 minutes. Old organelles are removed by a process known as **autophagy** (meaning "self-eating,") and are replaced by newly synthesized ones. Autophagic vacuoles form when an organelle is surrounded by a vacuole with which lysosomes then fuse. Lysosomal enzymes degrade the organelle and recycle its components in the cell. Autophagy is common in the cells of starving animals, which may continue to live for a limited time by consuming their own substance.

The Cytoskeleton. The **cytoskeleton**, consisting of various protein tubules and filaments in the cytoplasm, gives a cell its shape (Figure 2.20). Cytoskeletal elements may also be involved in providing the movement of amoeboid cells, the streaming of cytoplasm that can occur within cells, and the movement of nuclear components during cell division.

Microtubules, which are tiny hollow tubes of the globular protein **tubulin**, form part of the cytoskeleton. Microtubules are assembled near the center of the cell at a site called the **microtubule organizing center** (**MTOC**). The MTOC makes microtubules by polymerizing tubulin subunits. The cell can also disassemble polymerized tubulin; consequently, the cytoskeleton is a dynamic struc-

ture. In animal cells, a pair of organelles called **centrioles** occurs within the MTOC. Each centriole consists of nine sets of three closely pressed microtubules arranged in a circle around a central axis. As discussed later (p. 46), the centrioles seem to play a role in organizing microtubular structures during cell division. **Cilia** and **flagella**, locomotor organelles of cells, are also formed of microtubules. Both of these organelles arise from specialized centrioles called **basal bodies**, which are associated with the cell membrane. The microtubules of cilia and flagella are usually arranged in nine doublets around the periphery plus one pair in the center (Figure 2.21). (The structure and locomotor roles of cilia and flagella are discussed in Chapter 9, p. 197.)

Microfilaments, another type of cytoskeletal element, are long strands of protein, typically **actin** and **myosin**. In muscle cells, filaments of these proteins are arranged in a highly ordered pattern to form the contractile mechanism. (See Chapter 9 on the sliding filament theory of muscle contraction, p. 204.) In most other cell types, microfilaments occur throughout the cytoplasm and may participate in amoeboid movement and cell division.

Another component of the cytoskeleton, the **microtrabecular lattice** appears to be a spongelike

(a)

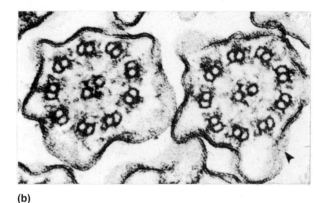

(b)

Figure 2.21 Cilia. (a) Cilia in the trachea of a rat, as seen with a scanning electron microscope (4560×). (b) The internal "9 + 2" microtubular arrangement of cilia can be seen when they are sectioned and viewed with a transmission electron microscope. The function of cilia in locomotion is discussed in Chapter 9.

network of fine fibers that connect microtubules and microfilaments throughout the cytoplasm. The lattice is thought to support organelles and divide the cytosol into a protein-rich region (the lattice itself) and a water-rich region among the fibers of the lattice. The lattice may also provide binding sites for cytosol enzymes, so that enzymes with coordinated functions would be located in one place. In this way, product molecules from one reaction would have to move only a short distance to another site where other enzymes would modify them further.

The chemical nature of the microtrabecular lattice is still not completely known, and some researchers still question its existence. They argue that the lattice appears as a result of the process involved in preparing cells for viewing with the electron microscope. As discussed in Chapter 1, this type of controversy is an essential ingredient in science, stimulating thought and often speeding the rate of discovery.

The Nucleus

Because the **nucleus** contains a cell's hereditary information, every animal cell has one at some

stage in its life cycle. Certain kinds of cells, such as mammalian red blood cells, lose their nuclei as they mature, and as a result, they have only limited repair capabilities and usually die if damaged. Other cells have more than one nucleus. Vertebrate liver cells often have two nuclei, and some tissues consist of multinucleate cells called **syncytia**. A syncytium forms when many cells fuse, the membranes between adjacent cells disintegrate, and the nuclei of the cells persist. Examples of syncytia are skeletal muscle cells and the body coverings of certain sponges and flatworms.

A **nuclear envelope** composed of two membranes forms the outer boundary of the nucleus. About 3000 to 4000 pores in the nuclear envelope regulate the passage of materials back and forth (Figure 2.22). Small molecules such as nucleotides, used in the synthesis of nucleic acids, pass readily into the nucleus, as do molecules that regulate gene activity. Meanwhile, the RNA molecules and ribosomes synthesized in the nucleus pass to the cytoplasm via the pores. The selectivity of the nuclear envelope prevents many other molecules, such as cytoplasmic enzymes, from entering, and prevents DNA from leaving.

Figure 2.22 The nuclear envelope. The surface of a cell nucleus as seen in a freeze-fractured cell with the scanning electron microscope (14,600×). Note that the nuclear envelope consists of inner and outer membranes (A, B), penetrated by many nuclear pores (NP). A mitochondrion (M) can also be seen.

Chromosomes. Inside the nucleus are structural units called **chromosomes**, which contain DNA. During and just prior to cell division, the chromosomes appear as distinctly separate, rodlike structures; when the cell is not dividing, they form extended, threadlike filaments, collectively called **chromatin**. Each chromosome is a single double helix of DNA wound in a regular pattern around globular proteins called **histones** (Figure 2.23). Because they are positively charged, histones bind to the negatively charged DNA, forming repeating structures called **nucleosomes**. Each nucleosome consists of a core of eight aggregated histone molecules with DNA wrapped around the outside. Just as a long thread becomes easier to manage after it is wound around a spool, so DNA is more conveniently packaged when it is coiled around the histones. The effectiveness of this packaging is indicated by the 46 human chromosomes. Each contains 2–8 centimeters (cm) of DNA, for a total length of over 1 meter (m) of DNA in each human cell nucleus—which is only 0.01 millimeter (mm) in diameter!

Nucleoli. One or more **nucleoli** are found in every cell nucleus (see Figure 2.17). Each nucleolus is composed of DNA plus RNA and protein. The first steps in the production of ribosomes occur here, and much of the nucleolus consists of ribosomes in various stages of assembly. Ribosomal subunits pass out of the nucleus and into the cytoplasm, where they join together to form complete ribosomes that can then begin the process of protein synthesis.

THE CELL CYCLE

A cell spends most of its life growing and maintaining its molecules and organelles. Cells that divide must double their contents during this growth period, or they would become progressively smaller over several divisions and eventually lose key components. This growth period is called **interphase**, and it lasts from 10 hours to an animal's lifetime, depending on the type of cell (Figure 2.24).

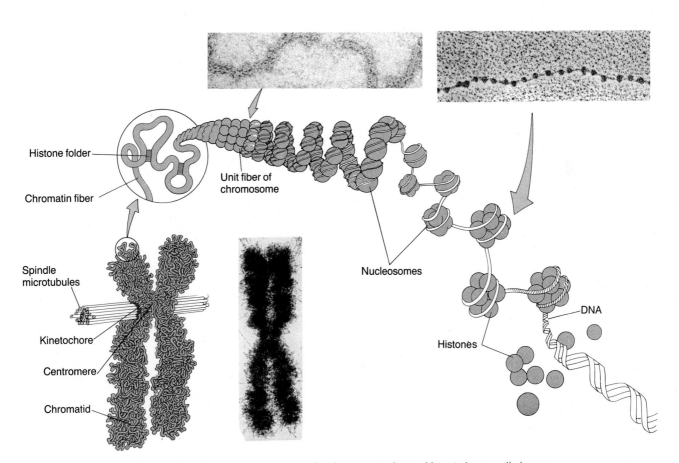

Figure 2.23 Chromosome structure. The chromosome depicted here is from a cell about to divide. A single long molecule of DNA occurs in each of the sister chromatids. The DNA wraps around histone molecules to form units called nucleosomes, which pack together to form the compact chromosome.

Histone folder

Chromatin fiber

Unit fiber of chromosome

Spindle microtubules

Kinetochore

Centromere

Chromatid

Nucleosomes

DNA

Histones

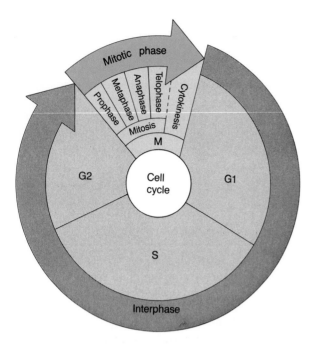

Figure 2.24 Stages of the cell cycle. For cells that regularly divide, the events in their life span are cyclic. DNA synthesis occurs during interphase, as do other growth events. Cells that no longer divide are thought to be physiologically equivalent to cells in early G1. In a typical cell cycle, mitosis may last about 2 hours. The G1 phase varies in length; S phase generally lasts about 8–10 hours, and G2 about 2–4 hours.

Because DNA is the hereditary material, its duplication is a critical interphase activity for potentially dividing cells. In early interphase, called the **G1** ("gap 1") **phase** of the cell cycle, no DNA is made in most cells. Then suddenly, as if on cue, DNA synthesis begins at several sites on each chromosome and continues for several hours, stopping when a copy of each chromosome has been made (DNA replication is discussed in Chapter 15, p. 362). This period of DNA synthesis is known as the **S phase**, and although highly variable, it typically lasts about 8–10 hours. When the S phase is over, each chromosome consists of two identical DNA molecules and associated proteins organized into two **sister chromatids**. Sister chromatids, each composed of a DNA molecule, are attached to each other by a **centromere**, a region of DNA that the chromatids share. Following the S phase, there is another gap period, the **G2 phase**, in which proteins that have a role in cell division are produced. Some of these proteins cause the chromosomes to condense into compact units, while others are involved in the breakdown of the nuclear envelope or in spindle formation.

The division period in the cell cycle consists of two processes: **mitosis**, or nuclear division, followed by **cytokinesis**, or cytoplasmic division. These processes occur in animals as they develop from fertilized eggs and as they replace worn out tissues. In the ovaries and testes, cells that will become eggs and sperm undergo a different form of nuclear division called **meiosis**. (This process, which reduces a cell's chromosome number by half in gamete formation, is covered in Chapter 14, p. 333.)

Mitosis

Cell biologists divide mitosis into four stages, defined by the shape and location of the chromosomes (Figure 2.25). During **prophase**, the first stage, the extended chromatin fibers gradually compact by coiling into well-defined chromosomes. At this point, each chromosome consists of two sister chromatids joined at their common centromere. As the chromosomes condense, the nucleolus breaks down and ceases to produce ribosomes, and the nuclear envelope and the endoplasmic reticulum fragment into small membranous vesicles that scatter throughout the cytoplasm. At the same time, the centrioles migrate to opposite ends, or **poles**, of the cell. The cytoskeletal microtubules dissociate into tubulin subunits and then reassemble into microtubules radiating from the newly positioned centrioles, forming a new structure called the mitotic **spindle**. Spindle microtubules extend into the disintegrating nucleus. On each centromere, proteins form a platelike structure, the **kinetochore**, which links the spindle fibers to the chromatids. The chromosomes are aligned along a plane halfway between the two poles of the spindle under the influence of the spindle fibers radiating from opposite poles. **Metaphase** is the mitotic stage at which this alignment is completed. In many animal cells, mitosis lasts for 1–2 hours, of which prophase and metaphase take about 45 minutes.

Anaphase begins when the spindle microtubules begin pulling the sister chromatids toward opposite poles of the cell, and the cell elongates as its poles move farther apart. This stage of mitosis lasts only a few minutes. **Telophase** begins when the chromatids arrive at the poles. At this time, new nuclear membranes begin to form, and the chromatids—now called daughter chromosomes—begin to uncoil and form chromatin. Thus, new nuclei form, and at the end of telophase, nucleoli appear in the new nuclei, and the spindle breaks down, marking the end of mitosis.

What has occurred in this process? One nucleus

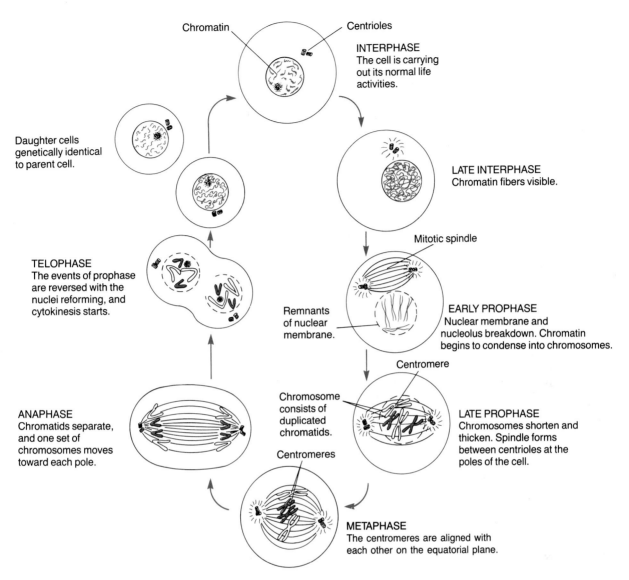

Chromatin

Centrioles

INTERPHASE
The cell is carrying out its normal life activities.

Daughter cells genetically identical to parent cell.

LATE INTERPHASE
Chromatin fibers visible.

Mitotic spindle

TELOPHASE
The events of prophase are reversed with the nuclei reforming, and cytokinesis starts.

Remnants of nuclear membrane.

EARLY PROPHASE
Nuclear membrane and nucleolus breakdown. Chromatin begins to condense into chromosomes.

Centromere

ANAPHASE
Chromatids separate, and one set of chromosomes moves toward each pole.

Chromosome consists of duplicated chromatids.

LATE PROPHASE
Chromosomes shorten and thicken. Spindle forms between centrioles at the poles of the cell.

Centromeres

METAPHASE
The centromeres are aligned with each other on the equatorial plane.

Figure 2.25 Mitosis and cytokinesis. Cell division, including mitosis (nuclear division) and cytokinesis (cytoplasmic division), is a continuous process, but biologists recognize several discrete stages as a convenience for studying the overall event.

has produced two nuclei, each of which contains a complete set of chromosomes. The cytoplasm of the cell now divides in such a way that each new cell will contain one nucleus and a complete complement of the cytoplasmic contents.

Cytokinesis

Cytoplasmic division occurs at the end of telophase. The plane of division is at right angles to the long axis of the spindle and passes through the middle of the spindle. A ring of contractile protein filaments (actin and myosin) encircles the cell midway between the poles and is attached to the inside of the cell membrane. The filaments slide past one another, drawing the cell membrane inward to form a furrow. Inward movement continues until the furrow contacts the spindle fibers, often making the cell look like an hourglass with a nucleus located at each end. The thin midbody then breaks, yielding two independent offspring cells.

The Importance of Cell Division

Some animals, such as crustaceans and fishes, increase in size throughout their lives, whereas most animals achieve a mature size and then stop growing. Cessation of body growth, however, does not mean that most cells cease to divide or that replacement of damaged molecules and organelles comes to a halt. Blood-forming tissues, the tissues that line the digestive system, and the skin of many animals are constantly being replaced through cell division. In fact, it is estimated that of the 60 trillion cells in a human adult, several million cells divide each second. (Other highly specialized cells, such as those found in vertebrate muscle or in the nervous system, rarely divide in adult organisms.)

Essay: How Cell Parts Are Isolated for Study

Much of what we know about cell function has come from research on isolated cellular components. The study of processes occurring in cells is sometimes greatly simplified when a component can be studied independently of other cellular molecules and organelles. Three separation techniques often used in cell and molecular biology are centrifugation, chromatography, and electrophoresis.

Centrifugation has been used to isolate cellular organelles since the 1940s. In *differential* centrifugation, a slurry of broken cells is placed in a plastic tube and spun at different speeds in a rotor for varying periods of time, from a few minutes up to 24 hours (Figure E1). As the speed of rotation increases, centrifugal force (measured as the force of gravity in g units) away from the center of rotation increases. Dense organelles such as nuclei quickly settle to the bottom of the tube and form a sediment (pellet), but lighter organelles such as mitochondria stay in suspension or settle on top of the nuclei. The effects are cumulative, so that rotation at 20,000 g for 10 minutes yields the same result as rotation at 2,000 g for 100 minutes.

Differential centrifugation alone rarely yields pure organelle preparations because other particles are trapped by the settling nuclei or by some other relatively heavy material. The purity of an organelle sample is improved by using *density gradient–equilibrium sedimentation* techniques. Instead of the centrifuge tube being filled only with a cell slurry, about two-thirds of the tube contains layers of sucrose solutions differing in concentration and density.

When a slurry of broken cells is added to the top of this gradient and the tube is spun in a centrifuge, cell organelles settle through the sucrose layers until they encounter a layer with a density equal to their own. Since few organelles have the same density, each layer represents a relatively pure fraction of a single type of organelle that can be collected for further study.

Small molecules such as sugars, amino acids, and nucleotides cannot be separated by centrifugation because they are in true solution in the cytosol. However, such molecules can be isolated by **chromatography**, a technique originally used to separate plant pigments. Molecules can be separated chromatographically because they have different polarities and thus different solubilities and charge attractions. Chromatographic systems contain a mobile phase that flows over a stationary phase. The mobile phase may be a gas, an aqueous solution of ions, or an organic solvent. The stationary phase can be a piece of coarse filter paper, a mass of polysaccharide beads, or minute glass beads perforated by hundreds of pores and channels. Frequently, the stationary phase is placed in a long tube, and the mobile phase flows down through the column—hence the name *column* chromatography. Figure E2 shows a mixture of neutral and charged molecules being added to a chromatographic column containing stationary polysaccharide beads with a positive surface charge. As the mixture passes through the column, negatively charged molecules are retained while all others flush through. The

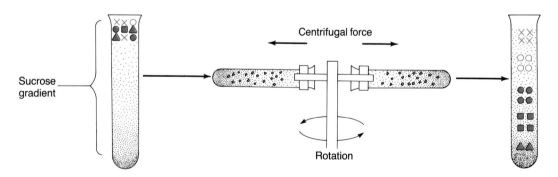

Figure E1 Centrifugation. When suspensions of particles are rotated at high speeds in a centrifuge, the particles can be separated by their speed of sedimentation or by their density.

Control of Cell Division

Cell division in tissues that are growing and in those that continually renew their cell populations must be controlled so that cells are produced at certain rates. For growing tissue, cell division must exceed cell loss, but growth must be consistent with the animal's overall growth rate. For cell renewal, cells must be produced at the same rate that they are lost.

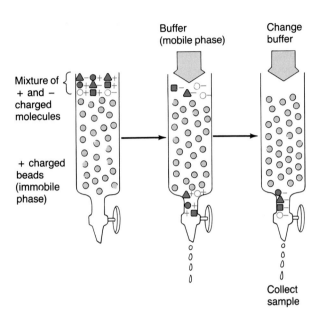

Figure E2 Column chromatography. In this procedure, molecules can be separated from complex mixtures by their attraction to charges on the column matrix. Changing the buffer can lead to loss of charge on the matrix or the retained molecules, allowing the molecules to be harvested.

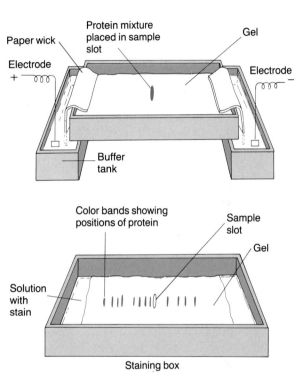

Figure E3 Electrophoresis. This technique is used to separate one type of protein or nucleic acid from another. When placed in an electrical field, large charged molecules will migrate toward one of the electrodes at a rate that is directly proportional to the charges on the molecules and indirectly proportional to their molecular weights. In many electrophoretic setups, the gel is poured as a density gradient and acts as a sieve to aid in separation.

retained molecules can be harvested by passing a buffer that causes the molecules to lose their charge through the column. By experimenting with different types of stationary and mobile phases, systems can be designed to separate virtually any kind of molecule from an extract of cells or biological fluids.

A third separation technique, **electrophoresis**, is often used to isolate and purify electrically charged molecules, especially proteins and nucleic acid. This technique separates molecules based on two properties: molecular weight and electrical charge. A typical electrophoretic apparatus is shown in Figure E3. Any charged molecules introduced into the electrical field between the electrodes will migrate toward the electrode having an opposite charge. A slab of polysaccharide gel (similar in consistency to gelatin) is placed between the electrodes so that the charged molecules must migrate through the gel. Large molecules will move more slowly than smaller ones. A complex mixture of proteins, for example, can be separated into its components, because different proteins usually have quite different charges and sizes. The gel is then stained to show the location of different molecules, and the molecules can be extracted for further testing.

To prepare very pure materials, a combination of separation techniques is often used. For example, to obtain nuclear proteins, nuclei are first isolated from a large number of cells by centrifugation. The nuclei are then chemically extracted, and the extract is passed through a column to separate proteins from nucleic acids. Finally, the protein mixture is run through electrophoresis, and the proteins are differentiated by staining the gel.

Control of division often involves communication among cells. Many cells produce **chalones**, chemicals that inhibit division in other cells of the same type when a certain population density (and hence inhibitor concentration) is reached. If the cell number decreases as a result of either injury or normal wear and tear, then the chalone concentration decreases, releasing the cells from inhibition and allowing them to divide. Other substances, called **growth factors**, directly stimulate cell division. Growth factors generally affect only the cells of their target tissues. Finally, cell division in some tissues may be inhibited by contact with neighboring cells, a phenomenon called **contact inhibition**. Cancer cells are not sensitive to contact inhibition, even though the cell types from which they are derived are sensitive to this check on growth. The control of cell division is one of the most active and interesting areas of research in cell biology today.

ORGANIZATION BEYOND THE CELLULAR LEVEL

Tissues

Most animals are made up of an assortment of **tissues**, which are highly organized groups of cells adapted to perform specific functions. Individual cells arise by mitosis, and each has a complete set of hereditary information in its nucleus. However, cells destined to become different tissues express different parts of that information and thus become specialized in form and function. The four basic types of tissues found in animals are epithelial, connective, muscle, and nervous tissues. **Histology**, the study of tissues, has revealed many subcategories of these four basic types. Mammals, for exam-

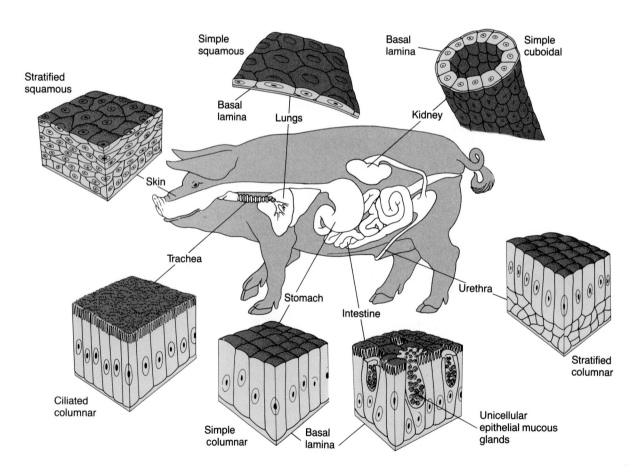

Figure 2.26 Epithelial tissues. Epithelial tissues line various cavities and tubular structures in animals.

ple, have about 200 tissue subtypes. In this section, we describe a few tissue types to show how animals are organized beyond the cellular level.

Epithelial Tissue. Various types of **epithelial tissue** cover external body surfaces and line internal cavities and tubular structures. Epithelial cells also form glands within the skin and digestive tract and make up certain hormone-secreting glands. An epithelium frequently consists of only a single cell layer, but it can also be composed of several layers, forming a **stratified** epithelium. Epithelial cells may be **squamous** (flat), **cuboidal**, or **columnar** in shape (Figure 2.26). A **basal lamina** usually occurs at the interface between epithelia and other types of tissue. Basal laminae consist of thickened regions of the epithelial cell plasmalemmas and of fine fibers (often of the protein collagen) in a gelatinous noncellular material secreted by the epithelial cells. Junctional complexes and basal laminae provide structural support for epithelia. Depending on the type of tissue, the laminae may also act as molecular or cell filters, as organizing sites where epithelial layers can regenerate after injury, and as pathways through which amoeboid cells can migrate between tissues. Because they are exposed to surface and internal wear, epithelial cells are constantly eroded and sloughed off. Thus, epithelial tissues are maintained by cell division.

Epithelial tissues have many functions. Ciliated epithelial cells lining internal passageways or forming external coverings help transport materials over the epithelial surface. Secretory epithelia forming the skin of flatworms, earthworms, and snails produce mucous lubricants that ease movement over and through dry surroundings. In other animals, such as jellyfish and hydras, secretory cells produce adhesive mucoproteins (see p. 35) that aid in capturing prey and in attaching the animal to substrates. In nematodes, annelids, and arthropods, secretory epithelial cells produce the nonliving cuticle that surrounds the organism. Some epithelial cells secrete irritants or toxins that protect animals from predators. In land vertebrates, epithelial cells on the body surface produce the protein **keratin**, which accumulates in scalelike cells and reduces water loss. The reproductive tissue that forms eggs in the ovaries and sperm in the testes is a special kind of epithelial tissue called germinal epithelium. (The structure and function of ovaries and testes are described in Chapter 14, p. 330.)

Connective Tissue. **Connective tissue** consists of either fixed or wandering cells embedded in material called **extracellular matrix**. Connective tissue cells called **fibroblasts** secrete much of the extracellular matrix, which consists of fibrous proteins (usually collagen) surrounded by proteoglycan molecules (cell glues) and other materials, depending on the specific type of tissue. The extracellular matrix provides a supportive framework and a medium of communication for the cells. Connective tissues support animal bodies by binding other tissues in place and giving form to organs. They also store fat, produce blood cells, and function in digestion and excretion in some invertebrates. Connective tissues are often the source cells for the development of muscle and skeletal elements in animals that regenerate lost body parts.

Connective tissues are classified according to the nature of their extracellular matrix (Figure 2.27). **Loose connective tissue** consists of separated, mobile cells (mostly fibroblasts), collagen fibers, and watery, gel-like proteoglycans. Loose connective tissue gives shape to most organs and strengthens the skin in vertebrates. Organs that stretch, such as the vertebrate skin, lungs, and urinary bladder, contain loose connective tissue with large amounts of the elastic, fibrous protein **elastin**. **Adipose tissue** (fat) is a specialized type of loose connective tissue in which cells store triglycerides as large droplets in their cytoplasm. In vertebrates, adipose tissue is found beneath the skin, in the spaces between muscle cells, and on the surfaces of various internal organs.

Dense **fibrous connective tissue** consists of many closely packed parallel arrays of collagen fibers, forming tendons or ligaments. **Rigid connective tissues** include the cartilage and bone of vertebrates, in which structural proteins (in cartilage) or calcium salts (in bone) are deposited in the extracellular matrix. (See Chapter 8 for more on the structure of skeletal tissues, p. 188.) Blood is a specialized type of connective tissue in which the matrix is the fluid plasma. (Table 6.1 lists the composition of human blood, p. 137.)

Muscle Tissue. Muscle cells, called **muscle fibers**, contain large amounts of the contractile proteins actin and myosin arranged as parallel filaments in the cytoplasm. These proteins allow muscle fibers to contract and exert force on points to which they are attached. (The structure and function of muscle tissue are covered in Chapter 9, p. 200.) Two general types of muscle tissue are **striated muscle**,

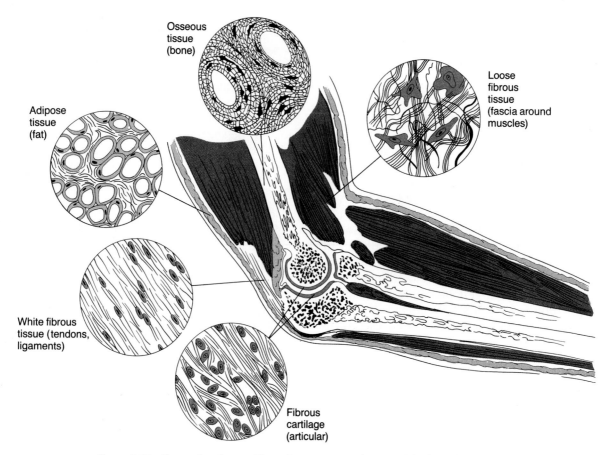

Osseous
tissue
(bone)

Loose
fibrous
tissue
(fascia around
muscles)

Adipose
tissue
(fat)

White fibrous
tissue (tendons,
ligaments)

Fibrous
cartilage
(articular)

Figure 2.27 Connective tissues. These tissues support the animal body and its organs. Some of the connective tissues in a mammalian forelimb are shown here.

whose fibers have alternating transverse stripes (striations) visible under a microscope, and **smooth muscle**, which is not striped. Both types produce locomotor movements in many invertebrates. In vertebrates, however, only striated muscle is involved in locomotion; smooth muscle moves materials through the digestive and reproductive tracts and through blood vessels. There are two types of striated muscle in vertebrates. **Skeletal muscle**, consisting of multinucleate fibers (actually syncytia), facilitates body movements and locomotion. **Cardiac muscle**, as its name implies, is found in the heart; it consists of uninucleate striated fibers.

Nervous Tissue. Two types of cells characterize nervous tissue. Nerve cells, or **neurons**, are specialized for communicating information along cytoplasmic processes called **axons** and **dendrites** (collectively called **neurites**). Information travels along an axon or dendrite as a momentary electrochemical change (nerve impulse) caused by the movement of sodium and potassium ions across the cell membrane. Nerve impulses pass from one neuron

to another when chemical transmitters are secreted into intercellular gaps called **synapses**. Some neurons also secrete hormones. A second type of cell in nervous tissue, the **glial cell**, nourishes the neurons. Specialized glial cells often surround and insulate axons and dendrites, promoting faster impulse transmission. (The structure and ◁ function of neurons and glial cells are covered in Chapter 10, p. 220.)

Organs and Organ Systems

In most animals, **organs** consisting of several types of tissues are adapted to perform specific functions. The vertebrate heart, for example, consists of epithelial, connective, muscle, and nervous tissues organized as an efficient, durable pump. Epithelial tissue lining the heart chambers prevents leakage and provides a smooth, moist surface over which blood can pass with little friction. Connective tissue gives the heart form and strengthens its walls and valves. Nerve cells direct the rhythmic contraction of the cardiac muscle. The heart, like all organs, represents a higher level of organization than the tissues composing it, and none of its component tissues alone can perform the heart's pumping function.

In most animals, specialized organs work together as parts of a yet higher level of organization, the **organ system**. The circulation of blood throughout the vertebrate body, for example, could not be accomplished solely by the heart. As part of the circulatory system, the heart is physiologically and anatomically dependent on the blood vessels. Other organ systems include the respiratory (gas exchange), digestive, and excretory systems. All these systems must work together to maintain homeostasis in an animal. The skeletal and muscular systems provide support, protection, and movement. Making the animal more than the sum of its parts, the nervous and endocrine (hormone) systems coordinate the activity of other body systems. How organ systems function in various animals is the subject of the remaining chapters in this unit.

☐ SUMMARY

1. Elements are composed of atoms, the smallest quantity of an element having that element's properties. Atoms are composed of protons, electrons, and neutrons. An atom's electrons, especially those in its outer shell, determine its chemical properties.

2. Ionic chemical bonds are formed when atoms donate or receive electrons from atoms of another element. Covalent bonds are formed when two or more atoms share electrons. Carbon, nitrogen, hydrogen, oxygen, and phosphorus readily form covalent bonds.

3. Water is the most abundant substance in cells. Its molecules are polar and readily form hydrogen bonds with one another.

4. Cells contain four main types of large (macro-) molecules. *Proteins*, large polymers of amino acids united by peptide bonds, serve as enzymes and structural components. *Lipids* include important membrane components (phospholipids) and energy reserve molecules. *Carbohydrates* are energy sources and starting materials in the synthesis of other cellular compounds. Glycogen is a storage polymer of the simple sugar glucose. Sugars are often bonded to proteins to form glycoproteins. DNA, the hereditary material, and RNA, which functions in protein synthesis, are *nucleic acids*.

5. Cell membranes consist of two layers of phospholipids and embedded proteins. Materials pass into and out of cells by diffusion directly across the cell membranes or through protein transport channels. Osmosis is the movement of water across a differentially permeable membrane. Using active transport, cells can move substances in or out against concentration gradients.

6. Cells contain cytoplasmic organelles, including mitochondria, where cellular respiration occurs; the endoplasmic reticulum, which assembles membranes and synthesizes proteins; the Golgi apparatus, which packages proteins; lysosomes, where materials ingested by cells are digested; and ribosomes, which are involved in protein synthesis. Microtubules, microfilaments, and the microtrabecular lattice form the cytoskeleton that gives cells shape and allows movement.

7. The nucleus is bounded by a double-membraned, porous nuclear envelope. Chromosomes in the nucleus contain genetic information as DNA. In cell division, the nucleus divides by mitosis after the DNA has replicated. Prophase, metaphase, anaphase, and telophase are the four stages of mitosis. Cytoplasmic division (cytokinesis) follows mitosis.

8. Animals have four basic tissue types. Epithelial tissues cover surfaces and line cavities, forming absorptive and secretory surfaces or barriers. Connective tissues give form to organs and form part of the skeleton. Muscle tissues contract and provide locomotion. Nervous tissues are specialized for communication.

9. Most animals consist of cells organized into tissues, tissues into organs, and organs into organ systems.

☐ FURTHER READING

Alberts, B., D. Bray, J. Lewis, M. Raff, K. Roberts, and J. D. Watson. *Molecular Biology of the Cell.* New York: Garland, 1983. *An exceptional cell biology textbook that thoroughly explains the structure and function of cellular organelles and molecules.*

Bretscher, M. S. "The Molecules of the Cell Membrane." *Scientific American* 253(1985): 100–109. *Describes the structure of the cell membrane and how it regulates the passage of materials.*

Dustin, D. "Microtubules." *Scientific American* 243(1981): 66–76. *Describes the synthesis and function of microtubules in several types of cells.*

Rothman, J. E. "The Compartmental Organization of the Golgi Apparatus." *Scientific American* 253(1985): 74–89. *Describes how the Golgi apparatus modifies proteins, sorts them, and packages them for secretion.*

Weber, K., and M. Osborn. "The Molecules of the Cell Matrix." *Scientific American* 253(1985): 110–121. *Explains how the various proteins that give form to a cell are studied.*

Weinberg, R. A. "The Molecules of Life." *Scientific American* 253(1985): 48–57. *An introduction to some of the remarkable recent findings of molecular biology.*

Bioenergetics and Metabolism

Animals use energy in many different ways. These leaf cutting ants (*Atta cephaloides*) are expending energy as they transport pieces of leaves to their nest.

The cells that make up an animal must obtain energy to synthesize molecules, grow, reproduce, and perform other vital activities. Thus, to understand animals, one needs some knowledge of **bioenergetics**, the use and transformation of energy by living systems, and **metabolism**, the chemical reactions that occur in cells. This chapter emphasizes the interrelation of bioenergetics and metabolism. We look at bioenergetics from the standpoint of what forms energy takes, whether it is required or released in a process, and how it is used physiologically. We discuss metabolism in terms of what types of molecules are necessary to start chemical events and what types of molecules are produced in selected metabolic reactions.

As you read this chapter, it might be helpful to continually ask yourself these questions: How is energy involved in the metabolic process being discussed? What are the starting materials and products of the metabolic reactions? How does a metabolic process benefit the animal, allow it to perform certain tasks, and survive in a particular environment? Answers to these questions address zoology at its most basic level.

WHAT IS ENERGY?

Energy, like the concepts of space, mass, and time, does not lend itself to simple definition. According to the classic definition, **energy** is the capacity to do work; work, in this context, means moving matter against a force that opposes movement. Energy may be either kinetic or potential, and the familiar types of energy—mechanical, electrical, chemical, solar, atomic, and heat—can take either of these

forms. **Kinetic energy** is the energy of motion, as when a horse canters across a field or a sodium ion moves across a cell membrane. Energy in this form is performing actual work. **Potential energy** is stored energy with the *capacity* to do work. When an osprey perches high in a tree by a lake, energy is stored in proportion to the bird's mass and its height above ground level. If the osprey launches from its perch and plummets to the lake to seize a fish, the stored potential energy becomes kinetic energy. Potential energy can also be stored in the chemical bonds of molecules and released as kinetic energy when the bonds are broken in chemical reactions. Bioenergetics, at the molecular level, is the study of the making and breaking of chemical bonds.

Biologists often measure energy in units called calories (cal), although the joule (J) is the Standard International Unit (1 cal = 4.17 J). A **calorie** is the

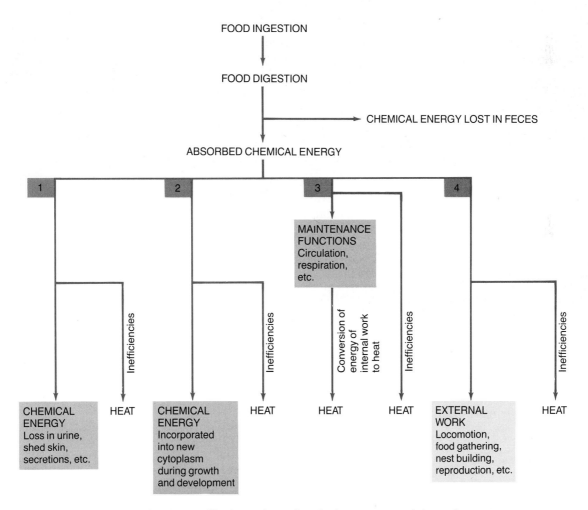

Figure 3.1 Major energy utilization pathways in animals. Energy expended in pathways 1 through 3 represents the basal metabolic rate of an animal. An active animal expends this energy plus whatever more is necessary for its activities. Note that some energy is lost as heat in every pathway.

amount of energy required to raise the temperature of 1 g of water 1 degree on the Celsius scale (by definition, from 14.5°C to 15.5°C). The kilocalorie (1 kcal = 1000 cal), also called a **Calorie** (with a capital C), is the unit used to express the energy value of foods. Units of energy are defined on the basis of heat energy because all forms of energy can be completely converted to heat.

The amount of work that animals must perform to stay alive determines their energy requirements. Figure 3.1 shows how an animal's energy intake can be directed into four work (energy) pathways. The energy required in the first three pathways is that needed for **basal metabolism**, that is, the energy supporting the minimum functions necessary to sustain life in a resting condition. The fourth pathway represents the amount of energy available for activities such as food gathering, reproduction, or migration. All four pathways characterize the active animal over long time periods.

THERMODYNAMICS

Energy flow is the movement of energy from one location to another or the conversion of energy from one form to another. The study of energy flow in physical or biological systems is called **thermodynamics**. Energy flow occurs when one animal preys on another or when molecules are transported in the circulatory system from one location to another within the same animal. Flow also occurs when potential energy is converted to kinetic energy during muscle contraction or nerve impulse conduction.

Thermodynamics deals with the exchange of energy between a system and its surroundings. A biological system can be an entity such as a single chemical reaction, a mitochondrion, a cell, a whole animal, or an ecosystem. The surroundings are everything in the universe other than the system.

The Laws of Thermodynamics

The **first law of thermodynamics**, also called the law of conservation of energy, states that energy can neither be created nor destroyed. Energy can move between locations or be converted from one form to another, but when a decrease in energy of one type occurs, an equivalent increase in some other type of energy must take place. This law tells us that when we examine a biological system, whether it be a food chain in a mountain lake or the

events of carbohydrate metabolism in a cell, all energy can be accounted for. The first law refers only to energy conservation, and it gives no information about how or where energy will move during a conversion or whether the energy will do work.

The **second law of thermodynamics** states that changes in energy always proceed in a direction that dissipates energy concentrated in one location. Energy movement tends to continue until the energy is uniformly distributed, that is, until an equilibrium is reached. A turtle basking in the sun accumulates heat energy. When it slides into cool pond water, the turtle will lose heat, not continue to gain it, and will cool only to the surrounding water temperature, not to a lower temperature. Likewise, when a chemical compound with a high energy content participates in a reaction, some of the potential energy in its chemical bonds will dissipate.

The second law tells us that the efficiency with which energy flows from an energy-releasing event to an energy-requiring one is always less than 100%. Some energy always escapes from a system as random molecular motion. The amount of energy lost depends on the temperature of the system and its surroundings when an energy transfer occurs. When a chemical reaction occurs, the energy it releases can take one of two forms: **free energy**, the energy available to do work, or **entropy**, the energy that dissipates as heat into the surroundings and is not available to do work. In the formal notation of thermodynamics, the relationship between entropy and free energy in a chemical reaction is expressed as

$$\Delta G = \Delta H - T\Delta S$$

where Δ ("delta") represents a change in the factor following it, G is the free energy that can perform work (in units of calories per mole), H is the energy of reaction (also in calories per mole), calculated as the potential energy of the products minus the potential energy of the reactants, T is the absolute temperature (in degrees Kelvin, where K = °C + 273), and S is entropy, the amount of energy dissipated throughout the universe (in calories per mole per degrees Kelvin).

In chemical reactions, there is an energy change whenever electrons change orbitals or chemical bonds are rearranged. The ΔG specifies the amount of energy released or absorbed when a chemical reaction occurs. A positive ΔG means that energy is absorbed from the surroundings and stored in the newly formed chemical bonds. Because the energy of the reaction products is greater than that of the reactants, energy must be supplied for the reaction to occur. Reactions requiring energy are said to be **endergonic**. A negative ΔG means that energy stored in the chemical bonds of the reactants is released. Reactions that release energy are **exergonic**. Because the energy of their products is less than that of the reactants, exergonic reactions proceed without energy input. For work to be obtained from an exergonic reaction, the reaction must be linked to a process that requires energy. If no linking mechanism is present to transfer the released energy, the exergonic reaction still proceeds, but its work potential is lost as heat. Heat produced in this way increases the kinetic energy of molecules in the immediate vicinity of the reaction. This kinetic movement is transferred to surrounding molecules by elastic collisions, thereby spreading gradually throughout the universe.

The implications of the second law of thermodynamics are far-reaching. Since all life depends on interactions of energy, and each interaction results in a loss of free energy, life on earth would cease if energy supplies were not continuously renewed. Only because photosynthetic microorganisms and plants continuously capture energy from the thermonuclear reactions of the sun can most biological systems continue operating. (Ocean vent communities are an exception; see the Essay on this subject in Chapter 25, p. 606.)

CHEMICAL REACTIONS IN LIVING CELLS

Several thousand different chemical reactions can occur simultaneously in a cell, and with up to about 36 million reactant molecules being converted to products each minute, it is difficult to imagine how the correct reactions take place. As you will see, a highly sophisticated system has evolved that ensures that the proper reactions occur.

Types of Chemical Reactions in Cells

Despite the great numbers of reactions that occur in cells, there are relatively few general types of reactions. Among the most important are the **oxidation-reduction** reactions, also called **redox** reactions. A molecule is *oxidized* when it loses an electron and *reduced* when it gains an electron. As one molecule is oxidized, another one must be reduced. Cells acquire energy by oxidizing organic food molecules while simultaneously reducing other molecules. As the term *oxidation* implies, oxygen often acts as an electron acceptor (an oxidizing agent). However, oxygen is not required for oxidation to occur; other molecules that can accept electrons more commonly serve as oxidizing agents in living cells.

One very common type of reaction in animal cells is **dehydrogenation**, the removal of hydrogen. This is actually an oxidation reaction, because when a hydrogen atom, which consists of one proton and one electron, is removed from one compound, it must be accepted by another, which is thus reduced. For example, when methane (a component of natural gas) burns, carbon is oxidized and oxygen is reduced because the oxygen serves as the hydrogen acceptor, as shown in the following equation:

$$CH_4 + 2\,O_2 \longrightarrow CO_2 + 2\,H_2O$$

In addition to redox reactions, three other kinds of chemical reactions frequently occur in animal cells. When food is digested, hydrolytic reactions break down large molecules such as proteins and starches into their component monomers. In **hydrolysis**, water is added to a molecule at a particular place, such as the peptide bond linking two amino acids. In the resulting rearrangement of chemical bonds, the molecule is cleaved into two smaller molecules (Figure 3.2). Whereas hydrolysis breaks molecules down, cells use an opposite type of reaction known as **condensation** (or dehydration) to synthesize large molecules from smaller ones. In

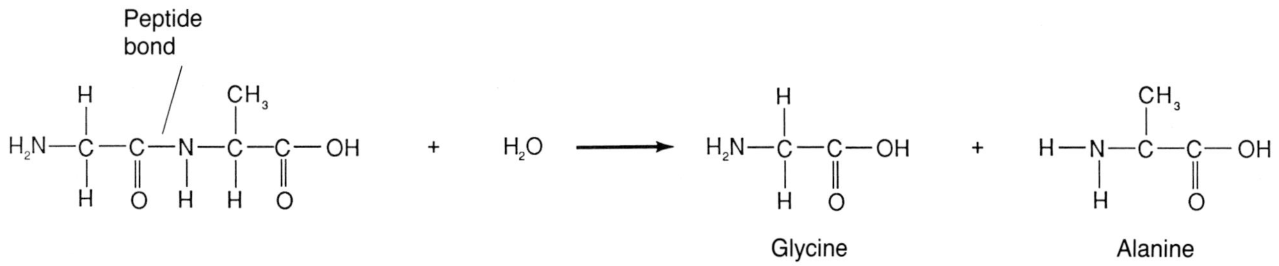

Figure 3.2 A hydrolytic reaction. A dipeptide is shown here reacting with water. When the peptide bond is broken, a hydrogen atom from water attaches to one amino acid (in this case, alanine) and an —OH group is added to the other amino acid (glycine).

condensation reactions, water molecules are produced as small molecules join to form larger ones. In another common type of chemical reaction called **isomerization**, the structure of a complex organic molecule is rearranged so that new internal chemical bonds are formed without the loss or gain of any atoms. The sugars fructose and glucose, for example, are isomers, for they contain the same number and kinds of atoms and have the same molecular formula ($C_6H_{12}O_6$) but have different three-dimensional structures (see Figure 2.12). The conversion of glucose to fructose or vice versa involves an isomerization reaction.

How Chemical Reactions Occur

A chemical reaction involves the physical interaction of reactants. According to the **collision theory** of chemical reactions, whenever reactants collide with sufficient velocity and at an appropriate orientation, they will react. The rate of the reaction is determined by the frequency of effective collisions. Since the velocity of molecular movement increases with an increase in temperature, chemical reactions usually proceed faster when heated.

Many chemical reactions occur very slowly or not at all at room temperatures. For example, glucose, a common energy source in cells, is quite stable and will not combine with oxygen unless heated. Observations of this sort for thousands of different compounds have led to the concept that before a chemical reaction can start, an energy barrier must be overcome. The barrier is probably a lack of sufficient kinetic energy to produce effective molecular collisions. **Activation energy** is the amount of energy needed to overcome this barrier.

Once its reactants are activated, a chemical reaction proceeds in accordance with the laws of thermodynamics. As reactants are consumed, their concentration decreases and the frequency of collisions among reactant molecules is reduced. At the same time, the concentration of products increases, and increased collisions of the product molecules may result in a reverse reaction, re-forming the reactants. Eventually, the forward and reverse rates of reaction become equal, reaching a **dynamic equilibrium**, where the reaction in effect stops. At this point, the free energy of the reaction is zero because energy is randomly distributed between reactants and products. No work can be done by a reaction in this state. The amount of reactant does not have to equal the amount of product for an equilibrium to exist; only the rates of the forward and reverse reactions must be the same.

How Catalysts Influence Chemical Reactions

Cells cannot survive at the temperatures that would be required for most cellular chemical reactions to proceed spontaneously. However, living organisms have the capacity to synthesize **catalysts**, substances that lower activation energy requirements. Catalysts allow chemical reactions to proceed at lower temperatures and faster rates, but are not consumed in the reactions they promote. Chemical reactions in cells are promoted by protein catalysts called **enzymes**. A typical animal cell contains several thousand kinds of enzymes, each promoting a specific chemical reaction. Each enzyme consists of a unique linear sequence of amino acids determined by the genetic information in the cell. By controlling which types of enzymes are present or activated, a cell can regulate which types of metabolic reactions occur.

How Enzymes Function

Figure 3.3 shows the energy changes associated with the breakdown of hydrogen peroxide by the enzyme catalase. Hydrogen peroxide is produced in certain metabolic processes and can destroy cellular structures if allowed to accumulate. Catalase acts as a detoxifying agent. Note that this enzyme reduces the required activation energy from 18,000 cal/mol to 5,500 cal/mol. The free energy (ΔG) and the entropy (ΔS) of the reaction are not changed by the enzyme; only the activation energy is affected.

The molecules that enzymes act on in promoting reactions are called **substrates**. Enzymes bind with their substrates to form **enzyme-substrate complexes**. A specific enzyme usually binds with only one type of substrate. This binding specificity seems to be based on complementary shapes of the two molecules. Just as a key will fit only one particular lock, so substrates fit only certain enzymes. The site on a large enzyme molecule that binds the smaller substrate molecule is called the **substrate-binding site**. This site may contain the **active site**, a specific chemical group actually involved in catalyzing the reaction. Binding of the substrate apparently causes a change in the enzyme's shape; the sketch in Figure 3.4 shows an enzyme before and after it binds to its substrate. The phenomenon of active change in an enzyme's shape following substrate binding is called **induced fit**.

Although the exact mechanism by which enzymes promote chemical reactions is not completely understood, three factors are thought to be involved. First, when a reaction involves two or

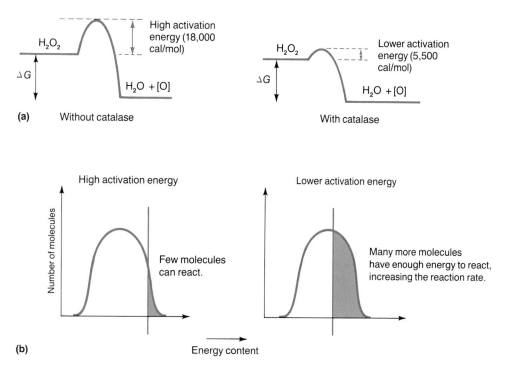

Figure 3.3 How enzymes function. (a) Catalase lowers the activation energy required to convert hydrogen peroxide to oxygen and water. It does not change the amount of free energy released in the reaction. Though the data given are for catalase, the principle is the same for all enzymes. (b) The overall effect of an enzyme is to allow more molecules to meet the activation energy requirement without increasing their kinetic energy.

more substances, the enzyme may hold the reacting molecules in the correct positions relative to each other for making or breaking bonds. Second, the enzyme's active site may contain charged regions that subject the orbitals of electrons in the substrate's covalent bonds to strain and distortion. The large size of most enzyme molecules may be

![Glucose, Hexokinase, Hexokinase-glucose complex]

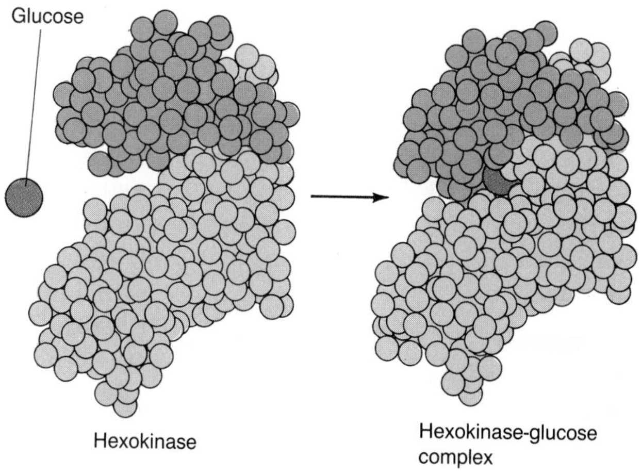

Figure 3.4 An enzyme-substrate complex. Model of the enzyme hexokinase before and after binding to its substrate, glucose. Note the change in the shape of the bound enzyme.

necessary to ensure that the proper strain is exerted on substrate molecules to break their chemical bonds. Third, some enzymes may promote reactions by directly acting as proton donors or acceptors.

After an enzyme has catalyzed a reaction, the former enzyme-substrate complex becomes an enzyme-product complex, which dissociates into product plus free enzyme. The free enzyme may then bind with more substrate and promote more of the same reaction. The formation of the enzyme-substrate complex, catalysis, and dissociation take only a few microseconds. A single molecule of catalase, for example, converts about 5,600,000 molecules of hydrogen peroxide to water and oxygen per minute at 20°C.

Cofactors and Coenzymes. Enzymes often cannot work alone and may require a **cofactor** to promote a reaction. Metallic ions derived from certain minerals essential in an animal's diet often serve as cofactors. Ions such as magnesium (Mg^{2+}), copper (Cu^{2+}), zinc (Zn^{2+}), and iron (Fe^{2+}) are bound and held in the active site of certain enzymes and presumably participate in the catalytic process.

Vitamins are organic molecules required in small amounts in animal diets to maintain growth and normal metabolism. (The roles of vitamins are listed in Table 4.1, p. 86.) Their main function is the regulation of physiological processes, and some of the B vitamins are essential components of nonprotein organic molecules called **coenzymes**. Coenzymes associate with enzymes and are necessary for their catalytic activity. Most enzymes that catalyze redox reactions require coenzymes called **electron carriers**. These carriers can either donate or accept two hydrogen atoms (along with their electrons) to reduce or oxidize the enzyme's substrate. The compound **nicotinamide adenine dinucleotide** (**NAD⁺**) is an important electron carrier involved in many metabolic redox reactions (Figure 3.5). When an enzyme catalyzes the oxidation of a substrate molecule, it binds the substrate at one site and NAD⁺ at a secondary site. Two hydrogen atoms then pass from the substrate to the NAD⁺ molecule, reducing it as shown:

$$NAD^+ + 2 H \longrightarrow NADH + H^+$$

Oxidized   **Reduced**

(The notation **NADH + H⁺** indicates that one of the hydrogen atoms is in the ionized form.) NADH + H⁺ is used as a source of electrons for chemical reduction in other metabolic reactions. Since NADH + H⁺ contains 52 kcal more energy per mole than NAD⁺, its chemical bonds are a potent energy resource in animal cells.

Enzyme Teams. As metabolism proceeds, reaction products seldom accumulate because most enzyme-catalyzed reactions are linked to other reactions. Food compounds such as glucose, for example, are degraded in stepwise fashion by teams of enzymes, each enzyme catalyzing a single chemical step. The equation for glucose metabolism in animal cells is often written as

$$C_6H_{12}O_6 + 6 O_2 \longrightarrow 6 CO_2 + 6 H_2O + energy$$

Although this equation is chemically correct, it is actually a summary that represents about 25 chemical reactions. No cell oxidizes glucose to carbon dioxide and water in one step. Instead, glucose is converted to CO_2 and H_2O in such a way that the potential energy stored in its chemical bonds is released in small quantities at each of several steps. The cell can harvest these small quantities of energy more efficiently than it can a single large quantity, but some energy is still lost as heat (entropy) at every step in the reaction sequence.

An enzyme team that sequentially breaks down a molecule is characterized by a simple property: The product of the first enzyme-catalyzed reaction

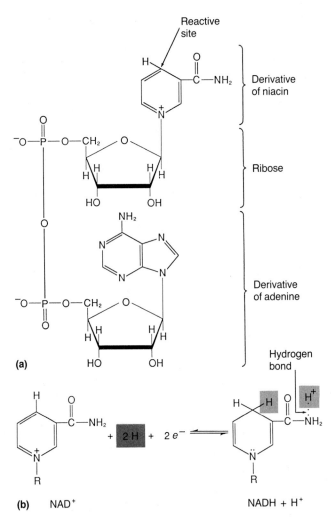

(a)

(b)   NAD⁺   NADH + H⁺

Figure 3.5 Structures of the coenzymes NAD⁺ and NADH + H⁺. These compounds are synthesized from the vitamin niacin, the organic base adenine, and the pentose sugar ribose. (a) Complete structure of NAD⁺, the oxidized form of nicotinamide adenine dinucleotide. (b) Reversible transformation of NAD⁺ to NADH + H⁺. R stands for the part of the molecule not directly involved in the reaction (i.e., the ribose and adenine derivatives).

serves as the starting material for the second reaction, and so on, linking several enzymes together in a **metabolic pathway**. In living cells, there is often continual flow of material through metabolic pathways because products are used as fast as they are produced.

Two general types of metabolic pathways are recognized. Those that break molecules down are called **catabolic** pathways. In **anabolic** pathways, more complex molecules are constructed from simpler molecules. The distinction between these two types of pathways is not always clear, because often a molecule is partially broken down and the breakdown products are used to synthesize a different type of molecule.

Energy Flow in Metabolism

All organisms face a significant metabolic challenge: to capture needed energy without destroying themselves with heat. Cells accomplish this by coupling exergonic reactions with endergonic reactions, so that energy flows in small increments to where it is needed, when it is needed. When an animal obtains food, its cells shuttle energy from one cellular reaction to another through a common intermediate, that is, a compound that participates in both reactions. The most common energy intermediate in cells is **adenosine triphosphate (ATP)** (Figure 3.6). Energy is built into this molecule by a condensation reaction and can be readily retrieved by hydrolysis for energy-requiring chemical reactions in the cell. **Adenosine diphosphate (ADP)**, a closely related compound, differs from ATP in that it has two, rather than three, phosphates. Both ATP and ADP are high-energy compounds. When a mole of ATP is hydrolyzed to ADP, about 7300 cal of free energy are released, depending on the internal conditions of the cell. Likewise, when ADP is hydrolyzed to adenosine monophosphate (AMP), which has one phosphate, an additional 7300 cal can be released. The covalent bonds that bind the second and third phosphates in ADP and ATP, respectively, are sometimes called "high energy bonds," but the high energy of the molecules is not really concentrated in these bonds. What is important in bioenergetics is not the amount of energy in any particular covalent bond, but the **free energy of hydrolysis**—the amount of free energy released and available to do cellular work when ADP or ATP is hydrolyzed. The pool of ATP within a cell can be thought of as a checking account in which energy is kept for rapid and convenient access. Each cell must make its own ATP, for ATP is not made in one cell and transported to another.

To make ADP or ATP, a cell must provide more than 7300 cal/mol because of the entropy factor. Let's look at how ATP can work as an energy shuttle between exergonic and endergonic reactions in a hypothetical set of metabolic reactions. For convenience, let us assume that a mole of reactant is converted to a mole of product and that the temperature factor in the energy equation is negligible. Assume that sugar X has a phosphate group covalently bonded to it (X–P). When X–P is hydrolyzed to X + P, 10,000 cal of energy per mole become available to the cell. If the enzyme that catalyzes hydrolysis is present, then X–P and ADP will be bound at its active site, and the cell might couple ATP synthesis to the X–P hydrolysis as follows:

Reaction	Energy Change
X–P $\longrightarrow$ X + P	−10,000 cal (ΔH)
ADP + P $\longrightarrow$ ATP	+ 7,300 cal (ΔG)
X–P + ADP $\longrightarrow$ X + ATP	− 2,700 cal (ΔS)

In this set of reactions, the exergonic hydrolysis of X–P yields 10,000 cal of energy; 7,300 cal are trapped in the synthesis of ATP from ADP, while

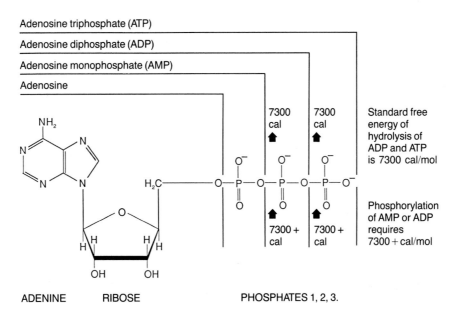

Adenosine triphosphate (ATP)
Adenosine diphosphate (ADP)
Adenosine monophosphate (AMP)
Adenosine

7300 cal   7300 cal   Standard free energy of hydrolysis of ADP and ATP is 7300 cal/mol

7300 + cal   7300 + cal   Phosphorylation of AMP or ADP requires 7300 + cal/mol

ADENINE   RIBOSE   PHOSPHATES 1, 2, 3.

Figure 3.6 Structure of ATP, ADP, and AMP. All these compounds consist of the purine adenine covalently bonded to the phosphorylated sugar ribose. ADP and ATP are unstable, high-energy molecules; they can be readily hydrolyzed to yield free energy for cell work.

2,700 cal escape as heat, contributing to the entropy of the universe.

Now assume that in another part of the cell, the compound Y–P is being formed at an energy cost of 5000 cal. If the proper enzymes are present, the following reactions could occur:

Reaction	Energy Change
ATP $\longrightarrow$ ADP + P	−7,300 cal (ΔH)
Y + P $\longrightarrow$ Y–P	+5,000 cal (ΔG)
Y + ATP $\longrightarrow$ Y–P + ADP	−2,300 cal (ΔS)

In this set of reactions, the endergonic process of forming Y–P is coupled to the exergonic breakdown of ATP.

Energy transfer by the donation of a phosphate group is called **phosphorylation**. After forming a chemical bond with a phosphate group, a molecule exists in a higher energy state than it did before.

Cellular Metabolism

In any cell, metabolism involves many different chemical reactions and reaction sequences, yet all living organisms have certain metabolic processes. The occurrence of the same types of energy-harvesting processes (described in this section) in virtually all forms of life suggests that these metabolic systems evolved very early in the history of life and that metabolic similarities reflect a common ancestry of all living organisms.

Cellular Respiration

Cellular respiration refers to all metabolic reactions that release energy for use in cellular work. (It should not be confused with breathing, also called respiration, which is a mechanism for ventilating the lungs or gills of animals.) In addition to generating energy, the reactions of cellular respiration also generate small molecules that can be used to synthesize larger molecules, such as lipids, proteins, and nucleic acids. To illustrate the principles of respiration, we emphasize the catabolic metabolism of glucose, a sugar frequently utilized by animals.

Glucose, a six-carbon sugar found in most foods, can also be produced by many cells from amino acids or other molecules. When excess glucose is available, body cells (especially in the liver and muscles) form glycogen, a storage polymer composed of glucose units (see Figure 2.13). Later, when energy is needed, the glycogen stores are bro-

ken down in hydrolysis reactions to yield glucose. Glucose may be catabolized by the cells that are storing it, or it may be released into the blood to supply other tissues. Marathon runners sometimes take advantage of this process and store energy by "carbohydrate loading" before a race. Eating foods high in carbohydrates may lead to glycogen synthesis in muscles and the formation of an energy reserve. The sudden exhaustion or "wall" that many runners experience about three-quarters through a marathon may represent the point where their muscle glycogen is depleted.

When glucose is broken down in respiratory reactions that require oxygen, the process is called **aerobic respiration**. **Anaerobic respiration** is the breakdown of glucose and other substrates in the absence of oxygen. The 23 enzyme-catalyzed steps in the aerobic oxidation of glucose can be grouped into four sequential metabolic pathways: glycolysis, pyruvate oxidation, the Krebs cycle, and the electron transport system. These metabolic pathways are common to almost all animals, and although detailed study of the pathways would reveal subtle biochemical differences among species the broad scheme is constant across the animal kingdom.

Glycolysis

Glucose is an energy-rich molecule that can be oxidized to yield 686 kcal of free energy per mole. **Glycolysis** is the first catabolic pathway by which this energy is released in cells. It occurs in the cytosol in many different cell types in any one animal. During this process, a molecule of glucose passes through a sequence of nine enzyme-catalyzed reactions and is converted into two smaller molecules of **pyruvate** (Figure 3.7). Throughout this chapter, we use the term *pyruvate*, which refers to the ionized form of the organic acid pyruvic acid. Later in the chapter, we will discuss other organic acids that, like pyruvic acid, are dissolved in cells in the ionized salt form. The names of all such compounds end in the suffix *-ate*.

The initial reactions of glycolysis activate glucose by phosphorylating it and rearranging its chemical bonds. The glucose-derived molecule is then split into two three-carbon compounds; these are isomers and can be interconverted, so the cell in effect produces two molecules of phosphoglyceraldehyde (PGAL) at this step. Both PGAL molecules go through the next five reactions, during which (1) inorganic phosphate in the cell combines with PGAL, (2) PGAL is oxidized as hydrogen atoms are removed from it and transferred to the elec-

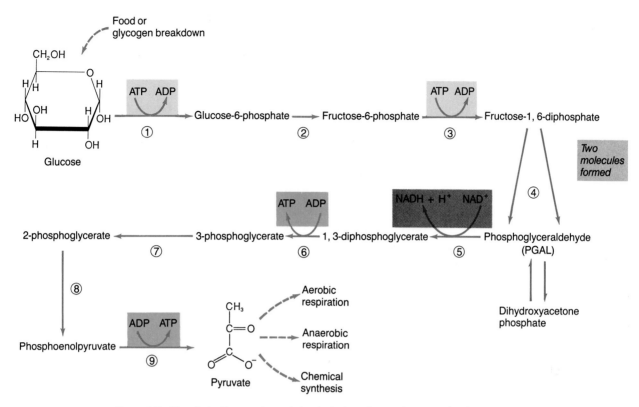

Figure 3.7 Glycolysis. The reactions of glycolysis degrade one glucose molecule to two pyruvate molecules, with a net production of two molecules of NADH + H⁺ and two molecules of ATP. (Notice that two ATP molecules are needed to convert glucose to fructose 1,6-diphosphate, and keep in mind that the pathway from PGAL onward is taken by *two* three-carbon compounds.)

tron carrier NAD⁺, producing NADH + H⁺, and (3) phosphate groups are transferred to ADP, producing ATP. It is important to realize what the cell is accomplishing by degrading glucose to pyruvate. Some of the free energy released during the reaction sequence is used to synthesize the high-energy compounds NADH + H⁺ and ATP. Thus, energy is acquired in a usable form to support the energy-requiring processes of life, and small molecules are obtained to build larger ones.

For every glucose molecule that undergoes glycolysis, four ATP, two NADH + H⁺, and two pyruvate molecules are produced. Since two ATP molecules are used in the early activating steps, the net ATP yield from glycolysis is only two molecules. The ATP molecules can be used to drive endergonic reactions. The NADH + H⁺ molecules can be a source of hydrogen for the chemical reduction of compounds in anabolic reactions, or they can be oxidized to release energy. The pyruvate molecules can be broken down further or converted into other necessary compounds, such as amino acids.

Energetics of Glycolysis. An energy analysis of glycolysis shows the changes in chemical potential

energy that occur through the reaction sequence (Figure 3.8). Of the 686,000 cal of free energy available from the breakdown of glucose, glycolysis produces a net harvest of 14,600 cal in the form of two ATP molecules. The rest of the energy is either contained in the two pyruvate molecules and the two NADH + H⁺ molecules, or has escaped as heat. Simple division of the energy captured in ATP by the amount of energy present in glucose shows that glycolysis captures only about 2% of the available energy. Any animal that can metabolize pyruvate or NADH + H⁺ further has an advantage because it does not require as much food to supply its energy needs.

Pyruvate Oxidation

When oxygen is available, most animal cells will catabolize pyruvate further, breaking it down completely to carbon dioxide and water in a series of pathways known as pyruvate oxidation, the Krebs cycle, and the electron transport system. However, oxygen doesn't actually enter the process until the last step of the electron transport system. Cells that have a source of oxygen and the many enzymes

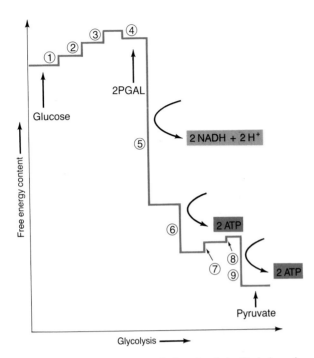

Figure 3.8 Free energy changes during glycolysis. Circled numbers correspond to the circled reaction steps in Figure 3.7. The cell invests energy during steps 1 and 3 and releases it during steps 5, 6, and 9 in amounts that the cell can trap in ATP or electron carriers.

needed for these metabolic reactions can obtain much more energy from glucose than cells that have only those enzymes needed for glycolysis.

A mole of pyruvate can be oxidized to release about 271,500 cal of free energy. To obtain this energy, cells first convert the two pyruvate molecules produced by glycolysis into two molecules of **acetate**. This conversion, **pyruvate oxidation**, occurs inside mitochondria and involves the removal of hydrogen atoms and one of the three carbon atoms from the pyruvate molecule. The carbon is released as CO_2 and diffuses out of the cell as a waste product, while the hydrogen atoms are captured in $NADH + H^+$. The pyruvate oxidation reactions are catalyzed by a complex of three different enzymes and several coenzymes acting sequentially on the pyruvate. One of the coenzymes, NAD^+, picks up the hydrogen atoms removed from the pyruvate; a second called **coenzyme A** (**CoA**), joins with the acetate and carries it into subsequent reactions. Coenzyme A is derived from pantothenic acid, one of the B vitamins.

Many different types of food derivatives, including fatty acid breakdown products, a number of amino acids, parts of nucleic acids, and most sugars, can be converted to acetate. In turn, acetates bind with CoA to form **acetyl-CoA**, the starting compound of the Krebs cycle. Acetyl-CoA may

consequently be viewed as the intersection where acetates coming from several side streets merge into a common metabolic highway.

The Krebs Cycle

The eight steps of the **Krebs cycle** (also called the citric acid cycle) were described in 1937 by Sir Hans Krebs, who subsequently received a Nobel Prize for his discoveries. The enzymes involved in this cycle are located in the mitochondria along with those that catalyze the formation of acetyl-CoA. Some of the enzymes are attached to the inner mitochondrial membrane, while others occur free in the mitochondrial matrix. At the start of the cycle (Figure 3.9), acetate leaves acetyl-CoA and

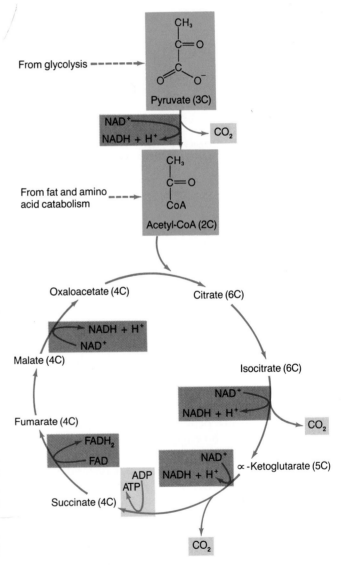

Figure 3.9 Pyruvate oxidation and the Krebs cycle. The reactions of the Krebs cycle break down acetyl-CoA to CO_2 and hydrogen atoms. Hydrogen atoms are picked up by the electron carries NAD^+ and FAD. Note that ATP is also made at one step.

combines with oxaloacetate, forming citrate. Citrate then undergoes a series of enzyme-catalyzed molecular rearrangements: One of its carbon atoms is lost as CO_2, and two hydrogen atoms are removed and passed to NAD^+. The five-carbon compound α-ketoglutarate results. This, in turn, is oxidized to succinate in such a way that an additional carbon atom is lost as CO_2, two more hydrogen atoms pass to NAD^+, and one ATP molecule is made using the energy released. Succinate is then oxidized to fumarate by removal of two more hydrogen atoms that are passed to a second electron carrier called **flavin adenine dinucleotide (FAD)**, reducing the FAD to $FADH_2$. FAD consists of the vitamin riboflavin (B_2) plus a derivative of ATP, and has a role in the cell similar to that of NAD^+. Fumarate then undergoes a series of molecular rearrangements that eventually lead to its conversion to oxaloacetate, which can combine with more acetyl-CoA so that the cycle continues.

Overall, what happens in the Krebs cycle is that acetate is completely disassembled into carbon dioxide molecules and hydrogen atoms that are attached to electron carriers. During two turns of the cycle (one for each acetate molecule from the original glucose molecule), molecules of organic acids are oxidized in a step-wise fashion, allowing the construction of energy-containing molecules that the cell can use. The energy released at one step is enough to generate one ATP molecule. The carbon dioxide produced diffuses out of the cell as a waste gas, and the reduced electron carriers that bear the hydrogen atoms enter another sequence of reactions, the electron transport system.

Energetics of Glucose Breakdown. Before looking at the electron transport reactions, let us take a quick glance at the balance sheet given in Table 3.1 to see what has happened so far in glucose catabo-

lism. These calculations are based on the energy content of a mole (6.023×10^{23} molecules) of glucose. If a cell oxidizes 1 mol of glucose, it gains energy in the form of 4 mol of ATP, 10 mol of $NADH + H^+$, and 2 mol of $FADH_2$. Because a mole of $NADH + H^+$ has a free energy content of 52 kcal owing to the hydrogen atoms it obtained from glucose, a cell has a great amount of potential energy stored in these molecules. In 10 mol of $NADH + H^+$ there are 520 kcal, or about 76% of the 686 kcal originally present in the mole of glucose. Through the electron transport system, the cell can harvest much of this energy as well as that contained in the $FADH_2$ molecules.

The Electron Transport System

The **electron transport system** consists of several kinds of electron carriers associated with proteins embedded in the membranes of the mitochondrial cristae. These carriers include flavin mononucleotide (FMN), coenzyme Q, and the cytochromes, which are positioned in the membrane in a series corresponding to their increasingly greater electron affinities. When electrons bind to a carrier with a weak affinity for electrons, they are removed from that carrier by another having a greater affinity. Thus, these carriers form a structural oxidation-reduction series. As the electrons pass from carrier to carrier, they release energy that is used to synthesize ATP. The process of using energy released in electron transport to generate ATP is called **oxidative phosphorylation**.

Electron (hydrogen) carriers ($NADH + H^+$ and $FADH_2$) reduced during glycolysis and the Krebs cycle pass their electrons (with associated hydrogen ions) to the electron transport system. The carriers in this series pass the electrons along to oxygen at the end of the chain, forming **metabolic**

Table 3.1 Balance Sheet for Glucose Oxidation*

	Glucose	Pyruvate	Acetate	CO_2	ATP	$NADH + H^+$	$FADH_2$
Glycolysis	1	**2**			2/**4**	**2**	
Pyruvate oxidation	2	**2**	**2**		**2**	
Krebs cycle			**2**	**4**	**2**	**6**	**2**
Total used/ produced	1	0	0	**6**	**4**	**10**	**2**

*Molecules used are indicated in regular type; molecules produced are in boldface.

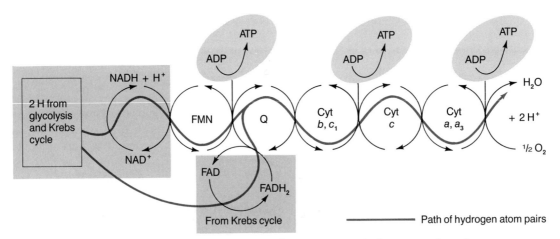

Figure 3.10 The electron transport system. This system accepts hydrogen atoms from electron (hydrogen) carriers reduced during glycolysis and in the Krebs cycle. Pairs of electrons are separated from hydrogen ions (protons) and travel through the system in a series of exergonic oxidation-reduction reactions. Three molecules of ATP can be synthesized for every pair of electrons that enters the transport chain on NADH + H$^+$; two molecules of ATP can be synthesized for every electron pair that enters on FADH$_2$. Cytochrome molecules (cyt) have complex structures. They contain a protein and an iron-containing heme group that is active in electron transport. Metabolic poisons such as cyanide and carbon monoxide exert some of their effects by binding with the iron, preventing electron transfer from taking place.

water (Figure 3.10). The flow of electrons through the transport system from the carriers to oxygen is a regeneration process for the carriers. This "recharging" of electron carriers such as NAD$^+$ is necessary because cells contain only a small amount of NAD$^+$, and without a regeneration mechanism, all NAD$^+$ would quickly be reduced to NADH + H$^+$. If this happened, glycolysis and the Krebs cycle would stop because no oxidized electron carriers would be available to accept the hydrogen atoms being removed from substrates.

Energetics of Electron Transport. Electron flow through the electron transport system is strongly exergonic. At three of the steps along the way, ATP synthesis is coupled to the energy release occurring. Since the aerobic respiration of 1 glucose molecule produces 10 NADH + H$^+$ molecules, 30 ATP molecules can be synthesized via this route for each original glucose molecule. When electrons enter the system from the two FADH$_2$ molecules produced in the Krebs cycle, only two ATPs are made from each molecule because these electrons were originally captured at a lower energy level. The electron transport system allows the synthesis of large quantities of ATP, 34 molecules in most cells, for every molecule of glucose aerobically oxidized. (Sometimes, 32 ATPs are made instead of 34; this depends on the mechanism employed to move the NADH + H$^+$ molecules produced in pyruvate oxidation into the mitochondria.)

Efficiency of Aerobic Respiration

Table 3.2 gives an accounting of how much energy from one glucose molecule is trapped by electron carriers and by ATP. Figure 3.11 reviews where these products are made during all phases of aerobic metabolism. Overall, the amount of energy captured as ATP from a mole of glucose by aerobic respiration equals the net moles of ATP produced (38) times 7300 cal. If we divide this number by the caloric content of a mole of glucose (686,000), we can

Table 3.2   ATP Produced by Complete Breakdown of One Glucose Molecule

	NADH + H$^+$	*FADH$_2$*	*ATP*
Glycolysis	2		4 (2)*
Krebs cycle	8	2	2
Electron transport system			
2 NADH + H$^+$ from glycolysis, × 3			6
8 NADH + H$^+$ from Krebs cycle, × 3			24
2 FADH$_2$ from Krebs cycle, × 2			4
Net ATP molecules per glucose molecule			38
% ATP from glycolysis			5%
% ATP from Krebs cycle			5%
% ATP from electron transport system			90%

*Two ATPs are used in the initial reaction steps, so the net total is two ATPs produced in glycolysis.

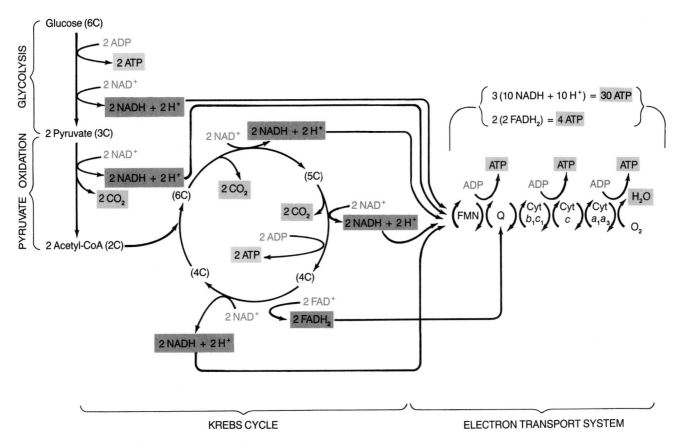

Figure 3.11 Summary of aerobic respiration of glucose. Glycolysis, pyruvate oxidation, and the Krebs cycle disassemble glucose into CO_2 and hydrogen atoms that are bound to electron carriers. Some ATP is produced in glycolysis and the Krebs cycle, but most is made in the electron transport system, where electrons pass down a redox gradient and eventually combine with oxygen to form metabolic water.

see that aerobic respiration is about 40% efficient; about 60% of the energy escapes from the animal as heat.

How ATP Is Made

When ATP is produced directly in glycolysis or in the Krebs cycle, a phosphate group is transferred from the sugar being dismantled to an ADP molecule. The phosphorylation process is more complex when it occurs in the electron transport chain. In 1978, the British biochemist Peter Mitchell was awarded the Nobel Prize for his **chemiosmotic hypothesis** of how mitochondria carry out oxidative phosphorylation. His hypothesis is based on the structural arrangement of enzymes and other compounds in the mitochondrial membranes, as well as on the chemical reactions the enzymes catalyze. Recall from Chapter 2 that the surface area of the inner mitochondrial membrane is greatly increased by folds called cristae. The enzymes in-

volved in electron transport are in two locations in the mitochondria: embedded in the membranes of the cristae and in the central matrix.

According to Mitchell's model (Figure 3.12), the components of the electron transport system are arranged on the inner membrane in such a way that as hydrogen atoms are removed from reduced $NADH + H^+$ molecules, the electrons are separated from the hydrogen nuclei, creating free electrons and hydrogen ions (H^+). As electrons move from molecule to molecule in the electron transport system, hydrogen ions are moved into the space between the inner and outer membranes of the mitochondria. Accumulation of hydrogen ions in this space creates a chemical energy gradient as a result of hydrogen ion (pH) differences across the membrane.

If a mitochondrion is examined by high-resolution electron microscopy after being specially stained, lollipop-like structures can be seen on its inner membrane (Figure 3.12c). Mitchell's model

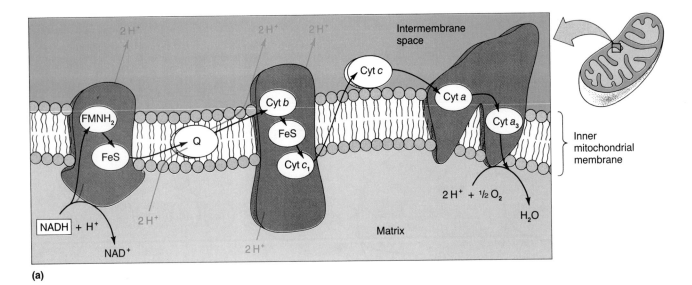

(a)

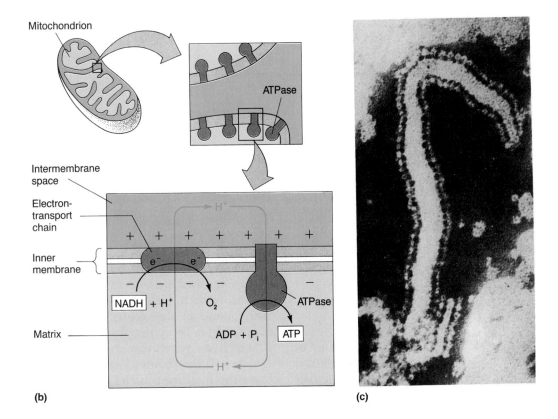

(b)   **(c)**

Figure 3.12 The chemiosmotic hypothesis of how ATP is made in the electron transport system. (a) As electrons pass from one component of the electron transport system to the next, they move back and forth from the inner to the outer face of the inner mitochondrial membrane in zigzag fashion. When the negatively charged electrons move outward to the carrier $FMNH_2$, the positively charged hydrogen ions follow and pass into the intermembrane space (between the inner and outer membranes). The electrons then move inward, passing to the next electron carrier in the series (FeS), but the hydrogen ions cannot follow because the outer face of the membrane is not permeable to them. As electrons move to the next carriers, they again move to the inner membrane face, where hydrogen ions from the dissociation of water in the matrix are attracted to the electrons by their negative

charge. The electrons with attracted hydrogen ions are then passed to cytochrome c (cyt c) , which is on the outer membrane face. The hydrogen ions remain in the intermembrane space when the electrons are again transported inward. The overall effect is that the electron transport system acts as a hydrogen ion pump, removing hydrogen ions from water in the matrix and trapping them in the intermembrane space. (b) This results in a hydrogen ion and electrical charge gradient across the inner membrane. Hydrogen ions pass back into the matrix only through the "lollipop" structures, which somehow use the energy of passing hydrogens to add a phosphate (P_i) to ADP, making ATP. (c) The structure of fragments of the inner membrane of a mitochondrion as seen with the transmission electron microscope.

proposes that the lollipop-like structures on the inner membrane are hydrogen ion reentry ports and that only at these sites can hydrogen ions diffuse back into the mitochondrial matrix, reducing the concentration difference. As the ions move through these channels, energy is released and is used to make ATP from ADP and inorganic phosphate. Enzymes at the entry ports catalyze the reaction forming ATP. The process can be thought of as active transport in reverse. Rather than using ATP to pump ions against a gradient, the cell uses a gradient to make ATP.

Mitchell originally published his hypothesis in 1961, but it took many years of testing before the majority of biologists became convinced of its validity. The model is now generally accepted, and current research is focused on determining how the energy from the hydrogen ion gradient is used to add a phosphate group to ADP in oxidative phosphorylation.

Catabolism of Fats and Proteins

Animals regulate their acquisition, storage, and release of energy. If they cannot obtain food with a caloric value equivalent to their energy requirements, they will catabolize their own tissues to make up the deficit. In most animals, fats are energy reserves laid down when caloric intake exceeds the caloric demand. Many migrating birds, adult salmon on spawning runs, and hibernating mammals survive only on energy supplied by stored fats. Fat has the advantage of being light and compact to store and carry. It contains about twice as much energy per mole as glycogen, and glycogen has the disadvantage of having to be stored with water. (About 3–5 g of water are required to store about 1 g of glycogen.)

As noted in Chapter 2, fats consist of up to three fatty acids covalently bonded to glycerol, forming a triglyceride (see Figure 2.11). Triglycerides can easily be hydrolyzed into separate fatty acids and glycerol. Glycerol can then be converted into pyruvate, which may be oxidized and subsequently enter the Krebs cycle. Fatty acids are long-chain molecules often consisting of 18 carbon atoms, each of which can be saturated with hydrogen atoms. These hydrogen atoms, if removed and bound to electron carriers such as NAD^+, produce significant amounts of ATP through the electron transport system.

The process of metabolic degradation is similar for all fatty acids. Coenzyme A binds with the carboxyl group of the fatty acid, and an enzyme then

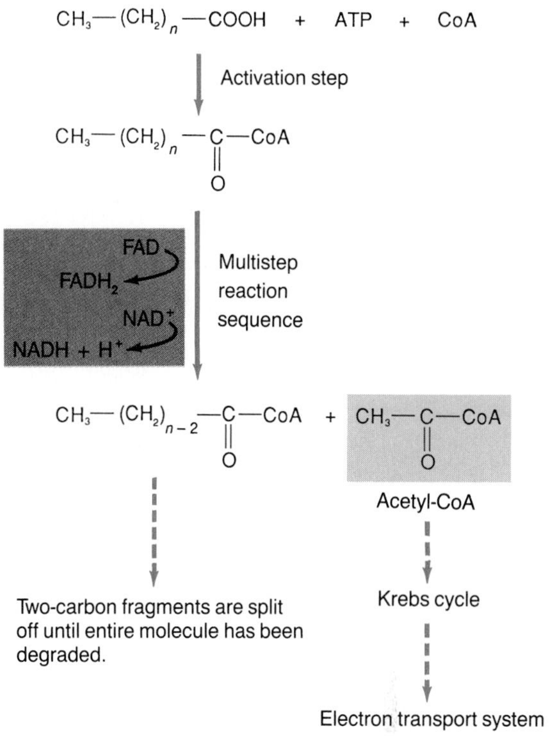

Figure 3.13 Oxidation of a fatty acid. Fatty acids are oxidized and broken into two-carbon acetyl fragments that enter the Krebs cycle.

cleaves the carbon chain so that the CoA is left attached to a two-carbon acetate fragment that enters the Krebs cycle (Figure 3.13). This process is repeated several times as the fatty acid is broken down two carbons at a time.

The catabolism of fatty acids illustrates an important aspect of metabolism. The only unique enzymes in this process are those associated with forming acetyl-CoA. The remainder of the enzymes are the same as those involved in glucose metabolism. The same is true when amino acids are catabolized. Proteins are first hydrolyzed to yield amino acids. The nitrogen-containing amino group of individual amino acids is cleaved off and escapes from the cell as ammonia, a toxic waste product. The remainder of the molecule is catabolized by glycolysis or the Krebs cycle, with hydrogen atoms being passed to NAD^+ and then entering the electron transport system. Having metabolic pathways perform double or triple duty reduces a cell's investment in scarce energy, materials, and space (Figure 3.14).

Anaerobic Metabolism

During aerobic respiration, NAD^+ is regenerated from $NADH + H^+$ by the electron transport sys-

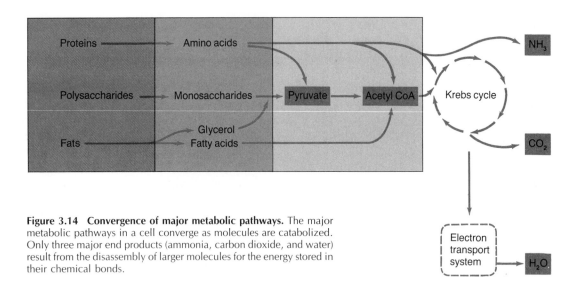

Figure 3.14 Convergence of major metabolic pathways. The major metabolic pathways in a cell converge as molecules are catabolized. Only three major end products (ammonia, carbon dioxide, and water) result from the disassembly of larger molecules for the energy stored in their chemical bonds.

tem, with oxygen acting as the terminal electron acceptor. Many animals, however, live in or sometimes experience conditions where sufficient oxygen is not available for aerobic respiration. Such animals can obtain energy by **anaerobic metabolism**, a form of cellular respiration in which glucose is oxidized without the use of oxygen.

A classic example of anaerobic metabolism occurs in vertebrate muscle. The cheetah in Figure 3.15 can sustain short periods of intense muscular work while running down prey or fleeing danger. But during these times, its circulatory system may not be able to supply the muscles with sufficient oxygen to keep the electron transport system running. When this happens, the muscle cells continue to perform work, but they do not derive energy from aerobic respiration. They continue to hydrolyze glycogen to glucose and break glucose down by glycolysis (see Figure 3.7), but without oxygen, pyruvate (the end product of glycolysis) cannot be

oxidized further. Recall that glycolysis yields two ATP molecules and two NADH + H$^+$ molecules per mole of glucose. Without oxygen, the energy in the NADH + H$^+$ molecules cannot be harvested, but muscle cells have evolved a way to regenerate (oxidize) these molecules, allowing them to be used over and over in glycolysis. Muscle cells use the enzyme **lactate dehydrogenase** to catalyze the transfer of hydrogen from reduced NADH + H$^+$ to pyruvate, converting pyruvate to **lactate** (Figure 3.16).

Figure 3.15 Intense animal activity leading to anaerobic respiration. A cheetah can attain speeds of 120 km/hr. However, limitations on its energy supplies restrict such activity to a distance of only several hundred meters.

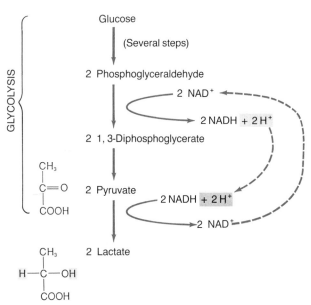

Figure 3.16 Anaerobic respiration. In this process, hydrogen atoms from NADH + H$^+$, generated in glycolysis, are passed to pyruvate, forming lactate and NAD$^+$. The NAD$^+$ can cycle back and accept more hydrogen atoms. In glycolysis, only two ATP molecules are generated per glucose molecule.

In the absence of oxygen, lactate is not metabolized further and therefore represents a metabolic dead end, a storage compound for hydrogen atoms from NADH + H$^+$. The pH of the body fluids can become acidic as lactate, an ionized organic acid, accumulates in the blood and around muscles, causing muscle fatigue. When the animal rests, muscle cells and liver cells remove hydrogen atoms from lactate and convert the lactate back to pyruvate, which can then be metabolized via aerobic respiration or converted to glycogen. The oxygen used in this recovery phase represents the animal's **oxygen debt**, the amount of oxygen needed to oxidize accumulated lactate.

Because anaerobic respiration consists only of glycolysis and an NAD$^+$ regeneration reaction, the efficiency of energy harvest is low (about 2%). Although this seems to be a disadvantage, anaerobic metabolism allows some animals to live in anaerobic environments and others, such as the cheetah, to be independent of oxygen periodically.

Alternative to Lactate Production. Many animals depend on anaerobic metabolism to supply most of their energy needs. For example, intestinal parasites live in surroundings where little oxygen is available. (Alternative metabolic pathways in parasitic flatworms are discussed in Chapter 22, p. 520; Figure 22.10.) And intertidal mussels at low tide close their shells and are anoxic (without oxygen) for several hours each day; some species of snails can survive in an experimental atmosphere of pure nitrogen for five days. Biochemists have found that these organisms use a form of anaerobic metabolism that differs from the lactate pathway of vertebrate muscle. Glycolysis still occurs, producing ATP and reduced NADH + H$^+$ that must be regenerated, but hydrogen atoms are not passed to pyruvate. Instead, they are passed to oxaloacetate, producing malate, which enters the Krebs cycle and is metabolized when oxygen becomes available. If you analyze the metabolic strategy here, you will see that although the mechanism is different from that of the lactate pathway, the problem being solved is the same. Oxidized NAD$^+$ is regenerated from reduced NADH + H$^+$, so that glycolysis continues to supply the animal's energy needs in the absence of oxygen.

Control of Metabolism

The production of ATP in aerobic or anaerobic metabolism is correlated with an animal's needs. Product formation is increased or decreased, sometimes even stopped, depending on the cell's requirements. By controlling its metabolism, a cell conserves energy and raw materials by building molecules only when needed.

Cells have several mechanisms for regulating the rate of metabolic product formation. Figure 3.17 shows a hypothetical three-step biochemical pathway in which compound A is converted to compound D. Clearly the supply of reactants will influence whether this pathway operates. If compound A must be transported into the cell, the reaction pathway will not operate until this transport has occurred. By regulating the rate of transport, the cell regulates the rate of formation of product D. The slowest reaction in a sequence of reactions is the pacemaker, or **rate-limiting step** for that pathway; in this example, the rate-limiting step could be the transport event (and/or some later reaction). By controlling the amount of substrate or enzyme available for any single reaction, the cell controls the entire pathway.

Once enough substrate molecules and enzymes are available, the only way the cell can regulate product formation is by modulating the catalytic activity of the enzymes. Three factors are known to be involved here. First, enzymes may be chemically modified so that their activity increases or decreases. This usually involves adding a phosphate to or removing a phosphate from the amino acid serine in the enzyme molecule. This leads to a

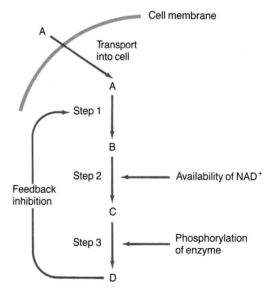

Figure 3.17 Regulation of a metabolic pathway. The cell can regulate this hypothetical biochemical pathway in several ways: by governing the rate at which compound A is transported across the cell membrane; by controlling the supply of NAD$^+$ necessary for the function of the step 2 enzyme; by phosphorylating the enzyme that catalyzes step 3; and by using the final reaction product D to regulate the enzyme that catalyzes step 1. Many other examples and combinations of regulatory mechanisms are possible in cells.

change in the enzyme's shape and consequently to changes in its catalytic abilities. Enzymes called **kinases** add phosphates to other enzymes, and **phosphorylases** remove phosphates. Many hormones change the rate of metabolism by affecting the phosphorylation of enzymes in cells. Epinephrine (adrenaline), for example, causes the phosphorylation and subsequent activation of the enzyme that hydrolyzes glycogen in the liver. When a mammal is stressed, epinephrine is released from the adrenal glands, causing glucose to enter the bloodstream and pass to cells throughout the body.

A second factor affecting enzyme activity is the availability of NAD^+, ADP, ATP, inorganic ions, and vitamins. If coenzymes or cofactors required for enzyme action are present in limited amounts, enzyme activity will also be limited.

A third important factor is the **allosteric regulation** of enzyme activity by small molecules called **modulators**. Many enzymes have two or more binding sites in their structure. One, the catalytic site, binds the substrate; the other, the allosteric site, binds the modulator molecule that regulates the enzyme's activity by causing changes in its three-dimensional structure (Figure 3.18). Some modulators make the enzyme more effective, while others

make it a less effective catalyst. In some metabolic pathways, for example, the product of the last reaction accumulates when it is not being used by the cell. When excess product accumulates, it combines with the first enzyme in the pathway and prevents it from functioning. This type of allosteric regulation, called **feedback inhibition**, prevents the synthesis of excess product by shutting down the entire pathway.

ENERGY REQUIREMENTS OF ANIMALS

Animal growth requires the synthesis of new molecules and cells. Even when an animal is not growing, the constant turnover of materials requires substantial metabolic activity, and cells must constantly furnish the necessary energy as ATP to make and replace proteins, phospholipids, nucleic acids, and polysaccharides. Cells may also perform mechanical work, such as amoeboid movement, ciliary or flagellar action, and muscle contraction. Energy is also required for active transport, when ions or molecules are moved against a concentration gradient. Cells perform this type of work when they draw food from surrounding fluids and maintain the proper ion balance.

Comparison of Energy Requirements

Figure 3.19 shows the energy requirements of several animals. It illustrates that, as might be expected, smaller animals require less total energy (in kilocalories per day) than do larger animals to

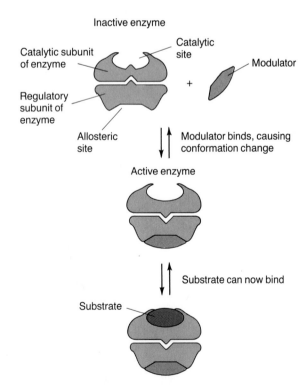

Figure 3.18 Allosteric regulation. Many enzymes have allosteric (regulatory) sites in addition to their catalytic sites. The binding of modulator molecules causes changes in enzyme shape, sometimes enhancing and sometimes inhibiting the efficiency with which the enzymes bind to their substrates.

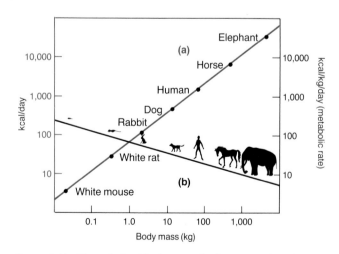

Figure 3.19 The relationship between body mass and energy requirements. Curve *a* (actually, a straight line) shows that the energy requirements of animals (kcal/day) increase with increasing body size; however, as curve *b* shows, on a unit weight basis (kcal/kg/day), the energy metabolism of large animals is lower than that of small animals.

Essay: Temperature and Animal Life

Temperature is one of the most influential environmental variables determining worldwide animal distribution. Temperatures in terrestrial environments can vary from 60°C in low-latitude deserts to −65°C in the Antarctic, and large diurnal and seasonal temperature changes may occur in temperate climates. Temperatures in aquatic environments are less variable because of water's high heat capacity, but can nevertheless fluctuate between −2° and 40°C. Temperature affects the fluidity of lipids in cell membranes and many membrane functions. It also influences the rates of metabolic chemical reactions. Animals that can maintain nearly constant (homeostatic) body temperatures achieve an independence of the environment, and they can colonize areas that others cannot.

An animal's body temperature results from a balance between heat gain and loss. Heat is generated by metabolism and may be gained from the environment. Heat escapes from an animal's body by radiation as infrared rays, by convection into air moving close to the body, by conduction to objects that contact the body, and by evaporation of water from the body surfaces. A high rate of metabolism and the slowing of heat loss by feathers or hair allow birds and mammals to maintain nearly constant warm body temperatures independent of environmental conditions. Because of their body temperature conditions, mammals and birds are often said to be "warm-blooded." In contrast to birds and mammals, most animals are not well insulated, and their metabolic heat rapidly dissipates into the environment. Commonly, all animals except mammals and birds are said to be "cold-blooded" because they do not maintain a constant warm body temperature. It is unfortunate that the terms *warm-blooded* and *cold-blooded* are so widely used, for they are misleading. Certain so-called cold-blooded animals can sometimes maintain a body temperature that is quite different from their environment. A swimming tuna can generate enough heat in its active muscles to increase its body temperature 12°C above that of its surroundings. (See the Chapter 29 Essay, "Not All Fishes Are Cold-Bodied," p. 717.) And an insect sunning on a hot rock may have a higher body temperature than mammals and birds. Cold-blooded animals that live where environmental temperatures remain nearly constant have relatively constant body temperatures.

Because the words *cold-blooded* and *warm-blooded* are inadequate, physiologists use four terms to describe temperature conditions in animals. Animals (mainly birds and mammals) that maintain a nearly constant body temperature are said to be **homeothermic**. Those that maintain a warm body temperature by means of internal (metabolic) heat production are **endothermic**. Animals with highly variable body temperatures are **heterothermic**. They may have a warm, relatively constant body temperature some of the time, but one that fluctuates with the environment at other times. Animals that derive body heat from the environment are **ectothermic**.

Most ectotherms are heterotherms, and most endotherms are homeotherms, although there are many excep-tions. Camels are considered endotherms, but their body temperature rises several degrees as they gain heat from the desert sun and then drops to "normal" at night. Many ectotherms also produce significant amounts of internal heat. Some insects fan their wings before flight in the early morning hours to warm the flight muscles to a point where they function optimally. The hairlike scales that cover the bodies of many nocturnal moths act as insulation that retains heat from muscle contraction, allowing these insects to be active during cool nights.

Homeothermic endotherms must regulate both heat production and dissipation. The center of temperature regulation in vertebrates is the hypothalamus, a brain region that also regulates sleep and blood pressure. Secretions from the hypothalamus regulate secretion of hormones by the adrenal and thyroid glands. The hormones epinephrine and thyroxin regulate metabolism, thus increasing or decreasing heat production. Behavioral adaptations also affect heat gain or loss (Figure E1). Shivering increases heat production. Animals may move into or out of the sun or seek sheltered areas. Heat escape can be reduced by elevating hairs or feathers to trap warm air next to the body, by regulating the amount of blood circulating through blood vessels close to the skin surface, or by sweating and panting.

Endothermic animals pay a high price to maintain their body temperatures. As much as 80% of their basal metabolic rate may be devoted to **thermoregulation**. Some birds and mammals lower thermoregulation costs by decreasing their body temperatures during periods of inactivity: Hummingbirds save energy by becoming metabolically inactive (torpid) at night; certain bats have lower body temperatures while asleep during the day; and winter hibernation in certain mammals reduces energy expenditure when food is not available. In addition, many species of birds migrate, thereby avoiding cold seasons.

Figure E1 Behavior related to body temperature regulation. An Indian elephant (*Elephas maximus*) seeks relief from heat.

sustain their life functions. A more subtle concept is revealed by dividing the caloric requirement of each animal by its weight in kilograms (kg). This calculation gives the **metabolic rate** of the animal, that is, the rate at which a kilogram of tissue from each animal would convert chemical energy to work and heat per day (kcal/kg/day). By calculating metabolic rates, we can dispense with the effect of size on total caloric requirements and show that as the size of the animal increases, the metabolic rate decreases. This produces what physiologists sometimes call the "mouse to elephant curve." What this means is that it takes more energy to keep a gram of small animal functioning than it does to keep a gram of large animal functioning.

Mammals (such as those shown in Figure 3.19) and birds have higher metabolic rates than do other animals. Mammals and birds are **endothermic**, meaning that they maintain a warm body temperature independent of the environmental temperature. Typically, other vertebrates (fishes, amphibians, and reptiles) and invertebrates are **ectothermic**, meaning that their body temperature depends on that of the environment. (See the Essay on temperature and animal life.) Despite these differences, results similar to those for mammals, such as those shown in Figure 3.19, are obtained for other groups of animals as well. In the next chapter, we will see how animals acquire energy for metabolism and other uses.

SUMMARY

1. Several basic energy concepts apply to metabolism. According to the laws of thermodynamics, energy is neither created nor destroyed when changed in form, so that all energy can be accounted for in a chemical reaction; and chemical reactions always proceed in a direction that dissipates energy as heat, thus increasing universal entropy.

2. Biochemical reactions (including oxidation-reduction, hydrolysis, condensation, and isomerization) are catalyzed by enzymes that change the rate but not the energy yield of reactions. Enzymes lower activation energy, so more molecules per unit time can enter the reaction at normal physiological temperatures. Within cells, many enzymes work in sequence to modify compounds in metabolic pathways.

3. Cellular respiration is the sum of the energy-yielding reactions in a cell. Aerobic respiration occurs in mitochondria and includes the reactions of glycolysis, pyruvate oxidation, the Krebs cycle, and the electron transport system. Oxygen serves as a final electron acceptor. Anaerobic respiration occurs without oxygen and involves only glycolysis as an energy-yielding process.

4. As food is catabolized (broken down), some of its potential energy is captured in the high-energy molecule ATP. Cells use ATP for chemical and mechanical work. Energy from ATP hydrolysis can activate molecules, allowing them to either enter into a reaction or change their shape. The synthesis (anabolism) of large molecules in the cell is an energy-requiring process, and large amounts of ATP are used to fuel growth and the replacement of cell components.

5. Energy requirements are related to body size. Small animals have higher metabolic rates than large animals. Moreover, basal metabolic rates are higher in endothermic animals than in ectothermic ones because energy is expended in temperature regulation.

FURTHER READING

Becker, W. M. *Energy and the Living Cell: An Introduction to Bioenergetics.* Philadelphia: Lippincott, 1977. *A general introduction to the concepts of bioenergetics at the cellular level.*

Brafield, A. E., and M. J. Llewellyn. *Animal Energetics.* Glasgow: Blackie and Son, 1982. *Considers cellular as well as ecological energetics.*

Crawshaw, L. I., B. P. Moffit, D. E. Lemons, and J. A. Downey. "The Evolutionary Development of Vertebrate Thermoregulation." *American Scientist* 69(1981): 543–550. *A review of some current hypotheses on the development of endothermy.*

Heinrich, B. "The Regulation of Temperature in a Honeybee Swarm." *Scientific American* 244(1981): 146–161. *Discusses how bees use social interactions to regulate their body temperature.*

Hinkle, P. C., and R. E. McCarty. "How Cells Make ATP." *Scientific American* 238(1978): 104–123. *Describes the chemiosmotic hypothesis of how cells make ATP using electron transport mechanisms.*

Lehninger, A. L. *Principles of Biochemistry.* 2d. ed. New York: Worth, 1982. *An excellent biochemistry textbook.*

Shulman, R. G. "NMR Spectroscopy of Living Cells." *Scientific American* 248(1983): 86–93. *Describes how a new technique is used to measure the synthesis of ATP in whole, functioning cells.*

CHAPTER 4

Feeding and Nutrition

As discussed in the previous chapter, living involves an expenditure of energy. A bird's wings, for example, perform mechanical work as they flap. To perform this work, the bird's flight muscles use electrical energy every time a nerve impulse is conducted, and chemical energy in the form of ATP to fuel nervous and muscular activity. Each hour, an active hummingbird may use about 0.2 kcal of en-

ergy for each gram of its body weight. Freshwater protozoa and flatworms must expend energy to expel the water that enters their bodies by osmosis. Other animals expend energy to construct burrows, chase down prey, court mates, and protect their offspring.

Almost universally, the immediate source of energy is ATP (Chapter 3 describes its structure and synthesis, p. 61, Figure 3.6), but organisms use two fundamentally different methods to obtain the raw materials to synthesize ATP. **Autotrophs** ("self-feeders") synthesize organic materials from simple

---

Animal diets are highly varied. The praying mantis, a carnivorous insect, eats a variety of other insects and is sometimes cannibalistic, eating members of its own species.

substances, using energy from an external source. The most abundant autotrophs are the photosynthetic protozoa and plants. Using sunlight as the energy source, these chlorophyll-bearing **phototrophs** synthesize sugar from CO_2 and H_2O. **Heterotrophs** ("other-feeders"), by contrast, must obtain already-synthesized organic materials from the environment. Animals and many protozoa are heterotrophic. Most feed on other organisms, ingesting fluids or solid organic food particles (**holozoic nutrition**), but some absorb nutrients from their surroundings (**saprozoic nutrition**). For example, the protozoan parasites that cause sleeping sickness absorb nutrients from the bloodstream of various vertebrates, whereas tapeworms, which are flatworms without a mouth or digestive tract, absorb food from the partially digested material in their host's intestine. (*Trypanosoma* life cycles and diseases are discussed in Chapter 19, p. 447; absorption of food across the body surface of tapeworms is covered in Chapter 22, p. 520, Figure 22.8.)

The metabolic fuel value, or "richness," of an animal's food depends on the amount and proportions of certain organic molecules, chiefly sugars, fats, and proteins, and on an animal's ability to digest particular substances. Moderately fat, protein-rich beef muscle has a fuel value for humans of about 2.9 kcal/g, more than ten times the energy value of plant material of the same weight. Plant matter has a higher water content than muscle, is much lower in fats and proteins, and is rich in cellulose, an abundant plant polysaccharide that humans and most other animals cannot digest.

Evolution has provided animals with a great variety of adaptations for exploiting food of different quality and form. Thus, plant material has a high fuel value for termites, deer, and other animals with adaptations for harboring microorganisms that can digest cellulose; many insects have structures for sucking blood and plant sap; sponges are adapted to filter microscopic particles from water; and many meat eaters are adapted for rapid pursuit and quick ingestion of prey. Much of the diversity among animals is an evolutionary consequence of the enormous diversity of available food. By exploiting different kinds of food, animals reduce competition for resources.

THE SEQUENCE OF EVENTS IN FEEDING

Like people who spend money to make money, animals must expend energy to obtain energy. Feeding usually begins with the search for food (Figure 4.1).

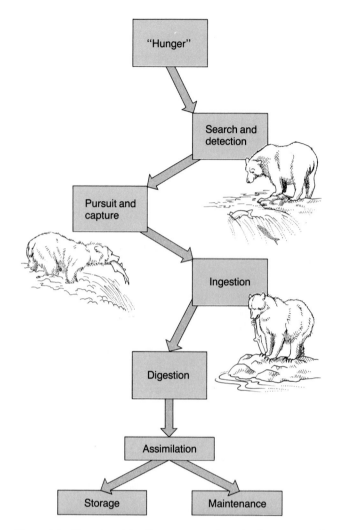

Figure 4.1 The general feeding sequence. An animal must expend energy at each step of the sequence, and the energy cost of feeding is part of the maintenance cost of the animal. To survive and reproduce, an animal must obtain more energy in its food than it expends in maintenance. Stored energy allows discontinuous feeding, along with a greater range of nonfeeding activities.

A meat eater such as a lion spends a major part of its feeding energy on the complex preliminaries of search, pursuit, and capture. But even animals that let their food come to them must spend energy in feeding. Spiders, for example, invest time and energy spinning intricate webs that trap the insects they will feed on. After food has been caught, it is eaten, or **ingested**. Then, before its energy can be used, the food must be broken down, or **digested**, into units that can be absorbed by an animal's cells. Generally, this means that food particles are physically and enzymatically broken down to relatively small molecules, primarily monosaccharides, amino acids, and fatty acids that the animal can use as energy sources or to build its own molecules. Following digestion, the feeding sequence concludes as food molecules enter and are incorporated into an animal's cells, a process called **assimilation**. Only when food molecules take part in a cell's metabolic reactions does the animal begin to gain energy for the energy it has expended in feeding.

The amount of food energy actually assimilated by an animal depends on the nature of the food. Hummingbirds feeding on sugar water are able to assimilate a remarkable 97%–99% of the food's energy. Most animals lose substantial amounts of energy in feces and urine; a mammal loses about 1.25 kcal from each gram of protein eaten (a gram of protein has an average fuel value of about 4.4 kcal for humans) because metabolism of proteins includes formation of urea, an energy-containing waste product excreted in the urine. Other organic compounds in urine and feces may add to the energy loss. Thus, many plant eaters discharge cellulose in the feces because they cannot digest it.

Food energy assimilated by an animal's cells is expended on general body maintenance, active transport across membranes, locomotor mechanisms, nervous system activity, and other ongoing functions. By feeding more or less continuously, some animals achieve a rough balance between daily energy expenditure and energy input. However, many animals have adaptations that permit intermittent, or discontinuous, feeding. In these animals, energy input during feeding periods exceeds energy expenditure. The resulting **net energy**, often stored in the tissues as glycogen, fats, or blood sugars, may then be used when the animal is not feeding. The ability to store energy has significant survival value. A nonfeeding animal can remain inactive for long periods of time, thereby avoiding the kind of exposure to predators that continuous feeders must risk. Stored energy also allows animals to care for their offspring, and the use of stored energy during hibernation or migration ensures winter survival of many species, including various snakes, squirrels, birds, and insects. Moreover, many insects feed only as larvae, storing sufficient energy for the short-lived adults to survive and reproduce without feeding.

NUTRITIONAL MODES

Animals can be grouped according to the broad categories of food they eat. Those that consume plants are called **herbivores**; those that consume other animals are called **carnivores** (meat eaters);

and those that eat both plants and animals are **omnivores** (Figure 4.2). The sequential passage of organic matter from plants to herbivores and to carnivores and omnivores is called a **food chain**.

(a)

(b)

(c)

Figure 4.2 Nutritional types among animals. (a) The koala of Australia, a herbivore, feeds on plant material (eucalyptus leaves). (b) Carnivores, such as this barn owl with a vole, eat other animals. (c) Omnivores, including cockroaches, many bears, and humans, eat both plant and animal material.

Herbivores are the key link in a food chain, for as they feed, solar energy captured by plants during photosynthesis is transferred to the animal kingdom. In aquatic environments, photosynthetic algae are the mainstay of the food chain and of herbivores. On land, animals may eat shoots, leaves, fruits, seeds, and nectar. Dietary specialization can be extreme: The koala *(Phascolarctos)* of Australia eats only eucalyptus leaves, and the silkworm caterpillar *(Bombyx)* requires mulberry leaves. But most herbivores have some flexibility in their feeding, and hunger sometimes forces them to accept foods they would ordinarily reject.

Many mammals, including dogs, walruses, and toothed whales, are carnivorous, as are frogs, hawks, snakes, and many fishes. There are also many invertebrate carnivores: Squids and octopuses eat crustaceans and fishes, and countless predatory insects eat other insects. Carnivores are rewarded with food that is especially rich in protein and fat, is compact, and is easy to assimilate. However, a great deal of energy must often be expended in catching and subduing mobile prey. Because small prey may not be economically worth catching, a carnivore's food is frequently bulky.

Carnivores overcome food bulk in one of three ways: by having a mouth and a digestive tract distensible enough to swallow prey whole, as in snakes; by reducing the food to a liquid broth with enzymes acting outside the body, a method used by many spiders and sea stars; or by having teeth for cutting and grinding food into smaller pieces.

Rats, raccoons, many bears, and people are examples of omnivorous vertebrates, which eat food of both animal and plant origin. There are also invertebrate omnivores, including cockroaches and certain snails. Are there adaptive advantages or disadvantages in being omnivorous? Omnivores avoid the hazards of overspecialization; imagine the fate of the koala should the eucalyptus disappear. However, because omnivore feeding structures are functional compromises, omnivores are less efficient at handling certain kinds of food. Compare our molars with those of a deer, or our incisors and canine ("eye") teeth with those of a polar bear, and you will see the difference (Figure 4.3). Our molars cannot grind vegetation as well as the deer's, and our canine teeth cannot tear meat as well as a polar bear's, but we can chew both plant and animal material moderately well.

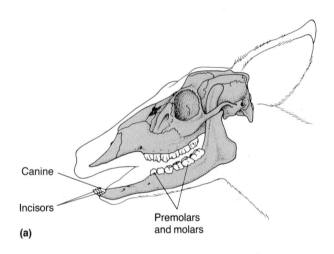

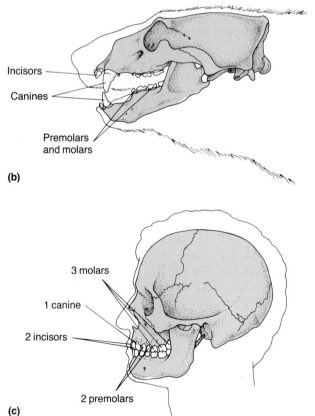

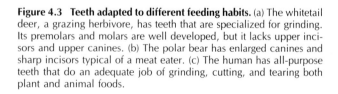

Figure 4.3 Teeth adapted to different feeding habits. (a) The whitetail deer, a grazing herbivore, has teeth that are specialized for grinding. Its premolars and molars are well developed, but it lacks upper incisors and upper canines. (b) The polar bear has enlarged canines and sharp incisors typical of a meat eater. (c) The human has all-purpose teeth that do an adequate job of grinding, cutting, and tearing both plant and animal foods.

FEEDING TYPES

Animals are often grouped according to the type of food they eat and how they obtain it. **Fluid feeders** suck or lap up fluids such as blood, plant sap, or nectar. Animals without mouthparts or a digestive tract, such as the tapeworms, could also be considered fluid feeders because they absorb nutrient-rich fluids from their hosts. **Filter feeders** are aquatic animals that strain or filter organic matter suspended in the water. Another group of animals are **substrate feeders**, eating materials on which they live. Substrate feeders include **deposit feeders**, which eat sediments such as soil or material at the bottom of a lake or ocean, and **substrate scrapers**, which eat algal and bacterial films on rocks and other surfaces. Finally, a large number of animals feed on active prey and on relatively large food masses. These feeding categories are convenient in comparing how and what animals eat, but not all animals fit into any one of these groups. Some animals change their feeding type as they mature. Fluid-feeding butterflies, for example, sip nectar, while their larvae (caterpillars) eat leaves.

Another way to group animals is by the relative size of the foods they eat. Those animals (including many filter feeders) that eat microscopic food such as bacteria, single-celled algae, and minute organic particles, are referred to as **microphagous**. Those that eat bulkier food (food masses) are **macrophagous**. Again, while these groupings are useful, it is not possible to draw sharp lines between microphagous and macrophagous animals. Many filter feeders, for example, are microphagous, while others are macrophagous; and some are both. In the following section, we will look at several types of feeding.

Fluid Feeding

Biological fluids are generally a rich source of nutrients, although certain ones are more accessible and more suitable as food than others. In mammals, blood flowing in vessels near the body surface is accessible to many bloodsucking arthropods, but any fluid deep in the body cavity is not. A suitable fluid provides a balanced diet containing the essential nutrients. Blood and plant sap supply carbohydrates, proteins and/or amino acids, vitamins, and minerals, but these are complete foods for only certain animals. Some fluid feeders drink liquids that are nutritionally incomplete or are very dilute. Many moths and butterflies drink nec-

tar, for example, but primarily for its water; they have already built up food reserves of fats and proteins during their larval (caterpillar) stage.

Many fluid feeders have similar types of adaptations. Blood-feeding ticks, mosquitoes, and vampire bats, for example, exhibit behaviors that make theft of a blood meal more likely to succeed, and they are often equipped with an anticoagulant that facilitates blood flow. Furthermore, animals that feed on blood are usually quite small compared to the host, a trait that may spare the host severe injuries, enabling it to remain available as a source of food.

The most remarkable adaptations for fluid feeding are modifications of mouth structures. Of the relatively few vertebrate fluid feeders, the sea lamprey, an eel-like jawless animal, has a funnel structure surrounding its mouth (Figure 4.4). The funnel

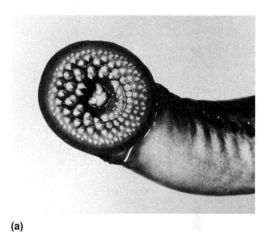

(a)

(b)

Figure 4.4 The sea lamprey, a fluid feeder. (a) The sea lamprey *(Petromyzon)* has an oral funnel with many teeth and a rasplike tongue at the center. (b) A sea lamprey attached to a carp. Sea lampreys will attack virtually any fish, but they prefer larger species, such as mackerel and cod in the ocean and lake trout or whitefish in the Great Lakes.

is lined with about 250 rasping teeth, and the mouth itself contains a rasplike tongue. The lamprey uses the funnel like a suction cup, gripping a fish tightly while the tongue rasps a hole in the body wall of the victim. The lamprey then sucks blood and body fluids from the wound, usually killing the fish. Several decades ago, lampreys virtually destroyed the commercial fishing industry in the Great Lakes. (See the Chapter 28 Essay on lampreys, p. 685.)

The only mammals that drink blood are the vampire bats *(Desmodus)* of tropical South and Central America (Figure 4.5). Vampire bats attack horses and cattle, using knife-sharp front teeth to reach the surface blood vessels. By measuring body weights of bats before and after the blood meals, zoologists have found that vampires can drink about 50% of their own body mass in blood each night. This is possible because part of a vampire bat's stomach is an expandable sac in which large quantities of blood can be stored prior to digestion in the intestine.

Plant lice, or aphids, have remarkable feeding adaptations. Some have mouthparts forming sharp, hair-thin tubes (stylets) that pierce the plants, aided by a salivary enzyme that loosens the plant cells. An aphid pumps sap into its gut, concentrating it to facilitate absorption of amino acids by expelling excess water and sugar from its anus as a sweet fluid called honeydew (Figure 4.6). Comparable "piercing-sucking" adaptations are found among mosquitoes, bed bugs, and many other bloodsucking insects.

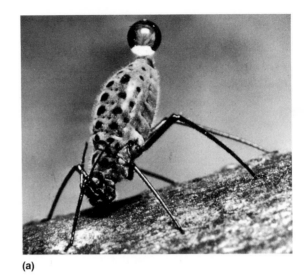

(a)

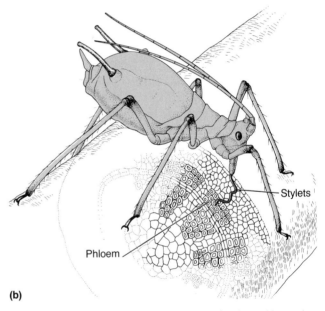

(b)

Figure 4.6 Fluid feeding by insects. (a) Note the drop of honeydew forming at the anus of this aphid as its deeply buried mouthparts (stylets) draw sap from the plant on which it feeds. (b) The stylets penetrate deeply into the food-conducting tissue (phloem) of the plant.

Figure 4.5 The vampire bat. The vampire bat *(Desmodus)* of South America is the only blood-drinking mammal. It uses razor-sharp teeth to cut the skin of sleeping horses, cattle, and birds, and then laps at the oozing wound.

Many fluid feeders are economically and medically important. Aphids and certain nematodes (roundworms) suck fluids from plants, causing serious crop losses. Bloodsucking insects transmit a number of serious diseases. Mosquitoes transmit malaria, and tsetse flies transmit certain flagellated protozoa *(Trypanosoma)* that cause sleeping sickness. Bubonic plague, famous as the Black Death of the Middle Ages, is a bacterial disease of rodents transmitted to humans by bloodsucking fleas. Plague is still a potentially dangerous disease; cases continue to be reported each year, including several in the United States. Bloodsuckers can also cause serious disease because of the amount of fluid they remove. Hookworms (Phylum Nematoda) pump blood from wounds they make in the intestinal lining of humans and dogs. Blood loss can be significant enough to cause severe anemia (reduced numbers of red blood cells).

Filter Feeding

Only animals with special adaptations can extract suspended material from the surrounding environment. Filter feeding requires a filtering apparatus and a way to make water flow through the filter, but it enables sessile (immobile) and slow-moving animals to exploit a vast supply of food not generally available to other animals. The abundance and food value of suspended food is indicated by the great diversity of filter-feeding animals. Most animal phyla have members that obtain food this way, and filter feeders occupy important positions in aquatic food chains. By eating nonliving organic particles, filter feeders make some of the energy from this food resource available to other animals. Living organisms suspended in water are collectively called **plankton** (from the Greek *planktos,* "wandering"). Planktonic organisms drift with water currents and waves, for they are unable to control their general course. Filter feeders may select **phytoplankton** (suspended algae and bacteria), **zooplankton** (suspended animals), or all these organisms. Phytoplankton are the major producers of organic matter in aquatic environments, and filter feeders that eat suspended algae and bacteria are an important link between the producer and higher consumer levels in aquatic food chains. The largest and many of the smallest animals are filter feeders (see the Essay on the next page).

Many filter feeders use mucus as a food trap, usually in conjunction with cilia or other structures for moving mucus to the mouth. *Chaetopterus,* a tube-dwelling marine annelid, uses flaplike structures called parapodia to pump water through its burrow and through mucus food traps (Figure 4.7). Periodically, a small ball of food and mucus forms and is passed to the mouth.

Other filter feeders that rely on cilia and mucus to trap food particles include many jellyfishes, corals, sea anemones, bivalve molluscs, and certain echinoderms. (Sea lilies and feather stars are described in Chapter 27, p. 657; Figure 27.6.) Clams and oysters have gills that are probably more important in food filtering than in respiration (Figure 4.8). Observations made through small glass win-

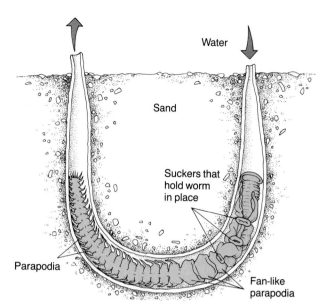

Figure 4.7 The marine annelid *Chaetopterus*. *Chaetopterus* pumps water through its U-shaped burrow with fan-shaped parapodia. Food particles trapped in mucus are rolled into a food ball and passed to the mouth.

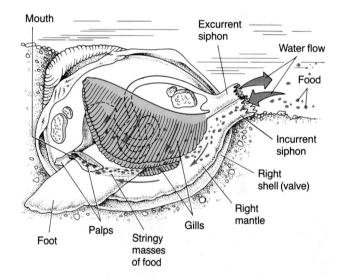

Figure 4.8 Bivalve molluscs. Bivalve molluscs are filter feeders. In this drawing, the left valve and left mantle have been cut away to show the feeding mechanism. Food particles deposited on the gills are moved by ciliary action to the gill margins (see arrows). The particles become trapped in a string of mucus that is moved anteriorly and guided into the mouth by the palps.

dows mounted in holes in oyster shells indicate that some sorting of particles occurs, with only the smallest particles being retained. Oysters appear to sample water with their gills, only intermittently secreting the mucus needed to trap food particles.

Not all filter feeders use cilia or mucus. Sponges use flagellated collar cells to draw in water and capture food particles. Barnacles, which are sessile

Essay: The World's Largest Animals Are Filter Feeders

The blue whale *Balaenoptera musculus,* reaching lengths of about 30 m and weights of about 102,000 kg, is the largest animal that has ever existed. This huge species and its main food illustrate the significance of filter feeding in ocean food chains. The blue whale is a filter feeder that subsists primarily on small, shrimplike planktonic crustaceans, commonly called **krill**. Krill are also filter feeders, straining smaller organisms from the sea.

The blue whale and other filter-feeding whales are toothless. They are called baleen whales because they have a brushlike series of 100 to 400 fringed plates of **baleen** (also called whalebone) located on each side of the upper jaw (Figure E1). The baleen is used in filter feeding. Blue and humpback whales represent one subgroup of baleen whales called the *finner whales. Right whales* constitute a second subgroup of baleen whales. Finner whales generally feed on krill, whereas right whales eat mainly smaller crustaceans (copepods) that congregate at the ocean surface. Methods of filter feeding differ somewhat between the two subgroups. In right whales, baleen fills the gap between the jaws even when the mouth is wide open. As a right whale swims, water streams into its mouth and out through the baleen. Minute animals entangled on the baleen are licked back into the gullet by the tongue. In contrast, finner whales gulp food and water into an expandable oral pouch (Figure E1a). When the mouth closes, the pouch muscles contract, forcing water out through the baleen and leaving a mass of food in the mouth.

Humpback whales use a remarkable behavioral adaptation for concentrating krill before filtering it from the water. Beginning about 15 m below the surface, a humpback swims slowly in an upward spiral, blowing bubbles from its blowhole. The quickly rising bubbles form a cylindrical screen, or *bubble net,* and krill and small fishes show an avoidance reaction that concentrates them at its center. The whale then swims with its mouth open up through the center of the bubble net, harvesting the catch. An adult humpback's stomach can hold up to 590 kg (1300 lb) of krill.

(a)

Baleen

Krill, fishes

Furrows

(b)

(c)

Figure E1   Baleen whales. (a) Illustration of a finner whale gulping food and water into its expandable pouch. When the pouch muscles contract, water is forced out through the baleen, leaving the food in the mouth. (b) The black right whale *(Eubalaena sieboldi)* feeds on copepods at the water surface. (c) The humpback whale *(Megaptera novaeangliae),* a finner whale, is seen here concentrating its food by means of a "bubble net." (Additional information on whales is given in Chapter 33.)

crustaceans, use bristly legs to filter food particles from the sea, while another crustacean, the mole crab, uses antennae in much the same way.

Macrophagous filter feeders include the blue whale, which feeds on relatively large (thumb-sized) crustaceans filtered from Antarctic waters. The American flamingo *(Phoenicopterus ruber)* is both macrophagous and microphagous; it uses its beak to filter macroscopic invertebrates and microscopic algae from muddy pond bottoms in the Florida everglades. Two species of African flamingos are able to coexist on the same lake without competing for food because one *(Phoenicopterus antiquorum)* has a coarse filter adapted for removing invertebrates from the bottom mud, whereas the other *(Phoenicopterus minor)* has a fine filter that retains microscopic algae. The whale and the flamingos show that filter feeding is not restricted to the sessile or sedentary members of the animal kingdom, but can be exploited by very active animals too.

Substrate Feeding

The food of substrate feeders is inactive and consists of material from the substrate or from subsurface tunnelings. The animal kingdom is rich in substrate feeders, including many marine worms and molluscs, as well as scores of insects that tunnel in plant leaves and stems.

Deposit Feeding. Earthworms are especially interesting representatives of this feeding type, and they have been studied intensively. In contrast to many *selective* deposit feeders that eat only organic material, the earthworm is *nonselective*, literally plowing its way through the earth while using a powerful muscular pharynx to suck in a mixture of inorganic soil particles, moisture, and partially decomposed organic matter. Nutrients are extracted, and the undigested residue, amounting to the bulk of what is ingested, may be expelled to the soil surface as castings. It has been estimated that the top 10 cm of a typical pasture passes through the gut of earthworms in about 12 years where the worm population is high. In this way, nutrients are recycled back to plants, and the soil is aerated and improved in texture.

Substrate Scraping. Many snails rasp food such as algae from surfaces using a filelike **radula** located just inside the mouth. (Snail radulae are considered in Chapter 24, p. 569; Figure 24.14.) Tropical-fish owners sometimes take advantage of these

feeding habits, using snails to clear algae from aquarium walls. Snails and other substrate scrapers may also eat films of bacteria that grow on decaying organic matter. Mucus and other organic molecules on animal feces are common substrates for bacterial growth.

The shipworm *Teredo*, a marine bivalve mollusc, is another type of substrate feeder. It lives in and eats wood. The shipworm's shell consists of tiny valves with cutting edges. The valves open and close like jaws, rasping into wood (Figure 4.9). Recent studies have shown that these molluscs have a unique gland (called the gland of Deshayes) that harbors bacteria that can digest cellulose. Thus, the shipworm eats wood, the bacteria digest it, and both the bacteria and the shipworm use the simple sugars that result. This is an example of **mutualism**, a relationship between two species that is beneficial to both. Research has also shown that the shipworm diet is deficient in amino acids and that its mutualistic bacteria produce nitrogenous compounds that the shipworm can use to make up the deficiency. Without this resource, the shipworm would be unable to synthesize proteins and nucleic acids.

Feeding on Active Food and Food Masses

A great number of animals and protozoa feed on active food and food masses. All are macrophagous. Two subgroups within this large category are organisms that ingest food without preliminary pro-

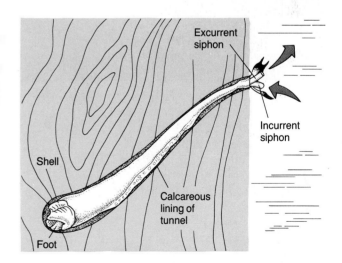

Figure 4.9 The shipworm *Teredo*. *Teredo* lives in a tunnel that it has bored in wood. The animal is attached to the inside of the tunnel by its small foot. Shipworms play an important ecological role in helping to decompose floating and submerged wood in the ocean, but are also quite destructive, weakening wharf pilings and the hulls of wooden ships.

cessing and those that cut, chew, or enzymatically reduce their food before swallowing it.

A wide variety of animals and protozoa ingest their food whole, and this has the advantage of speed. Many animals, safe from predators while concealed or motionless, bolt their food and return to cover to minimize danger. Rapid eating is also advantageous for animals whose food is small or fast-moving; a bird that catches insects in mid-air would soon starve if it ate slowly.

Amoebas use flexible extensions called pseudo-podia to engulf small organisms and particles, a process called **phagocytosis**. (Cell feeding by endo-cytosis is discussed in Chapter 2, p. 42.) An amoeba extends its pseudopodia toward the food object, enveloping it with the cell membrane (Figure 4.10). Food in contact with the amoeba's cell membrane for as little as 2 seconds stimulates engulfment, and a membrane-bounded **food vacuole** forms within a few minutes. Zoologists studying feeding by the

amoeba *Chaos carolinense* determined that it could form about 100 food vacuoles per day. They also calculated that these vacuoles had a total mem-brane area ten times that of the amoeba itself! The energy that *Chaos* expends in manufacturing mem-branes is thus a major part of its cost of feeding.

Many macrophagous invertebrates use a pump-ing and sucking action to swallow food whole. They usually kill or immobilize their prey before eating it, and muscles in the anterior part of their diges-tive tract provide the swallowing action. Many cnidarians, such as *Hydra*, are **tentacular feeders**. When small crustaceans or other suitable prey swim close to or touch a hydra's tentacles, nemato-cysts may discharge, entangling or impaling the prey. The tentacles then carry the whole victim to the hydra's mouth. (Nematocysts, stinging cap-sules that occur in cells called cnidocytes, are de-scribed in Chapter 21, p. 499; Figure 21.11.)

The majority of animals break their food down physically and/or enzymatically before eating it, and this has a number of advantages. Larger food items become accessible; a predator can kill prey larger than it could swallow whole, even larger than itself, then cut and chew the prey into pieces small enough to swallow. Subdividing also makes food more digestible, for it exposes more surface area on which digestive enzymes can act.

Animals that secrete enzymes into or onto their prey before swallowing it include many flatworms (planarians). The freshwater planarian *Dugesia* entangles small aquatic animals in mucus and then protrudes the anterior part (pharynx) of its gut and thrusts it through the victim's body wall. Once in-side the prey, the probing pharynx breaks up the tissues with the aid of enzymes, and the planarian

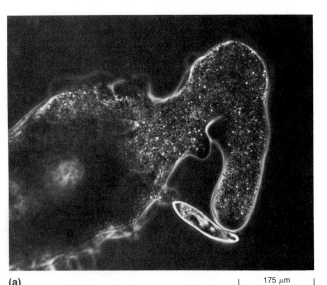

(a)

175 μm

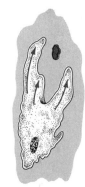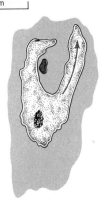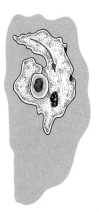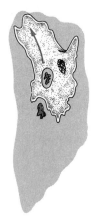

(b)

Figure 4.10   Phagocytosis. (a) The amoeba in the photograph is ingesting a *Paramecium* by phagocytosis. (b) These drawings show the sequence of pseudopodial formation and food engulfment during phagocytosis. Amoeboid blood cells and other amoebocytes in animals engulf particulate material and fluids in a similar way.

sucks the resulting juice. Many sea stars and spiders also predigest their prey with enzymes. Spiders inject paralyzing venom into insects caught in the web and then trickle digestive enzymes into the wounds made by the poison fangs. A spider can then readily draw up the victim's liquified body contents.

Most animals, including many vertebrates, use mouthparts to cut and chew their food before swallowing it. Digestive enzymes are often mixed with food as it is chewed. Many insects have chewing mouthparts, and many crustaceans have powerful, molarlike mandibles that crush and chew prey. Among the molluscs, squids and octopuses have mouthparts resembling a parrot's beak. Squids hold a fish with their tentacles while using their mouthparts to cut off the head and chop the body into bite-size pieces. Octopuses are too slow to catch most fishes, but fast enough to capture crabs and clams, kill them with poison from their salivary glands, and bite out the body contents.

DIGESTION AND ABSORPTION

Food is usually a complex mixture of many substances, including the nutritionally important carbohydrates, fats, and proteins. Carbohydrates provide energy and are a source of carbon atoms for use in many synthesis reactions. Fats also provide energy and play a key structural role as components of cell and organelle membranes. Proteins supply the amino acids that are rearranged to make most animal structures as well as the enzymes that catalyze metabolic reactions. Amino acids can also be used as an energy source through reactions involving the Krebs cycle (see Figure 3.9). The average fuel values of carbohydrates, fats, and proteins in human nutrition are about 4 kcal/g, 9 kcal/g, and 4.4 kcal/g, respectively, and in a well-balanced human diet, between 55% and 70% of the energy comes from carbohydrates.

Digestion prepares carbohydrates, fats, and proteins for assimilation into the animal by breaking them down into molecules small enough for **absorption** by the digestive tract. Most digestion is accomplished by enzymatically catalyzed hydrolysis. (Chapter 3 considers the hydrolysis of a dipeptide, p. 57; Figure 3.2.) Complex carbohydrates are hydrolyzed to monosaccharides, certain fats to glycerol and fatty acids, and proteins to their constituent amino acids. As mentioned in Chapter 2,

amino acids that an animal cannot synthesize and must obtain in its diet are called the **essential amino acids** (although the ones an animal can produce itself are no less essential for survival).

Most animals require carbohydrates, fats, and proteins, often called the **macronutrients**, as well as vitamins and minerals (the **micronutrients**) and water. Vitamins and minerals are needed for many metabolic functions. Minerals also occur in many body structures (e.g., in the bones and teeth of vertebrates), and some minerals are part of essential molecules (e.g., the iron in hemoglobin). In addition to carbon, hydrogen, oxygen, and nitrogen, which together form about 96% of the structural materials in living tissues, more than 20 other chemical elements seem to be essential (i.e., they must be supplied in the diet), in some cases in trace amounts. Tables 4.1 and 4.2 briefly describe the vitamins, the important mineral nutrients, and some of the trace minerals required by humans. Other animals have somewhat different needs; cats, for example, can produce their own vitamin C.

Anatomical Considerations

Through the course of evolution, methods for obtaining and processing food have generally become more elaborate. Protozoa such as *Amoeba* and *Paramecium* ingest food and then digest it in special food vacuoles. This is called **intracellular digestion**; it is the simplest and was no doubt the earliest type of digestion to occur within living organisms. Intracellular digestion is a relatively slow process, and only minute bits of food or dissolved substances can be processed in food vacuoles. Among animals, all of which are multicellular, digestion is exclusively intracellular only in sponges. Other animals are more active and complex, with most of their body cells specialized for specific functions other than digestion. Also, most body cells are not adjacent to cells that process food. Consequently, most animals require rapid digestion and rapid transport of digested nutrients to cells in all parts of the body. This efficient processing of food is achieved by means of various types of **digestive tracts**, (also called **guts** or **alimentary tracts**), adapted specifically for digestion and absorption. **Extracellular digestion** (i.e., digestion outside of cells) occurs in the gut in vertebrates and most invertebrates. Certain invertebrates use both intracellular and extracellular methods: Digestion begins in the gut and is completed by phagocytic cells lining the gut or by wandering amoebocytes (amoeba-like cells).

Table 4.1 Essential Vitamins

Vitamin	Physiological Function in Humans	Symptoms of Deficiency	Food Source
Fat-Soluble Vitamins			
Vitamin A	Part of visual pigments; maintains epithelia	Night blindness, infections	Vegetables, milk, eggs
Vitamin D	Absorption of calcium	Bone softness (rickets)	Liver oils, milk, eggs
Vitamin E	Not completely known; may protect cell membranes or aid electron transport	Sterility, muscular dystrophy, anemia	Green vegetables, seeds, oils
Vitamin K	Prothrombin synthesis during blood clotting	Hemorrhage (slow clotting)	Green vegetables, liver
Water-Soluble Vitamins			
Vitamin B_1 (thiamine)	Decarboxylation in Krebs cycle	Beriberi (neurological disorder)	Whole grains, meats
Vitamin B_2 (riboflavin)	Part of FAD^+	Eye and skin lesions	Whole grains, eggs, milk
Vitamin B_3 (niacin)	Part of NAD^+	Pellagra (skin and mental disorders)	Whole grains, meats, legumes
Vitamin B_6 (pyridoxine)	Amino acid metabolism	Nervous disorders, dermatitis	Whole grains, meats, vegetables
Pantothenic acid	Component of Coenzyme A	Nervous disorders, impaired antibody formation	Whole grains, meats, eggs
Vitamin B_{12} (cobalamin)	Nucleic acid metabolism	Anemia, nervous disorders	Eggs, milk, meats
Folic acid	Nucleic acid metabolism	Anemia (impaired red blood cell formation)	Whole grains, meats, vegetables
Biotin	Fat and amino acid metabolism	Dermatitis, muscle pains	Egg whites, meats, vegetables
Vitamin C (ascorbic acid)	Amino acid metabolism	Scurvy (breakdown of capillary walls)	Citrus fruits, fresh vegetables

The simplest gut has a single opening, the mouth, that admits food and expels wastes. Zoologists call this an **incomplete gut** because there is no anus. Cnidarians and many flatworms have an incomplete gut that is saclike in some cases, but branched in others. The branches increase the absorptive surface area and help distribute food. Most animals more complex than flatworms have a **tubular (complete) gut**; in this type of system, food enters the mouth and travels in only one direction through the gut, and indigestible matter passes out the anus. Evolution of a complete gut with its one-way passage of food allowed specializations to develop along the length of the gut; thus, the digestive tract of many animals includes regions adapted for food storage, grinding and mixing, digestion, absorption, and waste storage. The following sections describe these adaptations.

Physical Digestion

In most animals that eat solid food, the first step in digestion is physical, involving mechanical cutting and grinding. Mouth structures used to break up food are quite variable and correlate with the kind of foods that an animal eats. The mandibles of insects and crustaceans, the radulae of snails, the beaks of squids and octopuses, and the teeth and beaks of vertebrates are examples. Some animals soften and fragment their food internally, after they swallow it. Many animals have a saclike organ called the **crop** in the anterior part of the gut. Food

Table 4.2 Major Mineral Nutrients and Some Trace Minerals

Mineral	Physiological Function in Humans	Symptoms of Deficiency	Food Source
Calcium	Muscle, nerve, and blood function	Softened bones, muscle tetanus	Milk, eggs, fish, legumes
Chlorine	Acid-base and osmotic balance	Muscle cramps	Most foods, table salt
Cobalt*	Synthesis of hemoglobin	Anemia	Meat
Copper*	Synthesis of hemoglobin	Anemia	Meat, liver
Fluorine*	Inhibits mouth bacteria	Tooth decay	Milk, many toothpastes
Iodine*	Part of thyronines (hormones)	Goiter and cretinism	Iodized salt, fish
Iron*	Part of hemoglobin and cytochromes	Anemia	Liver, eggs, raisins
Magnesium	Nerve and muscle function	Muscle tetanus	Meat, milk, green vegetables
Manganese*	Enzyme cofactor	Reduced cell respiration	Whole grains, leafy vegetables
Molybdenum*	Part of xanthine oxidase	Impaired nitrogen metabolism	Milk, grains, leafy vegetables
Phosphorus	Constituent of bone, lipids, and nucleotides	Mineral loss from bones	Meat, milk, grains
Potassium	Nerve and muscle function, membrane potential	Changes in nerve and muscle action	Most foods
Selenium*	Part of glutathione peroxidase	Oxidative damage to membranes(?)	
Sodium	Osmotic balance, membrane potential	Kidney failure, impaired muscle action	Most foods, table salt
Sulfur	Part of many proteins	Reduced growth	Meats, eggs, milk
Zinc*	In some enzymes and insulin	Skin disorders, testis failure	Most foods

*Trace minerals.

is stored temporarily and softened here. Birds lack teeth and swallow food whole into a crop. They grind their food in a muscular **gizzard**, a part of the stomach lined with hard plates. Many bird species swallow pebbles or other gritty materials that mill the food as the walls of the gizzard move. As the grit wears smooth, it passes through the digestive tract and must be replaced. Without grit, even well-fed birds lack the means to digest the food they ingest and may starve. Many invertebrates also have a gizzard or similar internal grinding organ. Many insects and crustaceans grind food in a toothed gizzard. Earthworms have a gizzard that crushes and churns food before forcing it into the intestine.

Chemical Digestion

The second step in the process of extracting nutrients from food is **chemical digestion**, during which large food molecules are enzymatically hydrolyzed into smaller ones. The actions of some of the major digestive enzymes are summarized in Table 4.3. Animal foods are highly variable, and there is great variation in the difficulty of processing them. Vegetation presents the problem of cellulose, which few invertebrates and no vertebrates can utilize without the aid of cellulose-digesting microorganisms; blood and plant sap are nutritious but inconveniently dilute, and must be concentrated; other substances, such as honey, muscle, or egg yolk, provide both concentrated and relatively pure nutriment. Despite such differences in foods, and perhaps reflecting the common ancestry of animals, the general sequence of chemical digestion is similar throughout the animal kingdom.

In most animals with tubular digestive tracts, chemical digestion begins in the **mouth** (or **buccal**) **cavity**. Food then passes to the **pharynx** and is swallowed, reaching the stomach via a tube called

Table 4.3 Major Digestive Enzymes

Category	Action or Substrate	Products
Carbohydrases	Carbohydrate hydrolysis	
Amylases	Starch, glycogen	Maltose, glucose
Glucosidases	Disaccharide hydrolysis	
Sucrase	Sucrose	Glucose, fructose
Maltase	Maltose	Glucose
Lactase	Lactose	Glucose, galactose
Esterases	Fat hydrolysis	
Lipases	Neutral (true) fats	Fatty acids, monoglycerides
Proteases	Protein hydrolysis	
Exopeptidases	Break terminal peptide bonds	
Carboxypeptidases	Carboxyl end of chain	Free amino acids
Aminopeptidases	Amino end of chain	Free amino acids
Endopeptidases	Break peptide bonds within protein molecules	
Pepsin		Shorter chain lengths
Trypsin		Shorter chain lengths
Chymotrypsin		Shorter chain lengths
Dipeptidases	Split dipeptides	Free amino acids
Tripeptidases	Split tripeptides	Free amino acids, dipeptides
Tetrapeptidases	Split tetrapeptides	Free amino acids, tripeptides

the **esophagus**. In many animals, the walls of the pharynx and esophagus are lined with smooth muscles that assist swallowing and squeeze food toward the stomach. In some invertebrates, however, the gut lacks muscles and food is propelled through it by cilia on cells lining the gut cavity. The **stomach** is an enlarged, muscular part of the alimentary tract where food may be stored temporarily and where chemical digestion continues (or begins if it has not started in the mouth). After being partially digested in the stomach, the food passes to the **intestine**, where most chemical digestion takes place. In many animals, the intestine is subdivided; vertebrates have a small intestine and a large intestine, for example, and insects have an ileum and a rectum. The products of digestion are absorbed into the body fluids from cells in the intestine and transported throughout the body. In most animals, a circulatory system takes care of this transport, aided in vertebrates by the lymphatic system. (The lymphatic system is described in Chapter 6, p. 136; Figure 6.16.) In many invertebrates, the transport medium is fluid present in the body cavity. After the useful components of food have been absorbed, undigested wastes are moved to the posterior end of the digestive tract

and accumulated as **feces** in a **rectum**. At intervals, the feces are expelled through the **anus**.

Digestion in many animals depends on the actions or secretions of organs connected to the alimentary tract. Thus, digestive glands in lobsters and clams receive and digest fine food particles from the nearby stomach. Saliva from salivary glands moistens and digests food in many arthropods. Secretions from the vertebrate salivary glands and pancreas, as well as bile from the liver, flow into the alimentary tract and play important roles in digestion. The alimentary tract and its associated organs constitute an animal's **digestive system**. The organs and general functions of the vertebrate digestive system are summarized in Figure 4.11.

There are many adaptive modifications to the pattern just described. Some of the most remarkable adaptations are seen in herbivores. Compared to a carnivore's intestine, that of an herbivore tends to be relatively long, and passage through it is relatively slow, reflecting the difficulty of digesting plant material. In some animals, this difference can be seen in the life of a single individual. Tadpoles, for example, are alga-eating herbivores that develop into carnivorous (insectivorous) frogs. Corre-

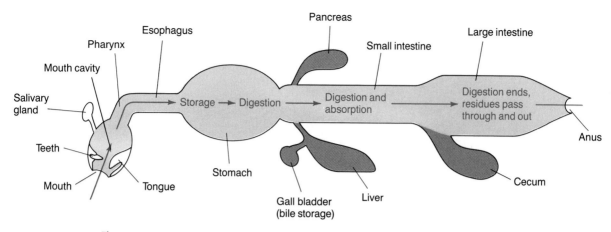

Figure 4.11 Sequence of functions in a typical vertebrate digestive tract. Without the teeth, tongue, pancreas, liver, and gallbladder, the drawing would apply to many invertebrates. A number of invertebrates have ceca in various locations, as well as organs comparable in function to the vertebrate liver.

lated with the difference between herbivory and carnivory, the intestine of a full-size tadpole is longer (both relatively and absolutely) than that of an adult frog.

Some of the most important digestive adaptations are those that maintain the mutualistic relationships between certain herbivores and cellulose-digesting microorganisms. The shipworm *Teredo* was mentioned earlier in the chapter. Among the insects, termites ingest wood that is then digested by microorganisms in their intestine. (Chapter 19 discusses the cellulose-digesting flagellates *Trichonympha*, p. 448; Figure 19.16.) Although termites can do severe damage to wooden dwellings, they are important in the ecology of tropical forests, for they help to recycle carbon and other elements present in dead and fallen timber. Among the vertebrates, some bacterial digestion of cellulose occurs in the large intestine of herbivores such as rabbits and horses. These and other herbivores and certain omnivores have one or more blind pouches called **ceca** that extend from the intestine. Ceca contain microorganisms capable of hydrolyzing cellulose. Rabbits have a particularly large cecum and a **vermiform appendix**. Both of these organs house cellulose-digesting bacteria, but they are located at the junction of the small and large intestine, beyond the point where most products of digestion are absorbed from the gut. Therefore, to absorb the material digested by the bacteria, rabbits must expel it as fecal fluid and reingest it, a process called **coprophagy**. Food may therefore be exposed to the rabbit's digestive processes twice. The human also has a cecum and a vermiform ap-

pendix (see Figure 4.15), but these organs are small and do not contribute to cellulose digestion. Perhaps the most remarkable adaptations for processing plant material are the multichambered stomachs of ruminant mammals such as cattle (see p. 91), sheep, deer, and camels.

FOOD PROCESSING IN REPRESENTATIVE ANIMALS

In this section, we illustrate some of the similarities and differences in digestive systems by comparing the digestive processes of four animals—the grasshopper and the cow (both herbivores) and the earthworm and the human (omnivores).

The Grasshopper

The digestive tract of a grasshopper is shown in Figure 4.12. Vegetation chewed by the mouthparts and moistened by saliva from the salivary glands enters the crop via the pharynx and esophagus. Food is further pulverized by the toothed gizzard and then passes into the stomach, where most of the chemical digestion takes place. With digestion complete, the food products are absorbed into the circulatory system. Using synthetic glucose and amino acids containing radioisotopic forms of carbon and hydrogen as tracers, researchers have found that food products in the stomach are moved forward into the gastric ceca and from there into

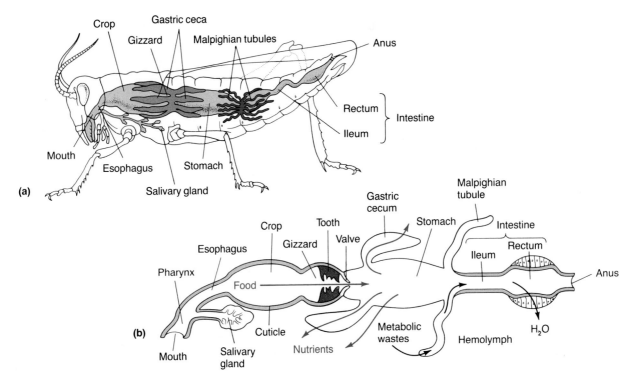

Figure 4.12 Digestive tract of a grasshopper. Food passes to the crop for storage, is ground in the gizzard, and is chemically digested in the stomach. Products of digestion enter the hemolymph by way of the gastric ceca and stomach. Metabolic wastes enter the gut through the Malpighian (excretory) tubules, and water and other valuable substances are recovered by cells of the rectum. (a) The anatomical location of digestive organs. (b) The paths followed by food and the products of digestion and metabolism.

the surrounding circulatory fluid (hemolymph). The undigested food that remains in the gut mixes with urine from the Malpighian (excretory) tubules and moves into the intestine as feces. In a very important last step, water is recovered from the feces and returned to the circulatory fluid by specialized cells in the wall of the rectum. Fecal pellets ranging from pasty to dry are then expelled from the anus. The feces are quite bulky because, as in other leaf-eating insects, the efficiency of food utilization by grasshoppers is poor. A juvenile grasshopper may digest and absorb only about 40% of the food it eats, and it assimilates an even smaller percentage of the food's caloric value.

The Cow

The digestive system of cattle and other **ruminants** is highly effective in extracting nutrients and energy from plant material. Digestion in their four-chambered stomach has been intensively studied (Figure 4.13). The **abomasum**, or true stomach, has digestive glands and opens directly into the intestine. In calves, the abomasum is relatively large

and provides a milk-coagulating enzyme called **rennin**. The other three chambers have a corneous lining and are derived from the esophagus. As cattle feed, they swallow masses of incompletely chewed grass into the **rumen** and **reticulum**. Here, billions of anaerobic microorganisms (cellulose-digesting bacteria and starch-digesting ciliates) reduce the grass to a pulp. The reticulum compresses the pulp into masses of cud, which a cow regurgitates and rechews at leisure (rumination), crushing the plant fibers. After reswallowing, the food passes to the rumen for further bacterial action and then passes, in succession, to the reticulum, **omasum**, and abomasum. Alternatively, the reswallowed food may bypass the rumen and pass directly to the reticulum and then to the omasum and abomasum. Gastric juices secreted in the abomasum complete the digestive process. In this mutually beneficial relationship, microorganisms have a warm, oxygen-free nutrient medium in which to live and reproduce, and the cow uses the cellulose breakdown products the microorganisms provide. Actually, the microorganisms themselves become a large part of the host's food; some of the

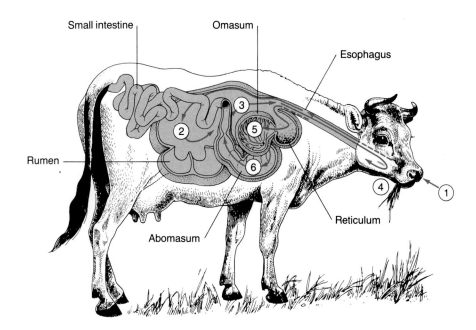

Figure 4.13 **Movement of food through the stomach of a cow.** Partly chewed food (1) enters the first two chambers, the rumen (2) and the reticulum, where bacteria and protozoa partly digest it. The cud (3) is returned to the mouth for further chewing (4) and then is reswallowed, passing to the omasum (5) and abomasum (6). After digestion with gastric juice, the material enters the intestine, where products of digestion are absorbed.

bacteria and protozoa that feed on bacteria are digested with the plant material, providing the host with important amino acids.

The Earthworm

The structure of the digestive tract of the earthworm is shown in Figure 4.14. Earthworms live in soil rich in decaying animal and plant matter. As an earthworm burrows, its powerful muscular pharynx, assisted by the liplike **prostomium** above the mouth, moves soil and organic material into the esophagus. The food passes from the esophagus into the crop, a storage organ, and then into the gizzard, where it is finely ground. After this phase

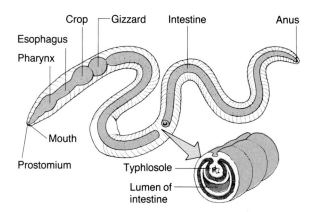

Figure 4.14 **Digestive tract of an earthworm.** A longitudinal view of a worm, showing the organs of the digestive tract. The cross section shows how the gut is invaginated to form the typhlosole, greatly increasing inner surface area.

of physical digestion, the food is forced into the intestine, which secretes enzymes that digest proteins, starches, fats, and possibly cellulose. (Reports differ on whether earthworms secrete the cellulose-digesting enzyme cellulase or rely on microorganisms in their intestine to digest cellulose.) Following digestion, nutrients are absorbed into the blood and body cavity fluid, which transport them throughout the body. Much of what an earthworm ingests is indigestible. Mixed with other wastes, this material passes out of the anus and, in some species, is piled at the burrow entrance as castings.

The earthworm intestine illustrates a general principle: Surface-dependent processes such as digestion and absorption are usually associated with extensive surface areas. Thus, the earthworm's intestine is long—roughly three-quarters the length of its body. The intestine is also deeply folded along its dorsal side, forming a **typhlosole** that increases the inner surface area. A variety of comparable adaptations increase the digestive and absorptive surface area in other animals. Internal folds and ridges are commonplace, as are the villi and microvilli characteristic of the human digestive tract.

The Human

The human digestive system, shown in Figure 4.15, consists of the alimentary tract (mouth, pharynx, esophagus, stomach, and intestine) and accessory digestive organs (salivary glands, liver, gallblad-

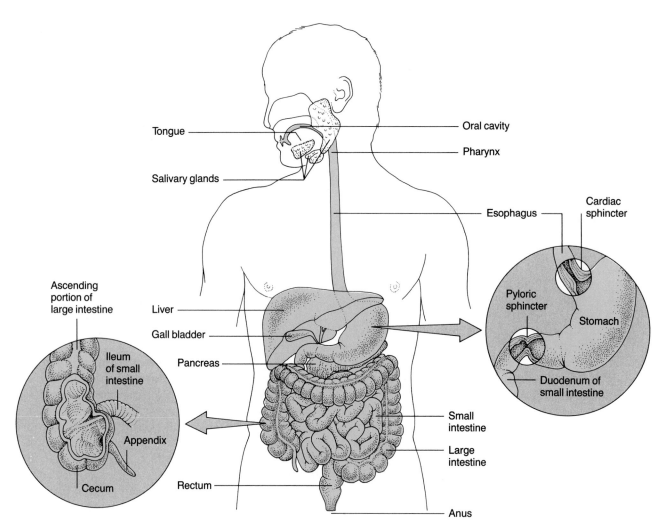

Tongue

Oral cavity

Pharynx

Salivary glands

Esophagus

Cardiac
sphincter

Ascending
portion of
large intestine

Liver

Pyloric
sphincter

Stomach

Ileum
of small
intestine

Gall bladder

Pancreas

Duodenum of
small intestine

Appendix

Small
intestine

Large
intestine

Cecum

Rectum

Anus

Figure 4.15 Principal components of the human digestive system.

der, and pancreas). The intestine is divided into a small intestine, into which the stomach empties, and a large intestine. The small intestine has three discrete regions: Its anterior 25 cm or so (the region closest to the stomach) is called the **duodenum**; its next 2.5 m is the **jejunum**; and the remaining 3.5 m is the **ileum**. The large intestine, also called the **colon**, is about 1.5 m long and extends from the posterior end of the ileum to the anus. The rectum is the highly muscular posterior region of the colon.

Chemical digestion begins in the mouth, where **salivary amylase** in the **saliva** catalyzes the hydrolysis of starch to the disaccharide maltose. Because starch is a polysaccharide formed in plant cells, the ability to digest it is particularly important for omnivores and herbivores. Saliva, which also contains water, salts, and the lubricating protein **mucin**, moistens the membranes of the mouth cav-

ity and makes food particles stick together in a mass called a **bolus**. The bolus is swallowed and then pushed toward the stomach by waves of circular muscle contractions called **peristalsis** passing down the esophagus. As the bolus passes along the esophagus, the lower esophageal **cardiac sphincter** relaxes, allowing the bolus to enter the stomach. (A sphincter is a ring of muscle capable of closing a tubular structure by constriction.)

Stomach Function. The human stomach performs several important tasks. First, it can store nearly a liter of food for several hours. Storage promotes efficient digestion by allowing the body to regulate the timing and amount of food that enters the small intestine. Second, the stomach mixes the food as part of the process of physical digestion. Gentle contractions of the stomach wall churn the food with digestive secretions. Third, the stomach plays an important role in chemical digestion.

Enzymatic digestion in the stomach is limited mainly to the hydrolysis of proteins by **pepsin**, an endopeptidase (see Table 4.3). Pepsin is a powerful protein-digesting enzyme. As a matter of self-protection, cells produce and release pepsin and other digestive enzymes in inactive forms called **zymogens**. Before they can catalyze the hydrolysis of foods, zymogens must be activated, usually enzymatically. The zymogen form of pepsin is called **pepsinogen**. Millions of gastric glands release some 2–3 L of **gastric juice** into the stomach every day (Figure 4.16). Gastric juice is a mixture of pepsinogen secreted by **chief cells**, hydrochloric acid secreted by **parietal cells**, and mucus secreted by **mucous cells**. Hydrochloric acid lowers the pH of the stomach environment to about 1.5. This helps break down proteins and kills some microorganisms. The stomach's acidity and previously activated pepsin convert pepsinogen to pepsin. The mucus forms a thin layer over the cellular lining of the stomach, protecting it from acid damage and from self-digestion by pepsin. The mucus layer also limits absorption from the stomach; only alcohol, various drugs including aspirin, and certain spices are absorbed here. Absorption of such substances from the stomach rather than from the intestine accounts for their rapid effects.

Chemical digestion in the stomach eventually results in a mixture called **chyme**, composed of gastric juice and semifluid food. Chyme is moved by muscular action toward the **pyloric sphincter** at the posterior end of the stomach. The sphincter relaxes at intervals, allowing brief gushes of chyme to enter the duodenum of the small intestine. Chyme entering the duodenum stimulates the release of hormones from duodenal tissue. Carried to the stomach by the circulatory system, these hormones suppress gastric secretion. Within 2–6 hours after a meal, nearly all ingested food has passed into the small intestine, and within 72 hours, about 70% of the undigested residue has appeared in the feces.

In pioneering studies in the early 1800s, William Beaumont, a U.S. Army surgeon, demonstrated the main features of human digestion. He had the unusual opportunity of observing digestive processes through a shotgun wound in the stomach and body wall of Alexis St. Martin, a Canadian trapper under his care. Beaumont determined how rapidly foods are digested and noted that emotions have a strong influence on the secretion of gastric juice. In the 1890s, the Russian physiologist Ivan Pavlov demonstrated that nervous and hormonal mechanisms regulate the release of gastric juice. The pertinent hormone is now known to be **gastrin**, which is synthesized by cells in the posterior region of the stomach, and which when released into the bloodstream, circulates to the gastric glands, which then release gastric juice.

The nervous system controls digestion in several ways. First, the smell or sight of food, and sometimes just the thought of food, stimulates brain centers that activate the gastric glands. In addition, food present in the stomach initiates nervous signals that travel from the stomach wall to the brain and back to the stomach again. Finally, local stimulation of the stomach by food triggers local secretions of gastric juice.

Intestinal Function: Digestion. Chemical digestion is carried out by enzymes embedded in the cell membranes lining the small intestine and by **pancreatic juice** from the **pancreas**. Chemical digestion of fats is aided by **bile** produced by the **liver** and stored in the **gallbladder** (see Figure 4.15). Digestion is generally virtually complete before chyme reaches the ileum.

Pancreatic juice and intestinal cell membranes contain enzymes that digest all major types of nutrients. Protein-digesting enzymes in pancreatic

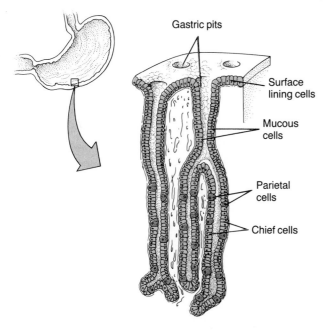

Gastric pits

Surface lining cells

Mucous cells

Parietal cells

Chief cells

Figure 4.16 Structure of a gastric gland. The gland is shown here in a longitudinal section. Millions of these glands are present in the human stomach lining, which is protected from its own secretions by mucus from mucous cells. The parietal cells secrete hydrochloric acid, and the chief cells secrete pepsinogen, an inactive form of pepsin.

juice are first released as inactive zymogens. **Trypsinogen** is activated to **trypsin** by an enzyme (enterokinase) in the intestine. Additional trypsinogen and two other zymogens are then activated to **chymotrypsin** and **carboxypeptidase** by trypsin itself. These enzymes function best in an alkaline medium; as the acidic chyme arrives in the intestine, it stimulates the release of bicarbonate ions from the pancreas, raising the pH. The combined actions of trypsin, chymotrypsin, and carboxypeptidase in the pancreatic juice, along with that of an **aminopeptidase** embedded in the intestinal cell membranes, complete protein digestion.

Pancreatic amylase completes the digestion of starch (begun by salivary amylase in the mouth) to the disaccharide maltose. **Maltase** then finishes the job by splitting maltose into two glucoses. Other enzymes embedded in the intestinal cell membranes act on sucrose and lactose, yielding glucose, fructose, and galactose, all relatively small molecules readily transported from the gut into the bloodstream.

Most fat reaches the human intestine unchanged, for the stomach digests only small amounts of it. Fat must be emulsified before **pancreatic lipase** can digest it. **Emulsification** is accomplished by bile, produced in the liver and stored in the gallbladder until needed in the intestine. Bile contains complex salts, comparable to those in a laundry detergent. The salt molecules have a fat-soluble part and a water-soluble part and are thus able to coat the fat globules, reducing surface tension. This, in turn, eases the breakup of the fat globules into smaller droplets, promoted by vigorous mixing movements of the intestine. Emulsification vastly increases the surface area of the fat globules and makes them more accessible to fat-digesting enzymes.

The digestive activities of the pancreas and liver are closely coordinated with those of the digestive tract. Chyme arriving from the stomach stimulates intestinal cells to release two hormones, **secretin** and **cholecystokinin** (**CCK**, also called **pancreozymin**). Secretin stimulates the release of bicarbonate ions and water by the pancreas, and bile by the liver. CCK stimulates the gallbladder to release bile into the intestine and triggers secretion of digestive enzyme precursors by the pancreas. In effect, each surge of chyme from the stomach is answered with a spurt of pancreatic juice and bile. This means that digestive enzymes will be available only when food is present, a safe and conservative way to use potentially harmful, energy-costly substances.

Intestinal Function: Nutrient Absorption. The small intestine is the chief organ of absorption, and it has several adaptations that increase its absorptive surface area. Its length alone (about 6 m in humans) provides an extensive internal area, and numerous circular folds and ridges extend around the inner surface (Figure 4.17a). Furthermore, every square millimeter of inner surface is covered with up to 40 fingerlike **villi**, each of which contains a network of blood vessels and a lymph duct called a **lacteal** (Figure 4.17b). Into these vessels and ducts nutrients are absorbed following digestion. Finally, the intestinal surface area is increased even more by a **brush border** on the cells of the villi. The brush border consists of minute projections called **microvilli**, visible with the electron microscope (Figure 4.17c). Contractile protein (actin and myosin) filaments within the microvilli allow movement, facilitating contact with nutrients.

Most nutrients are absorbed by the duodenum and jejunum, so that relatively little is left for the ileum to absorb. However, the ileum recovers about 94% of the bile salts released into the duodenum; these are carried back to the liver in the bloodstream for reuse. A combination of passive diffusion and active transport is responsible for moving nutrients across the cells lining the small intestine and into the bloodstream. (Diffusion and active transport are described in Chapter 2, p. 38.) Monosaccharides attach to protein carrier molecules for transport across the microvillar membrane, a process that requires the energy of ATP. They then diffuse passively into the blood vessels. Amino acids are transported into the villi and diffuse into the blood vessels in a similar way. Monosaccharides and amino acids are carried in the blood to the liver, where they may enter metabolic reactions in the liver cells or travel on to other organs. Lipid absorption is not as well understood. Some fatty acid and glycerol molecules follow the same path to the liver as the monosaccharides and amino acids do. However, after passive diffusion into the villi, most products of fat digestion appear to form protein-covered fat particles called **chylomicrons** that enter the lacteals (see Figure 4.17). Chylomicrons travel in the lymphatic system, bypassing the liver at first and arriving there by a more circuitous route.

Vitamins and minerals, the micronutrients, are also absorbed by the small intestine (see Tables 4.1 and 4.2). The water-soluble vitamins are quickly absorbed from the upper end of the small intestine as food is digested, while the fat-soluble vitamins are absorbed along with the products of fat diges-

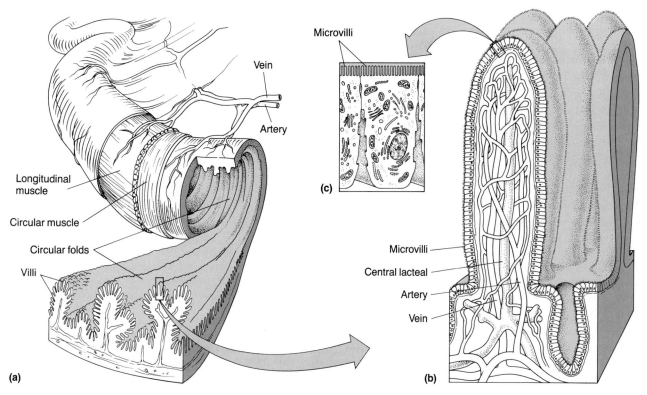

Figure 4.17 The lining of the human small intestine. (a) Note the many internal ridges and folds on which the villi form a dense covering of fingerlike projections. (b) Each villus contains a network of blood capillaries plus a lacteal of the lymphatic system and is covered with microvilli. (c) The microvilli further increase the absorptive surface area. The total absorptive area of the human small intestine is estimated to be about 250 m^2, or roughly the size of a tennis court.

tion. Both active transport and passive diffusion are important in mineral absorption. For example, ions of calcium, magnesium, iron, and sodium are transported actively from the small intestine, while chloride ions move passively, accompanying the active transport of sodium ions.

The Role of the Large Intestine. Material not absorbed from the stomach and small intestine enters the large intestine (colon), which performs several important functions. First, it is a major site of water absorption. In adult humans, the diet and digestive fluids add more than 8 L of water per day to the alimentary tract. About 90% of this water is absorbed as it passes through the small and large intestines. Second, the large intestine harbors large populations of bacteria that manufacture several micronutrients, including vitamin K and certain B vitamins (riboflavin, thiamine, and cobalamin). Production and absorption of such materials in the colon help meet the nutritional requirement for micronutrients. Third, water absorption by the large intestine and extensive bacterial action convert about 1 L of chyme per day into semisolid

feces. Feces are about 25% solid matter and 75% water. About 30% of the solid matter is undigested nutrients and inorganic material; about 30% is undigested roughage such as cellulose; and the remainder consists of bacteria, including harmless bacilli and potentially hazardous cocci. Feces entering the rectum stimulate a defecation reflex that generally results in their expulsion.

☐ S U M M A R Y

1. As heterotrophs, animals must obtain energy and organic nutrients, including essential amino acids, vitamins, and minerals, from their surroundings. Species that can store energy do not have to feed continuously and can spend more time on nonfeeding activities.

2. Animals can be grouped as herbivores (those that eat plants), carnivores (those that eat animals), or omnivores (those that eat plants and animals). Fluid feeders consume blood, plant sap, or nectar; filter feeders collect organic particles and/or small organisms from the surrounding water; substrate feeders eat part of the substrate on which they live; and many other animals catch and eat whole animals.

3. Food may be ingested with or without preliminary physical fragmentation or chemical digestion. Various kinds of beaks, teeth, and gizzards are used in physical digestion. Chemical digestion involves enzymatic hydrolysis of large food molecules into small ones.

4. Digestion in most animals is extracellular and intracellular. Only certain protozoa and sponges have entirely intracellular digestion. Cnidarians and some flatworms have an incomplete gut (no anus); most animals have a digestive system that includes a complete gut with a mouth and anus, organs that store, grind, digest, and absorb digested food, and associated glands that secrete digestive enzymes.

5. Most digestive tracts have adaptations that increase digestive and absorptive surface area. The intestine is often long and coiled and has various types of internal folds and outpouchings (ceca). Herbivores usually have longer intestines than do carnivores. A large cecum in rabbits and a four-chambered stomach in deer, cattle, and other ruminants house microorganisms that digest cellulose.

6. The sequence of digestion and absorption of nutrients is somewhat similar in animals with complete digestive tracts. Food may be pulverized by mandibles or teeth or ground in a gizzard, and it is often stored, at least briefly, in a stomach. Carbohydrates, proteins, and lipids are digested by similar enzymes in different species, and nutrients are absorbed from the alimentary tract into a circulatory system or fluid in a body cavity. In vertebrates, the liver, pancreas, stomach, and intestine secrete digestive materials, and the actions of these organs are coordinated by nervous and hormonal mechanisms.

☐ FURTHER READING

Hamner, W. M. "Krill—Untapped Bounty from the Sea?" *National Geographic* 165(1984): 626–643. *A beautifully illustrated portrayal of krill and feeding by humpback whales.*

Hoar, W. S. *General and Comparative Physiology.* 3d ed. Englewood Cliffs, N.J.: Prentice-Hall, 1983, Chap. 11. *Excellent coverage of the physiology of feeding and nutrition.*

Moog, F. "The Lining of the Small Intestine." *Scientific American* 245(1981): 154–176. *A well-illustrated article on the anatomy and fine structure of the vertebrate small intestine and its enzymatic functions.*

Morton, J. *Guts: The Form and Function of the Digestive System.* 2d ed. London: Edward Arnold, 1979. *A brief, readable text on animal digestive systems and processes, with an evolutionary perspective.*

Pivorunas, A. "The Feeding Mechanism of Baleen Whales." *American Scientist* 67(1979): 432–440. *A lucid account of filter feeding by various whales; includes a description of baleen.*

Spence, A. P., and E. B. Mason. *Human Anatomy and Physiology.* 4th ed. Menlo Park, Calif.: Benjamin/Cummings, 1987. *Includes a well-illustrated, readable chapter on the human digestive system, digestion, and absorption.*

Whitney, E. N., and E.M.N. Hamilton. *Understanding Nutrition.* 3d ed. St. Paul, Minn.: West, 1984. *Interesting, up-to-date, and enjoyable to read.*

CHAPTER 5

Respiratory Gas Exchange

As discussed in Chapter 3, most animals require molecular oxygen (O_2) to operate the metabolic machinery that supplies energy needed for growth, movement, and tissue maintenance. When oxygen is available, hydrogen atoms removed from organic

---

The killer whale *(Orcinus orca).* Although well adapted to aquatic life, these and other aquatic mammals must surface to obtain oxygen.

compounds during metabolism can move through the mitochondrial electron transport system, and ATP will be made. Since few cells can store oxygen, they must be continuously supplied with this gas or they cannot meet their energy needs. At the same time, carbon dioxide (CO_2) produced as waste in cellular metabolism must be released into the environment. The balanced exchange of gases between an animal and its environment is significant in the maintenance of homeostasis.

FACTORS AFFECTING GAS EXCHANGE

The relative amount of a gas in a mixture such as air is measured as the **partial pressure**, that part of the total gas pressure caused by a particular gas. Animals exchange gases with their environment by **diffusion**—in this case, the movement of gas molecules from a region of higher partial pressure to a region of lower partial pressure. Oxygen and carbon dioxide diffuse across an animal's respiratory surfaces, which are thin, moist cell membranes exposed to the environment. The respiratory surface must be wet, for only gases that are in solution will diffuse in or out of an animal. The physical factors that influence the rate of diffusion are described by **Fick's law**. In biological terms, this law states that the amount of a substance (in this case, a dissolved gas) crossing a surface per unit time increases (1) as the amount of surface area (respiratory membrane) becomes greater and (2) as the differences in partial pressures across the respiratory surface membrane increase. Fick's law also implies that the rate of diffusion will decrease as the thickness of the respiratory membrane increases.

Also affecting gas exchange is an animal's surface-area-to-volume ratio. Very small animals can rely solely on diffusion of gases across body surfaces because their surface-area-to-volume ratio is large. This means that all internal cells can be supplied with oxygen from surface diffusion. In contrast, larger animals generally have less body surface area per unit of body volume. Surface area increases as the square of an animal's dimensions, while volume increases as the cube of the dimensions. For example, consider two hypothetical animals that are roughly spherical. If the diameter of one animal is twice that of the other, then the larger animal has a surface area that is four times that of the smaller animal and a volume that is eight times larger. Thus, the surface-area-to-volume ratio of the larger animal is half that of the smaller animal. In biological terms, this means that oxygen will diffuse at half the rate per unit of metabolism into the larger animal compared to the smaller. However, as discussed in Chapter 3, larger animals have lower metabolic rates. This makes their oxygen demand per unit of metabolism lower than that of smaller animals, but it does not solve the gas diffusion problem. As a general rule, diffusion alone cannot adequately supply oxygen to animals greater in diameter than 1 mm, and the surface-area-to-volume relationship limits animal size.

Staying small is not the only way to solve the surface-area-to-volume dilemma. For example, various adaptive shapes can increase surface area (Figure 5.1). It is not surprising that animals are not spherical, because a sphere has the least surface area in relation to its volume of any geometric shape. Larger animals without respiratory organs have shapes that provide large surface areas and minimal diffusion distances between the body surfaces and internal tissues. Flatworms, for example, are elongate and flattened so that most of their cells are close to the outer surface. Many other animals are tubular, and though the body may be long, the diameter is small, resulting in a small diffusion distance to the innermost cells. Some animals **ventilate** their body surfaces so that a boundary layer devoid of oxygen does not develop at the critical environment/respiratory membrane interface. Surface cells on sponges, cnidarians, and many flatworms, for example, have flagella or cilia that create a water stream, and nematodes often thrash about in the water. This helps to stir internal as well as external fluids and maintains partial pressure gradients for gas exchange.

Larger aquatic animals have outgrowths of the body wall, called **gills**, that increase the surface area for gas exchange. Gills may also transport salt and serve in filter feeding, and they are the main sites of waste (ammonia) disposal in many aquatic animals. Gills are not common in terrestrial animals because exposed gill tissue quickly dries and collapses into a nonfunctional mass of cells. Instead, terrestrial animals generally draw air into internal tubes or sacs called **lungs**, which have large surface areas.

Animals with gills or lungs usually have circulatory systems, which further reduce the limitation that surface diffusion imposes on body size and shape. Circulatory systems transport oxygen from the gills throughout the body and carry carbon dioxide from the tissues back to the gills or lungs, thereby helping to maintain diffusion gradients as prescribed by Fick's law.

Other adaptations that promote gas exchange and delivery in animals are oxygen **transport pigments**, such as hemoglobin. These are unique proteins that increase the body fluid's capacity to hold oxygen. They greatly increase the respiratory system's efficiency by binding loosely with oxygen at the respiratory surface and then releasing the oxygen at oxygen-deficient tissue sites. Animals that have transport pigments need less respiratory membrane surface area than those that lack pigments. Most transport pigments also bind with car-

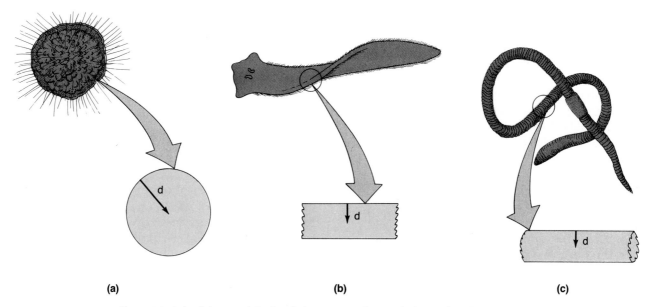

(a)   (b)   (c)

Figure 5.1 Animal shape and size in relation to gas exchange. The letter ''d'' in the illustrations represents the minimum diffusion distance from the surface. (a) Spherical organisms, such as this protozoan, must be small because their respiratory surface area per unit volume is very small. (b, c) Diffusion distance is minimized by the flattened shape of (b) a flatworm and by the small diameter of (c) a tubular animal. Drawings are not to scale.

bon dioxide, although this gas is more soluble than oxygen in body fluids.

In addition to the constraints imposed by Fick's law and surface-to-volume ratios, the physical properties of aquatic and terrestrial environments present unique challenges for animals.

Aquatic Environments

Aquatic animals live in a medium from which it is difficult to obtain oxygen. A liter of water saturated with air at 1°C contains only about 30–50 milliliters (ml) of oxygen, 5% of the amount found in an equivalent volume of air. Furthermore, water is dense (1 kg/L) and viscous compared to air. Aquatic animals that ventilate their respiratory surfaces must expend a considerable amount of energy to obtain this small amount of oxygen. A fish, for example, may use 20% of its basic energy output in moving water over its gills, compared to a mammal that may expend only 1%–2% of its energy output in breathing air.

The amount of oxygen dissolved in natural bodies of water is related to several factors. Oxygen enters water by diffusion from the atmosphere and from photosynthesis by aquatic plants. Increasing salt concentration decreases oxygen solubility, and at all temperatures, seawater contains somewhat less oxygen than fresh water. Water temperature

has a significant impact on oxygen solubility: Less oxygen can dissolve in water at higher temperatures, and this can cause considerable stress on ectothermic aquatic animals, because while oxygen solubility decreases, an animal's metabolism (and therefore its oxygen demand) increases with temperature. This effect is often pronounced in shallow ponds and streams, where temperature extremes can be common.

Various natural mixing actions distribute gases in bodies of water. Streams and rivers are aerated as water flows over rapids. Waves in larger bodies of water aerate surface layers, but wave effects are minimal below a few meters. Standing bodies of water, including lakes and oceans, mix by convection caused by daily solar heating, nocturnal evaporative cooling, and seasonal temperature changes. However, convection usually does not affect the bottom layers of deep, standing bodies of water, and oxygen content may vary significantly from the surface to the bottom. Decomposition of organic matter by microorganisms in sediments can remove all the oxygen and greatly increase the partial pressure of carbon dioxide in uncirculated water near the bottom. Darkness and reduced atmospheric exchange caused by snow and ice cover can cause oxygen levels in lakes to dip to zero. Without light for algal photosynthesis, oxygen is used by respiring organisms but not replaced.

"Winter kill," an extensive die-off of aquatic animals, may occur in some ice- and snow-covered lakes.

Carbon dioxide is about 200 times more soluble in water than is oxygen, and aquatic environments usually have higher carbon dioxide partial pressures than are found in the atmosphere. Once dissolved, carbon dioxide reacts chemically with water according to the following equilibrium reactions:

$$CO_2 + H_2O \rightleftharpoons \underset{\substack{\text{Carbonic} \\ \text{acid}}}{H_2CO_3} \rightleftharpoons H^+ + \underset{\substack{\text{Bicarbonate} \\ \text{ion}}}{HCO_3^-}$$

The bicarbonate ions produced in these reactions may then become chemically bound with various other ions in the water. Carbon dioxide is removed from water by photosynthesis of aquatic plants and by chemical precipitation as calcium carbonate.

Terrestrial Environments

In contrast to water, air is a rich source of easily obtained oxygen. A liter of air weighs only 1.19 g and contains 21% (210 ml) oxygen. The percentage of oxygen in the atmosphere and that of carbon dioxide (about 0.3%) remain relatively constant because of a balance between photosynthesis and aerobic respiration. Photosynthetic organisms use CO_2 and produce O_2, while organisms that respire aerobically use O_2 and release CO_2. In the past several decades, because of extensive burning of fossil fuels and the widespread clearing of vegetation from land, the amount of CO_2 in the atmosphere seems to be increasing, and there is concern that the overall atmospheric gas balance may be upset.

Although air may seem an ideal oxygen source, it is also relatively dry, thus challenging terrestrial animals to maintain moist respiratory surfaces and conserve body water. Air's ability to hold water is measured by its **relative humidity**, the percentage ratio of the actual amount of water vapor in the air to the maximum amount that the air can contain. Because relative humidity is temperature-dependent, air saturated with water vapor at a given temperature will absorb additional moisture when warmed. In fact, endothermic animals lose considerable amounts of water when they exhale cool air that has been warmed in their lungs. For example, a human breathing air at 23°C, with a relative humidity of 50%, loses about 350 ml of water per 24 hours through respiratory evaporation.

At sea level, air exerts a pressure of 1 atmosphere, which is sufficient to support a column of mercury (Hg) 760 mm tall in a barometer. The International System (SI) unit for pressure is the kilopascal, (kPa), which equals 0.133 mm Hg. The partial pressure of oxygen in air at sea level is equal to the percentage of the total pressure due to oxygen, or 0.21 × 760 mm = 160 mm Hg (or 21.3 kPa). At the summit of Mt. Everest, roughly 8840 m above sea level, air also contains 21% oxygen, but the total air pressure drops to about 240 mm Hg, and so the oxygen partial pressure is correspondingly lower (about 50 mm Hg).

The composition of air in restricted terrestrial environments can vary. In ant colonies or decaying logs, for example, oxygen partial pressures can fall to about 140 mm Hg because oxygen is consumed by respiration of decomposer organisms and because of poor atmospheric air circulation. Heavy rains may affect the gas composition in subterranean environments by temporarily blocking pores in the soil. As a result, the partial pressure of oxygen in soils can fall from 160 mm Hg to about 64 mm Hg, and the carbon dioxide partial pressure can rise to about 46 mm Hg from about 2.5 mm Hg. The sudden appearance of earthworms at the soil surface after a rainstorm is evidence of an escape response to such drastic soil gas changes.

In the remainder of this chapter, we will illustrate adaptations that several animals have evolved to carry out gas exchange. As you read about these adaptations, ask yourself the following questions: How does Fick's law apply here? How does this respiratory mechanism adapt this animal to its environment? What natural forces might have led to the evolution of this structure or process? Focusing on such questions makes it easier to perceive broad patterns in the variety of mechanisms that animals use for gas exchange.

GAS EXCHANGE IN AQUATIC ANIMALS

Most aquatic animals have a respiratory apparatus consisting of three components: a water-pumping mechanism, gills for gas exchange, and specializations of the circulatory system.

Invertebrate Mechanisms

Echinoderms.   Some of the simplest respiratory structures are the **dermal branchiae** ("skin gills") on sea stars (Figure 5.2). When observed in living

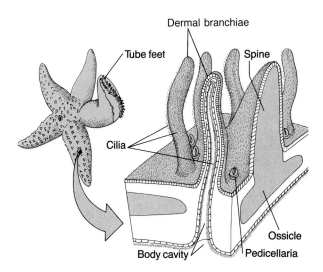

Figure 5.2 Dermal branchiae of a sea star. These simple gills are ventilated by cilia that stir fluids on the outside and inside of the animal, creating steep diffusion gradients for respiratory gases.

animals, these structures are thin, bubblelike projections of the body wall with hollow centers continuous with the body cavity. Cilia on the external surface of epidermal cells move water over the dermal branchiae, while cilia on cells lining the body cavity circulate body fluids inside the branchiae. These currents are driven in opposing directions (countercurrent), increasing the efficiency of gas exchange by maintaining maximum diffusion gradients for oxygen and carbon dioxide. (Countercurrent exchange in fish gills is discussed on p. 105.) Additional gas exchange may occur through the **tube feet**, thin-walled locomotor structures located on the undersides of the sea star's arms.

Annelids. Gas exchange in segmented worms occurs by diffusion across general body surfaces. Many polychaetes (marine annelids) have mobile lateral extensions of the body wall called **parapodia** that increase surface area and enhance gas exchange (Figure 5.3). Others, such as feather duster worms, have more specialized gill-like extensions of the body wall. Most annelids have a respiratory pigment in their blood that facilitates oxygen uptake and transport.

Tubifex worms, freshwater relatives of earthworms, live with their heads buried in oxygen-deficient muds on which they feed. Their bodies project upward into the water and are constantly moving back and forth. Blood vessels in their posterior segments are long, convoluted, and close to the body surface, an adaptation for obtaining oxygen and releasing carbon dioxide. Their anterior segments have a different circulatory pattern. Here the blood vessels are short and deep inside the body wall, not near the surface. This anatomical arrangement prevents oxygen obtained through the posterior segments from passing out of the anterior segments and into the anaerobic muds.

Molluscs. Gas exchange in molluscs occurs through gills called **ctenidia** and across the body surface. In bivalves, such as oysters and clams, ctenidia are used primarily in filter feeding, and gas exchange occurs mainly across the surface of the **mantle**, the sheetlike extension of the body surface (Figure 5.4). The mantle is richly supplied with circulating body fluids that contain a respiratory pigment.

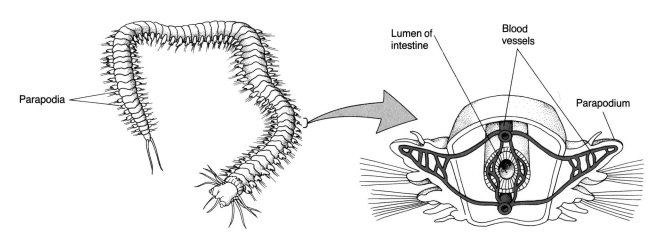

Figure 5.3 Parapodia of a polychaete annelid. These appendages function as gas exchange organs in some polychaetes. Blood circulates through closed vessels in each parapodium and in the body wall.

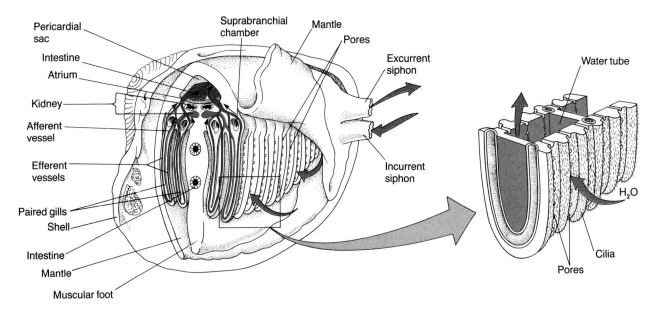

Figure 5.4 Gas exchange organs of a bivalve mollusc. Beating cilia on the gills (ctenidia) draw water into a clam through the incurrent siphon and drive it out through the excurrent siphon. Water passes over the ctenidia, but most gas exchange occurs across the mantle that lines the shell.

The ctenidia of marine and many freshwater snails are well-developed respiratory organs found in the mantle cavity (Figure 5.5). Each ctenidium has a featherlike shape consisting of a central axis with rows of filaments on one or both sides, an architecture similar to that in crustacean and fish gills. This shape provides a large surface area for a small amount of tissue. The ctenidial surface area of a 20 g whelk (a marine snail) approaches 160 square centimeters (cm^2), about three times greater than the external surface area of the whole animal. Some freshwater snails lack gills, but have a lung and breathe air by periodically coming to the surface. (Chapter 24 discusses pulmonate snails, p. 560; Figure 24.6.)

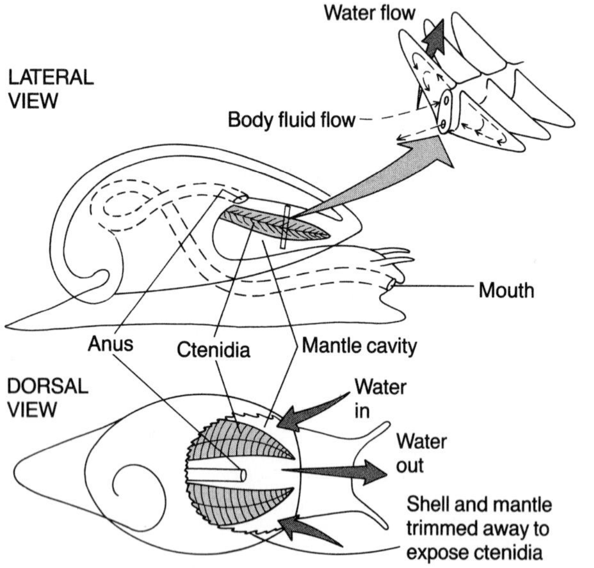

Crustaceans. Decapod crustaceans, a group that includes crayfish, lobsters, and crabs, have featherlike gills located laterally in a pair of **branchial chambers** covered by the hard outer exoskeleton (Figure 5.6). The gills provide about 7.5 cm^2 of gas exchange area per gram of body weight in marine forms, a value similar to that for the whelk. Water is drawn over the gills by the bailing action of a specialized appendage, the **gill bailer**, located near the mouth. The gill bailer beats up to five times a second, drawing water out of the branchial chambers. Water enters the chambers through openings located near the base of the walking legs.

Some crabs live on land and have gills adapted for exchanging gases with air. Their gills are smaller and have little surface area compared to those of aquatic crabs of similar size. Smaller gills reduce evaporative water loss and yet are adequate for gas exchange in oxygen-rich air. Coconut crabs of the South Pacific must live on land and will actually drown if forcibly submerged, because their small gills cannot extract sufficient amounts of oxygen from water.

Figure 5.5 Ctenidia (gills) of a marine snail. The ctenidia are located dorsally in the mantle cavity. Cilia on or near the ctenidia create a water stream over the gills. Most snails have only one ctenidium. The inset is a magnified view indicating the extensive surface area of a ctenidium.

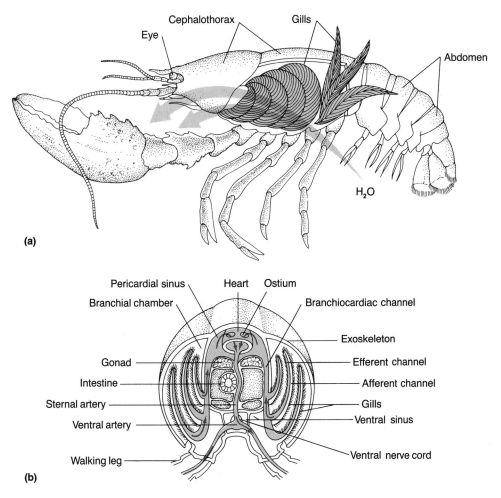

Figure 5.6 Gills of a lobster. (a) Cutaway drawing showing the path of water ventilating the gills. The sculling action of a body appendage, the gill bailer (not shown), drives the water. (b) Cross-sectional drawing of the gill region, showing major organs. Arrows indicate movement of circulatory fluid. Notice that circulatory fluids returning from the tissues enter the gills and exchange gases before entering the heart and being pumped throughout the body.

Fish Gills

All fishes have respiratory gills and most have a mechanism to pump water over them. Water enters a fish's mouth and is pushed out through gill slits, which open between four **gill arches** located in the sidewalls of the pharynx, or throat (Figure 5.7). On the outside of each gill arch are many pairs of featherlike **gill filaments**. Filament tips from adjacent arches interlock to form a sievelike screen through which water from the pharynx passes. Each filament bears many flat, platelike **lamellae** where gas exchange occurs. Special cells in the gill epithelium produce mucus that coats the gill surface and prevents invasion by microorganisms.

The lamellae greatly increase the surface area for gas exchange. Active fishes such as tuna have about 2 m² of exchange surface on their gills per kilogram of body mass, whereas less active fishes

have less lamellar surface. Smallmouth bass, for example, may have only about 0.13 m² of exchange surface per kilogram of body mass. The lamellar surface area in a particular fish is not fixed, however, and may change in response to oxygen availability. Catfish taken from oxygen-rich water and placed in oxygen-poor water may gradually double their gill surface area as they adapt to the new conditions.

Gill Ventilation and Circulation. The coordinated movements of the mouth and gill covers, or **opercula**, provide both suction and pressure to propel water over a fish's gills (see Figure 5.7a). As inhalation begins, the gill covers are closed, the mouth opens, and the mouth cavity expands. This creates a negative pressure in the mouth cavity, and water is sucked in. The mouth then closes, and a flap of

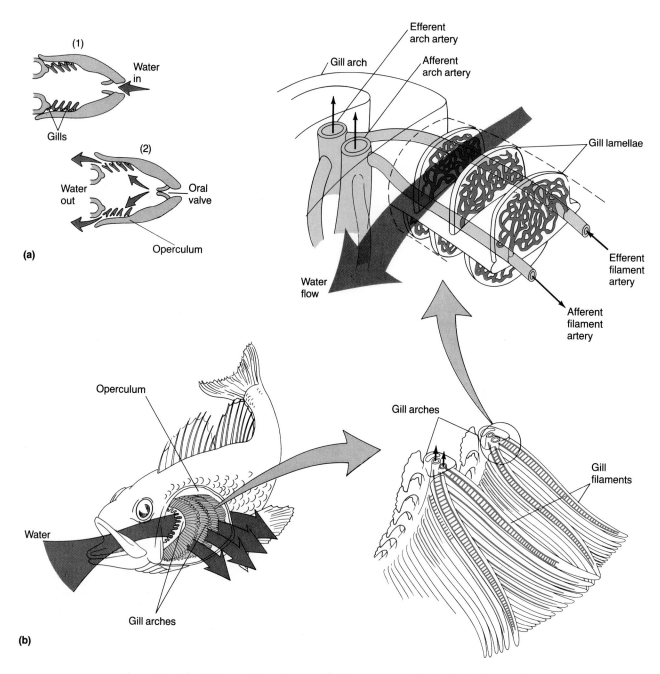

Figure 5.7 Gills of fishes. (a) The gills are located in a common chamber in the pharynx. The operculum covers the gills and, with the oral valve, helps regulate respiratory water flow. (b) This drawing shows the general pathway of water through the mouth, over the gills, and out the opercular openings. Magnified insets show the position of gill filaments during gas exchange and the direction of blood and water flow across the gas exchange surfaces, the gill lamellae. Blood flow in the lamellae is in a direction counter to the flow of water over the lamellae. Such countercurrent exchange promotes rapid and efficient uptake of oxygen from the water.

soft tissue, the **oral valve**, seals the inside of the mouth opening. The space between the gills and the opercula then expands, creating suction, and muscles constrict the mouth cavity, forcing water over the gills. Finally, water is forced out as the opercula rapidly open. When swimming fast, many fishes ventilate their gills passively by simply opening their mouths and opercula. Others, such as mackerel, lack a pumping mechanism and must swim continuously with open mouths to ventilate their gills. Such fish die when kept in small aquariums where swimming is restricted.

Gills are well supplied with blood vessels. Unoxygenated blood pumped from the heart enters each gill arch via a branchial artery that subdivides into afferent arch arteries serving each filament (see Figure 5.7). Small afferent filament arterioles carry blood to special flattened capillaries in the lamellae. The blood space in the flattened capillaries is kept open by regularly spaced **pillar cells**. Pillar cells are contractile, and regulate blood flow through the capillaries by closing certain capillaries periodically. Only about 60% of the lamellae carry blood in a resting trout, but during periods of activity, pillar cell relaxation may open all lamellae for gas exchange. The residence time of blood in a functioning lamella—the time during which gas exchange actually occurs—averages about 1 second. After blood passes through the lamellae, it collects in efferent filament arterioles, passes to efferent arch arteries, and flows into the dorsal aorta, which carries the oxygenated blood to the general circulatory system.

Countercurrent Exchange. Gas exchange in fish gills, as well as in the gills of many invertebrates, is greatly enhanced by **countercurrent flow**: Water flows *over* the gills in an opposite direction to blood flow *within* the gills (see Figure 5.7). The importance of countercurrent exchange is best illustrated by comparing it to the hypothetical alternative, **concurrent (parallel) exchange**, a situation in which flowing, oxygen-rich water is separated by

an oxygen-permeable membrane from oxygen-deficient blood flowing in the same direction (Figure 5.8a). As diffusion occurs, only half the oxygen in the water will enter the blood because an equilibrium will be reached where no diffusion gradient exists. In contrast, when blood and water flow in opposite directions, oxygenated blood about to leave the lamella encounters water that has just contacted the lamellar surface (Figure 5.8b). This water contains the maximum amount of oxygen available, and as the water passes along the lamella, it encounters blood with a progressively lower oxygen content. Thus, the diffusion gradient for oxygen is inward along the entire path of blood-water interaction, and oxygen continues to diffuse into the blood because no equilibrium is reached. About 80%–90% of the initial dissolved oxygen in water may be extracted in countercurrent exchange—far more than would be possible by a parallel flow arrangement, and with no greater energy expenditure.

GAS EXCHANGE IN TERRESTRIAL ANIMALS

Adaptations to air breathing occur in some flatworms, annelids, arthropods, snails, and verte-

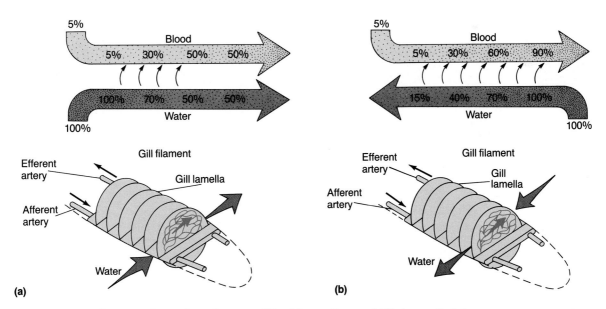

Figure 5.8 Comparison of water and blood flow patterns and diffusion gradients for oxygen movement. The figures indicate the change in oxygen content of both blood and water. (a) In concurrent flow, an equilibrium is reached in oxygen concentration between water and blood. Only a small amount of the total oxygen may be captured from the water; in this example, water leaving the gill retains 50% of its initial oxygen content. (b) In countercurrent flow, no equilibrium occurs between water and blood. More oxygen is obtained from the same amount of water without additional energy expenditure; water may leave the gill with only 10%–15% of its initial oxygen content.

brates. Terrestrial flatworms and annelids, such as earthworms and certain leeches, have no special respiratory structures. Gases are exchanged across the body wall, which is kept moist at considerable expense of energy. Mucus is secreted onto the surface by cells located over the entire body, and annelids release fluids from excretory pores in most of their body segments. Because of the need for a moist surface, terrestrial flatworms and annelids are behaviorally adapted to remain in moist places. Earthworms stay in damp burrows during the day, emerging only on humid nights.

Terrestrial arthropods, snails, and vertebrates have specialized respiratory structures and are able to inhabit dry as well as moist environments. Insects and certain other terrestrial arthropods have tubular gas exchange structures called tracheae that branch throughout the body. The primary gas exchange organs in vertebrates, snails and slugs, and some arthropods are lungs, internal cavities that are open to the atmosphere. There are three types of lungs. **Ventilation lungs**, found in terrestrial vertebrates, move air in and out via a pumping mechanism. **Diffusion lungs**, found in snails and slugs, exchange gases with the atmosphere only by diffusion (Figure 5.9a). **Book lungs**, actually a type of diffusion lung found in some arthropods, have their gas exchange surfaces arranged like the fanned-out pages of a book. The book lungs of spiders are located in the abdomen and open to the atmosphere on the ventral surface (Figure 5.9b). Book lungs are not ventilated other than by air flow caused by body movements in general activity.

Insects

Most insects are small terrestrial animals with a large surface-area-to-volume ratio. They are highly susceptible to drying, but are covered by a hard cuticle (exoskeleton) with a waxy outer layer that minimizes water loss. The tracheal tubes account for most gas exchange; the cuticle reduces surface diffusion of gases to an estimated 10% of the tissue demand.

The anatomy of a typical insect **tracheal system** is shown in Figure 5.10. Air enters the system through openings called **spiracles** in the lateral body wall. Hydrophobic hairs surrounding a spiracle prevent dust and water from entering. Spiracles open into **tracheae**, the largest tubes in the system. A thin lining of cuticle forming a helical ridge holds the tracheae open. Tracheae divide repeatedly, forming many fine **tracheoles** in intimate contact with the tissues. The tracheoles typically contain a small amount of fluid that may help oxygenate tissues when their demand for oxygen is high. An active insect requires more oxygen, and its tissues produce more metabolites than those of an inactive one. The metabolites raise the osmotic pressure of the intercellular fluids in contact with tracheoles; this draws the fluid out of the tracheoles, and air flows into the vacated space. As a result, the tissues are exposed to more oxygen. Tracheoles are most numerous in tissues with high oxygen requirements and often penetrate and end in muscle fibers. The tracheal system may occupy up to 50% of an insect's body volume, and it is so effective in delivering oxygen to tissues that the insect circulatory system plays little or no role in gas transport.

In small insects, diffusion of air into the tracheal system can provide sufficient oxygen to the tissues. In large insects such as grasshoppers and dragonflies, however, the distance from the spiracles to the deeper body cells is often too great for simple diffusion to supply the tissue needs. Large insects have a ventilation mechanism that circulates gases through the tracheal system. Many have **air sacs**, expansions of the tracheal system, located near large muscles. As the muscles contract during

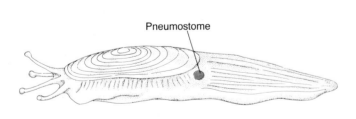

(a)

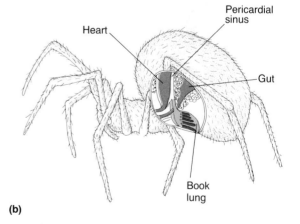

(b)

Figure 5.9 Invertebrate lungs. (a) Slugs and some snails have a diffusion lung formed from a region of the mantle that is well supplied with circulatory fluid. Gases enter and leave the lung through the pneumostome. (b) Spiders have book lungs.

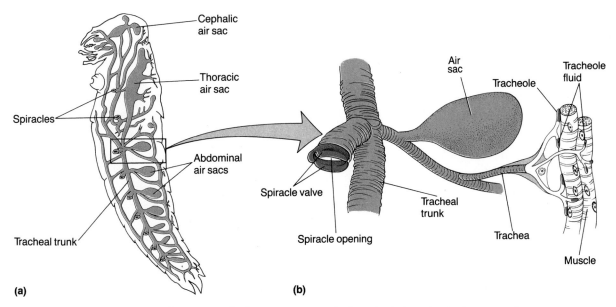

Figure 5.10 The insect tracheal system. Air enters the insect body through spiracles as a result of diffusion, gas pressure in the tracheal system being reduced below that of the atmosphere as tissues use oxygen, or muscles contracting around internal air sacs. Gases are exchanged through minute terminal branches called tracheoles that extend throughout the body.

body movement, they push on the sacs, forcing air out. Air is drawn into the system when the muscles relax. By using body movement to ventilate the tracheal system, an insect expends little, if any, additional energy in gas exchange.

Flow of air into the tracheal system also occurs by other means. As oxygen is used in the tracheoles, the gas pressure in these tubes can drop below atmospheric pressure when the spiracles are closed. The carbon dioxide produced in metabolism does not make up the difference in pressure because much, though not all, of it dissolves in the body fluids and escapes from the insect by diffusing through thin areas of the body covering. Consequently, a negative pressure approaching 4 mm Hg can develop in the tracheal system. When this happens, the spiracles open, and air rushes inward and supplies oxygen to the cells. The spiracles remain open for only a short time to minimize water loss. They are held open only if carbon dioxide levels in the body fluids reach a critical high. With the spiracles open, pressures are equalized, and carbon dioxide and water vapor diffuse outward while oxygen continues to diffuse inward.

Aquatic Insects. Some insects are able to live in aquatic habitats because their tracheal system has been adapted to obtain oxygen from water or air. Some aquatic insect larvae have **tracheal gills**, which are dense networks of fine tracheae in feath-erlike extensions from the body. Others, such as mosquito larvae, periodically rise to the surface and exchange gases with the air. (Other aquatic adaptations in insects are discussed in Chapter 26, p. 634.)

Amphibians

Although amphibians are aquatic and have external gills as larvae, most species lack gills and have air-breathing lungs as adults. Amphibian lungs are saclike with few infoldings, and they are not nearly as complex as those of reptiles, mammals, and birds. Amphibians ventilate their lungs by a mechanism called **buccal pumping** that is quite different from the vacuum mechanism used by reptiles, birds, and mammals. During inhalation, an amphibian lowers the floor of its closed mouth to draw air in through its nostrils. The nostrils are then closed, and the floor of the mouth is raised, forcing air back into the pharynx. Here the air enters the **trachea** (windpipe) through an opening called the **glottis**. The trachea divides into a left and right **bronchus**, each of which conveys air to the lungs. The lungs are filled, and air is held there for some time before it is exhaled. Contraction of body wall muscles and elasticity of the lung tissue causes exhalation (Figure 5.11).

In many amphibians, lung respiration is less important than **cutaneous gas exchange**, that is,

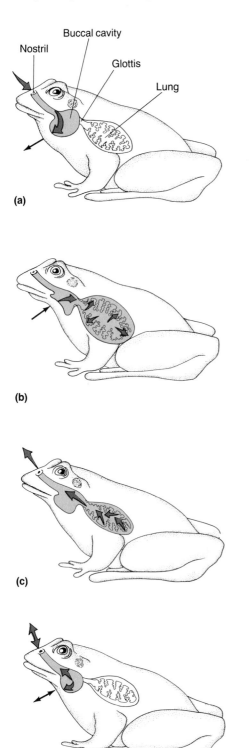

(a)

(b)

(c)

(d)

Figure 5.11 Respiratory movements in a frog. (a) Air is drawn into the mouth through the nostrils by expansion of the throat. (b) With the mouth and nostrils closed, elevation of the floor of the mouth and compression of the throat forces air into the lungs. (c) During exhalation, the glottis opens, and the elastic recoil of the lungs and contraction of the body wall muscles force gas out. (d) Substantial gas exchange occurs across the lining of the mouth when the throat flutters, drawing air into the nostrils and then pushing it back out.

Figure 5.12 The red-backed salamander *Plethodon cinereus.* A common amphibian in the Appalachian Mountains, this species represents a large family of salamanders that lack lungs. Gas exchange occurs across moist tissues in the mouth and across the general body surface.

respiration across the moist surface of the mouth or across general body surfaces. Many species move air in and out of the mouth while their lungs are inflated and the glottis is closed. Some of the oxygen in this air may diffuse across the lining of the mouth. Many amphibians also obtain oxygen and release carbon dioxide through their thin moist skin. In temperate climates, the relative roles of cutaneous gas exchange and gas exchange via the lungs may change with the season. In the summer, both lung and cutaneous exchange are important, but in the winter, when many amphibians are in mud burrows under water, all gas exchange occurs across body surfaces. One group of amphibians, the plethodontid salamanders, lacks lungs and gills and depends entirely on cutaneous gas exchange (Figure 5.12). Cutaneous respiration occurs to a lesser extent, or not at all, in reptiles, mammals, and birds because their skin consists of several layers of dry, dead cells that are effective in preventing moisture loss and also block oxygen uptake.

Mammals

As endotherms, mammals have a higher rate of metabolism and require more oxygen per unit of body weight to maintain internal homeostasis than do amphibians. Mammalian lungs are adapted to obtain more oxygen from air by having an efficient ventilation mechanism and much greater surface area.

Respiratory System Anatomy. Air enters the mammalian respiratory system through the nostrils, where large, airborne particles are filtered out by nasal hairs. It then passes into the nasal cavity and flows over a mucus-secreting epithelium covering three platelike bones, the **nasal conchae** (Figure 5.13). Here dust is trapped, the air temperature is adjusted to about 37°C, and the relative humidity of the air is increased, keeping the lung surfaces from drying. Air then flows to the pharynx, where it mixes with any air taken in through the mouth. The openings of the esophagus and the trachea are located at the posterior end of the pharynx. The **epiglottis**, a flap of tissue above the trachea, acts as a valve that covers the glottis during swallowing, preventing food from entering the respiratory system.

Air passing down the trachea travels through a specialized region called the **larynx** or voice box, which contains the vocal cords. As in amphibians, the trachea divides into left and right bronchi that enter the lungs. The trachea and bronchi have rings of cartilage in their walls that prevent collapse. In the lung, each bronchus subdivides to form thousands of **bronchioles**. Each bronchiole ends in grapelike clusters of very thin-walled **alveoli**, where gas exchange takes place. The two lungs in a 75 kg human have about 3×10^8 alveoli, providing a total gas exchange surface area of over 75 m^2, approximately 1 m^2 per kilogram of body weight. Compared to the less efficient frog lung, this is 15 times more surface area per unit of lung tissue.

To prevent clogging of the alveoli, the respiratory system has a self-cleaning mechanism. The

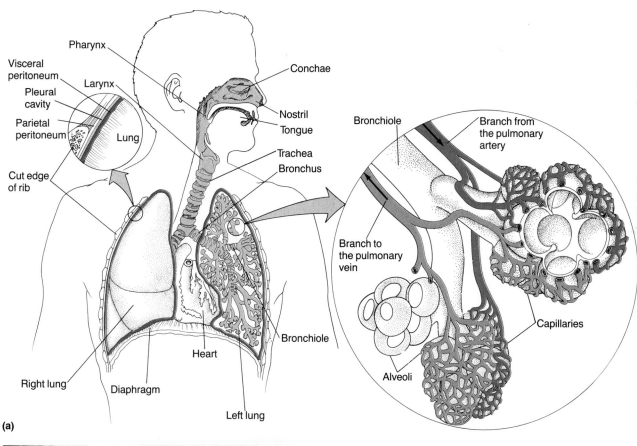

(a)

(b)

Figure 5.13 The mammalian respiratory system. (a) Air enters through the nostrils and passes over the conchae before entering the pharynx. From the pharynx, it passes through the glottis into the trachea. The trachea subdivides into two bronchi that branch into bronchioles conducting air to the alveoli. The inset shows the structure of several alveoli. (b) Alveoli in the lung of a mouse as seen with the scanning electron microscope (400×).

trachea and bronchi are lined by ciliated and mucus-secreting cells. Dust particles not filtered in the nasal cavity are trapped in the mucus and are moved toward the pharynx by the cilia. Mucus swept to the pharynx is then swallowed. If this **ciliary escalator**, as it is sometimes called, does not function properly, mucus and dirt can accumulate in the lungs, causing persistent coughs and a decrease in respiratory efficiency. Many irritants that we breathe, including tobacco smoke, impair the effectiveness of the ciliary escalator and may cause permanent clogging of alveoli.

Lung Ventilation. Each lung is housed in its own **pleural cavity**, which is a separate compartment of the thoracic (chest) cavity. Air moves into and out of mammalian lungs because changes in the volume of the thoracic cavity create either a positive or negative pressure. Thoracic volume is increased by expansion of the rib cage, which is accomplished by (1) contraction of the **diaphragm**, a large, domelike sheet of muscle separating the thoracic and abdominal cavities, and (2) contraction of

external intercostal muscles between adjacent ribs (Figure 5.14). Enlargement of the thorax creates a negative pressure of about 3 mm Hg. Air rushes into the respiratory system and expands the thin-walled alveoli until the intrathoracic pressure matches atmospheric pressure.

Inhalation requires the greatest amount of energy in the breathing cycle because the alveolar tissue is elastic and must be stretched, and because molecular attractions in the water film lining the alveoli exert a surface tension that resists expansion. Although surface tension may seem insignificant, the total force exerted by millions of alveoli can offer considerable resistance, as you can demonstrate by trying to pull apart two pieces of glass held together by a thin water film. Surface tension is important, for it provides some of the recoil force in the alveoli, helping the lungs expel air during exhalation. Without a counterbalance, however, the surface tension force would be strong enough to collapse the lungs and prevent them from expanding. Consequently, terrestrial vertebrates—in fact, perhaps all air breathers—have phospho-

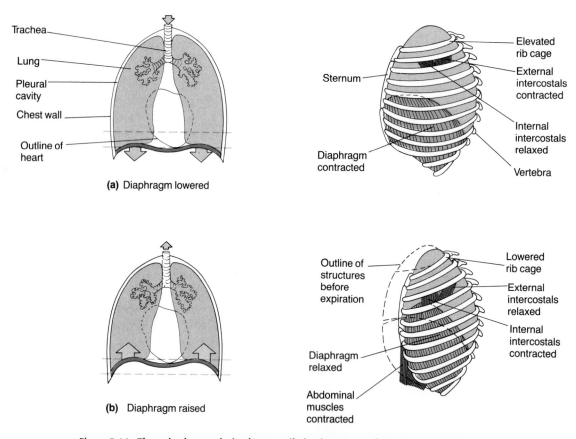

Figure 5.14 Thoracic changes during lung ventilation in a mammal. (a) During inhalation, air enters the lungs when the thorax enlarges as a result of rib cage expansion and by contraction (and consequent flattening and lowering) of the diaphragm against the abdomen. (b) During exhalation, gas is expelled when muscle relaxation reduces thoracic volume.

lipoproteins called **pulmonary surfactants** that reduce surface tension in the alveoli. When the alveoli are small and deflated, the surfactants are in high concentration relative to the surface area, and the alveoli do not collapse from surface tension. During inhalation, the surfactants help expand the small alveoli. Then, as the alveoli expand, the surfactants spread over a larger surface area and become less effective. Surface tension thus increases in enlarged alveoli, and they stop expanding. This allows air to flow to any unexpanded alveoli until all alveoli are inflated to a similar size.

Exhalation can be a passive or active process. Passive exhalation occurs when the rib cage is reduced as the intercostal and diaphragm muscles relax. Alveolar tissue elasticity, surface tension, and the weight of the rib cage then create a positive intrathoracic pressure of 3 mm Hg, pushing air out of the alveoli. Active exhalation occurs when the **internal intercostal muscles** and abdominal muscles contract and push the viscera forward against the diaphragm, increasing pressure on the lungs (see Figure 5.14). Thoracic pressure of up to 60 mm Hg is possible. Alveoli never completely collapse, however, even in forced breathing.

Each year, an adult human may inhale between 2 million and 5 million liters of air. A normal breath contains about 500 ml of air and is referred to as the **tidal volume** (Figure 5.15). Not all tidal air entering the lungs is fresh, however, because about 150 ml of exhaled gas (called gas because its composition is different from that of the atmosphere) fill the trachea and bronchial network. This "dead space" gas is the last to leave the lungs on exhalation and immediately reenters on inhalation. Therefore, each breath brings the lungs only about 350 ml of fresh air, which mixes with the **residual gas** that remains in the partially inflated alveoli between breaths. Since the lungs never receive pure air, the composition of alveolar gas always differs from that of the atmosphere: An alveolus usually contains about 15% oxygen and 5% carbon dioxide.

The maximum amount of gas that can be exhaled from the lungs after a forced inhalation is called the **vital capacity** (see Figure 5.15). During intense exercise, the lungs of the average human male may contain up to 6000 ml of gas, of which 4000–5000 ml are exchanged.

Circulation and Gas Exchange. The mammalian circulatory system is adapted to work in concert with the respiratory system. Oxygen-poor blood carrying carbon dioxide returns from the tissues and enters the right side of the heart. From there it is pumped through the pulmonary artery to the lungs, where it enters capillaries in the alveolar walls. About 75% of the surface of each alveolus is covered by capillaries, and most capillaries contact two adjacent alveoli, making rapid oxygen transfer possible. The diffusion distance between air and blood is less than 1 micrometer (μm). A given unit of blood stays in the lung capillaries for about 0.3–0.75 second, depending on the level of body activity. The rate of gas exchange changes according to the rate of blood flow and the rate of ventilation. When oxygenated blood leaves the pulmonary capillaries, it returns to the left side of the heart via the pulmonary veins and is then pumped to the tissues. The partial pressure of gases in the different components of the human respiratory and circulatory systems are shown in Figure 5.16.

Regulation of Respiration. Every 24 hours, a person normally inhales about 15,000 times. The necessary muscle movements are involuntary, though the frequency and depth of breathing can be consciously modified. Breathing is regulated by **respiratory centers** in the medulla oblongata and pons of the brain. Signals from these brain centers stimulate the muscles of the diaphragm and rib cage during inhalation and inhibit muscle activity during exhalation. Mammals are not highly sensitive to low levels of oxygen in the blood and depend mainly on elevated levels of carbon dioxide or hydrogen ions (lowered pH) in the blood to signal the need for increased respiratory movements.

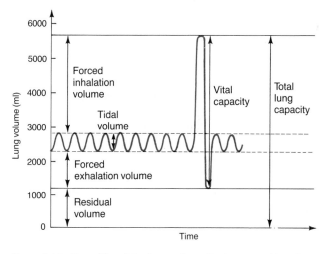

Figure 5.15 Capacities of the human lung. During normal breathing, a tidal volume of air enters and leaves the lungs. Forced inhalation can bring much larger quantities into the lungs, and forced exhalation can release some of the air normally kept in the lungs. A residual volume of gas remains trapped in partially filled alveoli despite the strongest exhalation.

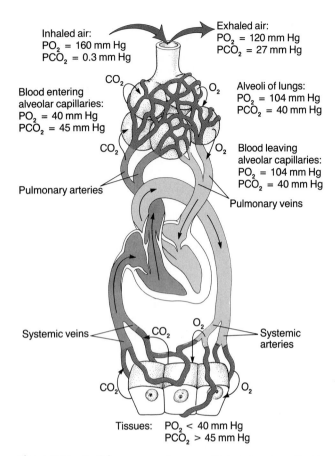

Inhaled air:
PO₂ = 160 mm Hg
PCO₂ = 0.3 mm Hg

Exhaled air:
PO₂ = 120 mm Hg
PCO₂ = 27 mm Hg

CO₂   O₂

Blood entering
alveolar capillaries:
PO₂ = 40 mm Hg
PCO₂ = 45 mm Hg

Alveoli of lungs:
PO₂ = 104 mm Hg
PCO₂ = 40 mm Hg

CO₂   O₂

Blood leaving
alveolar capillaries:
PO₂ = 104 mm Hg
PCO₂ = 40 mm Hg

Pulmonary arteries

Pulmonary veins

Systemic veins

CO₂   O₂

Systemic
arteries

CO₂   O₂

Tissues: PO₂ < 40 mm Hg
PCO₂ > 45 mm Hg

Figure 5.16 Partial pressures of oxygen (P$_{o_2}$) and carbon dioxide (P$_{co_2}$) in the respiratory and circulatory systems. This diagram shows the partial pressures of respiratory gases in the alveoli, blood, and tissues of a resting human.

Swimmers often take dangerous advantage of this CO₂-signaled regulatory mechanism to stay submerged longer than is safe. By taking several rapid deep breaths (**hyperventilating**) before submerging, a swimmer reduces the blood's carbon dioxide level but gains no more oxygen than is taken in by a few breaths. Carbon dioxide is produced as the person swims underwater, but because the starting level was low, blood CO₂ does not build up to a level that would stimulate an irrepressible urge to surface and breathe. Oxygen is depleted as the person swims, but because the system is not as sensitive to this, an underwater swimmer may be in trouble and not know it. With no urge to ventilate the lungs, the swimmer may lose consciousness for lack of oxygen and drown. (The Essay in this chapter discusses how aquatic mammals are adapted for diving.)

Birds

The gas exchange system may make up nearly 20% of a bird's body volume, compared to less than 10%

in mammals. The bird system is more efficient, extracting about 31% of the oxygen in inhaled air, while the mammalian system extracts about 24%. Only with such efficiency in gas exchange can birds fly at altitudes as great as 6000 m, heights at which humans must wear oxygen masks because the partial pressure of oxygen is insufficient to oxygenate the blood and supply tissue demands.

The gas exchange system of birds is structurally and functionally very different from that of mammals. Instead of having alveoli, bird lungs are composed of many parallel tubes called **parabronchi**, which have small, side-branching **air (gas) capillaries**. And unlike the expansive lungs of mammals, bird lungs have a fixed volume. The bird system also includes thin-walled gas storage pouches called **air sacs** in the posterior and anterior parts of the body cavity (Figure 5.17a). During flight, breathing movements are coordinated with wing beats, though not necessarily on a one-for-one basis. Some hummingbirds have ventilation rates of 330 breaths per minute, but their wing beat is even faster.

Unlike the tidal (in-and-out) flow of gases in mammals, the bird system maintains a more efficient, continuous *one-way flow* of gas through the lungs. To understand how gas flows through the bird system, it is helpful to think about movement of gas already in the system (from a previous breathing cycle) as well as about movement of a volume of new air into the system from the outside (Figure 5.17b). During inhalation, new air is drawn in through the nostrils or mouth and passes via a trachea and bronchi into posterior air sacs. Also during inhalation, gas already in the system moves ahead of the new air and passes from the lungs into the anterior air sacs. During exhalation, air flows from the posterior air sacs into the parabronchi in the lungs. Here oxygen diffuses into the air capillaries and carbon dioxide passes into the parabronchi. Also during exhalation, most of the gas from the anterior sacs is forced out of the system through the bronchi and trachea.

Gas exchange in the bird lung occurs by means of a **crosscurrent exchange** mechanism (Figure 5.17c). Blood capillaries intertwine closely with air capillaries, separated only by the one-cell-thick walls of the two types of capillaries. Any given blood capillary contacts air capillaries originating from only a short section of a given parabronchus. As air moves along a parabronchus, its oxygen content decreases and its carbon dioxide content increases as it contacts more and more blood capillaries. The blood, in turn, becomes enriched with oxygen and depleted of carbon dioxide to varying

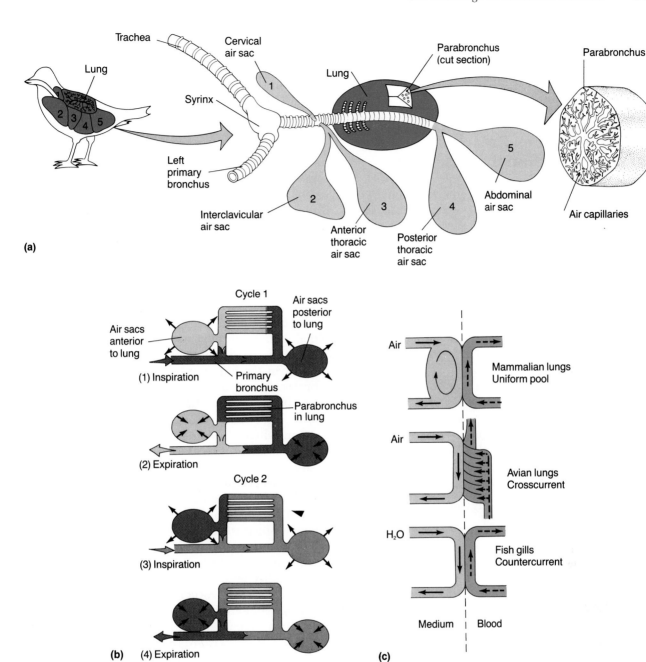

Figure 5.17 Respiratory system of a bird. (a) The system includes a pair of lungs and paired, thin-walled anterior and posterior air sacs. The inset shows the details of a parabronchus and the air capillaries that branch from it. The air capillaries are in turn surrounded by blood capillaries (not shown). (b) The path of air flow through the system. Gas flow is continuous, with one-way (posterior-to-anterior) flow through the lungs. During inhalation, the rib cage expands, and all air sacs enlarge and fill with gas. Newly inhaled air passes to the posterior air sacs. Most of the gas already in the system passes into the anterior air sacs from the lungs. During exhalation, the rib cage contracts, and gas in the anterior sacs passes to the outside. At the same time, gas in the posterior sacs flows through the lungs and toward the anterior air sacs. (c) The crosscurrent flow of air and blood in the avian lungs is contrasted to the countercurrent mechanism in fishes and the uniform pool of air employed by mammals.

degrees, depending on where it contacts the air capillaries along the parabronchial tube. Oxygenation of the blood will be greatest in those capillar-

ies nearest to sites where gas enters a parabronchus and lowest where it leaves. Blood with differing gas content therefore mixes in the veins draining blood from the lungs. The *average* oxygen content of blood in these veins can exceed the oxygen content in gas leaving the lungs.

GAS TRANSPORT IN BLOOD AND BODY FLUIDS

In animals with circulatory systems, the amount of oxygen delivered to the tissues per minute is a function of the rate of fluid flow through the tissues and the amount of oxygen carried per volume of body fluid. **Respiratory pigments** dissolved in blood or body fluid or contained in circulating cells increase the oxygen-carrying capacity of the circulatory system. These pigments have the capacity to bind oxygen at the respiratory surface and release it to the tissues. Five respiratory pigments have been identified in the animal kingdom, although some animals lack these molecules. Since hemoglobin, myoglobin, and hemocyanin are the most common pigments, they are the ones we will focus on in this section. Two other pigments, chlorocruorin and hemerythrin, occur in certain marine invertebrates. All respiratory pigments consist of special proteins complexed with iron or copper ions.

Function of Respiratory Pigments

The oxygen-binding and -releasing properties of respiratory pigments result from a delicate balance between a pigment's capacity to capture oxygen from the environment and its capacity to supply oxygen to the tissues. A pigment that binds oxygen strongly will capture large amounts of oxygen from the environment, but may not readily release it at the tissues. On the other hand, a pigment that binds oxygen weakly can release its oxygen rapidly at the tissues, but would not capture large amounts of oxygen at the respiratory surface. The evolution of respiratory pigments has been guided by these opposing forces as well as by the necessity of removing carbon dioxide from the tissues.

Hemocyanin is dissolved in the circulating body fluid of many species of animals in two phyla, Mollusca and Arthropoda. The protein portions of the molecule differ from species to species but are always large, having molecular weights of from 400,000 to 13 million. The molecules consist of several peptide chains, each associated with one copper atom. Two copper atoms and hence two peptides are required to bind one oxygen atom.

Hemocyanin is colorless when unoxygenated, but turns pale blue when oxygen is bound because the copper ion is oxidized from the cuprous (Cu^+) to the cupric (Cu^{2+}) form. The hemocyanins are efficient transport pigments, and they occur in such large and active animals as squids, octopuses, and the king crab.

Hemoglobins, the most studied respiratory pigments, are found in diverse animal groups, including vertebrates, annelids, some molluscs, and insects. Hemoglobins from different animal species differ by molecular weight and localization in the blood. Human hemoglobin consists of four polypeptide chains and four heme groups, each heme group containing one ferrous (Fe^{2+}) ion (see Figure 2.10d). The binding of oxygen to the ferrous iron in the heme group does not oxidize the iron to the ferric form.

Hemoglobins are transported in circulating red blood cells in vertebrates and some invertebrates. They may be dissolved in the fluid portion, or plasma, of the blood or body fluids in many invertebrates. Among adult vertebrates, only the Antarctic icefishes (Family Chaenichthyidae) are known not to have hemoglobin. The subfreezing conditions in which these fishes live allow enough oxygen to dissolve in the blood plasma to supply the tissue needs.

Each microliter of human blood contains about 5 million red blood cells, each carrying about 280 million hemoglobin molecules. This remarkable hemoglobin content greatly increases the blood's oxygen transport ability. Human plasma can carry only about 0.3 ml of oxygen per 100 ml of blood, considerably less than the amount needed to supply the average resting needs of human tissue. When hemoglobin is present in red blood cells at normal levels, each 100 ml of blood can carry 20 ml of oxygen, usually more than enough to supply tissue needs. Poisonous gases such as carbon monoxide have a higher affinity for hemoglobin than oxygen does. When a person inhales carbon monoxide, the sites of oxygen binding are gradually filled by carbon monoxide molecules, and oxygen is displaced, with potentially lethal consequences.

Some annelids, molluscs, and vertebrates have the oxygen-binding protein **myoglobin** in heart muscle and in red skeletal muscles. Myoglobin gives some muscles a red color. The dark meat in the breast of waterfowl and in the legs of turkeys and pheasants reflects their myoglobin content and the importance of these muscles as the primary means of locomotion. Myoglobin consists of one heme group containing iron, and one protein molecule about the same size as a single polypeptide

Essay: Diving Adaptations in Mammals

Although all mammals breathe air, certain species, including whales, dolphins, and seals, live in aquatic habitats. Several aquatic mammals can dive to great depths and remain submerged for long periods of time while foraging for food. The Weddell seal typically dives to 300–400 m for up to 15 minutes (Figure E1). Some whales submerge for up to an hour, and though most dive to less than 500 m, dives to 3000 m have been recorded. Two problems involving gas exchange confront these mammals. The need for oxygen is obvious, but equally important is the need to avoid a condition known as the **bends**.

Graphic accounts of the bends are found in stories of human deep-water diving. At 50 m beneath the surface, water pressures reach about 5 atmospheres. At this or greater pressures, large amounts of nitrogen from the air in the diver's lungs dissolve in the blood. If a diver whose blood is saturated with nitrogen at high pressure surfaces too quickly, the nitrogen forms bubbles in the blood at the reduced pressure. This is similar to the way bubbles form when a bottle of carbonated soda is uncapped, releasing the pressure inside. The bubbles in the bloodstream block or rupture capillaries throughout the body. The pain this causes can be excruciating, bending a person over and possibly causing a seizure if brain capillaries are blocked. Divers avoid the bends by ascending to the surface slowly, stopping at intermediate depths to allow nitrogen to leave the blood and reenter the lungs gradually.

Interestingly, unlike human divers, whales and seals do not ascend in stages from deep dives. Weddell seals routinely swim at depths of about 45 m under Antarctic ice and surface to breathe at "blowholes" that they keep open in the ice. How do these animals avoid the bends? Because air must be in contact with blood at high pressures for nitrogen to saturate the body fluids, one way to avoid the bends is to prevent high-pressure air-to-blood contact. Weddell seals accomplish this by exhaling instead of inhaling before diving. As the seal descends, increasing pressure collapses the trachea and lungs, and any remaining air is exhaled. In this way, no air contacts the blood when the seal is at great depths, and nitrogen saturation is avoided. However, it would appear that the seal's strategy would also cause an oxygen deficiency. How do these mammals cope with the oxygen demands of their tissues when submerged?

Figure E1 **The Weddell seal *(Leptonychotes weddelli)*.** A mother and her pup.

Deep-diving mammals have adaptations for storing oxygen. Generally, they have twice as much blood per unit body weight as other mammals. They also have a greater number of red blood cells per unit volume of blood. Thus, their blood serves as an oxygen reservoir. In addition, their muscle tissue contains myoglobin, a protein similar to hemoglobin, that allows oxygen storage at the main sites of consumption. Despite these adaptations for O_2 storage, sufficient oxygen is not available to support aerobic metabolism in all tissues during some extended dives. When oxygen reserves become depleted, physiological changes called the **diving response** occur. The heart rate slows to about one-half to one-tenth the normal rate, and blood circulation is restricted primarily to the heart and brain, tissues that cannot survive without oxygen. If the animal remains submerged long enough for the body muscles to use all the oxygen stored in myoglobin, the muscles can switch to anaerobic metabolism. This results in lactate accumulation in the muscles. When the animal surfaces, rapid breathing causes the lactate to enter the blood. Lactate is removed from the blood as oxygen becomes available. Some of the lactate is oxidized via aerobic respiration to carbon dioxide and water. Much of it, however, is converted into glycogen by the liver and muscle cells. (Aerobic and anaerobic respiration are discussed in Chapter 3, p. 62; Figures 3.11 and 3.16.)

subunit of hemoglobin. Each myoglobin molecule can reversibly bind one molecule of oxygen. Myoglobin's affinity for oxygen is slightly greater than hemoglobin's, and this promotes the diffusion of oxygen from the blood into the muscle cells. Oxygen stored locally in this way lies within easy diffusion distance of a muscle cell's mitochondria, supplying their oxygen needs for short times when the oxygenated blood supply is inadequate.

Factors Affecting Oxygen Binding to Hemoglobin. The amount of oxygen that will bind to hemoglobin depends on the partial pressure of the oxygen. Hemoglobin binds more oxygen when the partial pressure is high, as in the alveoli, than it does at the lower partial pressures found in the tissues. This relationship is illustrated by an oxygen equilibrium (dissociation) curve (Figure 5.18). This curve indicates how much oxygen is bound by hemoglo-

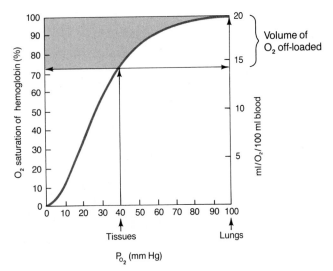

Figure 5.18 Oxygen equilibrium (dissociation) curve. At the partial pressures of oxygen (P_{O_2}) in the human lungs, hemoglobin is about 100% saturated with oxygen. At the partial pressures of oxygen found in the tissues, hemoglobin is only about 70% saturated. Therefore, about 30% of the oxygen will off-load to the tissues. This corresponds to about 6 ml of oxygen per 100 ml of blood.

bin as oxygen partial pressure increases and how much is released when partial pressure decreases. In the alveoli of human lungs, the partial pressure of oxygen is typically about 100 mm Hg. At this partial pressure, hemoglobin becomes fully saturated with oxygen to form **oxyhemoglobin**. In the tissues, the oxygen partial pressure is typically about 40 mm Hg, and some of the oxygen leaves the hemoglobin and diffuses to the tissues, where it is used in aerobic respiration. When all oxygen is

removed from hemoglobin, **deoxyhemoglobin** is formed.

Several other factors affect oxygen binding to hemoglobin. Once a single oxygen molecule binds to one of the heme groups in hemoglobin, the binding to the other three heme groups is enhanced in a progressive manner known as **cooperativity**. This helps hemoglobin to load with oxygen at the respiratory surface. Cooperativity also affects hemoglobin's capacity to release oxygen in the tissues. Once a single oxygen molecule diffuses from hemoglobin to the tissues, the remaining oxygen molecules are lost more readily in a domino-like fashion. As a result, hemoglobin molecules are usually found in a fully loaded or completely unloaded state.

Acidity also affects hemoglobin's oxygen affinity, and in this regard, temperature also plays a role in ectotherms, because water dissociates more at higher temperatures and thus becomes more acidic. If oxyhemoglobin is placed in an acid environment, the hydrogen ions alter the pigment molecule, lowering its oxygen-binding capacity. This effect is physiologically useful, for it enhances oxygen unloading in tissues. Because carbon dioxide forms carbonic acid when it reacts with cell water, high carbon dioxide concentrations will cause oxyhemoglobin to unload its oxygen. Active tissues, which need the most oxygen, also produce large amounts of carbon dioxide and thus facilitate unloading by oxyhemoglobin.

The effects of carbon dioxide on hemoglobin are evident from the oxygen equilibrium curves shown

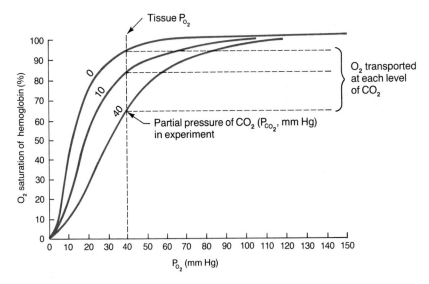

Figure 5.19 The Bohr effect. This graph shows that hemoglobin's affinity for oxygen is decreased by increasing levels of carbon dioxide. The Bohr effect occurs because carbon dioxide binds to hemoglobin and because the hydrogen ions from carbonic acid cause oxyhemoglobin to release oxygen.

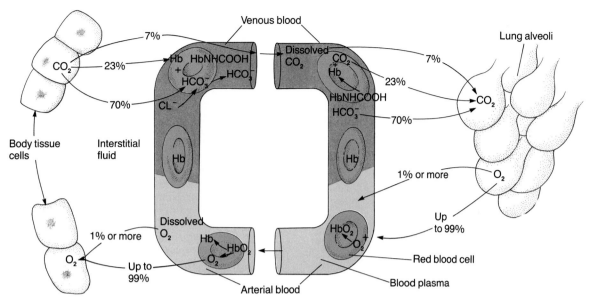

Figure 5.20 Transport and exchange of respiratory gases in the blood. This diagram summarizes the major events in gas exchange and transport of gases in the blood of a mammal. Percentages indicate the fractions of total O_2 and CO_2 carried in various ways in the blood. HbO_2 is oxyhemoglobin; HbNHCOOH is carbaminohemoglobin; and HCO_3^- is bicarbonate ion.

in Figure 5.19. The shift of the curves to the right with increasing carbon dioxide partial pressures shows that higher carbon dioxide levels are correlated with increased unloading of oxygen to the tissues. This phenomenon, which is actually acid sensitivity of oxyhemoglobin, is called the **Bohr effect**. At the respiratory surface, the Bohr effect disappears because carbon dioxide diffuses out of the blood. The overall effect is that oxygen loss is promoted in the tissues, but oxygen gain is promoted in the gas exchange organ.

Transport and Release of Carbon Dioxide. Carbon dioxide is transported in the blood of vertebrates in three ways. About 7% of the blood's carbon dioxide content is dissolved as CO_2 in the plasma. The rest is carried in some chemically combined form. About 23% binds directly to hemoglobin by combining with free amino groups of several amino acids to form carbaminohemoglobin (HbNHCOOH), and some 70% is carried as bicarbonate ion (HCO_3^-) formed by the action of carbonic anhydrase in red blood cells. Carbonic anhydrase catalyzes the formation of carbonic acid from carbon dioxide and water. Carbonic acid then dissociates into hydrogen ions and bicarbonate. Bicarbonate diffuses from the red blood cells into the plasma in exchange for chloride ions (Cl^-). Bicarbonate is important in transporting carbon dioxide and in buffering the blood against pH changes. Depending on the species, the pH of vertebrate blood ranges from about 7.3 to 8.2, but the pH of

any one species at a given temperature varies only about 0.2 unit of pH. Blood pH is maintained in the range of 7.35–7.45 in humans.

Carbon dioxide diffuses into the alveoli when free CO_2 is produced by the reversal of the bicarbonate reactions and when CO_2 is released from hemoglobin. Free carbon dioxide continues to enter the alveoli until the partial pressure of alveolar carbon dioxide nearly equals that found in the blood, roughly about 40 mm Hg. These and other phases of gas exchange at the respiratory surface and in the tissues are summarized in Figure 5.20.

☐ SUMMARY

1. Oxygen, needed to support aerobic metabolism, is captured from the environment, and waste carbon dioxide is voided, by diffusion across a moist gas exchange surface (cell membranes) in all animals. Gases diffuse down a partial pressure gradient. Depending on the animal, the gas exchange surface may be the outer body surface, gills, or lungs. Air breathers have access to plenty of oxygen, but must prevent their gas exchange surfaces from drying. Aquatic animals must work harder to obtain the relatively small amount of oxygen available in their environment.

2. Animals that live in moist habitats can use outer body surfaces to exchange gases. Animals that lack special gas exchange organs tend to be small because diffusion alone must meet their cells' demand for oxygen. Larger, more active animals have specialized gas exchange structures and a circulatory system that delivers oxygen to their tissues at a rapid rate.

3. Gills, which greatly increase external surface area for gas exchange, occur in many invertebrates, many larval amphibians, and fishes. Most gills are ventilated so that countercurrent flow of water and blood provides diffusion gradients that favor inward diffusion of oxygen and outward diffusion of carbon dioxide. Countercurrent mechanisms provide that water flows over the gills in a direction opposite to blood flow within the gills.

4. The insect respiratory apparatus consists of a tracheal system that branches throughout the body. Branches called tracheae and tracheoles carry oxygen directly to the tissues and dispose of carbon dioxide. The insect circulatory system plays little or no role in gas transport. Aquatic insects may have tracheal gills, or they may periodically rise to the surface to breathe air.

5. Most adult amphibians have simple, saclike lungs that are ventilated by a pumping action of the floor of the mouth. They may also obtain a significant amount of oxygen through the lining of the mouth and moist skin.

6. The mammalian respiratory tree consists of the nostrils, mouth, pharynx, epiglottis, trachea, bronchi, bronchioles, and alveoli. The lungs are ventilated by contraction of the diaphragm and rib muscles to increase the size of the thoracic cavity. Resistance to expansion due to surface tension of the alveoli is partially overcome by pulmonary surfactants. Gas exchange is regulated by nerve centers in the pons and medulla oblongata of the brain.

7. Bird lungs tend to have a fixed volume and consist of parabronchi and air capillaries. Gas flow through the lungs is continuous and unidirectional. Inhaled air passes first to the posterior air sacs. During exhalation, gas from the posterior air sacs flows through the lungs. With the next inhalation it passes into the anterior sacs and then out of the body during the next exhalation.

8. Several respiratory pigments, proteins that transport oxygen, occur in the animal kingdom. Hemocyanin, a copper-containing pigment, and hemoglobin, an iron-containing pigment, are the most common. All pigments can bind oxygen when its partial pressures are high and release it when partial pressures are low. High carbon dioxide concentrations and acidic conditions cause hemoglobin to release its oxygen in what is called the Bohr effect. This property ensures that active tissues receive oxygen from the blood.

9. Carbon dioxide is transported in the blood as a dissolved gas, as bicarbonate, and, in animals with hemoglobin, combined with hemoglobin.

☐ FURTHER READING

Feder, M. E., and W. W. Burggren. "Skin Breathing in Vertebrates." *Scientific American* 253(1985): 126–143. *Discusses extensive occurrence of skin breathing in vertebrates and circulatory adaptations in frogs.*

Nadel, E. R. "Physiological Adaptation to Aerobic Training." *American Scientist* 73(1985): 334–343. *Discusses the many physiological changes that occur in humans under aerobic training regimes.*

Perutz, M. F. "Hemoglobin Structure and Respiratory Transport." *Scientific American* 240(1978): 92–125. *Changes in hemoglobin associated with carbon dioxide and oxygen transport are discussed from a historical perspective.*

Rahn, H., A. Amos, and C. V. Paganelli. "How Bird Eggs Breathe." *Scientific American* 240(1979): 46–55. *Problems of gas exchange through egg shells are discussed.*

Randall, D. J., W. W. Burggren, A. P. Farrel, and M. S. Haswell. *The Evolution of Air Breathing in Vertebrates.* New York: Cambridge University Press, 1981. *Traces the anatomical and physiological changes associated with the evolution of air breathing.*

Schmidt-Nielsen, K. "Countercurrent Systems in Animals." *Scientific American* 244(1981): 118–128. *Discusses the role of countercurrent systems in respiration, excretion, and heat exchange.*

Schmidt-Nielsen, K. "How Birds Breathe." *Scientific American* 225(1971): 73–79. *Describes the anatomy and physiology of the avian respiratory system.*

Wells, R.M.G. *Invertebrate Respiration. The Institute of Biology's Studies in Biology No. 127.* London: Edward Arnold Publishers, 1980. *A general description of the respiratory mechanisms and structures in invertebrates.*

Internal Transport and Defense Systems

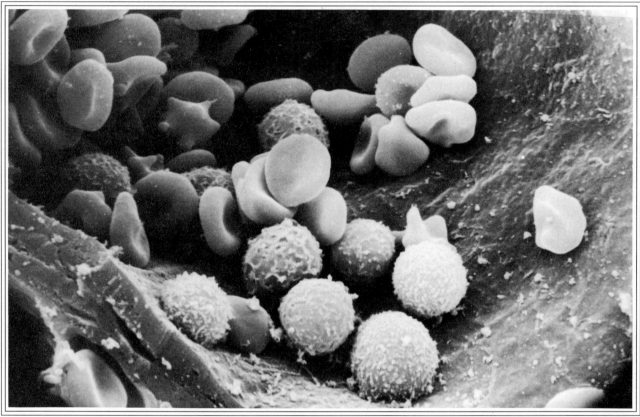

From *Tissues and Organs: A Text-Atlas of Scanning Electron Microscopy* by Richard G. Kessel and Randy H. Kardon. W. H. Freeman and Company Copyright © 1979.

Blood cells as seen with the scanning electron microscope. This micrograph shows erythrocytes (red blood cells) and leukocytes (white blood cells) of a mammal (4300×).

Fluid circulation, providing distribution of nutrients, respiratory gases, and wastes, is essential for the maintenance of homeostatic balance in all living organisms. Small animals with low rates of metabolism can rely on diffusion and membrane transport mechanisms to distribute materials throughout their bodies. Diffusion and membrane transport may be assisted by fluids surging back and forth with body movements. Larger, more active animals, especially those with separate organs for digestion, gas exchange, excretion, and hormone production, require a circulatory system to move their body fluids and dissolved materials from one organ to another. Circulatory systems generally consist of a pumping organ and a set of vessels that direct fluids along definite routes. Most circulatory systems also have built-in clotting mechanisms that stop loss of body fluids following minor injuries.

Circulatory fluids perform other functions in addition to transport. In many invertebrate animals, fluids provide support and hydraulic movement. Fluids contained in a limited body space that is acted on by muscles may serve as a **hydrostatic skeleton**. (Hydrostatic skeletons and hydraulic movement are discussed in Chapter 8, p. 176.) Circulatory fluids and cells also act as an immune system, removing pathogenic microorganisms and abnormal cells before they become established.

ORGANIZATION OF TRANSPORT SYSTEMS

Body fluids occur in functional compartments. All the space within an organism's cells constitutes the intracellular compartment, and the fluid within this compartment is called the **intracellular fluid**. About 70% of the weight of an adult vertebrate is water, and 45% of that is contained within cells. As discussed in Chapter 3, metabolic reactions occur in intracellular water, and the intracellular compartment is subdivided into smaller functional units. Protozoa have only an intracellular compartment, but animals may have several extracellular compartments, including (1) the *interstitial compartment* (all the spaces between cells), which contains **interstitial fluid**, (2) the *body cavity*, and (3) the *vessels* of the circulatory system. Some animals have additional minor fluid compartments. Vertebrates, for example, have extracellular fluids

in the gut, eye, nervous system, and urinary system. Fluids in these minor compartments do not circulate throughout the body or mix directly with circulating fluids. Materials are exchanged between fluids in the minor compartments and circulating fluids by diffusion and cell membrane transport. Collectively, the extracellular fluids make up about 20%–30% of the body weight of mammals. By contrast, molluscs, using body fluids for support and locomotion as well as internal transport, have about 35%–80% of their body weight in extracellular fluids.

Circulatory Systems

Circulatory systems generally consist of a pumping organ, or **heart**; a system of tubes, or vessels, collectively called the **vascular system**; and **valves** that prevent backflow of circulating fluid. Most valves are flaps of tissue projecting from the walls of the heart or vessels into the lumen (inner cavity) of the system. They are pushed aside when blood flow is in the proper direction, but are raised and block the vessel or heart passage when backward flow occurs.

There are two types of circulatory systems in the animal kingdom (Figure 6.1). One type, called the **closed circulatory system**, occurs in vertebrates and several invertebrates, including many annelids, squids, and octopuses. The vessels of such systems are continuous closed channels containing circulating **blood**; blood consists of a fluid called **plasma** in which circulating cells are usually sus-

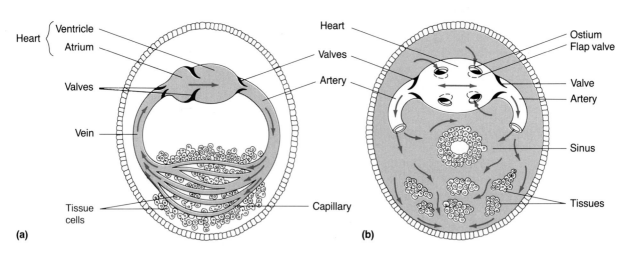

Figure 6.1 Types of circulatory systems. (a) In a closed system, blood flows within a circuit of arteries, veins, and capillaries. Exchange with interstitial fluids occurs by diffusion or filtration across the thin walls of the capillaries. (b) In an open system, hemolymph flows from arteries into a hemocoel composed of sinuses and interstitial spaces. It then percolates back to the heart and enters through openings called ostia.

pended. **Arteries** are vessels that convey blood away from the heart. They branch repeatedly, forming small arteries collectively called **arterioles**. In most closed systems, arterioles branch into **capillaries**, which are very small vessels with walls only one cell thick. The cell layer forming a capillary wall is called an **endothelium**. Capillaries branch repeatedly, extending into fine interstitial spaces. Networks of capillaries called **capillary beds** are the sites where oxygen and nutrients diffuse to the interstitial fluid and where wastes are collected. Blood remains within the vessels and does not bathe the tissues directly; but fluids from the blood filter through the capillary walls, carrying salts, dissolved nutrients, and various proteins into the interstitial spaces. Internal gas exchange also occurs across capillary walls. **Veins** are the vessels that convey the blood back to the heart. The small veins that actually collect the blood from the capillaries are called **venules**.

The second type of circulatory system, the **open circulatory system**, occurs in arthropods and most molluscs. The vessels are not continuous, and there are no capillaries. Fluid pumped by a heart is distributed by arteries to general body regions. The circulatory fluid flows out of the arteries directly into interstitial spaces, where exchange of gases and other materials with tissues occurs. Interstitial spaces are sometimes enlarged into saclike **sinuses**. Interstitial fluids drain back to the heart through collecting sinuses or veins. Because the circulatory fluid flows directly into interstitial spaces, it is not distinguishable from interstitial fluid. Thus, all extracellular fluids (including circulating and interstitial) in an animal with an open circulatory system are called **hemolymph** instead of blood. The sinuses and interstitial spaces are collectively called the **hemocoel**, meaning a cavity filled with hemolymph.

Circulation in open systems is generally more sluggish than in closed systems; and because hemolymph is pumped out into tissue spaces, fluid pressures in open systems are usually lower than in closed systems. The overall effect is that the time needed for hemolymph to circulate to all organs of the body may be longer and the composition of body fluids may vary more than in an animal with a closed system. Closed vessels, on the other hand, result in a less variable body fluid composition. Moreover, with a closed system, blood can be directed by dilation and constriction of vessels to specific parts of the body where it is most needed. Animals with open systems can do this by contracting certain body muscles, but fluid movements are usually not as precise.

Most very active animals have closed circulatory systems, but it should not be assumed that open systems are generally less effective or that animals with open systems are disadvantaged. Circulatory systems, regardless of the type, are efficient means of moving fluids from one body region to another. Open and closed systems represent alternative solutions to the problem of internal transport in large multicellular bodies. Actually, the distinction between closed and open circulatory systems is not always clear. Some annelids, for example, have typical closed systems; but many species have a well-developed system of arteries and veins, but have open channels or sinuses instead of capillaries.

Circulatory pumps, or hearts, are of two structural types (Figure 6.2). **Tubular hearts** are vessels with muscular walls that pulsate, pumping fluid through the rest of the system. In some tubular hearts, such as those of insects, fluids are driven by waves of contraction (peristalsis) sweeping from one end of the vessel to the other. Blood flows only one way, determined by the direction of peristalsis. Many invertebrates have tubular hearts that are nonperistaltic. Such hearts contract rhythmically, more or less all at once, rather than in waves.

The other type of heart is the saclike or **chambered heart**, characteristic of molluscs, crustaceans, and vertebrates. Chambered hearts have from one to four compartments. Most are two-stage pumps in which a thin-walled compartment, the **atrium**, receives fluid returning to the heart through veins. The atrium pumps fluid at low pressure into a second chamber, the heavily muscled **ventricle**, which then pumps the fluid at high pressure through the arteries.

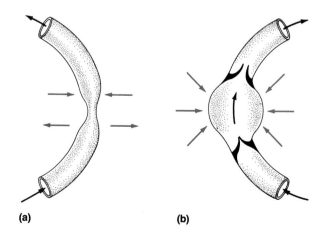

(a)   **(b)**

Figure 6.2 Types of circulatory pumps. (a) A tubular heart is a contractile portion of a major circulatory vessel. (b) A chambered heart is saclike. Valves in circulatory vessels prevent backflow of blood or hemolymph.

CIRCULATION IN PROTOZOA AND REPRESENTATIVE ANIMALS

Animal evolution has included a general trend toward an increasing complexity in circulatory mechanisms. The earliest animals were undoubtedly small and able to rely on the ebb and flow of fluids in interstitial spaces. As larger, more complex animals evolved, distinct circulatory systems also evolved, providing adequate internal transport. A look at circulation in some representative organisms illustrates the diversity of circulatory mechanisms and systems that have evolved.

Simple Circulatory Mechanisms

Protozoa.   Because protozoa are small, with high surface-area-to-volume ratios, they have sufficient surface area for gas, nutrient, and waste product exchange. In addition to diffusion, circulation in protozoa may also occur by **cyclosis**, the flowing of cytoplasm inside the cell, thought to be caused by microfilaments of the contractile protein actin. Cyclosis stirs intracellular fluid along the inside margin of the cell membrane, so that diffusion gradients are maintained. Digestive vacuoles also circulate by this mechanism, distributing nutrients throughout the cell.

Sponges.   These animals have many internal cells that are not equally exposed to oxygen in the external environment and that do not have equal opportunities for waste disposal. However, the body of a sponge is very porous, with canals passing from the body surface into the body mass. Flagellated cells called **choanocytes** line the internal cavities. Beating of the flagella draws water from the environment into the animal, thereby promoting waste and gas exchange. Furthermore, mobile cells called **amoebocytes** transport food, gases, and wastes to and from other body cells (Figure 6.3). Unlike most animals, sponges circulate the external environment through their bodies, instead of circulating an internal fluid.

Gastrovascular Cavities

As discussed in Chapter 1, invertebrate animals in several phyla lack a body cavity (either a pseudocoelom or a coelom). These animals also lack circulatory structures and rely heavily on diffusion of materials and gases across the body surface to pro-

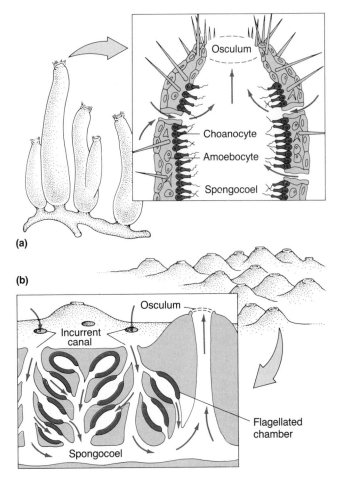

Figure 6.3   Circulation in sponges.   Flagella on choanocytes lining internal canals draw a constant stream of water into the animal through pores in the body wall. Arrows indicate the flow of water in (a) a simple vase-shaped sponge and (b) an encrusting sponge that has a thick, complex body wall.

vision their internal cells. Many such animals also have a **gastrovascular cavity**, which, as its name implies, functions in both digestion and circulation. Environmental water is taken into the cavity with food. Nutrients and dissolved oxygen diffuse from the water into the body, and wastes move in the opposite direction. Body movements and flagellated cells lining the cavity mix fluids and prevent the development of local accumulations of nutrients or wastes.

Cnidarians.   Hydras and several other cnidarians have a saclike gastrovascular cavity. In many jellyfishes and sea anemones, however, the gastrovascular cavity consists of a system of multibranched canals. In some jellyfishes, **radial canals**, which may be perradial, adradial, or interradial canals, extend from a central pouch to a **circular canal**

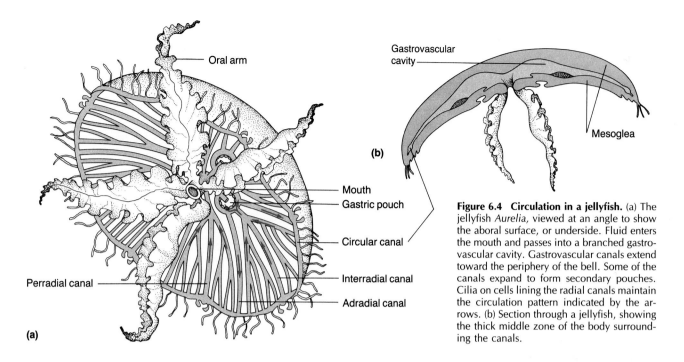

Figure 6.4　Circulation in a jellyfish. (a) The jellyfish *Aurelia,* viewed at an angle to show the aboral surface, or underside. Fluid enters the mouth and passes into a branched gastrovascular cavity. Gastrovascular canals extend toward the periphery of the bell. Some of the canals expand to form secondary pouches. Cilia on cells lining the radial canals maintain the circulation pattern indicated by the arrows. (b) Section through a jellyfish, showing the thick middle zone of the body surrounding the canals.

that borders the rim of the bell-shaped body (Figure 6.4). Cilia in the canals and periodic body contractions create currents that circulate water and materials throughout the cavity.

Flatworms.　Lacking a body cavity, a flatworm is a nearly solid mass of cells interspersed with fluid-filled interstitial spaces. Planarians and parasitic flukes have a highly branched gastrovascular cavity. Nutrients diffuse into the interstitial fluid from the gastrovascular cavity and are carried short distances to sites of use. The body surface is especially important in nutrient uptake and gas exchange in parasitic flatworms. In tapeworms, which lack a gastrovascular cavity, the body surface is the only site of uptake and exchange.

Circulation in Body Cavities

Most animals more complex than flatworms have a distinct body cavity between the gut and body wall. (Refer to the overview of the major phyla in Chapter 1, p. 6, and the Trends and Strategies section on body cavities, p. 531.) The body cavity is important in internal transport in many animals because fluids in the cavity bathe internal tissues and circulate as an animal bends or moves. Circulation in a body cavity is influenced by body organization. In nematodes, body cavity (pseudocoelomic) fluid may surge from one end of the animal to the other. In many annelids, however, cross

walls (septa) between each body segment restrict fluid movements to within individual segments. Coelomic fluid in annelids is circulated by cilia and body movements within each segment, but most transport is achieved via a circulatory system. Echinoderms have an extensive body cavity in which coelomic fluid is stirred by cilia on the peritoneal cells as well as by body movements (see Figure 5.2). In animals with closed circulatory systems, materials are exchanged between the blood and fluids in the body cavity. In open circulatory systems, hemolymph circulates back and forth between the vascular system and the hemocoel.

Open and Closed Circulatory Systems in Invertebrates

It is not possible to trace the evolutionary origin of open and closed circulatory systems, but the different systems we see today probably represent several lines of animal evolution. In some lineages, closed systems may have evolved from open ones. On the other hand, the open systems of arthropods most likely evolved from closed or partially closed systems similar to those of annelids, for it is fairly certain that arthropods evolved from annelid-like ancestors.

Earthworms.　The closed system of these annelids includes several **longitudinal vessels** that pass along the length of the worm dorsal to and ventral

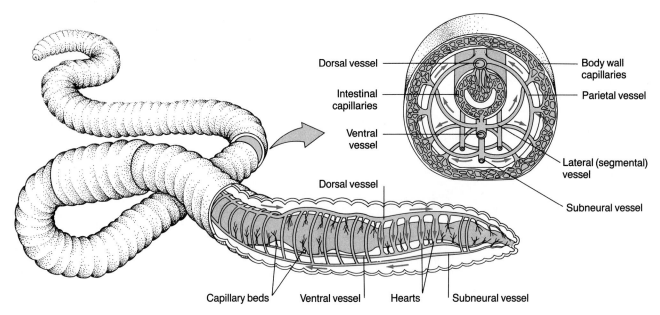

Figure 6.5 Circulation in an earthworm. Arrows indicate the main directions of blood flow. Segments anterior to the lateral hearts receive blood from small vessels off the large dorsal vessel. The diagrammatic cross section shows major vessels and capillary beds.

to the digestive tract (Figure 6.5). Smaller **lateral (segmental) vessels** branch off the longitudinal vessels in each segment and pass to capillaries in the body wall, the digestive tract, and other organs. Several parts of the system serve as peristaltic, tubular hearts. One of these is the dorsal vessel; peristaltic waves passing along its length from posterior to anterior drive blood forward. Also serving as hearts are five pairs of enlarged lateral vessels connecting the dorsal and ventral vessels in the anterior segments. Their contractions increase blood pressure in the system, driving blood into the ventral vessel. Valves in the dorsal and lateral vessels prevent backflow of blood. Blood flows from the ventral vessel into the lateral segmental vessels and then into capillary beds. Oxygen enters the blood by diffusion into the capillary beds in the body wall. Some blood from the capillary beds drains into the subneural vessel, which connects to the dorsal vessel via the parietal vessels in each segment. The rest of the blood drains directly into the dorsal vessel through small veins from the segmental capillaries. Annelid blood contains respiratory pigments (hemoglobin, chlorocruorin, or hemerythrin) usually dissolved in the plasma; the pigments increase the blood's oxygen-carrying capacity.

Insects. The open circulatory system of these arthropods has a single large vessel, the **dorsal aorta**

(Figure 6.6). The hemocoel is divided into several sinuses by longitudinal sheets of connective tissue called diaphragms. The three main body sinuses are a large **perivisceral sinus** surrounding most of the organs, a smaller ventral **perineural sinus** surrounding the ventral nerve cord, and a dorsal **pericardial sinus** surrounding the heart. Note in Figure 6.6 that part of the dorsal aorta is modified as a tubular heart. The heart is attached to the dorsal diaphragm by **alary muscles**. In many insects, contraction of the alary muscles expands the heart, sucking hemolymph into it through small circular openings called **ostia**. Flap valves of thin tissue inside the heart at each of the ostia are pushed aside when hemolymph flows in. When the heart itself contracts, the flaps block the ostia and prevent backflow. Contraction of the heart muscles drives hemolymph forward in the dorsal aorta. The rate of heart contraction varies with temperature and activity levels. When a sphinx moth is at rest, for example, its heart may contract about 40 to 50 times a minute, but during flight the rate can increase to 110 to 140 contractions per minute.

In most insects, the dorsal aorta ends abruptly below the brain, but in others it subdivides into small arteries supplying the eyes and appendages of the head. Fluid leaving the aorta or arteries percolates backward through the body sinuses to other regions of the body. Hemolymph enters the legs from the ventral sinus, and contractile vessels lo-

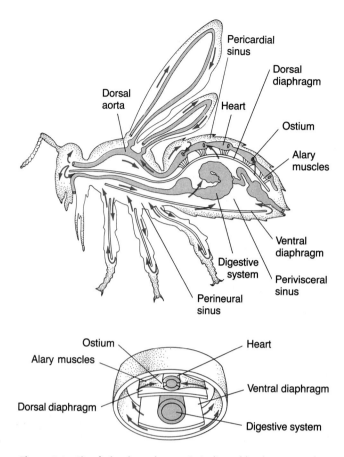

Figure 6.6 Circulation in an insect. As indicated by the arrows, hemolymph is pumped forward in the dorsal aorta. In many insects, as shown here, the dorsal aorta in the first nine abdominal segments forms the tubular heart. Diaphragms (shown schematically in the inset) divert hemolymph into sinuses. In the legs, hemolymph is divided by diaphragms into entering and exiting streams. The general pattern of wing circulation is indicated, but in the wings of most insects, several additional longitudinal veins and many cross-veins carry the hemolymph.

cated at the base of these appendages may boost hemolymph flow rate. Similar pulsatile vessels occur at the base of the wings, antennae, and neck in many insects. Hemolymph returning from the legs and ventral sinus enters the perivisceral sinus and moves slowly upward toward the heart. With few exceptions, insect hemolymph contains no respiratory pigments, and gas transport occurs by way of the tracheal system, not the circulatory system. (For details, see Chapter 5, p. 106; Figure 5.10.)

Crustaceans. The open systems of lobsters, crayfish, and other crustaceans include a single-chambered heart located in the thorax. Several large arteries distribute hemolymph to the hemocoel in all regions of the body (Figure 6.7). The hemolymph

then drains back into the ventral sinus. From here it flows via **afferent branchial vessels** to the gills, where gas exchange occurs. Oxygenated hemolymph leaves the gills through **efferent branchial vessels** and flows into the pericardial sinus. The crustacean heart acts as a suction pump, similar to the insect heart, but the sucking mechanism is somewhat different in crustaceans. When the heart contracts, it stretches the elastic ligaments that suspend the heart in the pericardial sinus. When the heart relaxes, elastic recoil of the ligaments expands the heart, drawing hemolymph from the pericardial sinus into the heart through paired ostia. Flap valves in the ostia and in the vessels prevent backflow during contraction.

Hemolymph may circulate quite rapidly in crustaceans. The heart contraction rate in lobsters is temperature-dependent, but is controlled by a cluster of nine nerve cells located on the heart's dorsal wall. In a 450 g lobster, the heart can contract about 100 times per minute. Each beat forces from 0.1 to 0.3 ml of hemolymph from the heart at a pressure of about 13 mm Hg, pumping about 10–30 ml of hemolymph per minute into the arteries. Because the hemolymph volume of a lobster this size is about 75 ml, all hemolymph may pass through the heart every 2–7 minutes.

Molluscs. Most molluscs, including snails, slugs, clams and other bivalves, and chitons have an open circulatory system (see Figure 24.2). Molluscan hearts have one, two, or three chambers. In bivalves, hemolymph returning to the heart first enters two expandable atria (Figure 6.8a). When stretched, the thin, muscular walls of the atria spontaneously contract, forcing hemolymph into a single highly muscular ventricle. Contraction of the ventricle forces hemolymph into one or (more typically) two arteries. One artery carries hemolymph anteriorly and the other carries it posteriorly into the hemocoel. During ventricular contraction, backflow of hemolymph into the atria is prevented by a valve at each atrioventricular junction. Hemolymph is oxygenated regardless of which arterial path it takes because it will pass through either the mantle or the gills, both of which are organs of gas exchange. From the hemocoel in the foot and viscera, hemolymph flows into the excretory organs. From here, veins carry hemolymph to the atria either directly or via the gills (Figure 6.8b). Unlike most molluscs, many bivalves lack respiratory pigments. Their extensive gill and mantle surfaces supply ample oxygen to support their rather sedentary existence.

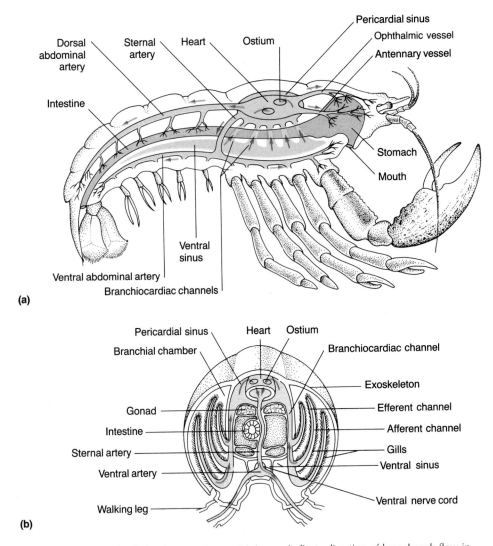

Figure 6.7 Circulation in a crustacean. (a) Arrows indicate direction of hemolymph flow in large vessels leaving the heart. Hemolymph enters the sinuses and percolates back to the gills (not shown here). (b) After passing through the gills, the hemolymph flows into the branchiocardiac channels and then into the pericardial sinus, from which it passes into the heart through the ostia.

Hemolymph is drawn into the atria in bivalves by a suction mechanism. A pericardial sac filled with fluid surrounds the heart. As the ventricle contracts within the pericardium, a negative pressure develops in the pericardial sac. Because the atria are thin-walled and partially located in the pericardial sac, this negative pressure causes them to expand, and hemolymph is drawn from the excretory organ to the heart. Contractile veins aid the process.

Squids and octopuses are unusual molluscs in that they have a closed circulatory system with three hearts. They have a three-chambered **systemic heart** (two atria and a ventricle) that receives oxygenated blood directly from the gills and pumps it through anterior and posterior aortas to capillaries throughout the body (Figure 6.9). In actively swimming squids, blood may leave the ventricle of the systemic heart at a pressure of about 40–60 mm Hg. This is about ten times the maximum pressure of the clam system. The systemic heart also beats at a relatively rapid rate (about 80 beats per minute when a squid is active, compared to a maximum of only about 10 to 20 beats per minute in bivalves). Squids and octopuses also have two single-chambered **branchial hearts** that receive blood through veins from the capillaries and pump it to the gills. These auxiliary hearts significantly increase the blood pressure and rate of flow in the gills. Also assisting oxygen transport is the

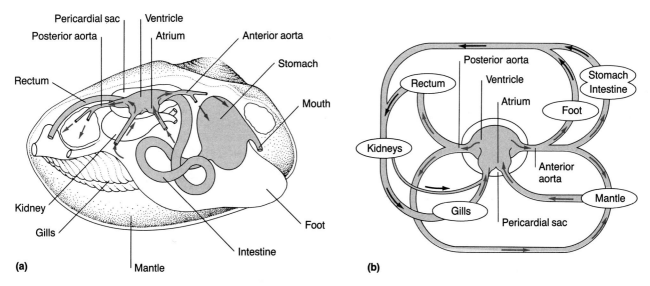

Figure 6.8   The open circulatory system of a bivalve mollusc. (a) The heart, which is three-chambered, is housed in a pericardial sac derived from the coelom. The ventricle is wrapped around the posterior part of the intestine. When the ventricle contracts, negative pressure in the pericardial sac causes the atria to expand and draw hemolymph in from vessels in the gills and mantle. (b) The general flow of hemolymph in the freshwater clam *Anodonta;* other bivalves, such as the marine mussel *Mytilus,* have only one vessel leaving the heart and a somewhat different circulatory pattern.

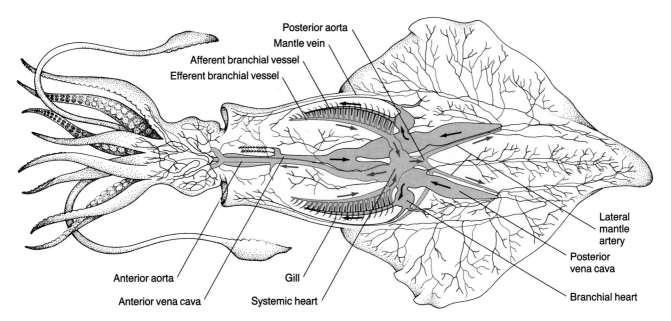

Figure 6.9   The closed circulatory system of a squid. A systemic heart pumps oxygenated blood through arteries to the body. Blood gathered in veins flows to the gills, where branchial hearts pump it through the gill capillaries.

respiratory pigment hemocyanin, which is dissolved in the blood plasma. Overall, this closed system distributes blood rapidly and under relatively high pressure, providing squids and octopuses with strong circulatory support for a generally higher rate of metabolism and higher activity level than other molluscs.

General Features of Vertebrate Circulatory Systems

The Vascular System.   All blood vessels in the closed circulatory systems of vertebrates are lined internally by an endothelium surrounded by a basal lamina (Figure 6.10). The endothelium forms

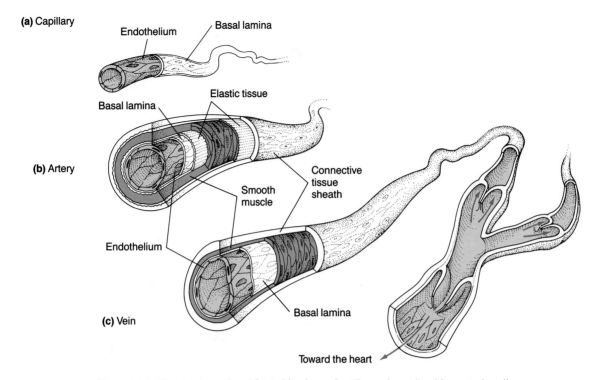

Figure 6.10 The structure of vertebrate blood vessels. All vessels are lined by a single-cell-thick endothelium. (a) Capillaries are formed of an endothelium covered by a basal lamina. Holes called fenestrae (not illustrated) between endothelial cells allow fluids to leave the capillaries. White blood cells may also leave the capillaries by squeezing through fenestrae. (b) Fluids do not leak out of the arteries because tightly knit muscle and connective tissue make up the arterial wall. (c) Veins have less muscular walls than arteries. Notice that flap valves allow blood to flow only toward the heart.

a nonabrasive internal surface, protecting blood cells as they tumble through the system. Small openings between adjacent endothelial cells in capillaries allow water and dissolved substances to pass into the interstitial spaces but prevent red blood cells and large proteins from leaving the vascular system. The endothelium and basal lamina of arteries and veins are surrounded by a zone of circular smooth muscle and elastic connective tissue fibers. Outside of this middle zone is a layer of collagen fibers running parallel to the vessel's long axis. Veins differ from arteries in having thinner, collapsible walls with less elastic tissue and circular muscle. The additional layers of tissues in larger vessels prevent materials from passing out of the vessels to the tissues. Therefore, the large vessels are conduits, and it is the capillaries that are most important in maintaining homeostasis in the tissues.

Circulatory Patterns. Correlated with the evolution of lungs for the breathing of air on land, the vertebrate circulatory system underwent some significant changes; consequently, the vascular system and heart of fishes are very different from those of terrestrial vertebrates. Bony and cartilaginous fishes have a single circulatory loop and a linear flow of blood through the heart. Blood flows from the heart to the gills and then directly to capillaries in body tissues. Other vertebrates have a double-loop circulatory pattern. One loop, called the **pulmonary circuit**, carries blood between the heart and the lungs; the other, called the **systemic circuit**, carries oxygen-rich blood from the heart to the body tissues and returns oxygen-poor blood from the tissues to the heart. Each circuit includes its own arteries, capillaries, and veins; and the heart pumps blood through both circuits. We are now ready to compare the main structural and functional aspects of the circulatory system of a bony fish, representing aquatic vertebrates, with those of a mammal, representing terrestrial vertebrates. (The evolution of vertebrate circulatory systems is discussed in Chapter 28, p. 696.)

The Circulatory Pattern in Bony Fishes. In fishes, blood returns to the heart from the liver via a large **hepatic vein** and from the rest of the body via large

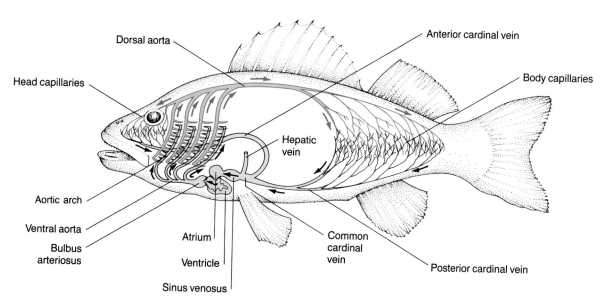

Figure 6.11 Circulation in a bony fish. Blood leaving the heart goes directly to the gills, where exchange of gases and wastes occurs. It then collects in the dorsal aorta and is distributed throughout the body. Cardinal and hepatic veins collect blood from the tissues and return it to the heart.

cardinal veins (Figure 6.11). The heart has four chambers. A posterior chamber called the **sinus venosus** receives blood from the veins. The sinus venosus is a thin-walled storage sac with little or no cardiac muscle. From it, blood flows into a muscular atrium, which expands and contracts, filling the thick muscular ventricle. Ventricular contraction forces blood at high pressure into the fourth heart chamber, called the **bulbus arteriosus**. Flap valves between the heart chambers prevent backflow of blood. Valves between the ventricle and the bulbus arteriosus develop from another part of the heart called the **conus arteriosus**. In bony fishes, this is represented only by the region of the valves, but in other vertebrates, the conus is a discrete heart chamber, and as you will see in Chapter 28, it was important in the evolution of terrestrial vertebrate circulatory systems.

Figure 6.11 shows the one-way path of blood through the four chambers of the fish heart. From the bulbus arteriosus, blood flows into the ventral aorta, which then divides into a series of right and left **aortic arches**. The aortic arches form **afferent branchial arteries** that break up into many tiny capillaries in the gill lamellae, where the blood is oxygenated. (See Figure 5.7 to review the process of gas exchange in fish gills.) Blood flows smoothly to the gills because the bulbus arteriosus is elastic. The bulbus expands when the ventricle contracts because the gill capillaries resist blood flow. Then, when the ventricle relaxes, the bulbus gradually

returns to its normal size. Without the action of the bulbus arteriosus, the ventricular contractions would force blood over the gills in spurts. From the gill capillaries, blood enters **efferent branchial arteries** that join to form the **dorsal aorta**. This large vessel conveys oxygenated blood backward along the length of the body, supplying the various organs and tissues. Smaller vessels, the **carotid arteries**, carry oxygenated blood to the head. From capillaries in the body tissues, blood collects in veins, and some of it passes directly back to the heart. Before returning to the heart, some of the blood in the posterior region passes first through a **portal system**, which is a system of vessels between two capillary beds. Blood returning from capillaries in the tail region flows through the **renal portal system** to the kidney, where wastes and salts are removed. Blood returning from capillaries in the intestinal region circulates through the **hepatic portal system** to capillaries in the liver, where many nutrients are absorbed and stored. Portal systems allow additional processing of the blood by the kidneys and liver, but they operate at low blood pressures. This means that blood flow is slow, and exchange with kidney and liver cells is by diffusion and secretion rather than by filtration through the capillary walls.

Compared to other vertebrates, fishes have a small blood volume. In bony fishes, the blood accounts for only about 5% of the body weight, while in mammals and birds, it is about 6%–10%. Note in

Figure 6.11 that all blood flowing to the tissues of a fish first passes through the gills. While this arrangement ensures that the blood headed for the tissues is highly oxygenated, it has a major disadvantage. A pressure drop occurs in the gills owing to the resistance to blood passage in the gill capillaries. Consequently, all parts of the circulatory system downstream from the gills operate at a lower pressure and flow than would occur if the ventricle pumped blood directly to the tissues.

The Circulatory Pattern in Mammals. The main features of the mammalian circulatory system are shown in Figure 6.12. The double-loop circulatory pattern can be envisioned by studying the mammalian heart and main vessels attached to it (Figure 6.12b). Mammals have a four-chambered heart (two atria and two ventricles) located in a separate portion of the body cavity called the **pericardial cavity**. The atria receive blood from veins, whereas the ventricles dispense blood to arteries. The right atrium receives oxygen-poor blood from capillaries of the systemic circuit via two large veins, the **anterior vena cava** and **posterior vena cava**. The right atrium pumps oxygen-poor blood into the right ventricle, which in turn pumps it through the pulmonary artery to the lungs. Oxygen-rich blood returns from capillaries in the lungs via the pulmonary vein to the left atrium. The left atrium pumps the highly oxygenated blood into the left ventricle. The left ventricle is the most muscular heart chamber. It pumps oxygen-rich blood to the systemic aorta, which connects the heart to the main arteries of the systemic circuit. The pattern of circulation in mammalian fetuses is different because the lungs are not yet functional. (Fetal circulation is discussed in Chapter 33, p. 808.)

Note in Figure 6.12 that blood flow through each side of the mammalian heart is unidirectional. **Atrioventricular (AV) valves** located between the heart chambers and **semilunar valves** at the junctions of the large arteries and the ventricles prevent backflow. Collapse of the atrioventricular valves back into the atria by the high pressures in the ventricles during contraction is prevented by the **chordae tendinae**, which are fibrous tendons attached to the valves and to **papillary muscles** on the ventricular wall.

Physiology of the Mammalian Circulatory System

Pressure Cycles in the Heart. The pumping actions of the mammalian heart can be explained in terms of the valve positions and pressures that de-velop during the **cardiac cycle**. A complete cycle involves a relaxation phase, or **diastole**, during which the heart chambers fill, and a contraction phase, or **systole**, during which the chambers empty. During diastole, the atria fill with blood, and the ventricles fill to about 70% capacity because the atrioventricular valves are open. Atrial contraction completes the ventricular filling, and with their walls fully expanded, the ventricles contract. Ventricular contraction increases blood pressure in the ventricles, and blood pushing toward the atria causes the atrioventricular valves to close. This allows ventricular pressure to further increase. When ventricular pressure is greater than the back pressure in the systemic aorta and pulmonary artery, the semilunar valves at the origins of these vessels open. Blood flows from the heart into the arteries so long as the ventricles continue to contract. Ventricular pressure during systole pushes blood into the systemic aorta against the resistance offered by the vessels. This stretches the elastic walls of the arteries, and when the ventricles relax during diastole, elastic recoil in the arteries squeezes against the blood, thereby maintaining pressure in the vessels, which in turn closes the semilunar valves, preventing backflow into the heart. The elastic recoil of the arteries also evens the flow of blood through the system. Without the elastic recoil, the intermittent heartbeat would cause the blood to flow in spurts. The pulse that can be felt in the arteries near the body surface is the expansion and recoil of the arteries, resulting from a pressure wave that travels through the blood with each heartbeat. Normal blood pressure in humans in the systemic circulation is about 120 mm Hg at complete ventricular contraction and 80 mm Hg during ventricular relaxation. The pulmonary circulation operates at a lower pressure, and the corresponding pressures are 25/10 mm Hg.

Rhythmicity of the Heartbeat. Rhythmic contraction is an inherent property of vertebrate muscle cells. This can be demonstrated by culturing cardiac muscle cells in the laboratory. About 1% of the cells will spontaneously contract and relax. When cultured cells contact each other, the entire cell mass contracts simultaneously, being paced by the most frequently contracting cell. The vertebrate heart contains **pacemaker cells** that have the inherent property of initiating contraction in themselves and neighboring cells. This autostimulation results from changes in electrical potentials in the muscle cell membrane. (Muscle activation is considered in Chapter 9, p. 207.) Vertebrate hearts are said to be **myogenic**, which means that contractions resulting

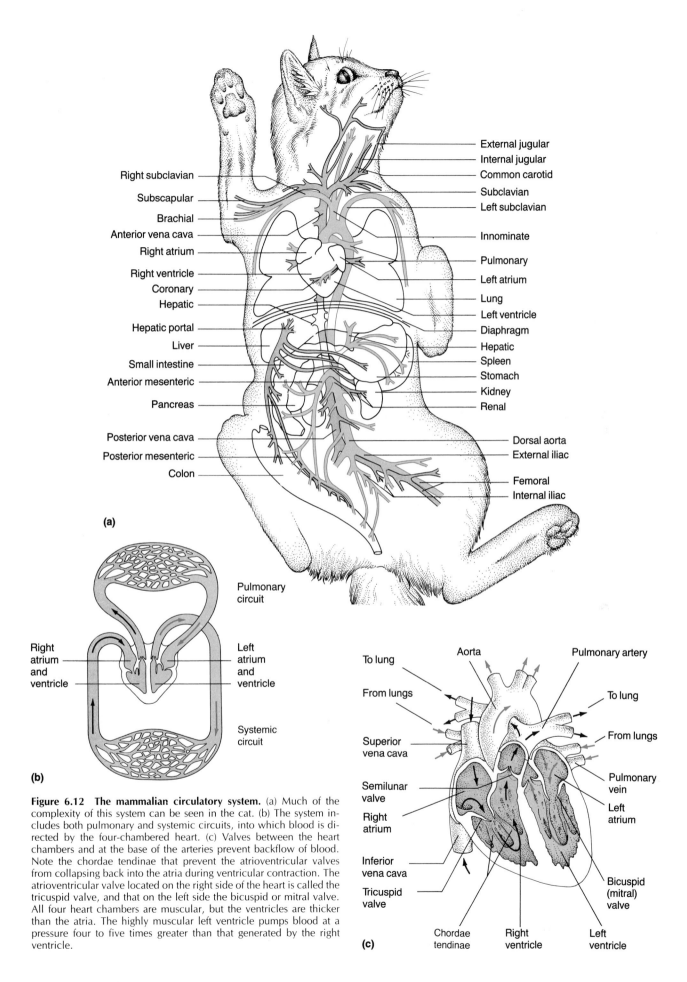

Right subclavian
Subscapular
Brachial
Anterior vena cava
Right atrium
Right ventricle
Coronary
Hepatic
Hepatic portal
Liver
Small intestine
Anterior mesenteric
Pancreas
Posterior vena cava
Posterior mesenteric
Colon

External jugular
Internal jugular
Common carotid
Subclavian
Left subclavian
Innominate
Pulmonary
Left atrium
Lung
Left ventricle
Diaphragm
Hepatic
Spleen
Stomach
Kidney
Renal
Dorsal aorta
External iliac
Femoral
Internal iliac

(a)

Pulmonary
circuit

Right
atrium
and
ventricle

Left
atrium
and
ventricle

Systemic
circuit

(b)

To lung
Aorta
Pulmonary artery
From lungs
To lung
Superior
vena cava
From lungs
Semilunar
valve
Pulmonary
vein
Right
atrium
Left
atrium
Inferior
vena cava
Bicuspid
(mitral)
valve
Tricuspid
valve
Chordae
tendinae
Right
ventricle
Left
ventricle

(c)

Figure 6.12 The mammalian circulatory system. (a) Much of the complexity of this system can be seen in the cat. (b) The system includes both pulmonary and systemic circuits, into which blood is directed by the four-chambered heart. (c) Valves between the heart chambers and at the base of the arteries prevent backflow of blood. Note the chordae tendinae that prevent the atrioventricular valves from collapsing back into the atria during ventricular contraction. The atrioventricular valve located on the right side of the heart is called the tricuspid valve, and that on the left side the bicuspid or mitral valve. All four heart chambers are muscular, but the ventricles are thicker than the atria. The highly muscular left ventricle pumps blood at a pressure four to five times greater than that generated by the right ventricle.

in the heartbeat originate in muscle cells. Some invertebrates have **neurogenic** hearts, in which the initiation of contraction is caused by nerve impulses.

In the intact mammalian and bird heart, pacemaker activity originates in the **sinoatrial (SA) node**, a patch of specialized muscle tissue (about 2 mm × 20 mm in humans) located near the site where the anterior vena cava enters the right atrium (Figure 6.13). A wave of excitation originating here travels across the atria to a similar node, the **atrioventricular (AV) node**, located near the wall between the two sides of the heart. Specialized muscle cells in the AV node delay the wave of excitation for about 0.1 second and then rapidly transmit the impulse to the ventricles via a thick strand of specialized cells, the **atrioventricular bundle**. The AV bundle branches into fine stems called **Purkinje fibers** that convey the excitatory impulses into the cardiac muscle. The short time delay followed by rapid conduction of the impulse to the tip of the ventricles is critical because it allows time for the ventricles to fill. The rapid conduction of excitation to the ventricular tip causes the ventricles to start contracting from the tip toward the openings of the arteries, so blood flow momentum is in the proper direction as blood exits the heart. In general, the heart rate is greatest in smaller animals, which have higher metabolic rates than larger animals. In elephants, the heart beats only about 25 to 40 times per minute, whereas mice have heart rates of 300 to 700 beats per minute. The fastest heart rates observed in mammals have been in flying bats, whose hearts can contract 1200 times per minute.

Cardiac Output and Its Control. **Cardiac output** is the amount of blood pumped by a heart ventricle per minute. It is equivalent to the **stroke volume**, or the amount of blood pumped by each ventricle per heartbeat, multiplied by the number of beats per minute. The stroke volume of the average human heart is about 75 ml. At a normal heart rate of 72 beats per minute, the human heart pumps about 5.4 L of blood per minute, a volume roughly equal to the total blood volume.

Stroke volume depends largely on the amount of blood in the ventricle prior to contraction. During exertion, an increased rate of return of blood to the heart can double the stroke volume. Up to a physiological maximum, the more the heart muscle is stretched by blood volume during diastole, the greater its force of contraction. This effect, called the **law of the heart**, is a major factor in regulating cardiac output in fishes. In mammals, the main effect of ventricular stretching is to ensure that the two ventricles pump an equal amount of blood in a given time period. Nervous and hormonal factors that influence heart rate are more important in regulating cardiac output in the mammalian heart.

In mammals, a **cardiac center** located in the medulla oblongata of the brain regulates the heart rate by influencing the pacemaker and the conducting system of the heart. The cardiac center receives information from receptors that sense the amount of carbon dioxide in the blood, the pH of the blood, and the blood pressure. Receptors in the carotid arteries in the neck, for example, respond when increased blood pressure stretches their walls. Receipt of impulses from these receptors causes the cardiac center to send inhibitory impulses to the sinoatrial node and conducting system of the heart. Impulses pass to the heart via the vagus nerves of the autonomic nervous system. (Organization of the vertebrate nervous system is covered in Chapter 10, p. 234; Figure 10.19.) The vagus nerves release a chemical called acetylcholine, which slows the heart rate. Conversely, when blood pressure falls, the cardiac center receives fewer impulses, and vagus output drops. At the same time, impulses along different nerves from the autonomic nervous system increase the release of a

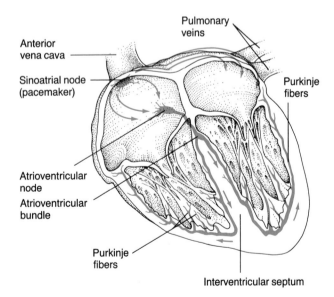

Figure 6.13 Excitation system of the mammalian heart. Arrows indicate how waves of excitation spread from the pacemaker (sinoatrial node) through the atria to the atrioventricular node. From the AV node, excitation rapidly passes via the atrioventricular bundle to the apex of the ventricles, where ventricular contraction is initiated, and then spreads back through the ventricular wall in the Purkinje fibers.

different chemical, norepinephrine, which causes the conducting system of the heart to increase the heart rate. Norepinephrine may also increase stroke volume by increasing the force of contraction. In times of stress, epinephrine (adrenaline), a hormone released from the adrenal glands, has effects similar to those of the nervous stimulation just described. Thus, cardiac output is controlled by the balance of excitatory and inhibitory signals that the heart receives.

Peripheral Circulation and Capillary Beds. The flow F of blood in circulatory vessels (or, for that matter, the flow of fluids in tubes of any type) is related to the pressure P and the resistance R to flow offered by the vessels according to the formula

$$F = \frac{\Delta P}{R}$$

where ΔP is equal to P_1, the pressure at the beginning of the vessels, minus P_2, the pressure at the end of the vessels. In terms of energy, the pressure produced by the right ventricle is expressed as the kinetic energy of the blood moving through the vascular system. This energy drives blood through the systemic vessels against a **peripheral resistance**, the resistance to blood flow in the vessels of the systemic circuit. Resistance results from friction as blood flows along the endothelial lining of the vessels. Friction produces heat, which represents degraded energy of movement of the blood. (Thermo-

dynamics and entropy are discussed in Chapter 3, ◁ p. 56.) Thus, the greater the resistance, the slower the blood flow and the greater the amount of energy the heart must expend to drive the blood. In vessels of any type, resistance to fluid flow increases with (1) increasing length of the vessels, (2) greater viscosity of the fluids, and (3) smaller diameter of the vessels. (A vessel's resistance is inversely proportional to the fourth power of its diameter.)

Considering that every cell in the body must have nearly direct access to a capillary, it is not surprising that most tissues have an enormous number of these exchange vessels. It has been estimated that the average human has about 96,000 kilometers (km) of capillaries with a combined surface area of over 0.6 hectare (ha), equivalent to 6000 m^2! Mainly because of the resistance resulting from this surface area and the small diameter of the capillaries, blood flow slows considerably in the capillary beds, allowing time for exchange of materials between the blood and the tissues.

Vessel diameter is the single most important factor affecting the flow in vascular systems because it can be altered readily. Small arteries, or arterioles, are extremely important in regulating blood flow into the capillary beds (Figure 6.14). Contraction and relaxation of **precapillary sphincters**, circular smooth muscles in the arteriole walls near the capillary bed, are controlled locally by the availability of oxygen and by the autonomic nervous system. When precapillary sphincters are closed, blood shunts around a capillary bed

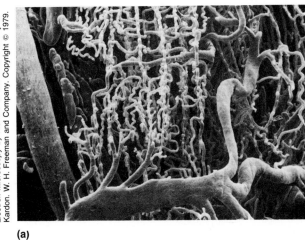

(a)

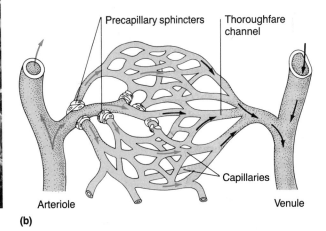

Precapillary sphincters | Thoroughfare channel

Capillaries

Arteriole

Venule

(b)

Figure 6.14 Capillary structure. (a) This scanning electron micrograph was prepared by injecting the circulatory system with plastic and then digesting away the surrounding tissue. Note the arterioles and venules and the convoluted capillaries (130×). (b) Blood flow through capillaries is regulated by precapillary sphincters in the arterioles. Blood may shunt past capillaries directly into venules when the sphincters are contracted.

through **thoroughfare channels**. If most arterioles are constricted, blood pressure throughout the body rises owing to increased peripheral resistance. Despite their number, the capillaries do not contain much blood. Only about 10% of the total blood volume is in the capillaries at any one time, because at any instant, only a small fraction of the capillary beds are open to blood flow. The arteries contain about 20% of the blood volume and the remainder is in the veins.

Fluid Exchange in Capillary Beds. Several factors influence the process of exchange between the blood and the surrounding interstitial fluids, and exchange in both directions constitutes a homeostatic mechanism that ensures overall balance in the transfer of materials between the circulatory system and the body cells. Water and small molecules readily cross a capillary's endothelial membrane by diffusion and endocytosis. The blood also exerts pressure on these materials, tending to make them filter from the capillaries into the interstitial space; this is called **pressure filtration**. As shown in Figure 6.15, several opposing forces determine fluid movements in capillaries. Hydrostatic pressure of the blood (HP_b) tends to force fluids out through small spaces between cells in the capillary walls. The interstitial fluid also has a hydrostatic pressure (HP_{if}), one that tends to force fluids back into the capillary; but as indicated in Figure 6.15, this is much less than the outward force of the blood. Proteins dissolved in the plasma and in the interstitial fluid also influence fluid movements in the capillaries. Solutes, including proteins, in a solution tend to draw water into the solution by osmosis. The **osmotic pressure** of a solution is the pressure required to prevent water from moving into a solution. Thus, osmotic pressure is a measure

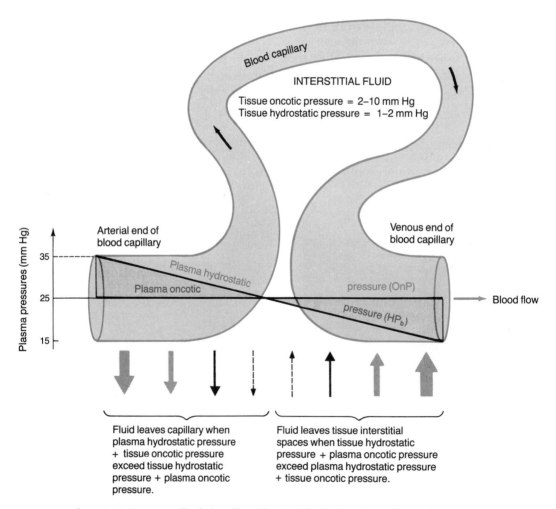

Figure 6.15 Pressures affecting capillary filtration. Fluid is forced out of the capillary into the interstitial spaces when outward pressures exceed inward pressures. Fluid is reclaimed by the capillary when inward pressures are greater than outward pressures.

Essay: How Some Animals Survive Extreme Cold

Birds and mammals that live in polar climates or spend significant amounts of time in cold water have a heat conservation problem. Blood circulating from the body core to the limbs, which have a high surface-area-to-volume ratio, quickly loses heat to the surrounding environment. Such losses are expensive for the animal, for heat production requires considerable energy.

As a mechanism to conserve heat, many birds that endure cold temperatures have countercurrent heat exchangers in their legs (Figure E1). Arteries that carry blood to the extremities are surrounded by veins that bring blood back to the body core. Heat passes from the warm blood entering the legs to the cold blood returning from the legs to the body core. Thus, heat is conserved and is not continually drained from the body.

Similar circulatory adaptations are found in many mammals, including caribou, foxes, and whales. A look at the blood flow to and from the flippers of marine mammals reveals some interesting aspects of temperature regulation. Some veins leaving the flipper are close to the arteries, while others are located farther away from them. If the animal is resting and needs to conserve heat, blood returns to the body through the veins next to the arteries. If the animal is active and producing excess heat, blood flows through the veins near the body surface and does not exchange heat with blood coming from the body core.

Maintaining body appendages at lower temperatures than that of the body core poses some problems. The temperature gradient in a gull standing on ice is substantial, from 38°C in the body core to 0°C at the surface of the feet. At such low temperatures, many of the lipids in the cell

Figure E1 The yellow-billed sheathbill *(Chionis alba)*. This bird thrives in the extreme cold of the Antarctic where it frequently scavenges food dropped by penguins.

membranes of the leg would be expected to solidify, with the effect that nerves and muscles would not function properly and cells would be fragile. But biochemical studies indicate that the membranes of cells from the extremities have a different lipid composition than those of cells in the body core. Phospholipids forming cell membranes in the extremities contain fatty acids that are less saturated than those in core cells. This means that the phospholipids and membranes in the extremities have lower freezing points, allowing the cells to continue functioning at very low temperatures.

of, and is directly proportional to, the tendency of a solution to take up water osmotically. Because they are unable to pass through the capillary walls, proteins produce *colloidal* osmotic pressures. The colloidal osmotic pressure of the plasma (about 25 mm Hg), due primarily to albumen (a protein dissolved in plasma) and called **oncotic pressure** (OnP), tends to draw fluids back into the capillaries. The pressure (about 2–10 mm Hg) produced by proteins in the interstitial fluid (IOP) has the opposite effect. When the outward-driving force (HP_b + OnP) is greater than the inward driving force (HP_{if} + IOP), the net movement of fluid is out into the tissue spaces. When the opposite condition occurs, fluid tends to move into the capillaries from the tissue spaces.

Two mechanisms may determine how fluid exchange occurs in capillary beds. One mechanism is illustrated in Figure 6.15. Fluids flow out of the capillaries and into the interstitial spaces on the arteriole side of the capillary bed and return to them on the venous side because the outward-driving force is high on the arteriole side and relatively low on the venous side. This difference results from a decrease in blood pressure as blood flows through the capillary bed. The other mechanism results from the action of precapillary sphincters. Fluids may pass *either* outward *or* inward through the entire capillary bed, depending on whether precapillary sphincters are open or closed. When the sphincters are open, HP_b is high and fluids flow out of the capillaries. When sphincters are closed, HP_b is low and fluid reenters the capillaries. The overall result in either mechanism is that about 99% of the fluid that leaves the capillaries reenters them.

Venous Return.   Blood enters the venules from the capillaries, and the veins from the venules, at pressures that are too low to force the blood back to the heart from the extremities of the body. Consequently, auxiliary pumping mechanisms drive the venous blood, and veins are equipped with flap valves that allow blood to flow only toward the heart (see Figure 6.10c). Movement of muscles in the limbs compresses the veins so that blood is forced past the valves but will not flow back when the muscles relax. Breathing movements provide an additional boost to venous blood flow. During inhalation, the negative pressure in the thoracic cavity causes expansion of the large veins and the atria, drawing blood toward the heart. When a person stands with the knees locked, blood collects in the legs because of the lack of movement and the shallow breathing. If blood return to the heart is inadequate, cardiac output falls and the person may faint from lack of adequate blood supply to the brain.

The Lymphatic System.   We have not yet discussed the fate of the 1% of the fluid that filters out of the blood capillaries but does not reenter them. If this fluid accumulates in the tissue spaces, it can cause swelling, or **edema**. Thus, the homeostatic mechanism providing exchange between blood and interstitial fluid involves one other major component—a fluid reclamation system called the **lymphatic system** (Figure 6.16). Intertwined with the vascular capillaries are small, blind **lymphatic capillaries** that take up excess fluid from the interstitial spaces (Figure 6.16b). Fluid in the lymphatic system, called **lymph**, is similar to interstitial fluid, but differs slightly in solute (mainly protein) content. Small amounts of protein that filter out of the capillaries from the blood are recovered by the lymphatic system.

The lymphatic system provides only one-way flow, back to the blood vascular system. Lymph capillaries merge into larger vessels that contain one-way valves. As fluids accumulate in the lymph

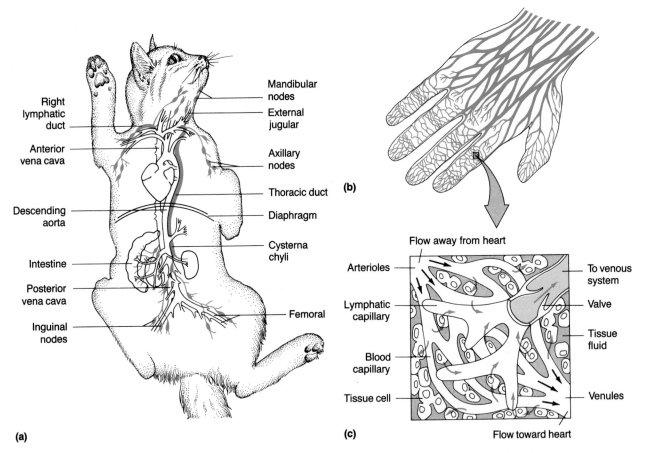

Figure 6.16   The mammalian lymphatic system. (a) Distribution of major lymphatic vessels and lymph nodes in the cat, and their positions relative to several of the major blood vessels. (b) The lymphatic vessels of the human hand illustrate the pervasive nature of the lymphatic system. (c) The relationship between the lymphatic capillaries and blood capillaries.

vessels, muscle contractions and breathing movements drive lymph from the lower part of the body and left side of the thorax and head into the large **thoracic duct**. In the cat, the thoracic duct empties into the left external jugular vein near its junction with the internal jugular vein (see Figure 6.12a). Lymph from the right side of the head and thorax is driven into the **right lymphatic duct**, which empties into the right external jugular vein. At various places in the lymphatic system, especially in the groin, arm pits, and neck, are **lymph nodes** containing cells called lymphocytes that are active in the vertebrate's system of internal defenses (discussed later in this chapter).

Composition of Vertebrate Blood. As shown in Table 6.1, mammalian blood is a combination of plasma (made up of water and a complex array of dissolved substances) and formed elements (cells and cell fragments). Formed elements are specialized for several functions. **Erythrocytes**, also known as red blood cells (RBCs), contain the hemoglobin so important in gas transport. RBCs are produced in bone marrow in terrestrial vertebrates and in the anterior kidney in fishes. In human adults, RBCs are produced in the bone marrow, function in the circulatory system for about 120 days, and then are removed by scavenger cells called **macrophages** located in the spleen, liver, and bone marrow. Mature RBCs in mammals lack nuclei, but those of other vertebrates are nucleated. White blood cells, called **leukocytes**, are a diverse group of cells produced in bone marrow and several other organs. Macrophages are a type of leukocyte, as are **lymphocytes**, and both play a role in the vertebrate body's defense against foreign materials. **Platelets** are small cell fragments that are important in blood clotting in mammals. Other vertebrates have whole cells called **thrombocytes** that perform the same function.

Table 6.1 Major Components of Human Blood

Plasma

Water (90%)
(at pH 7.35–7.45)
Dissolved substances

Plasma proteins	Major function	g/100 ml blood
Albumens	Maintain osmotic pressure	3.2–4.5
Fibrinogen	Clotting	0.2–0.5
Globulins	Immunity	1.4–3

	mg/100 ml blood
Glucose and other sugars	70–140
Amino acids	40
Lipids and cholesterol	700
Urea	14–40
Ions	mM
Na$^+$	142
K$^+$	4
Ca^{2+}	5
Mg^{2+}	3
Cl$^-$	107
HCO$_3^-$	27
HPO$_4^{2-}$	2
Gases	ml/100 ml blood
O$_2$	19 (arterial)
CO$_2$	52
N$_2$	0.9

Formed Elements	Number of cells/ μl blood
Erythrocytes (red blood cells)	4.3–5.8 × 10^6
Leukocytes (white blood cells)	5–10 × 10^3
50%–70% Neutrophils	
20%–40% Lymphocytes	
2%–8% Monocytes	
1%–4% Eosinophils	
0.1% Basophils	
Platelets (cell fragments)	2.5–4 × 10^5

HEMOSTASIS AND CLOTTING

The homeostatic internal environment that circulatory fluids provide for an animal's cells would be lost through minor surface wounds if animals did not have emergency leak-blocking mechanisms. **Hemostasis**, the arrest of circulatory fluid loss, is an important aspect of circulatory homeostasis. In the simplest of hemostatic mechanisms, found in many invertebrates, the musculature of the body wall simply contracts and closes off any wound until it heals. Such mechanisms are effective for soft-bodied animals, but they are not effective for hard-bodied animals or for those with high blood pressures. Many echinoderms, molluscs, and arthropods have a more complex hemostatic mechanism: Amoeboid cells in the hemolymph congregate at wound sites and plug the gap by **agglutinating** (clumping) with one another. In some animals, the amoeboid cells fuse, forming a large multinucleate cell that secretes a fibrous protein. This complex then entraps other cells, forming a plug, or **clot**, that stops fluid loss. In animals with clotting mechanisms, the circulatory fluid also has a mechanism that prevents clot formation under normal circumstances. Without a clot-preventing mechanism, drifting clots called **embolisms** could form in the circulatory system and block small arteries, causing damage to vital organs.

When a wound occurs in a vertebrate, damaged blood vessels constrict, reducing blood flow in the affected area. This process, called **vasoconstriction**, works with clotting to provide hemostasis in the vertebrates. In mammals, clotting involves about twelve different chemical factors. About 4%–6% of the protein in mammalian blood plasma is the soluble, nonfibrous clotting substance **fibrinogen** (see Table 6.1). When a wound occurs, a series of chemical reactions convert fibrinogen into **fibrin**, a fibrous, insoluble derivative. Figure 6.17 shows the major steps involved in this process. If the smooth endothelial surface of a blood vessel is disrupted, platelets adhere to the roughened surface. These fragile pieces of cells are easily ruptured and release small amounts of **thromboplastin** into the surrounding plasma. When calcium ions are present, thromboplastin converts another plasma protein called **prothrombin** into its active form, **thrombin**. Thrombin is a potent enzyme that cata-lyzes the conversion of fibrinogen to fibrin. As a fibrin meshwork builds up at a wound, red blood cells and damaged platelets are trapped in the fibers, and gradually the wound is plugged. Once formed, fibrin in the clot contracts and expels **serum**, a clear fluid component of plasma. Hemophilia, a hereditary disease in which the blood fails to clot, results from the absence of one of the clotting proteins.

BODY DEFENSE MECHANISMS

Animal bodies, and especially intracellular and extracellular fluids, make ideal culture media for infectious viruses, bacteria, fungi, and parasitic animals. Consequently, animals must have some means of defense to prevent harmful organisms from becoming established on or in their bodies. In addition, a defense mechanism is needed to remove cells that form tumors. These functions are performed by the **immune system**, whose products and cells are transported in the circulatory fluid. In invertebrates, the system is not as well known, and may not be as well developed, as it is in vertebrates.

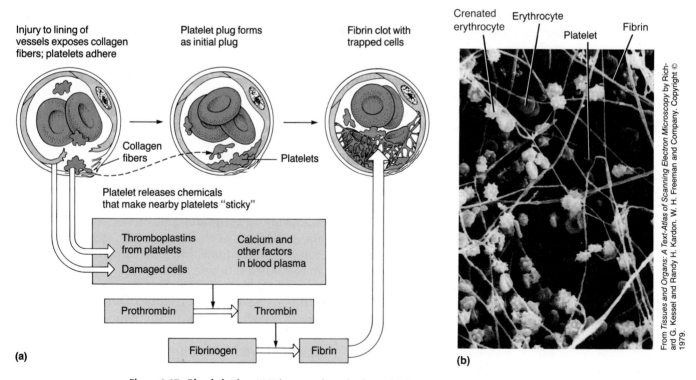

Figure 6.17 Blood clotting. (a) When vessels are broken, platelets aggregate and form a plug. A complex series of reactions is set in motion that converts soluble fibrinogen into insoluble fibrin, forming a meshwork of fibers that entraps cells. (b) Scanning electron micrograph of a clot. Red blood cells, many misshapen (crenated) in the abnormal environment of a clot, are trapped in a meshwork of fibrin threads (925×).

From *Tissues and Organs: A Text-Atlas of Scanning Electron Microscopy* by Richard G. Kessel and Randy H. Kardon. W. H. Freeman and Company. Copyright © 1979.

Three Levels of Defense

Surface Coverings. The first level of defense occurs at the body surface. The tough outer coverings of many animals are resistant to microbial penetration and may produce secretions that inhibit microbial growth. The mucous coats that cover many invertebrates, fishes, and amphibians continually slough off, preventing bacteria from establishing themselves on the body or respiratory surfaces. The surface mucus of many animals also contains certain antimicrobial secretions.

Nonspecific Cellular Responses. Most animals also have a second level of defense. Amoeboid "scavenger" cells wander about in interstitial spaces and in body cavities. These cells have the ability to recognize and engulf foreign materials, providing a nonspecific immunity that repels or kills invaders. Recognition usually involves detection of differences in the types of glycoproteins found on the surface membranes of foreign cells. These scavenger cells also remove dead host cells and other debris resulting from tissue breakdown.

The circulatory fluids of vertebrates and many invertebrates also contain cell secretions that lyse (break) foreign cells. Macrophages in vertebrates secrete an antiviral protein called **interferon**. And the body fluids of many invertebrates contain proteins called **nonspecific agglutinins** that cause various types of bacteria to agglutinate. The clumps are then engulfed by phagocytic cells. Or, if the foreign material is too large to be engulfed, amoeboid cells form an aggregate, **encapsulating** the foreign material and walling it off from the rest of the body. Such encapsulated masses may become surrounded with connective tissue and become calcified in the body, or they may gradually pass to the body surface, where they rupture and release their contents.

Inflammation, characterized by local redness, heat, and swelling, is a nonspecific response to a foreign substance or infection in vertebrates. Damaged body cells release chemicals (such as the tissue irritant **histamine**) that cause dilation of arterioles and increased capillary permeability, so that more fluid, including plasma proteins, passes from the capillaries into the interstitial spaces of the infected area. This causes the redness, heat, and swelling. Increased blood flow also brings more leukocytes into the area (at first mainly neutrophils; later, lymphocytes and monocytes; see Table 6.1). When capillaries are dilated, spaces between endothelial cells widen, allowing some leukocytes to migrate into the tissue spaces. These combat the infection by engulfing invading microorganisms and by secreting chemicals that attract other phagocytic cells. Fibrinogen can pass into the tissue spaces and be converted to fibrin, walling off the infected area. This localizes the infection in an abscess that may fill with **pus**, an accumulation of live and dead phagocytic cells (mainly neutrophils), tissue fluids, and usually microorganisms. Pain is often associated with inflammation because of the swelling and pressure on nerves and the irritating effects of cellular secretions on the tissues.

Another component of the nonspecific immune system in vertebrates is called the **reticuloendothelial system** (**RES**), which consists of phagocytic cells (mainly monocytes and macrophages) that remove foreign particles and cellular debris from body fluids. Monocytes, which are mobile phagocytic cells in the blood and lymph, become macrophages when they pass through the capillary walls and become fixed in various tissues. RES macrophages occur in many tissues and organs, including the liver, spleen, bone marrow, lungs, and brain. The effectiveness of the RES in removing particulate material from the circulatory system can be demonstrated easily. If particles, such as noninfective viruses, are injected into the vascular system of a laboratory mouse, 99.9% of them are removed from circulation in about 30 minutes.

Specific Immune Response. Aside from nonspecific defenses shared by nearly all animals, vertebrates have a third level of defense called the **specific immune response**. This involves several mechanisms that react to the presence in the body of specific foreign molecules called **antigens**. Antigens are large molecules, usually with molecular weights greater than 10,000, that elicit the specific immune response. They may be proteins, polysaccharides, complex lipids, or nucleic acids occurring as soluble molecules or on the surface of invading microorganisms. Certain small molecules collectively called **haptens** can also be antigenic if they are attached to large molecules. The small molecule formaldehyde, for example, may become antigenic when it chemically combines with certain proteins.

Many antigens elicit the synthesis of unique types of proteins called **antibodies**. Each type of antibody can bind with, and may inactivate, the specific antigen that elicited its synthesis. Antibodies, which are synthesized by lymphocytes, constitute one class of the plasma globulins (see Table 6.1). They may lyse bacterial cells, inactivate viruses, or detoxify bacterial toxins. Large particles, such as bacterial cells, may bind several antibody molecules, which in turn may bind with other bacteria, forming a clump. Such agglutinated bacteria may then be phagocytosed. Antibody function is enhanced by **complement**, a system of 18 plasma proteins that interact with antibodies, helping them to bind with antigens. Antibody-complement interaction often promotes antigen phagocytosis or foreign cell lysis.

A critical aspect of antibody production is **immunological memory**. When a vertebrate is first exposed to an antigen, development of the antibody response takes several days, and only a small amount of antibody is produced. This is called the *primary* antibody response. After the initial exposure, the eliciting antigen is remembered immunologically. Later exposure to the same antigen results in a *secondary* antibody response characterized by rapid production of the specific antibody (Figure 6.18).

Immunobiologists are interested in how the immune system recognizes, remembers, and reacts to millions of different antigens. Crucial to our overall understanding of the specific immune response system is how the system distinguishes between **self** (host cells) and **nonself** (foreign material). While our knowledge is far from complete, it is now possible to offer partial answers to these questions. Relatively little, however, is known about immunity in vertebrates other than mammals. Consequently, much of the following discussion is about mammals, especially humans.

Recognition of and Reaction to Antigens

Adult humans have about 10^{12} lymphocytes scattered through their circulatory system, interstitial

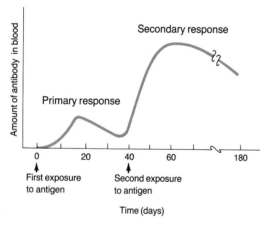

Figure 6.18 Primary and secondary antibody responses. When first exposed to an antigen, a mammal makes a smaller amount of antibody than when it encounters the same antigen a second or third time. Proliferation of memory cells (a specific clone of lymphocytes) accounts for the greater secondary response.

spaces, and lymph nodes. If collected together, these cells would have a mass about equal to that of the brain. Two types of lymphocytes are responsible for the specificity of the immune response. Both types originate from stem cells in the bone marrow in adult mammals. **B cells** pass from the marrow into the blood, lymph nodes, and spleen, where they function in **humoral immunity**, the production of antibodies that are released into the blood and lymph. Humoral immunity is most effective against bacteria, viruses, and soluble antigens circulating in the bloodstream. **T cells** are lymphocytes that pass from the bone marrow to the thymus, a glandlike organ located between the lungs. The thymus consists mainly of lymphocytes and is the site where T cells mature, become immunologically competent, and multiply. Competent T cells leave the thymus and take up residence in the lymph nodes or circulate in the blood or lymph. They participate in a variety of reactions collectively referred to as **cell-mediated immunity (CMI)**. CMI tends to be directed against fungi, protists, animal parasites, viruses already within host cells, tumor cells, and foreign tissue (such as an organ transplant). There are at least three types of T cells: Killer T cells bind with foreign cells and lyse them; helper T cells stimulate B cells to produce antibodies; and suppressor T cells stop antibody production in B cells after foreign materials are removed.

Humoral Immunity. Antibodies belong to a group of plasma constituents known as **immunoglobulins (Igs)**. They are usually called proteins, but they are actually glycoproteins, for they all contain some carbohydrate. Human blood contains five classes of immunoglobulins, the two most common being immunoglobulin G (IgG), a relatively small glycoprotein, and immunoglobulin M (IgM), a large glycoprotein. Research indicates that most vertebrates produce types of IgG and IgM molecules in response to bacterial and viral infections (see Figure 6.19). It is estimated that mammals can produce 10^7–10^9 different types of antibody molecules. Fishes and amphibians have a less extensive repertoire; researchers think they may produce only about 10^5 different types of antibodies and may rely more heavily on nonspecific defense mechanisms.

A typical antibody molecule consists of four polypeptide chains, two *heavy* (of higher molecular weight) and two *light* (of lower molecular weight), that associate into a flexible Y-shaped configuration (Figure 6.19a). As shown in Figure 6.19b, the human IgM molecule is a large complex of five

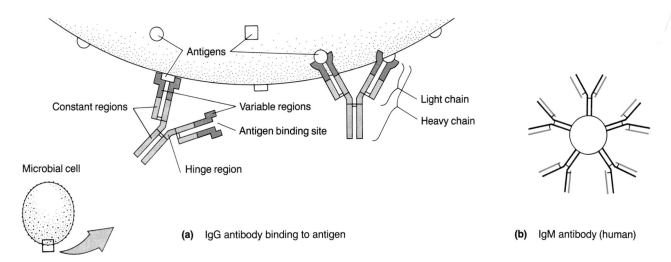

(a) IgG antibody binding to antigen

(b) IgM antibody (human)

Figure 6.19 Antibody structure. (a) IgG antibodies are proteins consisting of two heavy and two light polypeptides produced by different genes. The heavy chains consist partly of carbohydrate chains. Both heavy and light chains have constant regions and variable regions. The amino acid sequence in the variable region gives each antibody a unique affinity for only certain antigens. IgGs have two antigen-binding sites, each formed by the terminal regions of both the heavy and the light chains. A hinge region in each heavy chain gives flexibility during binding of antigen. (b) IgM antibodies are large, complex molecules with ten antigen-binding sites. They are usually produced during the primary antibody response. Disulfide bonds link the individual parts of the molecules.

Y-shaped subunits. In all types of antibodies, the amino acid sequence of each polypeptide chain has a *constant* and a *variable* region. The constant regions have the same amino acid sequence throughout a given class of antibodies, while the amino acid sequences of the variable regions are unique to specific types of antibody molecules. The variable regions give individual antibodies their specificity, causing them to bind with only one type of antigen.

Research indicates that antibodies specific to certain antigens are produced in the manner depicted in Figure 6.20. Soluble antigens that enter the body are carried via fluids or cells to the lymph nodes, where they come in contact with resident B cells that originally migrated there from the bone marrow. A small number of these cells have specific receptors on their surface that bind the introduced antigen. A B cell's receptor may be the specific antibody it can produce. Helper T cells in the lymph nodes are also involved in activating B cells. When all factors are present, B cells that can respond to the antigen do so by dividing. The offspring cells constitute a genetically identical clone with the ability to produce the antibody specific to the stimulating antigen. Some offspring B cells become activated and produce specific antibodies immediately. Other offspring B cells differentiate into long-lived **memory cells** that remain in the lymph nodes. If the same antigen is later encountered, these cells produce large amounts of anti-

body against that antigen. This constitutes the secondary antibody response characteristic of immunological memory (see Figure 6.18).

B cells that are synthesizing and releasing antibodies into the blood and lymph are called **plasma cells**. The structure of plasma cells reflects their activity (see the Figure 6.20 inset). Synthesis rates of 2000 antibody molecules per second per cell have been measured in mammals.

Immunological memory makes vaccination an effective measure against many microbial diseases. Viruses and bacteria are surrounded by protein and polysaccharide coats that an animal host recognizes as nonself, or antigenic. Some vaccines are prepared by inactivating viruses or bacteria with chemicals or heat so that they will not cause disease but will still have their antigenic properties. The noninfective vaccine is then injected into an animal, and the animal's immune system senses the antigen and produces memory cells against it. If the animal later encounters a live virus or bacterium with a protein coat similar to that of the microbe with which it was vaccinated, its memory cells produce antibodies rapidly and in quantities up to 100 times greater than the first time. This rapid defense mechanism may kill the invaders before they have a chance to establish themselves in the host. The animal is then said to be *immunized* against the disease organism. Animals that have had a disease and survived have, in a sense,

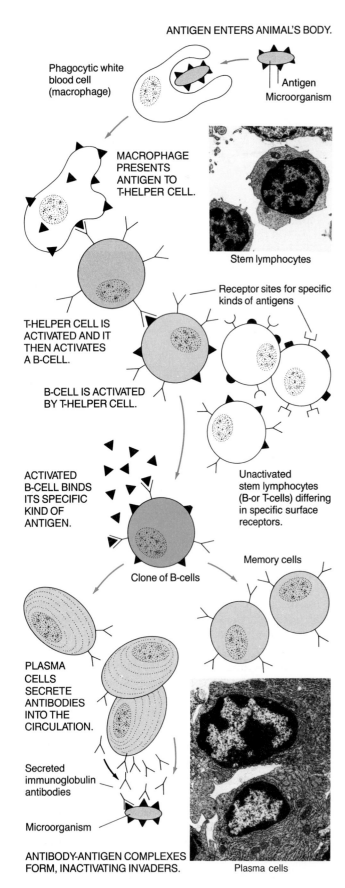

ANTIGEN ENTERS ANIMAL'S BODY.

Phagocytic white blood cell (macrophage)

Antigen Microorganism

MACROPHAGE PRESENTS ANTIGEN TO T-HELPER CELL.

Stem lymphocytes

Receptor sites for specific kinds of antigens

T-HELPER CELL IS ACTIVATED AND IT THEN ACTIVATES A B-CELL.

B-CELL IS ACTIVATED BY T-HELPER CELL.

ACTIVATED B-CELL BINDS ITS SPECIFIC KIND OF ANTIGEN.

Unactivated stem lymphocytes (B- or T-cells) differing in specific surface receptors.

Memory cells

Clone of B-cells

PLASMA CELLS SECRETE ANTIBODIES INTO THE CIRCULATION.

Secreted immunoglobulin antibodies

Microorganism

ANTIBODY-ANTIGEN COMPLEXES FORM, INACTIVATING INVADERS.

Plasma cells

been vaccinated naturally. As a result, they often have immunological memory of the disease organism and are immune to it.

Cell-Mediated Immunity (CMI). While B cells produce antibodies that are released into the circulatory system, T cells have specific antigen-binding components that remain attached to their surface membrane. Like B cells, T cells seem to respond to antigens by forming clones of cells, some of which become immediately active in CMI while others become memory cells (Figure 6.21). T cells can leave the circulatory system by moving between capillary endothelial cells into the interstitial spaces. In the tissues, they "examine" the cells they contact for "foreignness." Some T cells attach to foreign cells and lyse them; others secrete chemicals called **lymphokines** that have a variety of effects. Some lymphokines kill foreign cells directly, while others stimulate lymphocyte division or promote inflammation by attracting phagocytes to tissues and enhancing phagocytosis by macrophages.

It has been a major puzzle to learn how T cells can distinguish host cells from foreign cells. Vertebrate body cells have surface glycoproteins called **histocompatibility antigens** that give the cells of each animal a chemical identity. Each individual's immune system "learns" to recognize "self" early in life. One hypothesis proposes that this occurs in the fetus, when any T cells produced that react with fetal tissues are eliminated by some unknown mechanism. This process leaves only those T cells that will react with foreign antigens. Understanding how this mechanism operates is medically important, because in humans, many of the so-called autoimmune diseases (including arthritis, multiple sclerosis, and lupus erythematosus) result from the host's T cells destroying its own tissues.

The T-cell response can also cause the rejection of transplanted tissue. If tissues are introduced that have different histocompatibility antigens on the cell surface, the host T cells respond by attacking that tissue, causing cell breakdown and eventual

Figure 6.20 (Left) B cells: lymphocytes that produce antibodies. When antigens bind to surface receptors on certain B cells, these cells divide and increase in number. Some of the offspring B cells develop into plasma cells, which secrete antibodies. Others become memory cells that will secrete antibodies in response to a later exposure to the same antigen. Notice that before B cells are activated to produce antibodies, they are small stem lymphocytes with few organelles (top inset: a transmission electron micrograph of two lymphocytes from a rat lymph node, 2900×). Once activated, a B cell differentiates into a plasma cell characterized by prominent cytoplasmic organelles—ribosomes, rough endoplasmic reticulum, and Golgi apparatus—associated with protein (antibody) production and secretion (bottom inset: two plasma cells from a rat lymph node, 4100×).

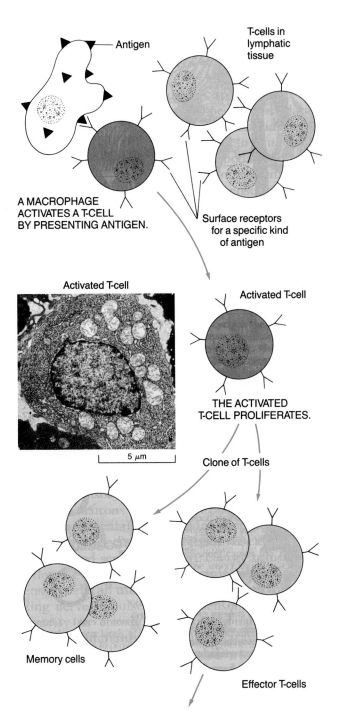

Antigen

T-cells in lymphatic tissue

A MACROPHAGE ACTIVATES A T-CELL BY PRESENTING ANTIGEN.

Surface receptors for a specific kind of antigen

Activated T-cell

Activated T-cell

5 μm

THE ACTIVATED T-CELL PROLIFERATES.

Clone of T-cells

Memory cells

Effector T-cells

EFFECTOR T-CELLS LYSE FOREIGN CELLS AND RELEASE LYMPHOKINES.

Figure 6.21   T cells: lymphocytes active in cell-mediated immunity. T cells, similar to B cells, respond to antigens by dividing to yield a clone of cells, some of which become active immediately while others become memory cells. The surface membrane of activated T cells binds antigens, and some activated T cells secrete chemicals called lymphokines. Activated T cells have a large number of ribosomes but lack the great development of rough endoplasmic reticulum and Golgi apparatus characteristic of plasma cells. Before activation, the T cell is a small lymphocyte, difficult to distinguish from those that form B cells (see the top inset in Figure 6.20).

transplant rejection. To minimize this response, donors and recipients are immunologically screened to match tissue antigen types as closely as possible. Immunosuppressive drugs may also be given to prevent the T-cell response, although these tend to make the patient susceptible to infections.

Overall, the specific immune response system is a highly effective homeostatic mechanism that provides vertebrates with protection against a vast array of potentially harmful disease organisms. As we have just seen, however, the system is not perfect. In addition to diseases resulting from T-cell problems, certain antibodies can produce harmful effects in the host. IgE antibodies in humans may elicit allergic responses, and in some diseases, the immune system "overreacts" to antigens of infectious agents, activating lymphocytes and producing antibodies in quantities that are harmful to the host. It is also important to realize that not all antibodies or CMI responses are protective against invasive organisms. The immune system produces many antibodies and cell-mediated responses that react with the specific provoking antigens but do not protect the host against a disease organism. Sometimes, lymphocytes or antibodies are not produced in sufficient quantity, or the antigens that the immune system reacts with are not vital to the invasive organism. An antibody reacting with a nonvital antigen may cause an invader little or no harm. Moreover, some disease organisms can elude the immune system by constantly changing their antigen makeup. (Surface coat alteration in ◁ *Trypansoma* is discussed in Chapter 19, p. 455.)

Recognition of Self in Invertebrates

Comparatively little is known about immune mechanisms in invertebrates, but they do seem to lack a complex specific response system comparable to that of vertebrates. Nonetheless, experimental studies indicate that many invertebrates—perhaps all animals—have certain immune capabilities, including the ability to distinguish self from nonself. Sponges, for example, can regenerate after being broken apart into individual cells. If cells from two or more sponges are mixed, cells from each of the former individuals recognize one another, become segregated from cells of other individuals, and reaggregate.

Tissue-grafting studies indicate that earthworms have both immunological memory and recognition of self. Coelomocytes (amoeboid cells in the body cavity) repel foreign materials in these animals. Grafts of body wall tissues made between individual worms from the same geographical population will survive for up to 250 days. If grafts are made between worms taken from geographically distant locations (such worms are presumed to be genetically very different), the grafts last for only about twelve days. Rejection occurs when coelomocytes surround and eliminate the grafted tissue. If a second graft from the same donor animal is made to a worm that rejected a first graft, the second graft is rejected in only five days, showing what seems to be a memory response. Furthermore, if coelomocytes collected from an earthworm that has rejected a graft rapidly are injected into an untreated new worm, the new worm will show the same accelerated rejection of a graft from the same donor. Recent research interest should provide important new information on invertebrate defense mechanisms in the near future.

☐ SUMMARY

1. Animals require circulatory mechanisms to maintain homeostasis. Body movements may circulate fluids in body cavities and interstitial spaces, but most animals have a circulatory system with vessels and either tubular or chambered hearts.

2. Squids, octopuses, vertebrates, and some annelids have closed circulatory systems, in which blood is pumped through arteries into capillary beds in the tissues. Exchange between the blood and tissue cells is by diffusion and filtration through the capillary walls. Blood returns to the heart through veins. Open circulatory systems, found in arthropods and most molluscs, consist of a heart that pumps hemolymph through arteries into body sinuses, collectively called the hemocoel. In the hemocoel, hemolymph bathes the tissues directly and then returns to the heart.

3. The vertebrate circulatory system underwent extensive evolutionary changes correlated with the evolution of the air-breathing lung. Fishes have a single blood circuit that includes the capillaries in the gills; blood flows through the heart and is pumped by the ventricle to the gills. Other vertebrates have two blood circuits. A pulmonary circuit carries blood between the heart and lungs. A systemic circuit carries blood between the heart and body tissues. The heart is partitioned to accommodate the two circuits and to separate oxygen-rich from oxygen-poor blood.

4. Circulation in vertebrates is controlled by nerve centers in the brain. Inhibitory and excitatory nerves control the inherent rhythmicity of heart contraction and the diameter of the arterioles in the vascular system. Heart rate and arteriole diameter regulate the flow of blood to the tissues.

5. In capillary beds, nutrients and oxygen are distributed to the tissues as fluid portions of the blood filter through the capillary walls. Fluid containing wastes is osmotically reabsorbed by the capillaries and by the lymphatic system.

6. Hemostatic mechanisms prevent loss of body fluids due to injury. Amoeboid cells may plug wounds, and some animals have clotting mechanisms that involve special proteins. In vertebrates, the soluble plasma protein fibrinogen is converted into its insoluble form, fibrin, by the action of platelets or thrombocytes at a wound site.

7. Most animals have defense systems that protect them against invasion by harmful organisms. These systems may involve nonspecific mechanisms or highly specific antibody or cell-mediated immune mechanisms. In vertebrates, special lymphocytes called B and T cells are responsible for immunity. Immune responsiveness depends on an animal's ability to distinguish self from nonself. The vertebrate immune system includes cells that are activated by an initial infection or exposure to an antigen. Clones of these cells respond more strongly and more quickly to a second antigen exposure.

☐ FURTHER READING

Cantin, M., and J. Genest. "The Heart as an Endocrine Gland." *Scientific American* 254(1986): 76–81. *Describes the role of the heart in secreting a hormone that helps regulate blood volume and pressure.*

Edelson, R. L., and J. M. Fink. "The Immunologic Function of the Skin." *Scientific American* 252(1985): 46–53. *Describes how the skin provides more than passive protection against foreign invaders.*

Hood, L. E., I. L. Weissman, and W. B. Wood. *Immunology.* Menlo Park, Calif.: Benjamin/Cummings, 1985. *A modern textbook describing many aspects of immunology.*

Jaret, P. "Our Immune System, the Wars Within." *National Geographic* 169(1986): 702–734. *Pictorial description of immune cells and their actions; superb scanning electron micrographs colored by special techniques.*

La Barbera, M., and S. Vogel. "The Design of Fluid Transport Systems in Organisms." *American Scientist* 70(1982): 54–60. *Engineering and design principles applicable to circulatory systems.*

Laurence, J. "The Immune System in AIDS." *Scientific American* 253(1985): 84–93. *Describes the immune system's reaction to viral infections and how the AIDS virus disrupts the system.*

Marrack, P., and J. Kappler. "The T Cell and Its Receptor." *Scientific American* 254(1986): 36–45. *Describes studies of cell surface proteins that account for the T cell's immune activities.*

Schmidt-Nielsen, K. *Animal Physiology: Adaptation and Environment.* 3d ed. New York: Cambridge University Press, 1983. *Includes two excellent chapters on circulatory systems, including a section on the physical principles of flow in tubes.*

Tonegawa, S. "The Molecules of the Immune System." *Scientific American* 253(1985): 122–131. *The incredible diversity of antibodies is described.*

CHAPTER 7

Internal Fluid Regulation and Excretion

The marine iguana *(Amblyrhynchus cristatus).* This unusual species inhabits the Galapagos Islands, living on land and feeding in the sea. Its body tends to dehydrate in either environment, and the marine algae that it eats are as salty as seawater. The animal's kidneys and thick skin reduce water loss, and special salt-excreting glands above the orbit of the eyes dispose of excess salts.

Regardless of where they live, animals must maintain constancy, or homeostasis, in their internal environment for their cells to function properly. As described in earlier chapters, digestive systems contribute to homeostasis by providing energy and nutrients; respiratory systems provide oxygen and dispose of waste carbon dioxide; and circulatory systems distribute these and other materials throughout the body. In this chapter, we will discuss how animals maintain constancy in their internal fluids, and we will examine the role of excretory systems in regulating the amount of water in the body, controlling the concentration of ions in body fluids, and disposing of certain waste products.

145

The three general types of external environments—marine, freshwater, and terrestrial—affect the internal fluids of animals differently. High salt concentrations in the ocean, for example, tend to increase salt concentrations in an animal's extracellular fluids and to promote the osmotic loss of water across gill and body surfaces. Freshwater environments present the opposite challenges. Animals in fresh water tend to lose ions and gain water. Terrestrial animals must conserve water *and* ions. This chapter explains how animals are adapted to these different environmental challenges.

Osmoregulation is the maintenance of a life-sustaining balance in water and solutes (dissolved substances) in body fluids. Animals gain water by drinking, by eating foods containing water, through metabolism, and through osmosis. They lose water by excretion, defecation, gas exchange (in terrestrial animals), and osmosis or evaporation. As described in Chapter 2, there is a net movement of water by osmosis across differentially permeable membranes from a region where solutes are low in concentration to a region where solutes are more highly concentrated (see Figure 2.16). Thus, if a 0.1 *M* solution of sodium chloride (NaCl) is separated from a 1 *M* NaCl solution by a membrane that is impermeable to Na^+ and Cl^- but that allows water to pass, more water will flow osmotically into the 1 *M* NaCl solution than will flow out of it. As a result, the 1 *M* solution will be diluted, while the concentration of the 0.1 *M* NaCl solution will increase. In time, if water movement is not regulated, the two solutions will come to equilibrium with equal NaCl concentrations, and net water movement will cease. At this point, the two solutions are said to be **isosmotic**, meaning they have the same solute concentration, usually called an osmotic concentration (from the Greek *isos*, "equal," and *osmo*, "a thrust"). As discussed in Chapter 6, the osmotic concentration or pressure of a solution is a measure of its solute concentration or a measure of its potential to draw water into it osmotically. Thus, isosmotic solutions have equal osmotic pressures. Osmotic concentrations of biological solutions are measured with an osmometer, which determines the freezing point of the solution. A solution's freezing point is lowered (depressed) by increasing the solute concentration, and freezing point is a direct measure of osmotic concentration or pressure.

Osmotic movement of water across body surfaces is not determined by the concentration of any one particular salt or organic compound. All of the salts, amino acids, sugars, proteins, and other molecules in solution contribute to the osmotic pressure. When there are differences in the total concentration of dissolved substances, and consequently a difference in osmotic pressure, between two regions, such as the external environment and an animal's body fluids, water tends to move into the region with the greatest osmotic pressure.

Seawater is isosmotic to the intracellular and extracellular fluids of most marine invertebrates; that is, the osmotic pressure of the body fluids is the same as that of seawater, so there is no net movement of water into or out of the animals. Freshwater animals tend to gain water because they are **hyperosmotic** to their environment (*hyper* means "greater than"); their body fluids have a higher solute concentration and a higher osmotic pressure than fresh water. Most marine fishes tend to lose water because they have body fluids that are less concentrated than seawater and thus are **hypoosmotic** to their environment (*hypo* means "less than").

Water is necessary for life, but water in its pure form can be harmful to animal cells. When placed in distilled water, an animal cell osmotically swells, its contents become diluted, and cellular metabolism is impaired. If swelling continues, the cell may rupture. By the addition of a single salt, such as sodium chloride, the external environment can be made isosmotic to the cell's cytoplasm and swelling will stop. However, most animal cells die in a medium containing only sodium chloride because life processes require a balanced mixture of dissolved solutes. For example, a frog's heart stops beating when placed in an isosmotic sodium chloride solution. The heart starts beating again if calcium ions are added, but the beat is irregular. With the addition of potassium ions, normal beating can be restored and will continue until the heart runs out of chemical energy. The ability of animals to regulate the ionic composition of their cells and extracellular fluids is a component of the overall process of osmoregulation and is called **ionic regulation**. Ions are lost with water in excretion and by transport across body surfaces. Animal cells constantly expend energy to obtain and retain certain vital ions at concentrations higher than the corresponding concentrations in the environment. They also expend energy to keep certain ions out of their body fluids.

The problem of maintaining water and ion balance in internal fluids is closely tied to the process of excretion. Cells release wastes into the extracellular fluids, and many animals excrete metabolic wastes as solutes in water. **Urine** is a general term for any type of nitrogenous metabolic waste produced by kidneys or other excretory structures. Many of the cellular mechanisms, tissues, and organs involved in osmoregulation also function in urine production and excretion. Thus, osmoregulation and excretion are closely interrelated aspects of homeostasis.

THE MAJOR TYPES OF ENVIRONMENTS

Aquatic Environments

There is a great variety of aquatic environments, but they are often grouped into three general categories, according to their relative salt (ion) concentration. Salt concentration can be expressed in several ways. Concentration of salts or gases dissolved in water is often measured as a fraction of the solution, either in parts per million (ppm), which is equivalent to milligrams (mg) per liter, or in moles (M) or millimoles (mM) of solute per liter of water. In zoology, the preferred unit of measurement for the *total* amount of dissolved solids in a solution is **salinity**, which is expressed in parts per thousand (symbolized by ‰).

Freshwater Environments. Generally, water whose salinity is less than or equal to 0.5‰ is considered **fresh water**. The amounts and types of salts dissolved in the water of lakes and streams are highly variable and depend on the chemical com-

position of a watershed's rocks and soils and on the leaching action of precipitation. Outflow streams from melting glaciers, where the water rarely touches the soil, have the lowest salinities. A few lakes, such as the Great Salt Lake and the Dead Sea, are not composed of fresh water, since they have salinities greater than that of the ocean. Such saline lakes have no outflow; water drains from the surrounding area into the lake bed and then evaporates, leaving leached salts behind. Hydrogen ion concentration, as measured by acidity and basicity, also varies widely in bodies of fresh water. Bogs can have a pH of 3.2, and alkaline desert pools can reach pH 10. In recent years, acid rains with a pH as low as 2.4, resulting mainly from the industrial production of sulfuric acid (H_2SO_4), have made many lakes in developed countries uninhabitable for fishes and many other aquatic animals.

Marine Environments. In contrast to lakes and rivers, the **seawater** of the open ocean is a relatively constant osmotic environment with a salinity between 34‰ and 36‰. Higher salinities (up to about 40‰) occur in equatorial oceans because of higher rates of evaporation. The concentrations of the principal ions in seawater are shown in Table 7.1. The pH of seawater is also relatively stable at about 8. Salinity is more variable in coastal areas than in the open ocean because of variable freshwater runoff from the land.

Brackish Water (Estuarine) Environments. **Brackish water** contains more salt than fresh water but less than seawater. It occurs at the mouths (estuaries) of large rivers, where river water meets and mixes with ocean tidewater. The Chesapeake Bay,

Table 7.1 Approximate Concentration of Some Common Inorganic Ions in Seawater, Fresh Water, and in the Circulatory Fluids of Selected Animals*

		Concentration (millimoles/liter)				
		Marine Animals			Freshwater Animals	
Ion	Seawater	Clam	Lobster	Fresh Water**	Crayfish	Trout
Na$^+$	480	475	460	0.7	200	160
K$^+$	10	12	9	0.1	9	5
Ca^{2+}	11	12	11	1.5	15	6
Mg^{2+}	55	54	5	0.3	2	1
Cl$^-$	560	555	500	0.5	180	120

*Intracellular fluid concentrations of these and other ions may be markedly different, especially in certain marine animals.

**These values give a general picture of the ionic content of fresh water, but the ionic composition of freshwater environments is variable.

the southern bayous of North America, and the Baltic Sea of Europe are examples of large estuaries. Many freshwater and marine animals are unable to tolerate the marked salinity changes in estuarine environments, yet brackish waters are among the most biologically productive of all aquatic environments. Species adapted to these transitional zones find them rich in nutrients and dissolved oxygen.

Terrestrial Environments

Terrestrial environments are osmotically harsh on animals because water is continuously lost from moist body surfaces by evaporation, and many animals use considerable amounts of water to dispose of wastes. Of the approximately 34 animal phyla, only 6 have managed to colonize land. Land animals gain water only periodically through drinking, feeding, or metabolizing, and they must store sufficient quantities to counteract continual losses. Because ingested water does not have an ionic composition that supports life, excess ions must be eliminated and scarce ions conserved.

Terrestrial environments are not homogeneous in the osmotic challenges that they present to animals. Sand dunes, Arctic ice fields, and wind-swept mountaintops are usually low in humidity and lead to rapid water loss. Animals that live in leaf litter, dense vegetation, or burrows encounter high humidities and do not readily lose water. In fact, many of the microscopic animals, such as nematodes, that live in the soil are actually aquatic, living in water films trapped in spaces between soil particles.

MEETING ENVIRONMENTAL CHALLENGES TO HOMEOSTASIS

Animals encounter different salt and fluid stresses, depending on their environment, but all animals must maintain a homeostatic internal environment for their cells. An animal's survival depends on its being able to balance water and salt gains with losses.

Adaptations in Aquatic Animals

Seawater contains enough vital ions to sustain life, and marine animals either conform osmotically to seawater or they are osmotic nonconformists. Their body fluids may be isosmotic, hyperosmotic, or

hypoosmotic to seawater. Freshwater species, on the other hand, must remain hyperosmotic to fresh water to sustain cellular metabolism. Marine animals whose body fluids remain isosmotic to seawater are called **osmoconformers**. Marine, freshwater, and estuarine animals that are hyperosmotic or hypoosmotic must expend energy to maintain osmotic concentrations different from that of the environment. Such animals are called **osmoregulators**.

Aquatic animals vary in their tolerance for changing salinities. Those that can tolerate only a narrow range of salinity are **stenohaline**. In contrast, **euryhaline** species, such as those living in estuaries, can tolerate significant changes in external salinity. Most, but not all, freshwater and marine species are stenohaline. Stenohaline and euryhaline species may be osmoconformers or osmoregulators.

Marine Invertebrates. Many marine invertebrates are osmoconformers (Figure 7.1). Unless their environment changes, they do not gain or lose water by

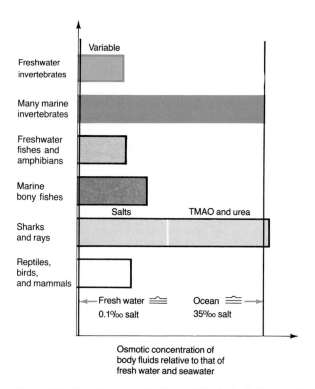

Figure 7.1 Osmotic concentrations of the body fluids of animals. Body fluid concentrations can vary considerably from the concentrations found in seawater and fresh water. In sharks and rays, a substantial part of the osmotic pressure in body fluids is due to urea and trimethylamine oxide (TMAO). Osmotic concentrations are measured as standard salinities symbolized by ‰ (parts per thousand).

osmosis. If the external environment becomes more concentrated than their body fluids, they lose water. If the external environment becomes more dilute, they gain water. Although most open-ocean species are stenohaline, they encounter little danger because the open ocean has a relatively constant salinity.

It is important to realize that being isosmotic to the surrounding water is not the same as having the same ionic composition as the surroundings. Thus, even though they are isosmotic to seawater, osmoconformers regulate the ionic composition of their body fluids—especially their intracellular fluids—and have concentrations of specific ions that differ from those in seawater (see Table 7.1). Because of these ionic differences, osmoconformers must expend a certain amount of energy to maintain fluid homeostasis. The cells of osmoconformers must cope with the problem of **volume regulation**. If a marine invertebrate encounters dilute seawater, its extracellular fluids become hypoosmotic to its intracellular fluids. This swells the cells and may disrupt their functions. Stenohaline osmoconformers have a poor ability to adjust to such changes; they swell and may eventually die. Euryhaline osmoconformers, on the other hand, adjust their intracellular solute concentrations so that their cells are isosmotic with extracellular fluids. In dilute environments, they may pump salts out of their cells, but more often, they reduce intracellular concentrations of nonessential amino acids, thereby lowering the osmotic pressure within the cells and reducing the cell's tendency to gain water. In hypoosmotic environments, euryhaline osmoconformers can increase intracellular solute concentrations.

The distinctions between osmoconformers and osmoregulators are not always clear-cut. Many marine animals conform over a range of external osmotic pressures but regulate beyond that range. This protects their cells from extreme osmotic changes while conserving energy when only minor changes occur in environmental salinities (Figure 7.2).

Some animals can avoid being exposed to an osmotic stress by changing their behavior. Marine invertebrates, such as oysters and certain barnacles that live in intertidal zones, are regularly exposed to flows of fresh water or to air. Under these conditions, the animals simply withdraw into their shells and seal themselves off from the environment. Some estuarine animals withdraw into burrows and stop ventilating their gills when fresh water passes overhead during low tide; these ani-

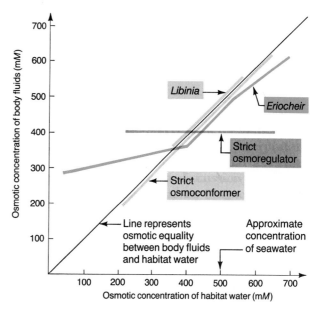

Figure 7.2 Relationship between environmental and body fluid osmotic concentrations. The body fluid osmotic concentration of a strict osmoregulator remains constant over a wide range of habitat salinities, whereas that of a strict osmoconformer corresponds to habitat salinity; these relationships are indicated by black lines in the graph. *Libinia* (spider crabs) are osmoconformers; with little or no ability to osmoregulate, they soon die in water having osmotic concentrations less than about 350 mM or greater than 550 mM. In contrast, the Chinese mitten crab *Eriocheir sinensis* is an osmoconformer at intermediate concentrations, but is an osmoregulator at high and low salinities. A native of Asia that was introduced into Europe, *E. sinensis* spends much of its life in rivers, but travels to the ocean to spawn. Its ability to migrate between freshwater and marine habitats depends on its osmoregulatory capability.

mals are able to avoid osmotic stress because salt water is denser than fresh water and therefore remains around them in burrows and sand.

Marine Bony Fishes. Body fluids in most marine bony fishes have salt concentrations that are about one-third of the concentrations found in seawater (see Figure 7.1), perhaps because some of these animals evolved in fresh water and later invaded the sea. Consequently, these vertebrates continually lose water by osmosis and gain salts by diffusion across the gills (Figure 7.3). To compensate for the water loss, they drink seawater. (If drinking is prevented experimentally, these animals dehydrate rapidly, losing about 7%–35% of their body weight as water in a day.) Water and many salts are absorbed by gut cells, and the excess salts must be eliminated. But the kidneys of marine bony fishes are not very effective in removing salt because they produce only a small amount of urine that is isosmotic to the body fluids. Then how do these fishes eliminate excess salts? The answer lies in the gills,

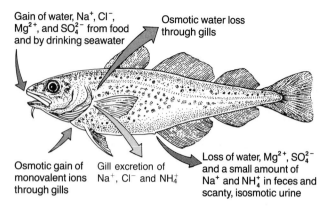

Gain of water, Na⁺, Cl⁻, Mg²⁺, and SO₄²⁻ from food and by drinking seawater

Osmotic water loss through gills

Osmotic gain of monovalent ions through gills

Gill excretion of Na⁺, Cl⁻ and NH₄⁺

Loss of water, Mg²⁺, SO₄²⁻ and a small amount of Na⁺ and NH₄⁺ in feces and scanty, isosmotic urine

Figure 7.3 Osmoregulation in marine bony fishes. Marine bony fishes such as this cod lose water by osmosis across their gills and also in urine and feces. They replace the water by drinking seawater. Cells in the gills excrete monovalent ions (Na⁺, Cl⁻). Most divalent ions are not absorbed and pass out in feces; others are eliminated in urine.

where special **chloride (ion)-excreting cells** actively transport chloride ions from the blood into the sea. Because sodium ions follow the chloride, the primary salt (NaCl) ingested with seawater is eliminated. Other salts are eliminated mainly in feces. As shown in Figure 7.3, the gills also excrete the nitrogenous waste product ammonia (as ammonium ions, NH₄⁺; discussed later in this chapter).

Marine Sharks and Rays. The body fluids of sharks and rays have only about half the salt concentration of seawater, and one would expect that, like bony fishes, they would lose water osmotically. However, sharks and rays maintain internal homeostasis very differently from the way bony fishes do and actually *gain* water osmotically. Sharks and rays retain the nitrogenous waste product urea at high concentrations (up to 2.5%) in their body fluids. They also produce and retain the compound **trimethylamine oxide (TMAO)** at concentrations ranging from 0.25%–1.43%. TMAO protects body proteins from the destabilizing effects of high urea concentrations. Together these substances make the osmotic concentration of shark extracellular fluids slightly higher than that of seawater (i.e., hyperosmotic to seawater; see Figure 7.1). Enough water enters sharks and rays osmotically that they do not need to drink seawater. Excess salts, such as those ingested with food, are eliminated by the kidney and by a **rectal gland** that excretes sodium chloride into the posterior part of the intestine.

Other Vertebrates That Osmoregulate Like Sharks and Rays. A few other vertebrates retain urea and TMAO as an adaptation to saline environments.

One of these, the bony fish *Latimeria chalumnae*, is the single surviving species of an ancient group of fishes called lobefins that gave rise to amphibians. (*Latimeria* and other lobefins are discussed in ◁ Chapter 29, p. 709; see Figure 29.7.) Commonly called a coelacanth, *Latimeria* is a bottom dweller in deep (60–600 m) oceans off South Africa. Another urea and TMAO retainer is the crab-eating frog, *Rana cancrivora*. This species inhabits the mangrove swamps of Southeast Asia and enters brackish water (with salinity concentrations up to 28‰) to feed. Its body fluids contain up to 0.45 M urea; thus, unlike other amphibians, which would die from dehydration in high salinity water, it does not lose water to the environment. Similar elevated urea and TMAO levels occur in the diamondback terrapin (*Malaclemys terrapin*), a turtle inhabiting brackish water marshes on the east coast of the United States.

Freshwater Animals. To achieve a medium suitable for cell activity, freshwater animals must maintain body fluid concentrations higher than that of fresh water; thus, they are all osmoregulators (see Table 7.1). They must compensate for a constant loss of salts by diffusion and a net gain of water across their gas exchange organs and general body surfaces.

Freshwater animals compensate for water gain and salt loss in three ways. First, some have low concentrations of solutes in their extracellular fluids, lessening the osmotic gradient between extracellular fluids and the environment. Freshwater clams, for example, have extracellular fluids that are only about 1/25 as concentrated as those of their marine counterparts. Second, the outer covering of many animals, such as that of crayfish and other crustaceans, is impermeable to water. This reduces the water-gaining surface area. Third, most freshwater animals expend energy to pump excess water into the environment from the body fluids and to actively transport salts from the environment into the body fluids.

Figure 7.4 illustrates how freshwater bony fishes overcome the osmotic challenges of their environment. Unlike their marine counterparts, these vertebrates do not drink water, but rather gain water by osmosis across the gills and by consuming food. Their kidney acts as a water pump, producing a daily volume of dilute urine equal to about one-third of the body weight. **Chloride (ion)-absorbing cells** in their gills replace salts lost in the urine, and vital salts are also obtained in food. Chloride-absorbing cells actively transport chlo-

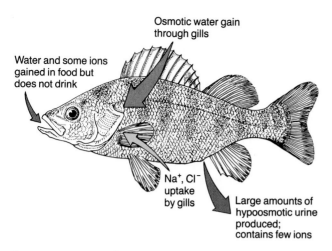

Figure 7.4 Osmoregulation in freshwater bony fishes. Freshwater fishes such as this perch gain water across their gills and the general body surface by osmosis. Excess water is eliminated in large quantities of urine that is hypoosmotic to body fluids. Chloride-absorbing cells in the gills actively transport salts into the body fluids.

ride and sodium ions from the environment into the body fluids. Such ion movements are linked to the elimination of waste products. As sodium (Na$^+$) moves inward across the cell membrane, for example, hydrogen (H$^+$) or ammonium (NH$_4{}^+$) produced in metabolism may move outward, maintaining electrical neutrality. Chloride (Cl$^-$) movement inward may be coupled to bicarbonate (HCO$_3{}^-$) movement outward. In these transport processes, the animal trades ionic forms of metabolic waste products for physiologically necessary ions.

Euryhaline Fishes. Many fishes live in estuaries, and several species migrate between fresh water and the sea at different times in their life cycle. Species such as salmon, steelhead trout, shad, and certain lampreys, which reproduce in fresh water, migrate to the sea to mature, and then return to fresh water to complete their life cycle, are called **anadromous** fishes. (See the Chapter 12 Essay on homing in salmon, p. 290.) **Catadromous** fishes, such as the freshwater eels of North America and Europe, feed and grow in fresh water but reproduce in the sea. (The life cycle of the eel is discussed in Chapter 29, p. 719; Figure 29.15.) Anadromous and catadromous fishes and estuarine species have highly adaptable osmoregulatory mechanisms that can adjust to the dehydrating saltwater environment as well as to the hydrating conditions of fresh water. When at sea, anadromous and catadromous species drink seawater, and their kidneys, chloride cells in the gills, and intestines conserve water. When they enter fresh water, they cease drinking,

and their kidneys produce large amounts of dilute urine while their chloride cells actively transport vital ions into the blood.

Adaptations in Terrestrial Animals

The single most important problem facing terrestrial animals is dehydration, although they also expend energy to maintain salt balance. The overall problem of balancing water and salt gains and losses is especially acute on land (Figure 7.5). Terrestrial animals obtain water and salts from food and drink, and certain amphibians, arthropods, and annelids can absorb some water and/or salts across the body and/or respiratory surfaces. Another important source of water is the **metabolic water** produced when food molecules are metabolized by aerobic respiration. The oxidation of 1 g of glucose and 1 g of fat yields 0.6 and 1.1 ml of water, respectively. (Reaction sequences that yield water are considered in Chapter 3, p. 67.) A person consuming an average American diet produces about 340 ml of metabolic water daily. Bears in temperate climates spend much of the winter in dens. They are in a drowsy state and do not feed. Recent research indicates that they depend on water from the oxidation of fats to replace the water that they lose by respiratory evaporation while in winter dens. About 46 kg of body fat are metabolized during a three-month period; this produces about 50 L of water, allowing a bear to emerge fully hydrated from a den. The desert kangaroo rat *(Dipodomys)*, a common inhabitant of the North American desert, can survive without drinking any water at all, rely-

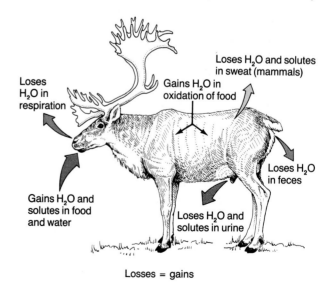

Figure 7.5 Major routes of water and salt gains and losses in terrestrial animals.

ing mainly on metabolic water. Kangaroo rats live on seeds and dry plant material, rarely eating living plants, and can live for months in the laboratory on dry food without water.

Terrestrial animals lose both water and salts in urine, feces, and glandular secretions. Water alone is lost by evaporation from respiratory surfaces. Birds and mammals lose considerable quantities of water by evaporation when they pant or sweat. Panting and sweating dispose of excess body heat by evaporative cooling. Panting results in the loss of essentially pure water, while both water and salts are lost in sweat.

Land animals have several types of adaptations that reduce water loss. A relatively impervious outer surface in many species helps prevent evaporative water loss. The shells of snails, the waxy coverings of insects, and the multilayered skin of vertebrates are examples. Generally, the most significant water loss results through evaporation from the gas exchange system. Having a vital need to conserve water, certain desert animals, including camels, the kangaroo rat, and the cactus wren *(Campylorhynchus)*, recover significant amounts of water from gas exhaled from the lungs. Their nasal passages are narrow and lined with extensive surface areas of mucous membrane where heat exchange occurs. When air enters the nasal passages, it is warmed and humidified as it passes over the mucous membrane. Evaporation cools these surfaces. When gas is exhaled, it passes back over the cool surfaces, and water condenses in the nasal passages. Up to 83% of the water in gas leaving the lungs can be reclaimed and used to humidify air inhaled in the next breath. Some amount of heat exchange and water condensation occurs in the same way in other animals with nostrils, but most animals lack sufficient surface area to recover significant amounts of water.

An animal's behavior may also help conserve water. Many terrestrial animals are nocturnal and thus avoid the heat and high evaporation rates of the day. Also, a number of mammals and birds sleep with their head tucked under a wing or between the hind legs. This conserves heat and traps moisture from exhaled gas in the fur or feathers. Air inhaled in the next breath is humidified as it passes over the trapped moisture.

For many terrestrial animals, urine and feces represent significant routes of water loss. We discuss the production of urine and related adaptations for water conservation in the next section. Water is lost in feces for several reasons. For one thing, moisture is needed to lubricate the intestine.

Also, few animals pass completely dry feces. Although many herbivores ingest food high in moisture content, much of the water is not absorbed and passes out in the feces. A cow, for instance, may pass 20–40 L of water daily in its feces. Seed-eating animals and carnivores generally have drier fecal material. When enough water is not reclaimed from the feces, diarrhea results and can cause severe dehydration and death. Animals with severe diarrhea can be treated with saline infusions to compensate for lost fluids and salts.

The extent of dehydration that terrestrial animals can tolerate varies considerably. Humans can lose only about 12% of their body water before dying. Camels can do much better; they can tolerate a loss of about 33%. No mammal can compare to the Florida spadefoot toad, a terrestrial amphibian that can tolerate a loss of 60% of its body water. Certain soil nematodes and a few other invertebrates hold the dehydration record. They can lose virtually all body water and become dormant when their environment dries up. (See the Chapter 23 Essay on suspended animation, p. 540.)

NITROGENOUS METABOLIC WASTES

Most of our discussion so far has been about ion and water homeostasis. We have seen that urine concentration varies, depending on the animal and its environment. In this section, we focus on the related topics of how nitrogenous metabolic wastes are produced and how they are excreted by animals in different environments. The kinds of nitrogenous wastes that animals excrete are related to the type of environment in which they live. Urine may be watery, semisolid, or solid, depending on the chemical form in which nitrogen is excreted. Waste excretion also illustrates that when the environment permits, animals are conservative in their expenditure of energy.

Ammonia

Most nitrogenous wastes result from the metabolic breakdown of amino acids derived by hydrolysis of proteins. An amino acid releases its amino group as **ammonia** in the reaction called **deamination** (Figure 7.6). Ammonia (actually NH_4^+ at the pH of body fluids) is a toxic compound because it tends to increase the pH of fluids (make them basic), and this can adversely affect cellular enzymes.

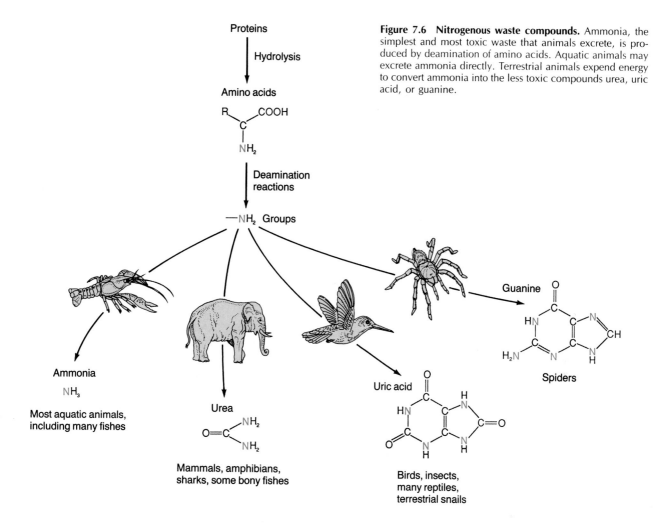

Figure 7.6 Nitrogenous waste compounds. Ammonia, the simplest and most toxic waste that animals excrete, is produced by deamination of amino acids. Aquatic animals may excrete ammonia directly. Terrestrial animals expend energy to convert ammonia into the less toxic compounds urea, uric acid, or guanine.

Animals dispose of ammonia in different ways, depending on their environment. Because ammonia is highly soluble in water and diffuses rapidly across cell membranes, aquatic animals usually dispose of it quickly across the gills or other body surfaces. Animals that excrete ammonia as their primary nitrogenous waste product are said to be **ammonotelic**. These animals expend little, if any, energy in excreting their nitrogenous wastes. Since they are not surrounded by water, terrestrial animals cannot similarly dump ammonia into their environment. Instead, they metabolically convert NH_3 into less toxic substances that can be concentrated, stored, and periodically excreted (see Figure 7.6). This process conserves water, which would otherwise be required to dilute the ammonia to less toxic levels, but it is an energy-costly process. Energy must be expended to chemically alter ammonia and to make enzymes that catalyze the conversion reactions.

Urea

Many terrestrial vertebrates and some invertebrates produce **urea** as their primary nitrogenous waste product. Theoretically, urea can be formed by combining two molecules of ammonia with one molecule of carbon dioxide:

$$2\ NH_3 + CO_2 \longrightarrow NH_2-\underset{\underset{O}{\|}}{C}-NH_2 + H_2O$$

Ammonia   Urea

However, its synthesis in living animals is not this simple and involves a biochemical pathway called the **ornithine cycle**. This cycle is fueled by ATP and occurs in the liver in vertebrates. It continuously converts ammonia to urea, which travels through the bloodstream to the kidneys. Animals that excrete most of their nitrogenous wastes as urea are said to be **ureotelic**.

Urea is highly water-soluble because it is a polar molecule. When dissolved, it forms hydrogen bonds with several water molecules, and it is toxic at high concentrations because its polarity can inactivate proteins. Consequently, urea must be kept dilute or its toxic effects must be countered by other means, such as those employed by sharks and the crab-eating frog described earlier. Ureotelic animals lose considerable amounts of water in excreting urea.

Some animals can switch between ammonotelism and ureotelism. When feasible, they save energy by excreting ammonia; but when conditions dictate, they expend energy to make the less toxic urea. Earthworms, for example, normally excrete ammonia; but when starved, their cells break down large amounts of protein and amino acids for energy, and they become ureotelic. Frog tadpoles are aquatic and ammonotelic; but adult frogs, which are semiterrestrial, excrete mainly urea. Certain aquatic snails normally produce ammonia; but when exposed to drying conditions, they burrow in mud, become dormant, and produce urea from ammonia. In addition to reducing waste toxicity, the increased urea concentration in their body fluids increases the osmotic pressure of their body fluids. This may decrease evaporative water loss and allow the snails to survive a certain amount of time without water.

Uric Acid

Many terrestrial animals have evolved a way to avoid the water loss associated with urea excretion. Most reptiles, birds, terrestrial snails, and insects are **uricotelic**, meaning that they excrete mainly uric acid. **Uric acid** is a water-insoluble compound, and one molecule is equivalent in nitrogen content to two urea molecules (see Figure 7.6). It is less toxic than ammonia, and because it is insoluble, it can be excreted as a suspension of crystals with little loss of water. The excretion of 1 g of uric acid by a bird requires only 1.5–3 ml of water, whereas the excretion of 1 g of urea by a mammal requires 60 ml of water. Chicken embryos excrete ammonia, urea, and uric acid, although they produce proportionately more uric acid as they grow in size. Because uric acid is virtually insoluble in water, it precipitates out of solution and does not kill the embryo as would ammonia or urea. A chicken embryo also minimizes nitrogenous waste excretion and increases metabolic water production by obtaining most (about 80%) of its energy from fats. By contrast, amphibian and fish embryos, which are

not encased by a shell and can dispose of ammonia in surrounding water, obtain over 70% of their energy from protein and only 30% from fat.

Uricotelism is a sound solution to the problems of ammonia toxicity and water conservation, and it is not surprising that so many land animals have evolved this method of waste disposal. Another large group of terrestrial animals, the spiders, have a similar excretory mechanism. Their nitrogenous waste product is guanine, which differs from uric acid only in that it contains an additional amino group. Both uric acid and guanine excretion have a higher energy cost than urea excretion, since ATP is used in the conversion of ammonia to uric acid or guanine.

EXCRETORY AND OSMOREGULATORY MECHANISMS

Having discussed the challenges that animals face in maintaining water and salt balance and in disposing of nitrogenous wastes, we now turn to some of the structures and mechanisms that maintain fluid homeostasis in protozoa and animals. Internal fluid homeostasis involves (1) compensating for the amount of water lost or gained in an environment, (2) regulating the amounts of dissolved salts and water needed in internal fluids for metabolism, and (3) eliminating metabolic wastes. These functions are performed by individual cell membranes at the body surface and/or by excretory organs. If excretory organs are involved, they process extracellular fluids and form urine.

Mechanisms in Freshwater Protozoa and Sponges

All protozoa and sponges exchange ions and wastes with their environment by diffusion and active transport through their surface membranes. Because marine protozoa and sponges are isosmotic to their environment, they experience no net gain or loss of water. Freshwater forms, however, constantly gain water by osmosis, and most have cell organelles called **water expulsion vesicles** (**contractile vacuoles**), that pump out excess water. At one time, it was thought that these organelles functioned in excretion, but four lines of evidence indicate that they function primarily in osmoregulation. First, marine and parasitic protozoa living in isosmotic environments lack water expulsion vesicles. Second, the rate of pumping by water expulsion vesicles is inversely proportional to environmental salinity. Pumping is greatest in distilled water and lowest in concentrated salt solutions. Third, if ATP formation is prevented by cyanide poisoning, pumping stops, and the cell rapidly expands as water continues to enter but is not removed. Finally, direct micropuncture and analysis of the vesicle contents indicate that the contents are hypoosmotic to the cytoplasm. Apparently, the vesicles are filled mainly with water, not salts or excretory products.

In *Amoeba* and many other protozoa, water expulsion vesicles form from the endoplasmic reticulum and are not permanent organelles. The ciliate *Paramecium* has two more-or-less permanent water expulsion vesicles located at the center of a complex system of tubules connected to the endoplasmic reticulum (Figure 7.7). Fluid is collected over the membrane surface of the endoplasmic reticulum. As it flows toward the water expulsion vesicle, the fluid probably becomes increasingly hypoosmotic because the endoplasmic reticulum actively transports salts and other molecules from it. How water expulsion vesicles discharge is not yet fully understood. They do not discharge under their own power; in at least some species, they seem to collapse and release their contents when squeezed against the plasma membrane by the surrounding cytoplasm.

Mechanisms Involving Body Surfaces

Cnidarians, echinoderms, and marine sponges lack special osmoregulatory and excretory structures. Cells on the body surface regulate water and salt content and eliminate wastes. Echinoderms are marine animals, and species that have been studied have body fluids that are osmotically virtually the same as seawater. Thus, the body surface of these animals does not seem to regulate extracellular fluids. Only the intracellular fluid seems to be regulated.

Marine cnidarians are isosmotic to their environment, but freshwater forms, such as hydras, are hyperosmotic. Membrane proteins in *Hydra* actively transport certain ions from the environment and excrete others. Recent studies indicate that hydras use their gastrovascular cavity to dispose of excess water. Fluids obtained from the gastrovascular cavity by micropuncture have a higher concentration of sodium ions than do the intracellular fluids or the surrounding water. Apparently, cells on the body surface actively pump sodium ions from the environment into the space between the outer and inner cell layers. The sodium ions then diffuse into the gastrovascular cavity, and water is thought to follow the ions osmotically. This expands the gastrovascular cavity, and water and salts are expelled through the mouth when the body contracts.

Tubular Excretory/Osmoregulatory Organs

Few animals rely entirely on their body cells to regulate internal fluids. Most bilaterally symmetrical animals have tubular organs that regulate water, salt, and waste concentrations in the body cavity, in interstitial fluids, and in circulatory fluids. These organs are generally called excretory organs, but in many animals they function primarily in osmoregulation. Tubular organs collect (filter) water, salts, and wastes from body fluids by one or more of three mechanisms: (1) Cilia beating within a tube create a negative pressure that draws body fluids into the tube through a filtering membrane; (2) active transport of salts by cells of the tubes leads to osmotic flow of water and certain solutes into the tubular system; and (3) positive blood or

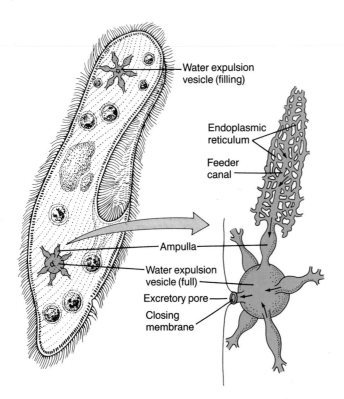

Figure 7.7 Osmoregulation in a freshwater protozoan. The water expulsion vesicle in *Paramecium* fills with fluid from the endoplasmic reticulum. When full, it releases its contents (which are hypoosmotic to the intracellular fluid) to the outside environment.

Water expulsion vesicle (filling)

Endoplasmic reticulum

Feeder canal

Ampulla

Water expulsion vesicle (full)

Excretory pore

Closing membrane

hemolymph pressures force body fluids into the tubular organ. In all cases, the composition of the fluid in the tubular organs is different from that of the body fluids, because certain materials, such as large proteins and cells, do not cross the filtering membranes into the tubes. The fluid in the tubular system is called a **filtrate** (it is not called urine or waste until it is excreted); excretory systems *concentrate* filtrates. Small molecules are reclaimed from the filtrate by a process called **selective reabsorption**. This may involve active transport, diffusion, and/or osmosis by the cells in the wall of the tubule. Tubule cells may also add excess salts and metabolic wastes to the filtrate from body fluids by **secretion**. Urine may contain waste products, other nonuseful compounds ingested with food, and any compounds not reabsorbed from the filtrate.

There are several types of tubular excretory/osmoregulatory organs. Many invertebrates have *nephridia*, which occur as two subtypes: *Protonephridia* are systems of fine, closed tubules that filter extracellular fluids; *metanephridia*, seen in most annelids, consist of open tubules that collect fluid via an opening from the body cavity. Arthropods have several unique types of blind, tubular systems that filter fluids from hemolymph. Crustaceans have *antennal* or *maxillary glands*. Insects and certain other terrestrial arthropods have *Malpighian tubules*. The vertebrate *kidney*, which contains many tubules called *nephrons*, is also a type of tubular organ.

Protonephridia. Protonephridia occur in a variety of invertebrates, including flatworms, rotifers, some annelids, and larval molluscs, as well as in the lancelet, which is an invertebrate chordate (see Figure 28.1). (Rotifers are discussed in Chapter 23, p. 545; Figure 23.9.)

The structure of a typical flatworm protonephridial system is shown in Figure 7.8. Two branching tracts of tubules ending in closed, hollow bulbs collect fluids from throughout the body and void urine through small openings called **nephridiopores**. Each bulb is formed by two cells. An expanded cap cell forms the bulblike blind ending of a tubule and contains cilia that project into the lumen (hollow center) of the system. The cilia beat with a flickering flamelike motion, and for this reason, early microscopists called the cap cells **flame cells**. Forming the terminal end of a tubule, each cap cell interlocks with a tubule cell. The area where the two cells interlock is thin and meshlike, with only a thin membranous covering separating the interstitial fluids from the lumen of the tubule.

Circumstantial evidence indicates that protonephridia act primarily as osmoregulatory organs (water pumps) in freshwater species. When freshwater flatworms are placed in a slightly hypertonic salt solution, ciliary beating in the flame cells, and filtration both stop. If the animals are then placed in distilled water, flickering starts again, and the urine produced is lower in Na^+ and K^+ than that produced in normal pond water. Protonephridia in parasitic flatworms, which are isosmotic to their environment, seem to function mainly in excretion.

The mechanism of urine formation by protonephridia has been studied for years, but is not yet fully understood. Researchers have learned from simple experiments with different-size molecules that protonephridia form urine by selective filtration of body fluids. Inulin (not to be confused with the hormone insulin) is a soluble polysaccharide molecule that does not enter cells or affect animals in any way. Because it is metabolically inert, it is a good marker for studying the movement of extracellular fluids. When radioactive inulin is injected into the body cavity of rotifers, it quickly enters the protonephridia and appears in the urine. When radioactively labeled larger-molecular-weight compounds are injected at the same time, they do not enter the protonephridia. Body fluid pressures do not seem to force fluids into the protonephridia, as evidenced by the fact that protonephridia isolated in a Petri dish continue to form urine. The best explanation is that the beating of the cilia creates a negative pressure that draws extracellular fluids through the thin, meshlike structure into the tubule lumen. The chemical composition of fluids in the lumen is then thought to be modified by reabsorption of sugars, amino acids, and some ions. Reabsorption has been demonstrated by injection of radioactive glucose into the protonephridia of tapeworms. Most of the radioactive glucose is reabsorbed from the tubules into the interstitial fluids, and little is voided in the urine.

Metanephridia. These tubular excretory organs occur in many annelids and in a variety of other invertebrates in several small phyla (described in Chapter 25, p. 603). Coelomic fluid enters metanephridia from the body cavity through a **nephrostome**, an open end of the metanephridial tubule. The tubule follows a twisting path to a storage **bladder**. Coelomic fluid is modified in the tubules by reabsorption of useful materials and secretion of excess salts. The bladder empties externally through a **nephridiopore**.

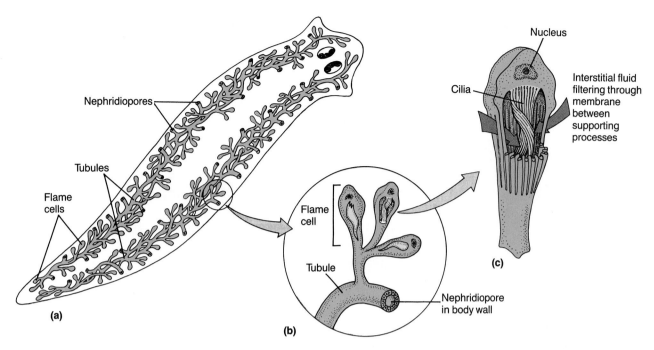

Figure 7.8 The protonephridial system in a freshwater flatworm. (a) Two tracts of blind tubules branch through the animal. Interstitial fluid is drawn into the system and refined in the tubules, and urine is excreted through the nephridiopores. (b) Ciliated flame cells form the blind ends of the tubules and collect fluid from the interstitial spaces. Note that the lumen of the tubule is continuous with a space surrounded by each flame cell. (c) Structure of a flame cell and its connection to an adjacent tubule cell as seen with the electron microscope. Although the mechanism of filtration is not fully understood, fluids apparently filter through a membrane supported by interlocking, rodlike processes of the flame cell and tubule cell.

Earthworms have a pair of metanephridia in each body segment (except the first few and the last one). Each body segment contains coelomic fluid formed as a filtrate from blood in the capillaries of the circulatory system. Each metanephridium extends through a septum (the division between two body segments), and the nephrostome collects fluid from the segment anterior to the one containing the remainder of the metanephridium (Figure 7.9). As the coelomic fluids pass through the tubules, salts and nutrients are reabsorbed and pass into capillaries surrounding each metanephridial tubule. Substantial amounts of fluid pass through the tubules. Each day, an earthworm may produce a volume of urine equal to 60% of its body weight. The metanephridial system primarily regulates salt and water concentrations. Ammonia, the chief nitrogenous waste product, diffuses across the body surface.

Recent studies indicate that the osmoregulatory function of metanephridia is regulated by hormones produced by the brain. If earthworms are placed in distilled water, they swell for about 2 hours but then return to normal size. If their front segments are removed, they lose the ability to regulate their fluid volume. However, injections of macerated earthworm brains restore volume regulation. The biochemical nature of the brain hormone and its precise effects on the metanephridia have yet to be discovered.

Crustacean Antennal and Maxillary Glands. These organs differ from each other mainly in that **antennal glands** have an excretory pore opening to the outside at the base of the antennae, whereas **maxillary glands** open at the base of anterior appendages called secondary maxillae. Crayfish, lobsters, and crabs have antennal glands, also called "green glands." Antennal and maxillary glands terminate internally in a bulblike **labyrinth** and a blind **end sac** that are bathed in hemolymph (Figure 7.10). Having a blind system such as this is advantageous to animals with an open circulatory system. If the end sac and labyrinth opened into the hemocoel, crustaceans would tend to lose hemolymph.

Fluid filters into an antennal or maxillary gland from the hemocoel through membranes of the end

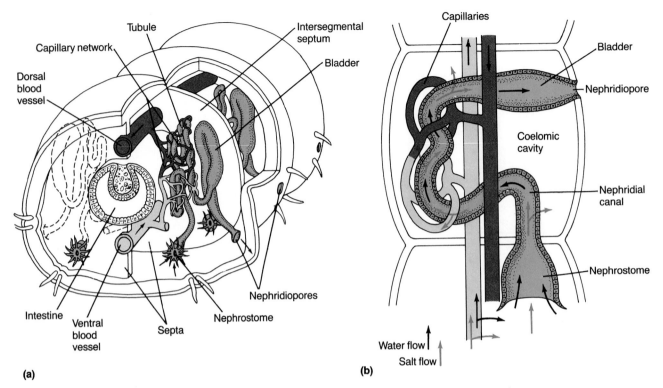

(a)

(b)

Figure 7.9 Metanephridia in an earthworm. (a) Cutaway drawing of an earthworm, showing two body segments as viewed from the anterior end. Only the left metanephridium is shown in each segment. Each segment bears a second metanephridium on the right side (indicated in outline). Coelomic fluids enter from the body cavity of one segment through the nephrostome and are modified as they pass through the tubules in the next posterior segment. Urine passes out through nephridiopores. (b) General organization of the metanephridial system, showing how salts are reabsorbed from the fluid in the metanephridial tubule into closely associated capillaries. During dehydration, water may also be reabsorbed via this route.

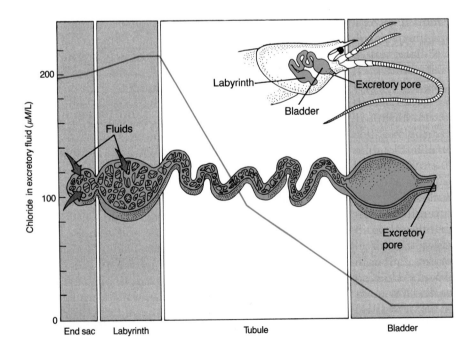

Figure 7.10 Structure and function of an antennal gland of a crayfish. A hemolymph filtrate is forced through the walls of the end sac and labyrinth. The graph indicates how chloride is reabsorbed into the hemolymph as fluid passes along the tubule and into the bladder. Sodium ions passively follow the chloride ions.

sac and labyrinth. Hemolymph pressure from the heart is the main driving force for filtration. The filtrate is refined as it passes along the coiled tubule. Marine crustaceans have a short tubule and produce urine that is isosmotic to their hemolymph. The tubule is proportionately longer in freshwater crustaceans, such as crayfish. The longer tubule allows more surface area for transport systems that reclaim salts from the tubular filtrate. Crayfish antennal glands dispose of excess water by producing urine that is hypoosmotic to the hemolymph. Notice in Figure 7.10 that the antennal gland also has a bladder. Additional ions may be absorbed from the urine that is stored here before it is voided through the excretory pore. Crustaceans are ammonotelic and eliminate ammonia mostly across the surface of the gills and gut. The gills also excrete or accumulate ions from the environment.

Malpighian Tubule–Hindgut Systems. In most terrestrial arthropods, the Malpighian tubule-hindgut system functions in excretion and osmo-

regulation (Figure 7.11). In most insects, Malpighian tubules are closed-end tubules emptying into the hindgut in the abdomen. They are bathed in hemolymph and form a filtrate by actively transporting K^+ (and sometimes Na^+) from the hemolymph into the tubules. Water follows osmotically, carrying with it small dissolved molecules, including wastes. Uric acid, the final nitrogenous waste product, passes into the lumen of the tubules as soluble salts called urates (mostly potassium urate). Large molecules such as proteins are excluded and remain in the hemolymph. Cells lining the hindgut reabsorb useful materials that pass into it from the Malpighian tubules as well as from the midgut. Water is removed osmotically from the hindgut when K^+ and Na^+ are actively transported into the surrounding tissue spaces. Materials not actively transported out of the hindgut are voided as a mixture of urine and feces. As fluid traverses the Malpighian tubules and hindgut, the pH drops and other changes occur, causing urates to form insoluble uric acid crystals that may be excreted with little or no water. Insects living in dry envi-

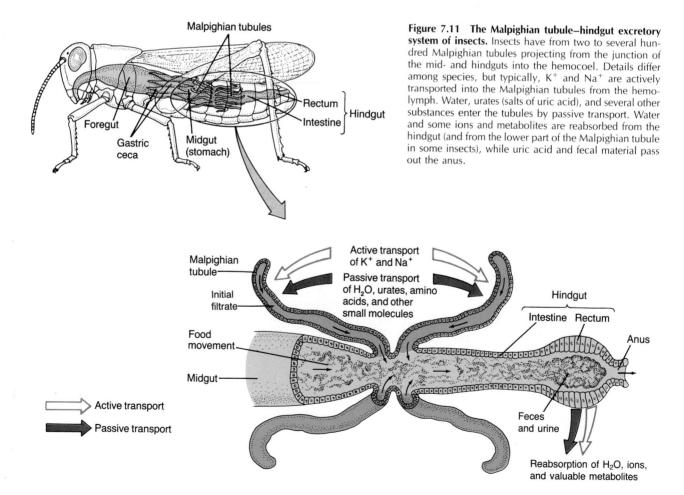

Figure 7.11 The Malpighian tubule–hindgut excretory system of insects. Insects have from two to several hundred Malpighian tubules projecting from the junction of the mid- and hindguts into the hemocoel. Details differ among species, but typically, K^+ and Na^+ are actively transported into the Malpighian tubules from the hemolymph. Water, urates (salts of uric acid), and several other substances enter the tubules by passive transport. Water and some ions and metabolites are reabsorbed from the hindgut (and from the lower part of the Malpighian tubule in some insects), while uric acid and fecal material pass out the anus.

Malpighian tubules

Foregut

Gastric ceca

Midgut (stomach)

Rectum

Intestine

Hindgut

Active transport of K^+ and Na^+

Passive transport of H_2O, urates, amino acids, and other small molecules

Malpighian tubule

Initial filtrate

Food movement

Midgut

Hindgut

Intestine   Rectum

Anus

Feces and urine

Reabsorption of H_2O, ions, and valuable metabolites

Active transport

Passive transport

ronments, such as certain flour beetles, never drink water and conserve water by producing a virtually dust-dry mixture of urine and feces. Any water lost is replaced metabolically.

Researchers have found that water balance in insects is regulated by a **diuretic** (urine-producing) hormone produced by specialized cells in the brain. When secreted into the hemolymph, this hormone reduces the active transport of ions out of the hindgut, thus also reducing osmotic reclamation of water. Such regulation is important. Locusts grazing on fresh grass or lettuce can consume 2 ml of water daily in their food. Since a locust's hemolymph volume is only 0.5 ml, the excess water must be voided. If the hindgut water recovery system operated without a control mechanism, much of this water would be absorbed, diluting the body fluids and causing tissues to swell.

Vertebrate Kidneys. Vertebrate **kidneys** are paired and located on the dorsal wall of the abdom-

inal cavity. As shown in Figure 7.12, mammalian kidneys are compact and bean-shaped; each is connected to a large duct called the **ureter**, which carries urine from the kidney to a temporary storage organ, the **urinary bladder**. The bladder empties to the outside via the **urethra**. The kidneys receive their blood supply from the renal arteries that branch from the aorta, and about 40 times as much blood per unit of organ weight may flow to the kidneys as compared to other organs. In the average human, this amounts to about 1100–2000 L of blood per day, or about 20%–25% of the total amount of blood pumped each day by the heart. From this quantity of blood, the kidneys form a filtrate of about 180 L/day, which is refined to about 1.5 L of urine excreted daily.

The functional units of the kidney are the tubular **nephrons**, which form a blood filtrate from the capillaries and modify the filtrate by selective reabsorption and secretion (Figure 7.13a). **Collecting ducts** receive filtrate from several nephrons and

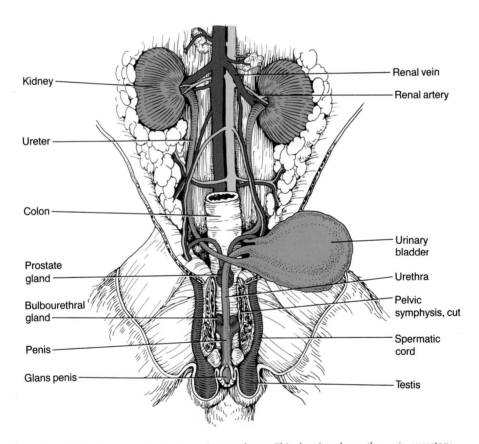

Figure 7.12 The urogenital system of a vertebrate. This drawing shows the main excretory organs, reproductive organs, and associated blood vessels of a male cat in ventral view. The system is somewhat different in other vertebrates. Reptiles and birds have a cloaca, a common chamber that receives the products of the urinary, digestive, and reproductive systems. In reptiles, the urinary bladder empties into the cloaca. Birds lack a urinary bladder and store urine temporarily in the cloaca. The kidneys and associated ducts are quite different in fishes and amphibians and are discussed in the context of vertebrate evolution in Chapter 28.

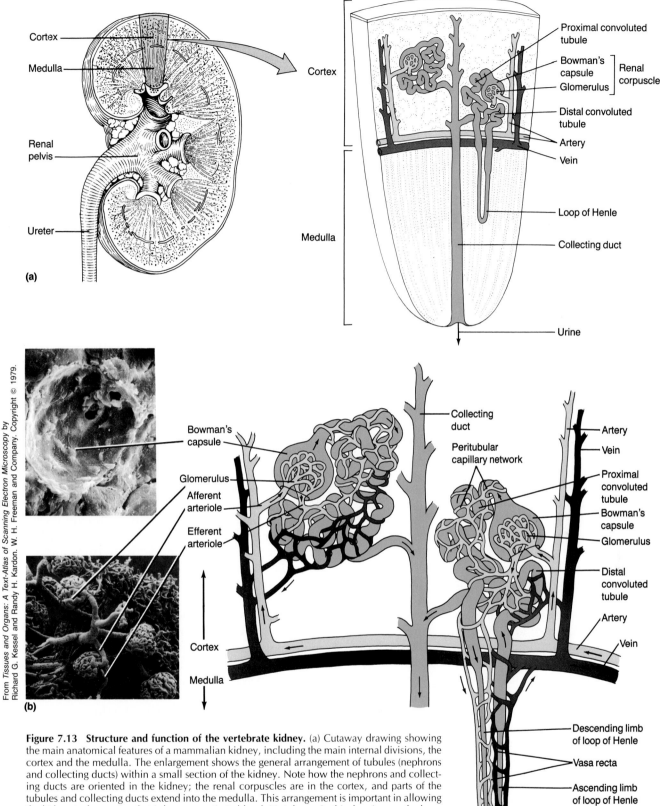

From *Tissues and Organs: A Text-Atlas of Scanning Electron Microscopy* by Richard G. Kessel and Randy H. Kardon. W. H. Freeman and Company. Copyright © 1979.

Figure 7.13 Structure and function of the vertebrate kidney. (a) Cutaway drawing showing the main anatomical features of a mammalian kidney, including the main internal divisions, the cortex and the medulla. The enlargement shows the general arrangement of tubules (nephrons and collecting ducts) within a small section of the kidney. Note how the nephrons and collecting ducts are oriented in the kidney; the renal corpuscles are in the cortex, and parts of the tubules and collecting ducts extend into the medulla. This arrangement is important in allowing the kidney to form urine that is hyperosmotic to blood. Not shown in this drawing are the large numbers of blood capillaries that envelop each nephron. (b) Two different types of nephrons and associated blood vessels in the mammalian kidney. Vasa recta surrounding the loops of Henle are important in water reabsorption. Arrows indicate direction of blood flow in these vessels. In humans, about 80% of the nephrons do not have well-developed loops of Henle and are located almost entirely in the cortex of the kidney as cortical nephrons. In the other 20%, which are called juxtaglomerular nephrons, the loops of Henle penetrate the medulla; juxtaglomerular nephrons are most important in water reabsorption. The scanning electron micrographs show the Bowman's capsule (above) and the microstructure of a glomerulus (below).

refine it further. The collecting ducts deliver urine (any materials not reabsorbed from the filtrate) to the ureter. The human kidney contains approximately one million nephrons, plus associated collecting tubules and blood vessels.

A nephron is made up of two main parts: a blood-filtering unit called the **renal corpuscle** and a **renal tubule** leading from the renal corpuscle to a collecting duct (see Figure 7.13a). The renal corpuscle consists of a tuft of capillary loops collectively called the **glomerulus** and a surrounding double-walled cup called **Bowman's capsule**. Bowman's capsule is formed by the expanded, closed end of the renal tubule. Renal corpuscles occur in the **cortex** (outer) portion of the kidney. Fluid forced out of the capillaries by blood pressure is collected in the lumen of Bowman's capsule, which is continuous with the lumen of the rest of the nephron.

The capillaries of the glomerulus have walls of thin endothelial cells, with small gaps in the junctions between adjacent cells. Specialized cells called **podocytes** surround the capillaries and with the capillary endothelial cells form a filtration membrane. Fluids crossing this barrier contain the small molecules and ions that occur in the blood. Blood cells and most proteins do not pass through the filter and are retained in the capillaries along with fluid that is not filtered.

The renal tubule has three structurally and functionally distinct regions that refine the filtrate obtained from the blood (see Figure 7.13a). The first region, which is adjacent to Bowman's capsule, is called the **proximal convoluted tubule** and is highly coiled. The second region, called the **intermediate tubule**, is quite variable in length, depending on the species and on the position of the nephron in the kidney. In some nephrons in the mammalian and bird kidney, the intermediate tubule assumes a long hairpin shape called the **loop of Henle** (Figure 7.13a and b). The loop of Henle helps conserve water and allows birds and mammals, alone among the vertebrates, to produce a urine that is hyperosmotic to the blood. The loop has a **descending limb** that may pass from the renal cortex deep into the **medulla** (central part) of the kidney. In the medulla, the loop of Henle reverses direction and returns to the cortex as the **ascending limb**. Downstream from the ascending limb is the third region of the renal tubule, called the **distal convoluted tubule**.

As shown in Figure 7.13b, the three functional regions of the renal tubules are surrounded by capillaries that branch from the efferent arteriole that leaves the glomerulus. **Peritubular capillaries** surround portions of the renal tubules in the cortex.

Capillaries called **vasa recta** surround the loop of Henle in the medulla. The filtrate in the lumen of the tubule is modified as materials are exchanged between the capillaries and the three tubular regions. In the exchange process, materials are reclaimed as they move passively, or are actively transported, first from the tubular lumen into the tubule cells, then from the cells into the interstitial fluid surrounding the tubule, and finally into the capillaries. The reverse is true for materials that are secreted into the filtrate. Impressive amounts of material are reabsorbed into the blood as the filtrate passes through the nephrons. The 180 L of glomerular filtrate formed daily by the two human kidneys contain about 1130 g of sodium chloride, of which 95% is reabsorbed; 99% of the approximately 450 g of sodium bicarbonate and about 145 g of glucose contained in the filtrate are reabsorbed at the same time. About 99% of the water is also reclaimed, as are substantial quantities of vitamins, amino acids, and other valuable nutrients.

Physiology of the Mammalian Kidney

Figure 7.14 shows the functional roles of the three regions of the renal tubule in the mammalian kidney, as well as the role of the collecting duct. Filtrate refining begins in the proximal convoluted tubule, where nearly 75% of the filtrate is reclaimed by the blood in the peritubular capillaries. Cells of the proximal tubule actively secrete hydrogen ions into the lumen and actively transport sodium ions out of the lumen into the interstitial fluid; water, chloride ions, and other solutes passively follow the sodium. The movement of hydrogen ions regulates the blood pH and makes the filtrate acidic. Cells of the proximal convoluted tubule may also produce ammonia, which diffuses into the filtrate. There it combines with hydrogen ions to form NH_4^+, which is excreted.

Notice in Figure 7.14 that the loop of Henle has thin- and thick-walled portions. The thin regions of the loop, including the descending limb and the lower portion of the ascending limb, do not actively transport ions. The descending limb is highly permeable to water, and there may be a net passive movement of some NaCl from the interstitial fluid into the filtrate in this portion of the tubule. The thin portion of the ascending limb is relatively impermeable to water, but is permeable to Na^+ and Cl^-. Consequently, there is a net movement of NaCl out of the filtrate in this portion of the ascending limb. The thick-walled portion of the ascending limb in the outer medulla has low permea-

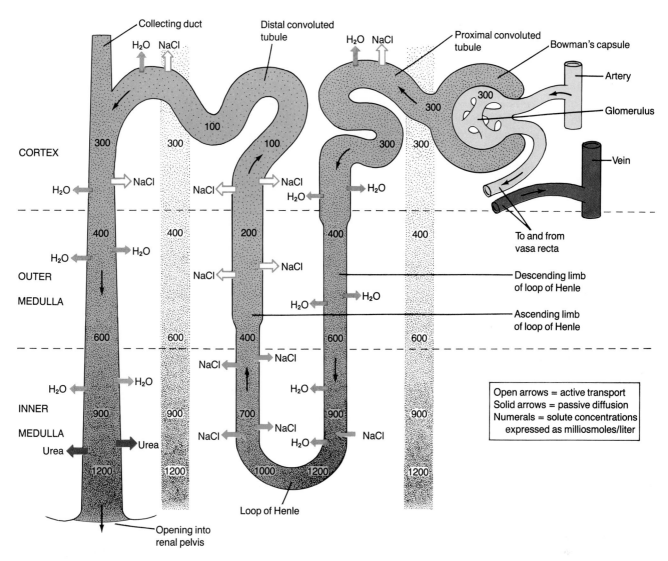

Figure 7.14 Functional zones in a mammalian nephron and collecting duct, and the mechanism of urine formation. This diagram shows the main activities of the three functional regions of a nephron with a loop of Henle, as well as the function of the collecting duct. Arrows indicate NaCl, water, and urea movements between cells lining the tubules and the interstitial fluid, and also indicate whether the movement occurs by active transport or by diffusion. Blood in the peritubular capillaries and vasa recta (shown in Figure 7.13b) exchanges materials with the interstitial fluid. The concentration gradients in the interstitial fluid of the cortex and medulla (indicated by the color gradient in the vertical bars and expressed numerically as milliosmoles of solute per liter) and in the filtrate within the loop of Henle (also indicated numerically and by color gradient) provide the conditions for osmotic removal of water from the collecting ducts in the medulla. These gradients develop largely because the ascending limb of the loop of Henle actively pumps NaCl out of the filtrate into the interstitial fluid and because urea moves passively out of the collecting ducts in the medulla into the interstitial space. As indicated in the diagram, some active transport of NaCl may also occur in the distal convoluted tubule and the collecting duct in the cortex.

bility to water; here, NaCl is actively transported out of the lumen. As a result, when the filtrate reaches the distal convoluted tubule, it is hypoosmotic to the interstitial fluid.

Ion and water movements in the distal convoluted tubule are similar to those in the proximal convoluted tubule. In addition to sodium chloride and water, bicarbonate ions (HCO_3^-) are reabsorbed in this region, and potassium ions are secreted into the lumen of the tubule and reabsorbed in the interstitial fluid, with net flow depending on the amount of potassium in the diet. From the distal convoluted tubule, the filtrate flows into the collecting duct, where more water is reabsorbed as the filtrate becomes urine and is conveyed to the ureter. Note in Figure 7.14 that urea moves passively out of the collecting duct in the inner medulla. As explained shortly, this is very important in assisting the nephron and collecting duct in forming concentrated urine.

Production of Concentrated Urine. The ability to produce urine that is hyperosmotic to body fluids is a significant adaptation for conserving water on land. In mammals, urine is concentrated mainly by the loop of Henle and the collecting duct. Notice in Figure 7.14 that there is a gradient of salt and urea concentration in the interstitial fluid of the kidney; salt and urea concentration is lowest in the outer cortex and gradually increases up to a maximum in the inner medulla. Notice also the position of the portions of the loop of Henle and collecting duct relative to the cortex and medulla. The salt concentration of the filtrate in the lumen of the loop of Henle shows a gradient similar to that of the surrounding interstitial fluid. These concentration gradients result from the difference between the salt and water permeabilities along the various parts of the loop of Henle and collecting duct and from active salt transport by the thick portion of the ascending limb of the loop of Henle. It is important to realize that the concentration gradients in both the interstitial fluid and the filtrate are maintained as the filtrate flows through the nephron. These solute concentrations largely account for the kidney's ability to produce hyperosmotic urine.

What actually happens in and around the loop of Henle? As the ascending limb of the loop of Henle actively transports NaCl from the filtrate into the interstitial fluid, it provides the essential driving force of the concentrating mechanism of the loop of Henle. The resulting high NaCl concentration in the interstitial fluid around the loop of Henle in the cortex and outer medulla causes a net flow of water out of the descending limb of the loop in these regions. Increasing amounts of water are drawn from the filtrate as the descending limb penetrates into the inner medulla, because urea from the collecting duct in this region increases the total solute concentration in the inner medulla. This results in a high solute concentration in the filtrate at the bottom of the loop of Henle (see Figure 7.14). As the filtrate flows through the thin portion of the ascending loop, NaCl flows passively out of the filtrate because its concentration is lower in the interstitial fluid in the inner medulla. Urea from the collecting duct plays a large role in maintaining high solute concentration in the inner medulla. Because the loop of Henle has low permeability to urea, the urea concentration remains high in the interstitial fluid of the medulla. Sodium chloride movement out of the filtrate continues in the outer medulla and cortex as a result of active transport

in the thick portion of the ascending limb of the loop.

As illustrated in Figure 7.14, the loop of Henle acts as a **countercurrent exchange mechanism**. (Similar mechanisms assist gas exchange between water and blood in gills, as discussed in Chapter 5, p. 105, and help regulate temperature in bird legs, as discussed in Chapter 6, p. 135.) Because the direction of flow of filtrate in the descending limb of the loop of Henle is opposite that of the ascending loop, the water loss and increasing concentration of the filtrate that occur along the descending limb are matched by the loss of NaCl along the ascending limb. Thus, as filtrate moves through the nephron, the osmotic gradient in the loop and that of the surrounding interstitial fluid are maintained.

The role of the collecting duct. Understanding the full effect of the loop of Henle and the gradients in the kidney requires a look at what happens to the filtrate after it leaves the loop of Henle. By the time the filtrate enters the collecting ducts, it has lost NaCl and water at several places along the nephron. At the same time, urea has been accumulating in the filtrate. Urea enters from the blood in the proximal convoluted tubule and the thin portions of the loop of Henle. Because virtually no urea passes into the interstitial fluid from the filtrate in the nephron, and yet NaCl and water are removed, the filtrate entering the collecting duct contains a high concentration of urea. The regions of the collecting duct in the cortex and outer medulla are not permeable to urea, but the collecting duct is permeable to water throughout its length. Thus, as the duct descends into the medulla, water flows osmotically from it. In the cortex and outer medulla, the NaCl gradient of the interstitial fluid (maintained mainly by the loop of Henle) draws water out of the collecting duct. This further increases the urea concentration in the filtrate. Finally, in the inner medulla, the collecting duct becomes permeable to urea, and some of the filtrate's urea flows into the interstitial fluid. (We have already mentioned how this movement of urea produces the high solute concentration around the lower loop of Henle in the inner medulla.) High urea concentration here provides the final step in making the urine hyperosmotic, for it draws still more water out of the filtrate in the collecting duct.

One additional aspect of kidney function enables the nephron to produce concentrated urine. Notice the vasa recta in Figure 7.13 and the direction of blood flow in these vessels. Blood flows from a glomerulus down into the medulla through vessels that lie close to those carrying blood back from the medulla to the cortex. This is another example of countercurrent exchange. Blood passing into the medulla gains salt and urea by diffusion from the increasing concentrations of these solutes in the interstitial fluid. Blood leaving the medulla loses these solutes as it passes through the opposite gradient in the interstitial fluid. The overall effect is that blood leaving the vasa recta carries little or no more salt and urea than it did upon entering these vessels. As a result, the tissues of the medulla are supplied with vital materials from the blood, but the blood does not interfere with the solute gradient on which water reabsorption depends. The vasa recta also help maintain the kidney's concentration gradient by carrying away much of the water that is reabsorbed by the loop of Henle and collecting ducts. If this water were not removed, it would dilute the solute gradient in the interstitial fluid.

Regulation of Kidney Function. Activity of the nephrons and collecting ducts is under hormonal and nervous control. A peptide hormone called **antidiuretic hormone (ADH)**, also called **vasopressin** in mammals or **vasotocin** in nonmammals, is produced by the hypothalamus of the brain and regulates water reabsorption by the collecting ducts. ADH formed in the hypothalamus passes via nerve fibers to the posterior pituitary gland, where it is stored until it is released into the blood. When a mammal is dehydrated, its blood osmotic pressure rises and causes **osmoreceptors** in the hypothalamus to trigger the release of ADH. Higher amounts of ADH in the blood cause an increased permeability of the collecting ducts. This results in more water being reclaimed by the salt and urea gradient mechanism (Figure 7.15a). Thus, a dehydrated mammal excretes a small amount of highly concentrated urine. If a mammal has excess water in its body fluids, little or no ADH is released into the blood, the collecting ducts become much less permeable to water, and a large amount of dilute urine results.

Another hormonal mechanism influences the kidney's rate of filtration of blood as well as its salt and water reabsorption (Figure 7.15b). This mechanism involves a group of cells called the **juxtaglomerular apparatus**, located in the afferent arteriole that supplies the glomerulus and the distal convoluted tubule. These cells produce an enzyme called **renin** when the blood pressure or blood sodium level drops, or when they are stimulated by certain nerves from the autonomic nervous system. When released into the blood, renin converts a blood glycoprotein called angiotensinogen to an active hormone called **angiotensin**. Angiotensin causes at least three things to happen:

1. Angiotensin causes the adrenal glands to release a steroid hormone called **aldosterone** into the blood. (The various roles of the adrenal glands are discussed in Chapter 11, p. 255.) Aldosterone promotes sodium transport out of the filtrate in the distal convoluted tubules. Because water follows the sodium, its rate of reabsorption also increases. Aldosterone also promotes secretion of potassium into the filtrate in the distal convoluted tubules and promotes sodium reabsorption from the large intestine, salivary glands, and sweat glands.
2. Angiotensin is carried in the blood to the hypothalamus, where it causes thirst sensations leading to drinking behavior.
3. Angiotensin raises the blood pressure by constricting the arterioles throughout the body.

The angiotensin mechanism is an adaptable homeostatic control that responds to changes in the salt/water balance of the body. Its activation by dehydration or blood loss following serious injury results in water and ion retention and in thirst, all of which lead to an increase in the volume of the blood. If blood pressure becomes too high, angiotensin levels decrease, and salt absorption and thirst are inhibited so that blood pressure is reduced and more urine is excreted.

Environmental Physiology of Vertebrate Kidneys

In this section, we will compare the kidneys of several groups of vertebrates, focusing on how kidney structure and function reflect the problems encountered in different environments. (This topic relates directly to the subject of kidney evolution, addressed in detail in Chapter 28.) As a group, vertebrates inhabit marine, freshwater, and terrestrial environments and therefore experience the full range of osmotic and ionic stresses.

The structure of the vertebrate nephron clearly reflects the challenges of the three types of environments. Most freshwater fishes, for example, have simple nephrons adapted to remove excess water from the body and to conserve vital solutes. There is a renal corpuscle (glomerulus plus Bowman's capsule) that filters water from the blood, plus a

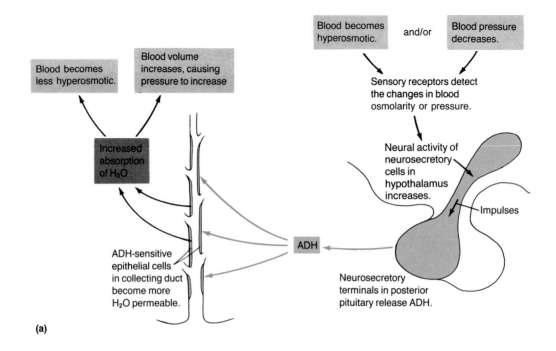

(a)

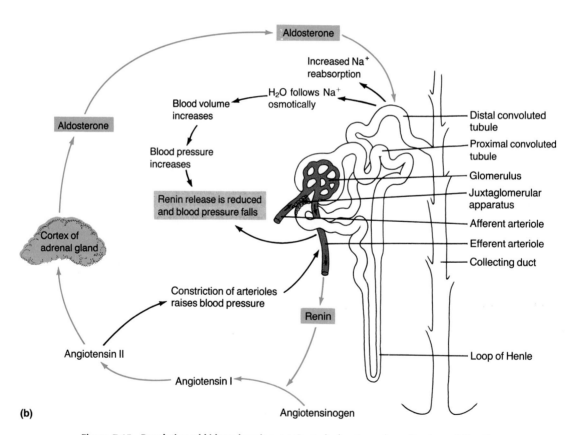

(b)

Figure 7.15 Regulation of kidney function. (a) Control of water reabsorption by antidiuretic hormone (ADH). (b) The renin-angiotensin control mechanism. Changing levels of renin, angiotensin, and aldosterone in the blood affect Na^+ and water reabsorption, rate of drinking, and the diameter of blood vessels.

Essay: Living at Sea and Solving the Salt Problem

Anyone lost at sea without a supply of drinking water must experience an extreme paradox. Although surrounded by water, the individual may die from dehydration. Most mammals, and indeed most vertebrates other than marine bony fishes, cannot alleviate dehydration by drinking seawater, because their kidneys cannot produce a urine that has a greater concentration of solutes than seawater. If they drink seawater, the excess salts are excreted with water in the urine, and more water is lost than was obtained.

Nevertheless, several reptiles, birds, and mammals are marine. And while little is known about how marine mammals cope with salt and water balance problems, birds and reptiles have been more thoroughly studied. Sea turtles, among the reptiles, and birds such as albatrosses and penguins remain at sea except to breed. Other oceanic reptiles are the highly venomous sea snakes *(Hydrophis),* which are totally marine, and the marine iguana *(Amblyrhynchus cristatus)* of the Galapagos Islands (see the photograph at the beginning of this chapter). The marine iguana spends much of its time on land but feeds in the ocean and eats marine algae, which have a salt content essentially that of saltwater. All of these seafaring vertebrates drink seawater, which has a salt concentration three times higher than that of their body fluids. Their kidneys are not markedly different from those of other reptiles and birds, and consequently, they cannot produce urine concentrated enough to remove such high levels of salt. How do these animals eliminate excess salts without losing too much water?

Oceanic birds and marine reptiles have special salt-excreting glands that remove excess salts from the blood while conserving water. In birds, salt-excreting (nasal) glands located near the eyes remove excess sodium chloride from the blood, leaving behind fresh water for the animal's needs. A highly concentrated salt solution drains from the glands into the nostrils (Figure E1). In marine turtles, the salt gland is a modified tear (lachrymal) gland, and in sea snakes, a salivary (sublingual) gland beneath the tongue excretes excess salt. Salt-excreting glands allow animals to desalinate seawater and survive without access to fresh water. Excretion of salt is not continuous and is regulated by the nervous system. High salt concentrations in the blood stimulate osmoreceptors thought to be located near the heart. These receptors send impulses to the brain, which in turn stimulates the glands to excrete salt until the salt concentration in the blood decreases to tolerable levels.

Salt glands are not limited to marine species. In fact, most birds have nasal (supraorbital) glands, but these are usually small and inactive. Active salt glands occur in desert reptiles and some birds that inhabit dry areas where there is little or no water. The diet of many animals in these environments is often high in potassium salts, and the ani-

mals tend to lose water by evaporation. Thus, the body fluids tend to accumulate salts. If excess salts were removed by the kidney alone, too much water would be lost. The salt glands of desert species, as in marine species, are able to remove salts dissolved in very little water and thus conserve this precious resource.

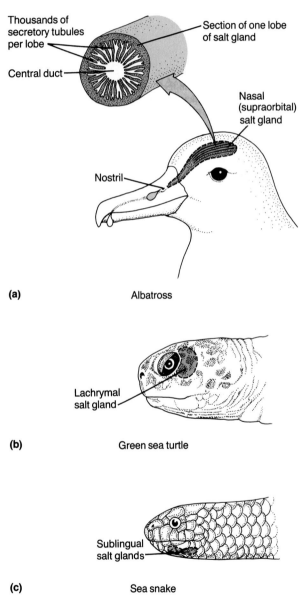

(a)   Albatross

(b)   Green sea turtle

(c)   Sea snake

Figure E1   Salt-excreting glands. These glands occur in a variety of animals, including (a) the albatross *(Diomedea),* (b) the Atlantic green turtle *(Chelonia),* and (c) the sea snake.

proximal and a distal tubule, both of which reabsorb essential solutes. The intermediate tubule is barely developed. Most marine fishes, which must conserve water and excrete salts, have a reduced renal corpuscle that removes little water from the blood. Some marine fishes lack glomeruli altogether and produce urine by the secretion of salts into the nephron. As discussed earlier, ionic regulation and excretion are largely functions of the gills in both freshwater and marine fishes.

The typical amphibian kidney functions mainly as a water pump. The nephron has a well-developed renal corpuscle, proximal and distal convoluted tubules, and a very short intermediate tubule. When in fresh water, amphibians actively accumulate ions through their skin. Excess water that enters the body osmotically is filtered by the renal corpuscle and excreted in watery urine. On land, where water loss becomes a problem, amphibians reabsorb water from the urine across their bladder wall. Australian aborigines are known to dig toads out of dried lake beds during droughts and squeeze them to obtain drinking water. ADH and aldosterone regulate the amphibian kidney as in mammals and influence the uptake of water and ions through the skin.

Unlike amphibians, reptiles and birds have thick, dry skin that reduces water loss. Uric acid excretion in reptiles and birds also conserves water. Moreover, many reptiles have reduced renal corpuscles, and some desert reptiles lack glomeruli. The wall of the cloaca also absorbs water from the feces and urine in reptiles and birds. It is probably because birds can produce uric acid that their loop of Henle is not as well developed or as effective in concentrating urine as it is in mammals. Many nephrons in birds have short loops or lack loops altogether. Marine and desert reptiles and birds also have special **salt-excreting glands** that dispose of large amounts (up to 60%–85%) of the salt ingested with seawater or food. (See the Essay on p. 167.)

There is an interesting correlation between the length of the loop of Henle and the habitat of certain mammals. Beavers, which are in little danger of desiccating, have short loops of Henle that do not dip into the medulla, and these animals produce a dilute urine. Desert rodents, which constantly face a water shortage, have very long loops of Henle and produce a highly concentrated urine.

Marine mammals, such as whales and seals, also face osmoregulatory problems. Their blood is hypoosmotic to seawater and they must maintain water in their body fluids. These animals obtain some fresh water from the oxidation of fats and carbohydrates but also consume large amounts of seawater and salts with their food. Few studies have been performed on kidney function in marine mammals, and the details of how these animals osmoregulate are not yet fully known.

□ SUMMARY

1. Osmoregulation is the maintenance of water and ion balance between an animal's body fluids and the external environment. Osmoregulation and excretion are often carried out by the excretory system. Ammonia is the primary nitrogenous waste product of most aquatic animals. Terrestrial animals excrete mainly urea, guanine, and/or uric acid.

2. Most marine invertebrates have body fluids that are isosmotic to seawater, although their specific ionic composition may be different. Marine bony fishes are hypoosmotic to seawater. Marine sharks and rays retain urea and trimethylamine oxide and are hyperosmotic to seawater. Stenohaline animals can tolerate only small changes in external salinities. Euryhaline species can tolerate greater changes and can live in estuaries.

3. Freshwater animals are osmoregulators, expending energy to combat the constant tendency of their body fluids to gain water osmotically and lose salts by diffusion. Their excretory organs promote water loss. Cells at body surfaces or on gills may accumulate ions from the environment.

4. Terrestrial animals have adaptations to reduce water loss. Body surfaces and respiratory systems conserve water. The kidneys of many birds and mammals conserve water by producing a urine that is hyperosmotic to the body fluids.

5. Freshwater protozoa and sponges have water expulsion vesicles that are mainly water pumps. Certain freshwater cnidarians osmoregulate by pumping salts into the gastrovascular cavity. Water follows osmotically and is expelled through the mouth.

6. Most invertebrates have tubular excretory/osmoregulatory organs. Flatworms and several other invertebrates have protonephridia that draw body fluids into tubules across a filtration membrane by the action of cilia. Materials are reabsorbed or secreted as urine is formed by the tubules. The metanephridia of annelids are tubular excretory organs that filter fluid collected from the body cavity. Fluids enter the open ends of the tubule and are modified and excreted as urine. Crustaceans have closed antennal or maxillary glands that filter fluid from the hemocoel. The Malpighian tubule–hindgut excretory system of insects conserves water by excreting uric acid. The hindgut reclaims water from the feces and urine, and some species produce dry feces.

7. Vertebrate kidneys contain many functional units called nephrons, which are tubules that collect and refine fluids from the blood. Each nephron consists of a renal corpuscle, proximal and distal convoluted tubules, and, in many species, an intermediate tubule. The renal corpuscle forms a filtrate from the blood, and the rest of the nephron refines the filtrate by reabsorption or secretion of water, salts, and organic molecules. The filtrate passes from the nephrons into collecting ducts, which form the final urine. Ureters convey urine from collecting ducts to a bladder or cloaca. In birds and most mammals, the intermediate tubule forms the loop of Henle, which allows these animals to reabsorb enough water to produce urine that is hyperosmotic to the blood. By acting as a countercurrent exchange system, the loop of Henle creates and maintains a solute gradient in the kidney that promotes water reclamation by the collecting ducts.

8. Vertebrate kidneys are regulated by hormones. Antidiuretic hormone, produced by the hypothalamus, promotes water reabsorption. Angiotensin, formed in the blood, stimulates release of aldosterone by the adrenal glands, promoting salt reabsorption by the distal convoluted tubules.

9. The structure of vertebrate kidneys correlates with the animal's environment. Marine fishes and reptiles have reduced glomeruli that minimize fluid loss. Freshwater fishes and amphibians have well-developed glomeruli. Only mammals and birds have the loop of Henle.

❏ FURTHER READING

Hoar, W. S. *General and Comparative Physiology.* Englewood Cliffs, N.J.: Prentice-Hall, 1983. *Contains an excellent chapter on osmoregulation.*

Little, C. *The Colonization of Land: Origins and Adaptations of Terrestrial Animals.* New York: Cambridge University Press, 1983. *Describes adaptations in the six animal phyla that have terrestrial species.*

Marsh, D. J. *Renal Physiology.* New York: Raven Press, 1983. *A concise description of kidney function written for medical students.*

Rankin, J. C., and J. Davenport. *Animal Osmoregulation.* Glasgow, Eng.: Blackie and Sons, 1981. *Describes osmoregulatory mechanisms in invertebrates and vertebrates.*

Smith, H. W. "The kidney." *Scientific American* 188(1953): 40–48. *A classic description of the structure, function, and evolution of the vertebrate kidney.*

Activity and Regulatory Systems

In the preceding chapters, our objective was to illustrate how the digestive, respiratory, circulatory, excretory, and osmoregulatory systems maintain critical differences between an animal's internal and external environment—that is, how these systems maintain internal homeostasis. We discussed these systems one at a time, yet in the whole animal, all systems work together and constantly influence one another, and an animal cannot be understood fully except as a whole, functioning unit. The remaining chapters in this unit focus on what might be called *activity and regulatory systems*—those that allow animals to move about and interact with their environment, and those that regulate and coordinate the activities of the whole body.

Mobility is one of the most fundamental and distinctive features of animals. Sessile (attached) animals move in place, while animals with locomotor ability can move from place to place. Being mobile increases an animal's ability to obtain food, escape enemies, and locate mates. Mobility may require moving parts, such as the wings of a bird, or it may require the participation of the whole body, as in a snake. Movement depends on several activity systems. Muscular or ciliary/flagellar and skeletal systems function together to provide movement. In many species, muscles and skeletal elements work together as a system of levers.

Most animals not only move, but do so in a directed, controlled manner. Directed movement requires a system for coordinating body actions. Consequently, most animals have a nervous system, which controls and coordinates muscular activity.

To move nonrandomly, and in a way that makes efficient use of energy to achieve such goals as feeding and reproduction, an animal also needs to obtain information about its environment. All animals have some type of sensory system, usually composed of various receptor structures that can sense different kinds of stimuli, such as light, sound, the presence of certain chemicals, or the existence of a magnetic field. As an animal's sense receptors detect various types of stimuli, its nervous system processes the information and then, often in concert with the chemical coordinating system (sometimes called the endocrine system), directs how the animal responds to the stimuli. The chemical coordinating system is composed of specialized glands and certain parts of the nervous system that secrete regulatory chemicals called hormones into body fluids. Body fluids distribute the hormones to sites

170

in the body where these chemicals can cause regulatory changes, such as ovulation in mammals or molting in insects. Both the nervous and the chemical coordinating systems exert control over other body systems. An animal's responses to environmental stimuli result from the interaction of all its activity and regulatory systems. The end result—the product of sensory reception, integration of information, chemical regulation, and response—is animal behavior.

The next five chapters examine the skeletal, muscular, nervous, chemical, and sensory systems of animals, and how these systems influence animal activity. The sixth and final chapter in the unit provides a look at the variety and scope of animal behavior.

Body Surface and Support Systems

An animal's outer covering, called its **integument**, is the interface between the internal and external environment. As such, the integument's primary function is to protect the body against mechanical injury, infection, and, in many animals, excess heat or cold. Depending on the animal and its environ-

ment, the integument may serve a variety of other functions. It may exchange gases, absorb nutrients, dispose of wastes, osmoregulate, sense environmental stimuli, and secrete chemicals that attract or repel other animals. In many animals, the integument also provides body support, in which case it acts as a component of the animal's structural framework, or **skeleton**. This chapter is about both integuments and skeletons. However, our emphasis is on body support, and so our discussion of integuments will focus mainly on how they provide skele-

The American flamingo (*Phoenicopterus ruber*). This species is found in the Caribbean area, including parts of Florida. It often rests with all of its weight supported by one of its long, thin legs.

172

tal support. The topic of skeletal support is closely tied to that of body movement, because an animal's skeleton does more than just provide body shape and protect the soft tissues and organs; it also provides a basis for movement. For an animal to move, its muscles, cilia, or flagella must be attached to, and apply force against, some type of skeletal part. Muscles, for example, can move an animal's body because they can exert force against each other and/or against the skeleton. (Transmission of force and movement are discussed in Chapter 9, p. 195.)

Integuments consist of one or more layers of living cells, usually covered by sloughed dead cells and/or cell secretions. The outermost layer (or layers), called the **epidermis**, is composed of epithelial cells. (For a discussion of types of cells and tissues, see Chapter 2, p. 50; Figure 2.26.) In some animals, the epidermis and mucous secretions produced by it constitute the entire integument; but in most animals, the epidermal cells secrete a nonliving surface layer called a **cuticle** that encases part or most of the animal. Cuticles are highly variable in structure. They are quite thin and elastic in some animals, while in others they are thick and rigid and provide strong support for the body.

There are three types of skeletons in the animal kingdom. In one type, the cuticle accounts for most, if not all, body support and is called the **exoskeleton**. The jointed, chitinous cuticle of insects and crustaceans, the calcareous cases of stony corals, and the rigid tubes of certain annelids are a few examples of exoskeletons. A second type of skeleton, called the **endoskeleton**, is internal. The cartilage and bone of vertebrates, which consist of living cells embedded in a matrix; the calcareous spicules of sponges; and the hard ossicles of sea stars are all endoskeletal elements. Exoskeletons and endoskeletons are relatively rigid frameworks, compared with the third type of skeleton. The **hydrostatic skeleton** consists of fluids held under pressure in a body compartment, the pressure being provided by the surrounding connective tissues and muscles.

Many animals have more than one type of skeleton, and some have elements of all three. Arthropods are supported mainly by an exoskeleton, but many species have their heads braced internally by cuticular structures. Vertebrates are supported mainly by an endoskeleton, yet most species have some exoskeletal elements, such as scales, shells in turtles, beaks, claws, and hooves. Many animals that have a rigid skeleton also have some hydrostatic support. In mammals, for example, erection of the penis results from blood engorging the spongy tissues in the penis. This compresses the veins in the penis, reducing the outflow of blood and making the spongy tissues rigid.

GENERAL PROPERTIES OF SKELETONS

Nearly all animals, whether they have exoskeletons, endoskeletons, or hydrostatic skeletons, derive considerable support from **connective tissues**. As mentioned in Chapter 2, cartilage and bony endoskeletons are specialized types of connective tissue. All types of connective tissue consist of cells in an extracellular matrix that the cells secrete (see Figure 2.27). Protein fibers in the extracellular matrix form reinforcing webs and meshworks throughout animal bodies. Fibrous connective tissues give shape to other tissues and organs, reinforce the body wall, and often hold organs in place.

Collagen, which may be thought of as molecular rope, is the most common type of reinforcing, fibrous protein in connective tissues. Collagen is composed of *tropocollagen* molecules, all of which are rich in the amino acids glycine, proline, and hydroxyproline. (Amino acid structure is covered in Chapter 2, p. 31.) Every third amino acid in a tropocollagen polypeptide is glycine, and because this amino acid has only hydrogen as a side group, it allows tropocollagen molecules to pack closely together and form collagen fibers. Tropocollagen molecules have three polypeptides bound to one another, forming a triple helix (Figure 8.1). In vertebrate collagen, the helical molecules assemble end to end in staggered parallel arrays to form collagen fibrils, which can be seen with the electron microscope (Figure 8.1 inset). Bundles of fibrils form collagen fibers, which are several micrometers thick. Collagen is an effective skeletal substance because it is not elastic and does not stretch. It is a major structural component in nearly all animals.

Endoskeletons and Exoskeletons

Endo- and exoskeletons are composite materials made of fibers embedded in mineral salts or organic polymers. Bone, for example, has calcium salts surrounding the collagen fibers in its intercellular matrix. By volume, about half of bone is collagen and half mineral matrix. X-ray diffraction studies of bone reveal that the mineral component

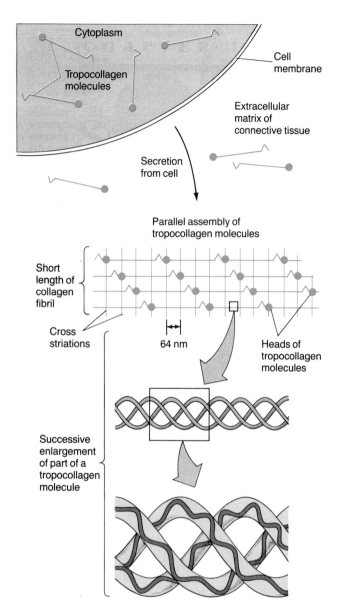

Figure 8.1 Structure of collagen, the most common fibrous material in connective tissue. Tropocollagen molecules secreted from cells spontaneously align and bond with one another, forming collagen fibrils. Note that each tropocollagen molecule consists of three polypeptide chains in a helical arrangement and that each polypeptide chain is itself a helix. The "heads" of the tropocollagen molecules are in register with each other at 64 nm intervals (1 nm = 10^{-9} m). The cross-striations evident in electron micrographs of collagen fibrils (inset, $34,200\times$) result from this orderly arrangement of tropocollagen "heads."

consists of needle-shaped crystals of **hydroxyapatite**, commonly called calcium phosphate and characterized by the formula ($Ca_{10}(PO_4)_6(OH)_2$).

As an animal moves, its skeleton sustains several different types of strains (Figure 8.2). In bearing loads, skeletal parts must withstand **compression**. This forces molecules closer together, and if the skeletal element is weak, it may crumble. **Tension** develops during pulling and tends to separate the molecules of a skeletal element. Force applied at an angle to a long bone's axis causes **shear**, which tends to force molecules to slide past one another; if shear is too great, a bone may fracture cleanly across its short axis. **Torsion**, or torque, results from a twisting mode of loading; if great

enough, the resulting shear may cause spiral fractures when the skeletal element breaks.

Reinforced concrete (concrete containing steel rods) is analogous to a hard animal skeleton. Concrete alone resists compression, but will break when subjected to shear or tension. Steel reinforcing rods make concrete resistant to shear and tension so that it can be used to build spanning structures, such as bridges. Likewise, each component of an animal's skeleton resists different types of strain. Collagen fibers in bone resist deformation caused by shear and tension, allowing bones to be quite long. (Some of the leg bones of a giraffe, for example, are 80–90 cm long.) Hydroxyapatite crystals resist compression, allowing bones to bear

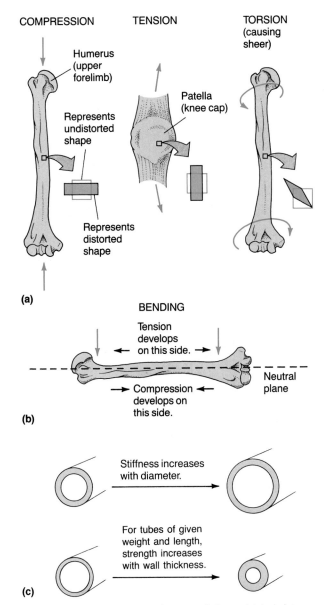

COMPRESSION

Humerus
(upper
forelimb)

Represents
undistorted
shape

Represents
distorted
shape

(a)

TENSION

Patella
(knee cap)

TORSION
(causing
sheer)

BENDING

Tension
develops
on this side. →

Neutral
plane

→ Compression ←
develops on
this side.

(b)

Stiffness increases
with diameter.

For tubes of given
weight and length,
strength increases
with wall thickness.

(c)

Figure 8.2 Types of forces acting on a skeleton. (a) A skeleton must withstand various kinds of loading, including compression, tension, and torsion, if it is to provide support for movement. (b) When a rod or tube, such as the vertebrate femur (thigh bone), is bent, the molecules on the outside of the curvature move farther apart as tension develops, while those on the inner curvature are compressed together. Molecules in the center (the neutral plane) are neither pulled away from each other nor compressed and therefore contribute little to resisting fracture. Tubes can be stiffer than rods of the same weight because the structural material is concentrated in the wall, where compression and tension develop, rather than in the neutral plane. (c) The stiffness and strength of a tube depend on diameter and wall thickness. Long bones, such as those in the leg or arm of a vertebrate, are typically thick-walled tubes.

heavy loads. The leg bones of an adult elephant, for instance, bear loads of about 7200 kg (8 tons)!

Stiffness in skeletal elements should not be confused with strength. **Stiffness** is the property of resisting bending, while **strength** reflects the load

that an element can bear before it breaks. Glass rods are stiff but not very strong, just the opposite of steel rods of the same diameter. When a skeletal element, such as a long bone of a leg, bends, two types of strain occur on the surface of the bend. Tension develops along the outside curvature of the bend and is greatest at the surface; molecules located here are displaced the greatest distance. Compression occurs along the inside curvature and is also greatest at the surface (Figure 8.2b). Molecules along the center line of a rod experience the least strain in bending and contribute the least resistance to bending; this "noncontributing" material is absent in a tubular long bone such as a femur, whose center is occupied by a nonrigid marrow.

The composite nature of bone makes it an excellent structural material. Calcium salts resist compression while the collagen fibers resist tension. Because of its fiber content, bone is also resistant to cracking. The mineral portion may develop small cracks, but these stop when they encounter fibers. Here, the energy forming the cracks dissipates along the length of the fiber; a crack crosses a fiber only if the energy forming the crack exceeds the fiber strength. Bone cells can repair small cracks in the hard matrix.

In general, skeletal shape and consistency represent a balance between strength and stiffness and the energy cost of carrying the weight of a skeleton. As already mentioned, most long elements of a skeleton are tubes, which are stiffer per unit weight than solid rods. This is because tubes have all structural materials at the outer edge, where compression and tension are greatest. For tubes, stiffness increases as the diameter increases for a constant wall thickness (Figure 8.2c). Strength also increases, but not in the same proportion; thin-walled, large-diameter tubes may be very stiff and not bend, but they quickly buckle rather than bend when they reach their load limit. Large-diameter tubes can be strengthened by adding to the wall thickness, but this adds weight. Nonetheless, long bones are usually thick-walled tubes.

Aquatic animals are supported to a large extent by the surrounding water; but the full weight of a terrestrial animal may be supported by the skeleton. The main skeletal elements that support much of a land animal's weight are subject to the scaling effect of geometry. If an animal's linear dimensions are increased, its volume increases as the cube of the change. Generally, a terrestrial animal's volume is directly proportional to its mass; thus an animal that doubles in linear dimensions requires a skeleton that can bear a body mass about eight times greater.

Cost-Benefit Analysis of Endo- and Exoskeletons. In general, exoskeletons are stiffer than endoskeletons, but not as strong. Their superior stiffness means that exoskeletons provide excellent support for muscle-based movement. On the other hand, exoskeletons have less load-bearing capacity than do endoskeletons. An exoskeleton can be strengthened by thickening, but this may increase weight to a critical level, and thick exoskeletons generally provide less internal space for an animal's soft tissues.

Exoskeletons also affect animal growth because there is a limited amount of space inside them. An animal with an exoskeleton can grow only if it can also increase the size of the exoskeleton (as in molluscs) or if it can shed (i.e., molt, as in arthropods) the old, smaller exoskeleton while replacing it with a new, larger one. If the exoskeleton is molted, the discarded organic material represents a major energy expenditure that is cast away. Molting also makes an animal vulnerable to predators. A new exoskeleton is usually soft, unable to provide firm protection, and muscle contractions may bend the soft skeleton rather than being translated into escape movements.

In contrast to exoskeletons, endoskeletons do not limit the space for internal organs and they can support greater weight. Thus, it is not surprising that the largest animals have endoskeletons. Endoskeletons also offer other advantages. For instance, the soft tissues surrounding an endoskeleton may prevent skeletal damage. Soft tissues absorb impact, and though they may be bruised by sharp blows, the resulting injury may be easier to repair than a broken skeleton. Endoskeletons usually include elements that surround and protect certain vulnerable organs, such as the heart, lungs, and brain of vertebrates.

Hydrostatic Skeletons

Body fluids within and between cells, in circulatory vessels, and within body cavities may contribute to hydrostatic support in an animal. Because water is incompressible, it will transmit pressure equally in all directions. If force is applied to the fluids in an animal's body, and if all parts of the body wall are equally strong because of a uniform network of collagen fibers or muscle layers, the body will be stiffened equally throughout. This is analogous to the way a soft hose with water flowing through it stiffens when the nozzle is closed. A hy-

drostatic skeleton manipulated by muscles may also provide **hydraulic movement**. Just as a water-filled balloon lengthens when squeezed at one end, if any part of an animal's body wall is more elastic than another, the pressurized fluid may expand that part of the wall. Certain aquatic organisms, such as hydras and sea anemones, rely almost entirely on hydrostatic support and hydraulic movement. As described in Chapter 24 (p. 567, Figure 24.12), bivalve molluscs use a hydraulic mechanism to extend their muscular foot for digging. Muscles force hemolymph into the tissues of the foot, causing it to expand and extend into the mud.

Many animals with hydrostatic skeletons have muscles in the body wall; some of these (circular muscles) encircle the body, while others run lengthwise in the body wall. When the circular muscles in a region contract, the body narrows, and the increased fluid pressure lengthens the animal if the body wall is elastic (Figure 8.3). Contrac-

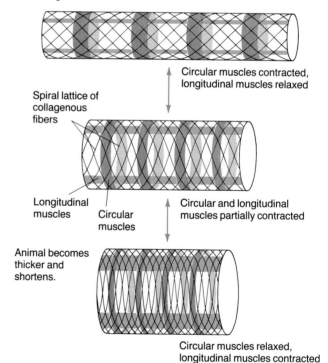

Animal becomes thinner and elongates.

Spiral lattice of collagenous fibers

Circular muscles contracted, longitudinal muscles relaxed

Longitudinal muscles

Circular muscles

Circular and longitudinal muscles partially contracted

Animal becomes thicker and shortens.

Circular muscles relaxed, longitudinal muscles contracted

Figure 8.3 Hydraulic movement in a tubular animal. These diagrams show the effects of circular and longitudinal muscle contraction on body shape in a generalized tubular animal. As the circular muscles contract and the longitudinal muscles relax, increased body fluid pressure causes the body to lengthen. Contraction of the longitudinal muscles and relaxation of the circular muscles shortens the animal as increased fluid pressure thickens it. The spiral arrangement of relatively inelastic collagen fibers in the body wall permits lengthening or thickening of the body without any change in body volume. If the collagen fibers were arranged rectilinearly, thickening or lengthening could not occur.

Figure 8.4 Peristaltic movement in a nonseptate worm. The many-eyed ribbon worm, *Amphiporus angulatus*, is a nemertine (Phylum Nemertea; discussed in Chapter 22). Unlike earthworms and other annelids, nemertines (and many other worms) are not segmented (i.e., not divided into segments by internal septa). These animals move as alternate contractions of circular and longitudinal muscles in the body wall pass in waves from anterior to posterior. Circular muscles contract first, constricting the anterior end of the animal and squeezing the front end forward. Contraction of the longitudinal muscles follows, shortening and expanding the body. As a result of the alternating action of the two sets of muscles, strong though irregular waves of peristalsis are transmitted from the front to the rear of the worm. In *Amphiporus*, the muscles exert force against a hydrostatic skeleton produced by the intercellular fluids in connective tissues between the gut and the body wall.

tion of the longitudinal muscles as the circular muscles relax has an opposite effect, shortening an animal with a simultaneous increase in body diameter. If both sets of muscles contract equally along the whole body wall, the body becomes more rigid. Alternate contractions and relaxations of circular and longitudinal muscles in backward-directed waves produce **peristaltic movement**, which is a type of hydraulic locomotion. Many worms with fluid-filled bodies can crawl or burrow by peristalsis. In this process, circular muscles contract and relax in a wave beginning at the anterior end. The circular wave pushes the front end forward. A wave of longitudinal muscle contractions follows the circular wave. This broadens the worm, making it touch and push against the substrate or anchor against the sides of a burrow (Figure 8.4).

A significant advantage of hydrostatic skeletons is that, unlike rigid skeletons, they add little weight to an animal, and little extra energy is needed to carry them around. However, hydrostatic skeletons also have some disadvantages. It is difficult for an animal with a hydrostatic skeleton to make precise local movements because fluids transmit pressures in all directions. Contraction of muscles in one body region may stretch muscles elsewhere, caus-

ing them to work against the increased pressure. This is not energy-efficient and tends to make movements based on hydrostatics rather slow. Furthermore, small wounds can lead to loss of fluid and impair locomotor capability.

SURFACE AND SUPPORT SYSTEMS

This section compares the structure and function of body surfaces and of rigid and hydrostatic skeletons as they occur in several types of animals. As you will see, animals have met the challenge of supporting and protecting their bodies in a wide variety of ways.

Animals Whose Main Support Is a Hydrostatic Skeleton

Cnidarians. Many of these animals are essentially a body wall only a few cell layers thick. A **gastrodermis** lines the gastrovascular cavity, and an epidermis covers the outside of the organism. In some cnidarians, such as *Hydra*, these cell layers are separated by a gel-like **mesoglea** containing collagen fibers that pass spirally around the body (Figure 8.5). (Details of the *Hydra* body wall are given in Chapter 21, p. 495; Figure 21.9.) Because it is contained by the body wall, fluid in the gastrovascular cavity provides hydrostatic sup-

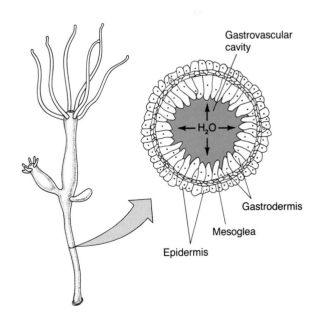

Figure 8.5 Organization of the body wall in *Hydra*. Fluids in the gastrovascular cavity act as a hydrostatic skeleton.

port. Moreover, the epidermis in some cnidarians secretes a delicate exoskeleton called the **perisarc** (see Figure 21.2a). The perisarc is composed of chitin, which is a class of structural carbohydrates common in exoskeletons (see arthropod chitin, Figure 8.8). Corals are cnidarians that secrete a massive calcareous exoskeleton. (Corals and coral reefs are discussed in Chapter 21, p. 496.)

Flatworms. Connective tissue and interstitial fluid are the main support elements in these acoelomate (cavityless) animals. Body movement results when body wall muscles contract, exerting forces on the interstitial fluids. Free-living flatworms are mobile bottom dwellers in aquatic habitats. Their body covering is a one-cell-thick epidermis anchored in a fibrous basal lamina that provides some body support (see Figure 22.7). Ventral cilia are also important in locomotion. The body covering of parasitic flukes and tapeworms is a complex syncytium called a **tegument**. The tegument provides some body support, but its main functions are nutrient ingestion and protection against digestion by host enzymes. (Tegument structure and function are dealt with in Chapter 22, p. 518; Figure 22.8.)

Nematodes. Roundworms have a one-cell-thick epidermis that secretes a multilayered cuticle consisting mainly of spiral layers of collagen fibers. (Structural and functional details of the roundworm cuticle are given in Chapter 23, p. 538; Figure 23.4.) The fibers are not elastic, but the layers are oriented at angles to each other so that when underlying muscles contract, the fiber layers shift over each other, allowing the body to bend. The body wall holds fluid in the body cavity (pseudocoel) under high pressures of up to about 120 mm Hg, and this provides strong hydroskeletal support.

Thrashing movements result from longitudinal muscles in the body wall working against the high pseudocoelomic fluid pressure. There are no circular muscles in the body wall.

Annelids. Segmented worms also have a flexible cuticle secreted by an underlying one-cell-thick epidermis. Bristlelike **setae** are produced by specialized cells in the epidermis. The cuticle is composed of crisscrossed layers of collagen fibers, but it is not as thick, relative to body size, as the nematode cuticle. And unlike nematodes, annelids have a segmented body wall, consisting of the cuticle and both circular and longitudinal muscle layers. Internal cross walls called **septa** divide the body cavity into compartments corresponding to the external annulations (Figure 8.6). Fluids in the partitioned body cavity provide hydrostatic support. In some annelids, especially marine burrowers and tube dwellers, the septa are not complete, and fluid can be forced from one body segment to the next. Waves of contraction of body wall muscles produce typical peristaltic locomotion involving the whole body. By contrast, earthworms (and many other annelids) have complete internal septa that largely prevent fluids from flowing from one segment to the next. The septa enable an earthworm to maintain different fluid pressures in different parts of its body; and with its own set of circular and longitudinal muscles and its own nerve supply, each segment or group of segments may function as a locomotor unit. Because of this ability to localize fluid pressure, earthworms have a more extensive range of movements than have unseptate worms (such as the nemertine in Figure 8.4), as anyone who has ever put an earthworm on a fish hook can appreciate. (For more on annelid movement, see Chapter 25, p. 591.)

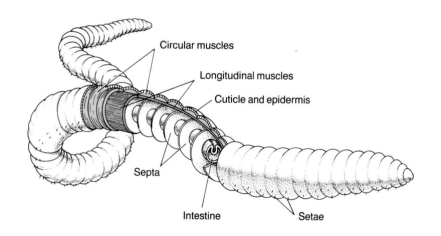

Circular muscles

Longitudinal muscles

Cuticle and epidermis

Septa

Intestine

Setae

Figure 8.6 Components of the hydrostatic skeleton in an earthworm. Longitudinal and circular muscles can exert forces on fluids in the body cavity (coelom). The coelom is divided into compartments by septa, corresponding to the body segments.

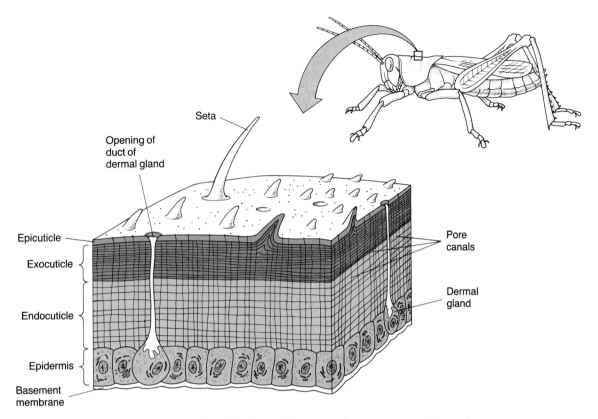

Figure 8.7 The arthropod cuticle. This multilayered exoskeleton consists of chitin and proteins secreted by the underlying epidermis. The thin epicuticle is impregnated with waterproofing waxes produced by epidermal cells. The epidermis also secretes a thin underlying layer, the basement membrane. [This is structurally equivalent to the basal lamina of many other animal tissues, but entomologists (insect specialists) prefer to use the term *basement membrane*.]

Animals Whose Main Support Is an Exoskeleton

Arthropods. A rigid, jointed exoskeleton is characteristic of this diverse group, which includes animals as different as crabs, millipedes, and insects. (In fact, the phylum name *Arthropoda* means "jointed foot.") The arthropod exoskeleton consists of a thick, multilayered cuticle secreted by a one-cell-thick epidermis. Special cells called **trichogen cells** in the epidermis produce bristlelike setae, and **dermal gland cells**, whose secretions have yet to be determined precisely, have ducts that pass to the body surface. Layers in the cuticle include a thick, innermost endocuticle, a somewhat thinner middle layer called the exocuticle, and a thin surface coat, the epicuticle (Figure 8.7).

About 30%–50% of the arthropod cuticle is made up of **chitin**, a long, flexible carbohydrate similar to cellulose (Figure 8.8). Chitin molecules combine to form fibrils that are laid down in oblique layers with different orientations, giving

the cuticle excellent impact strength. The **endocuticle** and **exocuticle** consist of a composite material analogous to fiberglass; chitin fibrils are embedded in a mass of several kinds of proteins. In the exocuticle, the protein molecules are chemically linked to one another. The epidermis produces and releases chemicals called *phenols* into the cuticle. The phenols become oxidized to *quinones*, which bond to protein molecules in the cuticle. By cross-linking neighboring protein molecules, the quinones convert the cuticle's soft protein matrix into the hard, often darkly colored exoskeleton. Cuticular hardness is directly related to the amount of protein cross-linking. Hardened plates of cuticle are called **sclerites**, and the hardening process is called **sclerotization** or **tanning**. In crustaceans, such as crabs and lobsters, protein hardening is augmented by calcium carbonate and calcium phosphate deposition in the endo- and exocuticle.

Different parts of an arthropod's exoskeleton contain varying proportions of chitin, hardened

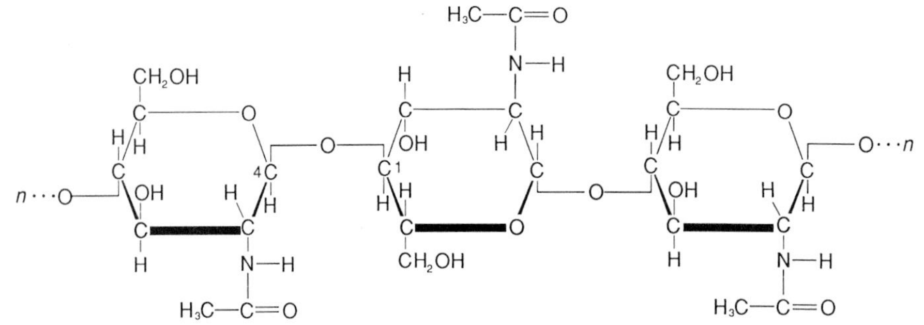

Figure 8.8 Chemical structure of chitin. This important constituent of arthropod cuticles consists of thousands of repeating units of *N*-acetylglucosamine covalently linked between carbon atoms 1 and 4, as shown. *N*-acetylglucosamine consists of a glucose backbone (making it similar to cellulose, the structural compound of plants) with a nitrogen-containing side group. These polymers form long fibrils.

(quinone-linked) proteins, and/or calcium salts. At moveable joints, the cuticle is flexible and contains little, if any, hard material. In contrast, cuticle of the head and mouthparts typically has a high proportion of sclerotized proteins and is quite rigid. In body areas where strength is required, cuticular regions may fuse, producing rigid sclerites. Some parts of hardened cuticle project into the interior of the body, forming knobs and platelike extensions, called **apodemes**, that serve as points for muscle attachment.

The **epicuticle** consists mainly of proteins, with water-repellent waxes in many arthropods. Many crustaceans also have calcium salts in this layer. A waxy epicuticle is an important adaptation in terrestrial arthropods. As small animals with high surface-area-to-volume ratios, they face constant water loss through the body surface. The waxy outer coat greatly reduces water loss and is one of the main reasons for the success of arthropods on land. Fine pumice powders are sometimes used as insecticides because they abrade the waxy layer of the epicuticle, so the insects dehydrate and die.

The arthropod cuticle is literally riddled with **pore canals** (see Figure 8.7). A cockroach cuticle, for instance, contains up to 12 million canals per square millimeter. Into each canal, a cytoplasmic strand projects from an epidermal cell, with each cell sending up 50 to 70 strands. Waxes and other materials synthesized by epidermal cells move to the body surface through these canals.

As mentioned earlier, arthropods solve the growth-limiting problem of the exoskeleton by **molting** a number of times as they develop into adults. (The process of molting, or shedding the exoskeleton, is also called **ecdysis**.) Growth thus appears to occur intermittently, rather than more or less continuously as in animals with endoskele-

tons. The number of molts varies with the species. Insects and many other arthropods stop growing and molting once they are sexually mature.

The molting process has been well studied in certain insects and crustaceans. During each developmental stage, the animal puts on weight faster and faster until body tissues fill the available space inside the exoskeleton. Then, just prior to molting, the animal stops eating. In preparation for a molt, the epidermis secretes a new epicuticle and an enzymatic solution called **molting fluid** under the old endocuticle. The epidermis secretes new cuticle, called **procuticle**, as the molting fluid dissolves the old endocuticle. The dissolved materials are absorbed and reused in the synthesis of new cuticle. After the old endocuticle has been dissolved, the arthropod swallows air or water, depending on the species. This causes splitting of the old exocuticle, and the animal moves forward out of the old exoskeleton. It is now encased in a soft, light-colored procuticle, covered by a thin epicuticle. The arthropod then swallows more water or air, expanding the new cuticle to create space for further growth. Deposition of new cuticle continues for several hours, during which the outer part of the procuticle is sclerotized to form the exocuticle. Growth then continues until the space within the new cuticle is filled (Figure 8.9).

Molluscs. A distinguishing characteristic of many, but not all, molluscs, is an external, rigid, calcareous shell that protects the body and provides attachment sites for muscles. The shell is produced by epidermal cells on the outside of the **mantle**, a living, sheetlike extension of the body wall. In many molluscs, the shell consists of three layers. The outer layer, called the **periostracum**, is composed of a tough protein called **conchiolin**. The

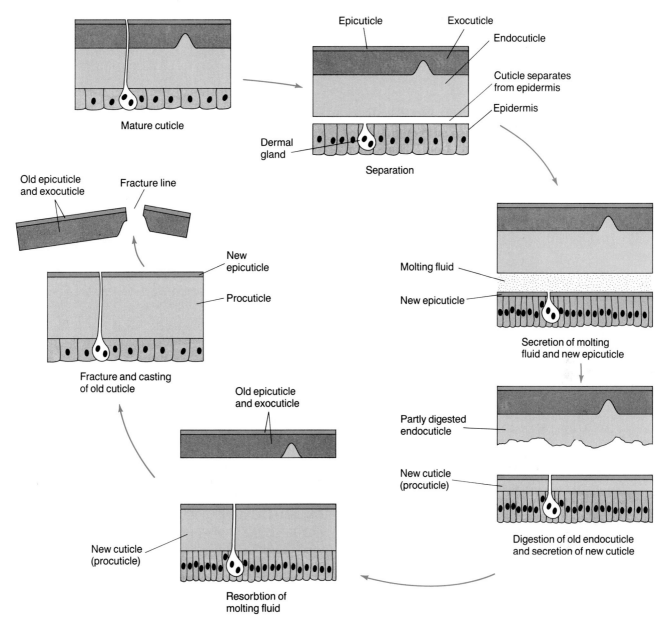

Figure 8.9 The major events in the molting process. Here the process is illustrated in a series of cross-sectional diagrams of the cuticle and epidermis of an insect. The first diagram, that of mature cuticle, is similar to the cuticle shown in Figure 8.7. As molting ensues, the old cuticle is digested from the inside out, and the new cuticle, at first called a procuticle, is secreted before the remnants of the old one are cast off. As a result, the insect is never without an exoskeleton.

periostracum protects the calcium carbonate of the underlying **prismatic layer** from erosion by acid in the water. Closest to the mantle is the lustrous **nacreous layer**, composed of alternating strata of calcium carbonate and proteoglycans (Figure 8.10). (Protein-carbohydrate complexes are discussed in Chapter 2, p. 35.) The proteoglycan layers form an organic matrix that resists tension in the same way that collagen fibers resist tension in bone. The shell grows in diameter as organic matrix and calcium carbonate are added to its outer edge. Growth in thickness occurs when material is added to the nacreous layer by epidermal cells located over the entire outer surface of the mantle. (See the Essay on p. 183 on calcification.)

Animals Whose Main Support Is an Endoskeleton

Sponges. The body of a typical sponge consists of two cell layers. A layer of flattened cells called **pinacocytes** covers the body and may line internal

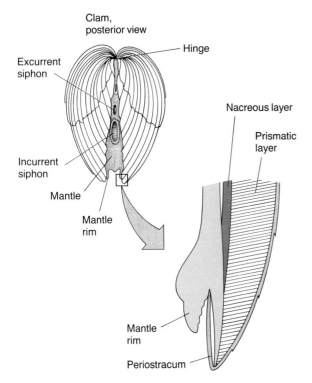

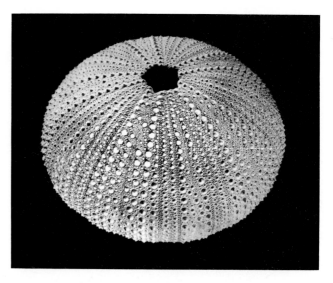

Figure 8.11 The test of a sea urchin. Echinoderms have an endoskeleton produced by the dermis of their skin. In sea urchins, endoskeletal plates form a globose test, or case, that is covered with a thin epidermis. The test contains all body organs. Movable spines (removed in this preparation) project from the endoskeleton.

Figure 8.10 Structure of a clam shell. The three primary layers of the clam shell are the periostracum, the prismatic layer, and the nacreous layer. The mantle, which produces the shell, lies between the shell and the main body mass. In clams and other bivalves, the mantle completely lines the inside of the shell except at the edges, where the valves meet.

canals through which water circulates. Internal chambers in which filter feeding occurs are lined by a layer of flagellated cells called **choanocytes**. A gelatinous zone called the **mesohyl**, located between the outer and inner cell layers, contains wandering amoebocytes. Some amoebocytes secrete endoskeletal elements, which are either hard **spicules** or soft, "spongy" protein fibers called ▷ **spongin**. (The structure and formation of sponge skeletons are discussed in Chapter 20, p. 477.)

Echinoderms. The integument of these animals consists of a thin, usually ciliated epidermis and an underlying connective tissue dermis. The dermis produces the skeletal elements, which are called **ossicles**. The ossicles are composed of magnesium carbonate and calcium carbonate crystals, and in many echinoderms, they are bound together by collagen fibers. Sea urchins have platelike ossicles that form a rigid, hemispherical endoskeleton called a **test** (Figure 8.11). In contrast, the ossicles of many sea stars are smaller, blocklike elements with movable articulations. Sea stars are able to bend and contort themselves into many positions

associated with feeding and locomotion. Both sea urchins and sea stars have spines that project from the ossicles. The spines of most sea urchins are set in sockets and are movable, whereas most sea star spines are extensions of the ossicles and are not movable (see Figure 5.2). Echinoderm spines are ◁ usually covered by the epidermis.

Chordates. A distinguishing characteristic of the Phylum Chordata is the development of an endoskeletal structure called the **notochord** at some stage in the animal's life cycle. The notochord is a semirigid rod that extends for most of the length of the animal beneath the dorsal nerve cord. In most chordates, the notochord is composed of large, fluid-filled cells and is encased in concentric sheaths of connective tissue (Figure 8.12). Rigidity results from the sheaths and from the hydrostatic pressure of the fluid in the inner cells. The notochord occurs in the early embryological stages of all chordates. Most likely, it was the main supportive element of early free-swimming ancestors of the phylum, but it persists as a continuous rod in the adults of only a few living chordates (e.g., lancelets, lampreys and other jawless vertebrates, and a few fishes). In most chordates, the notochord is replaced by the **vertebral column**, a segmented rod that stiffens the longitudinal axis of the animal, surrounds and protects the spinal cord, and allows a greater range of motion than the notochord.

The typical vertebrate skeleton has two main components: the **axial** skeleton, composed of the

Essay: Calcification—How Some Animals Make Hard Parts

A great variety of organisms, including plants, protozoa, cnidarians, molluscs, and vertebrates, synthesize calcareous skeletal materials, often of precise shapes and sizes (Figure E1). The process of biological calcification is com-

(a)

(b)

Figure E1 Complexity of skeletal structure. Many animals produce complex, ornately beautiful skeletal structures. (a) Top and side views of a snail *(Thaumatodon)* shell from the Fiji Islands. This species and many other small land snails that live in decaying leaves and in the upper few centimeters of soil have deeply sculptured shells. (b) The scanning electron microscope reveals the intricate microsculpturing between precisely spaced ribs of the shell of a closely related land snail *(Ptychodon)* from New Zealand. The ribs and microsculpturing between the ribs reduce the amount of area on the shell that has contact with the environment, helping to keep bits of wet debris from sticking to the shell.

plicated and requires considerable expense of energy. The organism must acquire calcium and other ions from the environment, often against a concentration gradient, transport them to the site of deposition, and provide an environment in which calcium carbonate ($CaCO_3$) and other calcium salts will precipitate. The process must also be regulated so that material precipitates at the proper rate and in the species-specific form of the skeleton.

Biological calcification is not well understood and is the subject of much current research. In recent work with molluscs, researchers have extracted the organic matrix of shells to study calcium carbonate deposition. Matrix is obtained by soaking shells in dilute acids to dissolve the calcium carbonate. By treating small pieces of isolated matrix with solutions containing calcium and carbonate ions at concentrations found in molluscan epidermal cells, researchers can, with the aid of a microscope, monitor the formation of calcium carbonate crystals in the matrix. Researchers postulate that the matrix plays several roles in the calcification process. It may provide the sites and the proper chemical microenvironment in which calcium carbonate crystals can form. And it may regulate the growth of crystals so that specific skeletal shapes result.

Although many questions remain, biochemical analyses, electron microscopy, and studies of cultured cells indicate that several events occur as $CaCO_3$ crystals form shell material in molluscs. First, epidermal cells in a groove at the edge of the mantle secrete proteins and proteoglycans. Some of these molecules form new periostracum (see Figure 8.10), while others (mainly proteoglycans) form a surface that attracts and/or binds calcium ions and provides loci or nucleation sites where crystallization begins. Recall from Chapter 2 that proteoglycans consist of a protein core from which carbohydrate molecules project. The specific structure of these molecules is significant, because calcium ions chemically bind to sulfate groups on the carbohydrates of the proteoglycans.

In addition to secreting proteins and proteoglycans, the epidermal cells also accumulate calcium ions from the aquatic environment and secrete them at high concentrations at the nucleation sites. (Land animals acquire calcium in food or in drinking water, and calcium ions are transported via the circulatory system to the site of skeletal deposition.) The epidermal cells also contain the enzyme carbonic anhydrase, which catalyzes the following reactions:

$$CO_2 + H_2O \underset{\text{Carbonic}}{\overset{}{\rightleftharpoons}} H_2CO_3 \rightleftharpoons$$

$$\overset{\text{anhydrase}}{H^+ + HCO_3^-} \rightleftharpoons 2\,H^+ + CO_3^{2-}$$

Carbonate

The carbonate ions react with the calcium ions bound to the proteoglycans, and $CaCO_3$ precipitates out of solution. Thus, the position of the proteoglycans in the matrix determines where calcium carbonate will be deposited and consequently the shape of the skeleton.

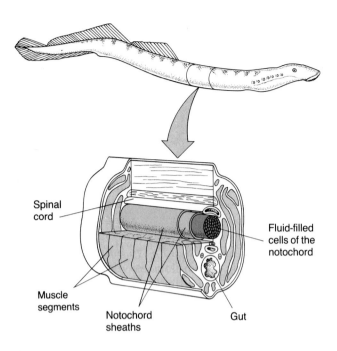

Figure 8.12 Structure of the notochord in a lamprey. The notochord is stiffened by connective tissue sheaths and by the hydrostatic pressure of the cellular fluids. The notochord prevents the animal from shortening when longitudinal muscles contract.

skull, vertebral column, and ribs, and the **appendicular** skeleton supporting the paired appendages. The **pectoral** (shoulder) **girdle** supports the forelimbs (fins, arms, or legs), and the **pelvic girdle** supports the hindlimbs. Paired appendages and the appendicular skeleton evolved in aquatic vertebrates as adaptations providing greater stability in the water. Because they are largely supported by the surrounding water, sharks, rays, and bony fishes have skeletons more adapted for locomotion than for support against gravity. Their paired fins are delicate structures used as steering and balancing organs, and they do not have to bear the load of the body's weight. In contrast, terrestrial vertebrates (amphibians, reptiles, birds, and mammals) have skeletons that hold the body up off land and provide support for various forms of locomotion (Figure 8.13).

Vertebrate skeletons are **articulated**; that is, they contain joints (also called **articulations**) between the bones or cartilages. Immovable joints, such as those between bones in the skull, are called **sutures**. There are several types of movable joints, as shown in Figure 8.14. Skeletal parts are held to-

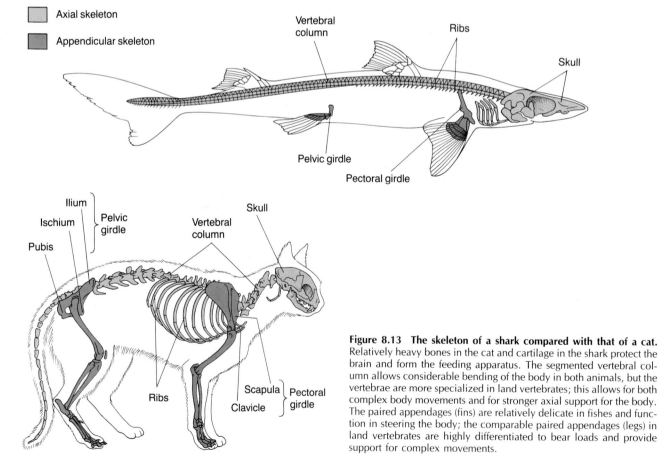

Figure 8.13 The skeleton of a shark compared with that of a cat. Relatively heavy bones in the cat and cartilage in the shark protect the brain and form the feeding apparatus. The segmented vertebral column allows considerable bending of the body in both animals, but the vertebrae are more specialized in land vertebrates; this allows for both complex body movements and for stronger axial support for the body. The paired appendages (fins) are relatively delicate in fishes and function in steering the body; the comparable paired appendages (legs) in land vertebrates are highly differentiated to bear loads and provide support for complex movements.

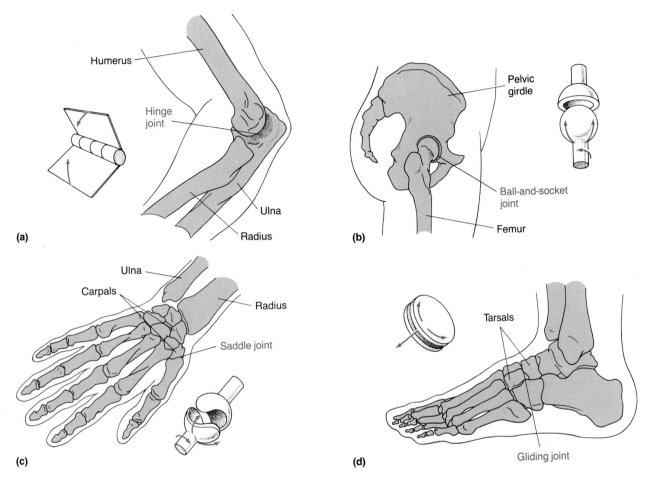

Figure 8.14   Types of joints in the skeletons of terrestrial vertebrates. (a) Hinge joints, found in the elbow and the knee, permit movement in one plane only. (b) Ball-and-socket joints, found in the shoulder and the hip, allow rotation and movement in several planes. (c) A saddle joint, which occurs in the human thumb, permits movement in two planes. (d) Gliding joints between tarsal bones in the ankle, carpal bones in the wrist, and vertebrae in the vertebral column permit rotation and movement in one plane.

gether at movable joints by **ligaments**, whereas **tendons** attach muscles to the skeleton. Joints capable of motion in more than one plane generally have several ligaments and muscles that span the joint and keep it aligned. Depending on where muscles attach relative to the joint, skeletal elements may move with a large force through a short distance or with less force over a longer distance. (For a discussion of levers in movement and locomotion, see Chapter 9, p. 211.)

The ends of bones in movable joints are modified to reduce frictional wear. The ends of articulating long bones, such as the femur, ulna, and radius shown in Figure 8.14, are expanded and covered with smooth **articular cartilage**. Joints are lubricated by a protein solution called **synovial fluid**. Exactly how this lubrication works is a subject of some debate. Two types of lubrication are possible. In *film* lubrication, fluid is trapped between the

two surfaces, keeping them apart. Friction is reduced because the surfaces do not touch. In *boundary* lubrication, the surfaces have lubricant molecules bound to them, keeping the surfaces apart. Joints are probably lubricated both ways, by film lubrication during light movement and by boundary lubrication during heavy loading. A lubricant bound or adsorbed to the bone surfaces is more efficient during loading stress because a film lubricant would be squeezed out from between the closely appressed surfaces.

The skeletal structure of appendages reflects an animal's mode of locomotion, and there are many interesting examples of alternative solutions to similar problems among the vertebrates. Bats and birds, for example, have wings. As shown in Figure 8.15, the bat wing is a modification of the vertebrate hand, while in birds the hand is much reduced, and it is largely the arm that supports the

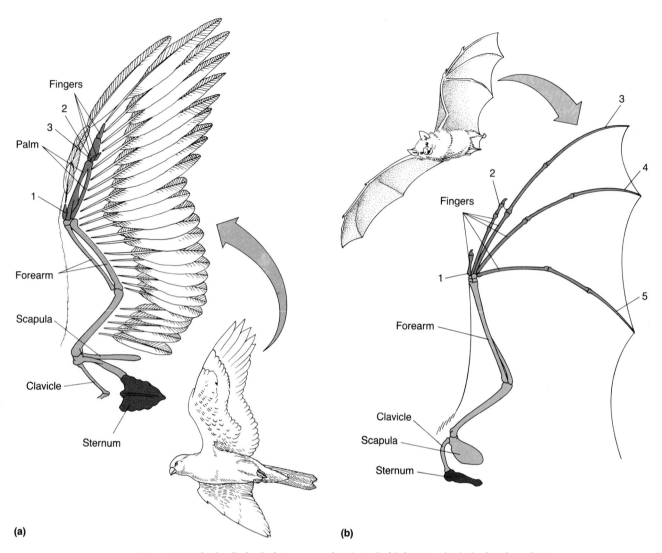

Figure 8.15 The forelimb of a bat compared to that of a bird. (a) In a bird, the first three digits (fingers) are reduced, and the fourth and fifth are absent. Most of the wing feathers are supported by the palm and forearm. (b) In bats, well-developed digits provide much of the membranous wing's structural support.

wing. The adaptive features of skeletons in various vertebrate groups are discussed in greater detail in Chapters 28 through 33.

Functions of the Vertebrate Skeleton

Supporting the Load of the Body. Architecture is important in determining how well a skeleton supports the body mass. The skeleton of amphibians and reptiles is not as effective as that of most birds and mammals in holding the body up off the ground. Amphibian and most reptile limbs project somewhat laterally from the body. In contrast, the legs of most mammals and birds provide much more effective support because they are positioned directly under the body. (Evolution of skeletal support and locomotor efficiency in the vertebrate classes are considered in Chapter 28, p. 689.)

For many centuries, bridge builders have recognized the value of the arch for supporting suspended loads. Likewise, the arched back of a cat or rat is well suited for supporting the weight of the viscera suspended below it (Figure 8.16a). But what happens when the arched vertebral column is made to bear loads in a vertical position, as in the human? Humans evolved from four-footed (quadrupedal) ancestors, and the human skeleton has adjusted to its upright position. In this position, the load is not evenly distributed along the vertebral column. Instead, the line of gravity is centered on the lower back and hip joints, and virtually all the weight is carried in this vertical plane. The imper-

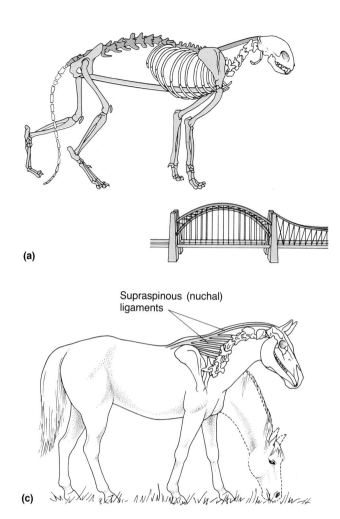

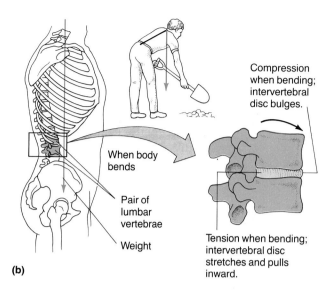

(b)

Figure 8.16 Architectural features of vertebrate skeletons. (a) The skeleton of a cat has features resembling the arched structure of a suspension bridge. The arched backbone increases the load-bearing capacity of the axial skeleton. The weight of the viscera tends to flatten the arched back and push the legs apart, but the abdominal muscles pull the ends of the arch together. This is an example of how the skeleton and muscles act together to bear loads. Note how the neck is held out in front of the body by muscles. (b) The upright position of the human does not allow the skeleton to function as effectively. The body weight is not distributed along the length of the arched back, but is borne in the vertical plane. Consequently, forces are concentrated on the lower vertebrae; when the body bends, as in the act of lifting an object, the cartilaginous discs between the lumbar vertebrae undergo compression and tension that may be great enough to cause injury. (c) Grazing mammals have nuchal ligaments (supraspinous ligaments) in the neck region that elevate the head in front of the body with little or no expenditure of energy.

fect adaptation of a vertebral column (originally evolved to bear loads suspended along its entire length) to the rather awkward upright position largely explains why so many people are troubled with lower-back problems (Figure 8.16b).

Quadrupeds typically hold their head out in front of the trunk. Carrying the head in this position, rather than vertically over the body, is adaptive for feeding. It also brings the main sense organs into contact with the environment ahead of the main body mass, but it may require considerable expense of energy. Many mammals, especially those with a long neck, have well-developed **nuchal (supraspinous) ligaments**, heavy, fibrous structures extending from the back of the skull to the dorsal tips of the cervical (neck) vertebrae. In grazing mammals, such as horses, cattle, and deer, the nuchal ligaments consist largely of elastic fibers (composed of the rubbery glycoprotein **elastin**) that return to their original length after stretching (Figure 8.16c). The head is thus held in front of the body by an elastic tissue spring. Muscle contraction

stretches the nuchal ligaments when the animal grazes; but when the muscles relax, the ligaments pull the head back to its original position with little additional energy expense.

In considering load bearing, it is necessary to consider muscles along with the skeletal system. Bones bear loads in compression and resist shear and torsion, while muscles bear loads in tension. Consider, for example, the loads borne by the femur, the upper leg bone of vertebrates. When a vertebrate runs and jumps, the head of the femur sustains considerable strain. In the human, the angle at which the femur joins the pelvis tends to make the loaded femur bow outward (Figure 8.17). If the femur were to bend too much, it would break. Bending is prevented by muscles and tendons that run from the ilium laterally along the femur to just below the knee. When these muscles contract, they tend to bow the femur inward, counteracting any outward bending caused by loading. Consequently, the bone remains relatively straight and there is little danger of breakage.

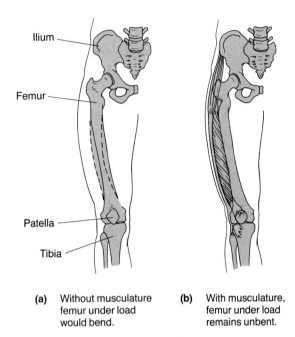

(a) Without musculature femur under load would bend.

(b) With musculature, femur under load remains unbent.

Figure 8.17 How bones and muscles bear loads together. (a) In humans, the weight of the body produces a bending load on the femur when the weight is supported by one leg only, as in jogging. Without musculature, the femur would bend outward. (b) Excessive bending of the femur is prevented by contraction of muscles that run from the ilium of the pelvis to the lower leg. The forces exerted by the muscles may even exceed the loads imposed by body weight.

Protection. There are numerous examples of this function of the vertebrate skeleton. You have already seen in Chapter 5 that bony fishes have a bony cover called the operculum that protects the delicate gills. The vertebrate skull forms a protective case around the brain and inner ears and partly encloses the eyes in deep-set sockets. The ribs of reptiles, birds, and mammals form a protective thoracic cage around the heart and lungs.

Mineral Storage. Bone is the major reservoir for calcium and phosphorus in the vertebrate body. Both of these minerals are essential for metabolism and are stored or released according to tissue needs. During times of mineral deprivation, calcium and phosphorus salts may be released from the bones and pass via the blood to cells throughout the body.

Two hormones maintain calcium salt homeostasis. In terrestrial vertebrates, parathormone, produced by the parathyroid glands, raises blood calcium levels by promoting the release of calcium from bone. Calcitonin, produced by the thyroid gland, has the opposite effect; it lowers blood calcium levels by promoting calcium deposition in bone matrix. In this way, bony matrix is resorbed and replaced throughout the lifetime of an individ-

ual. It is estimated that at any one time, about 3.5% of the bone in a human is undergoing replacement. (Calcium metabolism is discussed in Chapter 11, p. 258.)

Structure of Vertebrate Skeletal Materials

Vertebrate skeletons are composed of connective tissues, including loose and fibrous connective tissues, cartilage, and bone (see Figure 2.27). Cartilage and bone form most of the rigid parts of the vertebrate skeleton.

Cartilage. **Cartilage** is formed by connective tissue cells called **chondrocytes.** These cells secrete cartilage matrix, which consists of collagen fibers and huge molecular aggregates of proteoglycans. Each aggregate consists of 100 or more proteoglycans attached to a long central chain of the polysaccharide **hyaluronic acid** (Figure 8.18). Each proteoglycan consists of a protein molecule with about 100 to 200 polysaccharides of two types projecting from it. One of the side-projecting polysaccharides is **chondroitin sulfate**, and the other is **keratan sulfate** (not to be confused with keratin, the rigid protein in vertebrate skin). Chondroitin and keratan

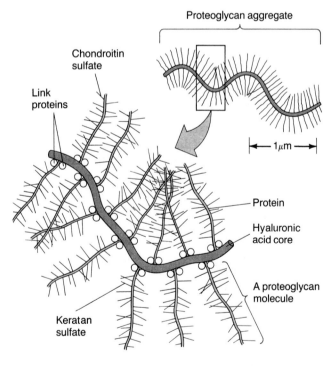

Figure 8.18 Structure of a proteoglycan aggregate in cartilage. These molecular complexes are sometimes as large as bacterial cells and have a molecular weight of over 100 million daltons! Note the hyaluronic acid core with proteoglycan molecules projecting outward. The chondroitin and keratan sulfates in the proteoglycans form hydrogen bonds with water molecules, making cartilage a gel.

sulfates are negatively charged and therefore attract and hold water molecules by hydrogen bonding. Each proteoglycan aggregate traps billions of water molecules, producing a gel-like physical state. Depending on the amount of proteoglycans in the matrix, cartilage can be amorphous or quite stiff. Because of the high water content of its matrix, cartilage is an excellent shock absorber and functions as such between many bones in the body.

When cartilage is compressed, the hydrogen bonds between the water and the chondroitin and keratan sulfates are broken, and the water is displaced. When the pressure is released, the hydrogen bonds are re-formed, restoring the cartilage to its original shape.

Bone. Figure 8.19 illustrates the structure of a long bone, such as a femur or ulna of a vertebrate

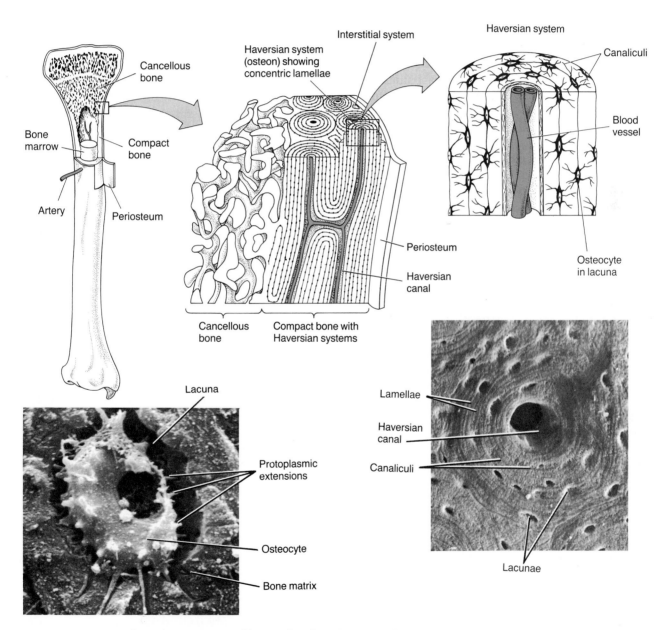

Figure 8.19 Structure of bone. A long bone from a vertebrate limb contains compact and cancellous bone. Successive magnifications show the microscopic structure of compact bone. The matrix of compact bone consists of concentric units called osteons (or haversian systems) fused together by nonconcentric interstitial systems. Osteocytes (bone cells) have cytoplasmic processes that extend into minute canals called canaliculi. The processes aid in the transfer of nutrients and wastes between the blood and the bone cells. The insets show scanning electron micrographs of (left) an osteocyte within a lacuna (5326×) and (right) a haversian system (587×).

limb. Such bones are hollow tubes with expanded hollow ends. The cavities are filled with **bone marrow**. Red marrow is the main site of blood cell production in terrestrial vertebrates. Yellow marrow consists mainly of lipid material. The outer surface of a bone is surrounded by a connective tissue sheath, the **periosteum**, to which tendons and ligaments attach. Cells in the periosteum secrete collagen and hydroxyapatite, increasing the bone's diameter. The shaft of a long bone is called the **diaphysis**, and the expanded, rounded ends are the **epiphyses**. The walls of the diaphysis, surrounding the marrow cavity, are composed of **compact** (dense) **bone** laid down in concentric layers. The epiphyses are composed of a surface layer of compact bone surrounding a central zone of **cancellous** (spongy) **bone**. Cancellous bone is latticelike in appearance, with a strong matrix of hydroxyapatite and collagen fibers.

Like cartilage, bone is a living connective tissue that grows and repairs itself when an injury occurs. Blood and lymph vessels and nerves penetrate bones through **nutrient canals**. Compact bone is arranged into functional subunits called **osteons**, or **haversian systems** (see Figure 8.19). Each osteon consists of a haversian canal containing a blood vessel surrounded by concentric rings of bony matrix. Tiny chambers called **lacunae** in the matrix contain the living bone cells, or **osteocytes**. Microscopic canals called **canaliculi** connect adjacent lacunae and allow nutrients to be distributed to, and wastes to be removed from, the osteocytes. Cancellous bone has a similar structure, but the matrix is arranged irregularly rather than in concentric rings, thereby providing maximum resistance to compression and torsion. It is the continued activity of osteocytes that makes both compact and cancellous bone living tissues.

Bone Formation

In addition to the compact/cancellous categories, vertebrate bone can also be classified on the basis of how it develops. Two developmental types of bone are synthesized by cells called **osteoblasts** that secrete collagen fibers and mineral components of the matrix. Osteoblasts eventually become surrounded by bony matrix and transform into osteocytes in lacunae. **Membrane bone** forms as osteoblasts in fibrous connective tissue secrete matrix materials. Small, irregular masses of mineral matrix eventually fuse to form flattened bones, such as the scales of fishes and certain bones of the skull. A second type of bone, called **endochondral bone**, forms as osteoblasts replace cartilage with bone

matrix. Most of the vertebrate skeleton, including all long bones, are primarily endochondral in origin. Long bones initially form as endochondral bone and grow in length by endochondral ossification, but membrane bone accounts for much of their increase in diameter. Except for their different modes of formation, endochondral and membrane bones cannot be distinguished.

Long bones grow in length at two cartilaginous growth centers called **epiphyseal plates**, located between the diaphysis and the epiphyses (Figure 8.20). On the diaphysis side, cartilage is resorbed and osteoblasts deposit bone. On the epiphysis side, chondroblasts deposit cartilage matrix, thus restoring the cartilage plate that is being consumed from the other side by the bone-forming cells. At the same time, other cells called **osteoclasts** resorb bone, making room for the advancing epiphyseal plate. In this way, the distance between the plates increases and the bone lengthens. At maturity, when bone growth ceases, the epiphyseal plate disappears and the epiphyses unite with the diaphysis.

The Vertebrate Integument

The vertebrate integument consists of two main layers, a basal **dermis**, composed of fibrous connective tissue, and an overlying **epidermis** of epithelial cells. This composite skin protects against abrasion, prevents invasion by microorganisms, blocks excessive fluid loss and gain, and, in many vertebrates, functions in gas exchange, ionic regulation, and excretion. Furthermore, the integument produces structures that serve exoskeletal functions. In terrestrial vertebrates, the epidermis produces hard scales, hooves, nails, beaks, feathers, and hair. Fish scales, which are bony plates, are formed by the dermis.

In all terrestrial vertebrates, living epidermal cells are covered by several layers of dead epidermal cells. The dead cell layers are said to be cornified, or keratinized, meaning that they contain the tough, water-impervious protein **keratin**. The number of cross-linking chemical bonds that occur between sulfur atoms in adjacent keratin molecules is directly related to the hardness of the keratinized layers. The closely packed keratinized layers slow moisture loss and are one reason why vertebrates have been able to colonize terrestrial environments. While the skin of amphibians is only a few cell layers thick and allows water to pass through, the skin of reptiles is heavily keratinized, impregnated with lipids, and covered with overlapping

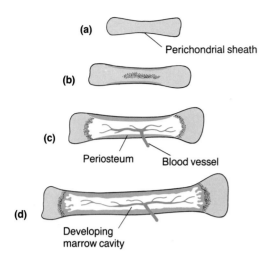

(a)

Perichondrial sheath

(b)

(c)

Periosteum Blood vessel

(d)

Developing
marrow cavity

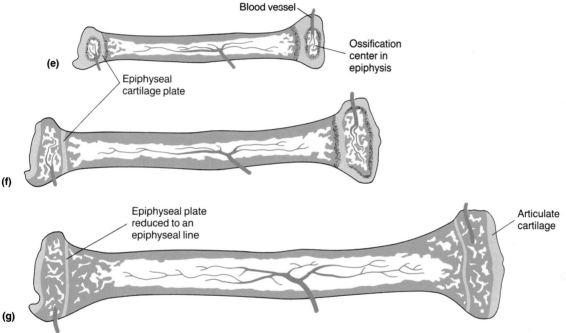

Blood vessel

(e)

Ossification
center in
epiphysis

Epiphyseal
cartilage plate

(f)

(g)

Epiphyseal plate
reduced to an
epiphyseal line

Articulate
cartilage

Figure 8.20 How a long bone grows. (a) In embryonic vertebrates, a long bone such as the humerus (upper arm) or tibia (lower leg) is at first a small cartilage model or precursor surrounded by connective tissue called the perichondrial sheath. (b) Cells at the center of the cartilage precursor begin to enlarge and die as the embryo grows. This leaves cavities whose thin walls calcify. (c, d) Next, blood capillaries invade the perichondrial sheath. In the process, the sheath is converted into periosteum, which becomes a source of bone-forming cells (osteoblasts). Osteoblasts lay down a collar of bone around the cartilaginous shaft, and the shaft grows thicker as bone is deposited on its outer surface. Meanwhile, bone-destroying cells called osteoclasts degrade bone and hardened cartilage on the inner surface of the shaft, creating a marrow cavity. (e) Other bone forming (ossification) centers appear in each epiphysis, separated from the shaft (diaphysis) by cartilage plates. (f) Cells on the diaphyseal side of the cartilage plate lay down bone, while those on the epiphyseal side make cartilage. This causes the bone shaft to grow lengthwise while it continues to increase in thickness. (g) As the animal reaches adult size, ossification of the diaphysis and epiphyses is completed, the epiphyseal cartilage plates are replaced by bone, and bone growth ceases.

protective, epidermal scales. In turtles and crocodiles, bony dermal plates underlie epidermal scales and stiffen the outer body wall into a protective casing. Birds also have epidermal scales on their legs, and their feathers are scalelike epidermal structures that provide protection, thermal insulation, and the airfoil surfaces necessary for flight. Mammals are characterized by hair, which helps to regulate heat exchange and protect skin cells from the damaging effects of ultraviolet light. (Adaptation of vertebrate integuments to aquatic and terrestrial environments is discussed in Chapter 28, p. 693. Also see the Chapter 3 Essay on thermoregulation, p. 73.)

The structure of mammalian skin is shown in Figure 8.21. The epidermis averages about 1 mm in thickness and consists of several discrete layers of epithelial cells. The outermost cornified layer, or **stratum corneum**, consists of dead, flattened cells filled with keratin. These cells slough off continuously and are replaced from below by division of cells in the germinal layer, or **stratum germinativum**. The stratum germinativum consists of cuboidal cells interconnected with desmosomes (see Figure 2.15a). Cells from the stratum germinativum synthesize keratin (which eventually fills their cytoplasmic space) and begin to flatten in an intermediate layer called the **stratum granulosum**.

Mammalian skin has an exceptional number of epidermal derivatives. Sweat glands, oil (sebaceous) glands, and scent (pheromone) glands are all derived from the epidermis, but penetrate into the dermis. Mammary glands and hair, both of which are distinguishing features of mammals, are also

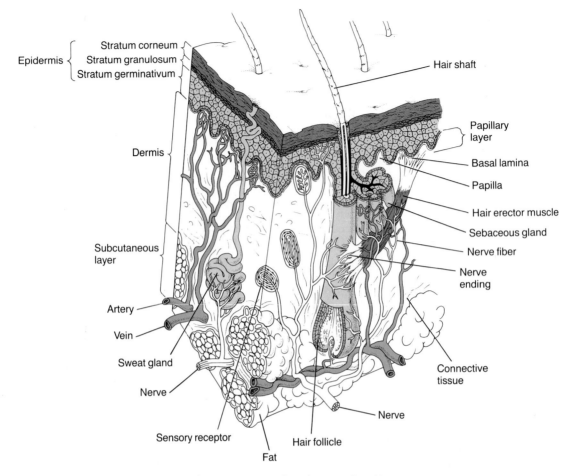

Epidermis {
Stratum corneum
Stratum granulosum
Stratum germinativum

Dermis

Subcutaneous layer

Artery

Vein

Sweat gland

Nerve

Sensory receptor

Fat

Hair follicle

Nerve

Hair shaft

Papillary layer

Basal lamina

Papilla

Hair erector muscle

Sebaceous gland

Nerve fiber

Nerve ending

Connective tissue

Figure 8.21. A section of mammalian skin.

epidermal derivatives. As shown in Figure 8.21, hair follicles penetrate deep into the dermis.

A thin basal lamina of collagen fibers synthesized by the basal cells of the stratum germinativum lies between the epidermis and the dermis. The outermost region of the dermis is made up of a layer of loose connective tissue called the **papillary layer**. Tufts (papillae) from the papillary layer bulge into the stratum germinativum and carry blood vessels close to the epidermis, which lacks blood vessels of its own. Many sensory structures, such as those detecting heat, touch, and pain, also occur in the dermis. Beneath the papillary layer is a layer of dense connective tissue containing closely appressed elastic and collagen fibers, giving the skin strength. Leather is made from the dermis of vertebrate skin by a tanning process that produces cross-links among the proteins in the dense fibrous layer.

Underneath the integument and attaching it to underlying muscles and bone is another layer of connective tissue called the **hypodermis** or **subcutaneous layer**. Fat is often deposited here, especially in aquatic birds and mammals. Subcutaneous fat is important as an energy reservoir and as a layer of thermal insulation.

COLORATION IN ANIMALS

Animals range in appearance from translucent, as in certain crustaceans, to opaque and brilliantly colored, as in butterflies and birds. Many animals are drably colored, a property that may help conceal them and allow them to attack prey or escape predators. In many species, bright colors develop with sexual maturity, and color patches play an important role in behavior associated with mating. Colors result from pigments (colored compounds) or from the light-refracting or -scattering properties of certain crystals. Pigments and crystals, sometimes in combination, are often contained in specialized color cells, called **chromatophores**, or they may be extracellular. Some chromatophores are color effectors; that is, they can change shape and consequently alter the appearance of the pigments or crystalline substances they contain.

Five general classes of coloring substances occur in chromatophores in the animal kingdom. Most animals have **melanins**, a group of brown-to-black pigments derived from the enzymatic oxidation of the amino acid tyrosine. Melanins occur in chromatophores called melanocytes, or melanophores. Melanocytes in the basal layers of human skin protect against the harmful effects of ultraviolet radiation in sunlight. Sunburn occurs when ultraviolet light damages cells in the lower layers of the skin. The damaged cells release histamine, a substance that causes local dilation of the capillaries. Skin tanning, very different from the process of leather tanning or the tanning (sclerotization) of the arthropod cuticle, results when sunlight stimulates melanocytes to synthesize more melanin and triggers the development of more melanocytes.

A second class of pigments, called **ommochromes**, vary from brown to red and yellow. They are derived from the amino acid tryptophan. These pigments are not confined to organelles but are dispersed throughout chromatophores. A third group of pigments, the **carotenoids**, are lipidlike and fat-soluble and occur in small droplets in the cytoplasm. Two types of carotenoids are the carotenes (yellow and reddish orange) and the xanthophylls (yellows). The red-yellow coloration of cnidarians, crustaceans, and goldfish result from these pigments. Animals do not synthesize carotenoids. Herbivorous animals obtain these pigments by eating plants, and carnivores obtain them secondarily by eating other animals.

Making up a fourth class of coloring substances are certain crystalline **purines** that yield yellow-white to silver colors. One of these is guanine, the nitrogenous base that occurs in nucleic acids and is an important metabolic waste product in many animals. (Nitrogenous wastes are discussed in Chapter 7, p. 152.) Color-yielding crystals of guanine and certain other purines, including hypoxanthine and uric acid, occur in chromatophores called iridophores. Refraction and scattering of light from these crystals yield the metallic blue and silver colors of many fishes, amphibians, and in-

sects. Animals that color their bodies this way are making use of some of their nontoxic metabolic wastes, rather than simply excreting them.

Structurally similar to the purines, a fifth class of coloring compounds called **pteridines** gives insect wings and eyes and the skin of many fishes and amphibians their yellow-white, orange, and red colors. Some pteridines are pigments, while others are colored or colorless crystals.

Some vertebrates, echinoderms, molluscs, arthropods, and annelids are able to change their colors according to environmental light conditions or as background colors change. Two mechanisms, both involving chromatophores, enable such changes. In molluscs, several small muscles surround each chromatophore. When the muscles are stimulated by nerve impulses, they contract, stretching the chromatophore and spreading the pigment over a larger area, darkening the animal. Chromatophores containing different types of pigments are controlled by separate nerves. Thus, precise color changes can be produced. The nerve-muscle mechanism is rapid, and many molluscs, especially squids, are noted for their ability to change colors quickly.

Color changes in crustaceans, fishes, and reptiles do not result from muscles stretching the chromatophores. In these animals, chromatophores are irregularly shaped cells with many highly branched cytoplasmic processes. Pigment-containing organelles and pigment granules are attached to elements of the cytoskeleton, and by a yet unknown mechanism, the pigments are moved through the cytoplasm of the cell. (The cytoskeleton is discussed in Chapter 2, p. 43; Figure 2.20b.) Pigment dispersal (but not retraction) requires energy, and color changes in animals with this type of chromatophore may take several minutes or longer. In crustaceans, pigment movement is controlled by a hormone produced in the eyestalk. In fishes, amphibians, and reptiles, pigment dispersal is caused by melanocyte-stimulating hormone from the pituitary gland, and adrenaline causes pigment aggregation and blanching. (Hormones and endocrine control mechanisms are discussed in Chapter 11.)

Many animals undergo long-term color changes by producing different pigments. Birds and some mammals molt seasonally and grow new feathers or hair, often of a different color. Tanning of the integument in arthropods and some vertebrates also causes color changes.

☐ S U M M A R Y

1. The epidermis of most invertebrates consists of a single layer of epithelial cells. Epidermal cells in many animals secrete a nonliving cuticle that serves as an exoskeleton, protecting the animal and translating the forces of muscle contraction into directed movements. The cuticle is most highly developed in the arthropods. In corals and molluscs, the epidermis forms a calcium carbonate exoskeleton.

2. Three types of skeletons occur in the animal kingdom. Exoskeletons and endoskeletons are rigid skeletons composed of cell products hardened by chemical reactions or by phosphate and carbonate salts of calcium. Hydrostatic skeletons result from muscles and connective tissues exerting pressure on body fluids. Most animals have more than one skeletal type.

3. Rigid skeletons consist of a matrix of protein and/or polysaccharide fibrils that resist tensile forces and a ground substance that resists compression. Skeletal elements in limbs are usually tubular, giving maximum stiffness per unit weight.

4. Animals with exoskeletons are limited in their growth to the space within the exoskeleton. They must either shed the skeleton periodically, as arthropods do, or increase the skeleton's size, as in the case of molluscs.

5. Sponges, echinoderms, and chordates have endoskeletons. The endoskeleton of terrestrial vertebrates, unlike that of aquatic vertebrates, supports virtually the full weight of the animal. The vertebrate skeleton consists of axial elements (skull, vertebral column, and ribs) and, in most vertebrates, appendicular elements, which support the paired limbs. Vertebrate skeletal materials are fibrous connective tissue, cartilage, and bone, all of which consist of living cells in a matrix. Cartilage matrix consists of collagen fibers and proteoglycan aggregrates; bone matrix consists of collagen and a mineral called hydroxyapatite organized into subunits called osteons. Osteocytes (bone cells) contained in minute chambers called lacunae in the matrix are active in mineral deposition and resorption.

6. The vertebrate integument consists of a multilayered epidermis and a connective tissue dermis. Integumental structures such as scales, feathers, and hair protect terrestrial vertebrates from thermal stress, ultraviolet radiation, abrasion, and water loss. Outer layers of closely packed, dead epidermal cells are filled with the tough protein keratin. Rigid structures derived from the epidermis, such as nails, hooves, beaks, and scales of terrestrial vertebrates, also contain this protein.

7. Animals have five types of coloring substances in cells called chromatophores. Pigments include melanins, carotenoids, ommochromes, and some pteridines. Color-yielding crystals include certain purines and certain pteridines.

☐ F U R T H E R   R E A D I N G

Caplan, A. I. "Cartilage." *Scientific American* 251(1984): 84–94. *Describes the role of proteoglycans in cartilage.*

Currey, J. D. *Animal Skeletons. The Institute of Biology's Studies in Biology No. 22.* London: Edward Arnold, 1970. *A short but excellent account of skeletal biology and mechanics.*

Edelson, R. L., and J. M. Fink. "The Immunologic Function of the Skin." *Scientific American* 252(1985): 46–53. *Explains the role of the skin in the immune response, as well as its role as a passive barrier in protecting vertebrates from invasive microorganisms.*

Hadley, N. F. "Surface Waxes and Integumentary Permeability." *American Scientist* 68(1980): 546–553. *Explains the role of lipids in protecting terrestrial animals and plants from desiccation.*

Hadley, N. F. "The Arthropod Cuticle." *Scientific American* 255(1986): 104–112. *Recent research on this complex, highly adaptable exoskeleton.*

Hildebrand, M., D. M. Bramble, K. F. Liem, and D. B. Wake, eds. *Functional Vertebrate Morphology.* Cambridge, Mass.: Harvard University Press, 1985. *Contains two authoritative chapters on skeletal systems.*

Kreighbaum, E., and K. M. Barthels. *Biomechanics: A Qualitative Approach for Studying Human Movement.* 2d ed. Minneapolis: Burgess, 1985. *Shows lines of force and skeletal positions in various athletic movements.*

Schmidt-Nielson, K. *Scaling: Why Is Animal Size So Important?* New York: Cambridge University Press, 1984. *Lucid discussion of how rates of processes must change in relation to size.*

Shipman, P., A. Walker, and D. Bichell. *The Human Skeleton.* Cambridge, Mass.: Harvard University Press, 1985. *A modern treatise on bone and the structure and function of the human skeleton.*

Vaughn, J. *The Physiology of Bone.* 3d ed. Oxford, Eng.: Clarendon Press, 1981. *Describes the various functions of bone other than in biomechanics.*

CHAPTER 9

Movement and Locomotion

Unusual locomotion in fishes. Airborne for short periods, flying fish skim above the ocean surface. A forceful flip of their tail propels them out of the water, and their large pectoral fins help them glide in the air.

Locomotion, the ability to move from one place to another, and the movement of body parts are so fundamental to the nature of animals that these two characteristics are often used to help define what an animal is. Body movement and various forms of locomotion, such as crawling, walking, swimming, and flying are important in reproduc-tion, protection, the search for food and shelter, and the distribution of species.

Although locomotor structures in animals vary, the mechanisms underlying locomotion and move-ment are similar: All convert chemical energy into mechanical energy; all use a skeleton to transmit the mechanical energy to a propulsive surface; and all use the propulsive surface to apply force against the environment. The nature of the environment determines the suitability of particular locomotor adaptations. Wings used for flying must be large enough to push effectively against air, a medium of

195

low density, while at the same time offsetting the force of gravity. The buoyancy of water makes gravity less of a problem for aquatic animals than it is for fliers, and water's higher density makes relatively small paddles, flippers, and fins ample means of locomotion. However, the viscosity of water impedes forward progress, so body streamlining is a common aquatic adaptation. Terrestrial animals such as the giraffe or kangaroo, however, would benefit very little from streamlining, since they push their bodies through the relatively unresistant air, but skeletal support against gravity is essential. The laws of motion together with the principles of hydrodynamics and aerodynamics determine what locomotor structures are adaptive for each kind of environment.

Newton's laws of motion govern the relationship between applied forces and locomotion. Newton's first law, stated in zoological terms, says that an external force must be applied to an animal to set it in motion. To move by its own efforts, an animal must use a "propellor" of some kind to develop force from the external environment. Octopuses and squids move by jet propulsion. An octopus squirting a jet of water from its mantle cavity is propelled by the outside force that the jet elicits from the surrounding water. Likewise, a hawk's wing elicits propulsive force (as well as buoyancy) from the air. Newton's second law is also relevant. If an animal is to change its speed or direction, an unbalanced force must be applied to the animal; force of a particular size will change the speed of a heavy animal less than it will a light one. Newton's third law, that every action must have an equal and opposite reaction, requires that forward-thrusting animals exert an equal force backward against the environment. An earthworm burrows by forcing its body against the walls of the tunnel as its anterior end presses forward. Similarly, a grasshopper exerts force against the ground each time it hops. Thus, an analysis of locomotion requires detailed information on animal structure, mass, and the size and direction of all forces.

This chapter first surveys locomotor adaptations of single cells and single-celled organisms and then considers muscle-based modes of locomotion.

MOVEMENT AND LOCOMOTION WITHOUT MUSCLES

Only animals have muscles. Protozoa move by means of cell organelles—pseudopodia, flagella, or cilia—and these locomotor structures also have critical locomotor and nonlocomotor functions in many animals.

Amoeboid (Pseudopodial) Movement

A **pseudopodium** is a cytoplasmic projection of a cell, and although movement by means of pseudopodia has been most extensively studied in amoebas, it is widespread in the animal kingdom. Mobile cells called **amoebocytes**, in the body wall of sponges and cnidarians and in the body cavity of many other animals, may perform a variety of functions, including phagocytosis of food particles, gas and nutrient transport, and waste disposal. White blood cells are amoebocytes that play key disease-combating roles in many animals. (The role of lymphocytes in vertebrate immunity is discussed in Chapter 6, p. 139.)

Pseudopodia form while a cell rests on a substrate such as underwater pond vegetation, the lining of a blood vessel, or the body wall of a sponge. Development of a pseudopodium begins with the formation of a smoothly rounded **hyaline cap** of clear, jellylike cytoplasm called **hyaline ectoplasm**. A stream of less viscous cytoplasm, called **endoplasm**, flows into the cap region through a tubelike passageway formed by **granular ectoplasm** beneath the cell surface (Figure 9.1a). The endoplasm carries with it a mixture of mitochondria, various small organelles, and crystals. When the endoplasm reaches the tip of the tube, it spills outward, like water from a fountain. The flaring endoplasm transforms into ectoplasm, which is then added to the front end of the tube. Eventually, the lengthening tube and surrounding plasma membrane become attached to the substrate, and the cell moves forward. At the base of the pseudopodium, ectoplasm transforms into endoplasm, maintaining a rough balance between the two kinds of cytoplasm.

Many questions remain to be answered about pseudopodial movement. Zoologists are still trying to determine what causes the motive force of pseudopodia and to define where the motive force is applied and how it is coupled to forward movement. An old idea that cytoplasm itself can develop motive force by contracting has recently gained experimental support. The electron microscope has revealed that amoebas contain filaments comparable in diameter to the contractile protein filaments of muscle. Biochemical research has confirmed that amoebas' filaments are composed of myosin and actin. It now seems likely that the contraction of cytoplasm in amoeboid movement is caused by

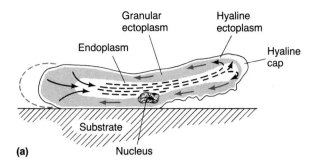

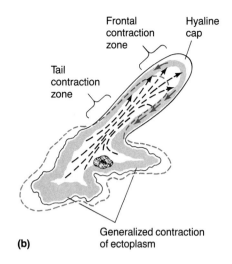

Figure 9.1 Formation of a pseudopodium. (a) Endoplasm flows forward into a forming pseudopodium; as the endoplasm nears the tip of the pseudopodium, the endoplasm transforms into gel-like ectoplasm. The ectoplasm then flows back peripherally toward the base of the pseudopodium and transforms back into endoplasm. (b) Regions where contraction would occur according to the frontal contraction, tail contraction, and generalized contraction hypotheses.

some process akin to the sliding of actin (thin) and myosin (thick) filaments that occurs in muscle contraction (described later in this chapter). This implies that the contraction mechanism in specialized muscle cells of higher animals is an evolutionary adaptation of "old" proteins to a new process and suggests an ancient kinship between animals and protists.

As to where the motive force of a pseudopodium is applied and how it is coupled to forward movement, there are three main hypotheses (Figure 9.1b). The *frontal contraction hypothesis* suggests that contraction of endoplasm at the front of the pseudopodium produces a force that pulls posterior endoplasm forward. The *tail contraction hypothesis* proposes that contraction of ectoplasm at the tail end of the organism, or at the base of a pseudopodium, pushes endoplasm forward into the forming pseudopodium. The *generalized contraction hypothesis* holds that ectoplasmic contraction over the whole body forces endoplasm into the new pseudopodia. Only more testing of these hypotheses can resolve which mechanism is actually involved in pseudopodial movement.

Flagella and Cilia

Locomotion by the bending and whipping movements of **flagella** and **cilia** is important in two major groups of protozoa, the flagellates and the ciliates. As with pseudopodia, these organelles also perform significant tasks in many different metazoans. Many flatworms and other small animals, for example, use cilia in locomotion. And food particles are carried into sponges in water moved by flagellated cells. Likewise, water and food particles are driven across the gills and into the mouths of clams and oysters, and mucus is

driven up the mammalian trachea, by the coordinated beating of cilia on epithelial cells. Because these cells are fixed in place, the forces they produce are transmitted to the surrounding medium. As components of internal organ systems or as means of locomotion, cilia or flagella occur in most animal phyla. They apparently are not prerequisites for success, however, for they are rare in two of the most successful phyla, the Nematoda and Arthropoda.

Flagella and cilia are very similar, although cilia are typically shorter than flagella. A flagellum or cilium consists of a shaft, the **axoneme**, extending from a centriole-like basal body within the cell. The axoneme is covered by the plasmalemma. As revealed by electron microscopy, nearly all axonemes consist of a **9 + 2 arrangement** of microtubules: Nine peripheral doublets of microtubules surrounding two single microtubules run the length of the organelle (Figure 9.2a). The individual microtubules consist of proteins called **tubulins** (Figure 9.2b). Figure 9.2a shows structures called **radial links** extending from the doublets toward projections on the central microtubules. Pairs of hooked arms extend toward neighboring doublets. The hooked arms contain **dynein**, an enzyme that releases energy for movement from ATP.

Given this structure, what causes the bending and whipping movements of cilia and flagella? Does bending occur as microtubule doublets on one side of the axoneme lengthen while those on the other side shorten? Compelling evidence favors another hypothesis—that bending occurs as microtubule doublets slide back and forth relative to each other while their lengths remain constant. This is called the **sliding tubule hypothesis**. In one set of experiments, axonemes were stripped of their outer membranes and any residual ATP. When the

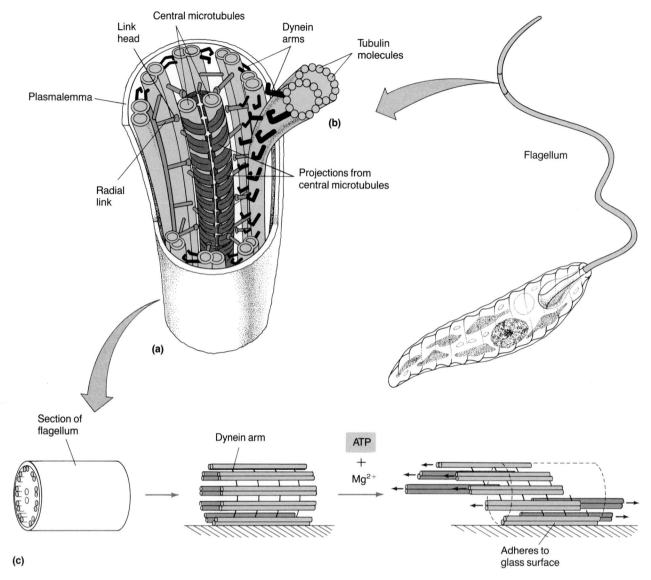

Figure 9.2   Flagellar structure as seen in the protist *Euglena*. (a) Three of the nine microtubule doublets have been cut away in the axoneme to show the central microtubules. Dynein arms form cross-bridges between adjacent microtubules and cause relative sliding of the microtubule doublets past one another. Radial links resist tubule sliding and cause the flagellum to bend. (b) An enlargement of one doublet, showing in cross section that it consists of molecules of the protein tubulin. (c) This diagram shows the microtubule doublets of a flagellum that has been treated with detergent and a proteolytic enzyme to remove the plasmalemma. The microtubules will adhere to glass, and the free ones can be seen to slide past each other when Mg^{2+} and an energy source (ATP) are added to the medium. In life, the flagellum does not adhere to a solid surface, and sliding causes the flagellum to bend. The relative sliding is thought to be caused by alternate formation and breaking of dynein cross-bridges between adjacent doublets.

naked axonemes were placed in a solution containing ATP, they beat normally. Rapid preservation "froze" the bend patterns, and electron microscopy showed that the microtubule doublets had slid past each other. The microtubule doublets on the outside of a bend underwent a greater amount of sliding than other doublets, while the doublets on the inside of a bend remained in place. Other research has shown that the dynein arms form cross-bridges between adjacent doublets. Alternate making and breaking of the cross-bridges is thought to pull one microtubule doublet past another, a process fueled by ATP and remarkably similar to the sliding filament mechanism described later in this chapter for muscles. The relative sliding of microtubules observed in experimental preparations is represented schematically in Figure 9.2c. Of course, sliding of the doublets alone would not cause cilia or flagella to bend or whip. Localized resistance to sliding occurs when groups of radial links bind or jam against projections on the central tubules; this resistance forces the sliding doublets to bend.

Although flagella and cilia share a common internal structure and mechanism of movement, the form or pattern of their beat on protozoan cells is generally different. A typical flagellum beats about 30 times per second. The beat is a flat or helical wave that travels the length of the shaft. As a familiar analogy, consider how waves travel along a rope when it is rapidly snapped at one end. At any given instant, several waves may be in motion along the rope, and if the rope is immersed in water, the waves displace water parallel to the rope's axis. A beating flagellum also displaces water; on a free-swimming protozoan, this generates thrust, or motive force, that propels the cell through the water. Figure 9.3 illustrates the motive forces produced by flagella of different lengths. In a long flagellum, the right and left lateral forces exactly balance, canceling each other; therefore, the entire motive force is straight along the flagellar

axis. Depending on the species of protozoan, the waves may travel from the base of a flagellum to its tip, or from the tip to its base, so flagella may either push or pull flagellates through the water. In short flagella, unbalanced lateral forces remain, so forward motive forces are reduced. This decreases the efficiency of the flagellum in moving a cell forward, but short flagella can move more rapidly than long flagella. Consequently, protozoa with short flagella generally swim faster and have more complex movements than do those with long flagella.

The fastest-moving protozoa are ciliates. In cilia, the lateral forces are entirely unbalanced, and fluid is moved at right angles to the shaft (Figure 9.3c). This is an essential feature of a cilium's action. A cilium has a characteristic cycle of action that includes a power stroke and a recovery stroke (Figure 9.4a). In the power stroke, the shaft is kept straight and stiff, bending only at the base; but on recovery, the cilium is held close to the cell surface. The rapid power stroke maximizes thrust, while the recovery pattern minimizes frictional resistance during return to the original position.

Cilia often occur in fields of thousands on a cell surface. Their beating is coordinated, producing a flow of fluid over the cell surface that propels a

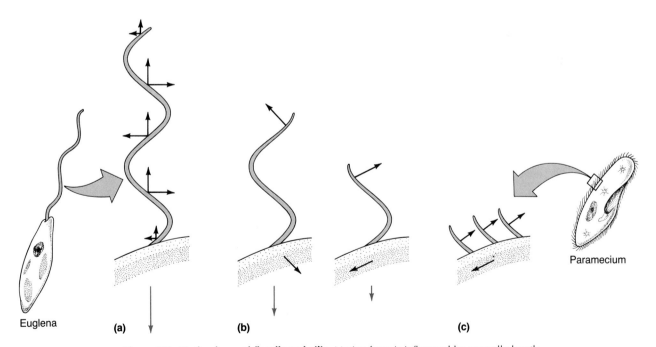

Euglena   **(a)**   **(b)**   **(c)**   Paramecium

Figure 9.3 Motive force of flagella and cilia. Motive force is influenced by organelle length and wave length. In (a), lateral forces balance, so motive force is entirely along the flagellar axis. In (b), lateral forces do not fully balance, so motive force along the axis is diminished, and a motive force at an angle to the axis is added. In (c), a cilium is shown moving water at right angles to the shaft axis because the motive force is limited entirely to unbalanced lateral forces.

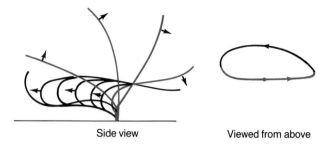

Side view   Viewed from above

(a)

(b)

Figure 9.4 Ciliary action. (a) The cycle of ciliary beating includes a power stroke (in color) and a recovery stroke. (b) This scanning electron micrograph of the surface of the ciliate protozoan *Paramecium multimicronucleatum* shows the waves of coordinated ciliary movement sweeping across the cell's surface (500×).

free-living ciliate in the opposite direction. The waves of synchronized beating can sometimes be seen with a microscope, spreading across a field of cilia like ripples crossing a pond (Figure 9.4b). Because the cilia lie so close together, fluid movement produced by one cilium triggers beating of adjacent cilia. Rows of cilia can become synchronized in this way, resulting in patterns of coordinated movement. The propulsive force of thousands of tiny coordinated cilia is much greater than that of a few long flagella, a factor in the evolutionary success of ciliates, which generally move faster, are larger, and are more diverse than flagellates.

Myonemes

Flagellar and ciliary movement, muscle contraction, and probably even pseudopodial locomotion all operate by a similar mechanism: microtubules or filaments actively sliding over each other. Movement caused by long contractile filaments called **myonemes**, however, involve a different mecha-

nism. Several protozoa are fastened to objects by contractile stalks that contain one or more myonemes. The freshwater ciliate *Vorticella*, for example, lives attached to plants by a thin stalk some 400 μm long (see Figure 19.11). Its stalk is extended for feeding, but a slight disturbance triggers rapid contraction of a myoneme in the stalk; this coils the stalk and pulls the cell down to safety.

Unlike sliding tubule or filament mechanisms, no ATP is needed for myoneme contraction. Where does the energy come from? The chemical potential of calcium appears to be the immediate source of energy. Protein molecules in the extended myoneme are held in an elongate shape by crosslinks that are stable at low calcium levels. When changes occur in the cell membrane, calcium ions rapidly enter from the surrounding water; calcium ions remove the cross-links, the protein molecules collapse into a shorter, more compact shape, and the stalk contracts. After contraction, the cell pumps calcium back out to allow re-formation of the cross-links and extension of the stalk, a process that *does* require ATP and considerable time. A fraction of a second is needed for stalk contraction, but its extension takes several seconds.

The chief adaptive advantage of the myonemal mechanism is its speed. A myoneme can shorten about 30% in 3 milliseconds (ms), four times faster than the fastest muscle. For this reason, the myoneme is probably the best possible escape mechanism for a stalked protozoan.

MUSCLE-BASED MOVEMENT AND LOCOMOTION

Muscles are contractile organs that move body parts. Muscles do not actively extend; they can only actively contract (shorten). Understanding muscle contraction and its control provides a foundation for understanding how animals move, how hearts pump, how food is moved through an animal's gut, and how an artist makes the delicate movements needed to paint a portrait or perform a piece of music.

Kinds of Muscle

Muscle is highly diverse, both structurally and functionally, and terms to describe its diversity are somewhat inadequate. Muscle terminology is also somewhat confused by the tendency in zoology to

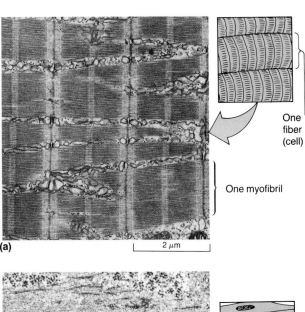

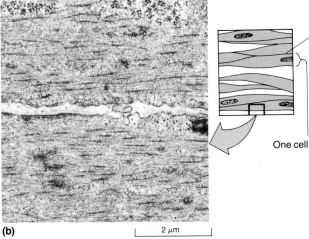

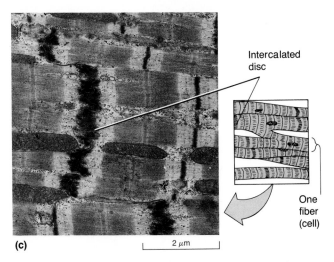

Figure 9.5 Types of muscle fibers. (a) Striated (skeletal) muscle. Note that the striations are in register throughout the fiber. (b) Smooth (visceral) muscle of vertebrates. Note the absence of striations, and note the single nucleus in the spindle-shaped cells (see inset). Many invertebrates have striated muscle in viscera. (c) Cardiac (heart) muscle. Note the striations. Intercalated discs form where the ends of fibers meet; they consist of the closely associated membranes of two adjoining fibers.

make generalizations based on the types of muscle seen in vertebrates. Muscle cells are called **fibers**, and zoologists generally recognize two kinds of muscle on the basis of fiber structure. **Striated muscle** consists of fibers in which the contractile proteins are arranged in repeating sequences, forming alternating dark and light bands called **striations** (Figure 9.5a). **Smooth muscle** has fibers that lack striations because their contractile elements are not organized into regularly repeating units (Figure 9.5b). A typical smooth muscle fiber is spindle-shaped and has a single nucleus. Vertebrates have smooth muscles (often called visceral muscles) in the walls of several viscera, including the stomach, intestines, urinary bladder and blood vessels.

Typically, smooth muscle contracts and relaxes more slowly than does striated muscle, and smooth muscle sometimes contracts and relaxes rhythmically. This is important in the alternating contractions and relaxations typical of peristalsis, the rhythmic process by which food is moved slowly but steadily along a digestive tract at a pace that facilitates digestion and absorption. Another example is the rhythmic contraction of uterine muscles as a mammal gives birth to offspring. Smooth muscle is also resistant to fatigue and may remain contracted for long periods with relatively low energy consumption. Slow and sustained contraction is important in the circulatory system, where muscle fibers in the walls of the arterioles often must remain contracted for long periods, an important element in maintaining blood pressure and regulating blood flow to the skin. In vertebrates, smooth muscle is sometimes called *involuntary* muscle because it is controlled automatically by the nervous system, specifically by nerves of the autonomic division of the central nervous system. (The autonomic nervous system is discussed in Chapter 10, p. 241.)

Nonstriated (smooth) muscle also occurs in several invertebrate groups. It is the predominant muscle type in many flatworms and occurs in the locomotor tube feet in echinoderms (see Figure 27.12). Smooth muscle also forms part of the large muscles that close the valves of clams and other bivalve molluscs.

Striated muscle is much more widespread and diverse than smooth muscle. Its fibers may be uninucleate or syncytial (multinucleate). Multinucleate striated fibers are large cells with nuclei often concentrated peripherally. In typical striated muscle, the alternating bands run from one side of the fiber to the other (see Figure 9.5a), but in some animals the striations are not perfectly aligned with one another. In bivalve molluscs, for instance, the striations follow a spiral course along the fiber. In vertebrates, striated muscle with multinucleate fibers is attached to the skeleton and is called **skeletal muscle**. Because they are not controlled automatically, vertebrate skeletal muscles are also called *voluntary* muscles. Unlike vertebrates, many invertebrates, such as arthropods, have striated muscle both in viscera and attached to the skeleton.

Striated muscle with uninucleate fibers is common in invertebrates but occurs in adult vertebrates only in the heart, where it is called **cardiac muscle**. As shown in Figure 9.5c, cardiac muscle has a distinctive appearance. Its fibers are branched, and where the ends of the fibers meet, the cell membranes mesh with each other to form **intercalated discs**. The cell membranes within an intercalated disc associate intimately with each other, forming two kinds of junctions. Desmosomes hold the cells together, while gap junctions allow electrochemical impulses to spread from one cell to adjacent ones. Cardiac muscle displays rhythmic spontaneous activity. Impulses passing through the gap junctions spread throughout the network of

cardiac cells, causing cardiac muscle as a whole to contract and relax spontaneously and rhythmically. The nervous system and hormones circulating in the blood regulate cardiac muscle's spontaneous activity. (Control of the heartbeat is discussed in Chapter 6, p. 132.)

Organization of Vertebrate Skeletal Muscle

Many years of research have produced a detailed understanding of the structure of vertebrate skeletal (striated) muscle and its mechanism of contraction. Less is known about most invertebrate striated muscle and about smooth muscle, although their mechanism of contraction seems to be similar to that of vertebrate striated muscle. Because vertebrate skeletal muscle has been studied so extensively, we will focus our attention on the organization and function of this type of muscle.

Figure 9.6 shows the anatomical arrangement of some of the major skeletal muscles in two vertebrates. The muscles are attached to the skeleton by tendons. Each muscle has two or more points of attachment: one or more **origins**, which are at the more proximal (closest to the trunk of the body)

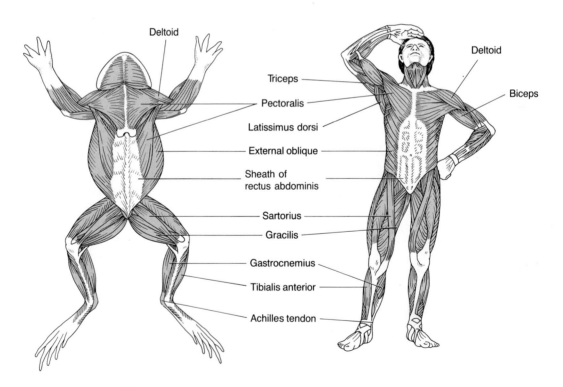

Figure 9.6 Major muscles of two vertebrates. Shown here are the major muscles in humans and frogs. Corresponding muscles, such as the sartorius of each animal, are homologous, meaning that they have similar embryonic and evolutionary origins. Homologies are thought to be an indication of an ancestral relationship.

and immovable ends of the muscle, and an **insertion**, which is at the more distal and movable end. For example, the gastrocnemius muscle is inserted on the heel via the Achilles tendon and has its origins on the femur (upper leg bone). Typically, the muscles are arranged in antagonistic pairs with opposing actions; thus, the triceps straightens and the biceps bends the forelimb at the elbow joint. As either muscle of a pair contracts (shortens), it extends the other one.

Figure 9.7 illustrates how a vertebrate skeletal muscle is organized. Each muscle is surrounded by a connective tissue sheath called the **epimysium** and is divided into bundles (fasciculi) of muscle fi-

bers. Each bundle, in turn, is sheathed by a connective tissue **perimysium**. Because the perimysium is inelastic, it helps prevent muscle from being overstretched; but the perimysium can fold, consequently it does not limit muscle shortening. Each muscle fiber in a bundle is separated from its neighbors by yet another connective tissue sheath, the **endomysium**, and by its own cell membrane, the **sarcolemma**.

Skeletal muscle fibers vary greatly in size, ranging from about 10 to 100 μm in diameter and, in some animals, up to many centimeters in length. The nuclei are scattered along the length of the fiber just beneath the sarcolemma. The cytoplasm

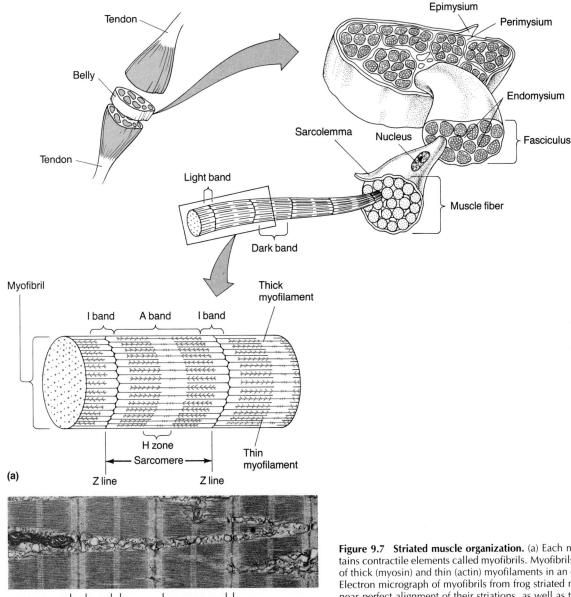

Figure 9.7 Striated muscle organization. (a) Each muscle fiber contains contractile elements called myofibrils. Myofibrils, in turn, consist of thick (myosin) and thin (actin) myofilaments in an orderly array. (b) Electron micrograph of myofibrils from frog striated muscle. Note the near-perfect alignment of their striations, as well as the dot-like granules of glycogen between the myofibrils (10,700×).

(also called the **sarcoplasm**) consists mostly of up to several thousand contractile **myofibrils** running lengthwise through the fiber. With a light microscope, alternate light and dark bands are visible along each myofibril. The striated appearance of the fiber as a whole results from the alignment of these bands. Many years ago, microscopists named the dark bands anisotropic bands, or **A bands**, and the light ones isotropic bands, or **I bands**, on the basis of differences in their microscopic appearance in beams of polarized light. The I bands are crossed by a dark line, the **Z line**; the Z lines divide the myofibril into a large number of functional units called **sarcomeres**. There is a lighter region in the middle of each A band that is called the **H zone**.

What is the molecular basis of the striations? Electron microscopy and biochemical studies have shown that the striations are composed of the protein molecules responsible for contraction. Each myofibril consists of longitudinally arranged **thick** and **thin myofilaments**. Figure 9.7a shows that only the thin myofilaments occupy the I bands, but both thick and thin myofilaments occupy the A bands, giving the A bands their darker appearance. In the lighter H zone, in the middle of the A band, there are only thick myofilaments.

Understanding muscle contraction requires a closer look at the molecular structure of the myofilaments. The thick myofilaments consist mainly of **myosin** molecules. Two polypeptide chains associate to form the elongated part of each molecule, which ends in two "heads." The heads form about 400 side projections from the shaft of each thick myofilament. As shown in Figure 9.8, thin myofilaments are more complex. Molecules of a globular protein called **G actin** polymerize to yield two beadlike strands that twist together, forming a shallow helix. Two other types of proteins, **tropomyosin** and **troponins**, occur in a regular pattern along the surface of each thin myofilament.

Contraction of Skeletal Muscle

The Sliding Filament Model. In 1954, in one of the great conceptual advances in animal physiology, British scientists Andrew Huxley (Nobel Laureate in 1963) and Hugh Huxley independently proposed that muscle contraction involves the sliding of thick and thin filaments (myofilaments) past each other within individual sarcomeres. In their **sliding filament model**, the overlap between the thick and thin filaments (myofilaments) increases as the muscle shortens (contracts), and decreases as the

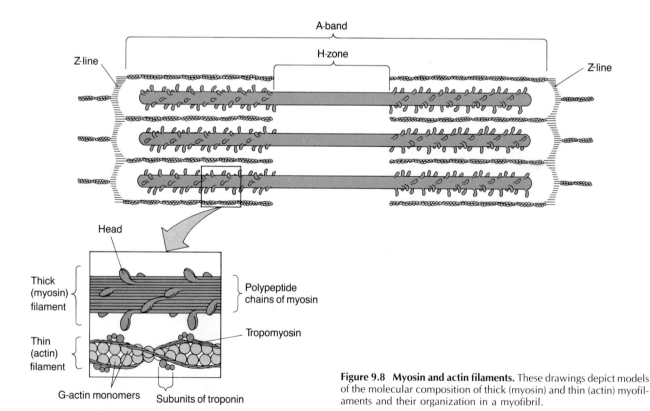

Figure 9.8 Myosin and actin filaments. These drawings depict models of the molecular composition of thick (myosin) and thin (actin) myofilaments and their organization in a myofibril.

muscle relaxes. Figure 9.9 illustrates these changes. As the thick and thin myofilaments slide, the I bands and H zones get narrower and the distance between the Z lines decreases, but the lengths of the individual myofilaments remain the same. Because each sarcomere shortens in this process, and a muscle consists of a great number of sarcomeres, the sliding filament mechanism can cause large movements.

In 1957, Andrew Huxley added an important feature to the sliding filament model when he postulated that sliding might result from the development of **cross-bridges** between the thick and thin myofilaments. A few years later, the existence of such cross-bridges was demonstrated. These structures are produced when the heads of the myosin molecules extend across to the adjacent actin filaments (Figure 9.10a). Biochemists discovered that bridge formation begins when ATP is hydrolyzed and its energy is transferred to the myosin heads. Magnesium ions (Mg^{2+}) and calcium (Ca^{2+}) are also required for muscle contraction. When these ions are available, the activated (energized) myosin heads can bind to the monomers of the actin filaments. In turn, when the high energy of the myosin heads is discharged, the heads bend back on the myosin molecule shaft. This bending exerts a pull

on the actin filaments, and sliding results. The cross-bridges break when the myosin heads bond with other ATP molecules, the head bends back to the original position, and the cycle of bridge formation begins again. Breaking of the old cross-bridges allows new cross-bridges to form, and it also allows for muscle relaxation. If a muscle's supply of ATP is depleted, the muscle cannot relax. For relaxation to occur, Ca^{2+} must be removed, but even without Ca^{2+}, a muscle's cross-bridges cannot detach from the actin filaments without ATP.

In vertebrate skeletal muscle, there are twice as many thin myofilaments as thick myofilaments, and since astronomical numbers of myosin heads are available for cross-bridge activity in the muscle as a whole, smooth and powerful contractions can result. Invertebrate striated muscle is more variable in structure; in some insect muscles, there are four to six times as many thin myofilaments as thick myofilaments, increasing the potential for cross-bridge formation and for powerful, sustained contractions.

The Role of Troponin and Tropomyosin. The troponin and tropomyosin constituents of the thin myofilaments play a vital role in regulating the sliding process (see Figure 9.8). Experiments have shown that in the absence of calcium ions (Ca^{2+}), tropomyosin molecules prevent cross-bridge formation by blocking the myosin head binding sites on the actin filaments. The myofilaments cannot slide, and thus the muscle cannot contract, unless the blockage is removed. The binding sites are unblocked by a sequence of events triggered by a nerve impulse arriving at a muscle fiber. The impulse causes Ca^{2+} to be released from storage sites elsewhere in the fiber (specifically in the sarcoplasmic reticulum, described later). Once released, Ca^{2+} attaches to the troponin molecules (Figure 9.10b). Although researchers are uncertain as to precisely what happens next, it is believed that the troponins change shape when bound with Ca^{2+}. Their new shape may force a shift in the position of the adjacent tropomyosin molecules, thus freeing the binding sites for cross-bridge formation.

Energy for Contraction. Energy for muscle contraction is derived from the enzymatic hydrolysis of ATP to ADP. Continuous regeneration of ATP is essential, because just a few seconds of intense activity can exhaust a skeletal muscle's supply of high-energy compounds. Biochemical studies have revealed that as ATP is used, a second high-energy compound in skeletal muscle—**phosphocreatine** in

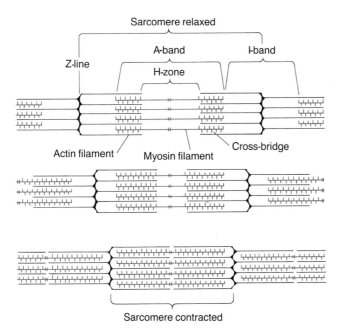

Figure 9.9 Contraction of a sarcomere. Thick myosin filaments and thin actin filaments of a sarcomere (contractile unit) of a myofibril slide past each other during muscle contraction and relaxation. As myofilament overlap increases during contraction, there is a correlated increase in the extent of cross-bridge formation. The lengths of the actin and myosin filaments typically do not change as contraction occurs.

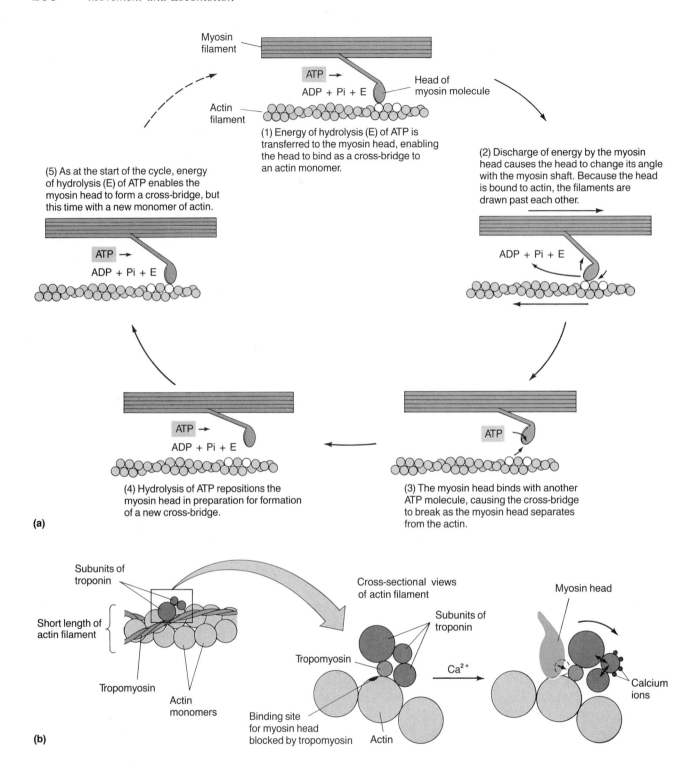

Figure 9.10 Model of the action of sliding filaments. (a) Myosin heads become cross-bridges that bind to actin, then change angle and shift the positions of the two myofilaments relative to each other. Binding and unbinding of the myosin head, and its change in angle, are dependent on ATP. (b) This short length of an actin filament shows the arrangement of the tropomyosin and troponin subunits. The cross-sectional view shows the position of the binding site for a myosin head. In the presence of Ca^{2+}, a change occurs in the spatial arrangement of the troponin subunits. The change allows tropomyosin to slip to one side so that a myosin head can bind to the actin, thus producing a cross-bridge.

vertebrates and **phosphoarginine** in many invertebrates—quickly transfers its phosphate group to ADP, producing ATP and creatine or arginine. Vertebrate skeletal muscle stores up to five times more phosphocreatine than ATP, enough to maintain the level of ATP during short periods of intense muscle activity. When the muscle is at rest, the store of phosphocreatine is regenerated by transfer of phosphate from ATP to creatine, and more ATP may be formed by cellular respiration.

Muscle Activation. Muscle contraction is the last event in an activation series that begins with the "command" to contract. The structural basis for activation is illustrated in Figure 9.11. Skeletal muscles receive the command to contract from motor nerve cells (also called motor neurons). These nerve cells carry impulses from the central nervous system (brain and spinal cord), activating the muscle fibers and setting the stage for contraction. The motor neurons branch as they contact the muscles. In the vertebrates, each branch usually serves a single muscle fiber, and each muscle fiber is innervated by only one branch of a motor neuron. **A motor unit**, or functional unit of vertebrate skeletal muscle, consists of one motor neuron and the muscle fibers that it innervates. In many invertebrates, such as crustaceans and insects, muscle fibers often receive branches from two or more motor neurons, an arrangement called **polyneural innervation**.

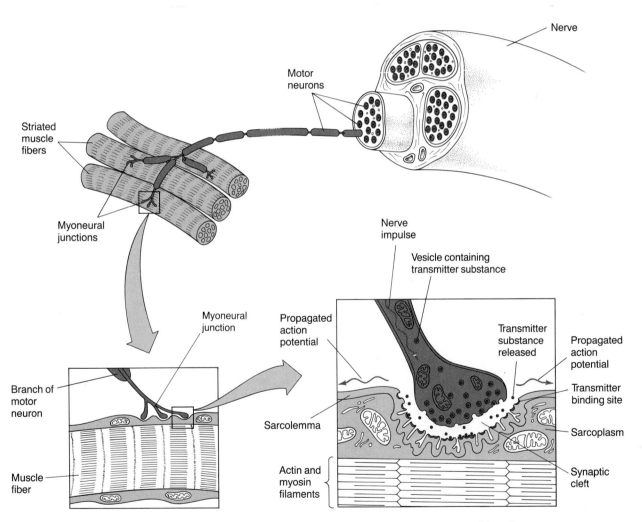

Figure 9.11 Innervation of a striated muscle fiber in a vertebrate. A motor nerve cell branches to serve a number of muscle fibers, forming a motor unit (i.e., a neuron and all the muscle fibers it innervates). Each muscle fiber typically has only one myoneural junction. Upon arrival of a nerve impulse at the myoneural junction, synaptic vesicles release a transmitter substance that diffuses to receptor sites on the sarcolemma. The transmitter substance changes the electrical potential of the sarcolemma, initiating a self-propagating action potential.

The termination of a neuron branch at a muscle fiber is called a **myoneural (neuromuscular) junction** or **motor end plate**. This is a splayed structure beneath which the membranes of a muscle fiber and neuron form a complex pattern of clefts, or spaces. The arrival of nerve impulses at a myoneural junction causes the release of a chemical transmitter from storage vesicles in the nerve cell. A chemical transmitter, such as acetylcholine, diffuses across the clefts and binds to receptor sites on the muscle fiber membrane, or sarcolemma. This increases the sarcolemma's permeability to Na^+ and K^+. The resultant change in ion distribution at the receptor sites causes a local change in the sarcolemma's electrical potential. If the change in electrical potential reaches a threshold level, an **action potential** is initiated. The action potential sweeps along the sarcolemma as a wavelike change in the membrane's electrical potential. (A more detailed description of action potentials is given in Chapter 10, p. 223.)

The action potential penetrates deep into the muscle fiber via a system of transverse tubules called the **T system**. Tubules of the T system are intimately associated with a complex network of tubes and sacs formed by the **sarcoplasmic reticulum**, a muscle cell organelle comparable to the endoplasmic reticulum of other cells. Rapid and uniform penetration of the action potential is facilitated by tubular **triads** located, in many vertebrates, on either side of the Z line (Figure 9.12). Each triad consists of a transverse (T) tubule flanked by two lateral sacs of the sarcoplasmic reticulum. As the action potential spreads into the fiber, it triggers the release of calcium ions from storage sites in the sarcoplasmic reticulum. As noted earlier, the Ca^{2+} interacts with troponin; the troponin forces the tropomyosin to shift its position, freeing actin filaments for cross-bridge formation; and the sliding of myofilaments ensues. The sarcoplasmic reticulum actively reaccumulates the Ca^{2+} during muscle relaxation, causing tropomyosin to block cross-bridge formation once again, and the myofilaments can then slide back to their original position.

Factors Influencing the Action of Skeletal Muscle

Several factors influence the kinds of movements and locomotion produced by skeletal muscle. These include the arrangement and kind of fibers, the fibers' rate of contraction, and the ability of the muscle to make gradual, or graded, responses.

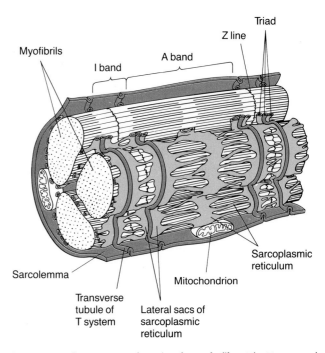

Figure 9.12 The structure of a striated muscle fiber. The T system of transverse tubules extends inward from the sarcolemma. Triads consist of a transverse tubule and two lateral sacs of the sarcoplasmic reticulum. Note how the sarcoplasmic reticulum surrounds the myofibrils and thus is in intimate contact with the functional contractile units, the sarcomeres. Mitochondria, which are the ultimate source of the ATP for contraction and relaxation, also lie in close contact with the myofibrils.

Fiber Arrangement.   The magnitude of force developed by a muscle is directly proportional to its cross-sectional area, and several different fiber arrangements have evolved that effectively increase this area (Figure 9.13). **Fusiform** muscles have fibers that run parallel to the length of the muscle. The thin, straplike sartorius and gracilis muscles of vertebrates are fusiform (see Figure 9.6). Extending from the hip and pubic regions to the tibia of the lower leg, these muscles, typical of the fusiform type, produce extensive motion (e.g., a variety of leg movements) but not much power. The opposite is true of **pennate** (featherlike) muscles, in which fibers are in diagonal arrangements. Many more fibers can be placed diagonally along a tendon than can be placed parallel to it, just as many more cars can be angle-parked along a street curb than parallel-parked. The pincers of crabs and crayfish can exert considerable force because they are closed by pennate muscles. A parallel-fibered fusiform muscle would close the pincer with much less force because it would have far fewer fibers. Among the most powerful muscles are those with the most complex fiber arrangements. In vertebrates, for

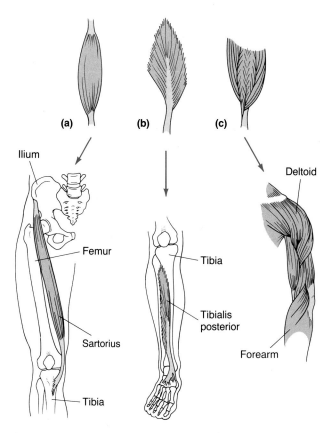

Figure 9.13 Muscle fiber arrangements. (a) A simple fusiform muscle, with fibers running parallel to the length of the muscle between two tendons. (b) A pennate muscle, with fibers inserted at an angle along the tendon. (c) A multipennate muscle, with a fiber arrangement conferring a large cross-sectional area and thus considerable power.

example, the **multipennate** deltoid muscle of the shoulder is a short, powerful muscle that rotates, flexes, and extends the forelimb.

Fiber Type and Rate of Contraction. Physiologists have shown that some muscle fibers fatigue quickly, while others are fatigue-resistant. Some fibers also contract more rapidly than others; the leg muscles of mice and the flight muscles of insects contract within a fraction of a second, whereas the body wall muscles of a sea anemone may take several seconds to contract. The type of fibers in a particular muscle is directly related to the tasks that the muscle performs.

Skeletal (striated) muscle fibers are classified into two types: tonic and phasic. **Tonic fibers** have a relatively slow rate of contraction and usually respond only to a succession of stimuli. Tonic fibers are relatively unimportant in locomotion, but are efficient in situations requiring sustained contrac-

tion and little length change. They allow animals to maintain posture and balance with relatively little expenditure of energy or fatigue.

Phasic fibers respond to an adequate single stimulus, or to multiple stimuli arriving simultaneously, with a rapid, twitchlike contraction and are far more important than tonic fibers in locomotor movements. Some phasic fibers, called *intermediate red fibers* or *slow oxidative fibers*, are rich in mitochondria and oxidative enzymes, contract slowly, and are slow to fatigue because they can make ATP oxidatively as fast as it is used. Fatigue-resistant fibers have high adaptive value in muscles that must make slow, repetitive movements, as in the tails of slow swimming fishes. Other phasic fibers, called *white fibers* or *fast glycolytic fibers*, have very high contraction rates and power outputs and occur where rapid movements are required, as in the tails of fast-swimming fishes and in the flight muscles of certain birds. A wild turkey, an ungainly, rather heavy bird that seems to prefer walking to flying, can accelerate rapidly to flight speeds of about 64 km/hr (40 mph), but can maintain such speeds only long enough to evade danger. As in the more familiar domestic turkey, most of its breast (flight) muscles are white. Because white muscle fibers are poor in mitochondria and oxidative enzymes, they quickly fatigue. When white muscle activity has ended, energy supplies are restored by glycolysis. A third type of phasic fiber, called *fast, oxidative fibers*, are red. Like white fibers, they contract rapidly; but like slow red fibers, they can make ATP quickly by oxidative phosphorylation, and therefore, they fatigue slowly. Fast oxidative fibers are adapted for rapid, long-term, repeated movements. They are common in the flight muscles of birds that spend much time in flight.

Several factors determine whether a particular skeletal muscle is adapted for tasks requiring great endurance, maximum power and speed, or some combination of these. Histological comparisons show that the proportions of slow and fast phasic fibers differ from muscle to muscle. Also, many animals have fiber types different from those just described. Sharks and other cartilaginous fishes have at least four types of phasic fibers, all with somewhat different contractile properties. Arthropods appear to have a continuous spectrum of types, ranging from fast to slow. The almost imperceptible leg movements of a praying mantis stalking a slowly walking grasshopper, and the explosively fast leap of the grasshopper as it escapes, indicate the contractile capacities provided by a range of fiber types.

Gradation of Response. When a skeletal muscle fiber is stimulated by a nerve impulse, it responds with a maximal contraction or no contraction at all, a phenomenon called the **all-or-none law**. A motor unit (all the muscle fibers innervated by one nerve fiber) also obeys the all-or-none law. Whole skeletal muscles, on the other hand, develop various degrees of force, and do so with smooth gradations. The nervous and muscular systems accomplish this **gradation of response** in several ways.

The number of motor units in a muscle varies greatly, from two or three in the leg muscle of an insect to hundreds in the leg muscle of a vertebrate. A graded increase in tension as a muscle contracts often depends on a gradual increase in the number of motor units brought into action, a process known as **recruitment**. Since recruitment takes time, the muscle as a whole contracts smoothly, even though its individual motor units are obeying the all-or-none law. The size of motor units also varies. Muscles involved in tasks requiring great precision, such as those of the fingers and eyes, have motor units with as few as 15 to 20 muscle fibers (and thus a high nerve-fiber-to-muscle-fiber ratio). This allows fine adjustment of force.

Recruitment applies mainly to the control of vertebrate skeletal muscles. Many invertebrates have skeletal muscles with few motor units, so there can be little recruitment. Graded muscle contraction in crustaceans and insects often depends on the capacity of individual muscle fibers to develop tension in proportion to the amount of transmitter released at the myoneural junctions. The amount of transmitter released depends, in turn, on the frequency (intensity) of nervous stimulation. As a result, the muscle as a whole may then exert forces ranging from the light grip used by an ant lifting immature offspring in its mandibles to the force required for the mandibles to cut and grind food.

An animal's nervous system may also cause graded muscular responses by varying the rate at which motor units are stimulated. This can be demonstrated by stimulating muscles electrically. If a single electric shock is applied to the nerve innervating a frog's gastrocnemius muscle (see Figure 9.6), the muscle responds with a single twitch (contraction). A muscle twitch has characteristic features that can be recorded and illustrated graphically. After a brief **latent period**, during which the nerve impulse is conducted to the myoneural junctions and muscle activation occurs, the muscle shortens during a **contraction phase**. Return to the resting muscle length then takes place during a **relaxation phase** (Figure 9.14a). The duration of the frog leg muscle twitch is about 100 ms. Representing a muscle's graded response capacity,

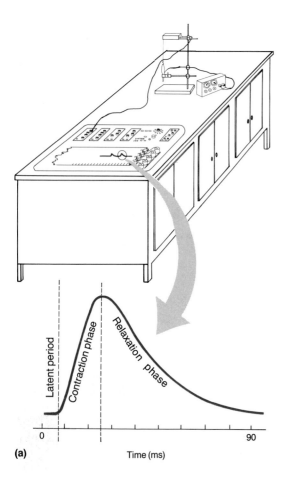

(a)

Latent period / Contraction phase / Relaxation phase

Time (ms)   0 ... 90

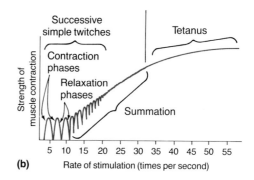

(b) Rate of stimulation (times per second)

Strength of muscle contraction

Successive simple twitches — Contraction phases — Relaxation phases — Summation — Tetanus

5 10 15 20 25 30 35 40 45 50 55

Figure 9.14 Responses of muscle to stimulation. (a) A recording of a single twitch in response to a single stimulus. The gastrocnemius muscle and sciatic nerve of a frog are attached through a force-transducing apparatus to a multipen recording device called a polygraph. When the nerve is stimulated electrically, the muscle contracts and exerts force on the transducer. The transducer and polygraph circuits cause a recording pen to move in proportion to the strength of the muscle's contraction. Oscilloscopes are often used instead of the ink-writing polygraph because their greater sensitivity enables them to display events of much smaller magnitude. (b) Development of tetanus in response to stimulation at different frequencies. Note temporal summation and incomplete tetanus at left and center.

Figure 9.14b illustrates what happens when a gastrocnemius muscle is given repeated electrical stimulations. Repeated small stimuli overcome the threshold value of some otherwise resting motor units, producing a greater shortening of the muscle (i.e., increase in tension) than would be produced by any of these stimuli alone. This phenomenon is called **temporal** or **wave summation**. Note in Figure 9.14b that partial relaxation phases are still detectable at low stimulation frequencies, but they disappear at high frequencies. As the frequency of stimulation increases, a graded increase in the tension exerted by the muscle occurs until all the motor units in the muscle are in a state of continued contraction called **tetanus**. (The word *tetanus* is also used to describe muscle rigidity and painful spasms caused by toxins of the bacterium *Clostridium tetani*. Involvement of the jaw muscles, causing lockjaw, and involvement of the respiratory muscles may lead to death.)

Efficiency in Muscular Movement and Locomotion

A major feature of animal evolution has been the development of more efficient ways to use muscle. In this context, efficiency has generally meant two things: (1) improving methods for returning contracted muscle to its original (noncontracted, extended) position and (2) enabling muscle to perform the greatest amount of work with the least amount of shortening. These two factors are often interrelated, and both are influenced by the architecture of an animal's body, especially that of the skeleton. Because muscles do not actively reextend themselves after contracting, they must be extended by some outside force. Muscles that pull against and displace certain skeletal parts may be returned to their original position, after they stop contracting, by the resilience of the skeleton. (See the discussion of the action of the nuchal ligaments in Chapter 8, p. 187; Figure 8.16c.) More commonly, muscles work as **antagonistic pairs**. Both muscles of an antagonistic pair are attached to and move a skeletal part, but only one muscle of the pair contracts at any one time. As it does, it pulls its antagonist back to its original, extended position. Antagonistic muscle pairs are often involved in moving body appendages that function as **levers**. Levers are simple machines that can increase the force being applied to some resistance or convert a small movement into a larger one. A bird's wing, for example, is a lever that converts the small distance the flight muscles shorten when they contract to the long distance traveled by the wing tip. Thus, one flight muscle pulls the wing up; then its antagonist pulls the wing down, and in so doing, extends the other muscle.

Figure 9.15 illustrates how levers are involved in movement. Every lever has three parts. First, there is a **fulcrum** (F); in animals, this is often the joint on which a lever of bone pivots. Second, an **effort arm** extends from the fulcrum to the area where the muscles are attached and apply **effort** (E). Third, a **resistance arm** extends from the fulcrum to the point where the load applies **resistance** (R). The three points F, E, and R may be in any of three anatomical sequences, corresponding to three kinds of levers called first-, second-, and third-class levers. Although examples of each lever class are found in animals, the most common are third-class levers, in which effort is applied between F and R.

The performance of a lever can be changed significantly by altering the relative lengths of the effort arm and resistance arm. For example, if we compare the forelegs of a horse and armadillo, we see that the same muscles and bones are present as third-class levers, but changes in the relative positions of the muscles adapt the forelegs to very different functions (Figure 9.16). After some 60 million years of evolution, the horse is well adapted for rapid running; it moves its legs rapidly in long strides without exerting unusually large forces. In contrast, the armadillo can apply large forces to the ground as it digs, but it runs slowly. As levers, how do their legs compare? The **mechanical advantage** (MA) of a lever, defined as the ratio of a system's output force to the force that must be applied to run the system, is expressed as

$$MA = \frac{\text{length of effort arm}}{\text{length of resistance arm}}$$

Eliminating obvious differences due to animal size, the mechanical advantage of the horse leg is about 1/13, compared with 1/4 for the armadillo. This means that for each unit of distance moved by the upper end of the horse's leg, the hoof moves 13 units in the same length of time. In other words, the force exerted on the upper end of the horse's leg is reduced 13-fold at the hoof; the comparable value for the armadillo is only 4-fold. Again making allowances for the difference in their sizes, the squat little armadillo is adapted for power and the long-legged horse for speed.

Animals with leverlike limbs far outnumber all other animals combined, testimony to the adaptive advantages of these structures. Attainment of rapid

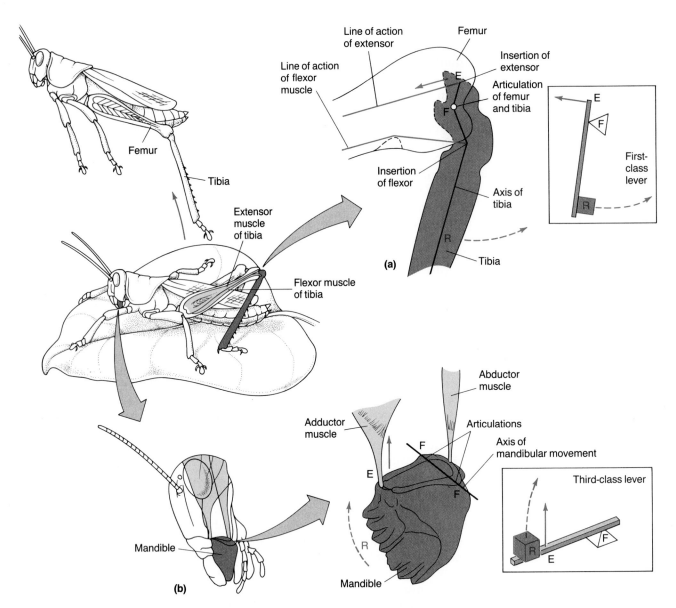

Figure 9.15 Musculoskeletal levers and antagonistic muscle pairs. (a) In first-class levers, the fulcrum (pivot point, F) is between the points of application of effort (E) and resistance (R). The hind tibia of a jumping grasshopper is an example of a first-class lever in action. The tibia extensor muscle and tibia flexor muscle, both located within the femur, act as an antagonistic pair as they alternately extend (straighten) and flex (bend) the insect's leg. At the beginning of a jump, the extensor muscle contracts, but movement of the tibia is at first prevented because the flexor contracts with the extensor; this causes distortion of the hinge between the leg joints. Then the flexor suddenly relaxes, and the energy released as the hinge returns to its normal shape helps to boost the grasshopper into the air by contributing to acceleration of the tibia. (b) In third-class levers, E is between F and R. The crushing action of a grasshopper's mandibles is based on a third-class lever. The adductor muscle has great power in closing the mandibles to grind food because the distance between E and R is much less than that between E and F. Another example of a third-class lever is the human arm; as the arm is flexed, E (the insertion of the biceps muscle on the forearm) is much closer to F (the elbow) than to R (the load-bearing hand).

locomotion with minimal energy input is the most common type of lever adaptation in animals. Lever principles have been a central feature in the evolu-tion of appendages and in the evolution of mecha-nisms as diverse as ciliary action and animal jaw movements.

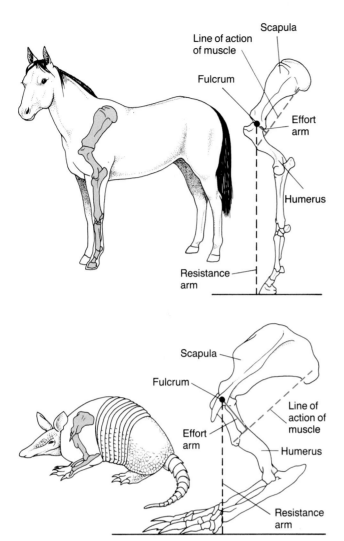

Figure 9.16 Comparison of the mechanical advantage in the legs of the horse and armadillo. Muscles actually pull at angles to the bones, so the effective length of the effort arm is not simply a measured length of bone. Instead, it is the perpendicular distance from the line of muscle pull to the fulcrum. The mechanical advantage for the horse is $1/13$ and for the armadillo $1/4$, indicating an adaptation for speed in the horse and for power in the armadillo.

Modes of Locomotion

The following section examines the principal methods of animal locomotion: walking and running, crawling and burrowing, swimming, and flying.

Walking and Running. Tetrapod (four-footed) vertebrates and hundreds of thousands of arthropod species use jointed, leverlike limbs for walking and running. Arthropods have taken two paths to greater locomotor rates. Some predators, like the spiders and centipedes, have long legs, and long strides can produce rapid locomotion (relative to body size), even though leg movement may be slow.

In contrast, the generally short-legged insects achieve relatively high speeds by means of very rapid leg movement.

Many tetrapod vertebrates achieve impressive speeds. The cheetah can run up to 113 km/hr chasing down its prey, while a gazelle manages about 96 km/hr to avoid becoming prey. The horse and red fox can race along at 64–80 km/hr, and a human competing in a 100 m dash does over 32 km/hr. Bringing up the rear, at only a few kilometers (or less) per hour are the tortoises, mice, and moles.

As discussed in Chapter 8, the skeleton is important in bearing the weight of a terrestrial animal and helping to maintain its body stability. Stability can be a significant problem, particularly for species with a small number of legs. Insects, which have only six legs, maintain stability by keeping three feet on the ground as they run. As two legs on one side of the body and one leg on the other side step forward, the remaining three legs form a stable tripod on the ground (Figure 9.17a). The two sets of three legs then alternate in forming the tripod. It is as if the insect were walking along on three-legged stools, never off balance, and generally able to keep from being tumbled by puffs of wind.

Some of the same features are found in four-legged walking by vertebrates. Some vertebrates use a moving tripod of three legs to maintain stability during slow walking (Figure 9.17b). At higher speeds, stability based on a tripod is lost because fewer feet generally contact the ground at any given moment. At a full run or gallop, all four feet of a wolf or zebra leave the ground simultaneously for part of each stride. (Two-legged, or bipedal, locomotion by humans is similar: One or two feet are always on the ground during walking, but both feet are off the ground simultaneously for part of each stride during running.) At high speeds, stability based on momentum replaces stability based on a moving tripod.

Crawling and Burrowing. These forms of locomotion often involve peristaltic movement. (Peristalsis is discussed in Chapter 8, p. 177; Figure 8.4.) Whether crawling or burrowing, an animal must be able to push forward against a substrate, usually because its body can anchor firmly in the substrate. An earthworm, for example, anchors by pressing its bristles into the soil, usually into the sides of its burrow. Because it is anchored, the worm is forced forward as it elongates. If it were not anchored, thrusting forces of its anterior end would propel

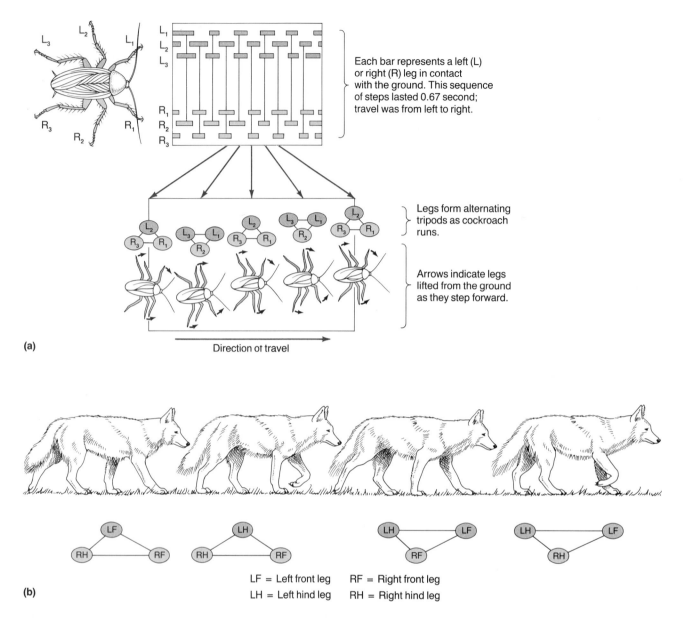

Each bar represents a left (L) or right (R) leg in contact with the ground. This sequence of steps lasted 0.67 second; travel was from left to right.

Legs form alternating tripods as cockroach runs.

Arrows indicate legs lifted from the ground as they step forward.

(a)

Direction of travel

(b)

LF = Left front leg   RF = Right front leg
LH = Left hind leg   RH = Right hind leg

Figure 9.17   Stepping sequences that help maintain stability on land. (a) A running cockroach and (b) a walking vertebrate keep three legs on the ground, thus maintaining a stable supporting tripod. The animal is most stable when its center of gravity is over the tripod. For many vertebrates, the tripod is lost at higher speeds; increased momentum then provides stability, as in a galloping horse.

the worm backward, as dictated by Newton's third law. Burrowing clams also illustrate the principle of anchoring. A clam digs its way into a tidal mudflat or river bottom with thrusting actions of its muscular foot. For the thrusts to be effective, the clam must anchor itself against backsliding by pressing its shell (valves) outward tightly against the surrounding mud. Once the foot is extended, the valves close slightly, releasing their grip. The tip of the extended foot expands, forming a termi-

nal anchor, and then pulls the shell deeper into the mud. (Bivalve digging is discussed in Chapter 24, p. 568; Figure 24.12.)

Swimming.   Most phyla include swimmers, and swimming takes many forms. Diving beetles and water bugs use their legs as oars; scallops, octopuses, squids, and some jellyfish jet along on bursts of water; and terrestrial vertebrates use a variety of arm and leg movements. The most versatile swimmers use undulatory movements to generate the necessary forces. Undulations may involve most of the body, as in many worms, eels, and snakes, or may be restricted to a posterior region, such as a tail fin, as in many fishes and aquatic mammals.

In most forms of undulatory swimming, waves of muscular contraction move backward along the animal's body as the animal moves forward. The body must be prevented from shortening as the muscles contract, but it must also remain flexible enough to undulate. This is accomplished by a relatively incompressible vertebral column in fishes, while many invertebrate swimmers manage with hydrostatic skeletons. The undulations that pass along the body (and tail fin of fishes) thrust against the water, producing forces that propel the animal forward. (Swimming in fishes is covered in Chapter 29, p. 715.) Streamlining, an adaptation for reducing resistance to movement through a medium as dense as water, helps to conserve energy in large, strong swimmers such as fishes, whales, seals, and penguins.

As shown in Figure 9.18, areas of muscle contraction in an undulating animal press against the

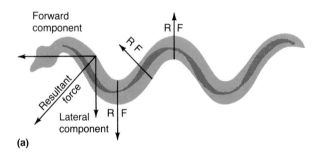

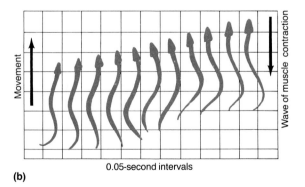

Figure 9.18 Undulatory swimming movements. (a) A swimming animal exerts forces that have both forward and lateral components that can be resolved to yield a resultant force (RF). The thick portions of the colored line indicate areas of muscle contraction. The lateral forces cancel each other, so if the forward component is greater than frictional resistance, the animal moves forward. (b) A swimming eel at successive 0.05-second intervals. Its swimming direction is opposite to the wave of muscle contraction.

water at angles ranging from fully lateral to straight back along the body axis. The opposing forces elicited from the water have *lateral* and *forward* components that can be resolved by simple vector analysis into **resultant forces** (RFs). Lateral RFs—those that alternate to the right and left sides of the animal—cancel each other, and there is no backward component. The animal will therefore slip ahead if the forward component is greater than the force of friction. The same principles apply to the undulatory locomotion of snakes moving through the grass and to the penetration of body tissues or soil by nematodes; their body thrusts are more effective, however, because they are moving through media that are firmer than water.

Flying. Flying is the fastest mode of locomotion. The peregrine falcon has been clocked at 350 km/hr (217 mph) during a dive at a 45° angle, making it the fastest animal on earth. The fastest bird in level flight, at 170 km/hr (106 mph), is the white-throated swift. Among the mammals, bats are relatively slow, with some species flying at about 25 km/hr. Among the insects, most dragonflies fly at 50–65 km/hr, and one species was clocked at 144 km/hr. Flight evolved in four evolutionarily independent lines of animals—insects, reptiles, birds, and mammals—indicating its adaptive advantages. Although flight is an energy-expensive form of locomotion, flying animals can cross greater obstacles and cover much more ground distance than can walkers or runners. (See the Essay on the energy costs of locomotion, p. 216.)

What factors make flight possible? When most objects are pushed through the air, the only force they develop is **drag**, the trail of turbulent air pulled along behind (Figure 9.19a). But objects with certain shapes may generate more than this. A person carrying a sheet of plywood on a windy day soon discovers that it develops **lift** as well as drag. A wing as flat as the board might glide steeply, but its drag would be so great compared to its lift that it could not sustain flight (Figure 9.19b). The key to successful flight is seen in the wing shapes of airplanes, birds, bats, and insects. All have the shape of an **airfoil**: The leading edge of the wing is thicker than the trailing edge, the upper surface is outwardly curved, and the lower surface is flattened or concave. Figure 9.19c shows airflow over an airfoil; notice that the airfoil reduces drag and increases lift. Because of the airfoil's shape, the air must travel farther, and therefore faster, over the wing than under it. As a result, air molecules are spaced farther apart above the wing and are closer to-

Essay: The Energy Costs of Travel

The energetics of locomotion and the comparative energy costs of different kinds of locomotion have interested zoologists for many years. Most is known about walking and running, the two forms of locomotion for which the human body is adapted, because it is easiest to obtain data on humans. Animals are not always cooperative in locomotor studies, and much less is known about swimming and flying.

Since no single species is adapted for all forms of locomotion, comparative energy costs for all forms of locomotion cannot be determined using any single species. This results in the problems of interpreting and comparing data from species to species. Although most animals are specialized for one form of locomotion, many animals are capable of, but not highly efficient at, a second form of locomotion. Many ducks and geese, for example, can swim in addition to flying, but they are far less efficient swimmers than the fishes. Humans are also capable of swimming, but they are so inefficient at it that they consume more energy in swimming than in any other form of locomotion. To compare the costs of locomotion, it is best to obtain data about a particular form of locomotion from animals that are adapted specifically for that form.

Walking, running, swimming, and flying have different energy requirements because animals on land, in the air, or buoyed by water are affected very differently by the forces of gravity and friction. A walking or running animal expends energy to counter both gravity and friction with the ground and to overcome the inertia involved in starting and stopping the limbs with each stride. A human spends less energy walking a certain distance than running it, but very rapid walking is more energy-costly than slow running. Because water is dense and supports a swimming animal's body weight, friction is generally a more important factor than gravity in swimming. Most swimming animals expend little, if any, energy in supporting their bodies, and body streamlining often reduces the energy that a swimmer must use to overcome friction. Compared to a swimming salmon or a running dog, a flying robin uses a greater percentage of its energy overcoming gravity than in dealing with other forces.

Several studies have shown that far more power (energy spent per unit time) is required for flight than for swimming or running, while swimming is least demanding (for animals adapted for it). But suppose the actual cost of transport is compared, thus taking into account the distance moved? Figure E1 illustrates, for a variety of animals (including running humans), how much work it takes to move 1 kg of body weight a distance of 1m. Keep in mind that each animal in this comparison is specifically adapted for the form of locomotion shown. When animals of similar weight are compared, flying proves to be less costly than running, with swimming least expensive of all. However, the data show that generalizing too broadly may lead to incorrect interpretations; a comparison of animals at the two ends of the weight range shows that a heavy terrestrial animal can cover a given distance with less energy cost than a small bird or fish.

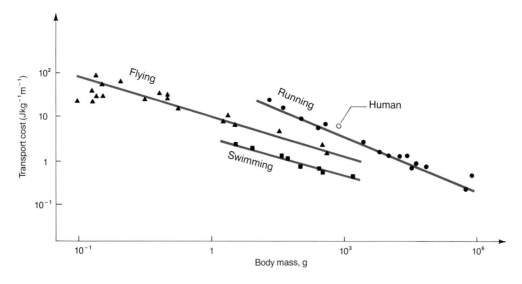

Figure E1 Cost of transport by running, flying, and swimming animals of various weights. Over part of the weight range, it is less energy-costly to fly than to run a given distance. (A running human is indicated by the open circle.) The vertical axis expresses transport cost in terms of work (joules) performed in moving 1 kg of body weight a distance of 1 m. The horizontal axis expresses body mass in terms of grams. Both axes are plotted on a logarithmic scale.

gether below it. In other words, the air pressure is greater under the wing than over it. Lift results as the wing is pushed from below toward the partial vacuum above it. By changing the wing's angle of attack to the flow of air, the animal controls how much of the resultant force is devoted to lift and how much to forward movement (Figure 9.19c). Wing flapping by birds, bats, and insects causes air to rush over their wings, creating lift because of the wings' airfoil. (Adaptation of bird wings for different types of flight are dealt with in Chapter 32, p. 776.)

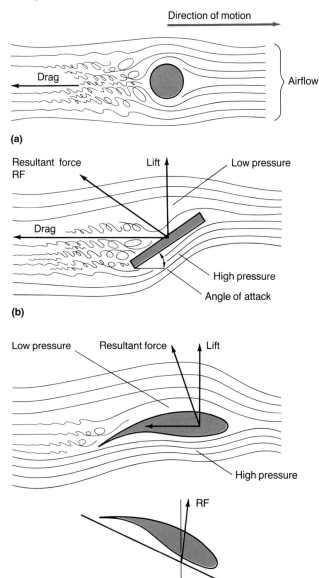

Figure 9.19 Influence of shape on drag and lift. (a) Most objects produce drag but no lift when they are placed in an airstream. (b) A flat board develops lift but also a great deal of drag because of turbulence. (c) An airfoil minimizes drag and maximizes lift. Glide angles are achieved by tipping the wing forward until the resultant force (RF) is directed slightly forward.

SUMMARY

1. Protozoa make the greatest use of pseudopodia, flagella, and cilia in locomotion, although some small animals and many larval forms move by means of cilia or flagella. The vast majority of animals have cells with pseudopodia, cilia, or flagella that move materials and/or fight infection within the body.

2. Formation of pseudopodia, the basis of amoeboid movement, is the least understood locomotor mechanism. The underlying motive force may depend on the relative sliding of protein molecules and the energy of ATP, a mechanism common to movement by cilia, flagella, and muscles.

3. Protozoan flagella propel water along the flagellar axis and either push or pull cells through the water. Cilia are smaller, and much like oars, they propel water parallel to the cell surface. In both cilia and flagella, microtubules in a 9 + 2 arrangement extend along the length of the shaft, and relative sliding of the microtubules produces the movement. Contractile threads called myonemes produce the sudden coiling of the stalks of certain sessile protozoa.

4. There are two general types of muscle: smooth and striated. Striated muscle cells (fibers) contain banded myofibrils and generally are larger than smooth muscle fibers. In the vertebrates, skeletal (locomotor) and cardiac (heart) muscles are striated, while smooth muscle occurs in the wall of the gut and many other viscera. Many invertebrates have striated skeletal and visceral muscle. When a muscle is stimulated by nerve impulses, myosin (thick) and actin (thin) filaments in the myofibrils interact with each other through cross-bridges. The cross-bridges draw the thin myofilaments past the thick myofilaments, shortening the myofibril and ultimately the muscle in a process known as the sliding filament mechanism of muscle contraction.

5. In vertebrates, a motor unit is composed of a neuron and all the muscle fibers that it stimulates.

6. Muscle activity is influenced by fiber type and arrangement, the rate of contraction of fibers, and the muscle's ability to produce graded responses.

7. The major modes of animal locomotion are walking, running, crawling, burrowing, swimming, and flying, all of which result from the contraction (shortening) of muscle cells. The evolutionary modification of skeletons and muscles as locomotor systems, including the development of leverlike appendages, has had a major impact on the morphological, ecological, and physiological diversity of animals.

❏ **FURTHER READING**

Alexander, R. M. "Walking and Running." *American Scientist* 72 (1984): 348–354. *Describes how legs and leg movements are adapted to minimize energy expenditure during locomotion.*

Alexander, R. M., and G. Goldspink. *Mechanics and Energetics of Animal Locomotion.* New York: Wiley, 1977. *Good integration of biomechanics and bioenergetics for the major modes of locomotion.*

Hildebrand, M., D. M. Bramble, K. F. Liem, and D. B. Wake, eds. *Functional Vertebrate Morphology.* Cambridge, Mass.: Harvard University Press, 1985. *Contains eight excellent chapters on locomotion written by experts in the field.*

Hoar, W. S. *General and Comparative Physiology.* 3d ed. Englewood Cliffs, N. J.: Prentice-Hall, 1983. *Includes a technical, up-to-date chapter on locomotor principles from a comparative perspective.*

Huxley, A. *Reflections on Muscle.* Liverpool: Liverpool University Press, 1980. *An interesting study of the development of the sliding filament model.*

Webb, P. W. "Form and Function in Fish Swimming." *Scientific American* 251 (1984): 72–82. *Correlates a fish's form with its swimming habits.*

Wilkie, D. R. *Muscle.* 2d ed. *The Institute of Biology Studies in Biology No. 11.* London: Edward Arnold, 1976. *A brief, readable account of muscle structure and function.*

Yates, G. T. "How Microorganisms Move Through Water." *American Scientist* 74 (1986): 358–365. *Explains, in terms of fluid dynamics, how cilia and flagella propel cells.*

Coordination: Nervous Systems

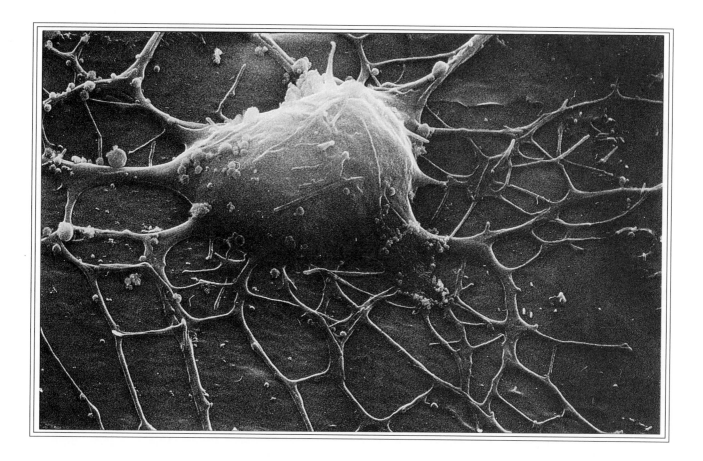

Coordination means harmonious, integrated action or interaction. An animal's survival depends on its ability to respond appropriately to stimuli from its environment. Virtually every action that an animal takes is adapted to its needs. The precision, direction, and speed of an animal's movements often

---

Nerve cells are called reurons. This photograph shows a neuron in a tissue culture established from brain cells of the housefly (*Musca domestica*).

determine whether food will be obtained, predators avoided, or mates found. Likewise, internal processes are precisely timed and scaled to meet specific needs within an animal. Digestive juices are released as food arrives in the digestive tract. Water is conserved or eliminated, depending on an animal's water balance. Blood flow through surface vessels increases with a rise in body temperature, cooling the body of a mammal or bird as excess heat radiates away. Such processes, which are important in homeostasis, are carried out by or-

219

gans capable of a great range of activity. A muscle may contract rapidly or slowly, and a gland may release a flood or a trickle of secretion. But the organs normally operate in harmony with one another and at levels appropriate to the task; this is what coordination means in the context of animal physiology. The two systems primarily responsible for coordinating physiological processes, the **nervous system** and the **endocrine system**, are discussed in this and the next chapter. An animal's activities in response to external stimuli and the maintenance of internal homeostasis result from the joint action of these two coordinating systems.

Nervous System Structures and Functions

The three major functions of the nervous system are (1) to detect and record changes in an animal's environment (internal as well as external), (2) to assess and integrate the incoming information, and (3) to respond by coordinating muscular or glandular activities that best serve the animal. The functional organization of the nervous system in most animals corresponds to these three primary activities.

Major Functional Subdivisions

The first major subdivision, the **sensory system**, samples the internal and external environments and receives information about pressure, temperature, light, sound, and chemicals; about the movement and position of the body and its parts; and about the status of internal fluids and organs, such as the concentration of O_2 in a mammal's blood or the fullness of an animal's stomach. (The sensory system is discussed in detail in Chapter 12.) The sensory system transforms these data into nerve impulses that are conducted along **sensory (afferent) nerves**; in most animals, these impulses enter a second subdivision, the **central nervous system (CNS)**. The CNS, which often consists of a brain and one or more nerve cords, assesses sensory input in terms of an animal's needs. Information from different sources—for example, the sight of food along with sensations of hunger and the odor of a hidden predator—is compared, weighed against information from previous experiences, and possibly stored as memory. Following this **integration** (processing and assessment) of informa-

tion, the CNS may trigger and coordinate activity in the **motor system**, the third subdivision. The motor system consists of **motor (efferent) nerves** that conduct nerve impulses to muscles and various other organs, causing suitable responses to sensory information. The part of the motor system controlling skeletal musculature is called the **somatic motor system** (or voluntary motor system). In contrast, the part controlling smooth (visceral) musculature and glands (i.e., organs over which an animal has little or no voluntary control) is called the **autonomic nervous system** (or involuntary nervous system). The sensory nerves that feed information into the CNS and the motor nerves that carry commands to the muscles and organs are sometimes collectively called the **peripheral nervous system** (**PNS**).

Cellular Organization

The basic structural and functional unit of any nervous system is the nerve cell, or **neuron**. Neurons transmitting nerve impulses from one part of an animal to another regulate muscles and organs, integrate information from different sources, and function in memory. Therefore, to understand how nervous systems operate at the level of the whole animal, it is necessary to look closely at organization and function at the neuronal level.

Neurons. Nerve cells consist of a cell body containing a nucleus, an endoplasmic reticulum, a Golgi complex, ribosomes, mitochondria, and one or more thin cytoplasmic extensions called **neurites** arising from the cell body. Neurites that normally receive nerve impulses from another neuron and carry the impulses *toward* the cell body are called **dendrites**; they are often short, numerous, and highly branched. Neurites that carry impulses *away* from the cell body are called **axons**, also referred to as **nerve fibers**. Axons in large animals may be more than a meter long and may extend, for example, all the way from the spinal cord to the foot. Axons commonly terminate by branching into many delicate filaments called **telodendrites**, each of which may have a buttonlike **synaptic knob** at its tip (Figure 10.1).

Each axon in the peripheral nervous system is surrounded by a sheath of flattened cells called **Schwann cells**. The outer membrane of the Schwann cells and their glycoprotein coats make up the **neurilemma**. Many axons are also enclosed by an insulating layer of protein and lipid called the **myelin sheath** (see Figure 10.1). The myelin

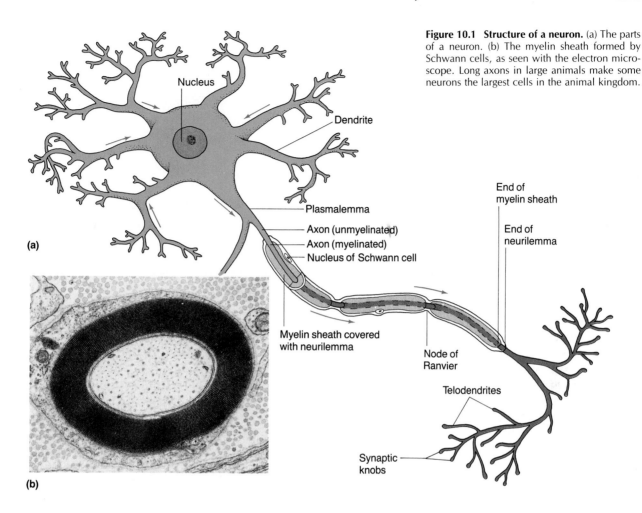

Figure 10.1 Structure of a neuron. (a) The parts of a neuron. (b) The myelin sheath formed by Schwann cells, as seen with the electron microscope. Long axons in large animals make some neurons the largest cells in the animal kingdom.

sheath forms during nervous system development as Schwann cells envelop an axon, wrapping it in concentric layers of their cell membrane, not unlike the way one wraps a cylindrical gift by rolling it up in tissue paper. The concentric layers become the myelin sheath. When the wrapping process is complete, the outermost membrane of the Schwann cells forms the neurilemma and the nucleus of each Schwann cell lies just under the neurilemma. Many Schwann cells are needed to form the myelin sheath along an axon. The small spaces between adjacent Schwann cells are represented by interruptions in the sheath called **nodes of Ranvier**. These nodes play a crucial role in the rapid conduction of nerve impulses in vertebrates.

Like the PNS, the CNS has both myelinated and unmyelinated axons, but it contains no Schwann cells. Instead, processes from cells called **oligodendrocytes** form myelin by encircling the axons. Myelinated axons in both the PNS and the CNS appear white because of the fatty material in the myelin sheath. Myelinated axons thus form the so-called white matter of the CNS, whereas unmyelinated

axons are part of the gray matter. The axons of invertebrates typically have thinner sheaths with little or no myelin and no nodes of Ranvier.

There are several types of neurons. **Sensory neurons** carry impulses from a sensory ending, perhaps a pressure sensor in an animal's skin, to the CNS. **Motor neurons** carry impulses from the CNS to the skeletal muscles and other effector organs. **Interneurons**, also called association neurons, serve as links between sensory and motor neurons in the CNS; interneurons sort out sensory input from various sources, channel information in the central nervous system, and inhibit or stimulate motor neurons. Sensory and motor neurons linked through interneurons may form **reflex arcs**, in which the neurons work together to produce **reflex acts**. The nerve impulses initiated by a sensory receptor in an earthworm's skin, for example, are carried by a sensory neuron into the longitudinal nerve cord (part of the earthworm's CNS), where junctions are made with one or more interneurons (see Figure 10.10). The interneuron, in turn, forms junctions with a motor neuron that carries impulses to the muscles, enabling the earthworm to respond with a reflex act. Many reflex arcs are even simpler, with the sensory neuron linked directly to a motor neuron. The reflex arc is a fundamental unit of nervous system function, but it is not the only simple type of neuronal system that produces motor responses. As you will see later, many types of motor activities are driven by neuronal systems that involve only interneurons and motor neurons. (See the discussion of central pattern generators, p. 232.)

Neuroglia. Although a nervous system's coordinating function depends on the neurons, supportive cells outnumber the neurons and account for more than half the weight of a vertebrate nervous system. Collectively, these cells are called the **neuroglia**, or **glial cells**. (*Glia* means "glue," reflecting an old idea that glial cells held, or "glued," the neurons together.) Schwann cells and oligodendrocytes are two types of glial cells. Several other cell types and their specific functions are now known: **Microglia** are small phagocytic glial cells responsible for removing tissue fragments and microorganisms at injury sites in the CNS. **Astrocytes**, which are star-shaped and very numerous, provide most of the support for neurons and blood vessels in the CNS; their close contact with blood vessels suggests that they may be involved in transferring nutrients to the neurons. Sheets of **ependymal cells** line the cavities in the CNS, including the spinal canal and the cavities (ventricles) of the vertebrate brain.

Nerves. A **nerve** is a bundle of nerve fibers in the PNS, supported by glial cells and tough connective tissue and often visible with the unaided eye. A comparable nerve fiber bundle in the CNS is called a **tract**. Microscopic examination of a vertebrate nerve in cross section reveals hundreds or even thousands of axons, each surrounded by a myelin sheath and a thin wrapping of connective tissue fibers called the **endoneurium**. Another wrap of connective tissue, called the **perineurium**, surrounds groups of axons; and a thick **epineurium** binds the axon groups together and surrounds the entire nerve (Figure 10.2).

Invertebrate nerves often consist primarily of parallel arrays of axons; there is relatively little connective tissue, and in tiny animals, such as water fleas or plant lice, the number of axons in a typical nerve is small.

Ganglia. **Ganglia** are an important feature of most nervous systems; they consist of aggregations of

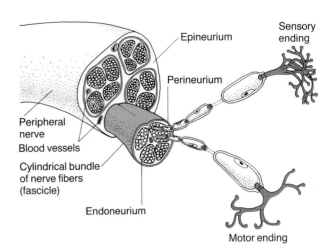

Figure 10.2 Structure of a nerve. The axons are grouped together into cylindrical bundles (fascicles) sheathed in connective tissue. Blood vessels among the bundles transport nutrients and oxygen to the axons and carry away metabolic wastes.

nerve cell bodies from which neurites arise and form nerves and tracts. In many animals, interneurons and junctions between neurons occur in the ganglia, making the ganglia centers for integration and coordination. Segmental swellings that occur along the ventral nerve cords of annelids and arthropods are ganglia (see Figures 10.10 and 10.12). In many animals, the anterior ganglia are enlarged and serve to process and act on sensory input from the major sense organs of the head. Enlarged anterior ganglia are called **cerebral (cephalic) ganglia** or, more commonly, the **brain**. Vertebrates have a well-developed brain and many ganglia elsewhere in the body.

Neural Junctions

Axons form junctions called **synapses** with other neurons. These are the sites at which nerve impulses are transmitted from one neuron, the **presynaptic neuron**, to another, the **postsynaptic neuron**. There are two general types of synapses. By far the commonest is the **chemical synapse**, in which the membranes of the presynaptic and postsynaptic neurons are separated by a **synaptic cleft** of 20–40 nanometers (1 nm = 10^{-9} m). A chemical transmitter substance released by the presynaptic neuron diffuses across the cleft and excites, or in some cases inhibits, the membrane of the postsynaptic neuron. Any one chemical transmitter has only an inhibitory or excitatory action at any given synapse. The other type of synapse, called an **electrical synapse**, has direct electrical coupling, with no intermediate chemical transmitter. Membranes of the two neurons at such synapses are less than 5 nm apart, and the nerve impulse, which is in the form of an electrical (ionic) current, flows directly from the presynaptic neuron to the postsynaptic neuron through intercellular channels.

As junctions in a complex communications network, synapses are control points where information may be passed, with or without modification, to the next neuron in the network. The effects of impulses arriving at synapses can be additive, or they can reduce or prevent initiation of impulses in the next neuron. Thus, the pattern of the nerve impulses traveling away from a synapse often differs from that of the arriving impulses. In effect, synapses are gates that determine what information is passed on. They are the basis of integration in the nervous system, determining whether or not a given environmental event will lead to changes within the body and ultimately to change in an animal's behavior.

The importance of synapses as control points is indicated by their appearance: large numbers of presynaptic neurons form scores of synapses with a single postsynaptic nerve cell body (Figure 10.3). The complexity of the resulting pathways can be enormous. There are several hundred million neurons in a typical fish brain, and the human brain consists of some 100 billion neurons, each having synaptic junctions with many others. Obviously, the task of tracing and determining the functions of neural pathways in complex nervous systems is extremely difficult, and is one reason the simpler nervous systems of earthworms, molluscs, and arthropods are often used in neurophysiological research.

Junctions do not only occur between neurons. **Neuroeffector junctions** occur between neurons and nonnervous structures, including unicellular and multicellular glands, chromatophores (color cells) in some species, and muscles. (Muscle-nerve neuroeffector junctions, called myoneural junctions, are discussed in Chapter 9, p. 208; Figure 9.11.) Neuroeffector junctions enable nerve impulses to activate or inhibit cells and organs directly.

Nerve Impulses and Their Transmission

The Nerve Impulse. Nerve impulses represent information being transmitted over a communication network of neurons. Therefore, one of the most basic problems of neurophysiology has been to learn what a nerve impulse actually is. The key to understanding the nerve impulse was the discovery, many years ago, that a difference in electrical charge exists across a cell membrane. Insertion of very fine electrodes into animal and plant cells revealed that the outer surface of a cell's membrane carries a higher (more positive) electrical charge than the inside (Figure 10.4). The difference in charge across the membrane represents an electrical potential that in nerve cells often amounts to about 70 millivolts (mV). Because this **membrane potential** exists when a cell is at rest (i.e., in the absence of a nerve impulse, in the case of nerve

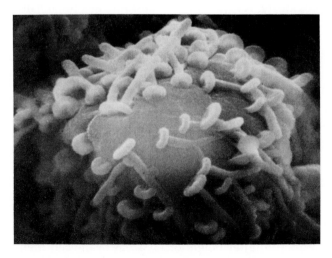

Figure 10.3 Synapses. There may be large numbers of synapses on the cell body of a neuron. This scanning electron micrograph of the cell body of a neuron from the mollusc *Aplysia* shows synaptic knobs of telodendrites on axons of other neurons. Each knob forms a synapse with the cell body (6000×).

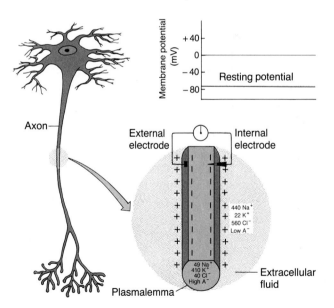

Figure 10.4 Membrane potential of a resting axon. The electrical potential (about −70 mV) of a resting axon's membrane is measured by insertion of a microelectrode through the membrane. The membrane is more positive outside than inside, owing to unequal distribution of cations (Na⁺ and K⁺), anions (Cl⁻), and protein molecules (A⁻). Ionic values are given in millimoles per liter of water.

cells), it is called a **resting potential**. A cell membrane having such a charge difference across its surface is said to be **polarized**.

Polarization results from an unequal distribution of ions across the cell membrane. The principal ions involved are the positively charged cations K^+ and Na^+ and the negatively charged anion Cl^- and proteins A^-. As indicated in Figure 10.4, Na^+ and Cl^- concentrations are higher outside the cell, while K^+ and A^- concentrations are higher inside. Two main factors account for this unequal distribution across the resting nerve cell membrane. First, the membrane is relatively impermeable to Na^+ and A^-, and second, the membrane allows free movement of K^+ and Cl^-. Consequently, while only a relatively small number of sodium ions pass inward, large numbers of potassium ions are able to pass outward toward the region of lower K^+ concentration surrounding the cell. This means that more positive ions are being transferred to the outside than to the inside, and therefore, the outside of the membrane becomes relatively more positive than the inside (see Figure 10.4). Eventually, the tendency of K^+ to diffuse outward is counterbalanced by the excess positive charges on the outside of the cell (because like electrical charges repel), and net transfer of K^+ ends. If there were no mechanism to prevent it, the continuing slow transfer of Na^+ into the cell would eventually lead to a new distribution of ions on the two sides of the membrane. (This would occur according to the conditions of the **Donnan equilibrium**: When two solutions are separated by a membrane permeable to some, but not all, ions in the solutions, a net movement of diffusible ions occurs until there is an equal number of positively and negatively charged ions in the cytoplasm on each side of the membrane.) However, the tendency for Na^+ to continue leaking slowly into the cell (and for K^+ to leak out of the cell) is countered by the active transport by the membrane of Na^+ out of the cell and K^+ into it. Ions such as H^+, Ca^{2+}, K^+, and Na^+ are often actively transported across cell membranes, but so general and important is the active transport of Na^+ and K^+ that the mechanism has been named the sodium-potassium pump (**Na^+-K^+ pump**). In fact, cells may use more than a third of their ATP to actively transport sodium and potassium ions. Thus, maintaining the nerve cell membrane's resting potential is an energy-costly process requiring ATP.

Because the resting potential, as we have just seen, depends on differences in the cell membrane's permeability to various ions, as well as on the Na^+-K^+ pump, if the permeability of the membrane is in some way changed, the electrical potential across it also changes. When a nerve impulse arrives at a region of the nerve cell membrane, a brief change in the membrane's permeability to Na^+ occurs. The subsequent inrush of Na^+ reduces the potential difference across the membrane because greater numbers of these positively charged ions enter the cell. Perhaps the simplest way to picture a nerve impulse (also called an **action potential** or **spike**) is as a narrow band of changed membrane permeability (and thus membrane potential) moving along an axon or dendrite (Figure 10.5a). If recording electrodes are placed at intervals along an axon, the changed membrane potential, called a **depolarization**, can be detected passing each electrode in succession. Such measurements show that the change actually does travel along the axon, in some cases at speeds greater than 100 meters per second (m/s)—well over 200 mph! Figure 10.5 shows several other properties of an impulse as well. First, the changes in potential are extremely brief, the membrane returning to its resting potential in about 1 ms. Second, the depolarization briefly overshoots to positive values, making the membrane temporarily more positive inside than outside. Third, depolarization is "slow" until a **threshold** of about -50 mV has been reached, and then it accelerates. And finally, as the membrane is being repolarized, the charge briefly undershoots the resting potential. The changes in electrical potential are related to shifts in the distribution of Na^+ and K^+, as indicated in Figure 10.5b. Although it is the rapid efflux of K^+ that restores the resting potential, an additional period of time is required before the Na^+-K^+ pump reestablishes the normal distribution of Na^+ and K^+ found in the unstimulated neuron. Before the resting potential is restored, the neuron is in a **refractory period**; that is, it cannot respond to stimuli of normal strength.

Axonal Transmission. As a narrow band of changed membrane potential traveling along an axon, the nerve impulse is not the same as an electrical current in a wire. What is passing along the axon is a change in electrical *potential*, not an electrical *current*. What, then, causes a change in membrane potential to sweep along an axon? The answer is that as soon as an action potential (nerve impulse) has been triggered, it causes local currents that alter neighboring areas of the cell membrane, making them more permeable to Na^+ and allowing an influx of these ions (Figure 10.6). Depolarization then takes place in a succession of neigh-

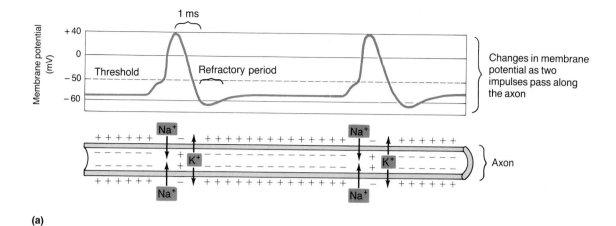

(a)

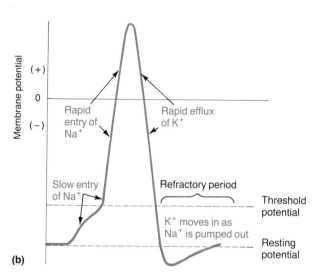

(b)

Figure 10.5 Characteristics of the nerve impulse. (a) Two nerve impulses moving along an axon. At the position of the impulses, the membrane potential is reversed (+ inside, − outside), owing to the influx of Na^+. The resting potential is restored as K^+ moves rapidly outward. During the refractory period, Na^+ is pumped out in exchange for K^+, restoring the ion distribution that is normal for the resting axon. (b) This graph is a summary of the ion movements associated with the major phases of the nerve impulse. Movements of ions are facilitated by the opening and closing of ion channels (gates) in the nerve cell membrane.

boring areas down the length of the axon. The action potential is thus *self-propagating*. An analogy can be seen in the burning of a firecracker fuse. The fuse burns in only one direction because it self-destructs, whereas the nerve impulse travels in only one direction because the membrane just traversed has a brief refractory period.

Propagation of nerve impulses along a myelinated axon is somewhat different from the process just described, because myelin is a good insulator with high resistance to local current flow and to changes in membrane potential. The nodes of Ranvier represent breaks in the insulation. They are sites where resistance is low and where action potentials are triggered progressively along the axon. Thus, impulse propagation occurs as a series of "jumps" from one node to the next, rather than as a continuous passing of the impulse from one area to an immediately adjacent area (see Figure 10.6).

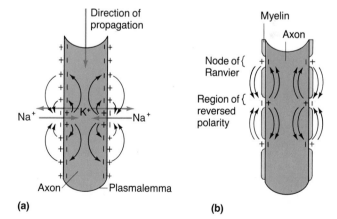

(a)   (b)

Figure 10.6 Propagation of a nerve impulse along an axon. (a) Depolarization of the axon membrane at the position of the nerve impulse produces local currents (movement of positive ions, as indicated by the black arrows) that, in turn, depolarize adjacent regions to the threshold level in those regions. However, the nerve impulse can proceed only in the region that is not refractory (in this case, the impulse is moving toward the bottom of the figure). (b) In myelinated fibers, the depolarizing currents are restricted to the nodes of Ranvier. Depolarization, and therefore the nerve impulse, "jumps" from node to node. Saltatory conduction is much faster than conduction in unmyelinated fibers, in which depolarization is necessarily continuous.

Saltatory (jumping) **conduction**, as this is called, is faster than nonsaltatory conduction, an advantage in large animals, where conduction distances may be great. It also saves energy because restoration of the resting distribution of ions by energy-consuming Na^+-K^+ pumps is limited to the node regions.

Impulses and Information.   Electrode studies have also revealed that all impulses that travel along a given neuron have the same size, or **amplitude**. If the threshold is reached, a full-sized impulse results, but no impulse occurs when depolarization fails to reach the threshold value. In short, an **all-or-none principle** applies, producing nerve impulses of constant amplitude. This means that the information content of nerve impulses cannot be based on their size. Instead, information must be conveyed by the **frequency** of impulses conducted along an axon.

Impulse frequency depends on the strength of the stimulus. Pressure sensors in an animal's skin, for example, generate more impulses per second in responding to a heavy touch than in responding to a light one. The same is true for other receptors, such as those sensing temperature or chemicals; they respond with greater frequency to more intense stimuli. At the level of the CNS, impulse frequency therefore provides information about the magnitude of environmental factors. Impulses travel along specific nerves, such as the optic nerve or auditory nerve, to specific parts of the brain, where their interpretation determines what an animal actually perceives (sound, vision, taste, etc.). Other methods used by the nervous system to code sensory information are described in Chapter 12 (see p. 271).

Synaptic Transmission.   Transmission of impulses across most synapses is a chemical event. A nerve impulse arriving at a synapse causes the synaptic knobs of the presynaptic neuron (the neuron carrying the impulse to the synapse) to release a **transmitter substance** (Figure 10.7a). The substance diffuses across a synaptic cleft to the postsynaptic neuron, where it may initiate or inhibit a nerve impulse. More than 50 transmitter substances have been identified, of which **acetylcholine** is probably the most common and the best understood. Acetyl-

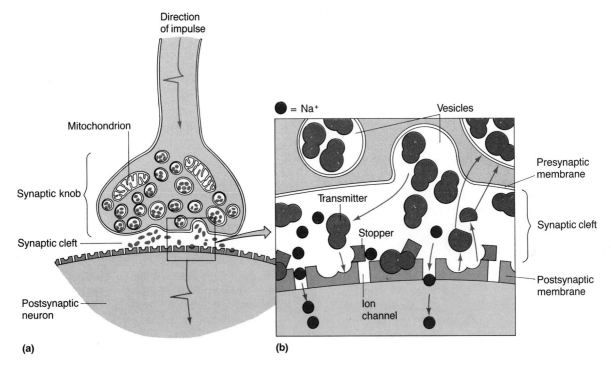

Figure 10.7   Events at a synapse. (a) At a chemical synapse, a synaptic knob of a presynaptic neuron is separated from the postsynaptic neuron by a synaptic cleft of about 20–40 nm. When a nerve impulse arrives, neurotransmitter molecules contained in presynaptic vesicles are released into the cleft. Neurotransmitter molecules diffuse across the cleft and bind to neuroreceptor molecules in the cell membrane of the postsynaptic neuron. Binding of the neurotransmitter depolarizes the postsynaptic membrane, initiating a nerve impulse. (b) A model of events at a chemical synapse, suggesting how the neurotransmitter, by unblocking ion channels, might alter the permeability of the postsynaptic membrane to Na^+, causing depolarization and initiating a nerve impulse.

choline is synthesized in the synaptic knobs of telodendrites and in the nucleus and stored there in synaptic vesicles. The arriving nerve impulse in the presynaptic neuron is thought to facilitate entry of Ca^{2+} into the synaptic knobs, and the Ca^{2+} then causes release of acetylcholine into the synaptic cleft.

Figure 10.7b shows a model of how synaptic transmission may occur. First, vesicles containing acetylcholine move to the presynaptic membrane and rupture, spilling acetylcholine into the synaptic cleft. After diffusing 20–40 nm across the synaptic cleft in less than a microsecond (μs), the acetylcholine molecules attach to receptor sites on the postsynaptic membrane. A receptor site and a transmitter molecule probably fit together chemically, somewhat like an enzyme and its substrate. The attachment of acetylcholine to the receptor sites changes the membrane's properties; the nature of the change depends on the nature of the postsynaptic membrane. In some cases, acetylcholine increases the postsynaptic membrane's permeability to Na^+, as in the model shown here. This causes an influx of Na^+ that depolarizes the membrane, and if the threshold is reached, an action potential results.

The synapse immediately prepares for the arrival of other nerve impulses by clearing acetylcholine from the receptor sites. The enzyme **acetylcholinesterase** does this within about 2 ms by splitting acetylcholine into acetate and choline. Acetate and choline are returned to the synaptic knobs, where acetylcholine is resynthesized. The pest control industry has taken advantage of this vital reaction in controlling insects: Organophosphate insecticides (and certain nerve gases stockpiled by military forces) are potent acetylcholinesterase inhibitors. These poisons interrupt the orderly transmission of nerve impulses by preventing the enzymatic breakdown of acetylcholine. Neural coordination of many physiological processes is disrupted, often resulting in death.

The events just described take place at an **excitatory synapse**, where the postsynaptic membrane is depolarized and an action potential results. There are also **inhibitory synapses**, in which the electrical potential across the postsynaptic membrane is increased rather than decreased, perhaps as a result of an increased permeability to K^+ or an influx of negatively charged chloride ions (Cl^-) instead of Na^+. A resulting increased electrical potential, or **hyperpolarization**, suppresses the formation of action potentials. Inhibitory and excitatory synapses working together determine how much and what kind of information crosses the synapses, which in turn determines whether the action of an effector organ is accelerated, slowed down, or stopped. As shown in Figure 10.3, a single motor neuron may receive inhibitory and/or excitatory signals from numerous presynaptic neurons. This provides a mechanism for variation in the type of response an animal may have to a stimulus. Moreover, we now know that more than one kind of transmitter may be released at the same synapse. (See the Essay on chemical messengers, p. 228).

Impulse transmission along a series of two or more neurons with chemical synapses is typically unidirectional, even though axons may transmit in either direction under experimental conditions. Synaptic vesicles usually occur only in the presynaptic neuron of each synapse, whereas receptor sites are confined to the postsynaptic neuron. Release and diffusion of the transmitter substance must therefore be toward the postsynaptic neuron. Yet, there are exceptions. Cnidarians, such as sea anemones, have synapses that are bidirectional because the adjacent neurons contain both presynaptic vesicles and receptor sites. Bidirectional impulse conduction may be important in helping radially symmetrical animals, which lack a central nervous system, to make general escape responses (such as overall shortening of the body) to stimuli that may be as narrowly confined as a touch to the body or as broad as a change in light intensity.

Animal Nervous Systems

The complexity of an animal's behavior is usually related to the complexity of its nervous system. Behavioral responses to environmental stimuli such as light, heat, or pressure may be limited to general increases or decreases in rate of movement or, at the other extreme, may involve the ability to reason. As the basis of animal behavior, nervous systems play a key role in enabling animals to exploit diverse environments. It is not surprising that animals in the phyla Arthropoda, Mollusca, and Chordata—the three most successful phyla (in terms of the capacity to exploit marine, freshwater, and land environments)—have highly complex nervous systems.

Major Evolutionary Trends

Major evolutionary trends in nervous systems correspond roughly with the evolutionary history of the animal kingdom. The general degree of com-

Essay: Nervous Systems Rely on Many Chemical Messengers

One of the most dramatic advances made in neurobiology during the past decade has been the discovery that there are large numbers of neurotransmitters, the messenger substances released from neurons at synapses and other junctions. Until about ten years ago, only five or six "classical" neurotransmitters were recognized, including acetylcholine, epinephrine, and norepinephrine. Now about 50 have been identified, and it is estimated that there may be as many as 200. Many of the newly discovered neurotransmitters are peptides, generally called **neuropeptides**.

Extremely sensitive antibody-coupled fluorescent-staining techniques and electron microscopy have made it possible to pinpoint the locations of many of the neuropeptides. Early studies suggested that each kind of transmitter occurred in a separate system of neurons in the brain. This would seem reasonable, given the astronomical numbers of neurons in the central nervous system of mammals. However, researchers soon found that a neuropeptide and a classical transmitter could occur as cotransmitters in the same neuron. The coexistence of two or more transmitters in a neuron now seems more the rule than the exception (Table E1).

What are the benefits of multiple messengers between neurons? Although researchers have few definitive answers, there are several hypotheses. For one thing, multiple messengers could carry more information than one. Thus, two transmitters could, more or less independently of each other, transfer different kinds of messages at synapses. While a classical transmitter might be involved in rapid transmission of nerve impulses, a neuropeptide could help regulate long-term intracellular events in an effector organ. Such a system, using one communications network (part of the nervous system) to produce several different effects, would be energy-efficient.

There is also mounting evidence that under some conditions, the classical transmitters and neuropeptides do not function independently at synapses. Instead, they may interact with each other to cause greater-than-usual activation of the postsynaptic membrane, particularly when neurons are already firing at a high rate. This may be a way of forcing through important messages when the nervous system's "lines are busy."

Research on multiple messenger phenomena is still in its infancy, but the discovery of cotransmitters is already influencing how neurobiologists view synaptic events. In a larger perspective, research on cotransmitters is also likely to influence our understanding of brain functions such as learning and memory and may help in the treatment of certain neurological disorders.

Table E1   Classical Transmitters and Peptide Transmitters Coexisting in Neurons

Classical Transmitter	Peptide Transmitter	Brain Region
Acetylcholine	Enkephalin	Cochlear nerves of guinea pig
	Substance P	Pons of rat
Epinephrine	Neurotensin	Medulla oblongata of rat
	Neuropeptide Y	Medulla oblongata of rat
Norepinephrine	Neuropeptide Y	Medulla oblongata of rat and human
5-Hydroxytryptamine	Enkephalin	Medulla oblongata and pons of cat
	Substance P	Medulla oblongata of rat and cat

plexity in body organization of animals is reflected in the nervous system, and evolutionary changes in nervous systems played a central role in increasing animal versatility.

Nervous systems are frequently categorized as *primitive* or *advanced*. In advanced nervous systems, nerves carry impulses into and out of a central nervous system. Primitive nervous systems lack the CNS. Their dominant feature is the arrangement of nerve cells into a **nerve net**. Impulses from certain sensory structures may be conducted along specific neuronal pathways in a nerve net, but there are no tracts or nerve cords as in advanced systems.

One highlight of animal evolution—the development of bilateral symmetry—closely paralleled the evolution of centralized nervous systems. Among the least complex animals, the cnidarians are radially (or biradially) symmetrical; cnidarian nervous systems are composed entirely of one or more nerve nets. Most other animals are bilaterally symmetrical, move forward through their environment, and have a CNS. For bilateral, forward-moving animals, there is an obvious advantage in having major sensory and feeding structures concentrated at the anterior end of the body, an evolutionary trend called **cephalization** (from the Greek *kephale*, "head"). Corresponding specialization and enlargement of the central nervous system serving the sensory and feeding structures produced cephalic ganglia, or a simple "brain."

Localization of most synaptic junctions in the CNS makes it the center for information processing and integration, and the integrative ability of the CNS is the most distinctive feature of advanced nervous systems. Many animals with advanced systems have one or more **nerve plexuses** (networks of small nerves) that resemble nerve nets. Nerve plexuses are part of the peripheral nervous system, which consists of nerves that serve the CNS. Consequently, nerve plexuses are subordinate to the CNS.

One of the most significant developments in advanced nervous systems was the evolution of faster nerve impulse conduction. We have already examined how saltatory conduction in myelinated nerve fibers increases impulse propagation speed in vertebrate nervous systems. Another effective development has been the evolution of giant nerve fibers in the CNS of many invertebrates. The diameter of a **giant axon** may be more than 30 times the diameter of a typical vertebrate axon. This allows extremely rapid propagation of nerve impulses because the electrical resistance of axonal cytoplasm to current (depolarization) flow is inversely proportional to an axon's cross-sectional area. One of several giant axons extending the length of the earthworm's ventral nerve cord conducts at 17–25 m/s, a thousand times faster than the 0.025 m/s conduction of ordinary thin fibers. Giant axons have great adaptive value, particularly in alarm or escape reactions. The sudden withdrawal of an earthworm retreating into its burrow, the rapid tail flip of a crayfish backing away from danger, and the darting movement of a squid as it jets after a school of fish are all responses mediated by giant nerve fibers. (The giant fiber system of squids is discussed in Chapter 24, p. 575; Figure 24.18b.)

Invertebrate Nervous Systems

This section briefly discusses the nervous systems of a series of animals with increasingly more complex nervous systems.

Cnidarians. Cnidarian behavior, which in some species includes swimming, somersaulting, capturing prey, and a variety of body shape changes, might seem almost too complex to be coordinated by one or two simple nerve nets of some 100,000 neurons. However, a honeybee with about 100,000 neurons has far more complicated behavior, including a complex social system and a unique body language. Viewed from this perspective, cnidarian behavior is indeed simple and reflects a fundamentally different way of arranging the same number of neurons. A closer look at nerve nets shows that their apparent limitations are offset by important advantages for radially symmetrical animals.

The cnidarian body wall consists of an epidermis separated from an inner gastrodermis by a region called the mesenchyme (or mesoglea). Figure 10.8 shows the pervasive nature of the nerve net in a sea anemone, with the neurons forming a meshwork in contact with the contractile parts of the epidermal cells and with sensory endings projecting from the body surface. A similar meshwork occurs at the base of the gastrodermis.

Neurophysiological studies with sea anemones such as *Metridium* (see Figure 21.6a) show how a nerve net can coordinate more than one behavior. If one of *Metridium's* tentacles is bitten by a fish or its body is prodded by an experimenter, the sea anemone does two things. First, it exhibits a closure reflex: A sphincter muscle around the mouth and retractor muscles contract, closing the mouth and withdrawing the tentacles. Then, some seconds later, muscle cells in the body wall contract, shortening the body. The response that will most likely protect the delicate tentacles and mouth occurs quickly, followed by a slower, general change in body shape. How are these two behaviors, one rapid and the other slow, mediated by a netlike arrangement of neurons?

The answer lies in the discovery that nerve impulses travel at different rates in various regions of the nerve net. This was demonstrated by giving *Metridium* electrical stimuli at selected points on its body and recording how long it took for the sphincter muscle around the mouth to respond by contracting. By far, the highest conduction rates were in the parts of the nerve net responsible for the closure reflex. The neurons in the fast-conducting parts of the nerve net are larger and tend to be aligned with each other and with the muscles they serve. Thick axons conduct impulses faster than thin ones, while their alignment gives the impulses a more direct path to follow. Thus, a nervous system without nerves, ganglia, or a brain can be par-

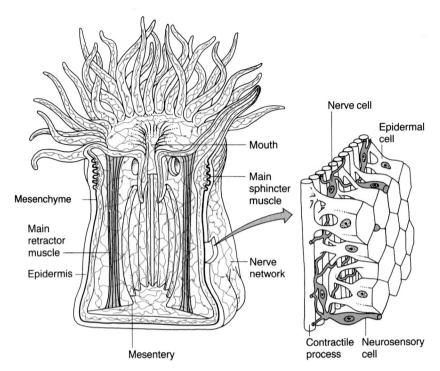

Figure 10.8 The nerve net in a sea anemone. The nerve net innervates the retractor and sphincter muscles and is in intimate contact with the contractile portions of the epidermal cells (inset).

titioned functionally by differential rates of impulse conduction.

Many cnidarians, including many sea anemones, have more than one nerve net, and each nerve net can perform several discrete functions. Studies with the free-swimming ephyra (immature medusa) of the jellyfish *Aurelia* (see Figure 21.16) show that the rhythmic, symmetrical contractions of the swimming beat are coordinated by a nerve net of large neurons servicing the muscle fibers involved in locomotion. A separate, more diffuse nerve net of smaller and very delicate neurons is distributed over the entire epithelium. This second nerve net coordinates feeding, allowing the ephyra to use one or more of its eight arms to sweep food toward its mouth. Swimming and feeding behavior can occur independently.

Flatworms. The typical flatworm nervous system has some of the features of an advanced system, though in an elementary form. As shown in Figure 10.9, the freeliving flatworm *Dugesia* (planarian worm) has longitudinal nerve cords that provide pathways for rapid transfer of information from one end of the animal to the other. Transverse nerves linking the longitudinal nerve cords help coordinate the activities of the animal's two sides. Photoreceptors in the eyespots and odor-sensitive

cells in the flaplike auricles send nerve fibers into a pair of cephalic ganglia located below the eyespots. The ganglia, though sometimes referred to as a brain, do little more than funnel nerve impulses from the sensory nerve fibers into the longitudinal nerve cords for transmission to the musculature and other organs. A planarian can detect and respond to chemical trails ("odors") of food in the water, and it can retreat from bright light to the safety of dimly lit cavities beneath rocks.

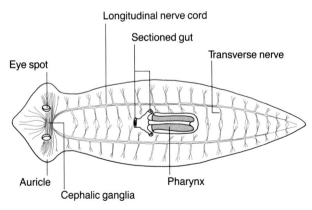

Figure 10.9 The nervous system of a flatworm. This dorsal view of a planarian worm (*Dugesia*), with most of the internal organs deleted, shows the ladderlike arrangement of the longitudinal nerve cords and the transverse nerves. A nerve net (not shown) is important in pharynx movements.

Annelids. The nervous systems of segmented worms, as typified by the earthworm, highlight the influence of body segmentation (metamerism) on nervous system structure and function. Figure 10.10 shows the central nervous system and nerves in the anterior body segments. Small cerebral ganglia serve as the brain. They are linked by circumpharyngeal connectives to subpharyngeal ganglia beneath the pharynx. The ventral nerve cord extends posteriorly from the subpharyngeal ganglia, forming a pair of fused ganglia in each segment. The ventral nerve cord gives off three pairs of lateral nerves in each segment posterior to the third segment. Each lateral nerve carries only a small number of sensory axons, even though there are thousands of sensory endings in the body wall; consequently, the CNS receives only broad patterns of information without precise detail, a factor limiting the variety of environmental information to which an earthworm might respond.

Several giant axons (described earlier) extend through the length of the ventral nerve cord. Because giant axons transmit nerve impulses very rapidly to locomotor musculature throughout the body, a worm can retract its entire body quickly, in contrast to its slow peristaltic locomotor move-ment. A quick retreat to the burrow by the usually slow-moving earthworm helps protect it from predators.

An earthworm's nervous system has many segmental reflex arcs; sensory and motor neurons in each body segment form synapses with each other or are linked through interneurons within the ganglia and nerve cord. The peristaltic locomotion of annelids may depend on both segmental reflex arcs and the conduction of impulses from one segment to the next by the ventral nerve cord. Simple surgical experiments show that if the cord is severed but the worm is otherwise left intact, normal peristaltic body movements occur anterior and posterior to the surgical site, but the movement of the two regions is not coordinated. However, if a worm is sectioned so that the two halves of the body are joined by the ventral nerve cord alone, normal peristaltic movements still occur in both regions. Moreover, the movements remain completely coordinated with each other. Only segment-by-segment conduction along the nerve cord can account for this coordination of the segmental locomotor reflexes.

These and other studies show that the ventral nerve cord and its ganglia act as centers for motor acts coordinated by central conduction. Whether

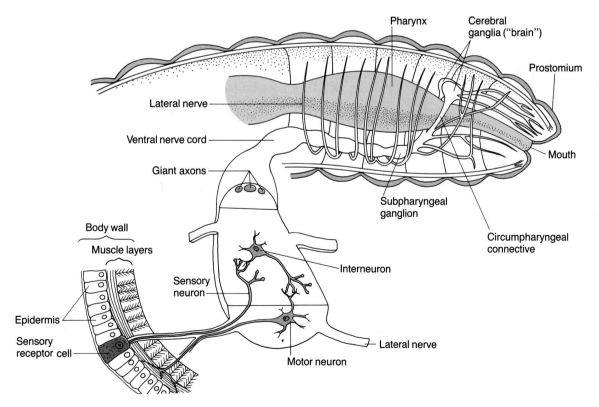

Figure 10.10 The nervous system of an earthworm. The system includes several anterior ganglia and a ventral nerve cord with segmental ganglia.

reflex arcs mediate body peristalsis remains a question. A growing body of data suggests that an alternative neuronal mechanism called a **central pattern generator,** or **central neuronal oscillator**, mediates peristaltic movements and other forms of rhythmic behavior. Central pattern generators (CPGs) consist entirely of interneurons that generate impulses that are transmitted through motor neurons, thus producing motor behavior without sensory input. CPGs can be detected by eliminating sensory input and monitoring rhythmic impulses generated by the CNS. Studies with leeches show that if all sensory neurons are cut, the CNS elicits rhythmic impulses matching the animal's normal undulatory swimming rhythm. (A CPG could be demonstrated directly if all sensory input were eliminated and the leech continued to swim. Unfortunately, to eliminate sensory neurons, the motor neurons must also be cut, and this immobilizes the animal.) Figure 10.11 shows the neuronal circuits that might function in a CPG-driven swimming sequence. Other research with leeches has shown that each ganglion in the ventral nerve cord has its own CPG and that undulatory swimming results from impulses generated by sequential firing of neuronally coupled CPGs along the ventral nerve cord.

One might ask what role the annelid brain plays in locomotion. Brain removal causes an earthworm to become restless and hyperactive; but the worm can still crawl normally, eat, and even copulate. Just the opposite occurs if the subpharyngeal ganglia are removed: The earthworm lapses into inactivity. Apparently, the subpharyngeal ganglia contain excitatory motor centers that the brain regulates by an inhibitory action. Thus, the brain seems to exert a moderating influence on the motor centers, increasing or decreasing their activity to produce corresponding changes in the activity of the worm as a whole.

Arthropods. The arthropod nervous system further demonstrates the diverse abilities of a segmentally arranged nervous system. Clearly, the great evolutionary success of arthropods is in part due to the localization of particular functions in specific body regions and the fusion of parts of the nervous system into large, complex nerve masses. In insects, for example, six segments are fused, forming a head bearing the major sense organs and feeding structures (Figure 10.12). The corresponding ganglia are enlarged and partly fused, forming the supraesophageal and subesophageal ganglia. The supraesophageal ganglion is generally called the brain because it is an integrative center and exerts considerable control over other parts of the nervous system. The three-segmented thorax usually has three pairs of legs and two pairs of wings, along with much-enlarged ventral ganglia coordinating locomotor movements. But in the abdomen, where

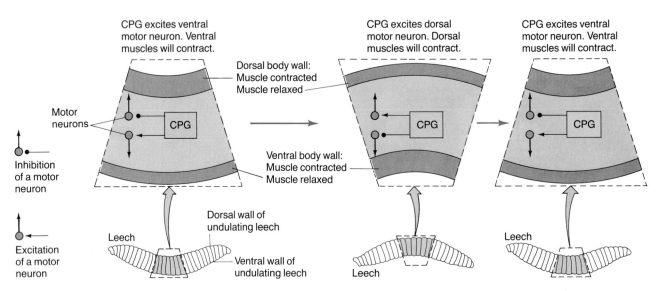

Figure 10.11 Control of rhythmic activity. Strong experimental evidence indicates that central pattern generators (CPGs) located in the segmental ganglia control undulatory swimming in leeches. This figure shows the possible action of a CPG mediating the body movements needed for undulatory swimming. Note that only interneurons and motor neurons are involved. Recent studies indicate that CPGs play an important role in mediating other rhythmic activities in annelids and other animals, and research on CPGs is one of the most active areas in neurobiology today.

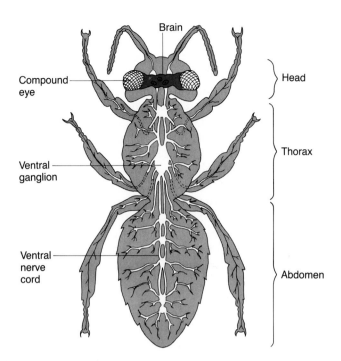

Figure 10.12 The nervous system of an arthropod. The arthropod nervous system resembles that of the annelids, but shows less uniform segmentation. Ganglia controlling the leg and wing musculature, for example, are typically enlarged and sometimes fused.

digestive and reproductive functions are controlled to a large degree by hormones, the ventral ganglia are usually small.

Locomotion, copulation, eating, and respiration are largely coordinated within or among the segmental ganglia. As in annelids, coordination may depend on chains of segmental reflexes. There is also strong evidence of the role of CPGs in controlling rhythmic activities, as in annelids, and the

ventral ganglia can exert considerable independent control. An insect leg surgically isolated with its ventral ganglion can make coordinated stepping movements when stimulated. In fact, such an isolated leg preparation can even "learn" *not* to step if it receives a small electric shock each time it does so.

Given the relative independence of the ganglia, what is the role of the brain? It functions mainly to integrate competing information flowing in from sense organs and to coordinate appropriate responses based on past experiences and an animal's physiological state. Though proportionately large by annelid standards, the arthropod brain is nonetheless very small, weighing less than a milligram in an insect the size of a honeybee. Functions that depend on very large numbers of neural circuits or connections can be developed only to a minor degree.

Molluscs. The general structure and some of the variety found in molluscan nervous systems are described in Chapter 24 (p. 575). Cephalopod molluscs, such as squids and octopuses, have highly complex systems. The octopus brain, composed of some 10^8 neurons, is proportionately larger in relation to body size and more highly organized than any other invertebrate brain. As shown in Figure 10.13, the octopus brain consists of a tight cluster of fused ganglia surrounding the esophagus. The ganglia are organized into highly specialized nerve tracts and lobes that form separate functional regions. Subesophageal lobes and basal regions of supraesophageal lobes are locomotor and postural centers, coordinating neuromuscular activity, especially that of the eight arms. The large optic

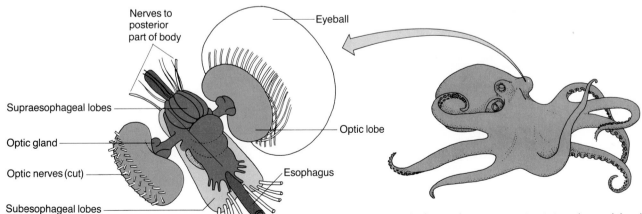

Figure 10.13 The brain of an octopus. Consisting of several fused anterior ganglia, this is the largest and most complex invertebrate brain. Compared to other invertebrates, octopuses are quite intelligent, relying extensively on learned responses in their everyday behavior.

lobes integrate information received from the animal's well-developed eyes. As demonstrated in numerous experimental studies, octopuses have a strong memory and a marked ability to learn rapidly. The dorsal regions of their brain are learning and memory centers. It is not surprising that learned responses are a significant component of octopus behavior. Many neurophysiologists believe that the complexity and integrative capacity of the octopus brain surpass that of certain vertebrates, such as lampreys and some fishes.

Vertebrate Nervous Systems

Compared to most invertebrate nervous systems (the cephalopod system excepted), vertebrate nervous systems are more complex and provide greater integrative ability. The vertebrate brain is also larger in relation to body size and provides a greater capacity for learning and memory. Relative size and integrative capacity are by far the greatest in the mammals. The brain of a large dog might weigh about 135 g. Still, this is small in relation to primate brains, which may contain billions of neurons and virtually countless synapses. A small gorilla, for example, and a human, both with body weights about equal to that of a large dog, have brain weights of about 430 and 1350 g, respectively.

The vertebrate central nervous system also has a different plan of construction than that of invertebrates. In contrast to the ventral, solid nerve cords of most invertebrates, the vertebrate nerve cord, or **spinal cord**, is dorsal, tubular, and single. The spinal cord is enclosed and protected by the vertebral column. Its anterior end is enlarged and specialized as the **brain**, which is encased and protected by a cartilaginous or bony brain case called the **cranium**. Both the brain and the spinal cord are enwrapped by one or more connective tissue membranes called the **meninges**. The **spinal canal** at the center of the spinal cord continues anteriorly as four **ventricles** (cavities) within the brain. **Cerebrospinal fluid** circulates through the spinal canal and ventricles, as well as in the space between the meninges and nervous tissue, where it transports nutrients and wastes, and serves as a shock absorber.

The vertebrate peripheral nervous system is also structurally unique. It consists of pairs of nerves that emerge from the brain and spinal cord and branch to various parts of the body. Reflecting the basic segmental arrangement of the vertebrate body, paired segmental **spinal nerves** emerge from the spinal cord through notches (foramina) on each side of each vertebra (Figure 10.14a). Each spinal nerve has a **dorsal root** and a **ventral root**, which unite a short distance lateral to the spinal cord. In mammals and birds, the dorsal root contains

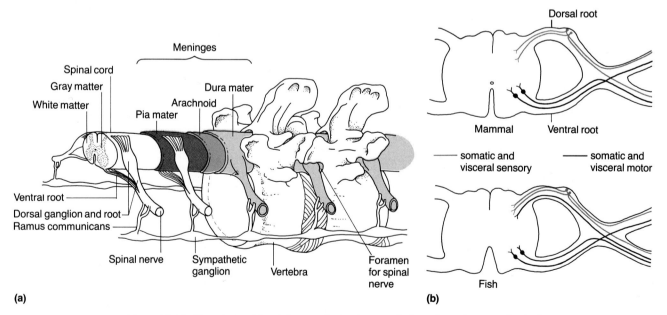

(a)

(b)

Figure 10.14 The relationship of the vertebral column, spinal cord, meninges, and spinal nerves. (a) Note the row of chain (sympathetic) ganglia connected to the spinal nerves by nerve branches called rami communicans. These ganglia are part of the autonomic nervous system, a major component of the peripheral nervous system. (b) The spinal nerves of a mammal and a fish, showing the distribution of visceral and somatic neurons in the dorsal and ventral roots.

mainly sensory (afferent) neuron fibers carrying impulses toward synapses within the spinal cord. Sensory nerve cell bodies are concentrated within a conspicuous swelling, the **dorsal root ganglion**, on each dorsal root. Sensory fibers that carry impulses from sense organs in the skin, muscles, and tendons are called **somatic sensory fibers**. Those carrying impulses from the digestive tract and other viscera are called **visceral sensory fibers** (Figure 10.14b). Motor (efferent) neuron fibers traverse the ventral root, carrying impulses out of the CNS to effector organs. **Somatic motor fibers** carry impulses to striated skeletal muscles. **Visceral motor fibers**

carry impulses to internal glands and smooth muscles of the gut, blood vessels, and other internal organs. Unlike the condition in mammals and birds, in many fishes and amphibians the dorsal root carries large numbers of visceral motor fibers in addition to sensory (somatic and visceral) fibers (see Figure 10.14b).

Paired **cranial nerves** arise from the brain. They are fundamentally similar to the spinal nerves but are not segmentally arranged and do not have dorsal and ventral roots. The cranial nerves are listed in Table 10.1 with a brief description of their sensory and motor functions. These nerves form a

Table 10.1 The Cranial Nerves of Vertebrates*

Cranial Nerve		Kinds of Fibers**	Function, Source, or Destination
Name	No.		
Terminal[†]	0	Sensory, S	Uncertain, possibly olfactory
Olfactory	I	Sensory, S	From olfactory epithelium
Optic	II	Sensory, S	From eyes
Oculomotor	III	Motor, S, V	To four of the six eye muscles
Trochlear	IV	Motor, S	To superior oblique eye muscle
Trigeminal	V	Sensory, S Motor, V	From head To lower jaw muscles
Abducens	VI	Motor, S	To lateral rectus eye muscle
Facial	VII	Sensory, S, V Motor, V	From lower jaw and taste buds To face muscles
Acoustic	VIII	Sensory, S	From inner ear
Glossopharyngeal	IX	Sensory, S, V Motor, S, V	From back of tongue and taste buds To throat, larynx, and salivary glands
Vagus	X	Sensory, S, V Motor, V	From taste buds and skin To visceral organs
Spinal Accessory[‡]	XI	Motor, V	To visceral organs; accessory to vagus
Hypoglossal[‡]	XII	Motor, S	To muscles of tongue

Key: Dark grey = somatic motor nerves. White = branchial nerves. Light grey = special sensory nerves.

*The distribution of somatic and visceral fibers in spinal and cranial nerves has important evolutionary implications. It is likely that the brain of ancestral vertebrates was very similar to the spinal cord. It is also likely that the cranial nerves were derived from segmentally arranged anterior nerves that emerged (much like spinal nerves) from the sides of the primitive brain. The dorsal and ventral roots of these primitive "cranial" nerves probably carried visceral and motor fibers in a pattern similar to that of modern vertebrates. Thus, the appearance of somatic and visceral fibers in the cranial nerves of modern vertebrates may indicate how the cranial nerves evolved.

**S = somatic, V = visceral. [†]Not in humans. [‡]Present only in amniotes (reptiles, birds, mammals); actually a motor portion of the vagus; exists as part of vagus in fishes.

complex, highly specialized part of the peripheral nervous system. But a pattern emerges if we consider which cranial nerves carry somatic fibers and which carry visceral fibers. As already mentioned, some vertebrates have visceral motor fibers in both the dorsal and ventral roots of spinal nerves. Somatic motor fibers, however, occur *only* in the ventral root of spinal nerves in all vertebrates. Four cranial nerves—III, IV, VI, and XII—constitute a group that carries somatic motor fibers. These nerves, which may be called **somatic motor nerves**, are believed to have evolved from the ventral roots of nerves that may have originated as anterior spinal nerves. Five other cranial nerves—V, VII, IX, X, and XI—contain sensory fibers and visceral motor fibers. This group is believed to have been derived from dorsal roots of anterior spinal nerves. Because they innervate structures derived from the gill (branchial) arches of vertebrates, they are called **branchial nerves**. (Gill arch evolution is discussed in Chapter 28, p. 689.) Finally, four cranial nerves— 0, I, II, and VIII—carry only somatic sensory fibers. These **special sensory nerves** carry impulses from the large sense organs of the head and are thought to have evolved uniquely with their sense organs; they may never have had distinct dorsal and ventral roots.

The Central Nervous System. Two important functions of the spinal cord in all vertebrates are the integration of reflexes occurring in the limbs

and trunk and the conduction of nerve impulses (information) to and from the brain. The brain, in turn, may make the spinal reflexes more complex. Both the brain and the spinal cord contain central pattern generators (CPGs) that control rhythmic activities without sensory input. Rhythmic activities known to be mediated by CPGs in vertebrates include breathing, heartbeat, peristaltic movements of the gut, and the various types of locomotion (swimming, walking, running, flying, and crawling).

Although rhythmic behavior seems to be mainly CPG-mediated, most rhythms are also influenced to some degree by sensory input. Thus, reflex arcs may have a significant moderating effect on CPG-controlled rhythms. Reflex arcs also mediate certain simple acts in vertebrates. Figure 10.15 illustrates how neurons are arranged in two kinds of reflex arcs. A simple, two-neuron reflex arc consists of sensory and motor neurons only. The knee-jerk, or extensor, reflex is a familiar example: Tapping just below the knee with a small hammer stretches stretch receptors in the extensor muscle, causing impulses to be sent along the sensory (afferent) neuron to a synapse with a motor (efferent) neuron in the spinal cord. Impulses then travel back to the leg and activate contraction of the extensor muscle, extending the leg. In a three-neuron reflex arc, a single interneuron links the sensory and motor neurons. Flexor reflexes with three neurons are common in the withdrawal of limbs from painful

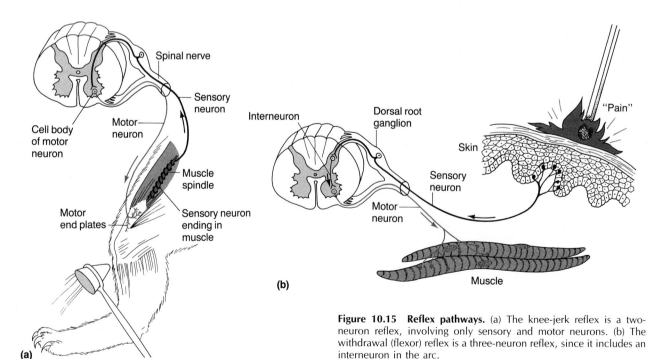

Figure 10.15 Reflex pathways. (a) The knee-jerk reflex is a two-neuron reflex, involving only sensory and motor neurons. (b) The withdrawal (flexor) reflex is a three-neuron reflex, since it includes an interneuron in the arc.

stimuli. More complex reflex arcs may involve interneurons crossing over to the opposite side of the cord or communicating with the brain. In the withdrawal from a painful stimulus, for example, it is the brain that introduces the sensation of pain—and, most likely, some verbal exclamation!

The brain exerts control over reflexes and central pattern generators. It may modify their activity in the light of other events and past experiences. Having several levels of control gives the vertebrate nervous system a remarkable complexity for integrating and coordinating incoming information and for eliciting appropriate behavior.

The vertebrate brain has three main parts, the **forebrain**, **midbrain**, and **hindbrain**, which together incorporate five structural divisions. Each division, in turn, consists of one or more functional components (Figure 10.16 and Table 10.2).

Looking first at the hindbrain, we see that its most posterior division is the **myelencephalon**, so named because it contains prominent tracts of myelinated neurons. Essentially an enlarged continuation of the spinal cord, the myelencephalon has a functional component called the **medulla oblongata**. Because it is well developed in all vertebrates, whereas other parts of the brain are not, the myelencephalon is considered the oldest part of the brain. The medulla oblongata has several important functions. Large white (myelinated) tracts of sensory and motor fibers in its walls transmit information and commands between the spinal cord and the rest of the brain. Cranial nerves VI through XII are associated with this region. Most important, the medulla oblongata contains control centers for such vital processes as breathing and regulation of heart rate. Even minor injury to the medulla may therefore have profound effects on the animal body.

Forming the anterior division of the hindbrain, the **metencephalon** includes the **cerebellum**, which coordinates neuromuscular activity and is sometimes called the body's "unconscious motor-coordinating center." The cerebellum is most developed in highly active vertebrates—that is, in those vertebrates in which locomotor activity requires a high degree of coordination, as in birds or mammals. In addition to coordinating movements, the cerebellum maintains body posture and equilibrium. Motor fibers from the cerebrum (which controls body movement) and sensory fibers from all of the sensory systems converge on the cerebellum, which integrates this input through its millions of inhibitory and excitatory synapses and neural circuits. The direction and rate of body movements already in progress are then modified as necessary to maintain posture, equilibrium, and coordinated actions.

In mammals and some birds, a transverse mass of fibers forms a prominent swelling called the **pons** on the ventral side of the metencephalon.

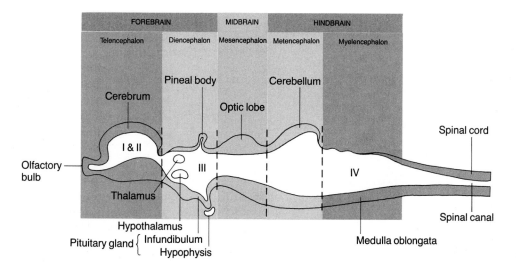

Figure 10.16 Vertebrate brain structures. Longitudinal section through the brain of a primitive vertebrate, showing, from anterior to posterior, the limits of the telencephalon, diencephalon, mesencephalon, metencephalon, and myelencephalon. Roman numerals I–IV indicate the ventricles, or cavities, of the brain. The relative positions of the thalamus and hypothalamus on either side of the plane of the section are approximate.

Table 10.2 Summary of Human Brain Structures and Functions

Division	Component	Function
Myelencephalon	Medulla oblongata	Controls heartbeat, respiration, defecation, blood vessel diameter
Metencephalon	Cerebellum	Coordinates muscular activity, equilibrium, posture, muscle tone
	Pons	Acts as fiber tract between cerebellum and cerebrum; acts as respiratory center, inhibiting inhalation
Mesencephalon	Superior colliculi	Act as visual center, controlling blinking, focus of the lens, and pupil aperture
	Inferior colliculi	Integrate auditory impulses, thus adjusting ear to sound volume
Diencephalon	Pineal body	Synchronizes rhythmic changes of body with light cycles; secretes serotonin and melatonin; possibly controls gonadal function
	Thalamus	Acts as central integrating centers for information passing to and from cerebral hemispheres; governs subjective feelings and primitive sensations
	Hypothalamus	Monitors blood pressure, body temperature, and heart rate; governs fat, sugar, and water metabolism and sleep rhythm; is source of releasing factors affecting anterior pituitary gland; is source of emotional reactions (fear, anger)
Telencephalon	Amygdaloid nucleus	Governs visceral activity; increases emotional intensity; keeps individual awake
	Corpus striatum	Has a stabilizing effect on voluntary movements
	Hippocampus	Has arousal effect on cerebral cortex; involved in recent memory ability; governs anger and fear responses of hypothalamus
	Cerebral cortex	Motor area initiates voluntary muscular movements; sensory area gives rise to senses of sight, smell, taste, hearing, touch, and equilibrium; association area involved with intellectual processes of memory, judgment, and reasoning

Though primarily a fiber tract, the pons contains a center concerned with regulating the inhalation of air into the lungs.

The **mesencephalon** (midbrain) in fishes and amphibians bears two prominent **optic lobes** that serve as visual centers. Though considered a primitive feature, optic lobes are also well developed in birds, which correlates with their exceptionally good visual system. In mammals, the midbrain has been overgrown by the cerebrum, and the integra-tion and coordination of visual impulses have largely been taken over by the cerebral cortex. The optic lobes in mammals have been transformed into the **superior and inferior colliculi** (Figure 10.17); the former control visual reflexes such as blinking, focusing, and adjusting the pupils to light intensity, while the latter have an auditory function, adjusting the ear to different sound levels.

The **diencephalon**, the posterior part of the fore-brain, is enveloped by the cerebral cortex in mam-

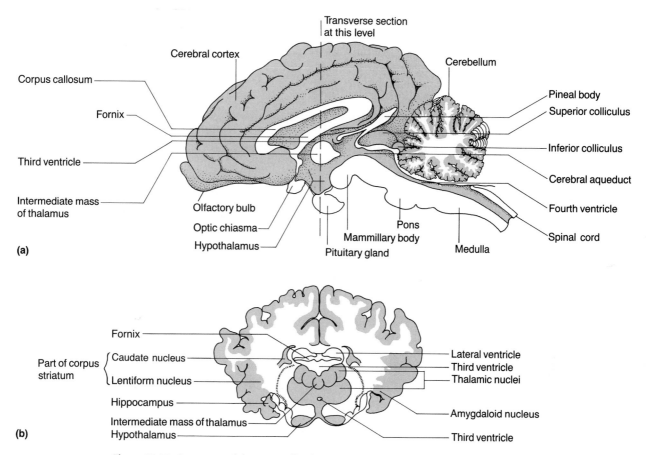

Figure 10.17 Structures of the mammalian brain. Major structures revealed in (a) sagittal and (b) transverse sections of the brain of a cow, including location of major basal nuclei deep in the interior.

mals and has several vital functions. The **pineal body** is responsive to changes in the intensity of light striking the eyes. Much like a time setter for a vertebrate's biological clock, it is involved in the synchronization of body rhythms with photoperiod, the daily cycles of light and dark. The pineal body also secretes regulatory substances, including the hormone **melatonin** (which affects distribution of pigment granules in the skin of amphibians) and **serotonin** (an inhibitory neurotransmitter substance). Moreover, the pineal body seems to regulate sexual cycles, perhaps by assisting the pituitary gland in controlling gonad function.

The thickened lateral parts of the diencephalon form the **thalamus**, a "relay station" that integrates all sensory impulses, except those for the sense of smell, before transmitting them to the cortex. In humans, the thalamus governs the intensity of sensations. Experimental stimulation of this brain region can make pain unbearable and pleasure ecstatic.

The floor of the diencephalon forms the **hypothalamus**, a composite structure that lies below the thalamus. Hypothalamic structures that are externally visible include the **mammillary bodies** (which integrate impulses for the sense of smell), the **infundibulum** (which connects to the pituitary gland), and the **optic chiasma** (where the optic nerves cross to opposite sides of the brain). Centers in the hypothalamus monitor fat and sugar metabolism, body temperature, water balance, heart rate, blood pressure, rhythm of sleep, and some genital functions. In fact, because it regulates so many vital internal processes, the hypothalamus is considered the body's **homeostatic center**, responsible for coordinating the many activities necessary to maintain the internal environment. Equally important, the hypothalamus is also the connecting link between the nervous and endocrine systems, the body's two coordinating systems. Much of the endocrine system is governed by the anterior pituitary, a gland that has sometimes been called the "master endocrine gland." However, the anterior pituitary gland is in turn governed by the hypothalamus. The important relationships between the hypothalamus, the pituitary gland, and vital body functions are discussed in detail in Chapter 11.

Adaptive development in the vertebrate brain is most apparent in the anterior division of the forebrain, the **telencephalon**. (The evolution of the vertebrate nervous system is discussed in Chapter 28, p. 700; Figure 28.17.) The **cerebrum**, which is divided into right and left hemispheres, becomes the dominant part of the brain, and in mammals, it overlies most of the other parts of the brain. Deep within the cerebral hemispheres, paired clusters of neuron cell bodies form the **basal nuclei** (equivalent to the **basal ganglia**). The most prominent of the basal nuclei are the corpus striatum and the amygdaloid nucleus.

The **corpus striatum**, a composite structure that includes several ganglia, is involved in the stabilization, or steadying, of voluntary movements. The **amygdaloid nucleus** is located below the thalamus and is attached to part of the corpus striatum. Experimental stimulation of the amygdaloid nucleus produces a variety of "visceral" responses, including salivation, gagging, defecation, and increased intestinal tract movement. Its destruction causes drowsiness, lack of interest, and loss of appetite, so it is possible that the amygdaloid nucleus may function to keep the individual awake and aware of the environment.

Another prominent, deep structure of the mammalian telencephalon, called the **hippocampus**, is intimately associated with the basal nuclei and cerebral cortex (see Figure 10.17). Sensory stimuli (e.g., odors and sounds) have an activating or arousing effect on the cerebral cortex via the hippocampus, and electrical stimulation of the hippocampus will awaken an anesthetized monkey. The hippocampus also appears to be important in recent, or short-term, memory. People with lesions (injuries) of the hippocampus may be unable to remember anything for more than a few minutes, though they may recall events that occurred years earlier. The hippocampus has an emotional function too; injury in this area may leave people unable to control fear and anger.

The **cerebral cortex**, as it appears in reptiles and mammals and has evolved to its highest level in primates, is the outer layer of the cerebrum. (Evolution of the cerebral cortex is discussed in Chapter 28, p. 702; Figure 28.17.) The cerebral cortex is 2–4 mm thick in humans, and it consists of several

billion neurons and synapses with interconnections far more complex than anywhere else in the brain. By stimulating small areas of the cortex with electrodes and noting the effects on movement or perception of sensations, physiologists have been able to map functional regions of the cortex. There are three general regions: **Motor areas** initiate contraction of the skeletal (voluntary) muscles; **sensory areas** receive impulses from the sensory receptors throughout the body; and **association areas** integrate and interpret sensory impulses. When the brains of various mammals are compared, an interesting relationship is evident: The percentage of the brain devoted to association corresponds with the adaptive complexity and integrative capacity of the brain. For example, most of a rabbit's cortex consists of sensory and motor areas, with relatively little devoted to integrative activity; this is less true of the cat, and even less so of the primates. The human brain has the greatest integrative capacity.

Figure 10.18 shows the distribution of sensory, motor, and association areas among the various lobes of the human cerebral cortex. The posterior region of the **frontal lobe**, along the **central fissure**, is called the somatic motor area because it initiates activity in the voluntary muscles. Just across the central fissure, in the **parietal lobe**, is the somatic sensory area, receiving impulses from taste receptors and sensory endings in the skin. The **temporal lobe** and the **occipital lobe** include the auditory and visual areas, respectively.

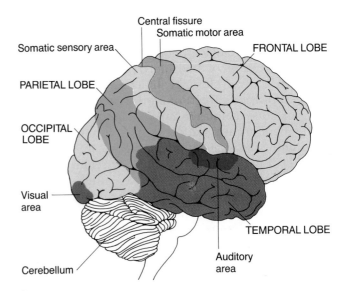

Figure 10.18 Functional regions of the human cerebral cortex. The central fissure separates the somatic motor area of the frontal lobe from the somatic sensory area of the parietal lobe.

The association areas—the parts of the cortex not occupied by the sensory and motor areas—include parts of each lobe. The occipital association area processes visual images. The temporal association area interprets new visual and auditory experiences in the light of remembered experiences. The parietal association area interprets the qualities of touched objects, such as their size, shape, and texture; it also interprets impulses from the inner ear, functioning in understanding spoken words. The association areas of the frontal lobe play a major role in defining one's personality. They determine the appropriateness of behavior by evaluating consequences, making considered judgments, planning for the future, and so forth. The fact that people differ in these traits is obviously a reflection of differences in the function of their frontal association areas.

The Peripheral Nervous System. The cranial and spinal nerves and their branches, making up the vertebrate PNS, are organized into the afferent and efferent divisions. The afferent (sensory) division consists of all the sensory nerves, whereas the efferent (motor) division consists of all the motor nerves and can be further divided into the somatic nervous system and the autonomic nervous system. The somatic nervous system is composed of motor nerves leading to skeletal muscle and skin. The autonomic nervous system, on the other hand, consists of motor nerves leading to smooth muscle in the gut, blood vessels, and other viscera, as well as to glands and to heart muscle.

The **autonomic nervous system (ANS)** provides nervous control of body systems that must operate automatically. Figure 10.19 shows the arrangement of the ANS in a mammal. Notice that the system consists of two divisions. Visceral motor portions of some of the cranial nerves (III, VII, IX, and X; see Table 10.1) and visceral motor portions of spinal nerves in the sacral region of the spinal cord form the **parasympathetic division** of the ANS. Vis-

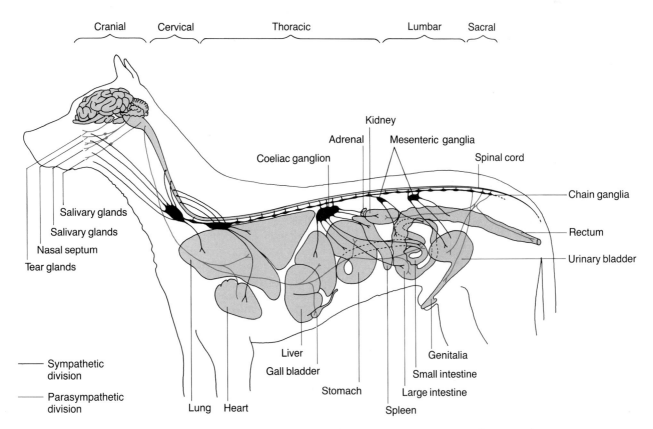

Figure 10.19 The autonomic nervous system. The autonomic nervous system of a mammal, viewed from the left side. Branches arising from four cranial nerves plus those from the sacral spinal nerves constitute the parasympathetic division of the autonomic nervous system. Branches arising from cervical, thoracic, and lumbar spinal nerves make up the sympathetic division. The chain, coeliac, and mesenteric ganglia contain nerve cell bodies and synapses of sympathetic fibers; parasympathetic fibers synapse in small ganglia close to the target organs.

ceral motor portions of spinal nerves located between the two parts of the parasympathetic division form the **sympathetic division**.

The key to control of body functions by the ANS is, first, that glands and visceral smooth muscles are innervated by *both* the sympathetic and the parasympathetic divisions and, second, that the divisions generally have *antagonistic* effects. The sympathetic division readies an animal for action (often for "fight or flight"). Sympathetic effects include constriction of blood vessels and therefore reduced blood flow to the skin and digestive tract; acceleration of heartbeat, thereby increasing blood flow to the skeletal muscles needed for fight or flight; contraction of sphincter muscles along the digestive tract, slowing food processing; cessation of secretion by salivary glands; dilation of the bronchi, thereby admitting more air to the lungs and increasing the rate of blood oxygenation; and dilation of the pupils of the eyes (to better see the threat). During a sudden fright, you have probably experienced these effects as your face paled, your heart pounded, and you inhaled air more rapidly. The parasympathetic division produces a set of essentially opposite effects.

The molecular basis for ANS function is indicated in Figure 10.20. At their first synapses (in either the parasympathetic or sympathetic ganglia), both the sympathetic and parasympathetic neurons release the transmitter substance acetylcho-

line. The parasympathetic fibers also release acetylcholine at their junctions with organs. However, a different substance, **norepinephrine** (also called **noradrenaline**), is released by the sympathetic fibers where they form junctions with the same organs. The two substances produce antagonistic effects on the innervated organs, and the response of the organ depends on the relative amounts of the chemicals. **Epinephrine** (also called **adrenaline**), a hormone secreted into the general blood circulation by the adrenal glands, has the same general effects on organs as does innervation by sympathetic nerve endings. (The adrenal glands are discussed in Chapter 11, p. 255.) Therefore, the sympathetic division is said to be **adrenergic** (after "adrenaline"), whereas the parasympathetic division is **cholinergic** (after "acetylcholine"). Under ordinary circumstances, the antagonistic effects of the sympathetic and parasympathetic divisions balance each other and provide a dynamic system of organ control.

☐ SUMMARY

1. Nervous systems consist of specific arrangements of neurons (nerve cells), nerves (bundles of nerve fibers), and ganglia (neuron cell bodies). The nerve impulse is a self-propagating wave of changed membrane potential traveling along an axon. Sensory (afferent) neurons carry impulses from sensory receptors toward the central nervous system (CNS); motor (efferent) neurons carry impulses from the central nervous system toward an effector organ. Interneurons link sensory and motor neurons, carry information from one part of the central nervous system to another, and integrate information from various sources.

2. Neurons form junctions (synapses) with other neurons and with effectors (glands, chromatophores, and muscles). Synapses between neurons and effectors are called neuroeffector junctions; neuron/muscle junctions are called myoneural junctions. Transmission across most synapses and neuroeffector junctions requires a chemical neurotransmitter substance, but some junctions have direct electrical coupling. Synapses are control points where information may or may not pass to the next neuron or may be modified.

3. Nervous systems can be classified as primitive or advanced. Primitive nervous systems have their neurons arranged in a nerve net. Animals with nerve nets can achieve complex behavior through a variation in their impulse conduction rate in different parts of the net and through the presence of different nerve nets that mediate different sets of behavior. Advanced nervous systems have a central nervous system that integrates and coordinates sensory input and motor output.

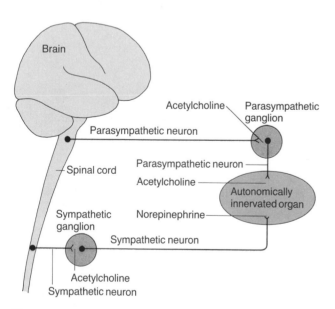

Figure 10.20 Parasympathetic/sympathetic antagonism. Comparison of innervation of an organ by parasympathetic and sympathetic fibers. The basis for the opposite effects of the two divisions is the release of different transmitter substances at the junctions with target organs.

4. Reflex arcs and reflex acts are fundamental functional units in nervous systems. Central pattern generators (nonreflexive neuronal mechanisms that operate without sensory input) mediate rhythmic activity. Reflexes and central pattern generators in segmented animals such as annelids and arthropods are organized on a segmental basis. The spinal cord of vertebrates functions as a reflex center and contains central pattern generators. The cord's actions include those that are simple and independent of the brain, as well as those that are closely allied to and highly modified by the brain.

5. The vertebrate central nervous system consists of the brain and spinal cord. The mammalian brain, the most complex in the animal kingdom, features an enlarged cerebrum with a well-developed cerebral cortex. Sensory information transmitted via the spinal cord to the brain is relayed by the thalamus to the sensory projection areas of the cerebral cortex. The motor area of the cortex is a command center for the body's muscles, with the cerebellum serving as a coordination center. Association areas of the cortex integrate and interpret sensory input and are involved in intellectual processes, including memory, judgment, and reasoning. The hypothalamus is the anatomical link between the nervous and chemical coordinating systems.

6. The vertebrate peripheral nervous system (PNS) consists of spinal and cranial nerves and their branches. The autonomic nervous system (visceral motor division of the PNS) regulates the activities of the heart, glands, and smooth muscles of the viscera and blood vessels. It operates on the principle of antagonistic but balanced sympathetic and parasympathetic divisions.

☐ FURTHER READING

Bullock, T. H., R. Orkand, and A. Grinnell. *Introduction to Nervous Systems.* San Francisco: W. H. Freeman and Company, 1977. *An excellent textbook covering invertebrate and vertebrate nervous systems.*

Camhi, J. M. *Neuroethology.* Sunderland, Mass.: Sinauer Associates, 1984. *An excellent synthesis of nervous system structure and function as they relate to behavior.*

Goslow, G. E., Jr. "Neural Control of Locomotion." Chap. 17 in *Functional Vertebrate Morphology*, ed. M. Hildebrand, D. M. Bramble, K. F. Liem, and D. B. Wake. Cambridge, Mass.: Harvard University Press, 1985. *An up-to-date look at the vertebrate nervous system in relation to its role in controlling muscle activity.*

Hubel, D. H. "The Brain." *Scientific American* 241(1979): 44–53. *One of eleven excellent, well-illustrated articles in an entire issue devoted to the nervous system.*

Keynes, R. D. "Ion Channels in the Nerve-Cell Membrane." *Scientific American* 240(1979): 126–135. *A well-illustrated discussion of how sodium and other ions pass through neuron membranes.*

Messenger, J. B. *Nerves, Brains and Behavior. The Institute of Biology's Studies in Biology No. 114.* London: Edward Arnold, 1979. *A brief, highly readable account of the relationship between nervous systems and animal behavior.*

Snyder, S. H. "The Molecular Basis of Communication Between Cells." *Scientific American* 253(1985): 132–141. *Emphasizes the interrelationship between nervous and hormonal systems of communication.*

Stevens, C. F. "The Neuron." *Scientific American* 241(1979): 54–65. *An excellent description of nerve cell structure and function.*

Wursig, B. "Dolphins." *Scientific American* 240(1979): 136–148. *A description of the nervous system and behavior of these highly intelligent mammals; interesting comparisons with humans.*

Coordination: Hormones and Endocrine Systems

Molting in insects is hormonally controlled. This scanning electron micrograph shows a cherry aphid *(Rhopalosium padi)* with its recently molted skin, feeding on a barley plant (71×).

In addition to having a nervous system that sorts and integrates sensory input, sends appropriate signals in the form of nerve impulses to effector organs, and participates in homeostasis, animals also have a second coordinating system whose signals are in the form of secreted chemicals called **hormones** (from Greek *hormon*, "to excite"). Compared to the fast-acting nervous system, where nerve impulses are generated rapidly and carried by nerve cell fibers more or less directly to the effector organs, the chemical coordinating system is fairly slow-acting. This is because hormones must be synthesized, released, and transported to the organs they regulate.

Given the remarkable versatility of the nervous system, one might wonder why animals need a second coordinating system. The reason seems to be that nervous control tends to be immediate and localized, whereas many body processes—including growth and metamorphosis, color changes, sexual maturation, and migratory and courtship behavior—involve complex responses

244

over a long term. These processes are seldom managed by the nervous system alone. Rapid control by a nervous system appears to be less suitable than a comparatively slow-acting, pervasive system of hormones more nearly matching the long-term nature of the processes they control. Hormonal control can also be more efficient than direct nervous control. A tiny quantity of hormone secreted by relatively few cells can govern enormous numbers of cells involved in such broadly based processes as growth and reproduction.

Traditionally, a hormone is defined as an internal secretion produced by a ductless gland and transported by the circulatory system to other parts of the body, where it regulates the function of a **target organ** or tissue. Because the hormone is secreted directly into blood capillaries within the gland (rather than through ducts), the gland is called **endocrine** (from the Greek *endon*, "within," and *krinein*, "to separate") or **ductless**. In invertebrates with open circulatory systems, the endocrine glands or their equivalents release hormones directly into the surrounding hemolymph.

While it is convenient to make the distinction between nervous and chemical coordinating systems, the two systems are not as distinct as was formerly thought. Many hormones are now known to be produced in the cell bodies of certain neurons and channeled through axons to a release point some distance away. Neurons that secrete hormones are called **neurosecretory cells**, and their hormones are called **neurohormones** (Table 11.1).

The nervous system uses neurohormones to control the production and release of **glandular hormones** by many endocrine glands. In many animals, the brain, via neurohormones, exerts control over at least some of the hormone-secreting glands. In vertebrates, for example, the hypothalamus of the brain controls the activity of the pituitary gland, which in turn regulates the secretion of hormones by many other glands. Of course, most neurons are secretory in the sense that they release neurotransmitter substances at synapses. However, neurotransmitters are short-lived compared to neurohormones, and they travel only the very short distances across synaptic clefts. Zoologists now recognize that the nervous and endocrine systems work together as a far-reaching integrative network, or **neuroendocrine system**, in which neurotransmitters, neurohormones, and glandular hormones are the main chemical messengers.

CONTROL OF HORMONE LEVELS: FEEDBACK LOOPS

The stimuli that trigger hormone release may be nervous or chemical or both. How does the arrival of a stimulus initiate hormone release? How is the hormone maintained in the blood or hemolymph at the concentration needed to control the target organ? Although the details differ from one gland

Table 11.1 Classification of Chemical Messengers

Type of Messenger	Produced by	Action	Examples
Neurotransmitters	Nerve cells	Transmit nerve impulses at synaptic and other neuroeffector junctions	Acetylcholine, serotonin, norepinephrine
Neurohormones	Nerve cells	Coordinate neuroendocrine functions; often transported long distances and stored before being released	Hypothalamic releasing factors, insect brain hormone
Glandular hormones	Ductless (endocrine) glands	Classic hormonal action; often transported long distances to target organs	Estrogen, insulin, thyroxines, ecdysone
Prostaglandins	Cell membranes (all tissues)	Local messengers with diverse effects; function within the tissues producing them	Various 20-carbon fatty acids
Enkephalins, endorphins	Brain tissue	Alleviate pain, influence memory, induce pleasure; natural opiates	Peptides, cleaved from β-lipotropin
Second messengers	All tissues	Intermediate chemical signals; convert certain enzymes to active form; regulate speed of many chemical reactions	Cyclic AMP
Semiochemicals*	Glands and organs secreting materials to the exterior of the body	Communication chemicals that affect behavior of other individuals after being released into surrounding medium	Insect sex pheromones, territory-marking odorants

*Includes pheromones, or chemical signals; discussed in Chapter 13, p. 309.

to another, most hormone release systems share certain basic features. When the appropriate stimulus arrives at the plasma membrane of an endocrine gland cell, the permeability of the membrane to calcium ions increases. The resulting rise in Ca^{2+} concentration within the cell stimulates transfer of the hormone molecules, often packaged in membrane-bounded granules, toward the plasma membrane. Expulsion of the hormone from the cell takes place by **exocytosis**, a kind of reverse pinocytosis. Another indication of the similarity between the nervous and chemical coordinating systems is that a similar process (involving Ca^{2+} and exocytosis) occurs at chemical synapses between nerve fibers.

Proper hormone levels are maintained in the circulatory fluid by means of **feedback mechanisms**, in which the product of a process controls the process. Maintaining a constant indoor temperature by means of a thermostatically controlled furnace is a familiar example of feedback. A furnace warms a house with its "product," heat energy. When the house temperature rises above a set point, a sensor in the thermostat turns off the furnace. Later, as the house cools, a circuit in the thermostat closes and the furnace turns back on. Regulation of this kind, originating from within the system, is a closed-loop control mechanism. Many biological systems also operate as closed loops. Thus, a hormone released into the blood by an endocrine gland triggers an action by its target tissue which, through feedback, inhibits or triggers further release of the hormone. Nearly all cases of hormonal regulation involve **negative feedback**, where the release of hormone is reduced or stopped. Just as negative feedback is used to maintain a house at a preferred constant temperature, so is it used to maintain an animal's biological systems at a preferred steady state, an aspect of the overall process of homeostasis.

Endocrine coordination by **positive feedback** also occurs, although it is rare. In this case, arrival of a hormone at its target causes the target to signal the release of still more hormone. The system rapidly becomes climactic or unstable, just the opposite of the homeostatic control provided by negative feedback. Positive feedback is important in such situations as the birth of a mammal, because it hastens the process to completion. During labor, the hormone oxytocin, produced by the hypothalamus, promotes contraction of uterine muscle. A positive feedback loop involving oxytocin and its target organ, the uterus, leads to the climax of birth.

MODES OF HORMONE ACTION

Unlike a motor nerve, which can directly stimulate a localized area, hormones are distributed by the circulatory fluid to all parts of the body. This broad distribution raises important questions concerning the relationship between hormones and their targets and about hormonal mechanisms at the cellular and molecular levels. Why, for example, does only a certain tissue or organ respond, while others ignore the hormonal messages? How is it that certain hormones reinforce each other, while others have conflicting actions? Why does a hormone exert its influence at only certain times in an animal's life, and how is it that a particular hormone is able to change functions during the life cycle? Is there a cause and effect relationship between hormones and gene transcription? Such fundamental questions about modes of hormone action provide the focus in modern research in endocrinology.

It is now known that hormones bind with specific **receptor molecules** associated with the cells of their target tissues. Thus, although both liver and muscle cells metabolize glycogen, the hormone glucagon stimulates only the liver to do so because liver cells have the necessary glucagon receptors and muscle cells do not. Once bound to its receptor molecules, a hormone acts by one of several mechanisms, depending somewhat on the chemical nature of the hormone.

Most hormones are either steroids, small proteins, peptides, or amino acid derivatives (Table 11.2). **Steroid hormones** are characterized by a relatively complex four-ring structure. They differ from one another in the kinds of side groups attached to the rings. In one widely accepted model, when a steroid hormone arrives at its target cell, it passes through the cell membrane, becomes bound to a specific receptor molecule in the cytoplasm, and is translocated into the cell nucleus (Figure 11.1a). There the hormone exerts its primary influence at the genetic level. It activates a specific gene that carries the genetic code for a specific protein. As a result of the gene's action, the cell begins producing the specific protein. The protein might be a key enzyme influencing the rate of a biochemical reaction, or it may be a structural protein required in a cell organelle. (Transcription and translation of the genetic code are discussed in Chapter 15, p. 366, Figures 15.11–15.13.)

Small-protein hormones and **peptide hormones** exert their influence through a **second messenger**.

Table 11.2 Chemical Classes of Animal Hormones

Chemical Class	Hormones	Examples of Chemical Structure
Steroids	Mineralocorticoids, glucocorticoids, testosterone, estrogen, progesterone, ecdysone	**Ecdysone**
Small proteins	Insulin, glucagon, parathormone, growth hormone, thyrotropic hormone	His-Ser-Gln-Gly-Thr-Phe-Thr-Ser-Asp Arg-Arg-Ser-Asp-Leu-Tyr-Lys-Ser-Tyr Ala-Gln-Asp-Phe-Val-Gln-Trp-Leu-Met Thr-Asn **Glucagon**
Peptides	Oxytocin, vasopressin, adipokinetic hormone, thymosin, releasing factors	Cys-Tyr-Ile-Gln-Asn-Cys-Pro Leu-Gly (NH$_2$) **Oxytocin**
Amino acid derivatives	Thyronines, thyroxines, epinephrine, norepinephrine	**Epinephrine**

When the hormone (the "first messenger") arrives at its target, it initiates formation of a second messenger, which then alters target cell function. The most common second messenger is **cyclic adenosine monophosphate (cAMP)**, a close chemical relative of the high-energy compound adenosine triphosphate (ATP). When the hormone binds to the target cell surface receptors, it activates **adenylate cyclase**, an intracellular enzyme that catalyzes the conversion of ATP to cAMP (Figure 11.1b).

The most important action of cAMP involves conversion of certain enzymes to their active form. This is done indirectly: First, the cAMP forms a complex with **protein kinase**, and then the cAMP–protein kinase complex activates the enzyme in a phosphorylation reaction requiring ATP hydrolysis. A cAMP-mediated hormone system functions by producing an amplifier-like effect. First, each hormone molecule triggers formation of *several* cAMP–protein kinase complexes. Each of these complexes, in turn, activates (phosphorylates) large numbers of enzyme molecules. Finally, the enzymes influence the rate at which various substances are produced within cells. The actions of some half a dozen invertebrate hormones are known to involve cAMP. (Vertebrate hormones that appear to act via cAMP are marked with an asterisk in Table 11.3.) In many cases, cAMP (and, indirectly, the hormones controlling its production) influence cell activity by regulating the rates of

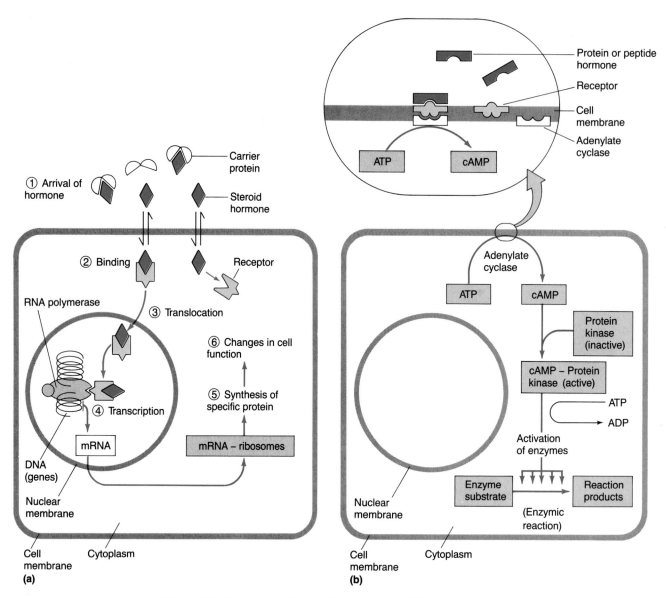

Figure 11.1 Models of hormone action. (a) A model for steroid hormone function. In this model, the steroid hormone, attached to a carrier protein, arrives at the target cell membrane (1). Once inside the cell, the hormone molecules are bound by specific receptor molecules (2). The receptors translocate the hormone into the cell nucleus (3), where the hormone influences the transcription of a gene (DNA) to messenger RNA (mRNA) (4). The mRNA then serves as a template for the synthesis of a protein product (5), such as an enzyme, which in turn alters cell function (6). (b) Small-protein and peptide hormone function. Cyclic adenosine monophosphate (cAMP), a second messenger substance, is formed from ATP in a reaction catalyzed by adenylate (adenyl) cyclase, an enzyme activated by a hormone (the primary messenger) binding to receptors on the cell membrane. The hormone thus exerts its effects indirectly, through cAMP.

enzyme-catalyzed metabolic reactions. Research on second messengers is an active area in endocrinology. Another second messenger, called cyclic guanosine monophosphate (cGMP), has already been identified, and it is likely that others will be found as well.

Amino acid-derivative hormones have several modes of action. For instance, epinephrine acts *via* the cAMP system, whereas thyronines may function directly at the gene level (as the steroid hormones do) or affect oxidative metabolism by mitochondria.

Not all hormones fit neatly into the four chemical categories summarized in Table 11.2. An important example among invertebrates is juvenile hormone in insects. Though not a steroid, juvenile hormone acts like the steroid hormones, influencing genes and causing the synthesis of specific proteins. More will be said about juvenile hormone later in this chapter.

OTHER CHEMICAL MESSENGERS

In recent years, a number of chemical messengers have been discovered that have some of the properties of hormones and neurotransmitters, but do not fit the customary definition of either one. Vertebrates have several types of hormonelike substances that are sometimes called "tissue hormones." Unlike true hormones, these substances elicit local changes in the tissues that produce them, are short-lived, and are not transported by the circulatory system. One such group, the **prostaglandins**, were so named because they were first discovered in the semen fraction produced by the mammalian prostate gland. Prostaglandins have been found in most mammalian tissues, and many physiologists think that these substances are synthesized in the membranes of virtually all cells. (See Table 11.1 and the Essay on prostaglandins.)

Another "tissue hormone," a substance called **histamine**, also has a local effect on tissues. Histamine is synthesized by connective tissue cells called **mast cells** from the amino acid histidine. Mast cells release histamine into extracellular fluid

---

Essay: Prostaglandins: Unique Chemical Messengers

About 16 different prostaglandins, a distinct class of cell regulators that serve as local messengers within the tissues that produce them, have been identified in vertebrates since the early 1960s, and many more may await discovery. These substances are synthesized from fatty acid chains derived from phospholipids in cell membranes. (The structure of phospholipids and cell membranes is discussed in Chapter 2, p. 37; Figure 2.14.) Cells secrete prostaglandins into interstitial fluid, but unlike true hormones, these substances are broken down so quickly or are taken up by neighboring cells so rapidly that they do not enter the circulatory system in amounts that can affect other tissues of the body.

Prostaglandins are known to regulate many cellular functions. In some cases, two or more different prostaglandins control cell activity by antagonistic action. Prostaglandin E (PGE), for example, promotes dilation of blood vessels and bronchial air passages leading to the lungs. As the vessels dilate, more blood can flow to the lungs for oxygenation by increased volumes of air or flow to the muscles with oxygen and nutrients. As a result, the body is better able to handle increased physical activity. Prostaglandin F (PGF), which differs slightly from PGE in chemical structure, constricts blood vessels and bronchial passages (Figure E1). The balance between PGE and PGF is thus important in maintaining normal levels of blood flow and lung ventilation. It is thought that homeostasis in general depends heavily on the balanced actions of prostaglandins.

Both PGE and PGF have other functions not necessarily involving mutual antagonism. For example, PGF causes degeneration of the corpus luteum in the ovary of many mammals, ending one ovarian cycle and setting the stage for the next (see p. 265 and Figures 11.14 and 11.15). And both PGF and PGE cause uterine muscles to contract, making them a useful aid in inducing labor during childbirth.

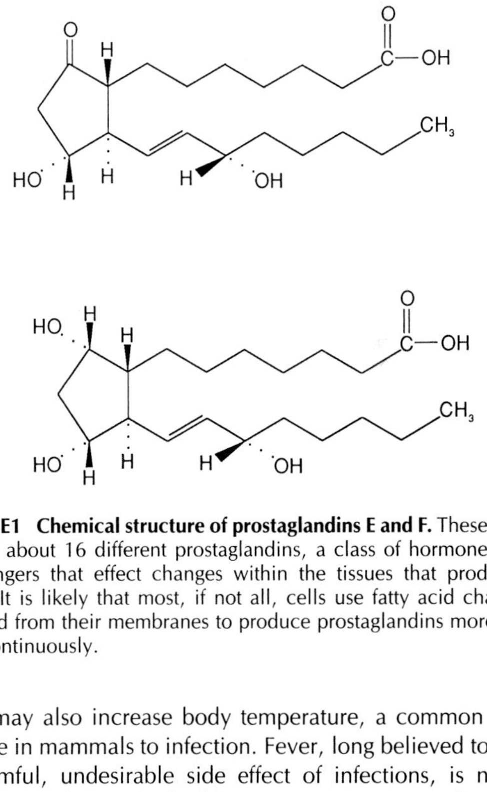

Figure E1 Chemical structure of prostaglandins E and F. These are two of about 16 different prostaglandins, a class of hormonelike messengers that effect changes within the tissues that produce them. It is likely that most, if not all, cells use fatty acid chains cleaved from their membranes to produce prostaglandins more or less continuously.

PGE may also increase body temperature, a common response in mammals to infection. Fever, long believed to be a harmful, undesirable side effect of infections, is now known to enhance the body's immune response to disease organisms. Aspirin's effectiveness in reducing fever is probably due to its ability to inhibit PGE synthesis.

We are only now beginning to appreciate the importance of prostaglandins and to understand how they affect cells. Continuing research will almost certainly add to the growing list of physiological processes influenced by these unique chemical messengers.

Table 11.3   Summary of Major Vertebrate Hormones

Producing Organ	Hormone	Target	Principal Effects
Hypothalamus	Releasing and inhibiting hormones* (see Table 11.4)	Anterior pituitary	Govern secretion of hormones by anterior pituitary
	Vasotocin (antidiuretic hormone, ADH)* (vasopressin in mammals)	Kidney tubules	Conserves water; controls muscle contraction in small-artery walls
	Oxytocin*	Uterus muscle, mammary ducts	Stimulates uterine contractions and release of milk
Pineal body	Melatonin	Hypothalamus	Mediates photoperiodic control over sexual cycles (mammals, birds)
Anterior pituitary gland	Thyroid-stimulating hormone (TSH)*	Thyroid gland	Controls synthesis and release of thyronines
	Adrenocorticotropic hormone (ACTH)*	Adrenal cortex	Releases hormones of the adrenal cortex
	Growth hormone (GH) (somatotropin)	Cells in general	Controls growth, protein synthesis, and size maintenance
	Follicle-stimulating hormone (FSH)*	Ovaries, testes	Controls follicle growth, estrogen production; spermatogenesis in males
	Luteinizing hormone (LH)*	Ovaries, testes	Controls ovulation, formation of corpus luteum; production of testosterone
	Prolactin (luteotropic hormone, LTH)	Mammary glands, other tissues	Controls milk production; osmoregulation in fish; tadpole growth; maintains corpus luteum
	Melanocyte-stimulating hormone (MSH) (intermedin)*	Chromatophores, melanocytes	Governs color changes by altering pigment granule dispersal in chromatophores in amphibians, reptiles; role in birds and mammals uncertain
Posterior pituitary gland	Stores and releases vasotocin (vasopressin; ADH) and oxytocin secreted by the hypothalamus (see target organs and principal effects under hypothalamus)		
Thyroid gland	Thyronines (triiodothyronine, thyroxine)	Cells in general	Stimulate oxidative metabolism; raise body temperature
	Calcitonin*	Bone cells	Governs calcium uptake into bones

when they are stimulated by the immune system, or when they are injured. Histamine promotes the inflammatory response, causing capillaries to dilate and assisting the immune system in fighting infection. (The role of inflammation in the vertebrate immune response is covered in Chapter 6, p. 139.)

More recently discovered than either prostaglandins or histamine are a group of peptides that function as natural painkillers. Researchers studying the effects of opiates (painkillers such as opium and morphine) discovered opiate-binding receptor sites on cells in the central nervous system. The existence of these receptor sites suggested that there might be naturally occurring opiates in the nervous system. In the mid-1970s, peptides were found that did, in fact, bind to the receptor sites. Called **enkephalins** (meaning "within the head") and **endorphins** (suggesting "internal morphine"), these

peptides are abundant in vertebrates. The powerful analgesic effects of the enkephalins and endorphins (they are even more effective than morphine) indicate that they may be important in pain perception. The discovery of natural opiates and their receptor sites will perhaps lead to important medical benefits. It may one day be possible to treat nervous system disorders with endorphins, enkephalins, and other peptides, and perhaps to use such substances as nonaddictive painkillers.

HORMONAL REGULATION

In the space of a single chapter, it is possible to discuss only a few of the literally hundreds of hormonal actions in animals. By far, most of what is

Table 11.3 *Continued*

Producing Organ	Hormone	Target	Principal Effects
Parathyroid glands	Parathormone*	Kidney, gut, skeleton	Governs absorption, excretion, and deposition of calcium
Thymus gland	Thymosin	Blood cells	Stimulates lymphatic immune response
Adrenal cortex	Mineralocorticoids (e.g., aldosterone)	Kidney	Govern water balance and mineral metabolism; promote inflammation
	Glucocorticoids (e.g., cortisol)	Cells of liver, muscle	Promote gluconeogenesis; antiinflammatory effect
	Androgens	Male fetus	Cause masculinization
Adrenal medulla	Epinephrine and norepinephrine	Most organs	Prepare body organs for flight or fight; same effects as sympathetic nervous system
Pancreas	Insulin	All cells except most neurons	Regulates uptake of glucose by cells, lowering blood glucose
	Glucagon*	Liver, fat	Converts glycogen to glucose, raising blood glucose
Testes	Testosterone and related metabolites	Cells in general	Develop and maintain male secondary sex characteristics and behavior
Ovaries	Estrogen	Uterus, ova	Regulates development of ova, uterine lining, secondary sex characteristics
	Progesterone	Uterus, mammary glands	Governs placental formation; sustains pregnancy
	Relaxin	Pubic symphysis	Softens pelvic ligaments prior to giving birth
Stomach	Gastrin	Stomach	Triggers the release of gastric juice
Intestine	Secretin	Pancreas, liver	Triggers secretion of pancreatic enzymes; bile formation
	Cholecystokinin (CCK) (Pancreozymin, PZ)	Pancreas, gallbladder	Triggers secretion of pancreatic enzymes; flow of bile from gallbladder
Activated in blood plasma	Angiotensins	Adrenal cortex, blood vessels	Stimulate release of aldosterone; vasoconstriction in blood pressure control; promote drinking

*Hormones that act through the cAMP mechanism.

known concerns the vertebrate hormonal system, but we also mention several invertebrate systems that have been studied in some detail. Animals as diverse as vertebrates, annelids, octopuses, and insects regulate their metabolism, skeletal and integumentary development, growth, and reproduction through hierarchical arrangements of endocrine structures and the nervous system.

Overview of the Vertebrate System

During vertebrate evolution, chemical regulation of many vital functions came to be based on a hierarchy involving the hypothalamus of the brain, the pituitary gland, and many of the endocrine glands. As an aid in keeping track of the "players" discussed in this chapter, some of the major vertebrate hormones, their interrelationships, and their sources are described briefly in Tables 11.3 and 11.4 and in Figures 11.2 and 11.3.

The **hypothalamus** can be considered a master control center that receives signals originating from sense organs sampling a vertebrate's external and internal environments. The hypothalamus responds to appropriate signals by releasing neurohormones that influence the **pituitary gland**. The pituitary gland, in turn, releases hormones governing the activity of many other endocrine glands (Figure 11.3). Thus, as a link between the central nervous system and much of the endocrine system, the hypothalamus enables factors such as odors, light conditions, blood osmotic pressure, and sexual activity to affect a wide range of physiological and behavioral processes. For example, a critical day length in the spring is the cue for the release of hormones that initiate seasonal breeding in many

Table 11.4 Hypothalamic Releasing and Inhibiting Factors (Neurohormones) That Regulate the Activity of the Anterior Pituitary

Neurohormone	Function
Thyroid-stimulating hormone–releasing hormone (TSH.RH)	Triggers release of TSH
Adrenocorticotropic hormone–releasing hormone (ACTH.RH)	Triggers release of ACTH
Growth hormone–releasing hormone (GH.RH)	Triggers release of GH
Growth hormone–inhibiting hormone (GH.IH)	Inhibits release of GH
Prolactin-releasing hormone (P.RH)	Triggers release of prolactin
Prolactin-inhibiting hormone (P.IH)	Inhibits release of prolactin
Gonadotropin-releasing hormone (G.RH)	Triggers release of follicle-stimulating hormone (FSH) and luteinizing hormone (LH)

mammals and birds, and nerve impulses from the uterus during copulation lead to release of eggs from a rabbit's ovary.

Figure 11.4 shows that the pituitary gland (also called the hypophysis) consists of two main parts.

The **posterior pituitary** (neurohypophysis) contains axons of neurosecretory cells whose secretory cell bodies reside in the hypothalamus. The posterior pituitary is a storage and release site for several neurohormones that these cells produce (see Table 11.3). The **anterior pituitary** (adenohypophysis) is a true endocrine gland in that it actually secretes several different hormones. The anterior pituitary is regulated by several releasing and inhibiting neurohormones secreted by the hypothalamus (see Table 11.4). These substances are transported from the hypothalamus to the anterior pituitary in a special system of blood vessels called the **hypothalamo-hypophysial portal system** (see Figure 11.4).

Control of Metabolic Processes

The regulation of metabolism in vertebrates depends directly on hormones from three endocrine glands: the pancreas, the adrenal glands, and the thyroid gland. It is important to keep in mind that the adrenal and thyroid glands are influenced by the hypothalamus-pituitary master control system (see Figure 11.3).

Pancreatic Hormones. Metabolic regulation in vertebrates depends to a large extent on an intricate array of hormones that govern blood sugar

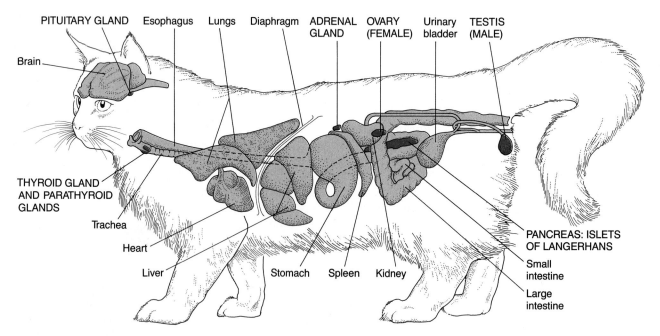

Figure 11.2 Mammalian endocrine glands. Location of the major endocrine glands relative to the other major organs in a cat. Not shown are the pineal body and hypothalamus (see Figures 10.16 and 10.17), both important neurosecretory structures of the brain.

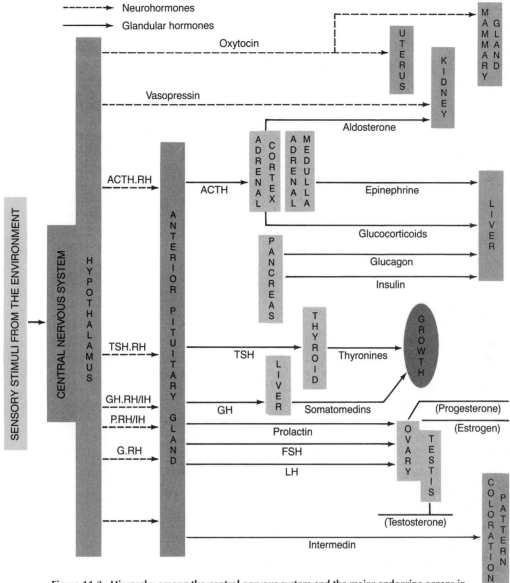

Figure 11.3 Hierarchy among the central nervous system and the major endocrine organs in vertebrates. The hypothalamus, as the primary control center, receives nervous input originating from stimulation of sensory receptors, both external and internal. The hypothalamus responds to these incoming nerve impulses by producing neurohormones that are carried to the pituitary gland. Hormones released from the pituitary influence other endocrine organs involved in regulating a wide range of processes, including reproduction, growth, and blood chemistry. Because the hypothalamus receives input from the external environment, it is possible for reproductive and other processes to be synchronized with seasonal changes. (ACTH.RH = adrenocorticotropic hormone–releasing hormone; ACTH = adrenocorticotropic hormone; TSH.RH = thyroid-stimulating hormone–releasing hormone; TSH = thyroid-stimulating hormone; GH.RH/IH = growth hormone–releasing hormone/inhibiting hormone; GH = growth hormone; P.RH/IH = prolactin-releasing hormone/inhibiting hormone; FSH = follicle-stimulating hormone; G.RH = gonadotropin-releasing hormone; LH = luteinizing hormone.)

concentrations. Two such hormones are secreted by groups of endocrine cells called the **islets of Langerhans**, located in the **pancreas**. The α cells in the islets produce the hormone **glucagon**, and the β cells produce **insulin** (see Table 11.2).

As a vertebrate eats food containing carbohydrates, the blood glucose concentration increases, and insulin is quickly released into the bloodstream. It was once believed that insulin is released in response to increased blood glucose following

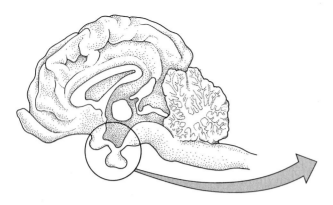

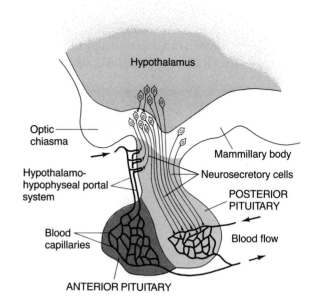

Figure 11.4 Pituitary gland regions. Note that the vessels of the hypothalamo-hypophysial portal system carry neurohormones from the hypothalamus to the anterior pituitary. The posterior pituitary stores and releases neurosecretions, such as oxytocin, received directly via nerve fibers from the hypothalamus. Both the anterior and posterior pituitary release hormones into the bloodstream.

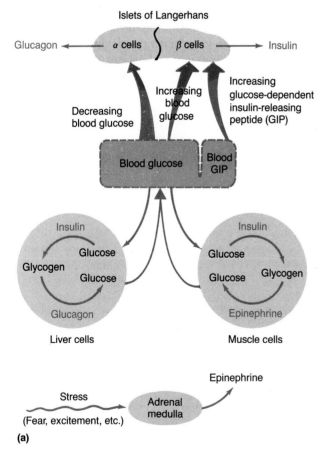

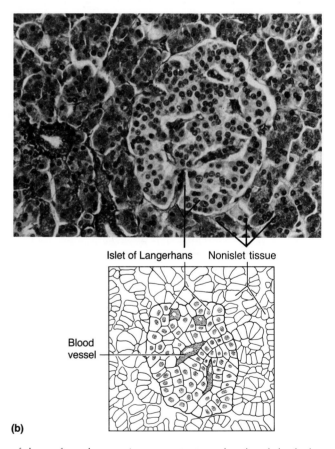

Figure 11.5 Regulation of blood glucose concentration. (a) After a meal, elevated blood glucose levels and glucose-dependent *insulin-releasing* peptide (GIP) from intestinal cells trigger the release of insulin from β cells of the islets of Langerhans in the pancreas. Insulin promotes glucose uptake and conversion to the storage polysaccharide glycogen by the liver and muscles. When blood glucose levels fall, glucagon released from the pancreatic α cells promotes mobilization of glucose from glycogen. In response to stress, the adrenal glands also produce hormones (epinephrine and glucocorticoids) that increase blood glucose levels. (b) This photomicrograph and sketch of pancreatic tissue show an islet of Langerhans. Nonislet tissue of the pancreas produces digestive secretions, while the islets produce insulin and glucagon.

glucose absorption from the intestine. It now appears that insulin release is stimulated by *glucose-dependent insulin-releasing peptide*, better known as GIP, a substance released from cells of the small intestine in "anticipation" of increased blood glucose (Figure 11.5). Insulin lowers blood glucose concentration by promoting glucose uptake by many tissues and by promoting glucose conversion to the storage polysaccharide glycogen. Insulin also inhibits glucose synthesis from amino acids. Because insulin lowers blood sugar concentration, it is a *hypoglycemic* agent.

In contrast, glucagon is released in response to a drop in blood glucose and is a *hyperglycemic* agent, raising blood glucose concentration by promoting the conversion of liver glycogen to glucose, which is released into the blood. Glucagon also stimulates the conversion of fats to free fatty acids, partly through its own action and partly by triggering the release of growth hormone from the anterior pituitary gland. Growth hormone helps mobilize fatty acids as an energy source in most cells. During periods of exercise and starvation, secretion of both glucagon and growth hormone increases while insulin levels fall, increasing the use of fat as an energy source and thus sparing glucose.

The delicate balance of these hormonal controls may sometimes go awry, with unfortunate consequences. Diabetes mellitus, which affects about 2% of the people in North America, is caused by a relative deficiency of insulin. In some diabetics, the β cells of the pancreas fail to produce enough insulin; in other cases, there is enough insulin circulating in the blood, but the target tissues either do not recognize it (because the cells have insufficient insulin receptors) or fail to respond to it. As a result, cells cannot take in and use glucose as fast as it enters the blood from the intestine, and blood glucose concentration rises. Excess glucose is excreted with large volumes of water required to produce the final urine. Severe dehydration may result, leading to tissue damage and possibly coma or death. Another potential hazard results from tissue response to deprivation of glucose as an energy source. Glucose-deprived tissues increase their metabolism of fats, and by-products of fatty acid oxidation called **ketone bodies** may then lower blood pH. The resulting acidity upsets respiratory and nervous functions and may also lead to coma or death. Fortunately, insulin treatments can restore the proper biochemical balance for many diabetics.

Adrenal Hormones.   Hormones from the **adrenal glands** of vertebrates are important in regulating metabolism. Adrenal glands consist of two discrete

tissues that develop from different embryonic tissues, are controlled by different mechanisms, and secrete different hormones. In mammals, the adrenal glands consist of an outer **cortex** of **interrenal tissue** surrounding an inner **medulla** of **chromaffin tissue**. In amphibians, reptiles, and birds, cells homologous to those of the mammalian adrenal cortex and medulla are closely intermingled, whereas in fishes they are in different regions of the body.

The adrenal medulla has direct nervous linkage with the sympathetic division of the autonomic nervous system. (The sympathetic division of the autonomic nervous system is discussed in Chapter 10, p. 242.) Its cells develop from the same embryological source that produces sympathetic neurons; thus, it is not surprising that the adrenal medulla produces the same hormones as those released by sympathetic fibers. (For a discussion of neural crest cells, see Chapter 16, p. 387.) **Epinephrine** (adrenaline, see Table 11.2) and, in lesser quantities, **norepinephrine** are released from the medulla upon arrival of nervous signals initiated by stress (excitement, fear, etc.). These hormones augment the effect of the sympathetic neurons in preparing the body to deal with stress in emergency fight-or-flight situations. One major effect of these hormones is in promoting the conversion of liver and muscle glycogen to glucose, which can be used as a source of energy by the muscles.

In contrast to the adrenal medulla, the adrenal cortex is under hypothalamus/pituitary control. The cortex is controlled directly by adrenocorticotropic hormone (ACTH) produced by the pituitary gland; secretion of ACTH, in turn, is mediated by ACTH-releasing hormone from the hypothalamus (see Figure 11.3 and Tables 11.3 and 11.4). The adrenal cortex produces several groups of hormones. Among these, the **glucocorticoids** directly affect metabolism; they help make glucose available by promoting its synthesis from amino acids. This process is called **gluconeogenesis** because new glucose is formed from noncarbohydrate sources.

Thyroid Hormones.   Vertebrate metabolism is widely influenced by **thyronines** from the **thyroid gland**. These hormones stimulate oxygen consumption and affect growth and metamorphosis (discussed later), ATPase activity, and RNA synthesis. The two main thyronines are **triiodothyronine** and **tetraiodothyronine** (also called **thyroxine**). The thyronines are bound to proteins for transport in the blood. Iodine is an important part of these hormones and thus is an essential nutrient.

Thyroid malfunction in humans can result in

severe and sometimes permanent detriment. Reduced thyroid function *(hypothyroidism)* in adults is associated with lowered metabolic rate and decreased body temperature. Other effects include lethargy, weakness, rough and dry skin, and slower mental activity. Administration of thyroid hormones is an effective treatment. Hypothyroidism in growing children can lead to abnormal bone development and, through effects on the nervous system, mental retardation. Impaired mental and physical development caused by hypothyroidism beginning in infancy is called cretinism.

Excessive thyroid function *(hyperthyroidism)* can also be a problem. Symptoms include elevated metabolic rate, body temperature, and heartbeat, and in some cases weight loss, nervousness, sleeplessness, and emotional instability. Drugs that inhibit thyroid function, radioactive iodine (which destroys thyroid gland cells when taken up and incorporated into the thyronines), and surgery are all effective treatments. In some cases, the thyroid gland becomes enlarged during either hyper- or hypothyroidism, a condition called **goiter**.

Metabolic Regulation in Invertebrates.   Many invertebrates are known, or believed, to have hormonal regulation of metabolism, but few systems have been thoroughly researched. Most is known about insect metabolism: Hormones controlling fat and sugar levels have been identified and, in one case, chemically characterized. In flies, **hypertrehalosemic hormone** (so named because insect hemolymph sugar is trehalose rather than glucose) raises hemolymph sugar levels to meet the enormous energy demands of the flight muscles. In flying locusts, **adipokinetic hormone**, a peptide of known amino acid sequence, spares blood sugar by promoting the oxidation of lipids by flight muscles.

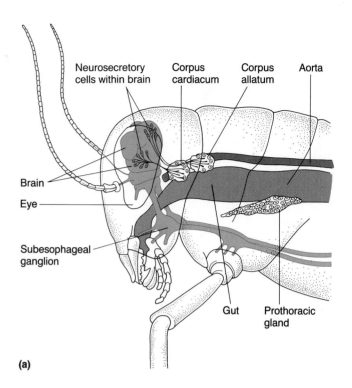

(a)

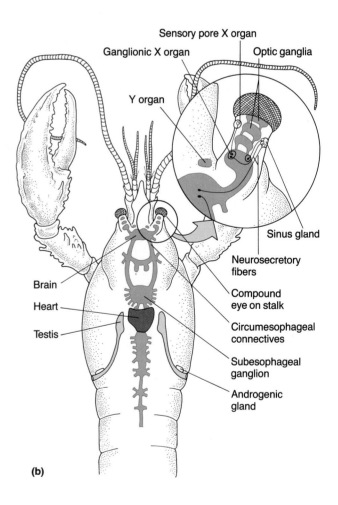

(b)

Figure 11.6 (Right)   Sources of hormones in insects and crustaceans. (a) In an insect, brain hormone, secreted by neurosecretory cells of the brain, is stored in and released from the corpus cardiacum. Brain hormone causes the prothoracic gland to release ecdysone, the molting hormone. Juvenile hormone released from the corpus allatum helps keep the insect juvenile; its absence permits maturation to the adult stage. (b) Crustaceans, such as this lobster, have a sinus gland in each eyestalk. The sinus gland receives, stores, and periodically releases neurohormones secreted by the X organ, which is a group of neurosecretory cells located in a ganglion in the eyestalk. One of the neurohormones, called molt-inhibiting hormone, inhibits release of ecdysone from a pair of endocrine glands called the Y organs. The X organ–sinus gland–Y organ complex regulates molting, growth, and sexual differentiation of the male.

Control of Integumentary and Skeletal Processes

Recall from Chapter 8 that an animal's body covering (integument) may have many functions, including gas exchange, excretion, osmoregulation, and temperature control. In addition, many animals are colored for showy display or camouflage, and some animals can even change color to make themselves less visible to predators. The integument may also be an animal's exoskeleton, as in the arthropods. Growth requires that arthropods periodically molt and replace their exoskeleton. The endoskeleton of most vertebrates is enlarged gradually by the addition of new skeletal material, and there is a dynamic relationship between bone minerals and their concentration in the blood. Hormones usually regulate or influence all these processes. This section briefly examines the role of hormones in molting and in the mineralization and demineralization of bone.

Molting. **Molting** encompasses a wide range of events. Amphibians, reptiles, and many invertebrates shed and reconstruct much of their integument at intervals, while the molting process of birds and mammals is seasonal (often occurring during breeding cycles) and usually restricted to changes in the feathers or hair.

Although many invertebrates cast their integument, detailed investigation of this process has been limited to exoskeletal molting in certain insects and crustaceans. The immediate molting stimulus is **ecdysone**, a steroid hormone released from paired **prothoracic glands** in insects and from paired **Y organs** in crustaceans (see Table 11.2 and Figure 11.6). The epidermis, as the target tissue, responds to ecdysone by secreting a new cuticle beneath the old one. Eventually, the partially digested remnants of the old cuticle split apart and are cast off. (The insect molting process is covered in Chapter 8, p. 180; Figure 8.9.) Release of ecdysone is controlled by neurohormones: Insect **brain hormone** (released from a storage organ, the corpus cardiacum) promotes ecdysone production and release; conversely, crustacean **molt-inhibiting hormone** (from the X organ–sinus gland complex in the eyestalks) inhibits ecdysone release (Figure 11.7). Thus, there is a close parallel between the control of molting in these two groups, but there is also a fundamental difference: Insects have a positive control system (neurohormones stimulate ecdysone release), whereas the crustacean system operates by negative control (neurohormones inhibit ecdysone release).

Among vertebrates, the timing of molts often correlates with environmental changes. When a mammal molts, the insulating properties and color of its coat may be drastically altered. Molts are often seasonal and prepare the animal for climatic changes. Generally, the winter coat insulates by trapping warm air in a dense layer of fine hairs next to the skin. The hairs of the summer coat are usually coarse and sparse, permitting free movement of cooling air. Seasonal color changes often

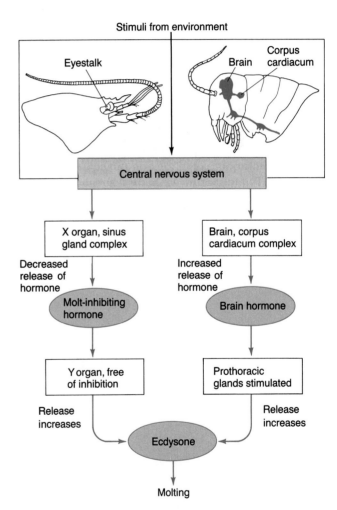

Figure 11.7 Control of molting in crustaceans and insects. Insects regulate molting by releasing a neurohormone (brain hormone) that promotes molting, whereas crustaceans regulate molting by releasing a neurohormone (molt-inhibiting hormone) that inhibits molting. This difference may reflect the long evolutionary independence of these two groups of arthropods. Depending on the species, various internal and external factors—ranging from day length and temperature to the nutritional state of the animal—suppress or stimulate release of the two neurohormones. The effect is to coordinate molting with the animal's state of growth and development, seasonal environmental factors, and the availability of resources such as food.

make the animal less visible; for example, the white winter coat of the snowshoe hare *(Lepus americanus)* matches a snowy background, while its brown coat helps it merge into the shadows of its summer habitat (Figure 11.8). In some mammals, hair loss is more or less continuous rather than seasonal, with molting passing successively from one area of the body to another. In rats, for example, molting starts on the abdomen and spreads slowly to the head and back. Humans have yet a different pattern; each hair follicle has a molt cycle independent of the follicles around it.

Birds molt completely each year, usually in late summer or fall after the nesting season, and many molt again in the spring before nesting begins. Feathers become tattered and even broken in the course of a year and must be replaced.

(a)

(b)

Figure 11.8 Molting in the snowshoe hare *Lepus americanus*. (a) Winter coat. (b) Summer coat.

Molting in mammals and birds is governed by some of the same factors that govern seasonal breeding cycles. Research with the field vole *Microtus,* a small rodent, has shown that the thyroid gland, adrenal glands, and reproductive organs most actively secrete hormones in the spring and slow down in the autumn. More hairs grow as levels of reproductive organ hormones decline; the hairs are finer because thyroid activity is reduced; and mature hair is retained because adrenocortical hormones are low. The vole then has a dense coat of fine hairs and is ready to face winter. In the spring, as hormone levels increase, the vole sheds mature hair and grows a thin summer coat of coarser hairs.

Calcium Metabolism. Controlled levels of calcium in the blood and intercellular fluids are important for many physiological processes, including muscle function, synaptic transmission in the nervous system, and cellular transport. Proper amounts of calcium are also needed to give the skeletons of molluscs, crustaceans, and vertebrates strength and hardness. Although crustaceans and molluscs are known to use hormones to regulate calcium metabolism during cuticle and shell growth, by far the best understood control of calcium levels is that of vertebrates.

Three hormones control the movement of calcium between the vertebrate intestine, kidneys, skeleton, and body fluids. First, **calcitonin** released from thyroid gland cells reduces the blood plasma calcium concentration by inhibiting bone demineralization and promoting bone mineralization by bone cells. (Bone structure and growth are discussed in Chapter 8, p. 189; Figures 8.19 and 8.20.) Second, **parathormone** (parathyroid hormone) released from the parathyroid glands in response to a decline in blood calcium promotes release of calcium from bones, reduces loss of calcium in the urine, and promotes absorption of calcium from the digestive tract. (Although mammals, birds, reptiles, and most amphibians have parathyroid glands, fishes do not; a hormone called prolactin may elevate blood calcium levels in fishes.) A third hormone, **1,25-dihydroxycholecalciferol (1,25-DHCC)**, increases the uptake of dietary calcium from the intestine into the blood. This hormone is a derivative of vitamin D, which is obtained in the diet and also formed in the skin by the action of the sun's ultraviolet rays—one of the few benefits of exposing the skin to the sun. The liver and kidneys convert vitamin D to the active hormone 1,25-DHCC when blood calcium and phosphate levels decline.

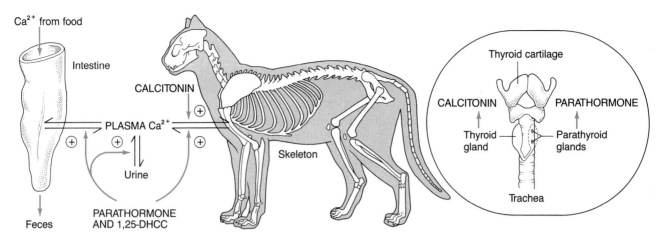

Figure 11.9 Regulation of calcium metabolism in a mammal. Calcitonin promotes bone mineralization, whereas parathormone has the opposite effect, as well as reducing calcium loss in the urine and promoting its absorption from the intestine. Meanwhile, 1,25-DHCC also promotes absorption of dietary calcium into the blood.

In summary, homeostasis in blood calcium levels in most terrestrial vertebrates is closely maintained by the interaction among parathormone, 1,25-DHCC, and calcitonin (Figure 11.9). Factors that reduce blood calcium level, such as inadequate dietary calcium or the drain on blood calcium during pregnancy due to development of the fetus, cause demineralization of the bones to make up the shortage. For this reason, women are often advised to increase their dietary calcium intake during pregnancy and while nursing offspring.

Control of Growth and Maturation

Hormones exert several effects on embryonic and postembryonic growth and differentiation. First, hormones can steer development into one of several alternative pathways. For example, embryonic tissues have the potential to differentiate into either male or female individuals, but sex hormones determine which sexual characteristics actually develop. Second, hormones can control growth. In so doing, they may determine the overall size of an animal and, to some extent, the proportions of its parts. Third, by activating various growth processes at specific times, hormones govern developmental schedules. The timing of the events of metamorphosis, a complex process that transforms a tadpole into an adult frog or a caterpillar into a butterfly, is an example.

Growth. As long ago as 1886, it was suspected that the vertebrate pituitary gland played a role in controlling growth. Excessive growth seemed to be

correlated with pituitary tumors, and certain kinds of dwarfism were associated with pituitary deficiency. In 1912, physiologists discovered that removal of the pituitary gland stopped or retarded the growth of puppies, a finding that was soon extended to other vertebrates, including humans. A few years later, in 1921, researchers found that repeated injections of pituitary extract during growth produced abnormally large experimental animals. Eventually, the active factor, named **growth hormone (GH)**, was isolated from the pituitary gland and identified as a protein consisting of 191 amino acids. The gene that codes for GH has been cloned and injected into the egg cells of certain mammals, where it can be incorporated into the host's chromosomes and expressed. Some researchers hope that this technique may someday be used to improve livestock production. (Gene expression through protein synthesis is discussed in Chapter 15, p. 362.)

GH exerts its effect on bone and cartilage indirectly by means of the **somatomedins**, a group of growth-promoting peptide hormones produced by the liver (see Figure 11.3). Somatomedin production is accelerated in the presence of GH, and the somatomedins circulating in the blood stimulate skeletal growth. GH also promotes protein synthesis in growing animals by facilitating the entrance of amino acids into cells.

Pituitary malfunction can have profound effects. Chronic (long-term) release of too much GH in growing animals accelerates the growth of most organs more or less proportionately, resulting in a condition known as *gigantism;* too little growth hormone slows growth and results in *dwarfism.* Because the brain grows independently of GH, failure of the body to grow when an animal is quite small can be disastrous. The brain keeps growing but the skull does not, and death may result from brain compression.

GH is not the only hormone that affects growth. Although the thyronines do not stimulate growth in the way that GH does, thyroid deficiency impairs growth by greatly reducing the growth-promoting effect of GH. Sex hormones from the reproductive organs also influence vertebrate growth in various ways. The female sex hormones generally are responsible for producing smaller bodies and earlier maturation. The male sex hormones are responsible for the adolescent growth spurt in human males and for the generally larger and heavier male frames. The sex hormones achieve these effects in part by promoting calcification of the growth centers (epiphyses) of long bones such as the femur and ulna, bringing growth to a halt somewhat earlier in females than in males.

Among invertebrates, there is some evidence that growth is also regulated by a general growth hormone. Formation of new segments in young sandworms (Annelida: *Nereis diversicolor*) is apparently controlled by a neurohormone secreted by the brain. When the posterior segments of the worm are lost accidentally or removed by a researcher, they are regenerated. Removal of the brain reduces the ability to regenerate new segments, but brain reimplantation restores this ability. The neurohormone has been called a growth hormone by some zoologists and a regeneration hormone by others. Shell growth in molluscs may also be under neurohormonal control; presumably, the shell-forming mantle is the target organ. In arthropods, the invertebrate group with by far the best understood endocrine system, there is no evidence of control by a growth hormone. Ecdysone, the molting hormone of insects and crustaceans, is sometimes called a growth hormone, but this is a misnomer because the hormone causes molting, not growth.

Metamorphosis. Many animals exist as two or more morphologically distinct stages during their postembryonic development. They transform from one body form (the *larva*) to another (the *adult*) by a process called **metamorphosis**. Metamorphosis enables different developmental stages of an animal to exploit different habitats, often using very different methods of locomotion, feeding, and behavior. Frequently, the larval form feeds, grows, builds up a store of fat, and then transforms into an

adult capable of reproducing. Metamorphosis is widespread in the animal kingdom, occurring in some vertebrates and in a host of invertebrates, including polychaete annelids, echinoderms, arthropods, and molluscs. The hormonal basis of metamorphosis is best understood in insects and amphibians.

Some insects undergo a series of molts to successively larger but essentially similar larval stages, variously called larvae, nymphs, or naiads, and a final molt to the adult stage. Often the young look and live much like the adult. However, in flies, beetles, butterflies, and many other insects, the larvae are not similar to the adults, and the final larval molt produces a **pupa**. Within the pupa, the larval tissues are broken down or replaced, and an adult—for example, a brilliantly colored butterfly—emerges from the pupal skin. The inclusion of a pupa in the life cycle allows greater adaptive divergence between the immature form and the adult. But what determines whether a given molt will be larva-to-larva, larva-to-pupa, or pupa-to-adult? In other words, what determines the nature of the molt?

Recall that insect molting itself is governed by brain hormone and ecdysone. Some years ago, insect physiologists discovered a third hormone that influences the nature of the molt. When the **corpora allata**, a pair of glands located in the head (see Figure 11.6), were removed from an immature insect midway through the larval period (e.g., from the third stage of a species that has five larval stages), premature metamorphosis to the adult form occurred. The larval period was cut short and a miniature adult developed, as if some "juvenilizing factor" had been removed along with the glands. Implantation of these glands into the final larval stage had quite a different effect: The larva molted into an extra larval stage instead of metamorphosing, and juvenility was retained. In some cases, the extra larval stage grew and later metamorphosed into a giant adult. Appropriately, the juvenilizing factor was named **juvenile hormone**.

Figure 11.10 illustrates how juvenile hormone interacts with ecdysone to determine the nature of a molt. A high proportion of juvenile hormone to ecdysone ensures that the molt will be accompanied by formation of larval structures; a small proportion of juvenile hormone to ecdysone triggers a pupal molt; and ecdysone acting alone promotes the differentiation of adult structures. It is possible that juvenile hormone, like ecdysone, exerts its influence at the level of the gene, determining whether genes for larval, pupal, or adult structures

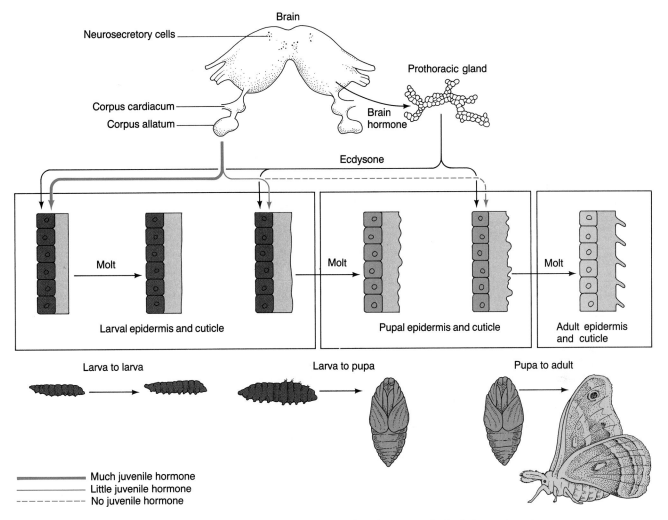

Figure 11.10 Endocrine control of insect growth and metamorphosis. Ecdysone from the prothoracic gland, released through the influence of brain hormone, acts on epidermal cells and initiates a molt. Juvenile hormone determines the nature of the molt: Abundant juvenile hormone causes a molt to another juvenile (larval) stage; less juvenile hormone causes a molt to a more mature stage, such as the pupa; absence of juvenile hormone allows molting to an adult.

are activated. In recent years, juvenile hormone has been chemically characterized. The hormone and its chemical cousins, or synthetic analogues, are being tested as possible insecticides, because spraying juvenile hormone on insects can kill them by disrupting their metamorphosis.

Among the vertebrates, metamorphosis is most conspicuous in frogs and toads, in which larval and adult life-styles are typically very different. The free-swimming, herbivorous, tailed, legless, gill-breathing tadpole of many frog species transforms into a four-legged carnivore with lungs. In most species, at metamorphosis, the tadpole loses most or all of its tail, changes its general body shape and means of locomotion, climbs onto land, and becomes terrestrial (Figure 11.11).

Metamorphosis in the frog is based on specifically timed changes in the concentrations of two hormones, prolactin and thyronine (thyroid hormone). Low thyronine levels and high prolactin levels in the young tadpole stimulate larval growth and prevent metamorphosis. As the hypothalamus and pituitary gland develop in the growing tadpole, thyroid-stimulating hormone–releasing hormone (TSH.RH) and prolactin-inhibiting hormone (P.IH) are released from the hypothalamus. Their release signals the anterior pituitary gland to release thyroid-stimulating hormone (TSH) and to withhold prolactin. As a result, blood thyronine levels rise and prolactin levels fall, triggering the onset of metamorphosis. Tail resorption and the other metamorphic changes follow.

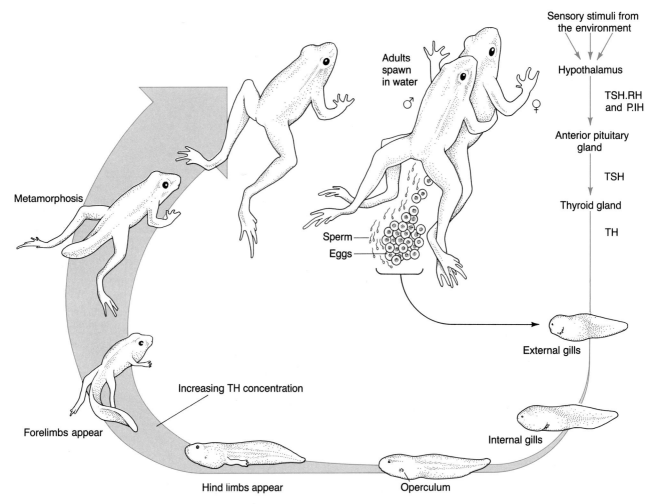

Figure 11.11 Frog tadpole metamorphosis. Tadpoles are aquatic and feed on algae and organic debris. Their metamorphosis into semiterrestrial or terrestrial adults is regulated directly by thyroid hormone (TH). Activity of the thyroid gland is regulated by thyroid-stimulating hormone (TSH), secreted by the anterior pituitary. Relative inactivity of the pituitary and thyroid keeps blood levels of TSH and TH low during the tadpole stage. During metamorphosis, blood levels of TSH and TH increase markedly, and cells in the animal become much more sensitive to TH stimulation. TH induces the drastic changes that result in the formation of a carnivorous adult frog. The time scale for larval development and metamorphosis varies greatly, depending on the species of frog and on water temperatures.

Control of Reproduction

The origin of experimental endocrinology can be traced to 1849, when the German physician A. A. Berthold discovered that a testis removed from a sexually mature rooster and transplanted into a castrated male (a capon) restored growth of the comb on its head. Berthold surmised that a factor released by the testis into the blood maintained the secondary sexual characteristics of males. Subsequently, many studies over a range of phyla have shown that reproduction in both vertebrates and invertebrates is under hormonal control.

As the following examples illustrate, hormonal regulation of reproduction is too diverse to allow much generalization. However, two principles seem to ensure reproductive success. First, release of neurohormones by neurosecretory cells synchronizes reproduction with suitable environmental conditions. By responding to stimuli such as day length, these cells ensure that the animal will make its reproductive effort at a time of year when the chances for success are greatest. Second, hormonal mechanisms usually bring body growth to an end before a period of reproductive activity begins. This increases the chances for successful reproduction by limiting reproductive effort to sufficiently mature animals.

Marine Annelids. Experimental studies show that in the marine polychaete *Nereis* (see Figure 25.1a), neurohormones that promote body growth inhibit reproduction; withdrawal of neurohormones when body growth ends allows reproduction to proceed. When experimenters cut *Nereis* worms in half, they discovered that sex cells matured only in the posterior half. Immature sex cells in the anterior half, which contained the worm's brain, failed to mature. Did the brain inhibit development? In many invertebrates, including annelids, it is possible to test this directly, because brains can be removed surgically without causing death, and brain transplants are performed routinely. When researchers removed the brain from *Nereis*, they found that the sex cells matured in the anterior half of the worm. Furthermore, when they implanted a brain into the brainless posterior half, maturation of the sex cells ceased. Evidently, the brain inhibited sexual maturation. The neurohormone responsible for the inhibitory effect has been named **annelid brain hormone**. The same hormone appears to promote body growth by stimulating development of body segments.

Several invertebrates in other phyla, including the freshwater cnidarian *Hydra*, have an inhibitory mechanism similar to that of *Nereis* to control and isolate the process of reproduction. Illustrating the flexibility of hormonal systems, certain nonnereid polychaetes make use of an opposite mechanism but arrive at the same result. In these annelids, the brain has a positive effect on sex cell maturation, and therefore its surgical removal is inhibitory. Reproduction is *stimulated* by a hormone produced only after body growth has ended.

Molluscs. There have been many studies on the hormonal control of reproduction in gastropod and cephalopod molluscs. Regulation of sexual maturity in the cephalopod *Octopus vulgaris* involves environmentally triggered inhibitory and stimulatory interactions between a brain center, an endocrine gland, and the gonads. In common with nonnereid annelids, reproduction is stimulated by a hormone after body growth has ceased.

As described in Chapter 10, the octopus brain is large and complex (see Figure 10.13). Optic nerves

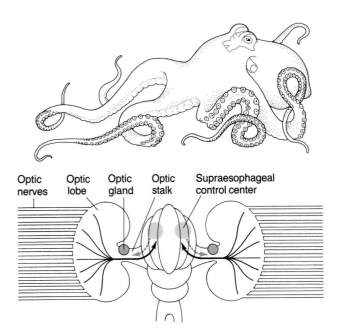

Optic nerves | Optic lobe | Optic gland | Optic stalk | Supraesophageal control center

Experimental procedures that resulted in gonadal development

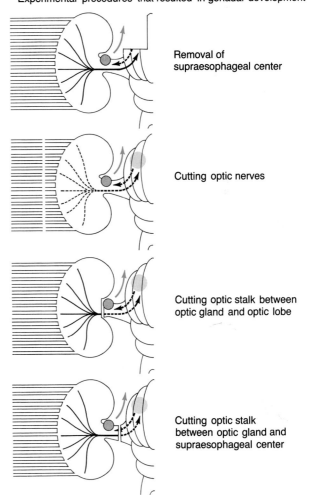

Removal of supraesophageal center

Cutting optic nerves

Cutting optic stalk between optic gland and optic lobe

Cutting optic stalk between optic gland and supraesophageal center

Inhibition of optic gland ----→   Stimulation of brain centers ——→
Stimulation of the gonads ——→   Nonfunctioning ------→

Figure 11.12 (Right) Experimental analysis of how sexual maturity is controlled in the octopus. Gonadotropic hormone from the optic gland promotes maturation of the gonads. The optic gland is inhibited by the supraesophageal control center in the brain. Because the brain center is affected by seasonal changes in day length, reproduction is timed to occur during favorable seasons.

pass from the eyes to large optic lobes located on either side of the brain. Optic stalks connect the optic lobes to the brain, and optic glands lie on each of the optic stalks (Figure 11.12). By blinding experimental animals, researchers found that they could induce the optic glands to secrete a gonadotropic hormone (a hormone that stimulates gonadal development or function). The hormone produces a dramatic increase in ovary size; before sexual maturity, the ovaries are about $\frac{1}{500}$ of the octopus's body weight, but after activation of the optic glands, the ovaries swell to $\frac{1}{5}$ the body weight, and egg laying begins. Additional experiments showed that a center in the brain called the supraesophageal control center exerts inhibitory control over the optic glands, preventing secretory activity and sexual maturation. This brain center is kept active by light stimuli transmitted via the eyes and optic nerves. Blinding the animal by cutting its optic nerves inactivates the brain center and eliminates its inhibitory effect on the optic glands. In fact, any surgery that removes the brain center or breaks its connections with the optic glands or eyes allows the optic glands to begin secreting gonadotropic hormone. Similar results have been observed in male octopuses. In nature, the brain center is activated and inactivated by changes in day length, a seasonal factor. As a result, reproductive activity is synchronized with the changing seasons.

Crustaceans and Insects. Many crustaceans are dioecious (i.e., they have separate sexes), and sex is under genetic control. However, male and female gonads and secondary sexual structures are not differentiated at the time of hatching, and several molts usually pass before males and females can be distinguished. Pioneering research on the control of sexual differentiation in crustaceans was carried out more than 30 years ago in France. The experimental animal was *Orchestia gammarella*, a common seashore-dwelling species in which the second pair of walking legs in the male has enormously enlarged claws that are used to hold down the female during copulation (Figure 11.13). Endocrine glands were discovered near the terminal ends of the male genital ducts. Eventually named **androgenic glands**, these remain small in females but enlarge in males, forming extensive strands of secretory cells that empty their contents into the hemolymph.

Extensive testing has shown that the androgenic glands have a masculinizing effect. Implanting them into a female transforms the ovaries into testes with sperm, and after a few molts, enlarged male claws develop. Recipient females even at-

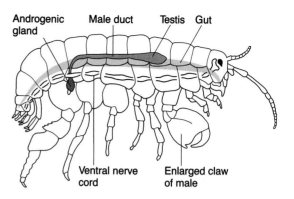

Figure 11.13 Development of male characteristics in the crustacean *Orchestia*. A hormone secreted by paired androgenic glands induces development of enlarged claws, which are male characteristics. Experimental removal of the androgenic glands causes regression of the testes and development of claws of the female type.

tempt to copulate with normal females, a futile effort because females lack male genital ducts. The opposite experiment, surgical removal of the glands from a young male, causes the gonad to produce egg cells instead of sperm, and oversize male claws fail to develop. Thus, sexual differentiation in the male seems to depend on **androgenic gland hormone**. The chemical nature of this hormone is not yet known, but it may be a steroid, because the vertebrate male sex hormone, testosterone, mimics its effects when injected into *Orchestia*, and testosterone is a steroid. What controls sexual differentiation in *Orchestia* females has not yet been resolved, although one or more hormones from the ovaries appear to be required.

Apart from the early sexual differentiation controlled by the androgenic glands and ovaries of *Orchestia gammarella* and other species that have been investigated, reproductive development in crustaceans is seasonal and under the influence of such environmental factors as day length and temperature. These factors operate through the X organ–sinus gland–Y organ complex, referred to earlier in our discussion of molting (see Figure 11.6b).

Sex differentiation in insects contrasts sharply with the situation in crustaceans. While crustaceans (and vertebrates) depend on hormones for this process, in insects it is under direct genetic control, with sex hormones apparently playing no role. (The single known exception occurs in the firefly *Lampyris*.) However, insect hormones take part in other reproductive processes, including synthesis of egg yolk protein and egg development.

Mammals. Although the primary function of the **ovary** is to produce eggs, the ovary is also an endocrine organ that synthesizes and releases the female sex hormones, **estrogens** and **progesterone** (Figure 11.14). The ovary undergoes cyclic events that are accompanied by cyclic changes in the uterus and in behavior in many mammals. In particular, during regular intervals, the female becomes more receptive to the male. She is said to be in "heat," or **estrus** (from the Latin *oestrus*, "frenzy, passion"), and the mammalian reproductive cycle is therefore generally called an **estrous cycle**. Humans and other primates lack a definite estrus, but in contrast to other mammals, have a **menstrual cycle** (from the Latin *mensis*, "month") in which the outer portion of the uterine lining increases in thickness and is shed periodically. **Menstruation**, discharge of the uterine lining, occurs at intervals of about 28 days in women and 35 days in chimpanzees, with shorter intervals for most other primates.

Estrous and menstrual cycles are governed by dramatic fluctuations in a neurohormone—**gonadotropin-releasing hormone (G.RH)** from neurosecretory cells of the hypothalamus—and in four glandular hormones—**follicle-stimulating hormone (FSH)** and **luteinizing hormone (LH)** from the anterior pituitary and estrogens and progesterone from the ovary (Figure 11.15).

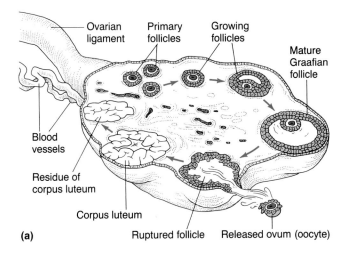

(a)

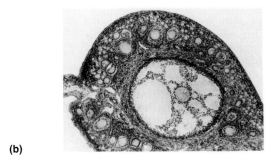

(b)

Figure 11.14 The mammalian ovary. (a) This drawing of a section of an active mammalian ovary shows various structures involved in egg production. (All the developmental stages shown here would probably not be present at the same time.) A primary follicle consists of a primary oocyte (a cell that develops into a second oocyte) surrounded by a single layer of follicle cells. Follicle cells support the oocyte and secrete female sex hormones. Depending on the species, one or more primary follicles develop into Graafian follicles, each containing an oocyte, during each reproductive period (estrus in most mammals). Ova (actually secondary oocytes) are released from the Graafian follicles during ovulation, and the remains of the ruptured Graafian follicle transform into a corpus luteum. If pregnancy occurs, the corpus luteum remains, and its cells secrete hormones that maintain the pregnant state. If pregnancy does not occur, the corpus luteum degenerates. These events, from primary follicle through pregnancy or corpus luteum degeneration, are repeated during each reproductive period. Collectively they are called the ovarian cycle. (b) Section through the ovary of a rat, showing several immature follicles and a mature follicle containing an oocyte.

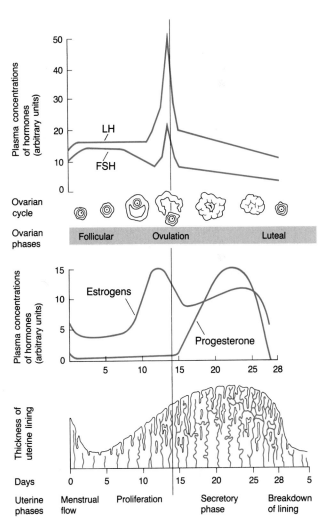

Figure 11.15 Phases of the reproductive (menstrual) cycle in the human female. The ovarian cycle illustrated in Figure 11.14 is correlated with cyclic events in the uterus (uterine cycle) and with cyclic changes in the blood concentrations of various hormones (LH, FSH, estrogens, and progesterone). The hormone levels regulate the ovarian and uterine cycles.

Pulses of G.RH from the hypothalamus stimulate the anterior pituitary to release FSH and LH, which are required for the ovarian cycle. The **primary follicles** in the ovary (see Figure 11.14) possess specific FSH receptors but no LH receptors. FSH promotes growth of a primary follicle. Then, about midway through the follicle's development process, FSH induces formation of LH receptors in the follicle cells, and LH causes the follicle to develop into a mature stage called a **Graafian follicle**.

The blood level of estrogens (produced by the follicle cells) is low during the early part of follicle maturation. At this time, the estrogens have a negative feedback effect on the hypothalamus and the pituitary, inhibiting the release of FSH and LH. But as the follicle approaches maturity, estrogen output increases sharply, soon reaching its maximum level (see Figure 11.15). The effect of estrogens on the hypothalamus now changes. At high levels, estrogens promote the release of G.RH and sensitize the LH-releasing mechanism of the pituitary. The G.RH can therefore promote a very sharp jump in the amount of LH in the blood. Because LH receptors are now present in the follicle cells, LH can exert its effects. First, LH promotes final maturation of the Graafian follicle, its rupture, and **ovulation** (release of the egg). Then LH causes the cells of the ruptured follicle to luteinize, meaning they change in appearance and form the **corpus luteum**, an endocrine gland capable of secreting large amounts of progesterone.

LH keeps the corpus luteum in an intact, active state. In humans, high levels of estrogens *plus* progesterone have a negative feedback effect, reducing the output of G.RH from the hypothalamus and LH from the pituitary. Consequently, the corpus luteum degenerates, and the levels of estrogens and progesterone (produced by the corpus luteum) fall sharply. With the negative feedback from these hormones removed, the hypothalamus releases G.RH, the pituitary releases FSH, and a new cycle of follicle formation begins. In some mammals, including domestic sheep, the corpus luteum degenerates through the action of a prostaglandin originating in cells of the uterus. (See the Essay on prostaglandins in this chapter, p. 249.)

The estrogens released from cells of the developing follicle have an extremely important task: the preparation of the uterus for pregnancy. Under the influence of estrogens, the uterine lining begins to thicken as nutrient-supplying glands and blood vessels proliferate. After ovulation, progesterone from the corpus luteum works with the estrogens, completing preparation of the uterus for the arrival and implantation of an embryo. Maintenance of the uterine lining requires continual stimulation by estrogens and progesterone. If there is no pregnancy, the corpus luteum degenerates, sex hormone levels plummet, and the thickened uterine lining deteriorates (see Figure 11.15). In humans and most other primates, this is accompanied by blood seeping from weakened uterine blood vessels to form the menstrual flow.

If a pregnancy does occur, the uterine lining becomes essential to the nourishment and protection of the embryo. Degeneration of the corpus luteum must therefore be prevented so that sufficient estrogens and progesterone are available to maintain the lining. In humans and other primates, the implanting embryo begins to secrete a hormone having properties like those of LH; called **chorionic gonadotropin**, this hormone maintains the corpus luteum from about the second week to the third month of a human pregnancy. The **placenta**, the organ developed from the embryo and uterine wall that nourishes the developing embryo, becomes an endocrine organ, too, secreting estrogens and progesterone and eventually assuming full responsibility for maintaining the uterus in a pregnant state. Only then does the corpus luteum regress. In mammals that use a prostaglandin to destroy the corpus luteum, the embryo somehow prevents release of the prostaglandin or makes the corpus luteum resistant to it. The nature of the signals from the embryo is a subject for further research.

Much as the ovary produces both ova and female sex hormones, the **testis** produces both spermatozoa and the male sex hormones, or **androgens**. The most prominent androgen is **testosterone**, which, like the female sex hormones, is a steroid (see Table 11.2). Testosterone is produced by **Leydig cells** surrounding the seminiferous (sperm-producing) tubules in the testis (Figure 11.16). The androgens effect a multitude of changes in males. For example, at sexual maturity, increased androgen levels bring about growth and secretory activity of accessory sex glands such as the prostate. Other androgenic effects include muscular growth, beard and other hair growth in humans, thickened vocal cords and a deeper voice, and behavioral changes, including courtship displays, sex urge, and sexual competence.

The pattern of testicular control has much in common with that of ovarian control. It involves the hypothalamus and anterior pituitary and even employs some of the same substances. G.RH from the hypothalamus stimulates LH release from the anterior pituitary. LH then promotes testosterone synthesis by the Leydig cells. But just how testosterone influences sperm formation is still uncer-

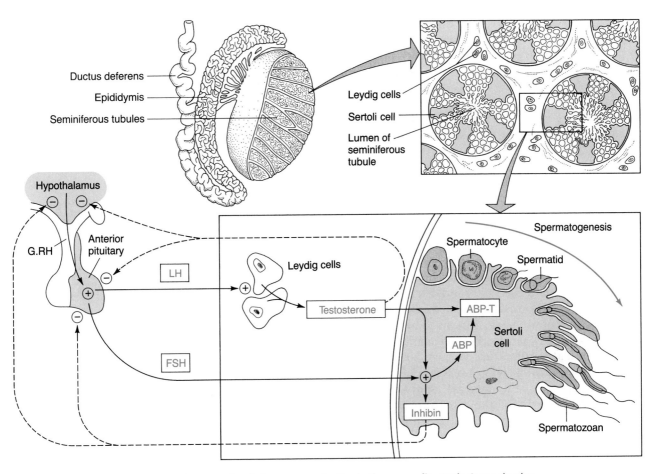

Figure 11.16 Sperm and male hormone production in the mammalian testis. Sperm develop in the seminiferous tubules of the mammalian testis. The Leydig cells produce the male sex hormone testosterone. The Sertoli cells support the germ cells and produce androgen-binding protein (ABP) and a protein hormone called inhibin. ABP binds to testosterone, forming ABP-T, which seems to promote sperm production. The hypothalamus and the anterior pituitary both have roles in promoting testosterone and inhibin formation. The interactions of all hormones and releasing factors are not completely understood, but as shown, the production of FSH, LH, and, indirectly, testosterone is controlled by negative feedback mechanisms.

tain. It is known that testosterone, together with FSH, causes **Sertoli cells** (which support the germ cells) to produce a protein called androgen-binding protein (ABP), which binds testosterone molecules (see Figure 11.16). The resulting testosterone-rich environment in the seminiferous tubules may be required for sperm formation. Testosterone and FSH also promote synthesis of **inhibin** by the Sertoli cells. Inhibin and testosterone modulate G.RH, LH, and FSH release by means of negative feedback loops to the hypothalamus and anterior pituitary.

Control of Other Processes

The roles of some additional hormones are described briefly in Table 11.3, and some are discussed in appropriate sections in other chapters.

The release of digestive enzymes as food passes through the digestive tract, for example, is regulated by the combined actions of the endocrine and nervous systems. (The secretion and function of gastrin, secretin, and CCK are discussed in Chapter 4, pp. 93–94.) Hormonal regulation of excretion and osmoregulation is discussed in Chapter 7. Aldosterone and a similar compound called **deoxycorticosterone** are **mineralocorticoids**, an important group of hormones produced by the adrenal cortex. (Aldosterone is discussed on p. 165.) In mammals, these hormones help maintain homeostasis by regulating ion uptake or secretion by the kidney, gut, salivary glands, and sweat glands. Mineralocorticoids and the other major group of adrenal cortex hormones, the **glucocorticoids**, are classified on the basis of their role in mammals. The same hormones often play quite different roles in other vertebrates.

Glucocorticoids, for instance, function in glucose metabolism in mammals, but regulate ion balance in many bony fishes.

As indicated in Table 11.3, mineralocorticoids and glucocorticoids also play important roles in regulating the immune response. While glucocorticoids such as cortisol, cortisone, and corticosterone generally depress the inflammatory response, the mineralocorticoids generally have the opposite effect. These opposing hormonal influences help maintain a dynamic balance in a mammal's response to infectious agents. (The inflammatory response, a component of the vertebrate immune response, is discussed in Chapter 6, p. 139.)

There is some evidence that hormones of the adrenal cortex are also important components of what is often called the vertebrate **stress response**. As first proposed in the 1940s and '50s by the Canadian researcher Hans Selye, the "stress hypothesis" holds that vertebrates—and perhaps all animals—respond to various stressors (such as infectious disease, extreme temperature, injury, or excessive crowding) in a similar way. A number of responses, collectively called the **general adaptation syndrome (GAS)**, are thought to constitute an animal's attempts to maintain homeostasis while confronting the stressor(s). GAS mainly involves a general increase in activity of the hypothalamus–anterior pituitary–adrenal cortex hormonal mechanism (Figure 11.17). The stress hypothesis holds that stressors stimulate the hypothalamus via nervous impulses or by increasing the body's output of epinephrine. (See the discussion of the sympathetic nervous system in Chapter 10, p. 242.) The hypothalamus, in turn, may induce the pituitary to secrete increased amounts of ACTH and decreased amounts of growth hormone and/or the reproductive hormones LH and FSH (see Table 11.3). Elevated levels of ACTH in the blood may cause the adrenal cortex to release excessive amounts of glucocorticoids into the blood. This would have the effect of reducing inflammation and maintaining blood glucose levels. It is thought that by reducing inflammation, GAS may prevent the body from harming itself. Thus, GAS may prevent the immune system from overreacting to stressors. The increasing glucose levels help maintain a higher rate of metabolism, helping the body adjust to the stressor.

Elevation of blood levels of ACTH and glucocorticoids during stress has been demonstrated experimentally. As suggested by Selye's hypothesis, these changes may represent attempts on the part of an animal's physiological mechanisms to compensate for stress. Such changes may not be harm-

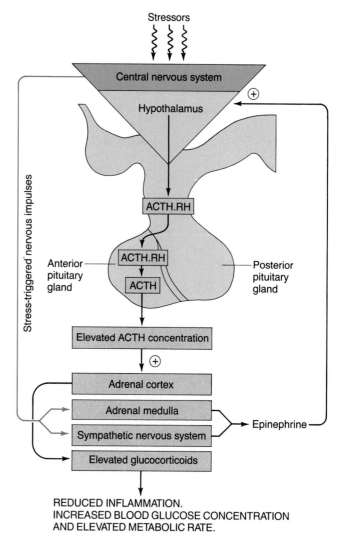

Figure 11.17 The stress hypothesis and the general adaptation syndrome (GAS). A wide variety of environmental stressors may cause the hypothalamus to stimulate ACTH production by the anterior pituitary. In turn, a hyperactive adrenal cortex may produce quantities of hormones that help the body compensate for the stress, but may have long-term detrimental effects.

ful in the short term. Depending on the stressor and the intensity of its effects, however, an animal may not be able to maintain homeostasis if stressed indefinitely. The stress hypothesis predicts that prolonged increase in hormonal activity and the increased body activities that it causes may eventually lead to the impairment of normal body functions.

The stress hypothesis is not universally accepted, and so far, its main contribution has been to stimulate research. It has been suggested that stress-related hormonal changes may decrease animal growth and reproduction and thus may have a regulating influence in certain animal populations.

☐ SUMMARY

1. Chemical messenger substances, called hormones, help to control and coordinate most life processes in animals. Hormones include neurohormones, produced and secreted by nerve cells, and glandular hormones, produced by the ductless endocrine glands. Hormones are usually distributed throughout the body by the circulatory system, reaching target organs with hormone-specific receptor sites. The nervous system, endocrine glands, and their biochemical products work together as a neuroendocrine system.

2. Synthesis and release of hormones are self-regulating processes usually involving negative feedback loops. As the amount of a hormone (or the amount of a substance regulated by a hormone) fluctuates above or below a required level, the gland producing or releasing the material is inhibited or stimulated. Positive feedback loops control the release of only a few hormones.

3. Most hormones are steroids, small proteins, peptides, or amino acid derivatives. Steroid hormones typically exert their effects by guiding the synthesis of mRNA. Protein and peptide hormones often influence the production of "second messengers" (commonly cyclic adenosine monophosphate, or cAMP), which then alter cellular functions.

4. Hormones regulate metabolism in vertebrates and in many invertebrates. In the vertebrates, pancreatic hormones regulate blood sugar; insulin acts to lower the level of blood glucose, whereas glucagon raises it. Thyronines from the thyroid gland influence metabolic rate, ATPase activity, and RNA synthesis in most cells, while growth hormone from the anterior pituitary gland helps to make fatty acids available as an energy source. Epinephrine (adrenaline) from the adrenal glands helps to make glucose available as an energy source during periods of excitement, fear, and other stresses, while glucocorticoids from the adrenal glands promote glucose formation.

5. Interactions among three hormones govern the distribution of calcium in the vertebrate body. Calcitonin from the thyroid gland reduces blood calcium levels; parathormone from the parathyroid glands has the opposite effect; and 1,25-dihydroxycholecalciferol facilitates absorption of dietary calcium from the gut into the blood.

6. Exoskeletal molting by arthropods is stimulated by ecdysone, a hormone released from the Y organs in crustaceans and the prothoracic glands in insects. Release of ecdysone is regulated by neurohormones: Brain hormone promotes release of ecdysone in insects, whereas molt-inhibiting hormone inhibits ecdysone release in crustaceans. Molting in vertebrates is also hormonally controlled.

7. Skeletal growth in vertebrates is stimulated by somatomedins, a group of peptides produced by the liver. Production of somatomedins is promoted by growth hormone (somatotropin) released from the anterior pituitary gland. Growth is also affected by thyronines and sex hormones.

8. Metamorphosis, the transformation from one body form to another, is governed by the relative amounts of ecdysone and juvenile hormone present at molting in insects and by changes in the amounts of thyronines and prolactin circulating in the blood in amphibians.

9. Many neurohormones and glandular hormones are involved in regulating reproductive processes. In general, hormonal controls gear reproductive activity to the time of year when reproductive success is most likely. Invertebrate reproductive hormones include annelid brain hormone (controlling sexual maturity of marine annelids), androgenic gland hormone (controlling male sexual differentiation in certain crustaceans), and gonadotropic hormone (controlling sexual maturity in cephalopod molluscs). In vertebrates, estrogens and progesterone (female sex hormones), testosterone (male sex hormone), and a number of hypothalamic neurohormones play key roles in reproduction.

10. Hormones produced by the adrenal cortex help regulate the immune response and may mediate a stress response called general adaptation syndrome. Various stressors may activate the adrenal cortex via the hypothalamus and pituitary gland.

☐ FURTHER READING

Berridge, M. J. "The Molecular Basis of Communication Within the Cell." *Scientific American* 253(1985): 142–153. *A discussion of the role of secondary messenger substances in intracellular communication.*

Buckle, J. W. *Animal Hormones. The Institute of Biology's Studies in Biology No. 158.* London: Edward Arnold, 1983. *A brief introduction to vertebrate and invertebrate endocrinology.*

Goldworthy, G. J., J. Robinson, and W. Mordue. *Endocrinology.* Glasgow and London: Blackie, 1981. *A comparative treatment organized by physiological process.*

Gorbman, A., W. W. Dickhoff, S. R. Vigna, N. B. Clark, and C. L. Ralph. *Comparative Endocrinology.* New York: Wiley, 1983. *A modern textbook of vertebrate endocrinology with coverage by gland.*

Raabe, M. *Insect Neurohormones.* New York: Plenum, 1982. *An excellent, brief review of insect endocrinology.*

Schally, A. V., A. J. Kastin, and A. Arimura. "Hypothalamic Hormones: The Link Between Brain and Body." *American Scientist* 65(1977): 712–719. *A good review of the master control activity of the hypothalamus.*

Snyder, S. H. "The Molecular Basis of Communication Between Cells." *Scientific American* 253(1985): 132–144. *A detailed but very readable discussion contrasting short-range with long-range chemical communication between cells.*

CHAPTER 12

Sensory Systems

The nervous and endocrine systems would not be able to coordinate an animal's vital processes without information about the internal and external environments. Control of urine production in many animals is based on information about blood chemistry, for example, while information about day length allows synchronization of reproduction with the seasons. Nervous and chemical coordination

usually requires more than one kind of information. A hungry predator's success in finding and running down a rabbit is a coordinated venture requiring information about the prey's odor and movements, the relative position of each of the predator's legs, and physical features of the trail. Animals receive environmental information through modified neurons called **sensory receptor cells**, which occur singly or in groups as part of **sense organs**. There are many kinds of receptor cells and sense organs. Some are specialized to detect chemicals in the environment, some respond to light, and still others respond to many kinds of

The tarsier (Tarsius syrichta). This rat-size primate lives in rain forests of the Philippines. Tarsiers are primarily nocturnal and insectivorous; their enormous eyes provide excellent night vision.

mechanical disturbances, including sound waves. Many animals also have receptors sensitive to thermal, electrical, and magnetic stimuli. In general, animal groups whose members have an assortment of well-developed sense organs are the most successful in terms of numbers of species and adaptability to diverse environments. Most animals in the three phyla most widely distributed in aquatic and terrestrial environments—Arthropoda, Mollusca, and Chordata—have a broad complement of sophisticated sense organs. Because sensory receptor cells and sense organs determine the amount and kind of information collected from the environment, understanding them is a major step toward understanding an animal's physiology, environmental relations, and behavior.

SENSORY MECHANISMS

Despite the remarkable diversity of sense organs, all sensory systems have certain important features in common. First, sensory receptors are strictly stimulus detectors; they do not interpret, or identify, stimuli. Stimulus interpretation, which involves distinguishing one type of stimulus from another (e.g., telling light from sound or odor), occurs as the nervous system receives information from the sensory system. Second, sensory cells are modified neurons that convert one form of energy (an environmental stimulus) into another (a nerve impulse). This energy conversion is called **transduction**, and sensory cells are energy transducers. Following stimulation by environmental stimuli, sensory cells all use the same mechanism for generating nerve impulses. Third, sensory cells have a capacity to encode sensory information by responding differently to stimuli of varying intensity. Finally, sensory cells have an ability to reduce their responsiveness to certain prolonged stimuli. This is called **physiological adaptation**.

Reception and Transduction of Stimuli

Because stimuli that affect receptor cells are actually various kinds of energy from the environment, energy transduction is a key concept in understanding sensory systems. **Photoreceptors** respond to light energy. **Mechanoreceptors**, such as the hairs on a spider's body, respond to mechanical energy associated with movement, touch, or sound. **Chemoreceptors** are organs of taste and smell that detect molecules in water or air. **Thermoreceptors** respond to differences in temperature.

Although different kinds of receptor cells are adapted to convert different kinds of energy, the underlying mechanism of transduction is similar in most cases. Like other neurons, a sensory cell membrane has an electrical resting potential, and reception occurs as stimulus energy causes the electrical potential to change. This usually creates a new potential called the **receptor potential** or **generator potential** (Figure 12.1). The generator potential results from a brief change in the membrane's permeability to sodium ions and, consequently, the passage of Na^+ into the receptor cell through ion gates in the cell membrane. The generator potential, in turn, produces localized electrical currents across neighboring regions of the cell membrane. These currents cause changes in the membrane's electrical potential which—if greater than threshold—become self-propagating nerve impulses. (Action potentials are discussed in Chapter 10, p. 224; Figure 10.5.) The impulses then travel along the cell's axon toward synaptic junctions with other neurons and become part of the information in the nervous system.

Sensory Coding

The generator potential must generate nerve impulses if information about a stimulus is to be conveyed to an animal's nervous system. A nerve impulse can inform an animal that a chemical is present, that the sky is light rather than dark, or that something is touching its body. But an impulse does not contain information about *how much* chemical or light or pressure there is, and such information is of vital importance in allowing animals to respond appropriately to environmental stimuli.

The key to providing quantitative information about environmental stimuli lies in the capacity of sensory cell membranes to respond *differently* to stimuli of different magnitudes. Depending on the strength of a stimulus, the size of the generator potential is greater or smaller (Figure 12.1). This is called a **graded response**, in contrast to the all-or-none nature of a nerve impulse. (The all-or-none principle is discussed in Chapter 10, p. 226.) As the size of the generator potential increases, the frequency of nerve impulses (number of nerve impulses generated per second) increases. Thus, a more powerful or larger stimulus produces larger generator potentials, and these generate nerve impulses at a higher frequency (see Figure 12.1).

The use of **impulse frequency** as an indicator of stimulus strength is one of several kinds of sensory

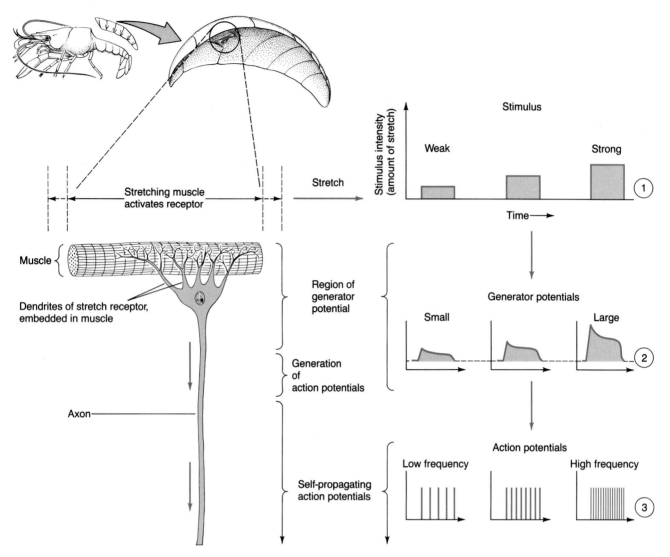

Figure 12.1 The relationship between strength (intensity) of stimulus, the resulting generator potentials, and nerve impulse frequency. Nerve impulse frequency reflects the intensity of stimulation, demonstrated here by the activity of a stretch receptor embedded in a muscle in a crayfish abdomen. When the crayfish arches its abdomen while swimming, the stretch receptors detect changes in muscle length. Stretching activates the receptors and leads to the generation of nerve impulses that represent information about the movement and position of the abdomen. Note that stretching the muscle by increasing amounts (graph 1) produces a corresponding increase in the magnitude of sensory cell generator potentials (graph 2). As the size of generator potentials increases, there is a progressive increase in nerve impulse (action potential) frequency (graph 3). In effect, increases or decreases in stimulus strength produce corresponding changes in the frequency of nerve impulses being transmitted toward the central nervous system.

coding. In a second kind of coding, receptors are connected to the central nervous system (CNS) in such a way that nerve impulses from a particular sense organ arrive in the CNS along a particular nerve, indicating a specific class of stimuli. In the parlance of modern neurobiology, this second type of receptor is "hardwired" to the CNS. For example, visual stimuli are signaled by impulses reaching the brain via the optic nerves, while impulses from chemical (odor) receptors travel along the ol-

factory nerves to a different region of the brain. Sensory coding may also involve specific kinds of receptor cells within a sense organ. For instance, about one-fourth of the receptor cells in a fly's eye are ultraviolet-light detectors, and impulses sent along fibers from these detectors are interpreted as ultraviolet by the CNS.

Coding can also involve a variation in the *total pattern* of nerve impulse output from a *group* of receptors, a process called **cross-fiber patterning**.

Coding by cross-fiber patterning is particularly important in the sense of smell because the receptor cells involved may be strongly stimulated by some chemicals, slightly stimulated by others, and even inhibited by still others. Because different receptor cells in a sense organ may respond differently to an array of chemicals, the total pattern of information passed to the central nervous system differs somewhat for each chemical. Cross-fiber patterning enables many animals to make fine distinctions among large numbers of scents. An example of cross-fiber patterning involving the sense of smell in an insect is described later in this chapter (p. 291; see Figure 12.16).

Physiological Adaptation

Physiological adaptation can be demonstrated by inserting recording electrodes into a receptor and then subjecting the receptor to constant, prolonged stimulation. The generator potential gradually decreases, reflecting the receptor's adaptation.

An example of physiological adaptation is our own reaction to certain strong odors—odors that seem to fade after first exposure. By "tuning out" unimportant stimuli, or stimuli that require no further action, physiological adaptation helps an animal detect and respond to new, possibly vital environmental information. Hearing the snap of a twig among other forest sounds might mean the difference between life and death for a bird being stalked by a cat. Some receptors adapt much faster than others. Many touch receptors adapt in less than a second, preventing an ongoing barrage of impulses from reminding us that we are wearing shoes or a wristwatch. In contrast, slow or minimal adaptation is common among receptors that provide ongoing information about the position of body parts and their movements, information useful in maintaining balance and posture. Providing more or less constant warning about potential injury, some pain receptors in humans also show very slow, even negligible, adaptation.

SENSES AND SENSE ORGANS

Light and Photosensitivity

Light is often described as wavelike, with each color having a different range of wavelengths. But light energy also behaves as discrete particles, which physicists call **photons**. So abundant are photons that on a bright day, billions of them enter the eyes of a large mammal every second.

Many organisms, including some that do not have specialized photoreceptors, are photosensitive; that is, they respond in some way to light. General **photosensitivity**—responsiveness to light without special light sensors—is uncommon in terrestrial animals, probably because their skin is relatively thick; but it is common among aquatic animals and protozoa, many of which have thin body coverings. In aquatic animals, nerves close to the skin surface may be stimulated by light, and in some animals, parts of the central nervous system are light-sensitive.

Photoreceptors

Photoreceptors, which are specialized structures that detect photons of light, vary from simple cell organelles or clusters of shaded cells that only give information about light direction to highly complex visual organs that can form images. Various terms (such as eyes, eyespots, and ocelli) are used to describe types of photoreceptors. Unfortunately, zoologists use these terms in a variety of ways. In this text, the word **eye** is used to refer to any type of multicellular photoreceptor. We use the term **eyespot** for simple light detectors (both cell organelles and eyes) that lack lenses and do not form images. The term **ocellus** usually refers to an eye that has a lens but has little focusing ability. Actually, however, this term is used so variously in biology that it has no concrete meaning except when applied to specific types of eyes in certain groups of organisms. Among the protozoa, for example, light-sensitive organelles are called eyespots or ocelli. Likewise, the simple eyes of flatworms (described shortly) are eyespots, but they are also called ocelli. At the other end of the spectrum, certain rather complex insect eyes, some of which can form images, are also called ocelli.

Despite the variety of photoreceptors, the fundamental process of light detection is universally similar in that it depends on a set of photon-induced chemical events called **photochemical reactions**. In most photoreceptor cells, the photochemical reactions involve a class of chemicals called **rhodopsins**, which are visual pigments related to vitamin A. In a photoreceptor cell, photons collide with and give up some of their energy to rhodopsin molecules, triggering a sequence of biochemical reactions that generate sensory nerve impulses. Arrival and interpretation of these impulses in visual centers of the nervous system produce the sensation of vision. Rhodopsin actually breaks down as it absorbs light energy. Specifically, light energy causes separation of rhodopsin's protein (opsin) and nonprotein (retinal, which is derived from vitamin A) components. Recombination of these components to form rhodopsin occurs rapidly in an enzymatic process in the vertebrate eye; a much faster photochemical reaction re-forms rhodopsin in octopus and insect eyes.

Photoreceptors That Do Not Form Images: Eyespots

Many photosynthetic protozoa (such as *Euglena*) and free-living flatworms have eyespots that are little more than light detectors, limited to giving information about the direction and brightness of light. These structures work on the principle of *shading*. The freshwater planarian worm *Dugesia*, for example, has two eyespots, also called **pigment cup ocelli**. Each of these structures is composed of a cup-shaped group of cells, with the mouth of one cup facing right and the other facing left (Figure 12.2). Light-absorbing pigment within the cells of each cup prevents light from passing through the cup's outer wall. Photoreceptor cells containing photosensitive pigment lie against the inside wall of each eye cup and are shaded except when light enters the open side of the cup. Dendrites, cell bodies, and axons of the photoreceptor cells extend ventrally to the planarian's cephalic ganglia, or "brain." Because light can reach the photoreceptor cells only through an eye cup's opening, light arriving from the left may enter only the left cup, while receptor cells within the right cup remain shaded. As it receives impulses from the photoreceptors, a planarian's nervous system compares information generated from the right and left and then, by turning its body, equalizes and minimizes the amount of light reaching the ocelli. This simple kind of sensory balancing enables a planarian to move directly away from the light toward the shadowy recesses beneath rocks, where it normally lives.

Image-Forming Eyes

For an animal to actually see objects, it must have eyes capable of casting images on a layer of photoreceptor cells called a **retina**. Animals in five phyla (Cnidaria, Mollusca, Annelida, Arthropoda, and Chordata) have image-forming eyes. As described in the following sections, the simplest type of image-forming eye has a retina, but lacks a lens. The great majority of animals have eyes with one or more **lenses** that collect and focus light, making bright, sharp images possible. Two very different

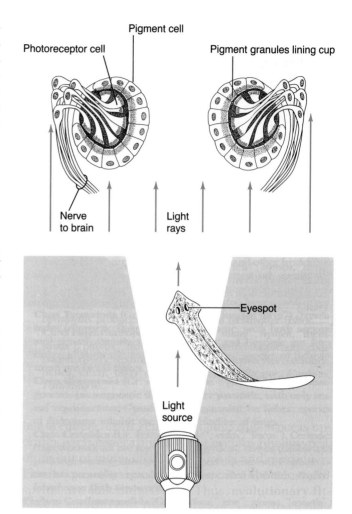

Figure 12.2 Light receptors and orientation in planarians. The eyes of planarians lack a lens or retina and are incapable of forming images. Each eye consists of a cup-shaped group of pigment-containing cells and a small number of photoreceptor cells arranged within the cup. The cup is oriented in such a way that its pigment shields the photoreceptor cells from light coming from behind or from the opposite side of the animal. A planarian may turn back and forth until nerve impulse output from the eyes is equal and minimal (i.e., until the eyes are equally and minimally illuminated); then the animal may crawl directly away from the light.

kinds of lens eyes have evolved. One type uses a single lens to form images, as in a camera, whereas the other, called the compound eye, has many lenses that form a composite, or mosaic, image.

Eyes Without Lenses: Pinhole Eyes. The eyes of the chambered nautilus, a cephalopod mollusc, form images by admitting light through a small opening, or pinhole (Figure 12.3). Light rays passing through the pinhole cast an inverted image on the retina at the back of the eye. Because of its small size, the opening produces a sharply focused image; but since very little light is admitted, the image is dim.

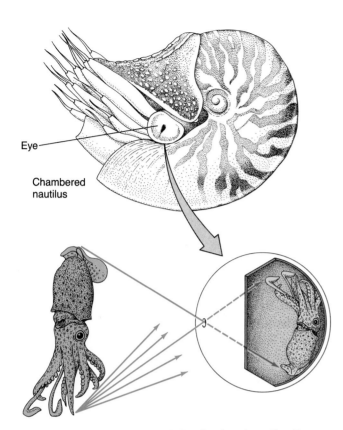

Figure 12.3 The pinhole eye of the chambered nautilus. Because little of the light scattered from an object passes through the pinhole into the eye, the resulting inverted image is dim.

Single-Lens Eyes. Certain jellyfish, polychaete annelids, spiders, many molluscs, and vertebrates have two or more single-lens, camera-like eyes. Spiders usually have eight, arranged in two groups of four near the anterior end of the body. Web spinners, which generally are sedentary spiders that wait passively for prey to blunder into their web, have eyes with little image-forming capability, whereas the eyes of hunting spiders can perceive objects and detect insect movements. Jumping spiders have truly sharp vision, an important factor in making accurate leaps. Many insects also have single-lens eyes called ocelli, although the main visual organs in adult insects and many immature forms are compound eyes.

The eyes of vertebrates are covered by a tough, protective layer of fibrous connective tissue (Figure 12.4a). The greater part of this layer forms the "white" of the eye, or **sclera**, while the transparent part forms the **cornea**, which admits light. In many vertebrates, the cornea and sclera are protected by upper and lower **eyelids**. Contraction of a muscle in each eyelid closes the eyes, and a levator muscle in the upper eyelids opens them. As the eyelids move,

mucous membranes on their inner surface moisten and cleanse the eyes.

After passing through the cornea, light crosses the fluid-filled anterior chamber and passes through the **pupil**. This is an opening in the center of a pigmented muscular diaphragm called the **iris**. In a process called **adaptation** (different from physiological adaptation, described earlier on p. 273), muscles in the iris control the amount of light that enters the eye by automatically changing the diameter of the pupil as light conditions change. In bright light, iris muscles encircling the pupil contract, narrowing the opening. In dim light, a set of radially arranged iris muscles contracts, pulling the pupil open so that more of the available light can enter (Figure 12.4b). The response of the iris muscles to changing light intensity is governed by the autonomic nervous system (described in Figure 10.19).

After passing through the pupil, light passes through the **lens**, which consists of cells containing large volumes of clear, fibrous cytoplasm. The lens is suspended by **suspensory ligaments** extending from its margin to the surrounding muscular **ciliary body** (see Figure 12.4a). Because the lens is biconvex, it converges light rays as they cross its front and back surfaces. This allows the lens to perform its chief function, which is to project and focus images of environmental objects onto the retina at the back of the eye. Because the cornea is convex in shape, it also bends light rays at angles that help the lens in focusing incoming light. The cornea's role in focusing is particularly important in terrestrial vertebrates; because light rays are strongly bent at the air/cornea interface, focusing can be completed by a relatively flat, weak lens. On the other hand, in most fishes, the cornea contributes little to focusing because the water/cornea interface bends light rays only slightly, making a more powerful, almost spherical lens necessary.

To focus images sharply, the eye must **accommodate** when viewing objects at different distances. The eyes of many fishes accommodate by moving the lens forward when viewing nearby objects and backward when focusing on distant objects. The human eye and that of many other vertebrates accommodate by changing the shape of the lens (Figure 12.5). When muscles of the ciliary body contract, the **choroid coat**, a vascular layer just behind the retina (see Figure 12.4a), is pulled forward slightly, reducing tension on the suspensory ligaments attached to the rim of the lens (see Figure 12.5). Through its own inherent elasticity, the lens then passively assumes a thicker, more spherical shape capable of focusing images of nearby ob-

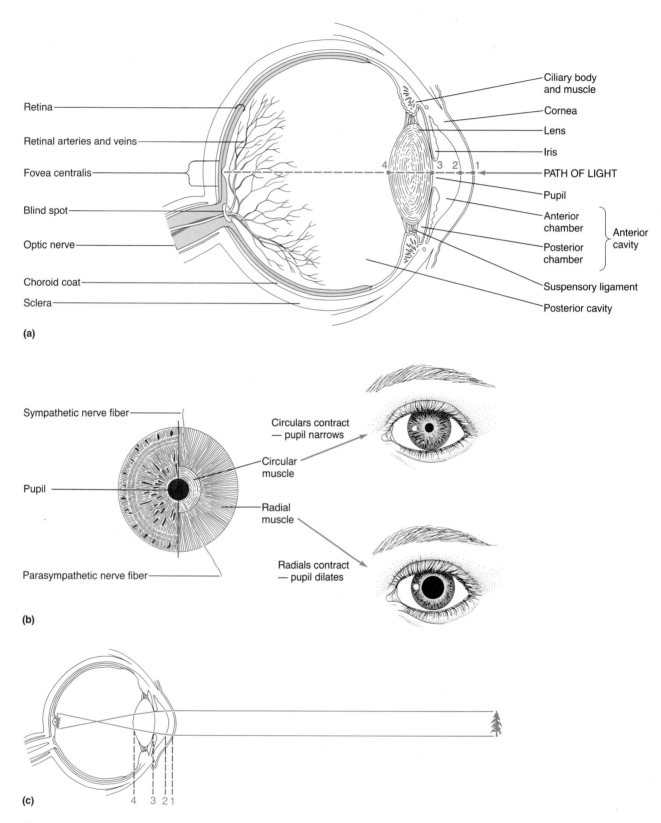

Figure 12.4 Structure of the mammalian eye. (a) The internal structures in a sagittal section of a human eye. (b) The diameter of the pupil changes as the eye adapts to changing light intensity. The pupil dilates in dim light as the radial muscles contract, and constricts in bright light as the circular muscles contract. Because the radial muscles are innervated by nerve fibers from the sympathetic division of the autonomic nervous system, the pupil tends to dilate and admit more light during periods of excitement, fear, and stress. (c) As light travels toward the retina, the light rays are bent (refracted) at four points: at the anterior and posterior surfaces of the cornea (1 and 2) and at the anterior and posterior surfaces of the lens (3 and 4).

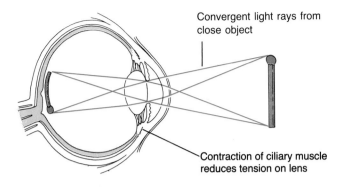

Convergent light rays from close object

Contraction of ciliary muscle reduces tension on lens

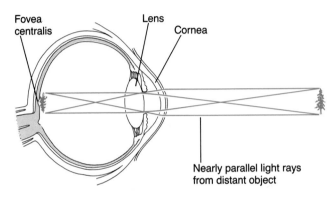

Fovea centralis

Lens

Cornea

Nearly parallel light rays from distant object

Figure 12.5 Accommodation in mammals. Accommodation is accomplished by changing the shape of the lens, which becomes flatter when viewing more distant objects. Note that the lens centers the image on the fovea centralis, the region of the retina having the greatest concentration of photoreceptor cone cells and therefore the region of greatest visual acuity. Birds accommodate by changing the curvature of the cornea as well as the shape of the lens.

jects. The lens gradually becomes less elastic with aging, thus losing some of its ability to focus on nearby objects. At about age 40, many people begin to compensate for this deficiency by wearing "reading" glasses. For focusing distant objects, the ciliary body muscles relax, restoring tension to the suspensory ligaments, which then pull the lens to a flatter shape. Thus, muscular effort is required for close viewing, but not for distance viewing.

The vertebrate retina consists of several kinds of nerve cells forming three layers. Light must pass through two transparent layers (the ganglion cell layer and the bipolar cell layer) before reaching photoreceptor cells in the third layer. Because the photosensitive layer is located in this position, under the other cell layer and nerve cell fibers, the retina of vertebrates is said to be **indirect**, or **inverted**, and the eye has a **blind spot** (Figure 12.6a). Photoreceptor cells are of two kinds called **rods** and **cones** because of their general shape (Figure 12.6b). When rods and cones are stimulated by light, their membranes generate generator potentials, which

trigger release of neurotransmitter molecules at synapses with the bipolar cells. In turn, the bipolar cells release a neurotransmitter onto the ganglion cells. Nerve impulses generated in the ganglion cells then travel along the **optic nerve** to the brain. In some vertebrates, such as frogs, a certain amount of integration of visual information takes place among the ganglion cells of the retina; but in mammals, most integration and interpretation occur in the visual cortex and visual association areas of the brain. (For a discussion of the functional regions of the brain, see Chapter 10, p. 237; Figure 10.17, 10.18.)

Many, though not all, vertebrates have both rods and cones. Humans, for example, have about 125 million rods and 6 million cones in each eye. Rods are extremely light-sensitive, responding even to dim light, whereas cones respond only to bright light and colors. Rods enable an animal to see in relatively dark conditions and to see shades of gray corresponding to light intensity, not unlike the shades seen on a black-and-white television set. At night, rhodopsin in the rods is reconstituted as fast as it is used, keeping the eyes maximally light-sensitive (**dark-adapted**); but in bright daylight, rhodopsin is depleted, and the sensitivity of the eye is reduced (**light-adapted**). It takes several minutes for the eyes to switch from the light-adapted condition to the dark-adapted condition, and vice versa.

The cones, at their best in bright light, and much less sensitive than rods in dim light, respond in several different ways because of differences in their opsins, the protein components of rhodopsin. Some cones are most sensitive to red light, others to green light, and still others to blue light. Combinations of these three populations of cones enable many primates (including humans), birds, reptiles, amphibians, and fishes to see gradations of many colors. Cones in birds and turtles have colored oil droplets that filter certain components of light, allowing separation of colors before light strikes the retina. Except for the primates, most mammals have few or no cones. Those without cones are completely color-blind, while those with some cones can see subdued colors. This is consistent with the fact that most mammals are active mainly at night, when the relatively light-insensitive cones would be of little value. It is also why bright red clothing doesn't make a hunter stand out to a deer.

Many vertebrates have a small area (or areas) of the retina called the **fovea centralis**, where focused light is centered (see Figure 12.5). In vertebrates with color vision, cones are highly concentrated in the fovea, making it the region of sharpest vision. In the human eye, the fovea has about 150,000

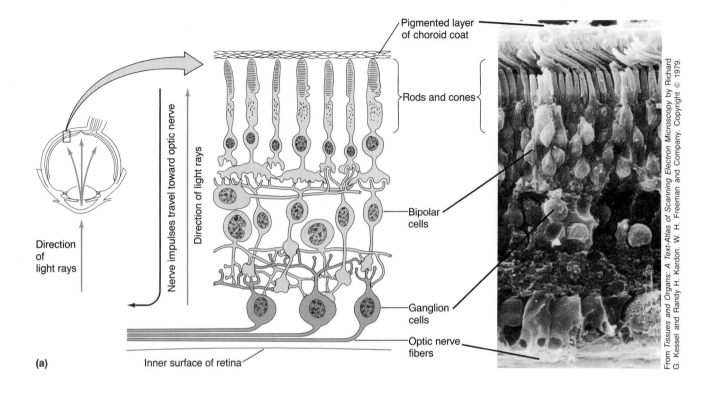

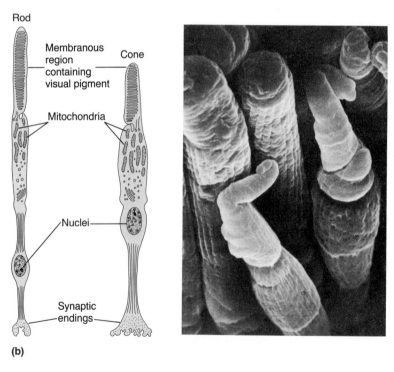

Direction of light rays

Nerve impulses travel toward optic nerve

Direction of light rays

Direction of light rays

Pigmented layer of choroid coat

Rods and cones

Bipolar cells

Ganglion cells

Optic nerve fibers

Inner surface of retina

(a)

From *Tissues and Organs: A Text-Atlas of Scanning Electron Microscopy* by Richard G. Kessel and Randy H. Kardon. W. H. Freeman and Company. Copyright © 1979.

Rod

Cone

Membranous region containing visual pigment

Mitochondria

Nuclei

Synaptic endings

(b)

Figure 12.6 Arrangement of cell layers in the vertebrate retina. (a) The retina as seen in the scanning electron microscope (1375×) and in an artist's interpretation. Note that light focused on the retina must pass through two transparent layers of nerve cells, first a layer of ganglion cells and then a layer of bipolar cells, before it can stimulate the photosensitive rods and cones. Nerve impulses generated by generator potentials in the rods and cones pass back to the ganglion cells via the bipolar cells. The axons of the ganglion cells come together to form the fibers of the optic nerve, which conducts the nerve impulses to the brain's visual centers for interpretation. Because the vertebrate retina is indirect, the eye has a blind spot where the optic nerve exits from the rear of the eye. (b) Rods and cones are named for their general shape. Note in the drawing that the elongate rods and somewhat squat cones have a membranous outer segment that contains visual pigment, a central region rich in mitochondria, and an inner end that forms synapses with bipolar cells. Because cones differ in their ability to absorb light of certain wavelengths, they can discriminate between wavelengths to provide color vision. The rods are more light-sensitive than the cones and are the basis for black-and-white vision in dim light. The scanning electron micrograph shows rods and cones from the retina of a mud puppy (an amphibian).

cones per square millimeter. Together with the excellent focusing power of the cornea and lens, this concentration of cones gives the eye good resolving power: At a distance of about 1 m from the human eye, two points as little as 1 mm apart can be seen as distinct entities. For comparison, the two points would have to be more than 20 mm apart to be resolved as separate entities by a honeybee's compound eyes. Some birds have well over a million cones per square millimeter, one reason why some hawks can see small rodents and other prey from high altitudes.

As with cones, the distribution of rods also corresponds to an animal's visual requirements. As the fovea evolved in nocturnal animals, space in the fovea that might have been occupied by cones, providing daytime color vision, was sacrificed in favor of the excellent night vision provided by tightly packed rods. Cats, which are most active at night, but have some cones and may see bright colors as pastels during the day, have many rods in the fovea, enhancing their night vision. Rods actually replace cones in the fovea of exclusively nocturnal animals, such as the owl monkey and tarsier. The very large size of a tarsier's eyes also enhances night vision; like binoculars with very large lenses, they are efficient light collectors (see chapter opening photograph). Rods are virtually absent from the human fovea, and this is consistent with the evolution of humans as fundamentally day-active animals. You may have noticed that it is often easier to see an object at night by looking at it from an angle of about 20°. Looking at the object at this angle allows the light from the object to strike the area of the retina having the greatest concentration of rods.

The complexity and capability of the vertebrate eye has few parallels. Yet, among invertebrates, one group, the cephalopod molluscs (squids and octopuses), have eyes that are structurally and function-ally very similar to vertebrate eyes. The squid eye is described in more detail in Chapter 24. As shown in Figure 24.18b, it has many of the same structures as the vertebrate eye. The lens is spherical and is moved back and forth in the eye for focusing, as in bony fishes. A muscular diaphragm controls the amount of light entering the eye. Quite different from the vertebrate condition, however, the cephalopod retina is direct; its photoreceptor cells lie in front of other nerve cells. As a result, light strikes the photoreceptor cells without passing through other nerve cells, and the eye has no blind spot.

Compound Eyes. Most arthropods (especially insects and crustaceans) and a few species of polychaete annelids have **compound eyes**. The compound eye is subdivided into an array of visual units called **ommatidia**, and each of these units gathers information from a small portion of the eye's field of vision (Figure 12.7). A single ommatidium is essentially a tube, with lenses at the outer end focusing light on photoreceptor cells deeper in the tube. Light entering neighboring ommatidia will almost certainly differ in brightness, so that the field of vision is seen as a mosaic image of dots of light differing in brightness.

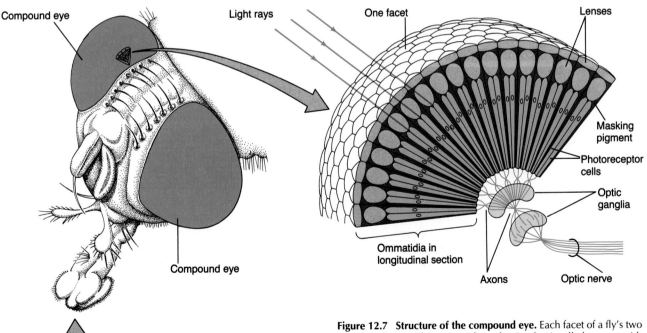

Figure 12.7 Structure of the compound eye. Each facet of a fly's two compound eyes is the outer surface of a visual unit called an ommatidium. Each ommatidium has lenses that transmit light to photoreceptor cells deeper within the ommatidium. Nerve impulses generated as these cells are illuminated are transmitted along axons to optic ganglia and then to the optic nerve for relay to interpretative centers in the brain. The combined action of the ommatidia produces a compound (mosaic) image.

The lenses of a compound eye produce coarse images; that is, they have poor **resolving power**. In a sense, the compound eye is an evolutionary compromise in which resolving power is sacrificed in favor of small size. A single-lens eye achieves its fine resolving power by being relatively large and heavy, luxuries the average arthropod can ill afford. A compound eye can fit on the head of a tiny insect and still have acceptable resolving power. A single-lens eye reduced to the same miniature dimensions would be considerably less functional. Of course, what an insect actually perceives also depends on how the visual inputs are processed by the neural circuits of the optic ganglia and brain.

The compound eye's mediocre resolving power is also balanced by its outstanding ability to detect motion, a feature especially useful to fast-flying insects, such as a dragonfly watching for flying prey or a bumblebee scanning for landmarks near its nest. Good motion detection requires that an eye be able to distinguish images arriving at closely spaced intervals; unusually rapid recovery of a compound eye's photoreceptor cells makes this possible. Some insects can distinguish images (flashes from a light source) arriving at a rate of 330 per second. Images tend to fuse at higher rates. In contrast, the human eye can distinguish only about 50 images per second, just low enough for image fusion of a 60-cycle incandescent light bulb. An insect viewing the same bulb would see 60 flashes each second.

Compound eyes have several other important capabilities. In some arthropods, for example, the compound eye adapts to changing light intensities. In bright light, masking pigment shields each ommatidium from its neighbors, so that light can enter an ommatidium only through its lenses (see Figure 12.7). The masking pigment probably has several functions. It allows formation of a higher-resolution mosaic image in bright light, and it may also prevent excessively intense light from reaching the visual pigments in the ommatidia. In dim light, the masking pigment moves out of the way, allowing light to enter the ommatidium from the sides as well as through the lenses. The dark-adapted eye can collect far more light, in effect becoming a better light detector. This feature is particularly useful to moths and other insects flying at dusk as they seek food, mates, or egg-laying sites.

Compound eyes may also provide for color vision. Many insects see ultraviolet, but shades of red appear black. Color vision is particularly important in day-flying, nectar-drinking insects such as honeybees. Foraging honeybees learn to recognize particular flowers by color, scent, and shape. Such

"bee flowers" generally have colored patterns, or "honeyguides," that direct the bee toward the center of the flower and the nectar stored at its base. Many flowers, or their honeyguides, reflect ultraviolet light, invisible to us but visible to honeybees. When a bee visits a red flower, what it actually is seeing is an ultraviolet blossom.

In addition to color vision, the geometric arrangement of visual pigment molecules in the photoreceptor cells makes the compound eye in some arthropods sensitive to **polarized light**. (In polarized light, all wave motions, or vibrations of light rays, are occurring in a single plane.) Blue sky has different patterns of polarized light, depending on the angle of the sun. This is somewhat difficult to imagine, because the human eye cannot see polarized light; but animals with polarized-light detectors can determine the position of the sun by the pattern of polarized light in the sky. Only a small patch of blue sky is necessary at any one time. The changing pattern of polarized light provides information about the animal's orientation relative to the sun throughout the day. Honeybees and certain species of ants use the blue sky's pattern of polarized light as a compass that points the way home when they are out foraging for food.

Because pioneering studies were done with bees, most is known about polarized-light detection by the compound eye of arthropods. But many other terrestrial and aquatic animals, including certain vertebrates, also have this capability and use it for orientation. Many birds and fishes and even certain mammals navigate long distances using patterns of polarized light.

Sound and Hearing

Although many animals make sounds incidental to burrowing, flying, chewing food, and other activities, only certain arthropods and vertebrates use sounds to communicate. Many remarkable sound-producing mechanisms have evolved. A bird forces air rapidly past two membranes in its **syrinx**, or voicebox, located at the posterior end of the trachea; the membranes vibrate with relatively pure notes. The tightness of each membrane can be changed independently by the contraction of associated muscles, allowing two pure notes to be made at the same time; by pulsing the notes and changing their pitch and loudness, a meadow lark can "compose" about 50 songs from some 300 notes. Some insects also sing by means of vibrating membranes. A few 17-year locusts can fill a woodlot with a loud, more or less continuous buzzing sound made by vibrating cuticular membranes. A mammal uses a somewhat different approach to sound production. Vocal cords in the larynx (see Figure 5.13) vibrate as air passes over them, something you can demonstrate by holding your fingers lightly against your Adam's apple as you talk. Unlike the relatively pure notes of a bird, the sounds made by a mammal have many overtones, or harmonics. Humans can "shape" sound by changing the size and shape of their throat and mouth cavities. We control the information content of our sounds more by this kind of sound shaping than by the production of pure notes.

Stridulation, another method of producing sound, is typical of many insects. A cricket, for example, passes a tiny file on one wing across a scraper on the other wing, making microscopic teeth on the file vibrate like small tuning forks. One pass of the file across the teeth produces the cricket's familiar chirp. Different patterns of chirping signal different kinds of behavior, such as courtship or aggression.

Understanding how animals hear sounds requires some knowledge of the nature of sound itself. Sound is generated when some structure, such as a guitar string or an animal's vocal cords, vibrates back and forth rapidly enough to send sound waves coursing through the medium in which the structure is vibrating, much as a stone splashing in the water sends ripples across a pond. Each time a vibrating structure moves back and forth, it makes a sound wave, and the number of such vibrations each second determines the sound's **frequency**, expressed as cycles per second, or Hertz (Hz). Frequency determines a sound's **pitch**, that is, how high or low the sound is. So far as we know, the range of frequencies that terrestrial vertebrates can hear extends from as low as 0.05 Hz in homing pigeons to as high as about 100,000 Hz in some bats. For humans, the audible range is about 20–20,000 Hz. Little is known about the frequency ranges of aquatic animals. Sound waves are also characterized by **amplitude**, which corresponds to a sound's loudness and is measured by the height of the wave peaks (Figure 12.8).

As sound waves move through a medium, molecules within the wave alternately move toward and away from each other, respectively producing regions of **compression** and **rarefaction** (Figure 12.8).

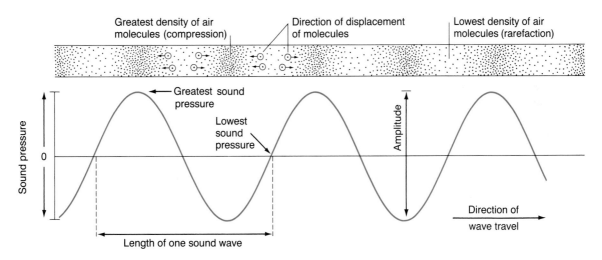

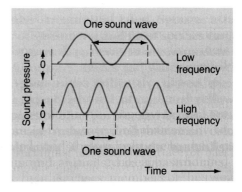

Figure 12.8 Sound wave characteristics. Sound waves are vibratory disturbances traveling in a medium such as air or water. A sound wave in air is characterized by pressure gradients caused by the displacement of air molecules into regions of high and low concentration. An ear receiving sound waves is subjected to both molecular (particle) movement and to alternating changes in pressure. Some ears are adapted to detect particle movement, while others respond to pressure changes. A sound's pitch, or frequency, is determined by the length of the sound wave; shorter waves have higher frequency (i.e., produce higher-pitched sounds), while longer waves have lower frequency. The inset illustrates the sound waves of two sounds differing in frequency; note the difference in the lengths of the sound waves. The loudness of a sound is determined by the amplitude of its wave; the greater the sound pressure, as indicated by the curve, the louder the sound.

Pressure rises as the molecules pack together more closely, and falls as they spread apart. Thus, an ear or any other object struck by a series of sound waves is subjected to a succession of high and low pressures and to molecules moving first one way and then the other.

Hearing organs are specialized mechanoreceptors. As discussed in the following sections, some

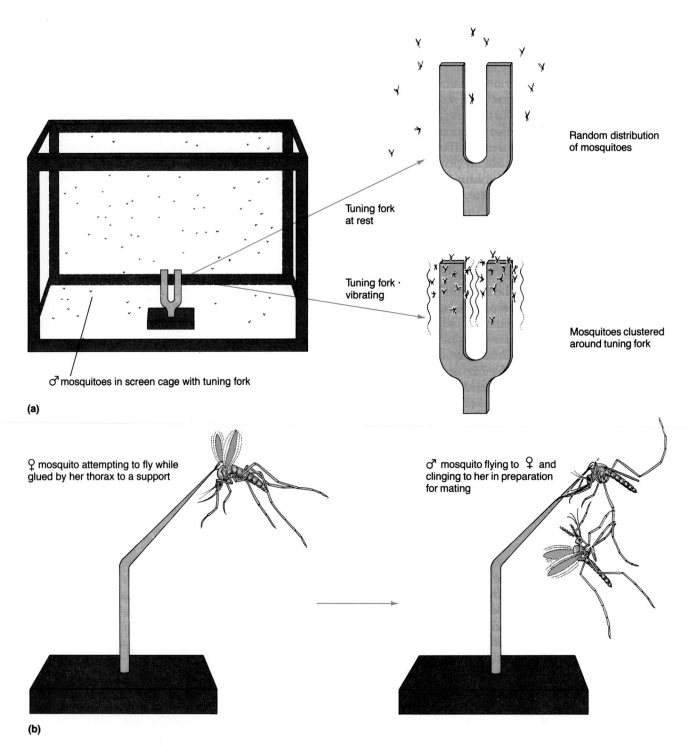

(a)

Tuning fork at rest

Random distribution of mosquitoes

Tuning fork vibrating

Mosquitoes clustered around tuning fork

♂ mosquitoes in screen cage with tuning fork

(b)

♀ mosquito attempting to fly while glued by her thorax to a support

♂ mosquito flying to ♀ and clinging to her in preparation for mating

Figure 12.9 Mosquitoes respond to sound. In nature, male mosquitoes locate females by the pitch of the sound produced by the female's beating wings. (a) In laboratory experiments, male mosquitoes are attracted to a tuning fork when it vibrates at the sound frequency characteristic of the sound made by a flying female's wings. (b) The wing sounds of a female attempting to fly while glued to a supporting needle also attract a male.

are responsive to molecular movement, while others detect pressure, and each type of organ has its advantages and disadvantages.

Mechanoreceptors That Sense Particle Movement

Air or water molecules rushing back and forth between high-pressure and low-pressure regions in sound waves can make delicate hairs vibrate at a frequency determined by the hair's thickness, stiffness, and length. Many insects have body hairs that vibrate at the same frequency as biologically important sounds. Certain caterpillars, for example, have body hairs that vibrate at the same frequency as the wing sounds of predatory wasps. When the sounds of a wasp's wings set an intended victim's body hairs in motion, warning signals (nerve impulses) are sent to the caterpillar's central nervous system. In a similar way, male mosquitoes detect

females by means of antennal hairs tuned to the frequency produced by the wingbeat of flying females (Figure 12.9).

Among vertebrates, the movements of many fishes and certain aquatic amphibians are guided by information from **neuromasts**, located on the body surface. Neuromasts are groups of cells bearing minute sensory hairs that are bent back and forth by the low-frequency vibrations of water flow and turbulence. In dark or muddy water, the hair cells help the animal to locate underwater obstructions, monitor the movements of other fish swimming nearby, or detect splashing waves. In many fishes and amphibians, neuromasts are localized within longitudinal canals of the **lateral line system** on each side of the body and head (Figure 12.10a). In common with other vertebrates, fishes also have two **inner ears** located just above and behind the brain (Figure 12.10b). Together, the lateral line system and inner ears form a hearing com-

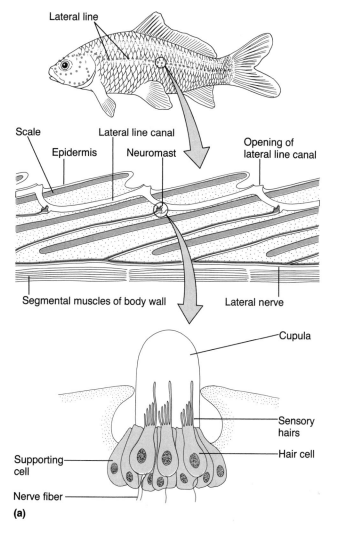

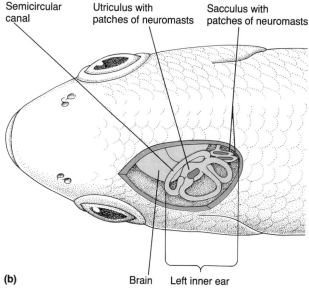

Figure 12.10 The auditory system of a bony fish. The lateral line system and inner ears constitute the acousticolateralis system in fishes. (a) Longitudinal section of a portion of the lateral line system of a carp. The lateral line system allows fishes to detect low-frequency vibrations. It consists of a row of neuromasts (clusters of sensory hair cells with hairs projecting into a gelatinous capsule, the cupula) set in a series of pits and canals along each side of the body. Movement of the neuromasts results in nerve impulses being sent to the brain via lateral nerves. (b) The left inner ear is illustrated here in a dorsal dissection. The inner ear functions primarily in detecting body position, but neuromasts in its sacculus and utriculus also respond to sound vibrations that pass through the skull.

plex called the **acousticolateralis system**. Groups of minute hairs within two inner ear sacs, the **utriculus** and **sacculus**, are moved back and forth by tiny pebblelike granules called **otoliths**. The otoliths are set in motion by vibrations from sound waves in the water conducted through the skeleton of the head. Nerve impulses generated by the sensory hairs of the inner ear and the lateral line neuromasts travel to the brain for interpretation. The inner ears extend the hearing of many fishes to higher frequencies. Other adaptations that enhance hearing in fishes are described in Chapter 29. (See the discussion of the Weberian apparatus, p. 718; Figure 29.14.)

Hearing based on the detection of particle movement has an important limitation. Sensory hairs vibrate best when oriented at 90° to the source of a loud sound, and are least responsive when pointed straight toward the source of a soft sound. Between these extremes, a sensory hair may vibrate with equal strength when it is broadside to the source of a soft sound and when it is pointed almost directly at the source of a loud sound. This means that an animal may receive conflicting signals about loudness and direction, a poor situation if what is being detected is a mate or a predator. Some animals circumvent this drawback by turning the detector (or the head or body) back and forth until reception is maximized or minimized, somewhat like rotating a radio direction finder.

Mechanoreceptors That Sense Pressure

The ears of most animals respond to sound wave pressure. In insects and in vertebrates other than fishes, sound is intercepted by a **tympanum** (also called a **tympanic membrane** or **eardrum**) stretched across the opening of a closed chamber filled with air. As sound waves arrive, the alternating high- and low-pressure zones within each sound wave move the membrane back and forth. The membrane vibrates at the frequency of the sound, and the extent of its movement corresponds to the sound's loudness.

Insect Ears. Depending on the species, insects have pressure-detecting ears on various parts of their bodies, including the front legs (Figure 12.11). Vibrations of the tympanic membrane stimulate sensory cells attached to the underside of the membrane, generating nerve impulses that travel to the brain for interpretation. Also depending on the species, insect ears respond to signals from below 1000 Hz to about 140,000 Hz, more than two octaves above the human range on the high side but well

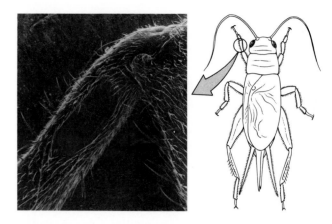

Figure 12.11 The tympanum (eardrum) of an insect. Photomicrograph of a tympanum on the front leg of a female cricket. A group of mechanoreceptor cells that are stimulated when sound waves cause the tympanum to vibrate are hidden from view beneath the tympanum.

within our range on the low side. Certain moths can hear extremely high notes, and this protects them against insect-eating bats. Many bats emit ultrasonic sounds for echo locating insect prey at night. As a moth's ears detect a bat's sound, the moth may make an abrupt change of course and often avoid the open mouth of the swooping bat.

Pressure-sensitive ears may also allow an animal to determine the direction from which a sound is being emitted. This is highly advantageous, particularly when it comes to pinpointing possible mates or predators.

Vertebrate Ears. Several major structural adaptations mark the evolution of vertebrate hearing. The lateral line system functions only in water, and was lost in the evolution of terrestrial vertebrates. Amphibians have the lateral line as tadpoles, but adults of terrestrial species do not. Aquatic reptiles and mammals, which arose from terrestrial ancestors, also lack the lateral line system. The inner ear has evolved as the sole hearing organ in land vertebrates.

Figure 12.12 shows representative stages in the evolution of the ear as seen in three groups of terrestrial vertebrates. The parts of the inner ear concerned with hearing in terrestrial vertebrates are housed in an outgrowth from the sacculus called the **lagena**. In the course of terrestrial vertebrate evolution, the lagena became adapted as the major receptor of airborne sound. The lagena is short in amphibians and reptiles, but in birds and mammals, it forms the elongate **cochlea**, containing the **organ of Corti** (the hearing organ). The cochlea is fluid-filled and is subdivided into longitudinal channels by several membranes. A **basilar mem-**

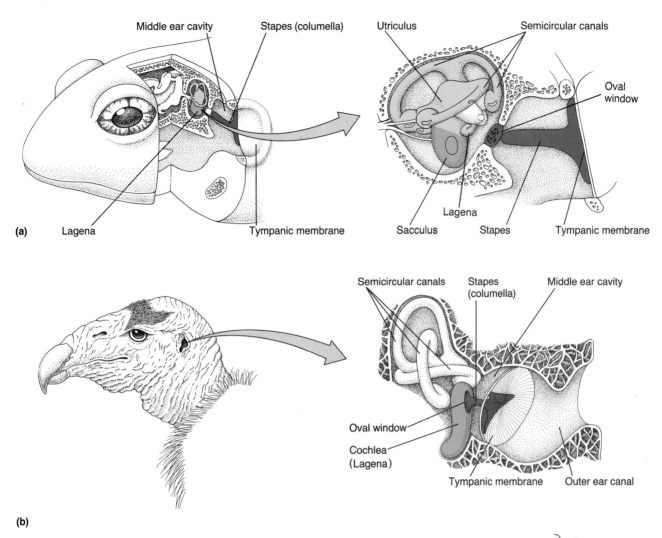

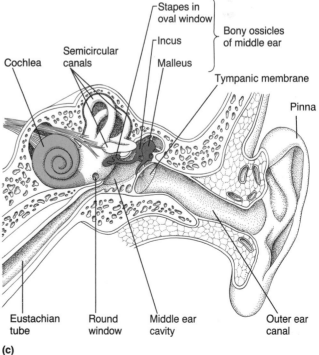

Figure 12.12 Representative stages in the evolution of ears in terrestrial vertebrates. The inner ear in all jawed vertebrates contains three fluid-filled semicircular canals and two sensory sacs, the utriculus and sacculus (also see Figure 12.10b). (a) Frogs and toads also have a middle ear with a single ossicle, the stapes (or columella), which transmits sound from the tympanic membrane (eardrum) to the membrane covering the oval window. Note the short lagena, which contains the main hearing organ of terrestrial vertebrates. In amphibians, as in fishes, neuromasts in the utriculus and sacculus may also respond to sound vibrations. (b) In birds, the lagena forms the elongate, fluid-filled cochlea, which houses an extensive hearing organ, the organ of Corti. As in amphibians, birds have a single middle ear ossicle, the columella. (c) In mammals, sound is transmitted by three bony ossicles (the malleus, incus, and stapes) to an elongate coiled cochlea (derived, as in birds, from the lagena). The coiled shape of the cochlea conserves space while housing a very long organ of Corti; the great length of the hearing organ provides for greater discrimination of sound frequencies.

brane bears hair cells along its length in the organ of Corti. The tips of some of the hairs of the organ of Corti are attached to the **tectorial membrane**, a relatively immobile structure that projects like a shelf out over the basilar membrane (Figure 12.13). As the hair cells bend, nerve impulses are generated and pass toward the brain along nerve fibers that are gathered together as the **cochlear nerve.**

Associated with the evolution of the inner ear as a pressure-detecting hearing organ, terrestrial vertebrates also evolved the **middle ear**, an air-filled cavity that is absent in fishes. The middle ear is separated from the inner ear by a membrane covering an opening called the **oval window**. The tympanum marks the outer limit of the middle ear. The tympanum is delicate enough to be moved by the relatively weak forces of airborne sound waves, yet large enough to respond to a broad range of frequencies. In amphibians and most reptiles, the tympanum is on the body surface, but in birds and mammals, it is recessed in an outer ear canal. The **eustachian tube** connects the middle ear to the pharynx, ensuring that air pressure is equal on either side of the tympanum. The middle ear houses tiny bones called **ear ossicles**, which transmit sound from the tympanum to the oval window. Amphibians, reptiles, and birds have a single ossicle called the **stapes** (or **columella**), whereas mammals have three ossicles called the **malleus, incus, and stapes** (see Figure 12.12). (The evolution of ear ossicles from pharyngeal and jaw structures is discussed in Chapter 28, p. 689; Table 28.1, Figure 28.10.)

Tracing the path followed by sound as it enters the ear of a bird or mammal allows us to see how the cochlea and other ear structures process sound. Sound waves enter the external auditory canal of the **outer ear** through an opening commonly hidden by feathers in birds and surrounded by the visible **pinna** in most mammals (see Figure 12.12c). The pinna is relatively small and ineffective in humans, but is a large "ear funnel" in many nocturnal mammals, efficiently gathering and concentrating the faint sounds made by predators or prey moving in the darkness. The outer ear canal directs the sound waves to the tightly stretched tympanum, which then vibrates, oscillating rapidly for high tones and slowly for low ones.

From the tympanum, sound waves are transmitted through the middle ear to the oval window by the ear ossicles. The oval window's membrane then sends vibrations into the fluid in the cochlea. The tympanum and ossicles do more than just transmit vibrations to the cochlea. Forces exerted by sound waves in the air are too small to make

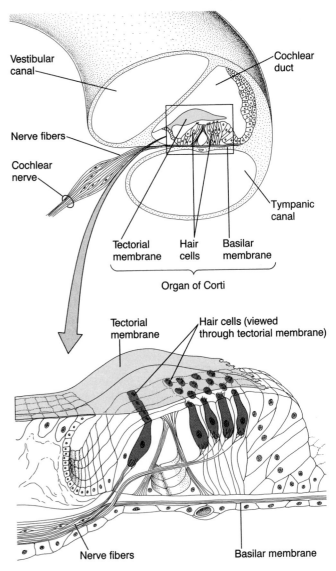

Figure 12.13 The organ of Corti. The organ of Corti, found in birds and mammals, consists of rows of mechanoreceptor hair cells supported by a basilar membrane. A long basilar membrane makes a lengthy array of mechanoreceptor hair cells possible, thus enhancing discrimination of different sound frequencies. The cross section through the cochlea shows its division into three canals separated from each other by membranes. The basilar membrane forms the floor of the middle canal (the cochlear duct). The tips of some of its hair cells are attached to the underside of the tectorial membrane. As vibratory disturbances travel through the cochlear fluid, the tectorial and basilar membranes move relative to each other, stimulating the hair cells. Nerve impulses generated by the hair cells travel toward the brain along the cochlear nerve.

cochlear fluid move unless they are first amplified. Through simple lever action, the bony ossicles approximately triple the sound pressure. More importantly, the tympanum concentrates pressure (via the ossicles) on the oval window, which has an area only $\frac{1}{30}$ that of the tympanum. As a result, pressure is multiplied proportionately, about 30-

fold, and together, the ossicles and tympanum multiply the pressure on the oval window some 90-fold, enough to set the cochlear fluid in motion. The stapes, which fits against the membrane of the oval window, has a pistonlike action that sends waves of amplified sound pressure coursing through the cochlear fluid, traveling outward in the **vestibular canal** and then back to the base of the cochlea in the **tympanic canal** (see Figure 12.13). The journey ends at a membrane covering the **round window**, another opening between the middle and inner ears. The membrane damps the vibrations, preventing potentially disruptive vibrations from traveling back toward the oval window. The waves vibrate the basilar membrane, causing a shearing action between its organ of Corti and the overlying tectorial membrane. The shearing action bends the hairs on the organ of Corti, producing generator potentials that trigger nerve impulses that travel along the cochlear nerve to the brain.

How does the organ of Corti distinguish between sounds of different frequencies—between the sound of a flying mosquito and the rumble of distant thunder? The answer is related to the structure of the basilar membrane, which is thin and tight at the end near the oval window but thicker and looser along its length. High notes make the membrane vibrate most vigorously where it is taut, and low notes where it is loose. In other words, the site along its length where the membrane vibrates most strongly depends on the frequency of the sound (Figure 12.14). An elongate cochlea is an adaptation permitting linear distribution of hair cells, the arrangement best suited for stimulation by sound waves of different frequencies.

Other Mechanical Senses

Mechanoreceptors perform a variety of functions in addition to detecting sound. Often, their locations allow them to detect deformations of the body wall and other structures. Vertebrates have many kinds of receptors in their skin. Many nerve fibers in the epidermis of a pig's snout, for example, end just beneath the surface as minute, highly touch-sensitive discs called Merkel's corpuscles; Meissner's corpuscles are touch-sensitive receptors in the epidermis of primates; and deeper in the skin of vertebrates, Pacinian corpuscles respond to pressures greater than a light touch. In humans, touch-sensitive receptors are particularly numerous in the lips and fingertips, accounting for the extreme sensitivity of these parts. Within the circulatory system of vertebrates, pressure receptors in the walls of certain blood vessels provide information necessary

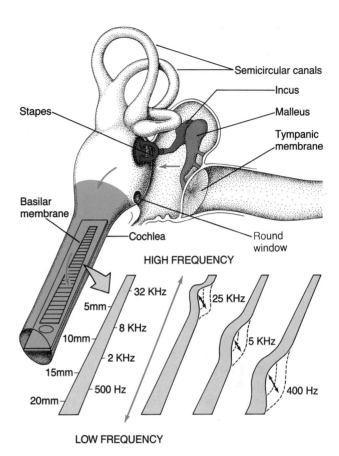

Figure 12.14 Response of the basilar membrane to different sound frequencies. Part of a mammal's ear, with the cochlea drawn as if straight rather than coiled. Vibratory disturbances are transmitted from the tympanic membrane to the oval window by the malleus, incus, and stapes. The stapes causes vibratory movement of the fluid in the cochlea, and this, in turn, causes the basilar membrane to resonate. The diagrammatic view of the basilar membrane shows that different frequencies of sound cause it to resonate at different places along its length. As a result, different sound frequencies cause different hair cells to be stimulated, leading to propagation of nerve impulses in different fibers of the auditory nerve.

for regulation of blood pressure. (Pressure receptors in carotid arteries and the cardiac center of the mammalian brain are discussed in Chapter 6, p. 132.)

Mechanoreceptors, including many touch- and pressure-sensitive receptors, also provide information about the position of limbs and other body parts. Such information is of key importance in coordinating locomotor and other movements. Also, by detecting constant or changing forces (including gravity and pressure), mechanoreceptors help an animal tell "which way is up," allowing an animal to right itself if turned upside down. Sense organs that provide postural information are sometimes called **proprioceptors**. Many of the setae and hairs on an animal's surface are proprioceptors. For example, setae on one joint of an arthropod's leg are bent by contact with the neighboring joint as the animal walks; the resulting nerve impulses represent information about the leg's position. Likewise, stretch receptors (sometimes called muscle spindles) within muscles or tendons are stimulated during muscular activity and provide positional information.

Many of the sensory structures important in proprioception are also **tactile**. The tactile sense, or sense of touch, allows animals to respond to physical contact and to detect movement of objects. Birds building nests, bees shaping waxen cells in their hive, a dog detecting fleas, and scores of other activities depend on this ability. Even detection of air or water flowing over the body surface is important to many animals, allowing them to compensate for the current by changing locomotor effort and direction.

Proprioceptors also help an animal maintain balance. In addition to its role in hearing, the inner ear of vertebrates functions as a complex organ of balance. Hair cells in the utriculus and sacculus (see Figures 12.10 and 12.12) are mainly gravity sensors that tell a vertebrate which way is up. Depending on the position of the head, gravity causes the otoliths in the utriculus and sacculus to touch different hair cells. The utricular and saccular hair cells also detect linear acceleration, allowing a vertebrate to know it is moving. The three **semicircular canals** in the inner ear augment the role of the utriculus and sacculus as organs of balance (see Figures 12.10 and 12.12). The three canals are attached at both ends to the utriculus, and they are arranged at right angles to one another, representing the three planes of space. Swellings called **ampullae** at the base of the semicircular canals contain patches of touch-sensitive hair cells. In contrast to the hair cells in the utriculus and sacculus, those of the semicircular canals detect angular acceleration. As a vertebrate turns its head, fluid within the semicircular canals is set in motion and deflects the hairs, generating nerve impulses. Because the three semicircular canals lie in three different planes, turning the head in any direction will cause movement of fluid and hair cell deflection in at least one of the canals. By constantly integrating data received as sensory impulses from the utriculus, sacculus, and semicircular canals, the brain maintains a complete picture of the body's position and direction of movement.

Many invertebrates have organs of balance that are remarkably similar to those of vertebrates. Gravity-sensitive organs called **statocysts** occur in a wide variety of animals from cnidarians (jellyfish) to molluscs and arthropods. (Actually, the vertebrate utriculus and sacculus are specialized types of statocysts.) All statocysts have a similar structure and function. Each consists of a multicellular sac containing neuromasts (mechanoreceptor hair cells) and from one to any number of hard granules called **statoliths** (Figure 12.15). As in the sacculus and utriculus, gravity pulls the statolith(s) against certain hair cells, depending on the position of the body, and this results in nerve impulses being sent to the central nervous system. The widespread occurrence of statocysts indicates that these structures may have been among the earliest types of sense organs to evolve in animals. It is likely, however, that statocysts evolved several times in the course of animal evolution.

Because organs of both balance and hearing operate on a similar basis—hairs on hair cells being displaced by hard substances or fluid—many zoologists think that there is an evolutionary link between these two types of organs. This is especially evident in the vertebrates, where the hearing and balancing organs work on a similar hair cell displacement principle and are intimately associated in the inner ear.

Chemoreception: Taste (Gustation) and Smell (Olfaction)

Chemoreception, the ability to detect and respond to certain chemicals in the environment, is a universal trait among animals and protists. Mate finding, recognition of territory, migration, and navigation often depend on the senses of taste and smell. (See the Essay on pathfinding in the ocean, p. 290.) Social life in the beehive or termite nest is strongly influenced by the detection of scents associated with the queen and other members of the group. Most animals, but particularly insects, use chemicals for communication. (Communication by pheromones is discussed in Chapter 13, p. 309.) The chemical senses may also be used in defense, as when skunks or certain insects produce noxious odors, or when an antelope breaks into a run after catching the scent of a lion.

Feeding activities are universally influenced by chemicals in the environment. In the cnidarian *Hydra*, feeding behavior is motivated by glutathione, a chemical released by wounded prey. Experimental addition of glutathione to the water induces swallowing movements and sometimes even causes nearby hydras to cannibalize one another. Planarians also detect chemicals in the water around them, using sensory cells concentrated on earlike auricles on either side of the head (see Figure 10.9).

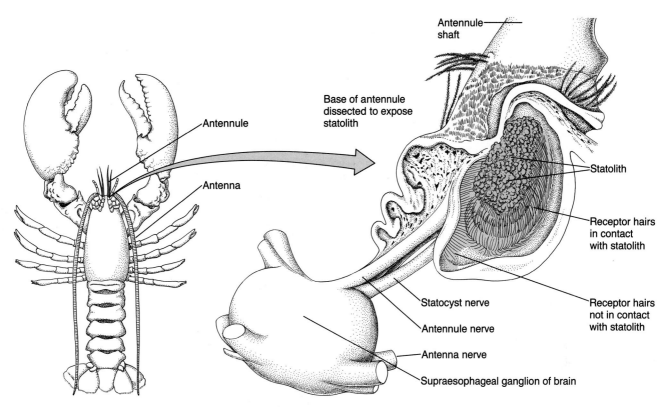

Figure 12.15 The statocyst in the antennule of a lobster. This diagram shows the basal segment of the right antennule cut open to expose the statocyst. Within the statocyst is a mass of fine sand particles cemented together to form a statolith that is cradled in a nest of receptor hairs. As the lobster moves about, the relatively heavy statolith shifts, stimulating some of the hairs more than others. The changing patterns of stimulation provide information about body position. A row of receptor hairs that is not in contact with the statolith responds to movement of fluid within the statocyst. As a lobster accelerates or slows down suddenly, the fluid moves relative to the receptor hairs, which bend as a result. Stimulation of these hairs gives the lobster information about acceleration and deceleration during locomotion.

Various molluscs, depending on the species, monitor water chemistry with chemosensory cells on the gills, on tentacles, in the mouth region, or in the mantle cavity where water is drawn in. An earthworm apparently can taste as it tunnels, sampling the soil with chemosensory cells scattered over its body and concentrated near its mouth. It goes without saying that taste and smell are important in our own feeding activity.

It is often difficult to distinguish between taste and smell because both senses may be involved in detecting a chemical. In vertebrates, however, smell usually means perceiving chemicals by means of the nose and its olfactory nerves, whereas taste typically involves oral sensory receptors and gustatory nerves. Moreover, chemicals that are smelled by terrestrial vertebrates and many invertebrates are volatile at relatively low temperatures and are therefore airborne, while those that are tasted are generally in solution when they contact

the taste receptors. Of course, in an aquatic environment, all chemicals detected are in solution, and there is really no distinction between taste and smell. The taste/smell distinction also breaks down in many invertebrates because they lack separate olfactory and gustatory nerves. Because most of what we know about chemoreception concerns insects and vertebrates, the following discussion is limited to those animals.

Taste and Smell in Insects. Insects taste by means of sensory hairs called **setae** on their mouthparts and tarsi (feet). Thus, an insect may identify its food simply by stepping in it. Insects use their sense of taste primarily in food selection and in the activation or inhibition of feeding responses, such as the extension of the mouthparts toward food. Female locusts (*Locusta*) make use of taste for a more unusual purpose. After digging a hole in sand, a female determines the salt content of the sand by

Essay: Pathfinding in the Ocean

Over the centuries, mariners have noted that many animals travel great distances to specific locations, as if they could navigate; yet the ocean appears to be a trackless expanse without guideposts. By marking animals for later recovery and tracking others fitted with ultrasonic transmitters, marine biologists have confirmed that tuna, green turtles, American and European eels, dolphins, and certain other animals can perform spectacular feats of navigation. Each December, for example, green turtles swim 2200 km from Brazil to Ascension Island, a 38 km² locus in the mid-Atlantic; and bluefin tuna tagged and released in the Gulf of Mexico swim at speeds up to 75 km/hr and soon arrive off the coast of Norway. We now realize that such animals are excellent navigators, using their senses as navigational instruments in an environment that is rich with pathfinding cues. Dissolved chemicals, water currents, temperature boundaries between water masses, the position of the sun, and the earth's magnetic field may all indicate direction for a marine animal. Some animals switch from one set of cues to another at different ages or in response to changes in weather, geographical location, season, or time of day.

The remarkable odyssey of the Pacific salmon (*Oncorhynchus*) illustrates oceanic navigation particularly well. When the time to breed arrives, several species of salmon ascend the Columbia River and other watersheds of northwestern Canada and the United States (Figure E1). After swimming hundreds of miles upriver, a female constructs a nest, a shallow depression on the bottom of a small tributary stream. The female then lays her eggs in the nest, and a male covers them with sperm. Adults die after spawning. Larval salmon soon hatch, leave their home stream, and drift to the ocean, the start of a journey to the Gulf of Alaska. Plankton feeders at first, the growing salmon gradually switch to feeding on herring and other small fishes, and school with other salmon north of the Aleutian Islands. Timing is somewhat variable, but often by the end of their third or fourth year, a remarkable thing happens. Responding to changing hormone levels, the salmon separate into groups of geographical origin and migrate back to the vicinity of the river system in which they were hatched, perhaps thousands of kilometers away. In their fifth year, some salmon find and ascend their home streams, where they spawn and die.

An intriguing pair of questions stem from this extraordinary migration. How does a salmon find its way back to the general vicinity of its home river system after spending years at sea far to the north, and how can a salmon distinguish one stream from all others and select the very one it left several years earlier? In the 1970s, A. D. Hasler of the University of Wisconsin discovered that a salmon uses its sense of smell to find its specific home stream and the upstream spawning ground. A unique combination of plants and soils seem to give each stream a unique scent that becomes imprinted on the nervous system of the young fish.

Figure E1 Adult salmon jumping a waterfall during their migration to spawning grounds in an Alaskan stream.

Arriving in the general vicinity of its home river system years later, a salmon swims along the coast until it detects the odor matching the imprint in its brain. Entering the river system, the salmon begins a struggle against currents, foaming rapids, and modern obstacles, such as hydroelectric dams, as it swims upstream to spawn.

Researchers are less certain how a salmon locates the general region of its home river system. Many marine animals have concentrations of magnetic crystals (magnetite) in their head, perhaps enabling them to navigate by detecting the earth's magnetic fields. However, adult salmon do not seem to have magnetic field detectors. Like most fishes, however, they do possess a sun compass. Given time in a local area, a fish soon learns to adjust its course to keep a constant angle between the sun and a selected direction. By making continual adjustments as it swims, and with allowances for the sun's changing altitude, a fish can navigate a straight course. A salmon probably relies on a sun compass as it navigates to the Gulf of Alaska and back to the neighborhood of its home river several years later. Then this resourceful navigator switches to its sense of smell and follows its nose home.

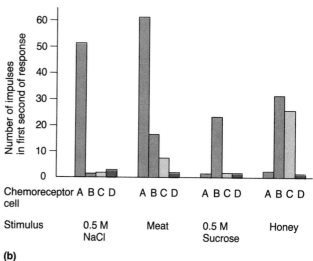

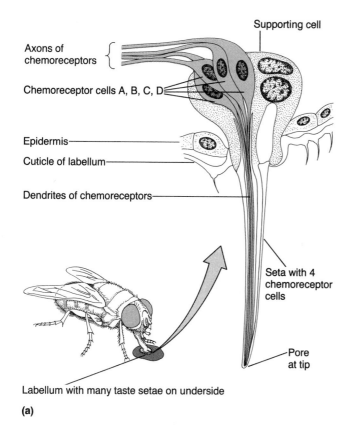

Figure 12.16 Responses of a blowfly's taste receptors. (a) Each of the many setae on one of the fly's mouthparts (the labellum) contains four chemoreceptor cells (A, B, C, and D). (b) Although the cells differ in being maximally responsive to different chemical compounds (e.g., A is a salt receptor most responsive to NaCl, whereas B is a sugar receptor most responsive to sucrose), they are not highly specific. Instead, each has a characteristic, broad spectrum of responses when exposed to natural foods; across all four cells, the response pattern differs for each food or compound. Such cross-fiber patterning enables an animal to distinguish between a very large number of tastes (or odors when the sense of smell is involved).

"tasting" it with chemoreceptor cells on her ovipositor (posterior egg-laying structure). She deposits her eggs in the hole only if the salt content is suitably low.

An insect's sensory setae are formed by the cuticle. The taste setae generally contain more than one receptor cell, each with a characteristic spectrum of responses. Each of the setae on the mouthparts of a blowfly *(Phormia regina)*, for example, contains four receptors (Figure 12.16). Each receptor is maximally responsive to either sugar, water, anions (negatively charged ions), or cations (positively charged ions). When exposed to natural foods, each receptor cell exhibits a broad spectrum

of response, and the pattern of responses of the four cells combined discriminates among different foods. Because the receptor cells are not precisely tuned to specific chemicals, responding instead to different spectra of chemicals, discrimination of individual substances depends on central nervous system integration of all the information available, rather than on simple receptor specificity. This is an example of sensory coding by means of cross-fiber patterning (described earlier, p. 272).

The sense of smell in insects is mediated by cuticular setae of various shapes and sizes on the antennae (Figure 12.17). Each of these olfactory setae usually has a few thousand to as many as 18,000

Figure 12.17 Olfactory receptors on insect antennae. (a) Chemosensory setae on the filaments of the antennae enable this male moth to detect and respond to minute amounts of an airborne sex attractant released by the female (6.5 ×). (b) A scanning electron micrograph of several filaments shows the enormous numbers of setae, one reason the male can detect extremely small amounts of attractant (70 ×).

pores through which scent molecules diffuse to underlying dendrites. Olfactory setae detect odors rising from chemical trails made by other insects of the same species. This allows ants, for example, to follow trails to food located earlier by other members of a colony. Flying insects, such as moths, may detect the odors of mates or food plants and in some cases may then turn and fly upwind toward the odor's source. In many species, females release a chemical that attracts males. So sensitive are the male's chemoreceptors that in some species, they may detect sex attractants at concentrations of only about 200 molecules per cubic centimeter of air.

Taste and Smell in Vertebrates.   Vertebrate taste receptors are located in **taste buds** that open to the external environment via a taste pore. Microvilli project from the taste cells into the taste pore at the surface of the bud (Figure 12.18). For chemicals to produce a taste sensation, they must be dissolved in water and then enter a taste pore. Then, by inter-

acting with the membranes of the microvilli, the dissolved chemicals produce a generator potential. Sensory nerve fibers forming synapses at the base of one or more taste cells carry complex patterns of nerve impulses to the brain.

Many fishes have taste buds covering their whole body surface, but in most terrestrial vertebrates, taste buds are confined to the tongue and other mouth surfaces that can be kept moist. It is customary to classify taste sensations as *salt*, *sweet*, *sour*, and *bitter*. "Tongue maps" show the boundaries of these sensations; in taste tests with humans, the anterior third of the tongue is found to be most sensitive to sweet and salty stimuli, the tongue margins to sour materials, and the back to sour and bitter substances (Figure 12.19). However, studies of generator or action potentials from receptors in these regions have failed to demonstrate a one receptor–one taste system. Instead, each receptor has its own pattern of response to several kinds of chemicals, but displays what might be called preferences. Vertebrate taste discrimination is thought

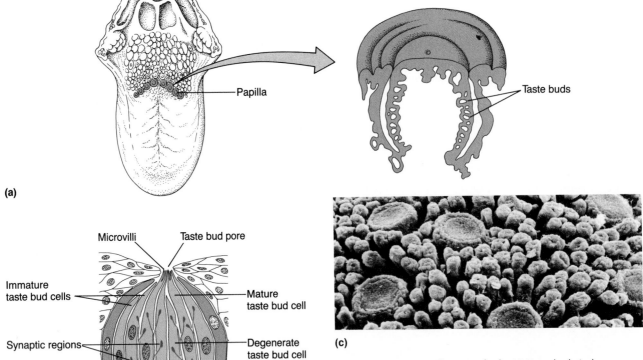

(a)

(b)

(c)

Figure 12.18 Mammalian taste buds. (a) Taste buds in humans are distributed over the tongue, soft palate (roof of the mouth), and parts of the pharynx, but are most abundant in papillae that extend down into the tongue's epithelium. (b) Taste bud of a rabbit. Microvilli from mature taste bud cells project into a minute taste bud pore. A taste bud also includes immature cells that have not yet developed microvilli, as well as degenerate cells that have lost them. (c) A scanning electron micrograph of the surface of a rat's tongue, showing sensory papillae surrounded by numerous smaller nonsensory papillae.

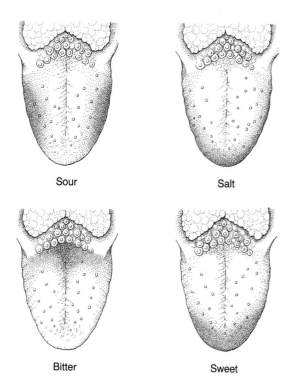

Sour

Salt

Bitter

Sweet

Figure 12.19 A "tongue map." This map shows the broad overlap of areas sensitive to sour, salt, bitter, and sweet substances in humans.

be extremely complex and to involve a sensory code based on cross-fiber patterning similar in principle to that of the blowfly (discussed in the previous section), only more complex. The total pattern of data (nerve impulses) sent to the brain probably differs for each chemical solution. The result is that many vertebrates can discriminate far more tastes than merely salt, sweet, sour, and bitter. Coding of innumerable combinations of these four fundamental sensations allows discrimination of a broad range of tastes. One further complication should be recognized: The strong interaction between the senses of taste and smell means that what is perceived as something tasted may actually be something both smelled and tasted.

Olfactory receptors and their supporting cells are part of a highly specialized tissue called **olfactory (nasal) epithelium**. Vertebrates generally have two olfactory organs with openings called **nostrils**, or **nares**, located on the head. In many fishes, the nostrils lead into dead-end pits or tubes containing olfactory epithelium. Water circulates in and out of the nostrils, and the olfactory epithelium monitors chemicals in the water as a fish swims forward. In terrestrial vertebrates, a nasal passage connects the nostrils with the mouth cavity. This allows a land vertebrate to breathe air without opening its

mouth and also to eat and breathe simultaneously. Having evolved from terrestrial vertebrates, whales, dolphins, and other aquatic mammals also breathe air. They surface regularly to quickly exhale and inhale through a nostril called the blowhole, which is located on top of the head.

Chemoreceptor cells in the olfactory epithelium bear cilia that project outward. Axons from receptor cells converge to form the **olfactory nerve**, which carries nerve impulses to the brain. In terrestrial vertebrates, the olfactory epithelium is located at the upper end of each nasal passage, so air is sampled for chemicals during breathing. Cilia on the receptor cells project out into a layer of mucus coating the surface of the olfactory epithelium (Figure 12.20). If individual molecules in the air passing through the nasal passage are to be detected, they must first dissolve in the mucus and then bind to receptor sites in the receptor cell membrane. The resulting generator potentials elicit nerve impulses that travel toward the brain for interpretation.

Vertebrate olfactory organs are highly sensitive, especially those of mammals. Sensitivity seems to be related to the number of receptor cells and therefore to the number of sites available for binding molecules. A cat or dog, with some 200 million receptors, responds to about 10,000 molecules per cubic centimeter of air of butyl mercaptan (thiobutyl alcohol), a foul-smelling chemical found in the odor of skunk and in small amounts in mammalian sweat. A bloodhound can follow a trail of human foot odor (mainly butyl mercaptan) that has passed through leather shoes to the ground. By contrast, a human, with only about 5 million receptors, responds only if the concentration of butyl mercaptan is at least a thousand times greater. Neural convergence contributes to olfactory sensitivity by amplifying neural signals: Millions of receptor cells converge their signals via synaptic junctions onto some thousands (45,000 in humans) of neurons to form the olfactory nerve.

Almost nothing is known about how some mammals discriminate thousands of different odors. According to one hypothesis, there are 30 or so different types of receptor sites in the membrane of each receptor cell. Each site might bind many different kinds of odor molecules, but would bind one type best. There may also be different numbers and concentrations of each type of receptor site on different receptor cells. The net result would be simultaneous but somewhat different rates of firing by many cells in response to a given substance. Other substances would produce a different pattern of responses, so the potential to discriminate chemi-

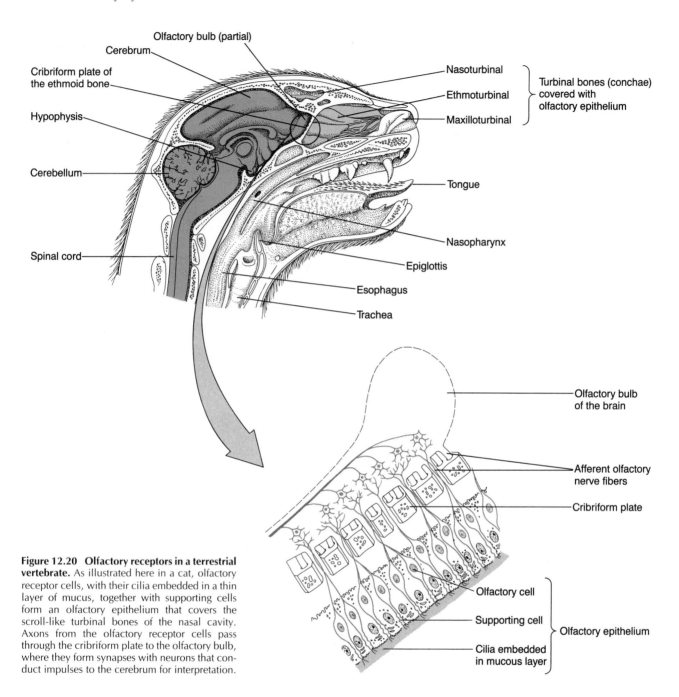

Figure 12.20 Olfactory receptors in a terrestrial vertebrate. As illustrated here in a cat, olfactory receptor cells, with their cilia embedded in a thin layer of mucus, together with supporting cells form an olfactory epithelium that covers the scroll-like turbinal bones of the nasal cavity. Axons from the olfactory receptor cells pass through the cribriform plate to the olfactory bulb, where they form synapses with neurons that conduct impulses to the cerebrum for interpretation.

cals on the basis of differential response patterns would be enormous. This explanation is hypothetical, and olfaction is an intriguing research area where great strides remain to be made.

Other Senses

A **thermal sense** plays a variety of roles in different animals. Many vertebrates respond to external temperature through several kinds of thermoreceptors in the dermis of the skin. In humans, for example, one kind of dermal receptor detects coldness between 12°C and 25°C, a second type detects warmth between 35°C and 45°C, and at intermediate temperatures, perception is determined by the relative activity of both receptor types. Some animals monitor their own internal temperature. In birds and mammals, certain neurons in the hypothalamus respond to changes in the temperature of the blood, acting as a biological thermostat in the regulation of body temperature.

Certain snakes use thermoreceptors in a particularly interesting way. Boas and the so-called pit vipers, including rattlesnakes, use extremely sensitive heat sensors, located on each side of the head, to track down birds or mammalian prey at night. They simply follow their victim's body heat to its source. If the environmental temperature is at least 10°C colder than the body temperature of a mouse, a rattlesnake can detect the presence and direction of the mouse from about 40 cm.

Virtually all invertebrates respond to environmental temperature changes. Many insects have specific temperature-sensing neurons incorporated into chemoreceptor organs, but some species have separate thermoreceptor cells. Thermoreceptors aid mosquitoes and other bloodsucking insects in locating mammalian or bird hosts; a bedbug is sensitive to temperature changes as small as 1°C, allowing it to follow a temperature gradient to a human or other mammalian host.

Electricity and **electroreception** are important in the life of many fishes and a few amphibians. Many fishes have electroreceptors set in pits on the surface of the head and body. Some species can detect extremely weak electrical stimuli (0.01 μV/cm). Certain large sharks, including the great white shark of current popular literature (see Figure 29.3a), can detect prey by sensing the extremely weak electric field produced by the prey's muscle activity. Long-distance navigation by Atlantic salmon and American and European eels may include responses to extremely small electrical currents induced by the flow of salt water through the earth's magnetic field. The electric eel (*Electrophorus electricus*), a large freshwater species found in South America, can generate charges up to about 600 V to kill or stun prey or ward off enemies. The eel's electrical organ consists of stacks of modified muscle cells on both sides of the body. Other species, such as the African electric fish (*Gymnarchus niloticus*), have an electrical sense based on electric fields that they generate about themselves. As an electric fish swims close to underwater objects, the lines of force in the animal's electric field are distorted. Electroreceptors along the sides of the animal detect the distortions. With information from its electroreceptors, *Gymnarchus* can find its way through murky waters, where vision is limited, and can detect predators and prey as they enter the electric field. Mild discharges may also serve as advertisement for mates.

Many animals have a **magnetic sense** that may help them navigate from place to place by detecting the earth's magnetic field. Crystals of magnetite (Fe_3O_4), also known as lodestone, have been found in the cells of many animals, including bees, some molluscs, dolphins, tuna, marlin, sharks, green turtles, migratory birds such as warblers and European robins, and homing pigeons. Each crystal is a tiny permanent magnet whose orientation is influenced by the earth's magnetic field. Although its exact mechanism is still unknown, the magnetic sense probably enables animals to use lines of magnetic force as navigational guides.

The ability to detect painful or noxious stimuli, such as excessive pressure, heat, or light, is yet another sensory adaptation in many animals. Receptors that detect potentially harmful stimuli often help keep an animal safe. Many species have unspecialized sensory nerve endings in the dermis of the skin that respond to extreme stimuli. Birds and mammals have special pain receptors (sometimes called **nociceptors**) that provide warning information about environmental events beyond the animal's normal operating range. Pigeons have thermal sensors that begin to generate action potentials in sensory nerves when the environmental temperature reaches the danger level of 42°C. A maximum rate of impulse generation occurs at 52°C.

❑ S U M M A R Y

1. Most animals are sensitive to light, certain chemicals, temperature changes, and various types of mechanical energy. Some also have sound, electrical, and magnetic sensors.

2. Stimulation of a sensory receptor cell produces a generator (receptor) potential, which generates action potentials (nerve impulses) that travel into the central nervous system. The magnitude of the generator potential corresponds to stimulus strength.

3. Large generator potentials elicit nerve impulses at a higher frequency than smaller generator potentials, allowing nerve impulse frequency to code for stimulus size. Other kinds of sensory coding include use of a particular nerve by nerve impulses from a specific kind of sense organ ("hardwiring"); the specific nature and capabilities of the sense organ being stimulated; and cross-fiber patterning.

4. Spiders, many molluscs, and vertebrates have single-lens eyes, while crustaceans, insects, many other arthropods, and a few annelids have compound (multilens) eyes. The compound eye has inferior resolving power and little or no ability to accommodate, but it is superior to the single-lens eye as a motion detector.

5. In general, hearing is limited to arthropods (mainly insects) and vertebrates, groups in which sound is used extensively in communication. In many insects, sound waves cause a tympanum to vibrate; the tympanum directly stimulates receptor cells attached to it. In vertebrates other than fishes, a tympanum intercepts sound waves and transmits the vibrations to one or three bony ossicles in the middle ear. Vibrations of the ossicles cause fluid in the cochlea of the inner ear to vibrate. Movement of the cochlear fluid bends hair cells, generating action potentials in the cochlear nerve.

6. Most animals have some type of gravity sensor that serves as an organ of balance. Statocysts, which are sacs containing hard granules and touch-sensitive hair cells, occur in a wide variety of animals.

7. Chemosensory ability is universal among animals and protists. Taste and smell involve sense organs that allow molecules access to specific receptor sites in receptor cell membranes. In insects, setae with one or more surface pores serve as detectors. In vertebrates, taste cells are grouped into taste buds, and olfactory receptors are part of the olfactory epithelium that lines the nasal passages.

❑ FURTHER READING

Ali, M. A., and M. A. Klyne. *Vision in Vertebrates.* New York: Plenum, 1985. *A brief yet comprehensive text on vertebrate photoreceptors.*

Dethier, V. G. "Other Tastes, Other Worlds." *Science* 201(1978): 224–228. *A highly readable description of the chemosensory world of insects.*

Dunbar, R. "How Animals Know Which Way To Go." *New Scientist* 101(1984): 26–30. *A nontechnical review of how animals navigate using such cues as the stars, the sun, magnetic fields, and smell.*

Gibbons, B. "The Intimate Sense of Smell." *National Geographic* 170(1986): 328–361. *A beautifully illustrated essay on vertebrate (especially human) olfaction.*

Hildebrand, M., D. M. Bramble, K. F. Liem, and D. B. Wake. *Functional Vertebrate Morphology.* Cambridge, Mass.: Harvard University Press, 1985. *Contains two chapters on sensory systems written by experts.*

Laverack, M. S., and D. J. Cosens. *Sense Organs.* Glasgow: Blackie & Sons, 1981. *A comprehensive, quite detailed description of the major sense organs.*

Michelsen, A. "Insect Ears as Mechanical Systems." *American Scientist* 67(1979): 696–706. *A readable, yet technical comparison of ears that detect particle movement and those that are pressure detectors.*

Stebbins, W. C. *The Acoustic Sense of Animals.* Cambridge, Mass.: Harvard University Press, 1983. *A clear, concise introduction to the science of hearing.*

Stevens, S. S., and F. Warshofsky. *Sound and Hearing.* Life Science Library. New York: Time Inc., 1965. *A superbly illustrated, highly readable description of the nature of sound and hearing.*

Zwislocki, J. J. "Sound Analysis in the Ear: A History of Discoveries." *American Scientist* 69(1981): 184–192. *Traces the major advances in understanding how the mammalian ear functions.*

CHAPTER 13

Animal Behavior

Social behavior in courtship and mating. Here a male booby *(Sula)* presents a pebble as a gift to a female. Many animals exhibit complex behavioral acts during reproductive periods. Such acts may be part of a complex mating process and may promote male and female pair bonding necessary for parental care of young.

Behavior in the context of zoology refers to the way an animal acts in response to stimuli. Behavioral science, or **ethology** (from the Greek *ethos*, "behavior"), is one of zoology's broadest subdivisions, and studying even a part of an animal's total behavior, such as how it feeds, how it finds suitable habitat, how it reproduces, or how it defends against natural enemies, may be a major undertaking involving specialists in many different fields. Yet, studying such questions is only part of the overall science of behavior. As is true for all fields of biology, ethology is also concerned with broader questions. In what way does an animal benefit by selecting a particular type of food or habitat, or by selecting a certain mate, or by defending the area in which it breeds? It is important to realize that behavior patterns, like all other features of animal species, are subject to evolution. Consequently, these broader questions can be answered only in an evolutionary context, that is, through analysis of the adaptive functions of behavior.

Modern ethology is an umbrella for many investigative techniques and viewpoints. It is also a robust and growing science with a rich ferment of hypotheses and theories stemming from diverse approaches. In general, there are two approaches to studying animal behavior today, and these are based on the two types of questions just mentioned. In testing hypotheses that attempt to answer the first type of question, one group of ethologists study the *physiological mechanisms* that underly animal behavior. The study of neural mechanisms that form the basis of specific behaviors, or the study of the connection between behavior and neural circuitry, is a rapidly growing subfield of modern ethology called **neuroethology**. Neuroethologists are interested in how animals sense environmental stimuli such as light and sound and how the nervous system processes such sensory inputs. To a large extent, their approach has grown out of the field of neurophysiology, and the tools of the neuroethologist are often the same as those used to study nervous system function. Because behavior, like other body functions, is influenced by chemical as well as nervous mechanisms, ethologists concerned with physiological mechanisms may also study the role of hormones and neurohormones in determining behavior. And because nervous and chemical coordinating systems result from inherited information encoded in an animal's genes, there is also much current interest in the genetic basis of behavior. In the current jargon of ethology, researchers studying the neural, hormonal, or genetic bases of behavior are concerned with the **proximate causes**—the mechanisms within an animal—that determine behavior.

The other approach in ethology is to develop and test hypotheses that focus on the evolutionary aspects, or **ultimate causes**, of behavior. In this approach, a researcher studies an animal's environmental relationships in terms of its behavior and attempts to determine how such phenomena as social behavior, migration, hibernation, and other behavioral adaptations affect survival of individual animals, populations, and species. The evolutionary approach to ethology has close ties with ecology, which is the study of environmental relationships of organisms. Both fields largely grew out of studies in natural history, which contributed much to the descriptions of animal behavior. Today, the field of **behavioral ecology** is a major area of re-

search in ethology. **Sociobiology**, the study of the evolution of social behavior, is one of its most active and controversial subfields.

The proximate and ultimate schools of ethology are both highly productive ways of looking at animal behavior, and these subdisciplines are not mutually exclusive. A meaningful explanation of an animal's behavior requires research in both. It is not possible in a single chapter to cover all the multifaceted aspects of ethology. This chapter, which follows chapters dealing with neural and hormonal coordination and sensory systems, first illustrates the connection between behavior and the physiological mechanisms that underly it—that is, behavior at the level of proximate causes. We then introduce certain ecological and evolutionary aspects of behavior.

THE UNDERLYING MECHANISMS OF BEHAVIOR

All but the simplest animal activity combines behavior that is fixed and predictable with that which may be different from time to time. Fixed, predictable responses to environmental stimuli are **innate** in that an animal performs them in a completely functional manner the first time it receives triggering stimuli. Innate responses take virtually the same form each time they occur, and they often persist more or less unchanged throughout an animal's life. Innate behavior includes reflexes and motor programs, both of which, to a large extent, result directly from instructions coded in an animal's genes. An animal's genes determine the specific types and arrangements of neurons that mediate the behavior. The environment affects the development of innate behavior in that it influences how genetic information is expressed as neural circuits. In contrast to innate behavior, **learned behavior** consists of responses that are relatively flexible; just how an animal acts in a specific situation depends on past experience. Everyone is aware that a circus animal or pet dog can be taught new routines through systematic use of training experiences. If the behavioral change persists, we say that the animal has learned. Consequently, learned behavior is often defined as *lasting modifications of behavior through experience*.

The basis of learned behavior is somewhat different from that of innate behavior. An animal's genes determine the overall characteristics of the nervous system that allows learned behavior to occur, but the specific neural circuitry underlying the behavior is not fixed; it may vary, depending on information stored in the nervous system from previous experience.

In the following sections, we examine the nature and adaptive advantages of innate and learned behavior. It is important to stress, however, that it is mainly for convenience that we discuss innate and learned behavior separately. Most animal activity is a combination of both of these types of behavior, and the distinction between innate and learned behavior is not always clear-cut. In the early days of ethology, it was customary to call innate behavior instinct, a label that tended to stifle further research into causes; instinct was thought to be caused by genes, and that was that. Eventually, the role of the environment in helping to shape innate behavior was recognized, and we now know that behavior that is mainly innate may be modified somewhat by an animal's experiences. Ethologists now realize that any particular behavior pattern may have innate and learned components and may vary to the extent to which it is controlled by genes and by the environment, including past experience.

Reflex Behavior

The eye-blink reflex and the knee-jerk reflex are familiar reflex acts. (The neural basis of reflexes are discussed in Chapter 10, p. 236; Figure 10.15.) **Reflexes** are simple acts because they typically involve coordination of relatively few muscles and little integration of information; they are usually rapid and automatic and without conscious control. A cat tumbling from a tree branch may avoid injury by flipping over to land on its feet, a **righting reflex** that begins the instant the falling cat turns upside down. Likewise, we are protected by a very rapid **withdrawal reflex** that lifts our hand from a hot surface before we are even aware of the temperature. The speed and automatism of reflexes are their chief advantage; they require no time-consuming thought or decisions.

Motor Programs

As discussed in Chapter 10, increasing evidence indicates that reflexes make up less of the behavioral repertoire of most animals than was formerly thought. In recent years, neurophysiologists and neuroethologists have discovered that much (and probably most) innate behavior results from motor programs that are "built into" an animal's nervous system during development. A **motor program** is a specific arrangement of neurons controlled by **command neurons** (interneurons in the central nervous system). As we shall see, activity based on a motor program is very often not displayed until some specific point has been reached in an animal's life cycle, when the behavior is "released" by some triggering factor. Therefore, it is appropriate to think of a motor program as preprogrammed behavior.

Advantages of Motor Programming. Motor programs benefit an animal in two basic ways. First, programming conserves space and material. Because programmed behavior results from particular arrangements of neurons established as the central nervous system develops (in the jargon of neurophysiology, the behavioral routine is "prewired"), few neurons are needed for transfer and integration of incoming sensory information. Motor programming is therefore especially advantageous for small animals such as arthropods; compared to most vertebrates, their nervous systems contain relatively few neurons and have low integrative capacity. Second, programmed behavior can usually be performed correctly the first time attempted, making an initial period of learning unnecessary. An insect, for example, can fly successfully on the first try. Motor programming is thus important in animals having no opportunity to see how tasks are done, let alone to practice them. Some animals are so short-lived (a few hours in the case of an adult mayfly) that a learning period would use up valuable time better spent in reproduction.

Fixed-Action Patterns. Both simple and complex behavior may result from motor programs. Among the simplest behaviors are the **fixed-action patterns**. In contrast to more complex behavior, a fixed-action pattern is strongly innate, and input from the senses has little effect on it. Once triggered by specific sensory impulses, certain neurons in the central nervous system cause a fixed-action pattern to run to completion in virtually unaltered form.

Fixed-action patterns are vital to many animals. Lobsters and certain molluscs (scallops and sea slugs) avoid harm with escape responses that are fixed-action patterns. A sea slug swims away from an encounter with a sea star with rhythmic arching movements made by alternately contracting dorsal and ventral muscles (Figure 13.1). Electrodes in-

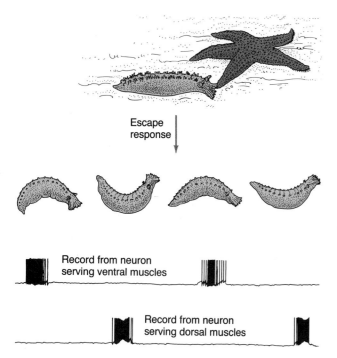

Escape
response

Record from neuron
serving ventral muscles

Record from neuron
serving dorsal muscles

Figure 13.1 Escape response of the sea slug *Tritonia*. The sea slug's escape from harm depends on a fixed-action pattern, displayed as a series of alternating dorsal and ventral arching movements caused by alternating contraction and relaxation of two sets of muscles. Electrophysiological recordings from two ganglion cells in the central nervous system, shown here in relation to body movements, are consistent with the idea that contact with a sea star activates a neuronal circuit (a central pattern generator, or CPG) in the slug's brain. The CPG functions as a switch, rhythmically triggering contraction of first one set of muscles and then the other.

serted into neurons serving the two groups of muscles reveal a corresponding alternation in nervous activity, as if an oscillating switch were turning one neuron off as it turned the other one on. Experimental evidence indicates that the switching mechanism is in the slug's brain; electrical stimulation of the brain triggers coordinated, rhythmic firing of brain cells that in turn excite the motor neurons serving the muscles.

In Chapter 10, we discussed a neuronal mechanism called the **central pattern generator (CPG)**, which functions as an oscillator switch. (See the coverage of structure and function of CPGs in annelids, p. 232; Figure 10.11.) It is likely that CPGs mediate many rhythmic fixed-action patterns, including the sea slug escape response. Once activated by a sensory stimulus (contact with a sea star, in the case of the sea slug), the CPG mediates the rhythmic fixed-action pattern with no further sensory input. Other examples of fixed-action patterns that are probably controlled by CPGs are the rhythmic wing and leg movements of insects. The

wings of a grasshopper, for example, once set in motion, continue to flap even when sensory nerve fibers from stretch receptors in the wing muscles are cut. Research has shown that the up-and-down flight pattern depends on a neuronal circuit switched on and off by rhythmic firing of command neurons in the insect's central nervous system.

Many fixed-action-pattern motor programs are nonrhythmic and therefore do not depend on oscillating on-off switches. Swallowing, gagging, and vomiting are commonly called reflexes, but they are actually nonrhythmic fixed-action patterns. In swallowing, stimulation of the upper throat by food triggers a motor program that coordinates the successive contractions of some eleven different muscles responsible for propelling food into the esophagus. According to our definition, a fixed-action pattern is automatically completed once it has started. Questioning whether swallowing meets this criterion, researchers showed that the swallowing program runs its course in dogs even when the food stimulus is immediately pulled from the animal's throat.

Complex Programmed Behavior. In contrast to simple fixed-action patterns, complex forms of **programmed behavior** may consist of a series of motor program subroutines, wherein the completion of one subroutine evokes the next, until the overall motor program has been completed. In addition, complex motor programs can often be tailored to fit changes or unique features in the environment.

The remarkable routine followed by a female digger wasp as she prepares to feed and shelter her offspring is an excellent example of complex motor programming (Figure 13.2). Using her legs and mouthparts, the wasp digs a tunnel with a brood chamber at its lower end. She then plugs the entrance with small pebbles and sand grains before flying off in search of a caterpillar with which to provision the nest. Upon finding a victim, the wasp seizes and stings it, paralyzing it with her venom. She carries the paralyzed caterpillar back to the burrow, temporarily laying it to one side while she unplugs the burrow entrance. The wasp then drags the paralyzed caterpillar to the bottom of the burrow and lays an egg on it. As the last act in her performance, the wasp climbs out of the burrow, replugs the entrance, and flies away. Soon she begins to prepare another burrow, repeating the entire process. In each burrow, a larval wasp eventually emerges from the egg and feeds on a diet of fresh caterpillar.

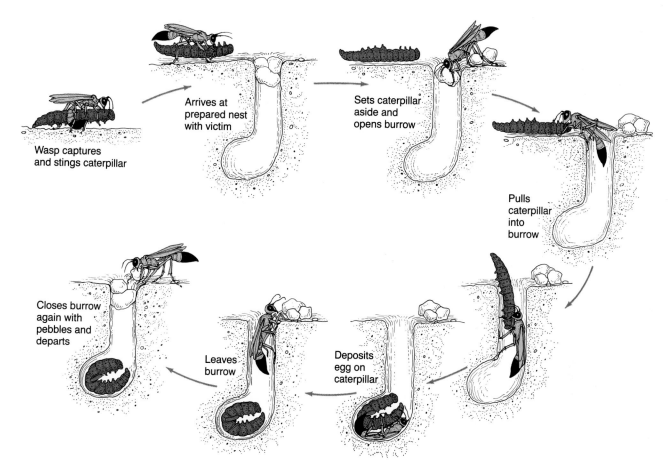

Wasp captures
and stings caterpillar

Arrives at
prepared nest
with victim

Sets caterpillar
aside and
opens burrow

Pulls
caterpillar
into
burrow

Closes burrow
again with
pebbles and
departs

Leaves
burrow

Deposits
egg on
caterpillar

Figure 13.2 Digger wasp nest-provisioning routine. A female digger wasp *(Ammophila campestris)* locates and stings a caterpillar, carries it to her previously prepared burrow, and leaves it to serve as food for the larva that soon hatches from her egg. The wasp's routine is an example of unlearned, innate behavior based on complex inherited motor programs that develop along with the wasp's nervous system.

The digger wasp's performance is an elaborate and immediately useful series of motor programs that illustrate the power of genetic coding. The sequence is mediated by neural circuitry that results directly from information encoded in the wasp's genes. Without ever having laid an egg or seen a larva, digger wasps make and stock their burrows correctly on the first try.

Arthropods do not have a monopoly on complex motor programming. Dam building by beavers and nest building by birds are largely genetically controlled feats, comparable to nest provisioning by digger wasps. A weaverbird, for example, has the innate ability to tie vegetation in about six different kinds of knots. Switching back and forth among knot-tying motor programs, it weaves a marvelous nest complete with an entrance that often foils snakes and other nest predators (Figure 13.3).

The Illusive Line Between Innate and Learned Behavior

From what we have said so far, it might seem that an animal's past experiences play no role in complex motor programs and that many animals are little more than preprogrammed automatons. Nothing could be further from the truth. To most modern ethologists, attempts to distinguish between many types of innate and learned behavior are fruitless, because all behaviors have both a genetic and an environmental component. Many behaviors illustrate this point. Humans are not born with innate motor programs for many tasks that, once learned, become virtually automatic. We type, play the piano, or tie a necktie only after a period of training. But a skilled typist can type and a pianist

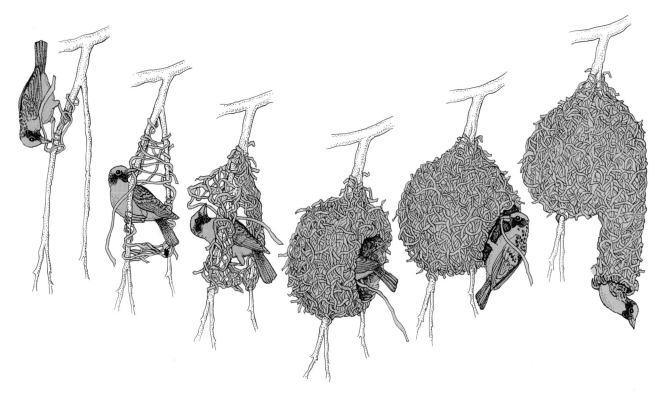

Figure 13.3 Nest construction by a weaverbird. This is one of the most complex examples of programmed behavior. With an innate (programmed) ability to tie several different kinds of knots, the bird switches from one knot-tying program to another as it weaves its remarkable nest from strands of vegetation.

can play without consciously selecting the keys, and they can even make the correct movements on an imaginary keyboard in the dark, where sensory input is minimal. After much time and practice, these tasks are performed so automatically that they actually come to be mediated by neural circuitry that is just as firmly programmed as an innate motor program. In this sense, they may be called **learned motor programs**. Of course, unlike an innate motor program, a learned motor program cannot be inherited, because it is not represented in the genes.

Adding to the difficulty in distinguishing between innate and learned behavior is the fact that many motor programs have an innate basis, but must be perfected through experience and practice before the behavior can be performed properly. For example, the singing behavior of most songbirds has both innate and learned components. Before a male song sparrow can sing, he must experience (hear) the song of his species. A male sparrow's innate programming restricts him to learning only

that song; when raised within earshot of several species of sparrows, the only song he acquires is that of his own species. The song is memorized early in the young bird's life, long before any attempt is made to sing it. A young male sparrow makes his first attempt to sing the following spring, in response to the release of specific hormones. Using a process of trial and error, the bird sings, and his vocal musculature becomes shaped, until, slowly but surely, vocal output comes to match the song-memory imprinted in his brain. It is likely that the neural circuitry directing the perfected pattern of muscle movements soon becomes incorporated into the brain as a motor program. Rapid learning of this kind, at a critical age and without any reward or punishment, is called **imprinting**. Because it is fast and is limited to a developmental period, imprinting reduces the probability of a learning error. Singing is a vital component of the reproductive cycle of songbirds, and a sparrow singing the wrong tune would probably be unable to hold a breeding territory and find a mate.

Behavior That Is Primarily Learned

As we have noted, modern ethology recognizes the importance of both heredity and environment as determinants of behavior. Learning is one of the ways the environment has a strong, direct effect on an animal's behavior. Animals vary a great deal in terms of the relative amounts of innate and learned behavior that they exhibit. There is also great variety in the learning abilities of different species and, indeed, of individual animals; but all animals, including humans, are limited in what they can learn. Each species' behavior patterns, including both innate and learned activities, have evolved as adaptive mechanisms that fit animals to their specific role in the environment. Let us now consider some of the various types of learning.

Habituation. **Habituation** is the simplest form of learning. Animals habituate as they learn *not* to respond to stimuli that are *not* followed by either punishment or reward. Even an animal with a relatively simple nervous system, such as a cnidarian or planarian, will learn to stop responding to harmless vibrations, and many insects will eventually stop trying to escape when joggled. Habituation allows animals to move about unperturbed by continuous low-level disturbances, while remaining ready to react to stimuli signifying food, mates, and danger.

Associative Learning. Some animals can learn to associate two or more unrelated stimuli with a reward or punishment. This process is called **associative learning**. In his now-classical pioneering studies, Ivan Pavlov, a Russian physiologist, demonstrated a type of associative learning called **classical conditioning**. In a series of experiments, Pavlov compared the amount of saliva produced by a dog (1) when meat powder was blown into its mouth, (2) when the dog was given meat powder and simultaneously stimulated by the sound of a bell or metronome or by the presentation of a light, and (3) when the dog was stimulated by a bell or light but not given meat powder. At first, the dog salivated only when given meat powder or meat powder plus another stimulus. But after eight to ten trials in which meat powder and another stimulus were presented together, Pavlov discovered that meat powder was no longer necessary; the dog had learned to associate taste with the accompanying stimulus and would salivate when it heard the bell or saw the flash of light. Salivation is normally an unlearned reflex stimulated by the smell, taste, or sight of food. Because the salivary reflex in Pavlov's experiments was conditional on sound or light, it is called a **conditional reflex**. Pavlov also demonstrated that a dog would lose a conditional reflex if exposed repeatedly to a light or sound without the smell or taste of food.

A second form of associative learning is important in nature. As animals move about, they encounter stimuli that they associate with various experiences. The associations then influence subsequent behavior. A fox or coyote hunting for food may learn by trial and error to associate certain odors, disturbed vegetation, sounds, and perhaps even footprints with the likely presence of a rabbit. As associations form (i.e., as learning progresses), the number of hunting errors decreases and the hunt becomes more efficient. Learning is accelerated (reinforced) by some form of reward and punishment; in the case of the fox, success is rewarded with food, and mistakes are punished with hunger. Associative learning involving trial and error is called **operant conditioning** to distinguish it from Pavlov's classical conditioning.

Latent Learning. In **latent learning**, animals store information obtained while exploring their surroundings and then use the knowledge to guide future behavior. Latent learning is important to most animals, including humans. A hiker may locate a pond or clearing in the woods because of a landmark noticed days earlier, and people in unfamiliar buildings often notice the location of fire exits and lavatories, knowledge they may use in the future. Experiments with the digger wasp conducted by the Dutch ethologist Niko Tinbergen are beautiful in their simplicity and illustrate the importance of latent learning in an insect (Figure 13.4). Tinbergen placed pinecones in a circle around the nest of a female digger wasp and then, after the wasp had visited the nest and flown away to seek food to provision the nest, he moved the circle of pinecones some distance to one side of the nest. When the wasp returned a short time later, she flew to the center of the displaced circle of pinecones, not to her nest. But was the wasp's response due to the circular arrangement of the pinecones or to the pinecones themselves? To answer this question, Tinbergen made a triangle of pinecones with the nest at the center of the triangle, and off to one side he made a circle of stones. This time the wasp flew to the circle of stones, not to the triangle of pinecones surrounding her nest. Thus, in her first flight,

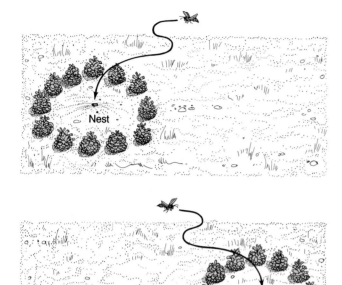

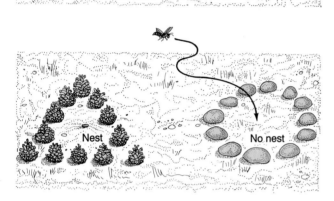

Figure 13.4 Latent learning in digger wasps. A digger wasp learns landmarks around its nest, later using the information to find the nest upon return from a foraging trip. The wasp locates the nest by recognizing the arrangement of landmarks (in this experimental case, circular versus triangular) rather than by the specific identity of a landmark (pinecones versus smooth stones).

the wasp seemed to have stored information about an *arrangement* of objects and later used the information to locate her nest.

Insight Learning. Insight learning, by far the most complicated form of learning, is the ability of some animals to adapt previous experiences to new situations. Faced with a new problem, the animal recalls past experiences and adapts parts of what it remembers to solve the problem. It does so even though the stimuli associated with the earlier experiences differ from the present ones. An animal

shows insight when it solves a detour problem, that is, when it determines a new route to some goal such as food or a den, on the first try. Taking even a simple detour to reach an objective is beyond the ability of most invertebrates and vertebrates. Despite their highly developed cerebral cortex, mammals such as dogs, cats, and raccoons can solve a detour problem only after repeated attempts, a matter of trial-and-error learning rather than insight. Insight is most highly developed and apparent in primates, especially humans. A chimpanzee experienced in handling boxes of various sizes shows insight when it stacks the boxes and uses them as a platform to reach a banana hanging above its head.

Learning and Memory

Earlier we defined learning as lasting modifications of behavior through experience. This implies corresponding changes in an animal's nervous system, producing what is commonly called **memory**. The most basic questions we can ask about learning mechanisms and memory concern the nature and location of the changes in the nervous system.

Although many unknowns remain, zoologists and psychologists are beginning to understand memory processes in humans and certain other mammals that have been studied extensively. Storage of information takes place in stages and involves several parts of the brain. First, the torrent of sensory information streaming into the brain in the form of nerve impulses is given a preliminary screening by the reticular formation, a network of neurons within the medulla oblongata (Figure 13.5). By acting as a filtering system that determines what information will or will not be passed along to higher brain centers, the reticular formation makes it possible for an animal's nervous system to ignore unimportant matters. Information surviving this initial screening passes to the amygdaloid nucleus and hippocampus for further processing.

The mammalian amygdaloid nucleus, one of several paired groups of nerve cell bodies concerned with fighting, fleeing, mating, and feeding, determines what information will be stored and what will not. By saving only potentially valuable data, the amygdaloid nucleus, like the reticular formation, keeps the brain from loading up with useless trivia. A malfunctioning amygdaloid nucleus may account for the "photographic memory" of rare persons who remember virtually everything they see. In humans, the amygdaloid nucleus also

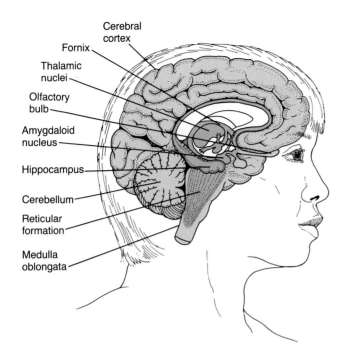

Figure 13.5 Brain structures involved in learning and memory. Incoming sensory information that passes an initial screening in the reticular formation of the medulla oblongata is transferred to the amygdaloid nucleus, where further sorting occurs and where "emotional labels" such as fear are linked to the information. The hippocampus then receives the information and holds it briefly as short-term memory before relaying it on to the cerebrum for more permanent storage as long-term memory.

"labels" information with the emotion of the moment. For example, a person who has an uncomfortably close encounter with a rattlesnake usually experiences fear. In the future, the individual will likely respond to similar sounds—even the harsh buzz of a grasshopper taking off—with a little jump and a moment of fear, because the amygdaloid nucleus has linked sound and fear. In this way, virtually all human behavior is influenced by emotion. Because we do not know how nonhuman animals register such phenomena as pain, fear, excitement, and other types of emotion, it is not appropriate to ascribe these traits to the behavior of other animals. In fact, most zoologists consider it essential to the study of behavior that any such tendency toward anthropomorphism be avoided. Clearly, however, the human stimulus/emotion linkage has strong survival value, and the nervous system of other animals may "label" potentially harmful stimuli in a similar way.

Information leaving the amygdaloid nucleus is stored briefly in the hippocampus and then relayed to the cerebral cortex, where it may become part of an animal's permanent or **long-term memory** (see Figure 13.5). It was once thought that the hippocampus was the actual site of memory, but in the 1950s, surgeons found that removing the hippocampus did not prevent people from remembering past experiences; however, postoperative events could not be remembered. Apparently, the surgery had destroyed the pathway to the site of long-term memory, not the site itself. In more recent experiments, old rats with degenerate hippocampi recovered the ability to form new memories following implantation of hippocampus cells from healthy young rats. As a result of such research, the hippocampus is now viewed as a **short-term memory** site that ultimately relays information to the cerebral cortex for permanent storage.

In addition to making strides toward understanding the role of different parts of the brain in learning, research is also advancing toward an understanding of the physical and chemical changes that take place in the brain during learning. Experimenters have found that the number of synapses in the hippocampus increases by as much as 40% when rats are trained to discern differences in light intensity. Other studies show that mastering a maze causes a marked reduction in the number of synapses in the rat hippocampus. These synaptic changes were temporary, but the studies indicate that certain types of learning may involve either enhancement or streamlining of neural pathways. (Nerve impulse transmission at synapses is discussed in Chapter 10, p. 226; Figure 10.7.) It is also noteworthy that some of these findings are applicable to both vertebrates and invertebrates. Conditioned learning in cats, for example, includes some of the same biochemical steps detected in the gastropod mollusc *Hermissenda*. (See the Essay on learning in molluscs, p. 306.)

CONTROL OF BEHAVIOR: DRIVE, MOTIVATION, AND RELEASE

Observations of animals under natural conditions suggest that readiness to undertake a behavioral routine fluctuates, as if the routine can be triggered only at certain times. For example, after passing the winter quietly feeding, apparently oblivious to any females present, a male redwing blackbird begins to interact aggressively with other males and to court females. Research shows that behavioral shifts of this kind result from changes in responsiveness to external stimuli.

Essay: Learning About Learning with Sea Hares and Sea Slugs

The relative simplicity of the nervous systems of the sea hare *(Aplysia)* and sea slug *(Hermissenda)* has made these molluscs attractive research models for neuroethologists attempting to describe the cellular and biochemical basis of learned behavior. Experiments on the withdrawal reflex of a sea hare and on learning in a sea slug reveal mechanisms that may represent universal elements in animal learning.

Aplysia has a gill on its right dorsal surface that is rapidly withdrawn into a protective cavity if the adjacent mantle tissue is touched, a reflex that can be modified by a simple kind of learning called **sensitization**. An electrical stimulation to the head or tail sensitizes the sea hare (Figure E1); then, when the mantle tissue is touched, gill withdrawal is more vigorous than before. The heightened sensitivity persists for only a few hours after mild stimulation, but may last for weeks after a series of stronger stimuli are administered. In other words, sea hare sensitization includes both short-term and long-term forms of memory.

The site of sensitization has been traced to the synaptic junctions between sensory and motor neurons. Sensitizing stimuli cause release of a chemical called serotonin at the synaptic junctions, the first step in a series of reactions that change the electrical properties of the synaptic membranes. As a result of the change, sensory impulses arriving at the synapses trigger release of greater amounts of neurotransmitter substance, generating a higher frequency of nerve impulses in the motor neurons. Consequently, the sea hare's gill is pulled in more vigorously following sensitization.

The long-term form of memory is not as well understood. One explanation may be that genes in the neurons are activated by the biochemical events resulting from relatively strong stimuli. Such activated genes might guide production of new structural proteins which, when incorporated into the synaptic membrane, could facilitate release of greater amounts of neurotransmitter substance. The presence of unusually large numbers of neurotransmitter release sites in the synaptic membranes of long-term-sensitized sea hares is consistent with this hypothesis.

Sensitization is a very rudimentary form of learning, and one might question whether the *Aplysia* mechanism applies to more complex learning. In an attempt to answer this, sea slugs were conditioned to associate a light being turned on with a sudden movement of their substrate, in the tradition of Pavlov's studies with dogs (see p. 303). Because the sea slugs learned to associate separate events, their learning was considerably more complex than *Aplysia's* sensitization. Nonetheless, *Hermissenda's* learning, like *Aplysia's*, is based on a change in the nerve cell membrane. An initial stimulus (light) followed closely by a second stimulus (substrate movement) initiated biochemical reactions leading to activation of an enzyme (protein kinase) that in turn altered the cell's membrane. The change in the membrane represents stored information (memory) about the association of two events, namely, that substrate movement follows light. Conditioned learning in certain vertebrates (e.g., cats) is also known to include some of the reactions discovered in *Hermissenda*, including the action of protein kinase. As a result of studies such as these with relatively simple invertebrate nervous systems and behavior patterns, researchers are making steady progress toward understanding the neural and biochemical mechanisms underlying learning.

An animal's responsiveness, such as a redwing blackbird's shifting into a reproductive mode, is governed by relatively long-lasting forces that some ethologists call drives. A **drive** is a physiological mechanism that modulates behavior or switches an animal from one major activity to another. Drives are often determined by hormones, and as discussed in Chapter 11, hormonal output effecting seasonal phenomena is often geared to changing day length. (Chapter 11 discusses the effect of day length on the hypothalamus in vertebrates, p. 251, and on the octopus brain p. 263.) Day length (photoperiod) increases to a maximum every spring and then decreases, providing precise, regular seasonal cues. Through mechanisms that are not fully understood, certain nerve cells in an animal seem to act as biological clocks, "measuring" day length (or perhaps night length) and triggering hormonal changes when the days or nights reach a critical length as determined by the animal's genes. Day- or night-length-initiated changes in hormonal output cause the seasonal switch to reproductive behavior in many vertebrates and invertebrates. Likewise, day or night length and hormones are often controlling factors in migratory behavior and in hibernation.

Of course, not all behavior follows a seasonal cycle, nor is all behavior regulated by hormones. Many animal activities result from a short-term version of drive called **motivation**. Being thirsty motivates an animal to seek water, hunger motivates it to search for food, and so forth. In essence, motivation is a frequently recurring need that can be satisfied by a behaviorial act—by the act of drinking in the case of thirst, or by the act of eating in the case of hunger.

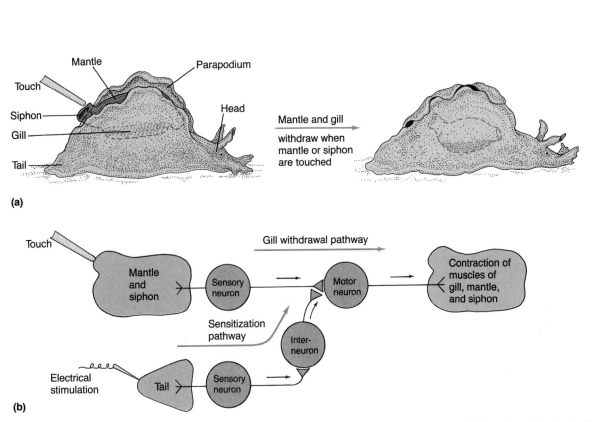

Figure E1 Sensitization in the sea hare *Aplysia*. (a) This mollusc withdraws its gill if its mantle is touched, a response that displays habituation (grows progressively weaker) with repeated touches. Release of progressively smaller amounts of neurotransmitter at the synapse between the sensory and motor neurons is responsible for the sea hare's habituation. (b) If the head or tail is electrically stimulated, the gill withdrawal reflex grows stronger through a process called sensitization. Sensitization occurs because the interneuron causes release of more transmitter molecules at the sensory-motor neuron synapse.

Drive and motivation put an animal in a state of readiness, like a rocket poised on its launchpad with all systems "go"; but for at least certain types of activity, something must still trigger the behavior. For example, the fixed-action patterns described earlier require an initial stimulus to become activated. Some 40 years ago, Niko Tinbergen and the Austrian ethologist Konrad Lorenz named these initial stimuli or sensory cues **releasers**. Tinbergen and Lorenz postulated that a releaser activates, by means of sensory impulses from a sense organ, a specific unit in an animal's brain called the **innate releasing mechanism (IRM)** and that the IRM, in turn, mediates the fixed-action pattern. To a neuroethologist, the IRM concept simply reveals how a group of neurons in the central nervous system (a CPG in the case of a rhythmic fixed-action pattern; see p. 300) responds to specific sensory impulses by commanding certain muscles to carry out an appropriate motor response.

Research with herring gulls illustrates the relationship between releaser stimuli, innate releasing mechanisms, and fixed-action patterns as perceived by Tinbergen and Lorenz. Motivated by hunger, a newly hatched herring gull signals to be fed by pecking at a red spot on its mother's bill. Pecking by the chick is a fixed-action pattern. In attempts to determine the releaser(s) for this behavior, Tinbergen exposed newly hatched chicks to models of adult gull heads differing in many factors, including presence or absence of a bill, shape, color, position of the red spot, orientation, and motion. As a result of Tinbergen's work and other studies, the most effective releaser stimuli proved to be a vertical bill (whether attached to a "head"

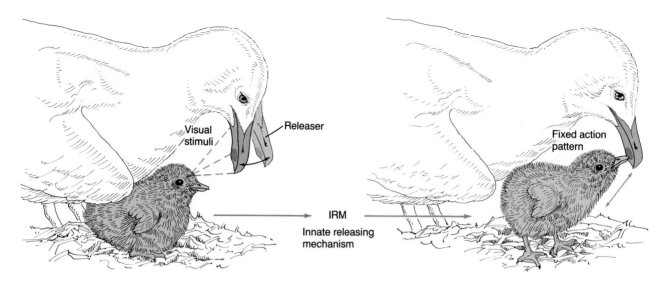

Figure 13.6 The relationship between releasers, innate releasing mechanisms, and fixed-action patterns. A hungry herring gull chick is innately sensitive to certain visual stimuli (releasers), namely, the vertical position of the mother's bill and the movement of its red spot as the bill is swung back and forth. The releasers activate an innate releasing mechanism (IRM) that links the releaser stimuli and the behavior pattern specifically associated with the stimuli. Theoretically, the IRM sets in action a motor program in which appropriate muscles perform a programmed series of contractions and relaxations. In the case of the herring gull and its chick, the resulting fixed-action pattern consists of a series of pecks by the chick at the red spot on the mother's bill. The mother will typically respond by offering food to her chick. Pecking behavior in herring gull chicks is released and guided by stimuli (releasers) for only a short time. A chick quickly learns what its parents look like, and within about two days, a learned mental picture of the parent is substituted for the releasers.

or not) swinging back and forth to give the red spot a pendulum-like motion. A parent gull with food moves its bill back and forth in front of a chick in the same way. The sensory impulses resulting from these releaser stimuli activate the IRM that releases the chick's pecking behavior (Figure 13.6).

BEHAVIOR IN THE LIFE OF ANIMALS

Thus far, we have considered some of the mechanistic causes of behavior. The remainder of this chapter is concerned more with behavior patterns as adaptive attributes of animal species. An animal's responses to external factors such as temperature and light, to thirst and hunger, and to reproductive drives make up much of its day-to-day activity. The adaptiveness of these behavioral responses affects an animal's fitness—that is, its ability to survive and reproduce. Our goal in this section is to give you a sense of behavior patterns as evolutionary adaptations.

We begin with a look at some of the simplest forms of behavior: nonoriented movements. We then look at several types of simple and complex orientation behaviors and then at the part played by communication in such activities as feeding, reproduction, and defense. To introduce the evolutionary ethologists' quest for ultimate causes of animal behavior, we conclude by examining several aspects of social behavior and sociobiology.

Nonoriented Movements: Kineses

Some animals and many protozoa respond to certain external stimuli simply by changing their rate of locomotion or by turning without actually orienting to the stimulus. Such activities are called **kineses**. Flatworms such as *Dugesia* (planarians), for instance, may speed up, slow down, or change course more frequently when they sense food molecules in the surrounding water. The human body louse *Pediculus humanus* moves about between the clothes and skin, a warm environment that averages about 30°C. Lice show a strong preference for this temperature. In circular test chambers with two halves at different temperatures, a louse lingers or makes few turns where the temperature is

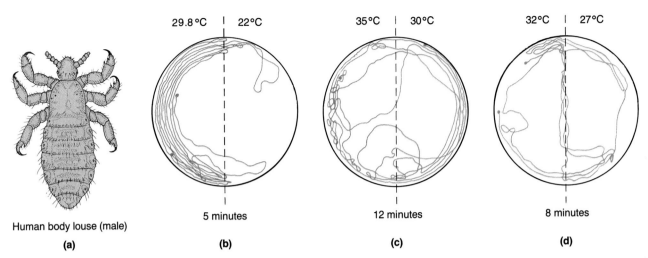

Figure 13.7 Kinetic and taxic behavior. (a) Dorsal view of a male human body louse, *Pediculus humanus*. In a circular chamber, the louse displays several kinds of behavior in response to temperature. In (b), the louse shows avoidance reactions to low temperature, spending most of its time at its preferred temperature of about 30°C; in (c), it makes relatively few turns at 30°C, but a large number of random turns at 35°C, an example of kinesis; in (d), the louse orients to the optimal temperature in the boundary region between 32°C and 27°C, an example of taxis.

optimal, but makes many turns where temperatures are lower or higher than 30°C (Figure 13.7b, c).

Despite their simplicity, kineses have definite adaptive value. With greater velocity and more frequent turning movements in a potentially harmful environment, an animal increases its chances of blundering into a more suitable one. Alternatively, decreased movement may allow an animal to remain in a favorable location.

Orientation

Orientation, which is the maintenance of a direction or spatial relationship in reference to environmental cues, is among the most important of all activities. To a large extent, an animal's ability to establish and maintain spatial relationships determines how successful it will be at practically everything else it does.

Taxes. Taxic behavior involves orientation to or away from one or more stimuli, such as different temperatures (Figure 13.7d). **Taxes** often involve a kind of sensory balancing act. An animal may turn until its bilaterally arranged sense organs are equally stimulated. Then it moves directly toward or away from the stimulus by maintaining balanced stimulation. Insects drawn to a lighted window or a street light show *positive phototaxis*, whereas a planarian moving into increasingly

darker surroundings in a stream may be displaying *negative phototaxis*.

Without experimental study, it is usually not possible to distinguish specific taxes. For example, does a planarian in nature really move away from light, or is it exhibiting *positive geotaxis* (orientation to gravity)? A planarian could even be orienting to certain chemicals *(chemotaxis)* on the bottom of a stream or responding to more than one type of environmental stimulus.

Taxic behavior is often complicated by simultaneous orientation to more than one stimulus. Fishes, for example, swim dorsal side up because they orient to light, but their phototaxis is influenced by a simultaneous response to gravity. Surgical removal of the inner ear eliminates a fish's sense of gravity and permits a simple taxic response to light. As shown in Figure 13.8, this produces calamitous consequences for a fish. (For a ◁ review of the inner ear, see the discussion on the semicircular canals, utriculus, and sacculus in Chapter 12, p. 288.)

Orientation in many animals is affected by chemicals, some of which are produced by the animals themselves. **Pheromones** are chemicals emitted by an animal to the external environment; they affect the behavior or development of other members of the same species in ways that generally seem adaptive for the emitter. Many species of insects intermix chemotaxis, other taxes, and

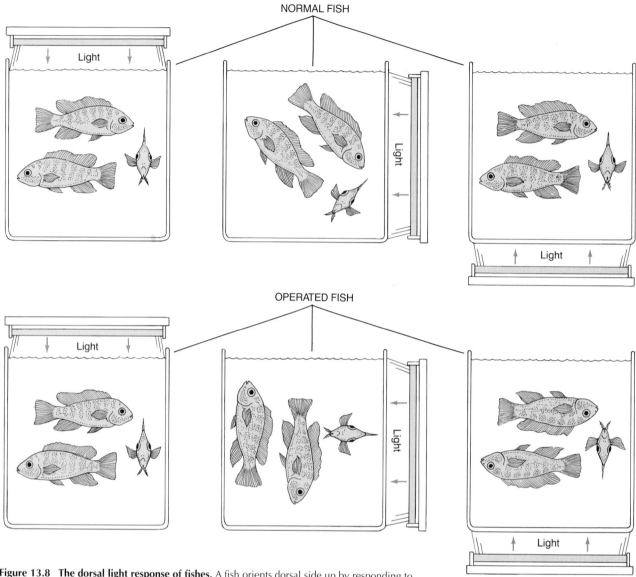

Figure 13.8   The dorsal light response of fishes. A fish orients dorsal side up by responding to light and gravity. In aquarium experiments, a normal, unoperated control fish was illuminated from above, from one side, and from below, producing the normal orientations illustrated in the upper three aquaria. The fish in the lower three aquaria was illuminated in the same way after part of its inner ear had been removed. Unable to sense the direction of gravity, the fish could orient only to light.

nontaxic behavior to locate mates. The female of certain moth species, for example, emits a sex pheromone that drifts downwind, forming an aerial odor trail. A flying male encountering the pheromone trail responds with an odor-triggered chemotaxic turn and commences to fly upwind (orientation to wind is called *amenotaxis*). The mechanism is incredibly sensitive; a female releasing as little as 0.01 micrograms (μg) of pheromone may trigger a response in males several miles away. The male moth adjusts his flight speed and upwind direction by detecting changes in the ground pattern passing beneath him. If a male hap-

pens to lose the odor trail, a zigzag motor program is triggered. Flying a crosswind zigzag pattern often brings the moth back into the odor trail, stopping the motor program and triggering a turn into the wind as before. Repeating the process as necessary, the male may eventually reach the female. As parts of a complex mechanism that ensures the reproductive success of these insects, all elements of this behavioral sequence—release of sex pheromone by the female, its detection by the male, odor-triggered upwind orientation, vision, and a central nervous system motor program for zigzag flight—are vital evolutionary adaptations.

Navigation. **Navigation** is the most complex type of orientation. In navigating, an animal not only orients itself in space, but also follows a course to a particular goal using specific environmental cues. Even if driven off course, a navigating animal is usually able to compensate and reorient toward its goal. Some animals can follow a course while maintaining an oblique angle with a source of stimulation. Many birds, fishes, and mammals make seasonal migrations to new habitats, in some cases traveling thousands of kilometers. On a smaller scale, ants and bees travel between their nests and sources of food. Depending on the species, animal navigators orient to the sun, stars, polarized light patterns in the sky, and even the earth's magnetic field.

Vertebrate navigation has been most extensively studied with migratory birds and homing pigeons. Caged homing pigeons fly toward home upon release. Likewise, when caged migratory birds are released, they tend to fly in the direction of migration. Species that migrate during the day use the sun as their compass. At least one night migrant, the indigo bunting, navigates using the stars. When researchers shift the position of the stars or sun (e.g., by using mirrors to reflect the sun's rays to the opposite side of indoor caged birds), the birds take off in the wrong direction. Because the sun and stars change position naturally as time passes, a bird using them as a compass must be able to compensate for their movement. "Clock-shifting" experiments show that day-migrating birds time-compensate by means of an internal clock. Groups of pigeons kept in a closed room with an artificial day/night cycle out of phase with the natural day/night cycle have their internal clocks reset; if the lights in the room go on and off 6 hours early, the birds' internal clocks indicate dusk when it is actually noon outside. With their sense of time running 6 hours fast, the pigeons set off for home 90° off course.

It is interesting that homing pigeons take off for home in the correct direction even on cloudy days, and most migratory birds can navigate perfectly well when the sky is overcast. This indicates that birds use more than just the sun's position to orient and navigate. There is evidence that many species can obtain compass bearing using the earth's magnetic field, which varies geographically. Most likely, they monitor magnetic field information along with solar information, but can rely solely on a magnetic compass when the sun is obscured. American ethologist William Keeton discovered the magnetic compass in a series of experiments with homing pigeons. Keeton found that pigeons with magnets fastened near their heads flew toward home when released on sunny days, but flew off in nonhoming directions when released on cloudy days. Control birds carrying nonmagnetic brass weights had no difficulty navigating on clear or cloudy days. Keeton surmised that pigeons carrying magnets were unable to navigate on cloudy days because the magnets prevented the birds from detecting the earth's magnetic field.

As yet, no one has been able to determine how animals detect magnetic fields. As noted in Chapter 12, however, many animals carry a crystalline substance called magnetite in their heads. (See the Chapter 12 Essay on pathfinding, p. 290.) Homing pigeons are among several bird species known to have magnetite. Among the insects, honeybees carry magnetite in their abdomens. Sailors formerly used magnetite as a compass, and it will be interesting to see if future research shows that other animals have put this substance to a similar use.

Communication

Broadly speaking, **communication** is any activity of one animal that changes the activity of one or more other animals. When a female moth announces her presence by releasing a sex pheromone, she is communicating with distant males. A baboon's snarl and fang-baring scowl is a message warning an approaching male to halt (Figure 13.9). Honeybees

Figure 13.9 Facial expression as a form of communication. The scowl of this baboon *(Papio hamadryas)* is a warning message that can deter an approaching male.

communicate when they perform an intricate dance that guides other members of the hive to distant food. And a skunk's message is clear as it advances with its tail held high, stimulating a bobcat to turn away. It is clear from these examples that communicative messages or signals may take several forms, including sound, light, chemicals, and touch, either alone or in combination. A moth may rely on a simple chemical signal, the baboon's warning is both visual and auditory, and the dance language of honeybees involves odor and body movement.

Communication has many functions in the animal world. It plays many vital roles in feeding, reproduction, and protection. As worker honeybees inform other workers about sources of food, they increase their hive's overall foraging effort. A moth locates a mate and reproduces. And a skunk's body position or scent may repel a predator. All forms of social behavior depend on some form of communication. The baboon's warning is one of many examples of the role of communication in maintaining social structure within animal groups. As a substitute for violent interaction among members of a social group, visual and auditory signals of social status or dominance maintain social order and often prevent severe injury or death.

Communication and Feeding. Because survival and the ability to reproduce depend on finding and identifying sources of food, it is not surprising that many species have evolved highly efficient methods of foraging. Currently, one of the most active areas in behavioral research is the study of foraging strategies. As discussed in Chapter 35, many animals seem to have evolved mechanisms that allow them to obtain maximum food with minimum effort. (Optimal foraging strategy is discussed in Chapter 34, p. 838.)

Foraging efficiency in many animals depends on communication with other members of the species. A gibbon's food-finding call announces the discovery of food to other gibbons, and a similar process occurs among some sea gulls. Food calls by a few gulls witnessing the dumping of garbage from a fishing boat may attract many other gulls, which soon become involved in a raucous, frenzied orgy of feeding. The circling flight of a vulture above a dying animal is also a food signal, sometimes attracting other vultures to the death scene from great distances.

The use of communication in feeding by honeybees is truly remarkable. Information about food quality, odor, distance, and direction is transmitted to "recruits" (uninformed workers) in the hive by "foragers" (workers that have just returned from gathering nectar or pollen). Upon receiving the message, the recruits fly directly to the flowers located earlier by the foragers. Aristotle thought that the foragers led recruits to the food, but tests made more than 2000 years later showed that recruits were able to fly directly to the food even after

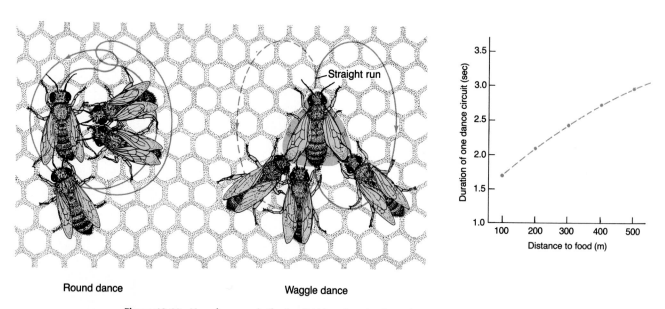

Round dance   Waggle dance

Figure 13.10 How far away is the food? When foraging honeybees return to the hive, they often perform one of two dances on vertical sections of honeycomb: a round dance, indicating that food is very near the hive, or a waggle dance, in which the number of waggles and duration of each dance cycle correlate with the distance of the food from the hive (see graph).

the foragers had been trapped while taking off on their second flight. In a now classical series of studies that began around 1915, Karl von Frisch and his colleagues in Germany deciphered the dance language of the bees.

Von Frisch discovered that returning foragers often perform either a round dance or a waggle dance on a vertical surface within the hive as potential recruits crowd around (Figure 13.10). Because the dances are generally performed in the darkness of the hive, the recruits have to follow the dance pattern by touch rather than by sight; as a forager dances, the worker recruits continually detect her position with their antennae. At first, von Frisch thought that the two dances indicated a two-word language, the round dance meaning "nectar" and the waggle dance meaning "pollen," but later he discovered the actual meaning of the dances. A honeybee doing the waggle dance vibrates (waggles) her body during the straight-run part of the dance pattern, and the number of waggles and duration of each dance circuit correlate with the distance to the food source (see Figure 13.10). The straight-run part of the dance pattern also indicates the direction of the food relative to the sun (Figure 13.11). A dance directed straight upward means that the food source is in the direction of the sun, whereas a dance directed straight downward means the opposite. A forager dancing at an angle to the vertical is informing recruits that the source of food lies at a corresponding angle to the sun. Thus, a dance at 60° to the left of vertical means that the food is 60° left of the sun.

After the recruits have monitored the waggle dance, they fly off in the general direction of the food source. Headed in the right direction, they are able to pinpoint the precise location of the food source by its odor. If the food source is a rich one, dancing may go on for long periods and involve many recruits. Because the sun changes position in the sky during the dance, how can the forager's directional information remain accurate? Like many other animals, honeybees have a built-in biological clock that compensates for the passage of time. As the sun's apparent position changes, the compensation mechanism causes the angle of the dance to change as well; consequently, flight directions remain accurate.

Given the great information content of the waggle dance, what is the purpose of the round dance? Von Frisch concluded that it represents a special case, indicating food too close to the hive for the dance to include any waggles. At close range, there is simply no need to give general directions, be-

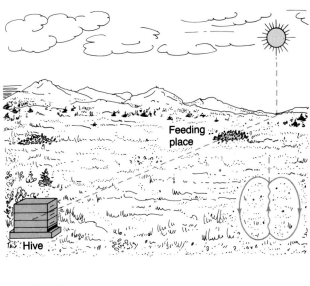

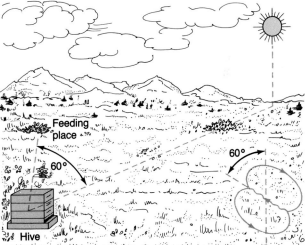

Figure 13.11 What is the food's direction from the hive? If the straight-run part of the waggle dance is straight up, food lies toward the sun; if this part of the dance is straight down (not shown), then food lies directly away from the sun; and if the dance is at an angle to the left or right of vertical, then food lies at the same angle to the left or right of a line between the hive and the sun.

cause the odor of the food close to the hive provides enough guidance for the recruits.

A particularly intriguing feature of honeybee communication is how a forager determines the food distances encoded in her waggle dance. How does a small flying animal measure distance? Von Frisch noted that foragers struggling against headwinds or flying uphill toward food typically overestimated distance in the waggle dance, while those coasting downwind or downhill underestimated it, as if they were equating distance with flight effort. To test the hypothesis that foragers determine distance by measuring effort, von Frisch fitted small flaps to some foragers to increase drag and attached bits of lead to others to increase their weight. In each case, greater flight effort was associated with a waggle dance message for increased distance. Although the dance language is precise in terms of sun and time, it evidently does not compensate for less constant factors such as wind and topography.

Communication and Reproduction. Communication in reproductive behavior typically occurs within a social system and often concerns possession and defense of territory. We will reserve discussion of social behavior until later and concentrate here on communication in the early part of the reproductive process, namely, species recognition and attraction of the sexes.

In many animals, one of the first steps in the reproductive process is species recognition. Accurate identification is important because animals can ill afford to spend time and energy attempting to mate without good prospects for offspring. Recognition may even mean the difference between life and death. For example, male jumping spiders are in imminent danger of being taken for prey by the larger female. To avoid this, the male spends considerable time waving his legs and dancing in front of the female until she recognizes him and indicates that she is receptive. Futile attempts at interspecific mating are largely prevented by species-specific morphology and signals.

To be effective, recognition signals must be specific. Chemical signals provide the necessary specificity in many species. Male silkworm moths respond to the female silkworm moth's sex pheromone, and male gypsy moths respond to the pheromone of the female gypsy moth. A pheromone released by female dogs in heat is specific for dogs, attracting males from substantial distances and acting to excite them. Basing recognition signals on chemicals has a number of advantages besides high specificity. Because chemical communication does not depend on vision, animals can recognize one another in the dark or while hidden from predators. Also, in contrast to many visual and auditory signals, odors are often effective over very great distances, as described earlier for moths.

Visual communication is often important in species recognition and in mate selection. In certain species of lizards, males identify and attract females by displaying head-bobbing movements. All males of a particular species perform virtually the same movements, but such sexual displays are quite different from species to species. In a similar way, male fiddler crabs identify themselves to females by waving the enlarged left claw (see Figure 26.7c) but the duration and frequency of the waving motion varies among species. The brightly colored plumage of some male birds (see Figure 32.3c) and the body markings in many other animals serve the same purpose. A few species of insects communicate by means of light that they produce themselves. Male fireflies fly about at night producing a pattern of luminescent flashes that is species-specific. A female firefly in the grass below recognizes the pattern and responds with a flash that guides the male to her. But in an area containing several species of fireflies, how does a male know which flash comes from a female of his own species? The time interval between the end of the male's flashing pattern and the female's response varies from species to species, and it is this interval that keys the male to a female of his species.

Sound is also a key feature in the reproduction of some insects and many vertebrates. It is especially effective in long-distance communication. The familiar chirping of male crickets (females are silent) are actually species-specific messages coded as patterns of sound pulses; changing the pulse pattern gives the chirps a different meaning. By using high-fidelity recorders and electronic sound-producing equipment, researchers have been able to manipulate cricket behavior and have made a start at analyzing the information content of the songs. Depending on the species, crickets sing four or five distinct songs. After females attracted by the *calling song* have arrived, the male cricket changes to a *courtship song* that often leads to mating. After copulation, the males of some species sing what some researchers have called a *postcopulatory* or *triumphal song*; it keeps the sexes near each other for a short period, perhaps allowing the male to "protect" his sperm investment by warding off other males. When two males meet, they sing a *rivalry*, or *aggression, song* that determines which male is dominant.

Among vertebrates, birds make extensive use of sound in species recognition and reproduction. Earlier we noted that songbirds imprint their species' songs as chicks, long before using them to attract mates. The unique character of a song is apparent to the experienced bird-watcher, who can often identify a species by the song alone; still, the human ear can detect only part of a typical bird song. (Spectrographic analysis of bird songs is discussed in Chapter 32, p. 787; Figure 32.15.) Males of many amphibians, mammals, and, to a lesser extent, reptiles also make specific sounds to attract and excite females.

Figure 13.12 Tail flagging by alarmed whitetail deer. It has not yet been determined whether this is a warning signal for other deer or an attempt to discourage a predator.

Communication and Protection. One of the benefits animals may gain by living in groups is mutual protection through alarm signals. But since an alarm becomes part of the environment of any animal able to detect it, animals of other species may also be affected by it. Therefore, there is often risk in giving an alarm signal, for it may attract predators. It is a challenging aspect of behavioral research to determine what a particular alarm signal actually communicates and, consequently, what adaptive role it plays. A frequently cited case in point is the tail-flagging behavior of the whitetail deer (Figure 13.12). Does tail flagging communicate alarm to other deer in a group, or is it mainly a signal to a would-be predator that its presence has been detected and further pursuit may be futile?

Many kinds of alarm signals have evolved, ranging from highly specific alarms to general warnings. Vervet monkeys of East Africa sound specific warning calls when they see leopards, pythons, and eagles—their chief predators—while ignoring commonly seen mammals, birds, and reptiles (Figure 13.13). On detecting an eagle, a monkey emits several staccato grunts that stimulate other vervets to look up in the direction of an attacking eagle. A python alarm, on the other hand, is a high-pitched call that stimulates vervets to look down. A third alarm, elicited by a leopard sighting, stimulates vervets to run toward the safety of trees. *Citellus beecheyi*, a species of North American ground squirrel, also makes three specific, acoustically different chirps to warn of either snakes, mammals, or hawks. At the opposite extreme, a dog barking at an unseen, unspecified threat is a familiar example of a general warning. Many social insects also sound a

Figure 13.13 Vervet monkey *(Cercopithecus aethiops).* These monkeys sound a different alarm for each of their major predators.

general alarm when disturbed or threatened. A warning signal, typically an alarm pheromone, stimulates some ants *(Lasius)* to scatter from their nests under rocks and logs until the disturbance has ended, and others *(Acanthomyops)* to rush toward the source of trouble to deal with it en masse.

Many types of alarms and warnings seem to operate mainly by communicating with members of other species. When threatened, many animals assume a more formidable appearance, a visual signal that may stop an advancing predator (Figure 13.14). The behavior of a cat in the presence of an aggressive dog is familiar. With its hair standing erect, legs stiff and straight, head high, and back sharply arched, the cat seems larger than usual and far more dangerous. Such behavior may prevent an injurious fight, because an experienced dog generally keeps its distance. Intimidation is not always visual, as exemplified by the awesome bellowing of a howler monkey.

Another way of thwarting predators is to display false clues. Certain fishes with camouflaged or inconspicuous eyes have a prominent eyelike spot on the tail. A predator attempting to head off fish may be deceived by the spot into attacking the tail end, allowing the fish to escape. Similarly, hairstreak butterflies have wing markings that may

Figure 13.14 A warning display by a reptile. Many animals warn or try to intimidate enemies by making themselves seem larger or more formidable than they really are. Here, an Australian frilled lizard *(Chlamydosaurus kingi)* expands its brightly colored frill in an attempt to frighten an approaching animal.

century English biologist), a harmless or palatable species (the *mimic*) is protected by its similarity to a harmful or noxious species (the *model*). The viceroy butterfly is palatable to birds, but is protected by its close resemblance to a distasteful species, the monarch butterfly (Figure 13.16a). Likewise, many harmless, palatable flies are mimics of venomous wasps and bees. After several unpleasant experiences, birds learn to recognize and avoid the wasps and bees, behavior that carries over to the harmless flies. A rare example of Batesian mimicry among mammals involves certain tree shrews *(Tupaia)* and squirrels *(Sciurus)* of Borneo. Predators learn by trial and error to avoid the tree shrews because of their obnoxious taste. The squirrels are perfectly palatable but look so much like the tree shrews that most predators avoid them as well.

In Müllerian mimicry (named for Fritz Müller, a nineteenth-century German zoologist), a different strategy is involved. Some unpalatable or poisonous species, including various butterflies, wasps, hornets, and certain birds and snakes are thought to have evolved common warning coloration. (See the discussion on convergent evolution in Chapter 17, p. 412.) Combinations of black, white, red, orange, and yellow in bold but simple patterns, as in the poisonous coral snakes, advertise the animal's presence to potential predators (Figure 13.16b). With few signals to recognize, a predator soon learns to recognize and avoid these dangerous animals. (Actually, because no specific model is being mimicked, Müllerian mimicry probably should not be called mimicry; but this name is firmly entrenched in zoology.)

direct attention away from the head and body (Figure 13.15). Moreover, some flying moths flash colorful wing markings that seem to disappear as the moth lands and folds its wings. A bird following the moth may continue to watch for bright markings and overlook the resting moth.

Mimicry is perhaps the most remarkable defensive strategy involving communication between two or more species. Broadly speaking, mimicry refers to the resemblance of an organism to another that is distasteful, a resemblance that confers protection from predators. In **Batesian mimicry** (named in honor of Henry Bates, a nineteenth-

(a)

(b)

Figure 13.15 Predator-foiling devices. (a) The tips of the wings of the Colorado hairstreak butterfly *(Hypaurotis chrysalus),* here shown resting on an oak leaf, mimic the head and antennae. The real head is at the right and is very inconspicuous; a predator might be attracted to the false head and wing tips, rather than to the more vital body parts. (b) When at rest with its head down, *Thecla linus,* a tropical hairstreak, moves its antenna-like wing appendages, helping to attract predators away from its head.

(a)

(b)

Figure 13.16 Mimicry. (a) An example of Batesian mimicry. The harmless, palatable viceroy butterfly *(Limenitis archippus)* (right) gains protection from predatory birds through its resemblance to the distasteful monarch butterfly *(Danaus plexippus)* (left), which is recognized and avoided by predators. (b) Müllerian mimics are thought to have evolved their color patterns through convergence to a common model. Some of the coral snakes are so similar that species can be distinguished visually only by simultaneous comparison. The snakes shown here are *Micrurus nigrocinctus* (top) and *M. fulvius* (bottom).

Social Behavior

The activities of many animals often depend on interactions with others. The resulting **social behavior** is qualitatively different from individual behavior and sometimes so complex that it almost defies analysis. One important difference between social and nonsocial animals is in the use of communication; nonsocial animals may or may not communicate, depending on the species, but communication is an essential ingredient of social behavior. Social relationships are usually between animals of the same species, but there are many exceptions. An aggressive young stallion raiding an aging stallion's harem of mares is displaying a form of *intraspecific* (within a species) social behavior. On the other hand, the relationship between a child and a pet kitten, or between ants and certain aphids, which the ants protect and obtain food from, is *interspecific* (between two species). The term *social* is also used to refer to **social systems,** in which groups of animals live together in some particular spatial arrangement. Social systems are often described by group composition—that is, the numbers, kinds and functions of individuals within a group.

Of the many kinds of social systems that have evolved, the least complicated is the **pair**. A parent-offspring pair, such as a horse and her foal, is a social system held together by relatively straightforward protective and nutritional needs. More complex interactions occur in other pairs, such as those formed by males and females during the breeding season. A social system may change during a species' life history. Animals that form pairs when breeding may aggregate when they are nonreproductive. Thus, many kinds of birds form flocks during the fall and winter, and many fishes and certain aquatic mammals, such as dolphins, school. Cattle, bison, horses, and many other hoofed species form herds, a social grouping that does not break down into pairs during the breeding season (Figure 13.17). Such aggregations of animals may be protective; a small fish in a large school may be safer than one swimming alone. (The adaptive significance of schooling behavior in fishes is discussed in Chapter 29, p. 720.) Birds and mammals in large flocks or herds may be protected from predators by their sheer numbers, as well as by warning signals given by sentinels or other members of the group. A social group is generally more successful than an individual animal in competing with other groups for resources.

Territoriality. That social animals often display territorial behavior was recognized hundreds of

Figure 13.17 Animal social groups. Aggregation into flocks, schools, and herds is a protective adaptation in many species. The bison *(Bison bison)* shown here are descendants of great herds, totaling more than 50 million individuals, that once roamed the western United States and Canada. Indiscriminate slaughter by hunters almost caused their extinction by 1900. Now, a few small herds are carefully managed under governmental protection, and some are raised for meat on ranches and farms.

(a)

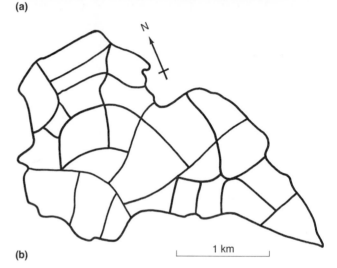

(b)   1 km

Figure 13.18 Territories of the tawny owl *(Strix aluco)* in Britain. (a) Paired for life, a male and female of this European species defend a territory, which is a well-defined, relatively permanent foraging area of woods. Each pair lives on wood mice and other rodents within its territory. (b) This area of habitat is divided into 25 territories, each supporting a pair of owls. The owls' rate of reproduction generally increases or decreases according to the number of rodents available as food. In a year when food is scarce, fewer eggs are laid, and energy is conserved as the owls raise fewer offspring. When abundance returns to the territory, the reproductive rate increases.

years ago, but not until about 1920 did scientists develop a strong interest in this form of behavior. Territorial behavior is now known to occur in many reptiles, mammals, and birds and in a few species of fishes, amphibians, and arthropods (Figure 13.18). No doubt, many more cases await discovery, especially among invertebrates. Broadly defined, an animal's **territory** is any defended area, and **territoriality** is the defense of the territory. Territory defense involves some type of contest, or **agonistic behavior**, between the territory holder and competitors. Often, the contest is a harmless **ritual** in which symbolic acts, such as teeth baring, snarling, or aggressive posturing, take the place of actual combat.

Territories and territoriality differ from species to species. In most species, the male is the defender, while in others, the female defends the territory, and in still others, a male and female defend the territory together. There are even cases in which all the males in a group cooperate in territorial defense, a practice found in some species of monkeys. Still another difference is in the resource being defended. Often, the resource is food, but territories based on protection of a mate, nest, or breeding site are also common. In general, territoriality can evolve only around resources that are defensible. A seal, for example, is highly territorial, but not over food. It cannot defend the great expanse of ocean in which it fishes. Instead, a seal's territory is a breed-

ing site on a secluded, relatively inaccessible beach. Likewise, an insect-eating swallow defends a territory around its nest, but not its large aerial hunting ground. In contrast to the territory, the general, undefended area over which an animal roams is called its **home range**.

Territorial behavior is essentially a strategy of exclusion. It denies an animal's competitors access to breeding and nesting sites and to limited resources such as food. Once territories have been established, they usually reduce the likelihood that time and energy will be wasted in dangerous fighting. In a typical territorial species, the most vigorous individuals successfully hold territory, mate, and transfer genes to the next generation. In effect, territoriality helps channel time, energy, and resources to reproductive activity.

Dominance Hierarchies. Many animals live in social groups organized into **dominance hierarchies** maintained by agonistic behavior. Both males and females may form dominance hierarchies, but only rarely does a single hierarchy include both sexes. In a typical dominance hierarchy, a single animal at the top of the hierarchy dominates all others, while the animal's immediate subordinate dominates all others except the one at the top, and so on. The animal at the bottom of the hierarchy is subordinate to all others. The result is a "peck order," so named because pioneering research was done with chickens, which express dominance by pecking their subordinates.

As shown in Figure 13.19 for the musk ox, males of many species sort themselves into reproductive hierarchies. Likewise, pairs of male bighorn sheep *(Ovis canadensis)* establish a hierarchy by repeatedly rushing full speed at each other and crashing their massive horns together until one sheep gives up or is seriously injured. The victor in a series of combats wins first mating rights to any sexually receptive female, while the second male in the hierarchy mates only when more than one female is

Figure 13.19 Agonistic behavior in a dominance hierarchy. Two musk ox *(Ovibos moschatus)* males engage in a battle consisting of a head-to-head pushing match in which their great horns serve as battering rams. Eventually, one of the combatants abandons the fight, becoming a subordinate in the hierarchy. The victor of a series of such fights dominates the hierarchy and has special breeding privileges.

receptive. Dominance hierarchies have important ecological and reproductive benefits. Once a hierarchy has been established, an animal's status is established, and consequently, less time is spent fighting and more time and energy are used for finding food, winning mates, and rearing offspring. Chances for survival also increase. The more dominant animals can take food with less fighting, while the more subordinate animals benefit by avoiding fights that they would no doubt lose.

In some species, hierarchies and territoriality exist at the same time. For example, bullhead catfishes *(Ictalurus)* may form a territorial hierarchy in which the dominant fish (established by fighting) takes the largest, most favorable area, while subordinates take less favorable areas. If an unknown fish approaches, the subordinates rush into the dominant bullhead's territory for protection. After battling the intruder, the bullhead chases his subordinates back to their own areas. Thus, in the bullhead hierarchy, the dominant member serves as group defender.

Introduction to Sociobiology and the Evolution of Social Behavior

It is necessary here to introduce some evolutionary concepts that are central to understanding all forms of social behavior. As recognized by Charles Darwin, **natural selection**, the driving force of evolution, operates through the reproductive process at the *individual* level. Natural selection provides the basis for the scientific analysis of social interactions. (For more discussion of natural selection, see Chapter 17, p. 402.) Using the logic of natural selection, sociobiologists formulate and test hypotheses to explain how a particular social activity improves an individual's fitness. By **fitness** is meant reproductive success, or the success one individual has relative to others of its species in passing its genes to the next generation. An alternative hypothesis considers the benefits that a group (the social group or the species group) derives from social interaction. It may seem logical that the main reason a social activity has evolved is that it benefits the group; but such "logic" is not consistent with the fact that natural selection operates at the individual level. Despite the superficial attractiveness of the group-selection hypothesis, repeated attempts to explain social behavior this way have not been fruitful, and ethologists generally reject group-selectionist ideas.

In light of evolutionary (individual selection) theory, social behavior may be defined as any activity by an individual (called the actor) that affects another individual (called the recipient). Sociobiologists study the evolution of social acts by measuring their effects on the reproductive success of the actor and the recipient; thus, the number of surviving offspring is a direct measure of reproductive success. Four fundamental categories of social activities may be recognized: **altruism**, in which there is cost (measured as decreased reproductive success) to the actor, and benefit (measured as increased reproductive success) to the recipient; **cooperation**, in which both actor and recipient benefit; **selfishness**, in which the actor benefits at the recipient's expense; and **spitefulness**, which results in detriment to both actor and recipient. Because the concept of altruism is at the core of modern sociobiology and may explain many social phenomena, we discuss it in this brief introduction. Of the other three categories, cooperation involves altruism, and selfishness and spitefulness may be important in many types of social interactions, but to a lesser extent than altruism.

Altruism. Because natural selection acts on individuals, it might seem that altruistic individuals would soon be selected out of a population and that altruism would be a rare type of social act. The opposite seems to be true, however, for altruism seems to be a very common form of social behavior. Individuals making alarm calls, for example, are probably performing altruistic acts. As we will see, the evolution of altruism may help explain some of the most complex forms of social behavior, such as that of the honeybee.

The key to understanding how altruism may result from natural selection is in considering natural selection at its most fundamental level. Ultimately, in selecting individuals, natural selection selects genes, which are passed via an individual's sperm or egg to the next generation. Individuals can be viewed, in this genetic or evolutionary sense, as carriers of genes. By focusing this way on gene selection, sociobiologists obtain a unique view of the cost/benefit analysis of altruism. What are the consequences of an altruistic act viewed from the level of an individual's genes? Suppose an actor performs an altruistic act and the recipient of the act is genetically related to the actor (i.e., the recipient has some of the same genes as the actor). Actually, this is common in nature, for many social groups are composed of closely related kin. An altruistic act controlled by some of the actor's genes may increase the genetically related recipient's chance of reproducing and passing these genes to

the next generation. The chances that altruistic genes will be selected are improved if the recipient and actor are closely related (share many genes). Consequently, viewed from the standpoint of an actor's genes, an altruistic act is potentially beneficial, and it is not surprising that genes for altruistic behavior seem to have been favored by natural selection. Natural selection of altruistic traits by this type of mechanism is called **kin selection**, and many types of social interactions may be explained on this basis.

Kin selection may have been a major force in the evolution of the intricate, tightly knit societies of insects such as honeybees, ants, and termites. In these animals, labor is divided among special kinds, or **castes**, of individuals. A hive of honeybees, for example, includes three castes, determined by genetic, nutritional, and hormonal factors (see Figure 14.13). The single *queen* in a hive is little more than a reproductive machine, laying several thousand eggs per day. By regulating sperm entry into her eggs, she governs the sexual balance of the hive. Female offspring develop from fertilized eggs and males from unfertilized eggs. The *drone* honeybee caste consists entirely of males, all of which lead idle lives. Their only function is to inseminate new queens. On a warm summer afternoon, when a queen makes her single nuptial flight, the drones fly from the hive to a predetermined drone congregation area, where they await the queen's arrival. Only about one male in a thousand actually inseminates a queen. When autumn comes, the drones are forced from the hive and soon die. *Workers*, the third caste, are females whose gonads never

Figure 13.20 The mating system of the sage grouse *(Centrocercus urophasianus).* This species, which inhabits high sagebrush plateaus of the North American Rocky Mountains, has communal display and breeding grounds called leks. Fifty or more males appear at a lek daily from February through May, strutting and displaying their bright feathers. Females come to the leks in April, select a male, copulate (usually only once), and leave. Only about 10% of the males copulate. As part of their courtship display, males form an upright fan with their tail feathers. Here we see a sage grouse male and a circle of hens.

develop. As the name suggests, the worker caste is the hive's labor force. Workers begin their careers as hive cleaners, and then for about three weeks they progress through a series of jobs—caring for larvae, building honeycomb, and guarding the hive from predators and invaders from other hives. If required to sting an intruder in defense of the hive, a worker dies in the act; its stinger remains in its victim, and most of the honeybee's viscera remains attached to the stinger. Surviving workers finally become foragers, seeking pollen and nectar for the hive. After foraging for several weeks, and flying perhaps a thousand miles, an exhausted worker drops to the ground and dies.

The honeybee society seems to rely on an extreme form of altruism. None of the workers reproduce; instead, they sacrifice all of their reproductive potential to the queen. Why would natural selection favor such supreme sacrifice? The answer is no doubt very complex, for the honeybee society has evolved as a result of many selective pressures. A now largely discredited idea that the mutual benefits provided to the group led to the evolution of honeybee and other insect societies is not consistent with the theory of natural selection. Worker sacrifice is, however, consistent with the concept of kin selection. Honeybee workers are sisters; that is, all are daughters of the queen in their hive. Consequently, in performing their work that benefits the queen, the workers are ensuring the passage of many of their genes to future generations.

Evolution of a Mating System. The concept of individual (gene) selection provides a theoretical basis for understanding many social phenomena. As a final note in this chapter, we will look at how this concept applies to the evolution of a complex **mating system**, the social organization of males and females involved in breeding.

During the breeding season, some animals have a social structure based on a combination of dominance and territorial behavior. Males congregate in a small area of the home range and wait there for the arrival of females. The **lek**, as the area is called, is generally open and flat, making it highly visible to females but nearly impossible for a predator to approach unseen. A lek may become a "traditional" area used year after year. Animals as diverse as grouse, certain sunfishes, reptiles, antelopes, and insects form leks. Attracted to the lek by the conspicuous displays of the males, the females generally move toward the lek's center, where the

more dominant males each defend a small territory. Depending on the species, the females may spend several hours or several days on the lek, observing the male displays. Eventually, a female selects a male, copulates, and then leaves the lek to reproduce without further participation by the male. Among sage grouse, which have been extensively studied, females generally select males that are dominant, gregarious, and attractive to other females (Figure 13.20).

How might this type of mating system have evolved? The key element in the system is the process of mate selection by the female. According to sociobiological theory, the lek mating system evolved as a mechanism that ensures that females choose and mate with males with the "best genes." The best genes in this case are those that have the greatest chance of ensuring the survival and future reproduction of the female's offspring. Thus, by choosing the male with the "best genes" and putting her genes with his in a fertilized egg, a female is giving her genes the best chance for future survival.

A Final Note. In closing, we would like to emphasize that sociobiology is relatively new and has a number of opponents as well as advocates. In fact, this is one of the most controversial areas in zoology today. While staunch advocates may hail sociobiology as one of the great revolutions in biological thinking, certain detractors argue that much of it may be an intellectual blind alley. Some of the strongest arguments against sociobiology arise when certain hypotheses are applied to the interpretation of human behavior. Detractors point out that much of human sociality has resulted from human culture, which develops so rapidly as to overwhelm the development of social behavior by the slow process of natural selection. The examples we have just discussed (honeybees and sage grouse leks) are less controversial, but it is important to realize that sociobiological interpretations of these and other behavioral phenomena are not beyond debate. On the other hand, there seems to be little doubt that sociobiology will continue to make enormous contributions to our understanding of animal behavior. In the past few decades, sociobiologists have generated a large body of strong hypotheses, and the field seems to have the momentum and intellectual attractiveness to enlist and hold the attention of an increasing number of biologists.

☐ SUMMARY

1. Ethology, the science of behavior, is concerned with describing behavior under natural conditions and with explaining behavior through studies of its physiological, developmental, and genetic foundations, its ecological significance, and its evolutionary history.

2. Reception of sensory information and its processing and integration by the nervous system—the underlying physiological mechanisms of behavior—constitute the field of study known as neuroethology.

3. Certain types of behavior are preprogrammed as discrete neural circuits (motor programs) in the nervous system. Motor programs develop as the nervous system develops, are inherited along with the animal's other traits, and provide the basis for innate behavior.

4. Motor programs may be relatively simple and inflexible, taking the form of fixed-action patterns, or they may be complex and modifiable through practice and learning. Some motor programs are triggered by a releaser stimulus, often a simple environmental cue "recognized" as a signal by an animal's neural circuits. An animal's responsiveness to a releaser and other stimuli fluctuates and is governed by such factors as hunger, thirst, and hormonal balance.

5. Behavioral patterns typically have both innate and learned components, both of which are influenced by genes. In habituation, the simplest kind of learning, an animal reduces its response to stimuli that are not followed by punishment or reward. Many animals learn to associate two or more stimuli with a particular reward or punishment, a process called associative learning. Latent learning occurs when an animal stores information and then uses it as a guide in later behavior. In insight learning, an animal's brain adapts remembered experiences to the solution of new problems.

6. Many animals respond to environmental stimuli by increasing their rate of random movement or turning, a phenomenon called kinesis. Animals also commonly display taxic (directed) behavior as they orient toward or away from environmental stimuli. Navigation, which is goal-oriented, is the most complex form of orientation.

7. Communication involving visual, auditory, tactile, or chemical signals and messages is a vital aspect of animal behavior. Social behavior involves communication between two or more animals of the same or different species. Social systems vary from pairs to large flocks, schools, and herds. Many social animals exhibit territory defense (territoriality) and/or dominance hierarchy (peck orders), thereby reducing the time and energy spent fighting for mates, food, and other resources and channeling efforts into reproductive activities.

8. Sociobiology, a significant and controversial area of research in ethology, applies the theory of natural (individual) selection toward developing an understanding of the evolution of social behavior.

☐ FURTHER READING

Alcock, J. *Animal Behavior, An Evolutionary Approach.* 3d ed. Sunderland, Mass.: Sinauer Associates, 1984. *A modern textbook covering the physiological, genetic, and evolutionary aspects of behavior.*

Anderson, E. W. *Animals as Navigators.* New York: Van Nostrand Reinhold, 1983. *A well-illustrated, highly readable look at navigational methods in a variety of animals.*

Bright, M. *Animal Language.* Ithaca, N.Y.: Cornell University Press, 1984. *A survey of major discoveries in animal communication compiled from interviews with working scientists.*

Camhi, J. M. *Neuroethology. Nerve Cells and the Natural Behavior of Animals.* Sunderland, Mass.: Sinauer Associates, 1984. *A clearly written introduction to the study of the neural mechanisms of behavior and to research in neuroethology.*

Dawkins, R. *The Selfish Gene.* New York: Oxford University Press, 1976. *A fascinating, nontechnical introduction to modern hypotheses of social evolution.*

Ferry, G. "The circuitry of learning." *New Scientist* 5(1984): 18–20. *An excellent account of ongoing research on relationships between simple learning (habituation and sensitization) and release of neurotransmitters at the synaptic junctions of the sea hare.*

Free, J. B. *The Social Organization of Honey Bees.* London: Edward Arnold, 1977. *An enjoyable, yet detailed account of honeybee behavior, including the language of the dance.*

Gould, J. L., and P. Marler. "Learning by Instinct." *Scientific American* 256(1987): 74–85. *An essay on the close interrelationship between learning and innate behavior.*

Griffin, D. R. *Animal Thinking.* Cambridge, Mass.: Harvard University Press, 1985. *A noted ethologist ponders whether animals have consciousness.*

Herbert, W. "Remembrance of Things Partly." *Science News* 124(1983): 378–381. *A highly readable account of forgetfulness in humans and monkeys and the anatomy of normal memory.*

Krebs, J. R., and N. B. Davies. *An Introduction to Behavioural Ecology.* Sunderland, Mass.: Sinauer Associates, 1981. *An interesting introduction to the ecological significance of behavior and to the study of how behavioral patterns have evolved.*

Trivers, R. *Social Evolution.* Menlo Park, Calif.: Benjamin/Cummings, 1985. *An introductory text on social behavior with background chapters on genetics and evolution.*

Wilson, E. O. *Sociobiology. The New Synthesis.* Cambridge, Mass.: Belknap Press of Harvard University Press, 1975. *A modern classic that set the stage for much of the research and popular interest in the evolution of social behavior; contains many interesting examples of invertebrate and vertebrate social behavior.*

Long-Term Survival and Evolution of Species

The next four chapters examine topics that differ in scope from those of the preceding chapters. In the last unit, we examined the mechanisms that animals use to solve the everyday (short-term) problems of existence, such as obtaining food, exchanging gases, maintaining salt and water balance, moving about, coordinating internal systems, and behaving appropriately in relation to the environment and to other individuals. In Unit II, we examine mechanisms that provide for the long-term survival of species, rather than the short-term survival of individuals; that is, we look beyond everyday problems and focus on mechanisms that provide for the continuity of animal species.

All the mechanisms that an animal brings to bear in solving short-term survival problems not only make its existence possible, but also promote its ability to reproduce. Unlike the physiological and behavioral mechanisms involved in everyday problem solving, however, a species' genes and reproductive mechanisms also participate directly in the evolutionary process. Genes provide the material on which evolution acts, because they are the hereditary entities that are passed from one generation to the next. As a species evolves, changes occur in the frequency and types of its genes, and these changes generally occur as a result of changes in the environment. Reproduction is the mechanism through which evolution operates, and evolution is the mechanism by which long-term change occurs.

"Survival of the fittest," a popular phrase used to describe evolution, actually refers to reproductive success, or, more specifically, to how many copies of an individual's genes are passed on to the next generation. Regardless of how adept an individual animal is at finding food, exchanging gases, escaping danger, and defending its territory, these abilities only count from an evolutionary standpoint if the individual reproduces and, in so doing, passes copies of its genes to the succeeding generation. If it does not produce offspring, its specific combination of genes ceases to exist when the individual dies. As we will see, all aspects of reproduction—how many times the animal reproduces in its lifetime, the number of offspring produced, the way the offspring develop, and the amount of parental care provided for the developing offspring—may affect the success an animal has in perpetuating its genes.

325

The first chapter in this unit examines animal reproduction, and the second describes the nature of genes and how they are transmitted from parents to offspring. The subject of the third chapter in the unit is how new animal bodies develop from directions contained in genes housed in single-celled fertilized eggs. The final chapter in this unit describes the process of evolution and how it occurs by genetic change through reproduction and natural selection.

Reproductive Mechanisms and Strategies

A copulating pair of damselflies. The male, in front, clasps the female behind the head with clasping organs on the tip of his abdomen. The female curls her abdomen forward, inserts the tip into a pouch near the anterior end of the male's abdomen, and receives sperm previously deposited there by the male.

Reproduction, the formation of new individuals that eventually replace their parent(s) in a population, is the only way a species can ensure its survival. Extinction is the result of reproductive failure. Unlike many of the life processes discussed in earlier chapters, reproduction is unique because it is not essential for individual survival. Whereas individual animals cannot survive without mechanisms that provide for nutrition, gas exchange, salt and water balance, and internal regulation, any animal can live its life without reproducing. In fact, the reproductive process may even put individuals at a disadvantage, because energy spent on such activities as dividing or producing buds, locating mates, defending breeding grounds, and producing eggs and sperm must be diverted from other activities that support individual survival.

Many different methods of reproduction have evolved, but all can be grouped into two fundamen-

327

tal types. **Asexual (vegetative) reproduction** is the production of offspring by parent individuals without the involvement of sex cells (sperm and egg). **Sexual reproduction** is offspring production involving **eggs** and **sperm**, which are collectively called **gametes**.

Although sexual and asexual reproduction are very different, both types usually produce more offspring than there are parents, thereby compensating for the failure of many offspring to survive the challenges of the environment. Typically, over a long term, either type of reproduction replaces parents rather than increasing the total number of individuals.

Asexual reproduction is a much simpler process than sexual reproduction. In fact, sexual reproduction requires so many complex and costly anatomical and physiological adaptations that one might wonder why sex ever evolved. As you will see in this chapter, however, sexual reproduction has strong survival value, and nearly all animals that have asexual reproduction also have some form of sex in their life cycle.

ASEXUAL REPRODUCTION

Asexual reproduction occurs when a parent splits, buds, or fragments to form pieces, each of which may develop into a separate animal. Asexually produced offspring, often called **clones**, are genetically identical to their parents and siblings because mitosis (nuclear division that produces identical offspring) is the basis of asexual reproduction. (Cellular reproduction is discussed in Chapter 2, p. 46; Figure 2.25.) Asexual reproduction occurs in several ways and is common among protozoa and many invertebrates, especially sponges, cnidarians, flatworms, annelids, and echinoderms.

Fission

Fission is reproduction by division of a parent body into two or more parts. **Binary fission**, the division of a parent into two more or less equal parts, is common in protozoa. The plane of cell division may be transverse, longitudinal, or asymmetrical, depending on the species. The planarian worm *Dugesia* and several other free-living flatworms reproduce by transverse fission, which is often called binary fission. Actually, however, the parent worm pinches into two unequal (anterior and posterior) parts that develop into two complete individuals (see Figure 22.12a). **Multiple fission**, the di-

vision of a parent into three or more parts that become new individuals, occurs in many protozoa and in several sea anemones and annelids.

Budding

In the form of asexual reproduction known as **budding**, offspring develop as small growths, or buds, from the parent. The buds may separate from the parent and become independent individuals, or they may remain attached and form a **colony**. Buds may form on the outside surface of the parent, as in many sponges and in the cnidarian *Hydra* (Figure 14.1a), or internally as cell aggregates. Internal buds in sponges are called gemmules. (See the discussion of gemmules in Chapter 20, p. 480; Figure 20.12.)

Fragmentation

Through **fragmentation**, many animals regenerate body parts that are lost through injuries, such as those sustained during encounters with predators.

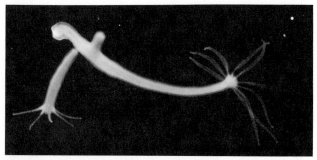

(a)

(b)

Figure 14.1 Examples of asexual reproduction. (a) Budding, as seen in this specimen of *Hydra oxyenida*, involves cells organizing into miniature individuals while still attached to the adult. Buds eventually break off and grow to adult size. (b) Asexual reproduction occurs following fragmentation in some animals. The small sea anemones *(Metridium senile)* in this photograph were produced by the large adult by pedal laceration, a type of fragmentation.

Sea stars and crayfish, for example, can cast off body appendages to escape the grasp of a predator or dispose of a badly damaged appendage. Such self-amputation is called **autotomy**. Crayfish can regenerate lost legs, but they do not reproduce in the process. By contrast, a lost arm of a sea star, if it is attached to a piece of the central disc, may regenerate into a new individual. (Asexual reproduction in echinoderms is discussed in Chapter 27, p. 667.) A similar type of asexual reproduction (fragmentation followed by regeneration) occurs in certain free-living flatworms and annelids (polychaetes) as a parent breaks or is broken into two or more pieces. In sponges and many sea stars, the entire parent body can be fragmented, and many of the pieces will regenerate. More often, small pieces break off from a large parent that remains intact. When food is plentiful, certain sea anemones reproduce by a process called **pedal laceration**. As a parent moves along the substrate, groups of cells break off from the base of the parent and eventually develop into new individuals (Figure 14.1b).

Polyembryony

This asexual process directly follows production of a fertilized egg. **Polyembryony** is the division of a fertilized egg into two or more identical cells that

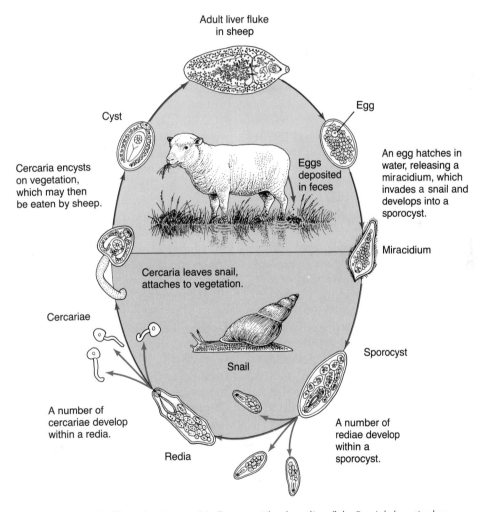

Figure 14.2 The life cycle of a parasitic flatworm. The sheep liver fluke *Fasciola hepatica* has both sexual and asexual reproduction in its life cycle. Fertilized eggs are released with the feces of an infected sheep. A ciliated larva called a miracidium hatches and invades a snail. In the snail, two stages of the parasite (sporocyst and redia) undergo a process of asexual reproduction called larval amplification. Many cercariae leave the snail and encyst on vegetation, forming metacercariae. If eaten by a sheep, a metacercaria will develop into a sexually mature adult. Larval amplification produces many individuals from one zygote and greatly increases the parasite's chances of long-term survival.

subsequently develop into a clone of genetically identical siblings. This process occurs in many species of wasps that lay their eggs in other insects. It also occurs in humans in the production of identical twins, as well as in the production of quadruplets in the armadillo.

Larval Amplification

In **larval amplification**, larval stages (rather than the fertilized egg, as in polyembryony) divide and multiply. This process occurs in the life cycles of many parasitic flatworms; it increases the chances that a single fertilized egg will result in at least one larval worm finding a suitable host (Figure 14.2). (Life cycles of other parasitic flatworms are discussed in Chapter 22, p. 525.)

Advantages and Disadvantages of Asexual Reproduction

Asexual reproduction seems to have several advantages over sexual reproduction. It eliminates the need for males and females to find each other during the sometimes brief period when both are sexually receptive, and it allows animals to direct energy that might otherwise be spent in sexual activities into producing young. As a result, asexually reproducing individuals in favorable environments can form large populations that rapidly exploit localized sources of food or other resources. Among the protozoa, many amoebas and flagellates reproduce only by asexual means, attesting the adaptiveness of asexual reproduction for certain organisms. The great potential for numerical increase by asexual reproduction can be seen in the marine annelid *Ctenodrilus serratus*, found along the California coast and in the West Indies. This species can reproduce generation after generation by transverse fission into six or seven pieces. Hypothetically, in just five generations, a single worm can form a clone of 7,776 to 16,807 individuals.

The potential disadvantage of asexual reproduction is that all offspring are genetically similar. The only source of genetic variability in asexually reproducing animals arises from mutations, and favorable gene mutations occur at very low frequencies. Whereas genetically similar individuals may thrive in a particular environment, if the environment changes to the extent that such individuals are not able to survive, the entire population may be eradicated.

SEXUAL REPRODUCTION

There are many types of sexual reproduction, but all involve the formation of gametes. Gametes are formed by special **germ cells**, which are usually organized into sex organs called **gonads**. Eggs are relatively large, nonmotile gametes that are often laden with energy reserves, called yolk. Sperm are small, motile gametes lacking extensive energy reserves. In any one species, fewer eggs are produced each generation than sperm, yet more energy is spent in egg production because of the energy it takes to produce fats and proteins contained in yolk. Sexual reproduction typically includes **fertilization**, which is the fusion of egg and sperm nuclei to form a **fusion nucleus** within a new cell called the fertilized egg, or **zygote**. The nucleus of a zygote contains twice the number of chromosomes as the gametes, because the zygote's DNA is derived from the combination of an egg nucleus and a sperm nucleus. (See the description of chromosomes in Chapter 2, p. 45.)

Many sexually reproducing species are **dioecious**, meaning they have separate sexes. Male individuals have gonads called **testes**, which produce sperm. Females have **ovaries**, which produce eggs. Ovaries and testes are the **primary sex characteristics** of female and male individuals. In addition, females and males of many species have **secondary sex characteristics**, which are all other features that function in sexual reproduction. In many animals, the gonads produce hormones that influence the development of these features. Secondary sex characteristics include body size, shape, and color, as well as other features that distinguish females from males. Species in which males and females appear different are said to be **sexually dimorphic** (Figure 14.3). Secondary sex characteristics also include various ducts and organs that convey gametes from the gonad to the site of fertilization. Gametes may leave an animal through a separate genital opening, but in many animals, they leave through the **cloacal vent**, a common opening, that also serves the urinary and digestive systems. Species in which eggs are fertilized externally have reproductive systems that simply convey the gametes into the surrounding water where fertilization occurs. Internal fertilization requires more complex anatomy, including special structures involved in transferring sperm from male to female and adaptations of the female system that make eggs available within her body for fertilization.

Sexual reproduction involving the fertilization of gametes in dioecious species is called **biparental reproduction**. Different from dioecious species are those in which each individual has both male and female sex organs. Such species are **monoecious**, and the individuals are called **hermaphrodites**.

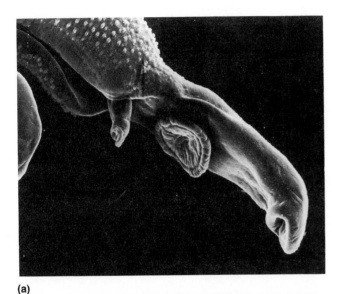

(a)

(b)

Figure 14.3 Sexual dimorphism. Many invertebrates and vertebrates, including humans, deer, and many birds, show obvious external differences (dimorphism) between males and females. Sexual dimorphism often plays a role in the courtship behavior that precedes copulation. (a) This scanning electron micrograph shows the anterior ends of a copulating pair of blood flukes *(Schistosoma mansoni)*. An adult male and a female copulate more or less constantly; the male, larger and broader than the female, has a ventral groove in which he holds the long, slender female. (b) Antlers are important weapons in the fights that determine dominance among male elk *(Cervus elaphus)*. Here, prior to copulation, a bull is stimulating a cow.

Advantages and Disadvantages of Sexual Reproduction

Why is sexual reproduction so common? How does it benefit a species, compared to asexual reproduction? Its chief benefits derive from the genetic variability that it provides each generation. Offspring resulting from zygotes contain a combination of genes from two parents and thus differ from one another and from their parents. The variability produced by such reshuffling of genes each generation makes sexual species adaptable to changing environments. Generally, there is a better chance that some of the variant offspring can continue to function and survive as an environment changes than there is if all offspring are the same. Genetic variability also provides a mechanism for increasing a population's overall fitness. That is, from a population of genetically diverse individuals, those individuals whose gene combinations enhance survival in the environment usually will produce more offspring (and consequently contribute more genes to the next generation) than those having less effective gene combinations. Over time, this mechanism generates populations composed mainly of individuals whose gene combinations provide a high degree of fitness for the environment.

The variability produced by sexual reproduction and the fate of varied offspring can be compared to a lottery where many different numbers are issued but only a few win. This is graphically illustrated in many fishes. A single pair of cod, for example, has the potential to produce approximately 8 million offspring in their lifetime. If the cod population is stable, however—that is, neither increasing nor decreasing in number over time—then on average only two offspring survive to adulthood. This means that 7,999,998 of a pair's offspring die because of predation, disease or lethal gene combinations. As in a lottery, most individuals are unsuccessful; but unlike a lottery, reproductive failure often is not random. Any combination of genes that promotes the survival of individuals increases the chances that offspring with these favorable genes will mature to produce future generations.

A hypothetical example comparing asexual and sexual reproduction will illustrate how gene reshuffling provides a long-term advantage in sexual reproduction (Figure 14.4). In an asexual species, if two beneficial mutations occur in two different individuals such that gene c mutates to C in one and gene d mutates to D in the other, there would be no way that both mutations could come together

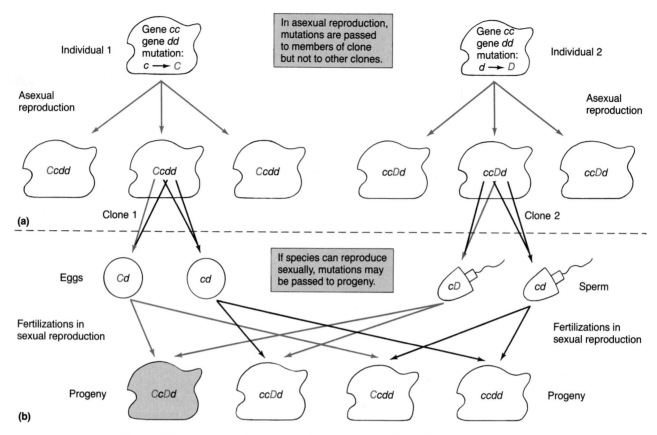

Figure 14.4 Comparison of the genetic effects of asexual and sexual reproduction. (a) When gene mutations occur in asexual species, the mutations are passed on only to members of the clone. (b) In sexually reproducing species, production of gametes by two parents allows mutations that occur in different individuals to come together in a new generation. Although mutations and genetic variability may have positive effects, there are also some potentially negative effects. Most gene mutations are not beneficial. In fact, many produce lethal effects. Also, many mutations are harmless alone, but may be lethal when they occur in combination with certain other mutations. In the case of lethal genes or gene combinations, there may be a greater chance that sexually produced individuals will be harmed by mutations.

in one individual. The CD combination would be found only when a second mutation occurred in one of the first individuals or their progeny. The chance of two favorable mutations occurring in a single individual is extremely slim. Through sexual reproduction, however, the CD combination is possible if the individuals carrying the mutations mate. If the CD combination is beneficial under certain environmental conditions, the combination will spread through the population in future generations as the CD offspring produce more offspring than other members of the population.

Advantages of Having Both Sexual and Asexual Reproduction

Many animals have life cycles that include both sexual and asexual phases, and there are distinct

advantages to this. Typically, in such species, asexual reproduction occurs when environmental conditions are relatively constant and favorable. At such times, asexual cloning allows rapid, energetically cheap population expansion and consequent exploitation of resources. On the other hand, sexual reproduction typically occurs when environmental conditions are deteriorating or are markedly different between two successive generations. The cooler temperatures and shorter day length that signal the onset of winter, or the drying of a seasonal pond, often are correlated with a switch from asexual to sexual reproduction. In such cases, the sexual process produces life cycle stages (usually dormant zygotes) that are resistant to unfavorable conditions.

Having two means of reproduction also may be advantageous for dispersal. Suppose, for example, that a species produces a dispersal stage, such as a zygote or a larva, by sexual means. If the dispersal stage can reproduce asexually, then once established in a new habitat, it can form an entirely new population by itself. When mature, its clones may in turn produce new larvae by sexual means and further extend the range of the species.

Chromosome Number in Gametes and Zygotes

Sexual reproduction requires that gametes have one-half the number of chromosomes found in the zygote. If gametes had the same number of chromosomes as the zygote, gametic fusion during fertilization would double the chromosome number each generation, eventually producing a mass of chromosomes too large to be contained in a single cell. A special type of nuclear division called **meiosis**, which reduces the chromosome number by half, is an integral component of sexual reproduction. Meiosis precedes, or is included in, the formation of gametes. The cells it produces are said to have a single **genome**, which is the **haploid (1n) number** of chromosomes. Gametes have a single genome, and when they fuse, the resulting zygote has two genomes, a **diploid (2n) number** of chromosomes.

In most animals, chromosome number reduction by meiosis occurs in the gonads in cells destined to form gametes. Zygotes develop into multicellular individuals by mitosis and cytokinesis (cell division). As a result, most animal life cycles include an alternation between haploid gametes and diploid individuals. (See Chapter 19 for a discussion of protozoan life cycle patterns, which are more varied than those of animals, p. 461.)

Key Aspects of Meiosis

The two genomes of the zygote and body cells of animals actually consist of pairs of matching chromosomes; one member of a pair is inherited from the female parent and the other is from the male (Figure 14.5). The chromosomes in a pair are said to be **homologous**, meaning that they carry genes for the same traits. One of the pairs of homologous chromosomes in humans, for example, carries the genes determining whether earlobes are attached or unattached, along with thousands of other genes. When meiosis halves the chromosome number, chromosomes from each homologous pair end up in different gametes. In the earlobe example, a gamete would carry only a gene for attached or

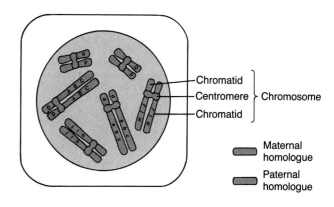

(a)

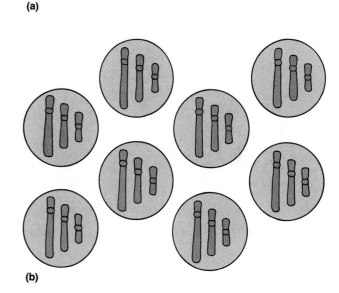

(b)

Figure 14.5 Effect of meiosis on chromosome number. Meiosis halves the number of chromosomes in a cell nucleus and ensures that zygotes receive genes from both parents. (a) This hypothetical cell is diploid (2n); it has three pairs of homologous chromosomes for a total of six chromosomes. Note that each chromosome has been replicated and consists of two sister chromatids joined by a centromere. The different colors of the homologous chromosomes in each pair indicate that they came from different parents (gray from the mother; brown from the father). Genes (represented by letters) that are on the same chromosome are said to be linked. (b) The types of gametes that can result from meiosis in an organism whose chromosome number is six. Note that the gametes are haploid (1n), meaning they contain one genome, or half the diploid number of chromosomes. Genes on different chromosomes assort independently of one another into the gametes. Independent assortment of paternal and maternal chromosomes during meiosis produces 2^n different gametes, where n is the organism's haploid chromosome number. In this example, $n = 3$; therefore, there are $2^3 = 8$ possible combinations of parental chromosomes and 8 possible kinds of gametes. In humans, $n = 23$.

unattached lobes; it would not carry genes for both. Consequently, children inherit one gene for the earlobe trait from each parent. In this way, meiosis ensures that zygotes contain genes for each trait from both parents. This mechanism for the mixing of genes from two parents produces much of the variation characteristic of sexual reproduction.

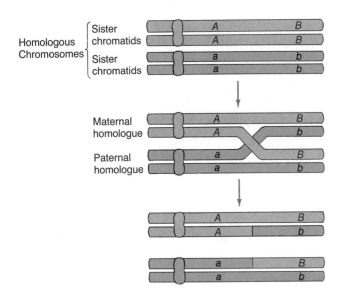

Homologous Chromosomes { Sister chromatids / Sister chromatids }

Maternal homologue

Paternal homologue

Figure 14.6 Crossing over. This process, which occurs during meiosis, produces chromosomes that differ from those of the parent cells. As crossing over occurs, homologous chromosomes reciprocally exchange segments of chromatids. This produces new combinations of linked genes (represented by the letters *A, b,* etc.) and new combinations of maternal and paternal genes (indicated by color). For simplicity in this illustration, only one pair of chromatids is shown crossing over, although several crossovers usually occur on each chromosome pair. In this way, many more different kinds of offspring can be produced than if crossing over did not occur.

Meiosis can also produce new combinations of genes. Genes that reside on the same chromosome are said to be **linked**. As a result, one might assume that all genes on any one chromosome would be passed as a group to a gamete. This would be true if it were not for another important aspect of meiosis. During the early stages of meiosis, homologous chromosomes pair with their partners and reciprocally exchange pieces of DNA in a process called **chromosomal crossing over**. During this process, the **sister chromatids** (halves of chromosomes held together by DNA and protein fibers at a centromere) break, and reciprocal pieces on homologous chromosomes rejoin to form different chromatids (Figure 14.6). Because the resulting chromatids have new combinations of genes (some genes from each parent), the rejoining process is called **genetic recombination**. Crossing over and genetic recombination augment the variability produced by the combining of genes from two parents during fertilization.

Similarities Between Meiosis and Mitosis

In halving the chromosome number and producing genetic variety in offspring, meiosis is a unique process; yet, it resembles mitosis in several ways.

(Mitosis is discussed in Chapter 2, p. 46; Figure 2.25.) Both processes are preceded by replication of DNA in the chromosomes. As a result, before a nucleus undergoes mitosis or meiosis, it contains twice the diploid amount of DNA, and each of its chromosomes consists of two identical sister chromatids. Also, immediately following meiosis or mitosis, and before DNA replication, the chromosomes consist of one chromatid. As either mitosis or meiosis commences, the chromosomes coil tightly and change from predominantly extended, thin threads to condensed, rodlike structures. Following this condensation process, the nuclear envelope disintegrates, and a spindle forms in the cytoplasm. Several spindle microtubules attach to the centromere of each chromosome as the chromosomes line up at the center of the spindle. However, the way that the chromosomes first become aligned at the spindle, as described in the following section, constitutes a major difference between mitosis and meiosis.

The Meiotic Process

For convenience, biologists divide meiosis into four stages—prophase, metaphase, anaphase, and telophase—which are named, as in mitosis, according to the structure and position of the chromosomes in the dividing cell. Unlike mitosis, which involves only one division and produces two cells (each with the same number of chromosomes as the parent cell), meiosis involves two sequential divisions, each consisting of the four stages, and can yield four haploid cells (Figure 14.7).

The first stage of meiosis is called **prophase I**. During this stage, the chromosomes condense and homologous chromosomes pair. As pairing occurs, the sister chromatids in each chromosome and the homologous chromosomes press closely together along their entire length in a process called **synapsis**. This mass of four tightly packed chromatids is called a **tetrad**, and it is during this stage that crossing over and genetic recombination occur. Following crossing over, the homologous chromosomes in each tetrad partially separate from each other, except at points called **chiasmata**, where crossing over has occurred. In some species, the division process may now be interrupted as long loops of DNA unravel from the chromosome. Genes in these DNA loops direct the synthesis of proteins necessary for further egg or sperm development (Figure 14.8). The duration of this interruption varies from species to species. It generally lasts a few days in sperm production and often much longer in

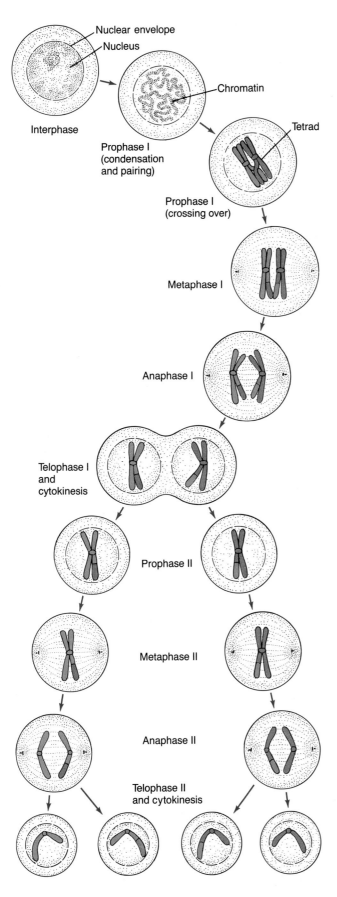

Interphase

Nuclear envelope
Nucleus

Prophase I
(condensation
and pairing)

Chromatin

Tetrad

Prophase I
(crossing over)

Metaphase I

Anaphase I

Telophase I
and
cytokinesis

Prophase II

Metaphase II

Anaphase II

Telophase II
and cytokinesis

Figure 14.7 (Left) The stages of meiosis. In this example, only one pair of homologous chromosomes is shown. Meiosis involves two divisions, each of which consists of four stages: prophase, metaphase, anaphase, and telophase. Prophase I is complicated and lengthy, in contrast to prophase II, which is short and fundamentally the same as in mitotic division. During meiosis I, homologous chromosomes first pair in prophase I; crossing over and genetic recombination occur in this phase. Homologous pairs then separate from each other into different daughter cells. During meiosis II, the two chromatids making up each chromosome separate from each other. The net result is that cells are produced having half the number of chromosomes of the original cell, and each chromosome in the gametes contains a mix of genes from both parents. In the production of sperm (spermatogenesis), each of the newly formed cells will become a sperm; in oogenesis only one egg results from the process (see Figure 14.12).

the formation of eggs. In the formation of frog eggs, the interruption may last several weeks. In the ovaries of a newborn human female, every egg is in the interrupted stage of prophase I and remains so until hormonally stimulated to resume development. Since no eggs are produced until puberty, and thereafter are produced (ovulated) at the rate of about one a month until menopause, this stage can last from about 13 years in some egg-forming cells (oocytes) to 50 or more years in others. Prophase I ends when the two chromatids of each chromosome condense into compact rods and the nuclear envelope breaks down.

Metaphase I occurs as the tetrads line up at the center of the spindle. This alignment of tetrads (rather than the single-file alignment of chromosomes that occurs in mitosis; see Figure 2.25) leads to the reduction in chromosome number. During metaphase I, spindle fibers from opposite poles of the cell attach to one or the other chromosome in each tetrad, but not to both (see Figure 14.7). At this stage in meiosis, the chromosomes are poised

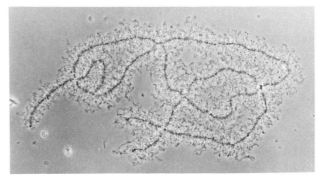

Figure 14.8 Lampbrush chromosomes. A pause in meiosis occurs at the end of prophase I, when the chromosomes uncoil slightly and send out loops of DNA that are active in RNA synthesis. (These chromosomes are called lampbrush chromosomes because of their similarity in appearance to the brushes used to clean the chimneys of kerosene lamps.) The chromosomes shown here are from the oocyte of a newt, *Notophthalmus viridescens* (240 ×), but similar types of chromosomes are found in many other animal species.

for the division that will yield haploid cells.

During **anaphase I**, the two homologous chromosomes in each tetrad separate from each other as they are pulled to opposite poles of the cell by the spindle microtubules. As a result, each pole will receive the haploid number of chromosomes.

Telophase I starts as the chromosomes arrive at opposite poles of the cell. Cytokinesis takes place, nuclear membranes re-form around the chromosomes, and two daughter cells are eventually formed. Figure 14.7 shows three important things about the chromosomes in each daughter cell. First, the number of chromosomes is half that in the parent cell. Second, each chromosome still consists of two sister chromatids. (Following telophase of mitosis, each daughter chromosome consists of a single chromatid.) Third, because of crossing over and genetic recombination, the chromatids in each chromosome consist of portions derived from both parents. A fourth point, not shown in Figure 14.7, is that each daughter cell contains a mixture of maternally and paternally derived chromosomes.

Just after telophase I or after a short interphase, during which the chromosomes may decondense and become somewhat threadlike, the nucleus in one or both daughter cells formed in meiosis I enters the second division of meiosis. (Both cells are involved in sperm formation; only one is involved in egg formation.) Meiosis II is mitotic, although it is not preceded by DNA replication. In **prophase II**, the nuclear envelope disappears, the spindle re-forms, and the chromosomes reattach to new spindle microtubules. **Metaphase II** starts when the chromosomes move to the center of this new spindle. As in mitosis, the chromosomes, each consisting of two sister chromatids, line up in single file. Metaphase II ends when the sister chromatids of each chromosome separate (see Figure 14.7). During **anaphase II**, the sister chromatids are pulled to opposite poles by the spindle microtubules, and each pole receives the haploid number of chromatids (now called **daughter chromosomes**). In **telophase II**, nuclear envelopes form around the daughter chromosomes, and cytokinesis occurs. If both cells formed from meiosis I complete meiosis II, as they do in sperm formation, four haploid cells are produced from the original diploid cell.

Formation of Gametes

The overall process of gamete formation, of which meiosis is a part, is called **gametogenesis**. Following chromosome reduction by meiosis, the resulting haploid cells differentiate into functional gametes. Formation of sperm, which occurs in the testis, is a specific type of gametogenesis called **spermatogenesis**. Egg formation, or **oogenesis**, occurs in the ovary.

Male Reproductive Systems and Spermatogenesis. In many invertebrates, the testes are simple sacs derived from the coelom. In most insects and vertebrates, however, the testis consists of a cluster of coiled **seminiferous tubules**. Figure 14.9 shows the anatomy of the testis and associated secondary sex organs in the human. Sperm are produced in the seminiferous tubules and then pass into highly coiled tubules of the **epididymis**, where they are stored temporarily. In humans and many other mammals, the two testes and epididymides are enclosed in a sac called the **scrotum**, which is suspended outside the main part of the body. Because the testes are housed in the scrotum, sperm-forming cells can be maintained at temperatures slightly below (as much as 5°C) that of the trunk of the body. At least in some species, including humans, sperm will not mature at the higher temperature. Some mammals (not including humans) have an opening between the abdominal cavity and the cavity of the scrotum, and during nonreproductive periods, they may draw the inactive testes up into the body cavity.

From the epididymis, sperm are collected by the **ductus (vas) deferens** and conveyed to the **urethra**, which passes through the penis. The urethra serves both the urinary and reproductive systems. It conveys urine from the bladder to the outside; or, when a male is sexually aroused and the penis is erect, **semen**, consisting of sperm and glandular secretions, can be ejaculated through the urethra into the vagina of the female. Three types of glands associated with the male reproductive tract secrete nearly all of the volume of the semen. Secretions from the paired **seminal vesicles** (which serve mainly in sperm storage in most animals), **prostate gland**, and **bulbourethral glands** (see Figure 14.9) supply nutrients to the sperm, lubricate the tubules for sperm passage, buffer residual acidity from the urine in the urethra, and contain prostaglandins that may stimulate vaginal and uterine contractions that help propel the sperm toward the uterus.

Figure 14.9 also shows the process of spermatogenesis in the seminiferous tubules. The sperm-forming cells in the walls of the tubules are progressively more differentiated from the outside to the inside. Relatively undifferentiated cells called **spermatogonia** line the rim of the tubule. These

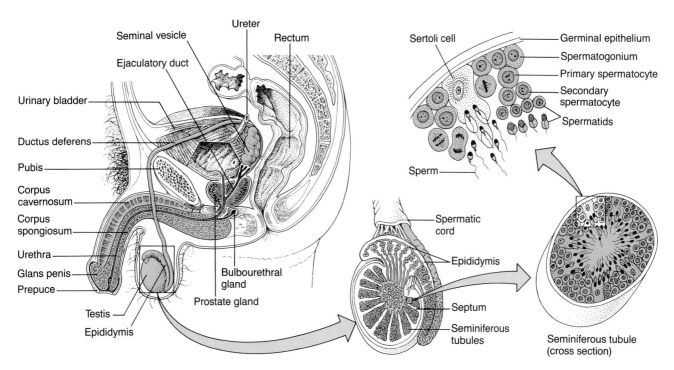

Figure 14.9 Anatomy of the human male reproductive system. A drawing of a section through the pelvic region, viewed from the left side. Note that the spermatic cord, consisting of the ductus (vas) deferens with nerves and blood vessels, loops upward into the body cavity from the testis to join the urethra. When men elect to have a vasectomy for permanent contraception, a physician cuts and ties off the vas deferens in the scrotum. This prevents sperm from entering the urethra. The cross section through a seminiferous tubule in the testis shows the location of spermatogonia and spermatocytes in the wall of the tubule; the sperm heads are embedded in the wall of the tubule during maturation. When maturation is complete, sperm pass to the epididymis and eventually to the outside through the ductus deferens and urethra. As discussed in Chapter 11, Leydig cells, located in the connective tissue among the seminiferous tubules, produce testosterone, which influences the development of secondary sex characteristics in the male (see Figure 11.16).

multiply by mitosis and become **spermatocytes**, which undergo meiosis. A **primary spermatocyte** undergoes the first division of meiosis to yield two **secondary spermatocytes**. Each secondary spermatocyte has a haploid number of chromosomes (each consisting of two chromatids). Each secondary spermatocyte then undergoes meiosis II, yielding four **spermatids**. Each spermatid also has a haploid number of chromosomes (each consisting of one chromatid). The spermatids attach to **Sertoli cells**, also found in the wall of the seminiferous tubules. The Sertoli cells nourish and regulate the differentiation of the spermatids into mature sperm. In a human, the development of mature sperm from a spermatogonium takes about 74 days.

The mature sperm of most animals are motile, flagellated cells, although roundworms and crustaceans have amoeboid sperm (Figure 14.10). Sperm shape develops after meiosis and involves changes in the cytoplasm and nucleus of the spermatids. Each flagellated sperm has a head containing a haploid nucleus. A fluid-filled, caplike **acrosomal vesicle**, at the tip of the head, contains enzymes that dissolve the protective coatings around eggs. Behind the head of the sperm, a midpiece contains mitochondria that supply energy to move the flagellum. A pair of centrioles are located at the base of the tail. Once released from the male, sperm are usually short-lived. Females of some species, however, have a special organ called the **seminal receptacle** or **spermatheca**, which stores sperm and releases them as eggs pass down the oviducts. Queen bees, for example, mate only once in their life, and they may store sperm and remain fertile for up to seven years.

Many animals release sperm in semen, but in a few vertebrates and in many invertebrates, sperm are packaged into fibrous capsules called **spermatophores** before they are released by the male. A single spermatophore may contain anywhere from hundreds of thousands to up to a billion sperm in a gelatinous mass. Spermatophores protect the

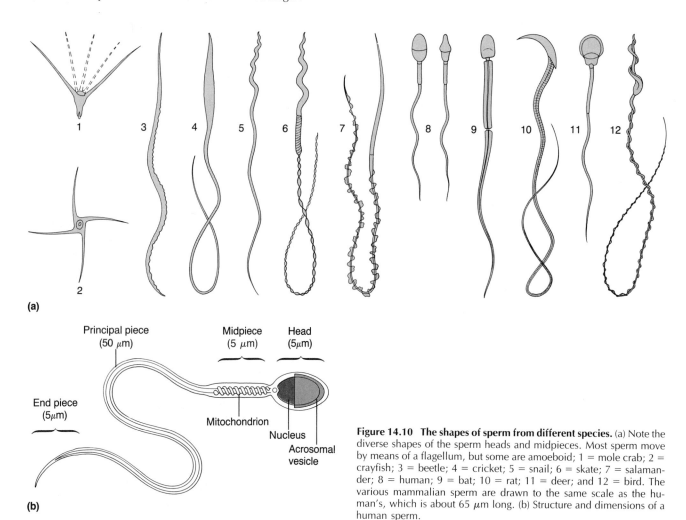

(a)

(b)

Figure 14.10   The shapes of sperm from different species. (a) Note the diverse shapes of the sperm heads and midpieces. Most sperm move by means of a flagellum, but some are amoeboid; 1 = mole crab; 2 = crayfish; 3 = beetle; 4 = cricket; 5 = snail; 6 = skate; 7 = salamander; 8 = human; 9 = bat; 10 = rat; 11 = deer; and 12 = bird. The various mammalian sperm are drawn to the same scale as the human's, which is about 65 μm long. (b) Structure and dimensions of a human sperm.

sperm from drying and assist in transferring a large number of sperm to the female. Male squids place spermatophores in the body of the female during copulation. In many other animals, the male lays spermatophores on the substrate. Airborne chemicals (pheromones) given off by the spermatophores may attract a female and may also act as sex repellents that discourage other males from copulating with females that have picked up the spermatophores. In many rodents, bats, snakes, and insects, gelatinous materials in the spermatophores or semen form a copulatory plug that physically blocks other males from copulating with a female and thus ensures the reproductive success of the first male.

Female Reproductive Systems and Oogenesis.
Ovaries are of two general types. Many invertebrates and bony fishes have tubular ovaries in which cells in the ovary wall produce eggs (ova). The pair of tubular ovaries in a single female

roundworm may produce 200,000 eggs per day. Other animals have more or less solid ovaries in which egg-forming cells reside in small cavities, or **follicles** (Figure 14.11). In vertebrates (except bony fishes), eggs (actually oocytes) are released when follicles mature and rupture through the ovarian surface. This process is called **ovulation**. Ovulated eggs enter the **oviducts** (called **Fallopian tubes** in mammals) and are conveyed to the **uterus** by cilia on cells lining the oviduct (see Figure 14.11).

Many aquatic animals, including most fishes, amphibians, and many invertebrates, have external fertilization. Eggs pass out of the female through a genital opening, and the male covers them with sperm. However, terrestrial animals require a more elaborate procedure to keep their gametes and zygotes from drying out; as a general rule, they copulate and have internal fertilization. Eggs are usually fertilized as they descend through the oviducts. Among the terrestrial vertebrates, birds and most reptiles produce zygotes that de-

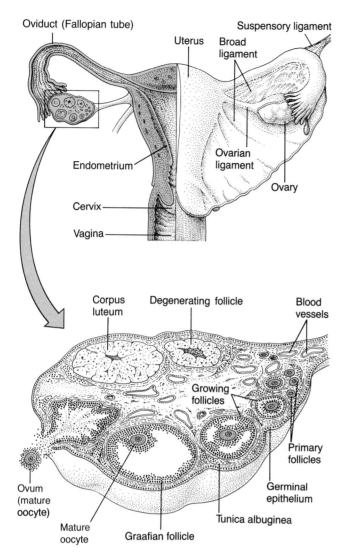

Figure 14.11 Anatomy of the human female reproductive system. The ovaries are paired, almond-sized organs suspended by ligaments in the lower body cavity. Within the ovary, follicle cells surround developing oocytes. At birth, the average human ovary contains about 400,000 primary follicles, each containing a primary oocyte that has entered prophase I of meiosis. Meiosis is arrested at this point until puberty. Usually only one primary follicle develops per month; consequently, only about 400 will develop during a woman's reproductive years. The rest degenerate before or at menopause. As a follicle develops, it enlarges and eventually ruptures, releasing an egg at the surface of the ovary in the process known as ovulation. The follicle then becomes a corpus luteum that secretes reproductive hormones until it disintegrates. Eggs are usually fertilized in the oviducts and pass to the uterus where they embed in the uterine endometrium for completion of development.

velop within shelled eggs outside the mother's body. In both terrestrial and aquatic mammals, on the other hand, fertilization and development take place internally. The mammalian uterus has thick muscular walls containing many blood vessels and a specialized lining, the **endometrium** (see Figure 14.11). In most mammals, zygotes attach to the endometrium some time after entering the uterus. The site of attachment develops into the **placenta**, an organ that allows nutrients to diffuse from the mother's capillaries to the embryo. The hormonal regulation of pregnancy is discussed in Chapter 11 (see p. 264 and Figure 11.15), and reproductive organs and patterns are described in more detail in Chapters 20 through 33.

Oogenesis differs from spermatogenesis in several ways, as we can see by looking at the process in vertebrates. The vertebrate ovary is covered by a thin sheet of tissue called **germinal epithelium**, which surrounds a layer of connective tissue called the **tunica albuginea** (see Figure 14.11). Egg-forming cells in the ovary are called **oogonia**. In some vertebrates, including humans, all oogonia are present in the ovary at birth; but in other vertebrates, such as many fishes and amphibians, which produce great numbers of eggs, oogonia develop in the mature animal by mitosis from germ cells of the germinal epithelium or tunica albuginea. Each oogonium becomes surrounded by a layer of supporting cells; together, the oogonium and supporting cells constitute a **primary follicle**. Under hormonal influence, primary follicles grow as the supporting cells form several layers and the oogonia enlarge to become **primary oocytes**. A primary oocyte undergoes meiosis I. During telophase I, the meiotic spindle is displaced from the center to the periphery of the oocyte. As a result, the plane of cleavage divides the oocyte into one large cell, called the **secondary oocyte**, and one very small nonfunctional cell, called a **polar body** (Figure 14.12). In most vertebrates, the secondary oocyte enters meiosis II, but the process stops in metaphase II. (The polar body may also undergo meiosis II, as indicated in Figure 14.12, but more often it disintegrates.) In response to hormonal stimulation, the follicle containing the secondary oocyte enlarges to form a mature follicle, called the **Graafian follicle**. At this stage, the secondary oocyte (containing the haploid number of chromosomes, each with two chromatids) is also called a mature oocyte. The mature oocyte (usually called an egg, or ovum) is ovulated as the Graafian follicle ruptures at the surface of the ovary. In most species, the oocyte remains in metaphase II until it is penetrated by a sperm cell. After penetration, the mature oocyte undergoes the final stages of meiosis II.

The hormonal regulation of follicle maturation and ovulation in the estrous and menstrual cycles of mammals is described in detail in Chapter 11 (see p. 265). In some mammals, eggs are ovulated

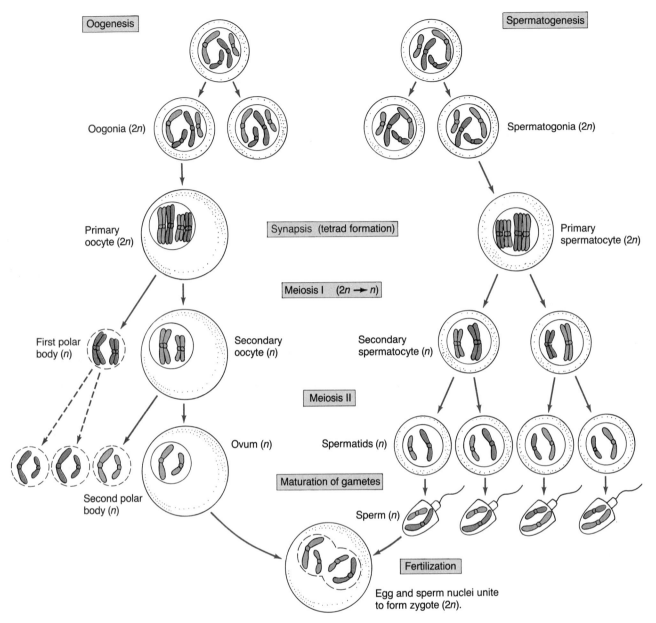

Figure 14.12 Comparison of oogenesis and spermatogenesis. Note how a single ovum is produced from one oocyte in the process of oogenesis, whereas four sperm are produced from one primary spermatocyte during spermatogenesis. Small, nonfunctional polar bodies are produced as the spindle is displaced to the periphery of the oocyte during meiosis. In humans, meiosis II, including the production of the second polar body, occurs after fertilization.

at regular intervals, whereas in others, release is under nervous control and may be stimulated by pheromones or by the appropriate sequence of courtship behavior patterns. In rabbits, cats, and ferrets, ovulation is stimulated by copulation.

An unanswered question concerning oogenesis in many vertebrates is what causes some oocytes to develop into eggs while others remain quiescent.

Why, as discussed in Chapter 11, do follicle-stimulating hormone (FSH) and luteinizing hormone (LH) from the pituitary gland cause only certain follicles to develop and mature, while most follicles remain inactive? It is known that hormonal control of follicles depends on the follicle cells' having specific hormone receptor sites. It is possible that these receptor sites become active only on those follicle cells that will develop. But even if this is true, what causes only certain receptor sites to become active at any one time? The answer is not yet known.

Eggs and Yolk. The eggs of most species contain **yolk**, which consists of substances (phosphoproteins, lipids, and glycogen) that supply energy and building materials for the developing embryo. In many species, yolk is not synthesized by the eggs. In fruit flies, for instance, it is made by nurse cells associated with the eggs in the ovary and passes directly through cytoplasmic bridges into the egg cytoplasm. Many other animals have a gland called the **vitellarium** that secretes yolk into the female's oviducts, and the eggs take up yolk as they pass along the oviducts. In vertebrates, a large molecular precursor of yolk called **vitellogenin** is synthesized by liver cells and secreted into the circulatory system. Vitellogenin travels via the blood to the ovary, where oocytes accumulate it and transform it into yolk. At the time that lampbrush chromosomes appear toward the end of prophase I of meiosis (see Figure 14.8), the oocyte undergoes tremendous growth. It is during this period that vitellogenin is synthesized in the liver and accumulated by the oocyte. At the same time, ribosomes, RNA, and proteins are synthesized and stored in the oocyte for use after fertilization.

Although producing only one gamete from meiosis may seem wasteful, if four eggs came from every cell undergoing meiosis, the amount of nutrient per cell might be insufficient to support embryonic development. With all yolk and other maternally contributed substances (ribosomes, RNA, and proteins) packaged into one cell, the embryo is ensured an adequate supply of nutrients.

Variant Forms of Sexual Reproduction

Several types of reproductive processes that involve sexual phenomena have evolved in addition to biparental reproduction.

Parthenogenesis. **Parthenogenesis**, the development to maturity of unfertilized eggs, has features that resemble sexual as well as asexual reproduction. Because parthenogenesis involves only eggs, some biologists call it "unisexual" reproduction. Others consider it a special type of asexual reproduction because it does not involve fertilization. We include parthenogenesis under sexual reproduction because the process often (though not always) involves the meiotic formation and development of eggs.

Parthenogenesis may have evolved from sexual processes that involved fertilization. Many invertebrates, including several species of flatworms, nematodes, crustaceans, insects, and annelids, and some vertebrates can reproduce parthenogenetically. Parthenogenesis has the advantage of saving energy and time by producing offspring without mating. Nonetheless, most parthenogenetic species also include sexual reproduction with fertilization at some point in their life cycle. Many animals reproduce parthenogenetically when environmental conditions are favorable and then reproduce sexually with the onset of drought or freezing temperatures. The resulting zygotes become dormant and remain so until favorable conditions return. (See the discussion of the life cycle of a rotifer in Chapter 23, p. 547; Figure 23.10.)

Parthenogenesis occurs in several ways because parthenogenetic eggs may be produced either mitotically or meiotically, and they may be either haploid or diploid. Whenever parthenogenesis involves meiotically produced eggs, genetic recombination occurs through crossing over. However, variability among the offspring is not as great as in sexual reproduction involving fertilization because only one parent, rather than two, contributes chromosomes to the offspring. Parthenogenesis that does not involve meiosis is clearly an asexual mechanism; in this process, the eggs develop into females that are all members of a clone.

In many social insects, including ants and certain wasps and bees, zygotes develop into females.

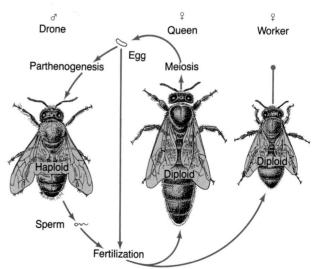

Figure 14.13 Parthenogenesis. In a honeybee *(Apis mellifera)* colony, workers and the queen develop from fertilized eggs and are diploid; males (drones) develop parthenogenetically from unfertilized eggs.

Unfertilized haploid eggs develop parthenogenetically into haploid males. In a honeybee colony, for example, the single queen (which produces all other individuals in the colony) and the many workers (which are sterile females) are diploid. The males, called drones, are haploid (Figure 14.13).

Parthenogenesis is much less common among vertebrates than among invertebrates, yet it is known to occur in one or more species in all vertebrate classes. Among the fishes, the live-bearing Amazon molly *(Poecilia formosa)* produces haploid eggs by meiosis. The haploid nucleus in an egg then replicates its hereditary material and divides, and the resulting two nuclei then fuse, making the egg diploid. The female is inseminated by a male, but sperm chromosomes do not contribute to the egg. As sperm bind to the egg, the egg is stimulated to develop, but the sperm nucleus does not enter the egg, and all chromosomes come from the female.

Hermaphroditism. Many sexually reproducing species are monoecious, with reproductive individuals (hermaphrodites) having both testes and ovaries. This phenomenon is called **hermaphroditism** or **monoecism**. Most species of flatworms are monoecious, as are several species of sponges, cnidarians, annelids, molluscs, arthropods, echinoderms, and chordates. Hermaphrodites also occasionally occur as abnormal individuals in dioecious species. Typically, hermaphrodites are incapable of self-fertilization because their anatomy prevents it, and copulation between two hermaphrodites usually must occur. Usually during copulation, both individuals release sperm, and eggs from both sexual partners are fertilized by sperm from the other partner. (Hermaphroditic land snails are discussed in Chapter 24, p. 578, Figure 24.6a; and copulation in earthworms is covered in Chapter 25, p. 600, Figure 25.13b.)

In many monoecious species, self-fertilization is prevented because individuals produce eggs and sperm at different times. Individuals undergo what is called **sex reversal**, being one sex early in the life cycle and the other sex later. There are two types of this **sequential hermaphroditism**: Hermaphrodites are either **protogynous** (first female) or **protandrous** (first male). Certain species of coral reef fishes called wrasses are protogynous. A single male lives with a harem of up to ten females. When the male dies or is removed, the largest female in the harem changes sex and takes over the group as a male. If a large female is transferred to a container with a small male, she still changes sex, indicating that size rather than gender of compatriots

influences the sex change. In nature, if a harem increases in size so that there are more females than can be held by a male, the largest female changes sex and starts a harem by leading some females away.

Simple logic allows us to predict which sex will occur first in sequential hermaphroditism. Because larger individuals are usually older, if larger size confers a reproductive advantage on one sex, the oldest individuals most likely will be of that sex. This is called the size advantage hypothesis. One would predict, for example, that wrasses are protogynous because large size gives males the advantage of being more effective in defending harems against intruders. On the other hand, if larger females produce more eggs, but large size does not give males any advantage, one would predict that older individuals would be females (in other words, the hermaphrodites would be protandrous). Protandry is common in bivalve molluscs, such as oysters. Because oysters are sedentary and do not defend territories or harems, size is not important in interacting with other members of the population. The size advantage in these animals is related to the energetically expensive task of egg production. Large individuals can produce more eggs than small individuals. Therefore, the probability of reproductive success is increased in oysters when small, young individuals are males and large ones are females.

REPRODUCTIVE STRATEGIES

Zoologists use the term *strategy* to describe various physiological, behavioral, and reproductive mechanisms that allow animals to survive and pass their genes on to the next generation. Use of the term is never meant to imply that animals have conscious strategies to solve problems. As discussed earlier, sexual reproduction has several potential advantages, but it also poses several basic problems that animals must solve. For example, both sexes must come into breeding condition at the same time, and the sexes must find each other and mate. Sexually produced offspring may also pose special problems. Typically, zygotes and developing young are helpless and require special care. In this section, we will look at strategies that have evolved for solving some of the problems inherent in sexual reproduction.

Reproductive Synchronization

In most sexually reproducing animals, neuroendocrine mechanisms synchronize gamete maturation and regulate mating behavior. Whereas animals living in nonseasonal environments such as tropical rain forests may remain in breeding condition throughout the year, it is not efficient for animals in seasonal environments to produce sperm and eggs continuously. Species in these climates produce young when environmental conditions most favor survival of the offspring. Cues for reproductive timing often come from one or more cyclic environmental changes, such as temperature, length of day (photoperiod), and availability of water and food. Photoperiod may trigger hormonal changes that influence the reproductive state of mature animals. (See the discussion in Chapter 11 about the hypothalamus-pituitary control of gonads, p. 251.) The piglike peccaries *(Tayassu)* breed all year round in arid regions and forests of South America, but produce the most young during the rainy season, when there is ample water for milk production. Amphibian breeding cycles are often keyed to warming temperatures and the availability of water in temporary ponds where they lay eggs. Reproduction in eider ducks *(Somateria)* that nest on arctic islands begins after ice linking the islands to the mainland melts. When isolated from mainland predators (mainly foxes), the birds can build their ground nests and lay eggs with less chance of being disturbed.

Moon cycles and tides may also provide reproductive cues in some species. The palolo worm, a bottom-dwelling annelid found on South Pacific coral reefs, has a precisely timed reproductive cycle (see Figure 25.12). Populations near Samoa breed on one night during a certain phase of the moon in late October or early November. On that night, the posterior half of the worm breaks loose from the rest of the body and writhes to the surface of the ocean. There the segments burst open and shed gametes into the sea, where mass fertilization occurs. If gamete release were not closely timed, the population might fail in its reproductive effort, because random release of segments would provide an insufficient number of gametes at any one time to ensure fertilization.

The reproductive cycle of the grunion, a small fish of California coastal waters, illustrates the importance of precise timing in the hatching process (Figure 14.14). The grunion spawns at night during periods of peak tides in the spring. Males and females are carried up on beaches by waves at

Figure 14.14   Breeding of the California grunion. This species, *Leuresthes tenuis,* breeds in wet sand during high tides. Mating and the early development of offspring depend on close synchrony with Pacific tides.

high tide. They deposit their eggs and sperm in holes in wet sand in positions that will not be washed out until the next peak tide reaches them, about two weeks later. After the two-week period, the eggs will hatch almost immediately when washed out of the sand, and the young grunions are carried out on the high tide.

Mate-Finding Strategies

Mating usually does not occur in animals with external fertilization. Adults simply release their gametes into the water at certain times, and fertilization occurs randomly. Exceptions include frogs, most of which have external fertilization involving male and female union, a process called **amplexus**. A male clings to a female, facilitates egg laying by applying pressure to her body, and covers the eggs with sperm as they emerge. (Amplexus in amphibians is discussed in Chapter 30, p. 742; Figure 30.14.) Species that mate usually have complex sensory and behavioral adaptations for finding mates, often over considerable distances.

Some animals congregate on or guard areas where there is a resource vital to members of the opposite sex. Dung flies lay eggs in fresh piles of cow manure. Male dung flies gather on fresh piles on the downwind side, where they are likely to encounter females that are following odors upwind to the dung. Males of certain nectar-feeding insects patrol areas near flowers and wait the arrival of females. Other animals produce visual, auditory, or chemical signals (pheromones) that attract members of the opposite sex.

Once potential mates, either males and females or hermaphrodites, have located each other, they must perform the specific mating activities of their species. Most vertebrates and many invertebrates perform specific **courtship rituals** that serve a variety of reproductive functions. Such rituals are species-specific and required prerequisites to mating. As a result, mating usually does not occur between individuals of different species, even if such individuals are capable of copulating. For many species, such as praying mantises, octopuses, and bears, physical contact between individuals is tolerated only at mating time. Courtship rituals in these species counteract aggressive responses from a sexual partner. Courtship behavior by males may also stimulate females to ovulate. In certain mice, for example, the male's urine contains pheromones that stimulate the female's estrous cycle.

Mate Selection

In those species that choose mates, some individuals in a population may mate frequently, while others fail to mate because they are rejected by members of the opposite sex. Because individuals differ in their ability to compete for mates, mate selection plays an important role in determining which individuals will pass genes to the next generation.

Competition for mates takes many forms. In many species, mate selection is based on sexual dimorphism. Males of many species are brightly colored or have pronounced body features. Such ornamentation may determine which males are dominant, and as components of courtship displays, they may determine which males will be chosen as mates by females. Male ornamentation, such as the peacock's tail and the antlers of male deer and elk, is of no direct benefit to a female and may even handicap the male. How did they ever become criteria that determine which males reproduce? The most widely accepted explanation is that in the early stages of evolution of a species' sexual behavior, male ornamentation may have had some benefit for the female. If most females then chose mates based on these male features, males with the features would have enjoyed greater reproductive success, and males without the features would not have passed their genes on to future generations. Such a system might have produced rapid evolution toward exaggerated ornamentation and pronounced sexual dimorphism. At some point, the handicap experienced by individuals with the ornamentation would have balanced any advantage gained, and the system would have come to equilibrium.

Social Organization Between Mates

The association of males and females for reproductive purposes involves social organization, or a **mating system**, among the participants. Fundamentally, mating systems involve either the association of a single male and female (**monogamy**) or the association of several reproductive individuals (**polygamy**). The biology of mating systems is an area of active investigation by behavioral ecologists who seek to understand how such systems evolved and what reproductive advantages they confer on species.

Monogamy. Monogamy is not a common mating system. Only about 5% of the mammal species are monogamous, yet more than 80% of the bird species mate this way. In some species, monogamy involves the formation of a pair bond between the male and female, and typically the male fertilizes all of the female's eggs. Pair bonding may last only through the breeding season, or, as in the Canada goose (*Branta canadensis*), a male and female may pair for life. When the sex ratio is 1:1 in a monogamous species, most individuals in the breeding population pair and mate. Pair bonding provides a mechanism for the care of the young, and monogamy is most common in species in which the need for extensive care of the young by both parents outweighs the advantages of either sex mating with others during a season.

Monogamy seems to be common in birds because incubating eggs and feeding more or less helpless hatchlings are energy-intensive jobs. By forming a long-term pair bond and sharing the duties, both parents may improve the chances that their genes will be passed to future generations. By contrast, most mammals reproduce in a way that makes the young entirely dependent on the female. She carries them in her uterus during development and supplies milk after they are born. For many male mammals, there is little advantage to forming a pair bond with a female because the male's role in rearing the young is limited. Probably as a result, males of many species are polygamous, increasing the chances that their genes will be perpetuated by mating with several females. In the few mammalian species that are monogamous, the male usually functions in territorial defense or in watching for predators or bringing food to the female or the young after they are weaned.

Polygamy. Many animal species are polygamous. In one form of polygamy, called **polygyny**, a single male mates with several females. Many mammals, for example, form breeding groups in which one male monopolizes several females; or the males may be promiscuous, mating with any receptive female encountered and forming no bonds. Polygyny is common, and its advantages seem to explain why. If a male invests the same amount of energy in gamete formation as does a female, he produces more than enough sperm to fertilize all the eggs of several females. For example, a male mammal may be able to ejaculate several times during each breeding season, and each ejaculum may contain over 100 million sperm. A female, on the other hand, produces at most about a dozen eggs in a season. By mating with several females, a male is taking fuller advantage of the energy he has invested in producing sperm. The more females he can make carry his offspring, the greater the chances that his genes will be perpetuated.

In populations where sex ratios are 1:1, polygyny means that there is intense competition between males for females. Some of the males will be successful, but most will not. The nature of the competition between males varies, as the following examples illustrate.

In *female defense polygyny*, a single male gathers a harem of females, including their dependent young offspring, and defends them against other males. In such a mating system, females often have opportunities to encourage male competition. Among elephant seals, for example, intruder males will constantly attempt to mate with members of a harem (Figure 14.15). When mounted, a cow will call loudly. On hearing the noise, the harem master will attempt to drive the intruder away. In this situation, females create a series of ongoing tests for a male's ability to dominate other males in the population. The female's behavior increases her chances of mating with the most fit male and so of combining her genes with those most likely to be perpetuated.

Another type of polygyny is *male defense polygyny*, in which breeding occurs in a specific area, called a **lek**. We looked at such a system in our discussion of the sage grouse *(Centrocercus)*, in Chapter 13 (p. 321). Females in lekking species do not enter the breeding condition synchronously, and their visits to the leks are spread over several weeks. During this time, there is fierce competition among the males for central positions on the lek, and throughout the breeding season, a dominant male's fitness for reproducing is constantly chal-

Figure 14.15 Polygyny in elephant seals. Among elephant seals *(Mirounga)*, a single male guards and mates with a harem of females.

lenged by other males. Arriving females show a preference for mating with a few dominant males at the center of the lek.

In a third type of polygyny, called *resource defense polygyny*, males monopolize or guard a resource needed by the females. The activity of male dung flies as they await females is one example. The marsh wren, a bird that breeds in marshes in North America, exhibits a resource-based mating system that involves a mix of polygyny and monogamy. The males defend their breeding territories, and as part of the courtship activity, each male builds numerous nests in his territory, far more than would be used in breeding. In one study, females were found to be attracted to the territories on the basis of the number of nests built. Males building an average of about 25 nests were polygynous (they had two mates), those building about 22 nests were monogamous, and those building fewer than 17 nests did not attract females to their territories. Presumably, in this species, the ability to build nests reflects food availability in the territory. Where food is plentiful, the males have more time and energy for nest building. As in other forms of polygyny, the female has a choice by which she tests males.

In contrast to polygyny, **polyandry** is a form of polygamy in which a female mates with more than one male. A female mallard duck in the breeding season is often mobbed by several drakes that copulate with her in rapid succession. Female cats and dogs in heat often attract and mate with several males. Polyandry is advantageous to females, but not in ensuring that a female's eggs are fertilized, for any one mating transfers more than enough sperm for that purpose. In mallards, cats, and dogs, polyandry may ensure that males with the most viable sperm (sperm that can swim fastest up her reproductive tract) fertilize her eggs.

Polyandry is common in invertebrates that have internal fertilization, especially those species in which the male contributes some type of mating "gift" to the female, such as food during the courtship ritual or as a result of mating. For example, male crickets and katydids transfer a large spermatophore, which is left protruding from the female's genital opening. After the male leaves, the female eats the spermatophore, bit by bit. The males that are successful in fertilizing her eggs are those that produce the largest spermatophores, for they take longer to eat. The longer the spermatophore is intact, the greater is the probability that some sperm will enter the female's reproductive tract and fertilize her eggs.

Strategies for Bearing Offspring

Three mechanisms for bearing offspring occur in the animal kingdom. No one mechanism is clearly superior to the others, and each represents a strategy for survival under a particular set of circumstances. In **oviparous** species, females release eggs into the environment, and young hatch in the environment. Fertilization may be internal or external. Most invertebrates and vertebrates (including most fishes, amphibians, and reptiles and all birds) are oviparous. Eggs laid in terrestrial environments have tough outer coverings that protect the developing embryo from desiccation and mechanical injury. Eggs laid in water are usually surrounded by a protective jelly coat. Some species shed their eggs indiscriminately, while others carefully choose sites for **oviposition** (egg laying). Among the insects, for example, walking sticks scatter their eggs in leaf litter, whereas flesh flies lay their eggs in raw meat, the only suitable habitat for larval feeding.

The alternative to oviparity is giving birth to young. Internal fertilization is required, and males of most species have a penis for transferring sperm. There are two types of animals that give birth to offspring. In **viviparous** species, embryos derive nourishment directly from the mother. Viviparity is characteristic of mammals, but also occurs in certain fishes, amphibians, lizards, snakes, and invertebrates. **Ovoviviparous** species also give birth to young, but the female retains the fertilized eggs in her reproductive tract, and the developing embryos derive most or all of their nourishment from the egg yolk. Ovoviviparity occurs in certain sharks and reptiles and in many insects.

Oviparity is considered the ancestral form of bearing offspring. Ovoviviparity probably arose when some females of certain species retained some of their eggs in their reproductive tract. By retaining eggs, a female may increase the chances that more of her offspring will survive. Not only are the eggs protected from predators, but they also are buffered from changes in the environment by the adult's well-developed homeostatic mechanisms. Viviparity may have evolved in certain ovoviviparous species as embryos drew some energy from their mothers before being released. Perhaps reflecting an early stage in the evolution of viviparity, in certain sharks, eggs hatch in the female, and early embryos are supported by a type of placenta in the uterus. Viviparity has the advantage of allowing the female to supply energy to developing embryos over an extended incubation time. This benefits both the female and her offspring. Rather than requiring (as in many cases of oviparity and ovoviviparity) that a female expend a massive amount of energy all at once in supplying eggs with yolk and other nutrients, viviparity spreads out the period of energy drain on a female, allowing her to feed and replenish her energy stores. The primary disadvantage of ovoviviparity and viviparity is that a female usually cannot mate again while carrying embryos in her reproductive tract.

Energy Budgeting in Reproduction

Two fundamental strategies of energy expenditure have evolved in the animal kingdom. One is to gamble on numbers. A great amount of energy is spent in producing massive numbers of offspring, but little or no energy is spent in caring for them. The numbers strategy accepts that most offspring will die, but out of the large numbers, a few offspring will survive and reproduce. In an extreme case of this, adult Pacific salmon use virtually all their energy reserves in migrating from the ocean to their home streams, where they die after spawning. (See the Chapter 12 Essay on pathfinding, p. 290.) Likewise, mayflies feed and grow only when immature. The aquatic mayfly nymphs metamorphose to flying adults, which breed, lay eggs, and die within a day or two (Figure 14.16). In contrast to these species, most animals reproduce more than once. Many fishes, for example, provide little or no parental care, but release prodigious numbers of eggs into the environment each year.

The alternative to gambling on numbers in reproduction is to produce relatively few offspring, but to invest much energy in caring for the young. Viviparity and ovoviviparity have evolved as alternatives to the numbers strategy. In addition, many oviparous species produce relatively few eggs but incubate eggs and care for newly hatched young. Among the oviparous fishes, male sea horses incubate fertilized eggs in a pouch on their ventral surface (Figure 14.17). Most birds spend a large portion of their life cycle rearing their young. (Nesting and care of the young in birds are discussed in Chapter 32, p. 789.)

In species that invest much energy in rearing young, the number of offspring produced at any one time generally correlates with the largest number of young that the parents can support. To produce fewer offspring does not realize the parents' full potential to reproduce. To produce more off-

Figure 14.16 Adult mayflies as reproductive "machines." These insects lack a digestive system and do not feed. Emerging from the water in which they have developed over the last year as larvae, mayflies may produce populations of adults so dense that the dead adults pile up on roads and bridges, creating slippery conditions that can become a traffic hazard. Here we see mayflies covering street lights and pedestrians in downtown Dubuque, Iowa.

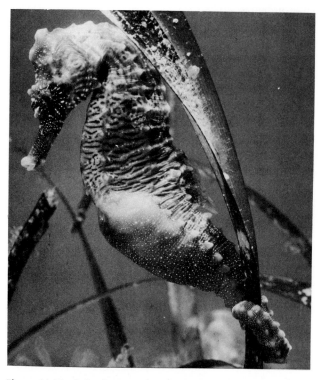

Figure 14.17 Oviparity among sea horses. Male sea horses (*Hippocampus*) incubate eggs in a ventral pouch. The spotted sea horse (*Hippocampus erectus*) shown here is "pregnant" with a bulging brood pouch.

spring invites loss of all the offspring; the female may become seriously weakened by reproducing, and even if the male participates in rearing young, he may not be able to care for them without equal participation by the female.

It is important to realize that the parental care strategy is not superior to the numbers strategy, or vice versa. Consider the very different strategies of a clam and a cardinal. A pair of clams may produce thousands of fertilized eggs each year, whereas a pair of cardinals may raise only a few offspring. In typical populations of these two species, the number of individuals remains about the same from generation to generation, meaning that despite how many or how few offspring are produced by a mating pair, only two individuals survive, replacing the parents in the population when they die.

SUMMARY

1. Asexually produced offspring are genetically alike. This allows populations of them to expand rapidly under constant environmental conditions but may put them at a disadvantage in changing environments. Sexually produced offspring are genetically variable, and as a result, they generally are able to adapt to new and changing environments.

2. Genetic variability arises during sexual reproduction because the genes get shuffled during the production of gametes by meiosis and because offspring usually inherit genes from two parents. Crossing over, which occurs during meiosis, produces gametes that have new combinations of chromosomes and genes. Meiosis is required in the life cycles of sexual species to keep chromosome number constant from generation to generation.

3. Meiosis occurs in gonads. Meiosis in the male results in four sperm cells; meiosis in the female results in only one functional egg. Females of some species reproduce parthenogenetically; that is, their eggs develop without being fertilized.

4. Dioecious species have separate sexes, that is, male and female individuals. Monoecious species have both male and female sex organs in one hermaphroditic individual, but hermaphrodites rarely are capable of self-fertilization.

5. Successful reproduction often requires that individuals come into breeding condition, find each other, and mate at a time that allows offspring to be produced when environmental conditions are optimal for their survival. In many species, there is intense competition among males or females for mating partners.

6. Sexual species are characterized by several types of mating systems. Polygyny, a type of polygamy in which each male has several mates, is the most common. Monogamy involves a permanent pair bond between the sexes for at least a breeding season.

7. Many animals produce large numbers of offspring, but provide minimal or no parental care; others spend much energy caring for a small number of offspring.

FURTHER READING

Bell, G. *The Masterpiece of Nature: The Evolution and Genetics of Sexuality.* London: Croom Helm, 1982. *Discusses the significance of sex and sexual reproduction and how these phenomena are maintained.*

Butron, R. *The Mating Game.* Oxford, England: Elsevier, 1976. *Well-illustrated book giving anecdotal examples about the beginnings of sex, fertilization, and sexual competition.*

Clutton-Brock, T. H. "Reproductive Success in the Red Deer." *Scientific American* 252(1985): 86–92. *The determinants of breeding success are reported from a twelve-year study of deer herds in a Scottish game preserve.*

Halliday, T. *Sexual Strategy: Survival in the Wild Series.* Chicago: The University of Chicago Press, 1980. *A brief, nontechnical treatment of mating systems and the nature of competition among members of a species during mating.*

Smith, J. M. *The Evolution of Sex.* New York: Cambridge University Press, 1978. *An authoritative work that discusses the advantages of sex, recombination, and breeding systems.*

Warner, R. "Mating Behavior and Hermaphroditism in Coral Reef Fishes." *American Scientist* 72(1984): 128–136. *Diverse mating systems, including sex reversal, are discussed for populations of coral reef fishes.*

Genetics

Normal and sickled human red blood cells. Sickle-cell anemia is a serious genetic disease that often causes early death in affected humans. Persons with sickle-cell anemia have abnormal hemoglobin and suffer great pain when misshapen red blood cells plug their capillaries and impair circulation.

The historical roots of **genetics**, the science of heredity, date back to the earliest attempts to produce desirable traits in domestic animals and plants by selective breeding. As a modern experimental science, however, genetics was born less than 150 years ago with the work of Gregor Mendel, an Austrian monk who studied inheritance patterns in garden peas. Mendel's results convinced him that heredity is based on discrete particles, which he called "factors" (and we now call genes). The prevalent, but erroneous idea of Mendel's era was that hereditary materials are fluids that blend together during fertilization. In a publication in 1866, Mendel, without any knowledge of meiosis, described how "factors" (genes) behave during meiosis and fertilization. He also described the phenomena of dominance and recessiveness and established genetics as a quantitative, predictive science by showing that inheritance patterns can be interpreted by the laws of probability and statistics.

Despite the significance of Mendel's work, his publication went unnoticed until other geneticists discovered it in the early 1900s. About the same time, biologists recognized the important connection between inheritance patterns and chromosome behavior during meiosis. Genetics quickly became a focal area in biology, and many inheritance patterns were studied in the ensuing decades. Since the 1950s and '60s, when the central role of DNA in carrying hereditary information was elucidated, genetics has been one of the most rapidly advancing areas in biology.

INHERITANCE PATTERNS

Mendel's discoveries and those made by later researchers allow us to understand and predict how genetic traits pass from one generation to the next. Understanding the patterns by which traits are inherited requires a firm understanding of how chromosomes behave during meiosis. (For a review of the effects of meiosis, see Chapter 14, p. 333; Figure 14.7.)

Inheritance Patterns Involving Single Traits

The fruit fly has been a favorite research animal in genetics since the early 1900s. These insects can be maintained in small containers on inexpensive food and will reproduce in about two weeks; even more conveniently, they have many sharply contrasting traits. Most fruit flies, for example, have normal wings, but some individuals, which survive only in laboratory cultures, have *vestigial wings* (Figure 15.1a). In this case, normal and vestigial are two contrasting variants of a single trait—wing shape. True-breeding stocks (pure-line individuals) of either normal- or vestigial-winged flies always produce normal- or vestigial-winged offspring, respectively. By crossing a true-breeding normal fly with a true-breeding vestigial individual, one can begin to study the inheritance pattern of the wing trait. Crosses such as this, which involve only one trait, are called **monohybrid crosses**. The two pure-line individuals are called **parentals** because they are the first to be crossed in the study. All their offspring, which are collectively called the **first filial (F_1) generation**, have normal wings. The inheritance pattern can be studied further by allowing the F_1 offspring to mate among themselves, producing the **second filial (F_2) generation**. In the F_2 generation, three-fourths of the offspring are normal winged, and one-fourth are vestigial winged.

Meiosis and the Principle of Segregation. Working with garden peas, Mendel explained results similar to those for the wing trait in fruit flies by his **principle of segregation**, also known as Mendel's first law. Mendel surmised that "factors" determining such traits occur in pairs and that one factor of a pair is **dominant**, while the other is **recessive**. When the factors are present together (as in the F_1 generation), the dominant factor masks the effect of the

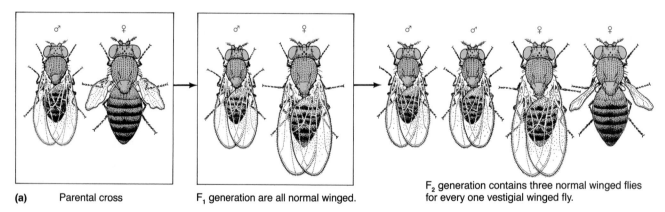

(a)   Parental cross F_1 generation are all normal winged. F_2 generation contains three normal winged flies for every one vestigial winged fly.

Figure 15.1 A fruit fly *(Drosophila melanogaster)* monohybrid cross. (a) This drawing shows the results from a cross between fruit flies differing in wing shape. In this case, parental types breed true, but when normal and vestigial-winged parents are crossed, all F_1 offspring have normal wings. When these F_1 flies are crossed, three-quarters of the F_2 offspring have normal wings and one-quarter have vestigial wings.

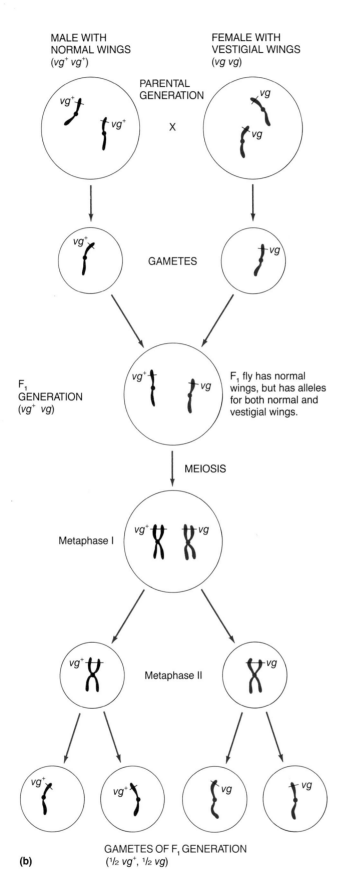

(b)

GAMETES OF F₁ GENERATION
(½ *vg*⁺, ½ *vg*)

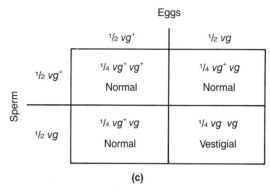

(c)

Figure 15.1 *Continued* (b) A meiotic model for the monohybrid cross in (a). Although fruit flies have eight chromosomes (four pairs), only one pair is shown here for simplicity. This model illustrates how Mendel's principle of segregation describes the meiotic process. Each homozygous parent has only one type of allele present at the gene locus governing wing shape. Consequently, meiosis yields only one type of gamete from each parent. Likewise, only one type of fertilization can occur between these gametes, and all F₁ offspring are the same. Each carries both types of alleles and is said to be heterozygous. Meiosis during gamete formation in the F₁ generation results in two types of gametes. As Mendel postulated (without knowing about meiosis), the two types of gametes form because the alleles (Mendel's "factors") segregate from each other during meiosis. (c) A Punnett square shows the four possible combinations of gametes from the F₁ generation in (a) and (b). Because gametes combine randomly at fertilization, the probability that a particular combination of genes will occur at fertilization is obtained by multiplying the gamete frequencies.

recessive factor. Most importantly, Mendel saw that the factors *segregate* into the gametes prior to reproduction; that is, each gamete contains only one factor of a pair.

Figure 15.1b illustrates how meiosis explains the fruit fly crosses just described. In modern terms, the gene for wing shape is said to have two alternative forms, called **alleles**, which produce different expressions of a trait (in this case normal and vestigial wings). Alleles for the wing trait occur only at one location on a chromosome, called a gene **locus**. Each chromosome is a collection of thousands of loci, governing the different traits that characterize an animal. As discussed in Chapter 14, chromosomes occur in matching pairs, called **homologous chromosomes**, with one member of each pair derived from the mother and the other from the father. Each chromosome of a homologous pair carries an allele for a trait. In fruit flies, the alleles for wing shape are represented by the letters *vg*⁺ (allele for normal wing, dominant) and *vg* (allele for vestigial wing, recessive). Note the distribution of these alleles in the flies and in the gametes in Figure 15.1b. Individuals having the same allele at a given locus on a pair of homologous chromosomes are said to be **homozygous** (*vg*⁺*vg*⁺

or *vgvg*, in this case, the parentals) for that trait. Individuals having different alleles at a given locus on homologous chromosomes are **heterozygous** (vg^+vg, the F_1 generation). Only the dominant allele (vg^+) is expressed in the F_1 flies; thus, all of these flies have normal wings. Heterozygous individuals are said to be **carriers** of the unexpressed recessive allele.

Figure 15.1 carries the wing trait inheritance pattern into the F_2 generation (produced by crossing two F_1 flies). Each heterozygous F_1 fly produces two types of gametes with equal frequency. A grid called a **Punnett square** illustrates the four possible combinations of the two types of gametes produced by each parent (Figure 15.1c). Because each type of sperm and egg is produced in equal quantity, each of the four types of sperm/egg combinations has an equal chance of occurring. As a result, the F_2 generation is expected to have three types of gene combinations—vg^+vg^+, vg^+vg, and *vgvg*—in a ratio of $1:2:1$.

Genotypes and Phenotypes. The combination of alleles in an individual is its **genotype**, represented by symbols such as *vg* and vg^+ or upper- and lower-case letters. In the fruit fly wing example, the $1:2:1$ ratio in the F_2 generation is called the **genotypic ratio**. In contrast to the genotype, an animal's actual appearance is its **phenotype**. Because the vg^+ allele is dominant, a $3:1$ **phenotypic ratio** (¾ with normal wings to ¼ with vestigial wings) is obtained in the F_2 generation (see Figure 15.1a). This ratio does not mean that when four offspring result from this cross, three will be normal and one vestigial. The ratios are **probabilities**, expressing the chance that the specified events will occur. Only in experiments involving hundreds of offspring are the actual ratios likely to be similar to those expected.

As the wing shape example illustrates, two alleles occurring at one locus can produce three genotypes in a population. If individuals in the F_2 population randomly mate, six possible kinds of crosses can occur. Table 15.1 shows these crosses and the expected genotypic and phenotypic ratios resulting from them.

The Test Cross. As illustrated by the preceding examples, the genotype of an animal with a dominant allele cannot be determined from its phenotype because the animal could be either homozygous or heterozygous. Geneticists use what is called a **test cross** to determine the genotype of an animal with a dominant phenotype.

Table 15.1 All Possible Crosses Involving Two Alleles at One Locus in Fruit Flies*

Cross No.	Parental Genotypes	Expected Genotypes of Offspring	Expected Phenotypes of Offspring
1	vg^+vg^+ × vg^+vg^+	All vg^+vg^+	All normal
2	vg^+vg^+ × *vgvg*	All vg^+vg	All normal
3	vg^+vg^+ × vg^+vg	½ vg^+vg^+; ½ vg^+vg	All normal
4	vg^+vg × vg^+vg	¼ vg^+vg^+; ½ vg^+vg; ¼ *vgvg*	¾ normal ¼ vestigial
5	vg^+vg × *vgvg*	½ vg^+vg; ½ *vgvg*	½ normal ½ vestigial
6	*vgvg* × *vgvg*	All *vgvg*	All vestigial

*vg^+ = normal wing, dominant; *vg* = vestigial wing, recessive.

Suppose we wished to determine the genotype of a normal-winged fruit fly. We know that it carries at least one dominant allele, but we do not know whether its second wing trait allele is dominant or recessive. To determine this, we would cross the fly (genotype vg^+?) with a homozygous recessive individual (crosses 2 and 5 in Table 15.1). If the fly in question carries two dominant alleles, all of the offspring will be normal. If the original animal is heterozygous, there is a 50% chance that some offspring will have the recessive phenotype. When a recessive phenotype appears among the offspring from a test cross, the original animal must have been a heterozygote. In this way, the genotype can be determined to be either homozygous or heterozygous dominant. Animal breeders and geneticists frequently use the test cross to establish pure breeding strains of animals. After identifying the genotype of an animal by test crossing, the breeder may either cull the animal from the breeding stock or use it for breeding purposes, depending on the breeder's interests. Because recessive alleles are hidden in a population, traits determined by them are much harder to eliminate than dominant traits.

Variations in Single-Trait Crosses. In heterozygotes, it is not always the case that one allele masks the other. In shorthorn cattle, for example, coat color results from two alleles (C^r and C^w) that do not exhibit typical dominance and recessiveness. Red shorthorns and white shorthorns are homozygous (red = C^rC^r; white = C^wC^w). Heterozygous shorthorns (C^rC^w) are roan colored. This type of inheritance pattern, in which neither allele is strong enough to mask the other and both alleles are expressed in heterozygotes (here producing the intermediate roan color), is called **codominance (incomplete dominance)**. In shorthorn cattle, a cross between two roan parents would produce three kinds of offspring in the following ratio: 1 C^rC^r (red): 2 C^rC^w (roan): 1 C^wC^w (white). Thus, in the case of codominance, the genotypic ratio is the same as the phenotypic ratio.

Multiple Alleles. So far, the inheritance patterns we have discussed involve genes with only two alleles. But many genes have more than two alleles (although any single individual has only two alleles for any one gene locus). Human ABO blood types illustrate the concept of **multiple alleles**. There are four blood types in the ABO system: A, B, AB, and O. Three alleles (I^A, I^B, and i^O) at a single gene locus produce these four blood types. Alleles I^A and I^B are *codominant;* when together, they produce the AB type. Allele i^O is recessive to I^A and I^B; consequently, type O blood occurs only in people homozygous for the i^O allele. Types A and B can both result from two genotypes (Table 15.2).

The ABO blood-typing system is based on differences between two types of glycoproteins present on the surface of red blood cells (RBCs). The glycoproteins, A and B, are determined by the allele combinations shown in Table 15.2. Because these glycoproteins act as antigens, they are often referred to as antigens A and B. Table 15.2 shows the distribution of the A and B antigens and the distribution of the corresponding antibodies in the four blood types. Because of the distribution of the two types of antigens and antibodies, only certain types of blood can be safely used for blood transfusions in humans. The safest method is to transfuse blood of the same type as the recipient's, but unless massive amounts of blood are to be given, it is also safe to transfuse blood from a donor whose RBC glycoproteins will not interact with the plasma antibodies

of the recipient. If the RBCs of the donor do interact, they will agglutinate (clump together), perhaps causing circulatory failure. For example, when type A blood (containing antigen A) is given to individuals with type B or O blood (both of which contain anti-A antibody), the donor's RBCs will agglutinate. On the other hand, type O blood can be given in moderate quantities to persons with any type of blood. It does not cause problems because the antibodies in its plasma are diluted by the recipient's blood before they can agglutinate the recipient's RBCs. (Antigen-antibody reactions are ◁ discussed in Chapter 6, p. 139.)

Harmful and Lethal Alleles. Many alleles, both dominant and recessive, are harmful and may even be deadly when they occur in homozygous combinations. The presence of lethal alleles in a population may be detected by unexpected ratios of offspring resulting from certain crosses. In chickens, for example, certain individuals are called creeper fowl because they have short, deformed legs. Creepers result from a dominant allele *(C)*, and such individuals are heterozygous *(Cc)*; normal chickens are homozygous recessive *(cc)* for this trait. Offspring of a cross between a creeper *(Cc)* and a noncreeper *(cc)* occur in an expected 1:1 ratio of creepers and noncreepers (Figure 15.2). However, offspring of a cross between two creepers *(Cc × Cc)*, which would be expected to yield the 3:1 ratio of creepers and noncreepers typical of a monohybrid cross, actually yield only two creepers per single noncreeper. The reason for this unexpected ratio is that individuals with the homozygous dominant genotype *(CC)* die before hatching.

Lethal genes have been identified in humans and in many domestic animals. Lethal dominant alleles are more likely to be eliminated from a population than are harmful or lethal recessive alleles. Under natural conditions, for example, creeper

Table 15.2   The Human ABO Blood Group System

Blood Type (Phenotype)	Possible Genotypes	RBC Glycoprotein (Antigen)	Antibodies in Blood Plasma	Safe Recipient of	Safe Donor to
A	I^AI^A, I^Ai^O	A	Anti-B	A, O	A, AB
B	I^BI^B, I^Bi^O	B	Anti-A	B, O	B, AB
AB	I^AI^B	A, B	None	A, B, AB, O	AB
O	i^Oi^O	None	Anti-A and Anti-B	O	A, B, AB, O

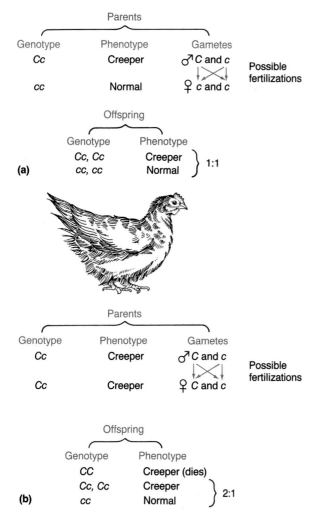

Parents

Genotype	Phenotype	Gametes
Cc	Creeper	♂ *C* and *c*
cc	Normal	♀ *c* and *c*

Possible fertilizations

Offspring

	Genotype	Phenotype	
(a)	*Cc, Cc*	Creeper	} 1:1
	cc, cc	Normal	

Parents

Genotype	Phenotype	Gametes
Cc	Creeper	♂ *C* and *c*
Cc	Creeper	♀ *C* and *c*

Possible fertilizations

Offspring

	Genotype	Phenotype	
	CC	Creeper (dies)	
(b)	*Cc, Cc*	Creeper	} 2:1
	cc	Normal	

Figure 15.2 Inheritance pattern of a lethal allele. In chickens, the dominant allele *C* is lethal in homozygous individuals. The heterozygous condition produces creeper fowl.

Inheritance Patterns Involving Multiple Traits

Inheritance in even the simplest animals involves the simultaneous passage between generations of thousands of genes. As we will see, the inheritance patterns of two or more traits are determined largely by whether the gene loci for the traits are located on the same or different pairs of homologous chromosomes. When loci are on different chromosomes, genes are inherited independently of each other. When they are on the same chromosome, genes are **linked** and tend to be inherited as a unit (see Figure 14.6).

Independent Assortment. Mendel described the results of several crosses involving the simultaneous inheritance of two traits in garden peas. In interpreting the results of these **dihybrid crosses**, Mendel formulated his **principle of independent assortment**, also called Mendel's second law. In modern terms, this principle states that genes for two or more traits are transmitted to offspring independently of each other when the genes are located on different pairs of homologous chromosomes.

In guinea pigs, coat color is determined by genes at a single locus; the allele *(B)* for black coat is dominant to the allele *(b)* for brown. Length of coat hair is determined by alleles at another locus on another pair of homologous chromosomes. The allele *(S)* for short hair is dominant to the allele *(s)* for long hair. If a guinea pig homozygous for both black and short hair is mated to a homozygous brown, long-haired guinea pig, all the F₁ offspring have black, short hair (Figure 15.3a). If these offspring are allowed to mate among themselves, their offspring (F₂) occur in a phenotypic ratio of 9 black, short hair: 3 black, long hair: 3 brown, short hair: 1 brown, long hair. All combinations of coat color and hair length appear, not just the two combinations found in the original parents (Figure 15.3b). As Table 15.3 shows and as Mendel surmised in stating his second law, the inheritance pattern of two independent traits such as these is a simple combination of two monohybrid crosses.

Independent assortment is explained by the way homologous chromosomes align during metaphase I of meiosis (Figure 15.4). Because alignment is random, all combinations of the alleles on separate chromosomes occur in gametes at equal frequencies. Therefore, an animal heterozygous at two loci on different pairs of homologous chromosomes produces four (2^2) types of gametes (as in the guinea pig model). For independently assorting traits, the maximum number of different kinds of gametes that can be produced by an individual is equal to 2 raised to the power of the number of homologous pairs of chromosomes in the individual.

Linked Genes and the Effect of Crossing Over. In fruit flies, the body color locus is on the same pair of chromosomes with the locus for wing shape.

chickens would most likely be killed by predators before reproducing; and animal breeders can easily remove such animals from breeding stocks. Recessive alleles, however, because their effects are hidden in heterozygotes, are much less likely to be eliminated by natural forces or by breeders.

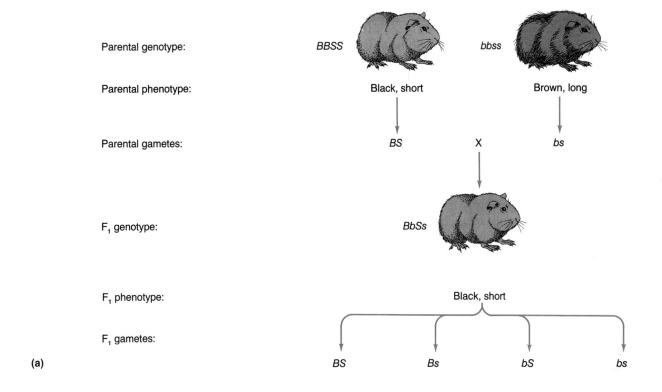

Parental genotype: *BBSS*   *bbss*

Parental phenotype: Black, short   Brown, long

Parental gametes: *BS*   X   *bs*

F₁ genotype: *BbSs*

F₁ phenotype: Black, short

F₁ gametes: *BS*   *Bs*   *bS*   *bs*

(a)

F₂ genotypes and phenotypes:

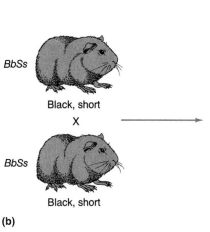

BbSs

Black, short

X

BbSs

Black, short

(b)

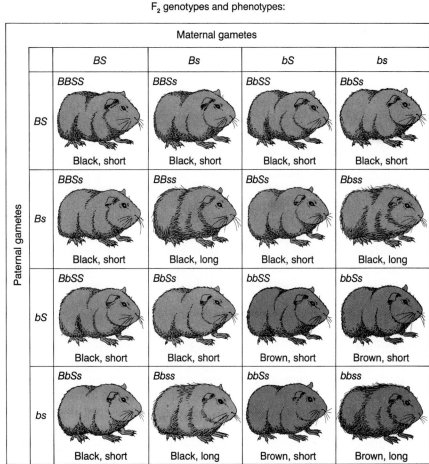

Figure 15.3 A guinea pig dihybrid cross. Here the two traits of interest are coat color and hair length. (a) If pure-breeding parents are crossed (*BBSS* × *bbss*), all of their F₁ offspring will be heterozygous at both loci (*BbSs*). Note that each F₁ individual can produce four types of gametes. (b) When the F₁ generation interbreed, 16 combinations of the four types of gametes are possible, giving the indicated genotypes and phenotypes in the F₂ generation. Note in Table 15.3 that the F₂ phenotypic ratio is 9:3:3:1.

Table 15.3   Short-Hand Method to Determine Genotypic and Phenotypic Ratios Expected from a Dihybrid Cross Between Heterozygotes for Two Independently Assorting Traits*

Actual Cross:	*BbSs*	×	*BbSs*
Two Equivalent Monohybrid Crosses:	*Bb × Bb,*		*Ss × Ss*
Genotypic Ratios in Offspring of Monohybrid Crosses:	¼ *BB*		¼ *SS*
	²⁄₄ *Bb*		²⁄₄ *Ss*
	¼ *bb*		¼ *ss*

How to Combine Two Monohybrid Crosses to Determine Genotypic Ratio:	¼ *BB*	×	{ *Genotypic Ratio* ¼ *SS* = ¹⁄₁₆ *BBSS* ²⁄₄ *Ss* = ²⁄₁₆ *BBSs* ¼ *ss* = ¹⁄₁₆ *BBss*
	²⁄₄ *Bb*	×	{ ¼ *SS* = ²⁄₁₆ *BbSS* ²⁄₄ *Ss* = ⁴⁄₁₆ *BbSs* ¼ *ss* = ²⁄₁₆ *Bbss*
	¼ *bb*	×	{ ¼ *SS* = ¹⁄₁₆ *bbSS* ²⁄₄ *Ss* = ²⁄₁₆ *bbSs* ¼ *ss* = ¹⁄₁₆ *bbss*

Phenotypic Ratios in Offspring:	¾ *B_* (black)	¾ *S_* (short)
	¼ *bb* (brown)	¼ *ss* (long)

How to Combine Two Monohybrid Crosses to Determine Phenotypic Ratio:	¾ *B_*	×	{ *Phenotypic Ratio* ¾ *S_* = ⁹⁄₁₆ *B_S_* (black, short) ¼ *ss* = ³⁄₁₆ *B_ss* (black, long)
	¼ *bb*	×	{ ¾ *S_* = ³⁄₁₆ *bbS_* (brown, short) ¼ *ss* = ¹⁄₁₆ *bbss* (brown, long)

*In this example, the traits are coat color and hair length. *B* = black coat (dominant); *b* = brown coat; *S* = short hair (dominant); *s* = long hair. Because the traits assort independently, the inheritance pattern is identical to that of two monohybrid crosses. Genotypic and phenotypic ratios (expressed here in fractions) in the dihybrid cross are obtained by multiplying all possible combinations of the monohybrid ratios together. In determining phenotypic ratios, a dash is used in genotypes with a dominant allele, because the second allele may be either dominant or recessive—it does not affect the ratio.

Gray body is dominant to black body. If a homozygous gray-bodied, normal-winged fruit fly mates with a black, vestigial-winged individual, all offspring (F_1) are gray with normal wings and heterozygous at both loci (Figure 15.5). Note that the alleles for gray body and normal wings are linked to each another, as are the alleles for black body and vestigial wings, in the parents, in their gametes, and in the F_1 generation. If we employ a test cross between F_1 flies and homozygous black, vestigial-winged individuals, we can see the effects of linkage. The test cross results in four types of offspring (see Figure 15.5a). These are the same four types of individuals that would be expected if the genes for body color and wing shape were on separate chromosomes and assorted independently. However, with nonlinked genes, the four types of F_2 offspring should occur in equal numbers, whereas the ratios obtained in this test cross (with linked genes) are quite different. About 40% of the offspring are gray, normal-winged flies; another 40% are black, vestigial-winged flies; about 10% are gray, vestigial-winged individuals; and 10% are black, normal-winged flies. How do we account for this difference?

Recall that as **crossing over** occurs during meiosis, homologous chromosomes first pair with each other and then reciprocally exchange parts. Conse-

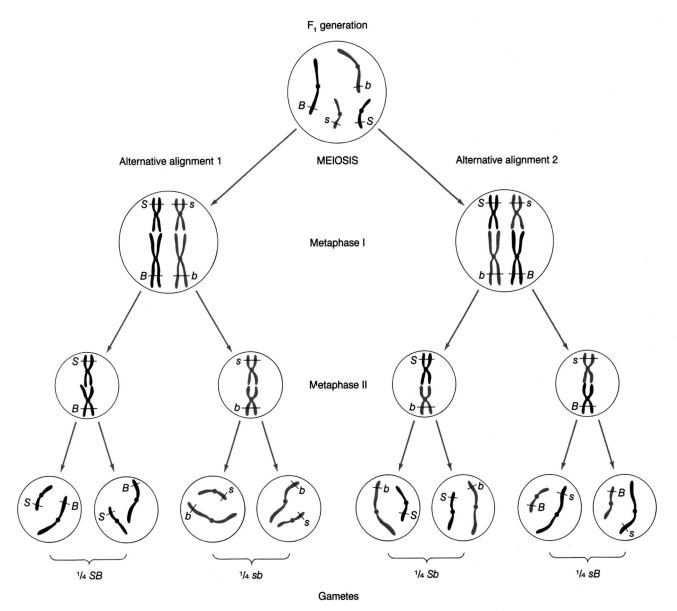

Figure 15.4   The meiotic production of four types of gametes by an individual that is heterozygous for two genes. Assume that the F₁ individuals are those shown in Figure 15.3. Because the two genes are on different chromosomes, they separate independently during meiosis. This separation process (independent assortment) and the inheritance patterns that can result from it were described by Mendel (without any knowledge of meiosis and gametogenesis). For simplicity, only two pairs of homologous chromosomes are shown in these cells. All F₁ cells are heterozygous, containing four different alleles for the two traits. During metaphase I of meiosis, chromosomes may align on the spindle in two alternative ways. Which alignment occurs in any particular cell is determined by chance. In a large population of gametes, all four allele combinations are present in roughly equal numbers.

quently, following meiosis, each chromosome is a mosaic of bits of itself and pieces of its homologue. When an individual is heterozygous at various loci on a pair of homologous chromosomes, crossing over creates new combinations of alleles that segregate into different gametes (see Figure 15.5a). If the gametes carrying chromosomes formed by crossovers participate in fertilization, offspring with new combinations of traits result. These are called **recombinant types** in contrast to the **parental types**, which carry the linkage groups of their parents.

PARENTAL GENERATION
Parental chromosome pairs, with
alleles for gray or black
color (*G*, *g*) and normal or vestigial
wings (*vg⁺*, *vg*):

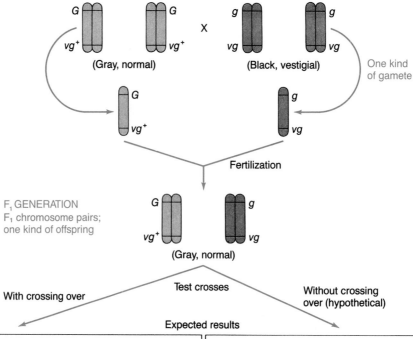

(Gray, normal)   X   (Black, vestigial)   One kind of gamete

Fertilization

F₁ GENERATION
F₁ chromosome pairs;
one kind of offspring

(Gray, normal)

With crossing over   Test crosses   Without crossing over (hypothetical)

Expected results

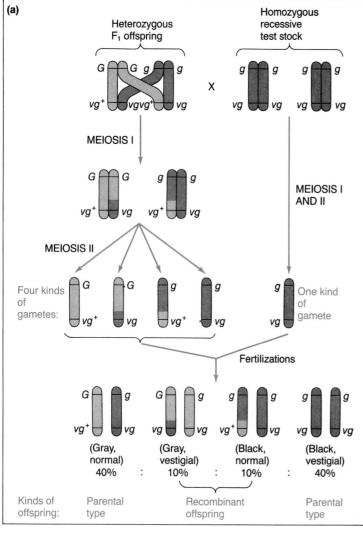

(a)

Heterozygous F₁ offspring   X   Homozygous recessive test stock

MEIOSIS I

MEIOSIS I AND II

MEIOSIS II

Four kinds of gametes: One kind of gamete

Fertilizations

(Gray, normal) 40% : (Gray, vestigial) 10% : (Black, normal) 10% : (Black, vestigial) 40%

Kinds of offspring: Parental type   Recombinant offspring   Parental type

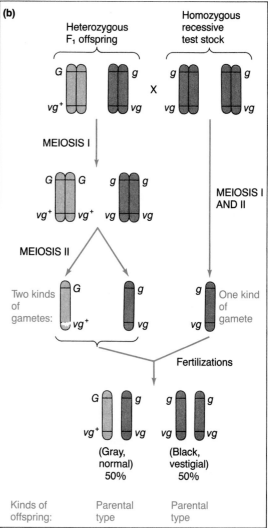

(b)

Heterozygous F₁ offspring   X   Homozygous recessive test stock

MEIOSIS I

MEIOSIS I AND II

MEIOSIS II

Two kinds of gametes: One kind of gamete

Fertilizations

(Gray, normal) 50%   (Black, vestigial) 50%

Kinds of offspring: Parental type   Parental type

Figure 15.5 (Opposite) A dihybrid cross involving linked genes. Inheritance of body color and wing shape is controlled by two linked genes in fruit flies. The two gene loci tend to be inherited as a unit because they are located on one pair of chromosomes, but crossing over during meiosis creates new combinations of alleles when gametes are produced from heterozygous individuals. Offspring of two homozygous parents are all heterozygous because each parent produces only one type of gamete. (a) Note that the heterozygous individuals can produce four types of gametes, two of which are new in that they result from crossing over. A test cross between heterozygous and homozygous recessive individuals produces four kinds of offspring. The proportions of parental type offspring to crossover types depend on the amount of crossing over that occurs between the two gene loci. This is the correct model for this inheritance pattern. (b) Note the results that would be expected from the same test cross if the genes were linked but no crossing over occurred: Only two kinds of offspring would result, both of the parental phenotype. This hypothetical model does not reflect the results of actual crosses because some crossing over always occurs.

The fruit fly body color/wing type model demonstrates the importance of crossing over in creating new combinations of alleles and therefore adding to the variability in the phenotypes resulting from sexual reproduction. If crossing over did not occur between the linked genes in the fruit fly model, only two types of offspring would result from the test cross. This result, which does not occur, is shown in Figure 15.5b.

Crossing over does not involve every chromosome locus in every meiosis. The closer together two loci are on a chromosome, the lower the probability that a crossover will occur in the space between the loci. Because of this relationship, geneticists can use the frequency of crossing over to measure relative distances between gene loci and can actually map the positions of genes on chromosomes. Crossover frequencies are measured by performing test crosses between homozygous recessive individuals and individuals heterozygous for the traits in question (as in the fruit fly model). By counting the number of parental and recombinant types of offspring, geneticists can calculate the percentage of recombinants among all the offspring. This percentage reflects the distance between gene loci on chromosomes in relative units.

Sex Determination and X Chromosome Linkage. Not all chromosomes in all species occur in homologous pairs. This is illustrated by chromosome patterns, or **karyotypes**, photographed during metaphase of mitosis. Figure 15.6 shows karyotypes for male and female humans. Note that all the chromosomes for the females can be arranged in pairs, but the male has two chromosomes that do not match each other. These nonmatching chromosomes are the **sex chromosomes**. The human sex chromosomes, and those of some other animal species, are called the **X** and **Y** chromosomes. All other chromosomes collectively are called **autosomes**.

Most **X-linked** genes—those on the *X* chromosome—have nothing to do with sex. In humans, for example, more than 200 genes have been identified so far on the *X* chromosome. Among these are the genes that determine the ability to see red and green. Figure 15.7 illustrates the type of inheri-

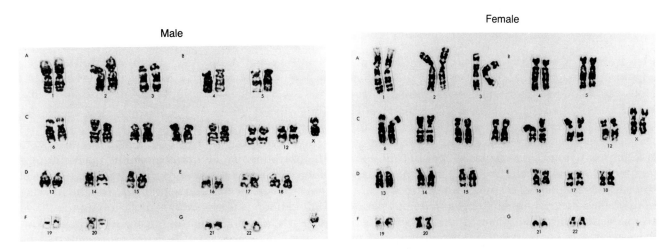

Figure 15.6 Karyotypes show the actual appearance of chromosomes during division. These photographs of chromosomes from a man and a woman show that there are chromosomal differences between the sexes. Females have two *X* chromosomes and no *Y* chromosome. Many other species show similar chromosome differences between the sexes. In birds, moths, and butterflies, the male has two *Z* chromosomes and the female only one. Turtles have no chromosomal differences between sexes.

Parents →		Normal ♀ $X^C X^C$		Carrier ♀ $X^C X^c$		Color-blind ♀ $X^c X^c$	
↓ Gametes →		X^C	X^C	X^C	X^c	X^c	X^c
Normal ♂ $X^C Y$	X^C	$X^C X^C$ Normal ♀	$X^C X^C$ Normal ♀	$X^C X^C$ Carrier ♀	$X^C X^c$ Carrier ♀	$X^C X^c$ Carrier ♀	$X^C X^c$ Carrier ♀
	Y	$X^C Y$ Normal ♂	$X^C Y$ Normal ♂	$X^C Y$ Color-blind ♂	$X^c Y$ Color-blind ♂	$X^c Y$ Color-blind ♂	$X^c Y$ Color-blind ♂
Color-blind ♂ $X^c Y$	X^c	$X^C X^c$ Carrier ♀	$X^C X^c$ Carrier ♀	$X^C X^c$ Carrier ♀	$X^c X^c$ Color-blind ♀	$X^c X^c$ Color-blind ♀	$X^c X^c$ Color-blind ♀
	Y	$X^C Y$ Normal ♂	$X^C Y$ Normal ♂	$X^c Y$ Color-blind ♂	$X^c Y$ Color-blind ♂	$X^c Y$ Color-blind ♂	$X^c Y$ Color-blind ♂

Figure 15.7 Inheritance pattern of an X-linked trait. In some species, genes that are carried on a sex chromosome (such as the X chromosome) are not inherited according to the same pattern as those on autosomes (nonsex chromosomes) because one of the sexes does not have an homologous pair of sex chromosomes. For example, human females are XX and males are XY. Alleles for normal vision and red-green color blindness in humans are X-linked, meaning that they are carried on the X chromosome as X^C or X^c. Color blindness is recessive to normal vision. Consequently, a woman heterozygous for this factor ($X^C X^c$) is a carrier, but does not have impaired red-green discrimination. A male, on the other hand, cannot be a carrier; his vision is either normal ($X^C Y$) or impaired ($X^c Y$). The Punnett square illustrates the kinds of offspring that can result from combinations involving these alleles for red-green discrimination. Note that of the 16 male and female offspring, 4 males but only 2 females have impaired vision. However, 4 additional females are carriers.

tance pattern that occurs with this particular X-linked trait.

Pleiotropy. In our discussion so far, we have assumed that each gene produces only one trait. However, there are many genes that influence several traits; such genes are said to be **pleiotropic** (from the Greek *pleio*, "more"). The *vg* allele in fruit flies, for example, influences several traits in addition to producing vestigial wings. And in rats, congenital deformities including thickened ribs, a narrowed trachea, nonelastic lungs, an enlarged heart, and blocked nostrils are all attributed to one pleiotropic gene. This gene has a wide-ranging effect because it carries the genetic information used to synthesize a connective tissue protein found in many different body parts. Pleiotropic genes typically exert their effect during an animal's early development, and in so doing, can influence the development of many structures derived later.

Inheritance Patterns Involving the Interaction of Genes at More Than One Locus

In contrast to the traits we have discussed so far, many heritable traits result from the interaction of genes at more than one locus. Such genes act in concert to yield a particular phenotypic trait.

Polygenes. Highly variable traits, such as the amount of butterfat in milk, the number of eggs laid per week, body weight, and body height, are traits of degree rather than of kind; that is, they are not the all-or-none expression of simple dominant and recessive alleles. Such traits may result from the *additive* effects of **polygenes** (many genes) at several loci. Table 15.4 presents a hypothetical model of the variation in human skin color that can result from the interaction of genes at three loci. In this model, the alleles *a*, *b*, and *c* each cause the synthesis of a small, constant amount of pigment, while the alleles *a'*, *b'*, and *c'* result in little or no pigment. The effect of all six of these alleles is equal (there is no recessiveness or dominance), and an individual's skin color results from the sum of the effects of the alleles at the three loci. Thus, individuals with genotype *aabbcc* have the darkest skin, and those with genotype *a'a'b'b'c'c'* have the lightest skin. As the table illustrates, 62 intermediate genotypes are possible, producing 7 different phenotypes for skin color. Note the relative frequency of genotypes producing the different phenotypes. From the model, one would expect greater numbers of individuals in a population to be medium to intermediate in skin color. This is generally the case in nature, although probably no natural system fits this idealized model exactly. Each allele would typically produce a somewhat different effect, and other factors, such as the environment and other gene interactions, might cause somewhat different ratios.

Other Types of Gene Interactions. In contrast to the polygene model, many heritable traits are produced by the combined but nonadditive action of genes at two or more loci. In a type of gene interaction called **epistasis**, one or more alleles (epistatic genes) at one locus *prevent* genes at another locus from being expressed. Epistasis is illustrated by the inheritance of coat color in guinea pigs. The expression of the coat color alleles (designated *B* and *b*; see Figure 15.3) is dependent on epistatic alleles at another locus that governs the production of an enzyme necessary in the early stages of pigment

Table 15.4 A Model Illustrating How Polygenic Inheritance Could Produce Skin Color Variation in Humans

Phenotypes

(Vertical label at left of genotype entries: GENOTYPES)

Darkest	Darker	Medium Dark	Intermediate	Medium Light	Lighter	Lightest
aabbcc	a'abbcc	a'a'bbcc	a'a'b'bcc	a'a'b'b'cc	a'a'b'b'c'c	a'a'b'b'c'c'
	aa'bbcc	a'ab'bcc	a'a'bb'cc	a'a'b'bc'c	a'a'bb'c'c'	
	aab'bcc	a'abb'cc	a'a'bbc'c	a'a'b'bcc'	a'a'b'bc'c'	
	aabb'cc	aa'b'bcc	a'a'bbcc'	a'a'bb'c'c	a'ab'b'c'c'	
	aabbc'c	aa'b'bcc	a'ab'b'cc	a'a'bb'cc'	aa'b'b'c'c'	
	aabbcc'	a'abbc'c	a'ab'bc'c	a'a'bbc'c'	a'a'b'b'cc'	
		a'abbcc'	a'ab'bcc'	aa'b'b'c'c		
		aa'bbc'c	aa'b'b'cc	aa'b'b'cc'		
		aa'bbcc'	aa'b'bc'c	aa'bb'c'c'		
		aab'b'cc	aa'b'bcc'	aa'b'b'c'c		
		aab'bc'c	aa'bb'c'c	aab'b'c'c'		
		aab'bcc'	aa'bb'cc'	a'ab'b'c'c		
		aabb'c'c	aa'bbc'c'	a'ab'b'cc'		
		aabb'cc'	a'abb'cc'	a'abb'c'c'		
		aabbc'c'	a'abb'c'c	a'ab'bc'c'		
			a'abbc'c'			
			aab'bc'c'			
			aabb'c'c'			
			aab'b'cc'			
			aab'b'c'c			
6*	5	4	3	2	1	0

*Number of dark alleles (those with no prime) in each genotype.

synthesis. If the epistatic alleles produce a non-functional enzyme, the coat color alleles cannot be expressed at all and the animal becomes an albino, lacking pigment.

Factors Affecting Gene Expression

As we have seen, genes represent the *potential* for traits to develop; they are not phenotypic traits themselves. Geneticists refer to the proportion of individuals that show the effects of a gene that they carry as the **penetrance** of the gene. They call the *amount* of phenotypic expression that a gene produces, in individuals that exhibit penetrance, the gene's **expressivity**. Penetrance and expressivity may depend on many types of interactions among genes, some of which we have just discussed, as well as on environmental conditions. There are many examples of environment influencing gene expression. Figure 15.8 shows one such example: the distribution of hair colors on Himalayan rabbits, which is partly a function of genetics and partly a function of body temperature.

Figure 15.8 Himalayan rabbit. These animals have dark extremities because one of the genes involved in pigment synthesis produces a temperature-sensitive enzyme that is active only at cooler temperatures. As a result, the animal's cooler extremities are dark, but hair on the warmer body is white. Siamese cats have dark extremities for the same reason. If a patch of white hair is shaved from a Himalayan rabbit's back and an ice pack is applied to the area, dark hair grows into the patch. If no ice is applied, the patch fills in with white hair.

THE MOLECULAR BASIS OF INHERITANCE

The discovery of the role of **deoxyribonucleic acid (DNA)** in inheritance was one of the great achievements in biological science. Research on DNA began in earnest in the 1950s and culminated in the realization that genes and chromosomes are composed of DNA and that this molecule is the carrier of hereditary information in all organisms. As described in Chapter 2, a chromosome contains one long DNA molecule. A gene is a discrete region of a DNA molecule that has a specific function, such as the production of a particular phenotypic trait.

DNA Structure and Function

A DNA molecule consists of two strands of linked monomers called **nucleotides** twisted together as a double helix. Each nucleotide has three components: the five-carbon sugar deoxyribose, a phosphate group, and a nitrogen-containing organic base (Figure 15.9). DNA contains four types of nucleotides, each with a different base. **Adenine (A)** and **guanine (G)** are relatively large bases collectively called **purines**; **cytosine (C)** and **thymine (T)** are smaller bases called **pyrimidines**. Different base sequences and ratios occur in the DNA of different individuals (except members of clones or identical twins), giving each individual a unique sequence of nucleotides along the length of its DNA. In all organisms, the amount of the purine A equals the amount of the pyrimidine T, and the amount of C equals the amount of G, because each A in one nucleotide strand hooks by hydrogen bonds only to a T in the other nucleotide strand, and G is always paired with C in a similar way (see Figure 15.9c). Because of this constancy in base pairing in DNA, the sequence of bases along one nucleotide strand is complementary to that of the other strand. **Complementarity** of base pairing provides the basis for DNA's two main functions: to make precise copies of itself and to serve as a template for the synthesis of **ribonucleic acid (RNA)**, molecules required in protein synthesis.

DNA Replication. DNA synthesis, called **replication**, always precedes mitosis and meiosis and results in two identical copies of an original double helix. During replication, DNA unwinds its double helix, separating into two strands of nucleotides. Each strand then serves as a **template** for the synthesis of a new complementary strand (Figure 15.10a). Free nucleotides having bases complementary to those of the template strands are hydrogen-bonded to the templates and covalently bonded to each other. The sequence of bases is copied faithfully, producing two DNA molecules, both of which are identical to the original molecule. Because each new DNA double helix is made from a template strand and a new strand of nucleotides, DNA replication is said to be **semiconservative**.

In a process called **bidirectional synthesis**, DNA replication actually begins at several sites called **replication origins** along a DNA molecule, where enzymes open the double helix. On either side of a replication origin, the enzyme, **DNA helicase**, breaks the hydrogen bonds between complementary bases, creating what is called a **replication fork** (Figure 15.10b and c). Replication forks move away from the origin in both directions. New strands of DNA form behind each advancing fork as the enzyme **DNA polymerase** binds to each template strand and catalyzes the synthesis of a complementary strand.

Each nucleotide strand in a DNA molecule has a deoxyribose sugar at one end (designated the 3' end) and a phosphate group at the other end (the 5' end). In any DNA double helix, the two complementary strands run in opposite directions; that is, one runs from 3' to 5', while the other runs from 5' to 3' (see Figures 15.9c and 15.10b and c). These directions are important, because DNA polymerase can construct new strands of nucleotides only in a 5' to 3' direction (referring to the new strands). Consequently, in a replicating fork, DNA polymerase catalyzes construction of one new strand, called the **leading strand**, *toward* the crotch of the fork, and the other strand, the **lagging strand**, *away from* the same advancing crotch (see the arrows in Figure 15.10b and c).

Bidirectional synthesis of DNA is known to occur in many eukaryotes, and until recently, it was assumed to be the only method of DNA replication in animals. Recent studies with the amphibian *Xenopus laevis* (African clawed toad), however, indicate that this model may not be universal. Replication forks are extremely rare in *X. laevis* embryos, but single-stranded DNA molecules are common, suggesting that in this species and perhaps in other animals, DNA molecules first unwind, and then replication occurs on single strands rather than at replication forks.

Genes Expressed as Specific Proteins. Many studies have shown that certain genes are expressed as

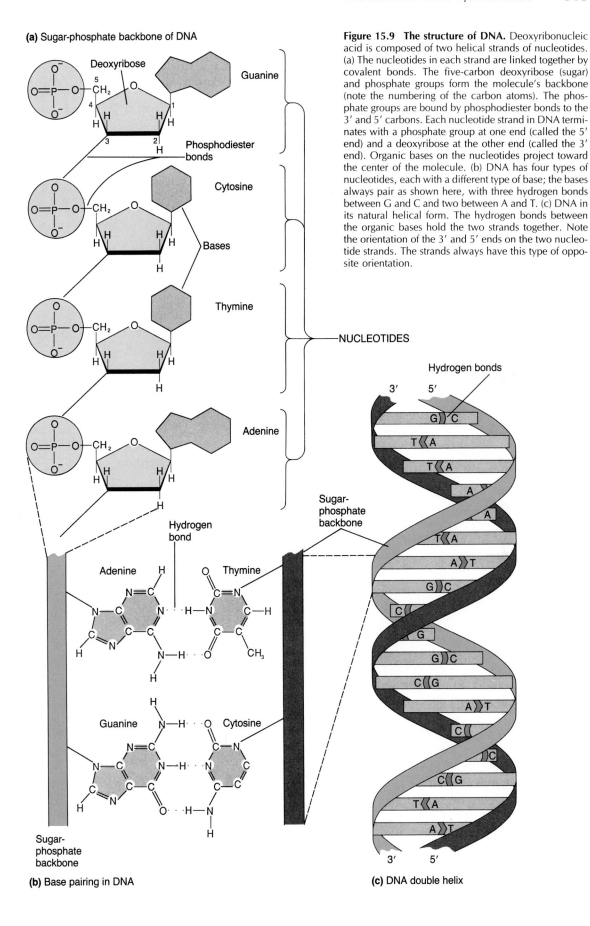

(a) Sugar-phosphate backbone of DNA

Deoxyribose

Guanine

Phosphodiester bonds

Cytosine

Bases

Thymine

Adenine

NUCLEOTIDES

Figure 15.9 The structure of DNA. Deoxyribonucleic acid is composed of two helical strands of nucleotides. (a) The nucleotides in each strand are linked together by covalent bonds. The five-carbon deoxyribose (sugar) and phosphate groups form the molecule's backbone (note the numbering of the carbon atoms). The phosphate groups are bound by phosphodiester bonds to the 3′ and 5′ carbons. Each nucleotide strand in DNA terminates with a phosphate group at one end (called the 5′ end) and a deoxyribose at the other end (called the 3′ end). Organic bases on the nucleotides project toward the center of the molecule. (b) DNA has four types of nucleotides, each with a different type of base; the bases always pair as shown here, with three hydrogen bonds between G and C and two between A and T. (c) DNA in its natural helical form. The hydrogen bonds between the organic bases hold the two strands together. Note the orientation of the 3′ and 5′ ends on the two nucleotide strands. The strands always have this type of opposite orientation.

Hydrogen bonds

Sugar-phosphate backbone

Hydrogen bond

Adenine   Thymine

Guanine   Cytosine

Sugar-phosphate backbone

(b) Base pairing in DNA

(c) DNA double helix

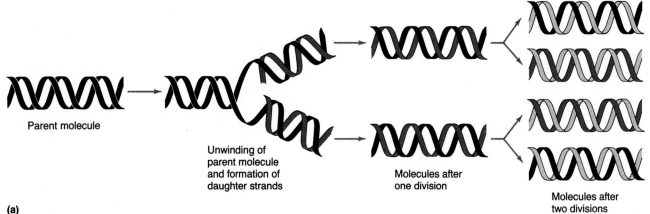

Parent molecule

Unwinding of parent molecule and formation of daughter strands

Molecules after one division

Molecules after two divisions

(a)

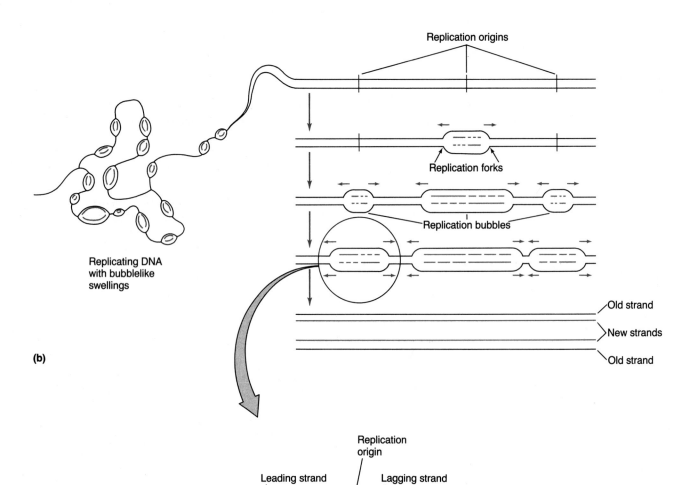

Replication origins

Replication forks

Replication bubbles

Replicating DNA with bubblelike swellings

Old strand

New strands

Old strand

(b)

Replication origin

Leading strand

Lagging strand

DNA $\{$ 5′ 3′ 3′ 5′

Organic bases

Lagging strand

Leading strand

(c)

Figure 15.10 (Opposite) Replication of DNA. Replication of DNA is semiconservative and bidirectional. (a) The double helix unwinds, and two daughter strands form on the two exposed nucleotide strands of the parent molecule. Note that the individual parental strands retain their identity through successive cell divisions. (b) Replication occurs simultaneously at many sites, which appear as bubblelike swellings on the DNA molecule. At each site, enzymes open the helix and create two replication forks, one at each end of the open portion of the molecule. (c) Replication at each fork occurs in opposite directions on the two template strands, but growth of the new strands always occurs in a 5′ to 3′ direction (i.e., from 3′ to 5′ on the template strands). Note that replication of the leading strand occurs toward the advancing crotch of each fork, while replication of the lagging strand occurs in a piecemeal fashion away from each fork. Research indicates that a short (about 10 nucleotides) piece of RNA called an RNA primer develops first on the exposed DNA templates. The RNA primer provides a 3′ sugar end on which DNA polymerase can add DNA nucleotides. Because of the 5′ to 3′ direction of synthesis, the two new nucleotide strands are synthesized somewhat differently behind a replicating fork. The leading strand forms continuously just behind the advancing fork. Synthesis of the lagging strand is more complicated. At first, RNA primer forms on the DNA template at the crotch. DNA polymerase then adds DNA nucleotides on the 3′ side of the primer (away from the crotch). As a result, the lagging strand is synthesized in separate pieces (each about 100 to 200 nucleotides long in animals). As the first piece is forming, a second one begins to develop on a second RNA primer at the crotch. The process continues, with DNA nucleotide fragments forming one behind the other. Each fragment grows until it runs into the RNA primer of the fragment just ahead of it. Repair enzymes remove the RNA primers and fill the gaps with DNA nucleotides. Eventually, the fragments are fused into a continuous strand of DNA nucleotides.

specific enzymes. Because enzymes are proteins, these studies illustrate the connection between gene action and protein synthesis. We also know that many phenotypic characters are determined by structural proteins. One of the central concepts in modern biology is that the genetic information stored in DNA directs the synthesis of thousands of different proteins in an animal's cells.

How Genetic Information Is Stored. To form the basis of gene expression, DNA must contain all the instructions determining how 20 different amino acids are put together to form an organism's diverse proteins. How can so much information be stored in an individual's DNA? The only aspect of DNA that is variable and that could provide an information bank vast enough for this purpose is the sequence of purine and pyrimidine bases in the nucleotides along the length of the molecule. But how does this information bank work, or stated another way, how is information prescribing protein structure coded in DNA molecules?

Because there are only four types of nucleotide bases in DNA, but 20 different amino acids in proteins, several bases together form a code for an amino acid in much the same way that two symbols—dots and dashes—can specify 26 letters and punctuation in Morse code. How many bases are

needed in each combination to code for the 20 amino acids? Combinations of two bases allow a code for only 16 amino acids (four different bases raised to the second power, or 4^2), but combinations of three bases allow a code for 4^3, or 64, amino acids, sufficient information for the approximately 20 amino acids found in proteins as well as signals that could control the start and stop of synthesis. In all organisms, from bacteria to plants and animals, genes in nuclear DNA have been found to be based on a code of nucleotide base triplets. This means that a gene coding for a protein containing 101 amino acids must contain at least 303 nucleotides. Actually, such a gene would be considerably longer, because some nucleotide triplets are also necessary to mark the beginning and end of the gene, and others serve as control elements that may turn the gene on or off or regulate events in protein synthesis after the gene has been read.

The Role of Ribonucleic Acid (RNA). DNA in animal cells is found in the nucleus and in small amounts in mitochondria, but none is present in the cytoplasm, where protein synthesis occurs. Consequently, special molecules of RNA convey information from the genes to the sites of protein assembly. RNA is similar to DNA in that it consists of four types of nucleotide monomers. It differs from DNA in having the sugar ribose in its nucleotides instead of deoxyribose; the pyrimidine base **uracil (U)** instead of thymine; and a single rather than a double strand of nucleotides.

Cells produce several types of RNA, each of which performs a specific function in protein synthesis. **Messenger RNA (mRNA)** is synthesized in the nucleus as a gene transcript, meaning a chain of RNA nucleotides complementary to those of the gene's DNA. The molecule is initially called a precursor because it includes unusable nucleotide sequences that are clipped out as the precursor is modified into mRNA and because some nucleotides are added after transcription. Messenger RNA eventually leaves the nucleus through the pores of the nuclear envelope and travels to the cytoplasm, where it can direct the synthesis of proteins. **Ribosomal RNAs (rRNAs)** are components of ribosomes. As discussed in Chapter 2, ribosomes are synthesized in the nucleolus of the nucleus. Each ribosome consists of a large subunit and a small subunit. The subunits are made in the nucleolus and move out to the cytoplasm, where they combine and become large multienzyme complexes that function as protein assembly sites. **Transfer RNA (tRNA)**, the smallest of the RNAs involved directly in protein synthesis, delivers amino acids to the ribosomes and is the connecting link between the mRNA and the appropriate amino acid.

Transcription. **Transcription** is the formation of an mRNA molecule as a complementary copy of a gene. This process begins when a region of DNA corresponding to a gene opens into two single strands (Figure 15.11). On one of the two strands, a short sequence of nucleotides called a **promoter** binds the enzyme **RNA polymerase**. This enzyme catalyzes the formation of mRNA by moving along a length of open DNA, reading the DNA nucleotides of one strand as a template. RNA nucleotides complementary to those on the DNA become aligned on the template and form hydrogen bonds with the bases of the DNA. RNA polymerase catalyzes the

formation of covalent bonds between nucleotides aligned on the DNA template. The action of RNA polymerase resembles that of DNA polymerase during DNA replication. It links RNA nucleotides in a 5′ to 3′ direction. Note that uracil (instead of thymine) always appears in the growing RNA strand across from adenine on the DNA template. At the end of a gene, a nucleotide triplet called a **terminator** (or "stop" triplet) causes the RNA polymerase to release the RNA and leave the DNA. Released polymerases may recycle to the same or other genes, and several RNA polymerases may read neighboring genes at the same time (Figure 15.11b). Overall, transcription results in the sequence of DNA nucleotides in a gene being rewritten in the complementary language of RNA. This

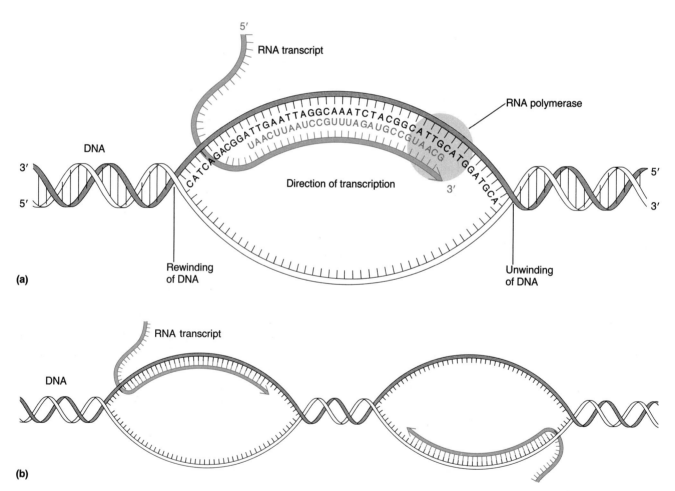

Figure 15.11 Transcription of the genetic code: mRNA synthesis. (a) A gene—a specific sequence of bases on one of the nucleotide strands in DNA—is transcribed when the DNA helix opens and a messenger RNA (mRNA) molecule is constructed on the gene. RNA nucleotides are joined in a sequence determined by the sequence of bases on the DNA template. Transcription resembles DNA replication in that enzymes add nucleotides to the mRNA molecule in a 5′ to 3′ direction. (b) Several regions (several genes) on DNA may be transcribed simultaneously and in different directions. However, for a given gene, only one DNA strand is transcribed.

Table 15.5 The Genetic Code*

Base in First (5') Position in Codon	Base in Second Position in Codon				Base in Third (3') Position in Codon
	U	C	A	G	
U	Phenylalanine	Serine	Tyrosine	Cysteine	U
	Phenylalanine	Serine	Tyrosine	Cysteine	C
	Leucine	Serine	STOP	STOP	A
	Leucine	Serine	STOP	Tryptophan	G
C	Leucine	Proline	Histidine	Arginine	U
	Leucine	Proline	Histidine	Arginine	C
	Leucine	Proline	Glutamine	Arginine	A
	Leucine	Proline	Glutamine	Arginine	G
A	Isoleucine	Threonine	Asparagine	Serine	U
	Isoleucine	Threonine	Asparagine	Serine	C
	Isoleucine	Threonine	Lysine	Arginine	A
	Methionine (START)	Threonine	Lysine	Arginine	G
G	Valine	Alanine	Aspartate	Glycine	U
	Valine	Alanine	Aspartate	Glycine	C
	Valine	Alanine	Glutamate	Glycine	A
	Valine	Alanine	Glutamate	Glycine	G

*The code consists of nucleotide base triplets written as mRNA codons that specify amino acids during protein synthesis. Except for the amino acids methionine and tryptophan, the code is redundant in that more than one codon specifies each amino acid. Most organisms that have been investigated have this genetic code, but there are a few exceptions. The codons UGA, UAG, and UAA specify different amino acids in certain protists and bacteria.

complementary copy, rather than the gene itself, is used in protein synthesis. Table 15.5 shows the genetic code written as triplets of organic bases, called **codons**, on mRNA molecules. Note that the codons are oriented with the 5' nucleotide on the left and the 3' nucleotide on the right.

Translation. **Translation** is the conversion of information in an mRNA molecule into an amino acid sequence forming a polypeptide. It involves the combined activity of mRNA, ribosomes, tRNA, and energy-containing molecules.

At least 31 different tRNAs occur in any given cell. Each tRNA has a unique sequence of 75 to 80 nucleotides and can bind covalently with only one of the 20 amino acids. A tRNA is said to be *charged* when an amino acid is covalently bound to it. The function of charged tRNA is to transfer amino acids to the correct place in a growing polypeptide. One loop in the tRNA molecule consists of a sequence of three nucleotides known as the **anticodon**. Note the orientation of the 5' and 3' ends of the tRNA molecule and that of the anticodon (Figure 15.12). This orientation is important in the process of translation.

Translation is initiated when a small ribosomal subunit binds with the 5' end of an mRNA molecule (Figure 15.13). The small subunit is moved along the mRNA until the subunit comes to the first set of three nucleotides (codon) to be translated. The start codon is AUG, the base triplet specifying the amino acid methionine. (In most proteins, this first methionine is later removed.) At the first codon, the small ribosomal subunit pauses and a large subunit associates with it. The resulting complete ribosome remains attached to the mRNA and "reads" along from the 5' to the 3' end. A methionine-charged tRNA whose anticodon ($^{3'}$UAC$^{5'}$) is complementary to the mRNA codon $^{5'}$AUG$^{3'}$ is part of the initial ribosome-mRNA complex. A ribosome has two adjacent sites (called sites A and P) where charged tRNAs can bind to it. The two sites span two codons on the mRNA molecule. A charged tRNA binds first to the mRNA at site A; then, as the ribosome moves along the mRNA to the next codon, this same tRNA becomes bound at site P.

In the example of Figure 15.13, as the ribosome moves to span the first two codons on the mRNA, the first (methionine-charged) tRNA becomes

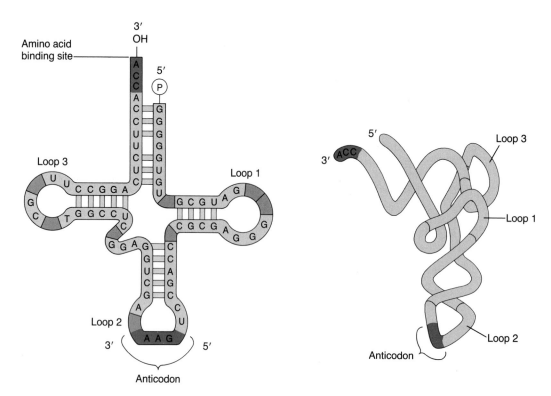

Figure 15.12 Models of transfer RNA (tRNA) molecules. The 3′ end of this relatively short molecule always has the nucleotide sequence CCA; this is the site that can bind an amino acid. A sequence of three bases near the middle of the twisted molecule serves as the anticodon. Each type of tRNA has a different anticodon and binds with a specific amino acid. The anticodon region forms hydrogen bonds with its corresponding codon, a triplet of complementary bases on an mRNA molecule, during the process of translation. The sketch on the left shows a schematic of a tRNA molecule, while its three-dimensional conformation is shown on the right.

bound to mRNA at site P on the ribosome. At the same time, a charged tRNA with an anticodon (GAC) complementary to the second mRNA codon (CUG) binds at the vacant A site. With the two tRNAs bound to the mRNA, two important events occur: (1) an enzyme catalyzes the formation of a peptide bond linking together the amino acids on the tRNAs at sites A and P, and (2) the methionine-charged tRNA releases the methionine, which remains attached (by the peptide bond) to the second amino acid (leucine); the uncharged tRNA for methionine then diffuses into the cytoplasm, where it will be recharged with free methionine. These events are repeated as the ribosome shifts from spanning the first and second codons to spanning the second and third codons. As a result, as the ribosome moves from the 5′ to the 3′ end of the mRNA molecule, it forms an increasingly longer polypeptide chain, one amino acid at a time.

Messenger RNA molecules end with a stop codon (UAA, UAG, or UGA). A stop codon binds an enzyme that reacts with the ribosome when it reaches the end of the mRNA, freeing the newly synthesized polypeptide from the ribosome and dissociating the ribosome into its two subunits. These subunits usually move on and bind to other mRNAs in the cytoplasm, reading their messages to produce other proteins. Because mRNA molecules are long, several ribosomes may read the same mRNA simultaneously and pass single file along its length. Several ribosomes in association with and simultaneously translating a single mRNA molecule are called **polyribosomes**, or **polysomes**.

Figure 15.13 (Opposite) Translation of the genetic code: protein synthesis. Charged tRNAs and ribosomes translate the genetic code contained in codons (triplets of nucleotide bases) along an mRNA molecule. Three stages are recognized: initiation, elongation of the polypeptide, and termination of synthesis (see text for details). Protein synthesis is an energy-consuming process. It has been estimated that synthesis of a polypeptide chain containing 100 amino acids requires an amount of free energy equivalent to about 400 conversions of ATP to ADP. Much of the free energy used in translation comes from the high-energy compound guanosine triphosphate (GTP).

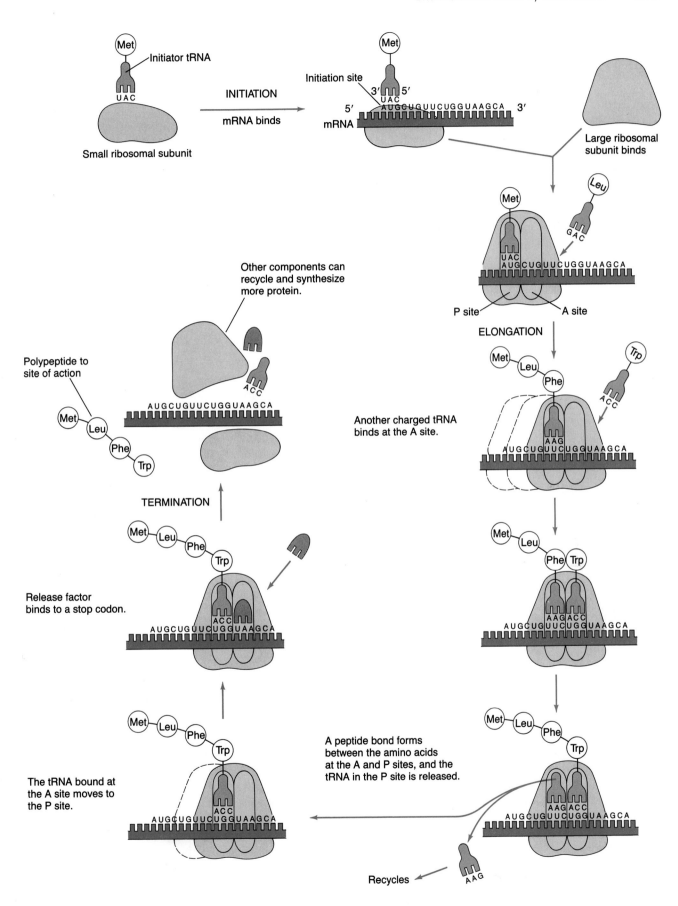

Essay: Making New Products by Genetic Engineering

A new, highly technical and commercial field of genetics has emerged since the late 1970s. Molecular geneticists can now manipulate genes in a variety of ways that can yield useful new products. Using one new technique, they can combine DNA from two different sources to form what is called **recombinant DNA**. In many cases, DNA that codes for a desired protein is combined with a small circular DNA molecule called a **plasmid**. Plasmids reside in the cytoplasm of most bacterial cells, where they replicate independently of the bacterial chromosome. To form recombinant DNA, plasmids are isolated from bacterial cells, specific enzymes are used to cut and splice the plasmids to the DNA that codes for a desired protein, and the resultant recombinant DNA is inserted into other bacteria. These bacteria can then be selectively cultured, and they will synthesize the protein. Thus, plasmids are used as vehicles to transfer DNAs from other sources into bacteria for mass production of useful proteins.

Several commercial uses of this genetic engineering technique have been developed. The genes for human insulin and growth hormone, for example, have been inserted into bacteria. The bacteria produce the hormones, which can then be isolated from bacterial cultures and purified. Hormones produced in this way are marketed and used to treat people with hormone deficiencies. This type of technology is replacing older, less efficient methods of extracting insulin from the pancreases of slaughtered cattle and pigs, and growth hormone from the pituitaries of human cadavers.

Genetic engineering techniques are also being employed to produce vaccines. Currently, most vaccines used to immunize people and domestic animals against viral diseases contain heat-killed or chemically killed viruses. The vertebrate immune system responds by making antibodies to the surface proteins of the viruses so that any subsequent infection by similar live viruses may be combated. Such procedures carry some risk because all of the viruses in a vaccine may not be killed when the vaccine is prepared. By isolating only the viral genes that code for surface proteins and inserting these genes in bacteria, researchers can create a bacterium that produces viral surface proteins. These surface proteins can then be isolated from the bacterial culture and used as a vaccine that is safe because it does not contain any live viruses.

Researchers are also trying to engineer organisms that would be useful in environmental applications. It may be possible to produce strains of corn and other grains with recombinant DNA that makes these plants more resistant to frost or able to produce their own nitrogen fertilizer. Moreover, patents have been issued for genetically engineered bacteria that break down oil spilled in shipping accidents, as well as for bacteria that are pathogenic to insect pests that feed on crops. There is some concern that creating new organisms by genetic engineering may cause future environmental problems, but the potential for increasing food production and fighting disease makes genetic engineering attractive to scientists and industrialists alike. **Biotechnology**, a broad field involving the commercial exploitation of genetically altered organisms, such as bacteria with recombinant DNA, promises to be a major industry in the near future. It is currently developing at an extremely rapid rate as researchers discover new ways to employ genetic engineering and as the products of recombinant DNA studies become useful and marketable.

Control of Gene Expression

Most genes are not constantly active. Some proteins are synthesized more or less continuously to fulfill a continuing need for metabolic enzymes or cell structures, but others are necessary only during embryonic development, and still others are needed only during reproductive periods. Differences in gene expression also occur among different types of cells. Pigment synthesis genes may be active only in skin cells, and antibody production genes only in lymphocytes. Gene regulation is one of the most active areas of research in genetics today. A complete understanding of how genes are regulated will provide a powerful tool for treating certain diseases and improving animal and plant breeds.

There are several types of gene control mechanisms. Controls may operate in the nucleus, where they affect transcription or mRNA formation from precursors, or they may act on the process of translation in the cytoplasm. In the nucleus, certain hormones may regulate target cell activity by causing specific genes to transcribe mRNA. (Steroid hormone action is discussed in Chapter 11, p. 246; Figure 11.1a.) Control can also be effected by the loss or gain of certain genes. In nematodes and certain flies, for example, portions of chromosomes are lost during embryonic development; consequently, mature cells do not contain the original complete complement of genes. In contrast, DNA in the salivary glands of larval fruit flies is replicated thousands of times during development to form giant chromosomes. This provides multiple copies of genes, RNA production may increase, and there is usually an associated rise in the rate of protein synthesis. Genes may also be regulated by chemical modification of DNA. Researchers have found that the presence of methyl (CH_3) groups on cytosine bases in DNA nucleotides is associated with certain types of gene inactivity in some animals.

In contrast to control mechanisms that affect genes directly, gene activity can also be controlled after mRNA has been transcribed. Research indicates that chemical factors in certain cell nuclei regulate which mRNA precursors become mRNA that will be translated. After mRNA is in the cytoplasm, the amount of mRNA that forms initiation complexes with ribosomes can determine how much of a given protein is actually synthesized.

Changes in DNA

Random chemical changes in the nucleotide bases of DNA occur at a low but constant rate in cells. Ultraviolet light, radiation, heat, and certain chemicals can speed spontaneous reactions, such as those that convert cytosine to uracil, those that cleave guanine from nucleotides, and others that produce cross-bridge formation between adjacent thymines. If these changes were not corrected, a substantial number of harmful genetic mistakes would accumulate during an animal's lifetime. Consequently, cells have enzymes that catalyze the repair of damage to their DNA. Repair enzymes can recognize changed nucleotides because such nucleotides contain bases not matching those on the opposite strand. The enzymes excise such bases and insert a matching nucleotide, correcting the mistake. One reason why DNA evolved as a double helix may have been to provide a reference template for these repair mechanisms.

Mutations are relatively stable, heritable changes in DNA. In the long term, mutations are vital, because they are the ultimate source of genetic variation that allows animal species to evolve in a changing environment. In the short term, however, most mutations are harmful to cells. Although there is always a small chance that a mutation will improve a gene's product, it is unlikely that any genetic change would be beneficial, because the genes that an organism possesses are the product of millions of years of evolution. Thus, it is vital to an individual organism that its DNA remain constant.

Gene Mutations. Failure to correct local alterations in the DNA nucleotide base sequence leads to changes at the gene level. Such gene mutations may be passed to future generations as newly formed alleles. Gene mutations are rare; they occur at a rate of only about one per 10^9 genes per DNA duplication. **Point mutations**, which are local changes involving one or a short sequence of nucleotide bases, occur most frequently. Point mutations may occur in three ways: by *addition* of a new base (or bases), by the *deletion* of an existing base (or bases), or by the *substitution* of one nucleotide base for another.

The addition or deletion of a single nucleotide can seriously disrupt protein synthesis by causing a condition known as a **frameshift**. Because an mRNA molecule is translated by having its information read as base triplets from beginning to end, the addition or removal of a base throws the reading process out of register. We can more easily understand a frameshift by considering a simple analogy in which three-letter words represent base triplets and the words go together to form a sentence, representing a gene:

THEMANSAWTHECATEATTHEBUG

When read as triplets from the beginning, this makes sense:

THE MAN SAW THE CAT EAT THE BUG

If the M is deleted but the message is still read as triplets from the beginning, a new "word" is formed at each position, and the message becomes nonsense:

THE ANS AWT HEC ATE ATT HEB UG

In this analogy, a frameshift, which destroys the message, occurs as the first A is read in place of the M. A similar frameshift in DNA might lead to the synthesis of a protein with different amino acids in all positions following the altered site. Such a protein may be completely nonfunctional.

The analogy also shows us that a base *substitution* may not affect the information content of the message as much as a deletion does. If B were to substitute for C, then the message would become:

THE MAN SAW THE BAT EAT THE BUG

In proteins, this type of change would cause only one amino acid difference between the proteins produced from normal alleles and those produced from mutant alleles. If such a change does not alter the function of the protein to the detriment of the organism, the change is called a **neutral mutation**.

In some cases, even a small base change can cause an abnormal phenotype. Sickle-cell anemia is an inherited recessive abnormality of human hemoglobin. A person who is homozygous for the abnormal allele has hemoglobin that causes malformation (sickling) of the red blood cells (see chapter opening photograph). Hemoglobin is a large protein composed of four polypeptides, two α globins, and two β globins (see Figure 2.10d). Analysis of the amino acid sequences of normal and sickle-cell hemoglobins has shown that only one amino acid out of 146 in the β subunit is different. The sixth amino acid in normal hemoglobin is glutamate, while in sickle-cell hemoglobin it is valine. Because the two codons for glutamate are GAA and GAG, and two of those for valine are GUA and GUG, the switch from normal to sickling can result from a single base substitution—GAA to GUA, or GAG to GUG (see Table 15.5).

Chromosome Mutations. This category of mutations involves large changes in a DNA molecule making up a chromosome, and often produces harmful or lethal effects. Sometimes, segments of a chromosome break loose and become independent. If such a chromosome fragment does not contain a centromere, it does not attach to the spindle during cell division. Consequently, the fragment ends up in only one daughter cell. The daughter cell lacking the genes contained on this piece of the chromosome is said to have suffered a **deletion**. In other cases, the fragment attaches to another chromosome, and DNA repair enzymes join it to the DNA of that chromosome in what is called a **translocation**. And sometimes, a fragment will rejoin its own chromosome after rotating 180°, causing what is known as an **inversion**. If an inversion occurs in the middle of a gene, the cell may produce an entirely new protein, because the order of codons in part of the gene is reversed.

☐ SUMMARY

1. Alleles are alternative forms of genes that occur at loci on pairs of homologous chromosomes. When the alleles are the same, an animal is homozygous. When they are different, the animal is heterozygous. In heterozygotes, the dominant allele is expressed and masks the presence of a recessive allele. Occasionally, alleles are codominant, with neither masking the other. Alleles segregate from each other during meiosis.

2. Genes located on sex chromosomes may show inheritance patterns that differ from the patterns of genes on autosomes (nonsex chromosomes), because one of the sexes may have only one sex chromosome while the other sex has two.

3. According to the principle of independent assortment, genes located on different pairs of homologous chromosomes are inherited independently of each other. Linked genes are on the same chromosome and tend to be inherited together. Crossing over between homologous chromosomes can break the linkage and create new combinations of alleles.

4. Genes are not traits; they merely represent the *potential* for traits to develop. The expression of genes is influenced by the environment and often by other genes. Genes are composed of portions of DNA, which is a long double helix of nucleotides. When cells divide, the DNA replicates as each old strand serves as a template for the synthesis of a new complementary one. Chromosomes consist of proteins and a single long molecule of DNA that contains the genes in linear order.

5. DNA replication and protein synthesis depend on the specificity of organic base pairing in DNA. In DNA, adenine always pairs with thymine, and guanine with cytosine. RNA contains uracil instead of thymine.

6. Genetic information is expressed as proteins. The information is stored in the genes as a series of triplets of nucleotide bases that specifies the order of amino acids in proteins.

7. There are several types of RNA in cells: mRNA is transcribed on DNA and carries the genetic information to ribosomes in the cytoplasm, where the information is translated into the amino acid sequence in proteins; during translation, tRNA delivers the correct amino acid to the mRNA at the ribosomes; and rRNAs occur in ribosomes. Translation of the genetic code into proteins depends on triplet code (codon-anticodon) interactions between mRNA and tRNA.

8. Gene expression is controlled by mechanisms that influence the availability of DNA for transcription and the interaction of mRNA with ribosomes.

9. Mutations are heritable changes that occur in DNA. Local mutations, involving additions, deletions, or substitutions of local areas of nucleotides in DNA are gene mutations. Chromosome mutations involve changes in large segments of DNA.

☐ FURTHER READING

Ayala, F. J., and J. A. Kiger, Jr. *Modern Genetics.* 2d ed. Menlo Park, Calif.: Benjamin/Cummings, 1984. *A modern textbook in genetics; contains a discussion of inheritance patterns integrated with molecular genetics.*

Croce, C. M., and G. Klein. "Chromosome Translocation and Human Cancer." *Scientific American* 252(1985): 54–60. *Describes how cancer-causing genes may be activated in cells of the immune system when segments of DNA are exchanged between chromosomes.*

Darnell, J. E. "RNA." *Scientific American* 253(1985): 68–87. *Describes how RNA may have been the first genetic material, even though in modern organisms it only serves as an intermediary in protein synthesis.*

Felsenfeld, G. "DNA." *Scientific American* 253(1985): 58–67. *Describes the interaction of DNA with various regulatory molecules.*

Zimmerman, B. K. *Biofuture.* New York: Plenum Press, 1984. *An introduction to genetic engineering and its implications for the future, written for the general public.*

CHAPTER 16

Animal Development

The diamondback terrapin (Malaclemys terrapin) hatching. This and other turtle species, many other reptiles, and all birds have shelled eggs. Turtles lay eggs in nests dug in soil or sand, and the parents do not care for the eggs. Hatching turtles cut through the eggshell using a cornified growth (egg tooth) on their upper jaw; they are well developed and self-sufficient when they hatch. The egg tooth is lost soon after hatching.

For centuries, biologists have been intrigued by how single-celled zygotes develop into animals. During the early eighteenth century, it was believed that eggs or sperm contained miniature adults that simply grew to mature size. Gradually, this idea gave way to the modern theory of **epigenesis**, which holds that an individual develops anew from a fertilized egg. We now know that a zygote, rather than carrying a miniature adult, carries a blueprint in the form of genetic material for the construction of an animal. A fertilized egg has the capacity to develop into a whole organism, which, depending on the species, may consist of from fewer than ten cells to over a trillion (10^{12}) cells.

The development of an animal from a zygote involves several processes. Increase in cell number and growth in size occur as a result of mitosis and cytoplasmic division. As cell numbers increase, the cells become specialized to perform specific functions. The process of cell specialization, or the generation of cell diversity, is called **differentiation**. Because nuclear divisions during development are mitotic, all cells of an adult are members of a clone (all are genetically identical). Understanding the molecular basis of how clonal cells become differentiated is a major research objective in modern biology. Another process of development, called **morphogenesis** (meaning the origin of form), occurs in animals: As cells differentiate, they become organized into cell layers, tissues, and organs.

The study of development from the zygote stage to birth (or hatching) is called **embryology**. Modern developmental biology encompasses much more than this, however, for developmental processes occur in adults as well as in embryos. Following embryonic stages, many animals undergo **direct development**; that is, their embryos develop into adultlike **juveniles**. Other animals undergo **indirect development**, meaning they have one or more **larval stages** that are distinct (often in form, function, and habitat) from the adult. Developmental biologists are especially interested in the complex series of changes in body organization that occur as larvae transform into adults. We examined some of these developmental events under the topic of metamorphosis in Chapter 11 (see p. 260 and Figure 11.11).

Today, developmental biology encompasses the study of all stages of an organism's life cycle. Cells in many tissues maintain and replenish themselves by continuing to divide and differentiate throughout life, and cells in adult animals that are capable of regeneration undergo growth, differentiation, and morphogenesis. Many of the diverse aspects of development are discussed in later chapters in sections titled "Reproduction and Development." In this chapter, we will concentrate on events in fertilization, early differentiation, and morphogenesis.

FERTILIZATION AND EGG ACTIVATION

This section considers the main events surrounding and including **fertilization**, the fusion of the genetic material (nuclei) of sperm and egg. We will first examine these processes in sea urchins and mammals, two well-studied animal groups. Sea urchins represent animals that have external fertilization, whereas mammals represent species in which sexual reproduction includes copulation and internal fertilization.

Sea Urchins

The eggs of sea urchins and many other marine animals are surrounded by two thin, gel-like glycoprotein coverings—an outer **jelly coat** and an inner **vitelline envelope** (Figure 16.1). Located just outside the plasmalemma, these coverings protect the egg against physical damage and invasion by microorganisms and play an important role in the interaction of sperm with eggs. A female sea urchin releases a yellow cloud of thousands of eggs into the sea. Simultaneously, she releases pheromones that induce nearby males to shed millions of sperm. At first, the sperm are inactive, but contact with seawater causes them to swim about randomly, finding the eggs by chance.

Sperm-Egg Interactions and Fertilization. Thousands of sperm may attach to an egg, but normally only one will penetrate the egg cytoplasm and fuse with the egg nucleus. As sea urchin sperm contact an egg surface, they undergo what is called the **acrosome reaction** (Figure 16.1b). A specific polysaccharide component of the egg's jelly coat elicits this reaction. First, the membrane surrounding the **acrosomal vesicle** in the sperm head breaks down, and enzymes released from the vesicle digest a passageway into the jelly coat. At the same time, the sperm's cap elongates into a slender **acrosomal process** that extends to the egg's vitelline envelope. A protein called **bindin** on the surface membrane of the acrosomal process attaches the sperm to receptor molecules (large glycoproteins) on the vitelline envelope. The bindin-glycoprotein reaction is species-specific; that is, bindin will adhere only to glycoprotein on eggs of its species. After the sperm is bound, enzymes from the acrosomal process digest an opening in the vitelline envelope, and the plasma membranes of the sperm and egg fuse. Microvilli on the egg surface near the fusion point then elongate, forming a **fertilization cone**. The fertilization cone contracts, pulling the sperm into the egg (Figure 16.1c). In the actual process of fertilization, the haploid nuclei of the sperm and egg (each called a **pronucleus**) fuse, forming a diploid **fusion nucleus**, or **zygote nucleus** (Figure 16.1d).

Prevention of Polyspermy. **Polyspermy**, which is fusion of more than one sperm pronucleus with an egg pronucleus, must be prevented; otherwise, the resulting triploid or greater number of chromosomes in the zygote nucleus may cause abnormal development. Two mechanisms activated by entry of a single sperm into an egg prevent polyspermy. By piercing sea urchin eggs with needlelike microe-

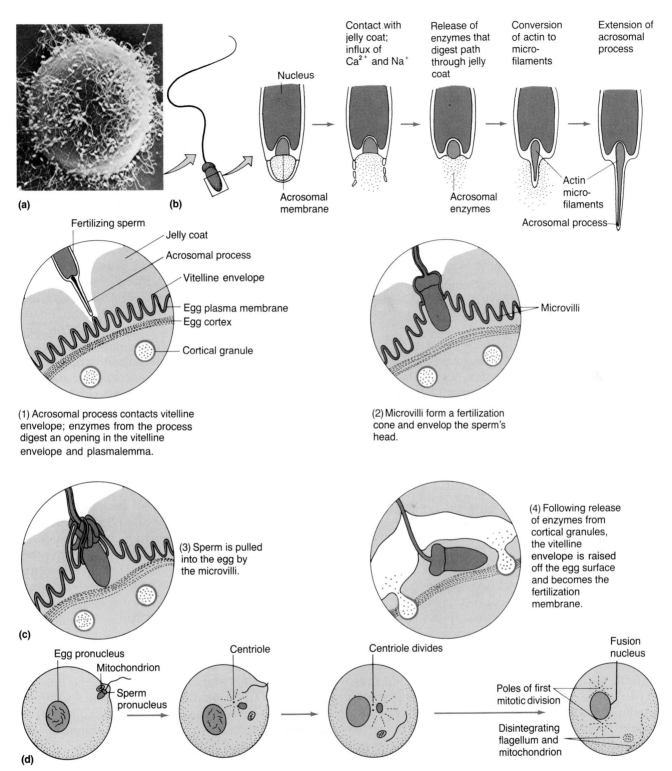

Figure 16.1 Fertilization of a sea urchin egg. (a) Thousands of sperm may bind to an egg but only one will fertilize it (775 ×). (b) The acrosome reaction. When a sperm contacts the jelly coat of an egg, sodium and calcium ions enter the sperm head and cause an increase in intracellular pH, which in turn causes extension of the sperm's acrosomal process. The acrosomal process contacts and binds with the vitelline envelope of the egg. (c) Microvilli that form a fertilization cone on the egg surface engulf the sperm head, and the sperm is then drawn into the egg. The fertilization membrane forms after a single sperm is brought into the egg. (d) Inside the egg, the sperm flagellum and mitochondrion disintegrate, and the sperm pronucleus fuses with that of the egg. The sperm's centriole divides and becomes the poles of the first mitotic division of the zygote.

lectrodes sensitive to certain ions, researchers can monitor changes in the ion permeability of the egg cell membrane. Within a tenth of a second after a sperm enters the egg cytoplasm, electrical changes in the egg cell membrane prevent other sperm from fusing with the membrane. Sodium ions flow into the egg; calcium (Ca^{2+}) is released into the cytoplasm from intracellular storage sites; the pH of the cytoplasm increases as hydrogen ions leave the egg; and the electrical potential across the plasmalemma switches from negative to positive. Sperm cannot fuse with the egg membrane unless its electrical potential is negative.

A second, slower mechanism prevents polyspermy by removing sperm that are bound to the vitelline envelope. Beneath the plasma membrane of a sea urchin egg are thousands of tiny sacs called **cortical granules**. As one sperm enters the egg cytoplasm, these sacs rupture through the egg surface and release enzymes between the plasma membrane and the vitelline envelope (Figure 16.1c). The enzymes digest proteins holding the vitelline envelope to the plasma membrane. Water then enters the space between the membranes, lifting the vitelline envelope, now called the **fertilization membrane**, away from the plasma membrane. Simultaneously, the cortical enzymes digest bindin on the sperm surfaces and also polymerize proteins in the fertilization membrane, making it impenetrable.

Mammals

Fertilization events in mammals are not as well understood as those in sea urchins. Studies are more difficult because fertilization occurs in the female reproductive tract (usually in the oviduct) rather than in the open sea. When ovulated, mammalian eggs are surrounded by a thick glycoprotein envelope called the **zona pellucida**, which is covered by a **corona radiata** formed of ovarian follicle cells (see Figure 16.5b). The zona pellucida is comparable to the vitelline envelope of sea urchin eggs.

To become capable of fertilizing an egg, mammalian sperm must reside for a period of one to several hours (depending on the species) in the female reproductive tract. Here they undergo a process called **capacitation**. Before capacitation, sperm cannot undergo an acrosome reaction; consequently, until sperm are safely inside the female, acrosomal vesicles will not rupture and release their enzymes. The acrosomal enzymes could cause serious damage if released in the male. Capacitation is not yet well understood, but it is known that the chemical environment of the oviduct causes the

process. Freshly ejaculated sperm undergo capacitation when they are placed in isolated fluids from the oviduct.

As an egg descends through the oviduct and contacts sperm swimming up the tract, acrosomal enzymes from a sperm digest the corona radiata, allowing sperm access to the zona pellucida. Glycoproteins on the surface of the zona pellucida recognize and bind the sperm. The binding reaction is usually species-specific, but species recognition at this level is not as critical in animals that have internal fertilization, because mating is unlikely between individuals of separate species. In common with sea urchins, ion movements in mammalian eggs may serve as a rapid block to polyspermy, and a cortical granule reaction modifies the zona pellucida, preventing further sperm binding.

Egg Activation

In sea urchins and most mammals, fertilization induces a marked increase in the metabolic activity of eggs. Fertilized eggs consume oxygen and synthesize DNA and proteins at much higher rates than unfertilized eggs. The activation process is not fully understood, but the egg's increased metabolic activity is correlated with the electrical potential changes, increase in intracellular Ca^{2+}, and rise in intracellular pH that accompany the reactions preventing polyspermy.

In many species, eggs do not become fully activated immediately after fertilization. Instead, fertilized eggs become temporarily dormant and serve as a resistant stage in the life cycle, allowing survival during unfavorable periods. Activation proceeds when favorable conditions return. (For example, see the discussion on the life cycle of rotifers in Chapter 23, p. 547; Figure 23.10.)

Yolk Distribution in Eggs

Eggs can be categorized according to the amount and distribution of yolk in the cytoplasm (Figure 16.2). Those with a small amount of yolk distributed uniformly throughout the cytoplasm are called **isolecithal eggs**. Embryos that develop from isolecithal eggs receive nourishment from their mothers or become able to feed themselves during early embryonic life. Most mammalian eggs, for example, are only about 0.1 mm in diameter and have little yolk, but most mammalian embryos are nourished in the mother's uterus. Sea urchin eggs are also isolecithal, but active, free-swimming larvae hatch rapidly from the eggs.

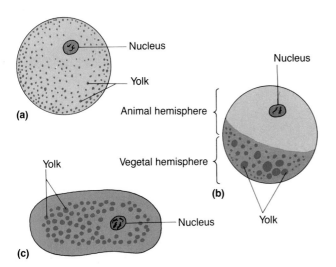

Figure 16.2 Distribution of yolk in isolecithal, telolecithal, and centrolecithal eggs. (a) An isolecithal egg has little yolk in its cytoplasm. (b) A moderately telolecithal (mesolecithal) egg has a moderate amount of yolk concentrated in its vegetal hemisphere. (c) A centrolecithal egg has much yolk, with the nucleus embedded in a small "island" of cytoplasm. During development, the yolk becomes surrounded by a peripheral layer of dividing cells.

globe of yolk; the yolk serves as a food source for developing embryos. The eggs of amphibians, squids, and octopuses are also telolecithal, although they are usually said to be moderately telolecithal, or **mesolecithal** because they have moderate amounts of yolk, and the entire egg divides. Most arthropods have unique **centrolecithal eggs** in which copious yolk is centrally located. The yolk becomes completely surrounded by the cytoplasm during development.

CLEAVAGE AND CLEAVAGE PATTERNS

Following fertilization and egg activation, the zygote begins to divide mitotically to form the multicellular embryo. The early divisions, which in most species involve no increase in size, are called **cleavage**. This process parcels the egg cytoplasm into increasingly smaller cells, called **blastomeres**. Throughout cleavage, the blastomeres are held together by glycoprotein molecules on the cell surfaces.

In contrast to the isolecithal condition, **telolecithal eggs** have much yolk, and it is concentrated in the half of the egg called the **vegetal hemisphere**. The half with the lesser amount of yolk is called the **animal hemisphere**. The eggs of birds, reptiles, and most fishes are extremely telolecithal in that a small disc of dividing cytoplasm sits atop a dense

Zygotes divide by specific, genetically determined **cleavage patterns**. The expression of these inherited patterns is influenced by the amount and distribution of yolk in the cytoplasm. There are two fundamental types of cleavage patterns. One type, called **complete** or **holoblastic cleavage**, occurs in isolecithal and mesolecithal eggs. Cleavage furrows

Table 16.1 Cleavage Patterns and Yolk Distribution in Eggs

Fundamental Pattern of Cleavage	Type of Egg	Specific Type of Cleavage	Animal Group
Holoblastic (entire egg divides)	Isolecithal (yolk scanty, distributed uniformly)	Radial	Sponges, Cnidarians, Echinoderms, lancelets*
		Spiral	Platyhelminths (many), Rotifers, Molluscs (most), Annelids (many)**
		Bilateral	Tunicates*
		Rotational	Placental mammals
	Telolecithal (mesolecithal; moderate yolk, concentrated in vegetal hemisphere)	Radial	Amphibians
		Bilateral	Molluscs (squids, octopuses)
Meroblastic (only part of egg divides)	Telolecithal (much yolk in vegetal hemisphere)	Discoidal	Fishes, reptiles, birds
	Centrolecithal (yolk in center of egg)	Superficial	Arthropods (most)

*Invertebrate chordates.

**Cleavage in nematodes resembles the spiral pattern.

pass completely through the yolk, and the entire egg divides. Typically, in isolecithal eggs, the sparse amount of yolk has little influence on holoblastic cleavage. By contrast, the polar concentration of yolk in mesolecithal eggs impedes cleavage in the vegetal hemisphere, producing fewer, larger blastomeres than in the animal hemisphere. A second type of cleavage pattern occurs in the extremely telolecithal eggs of birds, reptiles, and fishes and in the centrolecithal eggs of arthropods. The large amounts of yolk prevent the entire egg from dividing, and the eggs are said to undergo **incomplete** or **meroblastic cleavage**. Yolk actually prevents cleavage in the vegetal hemisphere in extremely telolecithal eggs or in the central zone in centrolecithal eggs. As Table 16.1 shows, there are several subcategories of cleavage within the basic holoblastic and meroblastic patterns.

Types of Holoblastic Cleavage

Radial Cleavage. In **radial cleavage**, the first two division furrows are said to be *meridional* because they form along meridians passing from pole to pole, as on a globe (Figure 16.3a). These first two furrows divide the zygote into four equal blastomeres containing similar amounts of cytoplasm and yolk. The third cleavage of each of the four blastomeres is usually at right angles to the first two and is therefore *latitudinal*. The third division furrow passes through what was the equatorial region between the animal and vegetal hemispheres in each of the four blastomeres to yield eight cells. From the eight-cell stage on, cell divisions alternate between meridional and latitudinal planes. The resulting cells are of equal size in many animals (see Figure 16.3a), but in others (e.g., sea urchins), the cells are of two or more sizes.

From about the 64-cell stage to the 128-cell stage in the development of many animals, a cavity called the **blastocoel** appears in the center of the embryonic cell mass and enlarges as the blastomeres continue to divide. Eventually, the embryo becomes a hollow ball of cells called a **coeloblastula** (commonly referred to as a **blastula**) (see Fig-

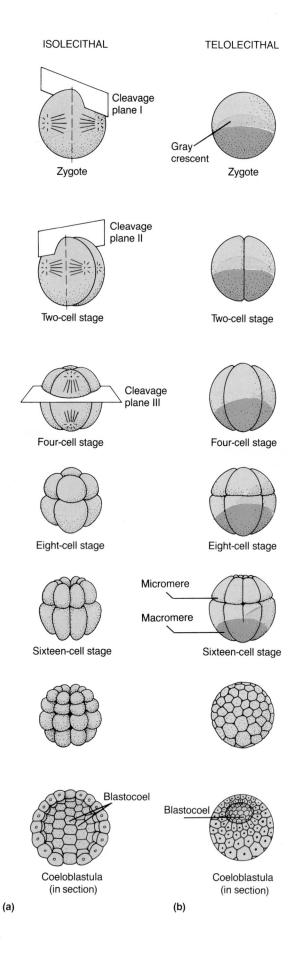

Figure 16.3 (Right) **Radial cleavage.** (a) In the isolecithal egg of a sea cucumber (an echinoderm), the first plane of division passes through the animal and vegetal poles. The second cleavage plane also passes through the poles but at right angles to the first. The third cleavage plane is perpendicular to the first two. Subsequent cleavage planes produce tiers of cells that lie directly on top of each other. Eventually, a hollow ball of cells, the coeloblastula, is produced. (b) In the telolecithal (mesolecithal) egg of an amphibian, the planes of division are as in echinoderms, except that in the vegetal hemisphere, progression of the cleavage furrows is retarded by the large amount of yolk.

(a)   (b)

ure 16.3). The blastomeres pump sodium ions into the blastocoel, and water follows osmotically, swelling the developing embryo. In many marine animals, the outer surface of the blastomeres bear cilia, which rotate the embryo inside the fertilization membrane. In sea urchins, enzymes secreted at about the 300-cell stage dissolve the fertilization membrane, and the coeloblastula becomes a free-swimming larva. Not all animals with radial cleavage form a coeloblastula. In some cnidarians, the blastula does not develop a cavity; instead, it remains as a solid body of cells called a **stereoblastula**.

Radial cleavage in the mesolecithal eggs of amphibians differs somewhat from the pattern just described. The large amount of yolk in the vegetal hemisphere retards the cleavage furrows, and division of the vegetal hemisphere lags behind that of the animal hemisphere. Also, because of the yolk distribution, the third cleavage is shifted upward toward the animal pole. As a result of this shifting and the more rapid division of the animal hemisphere, the fourth and successive divisions produce small cells called **micromeres** in the animal hemisphere and large cells called **macromeres** in the vegetal hemisphere (see Figure 16.3b). Amphibian eggs also have a pronounced pigmented area, called the **gray crescent**, between the animal and vegetal hemispheres. When the second cleavage furrows form in each of the two blastomeres at right angles to the first plane of division, they yield four blastomeres: two with and two without gray crescent material. When observed throughout the developmental period, blastomeres containing gray crescent become the left and right sides of the dorsal part of the animal, while the blastomeres lacking gray crescent become the ventral surface.

Spiral Cleavage. In **spiral cleavage**, the first two cleavages are meridional and produce four blastomeres of equal or nearly equal size, as in radial cleavage. However, the third and successive divisions are unequal, and the cleavage furrows are diagonal to the vertical animal-vegetal pole axis. During the third division, the four large cells give rise to four large macromeres in the vegetal hemisphere and four small micromeres in the animal hemisphere (Figure 16.4). Because of the diagonal

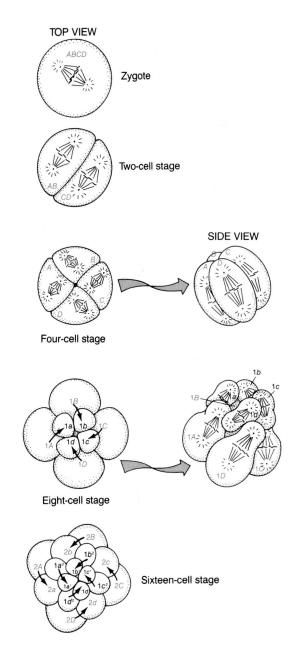

Figure 16.4 (Right) Spiral cleavage. This form of cleavage occurs in molluscs and annelids. The process viewed from the animal pole shows how the blastomeres are displaced into the cleavage furrows of underlying cells. Arrows indicate the source of the cells. Side views show the orientation of the spindles that displace the micromeres into the underlying cleavage furrows. Note that the first two cleavage divisions produce four large cells (macromeres), which, by convention, are designated A, B, C, and D. Four macromeres remain at the base (vegetal pole) throughout cleavage. The third division is latitudinal; it produces the eight-cell stage by separating a quartet of small cells (micromeres) from four macromeres. Thus, macromeres A, B, C, and D have divided to form macromeres 1A, 1B, 1C, and 1D and micromeres 1a, 1b, 1c, and 1d. Note (arrows) that the micromeres are displaced to one side (clockwise, when viewed from above) of the corresponding macromeres (see arrows). The fourth cleavage division, which is also latitudinal, produces the 16-cell stage. Macromeres 1A, 1B, 1C, and 1D divide, producing macromeres 2A, 2B, 2C, and 2D and a quartet of micromeres designated 2a, 2b, 2c, and 2d. As the macromeres divide, micromeres 1a, 1b, 1c, and 1d also divide, forming two daughter quartets ($1a^1$, $1a^2$, $1b^1$, $1b^2$, $1c^1$, $1c^2$, $1d^1$, and $1d^2$). Displacement of cells in the fourth cleavage division is counterclockwise. In subsequent divisions, alternation of clockwise with counterclockwise displacement continues, building up a blastula of macromeres and micromeres in which the path from one pole of the embryo to the opposite pole follows a spiral course, the feature that gives spiral cleavage its name.

cleavage, the micromeres do not sit directly on top of the macromeres as in radial cleavage; instead, they lie in the grooves between the underlying macromeres. In most species, when viewed from above, the micromeres resulting from the third division lie to the right of their sibling macromeres, and micromeres from the fourth division lie to the left of the macromeres. This alternating right and left displacement of micromeres continues through successive divisions, creating what appears to be a spiral arrangement of cells. As in the case of radial cleavage, spiral cleavage may yield either a coeloblastula or a stereoblastula.

Bilateral Cleavage. **Bilateral cleavage** occurs in several groups of animals, including some with isolecithal eggs and others with mesolecithal eggs (Table 16.1). In this type of cleavage, the early division planes establish a bilateral embryo, and only a single plane divides the embryo into symmetrical right and left halves. Cells on one side of the plane are mirror images of those on the other side. In certain tunicates (invertebrate chordates), color patterns in the cells reflect the bilateral symmetry.

Rotational Cleavage. Cleavage in placental mammals, called **rotational cleavage**, is unique. In the second division, one blastomere divides meridionally, but the other one divides at the equator, producing an asymmetrical four-cell stage (Figure 16.5a). (For comparison, note the symmetrical arrangement of cells in the four-cell stage of an echi-

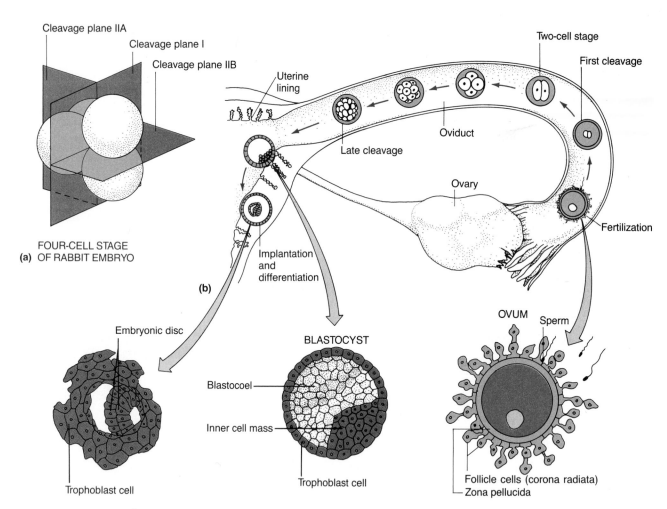

Figure 16.5 Rotational cleavage. This pattern occurs only in placental mammals. (a) Sketch of the cleavage planes in a rabbit embryo. Note the asymmetrical division of early blastomeres. (b) When ovulated, mammalian eggs are surrounded by a corona radiata composed of adhering follicle cells and a zona pellucida. Fertilization occurs in the oviduct, and cleavage proceeds as the embryo descends to the uterus. In the blastula (blastocyst) stage, trophoblast cells surround an inner cell mass, which develops into the embryo. The trophoblast develops into part of the placenta after the blastocyst implants in the uterine wall.

noderm in Figure 16.3a.) The blastomeres do not divide synchronously, and as a result, the embryos often consist of odd numbers of cells. Furthermore, at the eight-cell stage, the cells become compacted, forming a tight ball. In later stages, most of the blastomeres form an outer cell layer called the **trophoblast**, which surrounds a few internal cells called the **inner cell mass**. The blastula stage, called the **blastocyst**, consists of a rim of trophoblast cells around a fluid-filled blastocoel adjacent to the small inner cell mass (Figure 16.5b). The trophoblast cells eventually form the fetal layer of the placenta; the fetal layer attaches the blastocyst to the uterine wall of the mother. The inner cell mass develops into the embryo.

Types of Meroblastic Cleavage

Discoidal Cleavage. The eggs of fishes, reptiles, birds, and monotremes (egg-laying mammals, such as the duck-billed platypus) undergo **discoidal cleavage**. In this case, cleavage is confined to an island of cytoplasm called the **blastodisc**, situated at the animal pole of the cell (Figure 16.6a). The blastodisc divides repeatedly, with vertical cleavage planes passing down to the layer of yolk, but not into it. The resulting cell mass, called the **blastoderm**, is separated from the yolk by a **subgerminal space**. A blastocoel forms as some of the blastoderm cells move into the subgerminal space. The cell layer above the blastocoel is called the

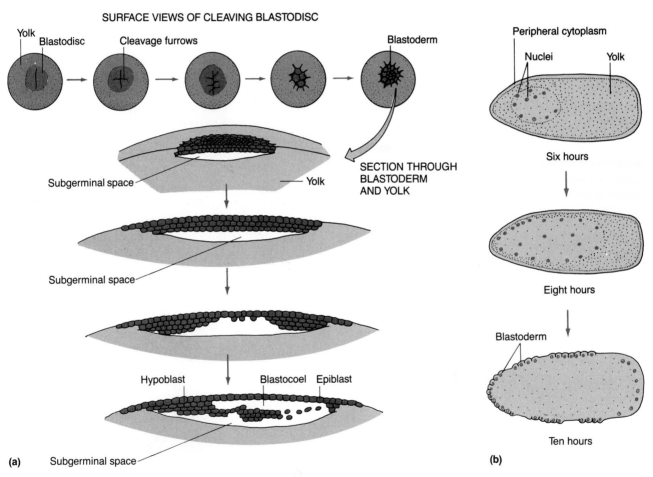

Figure 16.6 Meroblastic cleavage. (a) Discoidal cleavage in the highly telolecithal egg of a bird. Cleavage is confined to a small blastodisc on the surface of the yolk. The first several cleavage planes pass through the cytoplasm to the yolk, but not into it. Membranes separate the cells of the blastodisc from the yolk. Later division planes are parallel to the surface of the yolk, so that cell layers are built up. A blastula (discoblastula) develops as the layers separate, producing a blastocoel. (b) Superficial cleavage occurs in the centrolecithal eggs of insects. Numerous nuclei, each surrounded by a small amount of cytoplasm, are first produced in the depths of the yolk mass; the nuclei then migrate to the peripheral cytoplasm that surrounds the yolk and become separated from one another by membranes. The resultant cells form the blastoderm that differentiates into the embryo. Here we see the progress of cleavage and blastoderm formation in a moth egg 6, 8, and 10 hours after laying.

epiblast, and the lower layer, near the yolk, is the **hypoblast**. The blastula is thus a two-layered disc, called a **discoblastula**.

Superficial Cleavage. The centrolecithal eggs of most arthropods undergo **superficial cleavage**, a unique process in which cytoplasmic division occurs in a ring of cells surrounding the large yolk mass (Figure 16.6b). In an insect egg, rapid mitotic division of the zygote nucleus, located *within* the yolk, precedes cytoplasmic division. The resultant daughter nuclei migrate to the periphery of the yolk just under the egg cell membrane. Nuclei at one end of the egg later become the germ (reproductive) cells of the adult insect. Membranes enfold the remainder of the nuclei, forming a blastoderm that soon differentiates to form the embryo and its protective membranes.

GASTRULATION AND MORPHOGENESIS

The blastula stages resulting from cleavage are multicellular embryos, but otherwise they bear little or no resemblance to any adult animal. The generation of form, or morphogenesis, begins with the process of **gastrulation**, which is the production, by cellular movements and differentiation, of an embryo called a **gastrula** (Figure 16.7). The gastrula consists of an outer layer of cells, the **ectoderm**, and an inner layer, the **endoderm**. These are called the **primary germ layers** because they are the source of all cell layers and tissues of the adult animal. Most animals have a third germ layer called **mesoderm** that develops during or after gastrulation. Because

mesoderm develops from one or both of the primary germ layers, it is considered a **secondary germ layer**. Mesoderm derived from ectoderm is called **ectomesoderm**; that derived from endoderm is called **endomesoderm**. Most animals have all three germ layers and are sometimes said to be **triploblastic** (from the Greek, meaning "triple germ"). Cnidarians (e.g., hydras, jellyfish, and sea anemones) have a three-layered body, but their third layer, called mesenchyme or mesoglea, is fundamentally different from the mesoderm of other animals. Whereas cnidarian mesenchyme is derived from ectoderm, mesoderm of other animals is mainly endomesoderm.

In addition to forming the germ layers, gastrulation also performs the important function of placing the germ layers in positions corresponding to the location of tissues that the germ layers produce. The ectoderm, surrounding the gastrula, develops into the epidermis of the skin and into the nervous tissues. The mesoderm, located between the ectoderm and endoderm, forms many internal organs, as well as the muscles, skeleton, and blood (Table 16.2). The endoderm lines a blind gut tube called the **archenteron**, which develops into the adult gut. The archenteron has a single opening, called the **blastopore**. In annelids, arthropods, and molluscs, often collectively called the **protostomes** ("first mouth"), the blastopore becomes the mouth. By contrast, in echinoderms and chordates, collectively called the **deuterostomes** ("second mouth"), the blastopore does not become the mouth; in some deuterostomes, it becomes the anus, whereas the mouth develops secondarily from a distant part of the gastrula's wall. (Chapter 18 discusses the evolutionary significance of the blastopore, p. 427.)

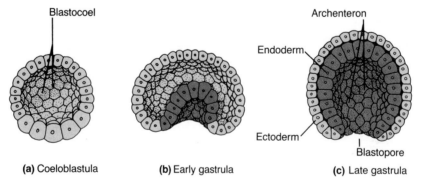

(a) Coeloblastula   **(b) Early gastrula**   **(c) Late gastrula**

Figure 16.7 Gastrulation in the isolecithal egg of the lancelet *Branchiostoma.* (a) Section of a coeloblastula. (b) This section of an early gastrula shows that invagination begins at the vegetal pole. (c) Section through a late gastrula stage, showing that the archenteron, formed by invagination, has obliterated the blastocoel.

Table 16.2 Derivatives of the Germ Layers in Vertebrates

Germ Layer	Adult Structures
Ectoderm	Epidermis and its derivatives; epithelial lining of mouth and rectum; sense receptors in epidermis; nervous system; adrenal medulla
Endoderm	Epithelial lining of digestive tract, except mouth and rectum; epithelial lining of respiratory system; liver; pancreas; thyroid; parathyroids; thymus; lining of middle ear; lining of urinary bladder
Mesoderm	Notochord; skeletal system; muscular system, including muscles of the gut; dermis of skin; circulatory system; coelom and peritoneum; excretory system; reproductive system, except the germ cells, which differentiate during cleavage before the three germ layers form

Gastrulation involves one or more of the five mechanisms shown in Figure 16.8. **Invagination** involves the infolding of cells of a blastula (usually at the vegetal pole); **epiboly** is the spreading of an outer sheet of cells around other cells; **involution** occurs as an outer cell layer turns in and expands beneath cells that remain on the outside of an embryo; **ingression** is the movement of separate outer cells into the blastocoel; and **delamination** is the formation, from a single layer of cells, of two or more sheets of cells.

Gastrulation of a coeloblastula by invagination produces the type of gastrula illustrated in Figure 16.7. This is called a **coelogastrula** because it contains an internal cavity, the archenteron; the blastopore marks the point of cell infolding. Gastrulation usually involves more than one of the five mechanisms, and not all species have a coelogastrula. Many species have a solid, two-layered gastrula called a **stereogastrula** that lacks an archenteron and a blastopore.

Echinoderms

In sea urchins and many other echinoderms, gastrulation begins with ingression; a few dozen micromeres, called **primary mesenchyme cells**, migrate into the blastocoel (Figure 16.9). At the same time, other cells in the vegetal pole invaginate, forming the blastopore and archenteron. This initial invagination results from contraction of actin

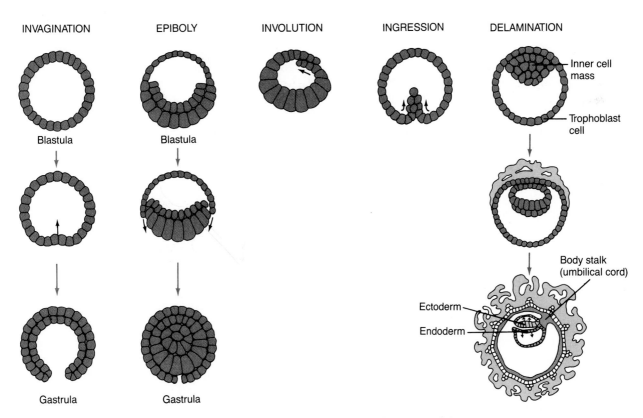

Figure 16.8 Five types of cell movements during gastrulation.

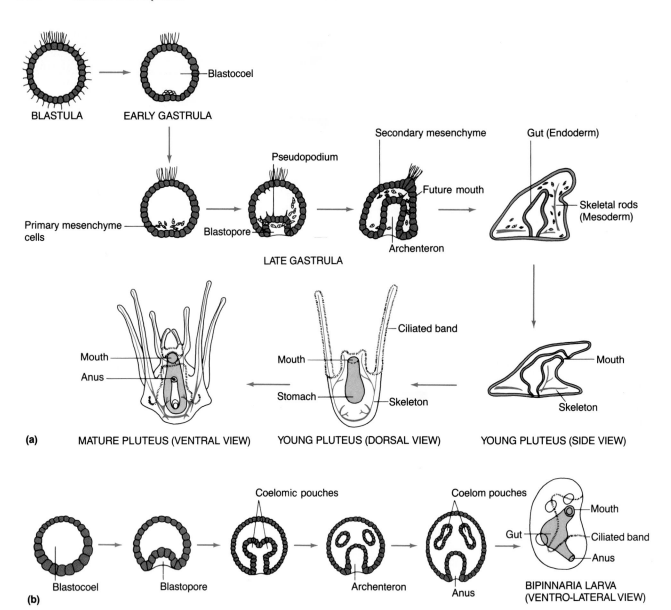

Figure 16.9 Gastrulation and mesoderm formation in echinoderms. (a) In many sea urchins, as shown here, as cells in the vegetal pole invaginate, some cells divide and give rise to primary mesenchyme, which eventually forms a transient larval skeleton of skeletal rods. Secondary mesenchyme, which forms later from the advancing archenteron, develops into mesoderm destined to become adult structures. (b) In other sea urchins and many other kinds of echinoderms such as sea stars (shown here), pouches that bud off of the archenteron form the mesoderm, which later produces adult structures. The coelom develops from the cavity in the pouches. Note the difference between the sea star bipinnaria larva and the pluteus of the sea urchin.

microfilaments in the cytoplasm on the outer surface of cells at the vegetal pole. A second phase of invagination ensues as cells called **secondary mesenchyme**, located at the advancing end of the archenteron, extend pseudopodia to the wall opposite the invagination. The pseudopodia attach to the wall and contract, drawing the archenteron across the blastocoel. The secondary mesenchyme becomes the adult mesoderm. In most sea urchins,

before the advancing archenteron contacts the wall of the gastrula, the secondary mesenchyme cells break free from the archenteron and migrate into the blastocoel (Figure 16.9a). Here, as the embryo develops, they proliferate and differentiate into mesodermal structures, including the coelom. In some sea urchins, and in many other echinoderms, such as sea stars, the mesoderm forms from one or two outpocketings, or pouches, of the archenteron

(Figure 16.9b); these pouches then bud off the archenteron and develop into mesodermal structures. The cavity within the pouches becomes the coelom. Because the coelom in these forms develops from sacs from the archenteron, echinoderms are said to be **enterocoelous**. Echinoderms share this type of mesoderm and body cavity formation with other deuterostomes. (Phylogeny of deuterostomes is discussed in Chapter 18, p. 435.)

The echinoderm mouth develops where the advancing front of the archenteron fuses with the gastrula wall. Typical of many deuterostomes, the anus develops in the region near the blastopore. As the mouth forms in the late gastrula of the sea urchin, bands of cilia develop on the gastrula's surface, and it becomes a **pluteus larva**, which filter-feeds on bacteria and algae (Figure 16.9a). The developing sea urchin is now self-sufficient and drifts with ocean currents, feeding and growing. Eventually, the pluteus will metamorphose into a juvenile sea urchin as it settles on the bottom. Gastrulation is similar in other echinoderms, but different types of larvae result (see Figure 16.9b).

Annelids and Molluscs

Most annelids and molluscs have spiral cleavage, which produces either a coeloblastula or a stereoblastula. Gastrulation by one or more of the five mechanisms produces either a coelogastrula or a stereogastrula. The blastopore, if present, forms the mouth. Typically, four macromeres and four micromeres resulting from spiral cleavage occupy the center of the annelid and mollusc gastrula, and division of these cells produces most or all of the endoderm (Figure 16.10a). In annelids and molluscs, one of the micromeres, actually the 4d cell in the lineage from the fertilized egg, divides into two **teloblast cells**, which then divide repeatedly to form two cell masses called **mesodermal bands** (see Figure 16.10a). Eventually, the bands split internally, forming pouches that develop into the coelom (Figure 16.10b). Because their mesoderm and coelom formation involves splitting of mesodermal bands, annelids, molluscs, and certain other protostomes are said to be **schizocoelous** (from the Greek *schizo*, "split"). (Phylogeny of schizocoels is discussed in Chapter 18, p. 435.) In many marine annelids and molluscs, the gastrula becomes a free-swimming, top-shaped **trochophore larva**, another hallmark of the protostomes (Figure 6.10a), quite different from the larvae of echinoderms and other deuterostomes.

Vertebrates

Vertebrates are deuterostomes and share certain developmental features with echinoderms; but ver-

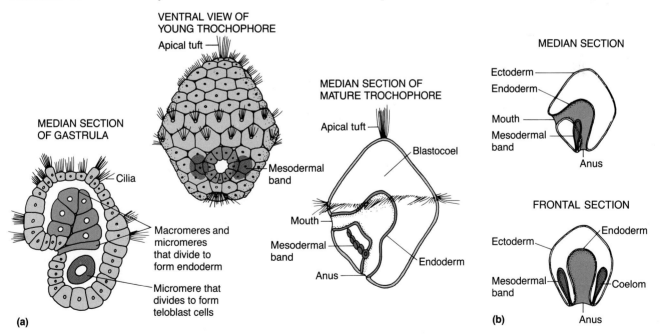

Figure 16.10 Gastrulation in a gastropod mollusc with a spirally cleaving egg. (a) Gastrulation of the ciliated blastula of a limpet (a gastropod mollusc) occurs by ingression. The teloblast cells give rise to bands of mesoderm. Gastrulation produces a free-swimming, ciliated trochophore larva. (b) The coelom develops from internal pouches in the mesodermal bands, as seen here in median and frontal sections.

tebrate gastrulation and the events following it have many unique aspects.

Gastrulation and Later Development in Amphibians.

Gastrulation in the yolk-laden coeloblastula of amphibians involves invagination, epiboly, and involution. Gastrulation begins in the area of the gray crescent near the equator as cells form a slitlike blastopore (Figure 16.11a). Involution then occurs as endodermal cells roll over the lip of the blastopore, proliferate into the blastocoel, and form the archenteron. Epiboly occurs as cells from the animal hemisphere migrate downward, converge on the dorsal lip of the blastopore, and involute. Inward movement of cells squeezes some of the yolk-laden cells into the blastopore, where they form the **yolk plug**. Cells that involute at the blastopore form both endoderm and mesoderm. Inside

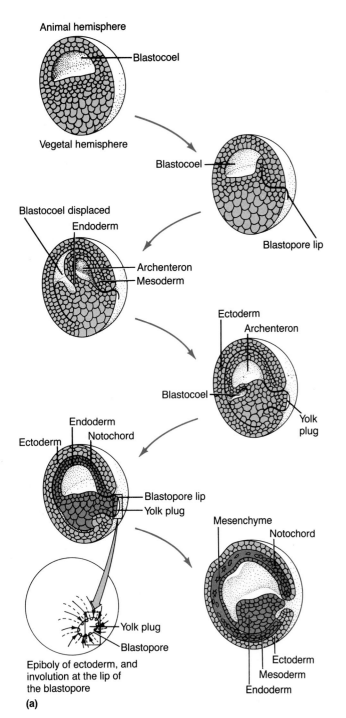

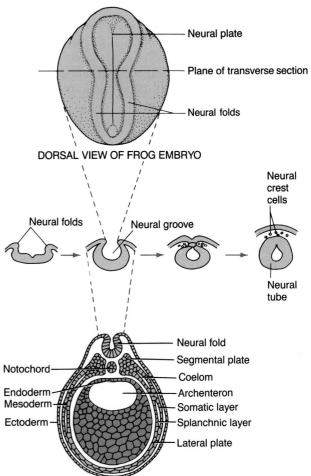

(b)   TRANSVERSE SECTION OF FROG EMBRYO

Figure 16.11 Gastrulation and ensuing morphogenetic events in amphibians. (a) Stages in gastrulation, shown in sagittal section. Note the progressive formation of the blastopore lips and the yolk plug in the blastopore. Also note the gradual elimination of the blastocoel as the archenteron increases in size. (b) Development following gastrulation. A notochord develops from the mesoderm dorsal to the archenteron. Neurulation occurs as ectoderm above the notochord rises up to form a neural plate. The center of the plate then sinks inward to form a neural fold. Eventually, the neural folds at the edge of the groove fuse, forming the neural tube beneath the outer covering of ectoderm. Mesoderm lateral to the notochord forms somites and lateral plates. Outer (somatic) and inner (splanchnic) layers of the lateral plates line the coelom.

the gastrula, mesoderm cells gradually fill the area between the archenteron and the inside of the ectoderm. As epiboly continues, involution occurs around the entire rim of the blastopore. Gastrulation is complete when all endoderm and mesoderm are inside the embryo and when the entire surface, including the blastopore with its yolk plug, is covered with ectoderm.

In later development, the mesoderm differentiates into the **notochord** and **segmental somites**, which will eventually form most of the skeleton and skeletal muscles, and **lateral plates**, which will form the coelom and urinary system. The ectoderm forms the skin, and ectoderm on the dorsal surface also gives rise to the nervous system (Figure 16.11b). Morphogenesis of the nervous system, called **neurulation**, involves an interaction between the mesoderm and the overlying ectoderm. The notochord and cells associated with it induce the overlying ectoderm to thicken, forming the **neural plate**. This type of influence of one tissue on another, called **embryonic induction**, is common in animal development. Neurulation continues as the central area of the neural plate gradually sinks inward and its edges arch over to form a **neural tube**. Cells of the neural tube and associated **neural crest cells** form all parts of the nervous system. The neural crest cells also form the pia mater (one of the membranes surrounding the brain), the Schwann cells that produce the myelin sheaths of nerves, pigment cells (chromatophores) throughout the body, and the adrenal medulla.

Gastrulation in Birds. The discoblastulas of birds (and those of fishes, reptiles, and egg-laying mammals) gastrulate by involution and epiboly. External cells turn under the surface layer and grow over the yolk. There is no archenteron or blastopore. The bird embryo develops entirely from the epiblast cells (see Figure 16.6). Supportive and protective membranes that grow around the developing embryo develop from the hypoblast and epiblast. Gastrulation begins with the formation of a thickened line of cells called the **primitive streak**, located in the epiblast. As the streak elongates, a depression called the **primitive groove** forms in it (Figure 16.12a). Surface cells involute through the groove and pass into the blastocoel. Thus, the groove is functionally similar to the amphibian blastopore. At the anterior end of the primitive groove, a circular cell thickening, called **Hensen's node**, functions like the dorsal lip of the amphibian blastopore. As in amphibians, epiboly causes surface cells to converge on and involute at the groove and node, forming endoderm and mesoderm cells that fill in the blastocoel. The primitive streak and groove gradually disappear after all endodermal and mesodermal cells have involuted. Beneath the disappearing primitive streak, mesodermal cells form a notochord (Figure 16.12b). The notochord then induces the overlying ectoderm to fold inward as a **neural fold** along the path of the regressing primitive streak. As the fold deepens, it gradually closes over to form a neural tube. Mesoderm on either side of the tube also organizes into somites and lateral plates, as in amphibians.

As a bird embryo gradually takes shape inside the eggshell, the hypoblast and epiblast produce four extraembryonic membranes: the **yolk sac**, the **amnion**, the **chorion**, and the **allantois** (Figure 16.12c). The yolk sac, actually an extension of the gut, surrounds the yolk and absorbs nutrients from it. The amnion forms a sac filled with fluid that protects the embryo by acting as a shock absorber and preventing desiccation. The chorion forms a continuous membrane that lines the inside of the shell and fuses with the allantois to form the **chorioallantoic membrane**. Blood vessels in the chorioallantoic membrane allow the developing embryo to exchange respiratory gases through pores in the calcified shell. Nitrogenous wastes, in the form of uric acid produced by the embryo, are also stored in this membrane and are left in the shell at hatching.

Gastrulation in Placental Mammals. Despite the lack of yolk in the embryo of placental mammals, the structures and behavior of cells during gastrulation closely resemble those of the yolk-laden eggs of reptiles and birds. Most likely, this is because birds and mammals both evolved from reptilian ancestors. The basic structures and mechanisms undoubtedly evolved in ancestral reptiles as adaptations that ensured development in eggs with large quantities of yolk. Mammals have secondarily adapted these features to development by a yolkless blastocyst in a uterus. (Vertebrate phylogeny is discussed in Chapter 28, p. 686; Figure 28.9.)

Once the mammalian blastocyst (see Figure 16.5b) has implanted in the uterus, it grows rapidly by cell division. Its inner cell mass reorganizes into an **embryonic disc** (Figure 16.13a). In a series of events very similar to those seen in bird development, gastrulation occurs by formation of a primitive streak across the hypoblast of the embryonic disc. As in birds, cells that will form endoderm and

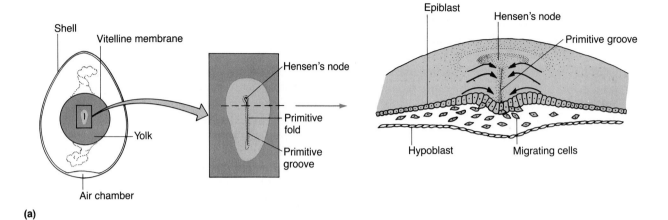

(a)

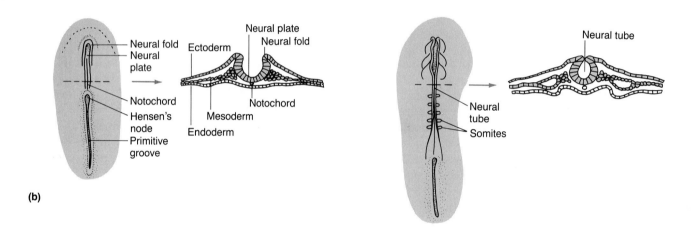

(b)

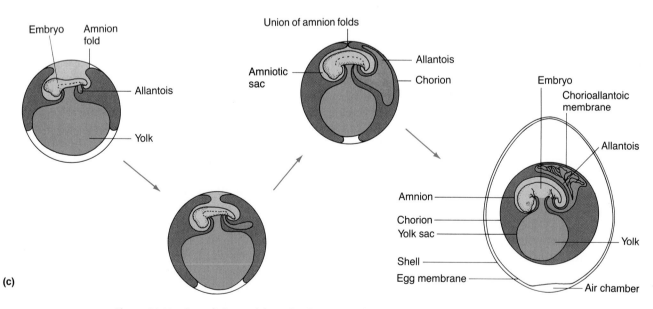

(c)

Figure 16.12   Gastrulation and later development in birds. (a) Formation of the primitive streak and groove during gastrulation. Cells from the epiblast migrate into the blastocoel to form the mesoderm. Arrows indicate epiboly and involution along the primitive groove. (b) Events in neurulation. (c) Formation of extraembryonic membranes surrounding the developing embryo.

Figure 16.13 Gastrulation and later development of the human embryo. (a) The embryo is implanted in the uterine wall. The trophoblast has become the chorion, and two membranes, the amnion and the yolk sac, have grown out from the embryo. (b) About four weeks after fertilization, the embryo is about 5–6 mm long and has a rudimentary digestive system and an actively pumping heart. (c) During the second month, rudiments of all major body organs form. Limb buds appear as paddlelike appendages without digits, and these eventually transform into arms and legs. (d) By week eight, the embryo is distinctly human in form and is called a fetus. During the remaining seven months prior to birth, the existing structures of the fetus continue to differentiate and grow. The sense organs and brain develop extensively during the second trimester (three-month period) of gestation. During the third (final) trimester, the embryo may gain up to 25 g/day. At birth, pressure caused by uterine contractions breaks the amniotic sac, and the amniotic fluid flows from the vagina. As a baby descends from the uterus and emerges through the birth canal, blood ceases to flow through the umbilical cord, and the rising carbon dioxide concentration in the baby's blood causes the brain's respiratory center to stimulate breathing movements. Uterine contractions continue after birth and expel the placenta and amniotic membranes.

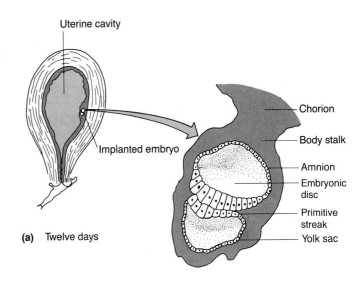

(a) Twelve days

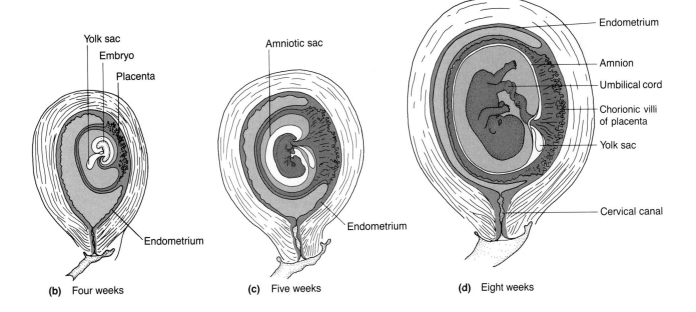

(b) Four weeks

(c) Five weeks

(d) Eight weeks

mesoderm migrate inward along the streak. Neurulation also occurs as the primitive streak fades. A yolk sac grows out from the embryonic disc beneath the embryo, and cells from the yolk sac form the germinal epithelium of the gonads and stem cells that develop into blood vessels. An **amniotic membrane** develops from the upper surface of the embryonic disc, and as in birds and reptiles, forms a sac full of shock-resistant fluid. An allantois also develops and becomes a storage sac for wastes. The **umbilical cord** contains blood vessels that circulate fetal blood to and from the placenta, where nutrients and wastes diffuse between the maternal and fetal circulatory systems (Figure 16.13b–d).

CONTROL OF DEVELOPMENT

Modern developmental biology is an active, exciting field. Much of the excitement stems from an anticipation that we may soon be able to fully understand the mechanisms that control cleavage, gastrulation, and cellular changes during differentiation and morphogenesis.

Maternal Versus Embryonic Control

Several lines of evidence indicate that the events of cleavage are programmed into an egg before fertilization and that an embryo's genes do not become active until after cleavage. Embryologists of the 1930s observed that when nuclei are removed from fertilized sea urchin eggs, the eggs still form mitotic spindles and undergo cleavage, despite the lack of chromosomes. Evidently, a nucleus is not needed to initiate development, and all information necessary for cleavage exists in an egg's cytoplasm.

These results have been confirmed in modern experiments. When eggs are treated with drugs that block RNA production by a zygote's genes, the zygote still develops to the blastula stage. However, if inhibitors of protein synthesis are administered, cleavage stops immediately. These results indicate that although the embryo's genes are not involved, proteins are being synthesized during cleavage. The messenger RNAs from which these proteins are produced must therefore be present in the egg's cytoplasm at the time of fertilization. Apparently, the embryo's genes become active only after at least some of these maternal RNAs are broken down.

Regulative and Mosaic Development

In many echinoderms and vertebrates, if blastomeres are separated at the two- or four-cell stage, each blastomere can develop into a complete embryo. This is called **regulative development**, meaning that each blastomere has the capability to regulate its own development and form an entire organism. Cleavage in eggs of regulative species is sometimes said to be **indeterminate** because the fate of the blastomeres is not fixed early in development. The fate of the blastomeres is determined after the four-cell stage, or as late as the 16-cell stage in some species, by cell-to-cell interactions among the blastomeres.

By contrast, removal of blastomeres from two- or four-cell stages of annelids or molluscs may result in abnormal or nonviable embryos. The fate of the blastomeres is established in the eggs of these animals, rather than by cell-to-cell interactions during cleavage. This type of developmental pattern is called **mosaic development**, because the eggs have regions of cytoplasm with fixed fates. When the egg divides, the resulting early blastomeres receive different types of materials, and each can produce only part of an embryo. Cleavage in which egg cytoplasm determines the fate of early blastomeres is sometimes said to be **determinate**.

The fate of various regions of egg cytoplasm can be traced experimentally by microinjecting small amounts of nonlethal dyes and observing which cells in later embryonic stages carry the stain. Such studies allow researchers to draw **fate maps** for eggs. These maps show which portions of an egg's cytoplasm give rise to specific organs in adults. Fate maps can be determined for both mosaic and regulative species. In many regulative species, the egg cytoplasm is regionalized, and if the early blastomeres all remain intact, the fate of the regions is predictable. The fate of the regions is not rigidly fixed, however, and if early blastomeres are removed, the remaining blastomeres can form the missing parts.

Genetic Regulation

As differentiation occurs in an embryo, cells undergo structural and functional changes and develop into different types of tissues. These changes result from the expression of genes as specific proteins. What occurs at the gene level in differentiation? It is known from nuclear transplantation experiments that cells do not lose genetic information during this process (Figure 16.14). If all the cells in an embryo are genetically identical but develop differently in different tissues, then certain genes must be used only in certain tissues. What causes some genes to be expressed in some cells but repressed in other cells? In other words, what causes differentiation?

Early in development, the cells of an embryo have the capacity to form several types of tissues. Ectoderm, for example, can form skin or nervous tissue. In the region where the eye develops, the ectoderm can have several fates, but gradually, these cells become less versatile and become committed to specific tasks. This progressive loss of versatility is called **restriction**, and the final commitment to a particular expression of genes is called **determination**. Restriction and determination occur continually and sequentially throughout development, but what is happening at the gene level to cause these phenomena? There is no firm answer as of yet, but several hypotheses have been proposed.

One hypothesis suggests that the availability of DNA as a template for RNA synthesis may be regulated by proteins that are bound with the DNA. Chromatin consists of DNA wound around cores of proteins, most of which are histones. (Chromosome structure is discussed in Chapter 2, p. 45; Figure 2.23.) Tightly packed regions of chromatin are not active in RNA synthesis; consequently, the degree of packing of DNA and histones could play a role in

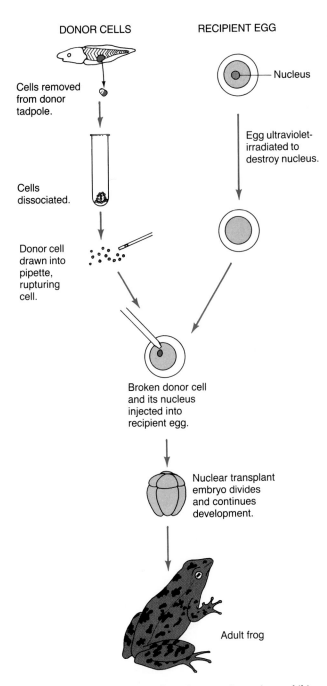

DONOR CELLS   RECIPIENT EGG

Nucleus

Cells removed
from donor
tadpole.

Egg ultraviolet-
irradiated to
destroy nucleus.

Cells
dissociated.

Donor cell
drawn into
pipette,
rupturing
cell.

Broken donor cell
and its nucleus
injected into
recipient egg.

Nuclear transplant
embryo divides
and continues
development.

Adult frog

Figure 16.14 A nuclear transplantation experiment in amphibians. Nuclei are removed intact from donor cells taken from a blastula, gastrula, or, as shown here, from the body surface tissue of a tadpole. Meanwhile, the nucleus of a recipient egg cell is destroyed with a laser microbeam or by ultraviolet irradiation. A donor nucleus is then injected into the egg. The egg then develops into a normal adult, showing that the transplanted nucleus has not lost any genetic information, even though the cell containing it was differentiated in the body surface of the tadpole donor.

regulating blocks of genes that are used only at specific times in development. The precise role of histones in differentiation is not known, but these proteins are generally thought to repress gene activity,

perhaps by limiting access of transcriptional enzymes to the DNA. Chromatin also contains a second group of proteins, called acidic nuclear proteins, that seem able to activate genes. It is possible that the interactions of histones and acidic nuclear proteins regulate at least some gene activity.

Recent studies have also indicated that gene expression during differentiation may depend on gene rearrangement. DNA contains so-called transposable elements, whole genes or segments of genes that are not expressed when they are in their original position in the chromosome. During development, these segments may move from their original position and be spliced into the DNA near a sequence of nucleotides called a promoter sequence. The promotor sequence enhances the binding of RNA polymerase with DNA, so that a gene or gene segment next to a promoter sequence is translated into messenger RNA and expressed. This type of mechanism has been described in genes coding for antibodies in mammals and in blood flukes. (Schistosomes are discussed in Chapter 22, p. 514; Figure 1.10b. For further discussion of differential gene expression, see "Control of Gene Expression" in Chapter 15, p. 370.)

Signals Regulating Differentiation

Whenever only certain genes are expressed in a cell, some type of regulatory mechanism must be activating these genes and repressing others, and such a regulatory mechanism must require some type of signaling device. In early development—perhaps through the gastrula stage—different chemicals in various regions of the egg cytoplasm may trigger gene regulatory mechanisms in cells that receive the egg cytoplasm.

The information content of the egg cytoplasm may be sufficient to direct gene expression in early embryonic stages, but it is likely that other signals become influential in the more complex stages of development. Much of development involves cell migration. For example, cells leave the mesodermal somites (see Figure 16.12) of vertebrate embryos and migrate to the limbs, where they differentiate into muscle cells. Nerve cells also grow out from the nerve cord to innervate the body. Such migrations require some type of signaling mechanism to tell cells where they are in reference to other cells. Such positional information originates in the interactions of cells, as in the induction of the neural tube by mesoderm of the notochord (discussed earlier, p. 387).

Cells inhabit a chemical and physical environment, called a **morphogenetic field**. In a developing embryo, chemical signals from this environment may determine which set of genes is expressed by a cell. Because new cell types appear throughout development, cells in an embryo reside in constantly changing morphogenetic fields. Chemicals called **inducers**, secreted into a morphogenetic field by certain cells, may play important signaling roles. As such inducers diffuse throughout the embryo, they create a highly varied intercellular environment. They establish chemical concentration gradients, and different chemicals may diffuse from several sources. Because cells in an embryo may vary in their response to inducers, the different concentrations and combinations of these chemicals may provide sufficient variety in gene expression to account for cellular differentiation into tissues during later stages of development.

The dorsal lip of the blastopore of amphibian embryos exerts a powerful directive force on morphogenesis and differentiation. When a dorsal lip is transplanted from one gastrula onto the ventral surface of a second gastrula, the lip induces the formation of a second gut and eventually the formation of Siamese twins (Figure 16.15). Induction of twinning takes place even when direct contact of the graft with the embryo is prevented by placing a porous filter between the host and the graft tissue. However, induction is blocked when an impermeable barrier such as metal foil is placed between the graft and the host. These experiments indicate that diffusable substances (inducers) trigger the expression of genes producing the twinning effect. Other experiments suggest that there may be universal signals that are recognized by cells during animal development. If rabbit mesoderm is transplanted into a chicken blastodisc, the mammalian mesoderm induces the formation of a neural tube in the bird embryo.

Cell Recognition and Adhesion

Chemical inducers alone do not account for all the events in the development of an animal's body form. If a sea urchin gastrula is gently agitated in calcium-free seawater, it dissociates into a suspension of single cells. If the cell suspension is transferred to normal seawater, a gastrula re-forms from the dissociated cells and continues developing into a normal larva. This experiment indicates that ectoderm, mesoderm, and endoderm cells convey positional information to one another and recognize one another so that only certain cells stick together. Similar experiments with adult sponges

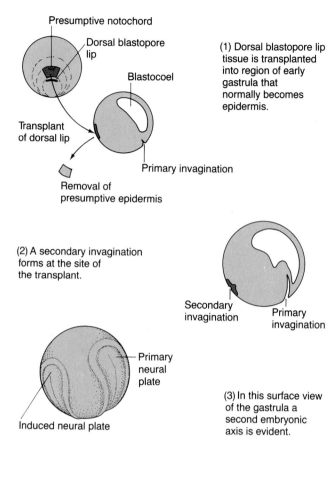

(1) Dorsal blastopore lip tissue is transplanted into region of early gastrula that normally becomes epidermis.

(2) A secondary invagination forms at the site of the transplant.

(3) In this surface view of the gastrula a second embryonic axis is evident.

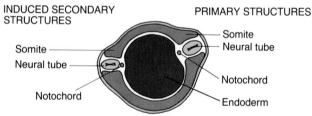

(4) The developing newt has both primary and induced secondary neural tubes and other structures.

(5) Eventually, Siamese twin newt tadpoles develop, each with gills, forelimbs, eyes, and other structures.

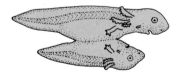

Figure 16.15 The organizer effect of the dorsal lip of the amphibian blastopore. If the dorsal lip of the blastopore is surgically removed from one embryo and added to a second, it induces the second embryo to develop into Siamese twins.

show that single-cell suspensions will reassemble into normal sponges. In sponges, the recognition signal is an acidic glycoprotein found on the cell surface.

Cell recognition and adhesion (attachment of one cell to another) are not confined to sea urchin gastrulas or to sponges; they are fundamental processes in development. One of the keys to understanding the overall developmental process may be in determining how membrane glycoproteins provide cell identity and cause certain cells to adhere to one another. Using highly specific antibodies coupled to fluorescent dyes, researchers have identified specific glycoproteins, called **cell adhesion molecules (CAMs)**, on the surface of developing embryonic tissues. In chick embryos, for example, the proteins on the surface of developing brain cells are not present on the surface of other tissues. Together with inducers, CAMs seem to be the signals that cells use to associate and develop into the unique three-dimensional arrangements of tissues that characterize organs and organ systems in multicellular organisms.

SUMMARY

1. Fertilization triggers the movement of ions into and out of cells, activating a zygote to start development and preventing polyspermy. Zygotes first undergo cleavage, which is cell division without increase in size. Cleavage results in the formation of a blastula. Eggs with sparse, evenly distributed yolk undergo holoblastic (complete) cleavage. In radial holoblastic cleavage, which occurs in echinoderms and many chordates, the cleavage planes pass either parallel or perpendicular to the animal-vegetal pole axis. In spiral cleavage, characteristic of annelids and molluscs, the cleavage planes occur at oblique angles to the animal-vegetal pole axis, so that the blastomeres are displaced as they are formed. Eggs with copious yolk undergo meroblastic (incomplete) cleavage; this includes discoidal cleavage in fishes, reptiles, and birds and superficial cleavage in the eggs of insects.

2. Gastrulation converts the blastula into a two-layered stage called the gastrula. The ectoderm covers the embryo; the endoderm becomes internal and is the precursor of the gut. Triploblastic animals have a third germ layer, the mesoderm, which forms tissues and organs between the other two cell layers. All adult tissues arise from these three germ layers. Gastrulation occurs by one or more types of cellular movement: invagination, epiboly, involution, ingression, and delamination.

3. Annelids and molluscs are called protostomes because their mouth develops from the blastopore, a region where invagination occurs during gastrulation. Echinoderms and chordates are called deuterostomes because their mouth develops as a secondary opening later in development, rather than from the blastopore.

4. Gastrulation in birds and mammals involves inward movement of endodermal and mesodermal cells along a primitive streak that forms across the surface of the blastodisc (blastula). Early development involves the formation of protective membranes around the embryo.

5. All cells in an embryo have the same genes, and differentiation of cells during development is due to differential gene expression. Egg cytoplasm influences cell activity in early development. Cells in an embryo may receive signals via chemicals that diffuse from organizing centers elsewhere in the embryo. Such chemical signals may determine which genes are expressed.

6. Development is characterized by many cellular migrations and specific associations. Cell adhesion molecules (CAMs) are glycoproteins that identify cells and allow cells to adhere to one another during morphogenesis.

FURTHER READING

Calow, P. *Life Cycles: An Evolutionary Approach to the Physiology of Reproduction, Development and Aging.* London: Chapman and Hall, 1978. *Life cycles are considered from evolutionary and ecological viewpoints.*

Edelman, G. M. "Cell-Adhesion Molecules: A Molecular Basis for Animal Form." *Scientific American* 250(1984): 118–129. *Describes the role of cell adhesion molecules in development.*

Gehring, W. J. "The Molecular Basis of Development." *Scientific American* 253(1985): 153–163. *Explains how a short section of DNA called the homeobox directs the basic development of the segmented architecture of a wide variety of animals.*

Gilbert, S. F. *Developmental Biology.* Sunderland, Mass.: Sinauer Associates, 1985. *An excellent modern textbook.*

Goodman, C. S., and M. J. Bastiani. "How Embryonic Nerve Cells Recognize One Another." *Scientific American* 251(1984): 58–66. *Reveals how, as nerve cells grow out from the central nervous system, they follow markers on the surfaces of other cells.*

Slack, J.M.W. *From Egg to Embryo.* New York: Cambridge University Press, 1983. *Discusses the mechanisms involved in development.*

CHAPTER 17

Evolution

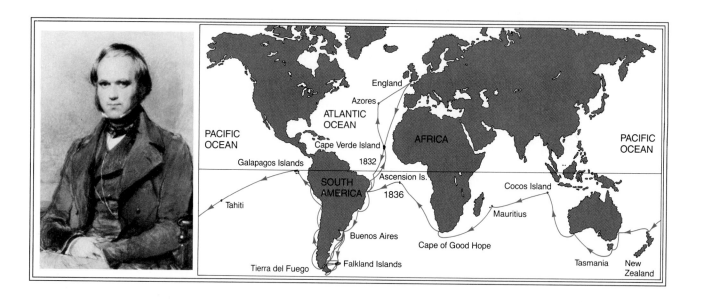

Evolution, the process of inheritable change arising in species over time, provides the basis of interpretation and unification of all fields of life science. Without an understanding of evolution, biology is a purely mechanistic science, able only to approach questions concerning how organisms are structured and function today. Knowledge of evolution allows us to understand present-day species as part of the overall history of life on earth. It allows a zoologist to think beyond the mechanistic level and to ask questions about how survival mechanisms and living species arose.

Evolution is sometimes depicted as merely a tenuous idea, but nothing could be further from reality. As discussed in Chapter 1, a theory in science is a well-established concept or principle that has withstood repeated testing. The theory of evo-

lution is no longer considered hypothetical. Most likely, there will always be a healthy scientific debate about *how* certain mechanisms of evolution operate, but there is no longer any serious question about *whether* organisms evolve.

THE DEVELOPMENT OF EVOLUTIONARY THEORY

The nineteenth-century English naturalist Charles Darwin is often thought of as the originator of evolutionary thought, but as Darwin himself recognized, this is not true. The idea that living species have arisen from other life forms or have changed over time can be traced back about 2000 years to the ancient Greeks. As with the development of many great concepts in science, a gradual increase

---

Charles Darwin and the voyage of HMS *Beagle* (1831–1836).

in scientific awareness and questioning set the stage for Darwin's contributions. Darwin and Alfred Wallace, another English naturalist, were the first to propose some of the fundamental mechanisms that explain how evolution operates. Darwin and Wallace were among the great "synthetic thinkers" of science, able to extract central, unifying principles from the more or less disjointed information and ideas that were available to them.

Despite the long history of evolutionary thought, the prevalent view of scientists and nonscientists alike in the centuries prior to Darwin was the belief, as expounded in Genesis in the Bible, that all species are immutable (unchangeable) and created in their present state (which was considered perfect) by a supernatural creator. The social and religious climate in Darwin's era (mid- to late 1800s) continued to support this view. Nonetheless, several facts and ideas had emerged in the eighteenth century that provided insight to Darwin and Wallace. European naturalists in the eighteenth century had begun to seriously question the biblical story of creation. The discovery of many fossil groups of animals with no living representatives posed the question, Were some of the creator's designs less than perfect? Similarities between fossils and living animals suggested to some naturalists (including Erasmus Darwin, Charles's grandfather) that species *had* changed over time. Then, in the late 1700s, French naturalist Jean Baptiste Lamarck made a truly significant contribution to evolutionary thought in postulating that animals evolved as a result of interacting with the environment. Lamarck expounded on his ideas in his *Philosophie Zoologique*, published in 1809, the year of Darwin's birth. Darwin himself gave much credit to Lamarck as an early proponent of species evolution. Unfortunately, Lamarck is remembered mainly for his erroneus ideas of how species evolve. Lamarck proposed that species evolve as characteristics acquired by an individual through use or disuse of its body parts are passed on to offspring. His idea that the giraffe acquired its long neck because its ancestors stretched higher and higher into the trees to reach leaves and that the increasingly stretched body parts were passed to offspring has become part of biological folklore. Nonetheless, despite his misinterpretations, Lamarck played a significant role in setting the stage for Darwin and Wallace.

In 1831, at the age of 22, Charles Darwin began a sea voyage that profoundly influenced his scientific thinking. He joined the crew of the surveying ship HMS *Beagle* and eventually became the ship's naturalist. Under command of captain R. H. Fitz-Roy, a devout fundamentalist Christian, the *Beagle*'s voyage lasted five years. While the rest of the ship's crew mapped coastlines, Darwin studied plant and animal diversity, experienced the force of earthquakes firsthand, collected and studied fossils, and had time to ponder why certain marine fossils could be found on the tops of mountains. Eventually, the *Beagle*'s voyage took Darwin around the world and provided him with a chance to document and contemplate the differences in plant and animal life on the major continents in the southern hemisphere and on sparsely inhabited oceanic islands. While on the voyage, Darwin read and was strongly influenced by a two-volume book entitled *Principles of Geology*, by the Scottish geologist Charles Lyell. First published in 1830 and 1832, these texts expounded on the concept of **uniformitarianism**, according to which natural forces producing changes on the earth's surface are the same, and occur at the same rate, today as in the past. In other words, modern geological principles could be applied to the study of earth history. Lyell's ideas were counter to the prevalent nineteenth-century view that earth history had included certain supernatural catastrophes and miraculous changes. Darwin's firsthand observations of geological phenomena during the *Beagle*'s voyage convinced him that Lyell's principles were correct. Darwin had learned that both earth and species history could be interpreted from natural events, without invoking the supernatural. It also became clear to him that the species composition of the earth had changed through geological time.

Another source of influence for both Darwin and Wallace was *An Essay on the Principle of Population*, written in 1799 by the English clergyman and economist Thomas Malthus. Malthus correctly perceived that species have an inherent capacity to multiply so rapidly that without natural checks to population growth, they would outstrip their food supplies in only a few generations. Malthus predicted grave consequences for the human species in light of what he saw as an ever-increasing, inadequately controlled population. (Exponential, ◁ or Malthusian, growth is discussed in Chapter 34, p. 830.) Darwin and Wallace read Malthus's essay in the 1830s, and both realized that overproduction of offspring each generation was a key ingredient in the mechanism of evolution.

In the mid-1850s, Alfred Wallace was doing field studies in the Malay Archipelago. He conceived the essence of the theory that species change by natural selection while in a feverish state brought on by an attack of malaria. Upon recovery, Wallace composed his thoughts and sent them to Darwin, who had already arrived at the same conclusions, but had been procrastinating about publishing them. Darwin was then compelled to publish, and did so in 1859 in the single volume *On the Origin of Species by Means of Natural Selection, or the Preservation of the Favoured Races in the Struggle for Life.* The basic tenets set forth in Darwin's text and in abstract form in Wallace's paper (published in 1858) still form the basis for our modern understanding of organismal change. In this and subsequent works, Darwin not only set forth evolutionary principles, but also contributed extensively to the mass of evidence that supports them. Because of the overwhelming contributions that Darwin made, his name has become almost synonymous with the theory of evolution.

Darwin's Theory

Darwin's interpretation of species changeability is based on three observations clearly set forth and exhaustively documented in his book. First, Darwin observed that individuals of a species are highly variable and that many variable traits are inherited. He pointed out that the many strains of domestic plants and animals that had been produced by artificial breeding represented offspring of variable populations whose inherited traits had been selected by breeders. Darwin also provided considerable evidence of variation among species living under natural conditions. Second, Darwin elaborated extensively on Malthus's observation of reproductive potential and perceived how it applies to the logic of evolution. Third, Darwin observed that a species' reproductive potential is seldom, if ever, realized. Because of limited resources, predation, disease, and so forth, most offspring die, and the sizes of most populations of organisms remain about the same from generation to generation. From these observations, Darwin concluded that in each generation, out of a variable population of offspring, the individuals possessing heritable traits that give them some advantage over other offspring survive and, most importantly, reproduce. Because only survivors produce offspring, advantageous traits would most likely be passed to future generations, while nonadvantageous traits would perish with the individuals that have them. Darwin called this concept **natural selection**. He presented strong evidence that natural selection, operating through differential reproduction, is the guiding force behind evolutionary change, and he

recognized that this force accounts for species diversity. Darwin also perceived that organisms living in the earth's varied environments would become highly diverse because different heritable traits would be advantageous in different environments.

As we will see, many aspects of evolutionary theory have been added to Darwin's basic model, but few contributions in science have explained as much, withstood the test of time as well, and stimulated as much other research as have Darwin's. The major weaknesses in Darwin's theory were that it did not propose mechanisms for the origin of hereditary variations or for the transmission of variations between generations. These problems were solved decades after Darwin's death.

The Castle-Hardy-Weinberg Principle

The genetic mechanisms needed to explain the processes underlying Darwin's theory were not understood until Gregor Mendel's work was rediscovered in the 1900s. With the increase in knowledge about inheritance patterns, mutations, and the events of meiosis, it gradually became clear that genes determine the variation documented by Darwin. But a debate soon began about whether Mendelian principles supported or refuted Darwin's theory. Certain biologists argued that variation should decline in populations because dominant traits would eventually replace recessive ones in individuals of a population. This suggestion was proved incorrect by an important mathematical model, now known as the **Castle-Hardy-Weinberg (C-H-W) principle** because it was developed independently by American biologist William E. Castle in 1903 and by English mathematician Godfrey H. Hardy and German physician Wilhelm Weinberg in 1908. Castle, Hardy, and Weinberg showed that the frequencies of alleles (defined in Chapter 15, p. 351) in a population remain constant from generation to generation if the following conditions are met:

1. The population is large.
2. All individuals in the population have an equal chance of mating with all others of the opposite sex.
3. No mutations occur, or a mutational equilibrium exists in the popoulation (i.e., the rate of mutation of allele *A* to allele *a* equals the rate of mutation of allele *a* to allele *A*).
4. No individuals migrate into or out of the population.

5. The population is not subjected to any kind of selection, so that all genotypes are represented equally among the surviving offspring.

The C-H-W principle can be demonstrated using two alleles, *A* and *a*, at one gene locus. In a hypothetical population, these alleles would be distributed in the three genotypes *AA*, *Aa*, and *aa*. If the frequency of each allele is expressed as a percentage, the frequency of allele *A* in the population could be represented as *p*% and the frequency of allele *a* as *q*%. Because these are the only two alleles at the locus, *p*% + *q*% = 100%. Suppose we consider a population of 100 individuals with the genotypic frequencies of ³⁶/₁₀₀ *AA*, ⁴⁸/₁₀₀ *Aa*, and ¹⁶/₁₀₀ *aa*. To obtain the allele frequencies, we first count the number of *A* and a alleles: for *A*, (2 × 0.36) + 0.48 = 1.20; for *a*, 0.48 + (2 × 0.16) = 0.8. This gives us a total of 2.00, so we then divide these values by 2 to obtain a total frequency of 1.00: for *A*, 1.20/2 = 0.6; for *a*, 0.8/2 = 0.4. Multiplying each of these proportions by 100 gives us the values *p* = 60%, and *q* = 40%. When this population reproduces, under the conditions set by Castle, Hardy, and Weinberg, will the frequency of alleles *A* and *a* change, and if so, what will the allele frequencies be in the next generation?

To answer these questions, we determine the types of gametes each genotype can produce and multiply that number by the frequency of the genotype in the population. Because 36% of the genotypes in the population are *AA* and produce only *A* gametes, 36% of the gametes will carry *A*. Because 48% are *Aa* and produce half *A* and half *a* gametes, 50% × 48% = 24% of the gametes will carry *A*, and 50% × 48% = 24% of the gametes will carry *a*. Because 16% are *aa* and produce only *a* gametes, 16% of the gametes will carry *a*. Summing the types of gametes, we see that 60% of the gametes will carry *A* and 40% will carry *a*. Note that these are the same frequencies as those of the alleles in the population.

We can extend our example to show that the genotypes in the new generation occur at the same frequencies as in the parental generation. The laws of probability tell us that *the probability of two independent events occurring together is obtained by multiplying their individual probabilities.* In the case of two simultaneous coin tosses, the probability that both tosses will yield a head (or a tail) is 0.5 × 0.5 = 0.25. Likewise, in the case of specific gametes combining at fertilization, the probability that any two gametes will combine equals the product of their two frequencies. In our example, the frequency of the *A*-bearing gametes is *p*. Thus, the probability that two *A* alleles will combine to form genotype *AA* is *p*². Likewise, the probability of *aa* is *q*². The probability that the heterozygote *Aa* (or *aA*) will be formed is *pq* + *qp*, which equals 2*pq*. If we substitute the values of 60% and 40% obtained previously for the frequency of *A* and *a* gametes, respectively, we see that the expected frequencies of genotypes in the next generation are the same as those of the parental generation (Figure 17.1). This demonstrates that if the C-H-W conditions are met, alleles are not lost from generation to generation, and variation is thus maintained.

The C-H-W principle demonstrated to the satisfaction of most scientists that Darwin's evolutionary theories were compatible with the principles of Mendelian genetics. Because the C-H-W principle shows what the genotypic frequencies will be when a population is not subject to influences that would change its allele frequencies, today's evolutionary biologists use C-H-W as a baseline against which to measure the amount of evolution in populations. If the C-H-W conditions are met, the frequency of genotypes is predicted by the relationship *p*² + 2*pq* + *q*² = 1. Therefore, if the value of *p* or *q* can be determined for a pair of alleles, the frequencies of genotypes in the next generation can be predicted. These predicted, theoretical frequencies can then be compared to actual field measurements of the frequencies of genotypes in a population. If predicted and actual genotypic frequencies differ, then

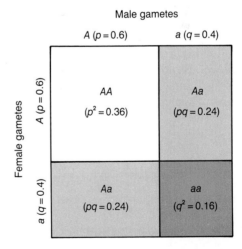

Figure 17.1 The Castle-Hardy-Weinberg Principle. This diagram shows that under certain conditions, the frequencies of genes in populations may remain the same from one generation to the next. Gametes bearing particular alleles are produced with the same frequency at which the alleles occur in a population. As this modified Punnett square shows, the frequencies of genotypes in the next generation are obtained by multiplying the frequencies of the gametes that combine to form zygotes.

one or more of the C-H-W conditions are not true for the population, and researchers can attempt to determine what factor or factors are changing the genotypic frequencies. For loci with only two alleles, allele frequencies may be determined by counting the homozygous recessive phenotypes in a population. The frequency of the recessive allele is equal to the square root of the percentage of the population expressing the recessive allele (because the frequency of *aa* is equal to q^2). Whenever there are only two alleles at a locus, $p + q = 1$. Thus, the frequency of p can be determined if q is known, and vice versa.

The Modern Synthesis

The study of allele frequencies in populations, including application of the C-H-W principle, is called **population genetics**. The merging of this and other aspects of genetic analysis with Darwin's concepts of evolution has greatly expanded our understanding of the process of evolution. The fusion of genetic and evolutionary concepts is sometimes called the *modern synthesis* or the *synthetic theory of evolution*. Reflecting this fusion, the species' variability documented by Darwin is now called **phenotypic variability**, and we know that phenotypic variability results from genetic variability.

It is convenient to study evolutionary change at one of two levels. Evolution at the species level and below is called **microevolution**. Microevolutionary analyses allow us to interpret Darwin's concepts of variation, differential reproduction, and natural selection in light of modern genetics. The other level, called **macroevolution**, is evolution above the species level. Biologists concerned with this level of biological change are interested in learning how major groups, such as animal phyla, originated or became extinct.

MICROEVOLUTION

Central to the modern concept of evolution is the concept of biological populations. A **biological population** is an assemblage of interacting individuals of a species. The individuals interact because they inhabit a particular geographical area. In sexually reproducing species, members of a population may interbreed. In this sense, interbreeding means the production of fertile offspring; consequently, gene mixing occurs in sexual populations. Biologists concerned with microevolution study mechanisms that generate changes in the gene frequencies of populations, mechanisms that can result in the formation of genetically different populations and new species.

Each population has its own assemblage of genes called a **gene pool**, which consists of all the alleles in all the individuals in a population. Thus, the amount of phenotypic variability that may occur in a population depends on the variability of the population's gene pool. It is at the level of the gene pool that evolution actually occurs; a gene pool evolves (changes) as various forces (those enumerated by the C-H-W principle) alter its allele frequencies.

Populations, Demes, and Clines

Large, widely distributed populations often consist of small geographical subpopulations called **demes**, which may be significant in evolution. Individuals in a deme often form a tight interbreeding unit, compared to the population as a whole. Some members of a deme may interbreed with members of other demes, and there may be migration among adjacent demes. In many cases, however, there is little interbreeding between demes and each deme may develop genetic and phenotypic uniqueness.

Even populations of highly mobile animals may consist of more or less unique demes. Although many species of birds fly thousands of miles on annual north/south migrations, most birds return to the area where they were hatched to mate and raise young. Consequently, the frequency of alleles in a local bird population will usually differ from the allele frequencies of other local populations of the same species. Small demes provide greater chances for **inbreeding**—mating between closely related individuals—and as a result, these local geographical groups tend to show a higher degree of homozygosity than the larger interbreeding groups.

In some populations, phenotypes will vary gradually and more or less continuously across the geographical range of the species. This type of geographical gradient in a trait (or traits) is called a **cline**. The coat color of deer mice *(Peromyscus),* for example, varies across their range, according to soil color. Mice that do not match their surroundings are usually eliminated from a population because they are easily seen by predators. Likewise, leg markings in zebras show clinal variation (Figure 17.2a).

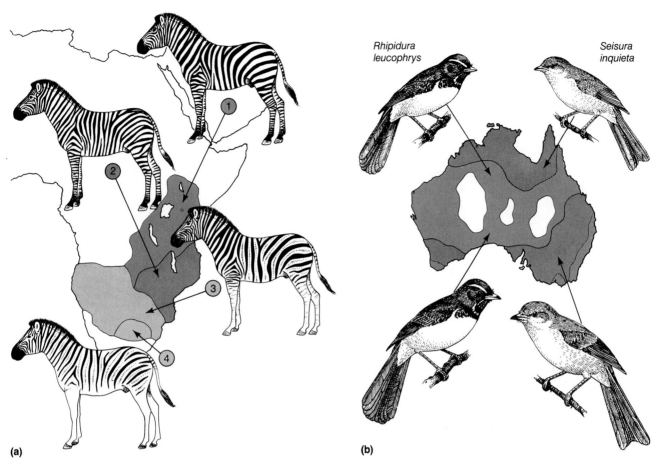

Figure 17.2 Species displaying clinal variation. Species often show gradual (clinal) phenotypic differences along a geographical range. (a) Zebras display clinal variation in the extent of striping on their legs. (b) Other endothermic animals, such as the Australian birds *Rhipidura leucophrys* and *Seisura inquieta*, tend to be larger and have shorter extremities in the colder parts of their range. The populations of *S. inquieta* are discontinuous, whereas that of *R. leucophrys* is unbroken. The gene flow that is possible in the unbroken population affects size differences: *R. leucophrys* individuals from the colder environment (at the southern end of the continent) are only 11% larger than those from the warmer (northern) part of the range, compared to a 22% difference between the largest and smallest *S. inquieta*.

Marked differences in physical traits or physiological processes are often seen between demes at opposite ends of a species' geographical range, because environmental conditions at such extremes are often very different. In many species of birds and mammals that are widely distributed across continents, members of demes in the colder part of the range are often larger and have smaller extremities than individuals in warmer demes (Figure 17.2b).

Maintenance of Variability in Populations

Genetic and phenotypic variability are of definite advantage to species. Variability can be viewed as a population insurance policy, for a variable population is more likely to include individuals that will be able to survive if the environment undergoes significant change. At the species level, loss of genetic variability may result in extinction.

Natural selection tends to reduce variability in a population's gene pool. It tends to eliminate those individuals whose genes provide them with fewer advantages than other individuals enjoy in a particular environment. Consequently, it is not surprising that many natural populations seem, on cursory examination, to be rather uniform. Yet, populations that have been thoroughly studied have been shown to be phenotypically and (to a much greater extent) genetically variable (Figure 17.3). In fruit flies, virtually all the alleles producing abnormalities that have been expressed in lab-

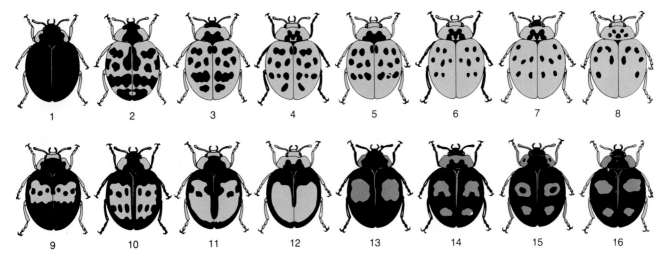

Figure 17.3 Phenotypic (morphological) variation in the ladybird beetle *Harmonia axyridis.* This species occurs in Siberia, China, Korea, and Japan. Color variations are characteristic of certain parts of the range. West-central Siberian forms are almost completely black (1); farther eastward, there is greater variability, and populations have a greater frequency of individuals with black spots on a yellow background (2–8) or with yellow spots on a black background (9–12). Individuals with red or orange spots on a black background (13–16) occur only in the Far East.

oratory cultures have been found to exist in nature in the heterozygous state.

Several factors contribute to and help maintain genetic variability in a population. New alleles in a gene pool originate from gene and chromosome mutations, which result from changes in DNA molecules. In sexually reproducing organisms, variation in a population's genotypes (and consequently phenotypes) is maintained largely by the reshuffling of alleles during fertilization and by genetic recombination during crossing over in meiosis. Also working in favor of genetic variability is the fact that natural selection is rarely, if ever, perfectly efficient. Selection cannot act directly on alleles unless they are expressed, so any effect that hides, or masks, an allele may help to maintain it in a population. For this reason, recessive alleles may not be affected by selective forces unless the alleles occur in the homozygous condition. Other factors that may maintain variability include various types of gene interactions and genetic control mechanisms that affect differential gene expression. (Inheritance patterns involving gene interaction and factors affecting gene expression are discussed in Chapter 15, p. 360.) Overall, the chances that recessive alleles (or other alleles that are not always expressed) will be completely eliminated from a population are slim.

Natural selection, although generally acting to reduce variability, may also have the opposite effect. Individuals are often subject to opposing selection forces, and evolution tends to proceed as a compromise between extremes. Because environmental conditions are generally quite variable, organisms that can tolerate changing conditions (such as varied weather, seasonal changes, fire, and earthquakes) are often favored by natural selection. Individuals phenotypically suited to tolerate such conditions often show a higher degree of heterozygosity than do extreme specialists. Thus, selection favoring generalist or "compromise" phenotypes tends to maintain genetic variability.

The Effect of Migration

The migration of individuals representing different phenotypes is often an important factor in changing and maintaining variety in gene pools. The exchange of genes between two or more populations is called **gene flow**. Unless a population is totally isolated from other populations of a species, gene flow may occur rather frequently. The overall influence of gene flow on gene frequencies is generally higher when the rate of migration is high, the receiving population is small and exhibits a high degree of homozygosity, and the contributing and receiving gene pools are very different.

Gene Pool Changes That Reduce Variability

Two evolutionary forces—chance and natural selection—tend to reduce genetic variability in populations.

Genetic Drift. Changes in a population's gene pool resulting from chance events are called **genetic drift**. Consider a population with the following frequencies of three genotypes: 64 *AA*, 32 *Aa*, 4 *aa*. A flash flood or other catastrophic event could result in the death of the four *aa* individuals. In this case, death would not be related to the genotype of the victims, but would be a chance event, like drawing an ace of hearts from a deck of cards. The effect of genetic drift is most profound in small populations, as illustrated both in nature and in model systems (Figure 17.4). Typically, evolution by genetic drift does not yield changes in a population's gene pool that produce phenotypes better suited to the environment, but there is always a possibility that this will occur.

Genetic drift may occur in populations when numbers decrease drastically during harsh times. This has been documented, for example, in California populations of the northern elephant seal *(Mirounga angustirostris)*. Unregulated hunting in the late 1800s reduced this species to about 20 individuals, creating what is called the **bottleneck effect** (Figure 17.5). Since about 1890, when this species became protected by law, their numbers have increased to about 30,000, which, by necessity, occurred largely through inbreeding. Electrophoretic studies of *M. angustirostris* blood samples taken in the mid-1970s showed that this species is extremely uniform genetically. Twenty-four genetic loci that were analyzed were homozygous. Because much more variability has been found in the closely related species *M. leonina* (southern elephant seal), which never sustained a drastic reduction, it is evident that the northern elephant seal's genetic variability was reduced when the species was decimated.

Genetic drift and the bottleneck effect are real possibilities for the world's many endangered species as their numbers decrease. The California condor *(Gymnogyps californianus)* now numbers fewer than three breeding pairs. If by chance all the offspring from these birds happen to be of one sex for several years, the species may become extinct. If the condor does manage to survive, its future populations will have resulted from inbreeding among so few survivors that its genetic variability, and probably its ability to adapt to changing environments, will most likely be minimal.

A special case of genetic drift is the **founder ef-**

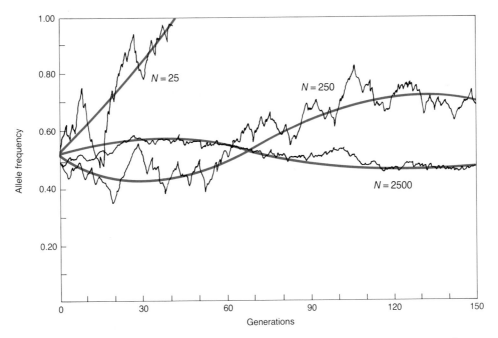

Figure 17.4 Genetic drift. This effect, a result of chance events, is most common in small populations. In a computer simulation, three populations (of a hypothetical species) varying in size from 25 to 2500 individuals were established. Each population had allelic frequencies equal to 0.5 for two alleles occurring at a single locus. With simulated random mating over several generations, gene frequencies changed by chance in the smaller populations more often than in the larger population. In one generation in the smallest population, all homozygous genotypes were produced, and the other allele was lost from the population, a situation known as fixation. Unless a mutation occurs, this population will exhibit only the homozygous genotype in future generations.

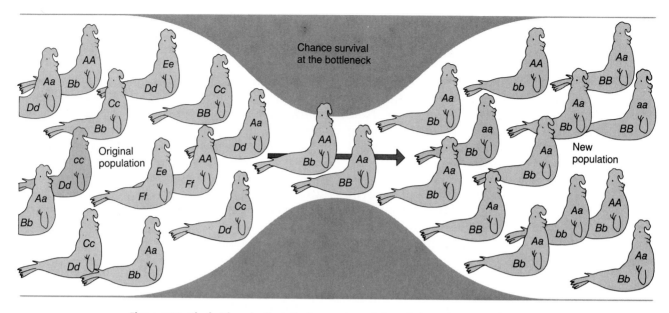

Figure 17.5 The bottleneck effect. If a large, substantially varied population is reduced to a small population, much of the variability is lost, even if the population later enlarges back to the original size. This may affect the species' ability to adapt. In the new population, note the increased number of individuals that are homozygous for recessive alleles.

fect, which occurs when a single gravid (pregnant) female or a small population of a species (the "founders") colonizes a new area. Because founders are few in number, they carry only a small sample of all the genes in the parent population. The founder gene pool may lack, or have a lower frequency of, alleles that are common in the parent population; or, conversely, rare genes may make up a much higher percentage of the founder gene pool. Because of inbreeding and loss of alleles through genetic drift, variability is typically low, and homozygosity tends to increase over time in founder populations. With these effects, more recessive traits are expressed, and because many of these are deleterious, founder populations may show decreased fertility, vigor, and fitness.

Artificial and Natural Selection. As Darwin pointed out, the development of domestic breeds with traits rarely seen in parent populations illustrates that species are highly variable. Selective breeding, also called **artificial selection**, mimics natural selection, and it is only a short conceptual jump from understanding how breeds of domestic dogs, farm animals, and grain crops have been developed to understanding selection in nature. As Darwin pointed out, artificial and natural selection operate through differential reproduction. Today, we recognize that natural selection changes the al-

lele frequencies in gene pools, and this process operates each generation as only certain phenotypes, often out of many offspring, survive and reproduce. Phenotypes that succeed in leaving the most offspring are called the fittest. Thus, **evolutionary fitness** is measured as the number of genes that a phenotype contributes to the gene pool of the next generation. Natural selection acting through differential reproduction produces future generations with a greater number of fit individuals than unfit ones. Actually, an individual's overall fitness or lack thereof may be determined by the alleles that provide the *least* fitness. In a variable population, an animal with the greatest number of characteristics that promote fitness may be the least fit overall if one allele in its genotype is more harmful than any others in the gene pool.

It is important to realize that natural selection acts on phenotypes; genes and gene pools are affected only indirectly by the process. As selection occurs, a population's gene pool loses the genes that were contained in phenotypes that did not reproduce. In favoring the most fit individuals, selection favors genes that provide reproductive advantage.

Patterns of Natural Selection. Three general patterns of natural selection may occur: stabilizing, disruptive, and directional selection. We can compare these three patterns by considering a model of phenotypic variation in a population. By plotting a measurable trait, such as tail length in mice (horizontal axis), against the number of individuals showing various phenotypes for that trait (vertical axis), we usually end up with a bell-shaped curve (Figure 17.6a). The reason for this is that most traits, such as length or height, are determined by several genes at more than one locus. Such genes often have more or less equal and additive effects, and the greatest number of genotypes produce intermediate phenotypes; in the tail length model, mice with extremely long or extremely short tails

are rare. (See the discussion on the inheritance of polygenes in Chapter 15, p. 360; Table 15.4.)

Stabilizing selection occurs when individuals exhibiting an extreme expression of a trait are selected against (Figure 17.6b). As a result, variation is reduced and fluctuates more narrowly about an average value. Stabilizing selection may have occurred in the selection for head size in human infants. Extremely small heads may have been selected against because they could not contain as much developing brain tissue as larger ones. And, at least prior to the development of reliable techniques for caesarian section, extremely large heads

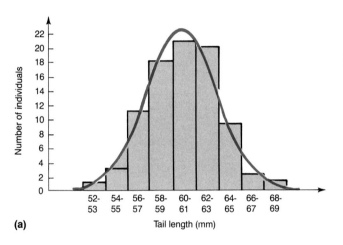

(a)

Figure 17.6 Variation in populations and natural selection. (a) A bell-shaped curve depicts variation in tail length in a deme of deer mice. The average value of the phenotypes is represented by the peak of the curve; note that the greatest number of phenotypes are intermediate (closer to the average value than to the extremes). (b) Three models illustrating possible patterns of natural selection. The gray areas represent animals against which selection is acting. Stabilizing selection against both extremes of a trait leads to a reduction in variability for that trait. It eliminates from later generations individuals that deviate extensively from the average condition. Disruptive selection acts against animals in the middle range of a trait's distribution, thereby favoring both extremes. By eliminating an intermediate expression of a trait, disruptive selection leads to diversification; consequently, the population develops two forms of the trait. Directional selection acts against one extreme, gradually shifting the population toward the other extreme of the phenotypic expression.

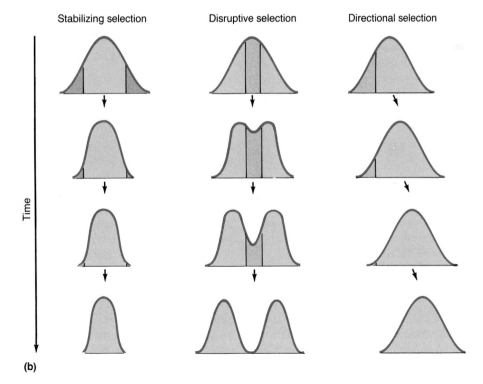

(b)

were probably selected against because babies having such heads could not pass through the mother's birth canal.

In **disruptive selection**, the intermediate phenotypes are selected against and the extremes are favored (see Figure 17.6b). Through this process, a population may be split into two distinct gene pools, and new species may form as a result, although many biologists question whether disruptive selection alone can give rise to new species. We will return to this type of selection in our discussion of speciation.

The two types of selection considered so far—stabilizing and disruptive—are sometimes called **nondirectional selection** because they do not change the average value of a phenotypic trait. By contrast, **directional selection** changes the average value by acting against individuals at only one of the phenotypic extremes. As a result, a population evolves toward one extreme or the other (see Figure 17.6b). A number of well-documented examples of directional selection have occurred among certain insect populations as a result of the extensive use of chemical insecticides since the 1950s. When new pesticides are first applied, they are usually effective in small amounts; but after continued use, many insecticides have become ineffective or are effective only in very large doses. The reason for this is that many (over 430 now known) insect species have developed insecticide resistance. The insecticides, acting as a selection force, have selected against susceptible phenotypes. As a result, out of variable insect populations exposed to insecticides, only phenotypes that are not susceptible to the poisons have survived. This has resulted in the rapid evolution of certain pest insects; and the populations that have been exposed to insecticides for an extended time now consist almost entirely of resistant phenotypes.

Adaptation

In the context of evolution, an **adaptation** is any inheritable characteristic that enhances an animal's fitness. Biologists also use the term *adaptation* to refer to the *process* whereby organisms become better fitted to their environment. In either sense, adaptations involve change, usually wrought by natural selection, in an animal's biochemistry, physiology, anatomy, or behavior and adaptations may occur at any level of organization. At the molecular level, sea anemones living along the Atlantic coast from Maine to Florida have enzymes adapted to function best at the local water temperatures. At the cellular level, a terrestrial vertebrate's skin cells occur in layers and contain large amounts of the protective protein keratin. At the organismal level, the shelled eggs of birds and reptiles are adaptations that protect the developing embryo and preserve its surrounding fluid environment while allowing respiratory gas exchange to occur.

The Niche Concept. Each species has a **niche**, which is defined as its total role, or everything it requires from and does in its environment. The niche includes all biotic (living) and abiotic (non-living) factors that make up the place, or **habitat**, where a species lives. Each species is molded to its niche by adaptations, and the ability of a species to adapt depends on its genes and on the nature of the selection pressures of its environment. If the genetic variation in the populations making up a species is not sufficient to allow the species to adapt to new environmental conditions, then the species may become extinct. As biotic and abiotic factors in an environment change and certain species die out, niches that these species occupied become vacant, constituting an opportunity for other species. Some biologists regard evolution as the exploitation of unoccupied or insufficiently used niches by existing species. In this sense, species may be viewed as participants in a game of adaptation, where both the game board and the players are constantly changing. This idea that successful species change constantly has been called the **red queen hypothesis** after the character in Lewis Carroll's *Through the Looking Glass* who had to keep running just to stay in the same place.

Industrial Melanism. Industrial development has produced many environmental changes, and it is not surprising that populations of certain species exhibit adaptations to these changes. Some of the best-documented changes are those involving insect populations that have changed from mostly light-colored individuals to mostly dark individuals as their environments have become darkened by soot. The most famous case of this phenomenon, which is called **industrial melanism**, involves the peppered moth (*Biston betularia*) in England (Figure 17.7). Studies of this and other species of moths in Europe and North America have shown that body color in moths is genetically determined and that predators (mostly birds) are the main selective agents. Dark (melanic) moths are better camouflaged in sooty environments; thus, they exhibit greater fitness and become more numerous than light-colored moths in industrial areas. The reverse

Figure 17.7 Color phases of the peppered moth, *Biston betularia.* Both photographs have both a light moth and a dark moth in view, but (a) has a soot-darkened background and (b) has a light, soot-free one. Predators are more likely to see moths that do not match their background.

(a)

(b)

situation holds in nonindustrial (non-soot-covered) areas. It is interesting to note how rapidly these types of population changes can occur. Populations of *Biston betularia* are known to have switched from the mainly light to the mainly dark condition in less than 50 years. With a recent trend toward cleaner air in some areas, dark populations have begun to change back to the light condition.

SPECIATION

The formation of new species from preexisting ones, called **speciation**, can result from the microevolutionary processes that alter gene frequencies in populations.

What Is a Species?

Biologists define a species as a group of populations consisting of organisms that interbreed or are capable of interbreeding, or producing fertile offspring, under natural conditions. Members of a species typically cannot reproduce successfully with members of other species; consequently, genes in one species cannot be passed to other species. Because each species is reproductively isolated from all others, a species is a discrete evolutionary unit with a defined gene pool.

The biological species is easy to define, but nature does not always acknowledge definitions. The central criterion of interbreeding does not apply, for instance, to species that reproduce only by asexual means. And even in the study of sexual species, breeding tests cannot always be performed to determine whether a group of organisms constitutes one or more species. Identifying fossil species is especially difficult, because one works with only impressions or mineralized remains in stone. Consequently, biologists and paleontologists (scientists who study fossils) usually use phenotypic traits to identify most species. Species identification is a challenging and ongoing process because species are variable, capable of change, and interrelated. (The problem of defining species is also discussed ◁ in Chapter 18, p. 438.)

Mechanisms of Speciation

For new species to arise, the first requirement is that a population must separate into two reproductively isolated groups. Two models have been proposed to explain how reproductive isolation can develop. When separation is initiated by a physical barrier that keeps parts of the population isolated in different geographical areas, the speciation that may result is said to be **allopatric** (*allo,* "different"; *patric,* "country"). This is probably the most common mechanism of speciation among animals. The second model, called **sympatric speciation**, is less common, and many biologists believe it has not been a significant force in animal evolution. The

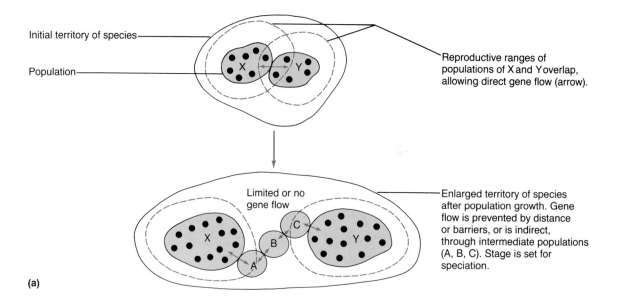

(a)

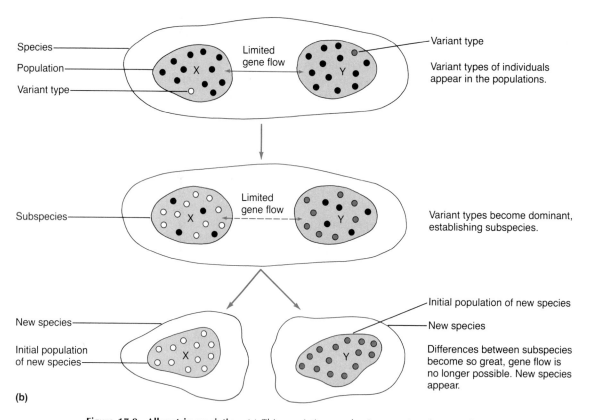

(b)

Figure 17.8 Allopatric speciation. (a) This speciation mechanism requires that populations become geographically separated by distance or by a barrier that reduces or prevents gene flow between the populations. (b) Each population's gene pool may then change uniquely because selection pressures may differ in different parts of the species' range. At first, the change may not be great enough to stop gene flow altogether, but as the number of variant types increases, separate species may be formed and gene flow ends. If the populations are reunited later and members of the two populations cannot interbreed, speciation has occurred.

sympatric model postulates that a population may develop into two or more reproductively isolated groups without first being split into geographically isolated groups. Disruptive selection (see p. 404) is considered capable of producing this type of speciation, and competition among members of a population for limited resources is often cited as a selective force that could disrupt (separate) populations. Among members of a population, any traits that reduce competition will be favored. Consequently, subpopulations (demes) could become uniquely fitted to obtain resources in ways that do not compete with those of other demes. If two or more demes develop uniqueness to the point of becoming distinct breeding units, sympatric speciation may result.

How Allopatric Speciation Occurs

When a species splits into two distinct populations that do not overlap geographically, gene flow between the populations is reduced or absent. Each gene pool may then accumulate unique combinations of genes and mutations without passing them to the other population. Eventually, genetic differences prevent individuals in the two populations from interbreeding, even if they are reunited (Figure 17.8). In certain cases, gene pools in newly isolated populations may already differ from each other. If either population has undergone genetic drift, or if it has resulted from founders, its gene

frequencies may be significantly different from those of the other population. Also, gene pools of demes at opposite extremes of a species' range are often markedly different. If such demes become geographically isolated, they are predisposed to evolve unique gene pools.

What constitutes geographical isolation depends on the species involved. For a bird species, geographical isolation might be caused by physical barriers such as deserts, mountain ranges, or oceans, though some birds can traverse even these. During migration, the smallest warblers and certain hummingbirds fly across the Gulf of Mexico, a distance of about 600 miles. It usually takes an extensive barrier to isolate populations of highly mobile species, but for some animals, barriers may be quite small. Earthworms, land snails, or millipedes may not cross a valley from one mountain range to the next. A well-documented example of geographical isolation involves two types of squirrels occupying two sides of the Grand Canyon. The Kaibab squirrel lives on the north rim of the Grand Canyon, whereas the morphologically distinct Abert squirrel occupies the south rim and parts of Colorado, New Mexico, and Arizona (Figure 17.9). Some researchers consider these two squirrels distinct species, but others believe they represent two distinct populations of one wide-ranging species, *Sciurus aberti*. It is postulated that a founding population of Abert squirrels invaded the north rim and gave rise to the Kaibab squirrels.

(a)

(b)

Figure 17.9 Squirrels of the Arizona Grand Canyon. (a) The Abert squirrel and (b) the darker-colored Kaibab squirrel are geographically isolated on either side of the Grand Canyon.

Allopatric separation into two populations may or may not result in speciation, and sometimes it is difficult to determine whether two types of apparently distinct organisms constitute one or two species. Prior to European settlement, the North American Great Plains had few trees and, correspondingly, few species of tree-nesting birds. Two types of orioles, the Baltimore oriole, inhabiting eastern forests, and the Bullock's oriole, inhabiting western woodlands, were separated by this vast, virtually treeless plain. Males of these two types of orioles have strikingly different plumages, and the two types were formerly thought to be distinct species (Figure 17.10). Within the last century, however, as settlers planted trees throughout the Great Plains, both types of orioles increased in number and spread onto the plains. A zone of hybridization about 320 km wide now extends from North Dakota to Texas. In the 1950s, about 5% of the orioles in this area were of one type or the other and the rest were fertile hybrids. Because they interbreed and produce fertile hybrids, the Baltimore and Bullock's orioles, formerly considered two species, are now considered one species, the Northern oriole (*Icterus galbula*). Evidence has been accumulating, however, that the two may be separate species after all, or they may be in the process of speciating. Hybrids seem to be less viable than nonhybrids, and the two distinct nonhybrid types are becoming more numerous than the hybrids in the overlapping populations.

Reproductive Isolating Mechanisms

During the time that populations are separated by geographical barriers, each population may develop **reproductive isolating mechanisms** that prevent its members from interbreeding with those of other populations. Isolating mechanisms may operate prior to or after zygote formation, and many species have more than one type of isolating mechanism.

Prezygotic Isolating Mechanisms. **Prezygotic isolating mechanisms** prevent zygote formation by preventing the sexes from coming into contact. Members of one population, for instance, may become nonreceptive to members of the opposite sex of the other population. Such mechanisms are often behavioral, so that sexual attraction between males and females is weak or lacking. Many species of birds, for example, have elaborate courtship songs and displays necessary for pair formation and successful mating (Figure 17.11). Partly because songs and displays of one species are not recognized by members of other species, interbreeding does not occur between members of separate species. Other types of species-specific breeding signals include the light flashes of fireflies and the pheromones (airborne chemical odors) released by many species of animals to attract members of the opposite sex. In most cases, these signals are species-specific, attracting only those members of the species that produces them.

Seasonal isolating mechanisms affect when males and females of a species come into breeding condition. A termite queen breeds once in a nuptial flight amid a swarm of males. In Florida, two closely related termite species are reproductively isolated because the nuptial flights of one species occur in the spring, while those of the other species occur in the fall.

(a)

(b)

Figure 17.10 Two types of orioles—two species or just one? (a) The male Baltimore oriole is markedly different from (b) the male Bullock's oriole.

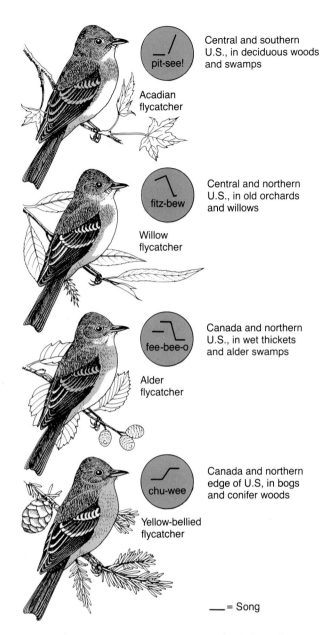

Acadian
flycatcher

pit-see!

Central and southern
U.S., in deciduous woods
and swamps

Willow
flycatcher

fitz-bew

Central and northern
U.S., in old orchards
and willows

Alder
flycatcher

fee-bee-o

Canada and northern
U.S., in wet thickets
and alder swamps

Yellow-bellied
flycatcher

chu-wee

Canada and northern
edge of U.S., in bogs
and conifer woods

———— = Song

Figure 17.11　Four species of closely related flycatchers (*Empidonax*). These species are similar in appearance and share a recent common ancestor. Males and females of different species do not interbreed because they occupy distinct niches and because male songs are species-specific (not recognized by females of other species).

Many closely related species breed during the same season, but maintain reproductive isolation by mating at different times of the day. Many species are active and breed only at night, whereas others are active only during daylight hours, and still others are crepuscular (active only during twilight hours).

Many species breed during the same season and in the same geographical area, but in separate regions of the habitat. This is called **ecological isola-** tion. The toad *Bufo woodhousei* inhabits the northern and midwestern regions of the United States; its range overlaps that of the closely related toad *B. americanus,* which occurs mainly in the eastern United States. Biologists have found that these species sometimes interbreed, but usually do not because one species breeds in the quiet waters of streams while the other lays its eggs in temporary rain pools.

Anatomical and physiological mechanisms can also prevent zygote formation. Sperm transfer is not possible between two internally fertilizing species if the copulatory structure of a female does not accommodate that of a male of the other species.

Prezygotic isolating mechanisms may also operate after mating. Sperm usually die when introduced into the female reproductive tract of a different species, and even if sperm remain alive, those of one species may not be capable of fertilizing the eggs of another.

Postzygotic Isolating Mechanisms. Even if mating can occur between members of populations that have been geographically separated, and even if zygotes form, the populations may still be reproductively isolated. They may have evolved **postzygotic isolating mechanisms** that prevent the zygotes from developing to maturity or from being able to produce offspring. The genes of two species may be incompatible, so that the zygote dies without developing; or **hybrids** may be formed by crossbreeding, but such hybrids may be sterile. The mule is a hybrid formed by mating a female horse with a male donkey. The horse has 64 chromosomes ($1n = 32$); the donkey has 62 ($1n = 31$). Mules, therefore, have 63 chromosomes and rarely can produce viable offspring because their chromosomes do not synapse properly to form tetrads during meiosis.

Outcomes of Geographical Isolation and Population Divergence

New species may or may not form as a result of geographical isolation and microevolutionary changes in populations. What happens if separated populations that were derived from a single species population come back together? Generally, there are three types of outcomes, depending on the amount of genetic divergence that the populations have undergone while separated. In some cases, members of each population may still be able to interbreed, and they simply merge back into one species population as genes are exchanged (Figure 17.12a). In other cases, members of the two popula-

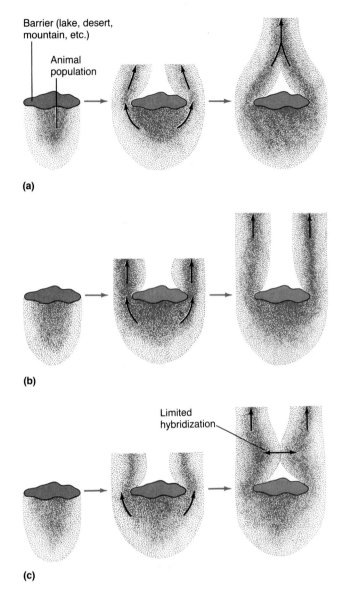

(a)

(b)

Limited
hybridization

(c)

Figure 17.12 Three possible outcomes of geographical isolation followed by reunion of populations. If speciation occurs, it usually involves three processes. First, a population must be separated into two or more populations that do not exchange genes. This is shown in all three examples here. Second, the separated populations must undergo genetic divergence owing to different selection pressures, founder effects, genetic drift, and new variations acting in concert to increase divergence. Third, if speciation is to occur, this divergence must proceed to the point where the populations become reproductively isolated from each other. (a) Two populations fuse back into one after being reunited; no speciation has occurred. (b) Speciation has occurred; reproductive isolation is complete. (c) Limited hybridization occurs after the barrier has been passed or removed, but hybrids are selected against; speciation occurs as a result of character displacement.

Barrier (lake, desert, mountain, etc.)

Animal population

tions will not be able to interbreed and will remain separate, in which case speciation has occurred (Figure 17.12b). In yet other instances, the populations may have developed significant genetic divergence, but there is some hybridization between

them (Figure 17.12c). Hybrids are often not as well adapted as nonhybrids, yet they must compete for the same resources as members of both parental groups. As a result, hybrids may be eliminated before reproducing, and in each population, natural selection typically favors phenotypes that cannot interbreed and produce hybrids with members of the other population. In both populations, natural selection favors any characters that reduce competition for shared resources. This phenomenon, called **character displacement**, has the overall effect of making populations undergo a rapid period of evolution, leading to speciation, once they come back together. Character displacement may be what is occurring in the oriole example discussed earlier, and it undoubtedly occurred among a group of birds referred to as Darwin's finches, which inhabit the Galapagos Islands.

Darwin's Finches. This group of 14 species of finches is believed to have evolved from a few immigrants that came to the Galapagos Islands from a single parent species on the South American mainland, in what is now Ecuador (Figure 17.13). The beaks of these birds are adapted for various feeding habits, and the 14 species can be grouped according to beak and feeding type. Of the seed-eating finches, three species occur on three of the Galapagos Islands, two species occur on one island, and only one species occurs on two islands. Two of the seed eaters, *Geospiza fuliginosa* and *G. fortis*, illustrate character displacement in beak size on islands where either of these species coexists with another species of seed eater. Different beak sizes allow these birds to consume different-size seeds, and character displacement occurs in response to competition. On islands with only one species of seed eater, there is no competition, and character displacement does not occur. Thus, where *G. fuliginosa* occurs alone, its beak size is quite similar to that of the larger *Geospiza fortis*. However, where both species inhabit the same island, *G. fuliginosa* has a smaller beak than *G. fortis*. Similarly, on islands inhabited by *G. fortis* and a larger, third species (*G. magnirostris*), the beak size of *G. fortis* is smaller than when *G. fortis* is alone or coexists with only *G. fuliginosa*.

MACROEVOLUTION

Macroevolution is the province of scientists who study the broad patterns of evolution. The fossil

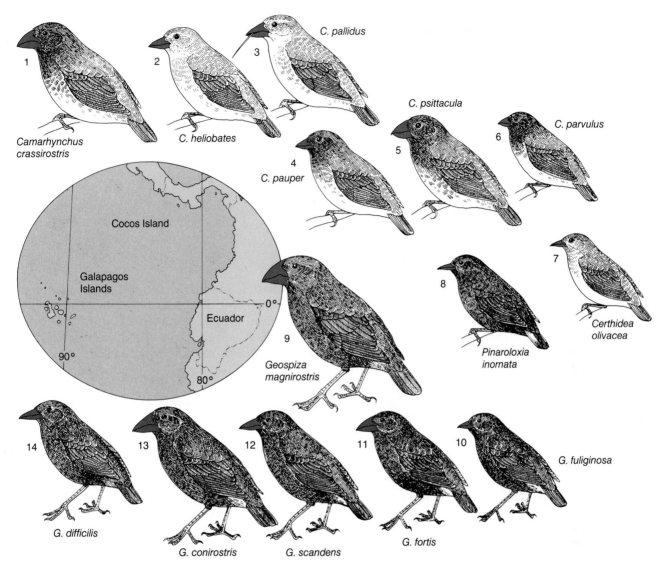

Figure 17.13 Darwin's finches. These 14 species are believed to have arisen by adaptive radiation from a small founder population of one species that colonized the Galapagos Islands. Diversity in beak size and shape reflects adaptive radiation (most of the 14 species have plier-like beaks, characteristic of finches). The 14 species are classified in four groups: tree finches (insectivores and vegetarians, genus *Camarhynchus*: 1,2,3,4,5, and 6); ground finches (cactus-flower eaters and seed eaters, *Geospiza*: 9,10,11,12,13, and 14); a tree-dwelling nonfinchlike species called the warbler finch (*Certhidea olivacea*: 7); and one species isolated on Cocos Island, north of the Galapagos Islands (*Pinaroloxia inornata*: 8). Several species coexist on the Galapagos Islands. Character displacement favoring reduced competition for shared food resources was important in the speciation of these birds.

record shows us that many more species have existed over geological time than are alive today, indicating that many extinctions have taken place in the past. It is also evident that many species alive today do not occur in the fossil record, implying that new species have evolved. The broad scope of macroevolution is concerned with extinction events as they fit into the study of **evolutionary lineages**, the sequences that have led from ancestral species to descendant species.

Patterns of Evolution

The fossil record indicates that there are five patterns of evolution (Figure 17.14). One of these patterns, divergent evolution, results in an increase in the number of evolutionary lineages. For any lineage of animals, different combinations of the five patterns may be discerned in the fossil record, reflecting the varying selection pressures that have acted on the lineage at various times.

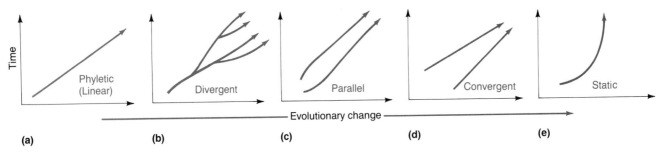

Figure 17.14 Patterns of change in evolutionary lineages. (a) Phyletic evolution leads to a linear sequence of speciations. (b) Divergent evolution occurs when one species diversifies into different groups that eventually become separate species. (c) Parallel evolution involves two species changing in a similar way. (d) Convergent evolution involves two different species developing similar characteristics. (e) In stasis, no evolutionary change occurs in a species for a long period of time.

Phyletic evolution, also called **anagenesis**, occurs when an evolutionary lineage follows a linear path. Along the path, one species gradually changes into another species. The second species replaces the first in geological time, so that at any particular time, only one species is present. Interbreeding is prevented by separation in time rather than by biological barriers. Paleontologists call these sequential species **chronospecies**. Phyletic evolution may occur as species change with gradual changes in their environment.

Divergent evolution, also known as **cladogenesis**, occurs when two or more lineages originate from one ancestral lineage. Biologists generally believe that the frequent recurrence of divergent evolution during the last half billion years has resulted in the great diversity of animals living today.

Convergent evolution, or convergence, occurs when two markedly different species or lineages undergo selection that yields organisms with similar physical features or physiological processes. Whales and dolphins superficially look like fishes, but these mammals and the fishes share only a very distant ancestor, and the mammals did not inherit their fishlike characteristics from that ancestor. Selection pressures of the aquatic environment operating independently in the mammal and fish lineages led to the evolution of fusiform bodies and fishlike appendages, characteristics that provide for efficient locomotion in water. The marsupial (pouched) mammals of Australia exhibit many examples of convergence with the placental mammals of other continents (Figure 33.3). Wolflike, catlike, and rodentlike marsupials have evolved, presumably as a result of adapting to niches in Australia similar to those on other continents.

Parallel evolution, or parallelism, is defined in several ways by different authorities, and some evolutionary biologists even question its conceptual value. As often defined, and as used in this book, parallelism refers to a similar pattern of evolution in two fairly closely related species or lineages over a long period of time. This is common within many groups of animals. Among the placental mammals, for example, rodents are the largest order, and species in separate suborders of the Order Rodentia have independently evolved similar body forms and many similar modes of feeding. Likewise, among the insects, many species of bees and ants (two closely related groups) exhibit parallelism in having complex social behavior.

In contrast to the preceding four patterns of evolution, a fifth pattern, called **stasis**, occurs when lineages exist for a long time without undergoing evolutionary change. Although patterns of evolutionary change occur repeatedly through the fossil record, they are not continuous. Many fossil lineages are known to have remained essentially the same for millions of years. So-called "living fossils" also exhibit stasis. The North American opossum (*Didelphis marsupialis*; see Figure 33.3a), the horseshoe crab (*Limulus polyphemus*; see Figure 26.2), and a bony fish known as the coelacanth (*Latimeria*; see Figure 29.7) are a few examples of living animal species that have changed only slightly in millions of years.

Adaptive Radiation

Lineages of animals that repeatedly diverge are said to undergo **adaptive radiation** (species diversification). Adaptive radiation occurs when a single ancestral species or lineage develops unique adaptations that allow it to invade a number of new niches. The ancestral lineage diversifies by speciating repeatedly and then specializes to fill various niches. To a great extent, evolution is a matter of chance and opportunity, and the lineages that radiate just happen to be in the right place at the right time with the right set of traits to take advantage of opportunities. Adaptive radiation is most easily studied on islands, as Darwin's finches clearly illustrate. On the Hawaiian Islands, birds called honeycreepers provide an even more striking example (Figure 17.15).

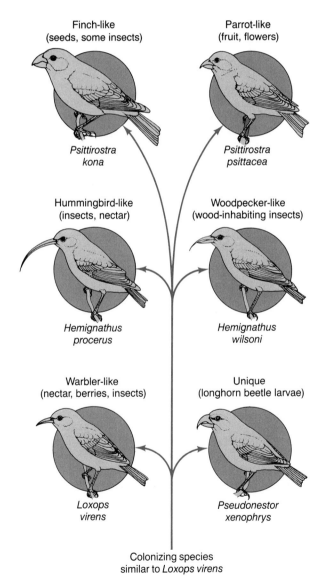

Finch-like
(seeds, some insects)

Parrot-like
(fruit, flowers)

*Psittirostra
kona*

*Psittirostra
psittacea*

Hummingbird-like
(insects, nectar)

Woodpecker-like
(wood-inhabiting insects)

*Hemignathus
procerus*

*Hemignathus
wilsoni*

Warbler-like
(nectar, berries, insects)

Unique
(longhorn beetle larvae)

*Loxops
virens*

*Pseudonestor
xenophrys*

Colonizing species
similar to *Loxops virens*

Figure 17.15 Speciation in Hawaiian honeycreepers. Striking variety in beak shape characterizes this complex group of species. All 15 species are believed to have arisen by adaptive radiation from a single species of finchlike birds that colonized the Hawaiian Islands.

Models of Adaptive Radiation

There are two schools of thought concerning how adaptive radiation proceeds. According to one view, called **gradualism**, evolutionary changes leading to adaptive radiation occur at a more or less steady, continuous pace through time (Figure 17.16a). This view was proposed by Darwin and has been widely accepted since then. The gradualist model assumes that the evolution of most species and lineages is mainly a result of phyletic evolution. Some gradualists also postulate that if the fossil record were complete, it would reflect the gradual evolutionary transitions from species to species in each lineage. Gaps—that is, missing transitional species—in the fossil record are interpreted as imperfections in the record, and it is assumed that fossils of transitional species simply have not yet been found.

A new school of thought concerning how adaptive radiation can occur has developed in recent decades. It is called **punctuated equilibrium**, and it holds that evolution may often occur in spurts rather than only as a series of gradual changes (Figure 17.16b). Thus, species may remain static for millions of years and then sustain rapid evolutionary changes in which the old species become extinct and new, related species fill the vacancies. By "rapid change," we mean time periods of 50,000 years or so—a mere instant in the perspective of geological history. Long-term stasis between periods of rapid evolution is thought to occur because mutations do not improve existing phenotypes. Gaps in the fossil record are interpreted as true reflections of organismal history. The punctuationist model predicts that gaps, rather than smooth transitions, would be the rule in the fossil record. Species would be extensively fossilized only after they became numerous and widespread. Consequently, one would expect that most fossils would represent the static periods of various lineages.

The punctuationist model has stimulated much controversy, and in the process, it has helped fuel interest and research in evolutionary theory. Both the gradual and the punctuationist models are consistent with microevolutionary theory, and many biologists now believe that macroevolution has involved both gradual and punctuationist mechanisms. There is general agreement that depending on the rate of environmental change, species change relatively rapidly or relatively slowly. The most realistic model of adaptive radiation is most likely one that includes elements of both gradualism and punctuationism.

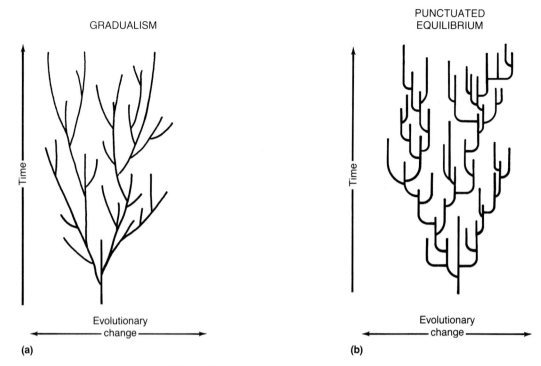

GRADUALISM

PUNCTUATED EQUILIBRIUM

Time

Time

Evolutionary change

Evolutionary change

(a)

(b)

Figure 17.16 Two models for adaptive radiation. These models represent alternative interpretations of how divergence from a common ancestor may occur. The degree of morphological change (horizontal axis) indicates the amount of evolutionary change that occurs in a lineage during geological time (vertical axis). Branches on the models represent divergent evolution. (a) In the model called gradualism, each branching line shows gradual morphological changes with time. (b) In the model called punctuated equilibrium, the lines show that the greatest morphological changes occur rapidly, just after divergence, and that most species spend long periods of time in stasis, undergoing little or no change.

Extinction and the Fossil Record

The fossil record shows that many species have persisted for several million years and then have become extinct. In some cases, a lineage has simply died out, as did many dinosaurs, but in other cases, the lineage continued to diversify even though the ancestral form is gone. The fossil record provides a great deal of information about animal history, but the record is incomplete. Only a few organisms become fossilized, and fossils of soft-bodied organisms are extremely rare. Moreover, as predicted by the punctuationist model, because many speciation events may have been restricted to small areas and small populations, few fossils would have resulted, and this may explain why there is rarely any trace of the origins of major animal groups.

If the ultimate end of a species is extinction, why is there such a diversity of organisms today? The answer is that as long as the rate of speciation exceeds the rate of extinction, diversity increases. Massive changes in the environment may lead to increased extinction at any particular time, but as extinctions occur, previously occupied niches are opened for exploitation by remaining species. The inherent variation in populations and the origin of new variations through mutation allow existing species to radiate and fill niches vacated by species that have become extinct.

EVIDENCES OF EVOLUTION

Paleontology, the study of fossils, provides irrefutable evidence that evolution has occurred. Fossils reveal that at no time have all the species of organisms that have lived on earth ever coexisted. Extinction has been a recurring feature of evolution. Animals and plants also appear in a chronological sequence in the fossil record, indicating that lineages have radiated through time. All of these facts are consistent with the concept of evolution and totally inconsistent with the idea of a single, recent, supernatural creation. Although the fossil record does not clearly indicate *how* living organisms have evolved, it is reasonable to postulate that the fossil record reflects the micro- and macroevolutionary mechanisms previously outlined.

Other Evidences

Studies in **comparative anatomy**, or **morphology**, provide strong support for evolution. Comparable structures in different animal species can often be traced to a common ancestral source, clearly showing that evolution has occurred. Such structures, which are said to be **homologous** and are called **evolutionary homologues**, may be either similar or very different in appearance and function. Thus, the flippers of whales and the forelegs of terrestrial mammals are evolutionary homologues. They show marked differences in structure and function, but fossil and anatomical evidence indicates that they evolved from a common ancestral body part, a paddlelike structure (Figure 17.17a). Homologous

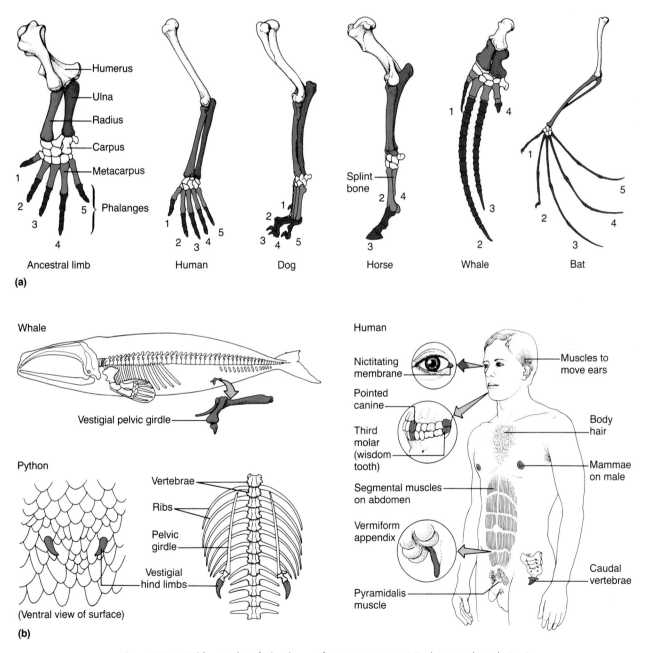

Figure 17.17 Evidence of evolution in vertebrate structures. (a) Evolutionary homologies in mammalian forelimbs. The basic skeletal plan was modified structurally and functionally as it became adapted to different environments. (b) Vestigial structures. Note also the vestigial splint bone on the horse limb in (a).

structures reflect the divergent pattern of evolution. By contrast, many similarities between animal species are **analogous**, that is, similar in structure and function but derived by convergence from different evolutionary origins. The fins of fishes and porpoises are examples of analogous structures, as are the wings of bats and butterflies.

In addition to functional body parts, many vertebrates have **vestigial structures**, which are more or less functionless organs that are homologous to well-developed structures in other species. In addition to the examples shown in Figure 17.17b, the limbs of horses also include vestigial parts (splint bones) that are functional as finger bones in many other vertebrates (see Figure 17.17a). (Other features of vertebrate anatomy that illustrate evolution are discussed in Chapter 28.)

Comparative embryology also offers much evidence of animal evolution. As described in Chapter 16, the early embryonic stages are similar in many animals and provide clues of evolutionary changes and of the relatedness of animal species and phyla. Vertebrate embryos are strikingly similar, and many of the embryonic structures indicate evolutionary relationships among all vertebrate groups. The human embryo, for example, shows evidence of pharyngeal gill structures and a well-developed tail. The "gill" structures develop into other features in the adult. (The evolution of vertebrate pharyngeal arches is discussed in Chapter 28, p. 689.) Other embryological evidence supporting animal evolution is discussed in Chapter 18.

Comparative biochemistry is yet another field that supports evolution. All organisms have DNA as their hereditary material and synthesize proteins by the same mechanism. Many aspects of metabolism, including glycolysis and the fundamental pathways of aerobic respiration, occur in most organisms. Recent work has shown similarities in the types of structural proteins, enzymes, and coenzymes that occur in cells of many organisms. All these factors point to a common ancestry, not only for animals, but for all organisms. Biochemical evidences that substantiate macroevolutionary trends in the animal kingdom are discussed in Chapter 18.

THE FOSSIL RECORD
AND GEOLOGICAL TIME

Biogeographers attempt to explain why species are distributed as they are and how their present and former distributions relate to prehistoric climates and habitats. Conditions on the surface of the earth have changed drastically over geological time. Continents that were once continuous drifted apart, isolating species from one another. As continents drifted, their climates changed as their position on the earth's surface changed. World climates also changed, causing sea levels to rise and fall. During warm periods, oceans covered land bridges between continents and invaded the interiors of the continents. During cool periods, large glaciers formed, lowering sea levels and covering much of the land masses. Mountain ranges gradually increased in height, only to be worn away by erosion. The oxygen content of the atmosphere increased from 0% to about 20%. An ozone layer also developed in the upper atmosphere, where it filters out much of the sun's mutagenic ultraviolet light. These changes have profoundly affected the evolution and distribution of living organisms.

The discovery of radioisotopes in the early 1900s led to a dating method for the earth's crust. At the time of formation, rocks contain a specific ratio of Uranium-238 (^{238}U) to Lead-206 (^{206}Pb). Over time, this ratio decreases because the ^{238}U (an unstable isotope) decays to ^{206}Pb (a stable isotope). Geologists have determined that the time required for half of a quantity of ^{238}U to decay radioactively is 4.5×10^9 years. This is called the half-life of ^{238}U; other unstable isotopes have different half-lives. Thus, if geologists measure the amount of ^{238}U and ^{206}Pb in a rock sample and find that the ratio ^{238}U : ^{206}Pb has decreased by one half from the time the rock formed, the rock can be dated at about 4.5 billion years old (the half-life of ^{238}U). Rocks can be dated reliably by independent measurements of several radioisotopes with different half-lives. If the rocks are sedimentary (formed by sediments that compact into rocks), some fossil remains may be trapped in the rocks by chance, and ages can thus be assigned to many fossils. Once dated, such fossils can be used as **index fossils** to date other sedimentary rocks in which they occur.

The Geological Time Scale

By correlating index fossils with radioisotopic dates, paleontologists have determined the order in which major groups of organisms appeared on earth. This information is summarized as the geological time scale shown in Figure 17.18.

The oldest rocks on earth were formed about 4 billion years ago. The oldest fossil cells—single-celled prokaryotes—are about 3.4 billion years old.

ERA	PERIOD	EPOCH	MILLIONS OF YEARS FROM START TO PRESENT	MAJOR EVENTS AND EXTINCTIONS
CENOZOIC	Quaternary	Recent	0.01	PLEISTOCENE EXTINCTION [Large birds and mammals extinct]
	Tertiary	Pleistocene	2	—*Homo* appears (1–2 million years ago).
		Pliocene	12	—Hominids evolve.
		Miocene	25	—Mammals, birds, lizards, and snakes diversify; insects abundant.
		Oligocene	36	—First monkeys and apes.
		Eocene	58	CRETACEOUS EXTINCTION [Ruling reptiles, many molluscs, and all ammonites extinct]
		Paleocene	65	
MESOZOIC	Cretaceous		135	—First placental mammals. —Continents extensively separated. —Dinosaurs numerous. —Mammals and birds appear. TRIASSIC EXTINCTION [35% of animal families extinct]
	Jurassic		180	—Continental drift under way. —First dinosaurs. PERMIAN EXTINCTION [50% of animal families extinct]
	Triassic		230	
PALEOZOIC	Permian		280	—Land masses fusing as Pangaea; reptiles radiate; amphibians decline; trilobites extinct. —First reptiles and insects; amphibians diverse.
	Carboniferous		345	DEVONIAN EXTINCTION [30% of animal families extinct] —First amphibians and terrestrial arthropods; diverse trilobites.
	Devonian		405	—First bony fishes, including dipnoi.
	Silurian		425	—First jawed fishes, the placoderms. —First vertebrates. CAMBRIAN EXTINCTION [50% of animal families extinct]
	Ordovician		500	—Most animal phyla established. Fossilized molluscs, crustacea, echinoderms, and pogonophora.
	Cambrian		570	
	Precambrian		1500	—Fossilized sponges, corals, jellyfish, annelid burrows, and early arthropods.
			3000	

Figure 17.18 The geological time scale. This scale illustrates the continuous series of major events in earth history. The scale is divided into main divisions called eras. Eras are subdivided into periods, and recent periods into epochs. These divisions are artificial because time is continuous, but they are useful in describing events and conditions. Six massive extinctions have been identified and have been superimposed on the geological time scale in this chart (indicated by diagonal lines). Recent work has suggested that extinctions may actually have a periodicity of about 26 million years.

Fossils of multicellular life first appear in rocks that were formed about 600 million years ago in the **Precambrian period**. The oldest-known fossils of animals include a few sponges, corals and jelly-fish, annelid worm burrows, and early arthropods dating from the late Precambrian. Undoubtedly, these fossils are scant remains of a much more di-verse fauna, but Precambrian rocks have under-gone such extensive changes that most of their fos-sil contents are no longer visible.

The Paleozoic (Old Animal) Era. The Precambrian period ended about 570 million years ago with the start of the "Old Animal" or **Paleozoic era**. The earliest part of this era, the **Cambrian period**, lasted for about 70 million years and was a time of great radiation of life forms. During the Cambrian, the earth's land mass probably consisted of two large continents and several small ones. Climates were relatively warm, and almost all life existed in vast ocean expanses. Most of the lineages that gave rise to modern animal phyla were already established by the mid-Cambrian.

The rest of the Paleozoic era is divided into five periods, during which life diversified along lines established during the Cambrian. The earliest vertebrate fossils date from the **Ordovician period**. In the **Silurian period**, fishes radiated from jawless vertebrates, and then later, in the **Devonian period**, arthropods appeared in considerable numbers on land. The end of the Devonian (about 345 million years ago) was marked by a mass extinction in which about 30% of the living families of animals were lost (see Figure 17.18). Life rebounded during the **Carboniferous period**, and large swamps and plant species that formed our modern-day coal deposits were prevalent. A diverse fauna of snails, scorpions, centipedes, and insects roamed the land, providing food for amphibians, which underwent an explosive radiation during the latter half of the Carboniferous period. The first reptile fossils also date from this period. The diversification of life on land generally continued through the **Permian Pe-**

Essay: How Did Life Begin on Earth?

"It is mere rubbish thinking at present of the origin of life. One might as well think of the origin of matter." So wrote Charles Darwin in the mid-1800s in response to questions about how life might have arisen from nonlife. Today, there are still more questions than answers about life's origins, but we have enough information to discuss the issue meaningfully.

Astronomers estimate that the universe began several billion years ago when matter exploded outward from a concentrated center of mass. Stars and planets, including the earth, condensed from gases formed by the explosion. A study of rocks brought back from the moon indicates that our solar system condensed about 4.7 billion years ago. Because the oldest known fossils of cell-like organisms are about 3.5 billion years old, life apparently originated during the billion or so years when the earth was condensing and cooling.

In the 1920s, the Russian chemist Alexander I. Oparin and British biologist John B. S. Haldane proposed that the early earth's atmosphere lacked free oxygen and consisted mainly of the gases water vapor (H_2O), methane (CH_4), ammonia (NH_3), and hydrogen (H_2). Oparin and Haldane suggested that molecules derived from this primeval atmosphere spontaneously assembled into more complex biological molecules. They postulated that the energy needed to assemble the molecules came from electrical storms and ultraviolet solar radiation. In the 1950s, Stanley L. Miller, then a graduate student working in the laboratory of American Nobel laureate Harold C. Urey, tested this hypothesis by mixing the four gases in a sealed flask containing no oxygen (Figure E1). The gases simulated the early earth's atmosphere. Some of the water vapor was condensed to simulate rain. Water at the bottom of the apparatus simulated the early ocean or pools of water, and electrical sparks were passed through the "atmosphere" to simulate lightning. After about a week, the water had turned color,

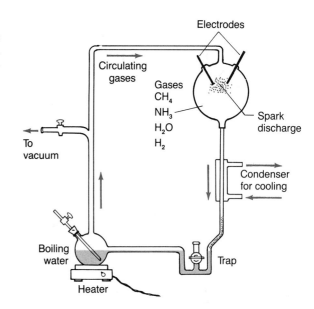

Figure E1 The Miller apparatus. Stanley Miller's spark-discharge apparatus simulated conditions on the early earth. Amino acids and other small organic compounds were generated spontaneously and dissolved in the fluid, or "chemical soup," at the bottom of the flask.

and analysis showed that it contained several kinds of small molecules, including some amino acids. Since the 1950s, many similar experiments, some using ultraviolet light or heat as the energy source, have shown that several chemical constituents of life can be generated under conditions similar to those thought to characterize the early earth. The nitrogenous bases adenine, guanine, cytosine, thymine, and uracil have been generated in laboratory simulations, as have certain nucleotides. These experiments show that many small molecules of which living cells are composed may have formed spontaneously in the absence of life.

riod. During much of the Permian, land masses were fused as **Pangaea**, the world continent. As continents collided, forming Pangaea, their edges were folded and forced upward, creating extensive mountain ranges. Mountains, including the Appalachians of North America, replaced many lowland coastal areas with higher, cooler environments. Much of the Pangaea supercontinent in the southern hemisphere was glaciated. At the end of the Permian (about 200 to 230 million years ago), another major extinction occurred in which about 50% of the animal families (some estimates say nearly 95% of all species) died out. Amphibian lineages were decimated, never to achieve significant numbers again. The Permian extinction may have

resulted from major changes in the continents, in climates, and in sea level.

The Mesozoic (Middle Animal) Era. The "Middle Animal" or **Mesozoic era**, also known as the age of reptiles, began following the Permian extinctions. Reptiles underwent great diversification during the first 35 million years of the Mesozoic, called the **Triassic period**. The oldest group of mammals (egg-laying mammals) evolved from one group of reptiles during the Triassic. The end of the Triassic period was marked by yet another major extinction, during which about 35% of the animal families, including about 80% of the reptiles, disappeared. The **Jurassic period** began about 180 to 190

It is one thing to synthesize small molecules such as amino acids and organic bases; it is another immense step to synthesize the complex polymers, or macromolecules (proteins, fats, carbohydrates, and nucleic acids) essential to life. Life as we know it cannot synthesize macromolecules without other macromolecules. Protein enzymes are required to form macromolecules, and nucleic acids are required to form proteins. How could macromolecules have formed before life existed? It is possible that some polymers formed without catalysts as small molecules became ever more concentrated in primeval seas, especially in backwater pools and in ocean foam. It is also likely that certain polymers formed from simple compounds that became exposed on drying clay surfaces. These conditions promote the kinds of synthetic reactions that form macromolecules. Without catalysts, chemical reactions that produced early macromolecules were undoubtedly random and slow; but it seems plausible that in a billion or so years, the early chemical mixtures could have given rise to a simple cell-like form of life.

Following up Miller's experiments, Sidney Fox and other American biologists found that if amino acid powders are heated in the absence of oxygen, polypeptides (long chains of amino acids) form spontaneously. Similar events may have occurred on the dry clay surfaces of the early earth. When Fox mixed the polypeptides with water, small cell-like vesicles called **proteinoid microspheres** resulted. These microspheres were able to accumulate sugars and amino acids from solutions and had crude enzymatic activity.

Although no one knows how the first cells actually evolved, Fox's microspheres provide a possible bridge between the chemical mixtures of the primeval oceans and the first cells. Given millions of years, microspheres or similar chemical aggregates may have developed a crude energy-yielding metabolism. Perhaps in the same time period, nucleotides within microspheres reacted together, forming nucleic acids. Because nucleic acids are self-replicating, once early cell-like entities acquired these molecules as well as a few proteins that may have acted as primitive catalysts, they could have evolved a genetic mechanism. If a cell-like entity with a crude energy-yielding metabolism developed a genetic system, it would have acquired the potential to reproduce and to have its metabolic and structural processes directed from within. Such an entity would have had the fundamental characteristics of life.

Most scientists believe that life had a common origin, because the genetic and metabolic systems of bacteria, protists, fungi, plants, and animals are all similar and based on similar molecules. It seems unlikely that these similarities would have developed if life had originated from different molecular mixtures.

The first cells were probably somewhat like present-day **prokaryotic cells**, the bacteria and blue-green algae. Prokaryotic cells lack a nucleus; their simple genetic system consists of a naked DNA molecule in the cytoplasm. By contrast, the complex **eukaryotic cells** of protists, fungi, plants, and animals have a distinct nucleus containing chromosomes made of DNA and proteins, and an extensive system of internal membranes, forming membranous organelles. Many of the organelles, as well as the nucleus, are thought to have originated as infoldings from the plasmalemma of primitive prokaryotic cells. Mitochondria and chloroplasts, because they have DNA that functions in organelle heredity, are thought to have been derived from prokaryotic cells that invaded and became internal parasites in other cells. Gradually, a mutually dependent metabolism may have arisen that made the parasites and their hosts incapable of existing independently. Other key events in cell evolution pertinent to animals are discussed in the Trends and Strategies sections in Unit III.

million years ago and lasted about 50 million years. During much of this period, despite changes in sea level, climates were much warmer than they are today. The poles were not ice covered and most land masses supported luxuriant plant growth. Tropical animals were distributed nearly worldwide. Reptiles, especially dinosaurs, continued to diversify, and birds and marsupial (pouched) mammals evolved from separate reptilian stocks. Coral reefs, which require tropical water temperatures, occurred at latitudes 10°–20° farther north and south than they do today. Toward the end of the Jurassic, large areas of land became covered by shallow seas. By this time, Pangaea had broken up into smaller continents that had begun drifting apart, and mountain building was common.

During the **Cretaceous period** (beginning about 135 million years ago), reptiles reached their zenith, dominating marine, freshwater, and terrestrial environments. Worldwide climates were warm and humid. The separated land masses that had once been Pangaea favored speciation of terrestrial animals. Late in the Cretaceous, new mountain ranges—the Rockies, Andes, and Alps—arose, and climates suddenly became cooler. The end of the Mesozoic (about 65 million years ago) was again marked by major extinctions; 50%–75% of all species disappeared, including most of the dinosaur lineages (see Figure 17.18). Fossil marine deposits indicate that plankton were decimated in late Cretaceous seas, suggesting that marine food chains were interrupted.

The Cenozoic (Recent Animal) Era. Following the Cretaceous extinctions, surviving lineages—and there were some in all animal phyla—underwent radiations that eventually filled vacant niches; consequently, species resembling many modern forms appear in the fossil record after the Cretaceous extinctions. The last 65 million years of geological time, called the "Recent Animal" or **Cenozoic era**, is divided into two periods. The **Tertiary period** lasted until about 1 to 2 million years ago, and the **Quaternary period** continues today. A general cooling of the earth has occurred during the Cenozoic. Estimates of temperatures in the northern Pacific region indicate that annual temperatures averaged near 20°C at the beginning of the Tertiary, but fell to lows of −2°–5°C in the early Quaternary.

During the mid-Tertiary period, climates were warm and wet enough to support vast forests over much of the globe. Among the animals that radiated to fill niches in these forests were the primates, especially monkeys and apes. At about the same time, continued drifting of the continents produced new upthrusts in the Rockies, Sierra Nevadas, Andes, Alps, and Himalayas, preventing much of the precipitation from reaching the interior of certain continents. Temperatures also continued to drop, forming polar ice caps, lowering sea levels, and producing drier conditions throughout the world. As a result, in the latter part of the Tertiary, many of the great forest expanses in continental interiors were replaced by grasslands and animals adapted to them. Since the beginning of the Quaternary, animal fossils became decidedly modern. In Africa, certain bipedal apes were speciating, eventually giving rise to our own species, *Homo sapiens*. (Human evolution is discussed in Chapter 33, p. 817.)

Meteors and Extinction. As this brief view of the fossil record indicates, the evolution of life has been interrupted by several mass extinctions. These events have had the important effect of opening niches into which surviving animals could radiate with little competition. But what could have caused these periodic extinctions? Recently, researchers have discovered on several continents that rock sediment dating from the Cretaceous/Tertiary boundary is rich in iridium, a rare element similar to platinum. Meteors are known to be rich in iridium, and several investigators have postulated that the Cretaceous extinction was caused by the collision of a large meteor with the earth. Such an impact could have vaporized the meteor and thrown up a dust cloud that would have blocked sunlight. All food chains would have been interrupted, perhaps for several years, and the temperature would have dropped as less solar radiation reached the earth's surface. This meteor extinction hypothesis is far from universally accepted, but it has started an intensive hunt among astronomers for mechanisms that might trigger periodic meteor bombardment of the earth.

☐ SUMMARY

1. Darwin and Wallace first proposed that evolution occurs by natural selection. They observed that all members of a species have heritable variations and that the reproductive potential of species is rarely, if ever, realized because environmental resources are limited. They theorized that natural selection occurs as only those individuals with traits that provide advantages in competing for resources with other members of the species may survive and produce offspring.

2. Principles of genetics provide an understanding of the hereditary mechanisms involved in evolution. Genetic variability results from mutations, which produce new alleles, and is maintained by recombination in meiosis and fertilization. Natural selection tends to reduce variability; it can eliminate deleterious dominant alleles from a population, but usually not deleterious recessive alleles, because the latter are not expressed in heterozygotes.

3. Castle, Hardy, and Weinberg demonstrated that Mendelian inheritance does not result in changes in allele frequencies, providing a population is large and randomly mating with no mutation, migration, or selection occurring. The Castle-Hardy-Weinberg principle is used as a baseline against which genetic change in natural populations can be measured.

4. A biological population is a group of interbreeding individuals of a species. Many populations are subdivided into local units called demes. Gene pools (all the alleles) in populations are the units of evolutionary change. A species is a reproductively isolated group of animals.

5. Microevolution is concerned with changes in population gene frequencies as they apply to evolutionary changes in populations and species. Macroevolution is concerned with evolution above the species level.

6. Genetic drift results from changes in gene frequencies (due to chance) in small populations. Inbreeding is common in small populations and often results in the expression of recessive traits.

7. Patterns of natural selection are disruptive selection, stabilizing selection, and directional selection. Natural selection produces structural, physiological, and behavioral adaptations that enhance an animal's fitness and mold an animal to its niche (the total role of a species in its environment).

8. Speciation may occur if one population separates into two populations that do not exchange genes. Different genetic variations may accumulate in each isolated population, producing reproductive isolating mechanisms. Two separated populations have become two species if their individuals cannot interbreed when reunited.

9. Patterns of macroevolution include phyletic (linear) evolution, divergent evolution, convergent evolution, parallel evolution, and stasis.

10. Adaptive radiation occurs when a species repeatedly speciates by divergence to fill unoccupied niches. Adaptive radiation may proceed gradually or in rapid spurts separated by long periods of stasis.

11. The fossil record, comparative anatomy, embryology, and biochemistry provide evidence of evolution. The geological time scale shows a sequential change in species over the last 600 million years. Many more species lived in the past than do at present. Mass extinctions have occurred at least five times in the last 500 million years.

☐ FURTHER READING

Ayala, F. J. *Population and Evolutionary Genetics: A Primer.* Menlo Park, Calif.: Benjamin/Cummings, 1982. *An introduction to the essentials of microevolution.*

Bambach, R. K., C. R. Scotese, and A. M. Ziegler. "Before Pangea: The Geographies of the Paleozoic World." *American Scientist* 68(1980): 26–38. *Reconstructs the movements of continents that occurred prior to about 240 million years ago.*

Cairns-Smith, A. G. "The First Organisms." *Scientific American* 252(1985): 90–100. *Develops the idea that the earliest "systems" to begin evolving may have been clay crystals.*

Calow, P. *Evolutionary Principles.* London: Blackie, 1983. *A brief treatise on micro- and macroevolution.*

Eisley, L. C. *Darwin's Century: Evolution and the Men Who Discovered It.* Garden City, N.Y.: Doubleday, 1958. *An interesting account of the social and scientific atmosphere of Darwin's era.*

Futuyma, D. J. *Evolutionary Biology.* 2d ed. Sunderland, Mass.: Sinauaer Associates, 1986. *A modern textbook in evolution.*

Lewin, R. *Thread of Life. The Smithsonian Looks at Evolution.* Washington, D.C.: Smithsonian Books, 1982. *A superbly illustrated introduction to evolutionary theory and diversity.*

Stebbins, G. L., and F. J. Ayala. "The Evolution of Darwinism." *Scientific American* 253(1985): 72–82. *Discusses how new studies and techniques in molecular biology are contributing to our understanding of evolution.*

Weaver, K. F. "Meteorites. Invaders from Space." *National Geographic* 170(1986): 390–418. *Illustrates the meteorite hypothesis of mass extinctions.*

Wilson, A. C. "The Molecular Basis of Evolution." *Scientific American* 253(1985): 164–173. *Describes how molecular biologists have gained a new understanding of evolution by studying mutations in DNA.*

UNIT III

THE ANIMAL KINGDOM

CHAPTER 18

Phylogeny and Classification

Modern study of evolutionary relationships among animals. Exhaustive museum collections, such as this display of moths and butterflies at the Smithsonian Institution, allow zoologists to compare morphological characteristics of species. Modern instruments, such as computers, help in the collection and analysis of data, and new techniques in molecular biology provide species comparisons at the biochemical level.

Animal diversity, which we examine in detail in this unit, is living evidence of the evolutionary processes discussed in Chapter 17. The sciences of **phylogeny**, or the reconstruction of evolutionary histories, and **classification**, the naming and logical grouping of organisms, make it possible to study this diversity in an organized way. Biologists have found that the most useful way to classify animals is to group them in ways that reflect the course of their evolution. Consequently, animal classification is based on our current understanding of animal phylogeny. Historically, classification and phylogenetic analyses were based primarily on comparative anatomy (studies of animal structure) and comparative embryology. These fields are still the foundation of classification of

425

most animal groups, but in recent decades, an influx of new analytical methods and data from disciplines such as physiology, biochemistry, genetics, and immunology have deepened our understanding of phylogeny. For this reason, classification is currently one of the most active and interesting research areas in zoology.

PHYLOGENY

A phylogenetic view of the animal kingdom considers the evolutionary history of the major animal groups. It postulates answers to macroevolutionary questions such as: In what general sequence did members of the animal phyla and their subgroups evolve? How are members of the phyla interrelated? What were common ancestors like, and to which groups did the common ancestors give rise?

Every species alive today is a blend of **ancestral characteristics**, which are unchanged traits from its remote ancestors, and **derived characteristics** that have evolved more recently (Figure 18.1). In discussing phylogeny, it has been customary to refer to animals with a higher proportion of ancestral traits as "lower" animals. Likewise, animals

with a high proportion of derived traits are often said to be "higher" or "advanced." Because the terms "lower," "higher," and "advanced" are not explicit and may imply a value judgment, there is a trend in zoology to abandon them. When we have occasionally used these terms in this book, we have tried to define or qualify them to prevent misinterpretation.

Sources of Phylogenetic Data

A long-standing hypothesis in zoology is that animals having the same or very similar developmental stages and having many body features in common had a common ancestor that possessed the same set of features. This hypothesis is applied in interpreting evolutionary trends discerned from morphological and embryological data.

In the early 1800s, the German embryologist Karl von Baer noted that very young embryos of so-called higher animals resemble those of lower animals and that young embryos resemble each other more than do older embryos. Somewhat later, the German evolutionist Ernst Haeckel elaborated von Baer's ideas and proposed that an individual's development, or **ontogeny**, repeats the phylogeny of the species or of a group, such as the phylum, to which the species belongs. Haeckel pos-

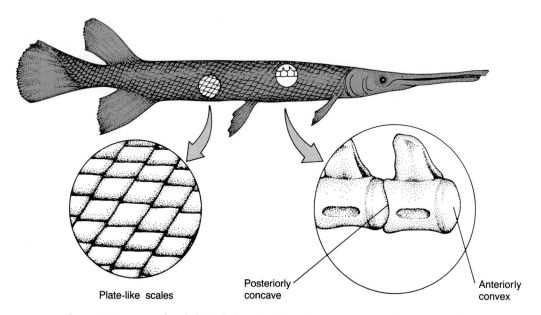

Figure 18.1 Ancestral and derived characteristics. Gars, a small group (seven species) of ancient bony fishes, have several ancestral (primitive) features. These include an armor of nonoverlapping diamond-shaped scales and a tail with slightly upturned vertebrae. These are homologous features shared by all gars and their common ancestor. By contrast, the derived features of gars have evolved since gars diverged from an ancestral group and are therefore unique to gars. Derived features include a long, thin snout and vertebrae that are anteriorly convex and posteriorly concave; other fishes have biconcave vertebrae.

Plate-like scales

Posteriorly concave

Anteriorly convex

tulated, for example, that the two-layered gastrula stage in the development of a bilateral animal represents a phylogenetic stage when the bilateral animal was gastrula-like as an adult. We now know that Haeckel overstated the case (ontogeny does not literally repeat evolutionary history), but his ideas stimulated thought and analysis of development and its relationship to animal evolution. It is now clear from studies of embryonic development in hundreds of animal species that embryos and larval stages change in the course of evolution. Each stage in the life cycle is subject to different selection pressures, however, and in most species, embryos and larvae change less drastically and at a slower rate than adult stages. Usually, as noted by von Baer, there is a greater similarity between the developmental stages of two different species than between the adults. Consequently, by analyzing developmental stages, biologists have been able to gain some understanding of relationships among living animals and to form hypotheses about what their common ancestors were like.

Some of the strongest evidence of evolutionary relationships between two or more animal groups comes from studies of evolutionary homologies, structural similarities that have evolved from a common ancestral structure. (Evolutionary homologies in vertebrate limbs are discussed in Chapter 17, p. 415.) Homologies in different animal species may be identified from studies of internal structure, embryonic development, and, when available, intermediate forms in the fossil record. It is likely that species sharing many homologous structures are more closely related than those sharing only a few. Many cases of homologous organs have been documented among the vertebrates, and as a result, vertebrate evolution is better understood than that of most invertebrates. (Homologous organs among vertebrates are discussed in Chapter 28, p. 689.)

Phylogeny and Embryonic Development. A popular model of animal phylogeny, and one that we present later in this chapter (see Figure 18.6), is based largely on evidence of animal evolution from comparative embryology and on von Baer's principle of embryonic similarity. Many zoologists believe that two large groups of animal phyla, the protostomes and deuterostomes, are distinguished on the basis of embryological criteria. These criteria were introduced in Chapter 16 and are summarized in Table 18.1. **Protostomes** ("first mouth," meaning the mouth is derived from the blastopore of the gastrula) include molluscs, annelids, and arthropods. As discussed in Chapter 16, arthropods have unique embryonic development, but they share several fundamental (ancestral) features with annelids. Their unique development is thought to be a derived rather than an ancestral characteristic, so they are usually considered protostomes. **Deuterostomes** ("second mouth," meaning the mouth is not derived from the blastopore) include the echinoderms (such as sea stars and sea urchins), hemichordates (a small, chordatelike group

Table 18.1 Embryological Evidence Delineating Two Large Groups of Animal Phyla

Embryological Feature	Protostomes	Deuterostomes*
Cleavage pattern	Spiral	Radial
Fate of blastopore	Mouth	Not the mouth
Mesoderm source	Single micromere in gastrula	Archenteron outpocketings or separating sheets
Coelom formation	Schizocoelous	Enterocoelous
Larva following gastrula**	Often a trochophore	Tornaria, bipinnaria, or others similar to these

*Among the chordates, placental mammals have rotational cleavage; and in most vertebrates, coelom formation more closely resembles the schizocoelous pattern than the enterocoelous pattern. (The schizocoelous and enterocoelous patterns are described in Chapter 16.) In other ways, vertebrate embryology closely resembles that of echinoderms and other deuterostomes. Consequently, it is assumed that the different vertebrate patterns were derived, rather than ancestral. The ancestral conditions of all chordates were most likely those of typical deuterostomes.

**The larval stages mentioned here are described in the next section of this chapter.

of marine animals discussed in Chapter 27), and chordates. As with the protostomes, not all animals in these phyla exactly fit the deuterostome mold. Placental mammals, for example, have rotational cleavage instead of the radial pattern typical of deuterostomes. Considering that placental mammals illustrate many derived characteristics but share ancestral features with other vertebrates, it may be assumed that they are deuterostomes with derived developmental patterns.

Larval Stages.   In many animals, the blastula or gastrula develops into a free-swimming larva. Undoubtedly, these larvae have changed considerably in the course of evolution, but comparing them across the major phyla reveals some interesting trends that may reflect animal history. Among the least complex animals, many sponges have a free-swimming flagellated **amphiblastula** that is unlike any other larval stage in the animal kingdom (Figure 18.2a). The amphiblastula eventually settles to the bottom and develops into an adult. Its outer flagellated cells come to line internal flagellated water chambers. Marine cnidarians typically have an ovoid, free-swimming larva called a **planula** (Figure 18.2b). The planula develops from a solid or hollow gastrula stage. It has a flattened mass or internal lining of endodermal cells surrounded by an outer layer of ciliated ectodermal cells. Planulae eventually settle, attach to the bottom, and grow into a hydralike polyp stage.

Among the flatworms (Phylum Platyhelminthes), many parasitic species have well-developed larval stages, but these are believed to be derived features, representing special adaptations for parasitism. Many free-living flatworms lack a larval stage; following gastrulation, the embryo usually flattens and develops into a small worm. Perhaps more revealing of ancestral relationships, several flatworms have a very simple body construction, and except for having mesoderm and rudimentary digestive and reproductive systems, such forms are not much more complex than the planula larvae of cnidarians. Although many flatworms have complex organ systems, some that do not are thought to resemble ancestral bilateral animals. (We will have more to say about the possible ancestors of bilateral organisms later in this chapter and in the Trends and Strategies section beginning on p. 485.)

Among the protostome eucoelomates, many molluscs and annelids have a ciliated, free-swimming larva called a **trochophore** (Figure 18.2c). Trochophores may represent ancestors of molluscs,

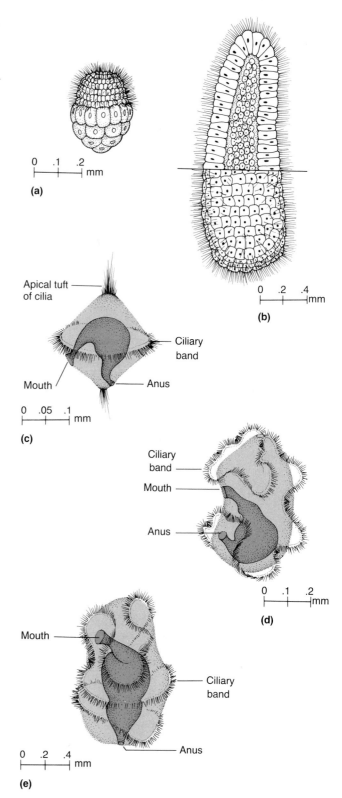

Figure 18.2   Free-swimming larvae of several marine metazoans. (a) Amphiblastula of sponges. (b) Partial longitudinal section of a planula (larva) of cnidarians; note its two cell layers. (c) The trochophore of an annelid resembles a spinning top; several molluscs have similar larvae. (d) The bilaterally symmetrical bipinnaria of sea stars resembles (e) the tornaria of hemichordates.

annelids, and other protostome phyla, but more likely they are living evidence that several protostomes had a common origin.

Among the deuterostomes, echinoderms and hemichordates have a free-swimming larval stage that is quite different from the trochophore. In these groups, the gastrula develops into a ciliated bilateral larva called the **auricularia** in sea cucumbers, the **bipinnaria** in sea stars, and the **tornaria** in hemichordates (Figure 18.2d and e). Similarity among these larval types may indicate a close phylogenetic relationship between echinoderms and hemichordates. In echinoderms, the larva eventually loses its bilateral symmetry and assumes the secondary radial symmetry of an adult.

Cellular and Molecular Biology. Several structural features of cells support the distinction between protostomes and deuterostomes. For example, ciliated epidermal cells are common in the animal kingdom, although two large groups (nematodes and arthropods) generally lack functional cilia. Ciliated cells are either monociliated (bear one cilium) or multiciliated. Flatworms have multiciliated cells, as do certain protostomes (molluscs and annelids). By contrast, many deuterostomes have monociliated epidermal cells. Some researchers believe the monociliated condition to be ancestral, because it is simpler and because some of the least complex animals (sponges and many cnidarians) have monociliated cells.

Protostomes and deuterostomes may also be distinguished by having different types of photoreceptors, as seen by electron microscopy. In certain echinoderms and chordates, photoreceptors are formed from cilia, and microvilli developed from the ciliary membrane increase the surface area of the photoreceptor cell. In certain flatworms and among protostomes (some molluscs, annelids, and arthropods), photoreceptors are specialized cell projections with microvilli, but they do not form from cilia.

At the biochemical level, the basic uniformity of chemical constituents (including amino acids, ATP, and common purine and pyrimidine bases in nucleic acids) of all organisms provides strong support for the theory that life had a common origin. Yet, despite a fundamental chemical similarity among all life forms, each species is chemically distinct from all others, and comparative studies of molecular similarities and differences among organisms can provide useful data in delineating animal groups. For instance, biochemical analyses of substances making up the skeletons and body coverings of animals provide some indications of broad phylogenetic trends. Chitin, a nitrogenous polysaccharide, is a common structural material among protostomes but is rare in deuterostomes. Echinoderms and chordates have mainly keratin, a tough protein, and bone or bonelike body supports.

Molecular and population genetics have also shed some new light on phylogeny. As species evolve, they become increasingly different genetically, so closely related species are genetically more similar than distantly related ones. It follows that if the genetic makeup of several species can be determined, the evolutionary relationships among the species may be better understood. Another factor makes genetic data uniquely useful in phylogenetic analyses. Compared with the evolution of major phenotypic features, such as larval or adult anatomy, evolution of genes is generally believed to be conservative; that is, genetic change is thought to be more directly dependent on time than is phenotypic change, and over a long period of time, genes may change at a more constant rate than do anatomical features. Thus, genetic change may serve as a type of "molecular clock" of evolution, and a species' genetic makeup may provide a relatively clear picture of its ancestry and evolutionary relationships. Also, because of its relative conservatism, genetic change is not likely to be affected by convergent evolution. (Convergent evolution is discussed in Chapter 17, p. 412.) While convergence of two distantly related species may make their adults and larvae look or act similar and thus appear more closely related than they really are, molecular change in the two species may continue to occur at a relatively constant rate, unaffected by convergence.

Because proteins are encoded by DNA, analyzing protein structure gives indirect evidence of genetic differences among species. It is thought that the degree of similarity in amino acid sequences in classes of proteins is directly related to the degree of genetic similarity between species. For example, the amino acid sequence of cytochrome *c*, one of the proteins in the mitochondrial electron transport system, has been determined in many animals (Figure 18.3). Compared to many proteins, cytochrome *c* changes very slowly and shows relatively few amino acid differences, even among rather distantly related organisms. By contrast, certain other proteins (such as the enzyme carbonic anhydrase) change rapidly, may vary greatly from one species to the next, and are more useful in assessing relationships among closely related organisms. These differences in protein changeability have foiled at-

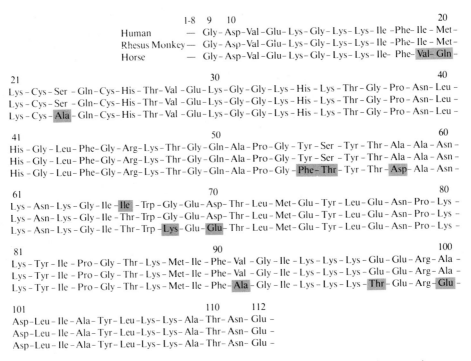

Figure 18.3 Amino acid sequences of cytochrome c in three mammals. In humans, rhesus monkeys, and horses, the molecule has 104 amino acids; in many other organisms (both plants and animals), positions 1–8 contain additional amino acids. These amino acid data reflect the relative phylogenetic distances between the three species shown. There is only one amino acid difference (color highlight at position 66) between human and rhesus monkey cytochrome c; horses and rhesus monkeys have cytochrome c molecules that differ at eleven amino acid positions.

tempts to establish a single molecular clock for evolution, but an overall average rate of molecular change can be calculated from studies of change at several genetic loci.

Electrophoretic and immunological techniques can also be used to study protein differences and hence phylogenetic relationships. Electrophoresis separates proteins according to how they migrate in an electric field. This technique is used widely in studies of genetic variability in species populations. In immunological studies, a protein such as serum albumin can be obtained from a particular species and used as an antigen: It is injected into a laboratory mammal, which then produces specific antibodies against it. Antibodies are obtained from the mammal's blood as antisera, which later react in various immunological tests with the albumin. Serum albumins are then obtained from other species and tested for reactivity with the antibodies. The strongest antigen-antibody reactions occur with the original albumin, but albumins from other species may also react with the antibodies. The strength of their reactivity with the antibodies is proportionate to how similar they are chemically to the original antigen, and this chemical similar-

ity may reflect the genetic similarity of the species tested. Table 18.2 presents data showing immunological differences among albumins of primates; Figure 18.4 illustrates the use of these data in reconstructing a set of phylogenetic relationships. Immunological and electrophoretic analyses are most useful in studying genetic similarities among closely related species. Their usefulness in studying relationships among species belonging to different phyla is limited because they detect only the molecular differences (not the similarities) among distantly related organisms.

Direct measurement of genetic relationships between species can be obtained by determining nucleotide sequences of genes or of messenger RNA transcribed from DNA. (Transcription and translation of the genetic code are discussed in Chapter 15, p. 366.) Unfortunately, it is difficult to isolate specific genes, and few such studies have been completed. More commonly used is a technique called *DNA hybridization*. In this procedure, DNA is separated (melted) into single strands by heating. Melted strands of one species are then allowed to react with those from other species. Upon cooling, if the strands of two species have similar base se-

Table 18.2 Immunological Differences (Estimated by Degree of Difference Between Albumins) Between Several Old-World (Eurasian and African) Primates*

Species Tested	Antiserum to:		
	Homo	Pan	Hylobates
Homo sapiens (human)	0	3.7	11.1
Pan troglodytes (chimpanzee)	5.7	0	14.6
Gorilla gorilla (gorilla)	3.7	6.8	11.7
Pongo pygmaeus (orangutan)	8.6	9.3	11.7
Symphalangus syndactylus (siamang)	11.4	9.7	2.9
Hylobates lar (gibbon)	10.7	9.7	0
Old-world monkeys (average of six species)	38.6	34.6	36.0

*Data represent the amount of reactivity between albumin of the species tested with the three antisera; the higher the number, the less the reactivity and the greater the immunological difference.

From F. J. Ayala and J. A. Kiger, Jr. *Modern Genetics*, 2d ed. (Menlo Park, Calif.: Benjamin/Cummings, 1984), p. 896. (Calculated from data in V. M. Sarich and A. C. Wilson, *Science* 158(1967): 1200.) Reproduced by permission.

quences, they will fuse together (hybridize), forming interspecific double helices. The extent of hybridization depends on how many base pairs match in the interspecific helices. This can be quantified and is a direct estimate of the degree of genetic similarity between the species tested. As with electrophoresis and immunology, DNA hybridization works best in studies of relatively similar species, for melted strands differing greatly in base sequences will not hybridize.

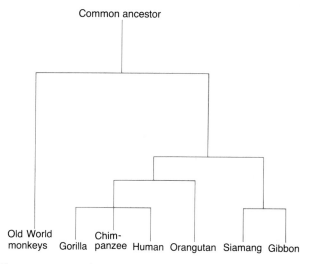

Figure 18.4 Use of immunological data to reconstruct phylogeny. This diagram shows evolutionary relationships among various old-world (Eurasian and African) primates as indicated by the immunological distinctiveness of their serum albumins (based on data presented in Table 18.2).

One other bit of new information from molecular genetics may apply to phylogeny at the phylum level and above. From studies of bacteria, certain nematodes, and fruit flies *(Drosophila)*, it had long been assumed that DNA replication occurs in all organisms by the formation of replication forks. (DNA replication is covered in Chapter 15, p. 362, Figure 15.10.) But new evidence indicates that DNA replication by forking is not universal, and in the animal kingdom, nonfork replication may be the rule in some deuterostomes (see *Xenopus*, p. 362). Additional research is needed to determine if this constitutes a basic evolutionary distinction between deuterostomes and other animal groups.

Interpreting Phylogenetic Data

From the kinds of data just described, zoologists develop models called **phylogenetic trees**, which are hypothetical reconstructions of animal evolution. Generally, the more uniform the group of animals, the easier it is to study and define phylogenetic relationships among its members. Relationships among the largest groups, the phyla, are the most difficult to define, and phylogenetic trees depicting the evolution of the entire animal kingdom can only postulate the major events in animal history.

Interpreting Phylogenetic Trees. Phylogenetic trees do not imply that any *living* species in any

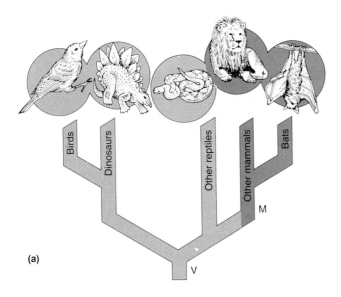

(a)

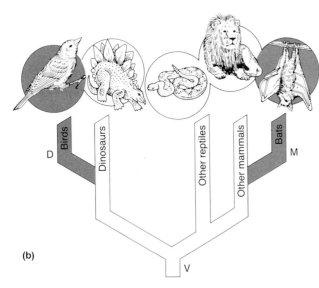

(b)

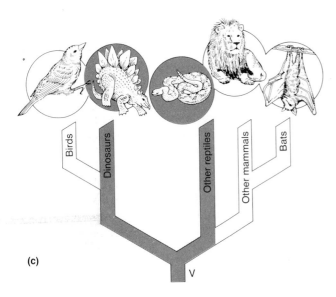

(c)

phylum is ancestral to any other species or group of animals. They simply represent interpretations of animal history constructed from evidence accumulated from living animals and fossils.

There are three types of phylogenetic groups of animals (Figure 18.5). A **monophyletic** group is one that includes a given ancestral form and all of its descendant species. Consider, for example, three groups of terrestrial vertebrates: the reptiles, birds, and mammals. Grouped together, as in Figure 18.5a, reptiles, birds, and mammals constitute a monophyletic group. They all evolved from a single, distant ancestral species that did not give rise to any other animals. Also, members of these groups are more similar to one another than they are to any other animals. A **polyphyletic** group, in contrast, includes descendants of more than one ancestral form. "Flying vertebrates" would be an example of such a group; bats and birds, though they share certain traits, including backbones and wings, did not arise from a common winged ancestor, and the common ancestor that flying vertebrates did share gave rise to many other types of vertebrates. Finally, a **paraphyletic** group includes the ancestral form and *some* of its descendants; an example would be the reptiles, from whose ancestral form mammals and birds also descended. We will have more to say about these distinctions among animal groups when we discuss animal classification. It is important to keep them in mind when interpreting the meaning of lines and branch points on phylogenetic trees.

A Phylogenetic View of the Animal Kingdom

A number of phylogenetic trees of the animal kingdom have been proposed. The one shown in Figure 18.6 postulates phylogenetic trends among the major animal phyla based largely on comparative anatomical and embryological data. The branches of this tree and the positions of phyla along them are meant to indicate the relative ages of the phyla and of ancestral groups, but not the ages or the

Figure 18.5 (Left) Three kinds of phylogenetic groups. (a) Monophyletic groups include an ancestor and all of its descendants. Here we see that one such group descended from ancestor *V*; a smaller monophyletic group descended from ancestor *M*. (b) A polyphyletic group (in this case, just birds and bats) has more than one ancestor; birds have ancestor *D*; bats have ancestor *M*. If the lineage of the polyphyletic group is traced to a distant common ancestor *V*, the ancestor also gave rise to other groups. (c) A paraphyletic group (in this case, just dinosaurs and other reptiles) has one ancestor *V*, but other descendants of that ancestor are not included in the group.

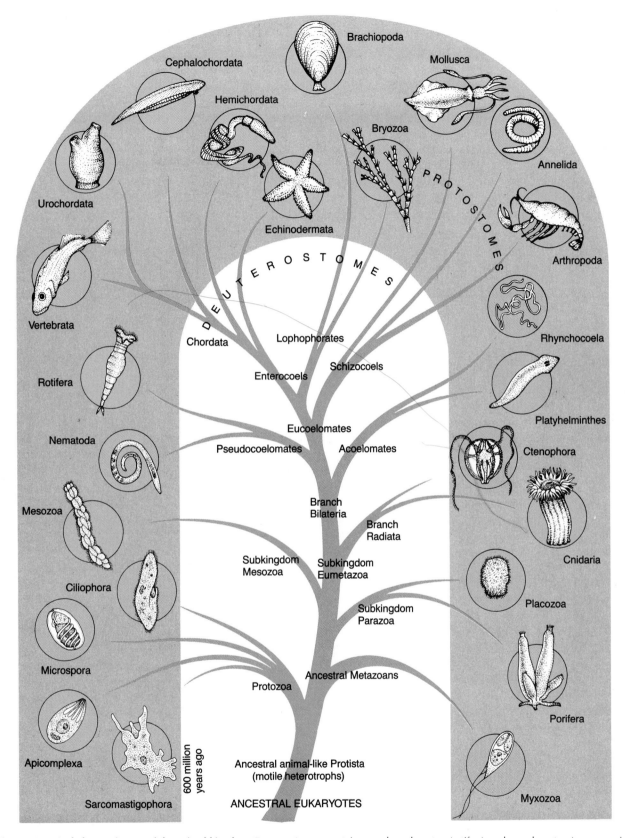

Figure 18.6 A phylogenetic tree of the animal kingdom. Four protist phyla are included, but many animal phyla with only a few species are not shown. Note that lineages of all modern phyla were probably established prior to 600 million years ago (before the Paleozoic era). Note that the pseudocoelomates and acoelomates are placed closer to the protostomes than to the deuterostomes. The reason for this is that certain pseudocoelomates (rotifers) and acoelomates (many marine flatworms) have protostome-like features, such as a spiral cleavage and a blastopore that forms the mouth. Some zoologists actually consider pseudocoelomates and acoelomates to be protostomes, but we have defined protostomes more narrowly, excluding animals without a mesodermally lined body cavity.

numbers of living species the phyla contain. Phyla whose members have levels of body organization or other significant characteristics believed to be ancestral are placed on lower branches of the tree. Phyla whose members show many derived features are placed on higher branches. For over 2.5 billion years (more than half of the earth's existence), prokaryotes—cells without nuclei—were the only life forms. A widely accepted hypothesis holds that ancestral eukaryotic cells, shown at the base of the trunk, evolved from symbiotic prokaryotes. (See the Chapter 17 Essay on the origin of life, p. 418.)

The first animal-like cells, which arose perhaps a billion years ago, were probably motile heterotrophs that may have been similar in body form to certain flagellate and amoeboid protozoa living today. All single-celled eukaryotic organisms, including algae and protozoa, are members of the Kingdom Protista. Several phyla of protozoa (Sarcomastigophora, Microspora, Apicomplexa, and Ciliophora) are on branches near the bottom of the phylogenetic tree. As a group, these animal-like organisms appear to be polyphyletic. Living species in these phyla may have evolved as recently as those in any other phyla, but their level of organization places them near the ancestral protists.

It is generally assumed that all multicellular life arose from unicellular forms. Since all members of the animal kingdom are multicellular, the origin of multicellularity was a major event in the evolution of animals, along with the associated specialization of cells to form tissues, division of labor among cells and tissues, and separation of reproductive from nonreproductive cells. (See the Trends and Strategies section beginning on p. 467.) Notice in Figure 18.6 that several animal phyla are thought to have branched off the main trunk from the ancestral metazoans. These groups—the Myxozoa, Mesozoa, Placozoa, and Porifera (the latter two constituting the **Subkingdom Parazoa**)—illustrate multicellularity in a very simple form and may represent several ancient ventures in the evolution of multicellular life. In reference to their simplicity as multicellular organisms, animals in these phyla are often called lower metazoans. Sponges (Phylum Porifera), for example, lack well-developed tissues and consist of aggregates of relatively few cell types. Unlike most animals with definite tissues and organs, sponge cells can reaggregate after being separated. As a group, the lower metazoans show little resemblance in embryology or body form to other animals. Most likely, these phyla constitute a polyphyletic group, and they are placed on separate side branches of the phylogenetic tree.

Back on the main trunk of the tree, metazoans with a more complex body form than that of lower metazoans are members of the **Subkingdom Eumetazoa**. There are several speculations about what the earliest eumetazoans were like. Some zoologists believe that the planula larva of cnidarians (see Figure 18.2b) represents the level of organization of an early eumetazoan and that a planula-like organism was ancestral to eumetazoans. The planula has two distinct cell layers and metamorphoses into an adult, yet it bears some resemblance to a ciliated colonial protist. But were planula-like organisms pivotal in the main line of animal evolution? Hypotheses about the origin of multicellular animals are discussed in more detail in the Trends and Strategies commentary on p. 468.

However the first multicellular organisms arose, there seem to have been two distinct evolutionary lineages of eumetazoans. Derived from one lineage, the **Branch Radiata**, are two phyla whose members are fundamentally different from other animals. They are radial or have a radial type of symmetry; most have unique eversible, threadlike cell organelles called nematocysts (see Figure 21.11); and unlike any bilateral animal, radiates lack true mesoderm. A third (middle) layer develops in most radiates, but it is not homologous with the mesoderm of bilateral animals. The middle layer in radiates is largely derived from ectoderm, rather than from endoderm. Because of their uniqueness, the radiate phyla are considered another side branch of the phylogenetic tree.

The other distinct lineage of eumetazoans, the **Branch Bilateria**, gave rise to the vast majority of animals. The bilateral lineage (considered to be two lineages by some zoologists) may have been initiated by a flatworm-like ancestor: a flattened, ciliated animal of three cell layers with head-to-tail orientation. (See the Trends and Strategies section beginning on p. 485.) Mesoderm in these animals and in all their evolutionary descendants were probably derived mainly from endoderm.

At least three distinct lineages seem to have been derived from ancestral bilateral animals. Representing the three lineages are three large groups of phyla: the **acoelomates** (having no body cavity), the **pseudocoelomates** (having an unlined body cavity, or pseudocoel), and the **eucoelomates** (having a mesodermally lined coelom). Some biologists believe that all these animals had a common ancestor, while other biologists hold that they arose from different bilateral organisms. (The evolution and significance of body cavities are discussed in the Trends and Strategies section beginning on p. 531.) Note the two branches emerging from the eucoelomates, one leading via the schizocoelous mode of coelom formation to the protostomes, and the other branch leading via enterocoelous coelom formation to the deuterostomes. Among the protostomes, molluscs are shown as having diverged relatively early from the schizocoelous branch. Molluscs show little or no evidence of body segmentation, while annelids and arthropods are markedly segmented. Even though some molluscs and annelids have trochophore larvae, the tree conforms with the hypothesis that the evolutionary lines of these phyla separated a very long time ago. On the other hand, the larvae and other features of echinoderms and hemichordates indicate that these phyla arose from fairly closely related ancestors. Deuterostome (especially chordate) phylogeny is discussed in some detail in Chapter 28.

Critique of the Model. Any phylogenetic tree depicting lineages for groups as broad and diverse as phyla is largely hypothetical. The fossil record is incomplete, and we can only postulate common origins and relationships among phyla on the basis of apparent common patterns in development and comparative morphology. Some groups pose greater problems than others. Relationships among the pseudocoelomate phyla, for example, are especially vague. Members of some nine phyla in this group have little in common except that most have a pseudocoel, and zoologists generally agree that these phyla constitute a polyphyletic group. (The nine pseudocoelomate phyla are discussed in Chapter 23.) Pseudocoelomates also have little in common with other animals.

Another criticism that many zoologists have of this model is that it relies so much on embryological data. A common objection is that in distinguishing protostomes and deuterostomes so conspicuously, this tree may overemphasize the significance of spiral and radial cleavage. These cleavage patterns seem fundamentally different, but many animals—both protostomes and deuterostomes—exhibit other types of cleavage patterns. Among the chordates, for example, tunicates and placental mammals have bilateral and rotational cleavage, respectively (Table 16.1). In defining deuterostomes, it is assumed that radial cleavage was ancestral and that the tunicate and mammalian cleavage patterns were derived from the ancestral radial plan. Likewise, among the molluscs, cephalopods (squids and octopuses) seem to have

derived patterns of development; their large, yolky eggs undergo bilateral cleavage, differing sharply from the spiral (apparently ancestral) plan of many other molluscs. It is likely that radial and spiral cleavage patterns are ancestral features of deuterostomes and protostomes, respectively, but there is no proof. Because of these and other assumptions that must be made to accommodate the protostome/deuterostome definitions, some zoologists would dispense with the use of protostome/deuterostome superphyla altogether. We have chosen to retain them mainly because they provide a focal point for discussion of animal phylogeny.

Without direct fossil evidence, it is impossible to paint an unassailable picture of the entire animal kingdom's phylogeny. In fact, doing so is not the goal of most zoologists interested in phylogeny. Phylogenetic trees of groups within each phylum can be more precise than a tree depicting all phyla. Nevertheless, despite its inability to accommodate all exceptions to general trends and despite the shakiness of some of its assumptions, the tree in Figure 18.6 serves a useful purpose. The importance of this model is to stimulate thinking and discussion about how the major groups of animals may have evolved and about which sources of data provide the best background for interpretations. The tree also provides an organized outline from which students can begin their study of animal diversity.

CLASSIFICATION

There are many ways to classify animals, but zoologists use a system standardized by the eighteenth-century Swedish naturalist Carl von Linné, more commonly known by his latinized name Carolus Linnaeus. Linnaeus and other prominent biologists of his time were concerned mainly with describing and naming species and with classifying them into groups that reflected their morphological similarities. Linnaeus, like nearly all scientists of his time, believed that species do not change; nevertheless, Linnaeus created a classification scheme that remains useful even today.

One of the most important features of the Linnaean system is that it is *hierarchical*. It consists of a series of groups within groups, or *nested sets*, from the all-inclusive kingdom down to the smallest subgroup, the species. Each group in the hierarchy is called a **taxonomic level**, or **category**, and there are seven major categories: kingdom, phylum, class, order, family, genus, and species. Taxonomic categories are not to be confused with **taxa** (sing. taxon), which are groups of real animals: Genus is a taxonomic category, whereas the Genus *Drosophila* is a taxon. Each species is given a **binominal**, or two-part name, in which the first word is the name of its genus (sometimes abbreviated) and the second is a descriptive specific term. Thus, the fruit fly *Drosophila melanogaster* belongs to the Genus *Drosophila* and has the specific epithet *melanogaster*. The species name, however, is *Drosophila melanogaster*. For classification purposes, the word *melanogaster* by itself conveys no meaning, and more than one species can have the same specific epithet, provided they are not in the same genus. For example, the species name of the Canada goose is *Branta canadensis*, and that of the beaver is *Castor canadensis*.

A second useful feature of Linnaean classification is that it can accommodate our expanding knowledge of animals without major rearrangements of the system. Today, zoologists may use as many as 29 taxonomic categories (many more than Linnaeus used) to classify certain species, especially those that are members of very large or complex phyla or classes. Each of the major categories may be subdivided into additional groups; even the species may be subdivided to classify recognizable subgroups within it. Table 18.3 shows the classification of the fish *Cyprinus carpio* and includes all categories of the classification system, even though some are not used in classifying this species.

A look at a section of the classification of birds illustrates more about how the system works (Table 18.4). All birds are classified in one large taxon, the Class Aves. Within this class, two similar birds, the wild turkey and the ring-necked pheasant, are classified in the same order (Galliformes), but the mallard duck and the bald eagle are placed in separate orders. Within each of these orders, several families may each contain several genera, and each genus may contain several species.

Taxonomy and Systematics

The terms *taxonomy* and *systematics* are often used interchangeably, although their meanings are not identical. **Taxonomy** is the theoretical study of classification, including its underlying principles and practical aspects. **Systematics** is the study of relationships and involves the application of taxonomic evolutionary principles to understanding phylogeny. Systematics is by definition comparative in its approach. Most taxonomists draw heav-

Table 18.3 The Modern System of Classification in Zoology

Taxonomic Categories*	Taxa of a Representative Animal, the Common Carp, *Cyprinus carpio*
Kingdom	Animalia (all animals)
Subkingdom	Eumetazoa (all animals with tissues)
Branch	Bilateria (all bilateral animals)
Grade	Eucoelomata (bilateral animals with a body cavity lined with mesoderm)
Division	Enterocoela (animals with the body cavity derived from the embryonic digestive cavity)
Subdivision	
Superphylum	
Phylum	Chordata (bilateral enterocoels with a notochord)
Subphylum	Vertebrata (chordates with a backbone)
Superclass	
Class	Osteichthyes (bony fishes)
Subclass	Actinopterygii (ray-finned fishes)
Infraclass	Teleostei (modern bony fishes)
Superorder	Ostariophysi
Series	Otophysi (with tiny sound-transmitting bones behind the head
Order	Cypriniformes (six families of carps, minnows, suckers, etc.)
Suborder	Cyprinoidea (carps and minnows)
Section	
Superfamily	
Family	Cyprinidae
Subfamily	Cyprininae
Tribe	
Supergenus	
Genus	*Cyprinus*
Subgenus	
Superspecies	
Species	*Cyprinus carpio*
Subspecies	
Variety	
Race	

*Boldface indicates the major taxonomic categories.

ily on the concepts of systematics in their work, and since most modern biological classifications are based on systematic studies, it is easy to see why the distinction between taxonomy and systematics is often blurred. Systematists attempt to distinguish, name, and group—that is, to *classify*—

Table 18.4 Partial Classification of Birds

Class Aves (birds)
  Order Galliformes (game birds: grouse, pheasants, turkey, and many others)
    Family Meleagrididae
      Genus *Meleagris*
        Species *Meleagris gallopavo* (wild and domestic turkey)

    Family Tetraonidae (grouse, ptarmigan)
    Family Phasianidae (quails, pheasants, etc.)
      Genus *Phasianus*
        Species *Phasianus colchicus* (ring-necked pheasant)
      Genus *Gallus*
        Species *Gallus gallus* (jungle fowl and domestic chicken)
      Genus *Pavo*
        Species *Pavo cristatus* (peacock, or peafowl)
      Genus *Lophortyx*
        Species *Lophortyx californicus* (California quail)
        Species *Lophortyx gambelii* (Gambel's quail)

  Order Anseriformes (screamers, ducks, geese, swans)
    Family Anhimidae (South American screamers)
    Family Anatidae (ducks, geese, swans)
      Genus *Anas*
        Species *Anas platyrhynchos* (mallard)
        Species *Anas acuta* (pintail)
      Genus *Aythya*
        Species *Aythya marila* (greater scaup)
        Species *Aythya affinis* (lesser scaup)

  Order Falconiformes (birds of prey: eagles, vultures, and many others)

animals so that the Linnaean hierarchy reflects the history of the animal kingdom. Obviously, the accuracy of phylogenetic classification depends on how much is known about the interrelationships of animals, and systematists frequently modify the classification system to reflect new knowledge about animals and their evolution.

Linnaeus's hierarchy provides an adaptable mold for evolutionary systematic classification. The higher categories, the phyla and classes, are defined more broadly than the lower groups, reflecting the fact that very basic ancestral traits are shared by all the animals in the higher categories. Theoretically, basic traits indicate older, more fundamental evolutionary relationships among animals than do the traits characteristic of the lower groups. All the animals in a class, for example, share several general features. Birds have feathers, are homeothermic, and communicate vocally; together, these general features define the Class Aves. Feathers and homeothermy are believed to have been among the distinguishing features of the first modern birds, and using these features to distin-

guish the class is an example of how the classification system attempts to both identify shared traits and reflect evolution. Within the Class Aves, lower taxa include the Order Anseriformes (see Table 18.4). This order contains several genera of ducks, all members of the Family Anatidae, which includes swans, geese, and ducks (Figure 18.7). *Anas* ducks, including the mallard *(A. platyrhynchos)* and the pintail *(A. acuta)*, frequent shallow ponds and marshes and feed from the surface by upending their bodies to grope for food on the pond bottom. They share certain features of the duck family (e.g., broad, flattened beak and webbed feet) with other genera of ducks, but differ in certain ways that reflect recent events in duck speciation. The so-called diving ducks of the genus *Aythya* frequent deep lakes and sea coasts. Unlike species of *Anas*, they dive and swim after food. The differences between these two genera are morphological and behavioral, but all ducks have many more similarities than differences. There are even fewer differences among species within these genera. Among diving ducks, *Aythya marila* and *A. affinis* are reproductively isolated but otherwise very similar. *Aythya affinis* is slightly smaller, and males are lighter with purple (instead of green) head feathers (see Figure 18.7b). Such minor differences probably indicate that these species are closely related and diverged more recently from a species of ancestral diving duck.

Naming and Classifying Animals

An internationally accepted set of rules called **The International Code of Zoological Nomenclature** ensures that each animal species is described and named according to the same standards. The code specifies methods for establishing taxa from the subspecies to the family level. Taxa above the family level are not regulated by rigid rules, but nonetheless are established as objectively as possible by application of the scientific method. The major points of the international code are as follows:

1. Species must be named and described in scientific journals so that other scientists can recognize them.
2. At the time a species is named, the preserved remains of a particular specimen (or group of specimens) must be designated as the **type specimen**. Each type specimen is deposited in a zoological museum, where it serves as an actual biological entity, or name bearer of the species, and is available for study by qualified scientists.

(a)

(b)

Figure 18.7   Four species of the Family Anatidae. All birds shown here are males. (a) Two closely related species of marsh ducks, a mallard *Anas platyrhynchos* (left) and a pintail *Anas acuta* (right). (b) Two closely related species of diving ducks, a greater scaup *Aythya marila* (left) and a lesser scaup *Aythya affinis* (right).

3. Binominals must be Latinized, with the name of the genus a capitalized noun. The specific epithet, a noun or an adjective apposing or modifying the generic noun, must begin with a lowercase letter. Both the genus and the specific epithet are italicized or underlined when in print. The domestic cat, for example, is *Felis catus*, from the Latin words *felis* ("cat") and *catus* ("sharp-eyed, keen").

4. Each binominal can be used to describe only one animal species. A genus name can be used to describe only one genus in zoology; a specific epithet can be used only once within each genus.

5. The binominal first given to a species holds priority over all others later used. The earliest names recognized are those published by Linnaeus in the tenth (1758) edition of his treatise on taxonomy, *Systema Naturae.*

6. The endings of certain taxa are standardized. Family names in zoology always end in *-idae;* subfamily names always end in *-inae.*

This process may seem cumbersome, but think for a moment about what would happen if zoologists used nonstandardized names for animals. Many species have several common names: The freshwater crayfish, *Orconectes virilis*, is also called a crawdad, a crawfish, and a freshwater lobster. Some common names are used for different species: A group of fishes and another group of mammals are both called dolphins. Binominals are in Latin, by convention a universal and unchanging language, and every scientist—whether writing in Japanese, English, or Russian—uses the same names.

Defining Species

In Linnaeus's era, biologists believed that since species were unchanging, they could be defined and described from one or a few specimens. To Linnaeus, a type specimen could represent a species perfectly. We now know that this approach is too simplistic and seriously misleading. Type specimens now serve as name bearers and sometimes as an aid to species recognition and comparison, but not as absolute species models. (Biological species definition is discussed in Chapter 17, p. 405.)

Linnaeus did not realize that each species has its own range of variation—its **intraspecific variation**—in phenotypic characteristics determined by its unique gene pool. This variation reflects the adaptation of species populations to changing environments, but it can also pose special problems for the systematist. The difficult and challenging task of modern taxonomists and systematists is to find diagnostic features that reflect the unique genetic differences between species and to estimate each species' range of phenotypic variability. Some species are easy to distinguish and define. Others—especially those that reproduce only asexually—cannot be defined without exhaustive study.

Since groups of individuals thought to be one species cannot always be tested for reproductive isolation, the systematist often must decide on the basis of other characteristics whether given groups constitute one or several species. Classically, the usual first step has been to look for distinctive morphological features, since the shape and size of animals are often easy to measure. Increasingly, however, systematists are turning to immunological and biochemical techniques that provide estimates of genetic variability. DNA base sequencing provides the most direct estimate of a species' genetic makeup, but so far, data have been generated for only a few species. Actually, body shape and size in adult and developmental stages, biochemistry, internal anatomy and physiology, behavior, and ecology may all provide information useful in species recognition. Systematists must often examine large population samples and make statistical comparisons of features that can be quantified to find distinguishing characteristics and estimate their range of variability.

Species definition is especially difficult when dealing with fossils. It is not possible to be certain what constituted a species of an extinct animal, and fossil species definitions typically are based on the appearance of related living species. Recently, however, modern analytical techniques have been applied to the study of fossils with encouraging results. In 1977, Rosalyn Yalow received the Nobel Prize for developing an extremely sensitive diagnostic technique called **radioimmunoassay** (**RIA**). Originally developed to detect minute quantities of insulin in blood, radioimmunoassay has several additional medical applications and has also been

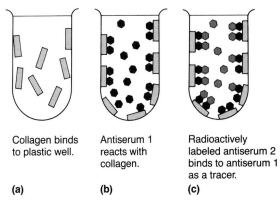

Collagen binds to plastic well.

Antiserum 1 reacts with collagen.

Radioactively labeled antiserum 2 binds to antiserum 1 as a tracer.

(a)   **(b)**   **(c)**

Figure 18.8 Radioimmunoassay. Radioimmunoassay (RIA) is a powerful diagnostic tool applicable to the study of fossils. The fibrous protein collagen, a major constituent of animal connective tissue, can often be recovered intact from fossils. Each species has its own specific type of collagen, and the RIA technique can identify tiny quantities of these species-specific molecules. First, antisera to collagen proteins thought to be related to those of the fossil are produced. RIA is then done in small plastic wells. (a) Collagen extracted from a fossil is placed in a well and some collagen molecules bind to the well. (b) A prepared antiserum (antiserum 1) is then added and allowed to react with the bound collagen. (c) Unbound antiserum is then removed, and a radioactive second antiserum (antiserum 2) is added to the well. The second antiserum reacts with the first antiserum molecules that reacted with the collagen. After unbound antiserum is removed, the amount of radioactivity detectable in the well is directly related to how much reaction took place between the collagen and the first antiserum. This gives an indirect measure of how similar the fossil collagen is to the collagen originally used to make the first antiserum. Human-type collagen has been identified by RIA from hominoid fossils nearly 2 million years old.

adapted for detecting minute amounts of species-specific protein in fossils (Figure 18.8).

The Process of Classification

The systematist's task is to construct classification hierarchies that reflect evolutionary events. Before an animal species can be placed in appropriate taxa, the available information about it must be scientifically assessed in light of evolutionary theory. The systematist attempts to ascertain which characteristics—morphological, biochemical, behavioral, and ecological—are shared by all members of a species and which features are unique to the species. Although phylogeny cannot be repeated in controlled experiments, systematists try to determine which characteristics were significant in a species' evolution. Taxa can then be established that reflect phylogenetic hypotheses.

Establishing higher taxa is especially difficult. Nature defines only the species (by reproductive isolation). All other taxa are human inventions, representing scientific hypotheses. It is not always obvious which characteristics of species should be used to distinguish classes, orders, or families, and different scientists may weigh evidence somewhat differently. Most higher taxa have been changed drastically since Linnaeus's time, reflecting not only advancements in evolutionary theory but also our ever-increasing general knowledge about animals. Sponges and sea anemones, for example, were formerly classified as plants because these animals appeared immobile and superficially plantlike; and echinoderms and many cnidarians were once classified together in the same phylum because they are radially symmetrical as adults. Taxa, like any scientific hypotheses, must often be reevaluated in light of new knowledge about evolution and characteristics of species; animal classification is not static, and taxonomic reassignments are made as necessary.

Traditional Evolutionary Systematics, Cladistics, and Phenetics.   There are three general approaches to classification today. Our discussion so far, including that of the phylogenetic tree (Figure 18.6), largely reflects the approach of **traditional evolutionary systematics**. This approach has dominated taxonomy and classification since evolutionary theory was first developed. Traditional systematists attempt to classify animals to reflect phylogeny and also the degrees of morphological difference between groups. Traditionalists may recognize paraphyletic groups as valid taxa and may use para- and polyphyletic groups in constructing phylogenetic trees. For example, the currently recognized Class Reptilia includes modern reptiles, dinosaurs, and ancestral reptiles; this is a paraphyletic group because it does not include birds and mammals, which were also derived from reptilian stocks. Likewise, among the protists, the Phylum Sarcomastigophora is undoubtedly a polyphyletic group, but it is currently recognized as a valid taxon more for convenience than for phylogenetic accuracy. Traditionalists also maintain that speciation may occur by anagenesis, or speciation ▷ within a single lineage, without branching. (Anagenesis is discussed in Chapter 17, p. 412; Figure 17.14a.) Each lineage in a phylogenetic tree drawn by a traditional systematist is meant to represent a sequence of speciation events; thus, each lineage includes several species. A branch point on a phylogenetic tree is meant to represent a major divergence from a lineage.

A relatively new and revolutionary school of modern classification is **cladistics** (from the Greek word meaning "branch"). Cladists (who prefer to be called phylogeneticists) are interested in devis-ing a classification that reflects only phylogeny. Their overall goal is to construct a system composed entirely of monophyletic taxa. Thus, each taxon in the hierarchy would consist only of an ancestral form and all its descendants. Cladists argue that the only scientifically testable aspect of phylogeny is the branching of one evolutionary line from another. They maintain that the tendency of traditional systematists to erect taxa and construct phylogenetic trees on the basis of *apparent* degrees of differences between organisms requires subjective judgment and is not scientifically testable. Thus, in the interest of making animal systematics more logical and objective, and more strictly reflective of phylogeny, cladists concentrate on trying to decipher major branch points in the evolutionary history of animal groups. Their phylogenetic trees represent only cladogenesis, or branching evolution. They reject as arbitrary and subjective the traditionalist tendency to define speciation events *within* unbranching lineages, and lineages on their phylogenetic trees do not represent anagenesis. In establishing a taxon, cladists stress the occurrence of unique, fundamental similarities that define the group, rather than differences between its individual members. If a group of animals shares unique fundamental features, cladists are not dissuaded from unifying the group, whereas traditional systematists may split groups on the basis of certain types of differences.

Figure 18.9 illustrates a simple cladistic model. The two diagrams, called **cladograms**, illustrate alternative hypotheses of phylogenetic branching patterns among four taxa. The hypotheses represented by the cladograms can be tested by comparing phenotypic features of the taxa. Taxon A, called the **outgroup**, is included because it appears to be only distantly related to the other three; as a result, it can be used to test whether features common to taxa B, C, and D are limited to these three taxa or are general features shared by a wide range of organisms. The key to testing cladistic hypotheses is determining evolutionary homologies, which are a direct reflection of the evolutionary branching pattern of taxa. Homologous features shared by closely related groups (taxa B, C, and D in our example), but not by the outgroup, are called **synapomorphies**. Synapomorphies are unique, unifying similarities (cladists call them "shared, derived characters") that members of a taxon have derived (inherited) from a common ancestor.

The following example illustrates the difference between the traditional view and the cladistic view. Traditional systematists hold that birds and

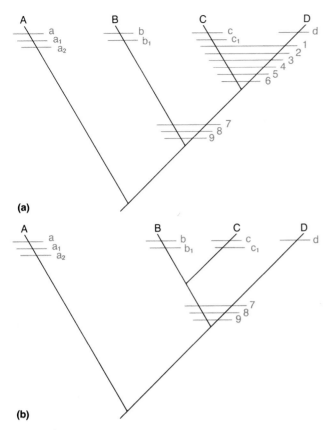

(a)

(b)

Figure 18.9 The cladistic approach. This approach allows for the formation of testable hypotheses about sequences of phylogenetic branching events. Here, two cladograms represent alternative hypotheses for branching patterns among four taxa (A–D). Unique features of each group are indicated by lines with lower case letters. Taxon A is the outgroup. Shared derived characters (synapomorphies) are indicated by horizontal numbered lines. Synapomorphies are the basis for testing alternative phylogenetic hypotheses. (a) One hypothesis proposes a branching pattern that is supported by six synapomorphies (those between taxa C and D); three additional synapomorphies support branch B–D. (b) This alternative hypothesis might be rejected on the basis of the data; it is not supported by any synapomorphies between taxa B and C; the B–D branch is the same, and is supported by three synapomorphies, as in the first hypothesis. Any features common to all taxa (A, B, C, and D) would support the A–D branch point; such features would probably be primitive (inherited from a distant common ancestor).

reptiles share a distant common ancestor but have evolved into unique groups of vertebrates. Modern birds, unlike any other known organisms, have feathers. In common with certain other vertebrates, they also fly, are vocal, are homeothermic, and lack teeth. By contrast, modern reptiles lack feathers, do not fly, are heterothermic, are generally nonvocal, and (with a few exceptions) have teeth. By stressing these apparent differences between birds and reptiles, traditional systematists place birds in the Class Aves, separate from the Class Reptilia.

Cladists, on the other hand, attempt to define the fundamental similarities between birds and reptiles and establish taxa without making intuitive judgments about the biological uniqueness of birds and reptiles. Although it is easy to form a mental picture of a bird and a reptile, and they are obviously different, try for a moment to be completely objective and not to view them as different unless they can be shown scientifically to be so. Now, from your neutral position (that of the cladist), consider the idea that birds may not have any traits that clearly distinguish them from *all* reptiles. Except for their feathers, fossil birds would be classified as reptiles; all of their other body features were markedly like those of certain small dinosaurs. Paleontologists also consider it likely that feathers were only rarely fossilized and that certain small birdlike dinosaurs (those with a high surface-area-to-volume ratio) had feathers or featherlike scales and that some of these reptiles were homeothermic. There is also evidence that some dinosaurs had birdlike behavior in nesting and caring for their young. Thus, none of the supposed unique features of birds seem to clearly distinguish them from reptiles. To have classification reflect phylogeny, cladists would dispose of the Class Reptilia, as traditionally defined, because it excludes birds and is thus paraphyletic. In a cladistic sense, birds and all reptiles belong together as a monophyletic taxon. Within this group, birds and their closest relatives, crocodiles and birdlike dinosaurs, would constitute a monophyletic subgroup. Figure 18.10 compares a cladogram of reptiles, birds, and mammals with a traditional phylogenetic tree. The cladogram is a phylogenetic hypothesis reflecting the logic just discussed; it can be tested scientifically by synapomorphic analysis, as outlined in Figure 18.9. The traditional tree, though used in this book and currently widely accepted, is more arbitrary than the cladogram.

Whereas the ultimate goal of traditional systematists and cladists is to describe the history of animals, a third approach to classification, called **phenetics**, is concerned with producing a practical, totally objective system of classification that reflects overall resemblances among taxa; phylogeny is not considered in phenetic classification. In attempting to distinguish two species, pheneticists apply a method called **numerical taxonomy**, which begins with the gathering of as much quantitative data as possible about the phenotypic traits of a species of interest. All data—all species characteristics that can be quantified—are treated as having equal weight. Species that share the greatest number of characteristics are placed close together in

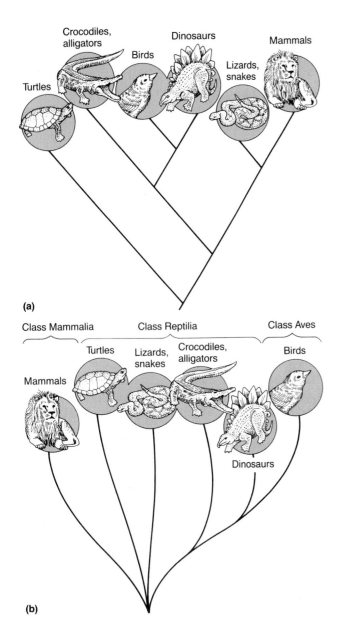

Figure 18.10 Cladistic versus traditional views of vertebrate phylogeny. (a) A cladogram depicting phylogenetic branchings. Birds are placed on a branch with dinosaurs and crocodiles; this branch would form a taxon, perhaps a class of vertebrates. Other reptile groups and mammals would form other classes. (b) A traditional phylogenetic tree of vertebrates, recognizing reptiles, birds, and mammals as three separate classes of vertebrates.

the system and separated from others with whom they share fewer characteristics.

Both cladistics and phenetics have brought systematics into the realm of true (testable) scientific inquiry. Most zoologists believe that phenetics is self-limiting simply because it fails to consider phylogeny. Traditional systematics is also limited as a science by its subjectivism, and among prac-

ticing systematists, the cladistic method is currently the most popular. There are rational arguments for the viewpoints of both traditional and cladistic systematics. To many, traditional classification reflects zoological practicality. In this sense, many zoologists would argue that due to present-day differences between birds and reptiles, it makes sense to classify and study these animals as separate groups. One could also argue that the subjectivity of traditional systematics is based on the educated opinions of generations of scientists; thus, it must have some truth and value. There is also general reluctance to accept sweeping changes in an established classification scheme that is familiar, generally workable, and used throughout the world for information storage and retrieval. The cladists argue that the traditional view is taken too much at the expense of logical consistency and that the rigorous scientific approach of cladistics has much more to contribute to our understanding of animal phylogeny. Simply in questioning many long-established axioms in classification, cladistics has significantly revitalized the field of systematics. It will be interesting to watch its influence on systematics and on established classification schemes over the next several decades. In *Systema Naturae*, Linnaeus classified all invertebrates except insects in the Class Vermes ("worms," or animals much longer than wide). Vermes is no longer used (except humorously) in any classification scheme. In a similar way, will, as some systematists predict, such taxa as Reptilia, Aves, and Mammalia eventually be discarded and replaced by more logical groups?

☐ SUMMARY

1. Animal diversity is the product of evolution. Phylogeny is evolutionary history. Zoologists attempt to devise classification schemes that reflect phylogeny.

2. Ancestral ("primitive") characteristics are those inherited from a distant ancestor; derived characteristics are those possessed by all species evolved from a common ancestor.

3. Phylogenetic trees are sets of hypotheses about animal evolution. Data supporting phylogenetic hypotheses may be obtained from virtually any field of zoology. Morphology and embryology have traditionally been the major data sources. More recently, molecular and cellular biology are making significant contributions.

4. A monophyletic group includes a given ancestral form and all its descendant species. A polyphyletic group includes descendants of more than one ancestral species. A paraphyletic group includes the ancestral form and some but not all of its descendants.

5. Metazoans (animals) most likely were derived from unicellular (protist) ancestors. Sponges and other "lower" metazoans are offshoots of the main line of animal evolution. Radiate animals (cnidarians and ctenophorans) are distinct from bilateral animals. Two commonly recognized subgroups of bilateral animals are deuterostomes (hemichordates, echinoderms, and chordates) and protostomes (molluscs, annelids, and arthropods).

6. Zoologists use a hierarchical system of classification consisting of a series of groups (taxonomic categories) within groups (nested sets); there are seven major categories from the all-inclusive kingdom down to the smallest subgroup, the species. Names of actual animals in taxonomic categories are called taxa. Each species has a binominal (a two-part name). Taxonomy is the science of classification. Systematics is the scientific study of evolutionary relationships.

7. Taxonomists attempt to define the range of phenotypic variation reflecting each species' unique gene pool and to find diagnostic features for each species.

8. Modern classification has three distinct approaches. Traditional evolutionary systematics and cladistics attempt to make classification reflect phylogeny. Cladists give strong emphasis to biological similarities to classify animals, and they construct phylogenetic trees using only monophyletic groups. Phenetics assigns equal weight to all characteristics and is purely taxonomic (does not consider phylogeny).

☐ FURTHER READING

Cracraft, J. "Cladistic Analysis and Vicariance Biogeography." *American Scientist* 71(1983): 273–281. *A strong argument for cladistic analysis.*

Duellman, W. E. "Systematic Zoology: Slicing the Gordian Knot with Ockham's Razor." *American Zoologist* 25(1985): 751–762. *A lucid review of the modern-day revolution in systematics research.*

Eldredge, N., and J. Cracraft. *Phylogenetic Patterns and the Evolutionary Process.* New York: Columbia University Press, 1980. *A full description of the logic and application of cladistic analysis.*

Gould, S. J. "Evolution and the Triumph of Homology, or Why History Matters." *American Scientist* 74(1986): 60–69. *A thoughtful essay on evolutionary methods and phylogeny.*

Lowenstein, J. M. "Molecular Approaches to the Identification of Species." *American Scientist* 73(1985): 541–547. *The application of immunological techniques to problems of fossils in systematics.*

Margulis, L., and K. V. Schwartz. *Five Kingdoms, An Illustrated Guide to the Phyla of Life on Earth.* San Francisco: W. H. Freeman, 1982. *A well-illustrated presentation of life's diversity.*

Mayr, E. "Biological Classification: Toward a Synthesis of Opposing Methodologies." *Science* 214(1981): 510–516. *Discusses problems and merits of three approaches in modern taxonomy.*

Sibley, C. G., and J. E. Ahlquist. "Reconstructing Bird Phylogeny by Comparing DNA's." *Scientific American* 254(1987): 82–92. *Application of the concept of the molecular (genetic) clock and the DNA-hybridization technique to avian systematics.*

Valentine, J. W. "The Evolution of Multicellular Plants and Animals." *Scientific American* 239(1978): 141–158. *Discusses the diversity of plants and animals through time and includes phylogenetic trees of invertebrates and vertebrates.*

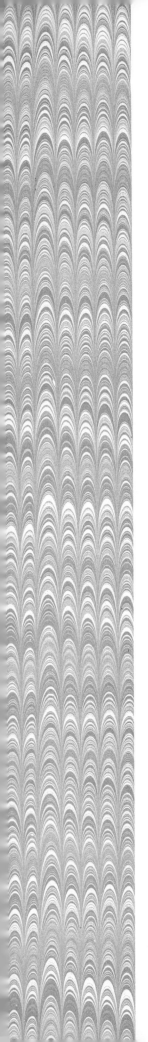

Animal Diversity

The previous units in this book examine the adaptations that have evolved in animals as solutions to major problems of existence. In this unit, our main goals are to introduce you to whole organisms, have you gain an appreciation for the great variety of animal life, and illustrate the major trends in animal evolution.

There are over 40 recognized phyla of animals and animal-like protists. All members of each phylum share a basic structural and functional similarity, and this fundamental architectural or adaptive "plan" of each phylum has been more or less successful as an evolutionary unit. The numbers of species and diversity of environments in which species are established are two measures of the adaptive success of a phylum. Phyla range in size from one consisting of only a single species to immense groups with many thousands of species. Twelve major phyla, containing about 99% of all living species, were briefly introduced in Chapter 1. One of them, the Phylum Arthropoda, contains nearly three times the number of species in all other phyla combined. Our discussion is a compromise: While we want to emphasize the major groups of animals, we also want to truly survey diversity; thus, this unit includes some discussion of all phyla.

We look first at the Protozoa, a diverse assemblage of animal-like protists whose single-celled bodies place them first in a phylogenetic sequence. It is appropriate to take protozoa as a starting point in studying animal diversity because these organisms resemble animals in their modes of nutrition and in being actively motile. Actually, there is no clear line of separation between unicellular and multicellular life, and animal-like protists undoubtedly gave rise to the first animals.

Protozoa

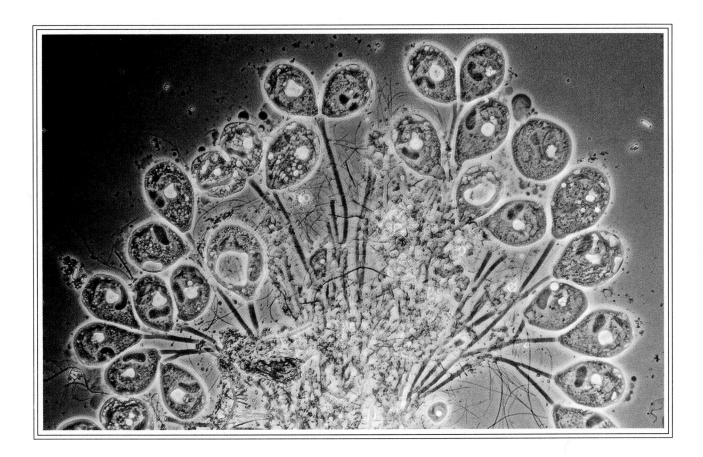

Kingdom Protista, Subkingdom Protozoa. *Vorticella*, which looks like an inverted bell on a long, contractile stalk, is a ciliate protozoan that often forms colonies.

In the 1670s, Antony van Leeuwenhoek, a Dutch merchant and lens maker, first observed and wrote about the single-celled, animal-like organisms we now call **protozoa**. Since then, about 30,000 fossil species and 35,000 living forms have been described. Most protozoa are free-living, and as a group, they occur throughout the world in all types of aquatic environments, including wet soil. About 10,000 living species are parasitic, and some of the world's most harmful diseases are caused by protozoa. As members of the Kingdom Protista, protozoa differ from multicellular organisms in that all their life-sustaining processes occur within a single cell. Protozoan diversity results mainly from specializations in cell membranes and organelles. Some protozoa are colonial, forming aggregates of virtually identical cells, but no protozoan has tissues.

Most protozoa are microscopic, but they range in size from about 0.005 mm to 5 mm. Only after

445

the electron microscope came into use were biologists able to observe much of the structural diversity and complexity of protozoa. Since the early 1960s, microscopic and experimental research on this group of organisms has blossomed. This chapter examines the four largest (of six) protozoan phyla.

PHYLUM SARCOMASTIGO- PHORA

The Phylum Sarcomastigophora consists of three types of protozoa. Those that move by means of cytoplasmic extensions called pseudopodia are called **amoebas**; those that have one or more flagella are called **flagellates**; members of a third group, the **amoeboflagellates**, have pseudopodia and flagella during different stages in their life cycles. The existence of amoeboflagellates suggests that amoebas and flagellates are closely related, and this is why all these organisms are grouped in a single phylum.

Amoebas

Species of *Amoeba* are common on the bottoms of freshwater ponds and lakes, where they move about and feed by pseudopodial action (Figure 19.1a). The active amoeboid stage, called the **trophozoite**, reproduces only by asexual means (binary fission); the life cycle also includes a dormant cyst stage that survives dry periods.

Amoeba and many other amoebas are covered by only a single cell membrane, the plasmalemma. By contrast, certain other freshwater amoebas, such as *Arcella* and *Chlamydophrys*, have an external case called a **test** surrounding the plasmalemma (Figure 19.1b). Tests are composed of hardened protein secretions, sometimes coated with fine sand grains or siliceous scales. The broad, spadelike or tubular pseudopodia of *Amoeba* and *Arcella* are called **lobopodia**. *Chlamydophrys*, by contrast, has tapered, threadlike **filopodia**.

Foraminiferans, a large group of marine amoebas, are very important ecologically. Most species secrete and live in multichambered calcium carbonate (limestone) tests and have thin, branching cell extensions called **reticulopodia** that form a net or reticulum (Figure 19.1c). The reticulopodium net allows foraminiferans to creep slowly over the

ocean floor or maneuver awkwardly as plankton, and it is coated with sticky mucus that traps microorganisms for food. Researchers estimate that a third of the world's ocean floor is covered with thick deposits of foraminiferan tests. Through geological time, fossilized tests have formed prominent limestone deposits throughout the world. The white cliffs of Dover, so much a part of British history, are composed of fossil foraminiferans.

Another group of amoebas are the **actinopods** (also called radiolarians and heliozoans), freshwater and marine protozoa that secrete complex skeletons (Figure 19.1d). Actinopods have tapered, spikelike pseudopodia called **axopodia** radiating from their spherical central cell body. The spherical shape of actinopods is supported by an *internal* skeleton; a rigid axis of internal microtubules supports the axopodia. Actinopods roll about slowly as plankton or on the bottom of lakes and oceans by retracting and elongating their axopodia.

Many amoebas are parasitic and pass from one host to the next as a dormant cyst. *Entamoeba histolytica* infects humans and may cause diarrhea and amoebic dysentery (bleeding intestinal ulcers). The life cycle of *E. histolytica* includes an active amoeboid stage that lives in the human intestine and a spherical cyst that passes out in feces. The disease is most common in the tropics, but may occur wherever sanitary conditions are poor; people become infected by swallowing the cysts. *Entamoeba histolytica* can now be grown on chemically defined media, and by using the large quantities of amoebas produced in culture, researchers are making progress in identifying the molecules that allow this important parasite to invade intestinal tissue.

Flagellates

In contrast to the simple plasmalemma of amoebas, flagellates are covered by a **pellicle**, a compound envelope of one or more cell membranes, often reinforced by underlying microtubules. This type of covering makes flagellates less flexible than amoebas, and whereas pseudopodia may bulge from any area on an amoeba's plasmalemma, flagellates usually hold a particular shape. *Euglena gracilis*, common in stagnant freshwater ponds, usually has an elongate body form with rounded anterior and tapered posterior ends (Figure 19.2a).

Many flagellates are plantlike autotrophs containing green, yellowish, or red organelles called plastids. Others are colorless, animal-like heterotrophs. A few, however, are normally autotrophic but also capable of heterotrophic nutrition. (This is

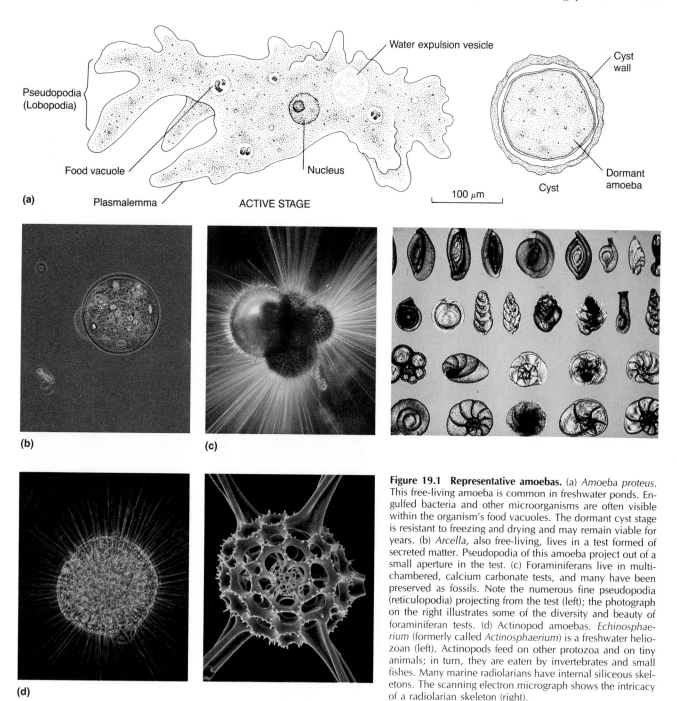

Figure 19.1 **Representative amoebas.** (a) *Amoeba proteus.* This free-living amoeba is common in freshwater ponds. Engulfed bacteria and other microorganisms are often visible within the organism's food vacuoles. The dormant cyst stage is resistant to freezing and drying and may remain viable for years. (b) *Arcella*, also free-living, lives in a test formed of secreted matter. Pseudopodia of this amoeba project out of a small aperture in the test. (c) Foraminiferans live in multi-chambered, calcium carbonate tests, and many have been preserved as fossils. Note the numerous fine pseudopodia (reticulopodia) projecting from the test (left); the photograph on the right illustrates some of the diversity and beauty of foraminiferan tests. (d) Actinopod amoebas. *Echinosphaerium* (formerly called *Actinosphaerium*) is a freshwater heliozoan (left). Actinopods feed on other protozoa and on tiny animals; in turn, they are eaten by invertebrates and small fishes. Many marine radiolarians have internal siliceous skeletons. The scanning electron micrograph shows the intricacy of a radiolarian skeleton (right).

why heterotrophic and autotrophic flagellates are usually classified together.) Cultures of *E. gracilis* can be "bleached" by growing them in the dark so that the cells lose their photosynthetic ability. If provided with dissolved organic nutrients, bleached individuals may survive and reproduce as heterotrophs for years. If such individuals are then placed in light, active chloroplasts will appear, and the organisms can once again become autotrophic.

Many flagellates are parasitic. Species of *Trypanosoma*, for example, infect the bloodstream of a wide variety of vertebrates (Figure 19.2b). The deadly human disease African sleeping sickness and the related Chagas' disease in Central and South America are caused by trypanosomes. Economic growth in large regions of Africa has been impaired by another type of trypanosome disease called nagana, which kills cattle and other domes-

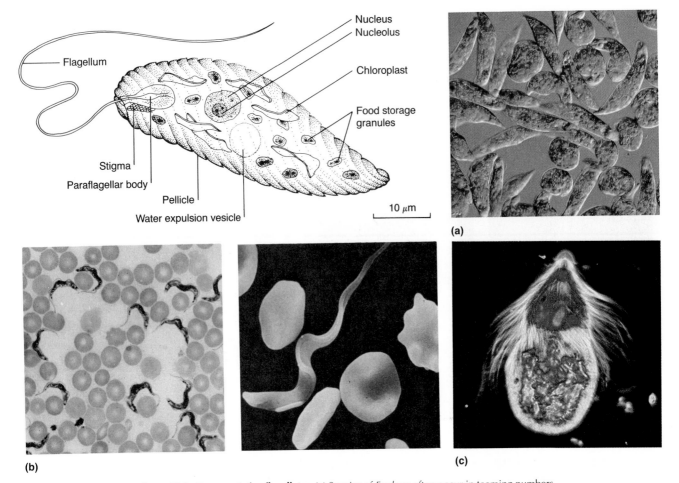

Figure 19.2 Representative flagellates. (a) Species of *Euglena* often occur in teeming numbers in stagnant freshwater ponds. Some species are capable of both photosynthesis and heterotrophic nutrition; this allows them to survive when light or organic nutrient levels are low. (b) Species of *Trypanosoma*, as seen in this scanning electron micrograph (right, 2300 ×), live in the bloodstream of many vertebrates. These parasites are about 20–30 μm long. As shown in the photograph on the left, they swim among red blood cells, absorbing nutrients (mainly glucose) from the plasma, and multiplying in the blood by binary fission. The life cycle includes development in an invertebrate (either a blood-sucking insect or a leech); trypanosomes are transmitted to a vertebrate when the invertebrate sucks blood. (c) *Trichonympha*, a cellulose-digesting flagellate (about 100–200 μm long) is found in termites and wood roaches. Note the numerous flagella that originate anteriorly on the cell, and also note the wood particles in the cytoplasm. Termites eat wood, but they cannot digest the cellulose in it. Fine wood particles in the termite gut are engulfed and digested in food vacuoles of *Trichonympha*. Simple sugars resulting from cellulose breakdown provide nourishment for these flagellates and their termite hosts. Neither the insect nor *Trichonympha* can survive without this mutually beneficial symbiotic association.

tic animals. African sleeping sickness and nagana are carried by tsetse flies, bloodsucking insects that may transmit infective stages of the parasites. Chagas' disease is transmitted to humans by bloodsucking insects called kissing bugs. Trypanosome diseases have been the subject of intense research throughout the world since these parasites and their insect vectors were identified in the late 1800s.

Unlike species of *Euglena* and *Trypanosoma*, which have only one flagellum, many flagellates have two or more flagella. Species of *Trichonympha*, which live in the gut of termites and wood

roaches and digest cellulose in wood eaten by their hosts, have many flagella (Figure 19.2c).

Many flagellates are colonial aggregates of a few to many thousands of cells. Colonies develop when cells divide but remain attached to one another. Species of *Volvox* are colonial autotrophs, consisting of bright yellow or green colonies large enough to be seen with the unaided eye (Figure 19.3a). *Volvox* colonies are common in shallow freshwater habitats. They rotate about slowly, propelled by the combined force of thousands of beating flagella (two on each of the cells). Specialized gametes (sex cells) develop in clusters distinct from the other

(a)

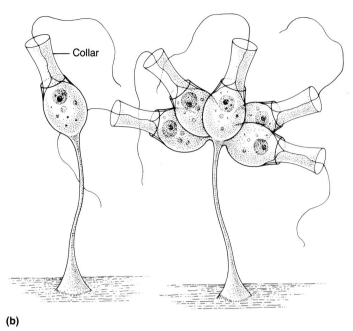

(b)

Figure 19.3   Colonial flagellates. (a) Species of *Volvox* are photosynthetic autotrophs that form brightly colored spherical colonies (up to about 0.5 mm in diameter) of up to 50,000 cells. Each cell has two flagella. Cells are held together in a single layer by secreted proteinaceous material; the interior of the colonies contains a secreted noncellular gelatinous mass. Here we see a number of colonies containing vegetative and sexual daughter spheres. (b) *Codosiga* and other choanoflagellates (about 12 μm long) live attached to the substrate in brackish, marine, and freshwater habitats. Their cells are structurally and functionally similar to the collar cells (choanocytes) that circulate water through sponges.

(somatic) cells in a colony; unlike multicellular organisms, however, there is no further differentiation or division of labor among cells.

Another group of flagellates, the **choanoflagellates**, are interesting from a phylogenetic viewpoint. Either solitary or colonial, their cells are usually stalked and have a transparent, meshlike collar surrounding a single flagellum (Figure 19.3b). The flagellum whips particles suspended in the water toward the collar. The collar may act as a sieve, trapping small particles that are ingested by the cell body. Structurally and in their mode of feeding, choanoflagellates closely resemble the collar cells (choanocytes) of sponges, and many zoologists believe that sponges arose from protozoa similar to choanoflagellates. (Choanocytes are discussed in Chapter 20, p. 478.)

Amoeboflagellates

Among the most common amoeboflagellates are members of the genus *Naegleria*, which inhabit moist soil and organically rich freshwater environments (Figure 19.4). They are sometimes found on the bottom of unsanitary swimming pools. One species, *N. fowleri*, may cause a deadly human disease called amoebic meningoencephalitis if it gains

access to the brain via the nostrils of unlucky swimmers. *Naegleria* species are usually amoeboid under favorable conditions, but flagellated stages

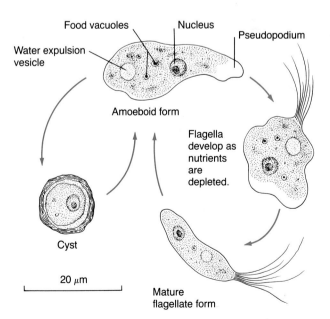

Figure 19.4   Amoeboflagellates. These sarcomastigophorans have both amoeboid and flagellated stages. This species, *Naegleria gruberi,* is common in organically rich soils and ponds. The dormant cyst is resistant to drying and freezing.

may develop when nutrients are scarce. Researchers speculate that flagella allow *Naegleria* to swim away from unfavorable habitats and toward favorable ones.

PHYLUM APICOMPLEXA

This phylum is well named, for its members have a complex of organelles located at the tapered end, or *apex*, of the cell (Figure 19.5a). All apicomplexans are parasitic, and the apical organelles probably allow them to penetrate through the cell membranes or tissue layers of their hosts. Although the apical complex can be seen only with the electron microscope, the general appearance of apicomplexan cells and their life cycle patterns also distinguish them. The life cycle of most apicomplexans

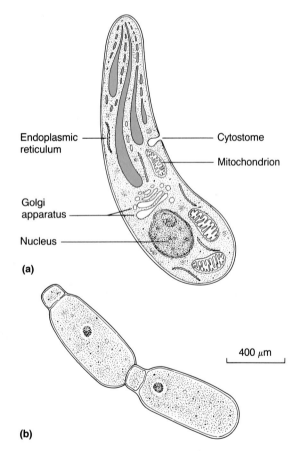

Endoplasmic reticulum

Cytostome

Mitochondrion

Golgi apparatus

Nucleus

(a)

400 μm

(b)

Figure 19.5 Members of the Phylum Apicomplexa. (a) These protozoa have unique apical penetration organelles that can be seen when cells, such as the one shown here, are sectioned and examined with the electron microscope. In this drawing, the apical complex is shown in color. (b) The gregarine *Gregarina* inhabits the gut of cockroaches and many other insects; as illustrated here, individual cells may stick together end to end prior to undergoing a sexual phase.

includes asexual multiplication by multiple fission (schizogony), a sexual phase called gametogony, and a second asexual phase called sporogony (see Figure 19.6 and Table 19.1). Most apicomplexans are members of the **Class Sporozoea**, which contains two subclasses.

Subclass Gregarinia

Gregarines (Figure 19.5b) are extracellular parasites that inhabit the digestive tract, body cavity, or other viscera of a wide variety of invertebrates. Generally, they cause no noticeable harm to their hosts.

Subclass Coccidia

Most **coccidians** are intracellular, and several genera cause significant diseases. Several species of *Eimeria* and *Isospora* that parasitize the digestive tract of vertebrates are of veterinary importance. Infected cattle, sheep, rabbits, chickens, dogs, and other domestic animals may develop the disease coccidiosis, characterized by severe diarrhea and dysentery. Drugs that prevent these infections are often contained in commercial feeds. Another important coccidian called *Toxoplasma gondii* is primarily a parasite of cats, but also infects a wide variety of other vertebrates. Humans may acquire an infection by eating infected beef that is not thoroughly cooked. *Toxoplasma* is generally harmless, except to the unborn; it can pass from an infected mother into a fetus and cause congenital brain damage.

From the standpoint of human medicine, the most important coccidians are species of *Plasmodium*, the organisms that cause **malaria**. The *Plasmodium* life cycle is complex and involves asexual reproduction—schizogony in vertebrates and sporogony in female mosquitoes—and a sexual phase (gametogony) with stages in both hosts (Figure 19.6). Schizogony in the vertebrate liver and red blood cells produces stages called **merozoites**. People with malaria suffer cyclic chills, sweats, and high fever and may die from toxins released when merozoites synchronously break out of red blood cells. After several phases of schizogony, gamete-containing cells called gametocytes are produced. If a mosquito ingests gametocytes with a blood meal, gametes mature and fuse in the gut of the mosquito. Sporogony then ensues in the mosquito gut wall to produce **sporozoites**, which are infective to the vertebrate. In addition to four species of *Plasmodium* that infect humans, more than a hundred other species infect other terrestrial vertebrates.

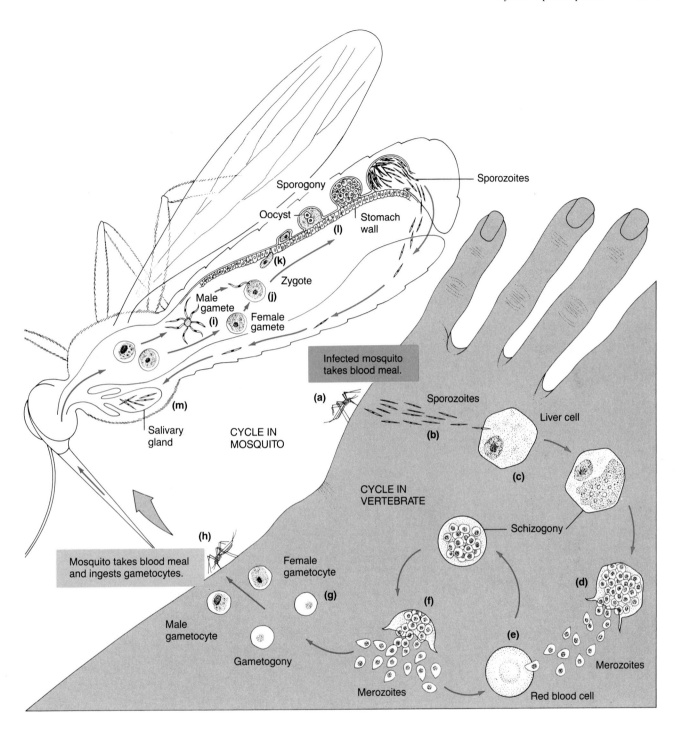

Figure 19.6 Life cycle of *Plasmodium*. (a) When an infected mosquito bites and obtains blood from a vertebrate, (b) infective sporozoites from the mosquito are inoculated into the vertebrate's bloodstream and (c) pass to the liver. Multiple fission (schizogony) then occurs in the liver cells, and (d) many merozoites are produced. The merozoites break out of the liver cells and (e) eventually enter red blood cells. Schizogony then occurs within the red blood cells. More merozoites are formed, break out of their red blood cells (f), and infect new red blood cells. (g) Eventually, some merozoites in the erythrocytes transform into sexual stages (gametocytes) to begin the sexual phase (gametogony). (h) For the life cycle to be completed, male and female gametocytes must by taken up by another mosquito. If this happens, (i) gametes develop from the gametocytes, and (j) fertilization occurs in the mosquito's gut. (k) The resulting zygote migrates to the outside of the mosquito's gut and develops into an oocyst. (l) Sporogony, another asexual multiplication phase, occurs in the oocyst. Many sporozoites are produced, and (m) these are carried in the insect's hemolymph to the salivary glands. From the salivary glands, sporozoites may be inoculated into another vertebrate when the mosquito feeds again.

PHYLUM MICROSPORA

Microsporans are intracellular parasites that form unique spores (Figure 19.7). As a group, they infect a wide variety of animals in most phyla, and many species are host-specific, infecting only a single host species. Tiny (5 μm) spores are the infective stage in the life cycle. Unlike nearly all other eukaryotic cells, microsporan cells contain no mitochondria; as intracellular parasites, they probably obtain the energy they need for growth from their hosts. Of several economically important microsporans, species of *Nosema* cause disease in honeybees and silkworms.

PHYLUM CILIOPHORA

Ciliates, which make up the largest phylum of protozoa, are distinguished by their cilia, by a sexual reproductive process called conjugation, by a unique multilayered pellicle supported by underlying fibrils, and by two different kinds of nuclei (one macronucleus and one or more micronuclei). The ciliate **macronucleus** is typically polyploid (i.e., it has multiples of the diploid chromosome number) and controls feeding, gas exchange, osmoregulation, and other nonreproductive functions; **micronuclei** are diploid and undergo meiosis prior to conjugation. During conjugation, haploid micronuclei are exchanged between two individuals.

Ciliates often occur in teeming numbers in stagnant pools of fresh water or seawater, as well as in moist soil. Several species are maintained in culture and have become important research organisms. Species of *Paramecium* are studied extensively by cell biologists, geneticists, and ecologists, and entire books have been written about them. *Paramecium* cells are slipper-shaped and covered with cilia; they show a distinct mouth region, the cytostome (Figure 19.8a). Another ciliate, *Didinium nasutum*, is a voracious predator of *Paramecium* (Figure 19.8b). Also popular in research, especially in molecular genetics and toxicology, is the small freshwater ciliate *Tetrahymena pyriformis*. *Tetrahymena* is about one-fifth the size of *Paramecium* and is tear-drop shaped. Like *Paramecium*, it feeds mainly on bacteria and is easy to maintain in culture. One of the most beautiful protozoa, the blue-tinted *Stentor coeruleus* is usually attached, but sometimes free-swimming (Figure 19.8c).

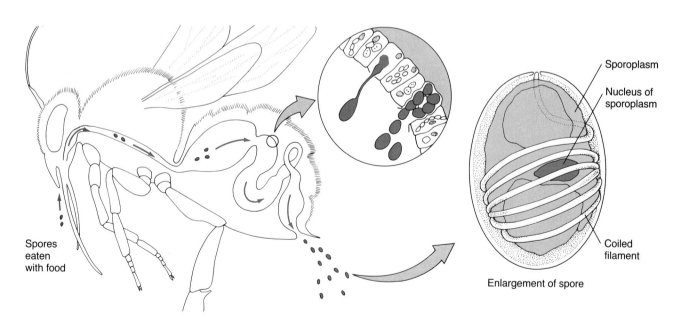

Spores eaten with food

Sporoplasm

Nucleus of sporoplasm

Coiled filament

Enlargement of spore

Figure 19.7 *Nosema apis,* **a typical microsporan.** This protozoan causes dysentery in honeybees and can kill entire colonies. A bee (shown here with a highly diagrammatic view of its gut and cells) becomes infected by eating the minute spores along with honey or pollen. The spore stage, as drawn from electron micrographs, contains an infective cell, the sporoplasm, and a coiled hollow filament, or tube. When a spore is eaten by a honeybee, its coiled tube extrudes and penetrates a host cell, and the sporoplasm is injected into the host cell through the tube. Once in a host cell, the parasite grows and multiplies by multiple fission. Eventually, many spores are formed, break out of the swollen host cells, and pass out with feces.

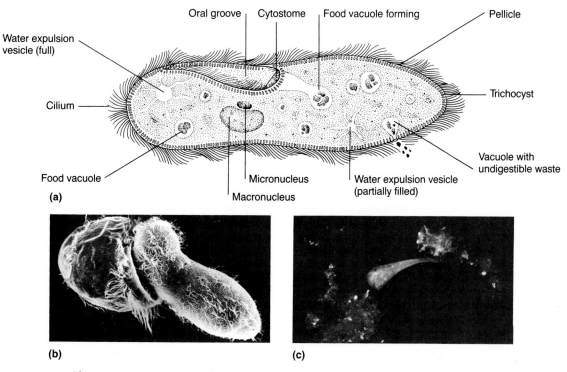

Figure 19.8 Representative ciliates. (a) *Paramecium,* one of the best known members of the Phylum Ciliophora, inhabits freshwater ponds and lakes, and some species live in brackish and marine environments. Notice its two types of nuclei. (b) The predatory ciliate *Didinium* (the globose cell with two rows of cilia) has paralyzed and is eating a *Paramecium* cell, as seen with the scanning electron microscope (350×). (c) *Stentor* is another common freshwater ciliate. Its macronucleus is beaded in appearance. Here we see *Stentor* and a rotifer among some green algae.

HABITATS AND ENVIRONMENTAL RELATIONSHIPS

Protozoa can grow and reproduce only in wet habitats because their surface membrane must remain moist to facilitate gas exchange and other vital functions, but they are found virtually anywhere there is water. Dense populations often appear in seasonally drying ponds, in tiny puddles that last for only a few days, and in surface scums on stagnant water. Some species are planktonic (suspended in the water), while many others live attached to plants, animals, and nonliving surfaces.

In wet soil, protozoa live in droplets and films of water in and around soil particles, microhabitats too small for most animals. Amoebas, with their elastic, maneuverable cell bodies, are often plentiful in soil, along with many small ciliates and flagellates. Estimates of total numbers of protozoa in fertile soils range from a few thousand to nearly a million individuals per gram of earth. Most soil protozoa have a dormant cyst stage in which they survive periods of freezing and drying; taking advantage of the short wet periods typical of soils, they usually can emerge rapidly from the cysts and begin reproducing soon after emergence. Cysts are often carried to new habitats by winds, water currents, and animals.

Role in Food Chains

Protozoan populations are important in aquatic food chains. At the bottom of a food chain, photosynthesizing by green flagellates often provides much of the organic energy available in freshwater and marine environments. Many amoebas, heterotrophic flagellates, and ciliates are predators, feeding on bacteria, yeasts, diatoms, algae, and one another. Other protozoa are scavengers, feeding on dead organic matter. Like bacteria, scavengers help decompose dead plants and animals and recycle nutrients back into food chains.

Bacteria are the most common food of protozoa. Species that feed on bacteria, and in turn are eaten by small animals or other protozoa, form a significant nutrient link between the decomposer bacteria and other organisms in food chains. Protozoa that eat bacteria often reach enormous population sizes, as do certain species that can carry out photosynthesis. Blooms, which are large, single-species populations of certain autotrophic flagellates,

Essay: Malaria—Hope for Future Control

Malaria—the term literally means bad air—is, and always has been, one of the most deadly human diseases. Despite massive control efforts and extensive research throughout the world, malaria still afflicts some 300 million people. Since the late 1950s, insecticide-resistant mosquitoes and drug-resistant strains of *Plasmodium* have evolved, and human populations have grown enormously in countries that cannot afford pesticides or drugs. About 3 million persons, mostly young children, die from malaria each year.

Today, research in immunology and molecular biology offers new hope that malaria may someday be conquered. Species of *Plasmodium* that infect humans can be grown and studied in certain monkeys, and much is being learned by studying species that parasitize mice, rats, and birds. Since the mid-1970s, stages of the most deadly species, *Plasmodium falciparum*, have been grown in red blood cells in the laboratory. Now mass-produced under computer control, such cultures are a ready source of *Plasmodium* organisms that can be used in many lines of research on malaria. A critical problem is that there are so many life cycle stages to study. One stage, the sporozoite, develops in mosquitoes and infects the human host. Other stages develop in liver cells and red blood cells (see Figure 19.6). To eliminate malaria, vaccines may have to be developed that produce immunity to several of these stages.

Some 100 laboratories around the world are now trying to develop vaccines against one or more of these life cycle stages. Researchers have recently cloned the gene that codes for a protein on the surface of the sporozoite of *Plasmodium falciparum*, making it possible to mass-produce this protein and use it in attempts to develop a vaccine against the sporozoites. Other researchers have identified a key protein that promotes the infection of human red blood cells. This substance, located on the surface of *Plasmodium* cells, may also be the source of a future vaccine. It can be extracted from cultured malaria organisms, and in the near future, it may be mass-produced from cloned genes. Another exciting line of research involves the identification and isolation of malaria proteins that allow the parasites to recognize red blood cells. If these molecules can be identified, they could also be used as vaccines. If injected into a host, they might stimulate the production of a host antibody that could render the malaria organism unable to recognize the host cell.

Although all these lines of research are developing rapidly, vaccination against malaria remains a distant hope. Not only must the vaccine be exhaustively tested for efficacy against the disease, but it must also be proved safe. Nonetheless, the possibility of a future worldwide malaria control program combining potent new vaccines with older methods of mosquito control and antimalarial drugs is cause for some optimism.

---

sometimes occur in lakes and oceans when optimal conditions promote rapid reproduction. Blooms may appear as large reddish or greenish areas on the water. Red tide, common on the east and west coasts of the United States, is caused by blooms of several flagellates, including one called *Gymnodinium brevis*. Toxins produced by these flagellates often cause widespread fish kills, and people have suffered temporary paralysis or died from eating toxin-contaminated seafood.

Parasitic protozoa are also important ecologically. Several protozoa parasitize other protozoa, and some infect plants. Others typically parasitize animals and are common on body surfaces, in the digestive tract, in many tissues and organs, and within cells. Species that infect internal organs of hosts live in body fluids and thus are aquatic, but they infect terrestrial as well as aquatic hosts. Those that parasitize their host's intestine usually pass from host to host as cysts, whereas those inhabiting blood and tissue fluids are usually transmitted to hosts by vectors, such as mosquitoes. In an ecological sense, parasites are consumers, draining nutrients and energy from their hosts.

COMPARATIVE STRUCTURE AND FUNCTION

Despite their variety, protozoa show certain fundamental similarities to one another and to the cells of other eukaryotes. Because their membranes are similar to those of animals, protozoa, especially *Tetrahymena* and *Paramecium*, are often used to study such universal phenomena as membrane transport, exocytosis, and hormone reception.

Body Support, Protection, and Locomotion

In conjunction with internal cytoskeletal elements, including microtubules and microfilaments, the body surface of protozoa, either a *plasmalemma* or a *pellicle*, supports and protects the cell, functions in locomotion, and forms a dynamic interface between the environment and the cytoplasm.

Pseudopodia and Amoeboid Movement. Movement by **pseudopodia**, called **amoeboid movement**, allows amoebas to move about freely on moist surfaces. As discussed in Chapter 9 (p. 196), amoeboid movement is not yet completely understood, and several species of *Amoeba* are maintained in laboratories and used in the study of this important means of cell locomotion.

Flagella and Cilia. These cell organelles are structurally identical, but may be distinguished by the way they move certain protozoan cells. (Flagella and cilia are discussed in Chapter 9, p. 197; Figures 9.2 and 9.3.) While flagella typically pull or push a cell, cilia essentially act like tiny oars, in effect, rowing a protozoan through the water.

Pellicles. **Pellicles** are essential for locomotion in flagellates and ciliates, for without a semirigid cell surface, the driving forces of flagella and cilia would be dissipated. The plantlike flagellate *Ceratium* has a rigid pellicle made of armorlike plates that form a solid protective covering similar to the cell wall of plant cells (Figure 19.9a). The pellicle is relatively flexible in species of *Trypanosoma*. It is

formed of a single membrane with an underlying row of microtubules adding some rigidity (Figure 19.9b). Covering the pellicle is a surface coat called the **glycocalyx**, which consists of mucus and other protective glycoproteins and which allows trypanosomes to survive in their vertebrate hosts. While in the vertebrate bloodstream, these parasites are under constant attack by the host's protective white blood cells and antibodies. The recent discovery that trypanosomes can elude their host's immune system by periodically changing their surface coats has been a major breakthrough in the effort to understand and control these parasites. Researchers are now trying to identify the specific chemical components of the glycocalyx that protect the trypanosomes; eventually, such chemicals might be used to develop a vaccine.

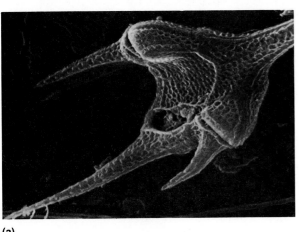

(a)

Figure 19.9 Pellicles of flagellates. (a) *Ceratium hirundinella,* as seen with the scanning electron microscope and in a drawing. The sculptured pellicle of this photosynthetic species consists of cellulose and protein plates lying between two cell membranes. *Ceratium* has two flagella, one posteriorly directed, the other wrapped transversely around the cell, causing it to rotate as it moves forward. (b) The pellicle of trypanosomes, interpreted here in longitudinal and transverse sections (1680×), consists of a single cell membrane with underlying microtubules. The pellicle covers the flagellum as well as the cell body. Next to the cell body, the flagellum is firmly attached to the cell surface by studlike structures. As the flagellum beats, the cell surface next to it ripples with the waves of contraction; the rippling cell surface and its attached flagellum are called the undulating membrane.

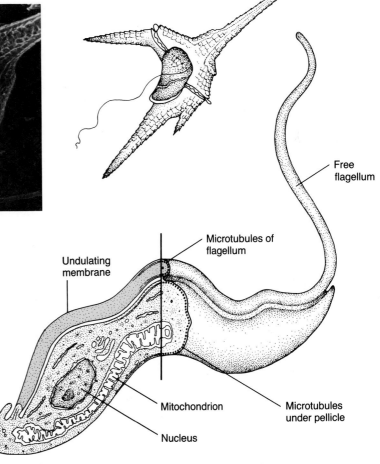

Free flagellum

Microtubules of flagellum

Undulating membrane

Mitochondrion

Nucleus

Microtubules under pellicle

(b)

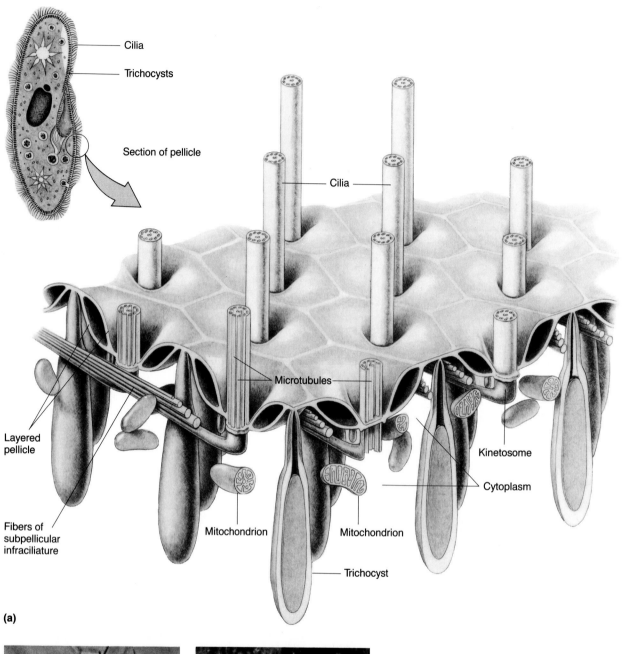

Cilia

Trichocysts

Section of pellicle

Cilia

Microtubules

Layered
pellicle

Kinetosome

Cytoplasm

Fibers of
subpellicular
infraciliature

Mitochondrion

Mitochondrion

Trichocyst

(a)

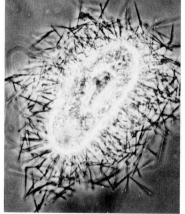

(b)

(c)

Figure 19.10 The complex pellicle of a ciliate. (a) As shown in this sketch of a section of a *Paramecium* cell, the pellicle is a complex structure supported by a microtubular and microfilament network called the subpellicular infraciliature. (b) A *Paramecium* cell after discharging its trichocysts, as seen with the light microscope. (c) The spearlike tips of discharged trichocysts, as seen with the electron microscope.

The pellicle of ciliates is a remarkably complex cell covering (Figure 19.10). It is composed of an outer membrane that covers the entire cell, including the cilia, and additional underlying membranes that do not extend over the cilia but form saclike spaces, or **alveoli**, in the pellicle. The alveoli form rows of tiny pits and ridges, providing flexibility at the cell surface. Each cilium arises from a basal body, or **kinetosome**, under the alveoli and projects from a pit in the cell surface. Cilia and their kinetosomes are arranged in longitudinal rows called **kineties**. Under the pellicle and associated with the kinetosomes is a complex array of microtubules and microfilaments called the **subpellicular infraciliature**, which provides support for ciliary locomotion.

Many ciliates also have unique flask-shaped organelles that project through the pellicle. Those of *Paramecium*, called **trichocysts**, may function in defense, for they discharge fine threads with spearlike tips in response to chemical, mechanical, and electrical stimuli (Figure 19.10b). Many predaceous ciliates, such as *Didinium* (see Figure 19.8b), have poisonous **toxicysts** used to capture and kill prey.

Myonemes. Many ciliates, such as species of the freshwater genus *Vorticella*, are sedentary and physically attached to the substrate by a stalk (Figure 19.11 and chapter opening photograph). Cilia in sedentary ciliates function mainly in feeding. Whole-body movements result mainly from the action of one or more contractile filaments called
▷ **myonemes**, which are contained in the stalk. (Myonemes are discussed in Chapter 9, p. 200.) If *Vorticella* is disturbed, its myoneme contracts and its stalk shortens, contracting like a spring. Species of *Stentor* also rely heavily on myoneme contraction for whole-body movement, even though they are covered with cilia (see Figure 19.8c).

Special Locomotor Organelles. Many flagellates and ciliates have special surface organelles that allow complex body movements. Among the flagellates, trypanosomes have an **undulating membrane** where the flagellum lies next to and makes intermittent contact with the membrane covering the cell body (see Figure 19.9b). As the flagellum undulates, the cell membrane at the contact points is forced to move with it, causing a rippling or undulating effect along the cell surface. This type of motion may help the trypanosomes to move about among the red blood cells in their vertebrate hosts. Many ciliates also have an undulating membrane, but it is formed from a single row of cilia that are so

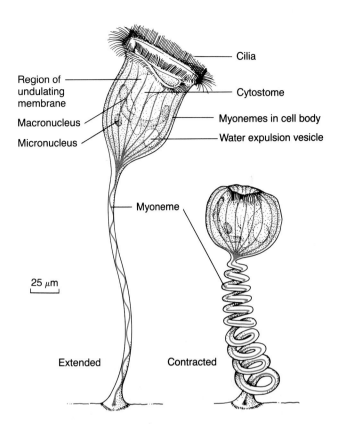

Figure 19.11 *Vorticella*. These stalked, freshwater ciliates, which are attached to submerged sticks and rocks, move in place by means of contractile myonemes. Their cilia function mainly in creating water currents for feeding. Note the undulating membrane in the cytostome, and the macro- and micronuclei.

close together that they seem to be fused. The cell membrane over the undulating membrane appears to undulate as this tight row of cilia beats.

Other compound locomotor organelles in ciliates are called **membranelles**. These consist of two or more rows of cilia clusters that form a fanlike organelle (Figure 19.12). In many ciliates, an undulating membrane and one or more membranelles are situated near the region where food is ingested. The feeding organelles of *Vorticella* consist of a large undulating membrane surrounding several smaller membranelles. In some of the more complex ciliates, such as species of *Euplotes*, membranelles and bristlelike clusters of cilia called **cirri** are the main locomotor structures (Figure 19.12a).

Other Protective and Supportive Structures. Many protozoa produce protective and supportive cell cases. **Tests**, as seen in the amoeba *Arcella* are cases that fit the cell surface closely (see Figure 19.1b). Cases called **loricae** fit loosely around the organism and allow considerable internal movement (Figure 19.12b).

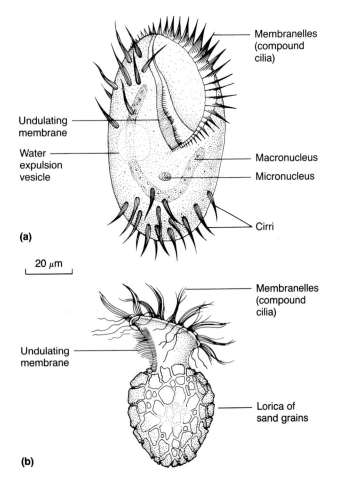

Membranelles
(compound
cilia)

Undulating
membrane

Water
expulsion
vesicle

Macronucleus

Micronucleus

Cirri

(a)

20 µm

Membranelles
(compound
cilia)

Undulating
membrane

Lorica of
sand grains

(b)

Figure 19.12 Ciliates with pronounced compound locomotor organelles. (a) Species of *Euplotes* are common in freshwater, brackish, and marine environments. They have only a few body cilia, but well-developed membranelles and cirri formed of clusters of cilia allow them to twist, turn, and swim rapidly forward and in reverse. Note also the small undulating membrane that functions in feeding. (b) Species of *Tintinnopsis* occur in deep marine environments and freshwater lakes. These ciliates inhabit a loose-fitting case, or lorica, formed of fine particles of sand attached to secreted material on the cell membrane; the compound cilia (membranelles) on *Tintinnopsis* are used in swimming.

Feeding and Digestion

Protozoa ingest some nutrients across their cell surface by diffusion and by membrane transport. Food may also be taken up by endocytosis. (See the discussions of phagocytosis and pinocytosis in Chapter 2, p. 42.) In protozoa with simple cell coverings, ingestion may occur at any site on the plasmalemma. Protozoa with pellicles often have a specialized **cytostome** where most or all ingestion occurs. Digestion occurs in intracellular **food vacuoles** as hydrolytic enzymes are added from circulating lysosomes. Nutrients are assimilated into the cytoplasm as digestion is completed, and undigested wastes are voided to the environment.

Circulation and Gas Exchange

Because protozoa are small and their surface-area-to-volume ratio is high, and because internal distances are small, the vital functions of circulation and gas exchange can be carried out without special organelles. Circulation is accomplished by intracellular diffusion of dissolved substances. The formation of pseudopodia in amoeboid protozoa involves streaming of cytoplasm, which also assists circulation. Gas exchange occurs as dissolved oxygen and carbon dioxide diffuse across the moist cell membrane.

Excretion and Osmoregulation

Protozoa, like most aquatic organisms, dispose of nitrogenous wastes as ammonia, which readily diffuses across the cell membrane. Freshwater sarcomastigophorans and ciliates have special osmoregulatory organelles called **water expulsion vesicles** (contractile vacuoles) that pump excess water out of the cell (Figure 19.8a). Because the cytoplasm of freshwater protozoa is hyperosmotic to the environment, water tends to flow into the cell and must be eliminated to prevent bursting. Accordingly, the water expulsion vesicles periodically fill with water from the cytoplasm, then expel it through the surface membrane. (Structure and function of water expulsion vesicles are described in some detail in Chapter 7, p. 154; Figure 7.7.) Protozoa living in brackish water are adapted to constantly changing salinity, and many species have water expulsion vesicles that are inactive in highly saline waters, but that actively pump water out when the environment is dilute.

Marine and parasitic protozoa are isosmotic to their environments. They maintain intracellular concentrations of compounds derived from their own metabolism at levels equal to the osmotic concentration of seawater or of host body fluids. Some of these protozoa have water expulsion vesicles, but the role of the vesicles is unknown.

Sensory Structures

The protozoan cell membrane is sensitive to many stimuli, including chemicals, light, electricity, temperature, and various mechanical forces. In some protozoa, certain regions of the cell membrane are especially sensitive to specific stimuli. The membrane covering the flagella and cilia often has a sensory function, and it is likely that some species have special sensory organelles yet to be discovered.

Many photosynthetic flagellates possess a pigmented **eyespot**, or **stigma**, which is thought to shade an underlying photosensitive zone. In *Euglena*, a reddish stigma seems to function as a light shade, allowing light from only a certain direction to strike an underlying swelling, the **paraflagellar body**, on the base of the flagellum (see Figure 19.2a). Experimental evidence indicates that the paraflagellar body is a type of photoreceptor that, when illuminated, causes the flagellum to move the cell toward the light.

BEHAVIOR

Behavioral responses in protozoa typically involve changes in movement or orientation. (See the discussion of kineses and taxes in Chapter 13, p. 308.) Chemotaxis is important in locating food, and parasitic protozoa often locate hosts and specific sites within hosts by following chemical gradients. Most protozoa exhibit negative phototaxis, although many autotrophs move toward light; photosynthetic flagellates in oceans and lakes move into upper, more lighted zones during the day and move downward at night. Most flagellates that have a stigma seem to detect and respond mainly to light direction, but *Euglena rubra* and species of *Volvox* respond to different light intensities. Low-intensity light increases flagellar beating in these autotrophs, and high-intensity light reduces it. In this way, the organisms increase their chances of finding a suitable light source and conserve energy upon locating it. *Volvox* colonies roll toward a light source because flagella on the side of the colony facing the light beat more slowly than those on the opposite side. In the laboratory, when all surfaces of *Volvox* colonies are exposed to equal light intensities, all flagella stop beating.

The so-called **avoidance reaction** of *Paramecium* is an example of a whole-body response pattern. It consists of a series of three negative taxes: rapid reverse, spinning, and forward movement in another direction (Figure 19.13). The reaction occurs when normal forward movement is impeded by mechanical obstructions, by threshold temperatures, or by critical levels of certain chemicals. *Paramecium* and many other ciliates also exhibit local ciliary movements in response to certain chemicals or in response to contact with other members of their species. Certain ciliates have a "mating ritual" that precedes sexual fusion (conjugation) of two cells. Before fusing, the cells spin, twitch, and touch each other.

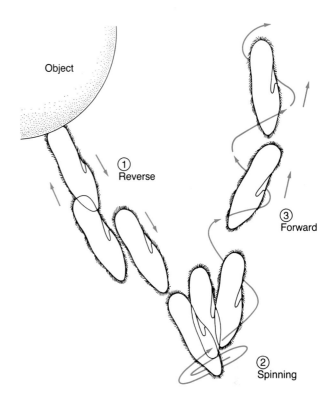

Figure 19.13 The avoidance reaction of *Paramecium*. When this ciliate swims into a rock or other solid object, its cilia respond to the impact by beating in reverse. This drives the organism backward a distance of several body lengths (1). Eventually, the ciliate spins on its longitudinal axis (2) and proceeds forward in another direction (3).

Some ciliates may even exhibit **habituation**, a form of simple learning. Attached individuals of *Stentor* react to rapid changes in water current by a quick withdrawal response mediated by contraction of their myonemes. This response may be lost if the stimulus is repeated several times.

REPRODUCTION AND LIFE CYCLES

The most common means of reproduction among protozoa are asexual (binary fission, multiple fission, and budding); and many flagellates and amoebas reproduce only asexually. The life cycles of many protozoa include one or more asexual phases, resulting in a great increase in numbers, and a sexual phase, which contributes to genetic variability (see Figure 19.6 and Table 19.1). The sexual phase may not result directly in reproduction. In most protozoa, the entire cell (individual) divides during reproduction. In a few protozoa, such as the colonial flagellate *Volvox*, special reproductive cells are differentiated from the vegetative (somatic) cells of the colony.

Table 19.1   Modes of Reproduction in Protozoa

Process*	Products
Asexual Reproduction	
Binary fission: equal division of a parent cell, either longitudinally (flagellates) or transversely (ciliates)	Two smaller daughter cells
Budding: unequal division of a parent cell	Offspring much smaller than adults
Multiple fission: division of a multinucleate parent cell	Many uninucleate offspring
(1) *Schizogony:* precedes sexual reproduction	Merozoites
(2) *Sporogony:* follows sexual reproduction	Sporozoites
Sexual Reproduction	
Gametogony: gamete formation	Gametes (specialized haploid sex cells)
Fertilization: fusion of gametes	Zygote (diploid offspring)
Conjugation: fusion of two cells like all others in the population	Synkaryon (diploid nucleus formed in both conjugating cells)
Autogamy: fusion of gametes or gametic nuclei produced by the same cell	

*Not all these processes occur in all species, although complex life cycles involving several stages are common, especially in parasites.

Asexual Reproduction

Binary fission occurs as an active, growing cell, often called a trophozoite, divides into two daughter cells of approximately equal size. In *Amoeba* (see Figure 19.1a), binary fission begins as locomotion ceases and the trophozoite compacts into a spherical shape. The amoeba then constricts centrally, and pseudopodia on the developing daughter cells pull the trophozoite apart to complete division. Most flagellates divide by *longitudinal* binary fission (Figure 19.14a). If cysts occur in the life cycle, then that is the stage in which fission often occurs. Ciliates typically divide *transversely* across rows of cilia (Figure 19.14b). Flagella and cilia do not divide, but new sets of these organelles are formed from kinetosomes duplicated prior to fission.

Budding, or unequal division, occurs when a portion of a parent cell (trophozoite) produces one or more smaller offspring. Budding often occurs in multinucleate amoebas as small portions of cytoplasm with a nucleus break off and form smaller cells. Budding is also common as a dispersal mechanism among sessile protozoa (Figure 19.14c).

Multiple fission (schizogony) is common among apicomplexan parasites and seems to be an adaptation to offset the odds against successfully passing from one host to another (see Figure 19.6 and Table 19.1). In schizogony, a uninucleate trophozoite undergoes multiple mitoses, producing a multinucleate cell. The multinucleate cell then divides to form many uninucleate **merozoites**. In **sporogony**, a uninucleate zygote forms many uninucleate **sporozoites**.

Sexuality in Protozoa

In many protozoan species, the sexual part of the life cycle is triggered by the onset of drought or low temperature, and **gametogony**, the production of gametes, often immediately precedes or is directly involved in dormant cyst formation. In the life cycle of *Volvox*, sexual stages usually appear when ponds begin to dry or when winter approaches (Figure 19.15). The resulting zygotes encyst and remain dormant until favorable conditions return.

Species of *Trichonympha* are so well adapted to their host termites and wood roaches that the sexual phase of these protozoa is triggered by increased levels of the molting hormone ecdysone in the insects. As the ecdysone level rises just before a molt, a *Trichonympha* cell secretes a protective cyst wall around itself. The cyst protects the protozoan while it is exposed during the molt. Gametes develop in the cysts (Figure 19.16).

There are three types of sexual processes in protozoa. The most familiar is **fertilization** by the

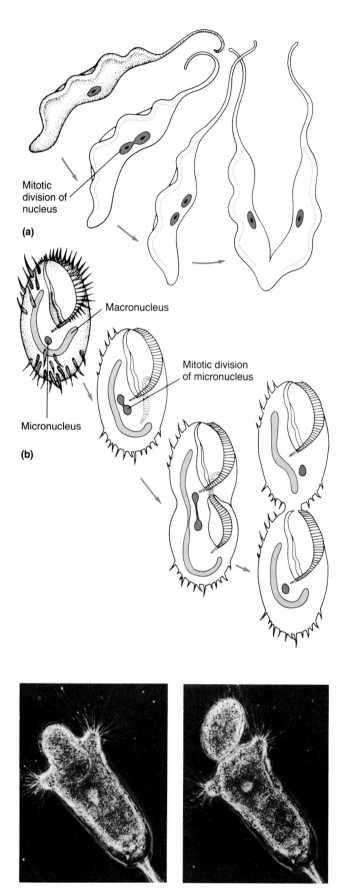

(a)

Mitotic division of nucleus

Macronucleus

Mitotic division of micronucleus

Micronucleus

(b)

(c)

Figure 19.14 (Left) Asexual reproduction in protozoa. (a) Flagellates, such as *Trypanosoma,* are characterized by longitudinal binary fission. (b) Ciliates, such as *Euplotes,* divide by transverse binary fission. The micronucleus divides by mitosis, while the macronucleus simply elongates and divides as the cell constricts. (c) Budding occurs in the sessile ciliate *Acineta tuberosa.* This species is a suctorian, a stalked ciliate that lacks cilia in its adult form. A small bud forms at the tip of the parent cell (left) and breaks free (right) to become a free-swimming ciliated swarmer. After dispersing from the parent, swarmers attach to a substrate, develop a stalk, and grow to adult size. Suctorians feed by capturing other protozoa on sticky cell projections called tentacles.

union of gametes. The *Plasmodium* life cycle includes this type of sexual process (see Figure 19.6).

A second type of sexual process occurs in a wide variety of flagellates, foraminiferans, gregarines, and ciliates. True gametes are not formed, but two cells called **gamonts**, which look like all the other individuals in the population, fuse, and the nuclei in the gamonts become gametic elements. Notice in Figure 19.16 that the two *Trichonympha* cells that fuse to form the zygote are identical and look like all the other cells in this flagellate's life cycle. In ciliates, this second type of sexual process is called **conjugation** (Figure 19.17). Two ciliate cells attach and their micronuclei undergo meiosis to form two or more **gametic nuclei**. One gametic nucleus from each cell then moves to the other cell. A diploid nucleus, or **synkaryon**, is then formed in each cell by fusion of the gametic nuclei. Thus, conjugation accomplishes genetic mixing. Reproduction is a separate process. After diploid nuclei are formed, conjugating cells separate, and then each reproduces asexually by binary fission. There are distinct varieties of certain ciliate species, and conjugation occurs only between members of the same variety. Within each variety, there are usually two or more cell types that differ in the chemical nature of their surface membrane. Conjugation occurs only between chemically different cells. There are at least six genetically isolated varieties of the ciliate *Paramecium bursaria.*

Autogamy, a third type of sexual process seen in some protozoa, is the fusion of gametes or gametic nuclei produced by the same cell. In many ciliates autogamy occurs within a single cell as two gametic nuclei fuse, producing genetic rearrangements without genetic mixing from separate cells—the ultimate in inbreeding.

Life Cycle Patterns and Chromosome Number

As a group, protozoa show four basic life cycle patterns: asexual, diploid, haploid, and haplodiploid.

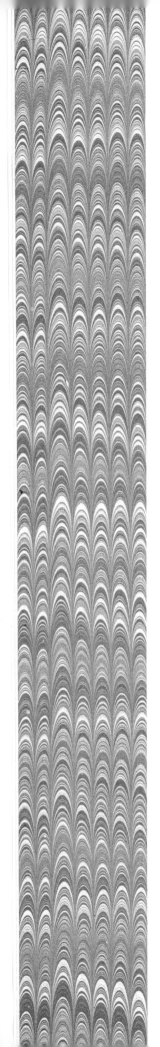

If separation of germ and somatic cells was the first step toward multicellularity, another early step toward interdependence and division of labor among cells undoubtedly would have been the specialization of some cells for maintaining homeostasis. An aggregation of cells in a multicellular body creates intercellular spaces, and it is highly advantageous to the cells that the fluid environment in these spaces be maintained in a balanced, steady state, distinct from the external environment. As some cells took on the task of maintaining the steady state of internal fluids, other cells would have gone on to specialize in other ways.

Several hypotheses have been proposed to account for the evolution of animals from unicellular life forms. One widely discussed idea, the **colonial hypothesis**, suggests that animals (metazoans) evolved from colonial flagellates. Several protozoan species, such as the photosynthetic flagellate *Volvox*, are colonial groups of cells with specialized eggs and flagellated sperm cells. Primitive animals might have evolved from heterotrophic flagellates with a colonial organization similar to that of *Volvox* as cell differentiation and division of labor gradually increased. The presence of motile, flagellated sperm in most animals and the occurrence of flagellated body cells in lower metazoans are cited as evidence for the colonial hypothesis.

An alternative idea, the **plasmodial hypothesis**, holds that animals arose from multinucleate ciliate protozoa that underwent internal cellularization. Many modern-day ciliates have several micronuclei that function in genetic exchange during the sexual process of conjugation. Perhaps some ancestral ciliates developed cell membranes around each of their numerous nuclei and gradually evolved into primitive multicellular forms. Ciliate micronuclei are diploid, like those of metazoan somatic cells, and this is cited as evidence supporting the plasmodial hypothesis. A tendency toward bilateral symmetry in certain ciliates may also indicate that bilateral metazoans arose via cellularization. Because ciliates lack flagellated sperm, the plasmodial hypothesis holds that differentiation of sex cells was not coincident with early multicellularization, but occurred later in metazoan evolution, after the lineage that led to modern ciliates split off. Likewise, the ciliate macronucleus, absent in metazoans, may have evolved later in the ciliate lineage.

Few biologists believe that all multicellular life evolved from a single ancestral lineage. Fundamental differences among modern multicellular life forms indicate that multicellularity evolved several times, in independent evolutionary lineages. The animal kingdom consists of several fundamentally distinct groups that probably represent independent evolutionary lineages (see Figure 18.6). As discussed in Chapter 18, radiate animals (cnidarians and ctenophorans) are fundamentally different from bilateral metazoans; thus, it is possible that radiates and bilateral animals had separate origins among protists. Perhaps radiate groups arose from colonial flagellates, while bilateral groups evolved via the plasmodial route from "cellularized ciliates."

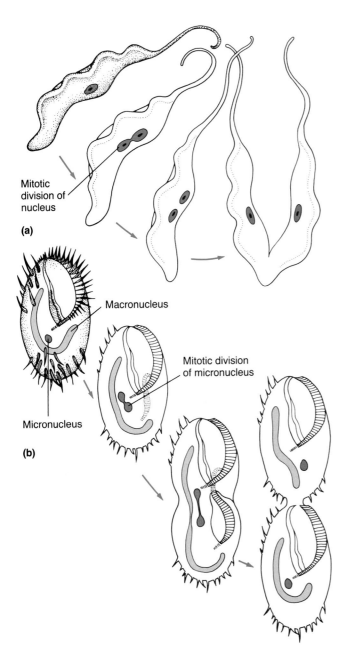

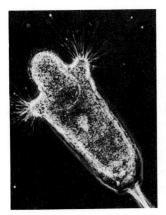

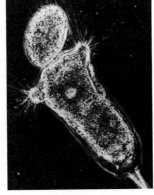

(a)

(b)

Mitotic
division of
nucleus

Macronucleus

Mitotic division
of micronucleus

Micronucleus

(c)

Figure 19.14 (Left) Asexual reproduction in protozoa. (a) Flagellates, such as *Trypanosoma,* are characterized by longitudinal binary fission. (b) Ciliates, such as *Euplotes,* divide by transverse binary fission. The micronucleus divides by mitosis, while the macronucleus simply elongates and divides as the cell constricts. (c) Budding occurs in the sessile ciliate *Acineta tuberosa.* This species is a suctorian, a stalked ciliate that lacks cilia in its adult form. A small bud forms at the tip of the parent cell (left) and breaks free (right) to become a free-swimming ciliated swarmer. After dispersing from the parent, swarmers attach to a substrate, develop a stalk, and grow to adult size. Suctorians feed by capturing other protozoa on sticky cell projections called tentacles.

union of gametes. The *Plasmodium* life cycle includes this type of sexual process (see Figure 19.6).

A second type of sexual process occurs in a wide variety of flagellates, foraminiferans, gregarines, and ciliates. True gametes are not formed, but two cells called **gamonts**, which look like all the other individuals in the population, fuse, and the nuclei in the gamonts become gametic elements. Notice in Figure 19.16 that the two *Trichonympha* cells that fuse to form the zygote are identical and look like all the other cells in this flagellate's life cycle. In ciliates, this second type of sexual process is called **conjugation** (Figure 19.17). Two ciliate cells attach and their micronuclei undergo meiosis to form two or more **gametic nuclei**. One gametic nucleus from each cell then moves to the other cell. A diploid nucleus, or **synkaryon**, is then formed in each cell by fusion of the gametic nuclei. Thus, conjugation accomplishes genetic mixing. Reproduction is a separate process. After diploid nuclei are formed, conjugating cells separate, and then each reproduces asexually by binary fission. There are distinct varieties of certain ciliate species, and conjugation occurs only between members of the same variety. Within each variety, there are usually two or more cell types that differ in the chemical nature of their surface membrane. Conjugation occurs only between chemically different cells. There are at least six genetically isolated varieties of the ciliate *Paramecium bursaria.*

Autogamy, a third type of sexual process seen in some protozoa, is the fusion of gametes or gametic nuclei produced by the same cell. In many ciliates autogamy occurs within a single cell as two gametic nuclei fuse, producing genetic rearrangements without genetic mixing from separate cells—the ultimate in inbreeding.

Life Cycle Patterns and Chromosome Number

As a group, protozoa show four basic life cycle patterns: asexual, diploid, haploid, and haplodiploid.

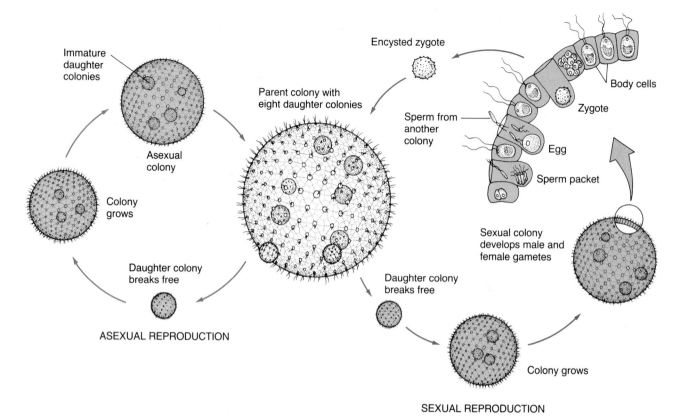

Figure 19.15 The life cycle of *Volvox globator*. Colonies reproduce asexually, producing small daughter colonies when conditions are favorable for growth. The daughter colonies eventually digest a passageway out of the parent colony, swim off, and grow. This cycle of asexual reproduction is repeated so long as conditions are favorable. Gametes are formed when the environment begins to dry or as temperatures and photoperiod decrease in autumn. Zygotes encyst as the overwintering stages, or as stages that withstand drought, later maturing into multicellular colonies.

Species that reproduce only asexually have the simplest type of life cycle, wherein only mitotic divisions occur, allowing no genetic exchange. The other three life cycle patterns occur in sexual protozoa and are categorized according to when meiosis and fertilization take place in the life cycle. Ciliates are diploid; the only haploid elements in their life cycle are the gametic nuclei resulting from meiosis. Meiosis that occurs at this position in the life cycle is called **gametic meiosis**. In contrast to ciliates, the flagellate *Trichonympha*, the apicomplexans, and actinopod amoebas have haploid life cycles. All life cycle stages *except* the zygote are haploid. The first division of the zygote is meiotic and hence referred to as **zygotic meiosis**, and gametes are produced mitotically. Foraminiferans display the haplodiploid life cycle, which involves a true **alternation of generations** of haploid and diploid individuals; this phenomenon also occurs in many plants and algae, but not in animals.

PHYLOGENY

The first animal-like eukaryotes probably were unicellular protists, and it is likely that early multicellular life arose from the same or similar ancestors that gave rise to the modern protozoan phyla. Unfortunately, we can only conjecture about the appearance of ancestral eukaryotes and about early protozoa because their soft bodies were not readily fossilized.

Zoologists generally agree that the earliest eukaryotes possessed flagella. Thus, the original ancestral stock of the modern protozoan phyla and multicellular life may have been "protoflagellates." The marked differences among the protozoan phyla suggest that the lineages of these phyla have been independent for so long (perhaps for over 1000 million years; see Figure 18.6) that they have little in common except their unicellularity. It is

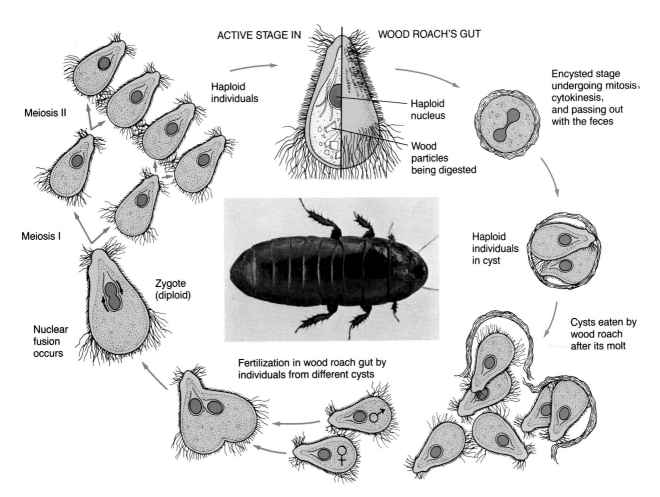

ACTIVE STAGE IN WOOD ROACH'S GUT

Haploid individuals

Meiosis II

Meiosis I

Haploid nucleus

Wood particles being digested

Encysted stage undergoing mitosis, cytokinesis, and passing out with the feces

Haploid individuals in cyst

Zygote (diploid)

Nuclear fusion occurs

Cysts eaten by wood roach after its molt

Fertilization in wood roach gut by individuals from different cysts

Figure 19.16   The life cycle of *Trichonympha.* Species of this flagellate live in the gut of termites and wood roaches, such as *Cryptocercus punctulatus,* shown here (3 ×). *Trichonympha* has a complex life cycle involving both asexual and sexual reproduction. It multiplies by binary fission except while its host insect is molting. As molting begins, the insect's molting hormone (ecdysone) causes the *Trichonympha* cells to encyst, and gametes develop in the cysts. When the insect sheds its exoskeleton, it also loses part of its hindgut-lining, which encloses the *Trichonympha* cysts and feces. After the molt is complete, the insects become reinfected by eating the cysts on contaminated wood. The *Trichonympha* gametes then break out of the cysts, and pairs from different cysts fuse in the host's gut, forming zygotes. Each zygote undergoes meiosis immediately. Thus, only zygotes are diploid; all other stages in the life cycle are haploid.

possible that the protozoan phyla arose from more than one stock of early protists; and their separate lineages may have originated among prokaryotes. If, as it seems likely, the flagellated condition was ancestral, the Phylum Sarcomastigophora is probably the oldest protozoan phylum. Within the Sarcomastigophora, species of *Euglena* that are capable of both photosynthesis and heterotrophy indicate that heterotrophic (animal-like) flagellates are rather closely related to many photosynthetic plantlike forms.

Of the other protozoan phyla, little can be said about phylogenetic relationships of the parasitic Apicomplexa and Microspora. It is possible to approximate the age of some groups in these phyla by estimating the age of their hosts, but in becoming adapted for parasitism, these protozoa have apparently changed so drastically that their relationships with other phyla cannot be discerned.

The ciliates constitute the best-defined protozoan phylum. All ciliates are diploid and have two kinds of nuclei. Three taxonomic classes of ciliates

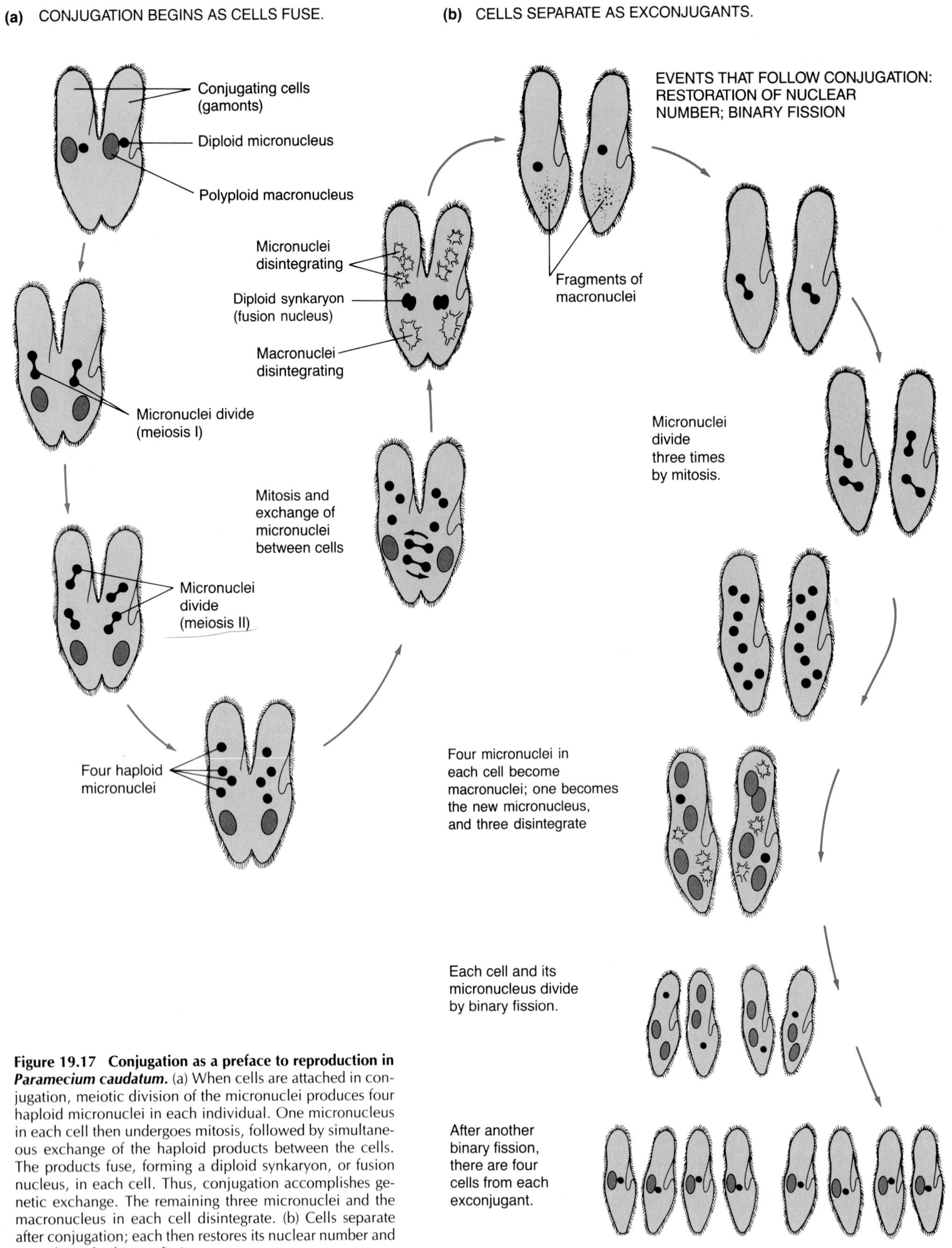

(a) CONJUGATION BEGINS AS CELLS FUSE.

(b) CELLS SEPARATE AS EXCONJUGANTS.

Conjugating cells (gamonts)

Diploid micronucleus

Polyploid macronucleus

Micronuclei disintegrating

Diploid synkaryon (fusion nucleus)

Macronuclei disintegrating

Micronuclei divide (meiosis I)

Mitosis and exchange of micronuclei between cells

Micronuclei divide (meiosis II)

Four haploid micronuclei

EVENTS THAT FOLLOW CONJUGATION: RESTORATION OF NUCLEAR NUMBER; BINARY FISSION

Fragments of macronuclei

Micronuclei divide three times by mitosis.

Four micronuclei in each cell become macronuclei; one becomes the new micronucleus, and three disintegrate

Each cell and its micronucleus divide by binary fission.

After another binary fission, there are four cells from each exconjugant.

Figure 19.17 Conjugation as a preface to reproduction in *Paramecium caudatum*. (a) When cells are attached in conjugation, meiotic division of the micronuclei produces four haploid micronuclei in each individual. One micronucleus in each cell then undergoes mitosis, followed by simultaneous exchange of the haploid products between the cells. The products fuse, forming a diploid synkaryon, or fusion nucleus, in each cell. Thus, conjugation accomplishes genetic exchange. The remaining three micronuclei and the macronucleus in each cell disintegrate. (b) Cells separate after conjugation; each then restores its nuclear number and reproduces by binary fission.

reflect a phylogenetic trend toward increasing complexity in the phylum. Members of the Class ▷ **Kinetofragminophorea** (*Didinium*, *Acineta*, and others; see Classification, p. 465 and Figure 19.8b) exhibit many features that seem to be ancestral: no compound ciliary organelles around the cytostome (cell mouth), and virtually no distinction between cytostomal and somatic (cell body) cilia. Members of a second class, the **Oligohymenophorea** (e.g., *Tetrahymena* and *Paramecium*) have cytostomal membranelles. Ciliates in a third group, the Class **Polyhymenophorea** (e.g., *Euplotes* and *Tintinnopsis;* see Figure 19.12), exhibit many derived features, such as pronounced compound feeding organelles and compound locomotor organelles that often replace body cilia.

Although no protozoan is multicellular, and none shows any evidence of tissues or organs, each individual is an integrated, complex living unit. Many protozoa have multiple nuclei, and ciliates have separate reproductive and somatic nuclei. The development of large colonies in several species reaches the point of separation of reproductive from somatic function. These features may indicate phylogenetic trends toward increasing complexity in body organization and may represent early stages in the continuum of evolutionary adaptations that led to multicellular life.

SUMMARY

1. Key Features of Protozoa. Protozoa, as members of the Kingdom Protista, are a diverse assemblage (probably several phylogenetic lineages) of unicellular or colonial eukaryotes whose diversity is based on structural and functional specialization of cell organelles. Free-living species are heterotrophic and/or autotrophic and are abundant in all types of aquatic environments, including moist soil; many are parasitic.

2. Cytoskeletal elements, a cell membrane, and sometimes a compound pellicle provide support and protection. Protective cases (tests or loricae) and resistant cysts are common. Locomotion is by pseudopodia, flagella, cilia, myonemes, or by compound ciliary organelles.

3. Feeding mechanisms include endocytosis, diffusion, and membrane transport. Digestion occurs in food vacuoles. Circulation is by diffusion and cytoplasmic streaming; gas exchange, osmoregulation, and excretion occur by diffusion across the cell membrane. Water expulsion vesicles expel excess water in freshwater and brackish-water species.

4. Sensory structures include the cell membrane, cilia, and flagella; some photosynthetic flagellates have a light-shading pigment spot with an underlying photosensitive zone. Behavioral responses include kineses, taxes, and simple learning.

5. Reproduction is typically asexual—by binary or multiple fission (schizogony or sporogony) or by budding. Some flagellates and amoebas are strictly asexual. In sexual species, meiosis may occur during gamete formation, just after zygote formation, or following asexual reproduction. Ciliates have separate reproductive and somatic nuclei.

FURTHER READING

Donelson, J. E., and M. J. Turner. "How the Trypanosome Changes Its Coat." *Scientific American* 252(1985): 44–51. *A lucid presentation of this important research finding.*

Farmer, J. N. *The Protozoa: Introduction to Protozoology.* St. Louis: Mosby, 1980. *Comprehensive coverage of the major groups of protozoa.*

Fenchel, T. *Ecology of Protozoa: The Biology of Free-Living Phagotrophic Protists.* Madison, Wis.: Science Tech Publishers, 1986. *Focuses on the important role of free-living protozoa in ecosystems.*

Gerster, G. "Fly of the Deadly Sleep. Tsetse." *National Geographic* 170(1986): 814–833. *Beautifully illustrated analysis of the impact of sleeping sickness in Africa.*

Godson, G. N. "Molecular Approaches to Malaria Vaccines." *Scientific American* 252(1985): 52–59. *Describes the major research advances that may lead to control of this deadly human disease.*

Jahn, T. L., E. G. Bovee, and F. F. Jahn. *How to Know the Protozoa.* 2d ed. Dubuque, Iowa: Brown, 1979. *Keys and brief descriptions of common species.*

Lee, J. J., S. H. Hutner, and E. C. Bovee, eds. *An Illustrated Guide to the Protozoa.* Society of Protozoologists. Lawrence, Kans.: Allen Press, 1985. *General introduction to all groups of protozoa written by experts.*

CLASSIFICATION

Kingdom Protista (Gr. *protistos*, "first of all"). Unicellular eukaryotes.

    Phylum Sarcomastigophora (Gr. *sarkodes*, "fleshy"; *mastigos*, "whip"; *phoros*, "bearing").

    Subphylum Mastigophora. Flagellates, including plantlike autotrophs ("phytomastigophorans"): *Ceratium, Gonyaulax,* and *Volvox;* several (e.g., *Euglena*) may be autotrophic and heterotrophic; strictly heterotrophic animal-like flagellates ("zoomastigophorans"): choanoflagellates (e.g., *Codosiga*); parasites include *Trypanosoma, Leishmania,* and *Giardia* in animals, *Phytomonas* in plants; *Trichonympha* is a cellulose-digesting mutualist in the gut of insects.

    Subphylum Opalinata (L. *opalus*, "opal"). Opalinids; common in the gut of frogs and toads; *Opalina.*

    Subphylum Sarcodina (Gr. *sarkodes*, "fleshy"; L. *ina*, "derived from"). Amoebas and amoeboflagellates.

        Superclass Rhizopodea (Gr. *rhiza*, "root"; *pous*, "foot").

        Class Lobosea (Gr. *lobos*, "lobe"). Amoebas with lobopodia:

Amoeba (free-living), *Entamoeba* (parasitic), *Naegleria* (amoeboflagellate), *Arcella*, *Chlamydophrys* (testate).

Class Granuloreticulosea (L. *granulus*, "small grain"; *rete*, "net"). Foraminiferans; amoebas with reticulopodia; mostly shelled, marine forms; many fossils; *Globigerina*.

Superclass Actinopoda (Gr. *aktinos*, "ray"; *pous*, "foot"). Amoebas with axopodia, usually with an internal skeleton of organic material, silica, or strontium sulfate.

Class Acantharea (Gr. *akantha*, "thorn"; L. *aria*, "like"). Acantharians; planktonic marine forms with a spiny strontium sulfate skeleton; *Acanthometron*.

Class Polycystinea (Gr. *poly*, "many"; *kystis*, "pouch"; L. *inus*, "like"). Radiolarians; planktonic and bottom-dwelling marine forms; most have a siliceous skeleton, some lack a skeleton; includes colonial forms (e.g., *Collozoum*, with colonies up to 10–20 cm).

Class Phaeodarea (Gr. *phaios*, "dark"; L. *aria*, "like"). Radiolarians; planktonic marine actinopods whose skeleton is composed of organic matter and silica; *Aulacantha*.

Class Heliozoea (Gr. *helios*, "sun"). Heliozoans (sun animals); common in fresh water; a few are marine, planktonic, or bottom-dwelling; *Echinosphaerium*.

Phylum Apicomplexa (L. *apex*, "point"; *com*, "together"; *plexus*, "interweaving"). Apicomplexans (perkinsids, gregarines, coccidians); parasites with apical penetration organelles; life cycles typically include asexual (schizogony, sporogony) and sexual (gametogony) phases.

Class Perkinsea. Perkinsids; parasites of oysters; life cycle does not include sexual reproduction; *Perkinsea*.

Class Sporozoea (Gr. *sporo*, "seed"; *zoon*, "animal"). Sporozoans; the zygote undergoes meiosis; sporogony forms infective sporozoites; gregarines and coccidians (*Eimeria*, *Toxoplasma*, *Plasmodium*).

Phylum Microspora (Gr. *mikros*, "small"; *sporo*, "seed"). Microsporans; intracellular parasites with no mitochondria; minute spores contain a germinal sporoplasm that is injected into a host cell through a coiled tube; *Nosema*.

Phylum Ciliophora (L. *cilium*, "hair"; Gr. *phoros*, "bearing"). Ciliates; diploid protozoa with two types of nuclei; cilia and/or compound ciliary organelles associated with a subpellicular infraciliature, which is present in all stages; mostly free-living.

Class Kinetofragminophorea (Gr. *kinetos*, "to move"; L. *fragmentum*, "a piece"; Gr. *phoros*, "bearing"). Ciliates with no compound ciliary organelles around the cytostome; *Didinium*, *Acineta* (freshwater carnivores); *Balantidium* (parasitic in the gut of pigs and humans).

Class Oligohymenophorea (Gr. *oligos*, "few"; *hymen*, "membrane"; *phoros*, "bearing"). Ciliates with compound ciliary organelles around the cytostome (often inconspicuous); *Tetrahymena*, *Paramecium*, *Vorticella* (stalked); *Ichthyophthirius* produces "ich" disease on freshwater fishes.

Class Polyhymenophorea (Gr. *poly*, "many"; *hymen*, "membrane"; *phoros*, "bearing"). Ciliates with conspicuous compound ciliary feeding and locomotor organelles (membranelles, cirri); body cilia often sparse; *Stentor*, *Euplotes*, *Tintinnopsis*.

The Evolution of Multicellularity

The success of unicellular organisms illustrates that multicellularity is not essential for survival, yet most of life's diversity is built on the multicellular plan, and only multicellular organisms have been able to adapt to and dominate dry terrestrial environments. The evolution of multicellular organisms was a major milestone in the history of life on earth.

Lacking fossil evidence, we can only postulate how multicellularity may have evolved, but two major developments seem to have been central to the process. True multicellularity does not exist without *division of labor* (interaction and interdependency) among cells. Division of labor, in turn, requires *differentiation and specialization* of cells to perform particular life functions. A cell specialized to perform only one or a few tasks is generally more efficient in its function than a cell that must do everything. No multicellular life form exists whose cells do not show some degree of differentiation, specialization, interaction, and interdependence; the more highly organized the animal, the more important these factors become.

Many zoologists postulate that the separation of sex (germ) cells from somatic cells was a crucial first step in the evolution of multicellularity. This is the most fundamental form of division of labor. Gametes are specialized for reproduction, and while developing, they are dependent on somatic cells. Because somatic cells are not involved in reproduction, they can specialize more efficiently in other activities. We saw in the last chapter that even colonial flagellates such as *Volvox* and the protozoa *(Plasmodium)* that cause malaria exhibit gamete–somatic cell differentiation. But most animals divide life's labors up more extensively: Their cells are specialized to perform one or only a few vital functions. Some cells are mainly structural and protective, others are involved primarily in feeding, others maintain internal homeostasis, and still others provide movement. Animals typically have a number of specialized cells grouped together into tissues that carry out these various functions. Cells making up tissues usually have gap junctions and other types of junctional complexes that bind adjacent cells and provide for chemical communication among them.

If separation of germ and somatic cells was the first step toward multicellularity, another early step toward interdependence and division of labor among cells undoubtedly would have been the specialization of some cells for maintaining homeostasis. An aggregation of cells in a multicellular body creates intercellular spaces, and it is highly advantageous to the cells that the fluid environment in these spaces be maintained in a balanced, steady state, distinct from the external environment. As some cells took on the task of maintaining the steady state of internal fluids, other cells would have gone on to specialize in other ways.

Several hypotheses have been proposed to account for the evolution of animals from unicellular life forms. One widely discussed idea, the **colonial hypothesis**, suggests that animals (metazoans) evolved from colonial flagellates. Several protozoan species, such as the photosynthetic flagellate *Volvox*, are colonial groups of cells with specialized eggs and flagellated sperm cells. Primitive animals might have evolved from heterotrophic flagellates with a colonial organization similar to that of *Volvox* as cell differentiation and division of labor gradually increased. The presence of motile, flagellated sperm in most animals and the occurrence of flagellated body cells in lower metazoans are cited as evidence for the colonial hypothesis.

An alternative idea, the **plasmodial hypothesis**, holds that animals arose from multinucleate ciliate protozoa that underwent internal cellularization. Many modern-day ciliates have several micronuclei that function in genetic exchange during the sexual process of conjugation. Perhaps some ancestral ciliates developed cell membranes around each of their numerous nuclei and gradually evolved into primitive multicellular forms. Ciliate micronuclei are diploid, like those of metazoan somatic cells, and this is cited as evidence supporting the plasmodial hypothesis. A tendency toward bilateral symmetry in certain ciliates may also indicate that bilateral metazoans arose via cellularization. Because ciliates lack flagellated sperm, the plasmodial hypothesis holds that differentiation of sex cells was not coincident with early multicellularization, but occurred later in metazoan evolution, after the lineage that led to modern ciliates split off. Likewise, the ciliate macronucleus, absent in metazoans, may have evolved later in the ciliate lineage.

Few biologists believe that all multicellular life evolved from a single ancestral lineage. Fundamental differences among modern multicellular life forms indicate that multicellularity evolved several times, in independent evolutionary lineages. The animal kingdom consists of several fundamentally distinct groups that probably represent independent evolutionary lineages (see Figure 18.6). As discussed in Chapter 18, radiate animals (cnidarians and ctenophorans) are fundamentally different from bilateral metazoans; thus, it is possible that radiates and bilateral animals had separate origins among protists. Perhaps radiate groups arose from colonial flagellates, while bilateral groups evolved via the plasmodial route from "cellularized ciliates."

In the next chapter we look at four phyla, collectively called the lower metazoans, that have only the simplest essentials of multicellularity. Some of the lower metozoans exist as only a small, fixed number of interdependent, highly differentiated cells. Others have large numbers of relatively undifferentiated cells. None of these animals have true tissues. Phylogenetically, the lower metazoans are often considered one or several side branches of the animal kingdom. It is likely that they arose from different protists than those that gave rise to other metazoans, either radiates or bilateral animals. Because there is virtually no evidence of affinities among any of the four lower metazoan groups, it is also possible that these animals represent four independent phylogenetic lineages.

CHAPTER 20

Lower Metazoans

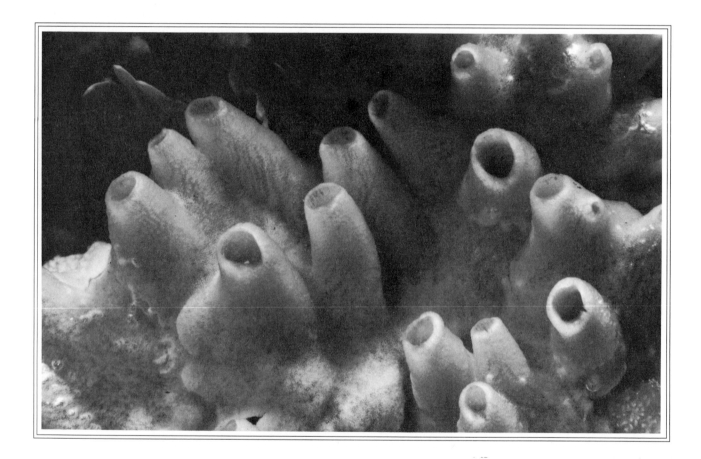

Halichondria panicea. This pastel-green breadcrumb sponge is common along the Pacific coast of the United States. It grows in large colonies that vary in shape according to the wave action in their habitat. Colonies in intertidal areas of heavy wave action are thick and stiff; those in relatively calm areas are thin and soft.

We use the term *lower* in referring to the animals in the four phyla discussed in this chapter because they have the least complex body construction of all metazoans. They consist of aggregates of cells that are more organized and differentiated than the cell groupings of colonial protists, but unlike other metazoans, they lack true tissues. Animals in these four phyla illustrate the fundamentals of multicellularity, and studying them may provide insights into how multicellular life arose.

Two of the phyla, the Myxozoa and the Mesozoa, consist entirely of parasitic species. Their adaptations to parasitic modes of life make deciphering their evolutionary background difficult. (Actually, many experts separate the Mesozoa into two phyla.) A third phylum, the Placozoa, contains only one species, a free-living marine form that may be the simplest living metazoan. The most studied group of the four is the Phylum Porifera, the sponges, sessile aquatic animals that bear little resemblance to any other metazoans.

PHYLUM MYXOZOA

Myxozoans infect fishes and a few amphibians, reptiles, and invertebrates. Of several hundred species in the phylum, only a few cause serious harm to their hosts. One species, *Myxobolus cerebralis*, causes a deadly disease of rainbow trout called whirling disease. The parasite destroys the cartilage supporting the central nervous system, causing the fish to lose nervous control of its body functions, tail-chase (whirl) spastically, and usually die. There is no cure, and whirling disease has deci-

mated trout populations in certain hatcheries throughout the world.

The myxozoan life cycle is not completely understood. In fish hosts, the myxozoan goes through a feeding stage called the **plasmodium**, which typically grows from a few micrometers to about 1 mm in length and which eventually forms many tiny **spores** (Figure 20.1). Both spores and plasmodia have separate somatic and reproductive cells, and somatic cells exhibit division of labor. Until the early 1960s, when electron microscopic studies showed that both the spores and the plasmodia were multicellular, myxozoans were thought to be unicellular, and even today, many zoologists classify them as protozoa. However, the myxozoan body organization—clusters of three to five types of cells specialized to perform different functions—is clearly metazoan. Also, some of the cells in spores are held together by junctions similar to the desmosomes between tissue cells in other metazo-

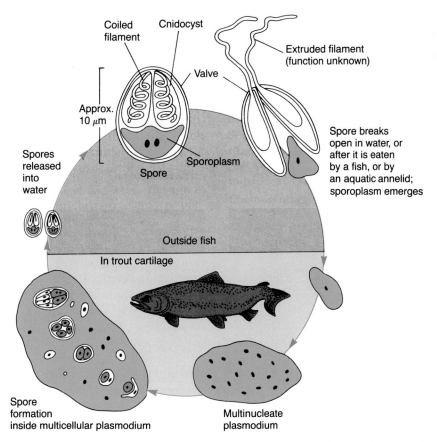

Figure 20.1 The myxozoan *Myxobolus cerebralis*. This organism is the causative agent of whirling disease in rainbow trout. Spores released into the water from an infected trout are tiny (about 10 μm long) and consist of five parts: two outside valves, two cnidocysts containing hollow coiled filaments of unknown function, and a single sporoplasm containing two nuclei. The sporoplasm infects new hosts when it is released from the spore. Recent evidence suggests that spores are infective to aquatic oligochaetes. Multicellular plasmodia grow in, and erode, cartilage in trout. Many spores develop in the plasmodium; each spore part develops from a separate cell. When an infected trout dies, spores are released into the water.

ans. Although little is known about the affinities of myxozoans with other phyla, one cell type in the myxozoan spore, the **cnidocyst**, closely resembles the cnidocytes of cnidarians (see Figure 20.1). (The cnidocytes of cnidarians are discussed in Chapter 21, p. 489.) This similarity may indicate that myxozoans are related to certain cnidarians.

Much research remains to be done on this group. Hopefully, future studies will answer questions about the life cycle and provide ways to control myxozoan diseases. The question of how the cell types in these organisms differentiate to function as simple multicellular units is basic to understanding the evolution of multicellularity.

PHYLUM PLACOZOA

The Phylum Placozoa was established in 1971 to accommodate a single marine species, *Trichoplax adhaerens*, which has puzzled zoologists since it was discovered crawling over the bottom of a saltwater aquarium in 1883. *Trichoplax* has now been cultured in marine aquariums for years. It is a flattened animal, constructed of two distinct cell regions (Figure 20.2). It moves slowly, propelled by flagella on its surface cells. Reproduction is by budding and fission, and researchers have seen

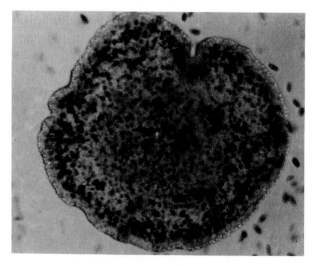

Figure 20.2 *Trichoplax adhaerens.* The only species in the Phylum Placozoa, *Trichoplax adhaerens* is small (0.5–3 mm) and has an irregular amoeboid surface. It consists of an outer covering of flagellated cells and an inner zone of loosely organized undifferentiated cells; cells on its ventral surface ingest fine food particles by endocytosis.

eggs developing in the inner cell zone. As with the Myxozoa, additional research is needed to better understand *Trichoplax* and its unique body organization. This animal illustrates certain fundamental metazoan features, and it may have much to tell us about the origin of metazoans. Its outer body layer is tissuelike, and with its separate somatic and reproductive cells, *Trichoplax* is a living model of simple multicellularity. Because *Trichoplax* is free-living, some zoologists believe that it, more than any other living animal, resembles ancestral metazoans.

PHYLUM MESOZOA

All of the approximately 50 species in the Phylum Mesozoa are parasitic in marine invertebrates. Two distinct subgroups of mesozoans are the **orthonectids**, parasites of clams, polychaete annelids, and many other invertebrates, and the **dicyemids**, inhabitants of the excretory organs of cephalopod molluscs (octopuses and cuttlefishes).

The mesozoan body consists of an outer layer of ciliated somatic cells surrounding one or more reproductive cells. Adult mesozoans are somewhat wormlike and range in size from about 0.5 mm to about 6.5 mm (Figure 20.3). Dicyemids typically have less than ten internal reproductive cells, called **axial cells**, whereas orthonectids have many. Both subgroups have sexual and asexual phases.

Many questions remain to be answered about mesozoans. Upon first infecting a host, dicyemids reproduce asexually; then they suddenly begin to reproduce sexually. Research indicates that high population density in the host or host reproductive hormones may trigger this switch, but results are not conclusive. Future research directed toward discovering the sexual trigger in these parasites could aid our understanding of how cells receive and respond to stimuli promoting their differentiation and how sex is determined at the cellular level.

Zoologists are uncertain about the phylogenetic position of mesozoans, and many doubt that the two mesozoan subgroups are even closely related. It is often suggested that mesozoans evolved their relatively simple form by degenerating from more complex ancestors, such as the flatworms (Phylum Platyhelminthes). An opposing idea is that mesozoans are living relics of an ancient group of metazoans that have not advanced beyond the cell aggregate stage of multicellularity.

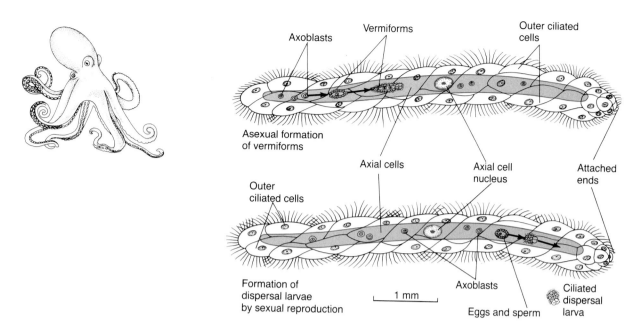

Figure 20.3 Dicyemid mesozoans. Adults consist of about 25 ciliated cells surrounding several elongated axial cells. Small somatic cells at one end of the body attach to the interior of the host's (octopus or cuttlefish) excretory organ. Adults reproduce as their axial cell nuclei divide and the daughter nuclei form special reproductive cells called axoblasts. Embryos called vermiforms, looking like tiny adults, develop asexually from axoblasts. Vermiforms emerge from the parent, remain in the host, and grow to adult size. Sexual reproduction occurs as axoblasts form eggs and sperm. Fertilization occurs in the parent, and zygotes develop into tiny, ciliated dispersal larvae. These pass out of the host in urine and may infect other hosts after a brief free-swimming period.

PHYLUM PORIFERA

Someone seeing sponges for the first time might have a difficult time deciding whether they are plants or animals; adult sponges are sedentary, and their animal-like features are not obvious. The name Porifera means "pore bearing," and sponges are porous filter feeders. They are distinguished from all other animals by having internal water chambers lined with special flagellated feeding cells called **choanocytes**, or **collar cells**.

A sponge is essentially a cluster of cells and tissuelike cell layers forming channels through which water circulates. Water and minute suspended particles are drawn in through many small pores in the body wall as a result of the beating flagella on the choanocytes. Many sponges also take advantage of external water currents to circulate water. Water circulates through the sponge, suspended particles are trapped by choanocytes, and water flows out through one or more large openings called **oscula**. Some sponges have a large internal water cavity called a **spongocoel**, but the interior of most species is a maze of small interconnecting

canals through which water flows to the oscula (see Figure 20.9).

As discussed in Chapter 8, sponges are different from most invertebrates in that they have an endoskeleton; their skeletal elements—hard siliceous or calcareous **spicules** or soft proteinaceous **spongin**, or both—are contained within the body. Besides getting support from the spicules and spongin, sponges also derive support from the water circulating through their internal canals. But they do not have a true hydrostatic skeleton because their porous body wall does not actively exert force on the water that they contain.

Most of the 4300 or so species of sponges live on the ocean floor; only about 150 species occur in freshwater streams and lakes. Sponges attach to a variety of substrates, and most species can assume different growth patterns to accommodate various types of bottom surfaces and water currents. Many marine sponges are brilliantly colored, their colors varying from bright red, orange, and yellow to various shades of blue, purple, and brown. Freshwater sponges are rather a drab brownish color, or a greenish color if they contain mutualistic algal cells.

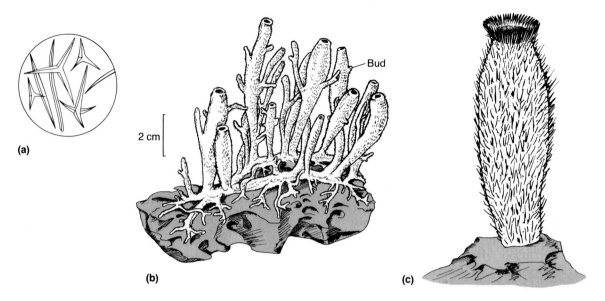

Figure 20.4 Calcarean sponges. (a) Calcium carbonate spicules, characteristic of sponges in the Class Calcarea, are unbranched or have three or four rays. (b) *Leucosolenia* attaches to rocks in shallow seas; it consists of small clusters of erect tubes interconnected at the base. (c) *Sycon* species, consisting of short (3 cm or less) vaselike individuals, also inhabit coastal waters.

REPRESENTATIVE SPONGES

There are four classes of sponges distinguished by the chemical nature and form of their internal skeletons. Members of three relatively small (i.e., consist of few species) classes, the Calcarea, the Hyalospongiae, and the Sclerospongiae, are exclusively marine. A fourth class, the Demospongiae, is by far the largest and includes marine and freshwater species.

Class Calcarea (Calcispongiae)

The **calcarean sponges** have a skeleton of generally unconnected *calcareous spicules* (Figure 20.4a). These animals are small, generally less than 10 cm high, and many exhibit radial symmetry. Most species occur in shallow marine habitats, from the intertidal zone down to about 100 m. Species of *Leucosolenia* and *Sycon*, common in coastal waters, are frequently studied in the classroom (Figure 20.4b and c).

Class Hyalospongiae (Hexactinellida)

Hyalosponges, or **hexactinellids**, are mostly deepwater forms, occurring mainly at depths from 500 m down to 5 km. They live in polar as well as tropical seas and are most common in the tropical west-ern Atlantic and near Japan and the Philippines. Most species are cylindrical or vase-shaped and range in size from a few centimeters up to about 1 m long (Figure 20.5).

The name Hyalospongiae refers to the *glassy (siliceous) spicules* characteristic of this group, and hyalosponges are often called glass sponges. The alternate class name Hexactinellida refers to the distinctive six-rayed shape of the group's main type of spicule. Most hyalosponges are not brightly colored, but they are the most elaborately structured of all sponges. Their body consists mainly of a network of fused spicules interwoven with a syncytial meshwork of amoeboid cells called **amoebocytes**. These elements form a cylinder surrounding a large central spongocoel. A **sieve plate** of fine spicules often covers the osculum. Whereas most sponges consist of three distinct cell layers (see p. 477), hyalosponges are unique in having no outer covering of flattened cells or any gelatinous middle layer. Their network of amoebocytes and fused skeletal elements take the place of these two layers. The body wall is riddled with choanocyte-lined channels that lead from the outside into the spongocoel.

Class Sclerospongiae

This small class of sponges was discovered during the 1960s in caves and crevices in Jamaican coral

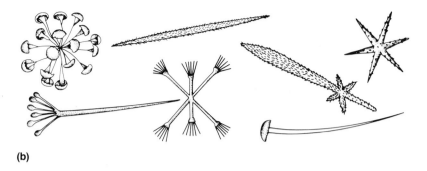

(b)

(a)

Figure 20.5 A hyalosponge. (a) *Euplectella aspergillum* (Venus's flower basket) and other members of the Class Hyalospongiae have been studied mostly from the remains of specimens dredged up from the ocean depths. Consequently, relatively little is known about the physiology of these sponges. Many, including *Euplectella,* are anchored to the sea floor by a tuft of rootlike spicules. (b) The spicules of the hyalosponges are siliceous, and many are six-rayed.

reefs. More recently, specimens have been found on reefs in tropical areas of the Pacific. **Sclerosponges** have a unique corallike skeleton consisting of a limy matrix in which siliceous spicules and spongin fibers are embedded. The living part of these sponges grows, veneerlike, over and down into irregular spaces in the skeleton.

Class Demospongiae

The **demosponges** include about 95% of all living sponges, and they are widespread in both shallow and deep seas and in both brackish and freshwater environments. Skeletons are varied in this group (Figure 20.6). Some consist of one- to four-rayed (but not six-rayed) siliceous spicules; others consist of a meshlike combination of siliceous spicules and spongin, or just spongin. A few species lack skeletal elements.

Demosponges show great variation in body form, taking advantage of available space in their habitats. Some are tubular or urn-shaped. Many others, including the freshwater sponges *Ephydatia* and *Spongilla,* and many marine demosponges, grow irregularly, encrusting substrates horizontally and vertically. Among the demosponges whose skeletons are made entirely of spongin, several marine species have been used for centuries as "bath sponges." These are collected by divers and then allowed to decompose so that only their proteinaceous skeleton remains (Figure 20.7).

Lime-boring demosponges of the genus *Cliona* create their own attachment sites by eroding and burrowing cavities in limestone. These sponges have specialized amoeboid "etching cells" that secrete an erosive substance that dissolves limestone. When an etching cell contacts a limy surface, such as a mollusc shell or a piece of coral, the cell secretes the chemical around its outer membrane. This dissolves (etches) a groove into the limestone

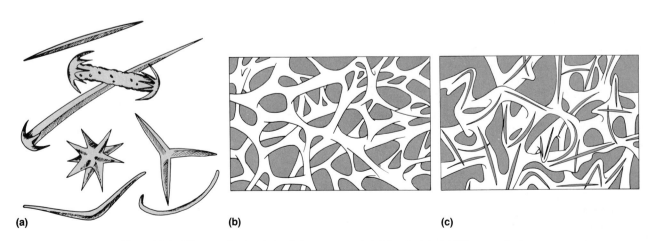

(a) **(b)** **(c)**

Figure 20.6 Microscopic structure of demosponge skeletal materials. (a) Siliceous spicules. (b) Spongin fibers. (c) Siliceous spicules embedded in spongin.

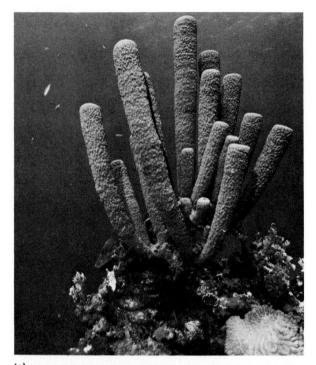

(a)

(b)

Figure 20.7 Demosponges. (a) *Verongia* is a tubular marine genus. Seen here is *Verongia lacunosa,* the purple tube sponge. Some species of *Verongia* grow to heights of about 1.5 m and often harbor dense populations of bacteria and blue-green algae, from which the sponge may obtain nutrients. (b) Skeletons of the marine "bath" sponges *Spongia* consist of only spongin fibers; more absorbent and resistant to wear than the more widely used synthetic "sponges," bath sponges are still harvested commercially and cultured in shallow tropical seas, especially the Mediterranean and Caribbean. Top (l. to r.): sheep's wool sponge; grass sponge; hard head sponge. Bottom: yellow sponge.

surface around the edge of the cell. Pseudopodia of the etching cell work into the etched groove, secreting more chemical and eventually undercutting a chip in the limestone. The chip flakes off, and the cell begins etching out another groove. Gradually, the main sponge body grows into the cavity that its

etching cells have created. Researchers have not yet identified the erosive substance, but postulate that it is either an enzyme, perhaps carbonic anhydrase, or an acid.

HABITATS AND ENVIRONMENTAL RELATIONSHIPS

Sponges have been prevalent in marine environments since Cambrian times, some 600 million years ago. Today, they are especially common in coastal seas, and many tropical species thrive on coral reefs. The distribution, abundance, and size of sponges with hard skeletons may be limited by the amounts of silicon or calcium in the environment. The freshwater species *Ephydatia muelleri* requires at least 1.6 mg of silicon dioxide per milliliter of surrounding water, yet other freshwater sponges seem to be less demanding and may simply develop less skeletal material in mineral-poor waters.

Wherever they live, sponges provide habitats for a wide variety of other animals. Small shrimps, barnacles, polychaete annelids, certain molluscs, and small fishes frequently live on or in the porous, water-filled bodies of sponges. Despite their hard spicules, sponges are eaten by several types of animals. A number of coral reef fishes feed extensively on them, as do several marine molluscs (sea slugs and limpets), a few polychaete annelids, and certain sea stars and sea urchins. The aquatic larvae of one family of insects, the spongilla flies, live in and on freshwater sponges, feeding more like parasites than predators by perforating cells and sucking out cytoplasm.

Photosynthetic algae and bacteria frequently grow between the cells of sponges. Little is known about the symbiotic relationship between these organisms and their sponge hosts, but the bacteria can be seen to multiply, are regularly engulfed, and are probably digested as food by sponge amoebocytes. Freshwater sponges commonly harbor green algal cells, whereas many marine sponges host blue-green algae. Studies of sponge populations on the Great Barrier Reef off Australia have shown that except for corals, sponges containing photosynthetic blue-green algae are the most prevalent organisms attached to the reef. The success of these sponges may be due to their ability to use nutrients produced by their algal symbionts. Sponges found on coral reefs typically have a thin, flattened body form, which provides greater surface area and greater exposure to light for algal photosynthesis and also makes the sponge less vulnerable to wave action.

Lime-boring sponges play a significant role in recycling calcium in the sea by helping to decompose mollusc shells, coral limestone, and deposits of certain calcareous algae. Unfortunately, the sponges also bore into the shells of live animals and may cause heavy losses of commercially important molluscs such as abalone, oysters, and scallops. The activity of boring sponges may also weaken sections of coral growth and promote underwater avalanches.

COMPARATIVE STRUCTURE AND FUNCTION

Sponges are composed of rather loosely organized cells and cell layers showing limited differentiation and division of labor. Their cells function more independently of one another than do the cells that form tissues in most metazoans.

Body Structure, Support, and Protection

Most sponges (except hyalosponges) have three tissuelike cell layers (see Figure 20.9). A skinlike layer, called the **pinacoderm**, formed of a single layer of cells called **pinacocytes**, covers most species. In some sponges, the pinacoderm is syncytial (con-

sists of a large multinucleate cell). A second tissuelike layer, called the **choanoderm** and formed of a single layer of choanocytes, lines certain internal water chambers. Internal areas that are not lined by choanocytes are lined by pinacoderm. The third layer, called the **mesohyl**, lies between the pinacoderm and the choanoderm. The mesohyl is a gelatinous matrix containing fibrous proteins and/ or spicules interspersed with a variety of amoeboid cells.

The sponge mesohyl is more like a region of loose cells and cell secretions than an organized layer. Its numerous wandering amoebocytes serve an amazing array of functions. Certain amoebocytes, called **spongocytes**, secrete spongin fibers, and others, called **sclerocytes**, secrete spicules (Figure 20.8). Undifferentiated amoebocytes called **archeocytes** build and repair the sponge body by differentiating into pinacocytes and choanocytes as needed. They may also protect the sponge by engulfing and disposing of any nonfood particles that enter the sponge and tend to clog its water canals. Amoebocytes can also become a special type of contractile cell, the **myocyte**; a cluster of myocytes often forms a smooth musclelike band around the osculum or around pores in the body wall. Located under the pinacoderm, the myocytes can regulate oscular or pore diameter and thus help regulate the flow of water through the sponge.

Sponge spicules, especially siliceous ones, provide protection as well as support. They often project from the sponge body, providing a spiny, predator-repellent body surface. Many sponges also produce certain organic chemicals (terpenoids and

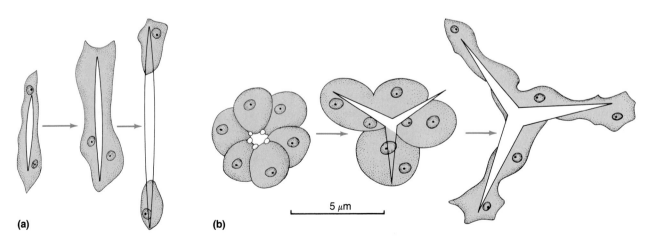

Figure 20.8 Spicule formation by sclerocytes. (a) A single-rayed spicule forms along a spindle fiber as a single sclerocyte divides. Eventually, the two sclerocytes disintegrate, leaving the spicule. (b) Multirayed spicules form in a similar way, but result from the coordinated activities of groups of sclerocytes.

benzoquinones) that are repellent or toxic to predators. These substances may be important in protecting the more vulnerable calcareous and soft sponges. Sponges are also known to secrete several bromine-containing organic chemicals that inhibit growth of neighboring corals and other sessile invertebrates, allowing the sponges to compete successfully for growing space.

Body Form in Relation to Water Circulation

Sponges have three basic body forms, which differ in the way their water-filtering system is organized and consequently in their efficiency of filter feeding. The most energetically efficient sponge bodies provide maximal contact between nutrient-laden water and choanocytes.

The simplest sponges, called **asconoids**, consist of clusters of small, upright, vaselike tubes (Figure 20.9a). The pinacoderm of these sponges is perforated by pores, each formed by a tubular cell called a **porocyte**. Water flows through the pores, into the central spongocoel lined by choanocytes, and out the osculum. Asconoid sponges are inefficient water filters; because of their body form, relatively little of the water in the large spongocoel is able to contact, and be filtered of its food particles by, the choanocytes. It has been suggested that this lack of efficiency is the reason the asconoid system occurs in only a few small species.

The other two structural types of sponges, called syconoids and leuconoids, have folded bodies that provide much greater surface area for water to contact the choanocytes. The spongocoel is small relative to the total cross-sectional area of the sponge (or even absent) and lined with pinacoderm (or amoebocytes in hyalosponges) instead of choanocytes. Water filtering is more efficient than in asconoid sponges because the choanocytes occur in many small chambers. This increases the choanocyte surface area in contact with water and slows the flow of water, allowing more nutrients to be filtered. In **syconoid** sponges, the body wall is uniformly folded, forming many separate choanocyte chambers called **radial canals** (Figure 20.9b). Water is drawn into each radial canal through pores or through a system of **incurrent canals** lined with pinacoderm.

The **leuconoid** form, the most complex and most efficient of the three, has an intricately folded body wall. Leuconoid sponges are essentially networks of water canals and many tiny choanocyte chambers (Figure 20.9c). The spongocoel does not exist as a single chamber but is subdivided into many incurrent and excurrent canals that conduct water to and from numerous choanocyte chambers. Leaving the choanocyte chambers, many small excurrent canals converge into one large one that conveys water to an osculum. Leuconoid sponges usually have several oscula (see the inset in Figure 20.9c).

Feeding and Digestion

Sponges do not have a digestive system, and food particles are ingested and digested by individual cells. In syconoid and leuconoid sponges, pinacocytes lining the internal water passages may engulf larger food particles (in the range of 1–50 μm), but the bulk of the food consists of minute (less than 1 μm) organic particles that are taken up by choanocytes. Bacteria often make up the bulk of the sponge diet, and organic chemicals produced by sponges have an antibiotic effect that may assist in filter feeding. These chemicals kill bacteria and induce bacterial clumping, allowing sponges to feed more effectively by trapping the bacteria in clusters.

Large marine sponges may filter and circulate enormous volumes of water. Studies of coral reefs have shown that along with coral animals, sponges can extract most of the suspended matter from seawater passing over them. Thus, sponges may contribute to making seawater between coral reefs and land less turbid than water at the outer edges of a reef.

The sponge choanocyte has three main components, each with a different function related to feeding (Figure 20.10). The first is a flagellum that propels nutrient-laden water through the sponge. The second is a mucus-coated collar around the flagellum, consisting of microvilli (fine cytoplasmic extensions of the choanocyte) and interconnecting microfilaments, forming a net with a mesh size of about 0.1 μm. Fine particles from this size up to about 1 μm may be trapped on the collar as water filters through it. The third component of the choanocyte is its cell body, which engulfs the food particles trapped by the net.

Experimental evidence shows that choanocytes do not digest the particles they engulf. Instead, food vacuoles move to the base of the choanocytes and pass their contents to amoebocytes in the mesohyl. It is possible to monitor intracellular digestion by using bacteria labeled with fluorescent dyes. When sponges are fed labeled bacteria and then killed at different time intervals after feeding, the appearance of fluorescence in the sponge cells

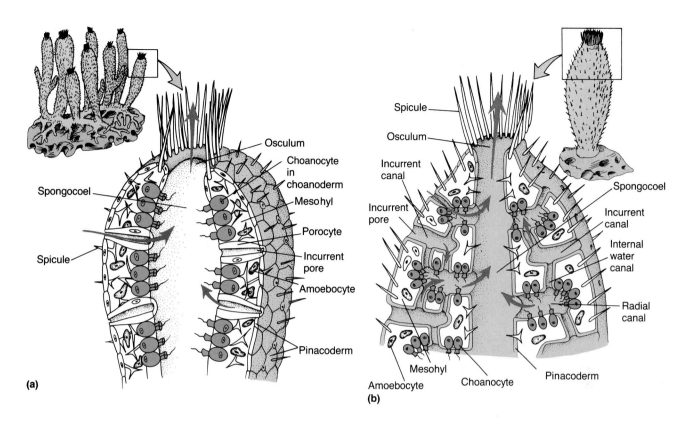

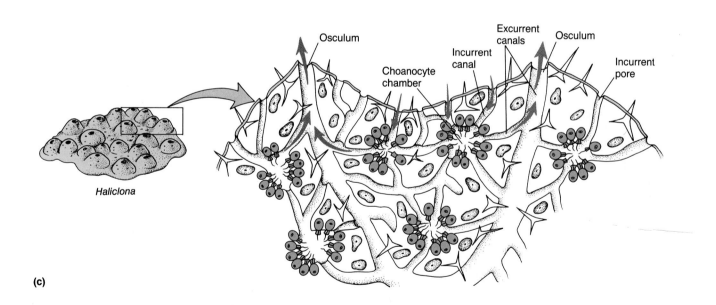

Figure 20.9 Sponge body forms. These cutaway drawings show the three body layers (pinacoderm, mesohyl, and choanoderm) and how these layers are arranged in the three basic body forms of sponges. Arrows indicate the flow of water through the sponge. (a) The asconoid form occurs in only a few small, calcarean sponges, including *Leucosolenia;* choanoderm lines the single large spongocoel. (b) The syconoid body form, seen in *Sycon* and many glass sponges, has a body wall containing pouchlike choanocyte chambers called radial canals. (c) The leuconoid form consists of a meshwork of choanocyte chambers with no spongocoel. Certain calcarean sponges and all demosponges are of this type. Notice that in syconoid and leuconoid sponges, internal water canals are lined by pinacoderm.

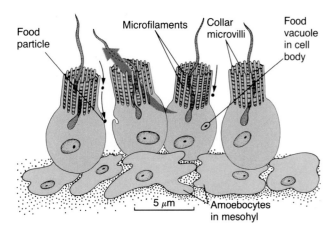

Figure 20.10 Choanocyte structure and function. Part of a choanoderm layer with four choanocytes. (The arrow indicates the direction of water current.) Notice the close association between choanocytes and amoebocytes in the underlying mesohyl; food particles trapped by choanocytes are digested by amoebocytes.

indicates what has happened to the bacteria. In the freshwater demosponge *Ephydatia fluviatilis*, undigested bacteria were moved from choanocytes to amoebocytes about 30 minutes after being filtered from the water current. About 4–5 hours later, digestion appeared to be complete, and fluorescing nutrients were dispersed throughout the sponge. Most, if not all, digestion occurred within the amoebocytes.

Other Life Functions

Most vital functions within sponges are performed by amoebocytes. These cells transport digested nutrients to other body cells. Because large volumes of water are constantly circulating through sponges, and because most of the cells are adjacent or close to water currents, wastes can diffuse from the cells and dissolved gases can be exchanged with little energy expense. Like freshwater protozoa, freshwater sponge cells have water expulsion vesicles that remove excess water. The leuconoid body form provides the greatest contact between cells and circulating water; thus, it is not surprising that the largest sponges have this body form.

Nervous Responses and Behavior

Sponges do not have specialized nerve cells or sensory structures, and behavior in adult sponges, like other vital functions, is largely a cellular phenomenon. In some species, pinacoderm cells respond to tactile and electrical stimuli, and when stimulated experimentally, these cells can conduct impulses to other regions of the sponge body. The velocity of impulse transmission is very slow compared with typical nerve impulse conduction, but it seems to provide a mechanism of simple response to external stimuli. Impulses may be relayed to effectors, such as myocytes or porocytes, which may regulate water flow in or out of the sponge. Choanocytes may also act as effectors. Certain demosponges and hyalosponges can increase or decrease water flow through their bodies by changing their rate of choanocyte flagellar beating. Choanocyte activity and water filtering may be arrested by touching or applying electric shock to the surface of these sponges.

Sponge larvae are free-swimming (see Figure 20.11), and their behavior significantly influences adult distribution and ecology. Swimming larvae are attracted to light and respond negatively to gravity. These responses are reversed as larvae settle to the bottom prior to attaching and developing into adults. What triggers these vital behavioral changes is not yet known. Other important questions awaiting answers are: How do larvae select certain types of water currents and substrates for attachment? And how is larval behavior mediated (for, like adults, larvae lack nerve cells)?

Reproduction and Life Cycles

All forms of reproduction in sponges involve archeocytes, or undifferentiated amoebocytes.

Asexual Reproduction

Budding occurs in some sponges as aggregates of archeocytes surrounding skeletal elements develop on the body surface. After reaching a certain size, the buds break free, attach to the substrate, and develop into adults. Colonies also form by budding: In colonial species, buds remain attached to, and grow with, the parent body. More common than budding is the formation of **reduction bodies**. These are aggregates of archeocytes that remain after a parent sponge disintegrates under unfavorable conditions. A single parent may leave many reduction bodies, each of which can grow into a new sponge when favorable conditions return.

Freshwater sponges and certain marine species form internal reproductive bodies called **gemmules**. These are small (up to 1 mm in diameter) aggregates of yolk-laden archeocytes covered by a thick protective layer of dead cells and spicules (see Figure 20.12). Gemmules may be produced throughout the growing season, but are often held within the parent sponge until the parent body disintegrates at the end of a growing season. Chemical inhibitors, photoperiod, or moderate temperatures may prevent earlier gemmule release. Gemmules of temperate-zone freshwater sponges can remain dormant and survive repeated freezing, thawing, and desiccation. When favorable conditions return, the mass of archeocytes emerges and develops into a new sponge. Of special interest to developmental biologists and endocrinologists is a cyclic AMP system that seems to be involved in regulating gemmule dormancy and germination. Levels of cAMP are high in dormant gemmules, but drop dramatically just prior to germination. (Cyclic AMP is discussed in Chapter 11, p. 247.)

Sponges have great powers of regeneration and often reproduce by this means. A minimum number of cells is usually required, but amoebocytes and even some pinacocytes and choanocytes of most species can reaggregate and form new sponges if they are separated experimentally. In nature, small pieces broken from parent sponges may form whole new individuals, and bath sponges are cultured from pieces cut from large individuals. Sponges living on coral reefs are often broken up by storms and undersea avalanches. After such events, fragment regeneration may account for significant sponge reproduction.

Sponges are excellent models for the study of basic cell aggregation and tissue-grafting mechanisms. Experiments in which separated cells of two sponge species are mixed have shown that the cells of each species can recognize and will reaggregate with one another, but not with those of the other species. Each species seems to have its own chemical recognition factor—a specific glycoprotein on its cell surfaces. Similar substances serve as identity factors on vertebrate cells.

Sexual Reproduction

The majority of sponges are hermaphroditic. Sexual reproduction generally occurs during environmentally favorable periods, while the sponges are growing. Gametes develop from archeocytes in the mesohyl or from choanocytes. Each individual produces sperm and eggs at different times, ensuring cross-fertilization. Sperm released into water currents passing out of one sponge are drawn into neighboring sponges, where they are ingested by choanocytes. Sperm-carrying choanocytes then lose their collar and enter the mesohyl as amoeboid cells; they carry sperm to the eggs, and fertilization occurs in the mesohyl. Zygotes develop into flagellated larvae.

Development and Life Cycles

Sponge development is unique. Most demosponges produce a larva called a **parenchymula**, which consists of an aggregate of archeocytes covered by flagellated cells (Figure 20.11a). The parenchymula resembles the stereoblastula (solid blastula) of many other metazoans. (The stereoblastula is dis-

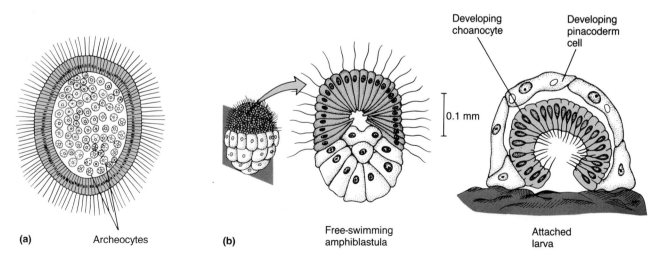

(a) Archeocytes

(b) Free-swimming amphiblastula

Developing choanocyte

Developing pinacoderm cell

0.1 mm

Attached larva

Figure 20.11 Sponge larvae. (a) The parenchymula of a demosponge. Its outer flagellated cells turn inward and become choanocytes. (b) An amphiblastula of a calcarean sponge, seen in surface view and in longitudinal section. This larva turns inside out as it attaches to the substrate and begins to develop into an adult sponge.

cussed in Chapter 16, p. 379.) As a parenchymula settles to the bottom and develops into an adult sponge, its outer flagellated cells turn inward and form choanocytes, while its archeocytes move outward and become the pinacoderm. This is different from the typical metazoan pattern in which the epidermis forms from the external cells of the blastula.

The larva produced by most calcarean sponges is a hollow **amphiblastula** with micromeres (small cells) bearing flagella (Figure 20.11b). The amphiblastula resembles the coeloblastula of other metazoans. (The coeloblastula is discussed in Chapter 16, p. 378.) In a pattern unique among metazoans, however, the macromeres of the amphiblastula grow around the flagellated micromeres, the latter becoming the choanocytes of the adult sponge.

Sponges are sessile as adults, but are dispersed to new habitats as gemmules, reduction bodies, or as larvae. The life cycle of the common demosponge *Spongilla lacustris* illustrates how this species is adapted to life in temperate freshwater environments (Figure 20.12).

PHYLOGENY

The Phylum Porifera is ancient. Fossil sponge spicules have been found in Precambrian rocks more than 600 million years old. The modern classes of sponges are well represented in the fossil record from the Cambrian, and it is evident that the main evolutionary lines within the phylum arose during Precambrian times. The antiquity of sponges, their unique embryology, and their pretissue level of body organization indicate that the sponge side

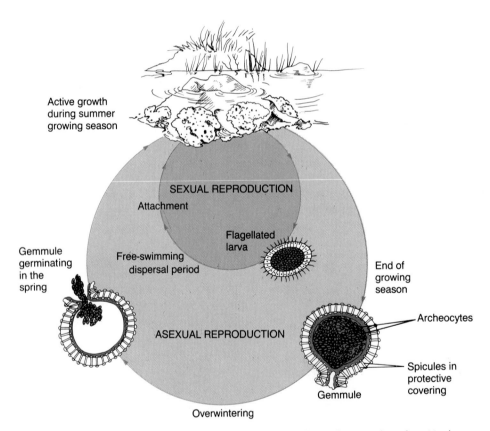

Figure 20.12 Life cycle of *Spongilla lacustris*. Common in lakes and streams throughout North America, this species reproduces sexually during warm summer months. The resulting flagellated larvae (about 0.2 mm in diameter) settle to the bottom and attach after a free-swimming dispersal period of several hours or days. Gemmules (about 0.5 mm in diameter) are released as overwintering bodies in the autumn. Each gemmule's internal mass of archeocytes develops into a new sponge after emerging in the spring. Only superficial parts of the parent body usually disintegrate at the end of a growing season. The rest of the sponge remains attached to the substrate and lies dormant, beginning to grow again when favorable conditions return. Thus, this species usually does not have to completely recolonize a suitable habitat at the onset of each growing season.

branch of the animal kingdom represents a unique evolutionary venture in multicellular body organization.

Sponge choanocytes are similar to the protozoan choanoflagellates, and many zoologists think that colonial protists similar to these flagellates were ancestral to sponges. (Choanoflagellates are discussed in Chapter 19, p. 449; Figure 19.3b.) Sponges resemble protozoa physiologically in that most of their vital functions are performed by individual cells. Their flagellated body cells and larvae also resemble colonial flagellates. However, sponges also show rudiments of division of labor and differentiation among cells. Specialized cells (choanocytes) have developed for feeding, and pinacocytes form the body covering. Nevertheless, most cells in sponges are relatively undifferentiated, and even choanocytes and pinacocytes do not become so highly differentiated that they lose their capacity to carry out diverse functions and reaggregate into new sponges if separated.

Sponges seem to have evolved mainly in the direction of increasingly more efficient water-filtering systems. With their simple body construction, the modern asconoids may be most similar to the earliest sponges. The most highly evolved sponges are the leuconoids, with complex, highly efficient water canals and numerous choanocyte-lined water chambers. Some leuconoid sponges have asconoid-like and syconoid-like developmental stages, and many zoologists believe that the leuconoid body form arose from simpler, less efficient asconoid and syconoid types. Evidence also suggests that the leuconoid condition arose more than once, that is, independently in several sponge lineages. By far, the greatest number of sponge species are leuconoids, a good indication of the relative success of this body plan.

☐ SUMMARY

1. Key Features of Lower Metazoans. Four phyla of animals exhibit multicellularity at the cell aggregate level of body organization. Their degree of complexity is between that of colonial protozoa and that of metazoans with true tissues. There is rudimentary cell differentiation and division of labor with separate somatic and reproductive cells, but there are no tissues or organs. Vital functions are carried out by individual cells. These phyla constitute evolutionary side branches of the animal kingdom; their affinities with one another and with other animals are obscure.

2. Members of the Phylum Myxozoa are parasitic with multicellular plasmodia and spores. The Phylum Pla-

cozoa consists of one free-living species, a flattened amoeboid aggregate of two cell layers. The Phylum Mesozoa is made up of parasites of marine invertebrates; mesozoans consist of an outer layer of ciliated cells surrounding one or more reproductive cells.

3. Members of the Phylum Porifera (sponges) are free-living, sessile, mostly marine filter feeders. Most have an outer covering (pinacoderm), a middle mesohyl with amoeboid cells and skeletal elements (hard spicules or leathery spongin), and one or many inner water chambers lined by collar cells (choanocytes). Flagella on choanocytes drive water through the sponge body.

4. Sponges have three body forms: asconoid types are vaselike porous sacs internally lined with choanocytes; syconoid types have a folded body wall containing separate choanocyte-lined water pouches; leuconoid types, the most common, have many separate choanocyte chambers.

5. Choanocytes filter food particles from water circulating through sponges; most other functions are performed by amoebocytes. Freshwater sponge cells have water expulsion vesicles that expel excess water.

6. Sponges lack nerve cells and sensory structures, but impulses conducted by body cells may effect changes in choanocyte activity and in water pore size. Flagellated larvae of sponges show various responses to light and gravity.

7. Sponges reproduce asexually by budding, by fragmentation and regeneration, and by formation of resistant stages (reduction bodies and gemmules). Most sponges are hermaphroditic; gametes are produced by special amoebocytes called archeocytes. Sponge embryology is unlike that of any other animal.

☐ FURTHER READING

Barnes, R. D. *Invertebrate Zoology.* 5th ed. Philadelphia: CBS College Publishing, 1987, Chap. 4. *A modern textbook account of the sponges; includes references to original research literature.*

Bergquist, P. R. *Sponges.* London: Hutchinson & Co., 1978. *A comprehensive account of modern sponge biology; includes some excellent electron micrographs.*

Lapan, E. A., and H. Morowitz. "The Mesozoa." *Scientific American* 227(1972): 94–101. *Lucid drawings and description of the dicyemid reproductive cycle and its potential for experimental research in developmental biology.*

Long, M. E. "Consider the Sponge." *National Geographic* 151(1977): 392–407. *Describes sponge diversity on a Caribbean coral reef; excellent color photographs.*

Pennak, R. W. *Freshwater Invertebrates of the United States.* 2d ed. New York: Wiley, 1978, Chap. 3. *Biology and classification of freshwater sponges.*

Simpson, T. L. *Cell Biology of Sponges.* New York: Springer-Verlag, 1984. *A modern, authoritative volume on poriferan biology.*

▢ CLASSIFICATION

Phylum Myxozoa (Gr. *myxa*, "mucus, slime"; *zoon*, "animal"). Myxozoans; parasites of fishes, reptiles, amphibians, and a few invertebrates; *Myxobolus*.

Phylum Placozoa (Gr. *plax*, "plate"; *zoon*, "animal"). One species: *Trichoplax adherens;* free-living, marine.

Phylum Mesozoa (Gr. *mesos*, "middle"; *zoon*, "animal"). Mesozoans (dicyemids and orthonectids); parasites of marine invertebrates; *Dicyema* infects octopuses and cuttlefishes.

Phylum Porifera (L. *porus*, "pore"; *fera*, "bearing"). Sponges.
  Class Calcarea (L. *calcis*, "lime"; Gr. *aria*, "like"). Calcarean sponges; marine asconoid, syconoid, and leuconoid species with one-, three- or four-rayed limy spicules; *Leucosolenia, Sycon.*

Class Hyalospongiae (Gr. *hyaleos*, "glassy"; *spongos*, "sponge"). Glass sponges; marine syconoid and leuconoid sponges with ornate, siliceous skeletons; six-rayed spicules, often fused to form a radially symmetrical latticelike skeleton; no pinacoderm or mesohyl; *Euplectella, Hyalonema.*

Class Sclerospongiae (Gr. *skleros*, "hard"; *spongos*, "sponge"). Corallike sponges; marine leuconoid sponges with a corallike limy covering; *Astrosclera.*

Class Demospongiae (Gr. *demos*, "people"; *spongos*, "sponge"). Demosponges; marine and freshwater leuconoid sponges with two- or four-rayed siliceous spicules and /or spongin; many marine genera, including: *Cliona* (lime-boring sponge), *Haliclona, Halichondria, Spongia* (bath sponges); freshwater: *Spongilla, Ephydatia.*

Symmetry and the Origin of Bilateral Animals

The concept of body symmetry was introduced briefly in Chapter 1, but this is such a fundamental aspect of animal architecture that we need to return to it as we begin our discussion of metazoans with true tissues. Recall that few animals are perfectly symmetrical, but an animal has symmetry if it can be divided into approximately mirror-image halves by one or more planes. A **radially symmetrical** animal is star-shaped, pie-shaped or cylindrical, with its body parts radiating from a central axis; it can be divided into similar halves by any plane passing longitudinally through its central axis (Figure Aa). A radial body, such as that of a sea star, hydra, or jellyfish, is adaptive for organisms that dwell on lake or ocean bottoms or that drift about in water currents. Radial animals can sense changes in the environment from all sides and can respond to stimuli from any direction.

A second type of symmetry, actually a specialized form of radial symmetry, is **biradial symmetry**. A biradial animal, such as a sea anemone, has radiating parts, but also has certain body structures that are not radially arranged, so that only two planes (perpendicular to each other) passing longitudinally through the animal's central axis divide the animal into mirror-image halves (Figure Ab).

Bilateral symmetry is the most common type of symmetry in the animal kingdom. A bilateral animal has mirror-image right and left sides and an anterior and posterior end. It can be divided equally by only a single longitudinal (midsagittal) plane passing from anterior to posterior and dorsal to ventral through the longitudinal axis (Figure Ac). The body parts of bilateral animals are adapted for moving head first through the environment. Accordingly, sensory and nervous structures are typically concentrated in the head, where they can detect and integrate environmental stimuli when the animal first confronts them. Feeding structures are also located on the head—in position to capture prey as the animal senses or pursues it.

If we consider body symmetry in a phylogenetic context, it is evident that the various forms of symmetry have arisen independently several times in different lineages. Among the protozoa, for example, many species are asymmetrical; others are radial, biradial, or bilateral; and still others, such as *Volvox*, are spherical. (A spherical organism has **universal symmetry**, for it is divided equally by *any* plane passing

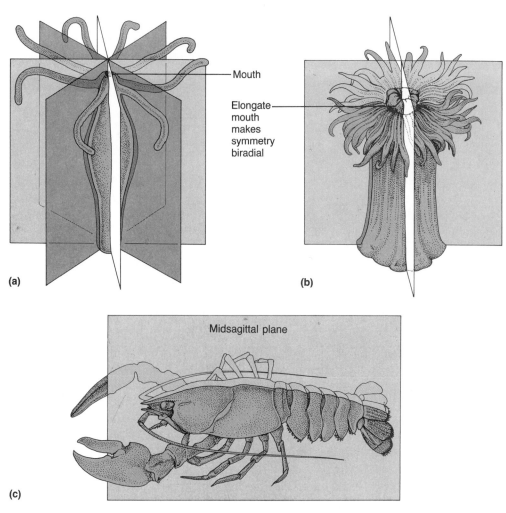

Mouth

Elongate mouth makes symmetry biradial

(a)

(b)

Midsagittal plane

(c)

Figure A Comparison of three types of animal symmetry. (a) Hydras are radially symmetrical. (b) Because of its elongate mouth, a sea anemone is biradially symmetrical. (c) Most animals, including this crayfish, are bilaterally symmetrical.

through its center.) Because it is unlikely that any of the lineages that gave rise to modern protozoan phyla also gave rise to animal phyla (see Figure 18.6), protozoan symmetry no doubt evolved independently of animal symmetry. Then, too, the lower metazoans discussed in Chapter 20 may be asymmetrical, radial, or bilateral. We have seen how different these animals are from the rest of the animal kingdom, and there is little doubt that their symmetry evolved independently of that of other animals.

As discussed in Chapter 18, most of the animal kingdom, namely, the Subkingdom Eumetazoa, consists of two large groups, the **Branch Radiata** and the **Branch Bilateria**, named according to the type of symmetry displayed. The Radiata, consisting of two phyla (Cnidaria and Ctenophora) of radial or biradial animals, show no evidence of bilateral symmetry. The Bilateria (bilateral eumetazoans) contains all other animals, and the origin of the bilaterian lineage was one of the great milestones in animal evolution.

What could the earliest animals in the Branch Bilateria have been like? Zoologists try to answer this question by first looking at living members of this huge lineage to determine which body features are common to all. It is likely that such universal features are ancestral in the lineage. Whether simple or complex, all bilaterians are bilateral, as least as larvae, and have a definite head-to-tail orientation. They also have true tissues, and their tissues develop from three germ layers: Ectoderm forms the epidermis and nervous system; endomesoderm (mesoderm that develops from endodermal cells) forms most internal tissues and organs, and endoderm forms the layer of cells lining the gut or a mass of internal cells involved mainly in ingesting and/or digesting food. Organs and organ systems (e.g., muscular, circulatory, gas exchange, and digestive systems) occur in most, but not all, bilaterians.

From these facts, it is possible to construct a hypothetical model of the earliest bilaterians. As indicated by Figure B, such animals were probably free-swimming or crawling, because their head-to-tail orientation would have been adaptive for forward movement. They probably lacked muscles, but were no doubt small and may have moved by means of external cilia. The body was most likely composed of three layers or regions, but cells in the regions may not have been highly differentiated, and the body was undoubtedly thin, allowing nutrients, gases, and wastes to diffuse to and from all cells. Also, there was probably no internal digestive cavity or body cavity. The anterior region may have housed some sensory cells and the components of a primitive nervous system, but otherwise was probably not highly differentiated.

This model, while hypothetical, represents a "best guess," based on available evidence, of a prototype bilaterian. As such, its basic traits, common to animals in several lineages, are considered *ancestral*, or *primitive*, to all bilaterian lineages. Prototypes are useful, for they establish hypothetical starting points that zoologists can use to compare divergent evolutionary trends between different lineages of animals.

The next chapter looks at the two phyla making up the Branch Radiata. Although the phylogeny of the Radiata is controversial, most zoologists believe that these animals were not derived from a bilateral ancestor and that their radial symmetry is primitive. In contrast, radial symmetry in an animal group such as the Phylum Echinodermata, in the Branch Bilateria, is undoubtedly a *derived trait*, characteristic of the echinoderm lineage. The remaining twelve chapters in this unit examine the diverse lineages of the Branch Bilateria.

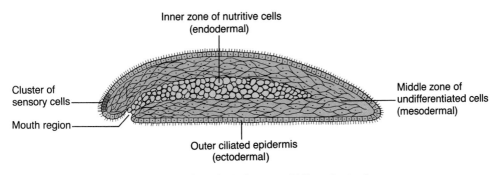

Figure B A hypothetical ancestral bilateral animal.

Radiate Animals

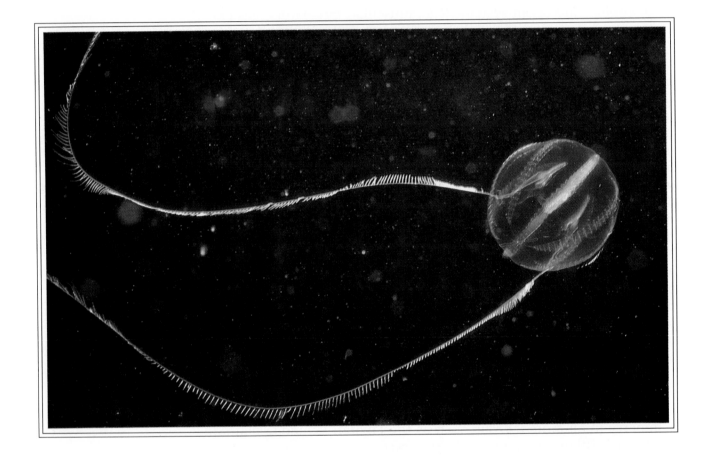

Two phyla of aquatic metazoans, the Cnidaria (nettle or stinging animals) and the Ctenophora (comb jellies), are called *radiate animals* because they are radially or biradially symmetrical. Cnidarians, including jellyfish, hydroids, sea anemones, and corals, are some of the most beautiful and ecologically important animals in marine and fresh waters.

Ctenophorans are much less numerous, and all are marine. The way in which all these animals move, capture food, and interact with their environment is strongly influenced by their body symmetry.

PHYLUM CNIDARIA

Nearly all of the approximately 11,000 living species of cnidarians are marine. They occur in all oceans, but abound in shallow temperate and trop-

---

A radiate animal. The sea gooseberry *Pleurobrachia,* a member of the Phylum Ctenophora (the comb jellies), is 2–3 cm in diameter. Sea gooseberries are planktonic and are found mainly in the northern Atlantic and Pacific oceans. Shown here in Atlantic waters is *Pleurobrachia pileus.*

ical areas. In fresh water, the phylum is represented by fewer than 50 species, including the familiar hydras of lakes and ponds and a few uncommon jellyfish.

Cnidarians have several distinctive features aside from their radial or biradial symmetry. The name Cnidaria refers to specialized cells called **cnidocytes**, present in all members of the phylum. A cnidocyte contains one or more eversible organelles called **cnidae**. Stinging cnidae, which can be discharged in defense or to help capture food and subdue prey, are called **nematocysts** (see Figure 21.11); other cnidae called **spirocysts** discharge adhesive threads that capture prey or adhere to substrates. Many cnidarians have two adult body forms: a cylindrical, bottom-dwelling **polyp** (as seen in sea anemones, corals, and hydras) and a bell- or umbrella-like free-swimming **medusa**, also called a jellyfish (Figure 21.1). Polyps reproduce asexually and sometimes sexually, and many are colonial. They are often attached to the substrate by a **pedal disc**. Medusae are usually sexual. Both forms have a ring of food-capturing **tentacles** surrounding their mouth. In some medusae, the mouth is at the end of a tube called the **manubrium**. The mouth leads into a **gastrovascular cavity**, so named because it serves both digestive and circulatory functions. The cnidarian gut is incomplete (i.e., it has no anus). As radially symmetrical animals, cnidarians do not have an anterior or posterior end. The surface bearing the mouth is called the **oral** surface; the opposite, the **aboral** surface.

The cnidarian body consists of three layers: an outer **epidermis**, a middle **mesenchyme**, and an inner **gastrodermis** lining the gastrovascular cavity. The epidermis and gastrodermis are epithelial tissue layers, each consisting of several types of differentiated cells among which there is considerable division of labor. The mesenchyme contains a high proportion of cell secretions (collagen fibers) that provide an elastic skeleton-like matrix called the **mesoglea**, in which epidermal and gastrodermal cells are anchored. The mesenchyme is an important third body layer, but it varies markedly in structure and function in different cnidarians. In hydras, the mesenchyme is essentially mesoglea, that is, a layer of noncellular material secreted by cells in the epidermis or gastrodermis; its only cells are a few wandering amoebocytes. In other cnidarians, most notably the sea anemones, the mesenchyme is highly developed with many different cell types.

Cnidarians are more diverse in body form and occupy a wider range of habitats than sponges or any other lower metazoans. Unlike the sessile filter-feeding sponges, many cnidarians are active, mobile predators. Also, because they have true tissues, cnidarians are structurally more advanced than lower metazoans. They are often referred to as tissue-level animals, because their vital functions

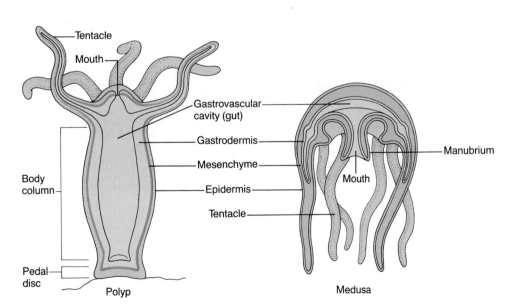

Figure 21.1 Cnidarian body forms. Jellyfish tentacles hang down in the water from the lower edge of the bell-shaped body, whereas the mouth and ring of tentacles are on the upper surface of the polyp. If the medusa settled upside down on the ocean floor and stretched its tentacles upward, it would resemble a polyp. Note the three body layers: epidermis, mesenchyme, and gastrodermis. The gastrodermis lines the saclike gastrovascular cavity.

are carried out mainly by tissues rather than by organs or organ systems. However, as you will see, this is an oversimplification, for cnidarians do possess certain organs and a nervous system as well. As a group, they are perhaps best viewed as structurally (but not phylogenetically) intermediate between sponges (and other cell aggregate, nontissue animals) and the more highly structured bilateral animals.

REPRESENTATIVE CNIDARIANS

Of four classes of cnidarians, only the Class Hydrozoa has both marine and freshwater species. The relatively small class Scyphozoa includes about 200 species of marine jellyfish. A third class, the Cubozoa, contains a few species of tropical, cube-shaped jellyfish whose stings are very dangerous. The fourth class, the Anthozoa, is the largest group, with over 6000 marine species of sea anemones, corals, and many others.

Class Hydrozoa

The majority of hydrozoans live as branching colonies of polyps, with free-swimming medusae in most species. Representatives of the marine genus *Obelia* are common in shallow coastal waters (Figure 21.2a). At least two kinds of individuals make up an *Obelia* colony. Feeding polyps are called **gastrozooids**. The less numerous reproductive polyps, or **gonozooids**, produce small free-swimming medusae by budding. Note in Figure 21.2a that each branch of a colony has both kinds of polyps. The epidermis of *Obelia* secretes a chitinous, proteinaceous exoskeleton, the **perisarc**, which surrounds and supports all parts of the colony except the freely moving tentacles. The layers of the body wall and the gastrovascular cavity are continuous, connecting all members of the colony, and branches of the colony are interconnected by rootlike extensions that anchor the colony to the substrate.

In contrast to *Obelia*, certain other colonial hydrozoans have medusae that remain attached to the colonial polyps, and in some hydrozoans, either the polyp or the medusa is reduced or absent. *Hydra*, for example, exists only as a solitary polyp (Figure 21.2b), whereas the marine hydrozoan *Gonionemus* has an active, free-swimming medusa and a small (less than 5 mm) polyp (Figure 21.2c).

Among the most bizarre and fascinating hydrozoans are *Physalia*, better known as the Portuguese

(a)

(b)

(c)

Figure 21.2 Class Hydrozoa. (a) A colony of *Obelia*. These small (2–3 cm high) colonial hydrozoans live attached to virtually any hard substrate. They are often mistaken for seaweed when they are pulled up with strings of algae on fish hooks or boat anchors. Note the feeding polyps (gastrozooids), reproductive polyps (gonozooids), and medusae budding from gonozooids. A chitinous perisarc surrounds and supports most of the colony. Also see Figure 21.15. (b) *Hydra*, common in fresh water, lacks a medusa. The solitary polyps often attach to bottom objects or plants, or hang upside down from the surface film in quiet pools. When fully extended, hydras are between 2 and 25 mm long (not including the tentacles). When disturbed, they can contract quickly to a short, stumplike form. (c) *Gonionemus*, a marine jellyfish, swims feebly or creeps along the ocean floor feeding on small invertebrates. *Gonionemus* and most other hydrozoan medusae have a velum.

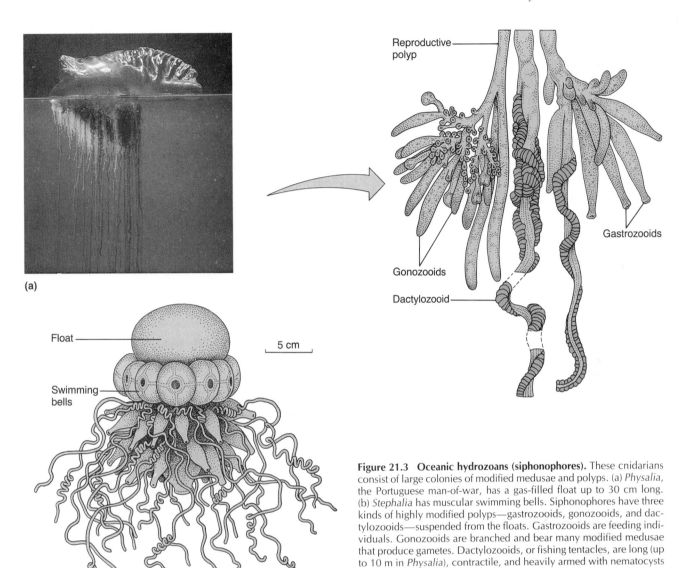

(a)

Float

Swimming bells

5 cm

(b)

Reproductive polyp

Gastrozooids

Gonozooids

Dactylozooid

Figure 21.3 Oceanic hydrozoans (siphonophores). These cnidarians consist of large colonies of modified medusae and polyps. (a) *Physalia,* the Portuguese man-of-war, has a gas-filled float up to 30 cm long. (b) *Stephalia* has muscular swimming bells. Siphonophores have three kinds of highly modified polyps—gastrozooids, gonozooids, and dactylozooids—suspended from the floats. Gastrozooids are feeding individuals. Gonozooids are branched and bear many modified medusae that produce gametes. Dactylozooids, or fishing tentacles, are long (up to 10 m in *Physalia*), contractile, and heavily armed with nematocysts for defense and food capture. They catch small fishes and other prey and pull them up to the gastrozooids for ingestion.

man-of-war, and *Stephalia* (Figure 21.3). Known collectively as **siphonophores**, or **oceanic hydrozoans**, these are large colonial predators, floating and swimming actively about in temperate and tropical seas. Their colonies consist of several kinds of tentacle-like individuals (actually modified polyps and medusae) hanging down into the water from a gas-filled float. The float, which may be a modified polyp, gives each genus a distinctive appearance and movement. *Physalia's* float projects above the water surface and acts as a sail. *Stephalia* has several muscular **swimming bells**, actually modified medusae, that form a ring around the base of the float and propel the colony by jetting water or pulsating. Some siphonophores can regulate the amount of gas in their floats and sink beneath the

ocean surface during stormy weather. Others regularly migrate up and down in the water, regulating their buoyancy to remain stationary at certain depths.

The nematocyst toxins of siphonophores can be dangerous or even fatal to humans, and experienced divers carry protection against them. Commercial meat tenderizers are often used, because they contain proteolytic enzymes. Since the toxins are proteinaceous, if tenderizer is rubbed on the skin just after a sting, its enzymes neutralize the toxins.

As a class, hydrozoans can be distinguished from other cnidarians by certain features: Their mesenchyme has wandering amoebocytes, but otherwise is noncellular; they have cnidocytes only in their epidermis; and most hydrozoan medusae (*Obelia* is an exception) have a **velum**, a shelflike extension of the body wall on the undersurface of the bell margin. Also, in most species, gonads develop in the epidermis.

Class Scyphozoa

Scyphozoans are sometimes considered true jellyfish; their medusae are large (usually 5–50 cm in diameter), whereas their polyp stage is small and inconspicuous. In contrast to most hydrozoans, scyphozoans typically have a cellular mesenchyme, a gastrodermis that bears cnidocytes, and medusae that lack a velum. Gonads in the scyphozoan medusae develop in the gastrodermis. Though the class is small, scyphozoans are abundant in all oceans. Certain species occur at great depths, while others, such as *Chrysaora*, the sea nettle, frequent coastal waters (Figure 21.4a). Also common in shal-

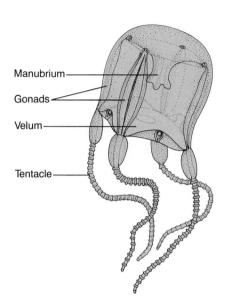

Figure 21.5 Class Cubozoa. Cubozoan medusae range in size from about 2 cm up to nearly 30 cm in total length. They are motile predators, feeding mostly on fishes in shallow tropical seas. Their nematocysts can inflict painful, dangerous stings. Over 50 people have died in Australian waters from multiple stings of the large species *Chironex fleckeri*, the sea wasp. Cubozoans have tiny polyps that transform directly into medusae.

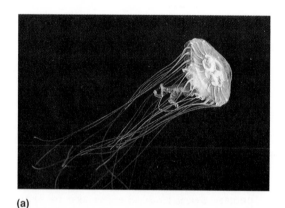

(a)

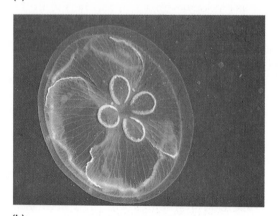

(b)

Figure 21.4 Class Scyphozoa. (a) Sea nettles, such as this species, *Chrysaora quinquecirrha*, are common in shallow temperate seas. Most species of *Chrysaora* are in the range of 3–5 cm in diameter, and can deliver an irritating sting. (b) The filter-feeding jellyfish *Aurelia*, ranging in diameter from about 5 to 25 cm, often appears in large groups in coastal waters. Shown here is the moon jelly, *Aurelia aurita*.

low seas, species of *Aurelia* are widely used as classroom examples of Scyphozoa (Figure 21.4b). Unlike most scyphozoans, which feed on small invertebrates and fishes, species of *Aurelia* are filter feeders.

Class Cubozoa

This class includes several marine jellyfish formerly classified with the scyphozoans. In common with scyphozoans, cubozoans have a reduced polyp stage, cellular mesenchyme, and gastrodermal gonads, but their cube-shaped medusae with four long fishing tentacles are unique (Figure 21.5). Cubozoan medusae also have a velum similar to that of hydrozoans.

Class Anthozoa

Anthozoans (sea anemones, corals, sea pansies, and sea fans) have no medusae, but their polyps are the most complex in the phylum. It is easy to understand why they were named anthozoans (meaning "flower animals"): Many are brilliantly pigmented and have a crown of tentacles (Figure 21.6). Anthozoans are characterized by a thick cellular mesenchyme, a gastrodermis that bears cnidocytes, and gonads that develop in the gastrodermis. The anthozoan digestive system is unique. As seen in a

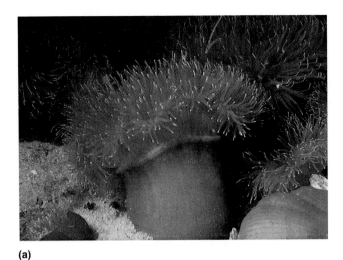

(a)

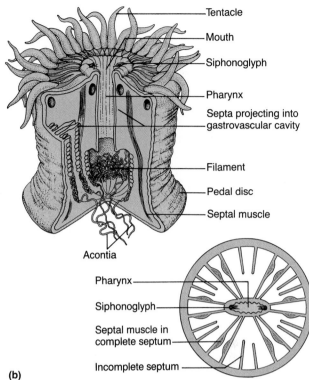

Tentacle

Mouth

Siphonoglyph

Pharynx

Septa projecting into gastrovascular cavity

Filament

Pedal disc

Septal muscle

Acontia

Pharynx

Siphonoglyph

Septal muscle in complete septum

Incomplete septum

(b)

Figure 21.6   Class Anthozoa (anemones). (a) Sea anemones, such as these *Metridium senile,* are solitary bottom dwellers in shallow seas and coastal tidepools. Most anemones feed on small fishes and invertebrates, but species of *Metridium,* common on both the Atlantic and Pacific coasts of North America, are filter feeders. Large individuals of *Metridium* may be 20 cm high; they are popular subjects of research, especially in neurobiology and behavior. (b) Sections of a sea anemone.

sea anemone, the mouth is usually elongate, making the animal biradially symmetrical. The mouth bears one or two folded clefts called **siphonoglyphs**. Cilia in the siphonoglyphs drive water into and out of a pharynx, an internal extension of the body wall. The pharynx opens into the gastrovascular cavity, which is divided by muscular septa called **mesenteries**. Some of the mesenteries extend from the body wall to the pharynx, and the partial compartments formed by these septa provide a hydro-

static skeleton that can be manipulated rather precisely by muscles. As a result, sea anemones can perform a range of surprisingly complex body movements. Notice that the mesenteries occur in multiples of six. (This is also true of the tentacles, although Figure 21.6 does not show it.) The free edge of each mesentery bears cnidocytes that can discharge and immobilize ingested live prey. In some species, these free edges of the mesenteries extend as cnidocyte-bearing **acontia** (Figure 21.6b).

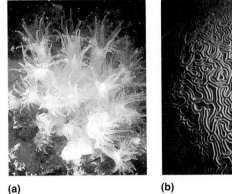

(a)

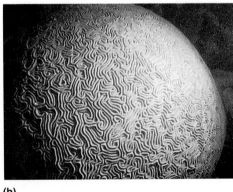

(b)

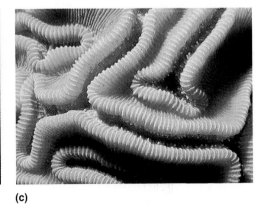

(c)

Figure 21.7   Class Anthozoa (stony corals). (a) Species of *Astrangia* are common in the cooler waters off the north and middle U.S. Atlantic and Pacific coasts. Shown here is the northern star coral, *Astrangia danae.* (b) Species of *Diploria,* such as *D. labyrinthiformis,* a brain coral, are important reef formers in tropical seas. (c) In this detail of *D. labyrinthiformis,* note the polyps, barely visible along the grooves.

Figure 21.8 Anthozoans with eight tentacles. Colonies of (a) the soft coral *Alcyonium digitatum* (dead man's fingers), (b) the organ-pipe coral *Tubipora musica,* and (c) the sea pen *Ptilosarcus gurneyi.*

Stony corals are similar to sea anemones in body form, except that coral polyps lack siphonoglyphs, they are mostly colonial, and they live in limestone exoskeletons (Figure 21.7). Polyps in a colony are interconnected, and the gastrovascular cavity is continuous throughout the colony. Corals are found in virtually all oceans, and several species form coral reefs in warm tropical seas. Reef-forming corals, such as *Montastrea, Diploria, Acropora,* and *Porites,* have enormous ecological impact in tropical oceans. Massive accumulations of their limestone exoskeletons form underwater reefs and islands, creating habitats for many animals, including many other anthozoans.

Both sea anemones and stony corals have septa and tentacles arranged in groups or multiples of six. But another subgroup of anthozoans are *octamerous*—that is, they have eight tentacles and eight septa. Octamerous anthozoans include the soft corals (such as *Alcyonium*), the horny corals, or gorgonians (sea fans, sea whips, and red corals) and the organ-pipe corals of the Genus *Tubipora,* all of which frequently live on coral reefs (Figure 21.8a and b). By contrast, sea pens and sea pansies typically inhabit soft substrates (Figure 21.8c).

ENVIRONMENTAL RELATIONSHIPS

Cnidarians are important members of marine food chains. They are typically carnivorous, although certain sea anemones produce cellulases and may be capable of digesting some plant material. Small crustaceans are a favorite food of many marine and freshwater species. The large oceanic hydrozoans, cubozoans, and certain anemones catch and eat fishes and larger invertebrates.

In fresh water, several flatworms and arthropods feed on *Hydra,* and certain marine fishes, flatworms, crustaceans, echinoderms, and molluscs (sea slugs) eat cnidarian polyps, especially small anemones and corals. On coral reefs, many fishes feed on the large amounts of mucus secreted by the corals. Some of the flatworms and sea slugs that eat cnidarians digest all but their victims' nematocysts. These pass from the predator's gut to its own outer surface and there serve as a defense mechanism for the predator!

Symbioses

Many cnidarians live in symbiotic associations with other organisms, and their body surfaces often harbor harmless protozoa, flatworms, and crustaceans. Cnidarians also have some harmful parasites, particularly protozoa. The freshwater amoeba *Hydramoeba hydroxena* often kills hydras and freshwater jellyfish by engulfing cells faster than its hosts can replace them.

Many sea anemones and corals live mutualistically with other animals. Brightly colored clown fishes commonly live among the tentacles of certain sea anemones (see Figure 29.12b). Upon encountering a suitable anemone, a clown fish darts rapidly in and out of the mass of tentacles. In so doing, the fish coats itself with the anemone's mucus; as a result, the anemone cannot recognize the fish and so does not discharge its nematocysts. In association with an anemone, clown fishes are protected from predators, and may in turn remove parasites and debris from the host.

Great numbers of cnidarians harbor photosynthetic protists or algae. The autotrophs are held in vacuoles in the cnidarian cells but are not digested. *Chlorohydra viridissima*, a common freshwater hydra, is green because it has symbiotic green algae in its gastrodermal cells. Likewise, the scyphozoan *Cassiopeia* is yellowish because its mesenchyme harbors dinoflagellates, photosynthetic protists often classified with the algae. These cnidarians are, in a sense, omnivorous: As herbivores, they consume some of the photosynthetic products of the algae, and as carnivores, they filter-feed or prey on other animals.

One of the world's most important symbioses is the mutualistic relationship between the reef-building corals and the dinoflagellates that live in their gastrodermal cells. Corals can grow without dinoflagellates, but they cannot form coral reefs without them. Research indicates that reef-building corals obtain more calories from their dinoflagellates than from their filter feeding. If symbiotic dinoflagellates are supplied with radioactive carbon dioxide in the laboratory, the radioactive CO_2 is used in photosynthesis, enabling researchers to trace the metabolic pathway of the carbon compounds through the dinoflagellate-coral system. Such research shows that the corals obtain some nutrients, probably in the form of glycerol, glucose, certain amino acids, and possibly some lipids, from their dinoflagellates. In turn, the protists may obtain metabolic water and useful animal waste compounds containing phosphorus and nitrogen from the coral animals. The protists may also produce certain nitrogen and phosphorus compounds that the coral can use; corals may rely on these compounds when they cannot obtain enough nitrogen or phosphorus by filter feeding. Dinoflagellates also obtain carbon dioxide from the coral, and in so doing may help the coral animals deposit limestone reef material. The following equilibrium reactions represent one of several hypotheses of how coral cells deposit limestone:

$$Ca^{2+} \quad + \quad 2HCO_3^-$$
(from sea water) (from CO_2 produced by coral cells)
$$\longleftrightarrow Ca(HCO_3)_2 \longleftrightarrow CaCO_3 \downarrow + H_2CO_3 \longleftrightarrow H_2O + CO_2$$

Notice that coral limestone ($CaCO_3$) precipitates when the reaction sequence proceeds to the right. The final product, CO_2, derived from the dissociation of H_2CO_3, must be consumed for the precipitation to continue. Two mechanisms ensure its steady removal. Coral cells convert the CO_2 to bicarbonate ions (HCO_3^-), and the dinoflagellates also consume it in photosynthesis.

COMPARATIVE STRUCTURE AND FUNCTION

With their two life stages, many cnidarians can take advantage of two different habitats in aquatic environments. Sessile or sedentary polyps are adapted to exploit *benthic* (bottom) habitats, while medusae can take advantage of *pelagic* (in the water column) zones. Thus, many species of cnidarians can exploit two different sources of food.

Body Structure, Support, Locomotion, and Protection

Cnidarian activities are determined by the structure and function of cells in the three body layers and by how these cells interact with the matrix of the mesenchyme and with the fluid in the gastrovascular cavity. In anthozoans, the epidermis of the body column consists mainly of columnar epithelial cells with amoeboid bases held tightly to underlying collagen fibers in the mesenchyme. By contrast, in hydrozoan polyps (and in the tentacles and mouth region of many other cnidarians), the epidermis is formed chiefly of a layer of **epithelio-muscular cells**, so named because they form the outer body covering (epithelium) and also a muscle layer. Note in Figure 21.9 that the bases of these cells stretch out into long processes containing contractile elements called myonemes. Collectively, the processes form a longitudinal muscle layer, which shortens the body and tentacles when it contracts (also see Figure 10.8). In the pedal disc and oral disc of many polyps, as well as on the undersurface of many medusae, epidermal cells form a radial muscle layer.

The gastrodermis of many cnidarians consists mainly of a row of flagellated **nutritive cells** that engulf food from the gastrovascular cavity. Flagella on these cells help circulate water and food in this cavity. Hydrozoan polyps and many anthozoans have **nutritive-muscular cells** that form a circular contractile layer in the gastrodermis (see Figure 21.9). Contraction of this layer decreases the diameter of the body. In anthozoans, the bulk of the musculature (often longitudinal and circular bundles) is derived from the gastrodermis, but the fibers are often located in the mesenchyme. Medusae generally lack gastrodermal muscles.

The epidermis and gastrodermis also have several other cell types (see Figure 21.9). Cnidocytes are common, and as mentioned earlier, their presence or absence in the gastrodermis helps distin-

Essay: Coral Reefs

Coral reefs are formed by the successive layering of limestone by many generations of stony corals and free-living red and green algae called coralline algae. Stony corals inhabit only the outermost layer of a reef, and each polyp may secrete several layers of underlying limestone. As one generation of polyps is replaced by another, layer upon layer of exoskeletal limestone gradually builds up to form a reef. Coralline algae spread over the reef surface and secrete calcium carbonate, which cements large portions of the reef together.

Coral reefs are among the most productive and diverse marine ecosystems (Figure E1). Most of a reef's energy is trapped by the corals' symbiotic dinoflagellates and by free-living algae, but other algae attached to the reefs contribute substantially to energy production. Studies indicate that about 1500–3500 g of carbon per square meter per year are incorporated into living tissue on coral reefs. Comparable estimates for open tropical oceans are about a hundred times less.

There are three types of coral reefs (Figure E2). **Fringing reefs**, the most common, form a shallow tidal area extending out from a continent or island. Many islands, including the Bahamas in the West Indies, bear fringing reefs. **Barrier reefs** differ from fringing reefs in that they are separated from a land mass by a lagoon. Extending for over 1800 km along the coast of northeast Australia, the Great Barrier Reef consists of both fringing and barrier reefs. The **atoll**, a third type of reef, develops on a submerged volcano and forms a ring of islands around a central lagoon. The Marshall Islands, including Bikini and many others in the Pacific, and the Maldives in the Indian Ocean are a few of the thousands of atolls in tropical seas.

All reefs have a similar profile with three main zones (Figure E3). The **reef flat** is variable in depth and provides diverse tidal pools and channels in which many different kinds of corals, other animals, and algae thrive. The **reef crest** is a zone of heavy wave action, where only those species adapted to withstand continuous pounding survive. Some of the common corals in Caribbean and Indo-Pacific reef crests are species of *Acropora*, the staghorn corals. Red coralline algae are also common on reef crests and are usually important in forming and holding this part of the reef together. Red algae and other species would probably overgrow many reefs if they were not grazed continuously by herbivorous fishes and invertebrates. Beyond the crest, the **reef front** slopes irregularly into the sea. The upper portions of the reef front are often the most productive areas of coral growth.

Coral reefs grow only in tropical seas where ocean temperatures average at least 20°C. The water must be clear and usually no deeper than 50 m to allow light penetration for photosynthesis by algae and by the corals' symbiotic dinoflagellates. Filter feeding by corals and multitudes of other reef animals helps keep the water over the reefs clear, promoting light penetration. Today, active reefs occur in the West Indies and in tropical areas of the Pacific and Indian oceans. Fossil evidence indicates that reefs were more widespread during warmer geological times (40 to 50 million years ago).

Coral reefs constantly change in response to water temperature, ocean floor subsidence, and tropical storms. Many reefs have died out completely after drastic changes in sea level during certain geological periods. Predators can also seriously affect reefs. Lime-boring sponges have a major impact on reef decomposition. The sea star *Acanthaster planci,* known commonly as the "crown of thorns" (see Chapter 27 opening photograph), eats coral polyps and has devastated several coral reefs off Japan and Australia since the early 1960s. Little is known about how and why *A. planci* appears on these reefs in such harmful numbers, but marine biologists suspect that its recent increase has resulted from human disruption of habitat and consequent decrease in populations of certain molluscs (conchs and sea slugs) that prey heavily on this sea star.

Human use and abuse of coral reefs have also increased in recent years. Scuba diving is becoming increasingly popular, and a large tourist trade centers around reef diving in the tropics. Underwater parks have been established in some areas in attempts to regulate overuse of reefs. Large-scale mining of coral limestone for use in road and house construction can remove reef material in excess of its replacement rate. Increased silting on many coasts from construction and agriculture has significantly reduced reefs in local areas. Oil spills and routine cleaning of oil tankers at sea are among the most serious modern threats to reefs.

Figure E1 Diversity on a coral reef. Reefs support a vast number of marine organisms. Except for the corals, sponges are often the most numerous filter feeders on reefs. Other abundant reef organisms include sea anemones, gorgonian sea fans, soft corals, flatworms, molluscs, annelids, crustaceans, and echinoderms. Chordates are often represented by the sedentary sea squirts (tunicates), as well as by the many species of fishes living in crevices and caves in the coral limestone or browsing in large schools over the reef surface. Certain large fishes periodically visit the reefs to have their body surfaces groomed by shrimps and small reef fishes (also see Figure 29.12c).

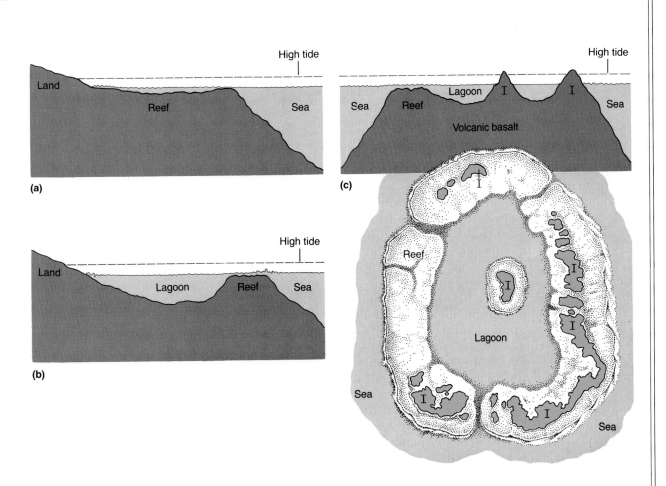

Figure E2   The three kinds of coral reefs. (a) A fringing reef. (b) A barrier reef. (c) An atoll. In (c), note the coral islands (I) projecting above the water surface. One hypothesis is that barrier reefs and atolls develop from fringing reefs around islands. If an island with a fringing reef subsides slowly, so that reef growth can keep up with the subsidence, a barrier reef may first form as a lagoon develops around the subsiding island. Eventually, the island may disappear, leaving an atoll.

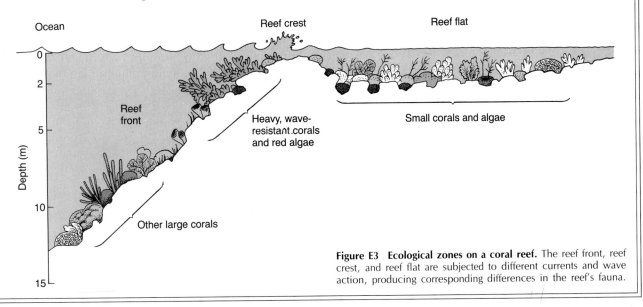

Figure E3   Ecological zones on a coral reef. The reef front, reef crest, and reef flat are subjected to different currents and wave action, producing corresponding differences in the reef's fauna.

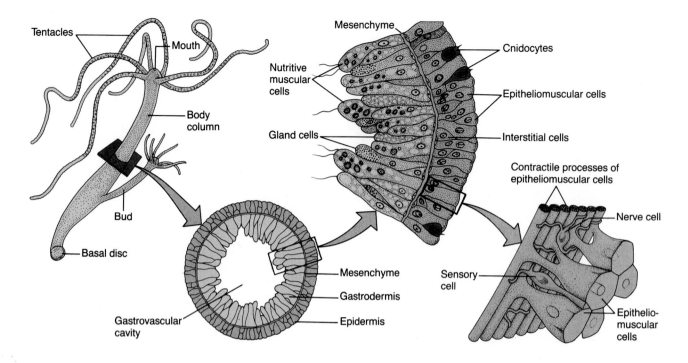

Figure 21.9   The body wall of a hydrozoan. This cross section of a hydrozoan polyp shows the different types of cells in the epidermis and gastrodermis, the relationship between the contractile and noncontractile parts of the epitheliomuscular cells, and the location of the nerve net.

guish the cnidarian classes. Gland cells in the gastrodermis secrete mucus and digestive enzymes. Similar cells, concentrated in the epidermis around the mouth, secrete mucus that lubricates food as it passes into the gastrovascular cavity. Mucus also helps protect intertidal species from drying out when exposed to air at low tide. Gland cells in the pedal disc of polyps secrete an adhesive substance that helps attach the animal to bottom substrates. Both body layers also have elongate nerve cells, some of which are sensory in function. There are also unspecialized interstitial cells that can differentiate into other cells as needed.

Most cnidarian polyps are supported mainly by a hydrostatic skeleton. (Hydrostatic skeletons are discussed in Chapter 8, p. 176.) Contraction of muscles in the mouth region holds water in the gastrovascular cavity, and muscles in the epidermis and gastrodermis exert pressure on the water. Held as a rigid column, the water provides support and can be manipulated by the muscles to produce various body movements. Body size can be altered by changing the amount of water in the cavity. Hydras and sea anemones can extend and shorten their bodies to capture prey or avoid danger while they are attached to a substrate. Hydras can also bend,

hold on to the substrate with their tentacles, detach their pedal disc, and somersault from place to place. They can also secrete gas bubbles and float with water currents. Sea anemones can glide along the substrate on their pedal disc, and some can detach and pull themselves along the bottom or swim by body flexure and tentacle movements.

Cnidarian medusae are supported mainly by fibers and water in their mesenchyme, which occupies most of their body mass. The umbrella-like shape of medusae is an adaptation for hanging and moving in the water column. Medusae may be moved about by winds and water currents, but they can also propel themselves for short distances. Most medusae sink slowly because they are slightly more dense than seawater. As a medusa sinks, circular muscles in its outer rim contract, compressing the umbrella and forcing water out from its undersurface. This creates a mild jet-propulsive force, driving the animal upward. Most medusae have radial muscles that act antagonistically to the circular muscles and serve to "open the umbrella." In hydrozoan and cubozoan medusae, the velum reduces the passageway for water being driven out of the bell and thus increases the jet-propulsive force.

In contrast to the motile hydras, certain anemones, and medusae, the sessile colonial polyps are supported mainly by a rigid exoskeleton. The chitinous perisarc of colonial hydrozoans and the calcareous exoskeleton of stony corals are examples. Individual coral polyps look like small (1–5 mm) sea anemones, but both the polyps and their exoskeletons are interconnected. The epidermal cells of each polyp secrete skeletal limestone in the form of a cup with a folded base. Each polyp fits tightly into its own cup and can contract so that all but its top surface lies within the exoskeleton (Figure 21.10).

Cnidocytes. These highly specialized cells contain the cnidarian's main offensive and defensive weapons, the dischargeable cnidae (nematocysts and spirocysts). Each cnida is a capsule containing a hollow coiled thread (Figure 21.11). Some cnidocytes are apparently under direct nervous system control, while others seem to be independent effectors in that their discharge is self-induced and not mediated by the nervous system. A bristlelike structure, the **cnidocil**, projects from many cnidocytes and may elicit discharge when stimulated by chemicals or mechanical forces. Other cnidocytes have sensory microvilli or cilia instead of a cnidocil.

Researchers do not yet understand exactly how the cnidae discharge. Nematocyst discharge has been studied the most and is a truly remarkable process. Just prior to discharge, an increase in water permeability of the nematocyst capsule causes the tip of the cell to open, and fluid pressure in the capsule apparently forces the nematocyst thread to evert, much the way air pressure can evert the fingers of a rubber glove that have been turned outside in. Through high-speed cinematography (40,000 frames per second), it has been determined that nematocyst discharge in *Hydra* is among the fastest cell activities known. The entire discharge takes only 3 milliseconds (ms), and the tip of the thread travels about 20 μm at a velocity of 2 m/s (this is about 40,000 times the acceleration caused by gravity). Once discharged, nematocysts are not retrievable, and the empty cnidocytes disintegrate. New cnidocytes differentiated from interstitial cells replace the spent cells in one or two days.

Hydras have four types of nematocysts. **Stenoteles** are specialized for capturing prey and for protection (see Figure 21.11). Stenotele threads

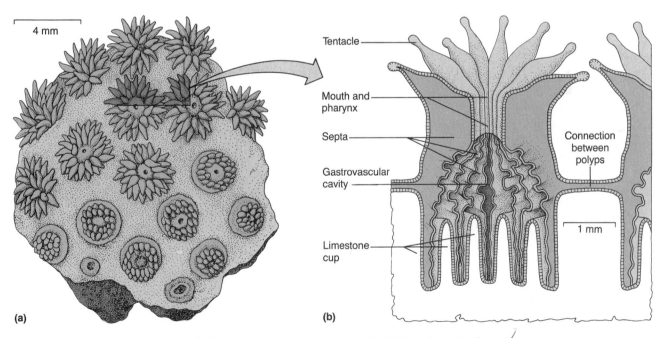

(a)

(b)

Tentacle

Mouth and pharynx

Septa

Gastrovascular cavity

Limestone cup

Connection between polyps

Figure 21.10 The limestone exoskeleton of stony corals. (a) View of a coral colony with some individuals withdrawn into the exoskeleton. Contracted coral polyps are safe from most predators, yet their symbiotic algae are still exposed to light and can photosynthesize. Most reef-forming corals remain in this protected position during daylight hours, extending their polyps and tentacles to filter-feed only at night. (b) Longitudinal section of an individual in its limestone cup. Note that the polyp is connected to its neighbors and that each cup in a colony is fused to adjacent cups.

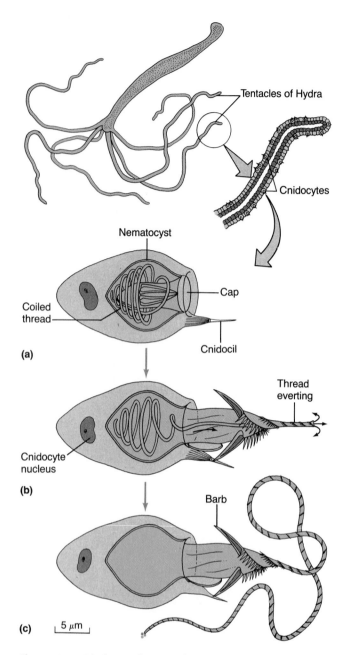

Figure 21.11 Discharge of a stenotele nematocyst. (a) Undischarged nematocyst. (b) Discharging nematocyst. (c) Discharged nematocyst. Note the droplets of toxin at the tip of the discharged thread, which is about nine or ten times as long as the cell when fully everted.

have an expanded base, are armed with spines and toxins that can damage nerves or muscles or destroy blood cells, and are discharged with enough force to penetrate the exoskeletons of small crustacean prey. **Desmonemes** have a thread that entangles prey. Two types of **isorhizas**, lacking the expanded thread base of the stenotele, function in defense or in attachment to substrates.

Feeding and Digestion

Among cnidarians, feeding has been studied most extensively in *Hydra*. Suitable prey, mainly small crustaceans and aquatic insects, trigger nematocyst release when they contact the tentacles. Most victims are paralyzed and killed within seconds by the nematocyst toxins. Contact with prey and certain small molecules that potential prey release into the water (especially the reduced form of the tripeptide glutathione and certain amino acids) trigger a series of reflexes known as the **feeding response**. As a hydra captures prey, it writhes about slowly, bends its tentacles toward its mouth, opens its mouth, and deposits the prey in its gut. The same or a very similar series of responses has been documented in over 20 other kinds of cnidarians.

Cnidarians digest food both extracellularly and intracellularly. Extracellular digestion in the lumen of the gastrovascular cavity includes protein hydrolysis and probably some carbohydrate breakdown; food is also softened and fragmented here, so that it can be taken up by phagocytosis. Digestion is completed in food vacuoles in the gastrodermal cells. Undigestible matter passes back out through the mouth.

Circulation, Gas Exchange, Osmoregulation, and Excretion

These processes occur by diffusion and cell transport. Wandering amoebocytes are especially important in nutrient and nitrogenous waste (ammonia) transport. The gastrovascular cavity is also important in circulation and other vital functions. As body movements and flagella circulate fluids in the cavity, nutrients and dissolved gases in the fluid are brought into contact with inner cell surfaces. Materials may circulate among all individuals in cnidarian colonies because the gastrovascular cavity is continuous. Gastrovascular circulation is especially important in medusae because their mesenchyme is thick, and proportionately less surface area is exposed to the outside. This poses a problem in provisioning inner cells, but most medusae have a multibranched gastrovascular cavity, often including a ring and several radial canals, that provide greater fluid contact with inner cell surfaces (see Figure 6.4).

Nervous and Sensory Structures

Cnidarians have a diffuse, noncentralized, weblike nervous system consisting of one to three **nerve**

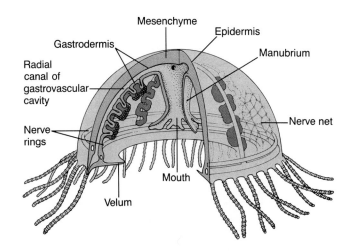

Figure 21.12 **Nervous and other structures of medusae.** Nerve nets occur at the base of the epidermis and gastrodermis, and nerve rings circle the base of the umbrella.

nets constructed of neurons (Figure 21.12). The cnidarian nervous system is a clear reflection of radial symmetry. Although neurons are usually clustered around the mouth region, there are no ganglia or brain. Without these elements, the system lacks central control, and when any point on the nerve net is stimulated, impulses may be transmitted equally in all directions from the stimulus point. (A detailed discussion of the structure and function of cnidarian nerve nets can be found in Chapter 10, p. 229; Figure 10.8.)

There is a concentration of nervous elements in some cnidarians: Augmenting their nerve net(s), many medusae have one or more **nerve rings** that coordinate the complex muscle actions providing tentacle movements and jet propulsion (see Figure 21.12). A special cluster of neurons in the nerve ring in some medusae acts as a pacemaker, initiating the muscle contractions that drive rhythmic swimming pulsations.

Cnidarians have two structural types of neurons. Bipolar neurons have two main processes, or neurites; multipolar neurons have several processes and can transmit impulses in several directions. Cnidarian nervous systems also include chemical synapses between neurons. At some synapses, only one neuron secretes a transmitter chemical, passing impulses only one way across the synaptic junction. At other synapses, both neurons secrete transmitter molecules, allowing two-way impulse transmission.

Rodlike sensory cells responsive to chemicals, touch, and perhaps other stimuli are common in the epidermis, especially on the tentacles, and in the gastrodermis in both polyps and medusae. Polyps lack sense organs, but medusae have two kinds of sense organs that represent adaptations to their active life in the open ocean. Photosensitive **ocelli,** consisting of sensory cells and a pigment spot or cup, often occur around the bottom of the bell margin just below the nerve ring. Gravity-sensitive

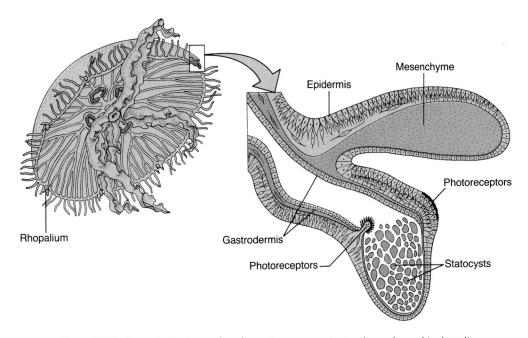

Figure 21.13 **Sensory structures of medusae.** Sense organs in *Aurelia* are housed in rhopalia located around the bottom of the bell margin. Each rhopalium contains photoreceptors, statocysts, and, in some cases, pacemaker neurons.

statocysts are situated in the bell margin at the base of the tentacles. Statocysts consist of a small calcium sulfate granule, the **statolith**, held in a sac lined with ciliated sensory cells. As a medusa is tipped by water currents, the statolith rubs against the cilia, stimulating sensory cells on the lower side of the statocyst. Motor impulses then stimulate appropriate muscle contractions to return the medusa to its upright position. Scyphozoans, such as *Aurelia*, have ocelli and statocysts housed in swellings called **rhopalia** (Figure 21.13).

BEHAVIOR

Despite their general lack of nervous system centralization, cnidarians show a remarkable range of behavioral responses. They respond to many different stimuli, and their responses are either local or involve whole-body movements, depending on the strength or type of stimulus. Hydras respond to touch, to sudden movement in surrounding water, and to noxious chemicals by rapidly contracting their tentacles or their entire body. They may also discharge nematocysts in response to touch. Hydras also exhibit a repeated sequence of activities: If undisturbed, individuals may stretch out and remain motionless for 5–10 minutes, then suddenly contract; the significance of this behavior is unknown, but hydras may perform this series of activities while monitoring changes in their surroundings.

Sea anemones exhibit specific avoidance responses when attacked by predators such as sea stars or sea slugs. They may contract tightly, or in some cases escape by burrowing into sandy substrates. Many species have pores in their body wall through which they extend their cnidocyte-armed acontia. Several species, including *Stomphia didemon*, respond to approaching predators or to extracts of certain predators by detaching and creeping, rolling, or swimming away, using tentacles or alternate side-to-side contractions of their longitudinal muscles (Figure 21.14).

Most cnidarians can detect light. Hydras show undirected movements in the dark, but become motionless in well-lighted areas. This may be a feeding adaptation, because small crustaceans commonly eaten by hydras often congregate in bright light, where algal food is abundant. Many sea anemones extend their tentacles to feed only during twilight hours or at night, but species with

(a)

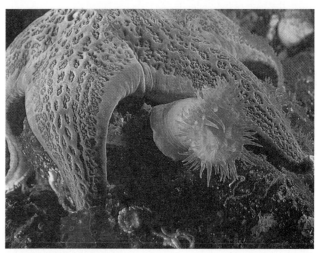

(b)

(c)

Figure 21.14 A sea anemone's response to a predator. (a) A leather sea star *Dermasterias imbricata* contacts the anemone *Stomphia didemon*. (b) The anemone contracts and detaches its pedal disc, then (c) rolls away.

symbiotic algae open their oral surface and spread their tentacles during part of the day. Medusae with ocelli can detect light direction, and they exhibit rhythmic behavior related to light-dark periods. Many are photopositive and spend daylight hours in the upper, well-lighted ocean zones; others spend time at the surface only during periods of moderate light; and still others are photonegative, sinking into deep water below the levels of light penetration during the day. Such movements may be adaptations to follow the movements of diurnal prey or to avoid predators.

Sea anemones in intertidal zones face desiccation when exposed to air during ebb tides. They solve the problem with a specific set of behaviors timed to occur during low tides. First, they expel most of the water from their gastrovascular cavity. Then they contract, pull their tentacles into their gut, and close their mouth tightly. This reduces exposure of thin body areas on the oral disc, which are subject to drying. Some species also protect their exposed surfaces by covering themselves with small pebbles.

REPRODUCTION AND LIFE CYCLES

Cnidarian life cycles typically include asexual and sexual phases. Meiosis occurs during gametogenesis; consequently, polyps and medusae are diploid. In species that lack the medusa stage, polyps usually reproduce both sexually and asexually.

Asexual Reproduction

Polyps reproduce asexually by budding, by longitudinal binary fission, and by regeneration after fragmentation. Hydras can re-form after being fragmented into separate cells. Even differentiated cells can participate in the process. Epidermal and gastrodermal cells may dedifferentiate and regenerate other cell types. Budding results in new polyps, medusae, or colonies. When a polyp bud develops in a colonial species, it remains attached to the parent organism. Fission and regeneration produce new polyps and are important means of reproduction in many cnidarians. Sea anemones often form clones by fission and **pedal laceration**. In pedal laceration, an anemone simply leaves portions of its pedal (attachment) disc behind as it moves to another location. Each piece then develops into a new anemone (see Figure 14.1b). Cloning allows species to exploit favorable habitats rapidly, and researchers have recently discovered that natural populations of *Metridium* and several other genera of anemones often consist of groups of clones.

Sexual Reproduction

Many cnidarians have separate sexes, but many others are hermaphroditic (individuals develop both male and female sex organs). Cnidarian gonads are not true organs, but consist of simple clusters of gamete-forming cells. Gametes are often released into the water, and fertilization is external. The zygote develops into a solid blastula (called the stereoblastula in soft corals) or, more typically, into a hollow coeloblastula. In hydrozoans, the coeloblastula becomes filled with cells and develops into a solid, ciliated **planula** larva. In many scyphozoans and most anthozoans, the planula is hollow. Anthozoan planulae often feed on microorganisms or are nourished by contained yolk. After swimming about for a short period, they attach to the bottom and develop into polyps.

Life Cycles

There are three basic types of cnidarian life cycles. In one type, which is rare, there is only a medusa stage. In a second, more common type, there is both a polyp and a medusa, although one stage may be less conspicuous than the other. The third type, characteristic of hydras and anthozoans, includes only a polyp stage.

All three kinds of life cycles occur in the Class Hydrozoa. Species of the hydrozoan jellyfish *Aglaura* are completely adapted to life in the open sea and have no polyp stage. Species of *Obelia*, with well-developed medusae and asexual colonial polyps, have the second type of life cycle (Figure 21.15). *Obelia* medusae are either male or female. Also representing the second type of life cycle, the freshwater jellyfish *Craspedacusta* and the marine hydrozoan *Gonionemus* have conspicuous medusae but small, inconspicuous polyps (see Figure 21.2c). The oceanic hydrozoans *Physalia* and *Stephalia* consist of floating colonies of asexual polyps and gamete-producing medusae (see Figure 21.3). Most members of the Class Scyphozoa, represented by *Aurelia*, also have the second type of life cycle (Figure 21.16). Two small polyp stages called the **scyphistoma** and the **strobila** reproduce asexually by budding. Budding from the strobila produces young medusae called **ephyrae**. The adult medusa reproduces sexually.

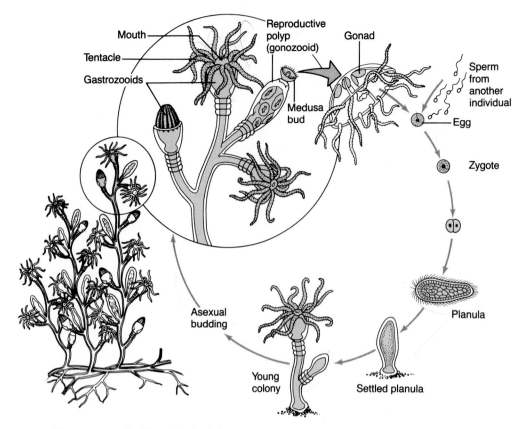

Figure 21.15 The life cycle of *Obelia*. This hydrozoan has distinct polyps and medusae. Its free-swimming medusae (about 1 mm in diameter) develop by budding from reproductive polyps called gonozooids. Medusae develop male or female gonads and release gametes that unite in the water. The zygote undergoes cleavage, a blastula forms, and a planula larva develops following gastrulation. After a short free-swimming period, the planula settles to the bottom, attaches to the substrate, and develops into a larval polyp that grows and develops by budding into a mature colony.

In the third type of life cycle, polyps reproduce both asexually and sexually. Hydras usually reproduce by budding during favorable growing periods, but produce male and/or female gametes at the onset of dry periods and in the autumn (Figure 21.17). In anthozoans, fertilization occurs externally or in the gastrovascular cavity. A resulting zygote produces a polyp, bypassing completely the medusa phase.

PHYLOGENY

There are several schools of thought about cnidarian phylogeny. The so-called **colonial hypothesis** proposes that cnidarians evolved directly from colonial protozoa, probably flagellates. (See the Trends and Strategies section beginning on p. 467.) Proponents of this idea point out that although cnidarian planulae are multicellular, they resemble flagellated colonial protozoa. The colonial hypothesis suggests that the cnidarian planula

is similar to some extinct species that was intermediate between protozoa and metazoans. Many proponents of the colonial hypothesis hold that **planuloids**, similar to cnidarian planulae, arose from colonial protozoa and gave rise to both the radiate and bilateral lineages of eumetazoans.

A second idea, the **plasmodial hypothesis**, suggests that the first eumetazoans (common ancestors of the radiate and bilateral lineages) were bilateral, perhaps somewhat like simple flatworms, and that these bilateral animals arose from multinucleate ciliates. Many zoologists discount this hypothesis because they do not believe that radiate animals could have been derived from bilateral organisms. Generally, the colonial hypothesis has been more popular among zoologists, but both hypotheses have been important in stimulating thought and discussion.

Whereas the colonial and plasmodial hypotheses postulate that the radiate and bilateral animals arose from a common multicellular ancestor, proponents of what might be called the **polyphyletic hypothesis** propose that radiate animals (cnidarians and ctenophorans) had a separate origin from

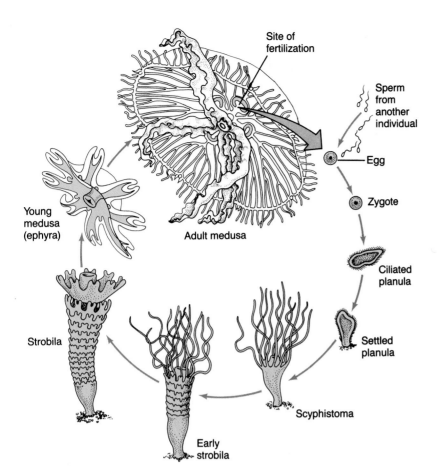

Figure 21.16 The life cycle of *Aurelia*. Adult medusae of this scyphozoan are dioecious, and fertilization occurs internally in the gastrovascular cavity of the female. Zygotes develop into free-swimming planulae that attach to the bottom and grow into a small polyp called the scyphistoma (5–7 mm high). The scyphistoma transforms into a second kind of polyp, the strobila (from a word root meaning pinecone; notice the resemblance). Larval medusae called ephyrae (about 5 mm in diameter) result from budding of the strobila.

that of all other animals. In other words, the radiate and bilateral lineages may have arisen from different types of multicellular ancestors or from different groups of protists. Several facts support the polyphyletic idea. First, only cnidarians have epitheliomuscular and nutritive-muscular cells, which perform several vital functions. Second, tissues in the mesenchyme of radiate animals develop mainly from ectodermal cells; all bilateral animals have true mesoderm that develops mainly from endodermal cells. Third, except for the myxozoan cnidocyst, there is no cell in any other animal that resembles the radiate's cnidocyte. Finally, some recent biochemical data show that certain hydrozoans, scyphozoans, and anthozoans have NADP-specific glutamate dehydrogenase, an enzyme that occurs in plants, bacteria, fungi, and at least one ctenophoran, but not in other animals. Cnidarians also have unique cell-to-cell junctions and produce mucus that is biochemically different from that of other animals.

Figure 21.17 (Right) The life cycle of *Hydra*. There is no medusa; polyps reproduce both asexually and sexually. Most species are dioecious. Eggs are held in epidermal cups at the polyp surface; once fertilized, they undergo cleavage and become surrounded by a protective chitinous capsule. The encapsulated embryos (0.5–1 mm in diameter), which eventually detach, can survive through dry periods and winters. Young hydras emerge when environmental conditions become favorable again.

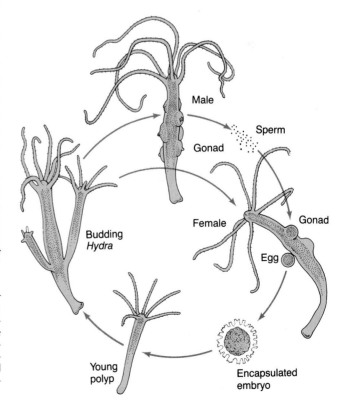

These facts clearly show that the radiate phyla should at least be placed on a side branch of the animal kingdom. If evidence continues to accumulate showing basic chemical dissimilarities between radiates and other eumetazoans, the assumption that radiates and bilateral animals had a common multicellular ancestor is probably incorrect. However, the colonial and plasmodial hypotheses are not necessarily invalidated by the polyphyletic hypothesis. As mentioned in the Trends and Strategies section on p. 468, they both may be valid, but in a more limited sense than often discussed. If radiate and bilateral groups are polyphyletic, perhaps radiates arose from colonial flagellates and bilateral groups arose via the plasmodial route from ciliates.

Fossil evidence indicates that the Phylum Cnidaria was established well before the dawn of the Paleozoic era (see Figure 17.18). Fossil jellyfish and and a few anthozoans have been found in Precambrian rocks. Among the cnidarian classes, the Hydrozoa is often thought to be the oldest. Scyphozoans and anthozoans may have evolved from ancestral hydrozoans during the Precambrian. Cubozoans, with their interesting mix of hydrozoan and scyphozoan features, seem to be intermediate between these two classes.

A challenging subject for debate is, Which evolved first in the cnidarian life cycle, the medusa or the polyp? Many zoologists would argue that since the medusa is the sexual stage, it is probably older. The polyp may have evolved secondarily as an asexually reproducing stage able to take advantage of bottom habitats; but there is no real evidence for this, and many zoologists believe that polyps preceded medusae. Among hydrozoans, there seems to have been an evolutionary trend toward emphasis of the polyp stage, and hydras may represent the zenith of this trend. Likewise, the simple polyp-only anthozoan life cycle may have evolved from more complex life cycles that included medusae. The complex, three-layered polyps of anthozoans seem to exhibit the greatest number of derived features of any cnidarians.

PHYLUM CTENOPHORA

There are about 100 species of ctenophorans, commonly called comb jellies; spherical species are also called sea walnuts or sea gooseberries. Several ctenophorans are brightly colored and luminescent, and many spherical species superficially resemble cnidarian medusae. Unlike medusae, however, ctenophorans are not umbrella-shaped, and they have paired body structures that make them biradially symmetrical, or else they are asymmetrical. Relatively little is known about the ctenophorans because many of them are rare and because most are fragile and difficult to collect and maintain in the laboratory. They are mainly free-swimming planktonic predators or filter feeders in the open sea (Figure 21.18), although some species frequent coastal waters and a few are bottom dwellers. There is no attached stage in the life cycle.

Species of *Pleurobrachia*, the sea gooseberries, range from about pea to walnut size and are common in temperate zones of both the Atlantic and Pacific oceans (see Figure 21.18b and chapter opening photograph). The name *Pleurobrachia*, meaning "rib arm," refers to two of the most distinctive features of these animals: their riblike ciliated **comb rows** and their long, armlike tentacles. There are eight comb rows, each row composed of ciliary plates that propel the animal through the water. A nerve net occurs under the comb rows. A single statocyst, located at the animal's aboral pore near the origin of the comb rows, serves as a balancing organ by controlling the beat of different comb rows. The tentacles, which arise through sheaths that extend deep into the body, contain muscles and bear adhesive food-capturing cells called **colloblasts** (**collocytes**) (see Figure 21.18b). A sea gooseberry feeds by extending its long tentacles to form a screen. Prey, mainly small planktonic crustaceans, are caught on sticky granules of the collocytes; then the tentacles contract toward the mouth, and the prey are wiped off into the mouth. There are no specialized structures for circulation or gas exchange. The branched gut serves to distribute nutrients and wastes. Cell clusters that surround an opening from the mesenchyme into the digestive tract may function in osmoregulation or excretion.

Ctenophorans have several traits that set them apart from cnidarians. They have a complete gut consisting of a mouth, a pharynx, a pouchlike stomach, biradially arranged digestive canals, and tiny anal pores. Their collocytes are functionally but not developmentally or anatomically similar to cnidarian cnidocytes. Also unlike cnidarians, ctenophorans do not have a planula larva. Their larvae are free-swimming, but resemble the adults.

Ctenophorans also have several traits that link

them to the Cnidaria. Despite their biradial symmetry, their basic body form resembles that of cnidarian medusae more than that of any other animal group. Also like cnidarians, ctenophorans have few organs (statocysts) and organ systems (digestive and sensory/neuromuscular systems); they are basically tissue-level animals. They have an epidermis, a gastrodermis, and a thick mesenchyme, similar to that of scyphozoans and anthozoans. Digestion, as in cnidarians, occurs both intracellularly and extracellularly.

With their mix of unique and Cnidaria-like traits, ctenophorans may have evolved from ancestors similar to medusoid cnidarians. It is also possible that the Cnidaria and Ctenophora had a common ancestor.

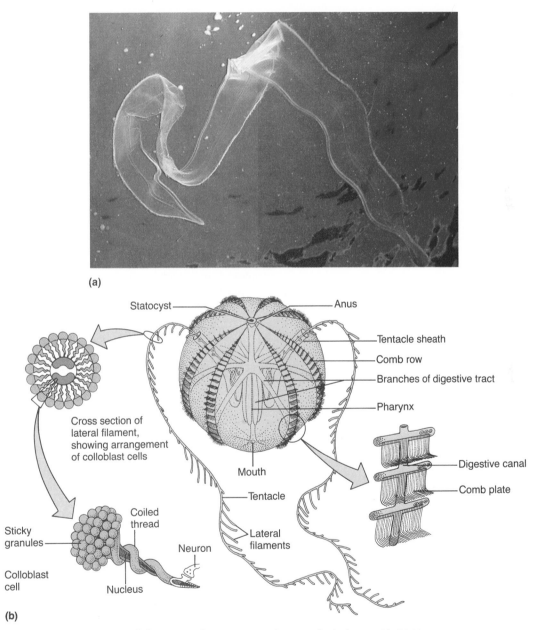

(a)

(b)

Figure 21.18 Phylum Ctenophora. Most species are spherical or ovoid. (a) Here we see a species that is flattened and beltlike, swimming in the open ocean. Muscles in the elongate body, which may reach lengths of 15–20 cm in some species, supplement the comb rows in providing an undulatory movement. (b) *Pleurobrachia*. The sheathed tentacles and internal branches of the complete gut give the animal its biradial symmetry. The enlargements show a comb plate and a colloblast from one of the tentacles. The colloblast's elongate filament is coiled within the cell and is tipped with numerous sticky granules; when the filament is discharged, prey are caught on the granules.

☐ SUMMARY

1. **Key Features of Radiate Animals.** Two aquatic phyla, Cnidaria and Ctenophora, are composed of radial or biradial eumetazoans. Most cnidarians and all ctenophorans are marine. The body wall consists of an epidermis and a gastrodermis separated by a mesenchyme, which is not homologous with mesoderm of bilateral metazoans. There are few organs and organ systems, but tissues are well developed.

2. The Phylum Cnidaria includes solitary and colonial carnivores with cnidocytes containing dischargeable cnidae. Many species have two adult body forms: a cylindrical, bottom-dwelling polyp and a free-swimming umbrella-like medusa (jellyfish). The gut is incomplete (the gastrovascular cavity lacks an anus); the mouth is circled by tentacles.

3. Many cnidarians have a hydrostatic skeleton and muscles in the body wall. Many colonial species are surrounded by a rigid chitin-protein exoskeleton called a perisarc, and corals secrete calcium carbonate exoskeletons.

4. Corals, several medusae, and some sea anemones are filter feeders. Digestion begins in the gastrovascular cavity and is completed within gastrodermal cells. Many species harbor photosynthetic protists from which they obtain some nutrients.

5. Polyps and medusae have one to three nerve nets at the base of the epidermis and sometimes the gastrodermis. Many medusae have a nerve ring, statocysts, and ocelli.

6. Polyps usually reproduce asexually, while medusae are typically sexual; many cnidarians are hermaphroditic. Some life cycles include both medusa and polyp stages; others have one stage reduced or absent. The zygote in many marine species develops into a ciliated, free-swimming planula larva, that develops into a polyp; in some species, the polyp produces medusae by budding.

7. Radiate animals constitute a side branch of the animal kingdom. The cnidarian planula larva resembles colonial protozoa and may be similar to the first eumetazoans or their ancestors. The Class Hydrozoa, with its diverse life cycles, may be the oldest of the four cnidarian classes.

8. The Phylum Ctenophora consists of biradial and predaceous or filter-feeding marine plankton. Ctenophorans have eight ciliary comb rows, a complete gut, and a mesenchyme with tissues. Ctenophorans lack cnidocytes, but have adhesive cells (collocytes) that function in food capture.

☐ FURTHER READING

Barnes, R. D. *Invertebrate Zoology.* 5th ed. Philadelphia: CBS, 1987, Chap. 5. *A good survey of diversity and functional biology of the radiate phyla.*

Brown, B. E., and L. S. Howard. "Assessing the Effects of "Stress" on Reef Corals." *Advances in Marine Biology* 22(1985): 1–63. *An authoritative review of changes in coral reefs, both natural and human-caused.*

Endean, R. *Australia's Great Barrier Reef.* New York: University of Queensland Press, 1982. *A discussion of the geology, ecology, and diversity of the largest groups of reefs in the world.*

Friese, U. E. *Sea Anemones.* Neptune City, N.J.: T.F.H. Publications, 1972. *A colorful, informative volume written in nontechnical prose.*

Summerhays, S. "Australia's Great Barrier Reef, a Marine Park Is Born." *National Geographic* 159(1981): 630–635. *Excellent color photographs.*

☐ CLASSIFICATION

Phylum Cnidaria (Gr. *knide*, "nettle"; L. *aria*, "like"). Cnidarians (coelenterates); aquatic, radial or biradial eumetazoans with cnidocytes.

Class Hydrozoa (Gr. *hydra*, "water serpent"; *zoon*, "animal"). Hydrozoans; most have polyp and medusa stages with noncellular mesenchyme; cnidocytes only in epidermis; medusa has a velum in most species; *Hydra* and *Craspedacusta* occur in fresh water; many marine genera include *Gonionemus* and the colonial forms *Obelia*, *Physalia*, and *Stephalia*.

Class Scyphozoa (Gr. *skyphos*, "cup"; *zoon*, "animal"). Scyphozoans; marine jellyfish; the medusa is dominant and lacks a velum; polyps absent or reduced; polyps produce ephyrae by budding; mesenchyme is cellular; cnidocytes occur in both epidermis and gastrodermis; *Aurelia*, *Chrysaora*.

Class Cubozoa (Gr. *kybos*, "cube"; *zoon*, "animal"). Cubozoans; marine cnidarians with a large, cube-shaped medusa with a velum and four long tentacles; polyps absent or reduced; polyps develop directly into medusae (without ephyrae); *Chironex*.

Class Anthozoa (Gr. *anthos*, "flower"; *zoon*, "animal"). Anthozoans (sea anemones, corals, sea pansies, sea fans); marine polyps (no medusae) with well-developed tissues in the mesenchyme; cnidocytes in both epidermis and gastrodermis; polyps of organ-pipe corals (*Tubipora*), soft corals (*Alcyonium*), sea fans, gorgonian corals, red corals (*Gorgonia*), and sea pansies (*Renilla*) have eight tentacles; sea anemones (*Metridium*, *Stichodactyla*) and stony corals (*Astrangia*, *Acropora*, *Montastrea*, *Porites*) have body parts arranged in multiples of six.

Phylum Ctenophora (Gr. *ktenos*, "comb"; *phoros*, "bearing"). Ctenophorans (comb jellies); *Pleurobrachia* (sea gooseberries), *Velamen*.

Symbiosis

Symbiosis literally means "living together," and in biology, a symbiotic relationship is any type of intimate association between individuals of two or more species. Many species have adopted symbiosis as their main survival strategy, and few animals exist in nature without serving as a habitat for thousands of microorganisms and often several metazoans. Organisms that live in or on individuals of another (the host) species are called **symbiotes**.

There is a broad range of symbiotic relationships. In the simplest type, called **phoresis**, an individual of one species may obtain a ride on an individual of another species. The rider is the symbiote, or more specifically, the **phoront**. A common example of phoresis is the accidental transport of cysts, eggs, or adults of one species on the body of another. In being dispersed, the phoronts may benefit from the ride, but they do not depend on it, and the host is unaffected. In other types of symbiosis, symbiotes are specifically adapted for exploiting their hosts and are partially or totally host-dependent. **Commensalism** (literally, "eating at the same table") resembles phoresis in that the host is unaffected, but the symbiote (commensal), although able to survive on its own, usually gains fitness by associating with the host. Typical commensals have behavioral adaptations that allow them to derive food and/or protection from their hosts. Many commensals eat food scraps left over by their hosts. Among the many cases of protective commensalism, several small marine fishes and crabs associate with sea urchins, hiding among the urchins' sharp spines when approached by predators.

Parasitism, a third type of symbiosis, involves greater symbiote and host adaptation and greater host exploitation than commensalism. Parasites benefit, while their hosts are adversely affected. *Obligatory* parasites are metabolically (usually nutritionally) dependent on their hosts and cannot survive without them. *Facultative* species can be parasitic if they gain access to a host, and free-living if they do not. Generally, parasites evolve to a state of not critically harming their hosts, for in destroying a host, a parasite is usually destroying its own environment. The host also evolves in the relationship, and in most cases, the protective responses of the host exert some degree of control on the parasite population. Parasitism is a common way of life. We have already seen that certain phyla (Apicomplexa and Microspora among the Protozoa,

---

Table A Animal Phyla with Parasitic Species

Myxozoa*	Rotifera	Annelida
Mesozoa*	Nematoda	Arthropoda
Porifera	Nematomorpha**	Pentastomida*
Cnidaria	Acanthocephala*	Chordata
Platyhelminthes	Mollusca	

---

*All species are parasitic.

**Only juveniles are parasitic.

and Myxozoa and Mesozoa among the lower Metazoa) are composed entirely of parasitic species. Fourteen of the 34 phyla of animals have some parasitic species, and 5 of these 14 phyla consist entirely of parasites (Table A).

Mutualism, a fourth type of symbiosis, involves mutual exploitation and a complex interdependence between individuals of two species. The relationship may be obligatory or facultative for either or both species, but both species derive benefit from it. Mutual associations often provide housing, food resources, protection, and other adaptive benefits unavailable to either species alone. For example, in the last chapter, we saw that the cellular association between coral animals and photosynthetic symbiotes results in the formation of coral reefs and mutual benefit for both creatures. **Symbiotic cellulose digestion** is another significant example. Few animal species can digest plant cellulose, but by providing a habitat in their intestine for cellulose-digesting bacteria and protists, many species exploit this readily available food resource. In Chapter 19, we described how this relationship works to the mutual benefit of termites and their intestinal flagellates. Likewise, cattle, goats, sheep, and many wild herbivores house cellulolytic microorganisms and exploit them in a similar way. Several other types of mutually beneficial symbioses include **protective mutualisms**, which often involve intricate behavioral interactions between species. One species may provide leftover food, while its mutual partner may give warning signals. Also involving complex behavior interactions, many animal species have become adapted to clean parasites and dead cells from the surface of other animals.

Phoresis, commensalism, parasitism, and mutualism are not always as easy to distinguish as their definitions might imply. It is often difficult to discriminate between certain forms of commensalism and facultative parasitism involving little or no visible effect on a host. Likewise, it is not always easy to distinguish between the one-way exploitation of parasitism and the two-way exploitation of mutualism. Host and symbiote physiology strongly affect these relationships. Ordinarily harmless parasites—even mutualists—may produce serious disease in undernourished or otherwise stressed hosts. A parasite may be harmful as its population expands soon after it infects a new host; then, as the host develops immune responses, parasite numbers may be reduced to a relatively benign level. Together, symbiotes and their hosts constitute coevolving units. Evolutionary adaptations allowing greater reproduction by a parasite, for example, may be followed by host adaptations providing greater resistance to the effects of the parasite. As coevolution occurs, one type of symbiotic relationship may evolve into another. Certain phoronts, for example, may gradually become more dependent on their hosts and eventually evolve into commensals or parasites; some parasites have probably evolved from commensal organisms; and hosts of certain parasites may be in the process of evolving ways to exploit their symbiotes, thus leaning toward mutualism. Symbiotic relationships are perhaps best viewed as frequently overlapping coevolving units, rather than as four distinct types of associations.

The next chapter is mostly about flatworms (Phylum Platyhelminthes), a group in which symbiosis, especially parasitism, is very common. One class of flatworms is largely free-living; the other three are parasitic. As you study and compare these animals, try to develop a sense of the various adaptations that allow parasitic species to exploit their hosts without causing serious harm. At the same time, try to weigh the benefits and potential liabilities of parasitism versus the "free life."

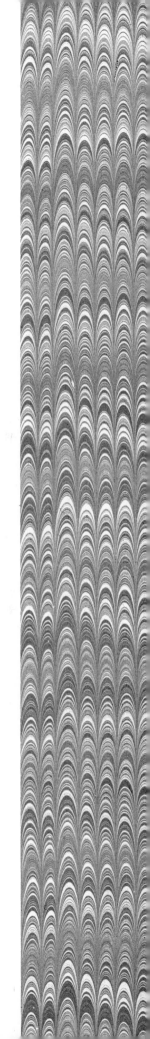

CHAPTER 22

Acoelomate Animals

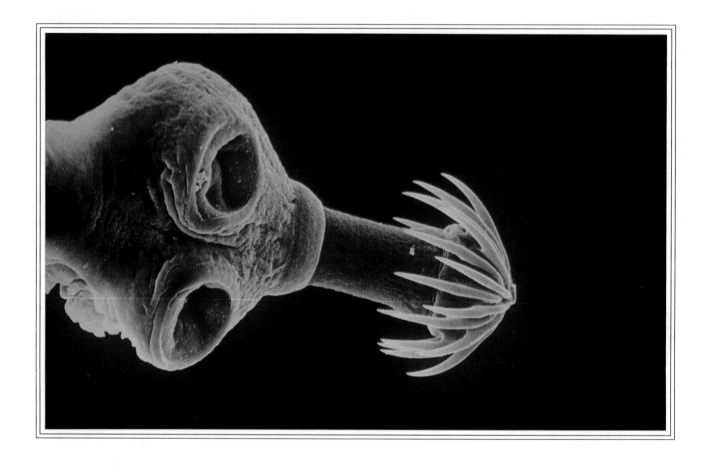

Three phyla of wormlike bilateral animals, the Platyhelminthes (flatworms), the Gnathostomulida (gnathostomulids), and the Nemertea (ribbon worms) are collectively called **acoelomates** because they lack a body cavity. Acoelomate body organization is more complex than that of radiate animals,

but simpler than that of most other bilateral metazoans. Acoelomates are triploblastic (they have three embryonic germ layers), and their mesoderm is derived mainly from endoderm, as it is in other bilateral animals. Most acoelomates have organs and several organ systems, and most (except ribbon worms) have an incomplete digestive tract (no anus). They are cephalized (have a head) and move headfirst through the environment. Many zoologists believe that the earliest metazoans were acoelomates, perhaps similar in body form to certain flatworms living today.

Phylum Platyhelminthes (flatworms). Many flatworms are parasitic. As seen with the scanning electron microscope, the scolex (anterior end) of this tapeworm *Acanthocirrus retrirostris* from a shore bird is adapted for attaching to the intestinal wall of its host (380×).

PHYLUM PLATYHELMINTHES

Flatworms live in marine, freshwater, and moist terrestrial environments, but out of about 15,000 species in the phylum, more than 85% (about 13,000 species) are parasitic. Several species infect humans and domestic animals, causing diseases of great medical and economic importance. Free-living flatworms include the familiar planarian worms of freshwater lakes and streams. Parasitic flatworms include the flattened, leaflike **flukes** and elongate **tapeworms** (see Figures 22.2, 22.3, and 22.4). The smallest flatworms are only about a millimeter long, but some tapeworms grow to lengths of 15–20 m. Several unusual reproductive features help distinguish the flatworm group as a whole. Most species have biflagellate sperm cells, and the sperm flagella have a 9 + 1 arrangement of microtubules. Most other animals have uniflagellate sperm with a 9 + 2 microtubule pattern. (The 9 + 2 arrangement is discussed in Chapter 9, p. 197.) Another unusual flatworm trait is that in many species, structures called **vitellaria,** separate from the ovary, produce yolk.

Flatworms lack rigid skeletal elements and have no circulatory and gas exchange systems. They either have an incomplete digestive tract or lack a gut altogether. Lacking a body cavity and a circulatory system, they have a limited capacity to service internal cells, and as a result, their body size and shape are limited. Flatworms attain sizes of more than a few millimeters only because their thin, flattened bodies provide short distances over which gases, nutrients, and wastes can diffuse to and from internal cells. Some zoologists speculate that the flatworm's lack of certain organ systems is a physiological handicap that has led to most species evolving as parasites. It is also possible, however, that some flatworms derived their simplicity in adapting to parasitism.

REPRESENTATIVE FLATWORMS

There are four classes in the Phylum Platyhelminthes. All free-living flatworms belong to the Class Turbellaria. There are two classes of parasitic flukes, the Trematoda and the Monogenea, and tapeworms belong to a fourth class, the Cestoidea.

Class Turbellaria

Except for a few commensal and parasitic species inhabiting invertebrates and fishes, **turbellarians** are free-living predators and scavengers. One group of free-living marine turbellarians called the **acoels** has the simplest body organization in the phylum (Figure 22.1a). Acoels are less than 10 mm long and are bottom dwellers in temperate and northern oceans. As their name implies, they lack any type of internal cavity, including a digestive tract. Their body consists of three cell regions: An epidermis surrounds a region of connective tissue cells, and an innermost region contains endodermally derived phagocytic cells that engulf and digest food. In some acoels, the phagocytic cells form a temporary digestive sac. Most acoels eat microorganisms, although some, including the common species *Convoluta roscoffensis*, may harbor intracellular green algae; little is known about this symbiosis, but *C. roscoffensis* can survive without algae. When the worm contains green algae, it loses its digestive cells, suggesting that it uses algal products for food. *Convoluta roscoffensis* frequents North Sea coasts, and its dense populations often give a green cast to tidal mud flats. The worms sometimes survive in anaerobic environments, perhaps by using oxygen produced by their algae.

Turbellarians other than acoels have some type of digestive tract. Members of a marine group called the **polyclads** have a highly branched gut (Figure 22.1b). Another group, called the **rhabdocoels,** is characterized by a simple, saclike digestive tube (Figure 22.1c). Rhabdocoels are common in marine coastal waters and freshwater lakes and ponds throughout the world. The best-known turbellarians are the freshwater planarians of the genus *Dugesia* (Figure 22.1d). The planarian digestive tract has a ventral mouth, muscular pharynx, and three branched internal pouches; planarians and other turbellarians with three gut pouches are collectively called **triclads.** Planarians feed on small invertebrates and dead animal matter. A few species are terrestrial, living under wet leaves and rotting logs in tropical rain forests.

Class Trematoda

Trematodes are called **digenetic flukes** because their life cycle is characterized by both sexual and asexual reproduction. With few exceptions, larval flukes, which undergo asexual reproduction, infect a mollusc (the **intermediate host**), and the adults, which produce gametes, parasitize a vertebrate (the **definitive host**). Adults of several species, in-

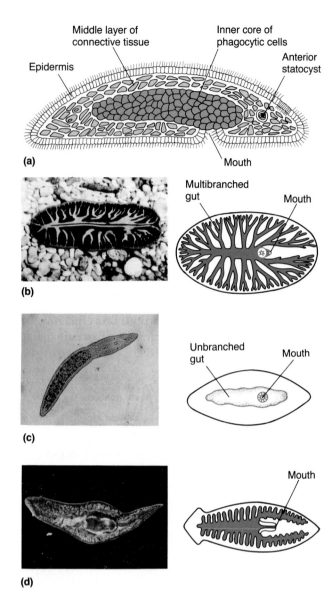

Figure 22.1 Class Turbellaria (free-living flatworms). (a) Body plan of an acoel flatworm. There are three cell regions, but the worm lacks a gut. (b) Polyclads are marine; many, such as this species (*Pseudoceros zebra*, from Hawaii), are brightly colored. The drawing shows the multibranched gut characteristic of polyclads. (c) The rhabdocoel *Stenostomum* (about 0.5 mm long) common in stagnant freshwater ponds, has an unbranched saclike gut (see drawing). Rhabdocoels inhabit coastal marine and brackish water lagoons, and all types of freshwater habitats. A few rhabdocoels are commensals, living on the body surfaces of marine and freshwater invertebrates and turtles. (d) *Dugesia* and other freshwater planarians often occur on the undersurfaces of rocks in clear streams. Common species are brownish black and are often lightly spotted. Planarians are triclads, their gut having three main branches (see drawing).

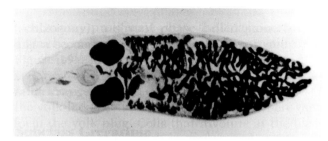

Figure 22.2 Class Trematoda (digenetic flukes). This species (*Platynosomum fastosum*) infects domestic cats. The hermaphroditic adult (shown here, stained and mounted for microscopic study) is about 8 mm long by about 2.5 mm wide; it lives in the bile ducts of the liver and may cause diarrhea and jaundice. Cats acquire the adult worms by eating lizards infected with encysted larvae. Early larval stages develop in snails.

severely impair host liver function, and drugs are not very effective against them.

From a medical standpoint, the most important trematodes are the **schistosomes**, the blood flukes. Three species of *Schistosoma* infect people, and blood fluke disease, or **schistosomiasis**, is one of the most significant human health problems. Unlike other members of the class, schistosomes are dioecious (they have separate male and female adults). The male and female worms live together in the host's blood vessels (see Figure 14.3a). The symptoms of schistosomiasis result mainly from the worms' eggs, which clog blood vessels, damage internal organs, and cause severe internal allergic responses, ulcers, and swelling. A new drug called praziquantel kills the adult worms, but nothing is known to remove the eggs in body tissues.

Closely related to the human schistosomes are other blood flukes whose definitive hosts are nonhuman mammals and birds. The adult worms cannot develop in humans, but if the larval flukes penetrate human skin, they cause a severe allergic skin reaction called swimmer's itch. The fluke larvae die in the skin and are eventually phagocytosed and digested by host cells. Outbreaks of swimmer's itch occur among people who swim in lakes containing snails infected with larval flukes.

Class Monogenea

Monogenetic flukes, or monogeneans, superficially resemble trematodes but have only one host species and reproduce only sexually. Nearly all of these worms infest the body surface, particularly the gills, of aquatic vertebrates. Also unlike the digenetic flukes, monogeneans have a very large, muscular attachment organ called the **opisthaptor**, with which they grip the body surfaces of their hosts (Figure 22.3).

cluding *Clonorchis sinensis* (which infects humans) and *Fasciola hepatica* (common in sheep and cattle), live in the bile ducts of their host's liver, feeding on tissue cells and blood. Figure 22.2 shows a species that infects the bile ducts of cats. Although liver flukes rarely kill their hosts, heavy infections

Essay: Schistosomiasis: An Age-Old and Modern Health Problem

Three species of schistosomes, the blood flukes, infect humans and cause the long-lasting and often deadly disease schistosomiasis. About 250 million people in 70 countries have this disease (Figure E1). *Schistosoma japonicum,* as its name implies, occurs mainly in Japan and Southeast Asia. Two other species, *S. mansoni* and *S. haematobium,* are widespread in Africa and the Middle East. *Schistosoma mansoni* is thought to have been introduced to the western hemisphere with the slave trade, and it now occurs in the West Indies and in tropical areas of South America. People in North Africa, particularly Egypt, have been afflicted with schistosomiasis since the dawn of civilization. Accounts of persons with the characteristic symptom of bloody urine have been found on ancient Egyptian scrolls, and schistosome eggs have been recovered from Egyptian mummies over 3000 years old. The first recorded cases of the disease in Europeans occurred in soldiers of Napoleon's army in Egypt around 1800, and during the mid-1800s, nearly 50% of the Egyptian population was found to be infected.

Schistosomes have several habitat requirements, including the availability of suitable snails in which to develop to the infective larval (cercaria) stage. Susceptible mammals must come in contact with cercariae in the water, and water with snails must be contaminated with urine or feces containing eggs. Proper disposal and treatment of human sewage reduces the incidence of blood flukes. The number of cases of schistosomiasis has decreased in Japan and Puerto Rico in the past 20 years, mostly because of improved sanitation and the reduction of snail breeding areas. In the same time period, however, Egypt's Nile River Valley has seen an increasing number of cases (see Figure E1).

The Aswan High Dam was constructed in the 1960s on the Nile River to increase food production. The dam provides irrigation water for desert areas and controls yearly flooding that formerly caused significant crop losses in the Nile Valley. Since Lake Nasser behind the dam was filled in the late 1960s, crops have been grown throughout the year in a 500-mile area along the Nile between Cairo and Aswan. Year-round cropping had previously been possible only in the Nile Delta area. Unfortunately, the agricultural benefits of the dam are being outweighed by the harmful influence of blood flukes. Since completion of the dam, the percentage of human cases of schistosomiasis in villages along the Nile between Cairo and Aswan and above the dam has increased from less than 10% to as much as 75% of the population. Schistosome-bearing snails thrive in the irrigation ditches created to carry water from the dam's lake into agricultural fields. Without proper sewage facilities in the area, contamination of the irrigation water with human waste provides ideal conditions for schistosomes. People frequently use the ditches for bathing and laundering and often come in contact with the infective cercariae.

Adding insult to injury, the productivity of the soil in the Nile Valley seems to have decreased since completion of the dam. For thousands of years, the yearly Nile floods had added new soil and nutrients to the area. Now, with the dam's flood control, this beneficial effect has been removed. Thus, as a result of its well-intentioned attempts to better feed its people, Egypt must now contend with an increased, but age-old health problem and the possibility of long-term crop failure.

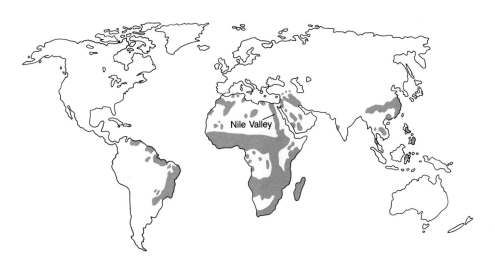

Figure E1 Distribution of schistosomiasis, human blood fluke disease.

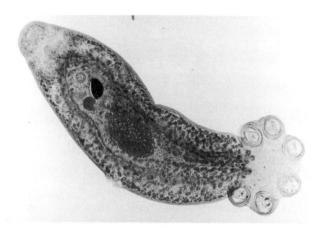

Figure 22.3 Class Monogenea. Most monogenetic flukes live on the body surfaces of fishes (both freshwater and marine), but a few live on frogs, turtles, and molluscs, and one even inhabits the eye of the hippopotamus, a semiaquatic mammal. Note the opisthaptor (posterior holdfast organ) of *Polystoma*.

Class Cestoidea

Tapeworms, or cestodes, have a tapelike body with an anterior attachment organ, the **scolex**, and a neck region attached to a chain of segments called **proglottids** (Figure 22.4). The proglottid chain, called the **strobila**, grows continuously by budding from the neck. Nearly all adult cestodes parasitize vertebrates, and most species inhabit the intestine. Larvae infect various tissues and organs in a wide variety of invertebrates and vertebrates. Tapeworms have no digestive tract; apparently, they lost it in becoming adapted to parasitism. The vertebrate intestine provides a ready supply of food, and tapeworms simply absorb nutrients already digested by their host.

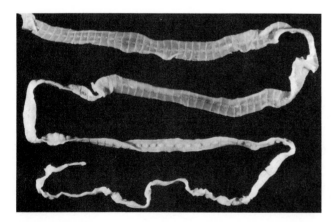

Figure 22.4 Class Cestoidea. Tapeworms, such as this specimen of *Diphyllobothrium latum* from a human, have a tiny scolex (head) that embeds in the intestinal wall of the host, as well as a chain (strobila) of segments containing mostly reproductive structures or mature eggs. The strobila can be many meters in length in some species, but less than 10 mm long in the smallest tapeworms (see Figure 22.16).

There are approximately 1000 species of tapeworms, and most cause little harm to their definitive hosts. In humans, adult worms may cause anemia and vitamin deficiencies, but infections can be rather easily diagnosed (by identifying eggs or proglottids in the feces), and chemicals are available that kill the worms. Larval tapeworms, on the other hand, often migrate in their hosts, and some grow to large size, causing serious tissue damage.

HABITATS AND ENVIRONMENTAL RELATIONSHIPS

The free-living flatworms and the parasitic flatworms form two profoundly different environmental groups.

Free-Living Flatworms

Free-living turbellarians occupy a wide variety of bottom substrates in all kinds of aquatic environments. Larger species, such as freshwater planarians and several marine polyclads, are often abundant on rocks, while the smaller acoels and rhabdocoels live mainly on mud or fine sand. The rhabdocoel *Gyratrix hermaphroditus* has an unusual tolerance for different salinities and occurs in marine, brackish, and freshwater environments. Freshwater turbellarians are influenced strongly by water temperature, dissolved oxygen, and current velocity. Species living in small, temporary ponds in temperate zones must be able to withstand marked daily changes in both temperature and level of dissolved oxygen. Many species produce resistant cocoons, which surround and protect their eggs from drying and freezing.

Few animals eat turbellarians, and some turbellarians secrete chemicals that repel potential predators. Several turbellarians are cannibalistic, eating their own kind as readily as other prey.

Parasitic Flatworms

The environmental relationships of parasitic flatworms are much more complex than those of free-living species. Parasites are adapted to two environments: the host, or **microenvironment**, and the general surroundings, or **macroenvironment** (Figure 22.5). Changes in either environment can affect the worms and can alter how the worms affect their host.

Figure 22.5 Factors influencing the relationship between a parasite and its hosts. During the course of a parasite (in this case a blood fluke) infection, the host-parasite relationship may change as a result of a number of variables affecting either the host, the parasite, or both. Host stress is especially important; as a host expends energy adjusting to stress factors such as temperature, pH, or salinity extremes, poor nourishment, or crowding, its susceptibility to infection and its tendency to develop serious disease may increase. Crowding of hosts can have a dual effect: It may cause host stress and it may provide greater opportunity for parasites to infect hosts. Many parasitic flukes and tapeworms have coevolved with their hosts to a level where the host's defenses regulate the parasites' growth and potential for causing disease. In such cases, the host-parasite relationship may be viewed as approaching a homeostatic condition.

Finding and infecting new hosts are significant challenges for parasites, and many species have increased their chances of survival by adapting to the habits and niches of their hosts. Some flatworms produce eggs or larvae only during seasons when hosts are available. Most trematodes and tapeworms have evolved to take advantage of the predator-prey relationships of their hosts; larval worms usually infect prey species, and adult worms usually develop in predators that commonly eat the hosts of the larval worms. Such life cycles require that the parasites be able to survive and reproduce in a great range of environments both within and outside hosts and that they be able to contact and infect new hosts. Typically, few offspring survive and complete the life cycle, but most species compensate for this by producing enormous numbers of offspring each generation.

Several studies have shown that the larvae of certain flatworms can alter their host's behavior, making the intermediate host more available to predators. For example, adults of the trematode *Brachylecithum mosquensis* inhabit the liver of the American robin (Figure 22.6). A robin becomes infected by eating a carpenter ant that carries larvae encysted on its brain. Uninfected ants are fast-moving and rarely sit out in the open; but infected ants become sluggish and spend considerable time during daylight hours on exposed rock surfaces, where they are much more susceptible to predators such as robins. It is the larval flukes that cause this behavioral change in the ants.

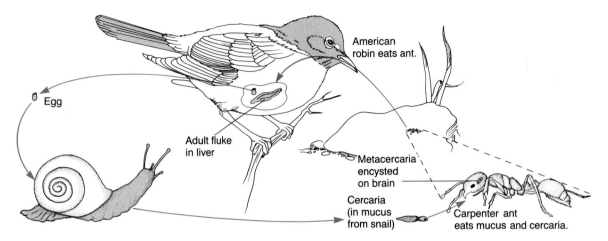

Figure 22.6 Behavioral change in a host caused by a parasite. In this example, the carpenter ant is infected with a larval fluke (metacercaria). Having the metacercaria encysted on its brain causes the ant to become more likely prey for the robin, which is the fluke's definitive host. The rest of the parasite's life cycle occurs in a land snail that becomes infected by accidentally eating the worm's eggs (passed in bird feces) on vegetation. Fluke larvae (cercariae) develop in the snail; infected snails periodically drop mucous "slime balls" containing the larvae, and carpenter ants become infected when they eat the slime balls.

COMPARATIVE STRUCTURE AND FUNCTION

In this section, we compare the body structures and functions of free-living turbellarians with those of parasitic flatworms. As you will see, there are profound differences, especially in the structure and function of body coverings. Flatworms have a living body covering that is relatively simple in free-living species, but exceedingly complex in parasites.

Body Support, Locomotion, and Protection

Flatworms are supported by a hydrostatic skeleton resulting from cell fluids in connective tissue surrounding their internal organs. A layer of fibers under the body covering provides additional support and attachment sites for underlying muscles.

Free-Living Flatworms.   Turbellarians have an epidermis consisting of a single layer of cells (Figure 22.7). Many cells, particularly ventral ones, are ciliated, and most turbellarians move by gliding, using their cilia and a slime trail secreted by epidermal gland cells. The worms' flattened shapes allow them to crawl over and press tightly to substrates. *Dugesia* and other planarians can also twist and turn because they have several layers (circular, longitudinal, and diagonal) of smooth muscle in

their body wall; similar muscles allow some polyclads to swim by undulating. Some turbellarians secrete adhesive substances that allow them to stick temporarily to rocks or sand grains when the worms are disturbed or in turbulent water.

Notice in Figure 22.7 that some cells in the turbellarian epidermis contain rodlike structures called **rhabdites**. Although little is known about the function of these structures, turbellarians often discharge rhabdites when disturbed, and the slimy sheath that results may protect the animal. Also, some turbellarians secrete chemicals that repel predators, and a few marine species that graze on cnidarians incorporate nematocysts into their epidermis. (Nematocysts are discussed in Chapter 21, p. 499.)

Parasitic Flatworms.   The living surface of parasitic flatworms is a continuous mass of cytoplasm (a syncytium) called the **tegument**, which extends from underlying cell bodies called **cytons** (Figure 22.8). The tegument provides body support, resists digestion by host enzymes, and absorbs small nutrient molecules. Rhabdites are absent, but in some species, the tegument contains chitinous spines. Much of the movement of adult worms is provided by attachment organs, which are special adaptations of the tegument and the well-developed underlying muscles. The suckers (holdfasts) of trematodes and the opisthaptors of monogenetic flukes can grip host surfaces and pull the worms over substrates (see Figures 22.2 and 22.3).

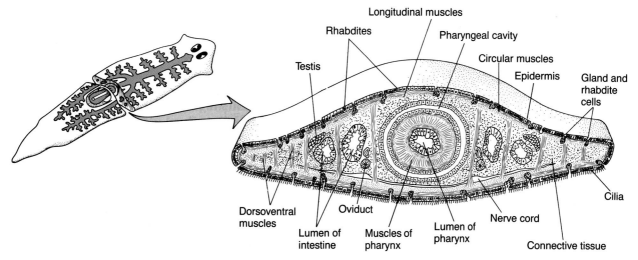

Figure 22.7 Transverse section of the freshwater planarian *Dugesia*. The single-cell-thick epidermis contains gland cells, sensory cells, and cells containing rhabdites. The body wall consists of circular and longitudinal smooth muscle layers separated by a diagonal layer. Thin bands of dorsoventral muscles extend from top to bottom.

One of the main challenges facing flatworms that live inside other animals is to keep from being digested or killed by the host's immune system. A fluke's or tapeworm's surface coat is its main line of defense against digestive enzymes in a host's intestine and against defensive cells and antibodies in the host's bloodstream. The tegument of flukes and tapeworms secretes a protective outermost coat called the **glycocalyx** (see Figure 22.8). This is a thin layer of molecules (mainly polysaccharides and glycoproteins) that are vital to the parasites' survival. Blood flukes are known to elude their

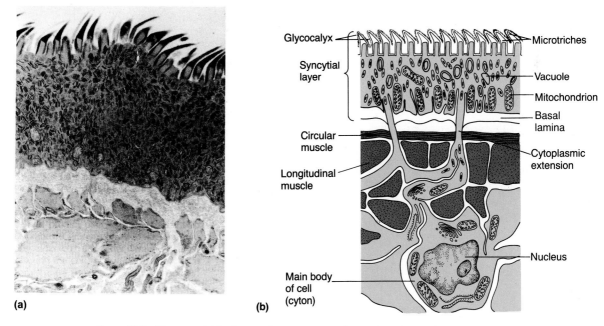

Figure 22.8 The syncytial body covering (tegument) of a tapeworm. (a) Section of the tegument as seen with the electron microscope. (b) Sketch of a section through a tapeworm tegument. Nuclei in the cell bodies (cytons) are protected from host enzymes and protective antibodies by their location deep under the layers of body wall muscles. Parasitic flukes (trematodes and monogeneans) lack the surface projections (microtriches); otherwise, their teguments are similar to the tapeworm's.

host's immune defenses by disguising themselves with molecules of the host. Their glycocalyx incorporates host antigens (glycoproteins or glycolipids), thereby thwarting the host's immune system. Since the immune system cannot distinguish the worms from host tissues, it cannot attack the worms. In this way, blood flukes can evade a host's defenses indefinitely. Because knowledge of the glycocalyx may provide keys to controlling important human diseases, research on tapeworm and trematode surface coats is an active area in parasitology today.

Feeding and Digestion

Turbellarians have a ventral mouth, often placed well back from the anterior end of the body. A pharynx is usually present, and in many species, it is muscular and can be protruded to manipulate or suck food (Figure 22.9). The branched digestive tracts of *Dugesia* and the polyclads provide a large internal surface area for enzyme secretion, nutrient absorption, and internal transport. Digestion begins extracellularly in the gut; cells lining the gut then endocytose the partially digested food, and digestion is completed within these cells.

In trematodes, virtually all regions of the gut and external body surfaces may be active in feeding, digesting, and absorbing food. By tying off the mouth of the sheep liver fluke *Fasciola hepatica*, researchers have found that this worm can survive and produce eggs by relying entirely on its tegument to absorb nutrients from the body fluids of its host. Having a gut merely allows the flukes to consume host tissue cells as well as liquid nutrients.

Lacking a gut, tapeworms rely entirely on nutrient absorption by their tegument. The tapeworm tegument is covered with projections called **microtriches** (sing. microthrix) (see Figure 22.8). Like the microvilli on the vertebrate intestine, microtriches greatly increase the absorptive surface area of the tapeworm tegument.

Many parasitic flatworms live in an oxygen-poor environment in their host; as a result, their metabolism is anaerobic. Few, if any, have a complete Krebs cycle or an oxygen-dependent electron transport system. Schistosomes, which live mainly in the oxygen-poor blood in veins, obtain most of their energy from glycolysis (see Figure 3.7). Many trematodes and tapeworms, living in the anaerobic contents of vertebrate intestines, have enzymes that carry respiration beyond glycolysis, perhaps allowing them to obtain more energy from organic molecules than can be gained from glycolysis alone (Figure 22.10). As more is learned about parasite metabolism, it may be possible to devise ways to control certain flatworm diseases. Parasites that use unique metabolic pathways may have enzymes not found in hosts, and drugs that inhibit such enzymes may kill the worms without affecting the hosts.

Circulation, Gas Exchange, Osmoregulation, and Excretion

Diffusion and cell membrane transport across external and gut surfaces accommodate most of the circulatory and gas exchange needs of flatworms. The branched gut of turbellarians and many flukes

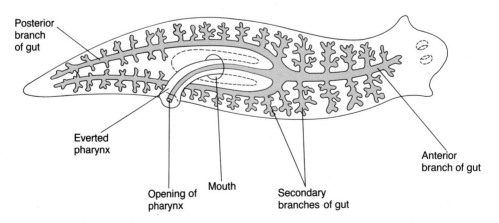

Figure 22.9 Ventral view of *Dugesia*. Note the three main branches (one anterior, two posterior) and many small secondary branches of the incomplete gut; the pharynx is eversible.

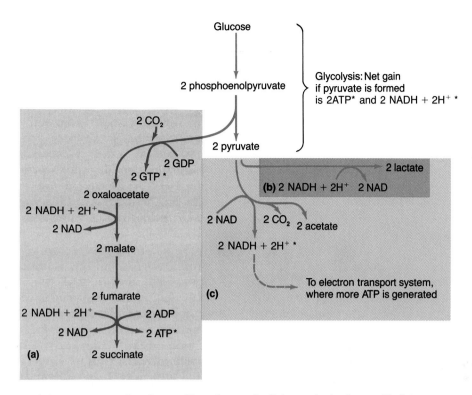

Figure 22.10 Examples of anaerobic pathways of cellular respiration in parasitic flatworms.
Molecules representing energy gain are marked with an asterisk. (For a review of the reactions of glycolysis, see Chapter 3.) (a) Many species have enzymes that convert phosphoenol-pyruvate, a precursor of pyruvate, to oxaloacetate. Carbon dioxide is fixed in this process, and two molecules of a high-energy compound, guanine triphosphate (GTP), are generated. Subsequent production of the organic compounds malate, fumarate, and succinate generates ATP in addition to that formed in glycolysis. Some flatworms can carry the reactions only as far as malate; others have enzymes that can carry them to succinate. Malate or succinate are then excreted as wastes. (b) Some flukes and tapeworms convert pyruvate to lactate (as do vertebrate muscles during periods of heavy activity). In so doing, they do not gain any more energy than that obtained in glycolysis, but hydrogen carrier molecules (NADH + H$^+$) produced by reduction during glycolysis are oxidized for future use; lactate is then excreted. (c) Conversion of pyruvate to acetate reduces more NAD to NADH + H$^+$ (in addition to that reduced in glycolysis). The reduced hydrogen carriers may then be oxidized in an electron transport system composed of cytochromes; if oxygen is not available, the final hydrogen acceptor may be an organic molecule, such as fumarate.

also aids in distributing nutrients, and circulation is enhanced as intercellular body fluids shift when flatworms move.

Many free-living and parasitic flatworms have **protonephridia** (flame cells) with interconnecting tubules (see Figure 7.8). As discussed in Chapter 7 (p. 156), these structures function mainly in osmoregulation, removing excess water from freshwater species, and they are absent from many marine turbellarians. Some wastes may be excreted via protonephridia, especially in parasitic flatworms; in turbellarians and in flukes that live on host body surfaces and are therefore exposed to water, ammonia is the main nitrogenous waste product, and

this readily diffuses into the aquatic environment from surface cells.

Nervous Systems and Sensory Structures

In conjunction with their bilateral symmetry, most flatworms have an anterior concentration of nervous structures. Most have a fused pair of "cerebral" ganglia, sometimes called the brain, but there are varying degrees of centralization and cephalization in flatworm nervous systems. The simplest type of nervous system in the phylum—in fact, the simplest type of nervous system in any bilateral animal—is the cnidarian-like nerve net of

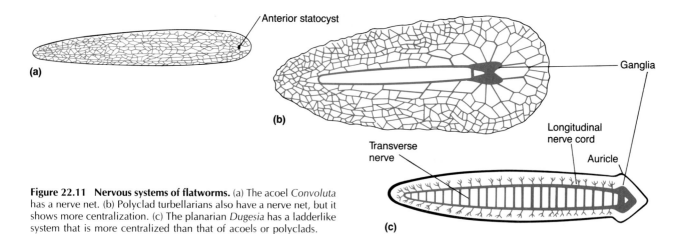

Figure 22.11   Nervous systems of flatworms. (a) The acoel *Convoluta* has a nerve net. (b) Polyclad turbellarians also have a nerve net, but it shows more centralization. (c) The planarian *Dugesia* has a ladderlike system that is more centralized than that of acoels or polyclads.

acoels such as *Convoluta* (Figure 22.11a). The only indication of cephalization in these animals is a slight anterior concentration of nervous elements surrounding a statocyst. Polyclads also have a nerve net, but are considerably more cephalized (Figure 22.11b). In planarians and in most tapeworms and flukes, the nervous system is more centralized, with one or several pairs of longitudinal nerve cords (Figure 22.11c). Transverse nerve trunks or rings that connect the longitudinal cords give the system a ladderlike appearance. Flatworms have sensory neurons, motor neurons, and interneurons similar to those of other bilateral metazoans.

Sensory structures typically include clusters of surface cells that are sensitive to chemicals, touch, water current, and light. *Dugesia* and other freshwater planarians have some of their sensory cells concentrated on earlike flaps called **auricles** (see Figure 22.11). Many turbellarians and larval flukes have two anterior **eyespots**. (Pigment cup ocelli are discussed in Chapter 12, p. 274; Figure 12.2.)

BEHAVIOR

Chemotaxis seems to be a major component of flatworm behavior. Many parasites find hosts by detecting host chemicals in water, and chemotaxis may play a key role in enabling trematodes and tapeworms to localize in specific sites within their hosts. Turbellarians also locate food by orienting toward a chemical source and moving in its direction. In laboratory experiments, when a freshwater planarian first senses mammalian blood, it raises and swings its head, presumably to monitor the

location, and then turns and crawls toward the source. If the chemoreceptors on one side of the head are removed by excising an auricle, the worm cannot orient and merely circles in place.

Touch reception plays a major role in dorsoventral orientation in many flatworms. The ventral surface of turbellarians responds positively to touch, but other body surfaces respond negatively. If flipped on its back, a planarian will twist and turn until some of its ventral ciliated surface contacts the substrate; then it will turn completely over and move away.

The ability to detect and react to light seems to be mainly a protective adaptation in free-living flatworms. Aquatic forms spend most of their time on the dark undersurfaces of rocks and other bottom objects. By contrast, larvae of the human blood fluke *Schistosoma mansoni* are photopositive, which may help direct them to areas frequented by their hosts.

Much research has been conducted to determine if, and to what extent, flatworms can learn. In one type of experiment, mild electrical stimuli have been used to train planarians to remain in bright light instead of moving away from it. The results of these and other studies have been conflicting, and in some cases where learning seemed evident, the worms lost their behavioral modification as soon as the invoking stimulus was stopped.

REPRODUCTION AND LIFE CYCLES

Many flatworms reproduce by both asexual and sexual means. Most species have well-developed reproductive systems, including organs for egg production and copulation. Parasites, especially, are highly prolific, and their life cycles may involve several asexual larval stages in two or more hosts.

Asexual Reproduction

Many turbellarians reproduce by transverse fission, forming new individuals in chains or in pairs (Figure 22.12). Some turbellarians have great regenerative ability, and if injured or broken apart, they can regenerate lost parts or produce whole new individuals. Pieces of *Dugesia* cut longitudinally will regenerate the missing right or left side. If a planarian is cut transversely, pieces anterior to its pharynx will form complete new worms, but posterior pieces usually lack a head if they regenerate at all. Many interesting questions remain about these phenomena. For example, what determines polarity in the transverse pieces so that their anterior cut surface regenerates a head and their rear cut surface forms a tail? Why is regeneration incomplete in posterior pieces? Zoologists speculate that an anteroposterior gradient in metabolic rate determines the animal's polarity. The head is most active metabolically, and the metabolic rate generally decreases from head to tail. Extracts of the head region inhibit head regeneration in transverse pieces, indicating that anterior regions are dominant over posterior regions. Recent studies indicate that neurohormones may make the anterior end

dominant. Because regeneration involves differentiation of specialized body parts from undifferentiated cells—virtually the same process that occurs in embryonic development—research on planarian regeneration may help explain certain aspects of animal development.

Trematodes and tapeworms lack regenerative ability, but their methods of asexual reproduction make these worms highly prolific. Trematodes undergo **larval amplification**, often increasing the number of offspring from each egg by more than a hundred. (Larval amplification is described in Chapter 14, p. 330; Figure 14.2.) Adult tapeworms bud continuously from the neck region just behind the scolex, but this is not asexual reproduction: As the oldest segments, or proglottids, at the posterior end of the worm become full of ripe eggs, they rupture or break off and are passed out with host feces; thus, the budding process is a component of sexual reproduction in tapeworms. In a few tapeworms such as *Echinococcus*, which has only a few segments per adult, the larval stages actually do reproduce asexually by budding (see Figure 22.16).

Sexual Reproduction

Most flatworms are hermaphroditic and have internal fertilization. Cross-fertilization generally occurs, and both partners receive sperm during copulation. Many tapeworms can self-fertilize, perhaps as a special adaptation for parasitism, allowing reproduction even if only a single worm infects a host. The unique schistosomes are dioecious, with the male permanently holding the female in a specialized ventral groove called the **gynecophoral canal** (see Figure 14.3a).

Male reproductive organs usually include a pair of testes, a copulatory organ called the **cirrus**, one or more sperm ducts, and a seminal vesicle (Figure 22.13). The female reproductive system includes one or more pairs of ovaries that are usually separate from the yolk-producing vitellaria. Yolk ducts convey yolk to the oviduct, where the yolk unites with the eggs. The eggs are fertilized in the oviduct or conveyed by oviducts to a seminal receptacle, where fertilization occurs. Drought-resistant shell material is added to the zygotes, which then pass to the outside via the uterus.

Life Cycles

Whereas the life cycle in turbellarians and monogenetic flukes is relatively uncomplicated, in trematodes and tapeworms it is typically complex.

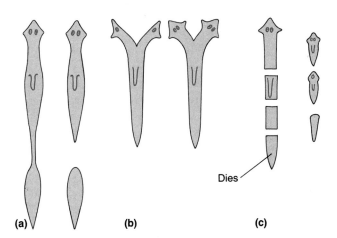

Figure 22.12 Asexual reproduction and regeneration in *Dugesia*. (a) In transverse fission, a parent worm produces two offspring by dividing posterior to the pharynx; regeneration of missing body parts occurs within about two months after separation, depending on temperature. (b) Regeneration of a two-headed worm following a partial longitudinal cut. Two whole worms can regenerate following complete longitudinal cuts. (c) Regeneration following transverse cuts illustrates the worm's polarity; when a worm is cut as shown here, only pieces anterior to the pharynx or those that include the pharynx can regenerate completely.

Dies

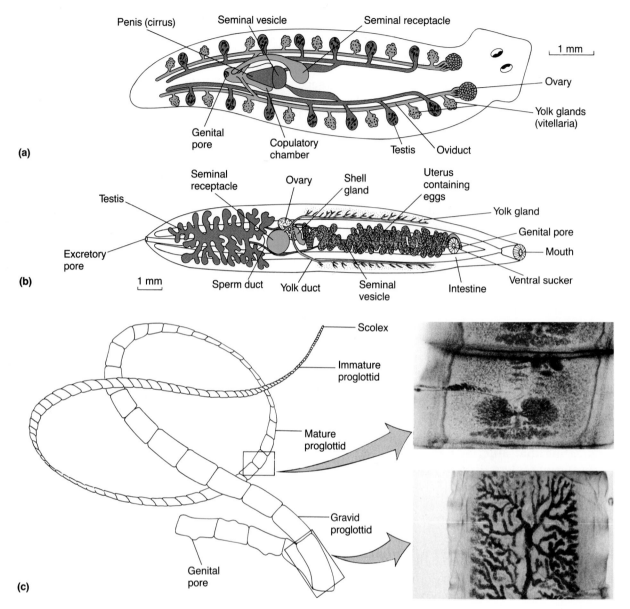

Figure 22.13 Male and female reproductive systems in hermaphroditic flatworms. (a) Planarians have a row of testes on either side of the body; note the pair of ovaries and separate yolk glands (vitellaria). (b) The trematode *Clonorchis sinensis*. Sperm from another adult are injected through the genital pore, swim up the uterus, and are stored in the seminal receptacle. Eggs are fertilized and receive yolk and shell material in a short oviduct (between the ovary and the uterus), then pass through the long coiled uterus and out through the genital pore. In the male system, sperm pass from the testes into sperm ducts, through the single seminal vesicle, and out the genital pore. (c) The beef tapeworm *Taeniarhynchus saginatus*, with adults up to 20 m long (though usually about 3 m), is the most common tapeworm in humans. The size and maturity of its proglottids increase from anterior to posterior. Each sexually mature proglottid, near the middle of the worm, has a complete set of male and female reproductive organs; fertilization occurs within each proglottid. Near the end of the worm, gravid proglottids have lost their reproductive organs and are full of fertilized eggs. Terminal proglottids break off or rupture, releasing eggs containing developing embryos into the host's feces. Infection with *T. saginatus* can be prevented by freezing raw beef for a week or more at −5°C or by cooking beef until it is no longer pink. The drug niclosamide, an inhibitor of anaerobic ATP production, kills the worms.

Free-Living Turbellarians. Many species alternate phases of sexual and asexual reproduction corresponding to environmental changes. A number of freshwater planarians living in temperate climates reproduce asexually during favorable periods; then, as temperatures cool and day length shortens in the autumn, or as temporary ponds dry up, they reproduce sexually, producing zygotes with a thick capsule that resists freezing and drying. The capsules hatch in the spring, or whenever conditions are once again favorable for growth. Some turbellarians have spiral or spirallike cleavage and a mouth that develops from the region of the blastopore, indicating that flatworms are more closely related to protostomes than to deuterostomes. Many turbellarians lack a larval stage, and their embryos look like tiny adult worms. In some marine polyclads, however, a stereogastrula develops into a free-swimming larva with four or eight ciliated lobes.

Monogenetic Flukes. The hermaphroditic adults of these parasites produce many fertilized eggs that are released into the water. Each egg hatches into a ciliated larva called an **onchomiracidium**, which swims about in search of a suitable fish or amphibian host. When it locates and attaches to one, the larva loses its cilia and transforms into an adult worm. In nature, most onchomiracidia run out of energy and die before finding a host, and few fishes are heavily infected. However, when fishes are crowded, as they often are in culture ponds, large numbers of monogenetic flukes may build up on the gills and result in serious disease, particularly if bacteria and fungi infect the wounds caused by the worms.

Digenetic Flukes. Trematodes may have as many as seven different life cycle stages: (1) adult, (2) fertilized egg, (3) **miracidium**, (4) **sporocyst**, (5) **redia**, (6) **cercaria**, and (7) **metacercaria**. Typically, stages 1, 4, 5, and 7 occur within hosts. Fertilized eggs, miracidia, and cercariae occur outside of hosts.

Miracidia hatch from fertilized eggs; they are ciliated and function in locating and infecting a suitable intermediate host. Cercariae, produced asexually in intermediate hosts, emerge and swim about in search of a definitive host (Table 22.1).

The life cycles of the Chinese liver fluke (*Clonorchis sinensis*) and the human blood fluke (*Schistosoma mansoni*) illustrate the basic patterns of reproduction and development in trematodes. *Clonorchis sinensis* has all seven stages in three host species (Figure 22.14). Humans become infected by eating infected raw fish.

In contrast to *C. sinensis*, *Schistosoma mansoni* has no redia or metacercaria in its two-host life cycle, and the adults are dioecious. Female worms deposit their eggs in the small veins around the host's large intestine. To leave the host, the eggs must pass through the blood vessel and surrounding connective tissue back into the intestine and then pass out with host feces. Considering this circuitous route, it is not surprising that few of the thousands of eggs produced ever leave the definitive host. Those that do leave hatch in water; miracidia emerge from these eggs, and if a miracidium penetrates a suitable snail, it develops into a sporocyst. The sporocyst produces daughter sporocysts, and after about a month, cercariae begin to emerge from the daughter sporocysts. Mammals, including humans, become infected with schistosomes by coming in contact with water containing the cercariae. The cercariae penetrate the skin and enter the circulatory system.

Table 22.1 Life Cycle Stages of Trematodes

Life Cycle Stage	Role in Life Cycle	Environment
Adult	Sexual reproduction	Definitive host (vertebrate)
Fertilized egg	Passes out of definitive host	Macroenvironment
Miracidium	Locates and infects intermediate host	Macroenvironment
Sporocyst*	Asexual reproduction	Intermediate host (mollusc)
Redia*	Asexual reproduction	Intermediate host (mollusc)
Cercaria	Locates and infects definitive host or second intermediate host	Macroenvironment
Metacercaria*	Nonreproductive encysted stage; develops into adult when eaten by definitive host	Second intermediate host or vegetation

*Absent in some species.

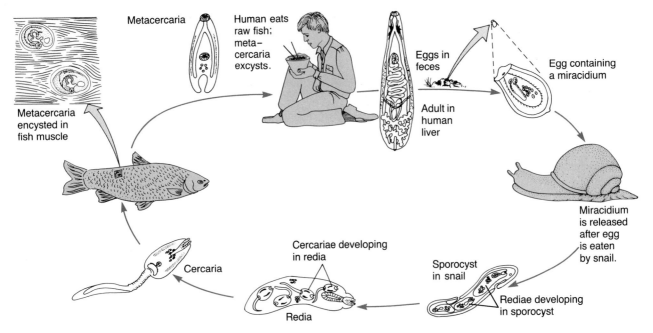

Figure 22.14 Life cycle of the trematode *Clonorchis sinensis*, the Chinese liver fluke. An egg containing a fully formed miracidium passes out of the definitive host in feces. If the egg is eaten by a certain kind of aquatic snail, the miracidium is released, penetrates internal tissues of the snail, and develops into a saclike sporocyst. Each sporocyst produces many rediae, and each redia buds internally to produce many cercariae. The cercariae escape from the snail and become free-swimming in the water until they either run out of stored energy or penetrate a fish. If a cercaria contacts a fish, it bores into it, loses its tail, and encysts in the muscle as a metacercaria. For the life cycle to be completed, a mammal must eat a fish infected with a metacercaria. Once in a mammal's intestine, the metacercaria emerges from its cyst, migrates up the bile duct to the liver, and develops into an adult fluke. Adult liver flukes may live and produce eggs for 20 to 25 years. Heavy infections in humans can seriously impair liver function, sometimes causing death. The disease is diagnosed by finding eggs in feces, but an effective treatment is not known.

Tapeworms. Unlike trematodes, many tapeworms do not reproduce as larvae. The fish tapeworm *Diphyllobothrium latum*, one of the largest parasites capable of infecting humans, has a partly aquatic life cycle with three larval stages (Figure 22.15). A free-swimming, ciliated larva called a **coracidium** hatches from the egg. If the coracidium is eaten by a copepod (a small aquatic arthropod), it loses its cilia and develops into a **procercoid** in the copepod's hemocoel. A third larval stage called the **plerocercoid** may develop in fish that eat infected copepods, and if a suitable mammal eats a fish with plerocercoids, adult tapeworms may develop in the mammal's intestine.

In contrast to tapeworms with aquatic stages, the beef tapeworm *Taeniarhynchus saginatus* has its entire life cycle in terrestrial animals (see Figure 22.13c). The adult worms live in the human intestine, and eggs containing a larva called an **oncosphere** pass out in the feces. The oncosphere is an important adaptation for life on land; it remains quiescent in the eggshell until eaten by an interme-diate host, and while quiescent, it is resistant to desiccation and is long-lived because it consumes little energy. A second larva, called the **cysticercus**, or **bladder worm**, develops in the body muscles of intermediate hosts (various herbivorous mammals, including cattle and sheep) that eat vegetation contaminated with eggs. Humans usually become infected with the adult worms by eating raw or rare beef containing bladder worms.

Among the few tapeworms that have asexual reproduction as larvae, one called *Echinococcus granulosus* has small adults (less than 10 mm long) that inhabit the intestine of carnivorous mammals (Figure 22.16). The adults cause little or no harm to their hosts. However, the larvae, called **hydatids**, which can infect many different mammals, including humans, can be truly dangerous. Hydatids reproduce by internal budding, often forming large cysts (10 cm or more in diameter) in the liver, lungs, or brain. The cysts can cause serious tissue damage, and to date the only effective treatment is surgical removal.

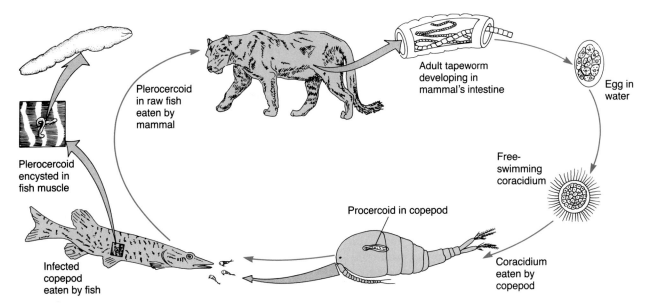

Figure 22.15 Life cycle of the tapeworm *Diphyllobothrium latum*. Hermaphroditic adults (up to 15 m long) in bears, dogs, skunks, seals, cats, and humans produce eggs that pass out with feces (also see Figure 22.4). The life cycle includes two intermediate hosts (a copepod and a fish), but no larval amplification. A ciliated coracidium, about 50 μm long, hatches if an egg is deposited in water. If a copepod eats the free-swimming coracidium, a second larva, the procercoid, develops to about 0.5 mm long in the copepod's hemocoel. Many small fish prey on copepods, and if a fish eats one that is infected, the procercoid emerges in the fish's gut and burrows through the intestine into the muscles. The third larval stage, the plerocercoid, develops to about 2–3 cm in length in fish muscle; if a mammal eats the infected fish, the life cycle is completed. In humans, the adults can cause anemia by absorbing large amounts of vitamin B_{12} from the contents of the intestine, but oral doses of niclosamide eliminate the worms.

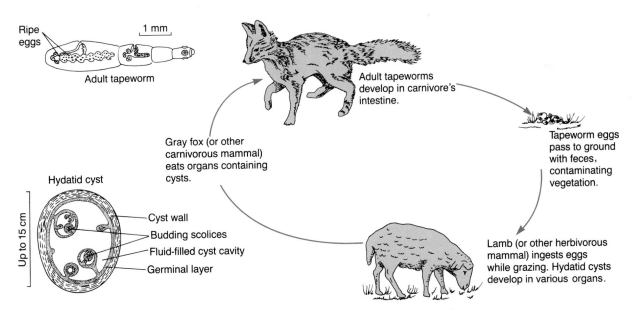

Figure 22.16 Life cycle of the tapeworm *Echinococcus granulosus*. Hermaphroditic adults live in the intestine of carnivores, particularly foxes, dogs, and coyotes. Herbivorous mammals are the most common intermediate hosts because they become infected by eating eggs on contaminated vegetation, but other mammals, including humans, may also harbor the larvae (hydatids). Hydatids reproduce by repeated internal budding to form tiny larval tapeworm heads (scolices). If a carnivore eats the hydatid in infected organs, each scolex can develop into an adult worm. *Echinococcus* species occur throughout the world. *Echinococcus granulosus* is especially common in the western United States, where large numbers of sheep are raised.

PHYLOGENY

Because flatworms are soft-bodied and seldom preserved as fossils, there is little evidence of how and when the Phylum Platyhelminthes originated. We must therefore look to the traits shared by all flatworms for clues about their early evolution. All flatworms are bilaterally symmetrical, and because they have the simplest body construction of all bilateral metazoans, we may assume that the origin of bilateral symmetry and the origin of the flatworm phylum were closely related if not simultaneous events in animal history. (See the Trends and Strategies section beginning on p. 485.) Many zoologists believe that flatworms originated as planuloid (cnidarian, planula-like) organisms. In support of this idea, there is some resemblance between certain acoel turbellarians and the free-swimming planula larva of some marine cnidarians. Many planulae and most acoels are bottom dwellers that move by means of surface cilia. They also lack a body cavity and typically have little differentiation of cells in their body layers. It is possible that ancestral planuloids gave rise to both radiate and bilateral animals. Some ancestral planuloids may have become sedentary and attached to the ocean floor, giving rise to the radiate lineage. Others—progenitors of the bilateral lineage—may have remained free-swimming and gradually evolved a head-to-tail orientation.

Not all zoologists believe that flatworms originated as acoels from planuloids. Some authorities postulate that the earliest flatworms were more like present-day rhabdocoels than acoels. There is also growing doubt that planuloids, or any animals with cnidarian affinities, were ancestral to bilateral animals. As discussed in Chapter 21 (p. 504), there is mounting evidence that cnidarians were not derived from stocks that also gave rise to bilateral animals. Thus, the cnidarian planula, or planuloid, may not be a good model for a bilateral ancestor. Some zoologists postulate that flatworms—either acoels or rhabdocoels—were derived directly from multinucleate ciliates, without the intervening planuloids.

It is generally believed that parasitic flatworms evolved from free-living turbellarians and that ancestral parasite stocks resembled rhabdocoels. Supporting this idea is the resemblance between certain living rhabdocoels, monogenetic flukes, and the redia larvae of trematodes (see Figure 22.14). Also, of the few living rhabdocoels that are parasitic, most infest molluscs, and trematodes have molluscs as one of their intermediate hosts. Parasitic flatworms have probably been coevolving with molluscan hosts and certain other invertebrates since the Precambrian, prior to 600 million years ago (see Figure 18.6). In most cases, vertebrates probably became secondary hosts of parasitic flatworms by eating molluscs and other invertebrates commonly infected with the worms.

PHYLUM GNATHOSTOMULIDA

First described in 1956, the Phylum Gnathostomulida contains only about 80 species of small (less than 1 mm long) marine acoelomates. **Gnathostomulids** inhabit spaces between sand grains in intertidal areas and are often numerous in wet, anaerobic sands (Figure 22.17). They have well-developed jaws with which they scrape microorganisms and fungi from sand particles. They have not been studied extensively, although their role in intertidal food chains and their metabolic adaptations to oxygen-deficient habitats would be fruitful research areas.

The phylogenetic relationships of gnathostomulids are enigmatic. Like flatworms, they have an incomplete gut, protonephridia, and an epidermis consisting of a single layer of cells. However, each epidermal cell bears only a single cilium, a rare condition among bilateral animals. The cilia provide a slow gliding movement that is enhanced by contractions of muscles in the body wall.

PHYLUM NEMERTEA (RHYNCHOCOELA)

There are about 800 species of **nemerteans**, or rhynchocoels (Figure 22.18). Most are intertidal marine predators or scavengers, although several species live symbiotically on marine clams and crabs, and a few others inhabit fresh water and moist soil. Nemerteans are flattened and range in length from less than a few millimeters up to 2 m. They are also called **ribbon worms** because of the ribbonlike appearance of certain elongate species.

The most distinctive feature of the phylum is a long hollow tube, the **proboscis**, held in a fluid-

Figure 22.17   Phylum Gnathostomulida. Members of this phylum, such as this species *Austrognathia christianae,* are often the predominant animals in muddy, anaerobic beach sand.

filled cavity called the **rhynchocoel**. The proboscis can be thrust out rapidly to capture prey or defend the nemertean, and in many species, it is armed with one or more sharp, piercing **stylets**. When it is inside the worm, the proboscis is inverted and held in the rhynchocoel by retractor muscles. It is extruded (thrust out) when contraction of body wall muscles increases the hydrostatic pressure in the rhynchocoel. Prey may be pierced by the stylet and encoiled by the proboscis, after which the retractor muscles pull the proboscis back into the rhynchocoel; the prey is then brought to the mouth and ingested.

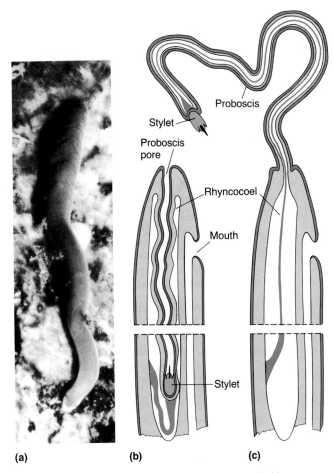

(a)   **(b)**   **(c)**

Figure 22.18   Phylum Nemertea. (a) The many-eyed ribbon worm *Amphiporus angulatus.* (b) Drawing of the anterior region showing the proboscis withdrawn into the rhynchocoel. (c) The proboscis is extended.

Nemerteans have traditionally been considered closely related to free-living flatworms. Like flatworms, they have protonephridia, their epidermis is ciliated (with rhabdites in the proboscis of some species), and except for the rhynchocoel, they seem to lack a body cavity. Unlike flatworms, however, they have a complete digestive tract, a closed circulatory system, and very simple reproductive structures. In contrast to the traditional view, a recent report suggests that nemerteans are eucoelomates and should be considered more closely related to annelid worms than to flatworms.

☐ SUMMARY

1. Key Features of Acoelomate Animals. Three phyla of bilateral, triploblastic metazoans have some organ systems but generally lack a body cavity.

2. The Phylum Platyhelminthes (flatworms) contains dorsoventrally flattened acoelomates with an incomplete gut (no anus) or no gut, and no circulatory system. Sperm cells typically have two flagella with a 9 + 1 arrangement of internal microtubules. Females have yolk glands (vitellaria) separate from the ovaries. Free-living flatworms (turbellarians) are mostly marine bottom dwellers; a few inhabit fresh water and moist terrestrial environments. Three flatworm classes are entirely parasitic.

3. Flatworms are supported by a hydrostatic skeleton and by internal connective tissues. Body wall muscles and surface cilia provide movement. Turbellarians have a single-cell-thick epidermis containing rodlike rhabdites. Parasites have suckers for attachment to hosts and a syncytial surface covering, the tegument, that absorbs food and resists digestion by host gut enzymes.

4. Most turbellarians feed on small invertebrates. Most parasitic flatworms ingest some food through the mouth, but most of the food (all, in the case of tapeworms, which are gutless) is absorbed by the tegument. Digestion is both extracellular and intracellular. The flattened body form and movement of materials in the digestive tract enable circulation and gas exchange to be performed entirely by diffusion and cell membrane transport. Some parasites live anaerobically in their host's digestive tract.

5. Freshwater flatworms have protonephridia that are mainly osmoregulatory; body fluids of many parasites are osmotically equivalent to those of their hosts. Nitrogenous wastes are excreted primarily as ammonia.

6. Flatworm nervous systems typically include anterior ganglia and longitudinal cords with transverse connectives; some turbellarians have a cnidarian-like subepidermal nerve net. Sensory structures are bilaterally arranged and concentrated anteriorly. Chemotaxis and phototaxis are important in body orientation in many turbellarians and in host location in certain parasites. Some flatworms may be capable of simple learning.

7. Turbellarians reproduce asexually by fission and/or fragmentation and regeneration; trematodes have several asexually produced larval generations. Adults are typically hermaphroditic and have internal fertilization. Many parasites have intricate life cycles involving one or several larval stages in more than one host species.

8. The earliest bilateral animals may have been flatwormlike. Turbellarians probably gave rise to parasitic flukes and tapeworms; trematode progenitors were probably parasites of molluscs.

9. Members of the Phylum Gnathostomulida are small marine invertebrates that feed on fungi and microorganisms and are often numerous in anaerobic beach sand.

10. The Phylum Nemertea consists of elongate, mostly marine worms that catch small invertebrates with a long eversible proboscis. They have a complete gut, a circulatory system, and a fluid-filled cavity (rhynchocoel) that houses the proboscis.

☐ FURTHER READING

Barnes, R. D. *Invertebrate Zoology.* 5th ed. Philadelphia: CBS, 1987. *Includes chapters on flatworms, rhynchocoels, and gnathostomulids.*

Dickson, S. "Biotechnology Is Deployed in the War on Parasitic Worms." *Genetic Engineering News,* September 1984: 18–19. *Describes modern research designed to find a control for schistosomiasis.*

McConnaughey, B. H., and R. Zottoli. *Introduction to Marine Biology.* 4th ed. St. Louis: Mosby, 1983. *Good discussions of ecology of marine acoelomates.*

Schmidt, G. D. *How to Know the Tapeworms.* Dubuque, Iowa: Brown, 1970. *Introduction and taxonomic keys to genera of the world.*

Schmidt, G. D., and L. S. Roberts. *Foundations of Parasitology.* 3d ed. St. Louis: Times Mirror/Mosby, 1985. *Includes ten chapters on parasitic flatworms.*

World Health. The Magazine of the World Health Organization, December 1984. *Entire issue devoted to worldwide efforts to control blood fluke disease.*

☐ CLASSIFICATION

Phylum Platyhelminthes (Gr. *platys,* "flat"; *helmins,* "worm"). Flatworms; bilateral acoelomates.

Class Turbellaria (L. *turbellae,* "a commotion"; *aria,* "like"). Turbellarians; most are free-living; a few are commensals on marine invertebrates; *Convoluta* and other marine acoels are gutless; *Notoplana* and other polyclads, which have a multibranched gut, are also marine; *Dugesia* and other freshwater planarians and a few marine and terrestrial genera are triclads (their gut is three-branched); *Stenostomum* and many other rhabdocoels, common in fresh water, have a saclike gut.

Class Trematoda (Gr. *trematodes,* "perforated form"). Trematodes (digenetic flukes); all are parasitic, with both asexual and sexual reproduction; *Clonorchis* and *Fasciola* are liver flukes of domestic animals and humans; species of *Schistosoma* are blood flukes.

Class Monogenea (Gr. *mono,* "single"; *genos,* "kind"). Monogeneans (monogenetic flukes); all are parasitic, with only sexual reproduction; *Gyrodactylus* is parasitic on fishes; species of *Polystoma* inhabit the urinary bladder of frogs.

Class Cestoidea (Gr. *kestos,* "girdle"; *eidos,* "form"). Cestodes (tapeworms); all are parasitic and gutless; *Taenia* (rabbit and pork tapeworms); *Taeniarhynchus* (beef tapeworm); *Echinococcus* has asexually reproducing larvae called hydatids; *Diphyllobothrium* (fish tapeworm).

Phylum Gnathostomulida (Gr. *gnathos,* "jaw"; *stoma,* "mouth"; L. *ulus,* a diminutive). Gnathostomulids; common in anaerobic beach sands; *Austrognathia, Gnathostomula.*

Phylum Nemertea (Gr. *Nemertes,* "the unerring one"). Nemerteans (rhynchocoels, ribbon worms); *Amphiporus* (intertidal); *Prostoma* (freshwater); *Geonemertes* (terrestrial).

The Significance of Body Cavities

None of the animals considered so far in this unit have a body cavity, that is, a fluid-containing space between the gut and the body wall. As indicated in Chapter 1, however, body cavity evolution has been of paramount importance in animal phylogeny, and most animals do have such a cavity.

Animals with a body cavity have many advantages. They have ample internal space where complex organs and organ systems can develop. They have a locomotor advantage in that their body is more flexible than it would be without a cavity. Fluid in a body cavity may provide hydrostatic support, and animals with a body cavity may be able to move by peristalsis. Fluid in a body cavity may also protect internal organs by acting as a shock absorber and as an insulation against marked temperature changes. Many invertebrates lack circulatory and respiratory systems and rely on fluid movement in the cavity for internal transport; nutrients, dissolved gases, and wastes may circulate in the cavity, and amoeboid cells can move about freely in its fluid, aiding in the distribution of materials. Free amoebocytes in the body cavity may also protect the body by phagocytosing foreign material or infectious agents that enter through wounds in the gut or body wall. In many invertebrates, metabolic wastes accumulate in the body cavity fluid and are filtered from it by the excretory system. Another important advantage is that with a body cavity surrounding the digestive tract, the gut can move independently of the whole body; peristaltic waves of muscle contractions can pass along the gut without involvement of the body wall. Most invertebrates and all vertebrates depend on this type of independent internal movement to propel food through the gut.

There are three kinds of body cavities. The **pseudocoel** of roundworms, rotifers, and other **pseudocoelomate** animals develops from the blastocoel, or the primary cavity of the embryo, and is not fully lined by mesoderm. In these animals, the muscles and other structures of the body wall and the internal organs are in direct contact with fluid in the pseudocoel. In contrast to pseudocoelomates, **eucoelomate** animals have a **coelom**, or true body cavity (the prefix *eu-* means "true"). The coelom is said to be "true" because it develops within mesoderm and is

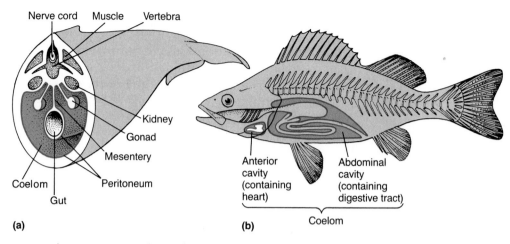

Figure A   Structure of a vertebrate, a representative eucoelomate. (a) The coelom is lined by mesodermal peritoneum that covers and suspends organs from the body wall. (b) In a fish, the coelom is divided into two parts, one containing the heart and the other containing the digestive tract and other organs.

usually lined completely by mesodermal tissue called **peritoneum** (Figure Aa). The peritoneum lines the inside of the body wall and surrounds the internal organs, including the gut. Thus, the organs are not free to move within the body cavity, but are bound in place and suspended from the body wall. This arrangement provides all the advantages of an internal cavity plus the advantage that the organs are held in place rather than being subjected to fluid movements in the cavity itself.

Three large groups of eucoelomates, the echinoderms, the annelids, and the vertebrates, have a spacious coelom. The coelom in echinoderms is important in virtually all vital functions. The annelid coelom is divided into fluid-filled compartments corresponding to body segments. As described in Chapter 9, this is important in annelid locomotion. As shown in Figure Ab, fishes have two separate compartments: an abdominal cavity, in which most of the digestive tract and associated organs are suspended, and a completely separate anterior cavity in which the heart is suspended. In birds and mammals, the lungs are suspended in a pair of additional coelomic cavities.

In the vast majority of eucoelomate species, the true body cavity is largely replaced during development by a third type of body cavity, the **hemocoel**. Like the pseudocoel of pseudocoelomates, the hemocoel develops from the blastocoel. Invertebrates with a hemocoel have an open circulatory system. Their circulatory fluid, or hemolymph, circulates in the hemocoel, which forms a network of interconnecting sinuses and channels throughout the body. In most molluscs, the coelom develops in the embryo, but remains in the adult only as small spaces around the heart and gonads and as parts of the excretory organs. In adult arthropods, the coelom persists only as gonadal spaces and, in crustaceans, as parts of the excretory organs.

The pseudocoelomates are the subject of the next chapter. There are many more species of these animals than there are acoelomates, but pseudocoelomates exhibit rather limited variety in body forms, compared to the enormously diverse eucoelomates.

CHAPTER 23

Pseudocoelomate Animals

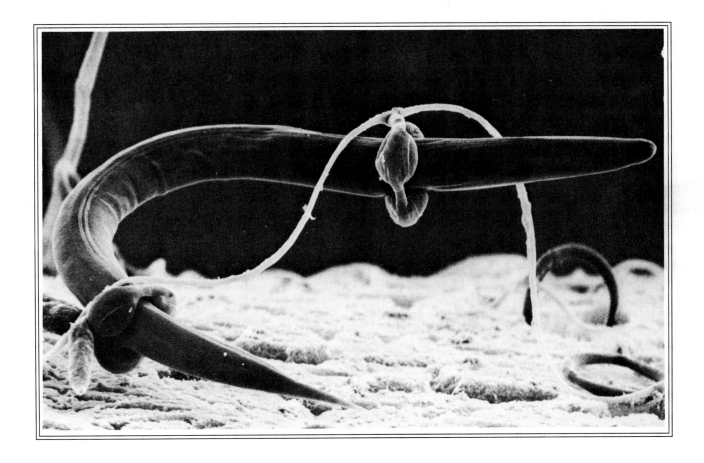

A soil roundworm (nematode) becoming food for a fungus. Nematodes, the largest group of pseudocoelomates, are numerous in wet soil, decaying wood, and rotting leaves. Here, the scanning electron microscope shows a rhabditoid soil nematode caught by a snarelike trap consisting of filaments of the fungus *Arthrobotrys anchonia*. The trap constricts as the nematode attempts to crawl through it. Fungal filaments then penetrate the captured worm and digest it.

Nine phyla of metazoans are collectively called pseudocoelomates, although they share only a limited number of traits and probably are not closely related. All are aquatic, and this includes parasites living in their hosts' body fluids and soil dwellers living in water droplets. None have organs specialized for circulation or gas exchange. Most have a fluid-filled **pseudocoel** surrounding a complete gut, and consequently, these animals have a tube-within-a-tube body plan. Internal organs are not suspended from the body wall, but are free in the pseudocoel. The body wall is covered by a **cuticle** that lies within a syncytial epidermis or is secreted by an underlying epidermis.

In many pseudocoelomates, mitosis ceases in somatic organs during embryonic development, resulting in individuals with a fixed, constant number of somatic cells (or nuclei if some tissues are syncytial). With this type of cell constancy, called **eutely**, the number of somatic cells in the adult is the same as that in the embryo when mitosis stops. Growth may continue after the full complement of cells is attained, but only because of an increase in cell size. As a result of eutely, each species may have its own cell number, and all members of a species have the same or nearly the same number. Actually, perfect eutely is rarely achieved, because certain somatic cells, particularly in the body covering and gut, continue to divide during growth in most pseudocoelomates. Nonetheless, compared to other metazoans, pseudocoelomates consist of a relatively uniform number of cells.

The nine pseudocoelomate phyla are quite distinctive. Two phyla, the Nematoda and the Rotifera, contain many species and are important in aquatic ecosystems. The other seven phyla have relatively few species, but are unique and interesting metazoans.

PHYLUM NEMATODA

The **nematodes**, or **roundworms**, constitute one of the most successful phyla on earth. With 90,000 or so known species, this phylum is larger than the other eight phyla of pseudocoelomates combined, and there are probably many thousands of nematode species yet to be discovered. Parasitic roundworms that infect humans, domestic animals, and crop plants have received a great deal of research attention; free-living species are more numerous and often beneficial, but are not as well studied. Nematodes are often the most abundant metazoans in moist soils, and they also frequent habitats as diverse as rainwater puddles, sewage ponds, water filters, deep marine and lake sediments, hot springs where temperatures exceed 50°C, and meltwater on polar ice sheets. Nearly 100,000 individuals have been counted in a single rotting apple. The so-called vinegar eel, *Turbatrix aceti*, thrives on bacteria and yeast cells in unpasteurized vinegar and on decomposing vegetables.

Despite the great number of roundworm species and the diversity of environments they inhabit, body forms are strikingly uniform in the phylum, and the phylum's success is largely a measure of the evolutionary adaptability of one basic architectural plan. The body is cylindrical and circular in transverse section—thus the common name round-

worm (Figure 23.1). Most roundworms taper to a point at their posterior end, but typically have a blunt anterior end with a mouth at its tip. Short bristles, sensory structures, and radially or biradially arranged mouthparts surround the mouth. The nematode digestive tube is straight and lacks muscles except in the pharynx. The body wall, covered by a secreted complex cuticle, has only longitudinal muscles and this limits the range of body movements. Affecting all body functions, the nematode pseudocoel is filled with fluid under high pressure. In common with arthropods, but not with other animals, nematodes lack cilia except in sensory structures. Also in common with some arthropods, nematode sperm cells are amoeboid.

REPRESENTATIVE NEMATODES

It is fairly easy to distinguish nematodes from other animals by their unique body form, but because of their uniform body plan, it is often difficult to tell one species of nematode from another. For the same reason, it is also difficult to classify nematodes into taxonomic categories above the species level. Two large classes are generally recognized. The Class Phasmidia includes roundworms that have a pair of unicellular structures called **phasmids** opening through two pores in the tail region (see Figure 23.1a). These structures seem to have sensory and secretory functions. A second class, the Aphasmidia, includes roundworms that lack phasmids.

Class Phasmidia

Most parasitic nematodes are members of the Class Phasmidia. Free-living phasmidians are numerous and important in freshwater habitats and in moist soils (see chapter opening photograph). The common genus *Rhabditis* and thousands of other small nematodes (a few mm long) are scavengers and bacteria feeders in the soil. They are so numerous in organically rich areas that their movements have a significant effect in mixing and aerating the soil. When observed with a microscope, their most prominent internal features are a bulblike muscular pharynx and an elongate reproductive system. Males are smaller than females and have a curled posterior end. One species of soil nematode, the phasmidian *Caenorhabditis elegans*, is widely cultured and used as a model research organism by

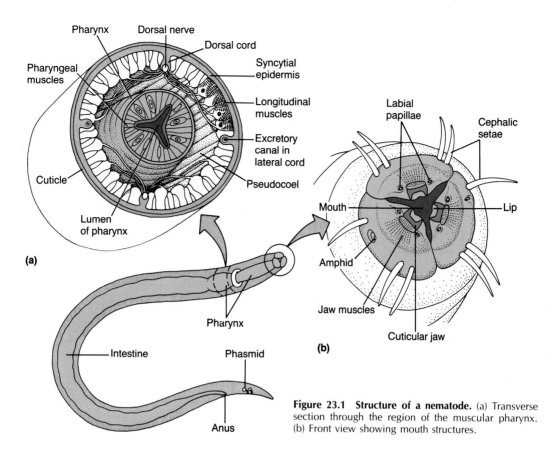

Figure 23.1 Structure of a nematode. (a) Transverse section through the region of the muscular pharynx. (b) Front view showing mouth structures.

geneticists and developmental biologists. (See the Essay on *C. elegans*, p. 537.)

Many phasmidian nematodes feed on plants and cause serious damage to agricultural crops. Many have spearlike mouthparts, or **stylets**, that pierce the walls of root cells and suck up plant juices. Species of *Heterodera* and *Meloidogyne* attack the roots of many crop plants. Juveniles of some species penetrate the roots and live inside the plants, which respond by growing rapidly around the worms, forming a gall. Annual crop losses to nematodes total several billion dollars, and in the United States alone, nearly 100 million dollars are spent each year on chemical nematicides. Crop rotation also helps reduce nematode damage. Moreover, bacteria and viruses that infect nematodes are being investigated as potential biological control agents for these pests.

Several species of phasmidians infect humans, who acquire the worms by swallowing infective eggs. The adult worms live in the human intestine. *Enterobius vermicularis*, the human **pinworm**, is common throughout the world, especially in temperate climates. As many as 30% of the college and university students in the United States have been or are infected. Pinworms are dioecious and resem-

ble typical soil nematodes in general body form. Their tiny eggs may be airborne and are frequently passed from person to person on soiled bedding. Several oral drugs are effective against the adults.

Nematodes of the genera *Ascaris* and *Toxocara* (intestinal roundworms) are quite large. *Ascaris* females may be up to 50 cm long and males up to 30 cm long. Species of *Toxocara* are similar to *Ascaris*, but are considerably smaller (Figure 23.2a). *Ascaris lumbricoides* lives in the human intestine; a very similar species, *Ascaris suum*, infects hogs and is often studied in zoology laboratory courses. Infection rates of *A. lumbricoides* exceeding 15% are not uncommon in rural areas in the southeastern United States, and much higher rates occur in poverty-stricken regions of the Orient, where sanitation is poor. The adult worms feed on intestinal contents and often cause anemia in children. In heavy infections, masses of worms can cause intestinal blockage, and adult worms may respond to crowding by wandering about in the gut, sometimes getting into and blocking liver or pancreatic ducts. Living in the anaerobic environment of the gut, *Ascaris* and other nematodes use pathways of metabolism that do not require free oxygen. Like certain flatworms, some are known to use fumarate

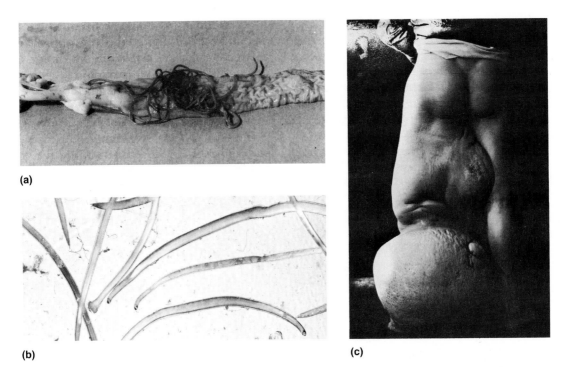

Figure 23.2 Nematodes of the Class Phasmidia. (a) Adults of *Toxocara cati* in a partially opened intestine of a domestic cat. This species infects both cats and dogs; adult females (6–15 cm long) and males (4–6 cm long) feed on partially digested food in the host's intestine. (b) The hookworm *Ancylostoma caninum* lives in the intestine of dogs, foxes, and cats and is similar to a hookworm that infects humans. Drugs such as tetrachloroethylene and mebendazole reduce hookworm numbers, but rarely kill all the worms. A relatively new drug named ivermectin is very effective against hookworms and other roundworm parasites, but so far, it is available only for certain verterinary uses. (c) Elephantiasis resulting from infection of the lymphatic system by phasmidian nematodes (filarial worms) of the genus *Wuchereria*.

instead of oxygen as a final electron acceptor in ATP synthesis (see Figure 22.10).

Unlike pinworms and *Ascaris*, **hookworms** attach to their host's intestinal wall and suck blood. They have special jawlike plates or teeth that grip the inner surface of the intestine, and they also have a muscular pharynx that pumps blood from host capillaries (Figure 23.2b). Two species, *Ancylostoma duodenale* and *Necator americanus*, infect humans. Recent studies indicate that about 500 mature *A. duodenale* worms attached to the human intestine can pump as much as 50 ml of blood per day from the intestinal wall, considerably more than the worms can use. The unused blood passes through the worms and into the intestine, and although much of the iron it contains may be reabsorbed before leaving the intestine, infected persons often develop severe iron and protein deficiencies.

Hookworm disease is most common in tropical and subtropical climates because the free-living juvenile worms require warm, moist soil conditions. The worms infect humans by penetrating the skin, usually through the feet, and wearing shoes greatly reduces a person's chances of becoming infected. Largely through educational efforts and sanitation programs, the number of human cases has been reduced tenfold in the southeastern United States, from nearly 50% of the population at the turn of the century to less than 5% of the population today.

Other phasmidian nematodes include the **filarial worms**, which are blood and lymph parasites in vertebrates. Species of the tropical genera *Wuchereria* and *Brugia* infect the human lymphatic system. The long, thin adults can block the lymph passages, causing inflammation, pain, and swelling. Long-term infection can result in gross swelling of large areas of the body, a condition called **elephantiasis** (Figure 23.2c). Another filarial worm, *Dirofilaria immitis*, is mainly a parasite of dogs, wolves, and foxes. It is common in the United States and causes heartworm. The adult worms lodge in the heart and large arteries of the lungs

Essay: The Nematode *Caenorhabditis elegans:* From the Soil to the Research Laboratory

This nematode, one of many that thrive in organically rich soil, is one of the most important research animals in the world, and probably more is known about its development at the cellular level than about that of any other eukaryote.

Researchers use *Caenorhabditis elegans* to study animal development and the genetic control of development for several reasons. To begin with, this species thrives in laboratory cultures seeded with bacteria for food and takes only two to three days at 25°C to complete its life cycle. Most important for genetic analysis, *C. elegans* is hermaphroditic and self-fertilizing (Figure E1); this allows researchers to maintain uniform genetic strains and study them through many generations. Also, developing embryos are easily obtained from gravid adults. Males, derived from nondisjunction of *X* chromosomes, are also cultured and used in certain genetic crosses. In addition, *C. elegans* has a relatively simple genetic system. Hermaphrodites have only six pairs of chromosomes, including two *X* chromosomes; males have a single (nondisjunct) *X* chromosome. Over 300 genes have been mapped on the six linkage groups.

As a typical pseudocoelomate, *C. elegans* exhibits eutely; adult hermaphrodites have 958 somatic nuclei; males have 1031. *Caenorhabditis* also has determinate cleavage, meaning that the egg cytoplasm determines the fate of early blastomeres. Together with eutely, this has made it possible for researchers to trace the derivation of all cells in both types of adults back to fertilized eggs, and studies of cell lineage in this species are contributing significantly to our general understanding of animal development. In one of the most promising lines of work, researchers use surgical lasers to analyze *C. elegans* development. At specific times in their division cycle, individual cells or nuclei can be destroyed with laser microbeams, and the resultant alterations in development indicate the normal role of the killed cells. Developing cells can also be laser-fused, and studies of this type indicate that the cytoplasm helps regulate normal development in *C. elegans*.

Studies of *C. elegans* have contributed much to our understanding of how cell-to-cell interactions influence cell differentiation (the development of cell specialization), but knowledge of the biochemistry of differentiation is far from complete. Likewise, as discussed in Chapter 16, our understanding of how genes regulate differentiation is still rudimentary. Nevertheless, because cell lineages are completely known in *C. elegans,* the timing of events in cell differentiation can be pinpointed to specific cells in the developmental sequence. These cells can then be isolated and studied intensively. Because many of the genes of *C. elegans* have been mapped, the effects of genes that control development can be studied sequentially through development. Researchers are also beginning to test the effects of injecting microquantities of DNA into developing cells.

Overall, research with *C. elegans* is very promising, and it is likely that future studies of this common soil inhabitant will help answer many central questions about the molecular and genetic basis of development.

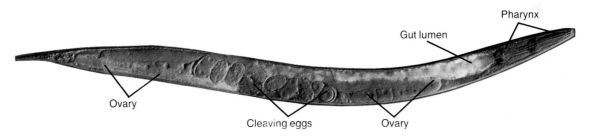

Figure E1 *Caenorhabditis elegans.* The lineage of every cell in the adults is known. This photograph shows a hermaphroditic adult. In the United States, the *C. elegans* Genetics Center at the University of Missouri maintains a bank of hermaphroditic and male strains, which are available to researchers.

and may cause heart failure in the animal. A dog may carry an infection for several months or years, until one day, while exercising heavily, it may have a sudden heart attack and die. Heartworms are difficult to eliminate once established, but dogs can be fed drugs that prevent infection.

Class Aphasmidia

Most aphasmidians are free-living, and several are abundant in freshwater and marine environments; but the group also contains some harmful parasites of plants and animals. Like phasmidians, species

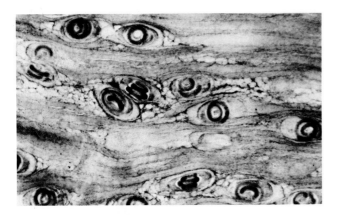

Figure 23.3   An aphasmidian nematode. The Class Aphasmidia includes the parasite *Trichinella spiralis* (the trichina worm), parasitic in many meat-eating vertebrates, including humans. This photomicrograph shows *Trichinella* juveniles (about 1 mm long) encysted in vertebrate muscle; hosts may carry from 1 to over 1000 juveniles per gram of muscle; more than 500 per gram can cause serious pain and muscle degeneration in humans. Adult males and females in the intestine can be removed by drugs, but a safe and effective drug for juvenile removal has not yet been found. Encysted juveniles may remain viable for months.

that attack plants have daggerlike stylets that pierce cell walls. Some aphasmidians transmit viral diseases to plants.

One notorious aphasmidian parasite, *Trichinella spiralis* (the trichina worm), infects carnivorous mammals, including humans, and is quite common in temperate and arctic areas. The adult worms live in the intestine and are quite small (1.5–3 mm long). The disease **trichinosis** results from tissue invasion by the females and juveniles. The females penetrate the intestinal wall and release juveniles into the bloodstream. Juveniles carried in the blood may penetrate many organs, causing severe trauma and death if large numbers invade the heart. After about ten days, the juveniles enter skeletal muscle cells and encyst (Figure 23.3). The next host becomes infected by eating the encysted juveniles in meat. Human infections are usually acquired by eating raw or undercooked pork.

HABITATS AND ENVIRONMENTAL RELATIONSHIPS

Many nematode species can thrive in habitats as diverse as soil, bottom sediments of oceans, freshwater lakes, and hot springs. As a consequence of this wide adaptability, many species are distributed worldwide. Many soil-dwelling species have

evolved drought-resistant eggs, and some can lose all body moisture and survive by becoming dormant when their environment dries up. (See the Essay on suspended animation, p. 540). Organically rich soils, bottom sediments, and sandy beaches may contain billions of nematodes per hectare (10,000 m²). The worms live mostly in the top 5–10 cm, where rotting organic matter is concentrated. Many species are scavengers, and their great numbers can profoundly increase the rate of decomposition of dead organic matter. Scavenging and herbivorous nematodes eat mainly bacteria, fungi, algae, and plants. Carnivorous species feed on a variety of small invertebrates, protists, and on other nematodes, including their own kind. Many organisms, particularly arthropods and some fungi that capture prey with sticky branches or snares, frequently consume roundworms (see chapter opening photograph).

As a group, parasitic nematodes infect virtually all plants and animals, and they may infect virtually all internal tissues and organs. A few parasitic nematodes infest the body surfaces of aquatic animals, but the group has not managed to exploit the body surfaces of terrestrial animals, probably because nematodes cannot tolerate dry conditions. Despite the bad reputation that a few dangerous species give the phylum, many parasitic nematodes cause little or no harm to their hosts, and certain parasitic species are being investigated as potential biological control agents for pest insects. One species of nematode holds promise in mosquito control, and others can transmit bacteria that kill insects destructive to corn, cotton, and cabbage crops.

COMPARATIVE STRUCTURE AND FUNCTION

The nematode body wall consists of a unique multilayered cuticle, an underlying epidermis, and longitudinal muscles (see Figure 23.1). The cuticle accounts for much of the adaptability of these worms. It is marked with surface rings, or annuli, but there is no body segmentation (Figure 23.4). Much of the cuticle consists of several strata of collagen, a tough, fibrous protein that maintains body shape and provides protection against mechanical damage. The cuticle and longitudinal muscles maintain unusually high fluid pressure in the pseudocoel (averaging 70–120 mm Hg in *Ascaris lum-*

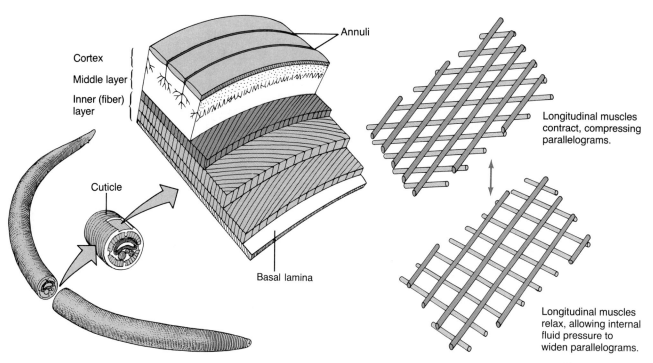

Figure 23.4 The cuticle of *Ascaris*. The cuticle consists of three main layers. An outermost cortex is laminated and covered with several layers of dead cell membranes. Branching canals extend through the cortex from external pores. In some species, the canals extend to the base of the cuticle and may be important in various functions, including gas exchange. A middle layer includes metabolically active enzymes and collagen. The cuticle can grow without molting, and the middle layer seems to be the site of biosynthetic reactions involved in growth. An inner (fiber) layer consists of three sublayers of collagen fibers. The diagonal arrangement of the fibers angled in opposite directions in successive layers produces a spiral basketwork in the body wall. The insets show how the fibers in these layers form "parallellograms" that allow body flexure. For simplicity, only two fiber layers are shown. Contraction of longitudinal muscles underlying the epidermis (not shown here) on one side of the body compresses the fiber parallelograms and shortens that side of the body; simultaneously, relaxation of muscles on the opposite side of the body allows internal fluid pressure to change the angles of the fiber parallelograms and extend the body.

bricoides). This provides strong hydrostatic support, and nematodes are quite resistant to crushing. However, with a relatively rigid body and a lack of circular muscles in their body wall, nematodes are capable of only simple body movements. Most can only undulate in the dorso-ventral plane. In *Ascaris*, three inner fiber layers of the cuticle form a basketwork-like arrangement that is important in body movement. Parallelograms formed by the fiber layers can change shape without altering the fiber lengths. Muscles on opposing sides of the body alternately contract and relax, causing the body to flex or undulate (see insets in Figure 23.4). Nematodes can swim slowly, but their type of motion is most effective when particles of soil or organic debris are next to the worms, providing good surfaces for leverage. When most nematodes are placed in a large volume of water, they thrash about and whip ineffectively; thus, in nature, nematodes are most abundant in soil or bottom sediments rather than in open water.

The nematode epidermis has several unusual features. It may be cellular or syncytial, and it extends into the pseudocoel as four cords (one dorsal, one ventral, and two lateral) (see Figure 23.1). The cords house epidermal nuclei, longitudinal nerves, and lateral excretory canals. Longitudinal muscle fibers extend between the epidermal cords, and unlike the usual condition (nerve fiber to neuromuscular junction), each muscle fiber sends out a thin arm that forms a synapse with either the dorsal or ventral nerve cord. (Nerve fibers and synapses are discussed in Chapter 10, p. 222.)

The nematode digestive tract is a straight tube that lacks muscles except in the mouth, pharynx, and rectum (see Figure 23.1). The mouth is lined

Essay: Prolonging Life by Suspended Animation

Many soil- and moss-dwelling nematodes, rotifers, and tardigrades have a remarkable capacity to prolong their lives. (Tardigrades are discussed in Chapter 26, p. 641; rotifers, p. 545.) They can survive in a state of suspended animation for decades or even centuries. Biologists call this phenomenon **cryptobiosis**, meaning "hidden life." Cryptobiotic animals are highly desiccated, and their metabolism is extremely low. Some researchers believe that metabolism may even stop completely, raising questions about how the cryptobiotic state can be distinguished from death. Cryptobiotic animals can withstand anaerobic and totally dry conditions for long time periods, and it is thought that their survival is prolonged if metabolism ceases. While cryptobiotic, they are much smaller and lighter than when active, and can be dispersed by winds to new, perhaps more favorable habitats. A few species have survived being held in a vacuum at temperatures as low as −270°C.

We are just beginning to understand cryptobiosis. Studies with nematodes indicate that two chemicals, glycerol and trehalose, a nonreducing disaccharide, are key compounds in cryptobiotic animals. Glycerol seems to protect dried body tissues against oxidation. Trehalose may act as an inert spacer between fragile protein molecules, preventing breakdown when the molecules are rehydrated, and may also prevent dried membranes from fusing together. To become cryptobiotic, a nematode, rotifer, or tardigrade must dry slowly. As the environment begins to lose moisture, the animals often aggregate into tight groups or twist into certain postures. Irregular surfaces, such as sand grains where pockets of water evaporate slowly, are ideal for slow drying. Rotifers reduce their exposed body surface by becoming ovoid. These activities slow the rate of drying and probably provide time for the synthesis of glycerol and trehalose.

The capacity to survive in a desiccated state is of great adaptive advantage for aquatic animals living in environments that periodically or irregularly dry out. Nematodes, rotifers, and tardigrades are all aquatic; that is, they require free environmental water for active existence. Yet, through cryptobiosis, they can live permanently in many habitats that would otherwise be lethal.

internally by cuticle and typically is surrounded by lips. Carnivorous nematodes usually have well-developed cuticular stylets, jaws, or teeth, but scavengers, bacteria feeders, and most parasitic species have relatively simple mouthparts. The highly specialized jaws and teeth of hookworms (see Figure 23.2b) and the oral stylets of plant-feeding nematodes are exceptions. The muscular pharynx of nematodes works in concert with the mouthparts to suck food into the gut against the internal fluid pressure. Lacking muscles, the intestine is collapsed by pseudocoelomic fluid pressure unless it contains food. Food is forced into the anterior end of the intestine by a pumping action of the pharynx, and an anterior intestinal valve prevents backflow. Food is moved posteriorly by the combination of pseudocoelomic fluid pressure exerted against the gut, body movements, and the pharynx forcing more food into the front of the intestine. Experimental evidence suggests that digestion begins in the intestinal lumen, but is completed within cells lining the intestine. Special muscles in the rectum periodically contract to open the anus for defecation. An *Ascaris* worm held in the air can propel its feces more than half a meter, a clear demonstration of its high internal fluid pressure.

Circulation and gas exchange in nematodes are aided by body movements, which help circulate the fluid in the pseudocoel. Cuticular pores, visible only with the electron microscope, may allow gases to diffuse more readily across the body wall. A type of hemoglobin has been identified in the pseudocoelomic fluid of a few species, but its significance is not known.

Osmoregulation and excretion in nematodes are not well understood. Unlike most other pseudocoelomates, roundworms lack protonephridia, but some have glandlike **renette cells** or variously shaped canals that may function in waste disposal (Figure 23.5). Nitrogenous waste products include ammonia, urea, and uric acid. At least in some species, the cuticle and digestive tube seem to function in osmoregulation.

The nervous system is centered in an anterior cluster of ganglia connected to a nerve ring around the gut (Figure 23.5). Nerve cords extend anteriorly and posteriorly from the anterior ganglia. Four longitudinal nerve cords (one dorsal, two lateral, and one ventral) unite at ganglia near the posterior end of the worm. The dorsal cord carries only motor fibers; the lateral and ventral nerve cords are ganglionated and carry both sensory and motor fibers.

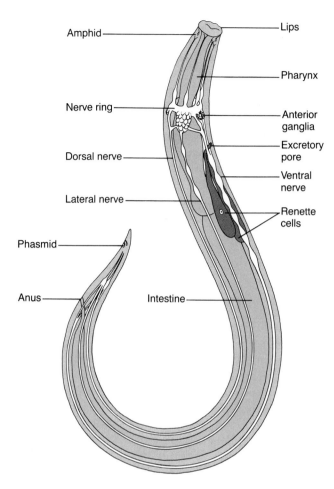

Amphid — Lips

Pharynx

Nerve ring — Anterior ganglia

Excretory pore

Dorsal nerve — Ventral nerve

Lateral nerve — Renette cells

Phasmid

Anus — Intestine

Figure 23.5 Nematode excretory, nervous, and digestive structures.

Nematode sense organs include hairlike **setae** set in pits in the cuticle and anterior and posterior cuticular thickenings called **papillae**, whose sensory elements are modified cilia. Setae and papillae are thought to be mechanoreceptors. Phasmids (in the Class Phasmidia) and similar structures called **amphids** near the anterior end of all nematodes seem to have both a chemosensory and a glandular function (see Figure 23.5). Both of these structures contain modified cilia as sensory elements. In addition to these sense organs, several free-living nematodes have photosensitive ocelli.

BEHAVIOR

Nematodes probably locate food, mates, and suitable habitats mainly by chemical signs and by touch. Females of several species produce male-attracting pheromones. (Pheromones are discussed in Chapter 13, p. 309.) Several free-living nematodes respond to temperature differences in the environment. Parasitic species may congregate in areas frequented by hosts. Juvenile hookworms, which must be on or near the soil surface to contact and penetrate a host, generally migrate upward when the soil is warm and moist, but move downward when the surface becomes too hot or dry. Juvenile filarial worms (e.g., *Wuchereria*, see p. 536) tend to congregate in the skin capillaries of vertebrate hosts during certain hours, usually at night. Here they can be ingested by mosquitoes during nighttime blood feeding.

REPRODUCTION AND LIFE CYCLES

Sexes are separate in most nematodes. The female and male reproductive tracts are long, coiled tubes that gradually increase in diameter from the gonad to the outside opening (Figure 23.6). Male systems may be paired or unpaired and include testis, sperm ducts, and seminal vesicles, which open into the posterior end (cloaca) of the intestine. The curved posterior end of the male has several cuticular spicules that aid in holding the female genital pore open during copulation. The female has a pair of ovaries, oviducts, and uteri, as well as a vagina that opens to the outside in a separate ventral pore near the middle of the body.

Nematodes have no larval stage. Juveniles are essentially small sexually immature adults, and there are four juvenile stages separated by three cuticular molts. The first two juvenile molts may occur before the egg hatches. A fourth and final molt precedes the adult stage, but the cuticle can continue to grow without molting during the adult stage. Nematodes exhibit eutely, and mitosis ceases in most tissues and organs by the time hatching occurs. (See the Essay on *Caenorhabditis elegans* in this chapter, p. 537.)

Molting seems to be controlled by neurosecretions from the anterior ganglia, and recent evidence suggests that the renette cells may secrete a constituent of the molting fluid. A fatty acid-like substance has recently been found that arrests development in *Caenorhabditis elegans*. When crowded, *C. elegans* produces this substance as a pheromone, causing late juveniles to become dormant and resistant to adverse environmental conditions.

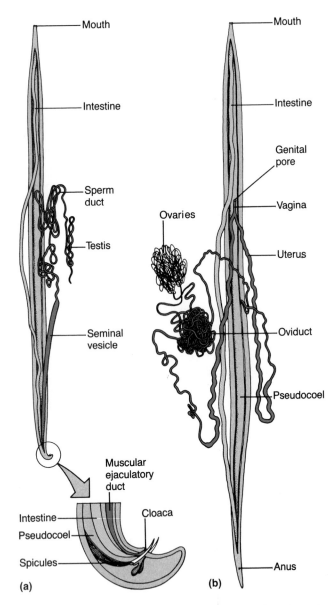

Figure 23.6 Reproductive systems of nematodes. Both (a) the male and (b) the female systems consist of long tubes that loop back and forth upon themselves, often forming tightly coiled masses. During copulation, the posterior end of the male adheres to the female's genital pore. Spicules on the male (see enlargement) hold the vagina open, and muscular contractions of the male's ejaculatory duct inject sperm into the female against the fluid pressure of her body cavity.

Most nematodes have simple life cycles. Following copulation, the amoeboid sperm migrate up the uterus and fertilize the eggs. Females either lay eggs containing developing juveniles or give birth to juveniles. Species inhabiting soil and temporary freshwater ponds produce eggs that can withstand desiccation and temperature extremes, often for years. The females of many plant parasites can transform into tough, cystlike cuticular sacs in which eggs or juveniles remain dormant but viable for long dry periods.

Most nematodes that infect animals have only one host in their life cycle, and often only the juveniles or adults are parasitic. *Enterobius vermicularis*, the human pinworm, has a simple one-host life cycle. Adults live in the human intestine, and females deposit eggs around the host's anus. If swallowed by the same or another person, the eggs hatch, releasing juveniles that develop into adults in the intestine. Species of *Ascaris* also have a single host, but the juveniles undergo a complex migration once inside the host's body. Adults living in the intestine produce eggs that pass out in feces. If swallowed by another host individual, the eggs hatch in the intestine; juveniles then penetrate the intestinal wall and enter the circulatory system. They are carried throughout the body in the bloodstream. Those that are carried to the lungs break out into air passages and migrate up to the throat. From there they must be swallowed and passed into the intestine to become adults. Hookworms follow a similar migration route after the juveniles penetrate the host skin.

The filarial worms (e.g., *Wuchereria* and *Dirofilaria*; see p. 536) have more complex life cycles involving two hosts: Adults and juveniles occur in vertebrates, and mosquitoes carry the juvenile worms from one host to another. The trichina worm *Trichinella spiralis* produces intestine-dwelling adults and muscle-inhabiting juveniles in the same host individual. To complete the *Trichinella* life cycle, however, a carnivorous host must eat meat containing encysted larvae.

PHYLOGENY

The affinities of nematodes with other phyla are vague. There is no other living group, including the other pseudocoelomate phyla, believed to be closely related to these worms. Several authorities postulate that nematodes first evolved in freshwater habitats and then colonized soils and oceans. Although certain parasitic species have evolved intricate life cycles, they appear to have arisen from free-living species without great modification in body form. In fact, many present-day free-living nematodes seem capable of becoming parasitic without substantial change. The nematode cuticle, feeding structures, and food habits appear in some ways to be "preadapted" for parasitism.

PHYLUM NEMATOMORPHA

The name Nematomorpha, meaning "thread form," describes the hairlike shape of these worms. **Nematomorphans**, or **hairworms**, constitute a little-known group (about 240 species) of mostly freshwater pseudocoelomates. Only one genus is marine. The free-living adults are gray, black, or brown and may be over 50 cm long, but are usually less than 2 mm in diameter (Figure 23.7a). Like nematodes, hairworms have no circular muscles, and their movements are limited to slow undulations or writhings. Sexes are separate, and males are generally smaller and more active than females. Sometimes, several mature adults can be seen writhing together at the edge of a lake or pond, and since many are thin enough to pass through water treatment facilities, they are occasionally reported in domestic tap water. It can be startling to find one of these creatures in the kitchen sink, but despite their unwholesome looks, hairworms do not infect humans or indicate water impurity. They merely show that the water comes from lakes or streams rather than from deep wells.

Internally, nematomorphans have a degenerate gut, and they do not feed as adults. The pseudocoel is nearly filled with a pair of elongate gonads and, in most, a mesenchymatous material. Nematomorphans lack organs for gas exchange, circulation, excretion, or osmoregulation. Their nervous system consists of an anterior ring around the gut; the ring is connected to a midventral nerve cord.

Unlike nematodes, nematomorphans have a distinctive larval stage, which is parasitic in terrestrial or aquatic arthropods or leeches. The larvae are superficially annulated and have a retractible **proboscis** (a snoutlike anterior projection) bearing three stylets and spines (Figure 23.7b). Lacking a gut, they feed by absorbing host body fluids across their body surface. A larva undergoes several molts and often grows to occupy most of the space within the host's cuticle. Hairworms that develop in terrestrial arthropods emerge only when the host is near water, and the worms' sex organs develop soon after emergence.

PHYLUM ACANTHOCEPHALA

The name Acanthocephala means "spiny head," and the most distinctive feature of the phylum, present in all species, is an anterior invertible pro-

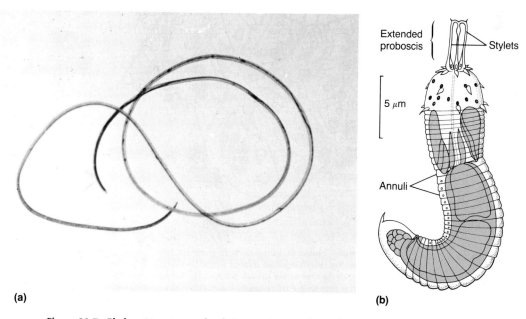

(a)   **(b)**

Figure 23.7 Phylum Nematomorpha (hairworms). (a) Males and females are free-living and resemble long coarse hairs; internally, they contain little except reproductive organs. (b) The larva is parasitic and feeds by absorbing nutrients across its body surface.

boscis with recurved spines (Figure 23.8a). There are about 700 species of acanthocephalans, and all are parasitic with two or more hosts in the life cycle. Adult males and females inhabit the intestinal tract of vertebrates and anchor to the gut wall by means of their proboscises. Larvae live in the hemocoel of arthropods (Figure 23.8b).

Acanthocephalans do not have a digestive tract. Like tapeworms, they seem to have derived their gutlessness in evolving as parasites. Nutrients from the host intestine are absorbed across the body surface, a living tegument that functionally resembles the teguments of parasitic flatworms. (The tegument of parasitic flatworms is discussed in Chapter 22, p. 518; Figure 22.8.) The acanthocephalan tegument has fluid-filled **lacunar canals** connected to two anterior sacs called **lemnisci** that serve as reservoirs for fluid used to hydraulically evert and expand the proboscis (Figure 23.8a). Fluid in the lacunae and in the pseudocoel provide hydrostatic support. A glycocalyx consisting of mucopolysaccharides and glycoproteins covers the tegument

and protects against host enzymes and immune defenses.

Many questions remain about the physiology of acanthocephalans. These parasites seem to have little capacity to osmoregulate. Their pseudocoelomic fluid is osmotically similar to that of the host gut contents, and the worms passively shrink in hypertonic fluids and swell in hypotonic fluids. Yet, in common with many freshwater flatworms and many pseudocoelomates, one group of acanthocephalans has protonephridia (of unknown function). Acanthocephalans have an anterior ganglion and several anterior and posterior nerve cords. Sensory structures seem to be poorly developed in adults, yet males locate females in the complex chemical environment of the vertebrate gut.

Acanthocephalans are most common in fishes, and in large numbers they can cause serious harm to their hosts. The proboscis can penetrate through the intestinal wall into the body cavity, causing bleeding and scar tissue formation, and the wounds often become infected with bacteria. Only a few

(a)

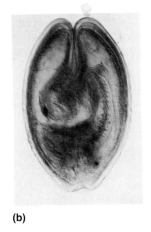

(b)

Figure 23.8 Phylum Acanthocephala (spiny-headed worms). These whitish or yellowish parasites range in size from about 1 mm up to 1 m in length. (a) This photomicrograph and diagrams show the anterior end of an *Acanthocephalus ranae* adult, a parasite of frogs. Note the lemnisci; with the proboscis pushed into the host's gut wall, muscles contract, forcing fluid from the lemnisci against the proboscis. This expands the proboscis and firmly anchors the worm in host tissue. Contraction of the proboscis inverter muscles pulls the proboscis, hooks inward, into the proboscis receptacle. (b) An acanthocephalan larva removed from an intermediate host; note its inverted proboscis (72×).

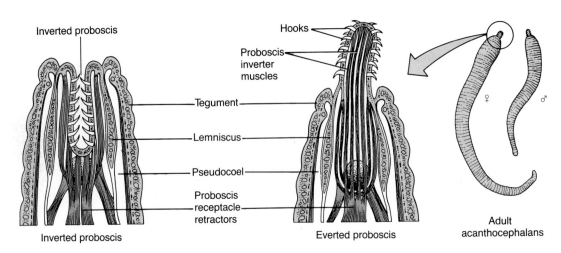

species occur in humans, and infections are rare because arthropods harboring larvae are seldom eaten.

PHYLUM ROTIFERA

Most rotifer species are less than 1 mm long and superficially resemble ciliate protozoa in their size and movements. More than 90% of the phylum's 1800 species live in freshwater lakes and ponds, where they are extremely common. Rotifers inhabit virtually all freshwater environments except fast-flowing streams; many thrive in moist soil, in water droplets on wet mosses, and in the upper few centimeters of wet sand on beaches. A few species are marine. Most rotifers move actively about over substrates and submerged plants in search of food, but some are attached, and a few are planktonic.

Rotifers have several distinctive features (Figure 23.9). On most kinds, a prominent anterior **corona** (crown) of cilia appears to rotate constantly and functions in swimming and feeding. Behind the corona, the pharynx is differentiated into a jawlike structure called the **mastax**. Unlike that of nematodes, the rotifer body is covered by a cuticle formed *within* a syncytial epidermis, and in many species the cuticle is thickened to form a **lorica**, or case. Body tissues have a relatively constant number of cells (or nuclei in the case of the syncytial epidermis), and although each species is different, rotifers generally consist of about 800 to 1000 somatic cells. Many species also have a tapered foot with movable toes and **pedal glands** that secrete adhesives, assisting in locomotion and in attaching to substrates.

Nearly all rotifers seen in nature are females. Males often occur only for short periods of time near the end of a growing season, and in many species they do not occur at all. Most rotifers develop parthenogenetically (i.e., individuals develop from unfertilized eggs). Soil rotifers typically have drought-resistant dormant eggs, and some, like many nematodes, can survive extended periods of desiccation. (See the Essay on suspended animation in this chapter, p. 540.)

REPRESENTATIVE ROTIFERS

The phylum is divided into two classes, the Digononta and the Monogononta. Females of the Class Digononta have two ovaries, while females of the larger Class Monogononta have only one.

Class Digononta

This class contains some of the most common freshwater rotifers. Species of *Philodina* (Figure 23.9a) are often abundant in water with a high content of organic material, such as sewage wastes, and often appear in old laboratory cultures of protozoa. The surface cuticle in *Philodina* appears segmented because it is superficially folded into annuli, allowing the body to move telescopically. Males are unknown in the genus. A single genus, *Seison*, is marine; males and females of *Seison* inhabit the body surfaces of certain crustaceans.

Class Monogononta

This diverse class contains over 90% of all rotifers, including the vast majority of freshwater species and several free-living marine species (see Figure 23.9b). Males occur in many species, but they are small, degenerate, and usually short-lived. Many monogononts are encased in a lorica. The class also has several sedentary members, such as *Collotheca*, which lives in secreted gelatinous tubes.

COMPARATIVE STRUCTURE AND FUNCTION

The cuticle and lorica provide protection and are the main body support elements, although fluid in the pseudocoel also provides hydrostatic support. Rotifers pull themselves through the water mainly by the action of coronal cilia. Planktonic species may swim continuously, never touching the substrate. Bottom dwellers swim and creep over surfaces using coronal cilia and circular and longitudinal muscle bands in the body wall (see Figure 23.9a). Muscles also retract the corona if the animals are disturbed. Species with a foot and pedal glands can move in an inchwormlike manner. With its toes glued to a substrate, a rotifer simply stretches its body forward, attaches at its anterior end, and then releases its foot.

Many rotifers feed on organic particles suspended in the water or on surfaces; others are active predators of protozoa and microscopic metazoans, including other rotifers. The rotifer gut typically consists of a mouth encircled by the corona, a pharynx, or mastax, an esophagus, a saclike

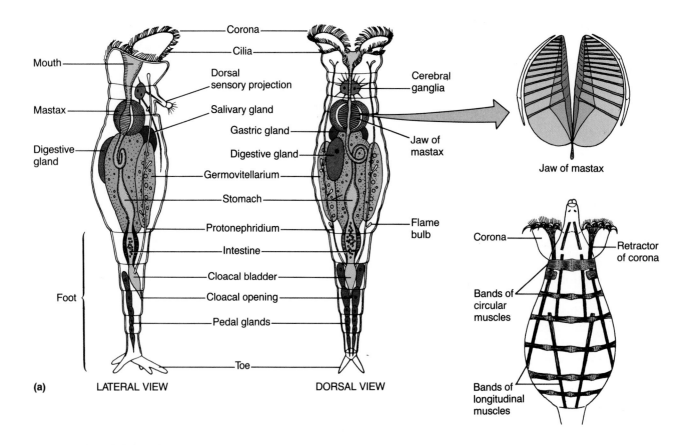

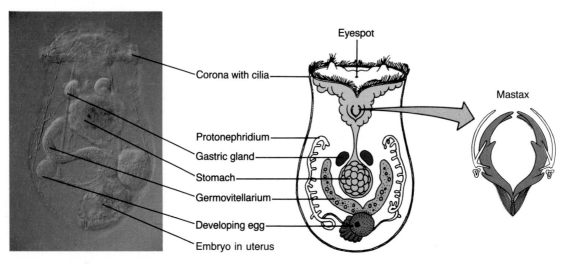

Figure 23.9   Phylum Rotifera. (a) The female of *Philodina*, a freshwater genus (Class Digononta), has a prominent corona. Early microscopists named these animals rotifers (meaning "wheel bearers") because of the corona's resemblance to a pair of spinning wheels. This rotifer's body is externally annulated; the elongate foot (extended in this drawing) consists of five cuticular units that can be moved in and out like a telescope. The enlarged view of the mastax shows two ridged plates that provide surfaces for grinding particulate food. Note that the circular and longitudinal muscles of rotifers are arranged in bands, different from the muscle layers typical of metazoans. The nervous system in most rotifers is highly ganglionated; the cerebral ganglia seen here are sometimes called a brain. (b) Photograph and sketch of a female of the predatory species *Asplanchna sieboldi* (Class Monogononta), common in freshwater plankton. Note that there is no intestine or anus; *Asplanchna* feeds by sucking fluids from its prey, ingesting little that cannot be completely digested and absorbed. In evolving as fluid-feeding predators, species of *Asplanchna* secondarily lost part of the complete gut of their ancestors. The mastax is pincerlike.

stomach, a short intestine, and a dorsal anus. Salivary and gastric glands secrete enzymes into the pharynx and stomach, and digestion is typically extracellular. Nutrient absorption occurs mainly in the stomach. Some rotifers lack an intestine and anus (see Figure 23.9b).

Structures of the corona and mastax reflect the feeding habits of rotifers and are useful in classification. In digonont rotifers, the mastax contains two prominent serrated plates effective in grinding minute plankton and particulate food (see the inset in Figure 23.9a). Monogonont rotifers have diverse feeding habits and a variety of mastax types; Figure 23.9b shows a pincerlike mastax.

Rotifers, like other pseudocoelomates, exchange gases and dispose of nitrogenous wastes by diffusion across body surfaces; pseudocoelomic amoebocytes and fluids moved by muscle contractions aid in circulation. Osmoregulation, at least in freshwater species, is carried out by protonephridia, with tubules extending longitudinally through the pseudocoel and emptying into a posterior cloacal bladder or into the gut (see Figure 23.9a and b). Protonephridia function mainly in the removal of excess water.

The rotifer nervous system consists of a cerebral ganglion that gives rise to several anterior and posterior nerves that are often ganglionated (see Figure 23.9a). Sensory structures include numerous ciliary clusters and sensory bristles concentrated on one or more short antennae or on the corona. One to five photosensitive eyespots may occur at the anterior end. In many species, a peculiar structure called the **retrocerebral organ** lies adjacent to the brain. It consists of a pair of glands and a sac containing symbiotic bacteria and may have a sensory function.

With their varied sensory structures, rotifers respond to a variety of stimuli, but their behavior has not been studied extensively. Planktonic species exhibit diurnal periodicity, moving downward during the day and upward at night. This seems to be mainly a response to light and may provide for concentrated nighttime feeding in productive surface waters and daytime safety from predators.

REPRODUCTION AND LIFE CYCLES

Except for parthenogenesis, rotifers have no asexual reproduction. When present, male rotifers are small and have a degenerate (probably nonfunc-

tional) digestive tract, but can swim rapidly, have well-developed sensory structures to locate females, and are sexually mature shortly after hatching. Depending on the class, females have one or two ovaries, an adjacent but separate vitellarium (yolk gland), and an oviduct that opens into the gut. The adjacent ovary and yolk gland are often called the **germovitellarium**. In species with both sexes, the male injects sperm into the female's cloaca or directly through her body wall, fertilizing eggs in the oviduct. Fertilized or parthenogenetic eggs or juveniles pass out through the anus.

Rotifers have no larval stages, and the cuticle is not molted during development. When mitosis stops, growth continues in females by increase in cell size. A few hours after hatching, females may be five to ten times larger than newly hatched juveniles. Despite eutely, individuals in large populations vary considerably in size. Species of *Asplanchna* can grow larger in direct proportion to the size of available prey. Researchers have recently found that a specific molecular form of vitamin E, α-tocopherol, induces growth in species of this genus. No metazoans are known to synthesize this compound; thus, it is assumed that *Asplanchna* obtains it indirectly from algae by consuming algae-eating prey. Presumably, as prey grow larger, they contain more α-tocopherol, and as females of *Asplanchna* consume greater quantities of this vitamin, the females produce larger offspring able to capture larger prey.

Rotifer life cycles range from simple to complex. In freshwater digononts, all individuals seem to be females that produce eggs mitotically. The eggs develop parthenogenetically, and species populations consist entirely of genetically independent clones. By contrast, monogonont rotifers have haploid males and two kinds of diploid females (Figure 23.10). So-called **amictic** females produce thinshelled diploid eggs that develop parthenogenetically. (*Amictic* means incapable of being fertilized.) Diploid eggs are also called summer eggs because they are produced when conditions are favorable, and they hatch within a few days. In this way, several generations of amictic females may develop by parthenogenesis, and the population increases rapidly. Monogonont rotifers have been cultured through hundreds of amictic generations in the laboratory. The other type of diploid individuals are called **mictic** females. For unknown reasons, diploid eggs from amictic females suddenly begin to develop into mictic females, which in turn produce haploid eggs by meiosis. Initially, there are no males to fertilize the eggs, but the haploid eggs

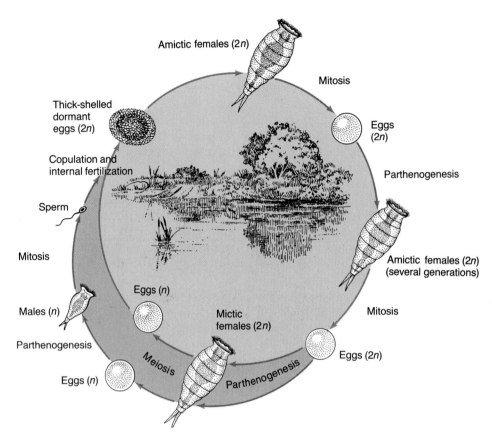

Figure 23.10 Life cycle of a freshwater monogonont rotifer. Parthenogenesis occurs in the development of diploid amictic females and haploid males. Development of both sexes is direct (there is no larva).

develop into males. Once males are in the population, they copulate with females and fertilize haploid eggs. The fertilized eggs develop a thick shell and can withstand harsh temperatures and desiccation; they remain dormant until favorable conditions return, at which time they hatch into amictic females.

This intricate life cycle, adaptive to life in seasonally changing environments, raises several interesting questions. What triggers the change from amictic to mictic generations, or, more specifically, what causes eggs from amictic females to develop into mictic individuals? Experimental and field studies indicate that different species are affected by different factors and that environmental rather than genetic factors cause the generation switch. Mictic females and males often appear in dense populations; thus, stress from crowding or food shortage or seasonal changes related to photoperiod, cold temperatures, or drought may be important, at least in some species.

PHYLOGENY

The affinities of rotifers with other phyla are obscure, but these pseudocoelomates have several features in common with certain acoelomates. The rotifer mastax resembles the jaws of gnathostomulids, but it is not known whether these structures are homologous. Moreover, the protonephridia of rotifers closely resemble those of freshwater rhabdocoels, and it is generally believed that the rotifers originated in freshwater habitats. (Gnathostomulids and rhabdocoels are discussed in Chapter 22, pp. 528 and 513.) Flatworms and most rotifers have separate ovaries and vitellaria, and the peculiar retrocerebral organ of rotifers may be homologous with a similar structure in certain acoel flatworms. Rotifers probably had their origins in the earliest acoelomates, or acoelomates and rotifers may have had a common ancestor among the earliest bilateral metazoans.

PHYLUM GASTROTRICHA

Gastrotrichs bear a slight resemblance to rotifers (Figure 23.11). They are of similar size (most are less than 1 mm long), but the body surface often bears scales or spines, there is no anterior corona, and the pseudocoel is reduced to tiny spaces or is totally absent. The ventral surface bears locomotor cilia, and in some species, each epidermal cell bears only one cilium. The cuticle of gastrotrichs is extracellular, secreted by an underlying syncytial epidermis more like that of nematodes than of rotifers. Many gastrotrichs have a forked tail with adhesive glands, and like many pseudocoelomates, they have protonephridia.

Only about 450 species of gastrotrichs have been described, but probably many more remain to be discovered. A few species inhabit moist soil, but most live on submerged plants and among sand grains in marine or freshwater habitats; they are rarely as abundant in fresh water as rotifers. Gastrotrichs feed on organic debris, bacteria, algae, and protozoa. Cilia and spines surrounding their mouth sort food particles, and the muscular pharynx provides a sucking force that helps pull food into the gut. A variety of organisms, including large amoebas, hydras, nematodes, annelid worms, and small arthropods, may consume gastrotrichs.

Gastrotrichs are hermaphroditic. Marine species typically copulate, ensuring cross-fertilization. It has long been assumed that most or all freshwater species reproduce entirely by parthenogenesis, for their testes do not seem to develop. However, recent discovery of sperm packets in certain freshwater gastrotrichs, previously thought to be parthenogenetic, indicates a need for further study of reproductive processes in these animals. Freshwa-

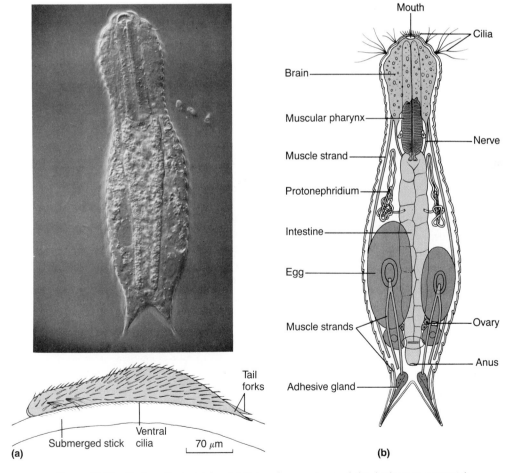

Figure 23.11 Phylum Gastrotricha. (a) External appearance of the freshwater gastrotrich *Lepidodermella squamata*. (b) Internal anatomy of a typical gastrotrich in dorsal view.

ter species produce two kinds of eggs, a thin-shelled type that hatches within a few days during favorable periods and a thick-shelled type that resists droughts and freezing temperatures. There is no larval stage.

Gastrotrichs show evidence of a distant relationship to acoelomates. Many of them lack a body cavity, and their ventral cilia may have been derived from the same ancestral source as those of turbellarian flatworms.

PHYLUM KINORHYNCHA

Only about 100 species of kinorhynchs are known, and all are small (less than 1 mm long) bottom dwellers in coastal seas (Figure 23.12). They feed on

fine organic matter and algae as they burrow through silt and mud. The name Kinorhyncha, which means "movable snout," refers to the spiny head that can be retracted inside the front end of the body. The animals burrow by pushing the head forward, planting the recurved spines in the mud and drawing the body up to the head.

Unique among pseudocoelomates, kinorhynchs show evidence of segmentation. Externally, the cuticle is divided into 13 units called **zonites**, each with articulating ventral and dorsal plates. The dorsal plates bear prominent spines. Internally, the muscles and nervous system also exhibit segmentation. A nerve ring encircles the pharynx and gives rise to a single ventral nerve cord with a ganglion corresponding to each zonite. This resembles the ventral ganglionated nerve cord of earthworms and arthropods. The digestive tube is straight with few specializations besides a muscular pharynx. Kinorhynchs have protonephridia similar to those in

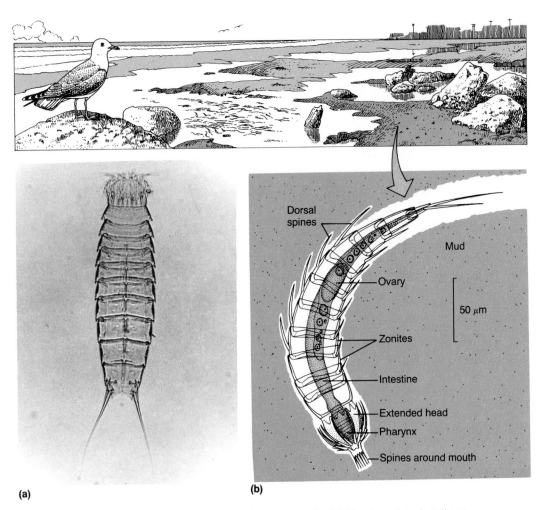

(a)
(b)

Figure 23.12   Phylum Kinorhyncha. (a) Photomicrograph of *Echinoderes dujardinii.* (b) Kinorhynchs live in wet mud and sand in intertidal zones.

other pseudocoelomates. Sexes are separate, but little is known about kinorhynch reproduction and development. There are conflicting reports of whether or not larvae occur, but most specialists hold that kinorhynchs lack larvae. Of the pseudocoelomates, these animals most resemble rotifers; their segmentation suggests some affinities with the metameric coelomates (annelids and arthropods), but without more knowledge of kinorhynchs, it is difficult to speculate about their phylogeny.

PHYLUM LORICIFERA

Many zoologists have the opportunity to discover and name new species, but it is a rare and exciting event to describe a new phylum. This most recently named phylum was first described in 1983 from specimens collected in marine sand (Figure 23.13).

Loriciferans are somewhat rotifer-like. The adults are covered with a heavily armored case, or *lorica*; thus the phylum name, which means lorica-bearing. Adults are mainly sedentary, adhering tightly to sand grains. The larva is more motile, with a thin lorica and a pair of toes used in swimming. Both stages have a flexible tubular mouth that can be retracted into the body. We know little about these animals, but they seem to be abundant and may play an important environmental role in wet beaches.

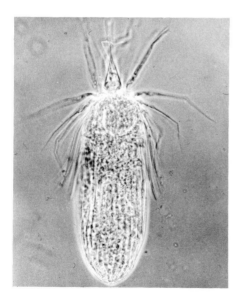

Figure 23.13 Phylum Loricifera. Loriciferans, including this species *Pliciloricus gracilis,* constitute a newly discovered phylum of marine sand dwellers.

PHYLUM PRIAPULIDA

There are only 13 species in this phylum, but **priapulids** are widely distributed from polar seas to the tropics and from tide waters to depths of 500 m. Ranging in length from 2 mm to about 8 cm,

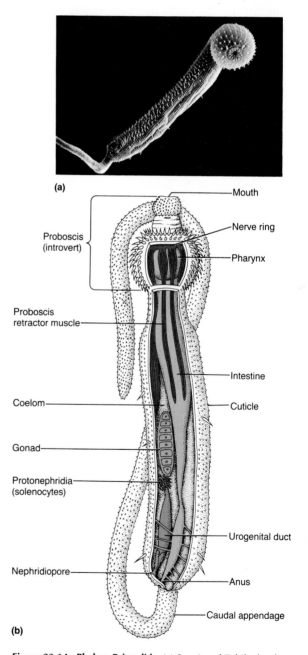

Figure 23.14 Phylum Priapulida. (a) Species of *Tubiluchus* burrow in intertidal mud; they lie in wait for prey with just their mouth exposed. This scanning electron micrograph shows *Tubiluchus corallicola* with its proboscis evaginated (50×). (b) This cutaway drawing shows the main internal structures, the protrusible proboscis, and the long caudal appendage. Other species have two short, multilobed caudal appendages.

priapulids are wormlike marine bottom dwellers that feed on small annelids and other invertebrates. Priapulids burrow by peristalsis and often sit upright in mud or sand, waiting for prey (Figure 23.14). The mouth, at the end of a bulbous **proboscis** (also called the **introvert**), is surrounded by spines that aid in grasping prey. Sensory papillae cover the proboscis. A spiny trunk and, in some species, one or two hollow **caudal appendages** that may function in gas exchange and sensory reception make up the rest of the body. The trunk bears superficial annuli, but there is no external evidence of segmentation. A flexible chitinous cuticle covers the body and is shed periodically as the animal grows.

Internally, priapulids have a straight gut, with an anus at the end of the trunk (Figure 23.14b). A muscular pharynx with recurved teeth pulls in food and masticates it. There are no specialized organs for circulation or gas exchange, although in some species amoeboid cells in the pseudocoel contain the red, oxygen-carrying protein hemerythrin. Protonephridia function in excretion. The nervous system, consisting of a nerve ring around the pharynx and a single ganglionated midventral nerve cord, resembles that of annelids. (The annelid nervous system is described in Chapter 10, p. 231; Figure 10.10.) Sexes are separate, but males and females look alike.

There are more questions than answers about priapulids. Debate over their phylogenetic position has waxed and waned since their discovery during the Linnaean era, and they have been classified as coelomates some of the time and as pseudocoelomates at other times. Because of incomplete knowledge of priapulid development, the nature of their body cavity remains in doubt, but the preponderance of evidence now indicates that priapulids are pseudocoelomates.

PHYLUM ENTOPROCTA

Phylum Entoprocta consists of about 100 species of sessile or sedentary filter feeders (Figure 23.15). Entoprocts are either solitary or colonial, and except for one freshwater genus, all species are marine, most inhabiting shallow coastal waters. One marine group is commensal on the body surfaces of various invertebrates. Entoproct colonies may form extensive matlike growths on rocks, but individuals are small (up to about 10 mm long). Each consists of a muscular stalk bearing a cup- or urn-shaped **calyx** with a crown of ciliated tentacles. A chitinous cuticle surrounds the stalk and calyx and provides some body support. The pseudocoel, which is filled with loose mesenchymatous tissue, extends throughout the body, and hydrostatic fluid pressure in the mesenchyme extends the tentacles

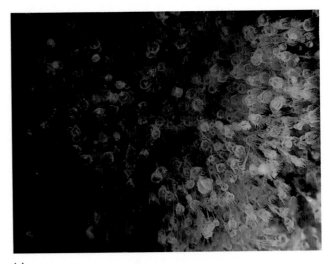

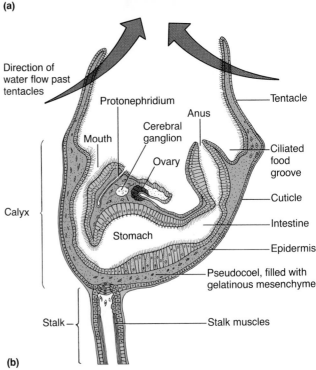

Figure 23.15 Phylum Entoprocta. (a) Portion of a colony of the marine species *Barentsia benedeni*. Muscles in the stalk allow entoprocts to bend and wave. Some species can creep slowly over the substrate, and a few solitary species can bend over and move by somersaulting on their tentacles. (b) Vertical section through one individual. Arrows indicate the flow of water between the tentacles.

(see Figure 23.15b). In feeding, cilia on the tentacles drive water upward, trapping food particles and sweeping them down along the inner surfaces of the tentacles to a ciliated groove leading into the mouth (Figure 23.15b). The gut lacks muscles, and food is moved through the mouth by cilia. Both the mouth and the anus are situated within the tentacular crown.

Most, if not all, entoprocts are hermaphroditic. Cleavage is spiral and leads to a free-swimming trochophore-like larva in some species. The phylogenetic position of the entoprocts is controversial. We have included them in this chapter mainly because they have a pseudocoel. They also bear some resemblance to sedentary rotifers and have protonephridia similar to those of flatworms and rotifers. In general appearance, however, entoprocts most resemble a phylum of coelomates called the Bryozoa (discussed in Chapter 27, p. 651). Accordingly, some zoologists consider the phyla Entoprocta and Bryozoa to be closely related. In placing the Entoprocta here, we are subscribing to the alternative view that the similarities between the two phyla probably resulted from convergent evolution.

flexing. Osmoregulation and excretion are carried out by cells of the gut and body surface and/or by renette cells or canals. The nervous system consists of a gut-encircling nerve ring bearing ganglia and connected to longitudinal nerves.

4. Most nematodes are dioecious; males are usually smaller than females and have posterior copulatory structures for internal fertilization. Many species develop parthenogenetically. There are no larval stages.

5. The other eight phyla of pseudocoelomates and their approximate number of species are: Nematomorpha (hairworms), 240; Acanthocephala (spiny-headed worms), 700; Rotifera (rotifers), 1800; Gastrotricha (gastrotrichs), 450; Kinorhyncha (kinorhynchs), 100; Loricifera (loriciferans), about 5 known, but probably many more; Priapulida (priapulids), 13; and Entoprocta (entoprocts), 100.

□ FURTHER READING

Barnes, R. D. *Invertebrate Zoology.* Philadelphia: CBS, 1987. Chap. 9. *An extensive chapter on the pseudocoelomates; these animals were formerly classified in one large phylum, the Aschelminthes, meaning "bag worms."*

Blonston, G. "To Build a Worm." *Science 84* 5(1984): 62–70. *Describes developmental and genetic research on Caenorhabditis elegans; includes color photographs.*

Gilbert, J. J. "Developmental Polymorphism in the Rotifer *Asplanchna sieboldi.*" *American Scientist* 68(1980): 636–646. *Describes the increase in size of this predatory species resulting from vitamin E consumption; includes color photographs of Asplanchna.*

Lewin, R. "New Phylum Discovered, Named." *Science* 222(1983): 149. *Interesting short summary describing the recent discovery of loriciferans.*

Moore, J. "Parasites That Change the Behavior of Their Host." *Scientific American* 250(1984): 108–115. *Explains how Acanthocephalan larvae, perhaps by biochemical means, make their hosts more vulnerable to predation by definitive hosts.*

Pennak, R. W. *Freshwater Invertebrates of the United States.* 2d ed. New York: Wiley, 1978. *Chapters on freshwater gastrotrichs, rotifers, nematodes, and nematomorphans.*

Poinar, G. O., Jr. *The Natural History of Nematodes.* Englewood Cliffs, N.J.: Prentice-Hall, 1983. *A well-illustrated introduction to the phylum.*

Walsh, J. "Rotifers, Nature's Water Purifiers." *National Geographic* 155(1979): 286–292. *Beautiful color photomicrographs of common free-swimming and sessile rotifers.*

□ SUMMARY

1. Key Features of Pseudocoelomate Animals. Nine phyla are grouped together as the pseudocoelomates, a diverse assemblage of aquatic eumetazoans. Most have a pseudocoel, a constant number of body cells or nuclei, protonephridia, and a complete gut with a well-developed pharynx. No organs are developed for gas exchange or circulation. The body is covered by a cuticle, and longitudinal muscles are often dominant in the body wall. Members of two phyla are capable of cryptobiosis. Phylogenetic affinities among the nine phyla and with other phyla are obscure.

2. Members of the Phylum Nematoda (nematodes, or roundworms) are cylindrical with radially or biradially arranged oral structures and with only longitudinal muscles in the body wall. The gut lacks muscles, except at the anterior and posterior ends. Muscle cells have projections that synapse with neurons. Cilia are absent except in certain sense organs. Most species are free-living in aquatic environments, including moist soil, but many others parasitize plants or animals. Nematode sperm cells are amoeboid.

3. Nematodes have a secreted multilayered cuticle; the body wall and pressurized fluid in the pseudocoel provide hydrostatic support. Movement is by whiplike body

□ CLASSIFICATION

Phylum Nematoda (Gr. *nema*, "thread"; *eidos*, "form"). Nematodes, or roundworms.

Class Phasmidia (Gr. *phasm*, "phantom"; *idia*, "tiny"). Phasmids (paired glandular/sensory structures) occur in the tail region; *Rhabditis, Caenorhabditis,* and *Turbatrix* are free-living; *Heterodera* and *Meloidogyne* infect plants and damage agricultural crops; *Enterobius* (human pinworm), *Ascaris* (intestinal roundworm), *Ancylostoma* and *Necator* (hookworms), *Dirofilaria* (heartworm), and *Wuchereria* (filarial worm) infect animals.

Class Aphasmidia (Gr. *a*, "without"; *phasm*, "phantom"). Phasmids are absent; most are free-living; *Dorylaimus* is common in fresh water; *Trichinella* (trichina worm) and *Dioctophyme* (kidney worm) are parasites of mammals; species of *Xiphinema* are plant parasites of economic importance.

Phylum Nematomorpha (Gr. *nematos*, "thread"; *morphe*, "form"). Hairworms (nematomorphans); adults are free-living; *Gordius* adults inhabit fresh water and moist soil; *Nectonema* is marine; larvae are parasitic in arthropods.

Phylum Acanthocephala (Gr. *akantha*, "spine"; *kephale*, "head"). Spiny-headed worms (acanthocephalans); adults have a spiny invertible proboscis and no gut; all are parasitic, mostly in the gut of fishes; larvae and juveniles infect arthropods; *Moniliformis* and *Macracanthorhynchus* occasionally infect people.

Phylum Rotifera (L. *rota*, "wheel"; *fero*, "to bear"). Rotifers; a ciliated corona surrounding the mouth and a muscular pharynx (mastax) with jawlike parts are diagnostic features; the cuticle is intracellular; parthenogenesis is common; freshwater and marine.

Class Digononta (Gr. *di*, "two"; *gonos*, "seed"; *ontos*, "a being"). Rotifers with two ovaries; *Philodina* and *Rotaria* are free-living; species of *Seison* are commensal on marine crustaceans.

Class Monogononta (Gr. *monas*, "single"; *gonos*, "seed"; *ontos*, "a being"). Rotifers with one ovary; *Asplanchna* species are planktonic predators; *Collotheca* is a sedentary tube dweller.

Phylum Gastrotricha (Gr. *gaster*, "stomach"; *thrix*, "hair"). Gastrotrichs; body is covered with spines and has two forked adhesive tubes at the posterior end; pseudocoel reduced to tiny spaces or absent; parthenogenesis common; *Chaetonotus* and *Lepidodermella* occur in freshwater and marine environments.

Phylum Kinorhyncha (Gr. *kineo*, "movement"; *rhynchos*, "snout"). Kinorhynchs; all are marine; evidence of segmentation in the spiny cuticle, muscles, and nervous system; the head is retractable; *Echinoderes*.

Phylum Loricifera (L. *lorica*, "clothed in armor"; *fero*, "to bear"). Loriciferans; marine sand dwellers; body mostly surrounded by an armored lorica; *Nanaloricus*, *Pliciloricus*.

Phylum Priapulida (Gr. *priapos*, "phallus"). Priapulids; bottom-dwelling marine worms with a barrellike anterior proboscis adapted for carnivory and deposit feeding; *Priapulus*, *Tubiluchus*.

Phylum Entoprocta (Gr. *entos*, "inside"; *proktos*, "anus"). Entoprocts; solitary or colonial; individuals are sessile and stalked; mouth and anus are situated within a crown of tentacles; many are colonial; there is one freshwater genus, *Urnatella*; *Pedicellina* and *Barentsia* are free-living and marine; *Loxosoma* is commensal on other marine animals.

Success of the Eucoelomates

Although we have now surveyed about half the metazoan phyla, we have yet to discuss the eucoelomates, a group that in terms of numbers of species and ability to inhabit diverse environments constitutes the most successful evolutionary lineage in the animal kingdom. Collectively, eucoelomates account for over 90% of all animal species. (This includes more than 1.1 million species in just two phyla, the Arthropoda and the Mollusca.) In terms of environmental diversity, only eucoelomate phyla (the Annelida, Arthropoda, Mollusca, and Chordata) and a few flatworms have both aquatic and truly terrestrial species. (By truly terrestrial, we mean able to reproduce and be active on land. We exclude nematodes here, because while numerous in soil, they generally require droplets of soil water to be active.)

Why has the eucoelomate lineage been so successful? No doubt the coelom, one of the few features that these animals have in common, provided adaptive advantages. But perhaps more to the point, the *combination* of a high degree of body organization (provided in part by the coelom), bilateral symmetry, and a three-layered body seems to have produced an immense adaptive potential not achieved by acoelomates or pseudocoelomates. This combination has allowed eucoelomates to evolve the most highly organized and efficient organ systems in the animal kingdom. Most eucoelomates have well-developed circulatory and gas exchange systems, while few acoelomates and pseudocoelomates have any structures specialized for these functions.

As described in Chapter 18, there are two evolutionary lines of eucoelomates, the protostomes and the deuterostomes, and both have produced immensely successful phyla. As shown in the phylogenetic tree in Figure 18.6, molluscs, annelids, and arthropods are protostome eucoelomates. Chordates, representing deuterostomes, are relatively few in number of species, but are highly successful in aquatic, terrestrial, and aerial environments. The fossil record indicates that phylogenetic lineages of these and other eucoelomate phyla diverged from ancestral eucoelomates between 600 and 1000 million years ago. Since then, the relative success of each phylum is a measure of the adaptability of the basic eucoelomate characteristics in combination with new features derived in each line. As you read about the eucoelomates, try to identify key features that appear to contribute to the relative success of each phylum.

CHAPTER 24

Molluscs

Phylum Mollusca. These shells, produced by species of the molluscan Class Gastropoda, illustrate some of the beauty and diversity of molluscs. Mollusc shells have been used as ornaments and currency since the dawn of human civilization.

The Mollusca is the second largest phylum in the animal kingdom, with about 110,000 species (including about 40,000 identified from fossilized shells). Although most molluscs are marine, many inhabit freshwater and terrestrial environments. The Latin word *molluscus* means "soft," and although many molluscs have a hard shell, several soft parts of the molluscan body are the distinctive features of the phylum (Figure 24.1). Most organs are located in a body mass called the **visceral hump**, which lies dorsal to a muscular *foot* used in locomotion. Many molluscs also have a *head* and a unique toothed, tonguelike feeding organ called a **radula** that probably functioned as the main feeding organ in ancestral species. In many living molluscs, the radula functions in rasping organic matter from rock surfaces, and this is probably the primitive mode of feeding in molluscs. The main body parts—the hump, foot, and head—are cov-

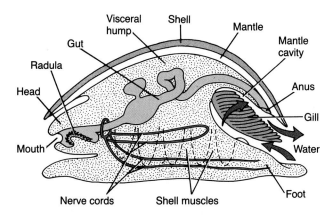

Figure 24.1 Diagnostic features of a mollusc. The earliest members of the Phylum Mollusca probably had these fundamental features and may have looked something like this generalized example. Of the main body features shown (foot, visceral hump, head, mantle, mantle cavity, and shell), nearly all living members of the phylum have at least a mantle, a mantle cavity, and some type of muscular foot. The visceral hump contains the heart, gonads, nephridium (kidney), and most of the gut. Arrows indicate the direction of water flow in the mantle cavity and over the gills (ctenidia).

ered dorsally by a cloak or sheetlike extension of the visceral hump, called the **mantle**. In most molluscs, the position of the mantle produces a space called the **mantle cavity** between the mantle and the main body mass. The mantle and the mantle cavity are key components of the molluscan body plan. The mantle produces the calcareous **shell**, which is usually attached to the mantle's outer edge. The mantle may also function in gas exchange, excretion, and sensory reception. In some molluscs, including squids and their allies, the mantle is muscular and has a locomotor function. A lung or gills often develop within the mantle cavity, which also may house the excretory and reproductive pores, and the anus. Water or air is circulated through the mantle cavity; consequently, soft tissues can be exposed to the environment and perform vital functions while they are protected by the overlying shell and mantle. The molluscan shell grows with the soft parts of the animal, and a mollusc does not shed its shell to grow.

REPRESENTATIVE MOLLUSCS

Molluscs are divided into seven classes, five of which are exclusively marine and contain fewer than a thousand living species each. Two large classes, the Gastropoda (snails and their allies) and the Bivalvia (clams and other bivalves), account for over 95% of all molluscan species. Both of these groups have freshwater and marine representatives, and some gastropods are terrestrial.

Class Monoplacophora

The fossil record indicates that the two genera and seven living species in the Class Monoplacophora are remnants of a group of marine molluscs that were plentiful 300 to 500 million years ago. Living monoplacophorans are small (less than 5 cm long), and most are members of the genus *Neopilina* (Figure 24.2a). They have survived with little change in body structure for hundreds of millions of years. Monoplacophorans live on rocky substrates at depths of 3000–6500 m.

Monoplacophorans have a mantle, a single dorsal shell, a ventral muscular foot, and a radula. Unlike other molluscs, however, several of their body parts are repeated. The foot is connected to the undersurface of the shell by eight pairs of retractor muscles; five or six pairs of gills project into the mantle cavity; the heart has two pairs of atria; and there are two pairs of gonads and six pairs of excretory organs. The nervous system is ladderlike with ten pairs of nerves innervating the foot. Repetition of body parts in these ancient molluscs leads some zoologists to postulate that ancestral molluscs were segmented, but this view is not widely accepted (see Phylogeny, p. 581).

Class Aplacophora

Aplacophorans (about 250 species) are wormlike, marine bottom dwellers, living mostly at depths of over 8000 m. The gills, mantle, and foot are either reduced or altogether absent. Most species have a radula but no shell, although shelllike calcareous granules may occur in the epidermis. Two subgroups, neither of which have been studied extensively, are browsers called **solenogasters** and elongate burrowers called **caudofoveates**.

Class Polyplacophora

The **Polyplacophora**, an ancient class composed of molluscs called **chitons**, includes fossil species dating from the Cambrian (over 500 million years ago). As the class name (meaning "many plates") implies, chitons have a unique shell composed of eight overlapping plates (Figure 24.2b and c). The shell plates are embedded in the mantle, which extends to the edge of the animal, forming a tough, pliable **girdle** that often bears calcareous bristles or spines.

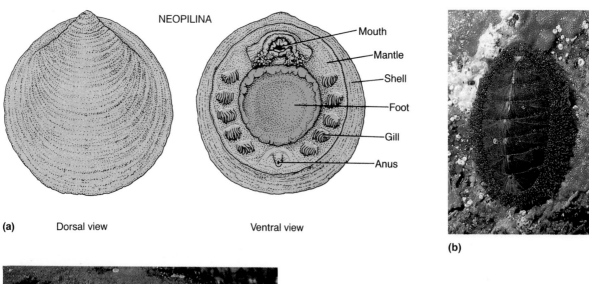

Figure 24.2 Members of two small classes of marine molluscs. (a) Class Monoplacophora. The "living fossil" *Neopilina.* (b) The chiton *Mopalia muscosa,* a member of the Class Polyplacophora. Note the eight exposed shell plates. (c) In *Cryptochiton stelleri,* another polyplacophoran, the shell plates are completely covered by the mantle.

All chitons are marine, and most of the 600 species in the class exhibit pronounced adaptations for life on intertidal rocks. The chiton body is flattened, providing protection against wave action. A large muscular foot grips tightly to rocks, and the overlapping shell plates and flexible girdle allow chitons to bend and fit snugly against hard substrates. Chitons have a radula, and when feeding, they move slowly about, scraping algae from rocks.

Class Gastropoda

Gastropods include the snails, slugs, limpets, and abalones, as well as many other closely related molluscs. There are about 40,000 living species of gastropods, and about 25,000 fossil species have been described, representing all geological periods back to and including the Cambrian. This is the only class of molluscs with terrestrial representatives.

Gastropods have several distinctive traits. Most have a spirally coiled single shell that rides on the body at an oblique angle. Most species also have asymmetrical body parts, and during development, gastropods undergo a peculiar body twisting called **torsion** (Figure 24.3). Torsion occurs during larval development. In some species, twisting occurs as muscles on the right side of the body develop before those on the left; this uneven growth causes the visceral hump to rotate, up to 180° above the foot and head. Before torsion, only the head faces forward; but after torsion, the gut and other organs are rotated so that the mantle cavity containing the anus, excretory pores, and gills lies over the head.

The adaptive advantages of torsion are not well understood, but since the gills face forward, they can be ventilated efficiently by undisturbed water drawn from in front of the animal. Also, torsion may provide some protection, for the head can be withdrawn into the shell before other organs, such as the foot; a snail might survive losing part of its foot (but not its head!) to a predator. Some zoologists postulate that torsion originated as a larval adaptation for protection of the head and that it provides little or no adaptive advantage to adults. In fact, having a torted body means that adult gastropods must overcome certain problems. Having the anus above the mouth and near the gills presents potential fouling problems, and much of the evolution of gastropods is related to solving problems created by torsion. In all torted gastropods, for example, the body is adapted so that fecal mate-

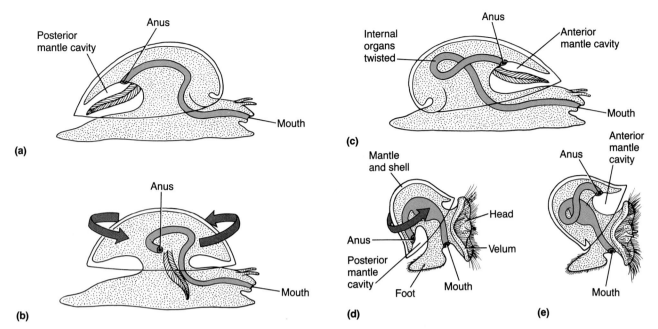

Figure 24.3 Body torsion in the Class Gastropoda. Color arrows indicate the direction of body twisting. (a) The position of the visceral hump, mantle, mantle cavity, and shell in an untorted gastropod is like that of a generalized mollusc (see Figure 24.1). (b) Torsion occurs as the body twists counterclockwise 180°. (c) The final position of the torted body above the foot. An adultlike animal is drawn in (a–c) to make the results of torsion easy to visualize, but torsion actually occurs in the larval stage, as shown in (d) before torsion, and (e) after torsion. Note the large velum (a ciliated swimming and filter-feeding organ) in the larva; the velum is lost in the radula-feeding adult.

Figure 24.4 Prosobranch gastropods. With over 30,000 species, this diverse group includes genera such as *Haliotis* (abalones) and *Patella* (limpets). Abalone shells, as seen in (a) *Haliotis cracherodi* (black abalone), have a shallow spiral, whereas limpets, such as (b) *Patella vulgata* (common limpet), lack a spiral altogether. Limpets are often attached alongside chitons on intertidal rocks. (c) The cowries, such as this species *Cypraea chinensis,* the Chinese cowry, from Hawaii, are among the most beautiful of the prosobranch gastropods. The mantle usually covers the outside of the cowrie's shell. (d) The highly mobile whelks, such as *Buccinum undatum* shown here laying eggs, prey on clams and other bivalve molluscs; whelks have an elongate siphon, an extension of the mantle, that functions as an incurrent water tube.

rial is swept away in water currents directed away from the mouth and gills.

Gastropods are classified into three large subgroups based on the structure and position of gas exchange organs. Making up the largest subgroup (about 30,000 species) are the **prosobranchs** (from the Greek, meaning "forward gills"); these gastropods have gills (a few have lungs) in an anterior mantle cavity. A few prosobranchs live in fresh water and on land, but most are marine. Herbivorous (algae-eating) prosobranchs include abalones and limpets, both of which have a shell (consisting of a single large whorl) and a foot adapted for gripping rocks (Figure 24.4a and b). Carnivorous prosobranchs include the cowries and whelks (Figure 24.4c and d). The foot of most prosobranchs bears a calcified thickening called the operculum; when the body and foot are withdrawn, the operculum seals off the shell opening.

A second subgroup of gastropods, the **opisthobranchs** (from the Greek, meaning "gills at the back"), was apparently derived from the prosobranchs. Opisthobranchs show evidence of having torted ancestors, but they exhibit "detorsion." Their body parts are partially or fully untwisted, and most are bilaterally symmetrical as adults. The mantle, shell, and gills are reduced or absent, and gas exchange occurs mainly across the body surface. Bubble shells, sea slugs (also called **nudibranchs**), sea butterflies, and sea hares represent some of the diversity in this mostly marine subgroup of about 3000 species (Figure 24.5). Bubble shells have an expanded head with a pair of flaps or tentacles bearing sense organs. The foot is large and used for burrowing or crawling over the ocean floor. Nudibranchs are among the most colorful marine animals. Many have projections on the dorsal body surface that assist in gas exchange. Sea butterflies are unusual in that they swim about in the open sea by means of broad, oarlike extensions of the foot. Sea hares *(Aplysia)* have earlike tentacles and resemble hares in the way they seem to hold and munch on large fronds of algae.

The third subgroup, the **pulmonates** (from the Latin, meaning "lung"), is made up of about 7000 species of gastropods. These molluscs lack gills but have a special adaptation for life on land—an internal air-breathing lung developed from the mantle and mantle cavity. Although pulmonates are common on land, many live in freshwater ponds and streams, and a few species are marine (Figure 24.6). Land slugs, also pulmonates, have only a small internal shell or have completely lost the shell, probably as an evolutionary adaptation to life in areas poor in calcium salts.

Class Bivalvia

Bivalves constitute a distinctive group of about 30,000 living species, including clams, oysters,

(a)

(b)

Figure 24.5 Opisthobranch gastropods. (a) Sea slugs (nudibranchs), such as this species *Dirona albolineata* (alabaster nudibranch), have no shell, mantle, mantle cavity, or internal gills; external projections function as gills. (b) Sea butterflies, also called pteropods (meaning "winged foot"), have either a lightweight shell and swim slowly about feeding on suspended particles or, like this specimen of *Clione limacina* (naked sea butterfly), lack a shell and are active predators; some pteropods occur in immense schools in northern oceans and are important in the diet of certain whales. Sea hares *(Aplysia)* are the largest opisthobranchs (up to about 40 cm long; see Figure E1 in Chapter 13).

(a)

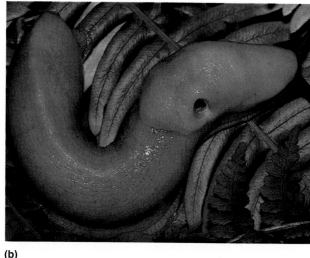
(b)

Figure 24.6 Pulmonate gastropods. (a) The large terrestrial snail *Helix pomatia* is eaten as the French delicacy escargot. This pair is mating. (b) Land slugs, including species of *Agriolimax* are common garden pests. Species of the related genus *Ariolimax*, giant land slugs (10–12 cm long), inhabit moist forests of the U.S. Pacific Northwest. This specimen of *Ariolimax columbianus* was photographed on the Oregon coast.

mussels, and scallops. All species are aquatic, and most are marine. The bivalve body is laterally compressed and housed in a shell of two (right and left) dorsally hinged valves (Figure 24.7a). The **hinge** is flanked on each side by an **umbo**, a raised area that is the oldest part of each valve. Concentric growth rings radiate from each umbo. Most bivalves burrow into mud at the bottom of lakes or shallow oceans; their main locomotor organ is a muscular, hatchet-shaped digging foot that can protrude from

between the valves. An extensive mantle lines the inside of the valves, and a spacious mantle cavity contains one or two pairs of large gills suspended on either side of the visceral hump. The gills are covered with cilia and are adapted primarily for filter feeding, but they are highly vascularized and also serve in gas exchange. Water circulates in and out of the mantle cavity and across the gills, usually through an **incurrent siphon** and an **excurrent siphon**, which are formed by the posterior edges of

(a)

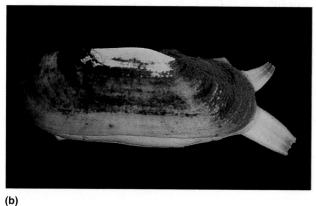
(b)

Figure 24.7 Class Bivalvia. (a) The freshwater mussel *Lampsilis ventricosa*, shown here from the right side, illustrates the basic features of the class. Note the broad, whitish foot extending ventrally from between the valves. The raised area surrounding the umbo is toward the top of the photograph. *Lampsilis* and many other freshwater mussels live in mud and sand at the bottom of lakes, rivers, and streams; with the dorsal (hinge) surface upward, the soft body of a clam is protected from predators above. (b) Marine bivalves are much more diverse: this razor clam *Tagelus plebeius*, normally buried completely in fine sand, has long siphons and a body shaped somewhat like the handle of an old-fashioned razor.

the mantle. As filter feeders, bivalves lack a radula; and with the entire body housed in the mantle cavity, they also lack a head.

Many bivalves lie almost completely buried in mud or sand, with only their incurrent and excurrent siphons exposed to the water. In this way, they can feed while hidden in the substrate. Razor clams have elongate, streamlined shells and tubular siphons that extend to the substrate surface to obtain water (Figure 24.7b). A large bivalve found on the U.S. Pacific coast, the geoduck (pronounced "gooeeduck"), lives as deep as 1 m in wet sand. Its siphons are fused together to form a long fleshy tube.

Several species of bivalves tunnel into rock or wood. Rock-boring bivalves resemble sand burrowers, from which they probably evolved, except that the anterior surface of the borers' shell is serrated. While the foot holds on to a rock surface, the serrated edge scrapes back and forth against it, and the rock is gradually abraded. The shell of *Teredo*, the shipworm (see Figure 4.9), is reduced to small cutting facets. While the animal's tiny foot adheres to the side of its burrow, the facets open and close like jaws, rasping into the wood. Shipworms are trapped in their burrows, with only their siphons (often more than 1 m long) forming a lifeline to the opening of the burrow.

Another group of marine bivalves do not dig or burrow, but live attached to rocks or wooden pilings. The mussel *Mytilus* is attached by fine **byssal threads**. Glands in the mussel's foot secrete a fluid that flows out to the substrate and then hardens, forming the byssal threads as the foot is retracted. Adult oysters lack a foot. An oyster larva, called a *spat* by oyster farmers, first attaches its right valve to a hard substrate by byssal gland secretions. The mantle edge then forms a permanent attachment by producing shell material that hardens and cements the valve to the substrate (Figure 24.8a).

The largest bivalves are the giant clams *(Tridacna)* of the Indo-Pacific, which sit with their valve hinge down in the ocean floor and their gaping valves and siphons upward (Figure 24.8b). Some specimens reach 1 m in length, not including the siphons, and with their heavy shells may weigh as much as 1200 kg. Giant clams are filter feeders, but they probably owe their great size to a mutualistic relationship with photosynthetic dinoflagellates housed in their mantle cells. The position of the clam exposes the mantle and the dinoflagellates to sunlight, and the clam may augment its diet by consuming some of the dinoflagellates' photosynthetic product.

(a)

(b)

Figure 24.8 Other marine bivalves. (a) Oysters (*Ostrea* and *Crassostrea*) are cultured for food throughout the world. They secrete extra shell material and cement their right valve to hard substrates. This photograph shows a pair of *Crassostrea virginica* from Maine. (b) The giant clam *Tridacna maxima* houses photosynthetic symbiotes in its mantle tissues.

Class Scaphopoda

All of the approximately 350 living species of the Class Scaphopoda, collectively called **tusk shells**, are marine and inhabit both shallow and deep oceans. Their whitish shell is an elongate toothlike or tusklike cone with a hole at both ends (Figure 24.9). Adhesive tentacles and a digging foot protrude from the anterior end of the shell. Tusk shells burrow headfirst through sand or mud, feeding on microscopic organisms trapped on the tentacles. Food caught on the tentacles is conveyed into the gut by a toothed radula. Gills are lacking, and gas exchange occurs across the extensive mantle surface as water is moved in and out at the smaller posterior end. Water is sucked into the mantle cavity when the foot is protracted, and is forced back out when the foot is retracted.

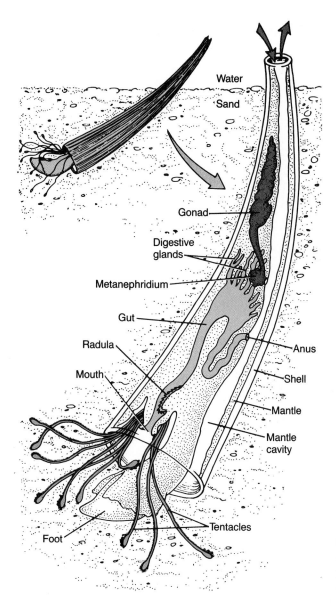

Figure 24.9 Class Scaphopoda. A tusk shell *(Dentalium)* burrowing through marine sand. Arrows indicate the movement of water in and out of the mantle cavity. Most scaphopods are less than 10 cm long, although a few are as long as 25 cm.

Scaphopods resemble bivalves in their mode of digging, in their poorly developed head, and in their developmental patterns. Also, in common with bivalves, fossil tusk shells date back some 500 million years.

Class Cephalopoda

Cephalopods, the nautiluses, squids, cuttlefishes, and octopuses (totaling about 600 living species and about 9000 fossil forms), are entirely marine. Nautiluses are mostly extinct, but a few species of the genus *Nautilus* live in the tropical Pacific, where they spend much of their time on the ocean floor. Compared to most cephalopods, nautiluses are rather slow swimmers; they live in a heavy shell that is cumbersome in the water (Figure 24.10a and b). Squids and cuttlefishes, on the other hand, have a streamlined body with an internal shell and are much more active swimmers (Figure 24.10c and d). Squids rank with fishes as rapid-swimming, successful predators of the open ocean. The giant squid *Architeuthis* (up to 450 kg and 18 m long with tentacles extended) is the world's largest living invertebrate. Giant squids live mainly in deep waters of the North Atlantic and they are sometimes eaten by sperm whales. Octopuses lack a shell and are mainly bottom dwellers (Figure 24.10e); when disturbed or in pursuit of prey, they can scuttle rapidly about over the substrate, or jet away by expelling water from their mantle cavity.

Cephalopods have the foot developed into a ventral muscular **funnel** that squirts water from the mantle cavity, providing jet propulsion. The mouth has chitinous, beaklike jaws and a radula and is surrounded by elongate appendages, called arms or tentacles. Cuttlefishes and squids, collectively called *decapods*, have eight arms and two long tentacles. Octopuses (or *octopods*), as their name implies, have only eight arms (no tentacles). All appendages of decapods and octopods bear suction discs. *Nautilus* is unusual in having up to about 90 arms (usually called tentacles) that lack suction discs.

HABITATS AND ENVIRONMENTAL RELATIONSHIPS

Molluscs have an enormous effect on aquatic environments. They live virtually everywhere in the oceans, from intertidal zones to the deepest

(a)

(c)

(b)

(d)

(e)

Figure 24.10 Class Cephalopoda. Cephalopods prey on fishes and invertebrates. (a) *Nautilus,* a remnant genus of a formerly large group, has a heavy shell. (b) This photograph shows the arrangement of internal chambers of the coiled shell. A coiled tube, the siphuncle, passes through the internal chambers. The animal occupies the outermost chamber. (c) *Loligo opalescens* and other squids are active open-water swimmers, but are most abundant in shallow, coastal waters; they can swim faster than any other invertebrates. (d) Cuttlefishes *(Sepia)* are common in shallow oceans. They are heavier, slower, and not as streamlined as squids, and spend most of their time on or near the bottom. (e) Species of *Octopus,* true bottom dwellers, are more sac-like; the body size of the largest species is about 35 cm, with arms up to 4 m long.

trenches, and it is likely that many marine food chains would collapse without them.

Molluscs in Marine Tidal Zones

Molluscs are numerous on sand beaches, tidal mud flats, and rocky seashores. Many species of clams lie buried in beach sand, filter-feeding on the rich plankton in tidal waters. Holes made by the tips of their siphons can often be seen as dark circles in the sand during low tides. Many gastropods prey on the bivalves.

As shown in Figure 24.11, distinct life zones occur on rocky seashores. Different organisms, dominant at different depths, form discrete bands on the rocks. Two types of gastropods (certain limpets and periwinkles) and some chitons are dominant in the upper intertidal zone; these molluscs are radular grazers, scraping algae from the wave-splashed rocks near the high-tide mark. Filter feed-

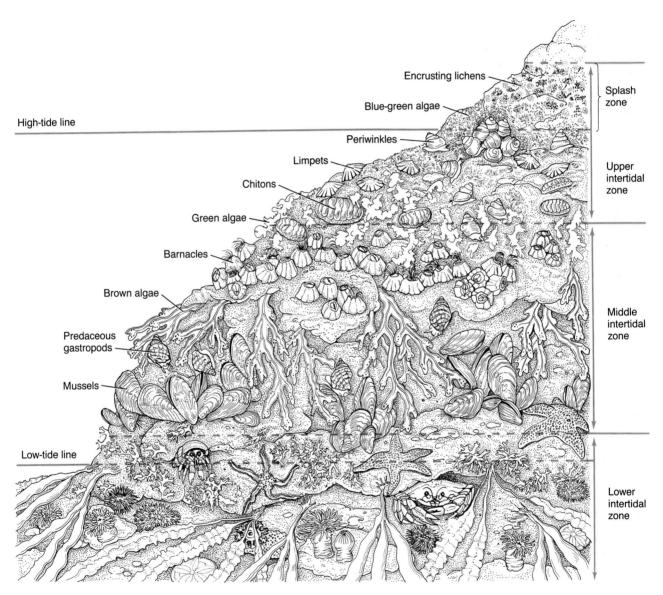

Figure 24.11 Profile of a rocky seashore. Note the three intertidal zones defined by the communities of organisms living in them. The upper intertidal zone includes a region on either side of the high-tide mark occupied by grazing snails and a few chitons. The grazing effect of these molluscs has been studied by removing them from certain areas of rock surfaces; a thick algal mat quickly grows over unoccupied surface. The middle intertidal zone, the area roughly between high and low tide, is dominated by attached filter feeders—bivalve molluscs and barnacles (crustaceans)—and gastropods that prey on filter feeders. The lower intertidal zone (at and just below the low-tide mark) is populated by a greater variety of species.

ers, such as oysters and mussels, which are permanently attached to rocks, cohabit the middle intertidal zone between high- and low-tide levels.

All intertidal organisms have some mechanism to prevent desiccation during low tides. Periwinkles have an operculum that seals off the shell aperture. Limpets and chitons fit tightly against rock depressions when exposed to air. Bivalves can close their shells completely and remain inactive for days if necessary. And certain mussels can respire anaerobically while closed during extended dry periods. (Anaerobic respiration is discussed in Chapter 3, p. 69.)

Freshwater Molluscs

Only two classes of molluscs—bivalves and gastropods—have colonized fresh water, but these groups are abundant and are important in the diet of many aquatic and terrestrial animals. Freshwater molluscs are also important as hosts of larval stages of parasitic flukes that infect vertebrates, including humans. (The life cycle of trematodes is discussed in Chapter 22, p. 525; Figure 22.14.)

An insufficient amount of dissolved oxygen is often a limiting factor for life in fresh water. Clams and gilled snails cannot survive where oxygen is depleted. They are also susceptible to heavily silted water, for their gills can clog and become nonfunctional if too many particles are trapped on them. On the other hand, many pulmonate snails live in fresh water and can tolerate highly polluted conditions with little or no dissolved oxygen, because they surface periodically and breathe air.

Terrestrial Molluscs

There are about 4000 terrestrial prosobranch snails; but with about 20,000 terrestrial species, the pulmonate snails and slugs are the most successful land molluscs. Pulmonates are more diverse on land than in freshwater environments.

Water is the most important limiting factor for land molluscs, and yet several species of snails inhabit deserts. Terrestrial species have special physiological and behavioral adaptations for water conservation. Land snails often aggregate in moist areas, and they can survive periods of extreme drought by withdrawing into their shell and becoming inactive until favorable conditions return. In this condition, their metabolic rate is greatly reduced. Species with an operculum may remain inactive in their sealed shells for months or even years. Similar to desert plants that flower and re-

produce in the few days when moisture is available, desert snails may be inactive for much of their lives. They may crawl about and feed only for a few hours or days when moisture is available, and some conserve water by producing only a few offspring each year.

The body surface of terrestrial gastropods, unlike that of most land animals, is moist and constantly losing water. Land slugs have especially wet, slimy skins, and lacking a protective shell, they are common only in moist areas and are active only when the atmosphere is humid. On dry days, slugs remain under leaves or other damp ground litter, often contracted into a tight ball to reduce surface area and water loss. In some ways it is surprising that there are as many as a thousand species of terrestrial slugs, and yet there are some adaptive advantages to not having a shell. Unlike snails, which require large quantities of calcium to build shells, slugs can live in habitats that are low in calcium. They can also squeeze into tiny holes and moist crevices. Compared to snails, slugs can survive greater loss of body water (more than half of the body weight in some species). By placing slugs that had lost much of their body water into containers of air saturated with moisture, researchers have demonstrated that the skin of these molluscs can absorb water from the air. In one set of experiments, nearly one-fourth of the body weight was restored within 2 hours by surface absorption.

Besides water loss, land animals must be able to withstand marked temperature fluctuations. Terrestrial snails avoid temperature extremes by moving under leaf litter or rocks. Certain slugs can moderate their body temperature for short periods in hot, dry air by evaporative cooling from their moist body surface.

COMPARATIVE STRUCTURE AND FUNCTION

This section concentrates on the functional anatomy of the main organs and organ systems in three prominent molluscan groups—the gastropods, bivalves, and cephalopods.

Body Support and Protection

The molluscan **shell** is protective and also provides attachment sites for muscles. In many molluscs, the shell gives little body support because the soft parts do not conform closely to shell contours. Support for the visceral hump and foot results more from a hydrostatic skeleton due to pressurized circulatory fluid. Lacking a shell altogether, octopuses and many slugs are supported entirely by a hydrostatic skeleton and by internal connective tissues.

Mollusc shells generally consist of three layers. ▷ (The structure of a clam shell is described in Chapter 8, p. 180; Figure 8.10.) The outermost layer, called the **periostracum**, is protective and is composed of **conchiolin**, a leathery protein. The middle **prismatic** layer and the pearly **nacreous** layer adjacent to the mantle consist of a conchiolin matrix impregnated with calcium carbonate. The periostracum and the nacreous layers are not always present; but either the nacreous or the prismatic layer provides muscle attachment sites.

The process of shell deposition in molluscs appears to be the same whether the shells are growing ▷ or being cemented to a substrate. (See the Chapter 8 Essay on biological calcification, p. 183.) Pearls are produced by bivalves in the same way. When a sand grain or some other foreign object lodges between the mantle and the inner surface of the valves, a pearl gradually forms as nacreous shell material is laid down around the foreign object. Although many bivalves can produce pearls, the finest ones are produced by pearl oysters of the tropical Pacific Ocean. Cultured pearls are produced commercially by placing tiny "seeds" (usually small pieces of mussel shell) under the mantle of a pearl oyster. Freshwater mussels are harvested in the United States, and their crushed shells are sold in Japan for seeding oysters.

The Shell of *Nautilus* and Other Cephalopods. The cephalopod *Nautilus* is unusual in that it lives in the outermost chamber of a multichambered shell (see Figure 24.10a and b). This arrangement is different from that of a snail, which occupies most of the space within a single-chambered shell. As a nautilus grows, it periodically forms a new shell chamber and seals off its old chamber, except for a small hole in the middle of the partition. An elongate cord of tissue called the **siphuncle** passes through the small holes and extends through all of the chambers. The siphuncle absorbs fluid from the empty chambers, which in turn fill with gas produced by the animal's respiration. The gas-filled chambers make the shell buoyant and allow the animal to remain upright and swim about.

Nautiluses are not found below 600 m because the water pressure at that depth (more than 40 times the atmospheric pressure) overwhelms their ability to equalize internal pressure, and their shells collapse. They often spend daylight hours on the seafloor (down to about 500 m) and swim up at night to feed near the ocean surface. As they swim to and from the ocean floor, they maintain neutral buoyancy at different depths by regulating gas volume in the shell chambers.

Except for *Nautilus*, cephalopods do not have a heavy shell. Cuttlefishes and squids have streamlined bodies and light internal shells and can swim faster than nautiluses. The cuttlefish *Sepia* is a relatively stiff-bodied cephalopod because its shell, or cuttlebone, is a rigid calcareous plate or rod. Cuttlebones are highly porous, and the pores contain nitrogen gas, providing buoyancy. In squids, such as *Loligo* and *Architeuthis*, the shell, called the **pen**, is thin and chitinous, rather than calcareous, and provides only minimal support. Cartilaginous plates and collagen fibers in the mantle and around the brain provide most of the body support in these squids.

Locomotion

The slow, gliding movement of a snail is typical of many molluscs that have a broad ventral foot adapted for moving over bottom surfaces. Speed is not essential, for the animal feeds as it moves and can retreat into its shell if attacked by predators. Muscle fibers in the foot run in several different directions and exert force against fluid in the circulatory system. The ventral surface of the foot, which is in contact with the substrate, is ciliated and secretes a mucous slime trail over which the animal glides. Locomotion may result from ciliary action or from waves of muscular contraction passing along the foot.

Most bivalve molluscs are efficient digging machines. A clam's valves slide easily into soft mud or sand as the ventral foot digs downward (Figure 24.12). During burrowing, the foot is forced into the substrate by contraction of **pedal protractor muscles** extending from the sides of the foot to the valves. The protractors compress the base of the foot, forcing it to extend. With the foot extended into the substrate (Figure 24.12b), **adductor muscles** contract, pulling the valves closer together, forcing water out of the mantle cavity, and increasing the flow of hemolymph to the foot. The ejected water loosens the sand around the animal. The end of the foot expands with increased hemolymph

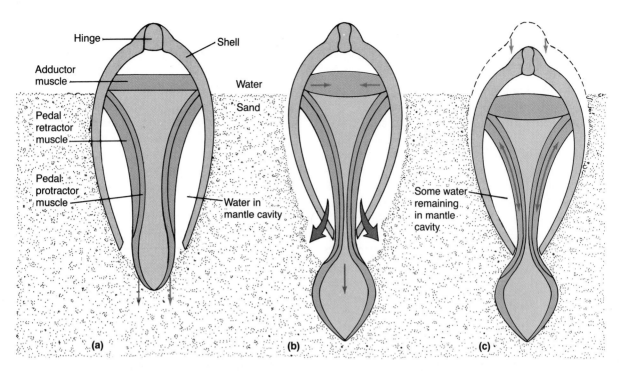

Figure 24.12 The digging mechanism of bivalves. Three sets of muscles are involved; color arrows indicate movements caused by muscle contractions; large arrows indicate water movement. (a) A clam partly buried in sand proceeds to dig in deeper; the elastic hinge ligament holds the valves open slightly, opposing the closing action of the large adductor muscles. (b) This position results from contraction of pedal protractor muscles (which extend the foot) and adductor muscles (which close the valves). (c) Contraction of pedal retractor muscles pulls the anchored animal downward.

flow. Finally, with the bulbous foot acting as an anchor, **pedal retractor muscles** contract, pulling the valves down into the sand behind the foot (Figure 24.12c). In many species, the anterior and posterior retractor muscles contract alternately, helping to rock the animal into the substrate.

Certain scallops (Family Pectinidae) are unusual bivalves in that they sit on the ocean floor but are unattached to bottom surfaces and do not dig or burrow (Figure 24.13). If disturbed, they can swim short distances by jet propulsion. Their foot is reduced to a small appendage around the mouth, but they have a very large adductor muscle (the edible portion of scallops). If touched by a predator, such as a sea star, a scallop can jet quickly away as striated fibers in its adductor muscle contract rapidly, clapping the valves together and squirting water out of the mantle cavity. Other muscles in the mantle edge direct the water jet. Slower-acting smooth fibers in the adductor muscle contract to hold the valves tightly closed for prolonged periods.

Cephalopods, the most active molluscs, also swim by jetting water from the mantle cavity.

Strong muscles in the mantle provide the locomotor force. As one set of muscles contracts, the mantle cavity takes in water; other muscles then contract, increasing water pressure in the cavity and

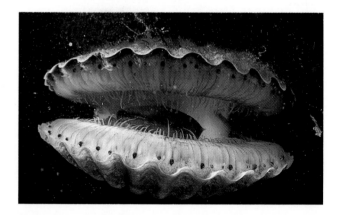

Figure 24.13 A nondigging bivalve. This close-up of the bay scallop (*Aequipecten irradians*) shows the row of eyes and touch-sensitive tentacles around the outside of the mantle; these sense organs can detect the approach of predators such as sea stars. Scallops live on the ocean bottom, but some species, including *A. irradians*, do not burrow into mud or attach to the bottom; instead, when disturbed, they move about by jetting water out between their shell valves and mantle.

sealing the mantle edges so that the water is expelled only through the ventral funnel. The animal propels itself backward or forward by directing the position of the funnel. Lateral fins in squids act as stabilizers for swimming.

Cephalopods use jet propulsion in a variety of ways. Species of *Loligo* can swim in bursts up to about 43 km/hr, faster than any other invertebrate, and they spend their entire life swimming and feeding in the open ocean. By jetting water at the ocean surface, the so-called flying squids (*Dosidicus*) can launch themselves into the air and glide for several meters at speeds up to 28 km/hr. Cuttlefishes are bottom feeders that use jet propul-

sion for hovering just above the seafloor. They also squirt jets of water at the bottom to disturb small prey such as crabs and shrimps. As bottom dwellers, octopuses only occasionally move by jetting water; more often, they pull themselves about over the ocean floor using the suction discs on their arms.

Feeding and Digestion

The typical, and probably ancestral, method of feeding in molluscs is radular scraping (Figure 24.14). The **radula** is a beltlike, toothed, chitinous membrane held in a **radula sac** in the mouth cavity.

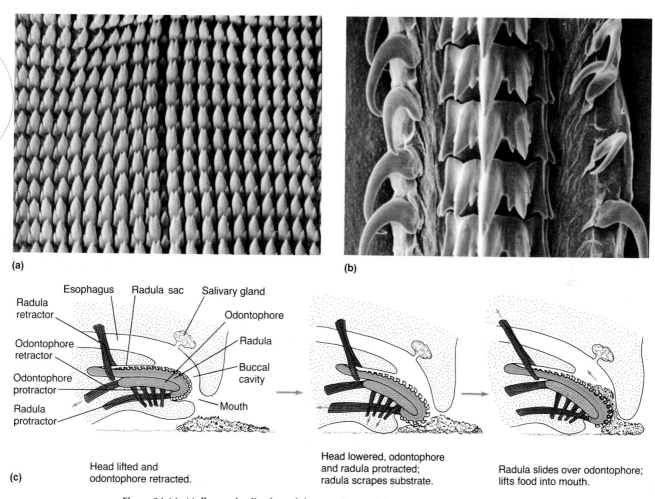

Figure 24.14 Molluscan feeding by radular scraping. Radular teeth are variable in number and shape, depending on the species, and are arranged in transverse rows. (a) The radula of an herbivorous species, the limpet *Siphonaria denticulata*, that scrapes algae and other food from rocks has many small teeth in each row (240×). (b) The radula of *Thais emarginata*, a marine carnivore, has a small number of strong teeth in each row (430×). (c) This diagrammatic longitudinal section through the head of a gastropod illustrates the scraping action of a radula. The radula retractor and protractor muscles slide the radula back and forth over the tip of the odontophore, a supporting cartilage pad. As the mouth opens, the odontophore protractor muscles thrust the odontophore forward against the substrate, and the radula scrapes food backward into the buccal cavity.

The radula grows continuously from the rear of the radula sac, so that the teeth and the radula membrane are replaced as they are worn down. Radular scraping resembles the licking action of the mammalian tongue. As the radula moves against the substrate, its teeth rasp off algae and bacteria and convey them into the back of the mouth. Many molluscs with a radula also have chitinous jaws that assist in holding or tearing food at the mouth opening. Food particles in the mouth are coated with mucus secreted by salivary glands.

The gut in many molluscs is only weakly muscled, and food is propelled through it mainly by ciliary action. In the ancestral condition, seen in some primitive living gastropods, cilia in the stomach create a rotating current that forms food particles into a mucous cord, the **style** (Figure 24.15). The food particles gradually fall out of the rotating style in the stomach as low pH reduces the stickiness of the mucus. The stomach wall bears chitinous plates that grind large food particles. Smaller particles are sorted by cilia into grooves in the stomach wall and in a cecum extending from the

stomach. Many species have digestive glands extending from the stomach; cells in the glands engulf fine particles, and much digestion occurs intracellularly. The digestive glands may also secrete enzymes into the stomach and cecum, where some digestion occurs extracellularly. Undigestible wastes are formed into compact fecal pellets in the intestine and egested from the anus into the excurrent flow of water.

Gastropods. Most gastropods are dietary generalists: Many are mainly algal scrapers or browsers and others are mainly carnivorous, but most species supplement their diet with dead plant or animal material when it is available. Most gastropods lack the ancestral, fine-particle sorting mechanism just described, and many have the radula, jaws, and anterior esophagus housed in a tubelike proboscis that can be extended to the substrate through the mouth. Probably as an adaptation for rasping through tough vegetation, radulae of herbivorous gastropods bear a great number of small teeth (over 700 per transverse row in some pulmonates). Also, as another adaptation for digesting plant matter, some herbivores have a gizzard, a muscular grinding organ containing abrasive sand grains, and unlike most metazoans, several herbivorous gastropods produce the enzyme cellulase and digest plant material completely.

Carnivorous gastropods generally have a larger radula with fewer teeth than that of herbivores. Whelks have an elongate proboscis effective for preying on bivalves (see Figure 24.4d). They can pry open bivalves by holding onto the shell with their foot and wedging their own shell between the victim's valves. Once a clam or oyster is partly opened, the whelk inserts its proboscis, rips the soft body apart, and eats it. Unique among carnivores, cone shells (genus *Conus*) harpoon small fishes or invertebrates by means of spearlike radular teeth containing a powerful nerve toxin (Figure 24.16a). The oyster drill *Urosalpinx cinerea* eats mainly small clams and oysters (Figure 24.16b). It first scores the shell surface of its prey by radular scraping and then secretes an acidic chemical that digests the shell at the scored site. About every half hour, the drill scrapes the shell again and applies more acidic material. It takes about 4 hours to drill through 1 mm of shell, or 15–20 hours to penetrate the average oyster shell. Once the hole is complete, the drill extends its proboscis into the oyster and consumes it.

Other means of feeding among gastropods include parasitism and filter feeding. Snails of the

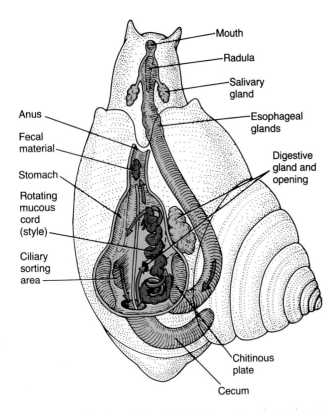

Mouth
Radula
Salivary gland
Esophageal glands
Digestive gland and opening
Anus
Fecal material
Stomach
Rotating mucous cord (style)
Ciliary sorting area
Chitinous plate
Cecum

Figure 24.15 The digestive tract of primitive gastropods. Food particles in the stomach are stirred into a mucous cord called the style. This type of mechanism is seen in primitive snails of the Order Archaeogastropoda (e.g., abalones and limpets) and was probably the ancestral molluscan condition.

(a)

Figure 24.16 Diversity in gastropod feeding mechanisms. (a) A cone shell can harpoon large prey with spearlike radular teeth. This photograph shows one cone, *Conus circumcisus,* being pursued by a second, *Conus marmoreus.* (b) The oyster drill *Urosalpinx cinerea* bores holes in the shells of other molluscs (such as the bivalve shown in the lower photograph) and consumes the soft parts. Like many carnivorous gastropods, *Urosalpinx* has an anterior proboscis housing its radula and other mouthparts.

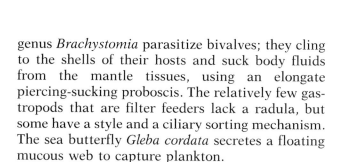

(b)

genus *Brachystomia* parasitize bivalves; they cling to the shells of their hosts and suck body fluids from the mantle tissues, using an elongate piercing-sucking proboscis. The relatively few gastropods that are filter feeders lack a radula, but some have a style and a ciliary sorting mechanism. The sea butterfly *Gleba cordata* secretes a floating mucous web to capture plankton.

Bivalves. Lacking a radula, bivalves have gills that form a large, mucus-coated, ciliated food sieve (see Figure 4.8). When a clam is feeding, its hinge holds the valves slightly apart, allowing water to be moved in and out. Water drawn in through the incurrent siphon passes between the gill filaments, where suspended particles are caught in the mucus or by cilia. The edges of the gills have ciliated food grooves that guide trapped particles toward the mouth. Particle sorting begins in the food grooves, where large unusable materials are rejected and passed out the excurrent siphon. A pair of **labial palps** collect particles from the food grooves, reject most inorganic matter, and pass suitable plankton and organic particles into the mouth.

The digestive system in bivalves contains a specialized type of style and ciliary sorting mechanism

(Figure 24.17). Food particles combine with mucus in the esophagus to form a **food string**. The stomach has an elongate outpocketing, the **style sac**, that secretes and houses a **crystalline style**, a proteinaceous rod impregnated with digestive enzymes and mucus. Cilia in the style sac rotate the crystalline style, pulling the food string into the stomach. As it rotates, the crystalline style rubs against a chitinous plate in the stomach. This action abrades the style, erodes digestive enzymes from it, and mixes them with the food string. Partially digested particles are gradually released from the food string as a result of low pH, and fall against the stomach's ciliated sorting region, where final particle sorting occurs. Undigestible material is passed into the intestine, while food is whipped into one or more digestive ceca, where absorption and some intracellular digestion occur.

Cephalopods. Unlike other molluscs, the jet-propelled cephalopods successfully pursue fast-swimming prey, mainly fishes and crustaceans. Prey are caught and held with the suction-cupped arms or tentacles and are sometimes showered with digestive enzymes or toxins before being stuffed into the mouth. Strong jaws in the mouth

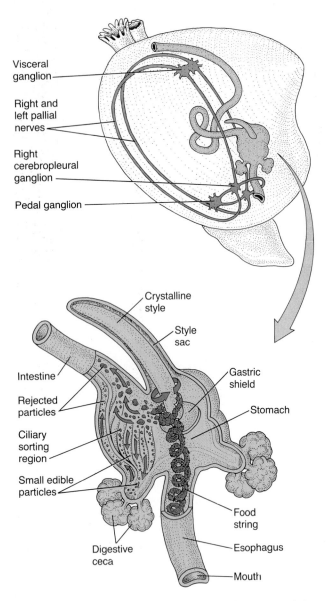

Visceral ganglion

Right and left pallial nerves

Right cerebropleural ganglion

Pedal ganglion

Crystalline style

Style sac

Intestine

Gastric shield

Rejected particles

Stomach

Ciliary sorting region

Small edible particles

Food string

Digestive ceca

Esophagus

Mouth

Figure 24.17 The digestive system of bivalves. The system includes a food string, crystalline style, and ciliary sorting mechanism (described in the text). The nervous system is included here to show its orientation with the digestive system. Note the lack of cephalization. Body functions are coordinated by fused pedal ganglia in the base of the foot, a cerebropleural ganglion on each side of the esophagus, and a pair of posterior visceral ganglia.

rip food apart, and a radula conveys it into the gut. Unlike typical molluscs, cephalopods have a strong muscular gut that moves food by peristalsis. Digestion and absorption begin in the stomach and are completed in a large cecum (Figure 24.18a). The stomach and cecum receive digestive enzymes from digestive glands sometimes called the liver and pancreas.

Gas Exchange and Circulation

Molluscs could not be as large and structurally complex as they are without having special gas exchange organs (mantle and gills or lungs) and a well-developed internal transport system.

Gills. Gills in the molluscan mantle cavity are called **ctenidia** (see Figures 5.4 and 5.5). Each ctenidium has many ciliated filaments, each of which is richly vascularized and supported by a stiff, chitinous rod. Ciliary beating draws water into the mantle cavity and directs it over the filaments, where gas exchange occurs.

As a highly diverse group, gastropods show many different types of breathing adaptations. Most have only one gill, rotated toward either the left or the right, at the front of the body. Water circulates across the gill from the right or left side (Figure 24.19a). Many snails have an elongate siphon through which water is drawn in and over the gill. The siphon is formed of extensions of the mantle and can be manipulated to control incurrent water flow. Nudibranchs (see Figure 24.5a) are exceptional in that they have no mantle, mantle cavity, or ctenidia. Instead, their body surface bears various types of projections that contain hemocoelic spaces and increase surface area for gas exchange. Many nudibranchs have branched, or club-shaped, dorsal projections called **cerata**. Species of the genus *Doris* have projections called secondary gills, encircling the anus.

In contrast to gastropods, monoplacophorans (*Neopilina*) and chitons have gills in a pair of ventrolateral furrows called **pallial grooves**. Each pallial groove lies under a shelf formed by the mantle and shell (Figure 24.19b). Water is drawn in anteriorly, passes along the grooves and over the gills, and exits posteriorly. A chiton can breathe as it grips a rock, because water circulates through its pallial grooves even when its outer mantle edge is tightly appressed to the substrate.

Gas exchange is also a function of the filter-feeding gills of bivalves (see Figure 5.4). As feeding organs, bivalve gills are larger and have more surface area proportionate to body size than do the gills of most molluscs. Gases also diffuse across the extensive bivalve mantle and general body surface. This can be especially important to intertidal species. It was recently discovered that when out of water during low tides, the intertidal mussel *Modiolus demissus* can obtain oxygen from the air. Enough oxygen can diffuse across its moist body

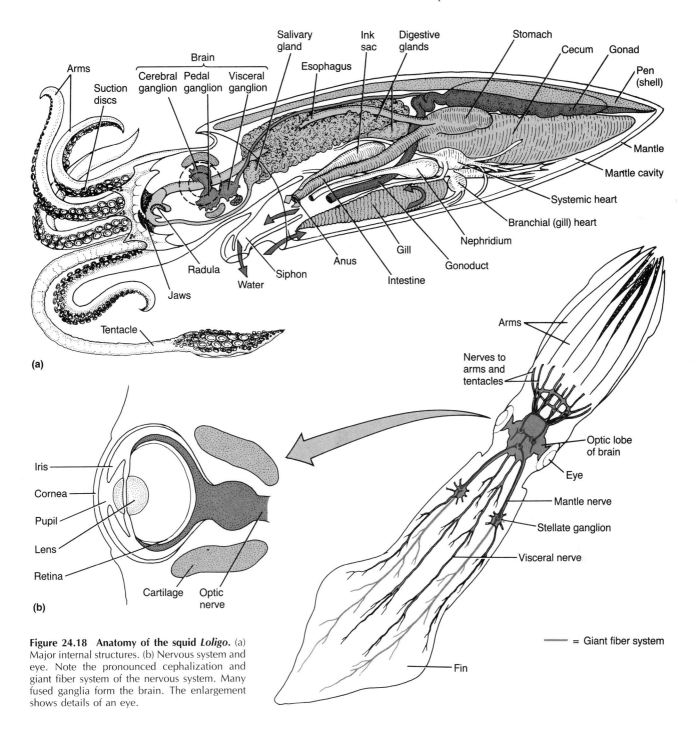

Figure 24.18 Anatomy of the squid *Loligo*. (a) Major internal structures. (b) Nervous system and eye. Note the pronounced cephalization and giant fiber system of the nervous system. Many fused ganglia form the brain. The enlargement shows details of an eye.

surfaces to provide for aerobic respiration in superficial tissues; the animal's deep tissues can respire anerobically.

Cephalopod gills are different from those of other molluscs in that the filaments lack cilia (see Figure 24.18a). The gills are large and efficient, however, and contractions of muscles in the mantle create a strong ventilating water current.

Lungs. The lung of terrestrial prosobranch snails and pulmonate snails and slugs is a single, highly vascularized internal sac formed from the mantle cavity. It is kept moist by mucous secretions, and it opens to the outside by a **pneumostome**, which, in many species, is located at the end of a breathing siphon. The lung, pneumostome, and siphon are all formed from the mantle. Aquatic pulmonates re-

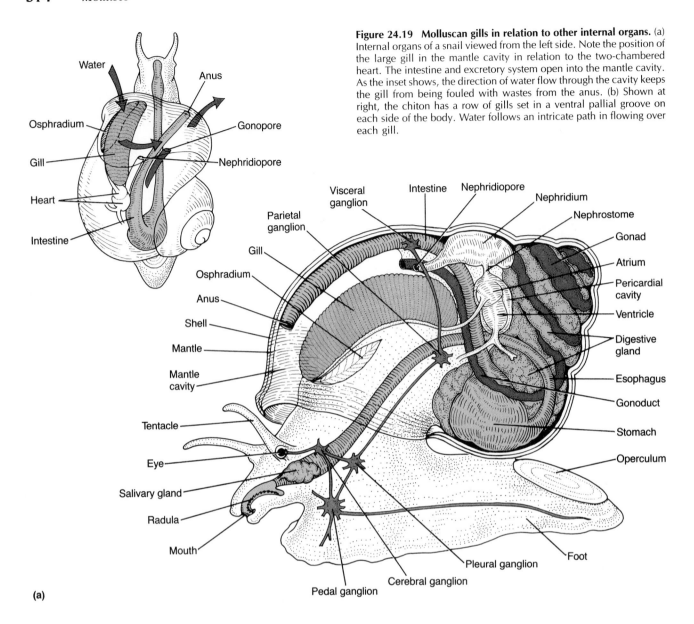

Water
Anus
Osphradium
Gonopore
Gill
Nephridiopore
Heart
Intestine

Visceral ganglion
Intestine
Nephridiopore
Nephridium
Nephrostome
Parietal ganglion
Gonad
Gill
Atrium
Osphradium
Pericardial cavity
Anus
Ventricle
Shell
Digestive gland
Mantle
Esophagus
Mantle cavity
Gonoduct
Stomach
Tentacle
Operculum
Eye
Salivary gland
Radula
Mouth
Foot
Pleural ganglion
Pedal ganglion
Cerebral ganglion

Figure 24.19 Molluscan gills in relation to other internal organs. (a) Internal organs of a snail viewed from the left side. Note the position of the large gill in the mantle cavity in relation to the two-chambered heart. The intestine and excretory system open into the mantle cavity. As the inset shows, the direction of water flow through the cavity keeps the gill from being fouled with wastes from the anus. (b) Shown at right, the chiton has a row of gills set in a ventral pallial groove on each side of the body. Water follows an intricate path in flowing over each gill.

(a)

spire by a variety of means. They may surface, open the pneumostome, and take air into the lung; or, in well-oxygenated waters, they may remain submerged indefinitely and allow gas exchange to occur across external body surfaces. Some species fill the lung with water and extract dissolved oxygen. A few species even have accessory gills developed from the mantle.

Circulatory Systems. Except for cephalopods, molluscs have open circulatory systems, and as explained in the Trends and Strategies section beginning on p. 531, body spaces in most noncephalopod molluscs are mainly hemocoelic. Cephalopods are exceptional in having a spacious nonhemocoelic body cavity. The cephalopod and

bivalve circulatory systems are described in Chapter 6 (see p. 125 and Figures 6.8 and 6.9). Both groups have a three-chambered heart (two atria, one ventricle), as is typical for molluscs with two gills. The accessory branchial (gill) hearts of squids are single chambered. Gastropods with a single gill have a two-chambered heart with one atrium and one ventricle (see Figure 24.19a). The gill and atrium on one (usually the right) side of the body are lost during torsion.

Osmoregulation and Excretion

Molluscs have one or a pair of excretory organs called **nephridia** (only one in most snails). Each nephridium consists of a coiled tube or sac near the

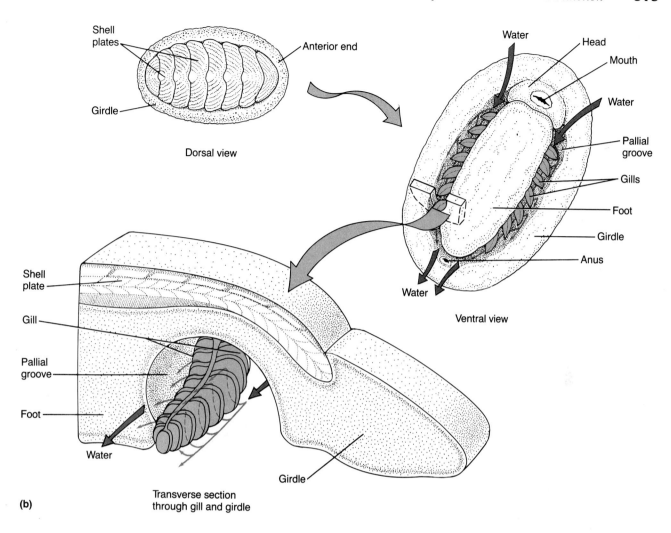

Shell plates

Girdle

Anterior end

Dorsal view

Water

Head

Mouth

Water

Pallial groove

Gills

Foot

Girdle

Anus

Water

Ventral view

Shell plate

Gill

Pallial groove

Foot

Water

Girdle

(b)

Transverse section through gill and girdle

heart (see Figures 24.18a and 24.19a). It usually extends from an opening called a **nephrostome**, located in the pericardial cavity (around the heart), to a **nephridiopore**, from which urine is discharged into the excurrent water passing out of the mantle cavity. Fluid accumulates in the pericardial cavity (which is derived from the coelom) from two sources. Some of it filters directly from the hemolymph through the wall of the heart and aortas; the rest is secreted by excretory glands in the wall of the pericardial cavity. The walls of the nephridia reabsorb usable substances from the fluid, leaving the rest to be excreted. Ammonia is the main nitrogenous waste in marine and freshwater molluscs, but many terrestrial gastropods and certain intertidal species produce uric acid as a means of conserving water. (Products of excretion are discussed in Chapter 6, p. 152.)

The nephridia of freshwater molluscs meet the osmoregulatory demands of the environment by disposing of excess water and conserving salts.

Freshwater gastropods and bivalves produce large amounts of urine with a lower solute concentration than that of their body fluids. Because their body fluids are approximately isosmotic to seawater, most marine molluscs expend little or no energy in osmoregulation. Squids, however, spend considerable energy in ion regulation. They remain buoyant in the water by maintaining high concentrations of light ammonium ions in their body fluid or in certain tissues.

Nervous and Chemical Coordination

Molluscan nervous systems vary from simple and relatively uncephalized systems to some of the most highly organized coordinating systems found in any invertebrates. Nearly all molluscs have a **nerve ring** around the esophagus and two pairs of longitudinal **nerve cords**: A pair of *pedal* cords innervates the foot, and a pair of *visceral* cords extends into the mantle and visceral hump. These components are probably primitive (ancestral) molluscan features. Chitons and monoplacophorans have these components organized into a ladderlike system with poorly defined (or no) ganglia and with transverse nerves connecting the main nerve cords (Figure 24.20).

Most other molluscs have a more centralized nervous system, with at least four pairs of **ganglia**. A pair of *cerebral* ganglia are located above the esophagus. *Pedal* ganglia are near the foot. A pair of *pleural* ganglia are associated with the mantle, but often lie under the esophagus, closely associated with the cerebral ganglia. *Visceral* ganglia lie in the visceral hump (see Figure 24.19a). Lacking a head, bivalves show little cephalization in their nervous system (see Figure 24.17). Their ganglia are of similar size, although the cephalic (cerebral) and pleural ganglia are fused. By contrast, gastropods and cephalopods show marked cephalization; in many species, several ganglia are fused anteriorly to form a distinct brain. They also have *parietal* ganglia in the viscera and *buccal* ganglia in the mouth region. Because of torsion, the nervous system of many gastropods resembles a figure eight twisted around the digestive system.

The brain of cephalopods, protected by a cartilaginous cranium, is larger and contains more ganglia than that of any other molluscs. (The *Octopus* brain is discussed in Chapter 10, p. 233; Figure 10.13.) Large cerebral ganglia, forming the dorsal part of the brain, connect to optic lobes and to large optic nerves serving the eyes. Beneath the cerebral region, the pedal ganglia send out nerves to the arms and funnel, and fused pleural and visceral ganglia give rise to paired mantle and visceral nerves. The mantle nerves extend to two large nerve centers, called the **stellate ganglia**, which control muscles active in jet propulsion. Squids also have a special system of **giant nerve fibers** extending from the brain into the arms and tentacles and posteriorly to the mantle muscles via mantle nerves and the stellate ganglia (see Figure 24.18b). These fibers have the largest diameter (up to 1 mm) of any known nerve cells, and as a result, they conduct impulses rapidly, triggering the fast attack or escape movements characteristic of squids. Squid giant fibers can be removed and studied in laboratory preparations and are used extensively in research on animal nervous systems.

Chemical Coordination. Hormonal regulation in molluscs has been studied most thoroughly in gastropods and cephalopods. Neurosecretory cell clusters occur in the main ganglia and in the walls of large blood vessels, and neurosecretions are known to regulate growth and reproduction, carbohydrate metabolism, and osmoregulation. Although little is known about the chemistry of the neurohormones, most appear to be small peptides.

In addition to neurosecretory structures, squids and octopuses have **optic glands** adjacent to the optic lobes of the brain (see Figure 10.13). As discussed in Chapter 11 (see p. 263), secretions from the optic glands influence gonadal development, and the cephalopod reproductive cycle is regulated by a complex interaction of seasonal environmental changes, neurohormones, and endocrine secretions. Similar to cephalopods, snails have one or two pairs of endocrine organs, called **dorsal bodies**, on the surface of the cerebral ganglia. Studies of the hermaphroditic land snail *Helix aspersa* indicate that the dorsal bodies are influenced by neurosecretions from the cerebral ganglia, and the dorsal bodies in turn secrete a hormone that promotes development of the female reproductive organs. The snail's testis is activated directly by neurosecretions.

Sensory Structures

Molluscs have a full complement of sensory structures. Chemosensitive and touch-sensitive struc-

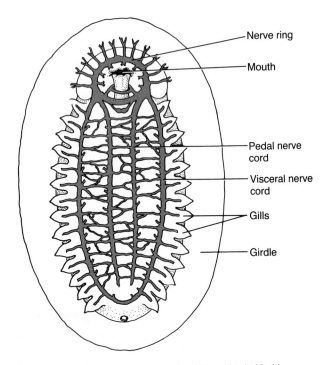

Figure 24.20 The nervous system of a chiton. This ladderlike system shows little cephalization.

Nerve ring

Mouth

Pedal nerve cord

Visceral nerve cord

Gills

Girdle

tures include many clusters of cells in the mantle, as well as an organ called the **osphradium**, located in the mantle cavity under the gills. The osphradium consists of many sensory cells in a folded region of mantle epithelium, and it seems to monitor chemicals in water ventilating the gills. Tentacles on the head of many gastropods (see Figure 24.19a) and the arms and tentacles of cephalopods also have many cell clusters sensitive to chemicals and touch. Other sense organs include gravity-sensitive **statocysts** adjacent to the brain or pedal ganglia. Cephalopods also have stretch-sensitive **proprioceptors** in body muscles. Squids can detect minute changes in their body position and make precise alterations in arm and sucker movements and in swimming direction.

Lacking a head, bivalve molluscs have many tactile and chemosensory cell clusters strategically located around the exposed outer edge of the mantle. The unattached scallops live a somewhat more precarious existence on the ocean floor than do bivalves that burrow into the substrate (see Figure 24.13). However, the scallop's well-developed eyes and chemosensitive and touch-sensitive tentacles on the mantle rim help offset its vulnerability to predators. Its eyes are complex; each has a cornea, lens and retina.

Cephalopod eyes superficially resemble those of vertebrates. Each has a transparent cornea, an iris diaphragm controlling how much light enters the eye, a lens, and a retina (see the enlargement in Figure 24.18b). The lens is rigid, and muscles focus images on the retina by moving the lens forward or backward. Unlike vertebrates, cephalopods have a **direct retina**, meaning that the photoreceptive cells of the retina face the front of the eye and receive light directly. The optic nerve forms from the back side of the photosensory cells. This means there is no "blind spot" where the optic nerve passes in front of the retinal cells, as in the vertebrate eye (see Figure 12.6).

BEHAVIOR

Molluscs, especially gastropods and cephalopods, show remarkably complex behavior, including homing, territoriality, elaborate mating activities, and learning. Homing and territoriality have been studied extensively in intertidal species. Many limpets and chitons spend their entire lives in a limited intertidal area, often moving less than a few centimeters per hour over rock surfaces. Several species graze over a **feeding territory** while immersed during high tide, then return to a specific rock depression (a **home site**) that closely fits their shell and mantle contours during low tide. Feeding territories rarely extend farther than 1 m from the home sites. Limpets and chitons may locate their home sites by chemical or tactile means. Some individuals seem to backtrack home over their own mucous trail, while others follow different routes and can find their way home even when researchers have placed roadblocks in their paths. Their ability to recognize the home depression can be eliminated if the edge of their shell or the depression is physically altered.

Cephalopods exhibit the greatest amount of behavioral diversity and complexity in the phylum. Their well-developed eyes and agility in the water enable them to be efficient predators and help them to elude danger. Also, as a protective device, most cephalopods have an **ink sac**, an extension of the gut near the anus, that produces dark pigment (see Figure 24.18a). The ink can be instantly expelled through the anus and exhalant water current, producing an ink screen that may decoy or confuse predators, or desensitize predators' chemoreceptors.

Behavioral experiments indicate that cephalopods (at least octopuses) cannot detect color, but can readily change body color. Their integument has color cells, or chromatophores, that are changed in shape by muscle contractions, thus altering body color. Chromatophores serve a variety of functions. They provide precise, split-second alterations in surface color patterns, perhaps confusing predators such as fishes. Cuttlefishes can virtually disappear by adjusting their body color to mimic a sandy seafloor. Other cephalopods may frighten predators by producing large, dark, eyelike spots on the dorsal surface. They can also flash their color cells on and off in various patterns to attract mates.

Octopus vulgaris thrives under laboratory conditions and is a model experimental animal for behavioral research. A few days after being captured, an adult octopus will take food in the laboratory, and it can thereafter be trained by means of food rewards or punishments. Octopuses learn quickly, and much has been discovered about the kinds of sensory stimuli they perceive. With taste receptors all over their body, particularly on the suction discs of the arms, they have a remarkable ability to discriminate chemicals in the environment. The octopus's perception of objects by touch is exceptional, but very different from that of most animals. Its suction discs are highly sensitive to touch and provide much tactile information. But the octopus probably perceives only the textures instead of the form of objects in its environment. Its body is almost entirely flexible, and without some rigid skeletal parts, the animal has no points of reference to compare the shapes, angles, and surface features of objects.

The octopus brain has about 30 functional centers. (See Figure 10.13.) Three dorsal brain centers function as memory banks, where sensory information from past experience is stored and retrieved as needed. These centers can be surgically removed without altering normal feeding or swimming activities. *Octopus* has both short- and long-term memory and can use its experience to develop new behaviors.

REPRODUCTION, DEVELOPMENT, AND LIFE CYCLES

Molluscs reproduce sexually. Individuals may have one gonad or a pair of gonads, which develop with the nephridia in small coelomic spaces associated with the pericardial cavity. The gonoduct typically develops from embryonic coelom. Most molluscs are dioecious, although certain oysters, scallops, and many freshwater bivalves and gastropods, including the pulmonate snails, are hermaphroditic. Cross-fertilization is the rule in hermaphroditic species, but a few snails self-fertilize, ensuring procreation when no mates are available.

Many molluscs, especially bivalves, lack copulatory organs and release clouds of sperm and eggs into surrounding water, where external fertilization occurs. Many species aggregate during breeding seasons, thereby increasing fertilization success. Gamete production and release are often synchronized with water temperatures, tidal changes, or seasonal photoperiods.

Gastropods

Most gastropods have copulatory organs and internal fertilization. There is only one gonad, located on the right side of the visceral hump (see Figure 24.19a). As a result of torsion, the right nephridium degenerates, and its duct becomes part of the gonoduct. In males, a single sperm duct opens on one side of the head, where a fold of the body wall extends out as a type of penis. The oviduct in the female often contains yolk glands and a seminal receptacle. Following internal fertilization, yolk is added to the zygotes, and they become free-floating (planktonic), or they are attached to the substrate in sticky gelatinous masses or hardened ropelike strings (Figure 24.21a).

Complex mating behavior often precedes copulation. A pair of terrestrial pulmonates may make intermittent oral and tentacular contact while circling each other, then twist closely together. Once twisted together, *Helix* snails stab (stimulate?) each other with calcareous darts produced in a muscular dart sac near the genital opening. Copulation follows the mutual exchange of darts. Being hermaphroditic, both partners receive sperm during copulation. Sperm are often transferred in **spermatophores** (capsules) that protect against drying.

Cephalopods

Cephalopod gonads are single sacs near the posterior (actually dorsal) end of the mantle cavity (see Figure 24.18a). Females have one or two short oviducts lying next to the rectum. The oviducts and a pair of adjacent **nidamental glands** secrete albumen, yolk, and protective membranes around the eggs. In males, sperm pass from a large testis to a seminal vesicle via a coiled sperm duct. Glands in the seminal vesicle secrete a spermatophore around a cluster of sperm (Figure 24.21b inset). The spermatophores are stored in a spermatophoric sac in the mantle cavity until copulation occurs. In squids, an elongate penis carries spermatophores from the sac to the anterior region of the mantle cavity near the male's funnel; the penis is not inserted into the female.

Before mating, a pair of squids may swim acrobatically around each other, waving their arms and tentacles and displaying body colors. During copulation, they position themselves head to head, their arms intertwine, and one of the male's arms, the **hectocotylus**, reaches through his funnel into the mantle cavity, picks up spermatophores, and thrusts them into the female's mantle cavity (Figure 24.21b). The spermatophores have a complex ejaculatory apparatus that everts when stimulated, releasing sperm into the female. A long thread attached to the spermatophore may trigger sperm release, or the spermatophore may evert when it swells with water.

The adults of many cephalopods die soon after spawning, but female octopuses typically care for their eggs until the young hatch. They attach their

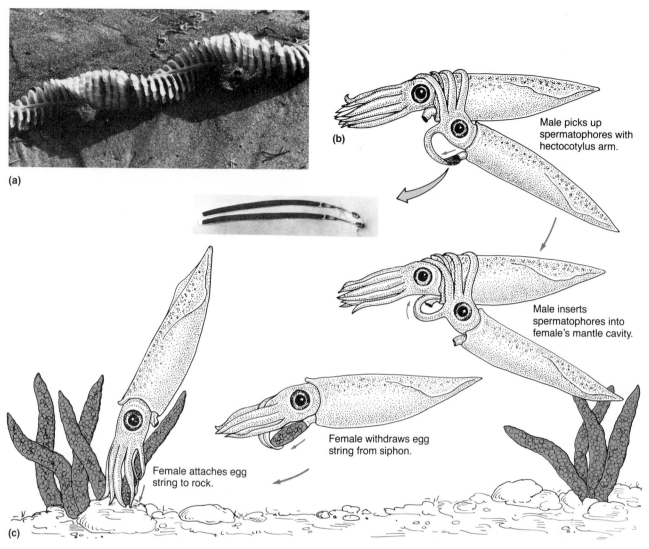

Figure 24.21 Reproductive adaptations. (a) Egg capsules of the whelk *Busycon canaliculatum*. (b) Copulating male and female squids. The male's hectocotylus arm transfers spermatophores (see inset photograph) to the female. As shown here, the hectocotylus arm does not look different from the other arms. Actually, it is modified, but rather subtly. In squids, it bears several small suckers for gripping spermatophores. In *Octopus*, its tip forms a depression for spermatophores. (c) Following copulation and fertilization, a female squid withdraws egg strings from her mantle cavity and attaches them to a rock or other hard surface.

egg clusters in rock crevices, guard them against predators, and regularly clean and aerate them by jetting water over them. Female squids use their dorsal arms to pull egg strings from their siphons; the egg strings are then attached to rocks or sandy seafloors, forming clusters called "dead man's fingers" (Figure 24.21c).

Development and Life Cycles

Most molluscs develop by spiral cleavage; cephalopods are the exception in having bilateral cleavage.

(Cleavage patterns are discussed in Chapter 16, p. 377.) A planktonic **trochophore** larva develops from the gastrula in some chitons, scaphopods, marine bivalves, and certain prosobranch snails. A unique **veliger** larva often develops from the trochophore (Figure 24.22). Veligers have a foot, mantle, and shell and also a large bilobed swimming organ called the **velum**. They are planktonic and subsist on yolk stores or filter-feed on suspended particles by means of the velum. Eventually, the veliger settles to the bottom, loses its velum, and develops into the adult. Many molluscs lack one or both free-

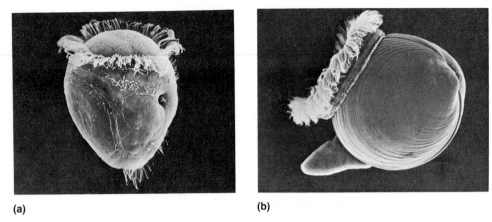

(a)   **(b)**

Figure 24.22 Molluscan larvae. (a) A scanning electron micrograph of a trochophore of the shipworm *Teredo navalis* (620×). (b) A scanning electron micrograph of a veliger of the shipworm *Lyrodus pedicellatus* (90×). Note the shell, foot, and ciliated velum. Propelled by their velum, veligers of many molluscs are planktonic.

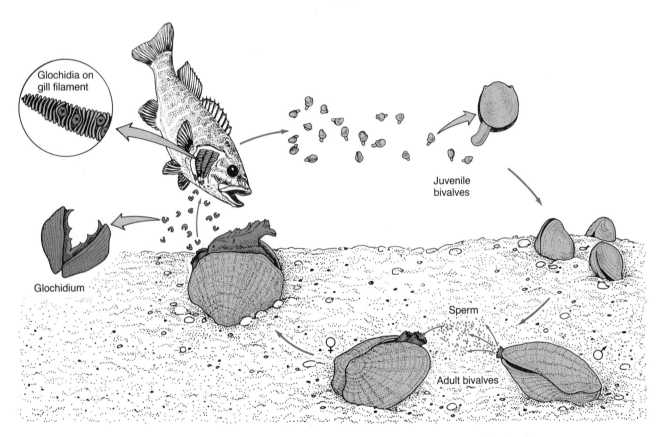

Figure 24.23 Life cycle of the freshwater mussel *Lampsilis*. Larvae develop in the female's gills and then on fishes. Following release of sperm by the male and internal fertilization in the female, several months of development occur in the female's gills. In the spring, a gravid female moves into shallow water and displays a remarkable adaptation: The edges of her mantle are modified to look like a small fish, serving as a lure for larger fishes. When a fish nudges or tries to attack the lure, the female releases a cloud of glochidia. Each glochidium has a pair of chitinous valves and looks like a miniature clam (0.05 to 0.5 mm in diameter). In response to chemicals in fish mucus, its valves clasp onto the fins or gills. A cyst forms around an attached glochidium, and after about a month in the cyst, a young mussel emerges (about the size of a glochidium), drops to the bottom, and grows to adult size. Glochidia do not usually cause significant harm to fishes, but large numbers of them in gill tissue may reduce gas exchange; and when the glochidia break out of the cysts, bacteria and fungi may invade the wounds.

swimming larval stages. In freshwater and land snails, neither larval stage is free-swimming, but they both develop within a protective egg capsule. Cephalopods have no larval stages.

Typical of freshwater organisms, and perhaps as an adaptation to living in currents, most freshwater bivalves lack planktonic larvae. Fertilization is also internal. During the summer, males release sperm into the water, and the sperm are drawn into neighboring females with incurrent water. Sperm unite with eggs in the oviduct or on the gills. Part of the female's gill then becomes a brood chamber for hundreds or thousands of developing eggs. The eggs hatch the following spring and are released as larvae called **glochidia**. Glochidia are specialized veligers, but they lack a velum and resemble tiny clams. When released by a female, they attach to gills and fins of fishes and live as parasites for several weeks. Attached to a fish, they are safe from being swept downstream and may be dispersed to favorable habitats. During the parasitic period, a glochidium transforms into a juvenile mussel, which drops off the host and then grows to an adult (Figure 24.23).

PHYLOGENY

Because of their hard shells, molluscs have left a rich fossil record. Nevertheless, we do not know what the earliest molluscs were like because the phylum's ancestors were not fossilized. The earliest molluscs most likely had most of the traits shared by living members of the phylum (see Figure 24.1), but probably lacked a shell.

Embryologically, molluscs seem closely related to annelids. Both phyla are eucoelomates and typically develop by spiral cleavage, and molluscan trochophores are very similar to those of many polychaete annelids. (Annelids are discussed in Chapter 25, p. 588.) It seems likely that lineages of these phyla evolved from common stock.

Repetition of certain body parts in the "living fossil" *Neopilina* (see Figure 24.2) has led some zoologists to postulate that the common ancestors of molluscs and annelids were metameric. It is more generally believed, however, that molluscan ancestors were not segmented, and the repeated body parts of *Neopilina* are not considered homologous with annelid segmentation. Both molluscs and annelids appear in the early Cambrian fossil record, indicating that they diverged as coelomates, perhaps from acoelomate stocks, more than 600 million years ago. While annelids apparently evolved mainly as digging and burrowing animals, the earliest molluscs probably became adapted as bottom creepers, scouring food from the substrate with a radula-like scraper. While annelids evolved an elongate metameric body with spacious, fluid-filled coelomic compartments—an efficient wriggling and peristaltic digging machine—early molluscs probably lost most of the coelom and developed a compact soft body adapted for bottom scavenging.

Among living classes, the monoplacophorans, gastropods, scaphopods, and cephalopods all have single-piece shells, indicating to some zoologists that they had a common origin. Perhaps the common stock of these groups resembled *Neopilina*. Chitons, with their multipiece shells, and the shell-less aplacophorans may have diverged from ancestral stock before the single-shell condition evolved. Bivalves may have evolved as specialized bottom diggers from single-shelled ancestors. Many zoologists postulate that bivalve ancestry can be traced to early Cambrian fossil species that had two valves as adults but single shells as larvae. The evolution of two valves attached by an uncalcified hinge was a significant adaptive milestone for an unsegmented digging animal. It allowed the animal to be protected dorsally by its shell while opening the valves ventrally to plow into the substrate.

Major adaptive radiations within the three main molluscan classes—the gastropods, bivalves, and cephalopods—are reflected in the different body forms and special adaptive mechanisms in these groups. Adaptations in feeding structures, routes of water circulation in the mantle cavity, and different degrees of body torsion account for much of the extreme diversity among gastropods. Development of a gill filter-feeding apparatus and adaptation of the foot as either a digging organ or a thread attachment device have been the major evolutionary trends among bivalves. Cephalopods, as a group, have evolved in the direction of increasing efficiency as agile predators with pronounced neurosensory capacity. Correlated with the development of their swimming ability has been a trend toward a lightening and streamlining of the body by reduction of the shell.

SUMMARY

1. Key Features of Molluscs. The Phylum Mollusca consists of seven classes. Five of six entirely aquatic classes are exclusively marine. One class, the Gastropoda, has aquatic and terrestrial members. A muscular foot is the main locomotor structure in most molluscs. The dorsal part of the body is usually surrounded by a mantle, which produces a calcareous shell in most species. The mantle typically forms a cavity, which houses ciliated gills or a lung between the mantle and the main body mass. The coelom is reduced in the adult to the pericardial cavity. Feeding is by means of a toothed rasping organ, the radula, by jaws, or by gill filter feeding.

2. Circulatory fluid in the hemocoel or in closed circulatory vessels (in cephalopods) provides hydrostatic support and hydraulic movement. Some scallops and most cephalopods move by jet propulsion.

3. Gills and mantle surfaces function in gas exchange in aquatic molluscs; some freshwater and terrestrial snails breathe air by means of a saclike mantle lung. Tubular nephridia function in excretion and osmoregulation. Aquatic species excrete ammonia; many terrestrial species excrete uric acid.

4. Molluscan nervous systems generally include a circumesophageal nerve ring, pairs of nerve cords to the foot, mantle, and viscera, and paired ganglia in various body regions. Cephalopods have a highly cephalized nervous system with giant nerve fibers. Sense organs include a pair of statocysts in the foot, chemosensory osphradia on incurrent gill surfaces, and photoreceptors ranging from simple ocelli to complex camera-type eyes, as in squids and octopuses.

5. Reproduction is entirely sexual, with external or internal fertilization. Most species are dioecious, but many gastropods are hermaphroditic. Cleavage is typically spiral, and many species have planktonic trochophore and veliger larvae; cephalopods have bilateral cleavage and no larva.

6. The similarity between molluscan and polychaete annelid development indicates that the molluscan and annelidan lineages diverged from a common stock, perhaps an acoelomate flatworm. The divergence probably occurred before metamerism evolved.

FURTHER READING

Abbott, R. T., and S. P. Dance. *Compendium of Seashells. A Color Guide to More than 4,200 of the World's Marine Shells.* New York: Dutton, 1982. *An authoritative color guide to mollusc shells.*

Boyle, P. R. *Molluscs and Man. The Institute of Biology's Studies in Biology No. 134.* London: Edward Arnold, 1981. *A brief, interesting introduction to the significance of molluscs as food, economic pests, and subjects of scientific research.*

Gosline, J. M., and M. E. DeMont. "Jet-Propelled Swimming in Squids." *Scientific American* 252(1985): 96–103. *Describes how thick collagen fiber layers in the squid's mantle provide strong jet propulsion.*

Roper, C.F.E., and K. J. Boss. "The Giant Squid." *Scientific American* 246(1982): 96–105. *Describes the anatomy and what little is known about the ecology of these deep-sea animals.*

Solem, A. *The Shell Makers, Introducing Mollusks.* New York: Wiley, 1974. *Emphasis on adaptive radiation, habits and body forms; includes excellent color photographs and scanning electron micrographs.*

Ward, F. "The Pearl." *National Geographic* 168(1985): 193–222. *Illustrates the process of pearl formation, and discusses the pearl industry in historical and modern times.*

Ward, P., L. Greenwald, and O. E. Greenwald. "The Buoyancy of the Chambered Nautilus." *Scientific American* 243(1980): 190–203. *Describes the function of the siphuncle in maintaining buoyancy as* Nautilus *grows.*

CLASSIFICATION

Phylum Mollusca (L. *molluscus,* "soft"). Molluscs.

Class Monoplacophora (Gr. *monas,* "single"; *plakos,* "flat plate"; *phoros,* "bearing"). Single-shelled marine species with several pairs of gills and nephridia; mostly Cambrian and Devonian fossils, but several species of *Neopilina* live in deep oceans.

Class Polyplacophora (Gr. *polys,* "many"; *plakos,* "flat plate"; *phoros,* "bearing"). Chitons; marine molluscs with a shell of eight dorsal plates; *Chiton, Cryptochiton, Katharina.*

Class Aplacophora (Gr. *a,* "without"; *plakos,* "flat plate"; *phoros,* "bearing"). Solenogasters and caudofoveates; wormlike molluscs without a well-developed mantle or shell; species of *Chaetoderma* and *Neomenia* live in deep oceans.

Class Gastropoda (Gr. *gaster,* "stomach"; *pous,* "foot"). Snails and their relatives; most are asymmetrical because their body undergoes torsion during development; marine gastropods include *Haliotis* (abalones), *Diodora* (keyhole limpets), *Busycon* (whelks), *Aplysia* (sea hares), and *Dendronotus* (nudibranchs, or sea slugs lacking a mantle and shell); freshwater, air-breathing snails include *Lymnaea, Physa,* and *Helisoma;* terrestrial snails include *Helix; Limax* species are terrestrial slugs.

Class Bivalvia (L. *bi,* "two"; *valva,* "leaf of a folding door"; *ia,* "state of being"). Bivalves (clams and their relatives) are gill filter feeders with no radula; the mantle is bilobed and produces two lateral shell valves; marine groups include *Mytilus* (mussels), *Pecten* (scallops), *Ostrea, Crassostrea* (oysters), *Tridacna* (giant clams), and *Teredo* (shipworms); freshwater bivalves include *Anodonta, Lampsilis,* and *Pisidium.*

Class Scaphopoda (Gr. *skaphe,* "something dug out, bowl"; *pous,* "foot"). Tooth or tusk shells; headless marine molluscs with a conical shell open at both ends; *Dentalium.*

Class Cephalopoda (Gr. *kephale,* "head"; *pous,* "foot"). Nautiluses, squids, octopuses; predaceous marine groups with a highly developed head; *Nautilus* is enclosed in a multichambered shell; the shell is reduced in *Sepia* (cuttlefishes); squids (*Loligo, Architeuthis*); *Octopus* lacks a shell.

Metamerism

In earlier Trends and Strategies sections, we examined the evolutionary significance of bilateral symmetry and body cavities. **Metamerism (segmentation)**, or subdivision of the body along its length into a series of **segments**, has also been an important element in animal evolution. Metamerism is a feature of annelids and of two of the most successful groups of animals in the animal kingdom, the arthropods and the chordates. In many annelids and arthropods, segmentation is obvious, because the segments are demarcated by external grooves that circle the body. Internally, parts of major organ systems are repeated in each segment. In the vertebrate chordates, segmentation often is not apparent externally; but internally, the serial arrangement of vertebrae, paired spinal nerves, paired kidney tubules, and repeating units of muscle (myomeres) signal metameric repetition of body parts. These segmentally repeated structures in metameric animals are called **serial homologies**. In annelids, metameric structures are generally quite uniform, whereas in arthropods and chordates, body parts that are serially homologous are often markedly different anatomically and may have different functions.

Comparative anatomy, embryological development, and the fossil record all indicate that segmentation evolved independently at least twice—once in the protostome (annelid-arthropod) lineage and again in the deuterostome (chordate) lineage. Because the relatively uniform segmentation of annelids is probably similar to the ancestral metameric condition in protostomes, hypotheses concerning the origin of metamerism often focus on the annelid body plan. Also, because most annelids have a well-developed segmented coelom, ideas about the origin of metamerism and the origin of the coelom are often intertwined. One school of thought, called the **gonocoel hypothesis**, holds that segmentation and a segmented coelom developed together in an ancestral flatworm or ribbon worm. The ancestor is assumed to have had serially repeated gonads, as seen in certain modern-day turbellarians and nemerteans (Figure Aa). (These animals are described in Chapter 22, pp. 513, 528; Figure 22.13.) According to this scenario, a space in each gonad remained after sex cells were ejected into the water, and the gonocoel hypothesis holds that the spaces would have evolved into a segmented coelom, as seen in annelids (Figure Ab).

Although superficially attractive, the gonocoel hypothesis is not supported by embryology. If the segmented coelom formed from gonadal cavities, it might be expected that embryonic development in modern

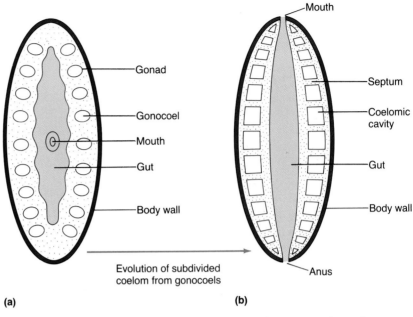

(a)   **(b)**

Figure A The gonocoel hypothesis. This hypothesis suggests that segmentation and a segmented coelom both developed in an ancestral acoelomate. In (a) we see a hypothetical acoelomate with an incomplete gut and serially repeated hollow gonads. Certain present-day acoelomate worms, such as the planarian *Dugesia* and some nemerteans (see Figure 22.18), which have serially repeated gonads, provide possible models for this hypothesis. (b) Development of a complete gut and ejection of sex cells into the water might have left a series of cavities, or gonocoels, which could have evolved into the subdivided coelom, as seen in annelids.

coelomates would reflect this—that gonadal development would precede and perhaps contribute to coelom formation. In fact, as described in Chapter 16, gonads develop *after* the formation of the coelom in coelomate animals, and gonadal tissue does not contribute to the peritoneal lining of the coelom.

An alternative idea, the **cyclomerism hypothesis**, postulates that metamerism originated in a radially symmetrical, sea anemone–like animal. The idea here is that the radial animal elongated and became bilateral with a mouth and anus; simultaneously, pouches of the gastrovascular cavity separated from the gut and developed into coelomic compartments (three at first—an anterior **protocoel**, a middle **mesocoel**, and a posterior **metacoel**) (Figure B). This is the condition of primitive deuterostomes. The protostome lineage may have evolved as the protocoel and mesocoel were lost. The unsegmented coelom of protostomes such as molluscs may have evolved from the metacoel, whereas the segmented coelom of annelids may have evolved from subdivision of the metacoel. The cyclomerism hypothesis is consistent with the enterocoelic (coelom from the gut) development of deuterostomes, but at least two lines of evidence speak against it. First, the hypothesis supposes—probably erroneously, as discussed in the phylogeny section of Chapter 21—that bilateral animals evolved from the radially symmetrical cnidarians. And second, the coelom in protostomes develops as a schizocoel, rather than from the gut, as would be expected if protostomes developed by cyclomerism.

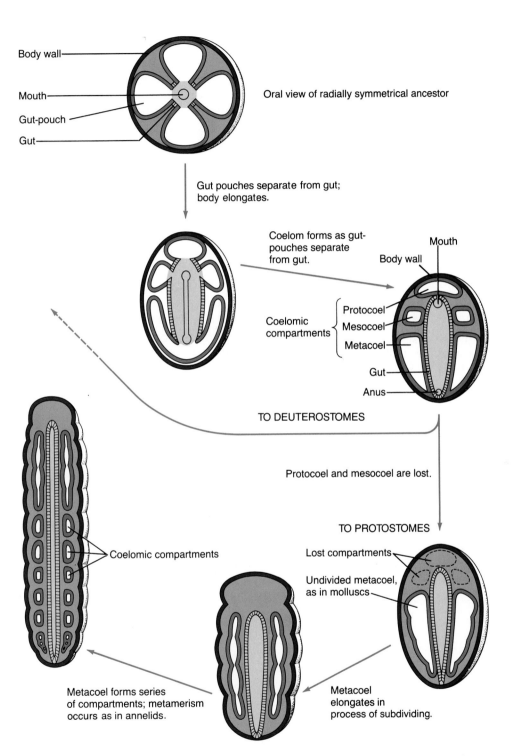

Figure B The cyclomerism hypothesis. This hypothesis postulates that metamerism originated in a cnidarian-like ancestor and that the coelom first appeared as pouches separating from the ancestor's gut. With the evolution of bilateral symmetry (indicated in these sketches by elongation of the body), the coelomic cavities became drawn out and subdivided along the animal's longitudinal axis. At first, three coelomic pouches developed. In the annelid lineage, a segmental series of coelomic compartments evolved from the original metacoel.

A third idea, which might be called the **locomotor adaptation hypothesis**, adopts a functional approach and is currently supported by most zoologists. The premise here is that the body cavity and metamerism were early adaptations that provided efficient locomotion in limbless animals. The coelom is believed to have evolved before metamerism. Thus, early nonsegmented animals with a fluid-filled body cavity and body wall muscles (but without limbs) would have been able to burrow in bottom muds more efficiently than those that lacked these adaptations. Such animals would have accrued additional locomotor advantages if their bodies (including the body wall musculature and fluid-filled coelom) became segmented. A segmented body cavity subdivides the hydrostatic skeleton into a series of partially independent chambers, facilitating peristalsis by allowing local or regional differences in hydrostatic pressure and muscle action. With these advantages, certain early coelomates (ancestral annelids) probably would have been able to do more than burrow in mud. They could have crawled efficiently on bottom surfaces by body undulations, as seen in many modern-day polychaete annelids.

Although the locomotor adaptation hypothesis focuses on the evolution of metamerism in invertebrates, vertebrate metamerism may also be viewed as mainly a locomotor adaptation. In fishes, for example, segmental blocks of skeletal muscle (myomeres) on each side of the body are positioned so that they exert local force on the notochord or vertebral column. Alternating waves of muscle contraction passing down the body produce undulatory swimming and crawling movements.

Annelids and Annelid-like Protostomes

This chapter examines the Phylum Annelida (segmented worms) and three small phyla (only about

A feather-duster worm, *Schizobranchia insignis*. This polychaete annelid, a member of the Family Sabellidae, feeds on plankton and organic particles filtered from seawater with ciliated tentacles (radioles) that it spreads apart like a Japanese fan. *Schizobranchia* and many other polychaetes live in sand tubes held together by mucous glue. When disturbed, they withdraw into their tubes.

550 species in all) of wormlike, marine bottom dwellers that bear some resemblance to annelids, namely, the Sipuncula, the Echiura, and the Pogonophora. Animals in all four of these phyla have a spacious coelom. They also have a multilayered body wall, with an outer cuticle secreted by an underlying epidermis, and with circular and longitudinal muscle layers. Coelomic fluid contained by the body wall provides hydrostatic support.

587

PHYLUM ANNELIDA

There are about 15,000 species of annelids divided among four classes. The **polychaetes (Class Polychaeta)** are aquatic (mostly marine), bottom-dwelling worms living at all ocean depths; the **oligochaetes (Class Oligochaeta)** include the earthworms, often abundant in damp soil and fresh water, and many marine species; **leeches (Class Hirudinea)** are most common in fresh water, although a few are marine, and some tropical species live in damp places on land. Terrestrial oligochaetes and leeches are restricted to moist habitats; since they lack lungs, their body surfaces must remain wet to facilitate gas exchange. Members of a fourth group, called the **branchiobdellids (Class Branchiobdellida)**, are symbiotic on the body surfaces of freshwater crayfish. Branchiobdellids are small worms (10 mm or less) that have at different times been classified as leeches and oligochaetes because they share certain features with both of these groups. So little is known about branchiobdellids that we only mention them here in passing.

The Annelida is one of only a few phyla with terrestrial, freshwater, and marine representatives, and this fact alone attests the adaptiveness of the annelid body plan. As discussed in the Trends and Strategies section beginning on p. 583, the segmented annelid body is an efficient locomotor machine, and segmentation has played a central role in determining annelid success.

REPRESENTATIVE ANNELIDS

All annelids are segmented, but as a group they display great diversity in appearance and habit. Annelids range in size from microscopic to several meters in length. Many swim or crawl about freely, often with considerable agility, while others pass their lives as sessile burrow or tube dwellers. They occur in a variety of colors; some are brownish, but many are brightly marked.

Class Polychaeta

With about 11,000 species, this is the most diverse annelid class. Polychaetes are often numerous on the ocean floor. Some species are voracious predators, actively foraging for other small invertebrates. Many others spend most of their lives con-fined to burrows or tubes, eating bottom deposits or particles filtered from the water. As their name implies, polychaetes (meaning "many chaetae") have many chitinous bristles, called **setae**. The setae project from paired, segmental appendages called **parapodia**, which aid in locomotion and gas

(a)

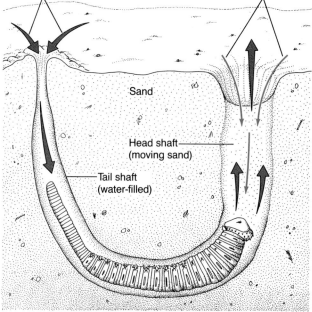

(b)

Figure 25.1 Class Polychaeta. (a) *Nereis vexillosa*, a clam worm, is common along the Pacific Coast of the United States. This intertidal species lives in a shallow burrow in the sand or mud. As a predator and scavenger, it has a pair of formidable chitinous jaws. Parapodia along each side of the body are used in crawling and swimming and also serve in gas exchange. Characteristic of the class, each parapodium bears many bristlelike setae. The closely related species *Nereis virens*, lives on the Atlantic Coast of North America. Species of *Nereis* have from 100 to 200 or more segments and may reach about 0.5 m in length. (b) *Arenicola*, a lugworm, lives in a U-shaped burrow. One arm of the U is loosely plugged with sand. Arrows indicate the direction of water flow and sand movement.

exchange and which distinguish the class (see Figure 25.5a). Some zoologists divide polychaetes into two broad nontaxonomic groups, the **errant**, or wandering, **polychaetes** and the **sedentary polychaetes**.

Clam worms are the most familiar of the errant polychaetes. People living near the ocean collect species of *Nereis* (Figure 25.1a) for fishing bait, much as inlanders collect earthworms. During the day, clam worms burrow in sediments, sometimes secreting mucus to form loose tubes. They feed at night by extending their body from the burrow or leaving the burrow altogether and seizing a variety of small invertebrates with their chitinous jaws. Clam worms have a distinct head, which bears several sensory organs, including five pairs of tentacles, other appendages called palps, and two pairs of eyes.

Sedentary polychaetes include *Arenicola,* a lugworm that lives in a U-shaped burrow in intertidal waters (Figure 25.1b). *Arenicola* looks superficially like an earthworm because its parapodia are reduced to low ridges. Water propelled by the worm's body movements flows into the end of the burrow nearest the animal's posterior end and exits through the head end. Lugworms eat bottom deposits that trickle into the head end of their burrow.

Remarkable feeding adaptations give some sedentary polychaetes a graceful beauty. Two groups, the sabellids and the serpulids, are known as fanworms or feather-duster worms. Sabellids such as *Sabella* and *Schizobranchia* (see chapter opening photograph) live on the seafloor in a permanent tube constructed of sand grains cemented together with mucus. Modified tentacles on the head can be extended from the tube in a fanlike array used in filter feeding. The brightly colored "fan" makes these polychaetes among the most beautiful of the annelids. Serpulid fanworms are similar but live in tubes of calcium carbonate secreted by the worm and cemented in clusters to rocks, shells, and other hard surfaces.

Class Oligochaeta

The Class Oligochaeta consists of about 3000 species. Terrestrial oligochaetes, or the earthworms, are often abundant in moist, nonacidic soil that is rich in organic material. Their burrowing helps condition the soil and greatly improves plant growth. In contrast to the polychaetes, oligochaetes ("few chaetae") have relatively few and inconspicuous setae, no parapodia, and a reduced head gen-

erally lacking appendages. They also have a conspicuous glandular zone, called the **clitellum**, covering several segments near the middle of the body. The clitellum secretes cocoons that house and protect eggs. The majority of oligochaetes are 10–30 cm in length, but tropical earthworms can be much larger. One collected in Australia was more than 3 m long. The common "night crawler," *Lumbricus terrestris,* is about 10–30 cm long, with its body subdivided into more than 100 clearly distinguishable segments (Figure 25.2a).

Freshwater and marine oligochaetes swim about freely, crawl over water plants, or are bottom burrowers. Most live in shallow water, at

(a)

(b)

Figure 25.2 Class Oligochaeta. (a) An earthworm, *Lumbricus terrestris.* Earthworms are highly vulnerable to water loss from their soft, moist body surface. They are protected from desiccation as long as they remain in their burrows in damp earth. Earthworms may leave their burrows on humid nights to copulate in dew-covered vegetation. (b) *Branchiura sowerbyi,* a member of the aquatic Family Tubificidae, lives head-down in mud on the bottom of ponds and lakes. Note the threadlike gills projecting from the posterior region of the worm. *Branchiura* waves its body and gills back and forth, presumably to increase the rate of gas exchange with the surrounding water. Most other tubificids, such as species of the very common genus *Tubifex,* lack threadlike gills.

depths of about 1 m, but some, such as *Tubifex* and *Branchiura* (Figure 25.2b), also thrive in organically rich muck containing little oxygen at the bottom of deep lakes and ponds. Dense populations (up to about 10,000 *Tubifex* worms per square meter) frequently occur in sewage lagoons. Many other oligochaetes live in marine intertidal zones, where they burrow in mud or live under rocks and masses of algae; still others live in deeper marine waters.

Class Hirudinea

Leeches are often thought of as bloodsuckers, although many of the approximately 1000 species are free-living predators that eat other annelids, snails, larval insects, and a variety of other small invertebrates. Most leeches live in the shallow margins of ponds, lakes, and slow-moving streams. The snail leech *Glossiphonia complanata* often lives in close association with its principal source of food, freshwater snails.

Parasitic leeches attach to the body surface of their hosts, and many are parasitic for only part of their adult lives. They suck blood or hemolymph, and most species prefer fishes, turtles, or snails. The European species *Hirudo medicinalis* (Figure 25.3), which attacks mammals, was once used in medicine to "let blood" from feverish patients, thereby supposedly withdrawing poisons and "excess" blood. "Medicinal leeches" are still used in some countries to remove blood from bruised skin.

Terrestrial leeches, some of which get to be 30 cm long, live on moist vegetation. Terrestrial species that are parasitic were a serious nuisance to soldiers in Vietnam.

Departing from the body plan of the earthworm or clam worm, most leeches are flattened dorsoventrally, body segmentation is less obvious, and setae are generally absent. Most leeches have an anterior sucker that surrounds the mouth, and a larger, cup-shaped posterior sucker. The suckers are used in looping locomotion (see Figure 25.6) and in adhering to hosts or prey. Leeches have 33 or 34 body segments, but the segments cannot be counted accurately from external observation because some are fused and form the suckers, while others are subdivided by narrow, ringlike external annuli. Internally, there are no septa and the coelom is greatly reduced. The only reliable way to count a leech's segments is to examine its nervous system; the number of pairs of segmental nerves equals the number of body segments.

HABITATS AND ENVIRONMENTAL RELATIONSHIPS

Because they are so numerous, annelids play an important role in aquatic and terrestrial food chains. Many invertebrates and fishes feed extensively on polychaetes and aquatic oligochaetes; many birds and rodents eat earthworms; and leeches are food for many fishes, reptiles, amphibians, certain birds, and even flatworms.

Only a few polychaetes live in fresh water, but these annelids are among the most widely distributed of all marine invertebrates. Some species thrive at great depths (see the Essay on ocean vent ◁ communities in this chapter, p. 606), while others inhabit intertidal sand or mud flats that are alternately flooded and exposed. Polychaetes are often very numerous along rocky shorelines, where they are able to withstand or behaviorally avoid extreme variations in temperature, light, and wave action.

The distribution of terrestrial oligochaetes is affected by the food value, water content, and pH of soil. Dense earthworm populations occur only in soils that contain a large amount of organic matter or a surface layer of humus. In addition, water content must not be too high. Water-logged soil reduces oxygen uptake by earthworms, causing suffocation or forcing the worms to the surface. Other respiratory problems occur if the soil pH is too low.

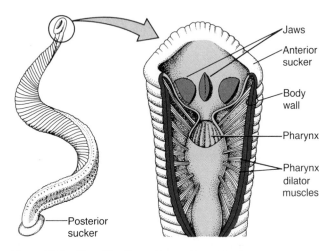

Figure 25.3 Class Hirudinea. The medicinal leech *Hirudo medicinalis* is a bloodsucker. It lives in shallow freshwater habitats and has a preference for mammalian blood. While gripping a victim with its anterior sucker, *Hirudo* uses its muscular pharynx as a pump to draw blood from cuts made by its three sharp jaws. Hirudin, an anticoagulant in the leech's saliva, keeps the blood from coagulating as the leech feeds.

Acidic soil generally has a low calcium content, a factor that allows carbon dioxide to accumulate in the soil. Excessive soil CO_2 impedes release of respiratory CO_2.

More than a century ago, Charles Darwin calculated that there were enough earthworms in an acre of pasture to ingest up to 18 tons of soil (dry weight) each year. More recently, biologists have shown that a pasture may support a greater weight of earthworms than of cattle! Pasture and woodland capable of supporting about 50 g of cattle per square meter contained up to 120 g of earthworms per square meter. Extensive burrowing by earthworms, to depths of several meters by larger species such as *Lumbricus terrestris*, aerates the soil and improves its drainage and fertility. Many earthworms carry ingested soil to the surface, defecating it as castings, which contain nutrients and minerals vital to plants. Bottom-dwelling oligochaetes and polychaetes have a similar "plowing" effect in aquatic environments.

Symbiosis

Bloodsucking leeches are only one of several types of symbiotic annelids. A number of polychaetes, oligochaetes, and leeches serve as intermediate hosts of parasitic flatworms. Several leeches transmit protozoan blood parasites (such as species of trypanosomes) to fishes, amphibians, and reptiles. (For a discussion of trypanosomes, see Chapter 19, p. 447.) Tube-dwelling polychaetes often serve as hosts to small commensalistic worms (including other polychaetes) and crustaceans. Other polychaetes live as commensals on the body surfaces of sea stars and other echinoderms.

COMPARATIVE STRUCTURE AND FUNCTION

Body Support and Locomotion

The annelid body is supported largely by the hydrostatic skeleton, a result of the coelomic fluid being held under pressure by the body wall (Figure 25.4). (Hydrostatic skeletons are discussed in more detail in Chapter 8, p. 176.) The body wall consists of a collagenous cuticle, a single-cell-thick epidermis, and at least two underlying sets of muscles (outer circular and inner longitudinal). The leech body wall is unusually thick, with a connective tissue dermis and several additional muscle groups.

In oligochaetes, all segments except the first and last bear four pairs of chitinous setae, each of which can be extended or retracted by its own musculature. Internally, septa between neighboring body segments divide the coelom into fluid-filled compartments corresponding to the segments. The septa are complete in species such as *Lumbricus terrestris*, but incomplete in many others. As discussed in Chapter 8 (p. 178), worms with complete septa have much greater ability to manipulate coelomic fluid, and thus have more body flexibility. Several distinctly different modes of annelid locomotion reflect differences in body wall and coelom structure.

Peristaltic Locomotion. Earthworms and many other annelids move by peristalsis. Rhythmic contractions and relaxations of body wall musculature pass wavelike along the body, with the circular and longitudinal muscles in each segment acting antagonistically. (Peristaltic movement is described in Chapter 8, p. 177.) In regions of the body where circular muscles are contracting, the body becomes longer and thinner and the relaxed longitudinal muscles are stretched; at the same time, regions where the longitudinal muscles are contracting become shorter and thicker, and the relaxed circular muscles are stretched. The earthworm's body travels forward as the wave of body thickening passes posteriorly. As an earthworm burrows, back slippage is prevented by the pressure of the thickened segments against the burrow wall and by setae inserted like barbs into the soil. Consequently, only segments in the thin regions of the worm actually advance.

Walking, Crawling, and Swimming. Clam worms such as *Nereis virens* can walk slowly, crawl rapidly, and swim. In comparing a clam worm's locomotion with an earthworm's peristaltic movement, we find two important anatomical differences. First, a clam worm's longitudinal muscles are arranged as four bundles (two dorsal and two ventral), not as two layers as in the earthworm. And second, clam worms have paired segmental parapodia bearing many setae (Figure 25.5a). Each parapodium (literally, "lateral foot") is divided into a dorsal **notopodium** and a ventral **neuropodium** and is supported by rodlike **acicula**. Oblique (acicular) muscles within the segment provide most of the parapodial movement.

A clam worm's parapodia provide most of the power for slow walking. During a backstroke (powerstroke), the setae and lower edges of a

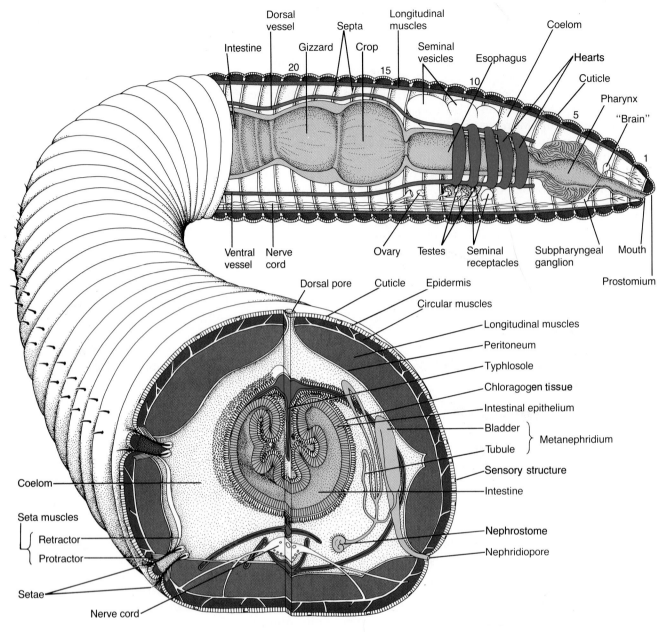

Figure 25.4 Anatomy of an earthworm. The longitudinal view shows the anterior structures from the right side; the digestive tract and body wall are cut medially. On the cross section, note that the section is divided, with the right and left halves offset from each other, to show a metanephridium and setae.

parapodium contact the sea bottom, and each parapodium acts like a lever, with rhythmic waves of parapodial strokes traveling along the body from back to front (Figure 25.5b). Rapid crawling depends primarily on the action of the longitudinal muscles. Alternating contractions of the left and right longitudinal muscles cause regions of the body to bend first to one side and then to the other. During the resulting writhing movements, the parapodia on each outside curve of the body act as levers pressing against the seafloor. Clam worm swimming is less efficient than crawling. In swimming, the parapodia are used as paddles rather than as levers. Their powerstrokes, together with an additional force against the water produced by waves of muscle contractions passing through the body, barely overcome the backward thrusts of the writhing body.

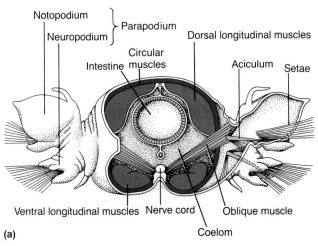

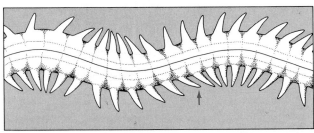

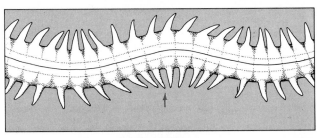

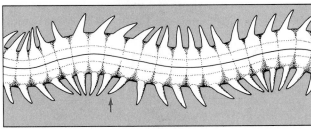

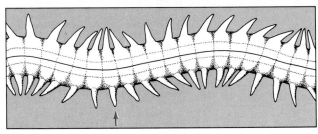

(a)

(b)

Figure 25.5 Clam worm locomotion and parapodial movement. (a) A cross section through the body and cutaway view of one parapodium shows the relationship among parapodia, acicula, setae, and musculature in one body segment. (b) In this series of sketches (from top to bottom), the arrow indicates successive positions of a parapodium as a clamworm crawls rapidly from right to left.

Locomotion in Leeches. The nonseptate body cavity of leeches is largely filled with connective tissue that forms a network of small, fluid-filled channels and sinuses. This arrangement is not conducive to peristalsis. However, with their complex musculature, leeches can move in a variety of ways. Many are graceful swimmers, propelling themselves with rhythmic vertical undulations of the body produced mainly by dorsoventral muscles. Leeches also use a looping action to crawl across underwater surfaces (Figure 25.6).

Feeding and Digestion

The earliest annelids were probably marine deposit feeders, burrowing in the bottom and eating a mixture of organic and inorganic particles. Other feeding methods in modern species reflect major adaptive radiations from the ancestral deposit feeder.

Deposit Feeding. Lugworms (see Figure 25.1b) and many other sedentary polychaetes are nonselective deposit feeders, ingesting a mixture of organic and undigestible inorganic material. Earthworms are also nonselective deposit feeders. Alternating dilations and contractions of an earthworm's muscular pharynx pump a mixture of inorganic particles, decaying vegetation, and animal matter into the mouth as the liplike **prostomium** acts as a guide (see Figure 25.4).

Many other annelids are selective deposit feeders, ingesting only organic material. The burrow-dwelling sedentary polychaete *Amphitrite* extends delicate tentacles that move over the seafloor by ciliary action. Tiny organic particles collected on the tentacles are channeled to the mouth along a ciliated groove (Figure 25.7).

Filter Feeding. Some of the most colorful annelids are filter feeders. The tube-dwelling fanworms and the sand-castle worm extend often beautiful, fan-shaped arrays of **radioles** from the tube entrance (see chapter opening photograph). Organic particles and minute organisms are filtered from the water as it passes between the radioles. A ciliated food groove in the axis of each radiole channels food toward the mouth. The highly specialized tube-dwelling polychaete *Chaetopterus* (see Figure 4.7) uses a mucous bag as a filter. As the bag fills with food, the food is rolled into a ball and carried by cilia along a groove to the mouth.

Predaceous Foraging and Parasitism. Many annelids actively seek, detect, capture, and then devour

Figure 25.6 Locomotion by leeches. This drawing shows the succession of body postures characteristic of looping locomotion. After a leech has attached its posterior sucker to a surface, as in (1), a wave of circular muscle contraction elongates the body, thrusting it forward (2 and 3). A leech then attaches its anterior sucker to the substrate (4) and detaches the posterior sucker (5). Next, the longitudinal muscles contract, thickening the body and drawing the posterior part forward (6 and 7). Leeches may also swim with smooth, dorsoventral undulations of the body (8).

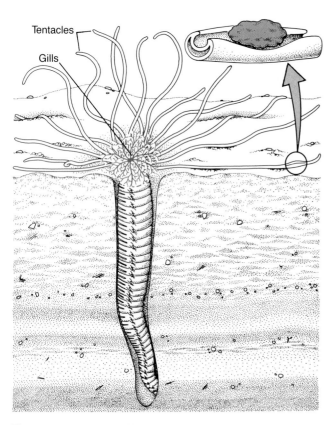

Figure 25.7 Deposit feeding by a polychaete. *Amphitrite*, shown here in a cutaway sketch of its burrow, is a sedentary polychaete that collects food particles on tentacles extended over the seafloor at its burrow entrance; the inset shows a small section of a tentacle and its food groove, along which organic particles are moved by ciliary action toward the mouth.

their prey. Burrow dwellers may use their burrow as a protective home base to which the successful forager returns with its prey. The burrowing polychaetes *Nereis* and *Glycera* seize prey with sclerotized jaws on a protrusible pharynx (Figure 25.8).

Leeches, both the predatory and the parasitic species, are mostly fluid feeders. Some species draw food in through a tubular proboscis extending from the mouth into the prey's body. The anterior sucker grips the prey as the muscular pharynx pumps the body fluids into the leech's mouth. If a victim is small enough (or the leech large enough), it may be sucked in whole. Bloodsucking leeches generally lack a protrusible proboscis and instead have three bladelike jaws in the mouth cavity (see Figure 25.3).

Digestion. Annelids typically have a muscular pharynx whose pumping action draws food through an anterior mouth into the buccal cavity and then forces it into the esophagus. Food may be moistened and softened by mucus in the pharynx. As in leeches, many polychaetes have a proboscis used to capture and then swallow small animals. The esophagus may be a simple tube, or it may be partly enlarged to form a **crop** in which food is temporarily stored. A highly branched, expandable crop enables the medicinal leech to drink up to ten times its own weight in blood, enough to provide

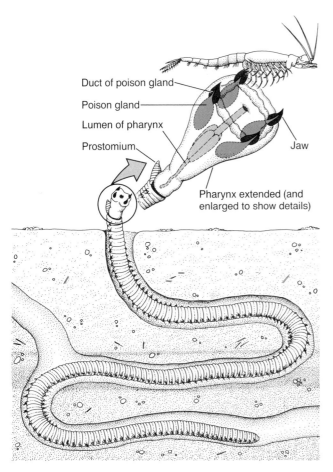

Duct of poison gland

Poison gland

Lumen of pharynx

Prostomium

Jaw

Pharynx extended (and enlarged to show details)

Figure 25.8 Extended pharynx of *Glycera*, a burrow-dwelling errant polychaete. When a small crustacean or other small invertebrate passes over *Glycera's* network of tunnels, pressure receptors on the predator's prostomium detect it. *Glycera* then creeps to a tunnel entrance near the intruder, very rapidly extends its protrusible pharynx (much enlarged in this diagram for emphasis), seizes the prey, and immobilizes it with poison from a gland at the base of each jaw.

sustenance for about nine months. The esophagus in some annelids also includes a grinding and mixing organ called the **gizzard**, which in the earthworm is heavily muscled and lined with cuticle. Earthworms and other deposit feeders have **calciferous glands**, which remove excess calcium from the blood (diets of deposit feeders are often very high in calcium) and secrete it as crystals of calcium carbonate into the esophagus. Ultimately, the crystals pass out with the feces.

Food leaving the esophagus enters the intestine, where digestion and absorption take place. The intestine extends for about 80% of the body length in an earthworm, providing considerable digestive-absorptive surface area. The earthworm also has a deep middorsal infolding called the **typhlosole** that expands its inner surface area.

Circulation and Gas Exchange

In many annelids, including *Nereis* and earthworms, blood flows through an intricate network of blood vessels and capillaries that form a closed circulatory system. (The major blood vessels and flow of blood in an earthworm are described in Chapter 6, p. 123; Figure 6.5.) In contrast to the closed circulatory system of earthworms and nereids, many other annelids have a partially or completely open circulatory system. Many polychaetes and some oligochaetes lack capillaries, and some have a large, open sinus around the digestive tract. In many leeches, the circulatory system is completely open, for it consists of the connective tissue network of sinuses and channels formed from the coelom.

The annelid circulatory system is important in gas exchange. Annelid blood is a relatively clear plasma that may be tinted red by a type of hemoglobin or green by chlorocruorin, both oxygen-carrying pigments. Earthworms and other annelids without gills depend on diffusion across the whole body surface for receipt of O_2 and elimination of CO_2. The total thickness of a typical earthworm's cuticle and epidermis is about 50 μm, but blood capillaries in the epidermis loop to within about 30 μm of the surface. Therefore, oxygen must diffuse only a short distance before it enters the bloodstream and is carried deeper into the body. Calculations indicate that distribution of oxygen by simple diffusion alone (i.e., without oxygen transport by a circulatory system) would limit an earthworm's diameter to about 1.5 mm.

Gills enlarge the gas exchange surface area of many polychaetes, but even gilled annelids rely heavily on diffusion of gases across the general body surface. In tube and burrow dwellers, parapodial or whole-body movement circulates water over the body surface. In earthworms, innumerable microscopic pores discharge mucus onto the body surface from unicellular glands in the epidermis. The mucus helps keep the surface soft and moist for gas exchange.

Osmoregulation and Excretion

As discussed in Chapter 7 (p. 146), annelids, like many other animals, combine the processes of osmoregulation and excretion. Typically, the excretory/osmoregulatory system consists of two **metanephridia** per segment. In most polychaetes and oligochaetes, a funnel-shaped, ciliated **nephrostome** at one end of each **metanephridial tubule** opens into the coelom of the preceding segment (see Figures 7.9 and 25.4). At the other end of the tubule, a **nephridiopore** opens to the exterior.

Our understanding of osmoregulation in annelids comes primarily from studies on earthworms and clam worms. Using earthworms (*Lumbricus*) from damp soil, one investigator compared the osmotic pressure of the urine with that of the coelomic fluid and found that urine sampled near the nephrostome has about the same osmotic concentration as coelomic fluid. However, as the urine passes along the nephridium, salt is reabsorbed by the tubule, and by the time the urine reaches the nephridiopore, its osmotic concentration has decreased approximately sixfold. Thus, the earthworm solves its chief osmotic problem—the net diffusion of habitat water into its body—by producing a very dilute urine.

Clam worms living in estuarine environments may experience great fluctuations in salt concentration, and they can control their water content by varying the concentration of their urine. Urine and coelomic fluid taken from a clam worm living in normal seawater have about the same osmotic concentration, but urine from a worm kept in very dilute seawater is only about 50% as concentrated as the coelomic fluid.

Ammonia is the principal nitrogenous waste product in most annelids, but terrestrial species may excrete mostly urea, which is less toxic than ammonia and does not have to be highly diluted. In its natural habitat of damp soil, where dilution of ammonia would be inadequate, *Lumbricus* produces urine rich in urea; but when placed in wet soil, it excretes mostly ammonia. Urea and ammonia are produced by **chloragogen tissue**, which, like the vertebrate liver, also synthesizes and stores fat and glycogen (see Figure 25.4).

Nervous and Chemical Coordination

As discussed in Chapter 10 (see Figure 10.10), the annelid nervous system is cephalized and segmentally arranged. Cerebral ganglia are connected to a ventral nerve cord, bearing one pair of ganglia per body segment. The role of reflexes and central pattern generators (CPGs) in mediating peristalsis and other rhythmic movements is also discussed in Chapter 10 (see p. 232; Figure 10.11). People who have collected night crawlers for fish bait know that an earthworm can retreat into its burrow in a fraction of a second. Similarly, a fanworm avoids predators by abruptly withdrawing its delicate food-gathering fan to the safety of its tube in response to a passing shadow or the slightest touch. Such escape responses are mediated by giant nerve fibers in the ventral nerve cord. (See Chapter 10, p. 229, for additional details.) Sensory stimulation, such as a flash of light, sudden darkness, or a touch, produces a barrage of impulses in the giant fibers; up to 500 impulses per second have been recorded. Their arrival at myoneural junctions causes an almost simultaneous contraction of all the longitudinal muscles in the body wall, shortening the worm with dramatic speed.

Coordination by Neurohormones. Neurosecretory cells, the source of neurohormones, have been identified in the central nervous system of polychaetes, oligochaetes, and leeches. Research involving surgical removal or transplantation of nervous structures has implicated the brain in neurohormonal regulation of many processes, including metamorphosis of immature polychaetes into reproductive forms, maturation of gonads and gametes, and regeneration of lost or damaged parts. The chief role of neurohormones in regulating reproductive processes is in providing synchrony. The mechanism is typically triggered by day length (or night length), a factor affecting the population as a whole. For example, synchronization of sexual maturity throughout a population of clam worms, followed by simultaneous spawning, promotes reproductive success by ensuring a high rate of fertilization.

Sensory Structures

Most annelids are very sensitive to chemical stimuli, suggesting that chemoreception is a particularly important source of environmental information. Sensory cell clusters that project through the cuticle in some earthworms seem to be chemoreceptors (Figure 25.9a), but in many annelids, specific chemoreceptors have not been identified. Specialized **nuchal organs** on the prostomium of many polychaetes are thought to be important in detecting food (Figure 25.9b); when these organs are destroyed experimentally, the worm no longer feeds.

Living in dark burrows or in the bottom mud of lakes and ponds, most oligochaetes lack distinct eyes, but they have photoreceptor cells scattered in the epidermis. Usually, these cells are concentrated at the anterior and posterior ends of the body. In earthworms, photoreceptors mediate attraction to weak light and repulsion from bright light, but are unable to form images (Figure 25.9a).

In keeping with their predatory (or parasitic) habits, leeches and many polychaetes have well-developed eyes. Leech eyes consist of a cluster of photoreceptor cells and a surrounding pigment cup

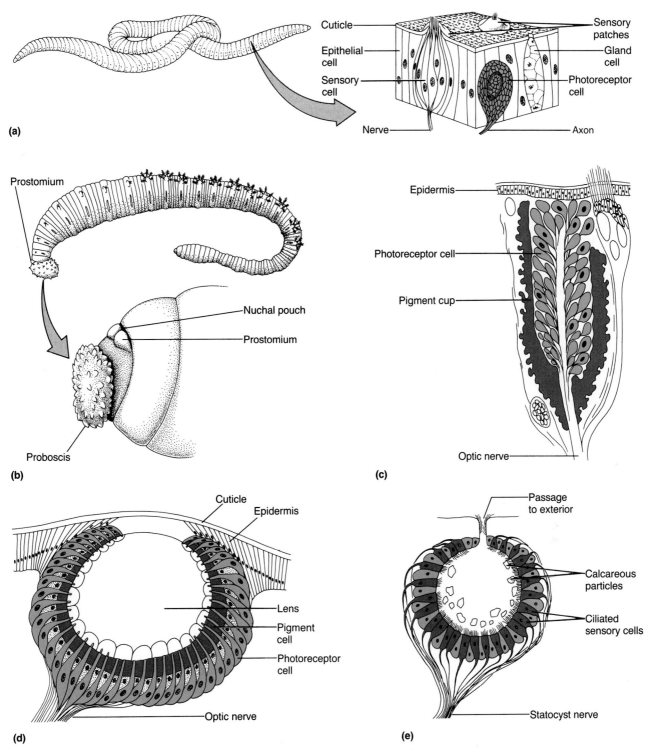

(a)

(b)

(c)

(d)

(e)

Figure 25.9 Annelid sensory structures. (a) A diagrammatic three-dimensional view of an earthworm's cuticle and epidermis, showing a photoreceptor cell, a sensory (probably chemosensory) cell cluster, and a gland cell. (b) The anterior end of the lugworm *Arenicola* (also see Figure 25.1b). Sensory cells in the nuchal pouch are probably chemoreceptors used in detecting food. (c) Section through one of up to ten eyes located on a leech's anterior segments (see Figure 25.3). Note that the eye lacks a lens for concentrating light or focusing images. Light striking the eye at an angle illuminates some of the photoreceptor cells, while other cells are shaded by the pigment cup; by turning its body one way or the other, a leech can increase or decrease the amount of light reaching the cells, thus detecting light direction. (d) Clam worms (see Figure 25.1a), as well as many other polychaetes, also have pigment cup eyes. Note that a lens rests in the cup formed by photoreceptor and pigment cells. Although the lens may increase the eye's light-gathering ability, it is probably not capable of forming clear images. (e) Lugworms and many other burrowing and tube-dwelling polychaetes commonly have a pair of statocysts near the anterior end of the body. Shown here in section is a statocyst from a fanworm, *Branchiomma*. As a worm changes its posture, calcareous particles shift position and stimulate different groups of sensory cells.

(Figure 25.9c). The pigment cup shades the photoreceptor cells, enabling the animal to orient toward or away from light. Leeches generally move away from light, but when hungry may move toward it to locate a host or a meal. Clam worms also have pigment cup eyes, usually occurring as two pairs on the prostomium (Figure 25.9d). Information obtained from two pairs of eyes doubtless helps the clam worm locate moving prey and detect potential predators. Certain planktonic polychaetes have very complex, image-forming eyes that can be focused by changing the shape and position of the lens.

Annelids also have a well-developed sense of touch. In polychaetes tactile cells are distributed over the entire body surface, but are most plentiful on the parapodia and tentacles. In addition, many free nerve endings projecting through the cuticle of earthworms and leeches are probably tactile.

Statocysts occur in many species of burrowing polychaetes, enabling them to orient to gravity as they burrow, head down in some cases but head up in others (Figure 25.9e). Statocysts also occur in fanworms, helping them in tube construction and in orienting their filter-feeding fan upward from the tube.

BEHAVIOR

Taxic behavior is an important feature of annelid life. Many tube-dwelling and burrowing annelids exhibit strong positive **thigmotaxis**, meaning they move toward and keep their body pressed against objects or surfaces. This response probably reduces the likelihood that a burrowing or tube-dwelling worm will be dislodged and carried away by wave action, water currents, or other animals. Thigmotaxis is sometimes powerful enough to override other responses; a worm may even hold its position in a crevice or burrow when other factors are harming it.

Exhibiting a type of phototaxis, many annelids are attracted to dim light but repelled by bright light. Zoologists have discovered that attraction to dim light (but not repulsion by bright light) is mediated by photoreceptors in the integument. Also, if the cerebral ganglia are surgically removed, the worms will move toward light of any intensity. This indicates that the cerebral ganglia are involved in triggering repulsion to bright light.

Sudden changes in light intensity also affects the behavior of many annelids. Those that extend part of the body from a tube or burrow while feeding usually retract quickly if a shadow is suddenly cast on them. In responding this way, they may often avoid approaching predators.

Most annelids are extremely sensitive to tactile stimuli, responding to touch by contracting longitudinal muscles that shorten the body. Recent research has shed some light on the neurophysiological basis of this response in leeches. Touching a leech's skin triggers nerve impulses in three kinds of sensory neurons: A slight touch stimulates T (touch-sensitive) cells; a somewhat heavier touch activates P (pressure-sensitive) cells; and an even more powerful contact triggers N (nociceptive) cells, which respond to noxious or painful stimuli. The cell bodies of the three sensory neurons have been identified in the segmental ganglia of the ventral nerve cord, where each forms a synapse with the motor neuron that activates the longitudinal muscles of the body wall. The longitudinal muscles contract to different degrees, depending on which of the synapses is transmitting information. As a result, the intensity of a leech's behavioral response is directly proportional to the strength of the touch on its body.

Learning

There has been considerable interest in the learning ability of annelids. **Habituation**, the simplest form of learning, has been studied in the rapid, giant fiber–mediated withdrawal response in fanworms. (Habituation is explained in Chapter 13, p. 303.) If fanworms are touched or exposed to sudden darkening, most of them withdraw into their tubes. After several repetitions of the stimuli, however, habituation occurs in most of an experimental group of fanworms, and only a few worms continue to withdraw (Figure 25.10). The cerebral ganglia have been identified as the site of this simple learning process. Such habituation allows an annelid to continue an activity, such as filter feeding, without wasting energy responding to repeated but harmless disturbances.

Earthworms and some polychaetes are capable of simple associative learning, in which an animal links, or associates, several kinds of stimuli. (Associative learning is discussed in Chapter 13, p. 303.) An earthworm can learn to enter one arm of a simple T-shaped maze and crawl toward a dark, damp shelter if the other arm leads to a sandpaper-lined area and an electric shock. Such learning is not rapid; between 20 and 100 trials are needed before a correct choice is made 90% of the time. But trained earthworms can learn a second maze with fewer trials, indicating that they have memory and

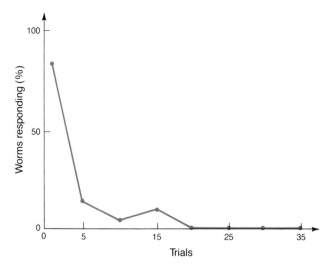

Figure 25.10 Habituation in a fanworm. Fanworms normally respond to touch by withdrawing abruptly into their tubes. However, in experiments with fanworms of the genus *Branchiomma*, repetitive touching produced habituation; the number of worms responding decreased, so that after only five experiences, fewer than 25% of the worms withdrew when touched.

can apply stored information to a new problem. Although earthworm memory seems to last only two or three days, their associative learning ability indicates that their central nervous system has considerable integrative capability. Associative learning is the dominant type of learning in vertebrates; among the invertebrates, annelids share this ability with certain arthropods and molluscs.

Figure 25.11 Development of the primitive polychaete *Polygordius.* Trochophores are top-shaped and are propelled by cilia. As the trochophore matures, a head develops and the lengthening body becomes segmented, gradually acquiring the features of the adult worm.

REPRODUCTION AND DEVELOPMENT

Many polychaetes and oligochaetes reproduce by both sexual and asexual means. Leeches have only sexual reproduction.

Asexual Reproduction

Polychaetes and oligochaetes have considerable regenerative power, with posterior parts of the body usually more readily regenerated than anterior parts. Consequently, loss of a few segments in an accident or to a predator is not necessarily fatal. Regeneration is not merely a damage control mechanism; many species reproduce by fragmentation followed by regeneration of missing parts. Asexual reproduction by budding is also quite common among polychaetes (see Figure 25.12a).

Sexual Reproduction and Development

Zoologists generally believe that ancestral annelids were dioecious and produced gametes in most of their body segments. Sperm and eggs were presumably expelled via gonoducts or nephridiopores into the ocean, where fertilization took place. Reproduction and development by many living polychaetes follow this ancestral scheme quite closely. However, in many species, the number of segments with gamete-forming ability is reduced, and gametes may be released into the ocean by rupture of the body. Following fertilization, which is usually external, the zygote undergoes spiral cleavage. Typical protostome development includes schizocoelous coelom formation, and some species have a

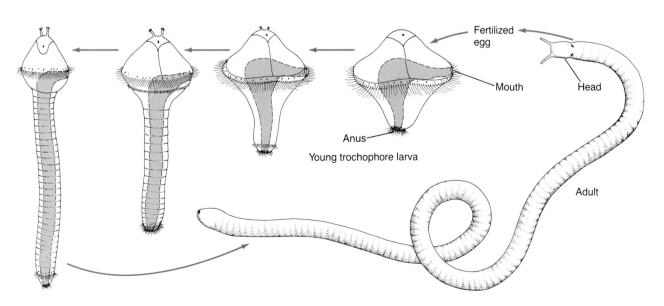

ciliated, planktonic trochophore larva (Figure 25.11). Many polychaetes lack a trochophore and instead have large yolky eggs that undergo direct development (no larvae).

The palolo worm *Eunice viridis* and a few other polychaetes undergo an unusual reproductive process (Figure 25.12b). The posterior half of the body in each sex enlarges and fills with eggs or sperm. On one night during October or November, the sexual segments of the worms break off and swarm to the surface, their astronomical numbers often clouding the water. A critical lunar photoperiod seems to determine the precise timing of these swarms by activating a neurohormonal mechanism. At the ocean surface, the sexual segments burst, spilling gametes into the sea. While the posterior pieces swarm, the anterior ends remain on the seafloor, each regenerating its lost end.

In contrast to most polychaetes, oligochaetes and leeches are hermaphroditic and have gonads in only a few segments, and most species copulate. In *Lumbricus terrestris*, each individual has both male and female reproductive systems located ventrally in segments 9 to 15 (Figure 25.13a). The male system includes two pairs of testes, one each in segments 10 and 11; two pairs of **sperm funnels** connected to a sperm duct (vas deferens) that leads to a male pore on segment 15; and two pairs of bulbous seminal vesicles that partially envelop the esophagus in segments 9 to 13 and often envelop the testes and sperm funnels as well. Immature sperm pass from the testes and complete their differentiation in the seminal vesicles. Then, during copulation, mature sperm enter the sperm funnels and pass along the vas deferens to the male pore.

The female system consists of a pair of ovaries

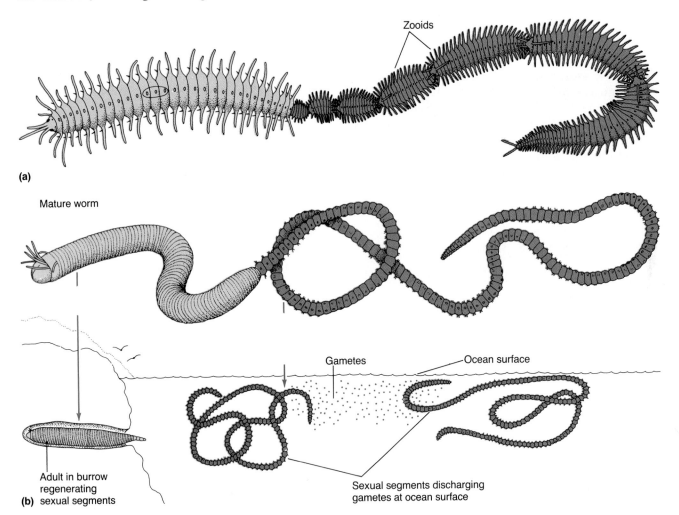

Figure 25.12 Reproduction in polychaetes. (a) Many marine polychaetes, such as this species *Autolytus purpureomaculatus,* reproduce asexually by budding offspring called zooids from the posterior end. (b) The palolo worm *(Eunice viridis)* of the tropical Pacific reproduces both sexually and asexually. The sexual segments break from the parent worms and swim to the ocean surface where they discharge sperm and eggs. Meanwhile, the parents begin the asexual regeneration of new sexual regions.

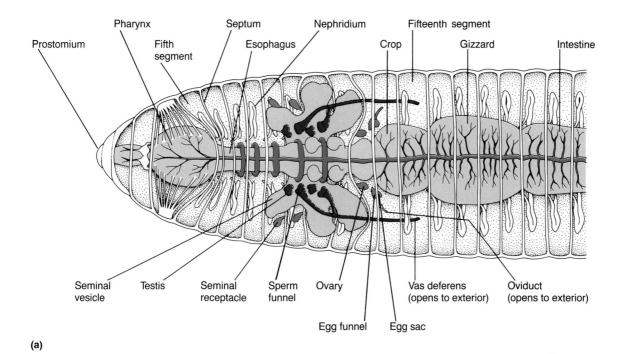

(a)

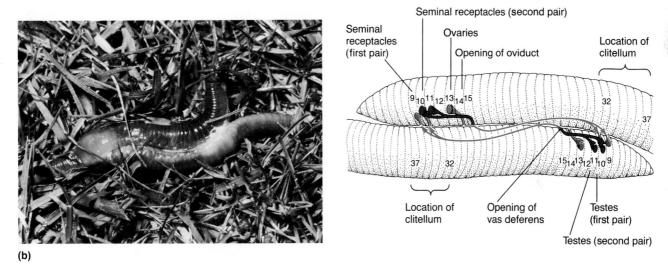

(b)

Figure 25.13 Reproductive system and copulation in earthworms. (a) Gonads and other reproductive structures are located in the anterior segments and are shown here in relation to the gut, metanephridia, and intersegmental septa. (b) Copulating pairs of *Lumbricus* can readily be seen with the aid of a flashlight on a wet night. The diagram shows the path followed by sperm as the worms inseminate each other. Sperm pass along a groove from the male genital pores (segment 15) to the seminal receptacles (segments 9 and 10) of each earthworm. Not shown is a slime tube that holds the worms firmly together in the region from segments 9 to 37.

in segment 13 and, just behind each ovary, an **egg funnel** with egg sacs. An oviduct leads from each egg funnel to a female pore on segment 14. During copulation, the female system receives and stores sperm in two pairs of seminal receptacles located in segments 9 and 10.

Earthworms copulate at night. Two worms, stretched partway out of their burrows, unite as shown in Figure 25.13b. The clitellum (segments 31 or 32 to 37) of each worm grasps segments 7 to 9 of the other, aided by special setae that actually penetrate the body wall. The worms are bound together by a "slime tube" of mucus secreted by the clitellum. In many oligochaetes, the male genital pores of one worm lie against the seminal receptacle openings of the other, facilitating sperm transfer;

some species even have a short penis. During copulation in *Lumbricus*, however, the openings lie some distance from each other. *Lumbricus* uses muscle contractions to form a ventral groove from the seminal receptacle openings of one worm to the male pores of the other. After each worm has deposited semen in the groove, wavelike pulses of muscle contraction push the semen along the groove and into the mate's seminal receptacles. Sperm transfer takes several hours to complete, and then the worms separate.

A few days after copulation, each *Lumbricus* deposits its eggs and its former mate's sperm in a cocoon secreted by the clitellum (Figure 25.14). The eggs are fertilized within the cocoon, and development begins. Ordinarily, only one embryo per cocoon develops, emerging two to three weeks later as a small but fully formed earthworm. Development is direct, with no trochophore.

Leech reproduction and development resemble the oligochaete pattern. Some species have a penis and ejaculate sperm directly from the male system

of one worm into the female system of another. In leeches without a penis, two worms intertwine in a tight embrace, and then each expels a spermatophore (a packet of sperm) against the body of the other. The spermatophore penetrates the body wall of the mate, releasing sperm that migrate to the vagina for storage. Following fertilization, the eggs are deposited in a cocoon secreted by the clitellum. Most leeches place the cocoon in a safe place in water or soil, but a few carry the cocoon with them. Development is direct, and young leeches emerge from their cocoons after a few weeks, becoming sexually mature in about a year.

PHYLOGENY

The evolutionary history of annelids is an important chapter in the phylogeny of the animal kingdom. As we mentioned in the Trends and Strategies section beginning on p. 583, ideas about annelid origins are intertwined with hypotheses about the origins of metamerism and the coelom, and zoologists generally agree that annelids and molluscs arose from a common protostome stock (see Figure 18.6). Most molluscs and annelids have spiral cleavage; and the mouth, mesoderm, and coelom develop according to the protostome pattern. In addition, some members of both phyla have a trochophore larva.

As to relationships within the phylum Annelida, polychaetes display several primitive features (e.g., metameric gonads and a trochophore) and are generally thought to resemble ancestral annelids more than oligochaetes or leeches do. Many zoologists believe that the oligochaetes diverged from ancestral polychaetes, whereas leeches probably arose from ancestral burrowing oligochaetes. An alternative view is that oligochaetes were ancestral to both polychaetes and leeches. The oligochaete body, with its lack of parapodia and few setae, is well adapted for burrowing. Conversely, the polychaete body, which has parapodia and is often capable of serpentine movement, is an efficient machine for errant surface dwelling, for swimming, and for circulating water in tubes. If the original annelid was a burrower (derived from a nonsegmented coelomate), then oligochaetes or oligochaete-like worms may have been the earliest annelids, and the polychaete and leech lineages may have been derived from these burrowing ancestors.

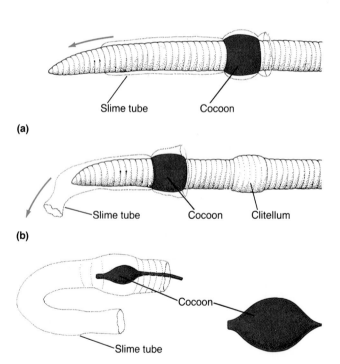

(a)

Slime tube   Cocoon

(b)

Slime tube   Cocoon   Clitellum

Cocoon

Slime tube

Figure 25.14 Earthworm cocoon formation. (a) In producing a cocoon, the clitellum first secretes a mucous "slime tube" around the anterior third of the body. Next, within the slime tube, the clitellum secretes the walls of the cocoon about itself. (b) After nutrients have been secreted into the cocoon by the clitellum, the cocoon slips forward and off the anterior end of the worm. As the cocoon passes over the female pores and seminal receptacle openings, eggs and sperm are deposited in it. The open ends of the cocoon seal shut as it slips free of the worm into the damp soil.

PHYLUM SIPUNCULA

Sipunculans, sometimes called "peanut worms" because of their appearance when disturbed and contracted, are whitish yellow or tan colored (Figure 25.15). Most of the approximately 350 species in the phylum inhabit sand and mud in intertidal zones and shallow seas, but several species live at great depths (down to about 7000 m). Sipunculans range in length from a few millimeters to about 75 cm. They feed mainly on bottom deposits or on organic matter suspended or dissolved in seawater. The burrowing activities of several species are important in mixing bottom sediments. Several sipunculans are common on coral reefs, where up to 700 individuals per square meter have been counted. They bore into coralline rock or live in tubes and shells vacated by other animals.

The sipunculan body consists of a saclike, unsegmented trunk and an anterior proboscis, called the **introvert**, that bears a terminal mouth (see Figure 25.15b). The introvert is extended by coelomic fluid pressure and withdrawn by retractor muscles. When sipunculans are undisturbed and feeding, the introvert slides in and out of the trunk continuously. The introvert bears sensory cell clusters, many short spines, and a terminal ring of hollow tentacles. Mucus-coated cilia on the tentacles collect food particles and convey them into the mouth. The digestive tube is long and coiled, with an anus located dorsally near the base of the introvert. Sipunculans lack gas exchange and circulatory structures; fluid in their coelom aids in internal transport, and they have red coelomocytes containing the oxygen-carrying pigment hemerythrin. A pair of metanephridia dispose of nitrogenous wastes collected from the coelomic fluid by amoebocytes, and also serve as the avenue of gamete release. The nervous system consists of ganglia above the esophagus attached by a nerve ring to a single ventral unganglionated nerve cord.

Most sipunculans are dioecious; gametes are produced in the coelom and then released into seawater, and fertilization is external. Cleavage is spiral, and development either is direct (no larva) or includes a free-swimming trochophore larva. The body cavity develops as a schizocoel. Sipunculans are not segmented at any time in their life cycle, but their developmental patterns indicate rather close affinities with other protostomes, including annelids. The sipunculan lineage probably diverged from early protostome stock before the segmented annelid body evolved.

(a)

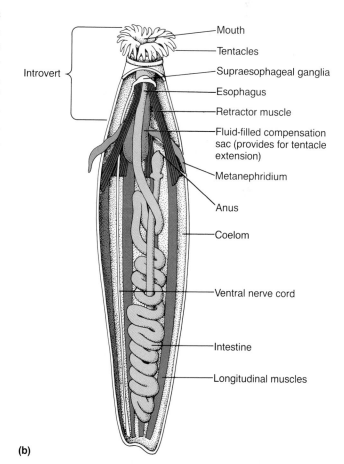

(b)

Figure 25.15 Phylum Sipuncula. (a) The proboscis (introvert) and branched tentacles are extended in this specimen of *Themiste lageniformis* photographed in a clump of oysters in a Florida lagoon. (b) A cutaway drawing showing the internal structures.

PHYLUM ECHIURA

Echiurans resemble sipunculans in general body form, size, and habitat. The phylum contains about 100 species, ranging in length from a few millimeters to about half a meter. A few species inhabit great depths (up to 10,000 m), but most live in sand burrows or rock crevices in coastal seas. The species *Urechis caupo* is cultured in the laboratory and studied extensively by developmental biologists.

The echiuran body has two parts, a cylindrical trunk containing internal organs and an anterior extensible proboscis (Figure 25.16). Echiurans are commonly called spoonworms because the proboscis is somewhat spatula-like, quite different from the introvert type of sipunculans. The ventral surface of the proboscis bears mucus-coated cilia and has infolded sides forming a gutter through which bottom deposits are conveyed backward into the mouth. Excretory organs include one or several pairs of metanephridia and a pair of **anal sacs** that pass wastes from the coelom into the rectum. The nervous system is like that of sipunculans, but most echiurans have a closed circulatory system similar to that of earthworms. Some echiurans also have coelomocytes that contain a type of hemoglobin.

Spoonworms are dioecious, and most species shed gametes via their metanephridia into the ocean, where external fertilization occurs. Species of *Bonellia* are unusual in having internal fertilization and marked sexual dimorphism; dwarf males live symbiotically within the females' metanephridia or coelom (see Figure 25.16). Lacking most organs, males apparently feed by absorbing nutrients across their body surface from the female's body. The sex an individual will become is determined hormonally. Some *Bonellia* larvae contact and enter the female and are induced by female hormones to become males. Larvae that do not enter adult females develop into females.

Like typical protostomes, echiurans have spiral cleavage, a schizocoel, and a trochophore larva. It seems likely that, in common with sipunculans, spoonworms evolved from early, nonsegmented protostomes.

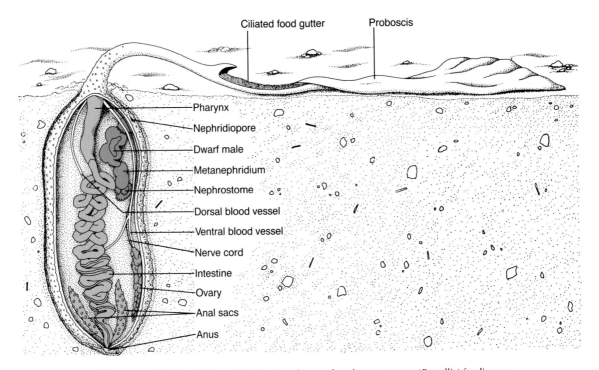

Figure 25.16 Phylum Echiura. This drawing shows a female spoonworm (*Bonellia*) feeding on organic deposits in sand. Note that her proboscis can be extended to several times the length of her body and has an expandable tip that increases surface area for food gathering. The trunk is cut away to illustrate internal structures; note the tiny male (about 1 mm long) inside the female's metanephridium.

PHYLUM POGONOPHORA

Pogonophorans, also called **beard worms**, are white or reddish with a crown ("beard") of pink or red tentacles (Figure 25.17). They live in upright, secreted, chitinous tubes at the bottom of deep oceans (over 100 m) and in certain areas are the

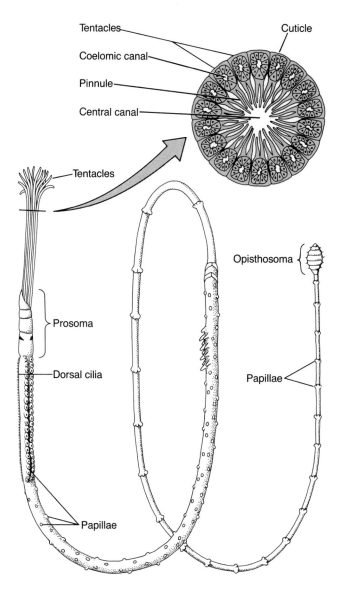

Figure 25.17 Phylum Pogonophora. *Lamellisabella zachsi* (removed from its tube) is only about 15 cm long. Note the three main body regions. The enlarged cross section shows structural details of the tentacles; note that the coelom extends into the tentacles. The pinnules have well-developed blood vessels.

most numerous animals at depths over a thousand meters. About 100 species are known, and all have been discovered since 1900. Prior to that time, they were often dredged up from ocean depths, but were not recognized as living organisms. These animals range in length from about 10 cm to over 2 m. The largest species, *Riftia pachyptila* (also called the vestimentiferan worm), is common at depths of about 2.5–6 km, where hot gases and molten lava escaping from the core of the earth through faults in the earth's crust create hydrothermal vents. *Riftia*, certain polychaete annelids, large bivalve molluscs, and several other types of animals thrive near these deep-sea vents. (See Figure E1b accompanying the Essay on ocean vent communities, p. 606.)

Pogonophorans are unusual among nonparasitic animals in that they lack a digestive tract. There is speculation that some of them feed by trapping bacteria and organic particles on their tentacles or by absorbing organic molecules dissolved in seawater. Tentacle cells bear many inward projections called **pinnules** that increase surface area. Cilia at the base of the pinnules may drive water through a central canal formed by the tentacles (see Figure 25.17), but there is no direct evidence that nutrients in the water are ingested. Electron microscopic and radiotracer studies show that internal tissues of *Riftia pachyptila* contain large numbers of symbiotic bacteria that supply the worms with energy-containing compounds.

The pogonophoran body consists of three distinct regions (see Figure 25.17). An anterior **prosoma** bears from one (spirally coiled) to about 200,000 tentacles, depending on the species. The middle part of the body, a long trunk, bears clusters of cilia and other surface projections called **papillae**. The trunk contains about 4 billion bacteria per gram in *R. pachyptila*. The third body region, the **opisthosoma**, was not discovered until the early 1960s, long after the phylum had been named and many species described. The opisthosoma is annelid-like in that it is segmented and bears paired setae that anchor the body to the tube. It often breaks off and remains in the tube when pogonophorans are collected.

Internally, the pogonophoran body is divided into three main coelomic compartments corresponding to the three external regions. The coelom in the opisthosoma is subdivided by septa between the segments. In common with most echiurans and many annelids, there is a closed blood vascular system. A type of hemoglobin dissolved in the plasma

Essay: Ocean Vent Communities

In 1977, the submersible research vessel *Alvin* was lowered into the Pacific Ocean near the Galapagos Islands, and biologists in the vessel descended 2500 m to explore an ocean vent (Figure E1a). Such vents exist along submerged mountain ranges where molten rock and hot gases pour out from the earth's interior along rifts between crustal plates in the ocean floor. Outpouring magma continuously creates new crust, and as a result, the seafloor spreads at these sites. Adjacent to the rifts, seawater seeps into porous rocks, becomes heated and mineral laden, and flows upward and jets out through tall, chimneylike structures called "black or white smokers." Although water from these chimneys may be 350°C, it is prevented from boiling by enormous water pressures (about 200–300 atmospheres) and is cooled rapidly by surrounding water (about 2°C).

Since 1977, several hydrothermal ocean vents have been studied, and a fascinating picture is emerging of life at these deep sites. Dense populations of giant tube worms (the pogonophoran *Riftia*; Figure E1b), 30 cm long whiteshelled clams, a few crabs, and polychaete annelids are among several animals that thrive here. All exist in total darkness, and no plants live here, for the vents are far below the level of light penetration.

Without plants to trap sunlight energy, what supports these animal communities? Little organic matter rains down from above, for overlying waters are not highly productive. Researchers found the answer in the porous rock crevices. Enormous populations of bacteria living in the crevices obtain energy (which they use to make organic molecules) by oxidizing hydrogen sulfide to elemental sulfur or sulfate. The hot, circulating seawater carries some of the bacteria out of the rocks and up into the surrounding water, and some of the vent animals filter-feed on the suspended bacteria. Symbiosis is also important in vent communities. Chemosynthetic bacteria housed in special tissues of the large pogonophorans produce organic molecules used by the worms as food. *Riftia*'s blood transports oxygen and carbon dioxide to the bacteria. A type of hemoglobin in the worm's plasma has a high affinity for oxygen and enhances oxygen transport. The bacteria use the oxygen to oxidize sulfur, and reduce the carbon dioxide to energy-containing organic compounds. Gill tissues of the large bivalves also contain symbiotic bacteria. Thus, unlike other environments on earth, which depend on photosynthesis by plants, ocean vent communities are supported by bacterial chemosynthesis.

So far, only about a dozen vent sites along the 73,000 km of deep-sea rifts have been explored. Over 250 strains of sulfur-oxidizing bacteria have already been identified from these sites, and no doubt other bacteria and new species of animals remain to be discovered. Many fundamental questions remain to be answered about the geology and

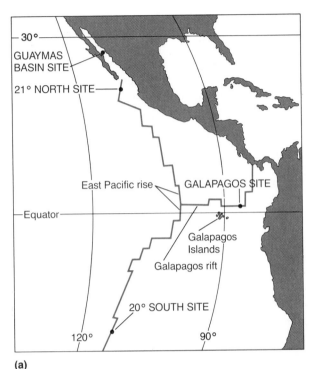

(a)

(b)

Figure E1 Life in ocean vent communities. (a) Map of known ocean vent sites. (b) A cluster of the giant pogonophoran *Riftia pachyptila* near a deep-sea vent.

biology of vent communities. The relationships between *Riftia* and its symbiotic bacteria are just beginning to be understood. Zoologists are also curious about how the vent animals can survive in concentrations of hydrogen sulfide that are toxic to other animals. Moreover, some of the vent bacteria have an incredible ability to survive in the superheated vent water. An understanding of their biomolecules could provide important information about biological heat tolerance.

makes the blood red. The pinnules of the tentacles contain blood capillaries and are probably involved in gas exchange, excretion, and perhaps in feeding in some species.

Most of the details of pogonophoran reproduction and development remain to be discovered. There are separate sexes, and sperm are packaged into spermatophores before they are released by the male. A solid blastula (stereoblastula) develops following bilateral cleavage. (Bilateral cleavage is described in Chapter 16, p. 380.) There are conflicting reports about whether the coelomic compartments are enterocoelous, schizocoelous, or both. Judging from their metameric opisthosoma, pogonophorans seem to be rather closely related to the annelids.

□ SUMMARY

1. Key Features of Annelids. Polychaetes, oligochaetes (earthworms and others), hirudineans (leeches), and branchiobdellids are metameric protostomes; most have a well-developed coelom.

2. Most annelids are free-living. Polychaetes are aquatic (mostly marine); oligochaetes and leeches occur on land and in freshwater and marine environments.

3. The annelid body is supported mainly by a hydrostatic skeleton. Many species live in secreted tubes or in burrows. Locomotion is by peristalsis, crawling, or swimming; many leeches move with a looping action.

4. Annelids have open or closed circulatory systems. Gas exchange occurs over the entire body surface, and many species have gills. Osmoregulation and excretion are carried out by paired segmental metanephridia.

5. The central nervous system consists of cerebral ganglia and a ventral nerve cord with segmental ganglia. Giant nerve fibers in the ventral nerve cord mediate rapid escape responses. Annelids possess photoreceptors, chemoreceptors, and mechanoreceptors. They commonly display negative phototaxis in strong light. Burrowers and tube dwellers are usually strongly thigmotaxic. Several annelids also exhibit habituation and simple associative learning.

6. Polychaetes are dioecious, whereas oligochaetes and leeches are hermaphroditic. Fertilization is external in most polychaetes. Oligochaetes and leeches copulate, but fertilization may be internal or external in a cocoon, depending on the species.

7. Polychaetes and oligochaetes probably evolved from burrowing marine worms; leeches most likely evolved from the oligochaetes.

8. The Sipuncula, Echiura, and Pogonophora are three small phyla of protostomes that resemble annelids developmentally, and in having a spacious coelom, multilayered body wall including a secreted cuticle, and a hydrostatic skeleton. All species (about 550 total) are marine bottom dwellers with an anterior proboscis.

□ FURTHER READING

Ballard, R. D., and J. F. Grassle. "Incredible World of the Deep-Sea Rifts." *National Geographic* 156 (1979): 680–705. *A pictorial description of the organisms and environments of deep-sea vents; includes color photos of giant pogonophorans and a description of the deep-sea submersible* Alvin.

Barnes, R. D. "The Annelids"; "Some Lesser Protostomes." In *Invertebrate Zoology*, 5th ed. Philadelphia: CBS College, 1987, Chaps. 10 and 17. *Detailed coverage of the annelids and annelid-like worms.*

Edmond, J. M., and K. Von Damm. "Hot Springs on the Ocean Floor." *Scientific American* 248 (1983): 78–93. *A detailed description of deep-sea vents as seen on several recent expeditions.*

Edwards, C. A., and J. R. Lofty. *Biology of Earthworms*. 2d ed. London: Chapman and Hall, 1977. *Modern, comprehensive account of earthworm biology.*

Leake, L. D. "The Leech as a Scientific Tool." *Endeavour* (New Series) 7(1983): 88–93. *Past and present use of medicinal leeches in medical practice, and their increasing use in neurophysiological research.*

Lee, K. E. *Earthworms. Their Ecology and Relationships with Soils and Land Use*. Orlando, Fla: Academic Press, 1985. *An authoritative treatise on the environmental relationships and importance of earthworms.*

" Tube Worm Nourished with Help from Within." *Science News* 120 (July 18, 1981): 38. *A short summary of research on symbiotic nutrition of giant pogonophorans.*

□ CLASSIFICATION

Phylum Annelida (L. *annelus*, "small ring"). The segmented worms; approximately 15,000 species.

Class Polychaeta (Gr. *polys*, "many"; *chaete*, "long hair"). Polychaetes; mostly marine, a few freshwater species; typically with numerous setae on segmentally arranged parapodia and a well-developed head; most are dioecious; errant polychaetes [e.g., *Aphrodite*, the sea mouse; *Nereis*, the clam worm; *Polygordius* (lacking setae and parapodia); *Glycera*, and *Eunice viridis*, the Samoan palolo worm] swim freely, burrow, crawl, or live in tubes. Sedentary polychaetes [e.g., *Arenicola*, the lugworm; *Sabella*, a fanworm; *Phragmatopoma*, the sandcastle worm; and *Chaetopterus*, a parchment tube worm] are primarily burrowers or tube dwellers.

Class Oligochaeta (Gr. *oligos*, "few"; *chaete*, "long hair"). Oligochaetes; terrestrial, marine, or freshwater; no parapodia and few setae; head reduced; monoecious, with a clitellum that secretes an egg cocoon; *Lumbricus* and *Eisenia*, earthworms; *Tubifex* lives in delicate tubes surrounded by mud.

Class Hirudinea (L. *hirudo*, "leech"). Leeches; freshwater, marine, or terrestrial; no parapodia or setae; anterior and posterior suckers; no internal segmentation except in nervous system; coelom reduced to network of small sinuses by musculature and connective tissue; monoecious, with a clitellum; *Hirudo* and *Haemadipsa* are bloodsuckers; *Glossiphonia* is predatory.

Class Branchiobdellida (Gr. *branchion*, "a fin"; *bdella*, "leech"). Branchiobdellids; commensals or ectoparasites of freshwater crayfish; a little-known group that appears intermediate between oligochaetes and leeches; *Cambarincola*.

Phylum Sipuncula (L. *sipunculus*, "a little siphon"). Sipunculans (peanut worms); marine, bottom-dwelling deposit feeders, with an extensible proboscis (introvert); *Sipunculus*, *Dendrostomium*, *Themiste*.

Phylum Echiura (Gr. *echis*, "serpent"; *oura*, "tail"). Spoonworms; marine, bottom-dwelling deposit feeders with a spatula-like proboscis; *Urechis*, *Bonellia*.

Phylum Pogonophora (Gr. *pogon*, "beard"; *phoros*, "bearing"). Beardworms; long, thin, gutless worms of deep oceans; the anterior end bears one or many long, filter-feeding tentacles; *Lamellisabella*, *Riftia*.

Arthropods and Arthropod-like Phyla

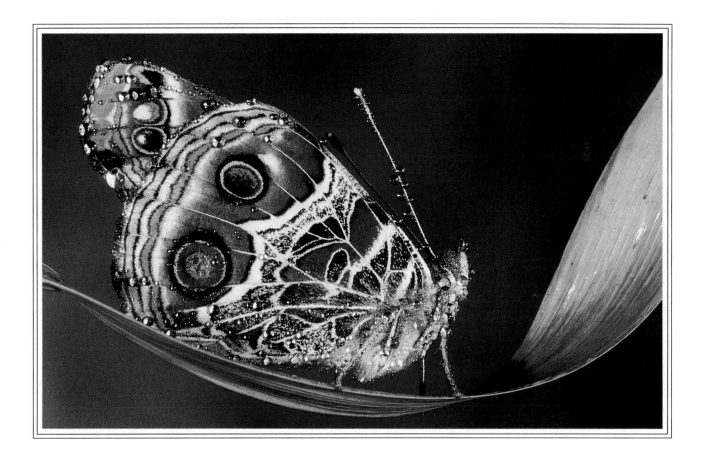

***Vanessa virginiensis,* the American painted lady.** This beautiful butterfly is one of approximately 120,000 kinds of moths and butterflies, making up the Order Lepidoptera of the arthropod Class Insecta. Of the million or so living species in the Phylum Arthropoda, the greatest number are insects.

Arthropods probably arose from annelid-like ancestors, and they are characterized by several annelid-like features. First, they are metameric, at least as embryos. Second, some of their body segments bear paired appendages reminiscent of, although probably not homologous with, the parapodia of polychaetes. Third, the arthropod nervous system is annelid-like in that it has a dorsal cerebral ganglion connected to a ventral, usually ganglionated, nerve cord.

Arthropods diverged from annelid-like ancestors in several important ways. Most significantly, they evolved a multilayered cuticular exoskeleton that protects internal organs and provides strong muscle attachments and support for the locomotor appendages. The cuticle is molted periodically as the animal grows. Second, the body segments in arthropods tend to be combined into functional

609

groups, or **tagmata**; the characteristic head, thorax, and abdomen of insects are examples. Third, arthropod appendages are jointed, articulating through hingelike connections that make complex limb movements possible. [The name Arthropoda is derived from the Greek words *arthron* ("a joint") and *podos* ("a foot").] Fourth, arthropods have a hemocoel as their main body cavity, rather than the metameric coelom found in most annelids.

In addition to the arthropods, three small phyla, the Tardigrada, Pentastomida, and Onychophora, whose members bear some resemblance to arthropods, are included in this chapter. These animals (about 670 species in all) lack jointed appendages, but have a secreted chitinous cuticle that is molted periodically, some evidence of body segmentation, an arthropod-like nervous system, and a hemocoel. Zoologists generally agree that these three phyla evolved from the line of metameric animals that gave rise to arthropods (see Figure 18.6).

PHYLUM ARTHROPODA

With astronomical numbers of individuals and about a million recorded species, the Arthropoda is the largest and most diverse phylum in the animal kingdom. The phylum has been prominent throughout much of animal history; primitive arthropods called trilobites were successful members of seafloor communities half a billion years ago, and for the last 300 million years, insects have been prominent on land.

REPRESENTATIVE ARTHROPODS

The Phylum Arthropoda is divided into four subphyla, three living and one extinct.

Subphylum Trilobita

Trilobites dominated the seafloors for millions of years and then became extinct toward the end of the Paleozoic era, about 280 million years ago. About 4000 fossil species have been identified, most of which were less than 10 cm in length. Typically, they had flattened bodies divided longitudinally into three lobes—the feature for which the subphylum is named—and transversely into anterior, central, and posterior tagmata (Figure 26.1). Each body segment had a pair of biramous (two-branched) appendages; one branch was probably used in walking, while the other branch, which was highly filamentous, may have been a gill. In addition, many species had prominent antennae and well-developed dorsal eyes.

Although some trilobites were apparently able to swim freely, and some were planktonic, most were probably bottom dwellers that used their shovel-shaped anterior end to dig food from the seafloor. No doubt, polychaetes and other invertebrates were dietary staples, although some trilobites probably digested organic matter eaten with the bottom muck.

Subphylum Chelicerata

There are approximately 65,000 identified living species of **chelicerates**, a group that is characterized by its first two pairs of appendages. The first are feeding organs called **chelicerae** (hence the name of the subphylum), and the second, adapted for a variety of functions (usually including feeding or sensory reception) in different groups, are called **pedipalps**. Also characteristic of the subphylum, chelicerates lack antennae, and their body is divided into an anterior **cephalothorax**, or **prosoma**, which bears the chelicerae, pedipalps, and walking legs, and a posterior **abdomen**, or **opisthosoma**.

Class Merostomata. This aquatic class was highly successful during the Paleozoic era, but most lineages became extinct by about 280 million years ago. **Merostomates** have a postanal tail spine, called a **telson**, and five or six pairs of abdominal **book gills**. One subgroup, the giant sea scorpions (eurypterids), some with bodies almost 3 m long, is extinct. A second subgroup, the horseshoe crabs (xiphosurans), has four living species. One of these, the largest living merostomate and one of the larger arthropods, is *Limulus polyphemus* (Figure 26.2).

Limulus polyphemus reaches about 60 cm in length, including its telson, and is common in shallow water along the western Atlantic coastline from New England to Yucatan. The body is covered dorsally by protective exoskeletal shields. The anterior shield, or **carapace**, is horseshoe-shaped and bears two prominent compound eyes, the only compound eyes known among the living chelicerates. *Limulus* bulldozes its way through sand and

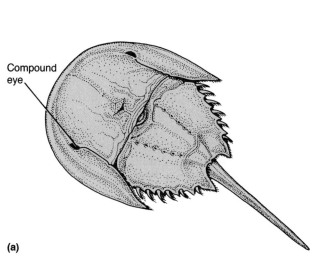

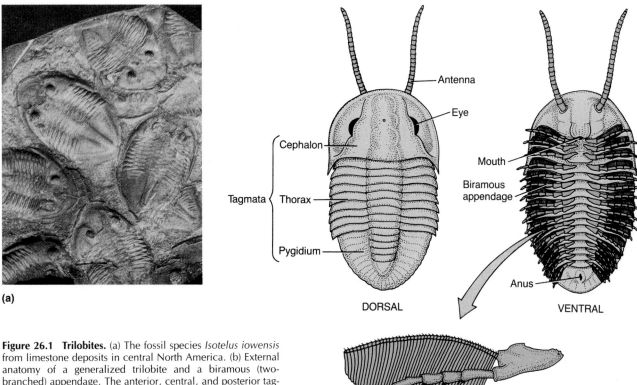

Figure 26.1 Trilobites. (a) The fossil species *Isotelus iowensis* from limestone deposits in central North America. (b) External anatomy of a generalized trilobite and a biramous (two-branched) appendage. The anterior, central, and posterior tagmata are called the cephalon, thorax, and pygidium, respectively.

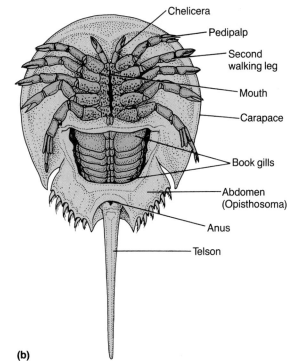

Figure 26.2 Class Merostomata: The horseshoe crab, *Limulus polyphemus.* (a) General view of the dorsal surface. (b) Anatomy of the ventral surface. The strong, spinelike telson at the posterior end of the body is used as a lever during locomotion and in flipping the body right side up. The pincerlike chelicerae are food-gathering organs, while the pedipalps are the first of the five pairs of walking legs. Most of the remaining appendages, those of the opisthosoma, are flattened, forming book gills.

mud in search of polychaetes, molluscs, and other small invertebrates that serve as food. Horseshoe crabs are widely used in research, especially in neurobiology and circulatory physiology.

Class Arachnida. The **arachnids** (approximately 65,000 identified species) include the scorpions, whip scorpions, spiders, pseudoscorpions, harvestmen, ticks, and mites. Arachnids probably originated in water, but living species are nearly all terrestrial. Adult arachnids have four pairs of walking legs.

Scorpions occupy habitats ranging from jungles to deserts in tropical, subtropical, and temperate zones (Figure 26.3a). Most are active at night, generally remaining hidden in burrows or under stones and plant litter during the day. The two features that impress people most about scorpions are their large pincers, which are actually pedipalps adapted for grasping prey, and the stinger at the posterior end of the body. Scorpions typically prey on insects and other small invertebrates. While holding an animal in its pedipalps, a scorpion may swing its abdomen and stinger forward to puncture the victim. Acting like a hypodermic needle, the stinger's barb injects venom into the victim. The venom is a neurotoxin powerful enough to kill or paralyze small invertebrates, but only a few species of scorpions are dangerous to humans.

Whip scorpions, or vinegaroons, typically occupy warm, humid habitats (Figure 26.3b). They are named for their thin, whiplike tail. Most species are about 2–4 mm in length, but the large American vinegaroon, *Mastigoproctus giganteus*, is about 7 cm long. Like the scorpions, whip scorpions hide by day and hunt at night, feeding on small invertebrates caught and crushed by the large pedipalps. Although they are not venomous, whip scorpions may repel opponents with a spray of concentrated acetic acid from glands near the anus.

Spiders include about 35,000 species and inhabit a wide variety of habitats on every continent except Antarctica. A spider's body has a narrow waist, or **pedicel**, separating the prosoma from an unsegmented opisthosoma (Figure 26.3c and d). Each chelicera is modified as a fang with a poison duct opening at its tip. Leglike pedipalps have sensory and reproductive functions. (A male uses his pedipalps to collect sperm from his gonopore and to place it into the female's gonopore.) The opisthosoma bears small appendages called **spinnerets**, which are the external openings of the silk glands.

Spiders are carnivores that prey mostly on in-sects. Hunting spiders, including wolf spiders, trap-door spiders, jumping spiders, and tarantulas, seize and immobilize prey with their fangs and pedipalps. Other spiders trap prey with strands or webs of silk.

Pseudoscorpions have large, pincerlike pedipalps similar to those of scorpions, but they are generally smaller than scorpions and have a flat, rounded opisthosoma without a stinger (Figure 26.3e). Their pedipalps contain venom glands. Pseudoscorpions prefer humid habitats, often living in great numbers under plant litter, where they feed mainly on small insects.

Harvestmen, known to many people as daddy longlegs, are sometimes mistaken for long-legged spiders (Figure 26.3f). Unlike spiders, they lack a pedicel, and most are omnivorous, eating a mixture of fruit, vegetation, dead organisms, and tiny insects. Harvestmen thrive in humid environments rich in organic matter and often occur in plant litter on the forest floor.

Ticks and **mites** may outnumber all other kinds of arachnids. Only about 25,000 species have been described, but there may be more than 100,000. Ticks are bloodsucking ectoparasites of many terrestrial vertebrates and are the vectors of many disease-causing microorganisms. Species of the genus *Dermacentor*, for example, transmit Rocky Mountain spotted fever, tularemia, and several virus diseases to humans (Figure 26.3g). Ticks cling to blades of grass and twigs along trails frequented by mammals, waiting for a host to pass. A tick's chelicerae form a stout shaft and blood-feeding tube that penetrates the skin of a host. Teeth on the chelicerae and barbs on the pedipalps anchor the tick to the wound (Figure 26.3h). Blood is required for egg production, and a female tick may engorge on blood until its body is greatly swollen and barrel-shaped.

Mites are generally less than 1 mm in length, and their small size allows them to occupy a great variety of microhabitats. Many are free-living scavengers, while others are parasitic, living in such diverse sites as mammalian hair follicles, insect tracheae, and the ear canals of many terrestrial vertebrates. One group, called the chigger mites, are parasitic only as larvae (Figure 26.3i). Some species attack humans, injecting digestive enzymes into the skin and sucking up digested cells. After the mite has dropped off, an aggravating itch may develop.

Class Pycnogonida. **Pycnogonids**, commonly called sea spiders, are often classified as

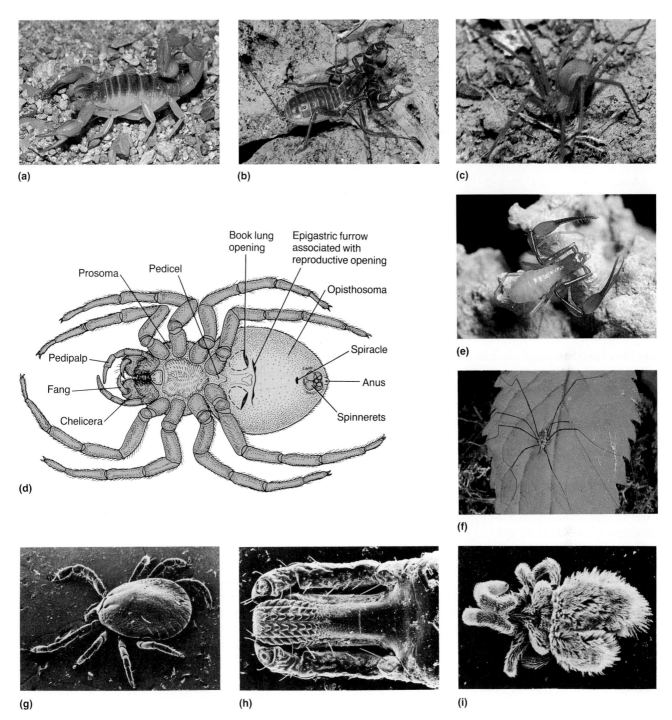

Figure 26.3 Class Arachnida. (a) A North American desert scorpion, *Hadrurus arizonensis*. (b) The whip scorpion, or vinegaroon, *Mastigoproctus giganteus*. (c) The brown recluse spider *Loxosceles reclusa*. The small, violin-shaped mark on the cephalothorax (prosoma) helps identify this dangerous species. (d) Diagrammatic view of a spider's ventral surface. Note the anus and spinnerets on the opisthosoma, as well as the openings of the respiratory system (book lungs and spiracle, leading to the tracheae) and reproductive organs. (e) A pseudoscorpion, *Vachonium*. (f) A harvestman, *Habrobunus maculosus*. (g) The American dog tick *Dermacentor variabilis*; this scanning electron micrograph shows an immature (nymphal) tick (20×). (h) A scanning electron micrograph of the chelicerae (forming the central projection) and pedipalps (on each side) of *Amblyomma americanum*, the lone star tick (46×). Note the teeth, which help to anchor the chelicerae in the host's skin as the tick takes a blood meal. (i) A chigger mite, *Trombicula alfreddugesi* (41×). Larvae of this and related species are skin parasites of vertebrates, including humans; *T. alfreddugesi* is found throughout the United States, except in mountainous areas of the west.

Figure 26.4 Class Pycnogonida. The sea spider *Nymphon*, seen against a background of sea stars. Most pycnogonids are between about 1 mm and 10 mm long, although some deep-sea species have a 6 cm long body with a 75 cm leg span.

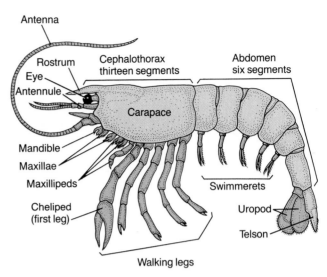

Figure 26.5 The crustacean body plan. The generalized crustacean body plan, illustrated by this shrimplike animal, has been adapted in various ways in the different crustacean lineages. Often, the head and thorax are fused as a cephalothorax covered by a shieldlike carapace, and the appendages are often adapted for tasks such as grasping food, chewing, swimming, and walking. In this crustacean, the first maxillae and mandibles and the five pairs of walking legs are secondarily uniramous, but the remaining appendages retain the biramous form.

chelicerates, but they are markedly different from other chelicerates, and many zoologists treat them as a separate subphylum. Of about 600 known species, all are marine, living on the bottom of virtually all ocean zones. Pycnogonids are carnivorous, eating a variety of small invertebrates, especially colonial cnidarians and bryozoans. The body bears four to six pairs of long walking legs, which are connected to a small segmented trunk, or prosoma (Figure 26.4). A headlike anterior region bears a pair of chelicerae (perhaps homologous with those of other chelicerates), usually a pair of small sensory pedipalps, and a proboscis. In front of the walking legs on males, a pair of unique **ovigerous legs** is used to bear developing egg masses. The opisthosoma is reduced to a small posterior projection bearing the anus.

Subphylum Crustacea

Crustaceans constitute a relatively small group (about 40,000 living species) among arthropods, but they are exceptionally important in aquatic environments. Crustaceans are most numerous in the sea, where their dense populations make up a large part of the total fauna and are a significant source of food for many other animals. Many freshwater species are also important in food chains. Relatively few species live on land.

The generalized crustacean body plan consists of a head and a trunk that is usually subdivided into a thorax and an abdomen. Appendages are segmentally arranged and typically biramous. Most crustaceans have gills, which are typically borne on the appendages. In most large species, the exoskeleton is calcified. Crustaceans exhibit enormous variety in body forms, but the head exhibits some uniformity throughout the subphylum; it usually bears a pair of compound eyes and five pairs of appendages. Crustaceans are distinguished by having two pairs of antennae. Thus, the first and second pairs of head appendages are **antennules** (first antennae) and **second antennae**. The antennae have various functions (e.g., sensory reception, locomotion, or feeding) in different crustacean groups. Behind the antennae, the third pair of head appendages are the **mandibles**, which usually serve as opposable mouthparts for chewing, grinding, or filter feeding. Two additional pairs of mouthparts, the **first** and **second maxillae**, are situated behind the mandibles (Figure 26.5).

Crustaceans probably arose from aquatic arthropods whose body segments and appendages, at least those of the trunk, were relatively uniform and unspecialized. Among living crustaceans, certain primitive species have uniform trunk segments and appendages, reflecting the ancestral condition. As a group, however, crustaceans exhibit marked regional specialization of body parts, with the number of segments and general appearance, especially of the thorax and abdomen, varying greatly among the crustacean groups. The head in many species is fused with the thorax, forming the cephalothorax. Appendages are typically specialized, and many have become secondarily uniramous (see Figure 26.5). In several groups, one or more anterior thoracic appendages are modified as feeding structures called **maxillipeds**. Of ten classes of living crustaceans, we describe five major ones, illustrating some of the specializations and variety in body forms in the subphylum.

Class Branchiopoda. Most abundant in fresh water, **branchiopods** include water fleas, fairy shrimps, and clam shrimps. Typically, the body is uniformly segmented, but some or all of the segments may be covered by a carapace. The carapace of a clam shrimp is bivalved, giving the shrimp the appearance of a tiny clam. Branchiopods generally have a large number of similar appendages used for a variety of purposes, including filter feeding, locomotion, and gas exchange. The water flea *Daphnia* (Figure 26.6a) and the brine shrimp *Artemia*, which is especially numerous in the Great Salt Lake of Utah, are common genera.

Class Copepoda. Most **copepods** are small, many less than 2 mm long (Figure 26.6b). There are only about 7500 species in the class, but copepods typically occur in dense populations, and they are among the most numerous animals in marine and freshwater environments. Copepods have a head fused with one or two thoracic segments. The head bears a single median eye and often very long antennules. The rest of the body consists of a tapered thorax (four or five segments) with appendages and a narrower abdomen without appendages. The first pair of thoracic segments are modified as maxillipeds. Most copepods are planktonic, swimming with their thoracic appendages and catching planktonic algae, tiny animals, and suspended particles (see Figure 26.18). About 1500 species of copepods are parasitic, found mainly on the body surfaces of fishes and marine mammals.

Class Ostracoda. Commonly called seed or mussel shrimps, **ostracods** look like tiny seeds or miniature mussels because the body is enclosed in a bivalved carapace (Figure 26.6c). Muscles close the carapace, and an elastic hinge opens it when the muscles relax. Somewhat more than half the living species are marine, a few are terrestrial, and the rest live in fresh water. A few predatory ostracods seize and eat tiny invertebrates, but most use setae on the mouthparts to filter organic particles from the water.

Class Cirripedia. Some 900 species of free-living and parasitic **barnacles** have been identified. Unusual among crustaceans, barnacles typically have a heavy calcareous shell, a feature that led early zoologists to classify them as molluscs (Figure 26.6d–g). The shell is actually the barnacle's highly modified exoskeleton. Free-living barnacles occur in two forms. Stalked (goose) barnacles are attached to rocks and other firm surfaces by an elongate stalk, whereas stalkless (acorn or volcano) barnacles are attached to the substrate directly by their calcareous shells. A stalkless barnacle's squat form helps it withstand strong currents and heavy pounding by waves. Aggregations of barnacles often live on wharf pilings, on ship hulls, and in wave-swept rocky intertidal zones.

Free-living barnacles are filter feeders, although some can capture small prey as well. With its shell open, a barnacle sweeps its jointed feeding appendages (cirri) through the water or holds them erect in currents, collecting plankton and other small food particles. Parasitic barnacles live in or on various marine crabs, echinoderms, and soft corals, and on whales; many parasitic species have lost the shell, appendages, and body segmentation.

Class Malacostraca. This class includes shrimps, crayfish, lobsters, and crabs, as well as a host of lesser-known animals, and accounts for about 70% of the crustaceans. **Malacostracans** are characterized by having from one to three anterior pairs of thoracic appendages modified as maxillipeds for feeding. Posterior to the maxillipeds, the thorax bears walking legs. The first pair of walking legs are often adapted for food capture and are called **raptorial legs**. In many species, such as crayfish and crabs, the raptorial legs are termed **chelipeds** because they are strong and stout and bear large terminal pincers called **chelae**. Typically, at least some of the thoracic appendages bear gills. Malacostracans are classified into more than a dozen subgroups, or orders; here we describe three major orders, constituting about two-thirds of the species in the class.

Order Isopoda. Isopods are predominantly marine, but include a number of freshwater, semiterrestrial, and truly terrestrial species as well. Most are about 1 to 3 cm long, but they range from less than a cm to 25 cm or more in length. Isopods have a single pair of maxillipeds, lack a carapace, and

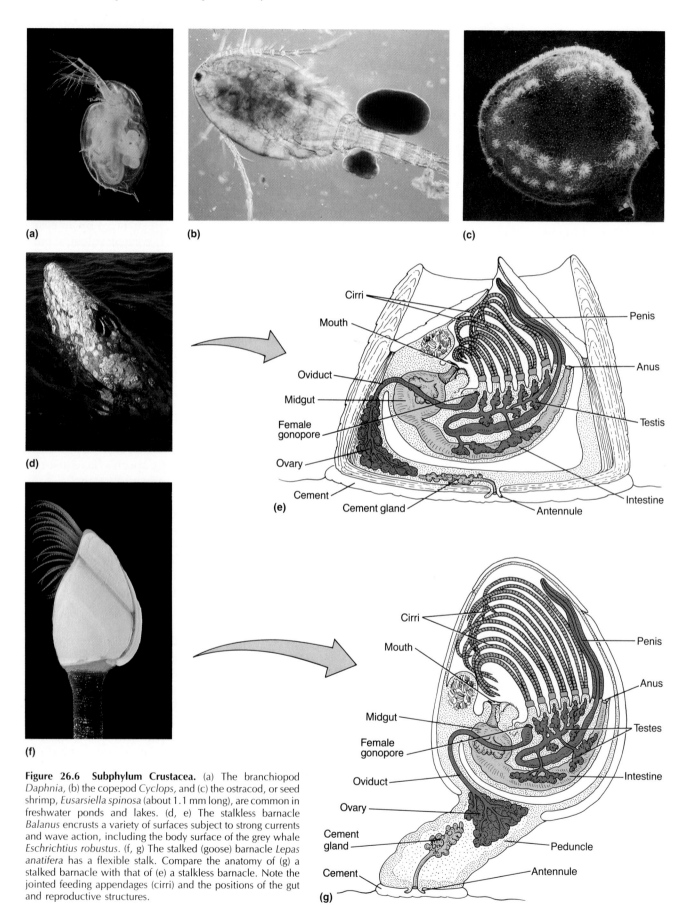

Figure 26.6 Subphylum Crustacea. (a) The branchiopod *Daphnia*, (b) the copepod *Cyclops*, and (c) the ostracod, or seed shrimp, *Eusarsiella spinosa* (about 1.1 mm long), are common in freshwater ponds and lakes. (d, e) The stalkless barnacle *Balanus* encrusts a variety of surfaces subject to strong currents and wave action, including the body surface of the grey whale *Eschrichtius robustus*. (f, g) The stalked (goose) barnacle *Lepas anatifera* has a flexible stalk. Compare the anatomy of (g) a stalked barnacle with that of (e) a stalkless barnacle. Note the jointed feeding appendages (cirri) and the positions of the gut and reproductive structures.

most are flattened dorsoventrally. Their gas exchange surfaces are part of the abdominal appendages. The group contains the most successful land-dwelling crustaceans—the pill bugs, or wood lice (Figure 26.7a). Generally restricted to very damp habitats by the need to prevent loss of body water, pill bugs can often be found under rocks or decaying wood, where they feed on living and dead plant and animal material. Several isopods are parasitic on fishes and on other crustaceans.

Order Amphipoda. **Amphipods** have much in common with the isopods. Like isopods, they are largely marine, but freshwater and semiterrestrial species also occur; amphipods also have one pair of maxillipeds and lack a carapace. Unlike isopods, however, most amphipods are flattened laterally rather than dorsoventrally, and they have thoracic rather than abdominal gills. Very dense populations of freshwater amphipods sometimes develop in ponds and streams. The most successful semiterrestrial amphipods are sand hoppers and beach fleas, which burrow on moist beaches or through damp forest litter, feeding on decaying organic matter (Figure 26.7b). On damp, overcast days, beach fleas may leave their burrows and hop about on algal masses washed in by the surf.

Order Decapoda. **Decapods** (ten-legged malacostracans) include the lobsters, crayfish, crabs, and shrimps (see Figure 26.7c–f; Figure 1.15d; Figure 26.12). With about 10,000 species, this is the largest crustacean order. Although most are marine, crayfish and a few species of shrimps live in fresh water, and some crabs and burrowing crayfish live on land. Crayfish, lobsters, and crabs generally move about by walking and crawling, whereas shrimps are adapted primarily for swimming. Decapods are distinguished from other malacostracans in having the first three pairs of thoracic appendages modified as maxillipeds, for food handling, and the last five pairs modified for walking (thereby giving these crustaceans ten walking legs). One or more pairs of walking legs are chelate (pincerlike), and the first pair are often developed as large chelipeds. The fused head and thorax form a cephalothorax covered by a carapace that encloses lateral gill chambers.

Lobsters live among rocks on the seafloor, walking slowly about scavenging for food. They often eat snails, clams, crabs, and even small fishes. The American lobster *Homarus americanus*, one of the largest living arthropods, may grow to over 50 cm in length and weigh about 20 kg. Crayfish some-

(a)   (b)   (c)

(d)   (e)   (f)

Figure 26.7 **Class Malacostraca.** (a) The Order Isopoda includes terrestrial species, such as the pill bug *Porcellio scaber.* (b) The Order Amphipoda includes the beach flea *Megalorchestia californiana.* Some beach fleas can hop more than 40 times their body length. (c–f) The Order Decapoda: (c) the fiddler crab *Uca minax* lives on tidal mud flats; (d) the hermit crab *Pagurus* inhabits discarded mollusc shells; (e) the Atlantic blue crab *Callinectes sapidus,* and (f) the shrimp *Pandalus platyceros* (up to about 15 cm long), are commercially important food species.

what resemble lobsters but are smaller, averaging about 10 cm in length.

Crabs are generally broad and flat, with a much-reduced abdomen flexed under the cephalothorax. This body shape puts a crab's center of gravity over its legs; well-balanced and unhindered by a dragging abdomen, many crabs are agile walkers. *Uca*, the fiddler crab, burrows in intertidal mud until the tide goes out, then emerges and scavenges for dead organisms and organic debris (Figure 26.7c). The male has an oversized cheliped that he waves or thumps as a lure to draw the female into his burrow. *Pagurus*, the hermit crab, lives in empty gastropod shells, periodically moving to a larger shell as it outgrows the old one (Figure 26.7d). Commercially important edible crabs include *Callinectes sapidus*, the Atlantic blue crab (Figure 26.7e), and *Cancer magister*, the Dungeness crab of the Pacific Coast. Also commercially important are many marine shrimps, including those of the genera *Penaeus* and *Pandalus* (Figure 26.7f).

Subphylum Uniramia

Uniramians are terrestrial or aquatic animals with uniramous (one-branched) appendages. The head appendages consist of one pair of antennae, one pair of mandibles, and one or two pairs of maxillae. The mandibles and maxillae are mouthparts. With nearly a million species of insects alone, the Uniramia is by far the largest subphylum in the animal kingdom. Insects constitute one of five classes in the subphylum (see Classification, p. 646). Two other well-known classes are the Diplopoda (millipedes, about 7500 species) and Chilopoda (centipedes, about 3000 species). We have not included a discussion of two small classes, the **Pauropoda** (about 120 species) and the **Symphyla** (about 400 species). Pauropods and symphylans are small (less than 10 mm) scavengers and predators of moist soil (see Classification section).

Class Diplopoda. The body of a **millipede** is essentially a multisegmented, many-legged cylinder ranging in length from about 2 mm to nearly 30 cm, depending on the species. The class name Diplopoda refers to the presence of two pairs of legs per **diplosegment** (Figure 26.8a). (Each diplosegment is derived from two original body segments.) With many of its numerous legs pushing at the same time, a millipede has the power to force its way slowly through forest litter and loose, rich soils as it feeds on decaying leaves and other plant matter. Although slow-moving and nonaggressive,

(a)

(b)

Figure 26.8 Subphylum Uniramia: Millipedes and centipedes. (a) A millipede (*Sigmoria aberrans*). The first segment behind the head is a limbless, collarlike structure called the collum, which, with the next three leg-bearing segments, forms the thorax. The remaining segments are fused in pairs called diplosegments (double segments), each of which bears two pairs of jointed legs. (b) A centipede's body segments are not fused into diplosegments; thus, there is just one pair of legs per segment. This species, seen here with a cockroach, is *Scutigera coleoptrata*, the house centipede.

millipedes are not defenseless. They usually roll into a tight protective spiral when disturbed, and in addition, many species produce an irritating fluid that large individuals can spray nearly 1 m. A bird hit in the eye with this secretion may become blind.

Class Chilopoda. **Centipedes** are flatter than millipedes and have one pair of legs per segment, all of which are single (Figure 26.8b). They are comparatively fast-moving carnivores, crawling over humid ground at night or through damp habitats under stones, logs, and forest litter, feeding on insects, earthworms, and snails. Large tropical species may even eat small toads and lizards. Large poison claws on the first body segment immobilize prey and transfer it to the mandibles for chewing. Many human dwellings shelter the house centipede, *Scutigera coleoptrata*, an exceptionally long-legged, fast-moving hunter. This centipede helps to control household insects and is harmless to humans.

Class Insecta. Most insects are less than 60 mm in length. Among the largest are certain beetles about the size of your fist and a species of walking stick from Borneo that can be as much as 33 cm long. At the other extreme are certain beetles less than 0.2 mm in length. Insects live in virtually every imaginable land habitat, from the tropics to the Arctic. Relatively few (some 20,000 species) are aquatic, yet all types of lakes, streams, and ponds support abundant insect populations, and some insects thrive in shallow marine habitats. Only the open ocean and deep-sea zones have been little exploited, reflecting the insects' terrestrial origin and the domination of these marine habitats by crustaceans. Insect success stems from a unique combination of factors. (See the Essay on p. 626.)

Despite their great diversity, insects share many fundamental features. The insect body consists of a **head**, **thorax**, and **abdomen** (Figure 26.9). The head is a protective capsule and a sensory and feeding center; it contains the cerebral ganglion (brain) and typically bears compound eyes, several simple eyes called **ocelli**, one pair of antennae, and the mouthparts. The thorax is an insect's locomotor center. Most adult insects have three pairs of legs and one or two pairs of wings, all borne on a three-segmented thorax. (Insects are the only invertebrates with wings.) Reflecting the diversity among insects, legs and wings have been adapted for a remarkable variety of functions. The abdomen consists of the last eleven body segments, although in many species some segments are greatly reduced or fused to one another. The abdomen contains the bulk of the internal organs, including the heart, excretory tubules, reproductive organs, and much of the digestive tract. Abdominal appendages are limited to highly modified terminal structures used in copulation and egg laying. In many species, the last abdominal segment bears a pair of mechanosensory structures called **cerci**.

Depending on the authority, the Class Insecta is divided into 25 to 30 orders, mostly on the basis of wing and mouthpart structure. The most prominent orders are described briefly in the remainder of this section.

Order Collembola. **Springtails** are wingless insects seldom more than 5 mm long (Figure 26.10a). Their wingless condition is considered primitive (in contrast to the fleas, for example, whose winglessness was derived from winged ancestors). Most springtails have a forked springing organ, held under the abdomen until its sudden release flings the springtail into the air for distances up to 7 or 8 cm. Springtails may be extremely abundant; one researcher estimated that an acre of meadowland had nearly 230 million individuals!

Order Ephemeroptera. **Mayflies** have four membranous wings, with hind wings much smaller than fore wings (Figure 26.10b). Immature stages live in streams, rivers, ponds, and lakes, where they

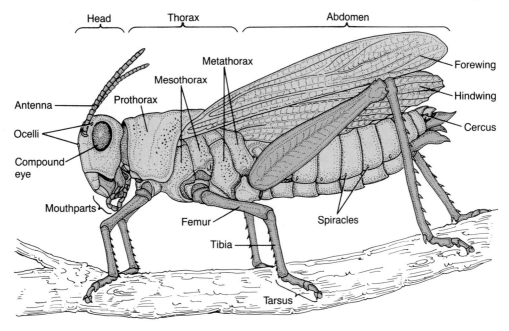

Figure 26.9 Class Insecta: External anatomy of a female grasshopper.

feed on vegetable matter. After a year or more of development underwater, mayfly larvae rise to the surface, molt, fly to the shore, rest for a time (often overnight) on vegetation, and then molt again to become adults. (Mayflies are the only insects to molt after their wings have become functional.) The adults lack a gut and do not feed; they mate and then die, usually within a day or two. Enormous numbers of mayflies may leave the water almost simultaneously, sometimes becoming a major nuisance as they accumulate along shorelines and on nearby highways (see Figure 14.16).

Order Odonata. **Dragonflies** and **damselflies** have long, narrow, membranous fore wings and hind wings; the wings are of similar size, and each has a rich network of veins (Figure 26.10c). Immature dragonflies and damselflies are aquatic and carnivorous, eating mainly small invertebrates, but larger species sometimes eat tadpoles and small fishes. Adult dragonflies are strong fliers, seizing and eating other insects on the wing. Damselflies are more slender than dragonflies and, unlike dragonflies, can fold their wings when at rest.

Order Orthoptera. **Orthopterans** (grasshoppers, locusts, cockroaches, crickets, katydids, praying mantises, and walking sticks) have membranous hind wings folded under straight, leathery fore wings (Figure 26.10d). In some species, wings are reduced or absent. Orthopterans are generally herbivorous (as in grasshoppers) or omnivorous (as in cockroaches), and their mouthparts are adapted for chewing. Praying mantises, however, are carnivorous, using their front legs to grip prey.

Order Isoptera. **Termites** are sometimes called white ants because after shedding their small, membranous wings at maturity, they superficially resemble wingless ants (Figure 26.10e). Termites differ from ants in having a wide connection between the abdomen and thorax, rather than a narrow "wasp waist." Depending on the species, termites live either in damp subterranean tubes and nests that may protrude above ground or in dry habitats above ground, for example, in stumps, posts, and wooden buildings. Termites eat wood, but depend on mutualistic protozoa in their digestive tract to digest the cellulose for them. Termites are social insects, living in highly organized colonies with several kinds of individuals, or castes, maintained by pheromones. Each colony has many sterile workers and soldiers, a single fertile queen (a large egg-laying machine), and fertile males (kings, if they leave the colony, mate with a virgin female, and establish a new colony).

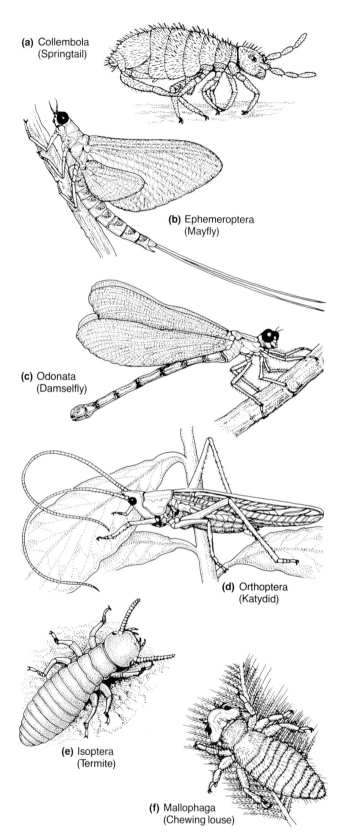

(a) Collembola (Springtail)

(b) Ephemeroptera (Mayfly)

(c) Odonata (Damselfly)

(d) Orthoptera (Katydid)

(e) Isoptera (Termite)

(f) Mallophaga (Chewing louse)

Figure 26.10 Subphylum Uniramia. Representatives of major insect orders.

Order Mallophaga. **Chewing lice** have mouthparts adapted for biting and chewing and are ectoparasites of birds and mammals (Figure 26.10f). They fasten to feathers or hair with legs adapted for clinging, and they eat particles of hair or feathers and bits of skin scraped from the host. None are known to attack humans, but they are sometimes serious pests of poultry and domestic mammals.

Order Anoplura. **Sucking lice** have piercing, sucking mouthparts and legs adapted for clinging to hair (Figure 26.10g). All members of this order are bloodsucking ectoparasites of mammals. Two species attack humans. *Pediculus humanus*, 2.5–3.5 mm in length as adults, consists of two varieties, the head louse and the body louse. *Phthirus pubis*, the crab louse, occurs mainly in the pubic area, but may also infest other hairy regions of the body. Adult crab lice are 1.5–2.0 mm in length.

Order Thysanoptera. **Thrips** are mostly small insects, about 0.5–5 mm long (Figure 26.10h). When present, the wings are very narrow and fringed with long, hairlike setae. Thrips are primarily plant feeders, sucking fluids from leaves, flowers, and other plant parts. If you look closely into the head of a dandelion or daisy, you will almost certainly see several thrips. They may be very abundant, sometimes becoming serious pests of cultivated plants and in some cases transmitting plant disease.

Order Hemiptera. Often called the **true bugs**, hemipterans include stinkbugs, bedbugs, plant bugs, back swimmers, water boatmen, water striders, and many others (Figure 26.10i). Their hind wings and the distal half of their fore wings are membranous, but the proximal half of the forewings is leathery (hence the name Hemiptera, meaning "half wing"). Some species, such as the bedbug, use piercing and sucking mouthparts to suck blood, while many other species ingest plant sap. Water striders walk on water by taking advantage of surface tension. Water boatmen and back swimmers swim by means of enlarged, oarlike hind legs and can fly when out of the water.

Order Homoptera. **Aphids** (plant lice), **leafhoppers, scale insects**, and other homopterans feed exclusively on plants, using piercing, sucking mouthparts (Figure 26.10j). Feeding by large numbers of individuals may cause severe injury to plants, and in addition, some homopterans transmit plant disease as they insert their mouthparts. It is not difficult to distinguish between homopterans and hemipterans. The fore wings of homopterans have the same texture throughout, and the wings

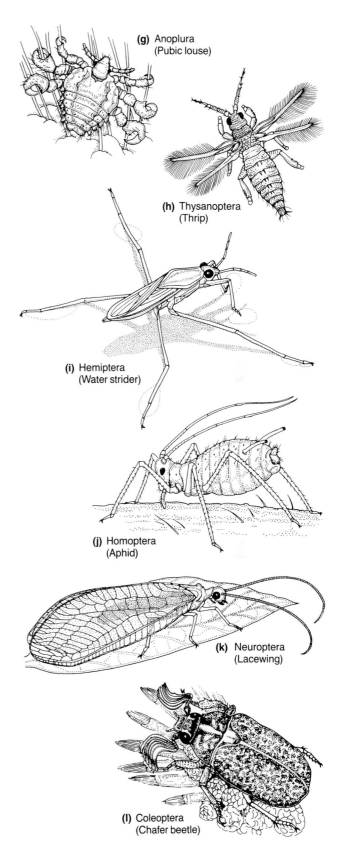

(g) Anoplura (Pubic louse)

(h) Thysanoptera (Thrip)

(i) Hemiptera (Water strider)

(j) Homoptera (Aphid)

(k) Neuroptera (Lacewing)

(l) Coleoptera (Chafer beetle)

Figure 26.10 *(Continued)*

of homopterans are held rooflike over the abdomen when at rest, whereas those of the true bugs lie flat.

Order Neuroptera. Lacewings are rather delicate, slender-bodied, weak fliers (Figure 26.10k). The front and hind wings, which are about the same size, are extensively veined, giving them a netlike appearance, and they are held rooflike over the body when at rest. The green lacewings *(Chrysopa)* are quite beautiful, with shining gold eyes and iridescent green bodies and wing veins. Because lacewing larvae are voracious predators of aphids, they are of considerable value to gardeners.

Order Coleoptera. Beetles constitute the largest order in the animal kingdom, with about 300,000 known species worldwide and some 30,000 species in the United States (Figure 26.10l). Varying in length from less than 1 mm to about 12 cm, beetles are readily recognized by their fore wings, which are hardened and thickened as protective covers, or **elytra**, for the membranous, locomotor hind wings. Beetles occur almost everywhere, from high mountains to the seashore, and their habits are extremely varied. Forests, streams, ponds, soil, dung, carrion, and plant material all support their quota of species. Most beetles are either carnivorous or herbivorous; a few species eat fungi, and some are scavengers. There is little exaggeration in saying that virtually anything organic is eaten by some kind of beetle.

Order Trichoptera. Caddis flies have hairy wings and bodies (Figure 26.10m). Their caterpillar-like larvae are aquatic, and depending on the species, catch waterborne particulate food in delicate nets, are carnivorous, or feed on aquatic plants and organic debris. The larvae of plant-feeding species build remarkable silk-lined cases within which they live. As silk is ejected from a spinneret on the floor of the larva's mouth, the tubular case is strengthened by the addition of bits of leaves, twigs, sand particles, or even tiny snail shells. Each species builds a characteristic case by selecting preferred materials, just as a bird does when it builds a nest.

Order Lepidoptera. Moths and **butterflies** have two pairs of membranous wings, the hind pair being smaller (Figure 26.10n). Typically, the wings and body are covered by scales (the dust you find on your fingers after you have held a butterfly). The mouthparts of adult moths and butterflies form a drinking tube that is coiled under the head when not in use. When the insect drinks nectar, the tube is uncoiled and extended deep into a flower.

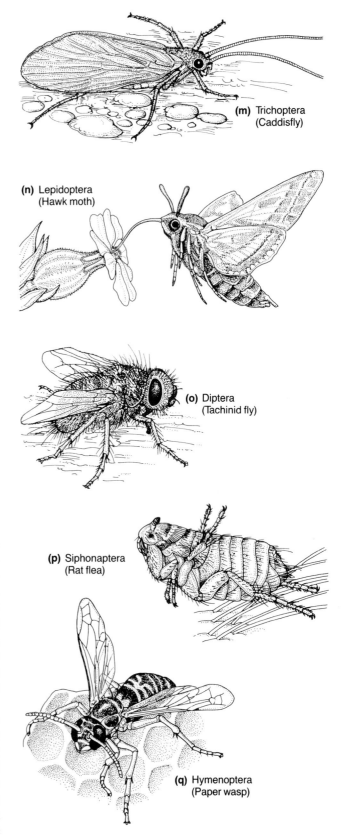

(m) Trichoptera
(Caddisfly)

(n) Lepidoptera
(Hawk moth)

(o) Diptera
(Tachinid fly)

(p) Siphonaptera
(Rat flea)

(q) Hymenoptera
(Paper wasp)

Figure 26.10 *(Continued)*

Larval moths and butterflies, called caterpillars, are mostly plant feeders, eating leaves or boring in stems, fruits, or other plant parts. As they feed and develop, caterpillars commonly make extensive use of silk. Tent caterpillars, for example, spin a massive shelter that may envelop an entire branch of a tree; and leaf rollers roll a single leaf into a tubular feeding shelter held together with silk strands. Many caterpillars spin silken cocoons within which they transform into adults.

Order Diptera. Dipterans, or **true flies,** include houseflies, fruit flies, mosquitoes, and horseflies, to name a few (Figure 26.10o). Unlike most flying insects, flies have only one pair of wings; their hind wings have evolved into small, club-shaped balancing organs called **halteres** that help keep flies on an even keel when in flight. The mouthparts are adapted for piercing and sucking, as in blackflies and mosquitoes, or for sponging, as in fruit flies and houseflies. Most adult dipterans feed on nectar, but many species prey on other insects, and many are bloodsucking. Mosquitoes, tsetse flies (of Africa), and other bloodsuckers may transmit diseases, such as malaria or African sleeping sickness, to their vertebrate hosts. Dipteran larvae live in a great variety of habitats. The larvae of mosquitoes and blackflies are aquatic; other larvae live within plant tissue, often as leaf miners, root borers, or stem borers.

Order Siphonaptera. **Fleas** are small, wingless insects with piercing and sucking mouthparts used to draw blood from birds and mammals (Figure 26.10p). Larval fleas have chewing mouthparts and typically feed on organic particles in the host's lair or nest. Adult fleas are good jumpers, their long legs giving great leverage; the human flea *Pulex irritans* can leap vertically more than 15 cm. Thus, fleas can readily jump from the substrate to a host or from one host to another. Movement between a host's hairs or feathers is facilitated by extreme lateral flattening of a flea's body. Aside from the irritation caused by their bites, fleas transmit a number of diseases, including the plague, or black death; and several may serve as intermediate hosts of tapeworms that can infect humans.

Order Hymenoptera. **Bees, wasps, ants,** and the many other members of this order have two pairs of membranous wings (Figure 26.10q). The leading edges of the relatively small hind wings are fastened with tiny hooks to the trailing edges of the larger fore wings; this improves flight efficiency by making the wings behave as single functional units. In most hymenopterans, the thorax and abdomen are separated by a narrow waist, which together with the four membranous wings, makes it easy to identify an insect as a member of this order. Many hymenopterans display complex behavior, including social organization. The order includes many predators and parasites of other insects, and certain hymenopterans, such as honeybees, are vitally important plant pollinators.

HABITATS AND ENVIRONMENTAL RELATIONSHIPS

Arthropods inhabit every continent, and of the many factors affecting their distribution, the most important are temperature, moisture, barriers to dispersal, and availability of suitable habitats or hosts. Physical barriers, such as mountain ranges and oceans, impede the spread of many arthropods, but such obstacles may not be a problem for some winged insects. The effect of habitat availability is clearly seen in a number of cases. Dog ticks occur only in grassy or brushy regions frequented by certain mammals. The silk moth *(Bombyx)* is limited by the distribution of mulberry trees on which its caterpillars depend for food and shelter. Human lice may occur virtually everywhere humans live.

Effects of Temperature

Most arthropods can live and multiply only within a fairly narrow temperature range. In most species body temperature fluctuates with changes in environmental temperature. As with other ectotherms, the rates of metabolic reactions and other temperature-dependent life processes typically decrease by about half with a 10°C drop in temperature or double with a 10°C rise. (For a discussion of ectothermy ◁ and other concepts in thermoregulation, see the Chapter 3 Essay, p. 73.) Like many other animals, most arthropods are active in temperature extremes for only short periods and soon die at temperatures above 35°–40°C and below 0°C. Exceptions include *Grylloblatta campodeiformis,* a small, wingless insect that lives at the edges of glaciers; this species is active between −3°C and 12°C. The firebrat *Thermobia domestica* is thermophilic; it is active between 12°C and 50°C and often occurs near bakery ovens and in steam tunnels.

Many arthropods have physiological or behavioral adaptations for surviving temperature extremes. Many temperate-zone insects avoid freezing by reducing the water content in their tissues as winter approaches and, in some cases, by producing glycerol or other substances that act as natural antifreezes. In addition, many arthropods can acclimatize to temperature; that is, the optimum temperature range of the species is not fixed, but may be shifted a few degrees by a period of "conditioning," or acclimatization. Thus, along the eastern seaboard of the United States, southern populations of some marine crustaceans have higher temperature optima than northern populations of the same species living in colder water. Comparisons of animals at the same latitude often show that a species has a higher temperature optimum in summer than in winter, again reflecting acclimatization. Acclimatization makes it possible for an arthropod to live at temperatures that would

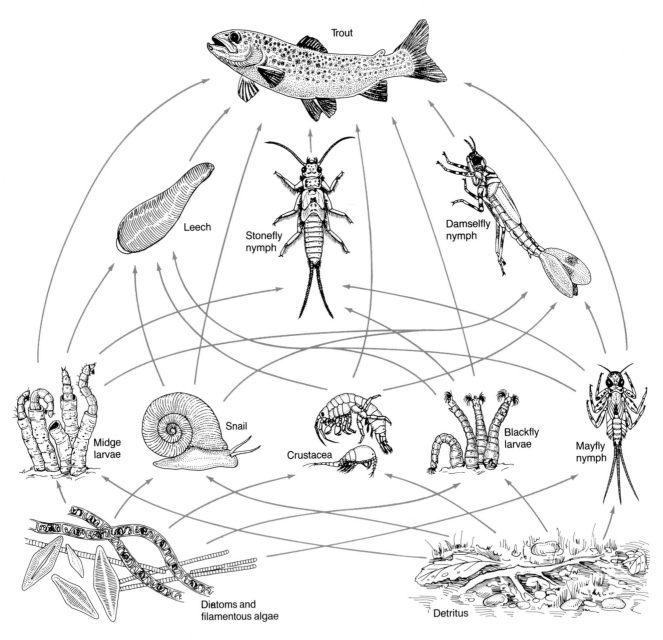

Figure 26.11 Feeding relationships in a freshwater community. Insects and crustaceans are among the most numerous organisms in most freshwater ponds, lakes, and streams. With other invertebrates, they form vital links between fishes and organic debris (detritus) and the solar energy–trapping algae.

otherwise be unsuitable or even lethal, and increases the likelihood that a species will be able to extend its geographical distribution.

Effects of Moisture

Water obviously governs the distribution of aquatic arthropods, and it also affects the distribution of terrestrial forms. Excessive soil moisture may limit the distribution of earth-dwelling arachnids and insects, because as air breathers, they cannot respire in water-saturated soil. Conversely, excessively dry soil may desiccate and eventually kill pill bugs, millipedes, centipedes, and those insects that lack a waterproofed (waxy) body surface.

Most crustaceans that have invaded land have behavioral adaptations that conserve water. Many crabs burrow into damp sand for protection against desiccation and predation. Many terrestrial crustaceans also have a peak of activity between dusk and dawn, taking advantage of lower temperatures and higher humidity.

Arthropods in Food Chains

Arthropods are significant components of food chains in virtually all environments. Chelicerates and insects are especially important on land; it is likely that without them as a food source, many species of terrestrial vertebrates could not survive. Likewise, insects and crustaceans are vital components of most freshwater food chains (Figure 26.11). Chelicerates and crustaceans are especially important as food in marine environments. Many smaller planktonic crustaceans, such as copepods and the shrimplike euphausiids (commonly called krill), form a vital link between planktonic algae and large carnivores in the sea. Copepods and krill eat drifting algae (phytoplankton) and are in turn consumed by other invertebrates, many fishes, and marine mammals. (See the Chapter 4 Essay on whales, p. 82.)

COMPARATIVE STRUCTURE AND FUNCTION

Support and Protection

The arthropod body wall, typically an armorlike, multilayered **exoskeleton**, or **cuticle**, affects almost all aspects of arthropod biology, including support, protection, locomotion, growth, and development.

We discussed the structure and function of the cuticle in considerable detail in Chapter 8 (see p. 179 and Figures 8.7 and 8.8). When first secreted by the underlying epidermis, new cuticle is soft and white, but part of it is soon strengthened by a hardening and darkening process called **sclerotization**. Sclerotization produces the brownish black body color typical of many arthropods. However, in some cases, sclerotization leaves the cuticle relatively clear, allowing pigments in the epidermal cells to show through as brilliant body colors and intricate markings. The black and yellow stripes of wasps and hornets result from bands of dark cuticle alternating with clear cuticle overlying epidermal cells containing a crystalline pigment called yellow pterine. Body colors also result from thin layers in the cuticle or from fine striations in its surface. These separate light into its component wavelengths, producing a rainbowlike array of colors that shimmer and shift in hue and pattern, like those of an oil film. Known as iridescence, this phenomenon gives many otherwise dull brown arthropods a colorful sheen. Color may be an important protective adaptation (see Figures 13.15 and 13.16).

Molting. Sclerotized cuticle is virtually inelastic. For an arthropod to grow, it must periodically shed (molt) part or all of its cuticle and secrete a larger one (Figure 26.12). Most arthropods molt and form new cuticle a specific number of times. For example, a housefly molts four times and a European corn borer moth molts six times, including the molt to the adult stage. Certain other arthropods,

Figure 26.12 A blue crab *(Callinectes sapidus)* molting. At genetically and physiologically determined intervals during development, the cuticle fractures along the middorsal line and the animal crawls free. Forces exerted by tissues and organs that had been compressed within the confines of the old exoskeleton cause the soft, flexible new cuticle to expand. Expansion is aided in many arthropods by the swallowing of air or water and by muscle contractions. Certain crabs double the osmotic concentration of their blood just before the molt. When the crab molts, osmotic uptake of water increases body volume by up to 40% and stretches the soft new cuticle before it hardens.

Essay: The Insects' Formula for Success

Because about 76% of all animal species are insects, it is only natural to ask what factors have made insects so successful, or, more precisely, how have insects been able to exploit so many niches and habitats?

A combination of factors accounts for insect success. One factor may be body size. A small body allows insects to occupy many habitats that would be inaccessible to larger animals. Examples include the leaf-mining insects that feed between the upper and lower surfaces of leaves and the parasitic insects that pass their immature stages within the eggs of other insects. Small size may also facilitate travel from one place to another. Fleas, for example, can travel on a dog or cat, moving with ease between the hairs, and tiny gnats can pass through a window screen.

Ability to fly is perhaps the most significant way in which insects differ from other arthropods, and many zoologists consider it the chief element in insect success. Food, habitats, and members of the opposite sex are seldom distributed uniformly in an environment, and flying back and forth over a considerable area increases the probability of finding them. Flight also increases the chances of escaping from enemies. In addition it allows a population to disperse more widely, broadening the species' distribution, and possibly reducing competition for food and space. Flight also allows long-distance migrations; populations of desert locusts, for example, may fly hundreds of miles to a fresh feeding area, and monarch butterflies fly south from Ontario and New England to Mexico as winter approaches.

A tough exoskeleton is a third major factor in insect success. It forms mouthparts that can pierce, cut, and grind a great variety of natural materials, including wood and even some metals. Shaped as arches, rings, and hollow cylinders, the exoskeleton gives armorlike protection and support for the body and its musculature. It makes the rough-and-tumble of insect life possible; mighty leaps, crash landings, heavy blows, occasional burial, and the stresses of flight seldom cause serious injury. In addition, most insects are waterproofed by a waxy layer on the surface of their exoskeleton. Without waterproofing, an insect's high ratio of body surface area to volume would result in very rapid water loss in dry air. Grasshoppers in a sweltering hot summer wheat field would quickly desiccate and die were it not for the waxy layer.

Metamorphosis, a fourth element in the insect success formula, allows most insects to have the best of two worlds as they grow and reproduce. A caterpillar on its food plant is little more than an eating machine, storing energy. Eventually, it metamorphoses to the adult butterfly, a winged form well suited for reproduction and dispersal of the species. The butterfly may sip a little nectar, but most of the energy needed for egg production and flight comes from energy stored by the caterpillar. Other insects, such as mosquitoes and dragonflies, exploit aquatic habitats as larvae and terrestrial habitats as adults.

A fifth element of success is rapid multiplication provided by a short life cycle and great reproductive potential. This allows many species to exploit briefly available food and other resources. Some mosquitoes complete a life cycle in temporary rain pools, and certain flies complete an entire generation in fresh cow manure before it becomes hard and dry. *Drosophila*, the fruit flies of genetic research, produce 25 generations per year under ideal conditions. Females lay about 100 eggs each, half developing into males and half into females. If the flies were allowed to reproduce for one year, with each female laying 100 eggs before she died, the number of flies produced by the end of the twenty-fifth generation would be about 10^{41}. If this astronomical number of flies were pressed together into a ball, with 1000 flies per cubic inch, it would extend from the earth to the sun!

Lastly, insects as a terrestrial group are dependent on land plants for food. Many insect species have coevolved with land plants, especially the flowering plants, or angiosperms. Most angiosperms depend on insects for pollination, and many insect species rely on nectar and pollen for food. As plants and insects have evolved adaptations that promote this mutually beneficial relationship, plant diversity has contributed to insect diversity, and vice versa.

such as the lobster, grow and molt at intervals throughout their lives. An arthropod is never without an exoskeleton; formation of the new cuticle precedes loss of the old one. Details of the molting process can be found in Chapter 8 (see Figure 8.9).

Locomotion

Arthropods have remarkable locomotor versatility; many species crawl, walk, swim, or fly, and some also run, jump, dive, or glide. This versatility is possible because of several important features of their limbs. First, limbs constructed of hard exoskeleton exploit the principles of leverage. An arthropod's leg is a lever whose free end moves a relatively great distance when relatively small muscular contractions move the pivotal end. (How limbs act as levers is discussed in Chapter 9, p. 211; Figure 9.15.) By shifting the position of the fulcrum, evolutionary forces have produced some limbs adapted for rapid locomotion, while other limbs are suited for the slower, more powerful movements of burrowing, lifting, or holding.

A second important feature of arthropod limbs is that the concentration of locomotor forces in the legs allows the animal to maintain continuous forward movement. This is in contrast to the peristaltic locomotion of an earthworm, in which each segment must come to a complete stop before moving again. A worm expends much energy overcoming inertia as it repeatedly accelerates and decelerates its entire body. The arthropod, once moving, need only accelerate and decelerate its limbs, a small mass compared to the whole body.

A third important feature is the jointed nature of the arthropod limb. Arthropod legs are divided into five or more hard parts linked to each other by flexible joints with hingelike articulations that usually restrict movement to a single plane. However, successive articulations along a leg are often set at angles to each other, an arrangement that allows the leg to undergo movements almost as freely as would a single ball-and-socket joint (Figure 26.13).

Arthropods differ greatly in the number of walking legs they have; insects have 6, spiders 8, decapod crustaceans 10, pill bugs 14, and centipedes and millipedes from 30 up to about 300. Whether an animal has 6 legs or 60, precise timing is required to keep the legs from tangling as each goes through its cycle of movement. Nervous coordination of the legs produces a pattern of action in which the legs step in orderly sequence, beginning with the posterior ones. A wave of leg action thus proceeds along the body from back to front in what is called a **metachronal rhythm**. This is easily seen in a slow-walking millipede, but leg movement is just a blur to an observer watching a ghost crab

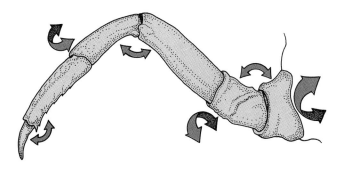

Figure 26.13 Movement of a crustacean walking leg. There are six or seven articulations along a typical crustacean walking leg, including the one with the body wall. Note that the leg joints have different planes of movement; consequently, the leg as a whole can move with great freedom.

scrambling along a beach at 160 cm/s (3.6 mph) or a cockroach traveling at 80 cm/s. (At this speed, a cockroach's leg takes almost 25 steps per second.) High-speed cinematography and frame-by-frame examination of the film make it possible to analyze the walking patterns of running insects (see Figure 9.17a). Such studies show that the basic metachronal rhythm is present even when the animal is walking or running at full speed.

Crawling—that is, locomotion with the body in contact with the substrate—is common among arthropods and is the only method available for legless larvae. Fly larvae (maggots) twist and turn but rely heavily on peristaltic locomotion as they crawl through decaying food. Horseshoe crabs use their legs and telson to crawl along the seafloor in search of worms or other food.

Swimming is an important means of locomotion in many aquatic arthropods, especially crustaceans. The earliest crustaceans were probably swimmers, with broad, biramous appendages adapted for paddling. Among living crustaceans, many swimmers are propelled by biramous abdominal appendages called **pleopods** or **swimmerets**; shrimps are highly adapted for swimming. For large-bodied arthropods with heavily muscled limbs, walking and crawling are more efficient than swimming; nevertheless, many large crustaceans can swim. Lobsters and crayfish generally walk on the bottom, but can use their flipperlike posterior appendages called **uropods** to dart backward out of harm's way; rapid flexing of the abdomen moves the uropods. Crabs are also mainly walkers, but some, such as the blue crab (*Callinectes*; see Figure 26.7e), can swim well enough to catch small fishes.

Many insects are accomplished swimmers. Some use their legs as oars (Figure 26.14). The surface area of the flattened legs is increased by long hairs (setae) that flare widely during the power stroke and then flatten against the leg during the recovery stroke. Other insects swim by undulations of the body; mosquito larvae, often called wrigglers (see Figure 26.21c), are a familiar example. Some insects even use jet propulsion. Dragonfly larvae draw water through their anus to ventilate rectal gills, but when startled can expel the water with enough force to jet their bodies forward several centimeters.

Insects fly by flapping their wings. Unlike the wings of birds and bats, which are modified limbs, insect wings are lateral expansions of the body wall. Thus, insects evolved their flight ability without converting any walking legs to wings. With the exception of the true flies (Diptera), which have

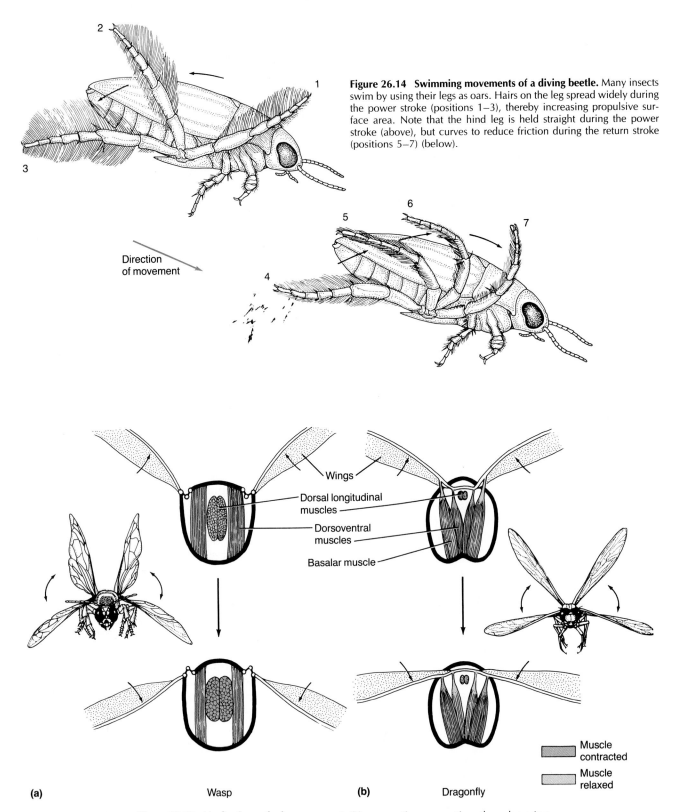

Figure 26.14 Swimming movements of a diving beetle. Many insects swim by using their legs as oars. Hairs on the leg spread widely during the power stroke (positions 1–3), thereby increasing propulsive surface area. Note that the hind leg is held straight during the power stroke (above), but curves to reduce friction during the return stroke (positions 5–7) (below).

Direction of movement

Wings

Dorsal longitudinal muscles

Dorsoventral muscles

Basalar muscle

Muscle contracted

Muscle relaxed

(a)   Wasp

(b)   Dragonfly

Figure 26.15 Mechanisms of wing movement. Diagrammatic cross sections through a wing-bearing segment of a wasp, an insect with a very high wingbeat rate, and that of a dragonfly. (a) A wasp's wings are moved by changes in the shape of the thorax. Note that the indirect flight muscles are attached to the exoskeleton of the thorax, not to the wings. (b) Dragonflies, which move their wings much more slowly, have direct flight muscles inserted at the wing base. In (a), the forewings and the hindwings are shown in the raised or lowered positions; in (b), only one set of wings is being followed up and down.

two wings, winged insects generally have four wings, a pair on each of the second and third thoracic segments.

Insect wings are complex levers that pivot on a fulcrum formed by the lateral thoracic body wall. Insects with high wingbeat rates move their wings by using **indirect flight muscles** to distort the thorax (Figure 26.15a). Once in motion, the muscle-thorax-wing system oscillates very rapidly, with only occasional nerve impulses needed to keep it operating. Oscillations have been recorded of up to 200 to 300 beats per second in many common flies, bees, and wasps and over 1000 beats per second in the tiny fungus gnat *Forcipomyia*. The wings of insects with relatively low wingbeat rates, including those of grasshoppers, dragonflies, and moths, are moved by **direct flight muscles**, and the wingbeat rate is synchronized with arriving nerve impulses (Figure 26.15b). Insect flight speeds are extremely variable. Mosquitoes average about 3 km/hr, honeybees about 22 km/hr, and dragonflies about 30 km/hr.

Feeding and Digestion

Much arthropod diversity is reflected in the variety of feeding methods and diets in the phylum. Among the Chelicerata, mites have the most varied food habits. Many species are herbivorous, and some mites cause severe damage to agricultural crops. Carnivorous mites eat a variety of small organisms, including crustaceans, nematodes, and insect eggs. Scavenging mites feed on a wide range of plant and animal materials. Parasitic mites include species infesting the gills of aquatic insects, crustaceans, and fishes. Some, such as chiggers, are parasitic only as larvae (see Figure 26.3i). Others, such as the mange mites, cause serious skin disease in humans and domestic animals, and controlling them on livestock is often costly (Figure 26.16). Compared to mites, ticks are quite uniform in their diets. They are primarily bloodsuckers, and as a group, infest nearly every kind of terrestrial vertebrate.

Spiders are liquid feeders that partially digest the body contents of their prey before ingesting it. Their chelicerae, modified as poison fangs, inject a paralyzing toxin into insects and other small invertebrate prey. Enzymes from a spider's mouth then soften and partially digest the paralyzed victim. The spider gut has a pharyngeal pump and a sucking stomach that ingest liquified food and pass it to the intestine and paired ceca (Figure 26.17). Products of digestion are absorbed by cells along the gut, and excess food materials may be stored in cells associated with the ceca. The ability to store

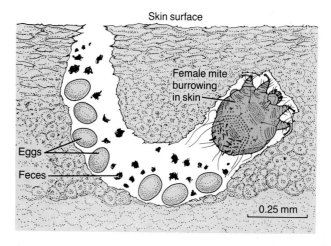

Figure 26.16 The sarcoptic mange mite, *Sarcoptes scabiei*. This species burrows and lays eggs in the skin of humans and domestic animals, causing severe itching and a condition called mange. This drawing shows a mite and its eggs in a skin burrow.

food allows many spiders to go for long periods between meals, and it also provides a stock of raw materials for silk production. Spiders are unique among arthropods in their use of both venom and silken webs for prey capture.

Many arthropods, especially certain crustaceans, are filter feeders. Planktonic copepods, for example, collect suspended algae and small zooplankton. From studies performed in the 1920s, it was assumed until recently that copepods used their feeding appendages to passively sieve, or strain, suspended food from the water flowing over them. Then, in the early 1980s, researchers using high-speed microcinematography discovered that copepods are active, rather than passive, filter feeders. Instead of acting as sieves, a copepod's feeding appendages act as paddles and food graspers. Some of the appendages drive water around the animal, creating feeding currents (Figure 26.18). At the same time, sensory receptors on the appendages sample the water for food, and a copepod's second maxillae actively capture suitable suspended material with small volumes of water. The first maxillae then push the food into the mouth.

Actually, because of the viscosity of water, appendages with setae as tiny as those on a copepod are coated with a thick, immobile layer of water, and it is impossible for them to act as particle strainers. Thus, it is likely that not only copepods, but many other small, filter-feeding invertebrates, capture particles actively, rather than passively straining them.

In contrast to filter feeders, many crustaceans seize relatively large, active prey or are scavengers.

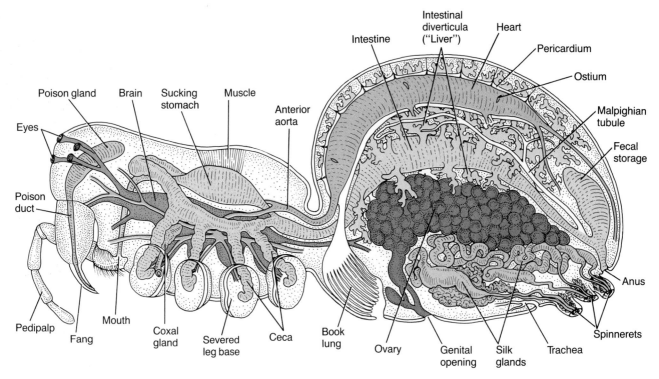

Figure 26.17 Internal anatomy of a spider. This cutaway diagram shows the main internal organs as viewed from the left side.

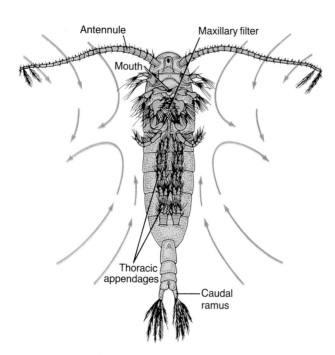

Figure 26.18 Filter feeding by a copepod. This ventral view of a planktonic copepod shows the feeding currents (arrows) produced by the appendages on the head and thorax. Note that the appendages bear many setae. The maxillary filter (first and second maxillae) actively captures suspended food particles from the water current, and pushes them into the mouth. The abdomen, narrower than the rest of the body, has only one pair of appendages, called caudal rami.

Some of their appendages are specialized for capturing, holding, and shredding animals. The large, claw-bearing legs (chelipeds) of crayfish, crabs, and lobsters are familiar examples of this specialization. Fragments of food are transferred to the mandibles, which crush and soften the food before it is swallowed.

Digestive Systems. The crustacean gut, typical of arthropods, is divided into a foregut, a midgut, and a hindgut (Figure 26.19). The foregut of larger crustaceans is often very complex, with specialized regions for grinding and straining food, which may have been only coarsely broken by the mandibles. Large malacostracans, such as crayfish, crabs, and lobsters, have a **cardiac stomach**, or **gastric mill**, where food is milled by cuticular teeth; a **pyloric stomach** strains the resulting particles between rows of fine setae (Figure 26.19b). Small particles from the pyloric stomach pass into midgut ceca. In malacostracans, ceca branching from the anterior end of the midgut form a large digestive gland, sometimes called the hepatopancreas or liver. Final digestion of food and absorption of nutrients occur in the digestive gland, while the intestine and anus serve primarily to discharge undigested matter.

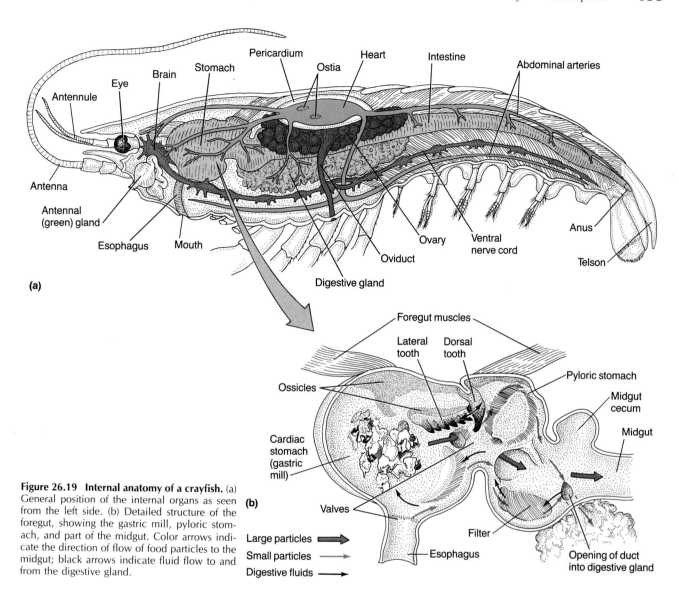

Figure 26.19 Internal anatomy of a crayfish. (a) General position of the internal organs as seen from the left side. (b) Detailed structure of the foregut, showing the gastric mill, pyloric stomach, and part of the midgut. Color arrows indicate the direction of flow of food particles to the midgut; black arrows indicate fluid flow to and from the digestive gland.

Large particles ➡
Small particles →
Digestive fluids →

The insect digestive system is described in Chapter 4 (see p. 89 and Figure 4.12). With their great diversity of feeding mechanisms, insects exploit an almost endless variety of foods. The chewing mouthparts of grasshoppers, carnivorous beetles, cockroaches, and thousands of other insects include stubby, leglike maxillae that hold the food while powerful, toothed mandibles cut and grind it (Figure 26.20a).

Insects ingest liquids in several ways. The piercing/sucking mouthparts of mosquitoes and aphids form needle-sharp stylets that pierce tissues, and channels through which host fluids are pumped into the gut (Figure 26.20b). The sponging mouthparts of houseflies soak up liquids from various surfaces (Figure 26.20c). Certain blood-feeding flies, such as blackflies and stable flies, combine piercing and sponging; they have stylets that puncture the host's skin, and blood from the wound is sponged up. Most bloodsuckers secrete anticoagulants into the wound to prevent clotting. The wound made by a blackfly continues to trickle blood for some time after the attack.

Many fluid-feeding insects obtain sugar-rich nectar from flowers. When a moth or butterfly lands on a flower, its coiled proboscis staightens out, probes the flower, and channels nectar into the digestive tract (Figure 26.20d). Honeybees and bumblebees also visit flowers and lap up nectar through tubular (though not coiled) mouthparts.

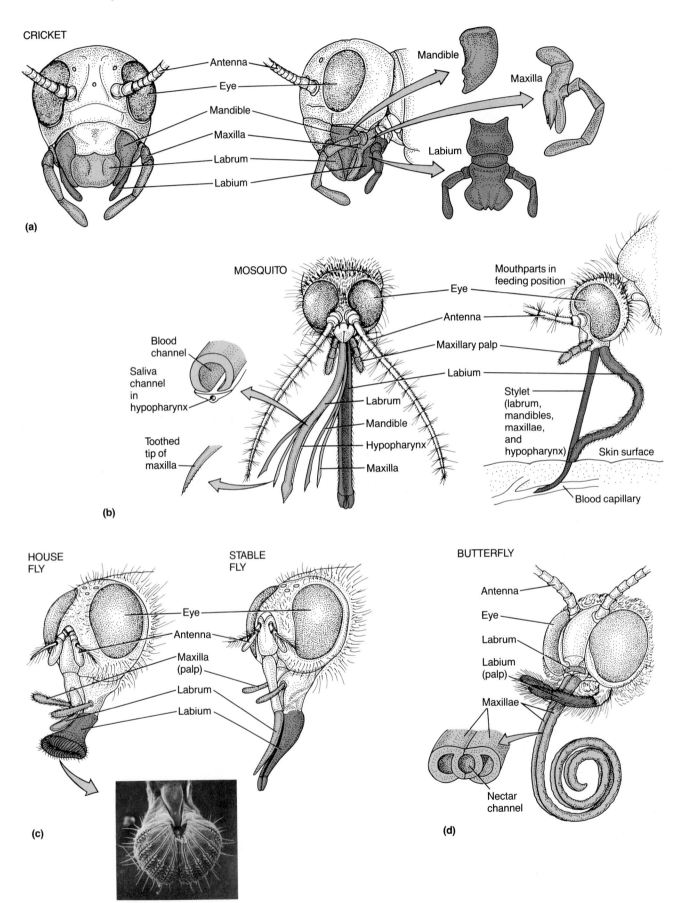

Figure 26.20 (Opposite) Adaptations of insect mouthparts. (a) The chewing mouthparts of a generalized insect such as a cricket or a cockroach represent what was probably the primitive condition. The mandibles and maxillae chew and manipulate food, and the flaplike labrum and labium help to hold food. Other types of mouthparts were probably derived from the chewing type. (b) In the mosquito, the piercing, sucking mouthparts are elongate, and the mandibles and maxillae are needlelike stylets. (c) The sponging mouthparts of a housefly, and the piercing-sponging mouthparts of a stable fly. The scanning electron micrograph shows the underside of the spongelike portion of the proboscis of a fruitfly, *Drosophila melanogaster*, with its numerous canals for channeling liquids. (d) The siphoning mouthparts of a butterfly.

Circulation and Gas Exchange

As seen in crustaceans and insects, arthropod circulatory systems are open but quite varied. (The circulatory systems of crustaceans and insects are discussed in Chapter 6, p. 124; see Figures 6.6 and 6.7 and Figure 26.19.) Hemolymph circulating in the hemocoel typically contains several kinds of amoebocytes called hemocytes. In many chelicerates and crustaceans, the hemolymph also carries a respiratory pigment, either the bluish, copper-containing protein hemocyanin or a type of hemoglobin. Insect hemolymph generally does not contain a respiratory pigment, as respiratory gases are distributed directly to tissues by the gas exchange (tracheal) system. In addition to its usual transport and protective functions, the circulatory system is also important in hydraulic movement. Arthropods commonly use hemolymph pressure to rupture their old cuticle prior to molting. Many spiders, lacking extensor muscles in their legs, use the hydrostatic pressure of hemolymph to extend these appendages, and newly emerged adult insects expand their wings hydraulically.

Arthropods are active, mobile animals with high oxygen requirements. Gram for gram, the muscles of a flying bee use 30 to 50 times as much oxygen as the heart or leg muscles of a human athlete at maximum activity, and during lift-off, a bee's oxygen demand is even greater. Accommodating these demands, arthropods have evolved highly efficient gas exchange systems that increase surface area.

Gills. The gills of aquatic arthropods are usually part of the segmental appendages. In many crustaceans, they are thin, often featherlike extensions of certain legs (see Figures 26.21a and 5.6). In many decapods, the gills are hidden and protected by the carapace. Specialized second maxillae, called **gill bailers**, help circulate water over a decapod's gills. A crayfish increases its oxygen uptake by moving its gill bailers more rapidly to direct more water

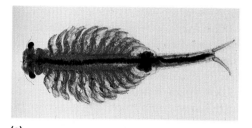

(a)

(b)

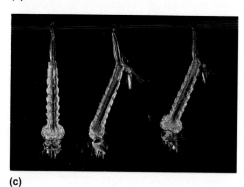

(c)

(d)

Figure 26.21 Gas exchange in aquatic arthropods. (a) The fairy shrimp *Eubranchipus*, a freshwater branchiopod, has many trunk segments bearing appendages that function in gas exchange, filter feeding, and locomotion. (b–d) Adaptations in aquatic insects: (b) damselfly nymphs have abdominal gills; (c) larvae of the mosquito *Culiseta*, breathe air through a tubelike siphon; (d) the whirligig beetle *Dineutes* has special water-repellent hairs that hold a bubble of air between its wings and abdomen. Species that use air bubbles or films are vulnerable to chemical wetting agents. Pollution of streams and ponds with detergents, for example, often spells disaster for these species, because with water-repellent hairs wetted, air bubbles and films are displaced by water.

over the gills when the water's oxygen content is low. By contrast, a lobster simply reduces its metabolic rate (and therefore its oxygen requirement) in poorly oxygenated water.

Although insects probably originated on land, some 20,000 species are aquatic, at least as juveniles. Sufficient oxygen may diffuse across the body surface of small aquatic individuals, but many larger forms supplement this route with gills (Figure 26.21b). The gills of many species are probably important only when oxygen is in short supply; in experiments involving surgical removal of the gills in damselflies, oxygen uptake over the general body surface could generally meet the oxygen demand.

Other Aquatic Adaptations. Many aquatic insects breathe air instead of relying on oxygen dissolved in the surrounding water. Mosquito larvae surface periodically and exchange gases by means of a posterior **breathing siphon** (Figure 26.21c). Many aquatic beetles and true bugs exchange gases by carrying bubbles or films of air next to their bodies (Figure 26.21d). As oxygen diffuses from a bubble or film into an insect's body, oxygen in the surrounding water may diffuse into the bubble or film. An air bubble usually shrinks as nitrogen diffuses out of it into the surrounding water, and this eventually forces the insect to surface for a new air bubble. Films of air held against the body are generally more long-lasting than bubbles, and some aquatic beetles retain the same air film throughout their lives.

Book Lungs. The paired **book lungs** of arachnids consist of numerous doubled sheets of thin tissue (lamellae), arranged somewhat like the pages of a book (see Figure 26.17). Each book lung in a spider is in a chamber that opens to the atmosphere via a narrow slit. One side of each lamella is exposed to air, while the other side is bathed by a flow of hemolymph circulated by the heart. Oxygen in the air diffuses across the lamella into the hemolymph, which transports the oxygen throughout the body. Hemocyanin enhances the hemolymph's ability to carry oxygen.

Tracheal Systems. Uniramians (insects, millipedes, and centipedes) have **tracheal systems** (described in detail in Chapter 5, p. 106). In addition, many spiders have tracheae (not homologous with those of uniramians) as well as book lungs. Tracheal systems typically consist of segmentally arranged **spiracles** opening to the atmosphere and a network of air-filled, cuticle-lined tracheal tubes (see Figure 5.10). After repeated branching, the smallest tubes merge with even smaller tubules called **tracheoles**. Often less than 1 μm in diameter, tracheoles may penetrate deeply into the tissues, especially skeletal muscle. The intimate association of tracheoles and tissue cells means that oxygen has to diffuse only a very short distance to reach sites of cellular metabolism. Carbon dioxide produced by the tissues moves in the opposite direction. This system of gas exchange is extremely efficient; even fully active insect flight muscles do not develop an oxygen debt.

Excretion and Osmoregulation

Arthropod excretion and osmoregulation are discussed in considerable detail in Chapter 7 (p. 157). In brief, aquatic arthropods, including the larvae of insects that are terrestrial as adults, excrete mainly ammonia. Many terrestrial arthropods conserve water by excreting wastes in nearly dry or pasty form. Most terrestrial arachnids excrete the purine guanine; land insects excrete mainly uric acid.

Osmoregulation differs greatly among arthropods. Marine species are typically osmoconformers (see Chapter 7, p. 148), but freshwater, estuarine, and terrestrial arthropods are osmoregulators. Freshwater crustaceans produce urine that is hypoosmotic to their hemolymph, and they actively absorb salt across their gills (see Figure 7.10). Also an osmoregulator, the brine shrimp *Artemia salina* (a branchiopod) is able to live in highly saline lakes because it produces hypertonic urine and secretes salt from its gills.

Excretory-osmoregulatory organs of arthropods include the **antennal** and **maxillary glands** of crustaceans (see Figures 26.19 and 7.10) and the **Malpighian tubules** of insects, millipedes, centipedes, and arachnids (see Figure 7.11). In addition to, or instead of, Malpighian tubules, many arachnids have **coxal glands** (see Figure 26.17). Antennal, maxillary, and coxal glands open to the outside via a pore located at the base of an antenna, maxilla, or coxa, respectively (A coxa is the basal segment of a walking leg.)

Nervous and Chemical Coordination

The arthropod nervous system is similar to that of annelids in being segmented, but it is considerably more complex and exhibits a marked tendency toward fusion of ganglia (compare Figure 25.4 with Figures 26.17 and 26.19). Many arthropods have a "brain" formed of fused anterior ganglia that encircle the esophagus. In insects, three ganglia in the thorax are partly or completely fused, providing greater local control of body appendages. In contrast to the vertebrate system, in which one neuron typically innervates a single muscle fiber, several neurons often innervate each muscle fiber in arthropods, and a single neuron may branch to serve several muscle fibers. Like annelids, many arthropods have giant neurons that provide rapid transmission of nerve impulses. Arthropod neurons may elicit fast, slow, or intermediate responses from muscle fibers, and they may be either excitatory or inhibitory. This high degree of control underlies an array of muscle responses as different as the steady, viselike grip of a crab's claw on prey, the superbly controlled, scarcely perceptible movements of a praying mantis stalking a grasshopper, and the fast, precise manipulation of silk threads by a web-spinning spider.

Chemical regulation in arthropods has been most studied in crustaceans and insects. The list of life processes known to be influenced by hormones has grown rapidly in recent years and includes reproduction, growth, molting, development, migration, osmoregulation, and hibernation. Neurohormones (hormones formed by neurosecretory cells of the central nervous system) are prominent in arthropods, and complex processes, such as molting and metamorphosis, may be governed by neurohormones in combination with hormones produced by nonneural endocrine glands (see Figures 11.6, 11.7, and 11.10).

Sensory Structures

Chemoreception, vision, touch, and hearing are important in the life of most arthropods. Chemoreceptors are generally most abundant on the antennae and mouthparts of crustaceans and uniramians and on the pedipalps and chelicerae of the chelicerates. They are also scattered over the body surface and legs of many insects. The housefly, for example, tastes its food with its feet, and many other insects taste the soil with their **ovipositors** (posterior egg-laying mechanisms) before laying eggs.

Most arthropods have well-developed eyes. The compound eyes of many species provide excellent perception of shapes and colors and are superb movement detectors (see Figures 12.7 and 26.9). Compound eyes are relatively large in fast-flying, hunting insects such as dragonflies, and in crustaceans such as crayfish and lobsters, which feed by seizing prey with quick, accurate thrusts of the claws. Certain immature insects, including grasshoppers, true bugs, mayflies, and dragonflies, also have compound eyes.

Arachnids, millipedes, centipedes, and many immature insects lack compound eyes. They have single-lens eyes (ocelli), often in clusters where they may function collectively, somewhat like compound eyes. Hunting spiders, for example, have two rows of four eyes each, located just above the fangs (see Figure 26.17). Most centipedes, active in a world of darkness, where vision has little importance, lack eyes or have only a few single-lens eyes. The house centipede *(Scutigera coleoptrata)*, a fast-running insect hunter, is an exception. It finds its prey with the aid of dense clusters of simple eyes.

Touch receptors and other mechanoreceptors occur on many parts of the arthropod body, often in association with chemoreceptors. Many of the setae on an arthropod's body are touch-sensitive. Other kinds of mechanoreceptors enable many arthropods to hear and to detect currents and pressure. (Pressure-detecting ears of insects are discussed in Chapter 12, p. 284; Figure 12.11.) Stretch receptors associated with the musculature monitor muscle tension and limb position. Working in conjunction with mechanoreceptor hairs at the leg joints, they provide the information required for balance, posture, and locomotor coordination. In addition, some crustaceans are kept informed about their posture and acceleration by statocysts located at the base of the antennules or elsewhere on the body.

BEHAVIOR

Arthropod behavior runs the gamut of complexity, from simple taxes to complex learning and intricate social behavior. Among the various forms of taxic behavior, phototaxis is probably of greatest importance to most arthropods. Millipedes and centipedes generally are negatively phototaxic, seeking shelter and moisture beneath leaf litter, rocks, and other objects. The same is true of many insects, although some fly toward light. Honeybees and some ants are able to orient at an angle to the sun, maintaining the angle and using the sun as a compass while traveling between food and nest (Figure 26.22). Among crustaceans, many planktonic copepods make daily vertical migrations in the ocean. At night they rise to the plankton-rich

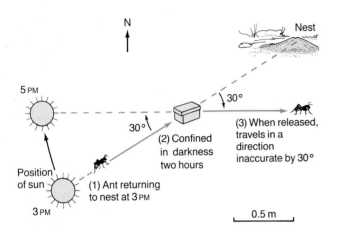

Figure 26.22 Sun-compass orientation. In an experiment demonstrating that ants navigate by means of sun-compass orientation, an ant *(Lasius niger)* was placed in a dark box for 2 hours and then released. Its direction back to the nest was inaccurate by the same angle (30°) traversed by the sun during the 2 hours.

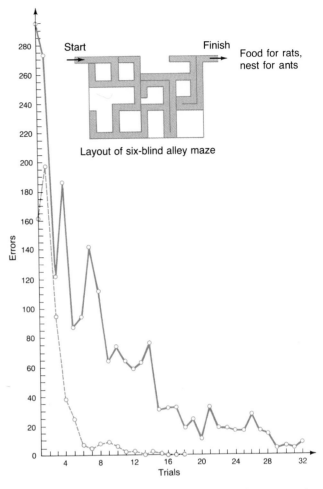

Figure 26.23 Maze learning by insects. Learning ability in insects has been demonstrated in maze-running experiments. In this graph, the performance of eight ants *(Formica incerta)* (solid line) is compared with that of eight rats (broken line). The ants learned to run a standard maze with six blind alleys (see inset), but they required many more trials than the rats to achieve a relatively error-free performance.

surface waters, where they feed under cover of darkness. Then, during the day, they sink away from the sunlight and visual predators.

Many of the adaptations that allow terrestrial and semiterrestrial crustaceans to live on land are behavioral in nature. Terrestrial crustaceans often follow a daily rhythm of activity. Peak activity at dusk, night, or dawn takes advantage of lower temperatures, higher humidity, and dew. Activity may also be timed to coincide with the presence of prey and absence of predators. The amphibious crabs *Uca* (see Figure 26.7c) and *Eriocheir* (mitten crab) excavate burrows and chambers in damp sand, behavior that protects against desiccation and predation.

Many arthropods are capable of learning. Habituation is commonplace; arthropods typically learn to ignore stimuli that are not followed by ill effects. Consequently, less time is lost from feeding and other activities because harmless disturbances are tolerated. More complex learning also occurs. Cockroaches and ants learn to run mazes (Figure 26.23), and learning is particularly important in the feeding activities of social insects. The foraging behavior of honeybees, for example, involves some learning. (This behavior is described in Chapter 13, p. 312.) Honeybees achieve a detailed knowledge of terrain during foraging flights. In the first few moments of a flight, a worker bee identifies landmarks in the vicinity of her nest and then recognizes these landmarks much later. Bees also learn to associate flower colors and odors with food.

As discussed in Chapter 13, many arthropods exhibit complex programmed behavior, which allows them to successfully perform intricate tasks without a long and potentially dangerous period of learning or practice. (See the discussions of digger wasps and honeybees in Chapter 13, pp. 303 and 313.) Some of the most awe-inspiring examples of programmed behavior among arthropods are the masterful uses of silk traps in prey capture by spiders. Orb weavers, such as the garden spider *Argiope aurantia*, spin especially beautiful webs (Figure 26.24). The silk threads are adhesive, so that when a fly or grasshopper flies into the web, it sticks. The spider quickly detects a victim struggling in the web, immobilizes it with venom and additional strands of silk, and then dines. Most orb weavers eat most of their old web each day, then spin a new one. This maintains the adhesiveness of the web and recycles the expensive silk proteins.

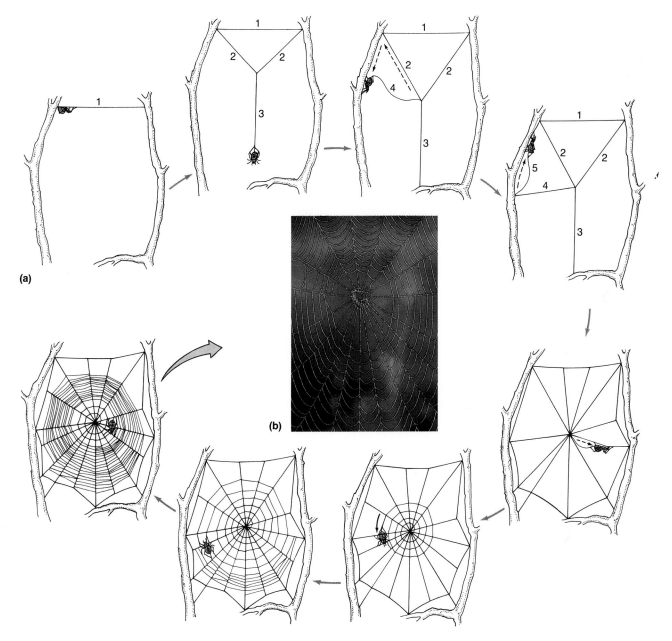

Figure 26.24 Spinning an orb web. (a) A garden spider begins the orb web by spanning the gap between two plant stems (or other objects) with several threads. It then drops a vertical thread from the center of the span to a lower branch. After a series of radii have been established, the spider joins them, first with a temporary silken spiral and then with a permanent one. The spider's nervous system can modify the spinning program to allow for differences in the arrangement of supporting plant stems and branches. (b) A completed orb web.

REPRODUCTION AND DEVELOPMENT

Arthropods are typically dioecious, the principal exceptions being barnacles and a few other crustaceans. Elaborate courtship displays by one or both sexes often precede copulation and insemination of the female. Some male spiders, after courtship, eject sperm onto a "sperm web." Using modified pedipalps, the male then picks up the sperm and transfers it to the female. In some spiders, the male's moment of glory ends as the female eats him; thus, the male supplies sperm and then, indirectly, nourishes the eggs. Male insects typically have a well-developed penis at the posterior end of the abdomen. Some crustaceans also have a penis, whereas others, such as crayfish and lobsters, transmit sperm via modified abdominal appendages (swimmerets). Fertilization generally takes place as eggs pass from the female.

A variety of adaptations provide protection for fertilized eggs and young. Female spiders deposit fertilized eggs in a silken cocoon, which, depending on the species, may be attached to the web or to a plant or may be carried by the females. Cocoons are often seen in human dwellings, hidden in basement corners or between wooden beams. Spiderlings hatch and undergo a short period of growth in the cocoon. After emerging from the cocoon, young wolf spiders ride on the mother's back for a period of time. Similarly, young scorpions (which hatch inside the female from eggs retained in the genital tract) cluster on the mother's back. Crustaceans typically protect their eggs, at least through part of the developmental period, by brooding them in a brood chamber or attaching them to special appendages. In crayfish, for example, the eggs are glued to the swimmerets with a sticky material. Insects deposit their eggs singly or in egg masses by means of ovipositors. Depending on the species, ovipositors are adapted for egg laying in soil, in or on plants, deep within tree trunks, as rafts of eggs floating on water, or in the eggs and tissues of other animals. Virtually every possibility is exploited.

Developmental patterns in arthropods are unique. Eggs in most species are centrolecithal (yolk is concentrated centrally), and cleavage is often superficial. (For a discussion of these developmental traits, see Chapter 16, p. 382.) The stage that emerges from the egg may be a miniature (juvenile) replica of the adult, but in most arthropods the hatchlings (larvae) differ markedly from the adult. Many crustacean and insect life cycles include a metamorphosis that converts the larva to the adult form.

Many crustaceans begin postembryonic life as a free-swimming **nauplius** larva (Figure 26.25). This tiny form has only three pairs of appendages, one eye, and no body segmentation. Segments and additional appendages develop during a series of molts. In most crabs and in the shrimp *Penaeus*, the later larva is called a **zoea**. Further development produces the **mysis** and **postlarva**, which, in many crustaceans, looks much like the adult. With the evolution of adaptations for brooding eggs and for terrestrial life, there has been a tendency to suppress larval stages. The nauplius stage is passed within the egg membrane in most decapods, and the zoea emerges at hatch. Crayfish, isopods, and amphipods brood their young throughout development, releasing them as postlarvae or juveniles.

Insects are often categorized by the nature of their metamorphosis (Figure 26.26). Primitively wingless species (Subclass Apterygota), such as silverfish and thrips, do not undergo metamorphosis. The hatchlings look like tiny adults, and after a period of growth involving several molts, they become sexually mature.

Simple metamorphosis is characteristic of grasshoppers and other orthopterans, lice, bugs, dragonflies, and a few other orders. Wings, when present, develop as external wing buds that become a little larger with each molt. Adult characteristics are attained gradually, so that with each molt, the growing juvenile becomes a little more like the adult. Typically, juveniles and adults both have compound eyes and similar mouthparts, and, most importantly, they often compete for the same kinds of food and space.

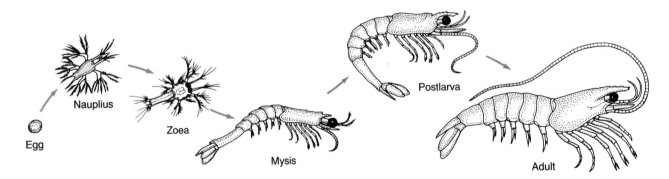

Figure 26.25 Crustacean development. Many crustaceans have an unsegmented larva, the nauplius, that hatches from the egg. The life cycle of *Penaeus*, a common shrimp in coastal waters of the Gulf of Mexico, includes a nauplius, a zoea, a mysis, and postlarval stages. These stages are not drawn to the same scale; the egg and nauplius are much enlarged relative to the later stages.

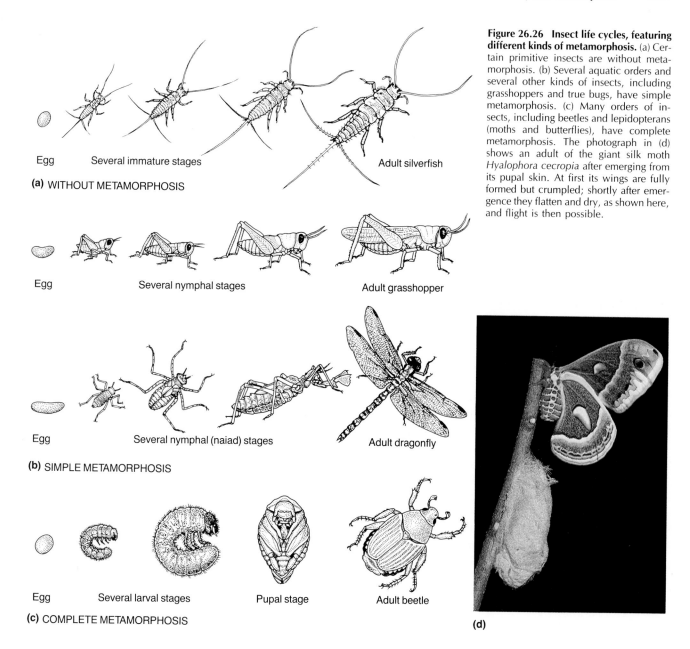

(a) WITHOUT METAMORPHOSIS

Egg Several immature stages Adult silverfish

(b) SIMPLE METAMORPHOSIS

Egg Several nymphal stages Adult grasshopper

Egg Several nymphal (naiad) stages Adult dragonfly

(c) COMPLETE METAMORPHOSIS

Egg Several larval stages Pupal stage Adult beetle

(d)

Figure 26.26 Insect life cycles, featuring different kinds of metamorphosis. (a) Certain primitive insects are without metamorphosis. (b) Several aquatic orders and several other kinds of insects, including grasshoppers and true bugs, have simple metamorphosis. (c) Many orders of insects, including beetles and lepidopterans (moths and butterflies), have complete metamorphosis. The photograph in (d) shows an adult of the giant silk moth *Hyalophora cecropia* after emerging from its pupal skin. At first its wings are fully formed but crumpled; shortly after emergence they flatten and dry, as shown here, and flight is then possible.

The great majority of insects have **complete metamorphosis**. Wings, when present, develop internally, and metamorphosis usually involves a drastic change in both body form and life-style. The magnitude of the change can be appreciated by comparing a leaf-eating caterpillar (a wormlike larva) to a nectar-drinking adult butterfly. The larva feeds, grows, and undergoes several molts as it passes through a succession of similar stages called larval **instars**. The last larval instar ceases feeding, becomes quiescent, and molts to the pupal stage. The **pupa** is usually an immobile stage in which larval tissues and structures are broken down and adult ones formed. Energy stored as fat by the larva is used in this transformation. The pupa is sometimes called a "resting stage," but actually it is a stage of intense metabolic activity and change. After adult structures have formed, the pupal skin splits at the anterior end, allowing the adult to emerge. The wings must dry and harden before the adult insect can fly. Complete metamorphosis permits feeding activity to be concentrated in one stage and reproduction and dispersal in another; the larvae crawl and eat, whereas the adults fly, reproduce, and may or may not eat. In species in which adults do feed, adults and larvae exploit different food sources and thus do not compete with the each other.

PHYLOGENY

Despite their unique development, arthropods are grouped with the molluscs and annelids as part of the large protostome lineage. The resemblance of the basic arthropod body plan to that of annelids suggests that arthropods evolved their cuticular exoskeleton, specialized body regions (tagmata), jointed legs, and complex organ systems by building on the metameric body plan pioneered by annelid-like ancestors. Most zoologists accept this broad picture of arthropod history, agreeing that arthropods and annelids probably shared a primitive metameric ancestor. There is also general agreement that the extinct trilobites and the chelicerates constitute two distinct subgroups of the phylum. There is some controversy concerning the affinities of other major groups; it was formerly assumed that crustaceans, insects, and other nonchelicerates were all closely related because they have antennae, mandibles, and similar compound eyes, but most zoologists now agree that the crustaceans are fundamentally distinct from insects and other groups, now called uniramians. Crustaceans are mainly aquatic and have two pairs of antennae and primitively biramous appendages; uniramians are mainly terrestrial, with one pair of antennae and primitively uniramous appendages. Also, studies comparing embryos and adult anatomy suggest that the mandibles of crustaceans may not be homologous with those of uniramians. The classification scheme followed in this text follows the general consensus that there are four fundamentally distinct subgroups of arthropods, which are consequently given subphylum rank (see Classification, p. 646).

The main controversy concerning arthropod phylogeny has to do with whether the phylum is monophyletic (had a single ancestor) or polyphyletic (had multiple ancestors). Did a single lineage of ancestral arthropods (protoarthropods) evolve from a primitive metameric ancestor, or did the major groups of arthropods arise as separate lines from more than one group of metameric ancestors (Figure 26.27)? There is a lack of fossil evidence from the Precambrian, when the major arthropod groups were evolving, and evidence from comparative anatomy and embryology has been interpreted to support both viewpoints.

The argument for monophyly is supported by the many unique synapomorphies (shared derived characters) possessed by all arthropods. (Synapomorphies are discussed in Chapter 18, p. 440.)

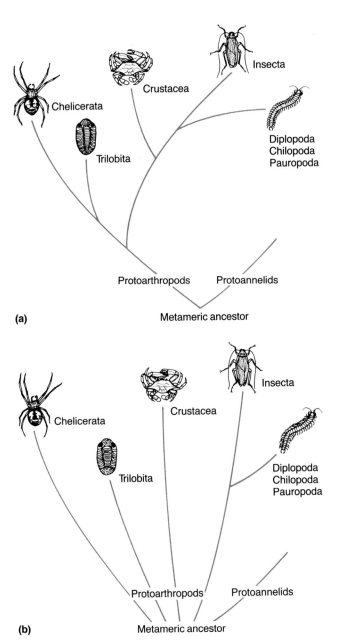

(a)

(b)

Figure 26.27 Arthropod phylogeny. (a) Pattern of arthropod phylogeny assuming a monophyletic origin. (b) Arthropod phylogeny assuming a polyphyletic origin.

Arthropodan metamerism and its functional tagmata—the chitinous multilayered exoskeleton, superficial cleavage, and jointed, paired appendages—unite all arthropods and set them off from all other animals.

The argument for polyphyly rests on somewhat shakier grounds. It holds that differences among the four arthropod groups are so fundamental that they could not have arisen from a single source. Uniramians, especially, are considered to have evolved separately from terrestrial ancestors, distinct from aquatic ancestors of other groups. The polyphyletic idea assumes that the evolution of the fundamental arthropod features ("arthropodization") occurred in two or more independent lineages and that modern similarities between the major groups resulted from convergent evolution. Although the polyphyletic hypothesis has attracted several strong advocates in recent years, most zoologists favor the monophyletic scheme.

PHYLUM TARDIGRADA

Tardigrades, commonly called **water bears** (about 500 species in all), live on wet mosses, on lichens, and among sand grains in ponds or on wet lakeshores and ocean beaches. These animals are yellowish, red-brown, or grayish blue. Their body is short and stubby, ranging in length from about 0.3 mm to 1.5 mm. Four pairs of stubby, clawed legs hold the body up off the substrate and are used in locomotion. (Figure 26.28). Thomas Henry Huxley, a contemporary of Darwin, first called these animals water bears because of the way they lumber over aquatic vegetation. Although tardigrades are rarely seen in great numbers, a handful of wet beach sand or moss will usually contain some. Several species thrive in the moss that grows on weather-beaten wooden roofs.

Water bears are aquatic. Even those species that live on terrestrial mosses and lichens are active only when the plants are wet. Like the rotifers and small nematodes that share their minute, often temporary aquatic habitats, tardigrades have the remarkable ability to become completely desiccated and undergo cryptobiosis, or suspended animation. (See the Chapter 23 Essay on cryptobiosis, p. 540.) Cryptobiotic tardigrades are resistant to extreme temperatures and drought. In laboratory tests, they have emerged from the cryptobiotic state and reproduced after exposure to temperatures ranging from about −250°C to 150°C and to the drying effects of ether and 100% ethyl alcohol. It is estimated that tardigrades have an active life span of 12 to 18 months, but may live for 50 years or more by interspersing active periods with months or years in the cryptobiotic state.

The tardigrade cuticle, much like that of arthropods, is multilayered and is molted anywhere from four to twelve times in a typical individual's lifetime. The cuticle consists of plates that resemble external segments. The only muscles are longitudinal and radially arranged bands, each composed of one or a few smooth muscle cells (see Figure 26.28b). Contraction of the longitudinal muscle bands arches the body, while alternating contractions of individual leg muscles produce a lumbering gait.

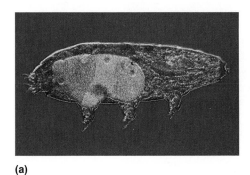

(a)

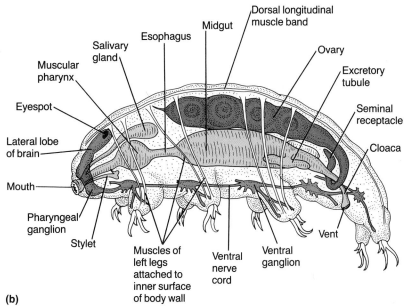

(b)

Figure 26.28 Phylum Tardigrada (water bears). (a) This species, *Milnesium tardigrada* (about 0.4 mm long), found throughout the world in aquatic and semiaquatic habitats, feeds mainly on other small invertebrates. (b) Drawing of a female tardigrade, showing the internal organs as they would appear if viewed through the body wall from the left side.

Tardigrades feed exclusively on fluids, mostly plant juices. They puncture individual plant cells or the bodies of small invertebrates, using a pair of retractable piercing stylets (see Figure 26.28b). Their straight digestive system includes a bulblike muscular pharynx and an expanded midgut. Digestive and excretory wastes as well as gametes pass out through a cloaca. Tardigrades lack circulatory and gas exchange organs.

The tardigrade central nervous system, like that of annelids and arthropods, consists of a multilobed cerebral ganglion ("brain") connected by a nerve ring to a double, ganglionated nerve cord. Sensory structures include tactile hairlike projections on the body surface, especially the head. Many species also have a pair of reddish or black eyespots over the cerebral ganglion.

Tardigrades are dioecious, and individuals have one large gonad (see Figure 26.28b). Fertilization may occur in the female's cloaca; or, in some species, the female deposits unfertilized eggs under her cuticle, and the male fertilizes them as she molts. Tardigrade populations usually consist mostly of females, and many species seem to consist entirely of parthenogenetic females.

From what little is known, tardigrade development appears to be unique. Cleavage is holoblastic, but otherwise unlike that of other animals. The coelom develops as an enterocoel, as in deuterostomes; but then typical of protostomes with a hemocoel, the coelom becomes reduced to small cavities around the gonads. In common with many pseudocoelomates, nuclear division in tardigrades ceases during development, resulting in a fixed number of nuclei and cells in adults. (See the discussion of eutely in Chapter 23, p. 534.) An adultlike juvenile hatches about two weeks after development begins.

Tardigrades pose many questions in phylogeny. A molted cuticle, legs held under the body, a metameric nervous system, and a hemocoel all indicate that tardigrades are related to arthropods. Yet, their enterocoelous development and resemblance to pseudocoelomates are puzzling. In common with arthropods and nematodes, tardigrades lack functional body cilia. They also have other features that may indicate an affinity with pseudocoelomates. Some zoologists consider the tardigrade body cavity a pseudocoel rather than a hemocoel. Also, tardigrade pharyngeal muscles are radially arranged, like those of nematodes, and the muscle bands of tardigrades are remarkably like those of rotifers (see Figure 23.9).

Do these features indicate a phylogenetic relationship among tardigrades, arthropods, and pseudocoelomates? To some zoologists, tardigrades represent a serious challenge to the usual grouping of phyla as pseudocoelomates, protostomes, and deuterostomes. Clearly, tardigrades merit more research attention, and future research on them should be rewarding.

PHYLUM PENTASTOMIDA

Pentastomids, commonly called **tongue worms**, are wormlike parasites of flesh-eating vertebrates. Although mainly found in snakes and lizards, about 10 of the 90 species in the phylum infect birds and mammals. Larval stages of several species may infect humans. The phylum is chiefly tropical, but certain species occur throughout the world.

Tongue worms attach to and suck blood from the nasal passages and lungs of their hosts (Figure 26.29). Their small head is adapted for gripping host tissue. It bears five processes: four claws or

Figure 26.29 Phylum Pentastomida. This scanning electron micrograph shows an adult female tongue worm with its anterior end embedded in lung tissue. Females range in length from about 1 cm to 15 cm, depending on the species. Males of most pentastomids are much smaller than females. Pentastomid infections may be symptomless or may result in bleeding, nasal discharge, excessive sneezing, and difficulty in breathing.

clawed legs and a snout bearing the mouth. The body wall (cuticle and underlying longitudinal and circular muscle layers) is segmented, but there is no internal segmentation. There are no circulatory, gas exchange, or excretory organs.

Tongue worms are dioecious and like many parasites are extremely prolific. Adults copulate in the host, and eggs are fertilized internally. Females may contain nearly half a million eggs that are gradually expelled from the host via the nasal passages or first swallowed and then passed out with feces. If embryonated eggs are eaten by a fish or small mammal, mitelike larvae hatch in the intestine, migrate into various host tissues, and encyst. If a suitable vertebrate eats an infected fish or mammal, the larval pentastomids excyst in the carnivore's gut, migrate up the esophagus, and enter the respiratory tract, where they develop into adults and complete the life cycle.

Zoologists generally agree that pentastomids are closely related to arthropods. The pentastomid cuticle, segmented body wall, hemocoel, and mitelike larvae all point to this conclusion. Moreover, recent studies indicating that pentastomid sperm are similar to those of certain crustaceans add support to this consensus.

PHYLUM ONYCHOPHORA

Onychophorans, also known as **walking worms** or **velvet worms**, are caterpillar-like, with many pairs of walking legs and with some segmentally arranged internal systems. The approximately 80 species in the phylum live in caves, under rocks, or in moist litter in tropical and subtropical forests of Central America and the southern hemisphere. The phylum's ancestry can be traced back to the Cambrian, over 500 million years ago. Fossil species were marine, yet they closely resembled living onychophorans, all of which are terrestrial. Zoologists postulate that ancestral marine species became adapted for life on ocean beaches and then spread inland in areas of abundant moisture.

Onychophorans are elongate, about 1.5–15 cm long, and covered by a chitinous cuticle that is rather soft, flexible, and unsegmented. The cuticle is molted, in patches, rather than all at once as in arthropods. The body wall has circular, longitudinal, and diagonal muscle layers composed of smooth fibers. Depending on the species, there are from 14 to 43 pairs of unjointed legs (Figure 26.30). A well-defined head bears two antennae and a ventral mouth flanked by a pair of shearing mandibles (probably not homologous with those of arthropods). An **oral papilla** on each side of the mouth bears an opening from an **adhesive gland**. The glands secrete a fluid that hardens rapidly, forming sticky threads when exposed to air. Onychophorans can spit the fluid as far as half a meter, foiling would-be predators or capturing prey (various small invertebrates) by entangling them in threads.

Onychophorans are a curious mix of annelid-like, arthropod-like, and unique features. Their appendages resemble annelid parapodia in being unjointed, but are arthropod-like in holding the body up off the substrate. Their excretory organs are metamerically arranged metanephridia, similar to those of annelids. A metanephridium, associated with each walking leg, has a ciliated nephrostome that collects fluid from a small coelomic sac and opens to the outside by a nephridiopore at the base of the leg.

Onychophorans have many arthropod-like features. Their esophagus and muscular pharynx are lined with chitinous cuticle. Their coelom is reduced to small cavities associated with the metanephridia and gonads. Their circulatory system, which is open and characterized by a dorsal tubular heart, plays only a minor role in gas transport. The hemocoel is partitioned by sheets of tissue extending longitudinally and dorsoventrally through the body. A tracheal system of air tubes delivers respiratory gases directly to and from body tissues. Tiny spiracles opening into the tracheal system are located over the whole body surface (see the inset in Figure 26.30). In contrast to the arthropod tracheal system, the onychophoran system is rather poorly adapted for land existence. The tubes are simple and unbranched, and the tiny spiracles cannot be closed to prevent water loss. To a large extent, this explains why onychophorans are restricted to humid environments.

The onychophoran nervous system resembles the annelid-arthropod type, but is also reminiscent of the ladderlike system of certain flatworms. A large bilobed brain above the pharynx connects to an unfused pair of ventral nerve cords. The ventral cords lack segmental ganglia, prominent in annelids and arthropods, but are connected by transverse cords. Sensory structures include touch receptors and chemoreceptors concentrated on the antennae, a pair of eyespots at the base of the antennae, and special moisture detectors, called **hygroreceptors**, on the body surface and antennae.

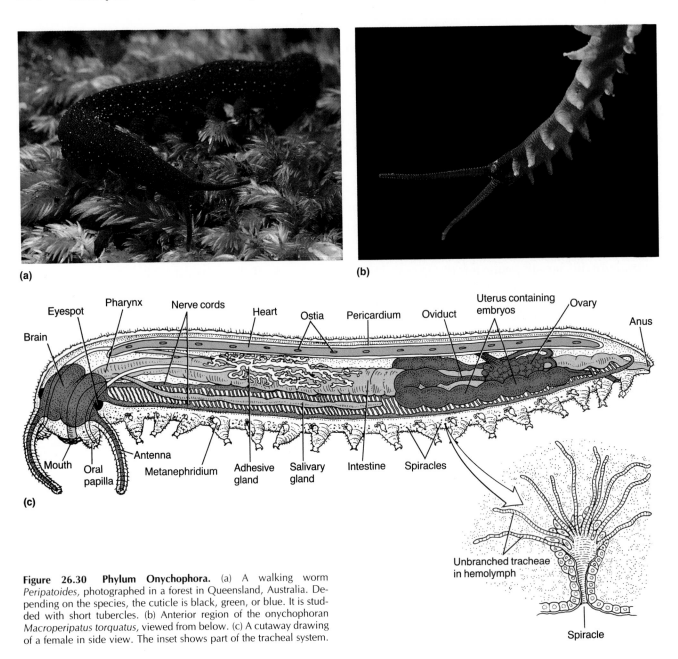

Figure 26.30 Phylum Onychophora. (a) A walking worm *Peripatoides,* photographed in a forest in Queensland, Australia. Depending on the species, the cuticle is black, green, or blue. It is studded with short tubercles. (b) Anterior region of the onychophoran *Macroperipatus torquatus,* viewed from below. (c) A cutaway drawing of a female in side view. The inset shows part of the tracheal system.

Onychophorans are dioecious, with dorsal gonads and a single posterior gonopore. Fertilization is internal and is followed by superficial cleavage, as in arthropods. The coelom develops as a schizocoel. Development is direct, and some species nourish their young by a placenta-like attachment in the female's uteri.

Because they are so much like arthropods and yet have some annelidan features, onychophorans have a special significance in animal phylogeny. They have sometimes been considered a link between the annelids and the arthropods. More likely,

however, onychophorans are descendants of a metameric group that was ancestral to arthropods. They may represent one of the earliest arthropod-like phyla that diverged from the stock of segmented protostomes that gave rise to annelids and arthropods (see Figure 18.6).

In addition to their importance in phylogeny, onychophorans have an interesting geographical distribution that tells us something about the earth's history. Closely related species of several genera occur on separate continents. Certain species of *Opisthopatus,* for example, are completely isolated from one another, some inhabiting moist tropical forests in Chile, others living in similar habitats in South Africa. This discontinuous distribution of closely related species is evidence that the African and South American continents were joined during former geological eras. (Continental drift is discussed in Chapter 17, p. 419.)

□ SUMMARY

1. ***Key Features of Arthropods.*** The Phylum Arthropoda (approximately 1 million recorded species) consists of segmented protostomes with paired, jointed, segmental appendages. Segments are often grouped into functional regions called tagmata. A chitinous exoskeleton supports and protects internal structures; it is molted periodically as a provision for growth. Eggs are centrolecithal, and cleavage is often superficial.

2. The four arthropod subphyla are the Trilobita (all species extinct), Chelicerata, Crustacea, and Uniramia. The chelicerates include marine and terrestrial classes; the crustaceans are primarily aquatic; and the uniramians are primarily terrestrial.

3. Arthropods have open circulatory systems. Gas exchange involves the circulatory system and filamentous gills (in crustaceans) or book lungs or book gills (in arachnids). In uniramians and many arachnids, a tracheal system carries respiratory gases to and from the tissues.

4. Antennal, coxal, or maxillary glands are the excretory and osmoregulatory organs in crustaceans and arachnids; insects have Malpighian tubules.

5. Arthropod nervous systems follow the annelidan pattern, but with extensive cephalization and varying degrees of consolidation of ventral nerve cord ganglia. Many body functions are chemically coordinated. Many arthropod adults and some juveniles have compound eyes. Simple eyes also occur in most arthropods and are the only visual organs in arachnids. Arthropods have many chemoreceptors, especially on antennae (when present) and mouthparts. Mechanoreceptors are widely distributed over the body, within the muscles, and at appendage articulations. Arthropod behavior includes combinations of taxes, motor programs, and learning.

6. Most arthropods are dioecious. Development frequently involves larval stages unlike the adult in form and life-style. The nauplius larva is characteristic of many crustaceans. During development, various kinds of metamorphosis convert larval arthropods to the adult form.

7. Arthropods probably arose from primitive annelid-like ancestors, and most likely their origin was monophyletic.

8. Three phyla—the Tardigrada (water bears), Pentastomida (tongue worms), and Onychophora (velvet worms)—contain metameric animals with arthropod affinities. Tardigrades cohabit water droplets on mosses and lichens and wet beach sands with rotifers and nematodes. They have several features (including cryptobiotic ability) in common with pseudocoelomates. Tongue worms are parasitic bloodsuckers, living in the nasal passages and lungs of carnivorous vertebrates. Onychophorans inhabit moist tropical forests; they may be similar to arthropod ancestors.

□ FURTHER READING

Amos, W. H. "Unseen Life of a Mountain Stream." *National Geographic* 151(1977): 562–580. *Illustrates the importance of arthropods in aquatic food chains.*

Cisne, J. L. "Trilobites and the Origin of Arthropods." *Science* 186(1974): 13–18. *Assesses the position of trilobites in arthropod evolution.*

Eisner, T., and E. O. Wilson. *The Insects: Readings from Scientific American.* New York: W. H. Freeman, 1977. *Well-illustrated articles on most aspects of insect biology.*

Evans, H. E. *Insect Biology.* Reading, Mass.: Addison-Wesley, 1984. *A modern entomology textbook.*

Foelix, R. F. *Biology of Spiders.* Cambridge, Mass.: Harvard University Press, 1982. *Probably the best modern treatment of this group.*

Hölldobler, B. "The Wonderfully Diverse Ways of the Ant." *National Geographic* 165(1984): 778–813. *A study of species and behavioral variety.*

Huffaker, C. B., and R. L. Rabb. *Ecological Entomology.* New York: Wiley, 1984. *Includes up-to-date descriptions of insects in food chains and webs.*

King, D. E. "Water bears." *Carolina Tips* 49(1986): 1–3. *An up-to-date summary of tardigrade biology; excellent photomicrographs.*

Manton, S. M. *The Arthropoda: Habits, Functional Morphology and Evolution.* Oxford, England: Clarendon Press, 1977. *Includes various ideas on arthropod origins and phylogeny, although heavily biased toward polyphyletic interpretations.*

Nelson, L. T. "Mosquitoes, the Mighty Killers." *National Geographic* 156(1979): 426–440. *A pictorial analysis of mosquitoes and their roles as vectors of human diseases.*

Prestwich, G. D. "Dwellers in the Dark, Termites." *National Geographic* 153(1978): 532–547. *A superbly illustrated essay.*

Rebach, S., and D. W. Dunham. *Studies in Adaptation: The Behavior of Higher Crustacea.* New York: Wiley, 1983. *Crustacean behavior from an evolutionary viewpoint.*

Resh, V. H., and D. M. Rosenberg. *The Ecology of Aquatic Insects.* New York: Praeger, 1984. *Thorough treatment of aquatic entomology.*

Ross, E. S. "Mantids, the Praying Predators." *National Geographic* 165(1984): 268–280. *A beautifully illustrated essay on praying mantises.*

Schmidt, G. D. and L. S. Roberts. "Phylum Pentastomida: Tongue Worms." Chap. 32 in *Foundations of Parasitology*, 3d ed. St. Louis: Times Mirror/Mosby College Publishing, 1985. *General biology of the pentastomids and their host-parasite relationships.*

☐ CLASSIFICATION

Phylum Arthropoda (Gr. *arthron*, "joint"; *pous*, "foot"). Arthropods; metameric protostomes with jointed appendages, a chitinous exoskeleton, and superficial cleavage.

Subphylum Trilobita (Gr. *tri*, "three"; *lobos*, "lobe"). Trilobites; extinct marine arthropods; the body was divided by two longitudinal grooves into three lobes and bore many uniform, biramous (two-branched) appendages; *Triarthrus.*

Subphylum Chelicerata (Gr. *chele*, "claw"; *keras*, "horn"). Chelicerates; body subdivided into an anterior prosoma and a posterior opisthosoma; usually with four pairs of walking legs; anterior appendages include chelicerae and pedipalps; no mandibles or antennae.

Class Merostomata (Gr. *meros*, "thigh"; *stoma*, "mouth"). Marine chelicerates with abdominal gills.

Subclass Eurypterida (Gr. *eurys*, "broad"; *pteron*, "fin, wing"). Giant water scorpions; extinct. *Eurypterus.*

Subclass Xiphosura (Gr. *xiphos*, "sword"; *oura*, "tail"). Horseshoe crabs; prosoma is somewhat horseshoe-shaped and broadly attached to an unsegmented opisthosoma bearing a stout, swordlike telson at its posterior end; possess book gills and compound eyes; *Limulus.*

Class Arachnida (Gr. *arachne*, "spider"). Scorpions, pseudoscorpions, spiders, harvestmen, mites, ticks, and other smaller groups; prosoma bears four pairs of walking legs; most are terrestrial.

Order Scorpiones (Gr. *skorpion*, "a scorpion"). Scorpions; possess large, pincerlike pedipalps; abdomen segmented and elongate, with a terminal stinger; *Centruroides, Hadrurus.*

Order Uropygi (Gr. *oura*, "tail"; *pyge*, "rump"). Whip scorpions or vinegaroons; possess large raptorial pedipalps; abdomen has a terminal whip; *Mastigoproctus.*

Order Araneae (L. *aranea*, "a spider"). Spiders; unsegmented abdomen is attached to the cephalothorax by a narrow waist; *Argiope, Pholcus, Latrodectus.*

Order Pseudoscorpiones (Gr. *pseudes*, "false"; *skorpion*, "a scorpion"). Pseudoscorpions; small, scorpion-like, with large raptorial pedipalps, but no stinger or whip on posterior end of abdomen; *Chelifer, Vachonium.*

Order Opiliones (L. *opilio*, "a shepherd"). Harvestmen ("daddy longlegs"); very long legs; small segmented abdomen is broadly joined to the cephalothorax; *Leiobunum.*

Order Acarina (L. *acarus*, "a mite"). Ticks, mites; small, with unsegmented abdomen broadly fused with the cephalothorax; *Dermacentor, Sarcoptes, Trombicula.*

Class Pycnogonida (Gr. *pyknos*, "compact"; *gonia*, "knee joint"). Sea spiders; prosoma is relatively large compared to minute opisthosoma; four to six pairs of long walking legs; *Nymphon, Pycnogonum.*

Subphylum Crustacea (L. *crusta*, "shell"). Crustaceans; possess biramous appendages, two pairs of antennae, and feeding appendages consisting of a pair of mandibles and two pairs of maxillae; most crustaceans are aquatic; the major classes are listed below.

Class Branchiopoda (Gr. *branchia*, "gill"; *pous*, "foot"). Water fleas, fairy shrimp, tadpole shrimp, and clam shrimp; flattened, seta-covered appendages used in filter feeding; mostly freshwater; *Daphnia, Artemia* (brine shrimp).

Class Copepoda (Gr. *kope*, "oar"; *pous*, "foot"). Copepods; most have a cylindrical body, usually long antennae, and a single median eye; both marine and freshwater species; *Cyclops, Calanus.*

Class Ostracoda (Gr. *ostrakon*, "shell"; *odes*, "like"). Seed or mussel shrimps; a bivalve carapace encloses the body; ostracods may be extremely abundant in both marine and fresh water; *Cypris.*

Class Cirripedia (L. *cirrus*, "curl"; *pes*, "foot"). Barnacles; body typically covered with calcareous plates; sessile as adults; entirely marine. *Balanus* (acorn barnacle), *Lepas* (stalked, or goose, barnacle).

Class Malacostraca (Gr. *malakos*, "soft"; *ostrakon*, "shell"). Shrimps, crabs, lobsters, crayfish, pill bugs, beach fleas; a 19-segment body typically divided into a head of 5 fused segments, an 8-segment thorax bearing the walking legs, and a 6-segment abdomen; includes orders Isopoda, Amphipoda, and the largest group of crustaceans, the Order Decapoda, with some 10,000 species; lobsters (*Homarus*), crayfish (*Cambarus, Astacus*), shrimps (*Penaeus, Pandalus*), and crabs (*Cancer*) are decapods.

Subphylum Uniramia (L. *uni*, "one"; *ramus*, "a branch"). Uniramians have unbranched (uniramous) appendages, one pair of antennae, and feeding appendages consisting of a pair of mandibles and one or two pairs of maxillae; most are terrestrial.

Class Diplopoda (Gr. *diploos*, "double"; *pous*, "foot"). Millipedes; elongate, many-segmented cylindrical uniramians with doubled segments, each bearing two pairs of legs; *Julus, Narceus.*

Class Pauropoda (Gr. *pauros*, "small"; *pous*, "foot"). Pauropods; small (0.5–2 mm long), millipede-like inhabitants of soil and leaf mold; *Pauropus.*

Class Chilopoda (Gr. *cheilos*, "margin"; *pous*, "foot"). Centipedes; elongate, many-segmented, dorsoventrally flattened uniramians with one pair of legs per segment; *Scolopendra, Scutigera.*

Class Symphyla (Gr. *syn*, "together"; *phyle*, "tribe"). Symphylans; small (2–10 mm long), centipede-like uniramians with twelve leg-bearing segments; occur in soil and decaying vegetation; *Scutigerella.*

Class Insecta (L. *insectum*, "cut into"). Insects; body is divided into a head, thorax, and abdomen; thorax bears three pairs of legs and, in the adult, usually two pairs of wings; the following synopsis omits minor orders.

Subclass Apterygota (Gr. *a*, "not"; *pterygotos*, "winged"). Proturans, silverfish, springtails; wingless insects thought to have evolved from wingless ancestors; no metamorphosis.

Subclass Pterygota (Gr. *pterygotos*, "winged"). Winged insects; wings secondarily lost in some orders through evolutionary adaptation; metamorphosis is either simple or complete; orders with more than 20,000 species are marked with an asterisk.

Order Ephemeroptera (Gr. *ephemeros*, "lasting only a day"). Mayflies.

Order Odonata (Gr. *odontos*, "tooth"). Dragonflies and damselflies.

Order Orthoptera* (Gr. *orthos*, "straight"; *pteron*, "wing"). Grasshoppers and allies.

Order Isoptera (Gr. *isos*, "equal"; *pteron*, "wing"). Termites.

Order Mallophaga (Gr. *mallos*, "wool"; *phago*, "to eat"). Bird lice and chewing lice.

Order Anoplura (Gr. *anoplos*, "unarmed"; *oura*, "tail"). Sucking lice.

Order Thysanoptera (Gr. *thysanos*, "tassel"; *pteron*, "wing"). Thrips.

Order Hemiptera* (Gr. *hem*, "half"; *pteron*, "wing"). True bugs.

Order Homoptera* (Gr. *homos,* "same"; *pteron,* "wing"). Aphids and allies.

Order Neuroptera (Gr. *neuron,* "nerve"; *pteron,* "wing"). Lacewings.

Order Coleoptera* (Gr. *koleos,* "sheath"; *pteron,* "wing"). Beetles.

Order Trichoptera (Gr. *trichos,* "hair"; *pteron,* "wing"). Caddis flies.

Order Lepidoptera* (Gr. *lepidos,* "scale"; *pteron,* "wing"). Skippers, moths, butterflies.

Order Diptera* (Gr. *di,* "double"; *pteron,* "wing"). True flies.

Order Siphonaptera (Gr. *siphon,* "a siphon"; *a,* "without"; *pteron,* "wing"). Fleas.

Order Hymenoptera* (Gr. *hymenos,* "membrane"; *pteron,* "wing"). Bees, wasps, ants.

Phylum Tardigrada (L. *tardus,* "slow"; *gradior,* "walk"). Water bears; aquatic herbivores showing evidence of metamerism and having affinities with pseudocoelomates, deuterostomes, and arthropods; nonchitinous exoskeleton molted periodically; cryptobiotic; *Macrobiotus, Echiniscus.*

Phylum Pentastomida (Gr. *pente,* "five"; *stoma,* "mouth"). Tongue worms; unsegmented parasites of the respiratory tract of vertebrates; body wall, including a cuticle, is metameric; mitelike larvae and sperm morphology indicate a close relationship to crustaceans; *Linguatula, Armillifer.*

Phylum Onychophora (Gr. *onychos,* "claw"; *phoros,* "bearing"). Velvet or walking worms; arthropod-like inhabitants of the moist tropics; unjointed legs and several internal systems are metameric; *Peripatus, Opisthopatus.*

Success of the Deuterostomes

So far, we have surveyed some 27 phyla of animals, making up about 96% of the species in the animal kingdom. The seven remaining phyla, collectively called the deuterostomes or deuterostome-like coelomates, consist of about 58,500 living species. This is a comparatively small number of animals; but nonetheless, deuterostomes constitute a highly successful assemblage. Four of the phyla—the Phoronida, the Brachiopoda, the Chaetognatha, and the Hemichordata—are small (about 500 living species altogether) and are entirely marine. Some of these phyla were much more diverse during former geological times. The "big three" of the seven deuterostome phyla are the Bryozoa (about 5000 aquatic, mostly marine, species), the Echinodermata (about 6000 marine species), and the Chordata (about 47,000 species). In terms of the variety of habitats colonized, the Chordata, which is represented by many species on land, in the sea, and in fresh water, is by far the most successful deuterostome phylum.

What factors contributed to the success of the deuterostome lineage? Answers to this question might be found by considering the ancestral features of deuterostomes. As discussed in Chapter 18, deuterostomes probably evolved from ancestors that had radial cleavage, enterocoelous body cavity formation, and a bilaterally symmetrical ciliated larva. Another ancestral deuterostome feature is a tripartite coelom: At least in the early embryo, the coelom is divided into an anterior **protocoel**, a middle **mesocoel**, and a posterior **metacoel**. (The tripartite coelom and the adaptive advantages of body cavities are discussed in the Trends and Strategies sections beginning on pp. 531 and 583.) All of these traits were undoubtedly important ancestral building blocks in the deuterostome lineage, but were they the most important keys to deuterostome success? Chordates, the most successful deuterostome phylum, lack several of the ancestral traits: They do not have a typical deuterostome larva and their patterns of cleavage and body cavity formation are somewhat different from those of typical deuterostomes. Thus, it is possible that chordate success, and perhaps the success of other deuterostome phyla, resulted mainly from adaptations derived *after* the individual phylum lineages diverged from the main stem of the deuterostome line.

If we were to look for the keys to success of just the chordates, we might look for features that these animals have in common with other highly successful animals. Based on species diversity and environmental distribution, the arthropods are the most successful phylum in the animal kingdom. As protostomes, arthropods have little in common with any deuterostomes. Yet, through convergent evolution, arthropods and chordates have come to share two very significant traits. For one thing, both groups are metameric (segmented), although arthropod and chordate metamerism evolved independently. As discussed earlier (see the Trends and Strategies section beginning on p. 583), metamerism has been a powerful force in animal evolution, and it is probably no coincidence that metameric animals constitute the largest and most ecologically diverse phyla in both the protostome and deuterostome lineages. In addition to their metamerism, arthropods and chordates also have rigid skeletal materials, and like many of their body systems, their skeleton is segmented. The arthropod exoskeleton is a complex cuticle, very different from the dermal endoskeleton of chordates. In both cases, however, metamerism and a rigid but segmented skeleton have been a winning combination. Primarily a locomotor adaptation, a segmented body is flexible, but if supported by hard skeletal parts, regions of it can be made rigid as needed for complex body movements. A segmented, strongly supported body is a highly adaptable locomotor machine, one that can support itself and move on land as well as in water. Many zoologists believe that being metameric and having strong skeletal support "preadapted" aquatic arthropods and chordates for colonizing land.

The following seven chapters on deuterostomes have three main thrusts. First, like all chapters in this unit, they survey diversity. Second, these chapters focus on key features that have allowed some groups to be more adaptable to diverse environments than others. Metamerism and skeletal features were central adaptations in deuterostome evolution, but other body systems played vital roles in making chordates one of the most successful phyla. Finally, these chapters reiterate threads—largely developmental patterns—that unite the deuterostome phyla. Echinoderm development, for example, illustrates the fundamental (perhaps ancestral) features of deuterostomes and sets the stage for tracing the main evolutionary trends in this important division of the animal kingdom.

CHAPTER 27

Lophophorates, Echinoderms, and Other Nonchordate Deuterostomes

Phylum Echinodermata. This sea star *Acanthaster planci*, the crown of thorns, illustrates the radial symmetry of adult echinoderms. Many other sea stars have five arms, or rays, projecting from a central disk, but as seen here, *Acanthaster* is multirayed. This species grows to about 67 cm in diameter; it eats coral and can be highly destructive to coral reefs. This specimen is grazing on coral in the South Pacific.

This chapter examines diversity among invertebrate deuterostomes other than chordates. Six phyla are discussed here; all are aquatic, and all but one are entirely marine. Of the six, the largest and ecologically most significant phylum is the

Echinodermata, but we look first at three smaller phyla—the Bryozoa, Phoronida, and Brachiopoda—collectively called lophophorates.

LOPHOPHORATE PHYLA

Lophophorates are nonmetameric, sessile filter feeders that live in a secreted chitinous or calcareous tube or exoskeleton. They are called lophophorates because their most prominent body feature is a circular or horseshoe-shaped crown of hollow tentacles called a **lophophore**. Actually an extension of the body wall, the lophophore stretches out for feeding, and cilia on its tentacles trap suspended particles and direct them toward the mouth. The lophophore encircles the mouth, but not the anus. In addition to its primary role in feeding, the lophophore also functions in gas exchange, and sensory cells receptive to chemicals and touch are concentrated on its tentacles. Both the lophophore and its tentacles contain coelomic spaces and are extended hydraulically as muscle contractions force coelomic fluid into these spaces.

The phylogenetic position of the lophophorates is controversial. We consider them deuterostomes because many of them show the radial cleavage and enterocoelous development typical of deuterostomes. However, because the larvae of some marine lophophorates bear slight resemblance to the trochophores of certain molluscs and annelids, and because, in a few types (i.e., phoronids), the blastopore becomes the mouth, some zoologists consider these animals phylogenetically intermediate between deuterostomes and protostomes.

PHYLUM BRYOZOA (ECTOPROCTA)

With about 5000 living species, the Phylum Bryozoa is actually among the ten largest phyla in the animal kingdom, and more than 15,000 extinct species are known, dating mostly from about 350 to 600 million years ago. Certain fossil species helped form ancient reefs.

Bryozoans are colonial, and most are marine, but about 50 species inhabit freshwater lakes and slow-moving streams. The prefix *bryo* is derived from the Greek word for moss, and these animals are often mistaken for aquatic fungi, moss, or seaweed. A growing number of researchers are now studying bryozoans, and the phylum may be the focus of greater interest in the future. One marine species has recently been found to contain a chemical substance called **bryostatin**, which is active in fighting a type of leukemia in mice.

Freshwater bryozoans grow horizontally or upright on shaded sides of rocks, plant stalks, or other solid substrates (Figure 27.1a). Several species are common, and some may cover more than 1 m² of substrate, providing attachment sites for many protozoa and small invertebrates. A few turbellarians, snails, leeches, and some arthropods eat bryozoans.

Marine bryozoans typically grow on solid substrates in shallow coastal waters. Like corals, several species help form extensive underwater reefs. Others colonize floating mats of algae that may be carried far out to sea, and a few live in deep oceans (up to 6000 m deep). Like barnacles, marine bryozoans may encrust boat hulls, and if not regularly removed, they can cause significant fouling.

Individuals in a bryozoan colony are called **zooids**. Each zooid consists of a **polypide** (the digestive tract crowned by the lophophore, muscles, and most other organs) and a **case** formed of the body wall including the exoskeleton (see Figures 27.1 and 27.2). All zooids in freshwater bryozoans look alike. In contrast, marine bryozoans may have several kinds of zooids. While most of a colony is composed of feeding individuals with a circular or horseshoe-shaped lophophore, some zooids are beaklike or bristlelike. These zooids defend the colony either by grasping or sweeping away other animals crawling over it, or by attempting to build tubes on it.

The bryozoan lophophore and its tentacles extend as muscle contractions decrease the volume of the coelom in the case. This causes coelomic fluid to push the polypide out the aperture and to engorge the hollow tentacles. Powerful retractor muscles extending from the body wall to the base of the lophophore pull the polypide inside the case. Freshwater bryozoans have a flexible chitinous exoskeleton. The adjacent sides of their zooids are open, and the coelom is continuous throughout the colony (see Figure 27.1). The lophophore extends when body wall muscles contract, compressing the sides of the case. Marine bryozoans lack body wall muscles, and each polypide resides in a complete, rigid case with calcareous or chitinous sides (Fig-

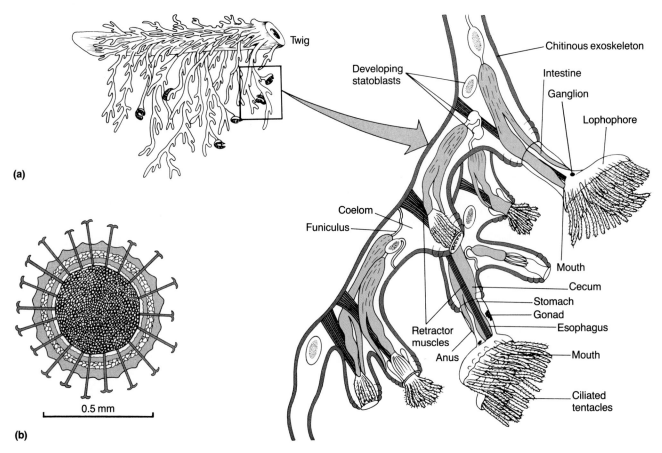

Figure 27.1 Phylum Bryozoa. (a) Sketch of a colony of the freshwater genus *Plumatella* on a plant stalk, enlarged about 1½×. A portion of the colony is enlarged still further to show internal structures. The body wall is covered with chitinous exoskeleton and lined internally with peritoneum. The spacious coelom is continuous throughout the colony and extends into the hollow tentacles. (b) The resistant stage, or statoblast, of *Cristatella mucedo,* a freshwater species, consists of flattened masses of germinal cells surrounded by a protective wall with spiny projections.

ure 27.2). Some species have pores in the sidewalls, but the pores are filled with cells. Contraction of special lophophore protractor muscles decreases the volume of each zooid's coelom by pulling a flexible membrane into the coelom.

When expanded, the bryozoan lophophore forms a particle trap and funnel. The tentacles have cilia on their inner and lateral surfaces, but not on the outside. The cilia drive water carrying suspended particles downward toward the centrally located mouth (see Figure 27.2). In most bryozoans, the tentacles remain stationary during feeding, but in a few species, they can close around small prey. Several species can rotate the lophophore, apparently to sample particles in the water, and some can select particles by flipping them toward or away from the mouth with a tentacle. An expansive, muscular pharynx helps suck food particles into the Y-shaped gut. Extracellular digestion oc-

curs in the stomach, and nutrients are absorbed by a large cecum and intestine. Because the anus is located on the outer surface of the lophophore, bryozoans are sometimes given the phylum name **Ectoprocta**.

Bryozoans have no respiratory, excretory, or circulatory organs; coelomocytes and coelomic fluid are circulated by cilia on the peritoneum and distribute nutrients, dissolved gases, and metabolic wastes. Nutrients may also be transported along a cord of tissue called the **funiculus**, which extends from the cecum to the peritoneum (see Figure 27.1). The bryozoan nervous system includes a ganglion dorsal to the pharynx, a circumpharyngeal nerve ring, and a subepidermal nerve net. In some species, small nerves extend through exoskeletal pores, allowing some coordination between adjacent zooids. Most bryozoans respond negatively to light and live in shaded areas.

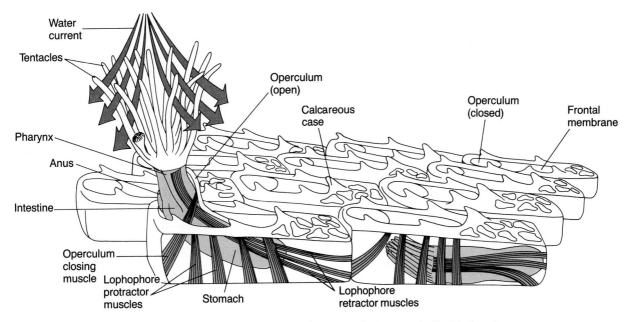

Figure 27.2 Part of a colony of the marine bryozoan *Electra*. Note the flexible frontal membrane over part of each zooid's case. The lophophore is extended when protractor muscles pull the frontal membrane down into the case; this decreases the coelomic volume of the case and increases coelomic fluid pressure, which forces the polypide out the aperture and extends the lophophore. Other muscles operate a trap-door-like operculum that closes behind the retracted lophophore. Arrows indicate the feeding mechanism. Cilia on the sides of the tentacles drive water downward from above the lophophore and out between the tentacles; suspended food, mostly algal cells and bacteria, is trapped by cilia on the inner surfaces of the tentacles and swept into the mouth.

All freshwater bryozoans and most marine species are hermaphroditic. Gametes are shed into the coelom, where fertilization usually occurs. In most species, sperm leave the coelom of one zooid and enter another zooid of the same colony through tiny pores at the end of the tentacles. Fertilized eggs are usually brooded in the coelom. Cross-fertilization among neighboring colonies occurs periodically in marine species, providing greater opportunity for genetic variability. Bryozoans have deuterostome-like radial cleavage, but additional research is needed to fully document their embryogeny. Some marine species have a free-swimming ciliated larva that bears some resemblance to the trochophore of molluscs and annelids. A bryozoan larva swims for anywhere from a few hours to several months, depending on the species, then sinks and attaches to the underside of a rock or other object and grows into a zooid. A colony is then formed by budding.

Freshwater bryozoans produce dormant stages called **statoblasts** (see Figure 27.1b). Statoblasts are budded from the funiculus and are usually released when the colony disintegrates at the end of a growing season. They can withstand freezing, drying, and extreme heat and will hatch only with the return of favorable water conditions. Statoblasts are often produced in great numbers and may be carried long distances by wind, water, or aquatic birds and mammals.

PHYLUM PHORONIDA

Phoronids are wormlike lophophorates, and with fewer than 20 species, this is the smallest lophophorate phylum (Figure 27.3). Phoronids range in length from a few millimeters to about 50 cm, and all are marine. They live singly or in large aggregates in secreted chitinous tubes buried in sand or attached to pilings, shells, or rocks in tidal areas and coral reefs.

Protected and supported by their tubes, phoronids have a thin, nonchitinous cuticle. The phoronid lophophore consists of two parallel folds of the body wall. Contraction of circular muscles in the body wall expands the lophophore hydraulically. Contraction of longitudinal muscles in the body wall shortens the body, pulling it and the lo-

(a)

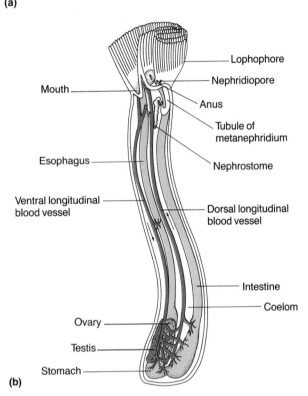

Lophophore

Nephridiopore

Mouth

Anus

Tubule of
metanephridium

Nephrostome

Esophagus

Ventral longitudinal
blood vessel

Dorsal longitudinal
blood vessel

Intestine

Coelom

Ovary

Testis

Stomach

(b)

Figure 27.3   Phylum Phoronida. (a) Individuals are pink, orange, or yellow. Their secreted tubes are often covered with sand grains. Unlike bryozoans, phoronids are not attached to their tubes and move freely within them. When disturbed, the entire body withdraws into the tube. This photograph shows a group of *Phoronis vancouverensis*, a subtidal species of the Pacific coast, with tentacles spread to gather food. (b) Cutaway drawing of a phoronid removed from its tube.

the anus. The nervous system includes a giant motor fiber that innervates the longitudinal muscles, allowing the animal to withdraw into its tube very quickly if disturbed.

Phoronids are hermaphroditic. Gametes pass from the coelom out through the metanephridial tubules and are then carried in an external groove to the space between the folds of the tentacles, where fertilization occurs. Cross-fertilization is the rule. Eggs of some species become planktonic, while others are brooded on the lophophore. Cleavage follows a unique pattern that seems to be a variant of the radial plan; coelom formation resembles the enterocoelous pattern of deuterostomes. Unlike the deuterostome pattern, however, the blastopore develops into the mouth. Following gastrulation, an elongate, ciliated larva, called an **actinotroch**, develops in both brooding and nonbrooding species. After a free-swimming period, the larva sinks to the bottom, undergoes metamorphosis, and secretes a chitinous tube. Budding is also an important means of reproduction in some phoronids, and one species lives in large clones generated by budding and transverse fission. Although phoronid development raises questions about whether these animals are deuterostomes or protostomes, phoronids share several fundamental traits with bryozoans, and it seems likely that phoronids and bryozoans arose from a common stock.

PHYLUM BRACHIOPODA

Brachiopods are commonly called lampshells because they resemble ancient Roman oil lamps. Having two shell valves, they also resemble bivalve molluscs, and until the middle of the last century, brachiopods were classified in the Phylum Mollusca (Figure 27.4). Actually, bivalve molluscs and brachiopods are fundamentally different, and their superficial similarity results from convergent evolution, probably as both groups became adapted to sessile filter feeding. The valves of bivalve mollusc shells cover the right and left sides of the animal (see Figure 24.7a), whereas lampshells have dorsal and ventral valves. Also, bivalve molluscs filter-feed with their gills, while brachiopods use a lophophore.

Brachiopods are exclusively marine, and although most species live from the tidal zone to the

phophore into the tube. Phoronid feeding is similar to that of bryozoans.

The phoronid body consists mainly of an elongate **trunk** that is expanded posteriorly, helping to anchor the animal in its tube (see Figure 27.3). Unlike other lophophorates, phoronids have a closed blood vascular system. Two large longitudinal vessels with muscular walls serve as pumps, and red blood cells contain a type of hemoglobin. A pair of metanephridia filters fluid from the coelom, and urine is excreted through two nephridiopores near

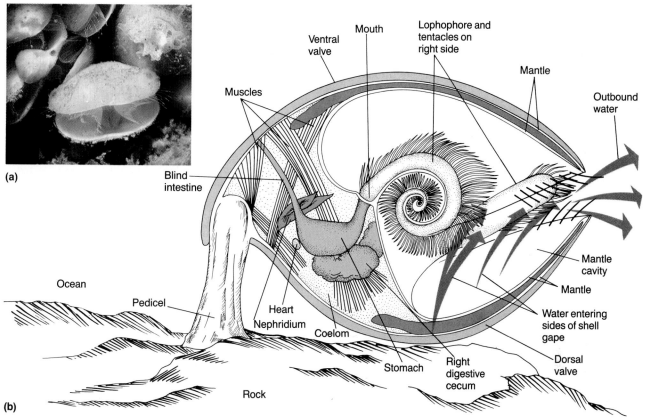

(a)

(b)

Figure 27.4 Phylum Brachiopoda. (a) A cluster of *Terebratulina septentrionalis.* Brachiopods have gray, yellowish, or reddish valves. (b) This cutaway drawing shows internal structures as viewed from the left, with the left side of the lophophore removed. Note how the multilobed lophophore projects into the mantle cavity. Arrows indicate the flow of water through the mantle cavity; water enters laterally between the slightly gaping valves and leaves through the gape at the anterior end.

edge of the continental shelves (down to about 200 m), a few species inhabit far greater depths (as deep as 6000 m). There are only about 300 living species in the phylum, but brachiopods were a dominant marine group during former geological times. Their hard shells readily fossilize, and 30,000 extinct species have been described. Living brachiopods are fairly small (5–80 mm in length), but certain fossil species were up to 300 mm long. Cambrian rocks estimated to be 600 million years old contain abundant brachiopod fossils, and it is likely that many Cambrian species evolved during the Precambrian. Brachiopods reached peak abundance and diversity between 300 and 400 million years ago, after which many species died out in a relatively short time. Perhaps, as some zoologists postulate, these lophophorates were unable to compete successfully as filter feeders with the more efficient bivalve molluscs. Living brachiopods are remarkably similar to certain fossil forms. *Lingula* may be one of the oldest living genera of animals

(Figure 27.5); the fossil record indicates that it has changed little since the Ordovician period, over 400 million years ago.

Brachiopod shells are calcareous and/or chitinous and are secreted by an underlying mantle. The valves are hinged in some species, but they are held together entirely by muscles in others. The ventral valve is often larger and is usually attached to a hard substrate, either directly or by a stalklike **pedicel.** Some brachiopods, such as *Lingula,* have a muscular pedicel used for burrowing and anchoring in mud or sand (see Figure 27.5).

The brachiopod lophophore is multilobed, and it occupies most of the mantle cavity between the valves (see Figure 27.4b). Contraction of paired **diductor muscles** opens the valves, pulling the soft body parts posteriorly and forcing coelomic fluid forward into the lophophore. In many species, the dorsal valve bears special grooves and projections that help support the expanded lophophore. Water is generally drawn in from the sides of the mantle,

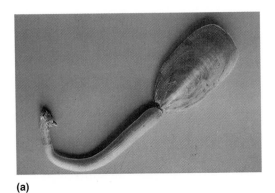

(a)

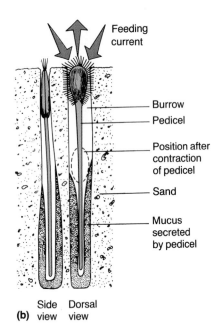

Feeding current

Burrow

Pedicel

Position after contraction of pedicel

Sand

Mucus secreted by pedicel

Side Dorsal
(b) view view

Figure 27.5 The brachiopod *Lingula*. (a) *Lingula* is a common intertidal inhabitant of the Pacific and Indian oceans. (b) Contraction of muscles in the pedicel pulls the animal into its burrow in sand or mud. Arrows show direction of water while the animal is feeding.

passes between the numerous tentacles of the lophophore, and is driven out the front of the animal. Suspended particles, trapped by cilia and mucus on the tentacles, pass to ciliated food grooves at the base of the lophophore and are swept into the mouth. Rejected particles pass into the excurrent water flow. The cilia can reverse water flow to cleanse the lophophore. Similar to bivalve molluscs, brachiopods have adductor muscles with striated "quick" fibers that contract rapidly to close the valves and smooth "catch" fibers that keep the valves closed for long periods of time. Other muscles allow the animal to turn on its pedicel.

The brachiopod digestive tract and other organs are located in a large coelomic space behind the mantle cavity. Both intra- and extracellular diges-

tion may occur in the stomach and in large digestive ceca. The intestine is a blind tube in many brachiopods, but others have a complete gut with an anus opening near the pedicel or into the excurrent water in the mantle cavity. Gas exchange occurs across the extensive surfaces of the mantle and lophophore, and coelomocytes carry the reddish respiratory pigment hemerythrin. There is also an open circulatory system (including a small heart near the stomach) and one or two metanephridia.

Most brachiopods are dioecious. Gametes are shed into the ocean via the metanephridia, and fertilization is external. Free-swimming stages—either a ciliated larva or a juvenile much like a miniature adult—develop following radial cleavage and enterocoelous mesoderm and coelom formation. The free-swimming stages eventually settle to the bottom and grow into adults.

PHYLUM ECHINODERMATA

Echinoderms (about 6,000 living species) occur virtually everywhere in the oceans of the world, from shallow coastal waters to abyssal depths (down to 12,000 m). Often they are the most numerous species in deep oceans. These deuterostomes have three distinctive features. First, they are bilaterally symmetrical as larvae, but are typically radially or biradially symmetrical as adults. As seen in most sea stars, the adult body tends to be **pentamerous**, meaning that it has five radiating parts. Second, the name *echinoderm*, meaning "like a sea urchin" or "spiny skin," refers to the calcareous spines or spicules embedded beneath the skin. The spines are actually components of a hard endoskeleton formed of uniquely structured calcareous ossicles. Third, echinoderms have a hydraulic locomotor system unlike any other system of movement in the animal kingdom. Usually called the **water vascular system** (although it is not primarily a circulatory system), this is a set of interconnected internal water tubes with many small external projections called **tube feet**, or **podia**. The water vascular system provides the slow, steady movement characteristic of sea stars and many other echinoderms, and it has a variety of other functions as well. Radiating bands or grooves of the echinoderm body housing parts of the water vascular system are called **ambulacra**.

REPRESENTATIVE ECHINODERMS

There are six classes of living echinoderms, one of which was described only recently (1986).

Class Crinoidea

Only about 625 living species represent the **crinoids**, an ancient group, which reached its peak about 350 million years ago, during the Carboniferous period. Many of the approximately 5000 fossil species have been described from thick beds of limestone, consisting largely of fossils formed when the seas inundated the central plains of North America. Unfortunately, less is known about crinoids than about any other group of echinoderms; many species occur only in deep oceans and are difficult to keep alive for study. **Sea lilies** are flowerlike crinoids permanently attached by a stalk (Figure 27.6a). The more numerous **feather stars** have stalks as larvae, but the adults are stalkless (Figure 27.6b).

Crinoids feed on suspended organic particles and plankton, using mucus-coated tube feet. It has been suggested that the water vascular system originated as a filter-feeding device, and crinoids illustrate how the system may have worked in ancestral echinoderms.

Class Asteroidea

Asteroids, the **sea stars** (about 1600 living species), have arms projecting from, but not sharply set off from, a central disc. Sea stars move by means of tube feet set in **ambulacral grooves** on the undersides of the arms (Figure 27.7). Many species have

(a)

(b)

Figure 27.6 Phylum Echinodermata: Class Crinoidea. (a) Only about 80 species of sea lilies remain today, and some zoologists believe that this group is becoming extinct. (b) Feather stars, such as this species *Nemaster rubiginosa*, often attach temporarily, but also swim or creep over the bottom. They are found at great depths as well as in coastal areas and on coral reefs. Many shallow-water species are beautifully colored in shades of crimson, orange, yellow, and purple.

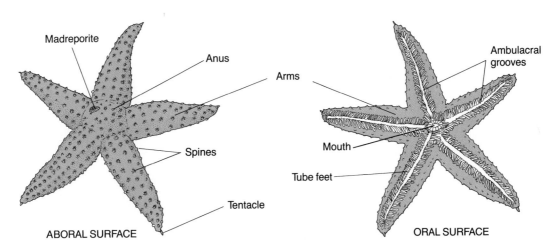

ABORAL SURFACE

Madreporite

Anus

Arms

Spines

Tentacle

Ambulacral grooves

Mouth

Tube feet

ORAL SURFACE

Figure 27.7 Class Asteroidea. Typical of the class, the five arms and central disc of this sea star are not clearly demarcated. The underside (oral surface) of the arms bears ambulacral grooves containing rows of tube feet.

the characteristic 5 arms, but others can have 6, 7, 11, or up to 50 arms (see chapter opening photograph). Some species are rather drab in color, but others are shades of bright blue, orange, or red.

Class Ophiuroidea

With about 2000 species, **ophiuroids** constitute the largest class in the phylum. Commonly called **brittle stars** and **basket stars**, these echinoderms are agile and often brightly colored. They are called ophiuroids, meaning snakelike, because they have long, solid arms that move in a serpentine manner, pulling the animals rapidly about. Brittle stars have long unbranched arms, whereas the arms of basket stars are dichotomously branched. In contrast to sea stars, both brittle stars and basket stars have arms sharply set off from the central disc (Figure 27.8). There are no open ambulacral grooves.

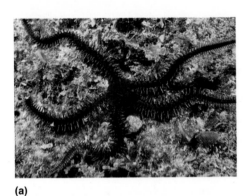

(a)

(b)

Figure 27.8 Class Ophiuroidea. Ophiuroids are generally smaller than sea stars and are often concealed in rock and coral crevices. Notice how much more clearly demarcated the arms and central disc are, compared to those of asteroids. (a) The brittle star *Ophiocoma aethiops* is a common inhabitant of shallow tidepools along the Pacific coast of the United States. This is a large species, with an arm span of up to 44 cm. Several other species of brittle stars thrive in the ocean abyss, feeding on organic debris that sinks from productive waters above. (b) The basket star *Gorgonocephalus caryi*, about 25 cm in diameter, catches plankton and sometimes small invertebrates and fishes by periodically extending its multibranched arms. Each of the five arms of this species branches about 12 times and can be rolled in toward the mouth.

The tube feet lack suckers and with few exceptions do not contribute significantly to locomotion. Instead, they function mainly in chemoreception and food capture. Ophiuroids are widely distributed in all oceans, and several species are numerous at depths down to 6000 m. They outnumber sea stars and other larger echinoderms on many seafloors. Populations with as many as 2000 individuals per square meter have been documented.

Class Concentricycloidea

This class consists of the single species *Xyloplax medusiformis* found at depths of about 1200 m off the coast of New Zealand. First identified in 1986, *X. medusiformis* is about 2–8 mm in diameter and lacks a gut and radiating arms (Figure 27.9). Unlike any other living echinoderm, its water vascular system forms two rings (others have one), and it has a single ring of tube feet that project from between ambulacral grooves. The ventral surface is covered by a thin, membranous **velum**.

Class Echinoidea

Echinoids (about 900 species) include the spiny, globose **sea urchins**, ovoid **heart urchins**, and flattened, disklike **sand dollars** (Figure 27.10). Like sea stars and brittle stars, these echinoderms frequent coastal areas as well as the ocean abyss (down to at least 5000 m) and are common in tropical as well as polar seas. Echinoids lack arms and have an armorlike **test** (not a shell because it is internal) formed of tightly knit skeletal plates. Tube feet bear

Figure 27.9 Class Concentricycloidea. The recently discovered *Xyloplax medusiformis* is disklike and, as its name implies, resembles a medusoid cnidarian.

(a)

(b)

Figure 27.10 Class Echinoidea. (a) The slate pencil urchin *Heterocentrotus mammillatus* uses its thick, blunt spines to wedge itself in crevices on coral reefs. The red spines have sometimes been used for writing on slate. (b) The sand dollar *Echinarachnius parma*. Sand dollars (and the less-flattened, more ovoid heart urchins) usually bury themselves under a thin layer of sand or mud. They have many small movable spines that are their main locomotor organs.

suckers and are the main locomotor organs in sea urchins. The tube feet project through the test in five ambulacral areas. Echinoids also have movable spines that assist in locomotion in sea urchins and are the main locomotor organs of sand dollars and heart urchins. Sand dollars and heart urchins are also called "irregular echinoids" because their tests are not radial. Irregular echinoids have anterior and posterior ends and are bilateral. This is tertiary bilateral symmetry, for as larvae, they are first bilateral, then radial, and finally bilateral.

Class Holothuroidea

The common name for **holothurians** is **sea cucumbers**, and they are well named, for many of them truly resemble cucumbers. They lack arms, and many species are rather drab in color, although some are bright red, orange, or purple (Figure 27.11). Most of the approximately 1100 species in this class are sluggish bottom dwellers, and several species burrow into mud and sand. The body wall is soft with small ossicles embedded in the dermis,

(a)

Figure 27.11 Class Holothuroidea. Sea cucumbers have anteroposterior orientation, with a crown of tentacles at the anterior end. The tentacles function in filter feeding or in sweeping bottom sediments into the mouth. (a) Species of *Cucumaria,* common on both sides of the Atlantic and on the U.S Pacific coast, have five ambulacra bearing tube feet. Ambulacra are equally spaced around the body. Their suckered tube feet allow them to anchor in rocky crevices and under·stones. This species, *Cucumaria miniata,* an inhabitant of rocky intertidal areas, reaches 20 cm in length. (b) *Stichopus californicus,* up to about 50 cm long and common on the Pacific coast of North America, is a deposit feeder, mopping the seafloor with its tentacles to obtain organic material. Only one surface has functional tube feet and contacts the substrate. This closeup shows the delicate branching of the tentacles around the mouth.

(b)

and there are no hard spines. Most sea cucumbers are about 10–30 cm long, but a few species are as long as 2.5 m when fully extended.

Sea cucumbers have locomotor tube feet with suckers, as well as modified tube feet that form feeding tentacles around the mouth. In many species, only one side of the animal has functional tube feet; this side is always applied to the substrate. Because of this feature, and because the mouth is at one end and the anus at the other, sea cucumbers tend to be more bilateral than radially symmetrical.

Holothurians inhabit all oceans of the world. Several species are common in shallow seas, including coastal tide pools, while others live at great depths, where they are often among the most numerous animals. Dredgings from depths as far down as 10,500 m in waters off the Philippines have yielded many specimens. Deep-sea species often have pronounced bilateral symmetry.

HABITATS AND ENVIRONMENTAL RELATIONSHIPS

Despite their considerable success in the ocean, echinoderms have not colonized fresh water, and only a few species can tolerate brackish waters. As discussed in Chapter 7, living in fresh water requires that the body be able to dispose of excess water, and echinoderms, unlike other animals, largely lack this ability. (See the discussion of freshwater animals in Chapter 7, p. 150.) It may be for this reason that the Echinodermata is the only major phylum (with over 4000 species) that does not include any freshwater species (see Table 1.3).

Great numbers of echinoderms live on coral reefs, where they play a significant role in tropical food chains. Despite their firm, often spiny endoskeleton, echinoderms are important in the diet of many animals, especially crustaceans, molluscs, and fishes. Brittle stars are often common on coral reefs, where they may occupy almost any small rock cranny or discarded shell. Sea urchins are important in controlling algal growth on coral reefs; die-offs of populations of algae-eating sea urchins in the Caribbean have proved harmful to several reefs. Algae, normally controlled by the sea urchins, overgrew the corals, blocking out light necessary for photosynthesis by the coral animals' algal symbiotes.

Dense populations of sea urchins may also be harmful. In recent years, large numbers of *Strongylocentrotus*, common algae-eating sea urchins, have destroyed extensive growths of kelp (multicellular algae) off the California coast. Kelp beds form important habitats for many marine animals. The number of sea urchins increased markedly when populations of one of its main predators, the sea otter, were greatly reduced by human activity.

Sea stars eat a wide variety of invertebrates, including other echinoderms, and some even eat small fishes. The so-called crown-of-thorns sea star, *Acanthaster planci*, eats coral polyps and has damaged extensive areas of reef in the tropical Pacific. (See chapter opening photograph and the Chapter 21 Essay on coral reefs, p. 496.) Species of *Asterias* thrive on bivalve molluscs and often congregate where oysters are cultured commercially in shallow, temperate seas. A single large sea star may consume as many as ten oysters a day. Oyster farmers control *Asterias* by dragging large mops of loose threads over oyster beds. The sea stars become entangled in the mops.

Many echinoderms form symbiotic relationships with other animals. Tiny brittle stars live as commensals in the bodies of sponges and corals; the brittle stars feed on fine particles swept in by the host. Pearlfishes, small tropical fishes of the genus *Carapus*, live in the hindgut of sea cucumbers during the day, swimming out through the anus at night to feed in the ocean. Many parasitic and commensalistic protozoa, flatworms, nematodes, and even gastropod molluscs live in and on echinoderms, but little is known about these host-symbiote relationships.

COMPARATIVE STRUCTURE AND FUNCTION

Despite the differences among the six classes, all echinoderms are fundamentally similar. Except for the sedentary stalked crinoids, most adult echinoderms are free-moving inhabitants of the seafloor. Because sea stars and sea urchins are the best-known members of the phylum, we will concentrate on their characteristics, mentioning other groups as appropriate.

Body Support, Protection, and Locomotion

Echinoderms have a multilayered body wall covered with a thin, often ciliated epidermis. Beneath

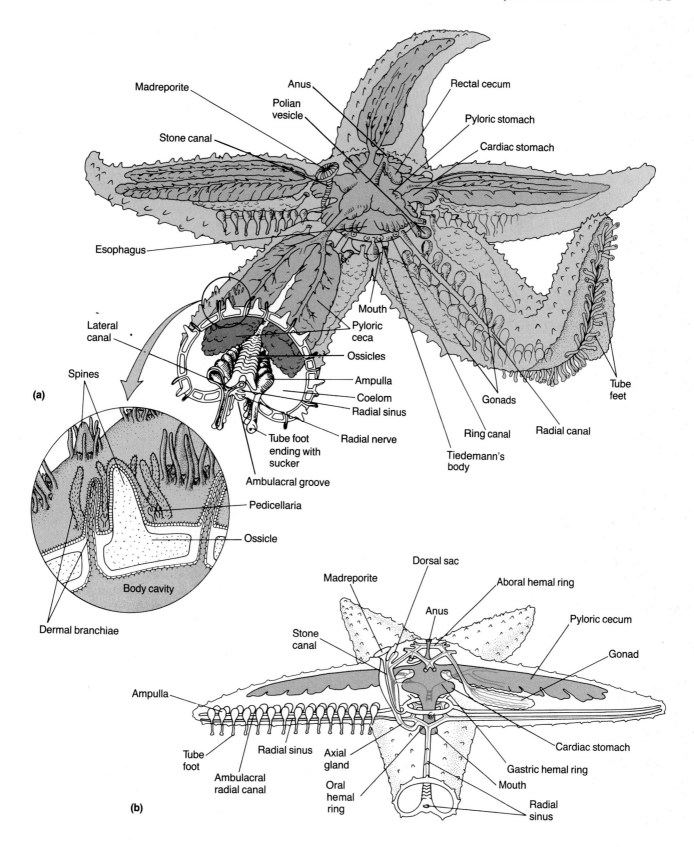

Figure 27.12 Cutaway drawing of a sea star. (a) Details of the endoskeleton, water vascular system, and other internal organs. (b) The hemal system in relation to the digestive and water vascular systems.

the epidermis, a connective tissue dermis forms the endoskeletal **ossicles**. The ossicles in sea stars, brittle stars, and most other echinoderms resemble small blocks or rods supporting the disc and arms, and the ossicles often extend into the epidermis as **spines** (Figure 27.12). Ossicles and spines are composed of calcium carbonate, and there is some evidence that each ossicle or spine consists of only one large crystal. Spines provide protection, mainly against large predators. The movable spines of sea urchins are not fused to the endoskeletal plates that form the globose test (Figure 27.13); the spines articulate with the plates by muscles, and some species have long, formidable spines.

In addition to spines, many sea stars and sea urchins also have special pincerlike structures called **pedicellariae** that constantly clean the skin surface. Often encircling the base of the spines, these structures resemble tiny jaws projecting from the body surface (see Figures 27.12 and 27.13). Pedicellariae wave about, snapping at surface contaminants such as minute, colonizing organisms and debris. They contain two or more hard ossicles and are opened and closed by antagonistic muscles. Experiments indicate that at least some pedicellar-iae are independent effectors; that is, they respond to stimuli without being activated by impulses from the nervous system. Although tiny, pedicellariae can be very effective weapons. Those of many sea urchins contain poison sacs that secrete paralytic and toxic substances. Pedicellariae in the sea star *Stylasterias* have recurved toothed spines that can trap small fishes.

Muscles of the Body Wall and Locomotion. Echinoderms typically have a thin outer circular muscle layer just under the dermis, as well as a thicker longitudinal layer bordering the peritoneum of the coelom. In sea stars, these muscles may flex and elevate an arm, helping the animal to right itself if flipped upside down. Having a rigid test without arms, sea urchins lack muscles in the body wall. In contrast, body wall muscles provide virtually all locomotion in brittle stars. The arms of both brittle stars and basket stars are much more flexible than those of other echinoderms. Each arm contains a row of ossicles, and longitudinal muscles between adjacent ossicles provide the serpentine movement typical of the group. When a brittle star moves, it lifts its central disc off the substrate, and one or

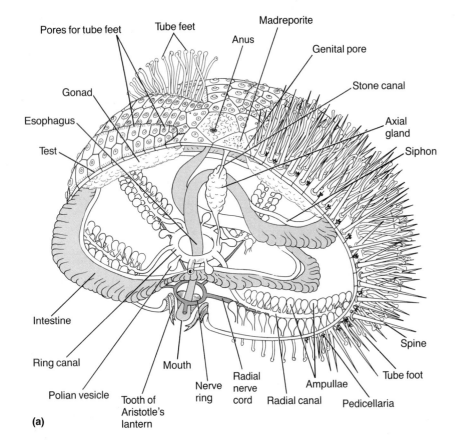

(a)

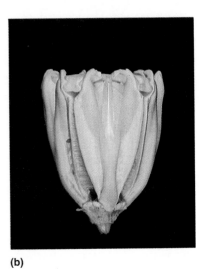

(b)

Figure 27.13 Anatomy of a sea urchin. (a) Note the arrangement of ambulacral areas composed of flattened plates (ossicles) bearing holes through which the tube feet protrude. The inner surface of each ambulacral area bears a radial canal and ampullae associated with the tube feet. Spines project from the ambulacral plates and also from plates forming the interambulacral areas. (b) The complex chewing organ, Aristotle's lantern, from *Echinometra lacunter*, from waters off Bermuda.

two of its arms pull forward. The other arms trail behind, assisting forward movement by pushing or rowing.

Movement in many sea cucumbers also results mainly from the action of muscles in the body wall. The body is highly flexible because its ossicles are tiny and widely separated. Also, there are no spines or pedicellariae. Many sea cucumbers burrow into bottom muds, and some can swim for extended periods of time using specially developed fins or using wormlike movements produced by muscles exerting force against the fluid-filled coelom.

Water Vascular System. This hydraulic system is similar in all echinoderms, and understanding how it works provides an appreciation for much of the uniqueness of echinoderms. The system is derived from the coelom and associated mesodermal tissues and consists of radially arranged interconnecting canals lined with ciliated epithelium. The fluid in the water vascular system is much like seawater, and studies with radioactive tracers show that, at least in some species, seawater may enter the system.

As shown in Figures 27.12 and 27.13, sea stars and sea urchins have a **ring canal** encircling the gut in the central disc. Pouches called **Tiedemann's bodies**, extending from the ring canal, produce coelomocytes. Bulblike **Polian vesicles**, apparently involved in regulating fluid pressure in the system, may also extend from the ring canal. The ring canal connects, via a calcified **stone canal**, to a furrowed skeletal plate, the **madreporite**, which is on the aboral surface (the top, or side opposite the mouth) in sea stars and sea urchins. The water vascular system communicates with the outside environment through pores in the epithelium covering the madreporite. The madreporite may allow equilibration of the hydrostatic pressure inside the water vascular system with that of the environment. In most sea cucumbers and crinoids, the madreporite is internal, and in these animals it may be involved in equilibrating pressures between the coelom and the water vascular system.

The remainder of the water vascular system usually consists of five (or multiples of five) **radial canals** projecting into the arms (in sea stars) or ambulacral radii of the body (in sea urchins). In sea stars and sea urchins, radial canals have side branches, called **lateral canals**, each with a bulblike muscular **ampulla** at the base of a hollow, blind tube foot, or **podium**. Some sea stars have up to 1200 tube feet projecting through spaces in or between ossicles lining the ambulacral grooves.

Contraction of muscles in the wall of an ampulla forces water into the attached tube foot, causing the foot to push against the substrate. A valve in each lateral canal prevents fluid in the tube foot and ampulla from passing into the radial canal. Tube feet of most sea stars and sea urchins have a terminal sucker, providing a gripping surface effective for holding onto and moving across rock surfaces. A tube foot is bent and turned by alternate contraction of longitudinal muscles in its walls.

Sea stars and sea urchins move as a result of the coordinated action of the tube feet. In a forward-moving sea star arm, the tube feet bend forward over the substrate, the suckers contact the surface, and an adhesive vacuum is created as the centers of the suckers are pulled in. The suckers are then withdrawn as longitudinal muscles shorten and bend the tube feet backward, forcing water back into the ampullae. Although the system is slow, it provides a steady, powerful locomotor force. Suckered tube feet are not very effective on soft substrates or in sand, and consequently, most sea stars and sea urchins live on hard surfaces. Those that inhabit mud or sand typically have suckerless, pointed tube feet.

Feeding and Digestion

As a group, echinoderms eat an enormous variety of foods, and most are opportunistic feeders. Many species are bottom scavengers, and even those that are chiefly carnivorous or herbivorous may supplement their diets with dead organic debris available on the seafloor. Brittle stars scavenge, but can also filter-feed by waving their arms about, collecting organic particles and fine plankton on mucous threads strung between their spines or tube feet. The tube feet accumulate particles and transport them to the mouth. Brittle stars and basket stars have an incomplete gut, with a stomach but no intestine or anus; little is known about digestive function in these animals.

Sea cucumbers have varied feeding habits. Some species are deposit feeders, ingesting mud as they burrow through it. Others extend their tentacles into the water to filter-feed, or sweep the tentacles over the substrate to pick up bottom deposits. Particles are trapped in sticky mucus on the tentacles, and then, one by one, the tentacles are swallowed, and the particles are removed in the pharynx. The tentacles are recoated with mucus as they are pulled out of the pharynx.

Crinoids (see Figure 27.6) are strictly filter feeders. Their feathery arms, equipped with tiny, mucus-coated tube feet, form an extensive sticky trap. As particles adhere to the mucus, a beating action of the tube feet flips the trapped particles into the adjacent ambulacral food groove. Cilia in the groove then drive the food particles toward the mouth.

Most sea stars are carnivores. Because they are incapable of catching rapidly moving prey, their diet consists mainly of slow, bottom-dwelling molluscs and other invertebrates that they encounter by chance. A few species have highly specific diets (one species eats only sea cucumbers). Conforming with the rest of the sea star's body, the digestive tract is radially arranged. It consists of the mouth; a short, vertical esophagus; a large, two-part (cardiac and pyloric) stomach; a short, vertical intestine; and the aboral anus, all of which lie in the central disc (see Figure 27.12). Food is moved through the system mainly by ciliary action. As in ophiuroids, a few sea stars lack an intestine and an anus. Radially arranged sacs called **pyloric ceca** extend into the arms from the pyloric region of the stomach. Digestion begins externally or in the cardiac stomach, and material is swept from the cardiac stomach into the elongate ceca, where further digestion, absorption, and storage occur.

Sea stars frequently eat bivalve molluscs. Upon encountering a clam, scallop, or oyster, a sea star extends its arms around the mollusc and grips both valves with its tube feet (Figure 27.14). The sea star's body wall muscles then contract, increasing pressure on its coelomic fluid. This pushes the cardiac stomach out through the mouth and into the minute space between the valves and mantle edges of the bivalve. The sea star may pull its prey's valves apart slightly, but there is usually enough space for the stomach to slip between the valves even when the mollusc's shell is tightly closed. A

crack as small as 0.1 mm may be wide enough to admit the stomach. Once inside the shell, a sea star's stomach is not injured even if the mollusc closes its valves as tightly as possible. The everted stomach exposes the soft parts of the mollusc to digestive enzymes. Soft tissues are broken down to the consistency of thick soup and eaten, while hard parts are simply left behind.

Sea urchins eat a variety of foods, but most are chiefly herbivorous. Most species scrape algae from rocks or cut fronds of kelp. Sea urchins and sand dollars have a complex feeding apparatus called **Aristotle's lantern** surrounding the mouth cavity and pharynx (see Figure 27.13). The lantern is radially symmetrical and has replaceable teeth that can be moved in and out of the mouth, allowing a scraping or biting action. Radially arranged sets of muscles move the various parts of the lantern. The rest of the sea urchin gut consists of a short esophagus leading to a long intestine that provides the necessary surface area for the digestion of plant roughage. A long **siphon** lined internally with cilia lies adjacent to most of the intestine. The siphon may function in extracting water from the intestine, concentrating the food, and aiding digestion. As in sea stars, the anus is located on the aboral surface.

Osmoregulation, Circulation, Gas Exchange, and Excretion

Echinoderms cannot regulate water flow across their body wall; consequently, the osmotic pressure of their coelomic fluid conforms to that of the environment. Some species have been shown to have a weak ability to regulate ion flow, and the concentration of certain ions in their coelomic fluid differs slightly from that of the environment. But in most echinoderms, the coelomic fluid is virtually identical to the environment (i.e., it has the same types and concentrations of ions as the environment). Because of the body wall's inability to osmoregulate, body cells bathe in internal fluids whose ionic makeup does not support cellular metabolism. As a result, ion regulation is mainly a cellular function in echinoderms, and individual cell membranes expend considerably more energy than those of most other metazoans in maintaining intracellular homeostasis.

Coelomocytes circulating in the water vascular system and in the coelom are important in internal transport. Dissolved gases and nitrogenous wastes diffuse through the coelomic fluid and body wall or are transported by the coelomocytes. Cilia on the peritoneal tissues lining the coelom and on the

Figure 27.14   A sea star feeding on a bivalve mollusc. In this process, the sea star inserts its everted stomach between the victim's valves.

inner lining of the water vascular system circulate fluids and coelomocytes. Sea stars and sea urchins also have epidermal cilia that maintain a flow of seawater over the body, creating a diffusion gradient that promotes exchange of dissolved gases and wastes.

Tube feet are covered with thin epidermis and are important in gas exchange and waste excretion. Experiments show that oxygen uptake across the surfaces of tube feet may exceed 50% of the total oxygen uptake in sea stars and may account for all gas exchange in crinoids. Sea urchins and other echinoids have suckerless tube feet specialized for gas exchange. These respiratory tube feet are divided internally by a longitudinal septum. Cilia in these tube feet drive water toward the tip of the foot on one side of the septum and in the reverse direction on the other side. In heart urchins and sand dollars, cilia on the outside of the tube feet drive water opposite the internal current, creating a countercurrent flow that promotes efficient gas exchange. (Countercurrent flow is discussed in Chapter 5, p. 105.)

Complementing the respiratory surface of the tube feet, sea stars also have projections of the body cavity called **dermal branchiae**, or skin gills (see Figures 27.12a and 5.2). Like tube feet, these structures are especially effective as diffusion sites because their walls, consisting of a thin layer of epidermis and the peritoneum, are all that separate the coelomic fluid from the seawater. Coelomo-cytes carry wastes to the branchiae for disposal, and gases readily diffuse in and out of the coelom across the branchiae.

Sea cucumbers are unique in having two clusters of branched tubes, called **respiratory trees**, that are the main gas exchange/excretory organs (Figure 27.15). The trees branch into the coelom from the cloaca, and seawater is pumped in and out of the trees by muscular action of the cloaca. Gases are exchanged between the coelomic fluid and the seawater in the respiratory trees, and nitrogenous wastes (mostly ammonia) diffuse into the trees from the coelomic fluid.

Hemal System. Although the coelomic fluid and water vascular system are important in circulation, many echinoderms also have an open (hemolymph-containing) **hemal system**, with main channels paralleling those of the water vascular system. In sea stars, two or three **hemal rings** encircle the gut, connecting to a radial sinus in each arm and to sinuses in the gonads and body wall. A vertical cluster of elongate sinuses, called the **axial gland**, connects the rings (see Figure 27.12b).

Little is known about the function of the hemal system, except in sea cucumbers, where it is intimately associated with the respiratory tree and plays a role in gas circulation and exchange. The hemal system does not seem to circulate gases in sea stars, for removal of parts of the system does not alter the rate of gas exchange. In sea stars, he-

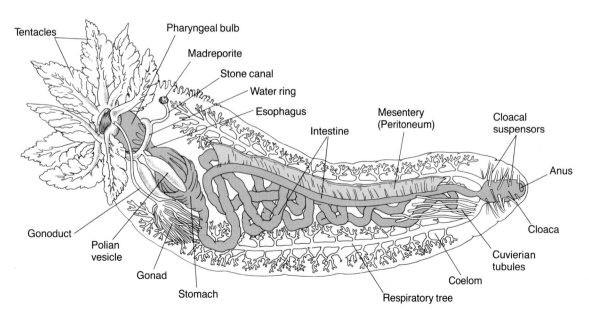

Figure 27.15 Internal anatomy of a sea cucumber. This drawing shows the relationship between the digestive system and the respiratory trees.

molymph may transport nutrients from the gut, or as recent evidence suggests, the hemal system may be a type of immune system. The axial gland in sea stars is the main source of amoeboid hemolymph cells, and these cells may help fight disease organisms or repair damaged tissues.

Nervous and Sensory Structures

Conforming with body symmetry, the typical echinoderm nervous system is radially arranged and not strongly centralized. A **nerve ring** usually encircles the gut and gives rise to **radial nerves** composed mainly of sensory neurons carrying impulses from the tube feet (see Figure 27.13). These main sensory elements lie just under the epidermis and connect to a subepidermal **nerve net** that innervates all parts of the body. One or two other nerve nets, sometimes with nerve rings around the gut, and associated nerves are mainly motor in function. A nerve net in the peritoneum and motor nerves lying in deep tissue carry impulses to muscles attached to ossicles and tube feet. Cell bodies of these motor neurons often form small ganglia at the base of the tube feet.

The main sensory elements of echinoderms are receptor cells scattered over the body surface. Echinoderms generally move away from strong light, and many sea stars have a modified tube foot that forms a photosensitive **eyespot** at the tip of each arm. Sea urchins and burrowing sea cucumbers also have **statocysts** that detect the position of the body relative to gravity. Associated with their tendency toward bilateral symmetry, sea cucumbers have concentrations of epidermal sensory cells at both ends of the body.

BEHAVIOR

Echinoderms respond to a variety of environmental stimuli, and their behavior has been the subject of much experimentation. As might be predicted from their noncentralized nervous system, much echinoderm behavior involves local body areas, with some spreading of responsiveness to other body parts. Whole-body movements requiring overall coordination of tube feet result from centralized integration that probably occurs in the nerve ring or radial nerves. The complex sequence of behaviors exhibited by a sea star feeding on a bivalve mollusc requires considerable integration of sensory and motor information.

Sea stars and brittle stars have "dominant arms" that tend to lead as the animal moves. Arm dominance is usually temporary and seems to depend on the animal's immediate response to an environmental stimulus. The ability to switch arm dominance is adaptive for radially symmetrical animals, and researchers postulate that sensory nerves in each arm contain special "dominance centers" that provide this capability. A dominance center might consist of a group of integrative neurons in a sensory nerve. Inhibition or facilitation of synapses at these centers may determine whether an arm is dominant. A dominance center in a leading arm probably exerts temporary control over the centers in other arms.

Most echinoderms can right themselves if flipped upside down, and their righting movements constitute a coordinated set of responses. Sea stars right themselves by twisting the tips of one or two arms so that some of the tube feet contact and adhere to the substrate. Muscles in the body wall accomplish the twisting. The rest of the animal can then flip over the adherent arms. If a sea urchin is placed on its aboral surface, it will extend its tube feet from one ambulacral area (see Figure 27.13). The feet attach to the substrate, and the animal begins to roll. As it rolls, the tube feet on the underside release their grip on the substrate, and others closer to the oral surface grip it. This wave of tube foot action continues to roll the animal until it is back to its correct position, with its mouth against the substrate.

Lacking protective spines, sea cucumbers have evolved behavioral means of self defense. Some have special defense organs called **Cuvierian tubules** attached to their respiratory trees (see Figure 27.15). If a sea cucumber is attacked, the tubules can be expelled from its anus. The tubules of some species are sticky and can entangle small would-be predators. Those of other species contain a toxic chemical called holothurin that can paralyze or kill a small fish. Many sea cucumbers will eviscerate themselves if disturbed. Their body wall contracts, rupturing the digestive tract behind the mouth or in the cloaca, and the gut and most other viscera are ejected through the mouth or anus. Since a sea cucumber can regenerate its lost parts within a matter of weeks, evisceration could serve as a defense mechanism; a predator might eat the ejected viscera and leave the rest of the animal. However, behaviorists are not convinced that evisceration is a defensive response. Other factors, including overcrowding and polluted water, can elicit the response, and for unknown reasons, certain species eviscerate seasonally. Additional research is needed to fully understand this curious behavior.

REPRODUCTION AND LIFE CYCLES

Many echinoderms reproduce both asexually and sexually. The ability to regenerate lost parts is universal in the phylum, and some species can fragment into pieces that regenerate clones of new individuals. If grasped by a predator, some sea stars and many brittle stars can cast off arms. In this process, all muscles at a particular site in the arm contract simultaneously, pinching off the distal portion of the arm. Many sea stars and brittle stars can also reproduce asexually by breaking into two approximately equal halves. Imperfect regeneration often follows, and individuals with more than the usual number of arms sometimes result. If a sea star is broken into several pieces, new individuals may regenerate from pieces that include a substantial part of the central disc. Before this fact was discovered, oyster growers trying to control predatory sea stars may have propagated them instead. The growers would trap the sea stars, chop them into pieces, and then throw them back into the ocean. Although most of the pieces died, a few would regenerate into whole new individuals.

Nearly all echinoderms are dioecious, although males and females generally look alike. Figures 27.12 and 27.13 show the elongate gonads, which parallel the ambulacral grooves in sea stars and sea urchins. At spawning time, adults of most species shed large numbers of gametes into the sea through gonopores, usually on the aboral surface, and fertilization is typically external. Adults may aggregate during the breeding season, and researchers postulate that chemical signals released from gametes or from a spawning adult trigger the release of sex cells by individuals of the opposite sex.

About half of all echinoderms brood their fertilized eggs, and brooding species produce fewer eggs (with more yolk) than other echinoderms. Fertilization in brooding species usually occurs on the body surface of the female, and eggs are held in specialized areas on her skin or in the cardiac stomach.

As typical deuterostomes, echinoderms have radial cleavage that yields a coeloblastula (see Figure 16.3). The blastopore forms the anus. Mesoderm and coelomic sacs develop enterocoelously, and the mouth forms late in gastrulation as a secondary opening.

Free-swimming larvae that develop from the gastrula are bilaterally symmetrical planktonic filter feeders. The early sea star larva, characterized by bands of locomotor cilia and a complete gut, is called a **bipinnaria** (Figure 27.16). A bipinnaria

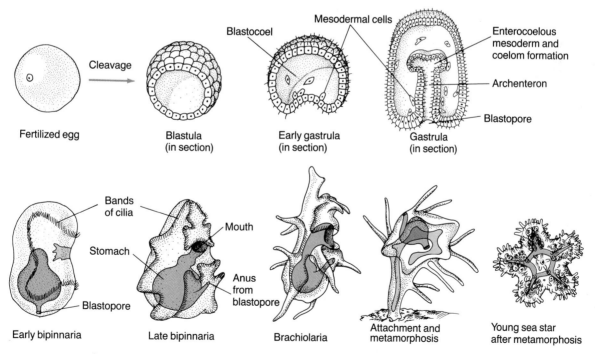

Figure 27.16 Development of a sea star. Gastrulas are ciliated and free-swimming. The cilia become restricted to bands as the gastrula transforms into the bipinnaria larva. Next, arms develop, but the brachiolarian's arms do not develop into the adult sea star's arms. Other classes of echinoderms have somewhat different larvae, but the basic developmental sequence is similar in all echinoderms.

eventually grows lateral lobes from its body wall, and when the lobes become armlike, the larva is called a **brachiolaria**. After a planktonic existence, a brachiolaria settles and attaches to the bottom, loses its larval arms, and metamorphoses into the radially symmetrical adult.

PHYLOGENY

Echinoderms are the most complex of the radially symmetrical animals. Unlike cnidarians, for example, they are coelomate, they have highly developed organs and organ systems, and they are actively motile. Because echinoderm larvae are bilateral, ancestral echinoderms are believed to have been bilaterally symmetrical, probably existing as bottom dwellers or free swimmers. Their primary bilateral symmetry and deuterostome features strongly suggest that echinoderms share a distant ancestor with chordates and hemichordates (see Figure 27.20). (Chordate evolution is discussed in Chapter 28, p. 681.) As part of the deuterostome lineage, echinoderms have little in common with the annelids, arthropods, and other protostomes.

Symmetry is a central aspect of echinoderm evolution. Ancestral echinoderms may have evolved their secondary radial symmetry as an adaptation to sedentary or sessile life. Perhaps later, as most echinoderms became free and motile,

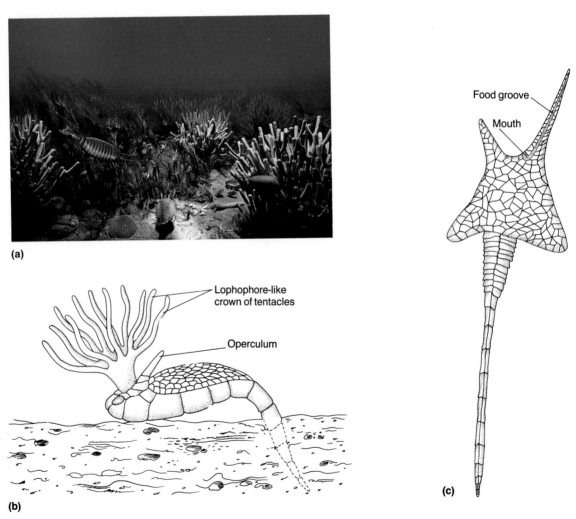

Figure 27.17 Fossil echinoderms. (a) This reconstruction of an early Cambrian ocean shows the type of habitat in which many early echinoderms thrived. (b) *Gyrocystis*, a carpoid, may have lacked ambulacra and used a lophophore-like crown of tentacles for filter feeding. (c) *Dendrocystites* is also a carpoid. It has been postulated that the stalk of forms similar to *Dendrocystites* was really a chordatelike tail, but this idea has not been generally accepted. The two types of carpoids shown in (b) and (c) may have used their stalks to anchor in mud.

they remained radial. Most echinoderms are rather slow and sedentary anyway and may have experienced little selective pressure to depart from the radial plan of their ancestors. The tendency toward bilateral symmetry in heart urchins and certain sea cucumbers is a tertiary adaptation, no doubt correlated with forward movement in these animals.

The echinoderm phylum is ancient, and the hard endoskeletons of these animals have left a rich fossil record. About 20,000 fossil species have been described in about 16 extinct classes. Echinoderms were diverse in the Cambrian (about 600 million years ago), and undoubtedly there were many species during Precambrian times. Some of the oldest fossils, dating from the early Cambrian (Figure 27.17a), were stalked or stalkless sessile forms called **eocrinoids**. These forms superficially resembled the crinoids of today, and like them, may have used a water vascular system for filter feeding. Thus, it is suggested that the water vascular system originated in ancestral echinoderms as a filter-feeding device; then, as motile echinoderms evolved, the system became adapted for bottom locomotion. Early sessile echinoderms may have evolved their hard endoskeleton as a support for the filter-feeding water vascular system.

The **carpoids**, another fossil group, were numerous from the Cambrian through the Devonian (Figure 27.17b and c). Many of these animals lacked ambulacra and were bilateral or asymmetrical; some had an elongate stalk that may have anchored them in mud. A number of carpoids may also have had a crown of tentacles, similar to that of the lophophorate animals, and some zoologists postulate that lophophorates and echinoderms had a common ancestor. Despite the wealth of fossils and a voluminous literature on the subject, the phylogeny and affinities of echinoderms remain very speculative and controversial.

PHYLUM CHAETOGNATHA

The Phylum Chaetognatha consists of about 70 species of common planktonic predators. **Chaetognaths**, commonly called **arrowworms** because of their dartlike appearance, are all marine (Figure 27.18). They are numerous in most oceans, forming an important link in the food chains between small herbivores and larger planktonic animals. In one of his lesser-known reports, Charles Darwin commented that chaetognaths are "remarkable for obscurity of their [phylogenetic] affinities." Darwin had been studying reproduction in arrowworms of the genus *Sagitta*, and when he wrote his report in 1844, chaetognaths had already been classified in Linnaeus's Class Vermes (worms) for 75 years, ever since their discovery in 1769. Today, there is general agreement that arrowworms are deuterostomes, but questions remain about their relationships with other animals. Fossil chaetognaths dating from the Cambrian, nearly 600 million years ago, show that the phylum is ancient. Fossils closely resemble modern species, indicating that the general body form has been successful and has not greatly changed for hundreds of millions of years.

Chaetognaths have what some zoologists call "a predaceous look." Ranging in length from about 1

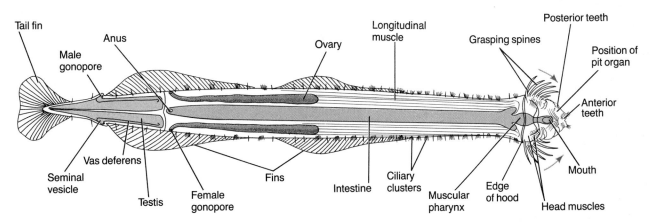

Figure 27.18 Phylum Chaetognatha. Dorsal view of a chaetognath. Internal structures are visible through the transparent body wall. Arrows on each side of the head indicate movement of spines to grasp prey.

to 15 cm, their dartlike bodies have a prominent head bearing two eyespots and a large central mouth flanked by rows of short teeth and long bristlelike grasping spines. Two **pit organs** near the teeth produce toxins that immobilize prey. An anterior extension of the body wall, the **hood**, covers the head while the animal is swimming, providing a more streamlined shape. Complex muscles that twist the head, move the spines and teeth, and open and close the mouth are visible through the transparent body covering. Fins on the sides of the body and tail provide stability and buoyancy in the water.

Chaetognaths are covered by a thin cuticle, and their body wall has only longitudinal muscles, providing jerky, dartlike swimming movements. The animals move in short spurts, then glide and drift, waiting for prey. They snap their grasping spines and teeth around small planktonic crustaceans, worms, and larval fishes. A muscular pharynx just behind the mouth pushes food into a straight intestine that ends in an anus on the ventral surface in front of the tail fin. There are no respiratory, circulatory, osmoregulatory, or excretory organs; fluid in the body cavity assists in circulation. The nervous system is strongly cephalized, with a cerebral ganglion and a nerve ring around the pharynx extending to lateral and ventral ganglia. In addition to eyespots, sensory organs of chaetognaths include fanlike **ciliary clusters** covering the body (see Fig-

ure 27.18). These structures detect vibrations in the water and may be important in detecting prey movements. In addition, a **ciliary loop** just behind the head may have a tactile or chemosensory function or may be involved in sperm transfer.

The entire life cycle of most chaetognaths is planktonic. Adults are hermaphroditic, and in species that have been studied, seminal vesicles package sperm into spermatophores that seem to rupture through the body wall and stick to the cuticle. Spermatophores are exchanged when two arrowworms touch; sperm enter the gonopore of the female system, and eggs are fertilized internally. Fertilized eggs pass out through the gonopore and either float about in plankton or stick to the outside of the adults while development proceeds. Development is deuterostome-like, but with some unusual events in the gastrula stage. Mesoderm and coelomic pouches develop by infoldings of the endoderm of the archenteron (rather than by the typical outpouching; see Figure 16.9b). The chaetognath coelom forms as mesodermal infoldings grow into the archenteron and divide it into two pouches (Figure 27.19). Also unique among deuterostomes, the body cavity in adult chaetognaths does not appear to be lined by peritoneum. Additional research is needed to determine if the peritoneum is lost during development or if the adult cavity is a pseudocoel. Juvenile arrowworms look like small adults, for there is no true larval stage.

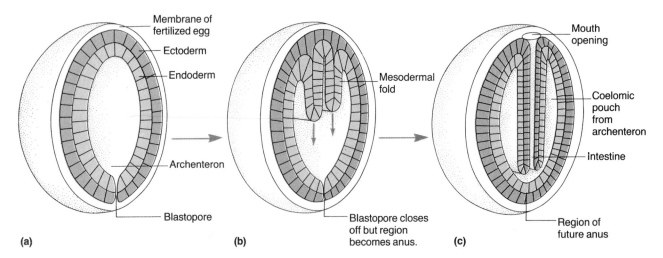

Figure 27.19 Early development in a chaetognath. All stages occur within the membrane of the fertilized egg. (a) Section through an early gastrula. As in many other deuterostomes, the region of the blastopore becomes the anus. (b) Formation of mesoderm and coelom by infoldings of the endodermal walls of the archenteron. (c) A late gastrula, in which the infolds have nearly fused with endoderm on the opposite (posterior) end of the gastrula. Two lateral coelomic sacs are formed from the archenteron, and the space between the two infolds becomes the intestine. Also at this stage, the mouth opening has formed at the anterior surface. Later in development, the coelomic pouches of the embryo become filled with cells. The adult body cavity (probably different from the larval cavities) develops after the juvenile is fully formed.

PHYLUM HEMICHORDATA

The phylum Hemichordata contains about 120 species of marine bottom dwellers. **Hemichordates**, as their name implies, bear some resemblance to chordates and were formerly classified in the Phylum Chordata; but since the 1940s, zoologists have generally agreed that hemichordates constitute a separate phylum. There are two distinct subgroups of hemichordates: the enteropneusts, which are elongate and free-moving, and the pterobranchs, which are vase-shaped and sedentary.

Enteropneusts, or **acorn worms** (about 100 species), are common inhabitants of tidal mud flats. Some live in U-shaped burrows, while others wriggle sluggishly about on mud surfaces (Figure 27.20a). A recently identified species inhabits the rocky edges of deep-sea vents in the Galapagos rift

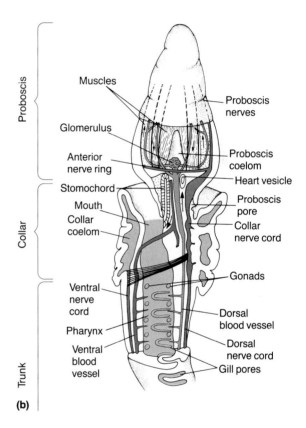

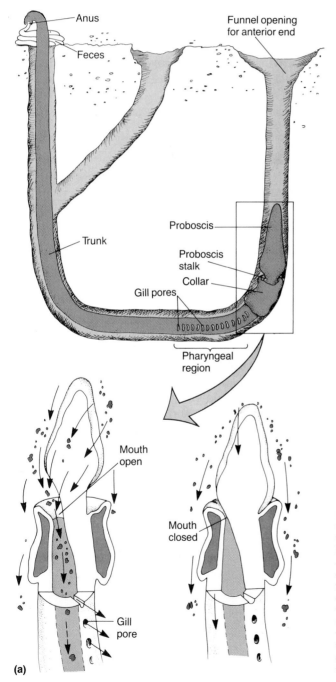

Figure 27.20 Phylum Hemichordata. (a) An acorn worm in its burrow. Acorn worms are characteristically sluggish, and the burrowing types may remain in their burrows for long periods of time. The proboscis is in almost constant motion, aiding feeding and gas exchange at the body surface. The arrows on the enlargement indicate the flow of water during filter feeding and while the mouth is closed. Notice that some particles do not enter the mouth, but are passed backward over the collar. Food particles that enter the mouth remain in the gut as water passes out the gill pores. (b) Cutaway drawing of the anterior region of an acorn worm, showing the digestive, circulatory, and nervous systems. Arrows indicate the direction of hemolymph flow. Note that the heart vesicle is not part of the circulatory system; it has muscles whose contractions help propel hemolymph in the adjacent central sinus.

▷ area. (The Chapter 25 Essay describes ocean vent communities, p. 606.) Different species range in length from about 10 cm to about 1.5 m. The body has three distinct regions: An anterior conelike **proboscis** is connected by a short stalk to a **collar**, which is fused to a long **trunk**. The anterior region of the trunk has a series of **gill pores** opening from the pharynx on each side. Depending on the species, there are from about 10 to over 100 pairs of gill pores.

The body wall of acorn worms is covered by a ciliated, mucus-coated epithelium. Unusual for coelomates, muscles develop from the lining of the coelom, and in adults, the body cavity is largely filled with connective tissue and muscles. The muscles of the proboscis and collar are much more strongly developed than those of the trunk. External cilia aid locomotion, but forward movement and burrowing are accomplished mainly by peristaltic contractions in the anterior body regions. The trunk is more or less passively pulled behind the advancing proboscis and collar.

Most acorn worms eat bottom deposits and also filter-feed. Inorganic matter passes through the gut and is defecated at the substrate surface (see Figure 27.20a). Recent studies indicate that acorn worms contribute significantly to the turnover of ocean sediments. In a study off the coast of Japan, densities of about 6×10^7 acorn worms per square kilometer were estimated to produce about 4×10^4 kg of fecal material per day.

When filter-feeding, acorn worms wave their proboscis about, bringing fine suspended particles and plankton into contact with the mucus and cilia on their body surface. Cilia drive food particles and water into the ventral mouth (see Figure 27.20a). The particles remain in the pharynx while cilia on the gills drive water out through the gill pores. An acorn worm can regulate its food and water intake by drawing the front edge of its collar over its mouth. In this way, the animal can close off its mouth completely, or it can close it off only partially to admit particles of a certain size.

Acorn worms have an open circulatory system with contractile dorsal and ventral vessels (Figure 27.20b). Oxygenation of the hemolymph occurs in the gills and skin. Hemolymph from the gills and other body tissues circulates forward in the dorsal vessel and backward in the ventral vessel. From a large central sinus (an expansion of the dorsal vessel), hemolymph is pumped into the **glomerulus**, a clump of sinuses bulging into the coelom at the base of the proboscis. Little is known about excretion in acorn worms, but the glomerulus may function as a type of kidney.

The nervous system of acorn worms consists of a nerve net under the epidermis connected to a dorsal and a ventral nerve cord (see Figure 27.20b). The dorsal cord extends into the collar and proboscis, where it forms an anterior nerve ring and many small nerves. In some species, the collar cord is hollow, resembling the dorsal hollow nerve cord of chordates. The collar cord also contains giant nerve fibers, providing rapid impulse conduction for muscles moving the collar edge over the mouth. This allows an acorn worm to quickly cover its mouth and prevent oversized particles or other harmful matter from disrupting the filter-feeding mechanism. Sensory cells are concentrated on the proboscis.

Acorn worms reproduce both asexually and sexually. If torn or pulled apart, they can usually regenerate lost parts, and whole worms may regenerate from pieces of the trunk. Adults are dioecious, and males and females look alike. Many separate gonads occur in rows on each side of the anterior trunk, and each one opens to the outside through a small pore. Gametes are shed into the ocean, and fertilization is external. Development follows the deuterostome pattern. There is no larval stage in some species, but many acorn worms have a ciliated larva called a **tornaria**, which closely resembles the bipinnaria larva of sea stars (see Figures 27.16 and 27.21). The tornaria is planktonic for a few days or months, depending on the species, and then settles to the seafloor and metamorphoses into the adult worm.

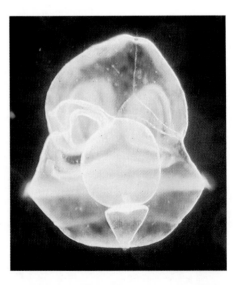

Figure 27.21 The tornaria larva of acorn worms. This bilaterally symmetrical larva, seen here from the right side, resembles the echinoderm bipinnaria. Details of the complete digestive tract are clearly seen in this dark-field photomicrograph.

Whereas larval acorn worms resemble echinoderms, the adults have several chordatelike features. The pharyngeal gills are similar to those of invertebrate chordates. (See the discussion of subphyla Cephalochordata and Urochordata in Chapter 28, p. 677.) The hollow nerve cord in the collar of some acorn worms has no counterpart in the animal kingdom except for the dorsal, hollow nerve cord of chordates. On the other hand, the rest of the nervous system is solid, and unlike the chordate nervous system, there is a ventral nerve cord. Also, the circulatory system, with hemolymph circulating forward in the dorsal vessel, is more like that of annelids than of chordates.

One other structure in acorn worms merits mention. Adults have a pronounced anterior projection from the mouth cavity called the **stomochord** (see Figure 27.20b). The walls of the stomochord are partly cartilaginous, and this structure was once thought to be homologous with the notochord of chordates. It was on this basis that zoologists of the late nineteenth century classified hemichordates in the Phylum Chordata, but it is now known that the stomochord is not homologous with the chordate notochord.

Pterobranchs (about 20 species), the second subgroup in the Phylum Hemichordata, superficially look more like lophophorates than acorn worms. Their most prominent feature is a filter-feeding apparatus consisting of two or more pairs of lophophore-like tentacled arms. Most pterobranchs live in small colonies of saclike individuals in interconnected gelatinous tubes (Figure 27.22). They are most numerous in fairly deep oceans (about 250–600 m), although some occur in tidal areas. No pterobranchs are common, and much research remains to be done on these animals.

Although modified for their sessile existence, individual pterobranchs have the same body regions as acorn worms (compare Figures 27.20b and 27.22). There is a muscular, shield-shaped proboscis covered with gland cells that secrete the tube. The proboscis grips the inner surface of the tube, allowing the animal to pull itself up or down inside the tube. The collar region of the body expands as the tentacled arms. Cilia on the tentacles direct water currents, moving fine suspended particles and plankton toward the mouth. The saclike pterobranch trunk bears a stalk extending to the inside of the tube or, as in *Rhabdopleura*, to a rootlike **stolon** joining other individuals in a colony. Contraction of muscles in the stalk retracts the animal into its tube.

Similar to many lophophorates, pterobranchs have a U-shaped digestive tract with the mouth and anus at the base of the collar. Species of *Cephalodiscus* have a pair of gill pores in the pharynx. *Rhabdopleura* lacks gills, but has two internal pharyngeal furrows reminiscent of gill pores (see Figure 27.22). Other organ systems and a tissue-

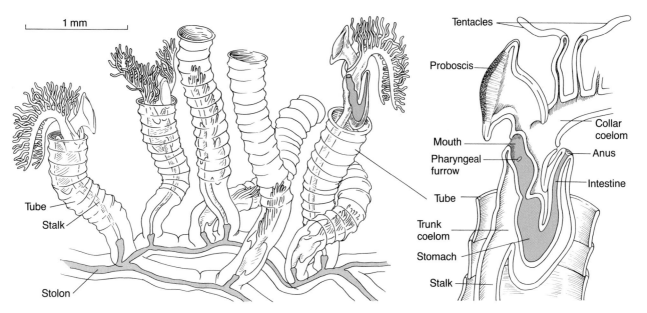

Figure 27.22 A colony of the pterobranch *Rhabdopleura*. Colonies are usually less than 10 cm across and consist of several interconnected individuals attached by a rootlike stolon to rocks or shells. Individuals are about 1–8 mm long, are usually brown, and have a pair of arms. In contrast to *Rhabdopleura*, individuals of the genus *Cephalodiscus* have five to nine pairs of arms, are not connected, and have two gill pores.

filled coelom are similar to those of acorn worms. Pterobranchs may be monoecious or dioecious. Some species are known to brood fertilized eggs in their tubes, but life cycles and development have yet to be thoroughly studied in any species. As typical deuterostomes, some pterobranchs have radial cleavage, enterocoelous development, and a free-swimming larva that resembles the tornaria of acorn worms. Upon settling, the larva metamorphoses into the saclike adult form, and colonies develop by asexual budding of the stalk or stolon.

It is unfortunate that pterobranchs have attracted so little research attention, for they may be of considerable phylogenetic significance. Some zoologists consider them relatives of both lophophorates and echinoderms. Their tentacled arms resemble lophophores, and their mode of feeding resembles that of both lophophorates and crinoid echinoderms (see Figure 27.6). On the other hand, pterobranchs may have evolved their saclike body and tentacled arms independently of lophophorates or crinoids as an adaptation for sessile life on the ocean floor. The pharyngeal gill pores or furrows in pterobranchs indicate a distant relationship to chordates. A hypothesis concerning how ancestral chordates may have evolved as sessile gill feeders, possibly from pterobranch-like ancestors, is discussed in Chapter 28.

SUMMARY

1. Three aquatic phyla (Bryozoa, Phoronida, and Brachiopoda) of nonmetameric, enterocoelous coelomates are collectively called lophophorates because they have a crown of ciliated tentacles called a lophophore. The lophophore contains fluid-filled coelomic channels and is extended hydraulically for filter feeding.

2. **Key Features of Echinoderms.** Echinoderms are marine deuterostomes, often with bilateral larvae. Adults have radial symmetry, but in some species, bilateral symmetry is superimposed on the radial plan. Body support is provided by an endoskeleton of unique calcareous dermal ossicles; dermal spines are protective and give the body surface a spiny appearance. The typical locomotor system is a coelom-derived hydraulic (water vascular) system with movable tube feet.

3. Osmoregulation in echinoderms is a cellular function. Circulation occurs via coelomic cells and fluid, and an open hemal system occurs in some species. Gases and wastes diffuse across the surfaces of the tube feet, dermal branchiae in sea stars, and respiratory trees in sea cucumbers.

4. The echinoderm nervous system consists of one to three nerve rings encircling the gut and radial nerves that connect to subepidermal nerve nets. Sensory cells are scattered over the entire body. Some species have statocysts and eyespots; some have pincerlike independent effectors called pedicellariae.

5. Nearly all echinoderms are dioecious with external fertilization. Typical deuterostome development often leads to a free-swimming, bilaterally symmetrical larva that metamorphoses into the radial adult; many species brood their embryos.

6. Echinoderms were probably derived from bilateral ancestors that developed radial symmetry and an endoskeleton as adaptations for sessile life. The unique water vascular system may have evolved first as a feeding system, becoming adapted secondarily as a locomotor system.

7. Two phyla, the Chaetognatha (arrowworms) and Hemichordata (acorn worms and pterobranchs), contain fewer than 200 species of unsegmented marine deuterostomes. Arrowworms are dartlike planktonic predators with unique developmental features. Hemichordates are bottom dwellers with three body regions (proboscis, collar, and trunk). The trunk bears pharyngeal gill pores. Acorn worms are elongate burrowers that feed on bottom deposits or suspended particles. Pterobranchs are saclike, sessile filter feeders.

FURTHER READING

Clark, A. M. *Starfishes and Related Echinoderms.* 3d ed. Neptune City, N.J.: T.F.H. Publications, 1977. *A concise, exceptionally well illustrated introduction to echinoderm biology.*
Ghirardelli, E. "Some Aspects of the Biology of the Chaetognaths." *Advances in Marine Biology* 6(1968): 271–375. *An authoritative review of research on chaetognaths.*
Jackson, J.B.C., and T. P. Hughes. "Adaptive Strategies of Coral-Reef Invertebrates." *American Scientist* 73(1985): 265–274. *Long-term analyses illustrate how bryozoans and other sessile organisms adapt to limited space available on coral reefs.*
Lester, S. M. "*Cephalodiscus* sp.: Observations of Functional Morphology, Behavior and Occurrence in Shallow Water Around Bermuda." *Marine Biology* 85(1985): 263–268. *One of the few studies of pterobranch hemichordates.*
Macurda, D. B., Jr., and D. L. Meyer. "Sea Lilies and Feather Stars." *American Scientist* 71(1983): 354–365. *Ecology and functional aspects of crinoids on coral reefs; excellent color photographs.*
Nichols, D. *Echinoderms.* 4th ed. London: Hutchinson University Library, 1969. *Includes classification, phylogeny, and major fossil groups.*
Richardson, J. R. "Brachiopods." *Scientific American* 255(1986): 100–106. *Diversity and physiology of these ancient animals.*
Woollacott, R. M., and R. L. Zimmer, eds. *Biology of Bryozoans.* New York: Academic Press, 1977. *Sixteen chapters on bryozoology by researchers.*

☐ CLASSIFICATION

Phylum Bryozoa (Gr. *bryon*, "tree moss"; *zoon*, "animal"). Bryozoans, or moss animals; freshwater and marine lophophorates forming encrusting, branching colonies with a calcified or chitinous exoskeleton; *Plumatella* (freshwater); *Bugula* and *Electra* are marine.

Phylum Phoronida (L. *Phoronis*, surname of Io, a woman in Greek mythology who was changed into a white heifer and wandered over the earth). Phoronids; marine, wormlike tube dwellers with a spirally coiled lophophore; *Phoronis.*

Phylum Brachiopoda (Gr. *brachion*, "upper arm"; *pous*, "foot"). Lamp shells; attached, often stalked marine lophophorates with dorsal and ventral calcareous valves secreted by a mantle; fossil species about 100 times more numerous than living forms; *Terebratula, Lingula.*

Phylum Echinodermata (Gr. *echinos*, "a sea urchin or hedgehog"; *derma*, "skin"). Echinoderms; marine deuterostomes with bilateral larvae and usually radial adults, a water vascular system, and a unique calcareous endoskeleton.

Subphylum Homalozoa (Gr. *homalos*, "level"; *zoon*, "animal"). Carpoids; four classes of asymmetrical fossil echinoderms (all extinct); *Dendrocystites.*

Subphylum Crinozoa (Gr. *krinon*, "lily"; *zoon*, "animal"). Crinozoans; radially symmetrical echinoderms with filter-feeding arms; includes several extinct classes.

Class Crinoidea (Gr. *krinon*, "lily"; *oideos*, "form of"). Crinoids (sea lilies and feather stars); mostly fossil species with a stalk composed of skeletal ossicles attached to a crown, bearing five or more featherlike arms; ambulacral grooves are ciliated and function in filter feeding; no spines or pedicellariae; *Antedon, Nemaster.*

Subphylum Asterozoa (Gr. *aster*, "star"; *zoon*, "animal"). Asterozoans; free-moving, star-shaped echinoderms with arms (rays) projecting from a central disc.

Class Asteroidea (Gr. *aster*, "star"; *oideos*, "form of"). Sea stars (asteroids); arms are not sharply set off from the central disc; open ambulacra usually have suckered tube feet; aboral surface usually bears pedicellariae; *Asterias, Crossaster, Acanthaster.*

Class Ophiuroidea (Gr. *ophis*, "snake"; *oura*, "tail"; *oideos*, "form of"). Brittle stars and basket stars (ophiuroids); arms are sharply set off from the central disc; ambulacra are closed, covered with ossicles, and contain tube feet without suckers; no pedicellariae; muscular articulated arms, rather than tube feet, provide locomotion; *Ophiothrix, Ophiocoma* (brittle stars); *Astrophyton* (basket stars).

Class Concentricycloidea (L. *concentricus*, "concentric"; *cyclus*, "ring"). Sea daisies; medusiform asterozoans with a double water-vascular ring and concentric skeletal elements; *Xyloplax.*

Subphylum Echinozoa (Gr. *echinos*, "hedgehog or sea urchin"; *zoon*, "animal"). Echinozoans; free-moving, armless echinoderms with closed ambulacra and suckered tube feet; includes several extinct classes.

Class Echinoidea (Gr. *echinos*, "hedgehog or sea urchin"; *oideos*, "form of"). Echinoids; globose (sea urchins) or flattened (sand dollars) echinoderms with a rigid test formed of fused platelike ossicles; possess movable skeletal spines and several types of pedicellariae; *Arbacia, Strongylocentrotus* (sea urchins); *Clypeaster* (sand dollars).

Class Holothuroidea (Gr. *holothourion*, "sea cucumber"; *oideos*, "form of"). Holothurians (sea cucumbers); cylindrical echinoderms with a leathery body containing microscopic ossicles, but no spines or pedicellariae; *Cucumaria, Stichopus.*

Phylum Chaetognatha (Gr. *chaite*, "long hair"; *gnathos*, "jaw"). Arrowworms; dartlike, predatory marine deuterostomes with unique enterocoelous development; *Sagitta.*

Phylum Hemichordata (Gr. *hemi*, "half"; *chorde*, "cord"). Hemichordates; bottom-dwelling marine filter feeders; many have pharyngeal gill slits; *Saccoglossus* (acorn worms); *Rhabdopleura, Cephalodiscus* (pterobranchs).

Introduction
to the Chordates

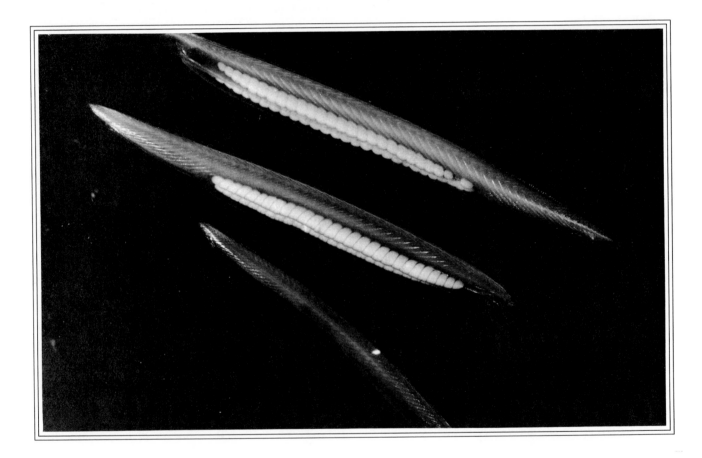

The lancelet *Branchiostoma floridae,* a member of the chordate Subphylum Cephalochordata. Lancelets (about 5–15 cm long) filter-feed while partly buried in ocean sand.

Chordates inhabit all types of aquatic and terrestrial environments. Of about 47,200 living species in the phylum, about 2,100 are filter-feeding marine invertebrates. The rest, typically having a bony or cartilaginous backbone, are vertebrates.

Chordates are bilaterally symmetrical, and most have a well-developed head and tail. Metamerism is evident in their body musculature and in the vertebrate skeleton. All chordates have four structures that are diagnostic of the phylum. An endoskeletal rod, the **notochord**, provides longitudinal support, at least during larval life. The notochord lies just under a **dorsal tubular nerve cord** with an anterior expansion, the brain. The pharynx bears lateral **gill slits** that function in filter feeding or gas exchange or that develop into other structures in land vertebrates. Finally, a **postanal tail** contains the posterior extensions of the notochord and nerve cord. With few exceptions, chordates also have a highly organized closed blood vascular system with a ventral heart that pumps blood posteriorly through a dorsal aorta.

The **Phylum Chordata** is classified into three subphyla. Invertebrate chordates are divided into two subphyla, the Cephalochordata (lancelets) and the Urochordata (tunicates). Besides lacking vertebrae, invertebrate chordates have a greatly reduced coelom or lack a coelom altogether, and their pharynx, perforated with many gill slits, forms a bag-like sieve that functions in filter feeding. All vertebrates are members of the Subphylum Vertebrata.

SUBPHYLUM CEPHALOCHORDATA

There are only two genera and about 30 species in this subphylum. **Lancelets** (*Asymmetron* and *Branchiostoma*, formerly *Amphioxus*) superficially resemble fishes (see chapter opening photograph). They inhabit coastal waters in tropical and temperate seas, spending most of their time partly buried tail downward in sand. Lancelets seem to prefer smooth-surfaced, porous sands with little organic decomposition, and they are not tolerant of oxygen-poor waters. In the feeding position, the anterior mouth and gill apparatus is exposed for filter feeding.

It would be difficult to imagine a better model than the lancelet to demonstrate the four diagnostic chordate features. A notochord lies under the dorsal nerve cord, and both of these structures extend the length of the body. The pharynx has about 100 gill slits on each side, and a prominent tail extends posterior to the anus (Figure 28.1). Much of the lancelet body consists of **myomeres** (segmental blocks) of longitudinal muscles, clearly visible through the thin epidermis. Connective tissue **myosepta** separate the myomeres, and when the animal is swimming or burrowing, myomeres on opposite sides of the body contract in alternate waves from anterior to posterior, providing side-to-side undulations. The notochord provides rigidity along the longitudinal axis. Prominent dorsal "fin rays" are actually food storage organs that accumulate energy reserves for use during reproductive periods.

Filter feeding involves all anterior parts of the digestive tract. The mouth region is adapted for sorting and collecting suspended particles. An **oral hood** has bands of cilia that propel water and suspended particles toward the mouth. Tentacle-like **cirri** projecting from the oral hood prevent large particles from entering the mouth. The **velum**, a membrane around the mouth, also has ciliated tentacles that strain particles. Cilia on the gills drive seawater and selected particles through the mouth and into the pharynx. Water passes out through the gill slits into a large internal space called the **atrium**. Muscle contractions periodically force water out of the atrium through the **atriopore**. Food particles are caught by the gill cilia and by mucus secreted by the **endostyle**, a ciliated ventral groove in the pharynx. The mucus and particles form a **food string** that passes into the intestine. Here, ciliary action rotates the food string and passes it into a large **hepatic cecum**, where both extracellular and intracellular digestion occur.

The lancelet circulatory system is closed and resembles that of vertebrates, especially fishes (see Figure 6.11). There is no heart, but the **ventral aorta** and small bulbs at the base of each gill are contractile and propel blood dorsally into the gill vessels. From the gills, blood passes posteriorly in dorsal vessels to capillaries in the intestine, skin, and other body organs. Oxygenation occurs in the skin and gills. **Cardinal veins** return blood to the ventral aorta, and there is a **portal system** carrying blood from capillaries in the gut to capillaries in the hepatic cecum.

Lancelets are unique chordates in that they have **protonephridia**, which collect fluid from small coelomic sacs at the top of the gills and pass urine into the atrium (see the inset in Figure 28.1). (Protonephridia are described in more detail in Chapter 7, p. 156; Figure 7.8.) The atrium occupies much of the body space. It develops from folds of the body wall and is not coelomic. Little is known about excretion and osmoregulation in these animals, but it is likely that the skin, gills, and protonephridia excrete ammonia and maintain the salt and water balance.

The lancelet nervous system includes segmentally arranged nerves branching from the dorsal nerve cord. There is no brain or anterior sense organs, but sensory cells occur over the body surface, especially in the oral hood and mouth region.

Lancelets are dioecious, and their sex organs are metamerically arranged. Between 20 and 25 pairs of gonads develop on the inner wall of the atrium. Gametes are released from the atrium, and fertilization is external. Lancelets are model deuterostomes, with holoblastic radial cleavage and enter-

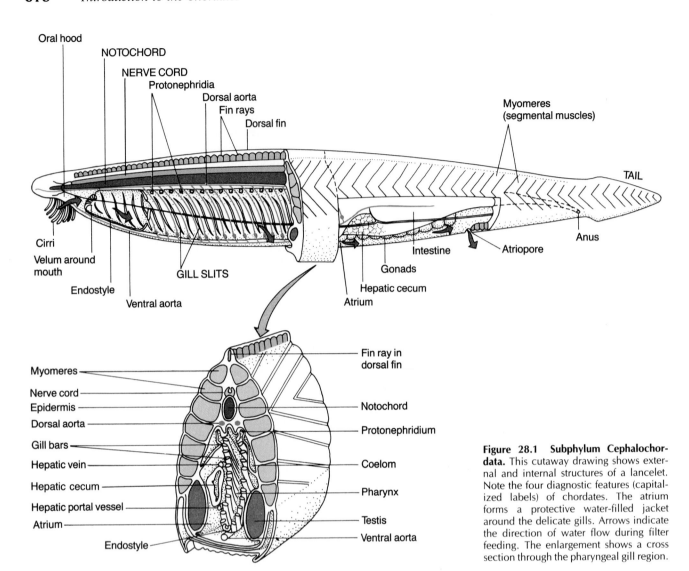

Figure 28.1   Subphylum Cephalochordata. This cutaway drawing shows external and internal structures of a lancelet. Note the four diagnostic features (capitalized labels) of chordates. The atrium forms a protective water-filled jacket around the delicate gills. Arrows indicate the direction of water flow during filter feeding. The enlargement shows a cross section through the pharyngeal gill region.

ocoelous mesoderm formation. Each has a free-swimming ciliated larva that spends several months as a planktonic filter feeder, then settles to the bottom, burrows into sand, and transforms into an adult.

SUBPHYLUM UROCHORDATA

The name Urochordata comes from the Greek word meaning "tail cord" and refers to the notochord in the larval tail. Unlike other chordates, members of this subphylum lack a coelom and do not show any signs of metamerism during larval or adult life. Urochordates are also called **tunicates** because some of them live in a secreted cloak, or tunic,

which contains a cellulose-like substance called **tunicin** and provides protection and support.

There are about 1400 species of urochordates occurring from the poles to the tropics, mostly in water less than 100 m deep. About 100 species live at greater depths (down to about 5000 m), and a few species inhabit wet beach sands. Of the three classes in the subphylum, the Ascidiacea is the largest, with about 1300 sessile species. Members of the classes Larvacea and Thaliacea are free-swimming plankton.

Class Ascidiacea

Members of this class are commonly called **ascidians** or **sea squirts**, because when disturbed, they squirt water out of an excurrent siphon on the side of their body. Adults are saclike or cylindrical, and many species are colonial (Figure 28.2a). They attach to hard substrates as well as to sand or mud. Except for a large gill apparatus, adult ascidians do not look much like chordates. Their chordate features appear in their free-swimming, planktonic larval stage.

In adult ascidians, the body wall, called the **mantle**, consists of an epidermis that secretes the tunic, a connective tissue dermis, and three muscle layers. The **excurrent siphon** opens through the mantle and tunic from an internal water-filled atrium. An **incurrent siphon** leads into the mouth. Both siphons can be closed by contraction of sphincter muscles.

The ascidian filter-feeding mechanism is similar to that of lancelets (Figure 28.2b). Cilia bordering the gill slits draw seawater and suspended particles in through the incurrent siphon and mouth. Tentacles around the mouth sort and reject oversized particles. Water passes out of the pharynx through the numerous gill slits, into the surrounding atrium, and is forced out the excurrent siphon by muscular contractions. The pharynx has a mucus-secreting endostyle. Food particles trapped in the mucus are propelled through the gut by cilia, and digestion occurs in the stomach. Wastes are egested from the anus into the excurrent water leaving the atrium. In many ascidians, much of the gut lies in the **visceral cavity**, separate from the seawater-filled atrium. The atrium and visceral cavity develop from folds of the mantle, not from a coelom.

In striking contrast to the typical chordate plan, the ascidian circulatory system is open, with a tiny tubular heart that pumps hemolymph in alternating directions. In each complete cycle, the heart pumps about 20 times; about 10 beats propel hemolymph out to body sinuses in one direction; then as hemolymph collects back into the opposite side of the heart, another 10 beats pump it in the reverse direction. The heart and main vessels lack valves, but body fluid pressure seems to close off the vessels at the nonpumping end of the heart. The hemolymph carries dissolved gases and wastes, and gas exchange and ammonia excretion occur across the pharynx and tunic. The hemolymph contains no respiratory pigments, but may contain unusually high concentrations of certain rare elements. Whereas seawater contains only trace amounts of the element vanadium, the hemolymph of certain sea squirts may have concentrations of nearly 2 g/L. Vanadium is concentrated by cells in the pharynx and may be used as a reducing agent as tunicin is formed in the mantle.

The nervous system of ascidians, consisting mainly of a ganglion in the mantle and a nerve net, is unlike that of other chordates, but similar to nervous systems of many other sessile animals. Sensory organs are lacking, but receptor cells are aggregated around the siphons.

Most ascidians are hermaphroditic, with single or paired ovaries and testes in the visceral cavity. Each gonad has its own duct that releases gametes into the excurrent water. Following external fertilization, ascidians undergo a developmental se-

(a)

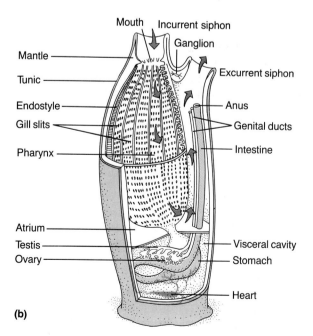

(b)

Figure 28.2 Subphylum Urochordata: Class Ascidiacea. (a) The colonial ascidian (tunicate) *Ciona intestinalis* lives on pilings and floats in shallow waters of the Atlantic coasts of Europe and North America and along the U.S. Pacific coast. Many other ascidians are solitary. Individual tunicates range in height from a few mm to nearly 40 cm; individuals of *C. intestinalis* are up to about 15 cm high. (b) Cutaway drawing of an individual adult ascidian, showing the direction of water flow through the gill apparatus and atrium.

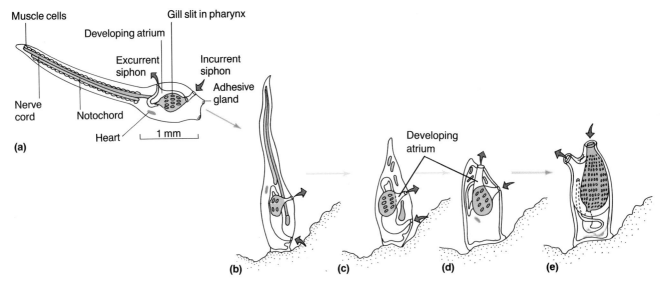

Figure 28.3 Metamorphosis of a larval ascidian. (a) A free-swimming, bilaterally symmetrical larva has a complete gut with an atrium developing around the pharynx. (b) The larva attaches to the substrate by adhesive glands, and metamorphosis begins. (c) The larval tail, notochord, and nerve cord are resorbed; the atrium expands around the pharynx, and more gill slits form. (d) Late in metamorphosis, the gut has rotated toward the adult position and the atrium continues to expand around the pharynx. (e) Metamorphosis to the sessile adult stage is complete; the water siphons and gut are in the adult position.

quence quite different from that of typical deuterostomes. Cleavage is bilateral, mesoderm forms from clusters of cells on the sides of the archenteron, and coelomic pouches do not appear. It is likely that the coelom was lost as the atrium evolved for filter feeding in these animals. (Bilateral cleavage is discussed in Chapter 16, p. 380.)

The ascidian larva, sometimes called a "tadpole" because it superficially resembles a frog larva, is elongate and bilaterally symmetrical (Figure 28.3a). It has a postanal tail with a notochord and nonsegmented muscles. A dorsal hollow nerve cord extends most of the length of the body. The larva does not feed, although it has a complete gut with a pharynx resembling that of the adult. Larvae are planktonic for a few hours or days, then attach to the bottom and metamorphose to the adult stage (Figure 28.3).

Class Larvacea (Appendicularia)

About 70 species of small (up to about 5 mm long), nearly transparent planktonic chordates make up this class. **Larvaceans** exhibit all fundamental chordate features and are free-swimming throughout life. The body is similar to the tadpole larva of ascidians, but is somewhat U-shaped with a long tail (Figure 28.4a). It is generally believed that lar-

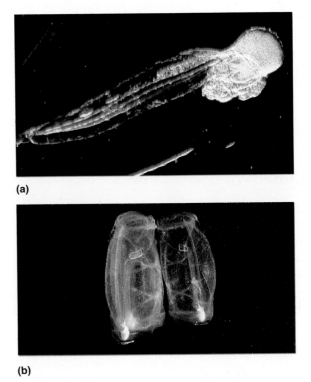

Figure 28.4 Planktonic urochordates. These animals are numerous, and many species are important in the diet of filter-feeding baleen whales and other plankton eaters. (a) Class Larvacea. These urochordates eat fine planktonic algae trapped in their proteinaceous houses. (b) Class Thaliacea. The salps, including these specimens of *Pegea confoederata*, are also filter feeders.

vaceans evolved from ascidian-like urochordates that lost the sessile stage in the life cycle.

Larvaceans have a unique method of filter feeding. They secrete a proteinaceous tunic called a **house**, which is often much larger than the animal. The house has incurrent and excurrent openings and serves as a plankton trap. A coarse fibrous screen over the incurrent opening prevents large particles from entering the house. The long tail whips back and forth, driving water through the house. Minute plankton that many other filter feeders cannot trap are caught on a fine mesh screen inside the house and are taken into the mouth. A valve at the excurrent opening intermittently releases jets of water, propelling the animal and its house slowly through the water. The house is a temporary structure. As its filters clog (often in a matter of hours or days), the larvacean simply exits via an escape door and secretes a new house.

Class Thaliacea

Members of this class (about 30 species), commonly called **salps**, range in length from about 10 cm to colonies that reach 4 m. They lack a tail as adults and are barrel-shaped, with a large water intake at one end and an excurrent opening at the other (Figure 28.4b). Muscle bands encircling the body contract, jetting seawater from the excurrent opening and moving the animal forward. The salp body is transparent but often brightly luminescent. Several species are numerous and may light up the ocean surface at night.

Chordate Evolution

Zoologists generally agree that the earliest chordates were invertebrates, perhaps gill filter feeders somewhat like urochordate larvae or cephalochordates. Unfortunately, few invertebrate chordates have been fossilized, and there is no direct evidence of the main events in early chordate evolution. Several cephalochordates and urochordate-like fossils have been found in Silurian rocks (about 425 million years old), but chordate origins undoubtedly predated any of these fossils by many millions of years. Actually, the oldest chordate fossils are the remains of fishlike jawless vertebrates called **agnathans** (from the Greek word meaning "without jaws"). Dating from the Cambrian period, about 525 million years ago, extinct agnathans ranged in length from about 0.1 cm to 1.5 m and had a head

and a filter-feeding gill apparatus. Many species were covered with an armorlike bony exoskeleton (see Figure 28.8). These chordates were probably preceded by less specialized, soft-bodied species that were not fossilized.

Central questions concerning chordate evolution are: What were ancestral chordates like? What group of nonchordate invertebrates gave rise to the chordate line? and If the first chordates were invertebrates, what invertebrate chordates were ancestral to the first vertebrates? A hypothesis developed in the 1920s by British biologist Walter Garstang is generally accepted as the best interpretation of early chordate evolution. Garstang postulated that ancestral chordates were sessile bottom dwellers with a filter-feeding gill apparatus similar to that of modern ascidian adults. Garstang further postulated that the notochord, dorsal nerve cord, and postanal tail appeared later, in a free-swimming larval stage that evolved as a secondary means of dispersal and that was added to the ancestral chordate life cycle. Garstang thought that such a larva and the phenomenon of **neoteny** (reproduction by a larvalike adult) probably played a central role in chordate evolution. He postulated that the free-swimming larva became the more important stage in the early chordate life cycle. The larva may have become longer-lived than the adult stage, eventually developing sex organs and reproducing. If this occurred, the sessile stage may have dropped out of the life cycle. Thus, a reproducing (neotenic) urochordate "larva" may have been ancestral to other chordates.

Support for Garstang's hypothesis is seen in the life cycle of the larvaceans. Lacking a sessile stage, these urochordates seem to illustrate neoteny and may be living representatives of the early evolutionary events that led to other chordates.

Elaborating on Garstang's ideas, Alfred Romer, an American paleontologist, developed the idea that the chordate body evolved from two sets of components. **Visceral components** include the digestive tract and other organs of the body cavity, as well as muscles, nerves, and skeletal parts derived from or associated with the gut or coelom. Thus, the gill apparatus, as a derivative of the pharynx, is a visceral component. **Somatic components** are elements of the outer body—the integument, the body wall, and their derivatives—all adaptations for active forward movement. Of the diagnostic chordate features, the dorsal nerve cord, notochord, and tail are all somatic components. Romer suggested that chordate evolution occurred in several visceral and somatic stages. In this sense, the earliest chordate, an ascidian-like bottom dweller, was essentially a nonmotile visceral animal. It had a large visceral (pharyngeal) component, but its somatic elements (probably a tunic and a nerve ganglion) had little or no locomotor function and were not like those of a typical chordate. Typical chordate somatic elements were probably evolved in the bilateral, free-swimming larva of the early chordates. With the evolution of the neotenic larva, visceral and somatic components may have become more nearly equal in importance: The ancestral larva had a large pharynx for filter feeding, but as a bilateral, free-swimming animal, it also had well-developed somatic components. Compared to larval urochordates, vertebrates have highly integrated visceral and somatic components, and Romer postulated that an increase in visceral and somatic integration and a trend toward greater development of somatic components were central features of early chordate evolution.

Assuming that urochordate-like visceral animals were ancestral chordates, what animals might have given rise to them? It would be logical to look for urochordate ancestors among other sessile filter feeders in the deuterostome line. Many zoologists consider the **pterobranchs**, sessile filter feeders of the Phylum Hemichordata, good models of urochordate ancestors (see Figure 27.22). Pterobranchs have external arms and tentacles rather than gills for filter feeding, but some species have gill pores in the pharynx. Perhaps the pharyngeal gill apparatus evolved in animals similar to pterobranchs as a more efficient feeding system and eventually replaced the external arms.

Filter feeding by means of tentacled arms is common in several phyla of animals. In their mode of feeding and sessile habit, pterobranchs resemble certain stalked echinoderms, such as the extinct eocrinoids or carpoids (see Figure 27.17). Also, both pterobranchs and certain carpoids bear some resemblance to lophophorates. (Lophophorate phyla are discussed in Chapter 27, p. 651.) Perhaps all these animal groups are distantly related and shared an ancestor with early chordates, but linking them is purely conjectural. It is also possible that they evolved their similarities independently through adaptation to sessile filter feeding.

Turning to the question of vertebrate origins, several threads link urochordate-like ancestors with early vertebrates. The extinct agnathans resembled invertebrate chordates in being filter feeders. It is also possible that nonfossilized agnathans resembled urochordate larvae more than any of the fossilized species. Among the invertebrate chordates, lancelets have both urochordate and verte-

brate features. Their gill apparatus is urochordate-like, but their circulatory system and metameric nervous and muscular systems resemble those of vertebrates. On the other hand, lancelets are specialized, and their protonephridia and metameric gonads do not have counterparts in any other chordates. For these reasons, zoologists generally agree that lancelets do not represent ancestral vertebrates. Lancelets could represent a stage in chordate evolution after the early neotenic larvae lost the sessile life cycle stage, but they probably constitute a side branch off the mainstream of chordate evolution.

Another bit of evidence linking invertebrate chordates and vertebrates is seen in the life cycle of lampreys, a group of living agnathans. Larval lampreys, called **ammocoetes**, have all the fundamental features of chordates and are remarkably similar to urochordate larvae and to lancelets (see Figure 28.7b). Much like lancelets, ammocoetes live in sand and filter-feed with gill structures. They also have vertebrate kidneys, a liver, and a pancreas, and their muscles are segmented.

Figure 28.5 summarizes some of the possible events in early chordate evolution. Based on similarities between pterobranchs and urochordates, it is postulated that chordates arose from deuterostomes that filter-fed by means of tentacled arms. Pterobranchs are on a side branch of the phylogenetic tree, but may represent ancestral stocks, as may certain filter-feeding echinoderms. Following the main trunk of the tree, gill filter feeding seems to have been on the mainstream of chordate evolution. If arm feeders were ancestral to chordates, their adaptation to gill filter feeding was a significant milestone in chordate evolution. Consistent with Garstang's hypothesis, urochordate-like bottom dwellers and their neotenic, free-swimming larvae are shown on the mainstream of chordate evolution leading to filter-feeding agnathans.

SUBPHYLUM VERTEBRATA

Vertebrates, the most successful group of chordates, are characterized by a **spinal column** of **metameric vertebrae** and a **skull**. These elements provide support and protection for the dorsal hollow nerve cord (spinal cord) and its anterior expansion, the brain.

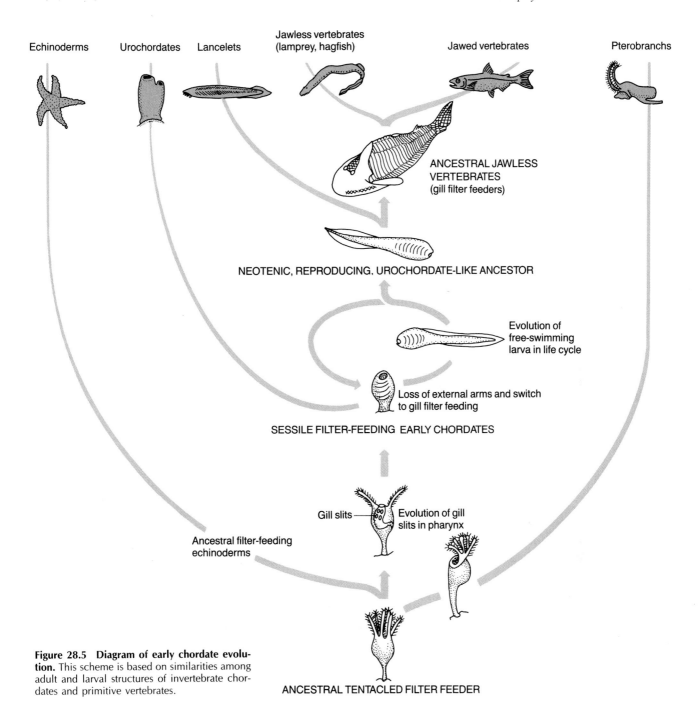

Figure 28.5 Diagram of early chordate evolution. This scheme is based on similarities among adult and larval structures of invertebrate chordates and primitive vertebrates.

REPRESENTATIVE VERTEBRATES

Vertebrates are classified as agnathans (jawless vertebrates), fishes, amphibians, reptiles, birds, and mammals. Agnathans are often considered fishes because they are aquatic and generally fish-like in body form, but they are very different from true fishes. As a group, agnathans constitute the Class Agnatha in the Superclass Agnatha. All other vertebrates are classified in the Superclass Gnathostomata (from the Greek word, meaning "jawed mouth").

Superclass Agnatha, Class Agnatha

There are only about 60 living species of agnathans, but from 400 to 500 million years ago, these were

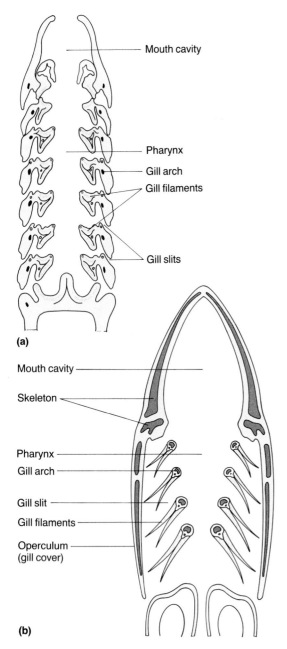

(a)

(b)

Figure 28.6 Two types of vertebrate gills. The gills of (a) an agnathan compared with those of (b) a bony fish. These views are from the ventral side of the body. Note the position of the gill filaments relative to the gill arches.

the most diverse vertebrates. Besides lacking jaws, agnathans have several features that distinguish them from other vertebrates. They have gills with soft tissues internal to the gill arches, and their gills project inward toward the pharynx; the gills of fishes project outward (Figure 28.6). Agnathans have only one or two semicircular canals in their inner ear; all other vertebrates have three. And unlike other vertebrates, living agnathans (and many

extinct species) lack paired appendages. Moreover, in common with only a few fishes, agnathans are primitive in having a notochord as adults.

There are two subgroups of living agnathans, the lampreys (about 40 living species) and the hagfishes (about 20 living species). Although only distantly related, animals in both of these groups have entirely cartilaginous endoskeletons, lack external armor, including scales, and are specialized predators or scavengers.

Lampreys spend their larval life (3 to 6 years) in fresh water (Figure 28.7a and b). Depending on the

(a)

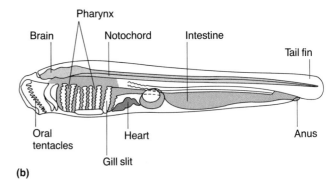

(b)

(c)

Figure 28.7 Living members of the Class Agnatha. (a) An adult least brook lamprey, *Lampetra aepyptera.* (b) The lamprey's ammocoete larva, whose internal structures are seen here from the left side, closely resembles cephalochordates. (c) An adult California hagfish, *Eptatretus stouti.* These agnathans are also called "slime hags," because when disturbed, they secrete great quantities of mucus from a row of skin glands on the sides of the body.

species, adults range from about 75 mm to 1 m long and may live in fresh water or in the sea. Some species are strictly filter feeders, feeding only as ammocoetes and living for only a few days as adults, just long enough to reproduce. Adults of several large-bodied species are predaceous (some zoologists say parasitic), attaching themselves to fishes with a toothed oral disc (see Figure 4.4). They rasp into the victim's body wall, inject an anticoagulant into the wound, and suck out body fluids. Predaceous lampreys remain attached to their hosts until full of blood, then drop off, leaving an open wound that may become infected with bacteria and fungi. (See the Essay on the destruction of the Great Lakes fishery, and Figure 34.15.) Adult lampreys have seven pairs of gills that unlike fish gills are not ventilated with water from the mouth. A lamprey can breathe while attached to a fish because its gills pump water in and out, working independently of the mouth.

Essay: Destruction of the Great Lakes Fishery

The sea lamprey *(Petromyzon marinus)* is a native inhabitant of the Atlantic Ocean and fresh waters of the eastern seaboard of North America. Adults are large (reaching about 50 cm in fresh water) and prey on large fishes. They migrate into freshwater streams to spawn, and some adults remain in large inland lakes with outlets to the sea. Adults have always had access to Lake Ontario via the St. Lawrence River, but the other Great Lakes had no sea lampreys until two artificial waterways, the New York State Barge Canal (connecting Lake Erie with the Hudson River) and the Welland Canal (around Niagara Falls) were opened, providing sea lampreys access to all of the Great Lakes (Figure E1).

Sea lampreys have an enormous reproductive capacity; although females spawn only once, they may produce 200,000 eggs. Populations began growing rapidly in most of the western Great Lakes starting about the 1920s. Only Lake Erie failed to develop dense lamprey populations, perhaps because it is too warm and has few streams suitable for the filter-feeding larvae.

Adult sea lampreys prefer to feed on larger fish species, such as lake trout and whitefish, which were commercially important in the Great Lakes. As a result of lamprey predation (probably coupled with a concurrent increase in commercial harvest and water pollution), in less than three decades (from about 1920 to 1950), the yearly catch of the Great Lakes' commercial fish species declined from nearly 12 million pounds to virtually nothing. As the commercial species declined, lampreys attacked other fishes, and by the late 1950s, the Great Lakes were mostly depopulated of large fishes. Without prey, lamprey populations also declined. Finally, in the early 1960s, after an enormous research effort, a chemical was discovered that is fairly selective in killing ammocoete larvae. By applying the chemical in streams where lampreys spawn, biologists have been able to control the lamprey populations, and other fishes have been successfully restocked in the Great Lakes.

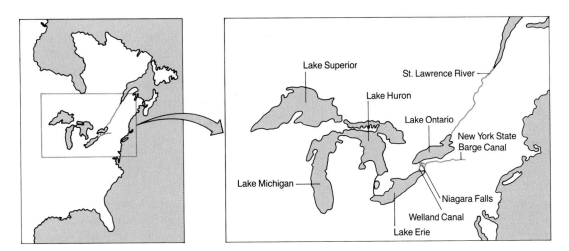

Figure E1   The Great Lakes and access to the sea. Niagara Falls (between Lake Ontario and Lake Erie) was once an effective barrier to lamprey dispersal. Note that the New York State Barge Canal (first opened in 1825) and the Welland Canal (opened in 1932) bypass Niagara Falls.

Hagfishes are believed to be degenerate descendants of extinct bottom-scavenging agnathans (Figure 28.7c). Although classified as vertebrates because of their general appearance, hagfishes are very different from all other living vertebrates because they lack vertebrae. They also have three accessory hearts, one in the tail and two in the venous system. Unlike lampreys, hagfishes have no larval stage and are exclusively marine. They have from 5 to 14 pairs of gills that in some species open through a single external pore. Hagfishes also lack the lamprey's sucking disc, but have teeth (not homologous with those of gnathostomes) and can open and close their mouth laterally. They eat small invertebrates or scavenge on the seafloor, frequently burrowing into dead or dying fishes and consuming the contents. Hagfishes are sometimes found wriggling inside the "husks" of fishes caught in fishing nets. Hagfish eyes are rather weakly developed and are covered with skin, but around the mouth are three pairs of touch- and chemical-sensitive tentacles, called **barbels**, that help the hagfish locate food.

Extinct agnathans are often called **ostracoderms** in reference to the bony armor of some species, but this name is no longer used in classification. Agnathans were diverse for nearly 100 million years, and different authorities recognize from 4 (used in this book) to about 20 major subgroups, dating mostly from about 375 to 450 million years ago.

The most diverse agnathans, also the oldest known vertebrates, are called **heterostracans** (Figure 28.8a). Heterostracans were numerous from the Ordovician to the Devonian Period (about 375 to 500 million years ago). A second subgroup of extinct agnathans, the **thelodonts**, had well-developed paired fins and were weakly armored with small scales (Figure 28.8b). Members of a third subgroup, the **osteostracans** (from the Greek word meaning "bony shell"), were encased in heavy bony plates and had a somewhat flattened head shield (Figure 28.8c). Paleontologists have been able to reconstruct precise details of osteostracan anatomy because internal head bones of several species were fossilized. Blood vessels of certain species resembled those of larval lampreys, perhaps indicating that lampreys and osteostracans had a common ancestor.

A fourth extinct subgroup of agnathans, the **anaspids**, were laterally compressed and generally less armored than the osteostracans. Some, such as *Jamoytius*, were covered with thin bony scales and had lateral fins (Figure 28.8d).

Superclass Gnathostomata

Jawed vertebrates—the fishes, amphibians, reptiles, birds, and mammals—are described in detail in subsequent chapters. They were also introduced in Chapter 1 (see Figure 1.18). Recall that there are two living classes of fishes. Sharks, rays, and related cartilaginous fishes constitute the **Class Chondrichthyes**. Bony fishes, by far the largest group of vertebrates with about 21,000 species, make up the **Class Osteichthyes**. Other vertebrates are collectively called **tetrapods** because they have four limbs or are descended from four-limbed ancestors. The **Class Amphibia** includes the frogs, salamanders, and a limbless, wormlike group called the **caecilians**. (See the discussion of the Order Gymnophiona in Chapter 30, p. 729.) As the name Amphibia, meaning "double life," implies, many of these animals spend part of their life on land, but most species must return to water to reproduce. Turtles, crocodiles, snakes, and lizards constitute the **Class Reptilia**, a group that is better adapted for life on land than are the amphibians. The reptilian body has evolved efficient water-retention capabilities, including a nearly impermeable integument. Ancestral reptiles gave rise to the other two classes of vertebrates, the **Class Aves**, or the birds, and the **Class Mammalia**. As discussed in Chapter 18 (see p. 441), birds are very similar to reptiles, but may be distinguished by their feathers and physiological maintenance of constant, warm body temperature (homeothermy). Mammals have hair, are homeothermic, and have milk-secreting mammary glands. Reptiles, birds, and mammals are collectively called **amniotes** because they all have special membranes that surround and protect the developing embryo. One of the membranes, called the amnion, forms a fluid-filled sac containing the embryo. (See the discussion of gastrulation in birds and mammals in Chapter 16, p. 387.) Agnathans, fishes, and amphibians lack the amnion and are collectively called **anamniotes**.

VERTEBRATE PHYLOGENY

Figure 28.9 presents a broad outline of vertebrate evolution. Notice the position of the agnathans. Modern lampreys and hagfishes represent two ancient lineages derived from the oldest known vertebrates, the extinct filter-feeding agnathans. Fossil lampreys about 275 million years old are remarkably similar to modern species.

Figure 28.8 Extinct agnathans in an artist's reconstruction of a Silurian seafloor. Lacking jaws, agnathans probably fed by stirring up mud, ingesting it, and trapping suspended particles on their gills. (a) The heterostracan *Anglaspis* (about 6 cm long), representing the most diverse group of agnathans, was covered with large, heavy scales. (b) The thelodont genus *Phlebolepis* (about 8 cm long) may have been a fairly agile swimmer; it was covered with light scales and had a pair of flaplike fins behind the head. (c) The osteostracan *Hemicyclaspis* (about 20 cm long) also had paired fins. Note the heavily armored, dorsoventrally flattened, shieldlike head; the dark band on the lower margin of the head is believed to have been a sense organ. (d) The lightly scaled anaspid *Jamoytius* (about 15 cm long) had a circular anterior mouth, paired lateral fins, and many gill openings. The scorpion-like arthropods in the drawing are eurypterids, probably important predators of agnathans during the Silurian period. Most eurypterids were about 5–30 cm long, but some were nearly 3 m long.

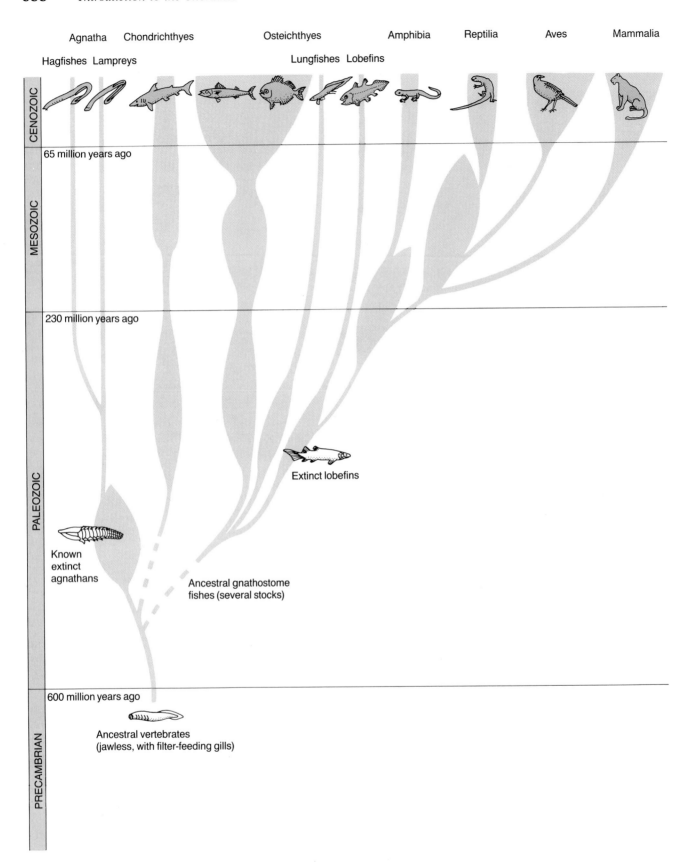

Figure 28.9 Phylogenetic tree of the living classes of vertebrates. The relative diversity of groups through time is indicated by the width of the lineages.

The oldest-known gnathostomes are fossils of armored fishes that lived about 450 million years ago. Because of the difference between agnathan and gnathostome gills, it is unlikely that gnathostomes were derived from agnathans of the types seen in fossilized form. Thus, Figure 28.9 shows ancestral gnathostomes as derivatives of jawless vertebrates (agnathans that were not fossilized) that predated the fossil agnathans. (It is also possible that gnathostomes and agnathans were derived from different ancestors.) Note that modern chondrichthyans and bony fishes (osteichthyans) are thought to represent separate lineages of ancestral gnathostomes.

The first tetrapods were amphibians that apparently arose from a group of bony fishes called **lobefins** (see Figures 29.19 and 30.16). Lobefins had paired fins with bony supports resembling those supporting tetrapod limbs. They also had a pair of lungs that could be filled with air via a connection between the external nostrils and the pharynx.

Reptiles probably evolved about 300 million years ago from certain stocks of amphibians. The earliest forms resembled amphibians, but unlike amphibians, they evolved a shelled egg that resisted drying on land. Birds and mammals, the phylogenetically youngest groups of vertebrates, evolved from two distinct groups of reptiles.

EVOLUTION OF VERTEBRATE STRUCTURE

Although members of the various vertebrate classes look quite different from one another and have become adapted for different ways of life, they all share a common body type with fundamentally similar organ systems. Vertebrates clearly illustrate evolutionary adaptation of homologous structures (discussed in Chapters 17 and 18, pp. 415 and 427).

The following discussion compares adaptations in organ systems of the vertebrate classes. It emphasizes evolutionary changes that allowed vertebrates to colonize land. The discussion also illustrates Romer's concept of the vertebrate body as an integrated somatic-visceral unit. Somatic and visceral components of the various organ systems are indicated in the following headings by the letters S and V.

The Skeletal System (S, V)

All vertebrates have an endoskeleton formed of fibrous connective tissue, cartilage, and/or bone. As discussed in Chapter 8 (see p. 190), vertebrate bone develops in one of two ways. **Endochondral bone** forms in three stages. Fibrous connective tissue develops first; this is replaced by cartilage, and finally the cartilage is replaced by osteocytes and bone matrix. **Dermal (membrane) bone** develops directly from fibrous connective tissue without the intervening cartilage stage. As you will see, these two sources of bone have phylogenetic significance.

The vertebrate skeleton is an integrated combination of visceral and somatic components. **Gill arches** (cartilage or bony supports for the gills) and structures derived from gill arches constitute a set of visceral elements. Two sets of somatic components are the **axial elements** (vertebral column, ribs, sternum, and skull, all providing central support for the body) and the paired **appendicular elements** (pectoral girdle, pelvic girdle, and limb supports, all supporting the body appendages).

Gill (Visceral) Arches. Vertebrate gills have undergone remarkable adaptations from their ancestral condition as filter-feeding structures. Gill pores or slits appear during some stage of development in all vertebrates. They form as invaginations from the outer body wall (ectoderm), grow inward, and fuse with outpouchings from the endoderm of the developing pharynx. Each point of fusion develops an opening, or gill slit, supported by a gill arch. Skeletal elements of the gills develop from ectoderm, unlike the two sets of somatic components (axial and appendicular elements), which develop from mesoderm. Bones of the gill arches and their derivatives are endochondral.

The evolution and embryological derivations of the gill slits and arches have been investigated in all vertebrate classes. Agnathan gills are not homologous with those of gnathostomes, but gill homologies in gnathostomes are a classic study of evolutionary adaptation. In the ancestral condition in gnathostomes, there may have been up to 10 to 15 pairs of similar gills, all functioning in feeding and respiratory gas exchange. Present-day gnathostomes have seven or fewer pairs of gills, some or all of which become nonrespiratory structures. Jaws and jaw supports first evolved from anterior gill arches, and in tetrapods the gill arch structures undergo remarkable developmental changes, transforming into parts of the skull, including tiny ossicles in the middle ear and cartilages supporting the larynx (voice box) (Table 28.1 and Figure 28.10).

Axial Elements. All axial components, except the skull, are formed entirely of notochord, cartilage, or endochondral bone. The **vertebral column** is

Table 28.1   Evolution of the Gill (Visceral) Arches in Selected Vertebrates

Arch No.	Sharks	Amphibians, Reptiles, Birds	Mammals
		Derivatives of Arches	
I	Upper jaw	Two skull bones	One skull bone and ear ossicle (incus)
	Lower jaw	One of lower jaw bones	Ear ossicle (malleus)
II	Five cartilages supporting jaw	Ear ossicle (columella = stapes) and hyoid apparatus* (supports lower jaw and tongue)	Ear ossicle (stapes) and hyoid apparatus* (tongue support)
III–VII	Respiratory gills	Hyoid apparatus* and cartilages of the upper throat	Hyoid apparatus* and cartilages of the larynx

*The hyoid apparatus consists of several elements and is derived from gill arches II and III.

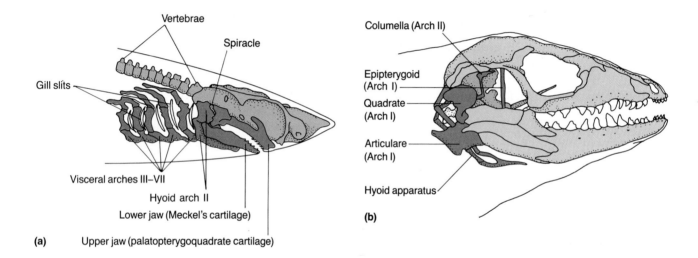

(a)

(b)

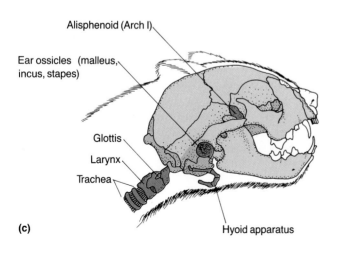

(c)

Figure 28.10   Evolution of the gill (visceral) arches. The skull and associated structures of representative gnathostomes illustrate the homologies of gill arch structures. Dark color: gill arches and their derivatives; intermediate color: somatic cartilages or somatic endochondral bones; light color: somatic dermal (membrane) bones. (a) The head of a shark illustrates the condition of early gnathostomes; the anterior gill arches have become adapted as jaws and jaw supports. Note the similarity between the jaw and jaw support cartilages and the gill arches. The gill slit between arches I and II has been reduced to a small hole, the spiracle. (b) Skull of a reptile, showing the derivatives of gill arches I and II. (c) Skull and associated visceral cartilages of a mammal.

prominent in all vertebrates except lampreys and hagfishes. In these agnathans, the notochord persists in adults and is the chief support element for the spinal cord. Similar to the vertebrae of fishes and tetrapods, small blocks of cartilage called **protovertebral blocks** provide auxiliary support for the notochord and spinal cord in lampreys.

Vertebrae in a fish are generally alike (Figure 28.11a), but in comparing the vertebral column through the tetrapod classes, from amphibians through reptiles, birds, and mammals, one sees a trend toward a greater modification and regional specialization of certain vertebrae. Birds and mammals have five distinct regions in the vertebral column. Vertebrae of the **cervical** (neck) region are distinct from those of the **thoracic** (chest), **lumbar** (middle back), **sacral** (lower back), and **caudal** (tail) regions (Figure 28.11b). Regionalization of the vertebral column provides support for specialized muscles, allowing complex movements in land vertebrates. Birds also have many fused vertebrae, providing added strength and stability for flight.

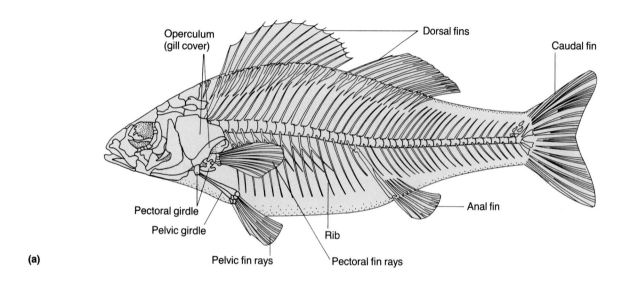

(a)

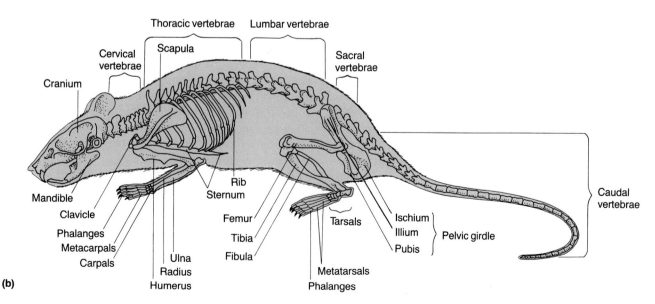

(b)

Figure 28.11 Regional specialization in vertebrate skeletons. Comparison of the skeletons of (a) a bony fish (a perch) with that of (b) a mammal (a rat).

Ribs in fishes are cartilaginous or bony extensions of vertebrae (see Figure 28.11a). They form in the connective tissue septa (myosepta) between muscle blocks. The ribs of amniotes enclose the thoracic region, protecting the heart and lung cavities and providing attachments for muscles active in breathing movements (see Figure 28.11b).

The **sternum** (breastbone) first evolved in amphibians. In reptiles and mammals, the sternum consists of a series of cartilages and/or bones articulating with the ventral ends of the ribs. The sternum is most developed in birds, where it forms a single large **sternal keel**, providing attachment sites for the massive flight muscles (see Figure 32.4).

Vertebrate **skulls** clearly illustrate Romer's principle of visceral-somatic integration. The skulls of hagfishes, lampreys, and chondrichthyans are formed entirely of cartilage. The agnathan skull is entirely somatic. Chondrichthyans show a rudimentary integration of visceral and somatic elements in having jaws, which are derived from the visceral (gill) arch skeleton, attached to a braincase formed of somatic cartilages. In bony fishes and tetrapods, the skull consists of a highly integrated unit of bones from three different sources: visceral elements and two groups of somatic bones, one endochondral and the other dermal (see Figure 28.10). Compared to fishes and amphibians, amniotes also have greater fusion of skull bones, generally fewer gill arch derivatives, and extensive development of the relatively few dermal bones forming the jaws and outer shell of the skull.

Appendicular Elements. The girdles lie within the trunk of the body and provide a base of support for the limbs. The **pectoral (shoulder) girdle**, supporting the forelimbs, includes both endochondral and dermal bones in bony fishes and tetrapods. The girdle is firmly attached to the skull in fishes, but is free of the skull and embedded in strong pectoral muscles in tetrapods. The tetrapod condition provides two important advantages for life on land: greater forelimb flexibility and greater head movement. Tetrapods also have fewer bones in the pectoral girdle than do fishes. Most mammals have only one or two pectoral bones: the **scapula**, or shoulder blade, and the **clavicle**, or collar bone. The clavicle provides support for muscles that pull the forelimbs and is especially well developed in monkeys that use their forearms to brachiate, or swing through trees. The clavicle is reduced or absent in mammals that are specialized for running, such as the horse. Birds have four bones, the scapula, coracoid, clavicle, and interclavicle, all providing attachments for large wing muscles. The clavicle and interclavicle are fused as the **furcula**, or wishbone (see Figure 32.4).

The **pelvic girdle**, also called the **pelvis**, is cartilaginous or endochondral. In contrast to the pectoral condition, the pelvic girdle is not attached to the vertebral column in fishes. But it is firmly attached to sacral vertebrae in tetrapods, providing strong support for the hindlimbs. Most fishes have only one pelvic element (bone or cartilage) per side, whereas tetrapods have three to five elements per side (see Figure 28.11b).

Vertebrate **limbs** are of two general types, reflecting the different demands of aquatic and terrestrial environments. Fishes have **paired fins**, reinforced in nearly all species by cartilaginous or bony **fin rays**. Supported by water, fishes use their fins for steering and balance and move foward by undulations of the body and tail fin. Only one group of fishes, the lobefins, have limb bones that are homologous with those of tetrapods (see Figure 30.16). In contrast to fish fins, tetrapod limbs evolved as adaptations for supporting and moving the body on land. Unsupported by water, terrestrial animals must counteract a greater effect of gravity. To move most effectively on land, they must hold themselves up off the substrate. The limbs of tailed amphibians (e.g., salamanders) and reptiles (e.g., lizards) support the body, but are held laterally. As a result, considerable energy is consumed in overcoming gravity to keep the body off the ground; locomotor movements in these animals are mainly undulatory, as in fishes. In contrast, the limbs of birds and mammals support the body from underneath and also provide most of the locomotor power. Birds and mammals do not undulate; their limb muscles provide locomotor power for walking, running, or flying.

Tetrapod forelimbs and hindlimbs contain two sets of bones of endochondral origin (see Figure 28.11b). One set, the **proximal elements**, consists of three bones per limb. Articulating with the girdles is a single element, the **humerus** in the forelimb and the **femur** in the hindlimb. The other two proximal elements are the **radius** and **ulna** in the forelimb and the **tibia** and **fibula** in the hindlimb. **Distal elements**, which make up the other set of limb bones, are numerous and have undergone greater evolutionary modification than have proximal elements. In the forelimb are **carpals**, **metacarpals**, and **phalanges** (**digits**); corresponding bones in the hindlimb are the **tarsals**, **metatarsals** and **phalanges**.

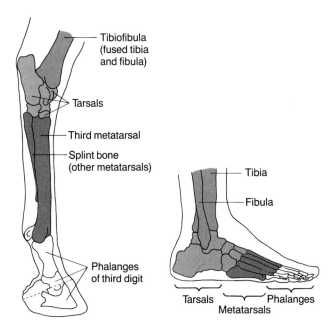

Figure 28.12 Comparison of the human hindlimb with that of a horse. The walking surface of the human foot extends from the phalanges to the tarsals (ankle bones). Human sprinters attain their greatest speed and efficiency by running on the tips of their toes. Horses represent the ultimate evolutionary solution to running efficiency. Their long legs, formed by elongate limb bones, provide excellent leverage. They run on the tip of a single phalanx (the distal phalanx of the third digit) covered by a cornified hoof.

The ancestral tetrapod limb was **pentadactyl**, meaning that it had five digits. This condition is evident in human hands and feet and in the limbs of several other mammals, such as bears. Derived limbs, seen in tetrapods specialized for certain activities, such as running or flying, may have fewer than five digits as a result of adaptive loss or fusions of bones. The horse is derived from a long lineage of runners, and its limbs are highly specialized for this function. Both its forelimbs and hindlimbs show fusions and elongations in proximal and distal elements (Figure 28.12). The effectiveness of horse limbs in running is discussed in Chapter 9 (see p. 211 and Figure 9.16).

The Muscular System (S, V)

The broad patterns of muscle evolution are similar to those of skeletal evolution, for the change from undulatory, body locomotion in aquatic vertebrates to nonundulatory, limb locomotion on land involved muscles as well as skeletal parts. Axial (trunk) muscles of the main body are distinct from limb muscles of the paired appendages. Axial and limb muscles are somatic, but muscles moving the gill structures are visceral. Jaw muscles and muscles of the tongue in gnathostomes are derived from gill arch musculature.

In the ancestral condition, as seen in most aquatic vertebrates, the great bulk of muscles are segmented axial bundles (myomeres) that power body undulations. The number of myomeres corresponds to the number of vertebrae. In the derived condition seen in most tetrapods, axial muscles account for much less of the total muscle mass, their segmentation is less pronounced, and limb muscles provide the main locomotor force.

Representatives of several vertebrate classes graphically illustrate the main evolutionary trends in musculature. Agnathans are the most primitive aquatic vertebrates. Except for small muscles in the head, their entire muscle mass consists of axial myomeres. The myomeres flex forward and backward, allowing the body to twist and turn; but without paired appendages, lampreys and hagfishes have rather limited swimming movements. The myomeres in chondrichthyans and bony fishes are complex and divided into dorsal and ventral groups by a sheet of connective tissue called the **horizontal septum**. These adaptations provide greater power and diversity in swimming movements (see Figure 29.13).

Limb muscles in fishes are quite small compared to those of tetrapods, and they serve in changing position of the fins, rather than in driving the body forward. Most tetrapods have massively developed limb muscles that power their strong and highly mobile paired appendages. Tetrapod axial muscles are not only reduced in mass relative to the limb musculature, but are also much more specialized than those of fishes.

The Integumentary System (S)

To illustrate the major differences between integuments adapted to aquatic systems and those adapted to terrestrial environments, we compare the integument of fishes, amphibians, and amniotes. This comparison shows that in adapting to land, the vertebrate integument, which consists of an epidermis and an underlying dermis, underwent a general thickening, layering, and moisture-proofing of dead cells at the surface, providing protection against drying and abrasion.

All vertebrates have a **cuticle**, the outermost layer of the epidermis, composed of sloughed dead cells and cell secretions that cover the living cells of the epidermis. The fish cuticle is very thin and consists of mucus, protective antibodies, and a few cells. This allows close interaction with the environment, for living epidermal cells, including mucus-secreting cells, are nearly in contact with the water. Pigment cells, or chromatophores, in the dermis can change body color to blend with the environment (Figure 28.13a). Fish scales, providing surface support and protection, are formed in, and remain embedded in, the dermis. Scales in bony fishes are overlapping plates of fibrous connective tissue impregnated with a bony matrix. Shark scales are very different (see Figure 29.2); toothlike, with a pulp cavity surrounded by dentine, they are homologous with shark teeth and the teeth of other gnathostomes. All types of fish scales grow with the animal.

Amphibian skin is probably most like that of the earliest land vertebrates (Figure 28.13b). It reflects the dual (aquatic and terrestrial) life of most am-

phibians and is not especially well adapted for life on land. The cuticle consists of a layer of dead cells called the **stratum corneum**. It is thicker than that of fishes, but thin compared with that of other tetrapods. The living layer of the epidermis is similar to that of fishes, but has large multicellular glands that secrete mucus or various toxic substances that repel predators. The stratum corneum cannot be too thick because the integument in most amphibians plays a large role in gas exchange. Thus, the skin surface must be kept moist, and large amounts of water are continuously lost from it during exposure to air.

Reptiles, birds, and mammals—truly terrestrial vertebrates—have a greatly thickened, keratin- and lipid-impregnated stratum corneum, providing an effective barrier to moisture loss (Figure 28.13c). Lacking glands, reptilian skin is virtually dry and is largely covered with scales formed from the thick stratum corneum. As reptiles grow, they periodically shed their skin, including the outer portion of their scales. Birds have one large oil gland, the **uropygial gland**, located dorsally at the base of the tail (see Figure 32.4). A bird uses its beak and neck to spread oil from the gland through its feathers. Beaks and claws of reptiles and birds are derived from thickenings in the epidermis. Scales on bird legs and feathers also develop from the epidermis and are homologous with reptilian scales. The mammalian epidermis is highly glandular (see Figure 8.21); mammalian claws, nails, hooves, and hair are derived from the epidermis.

The Digestive System (V)

Unlike other organ systems, the gut is remarkably similar in all vertebrates. It is suspended in the body cavity by the **dorsal mesentery**, which extends from the **parietal peritoneum** lining the inside of the body wall, and is continuous with the **visceral peritoneum** (Figure 28.14). Two or three layers of smooth muscle in the gut wall (two forming the tunica muscularis and sometimes a third called the muscularis mucosae) provide peristaltic movements, propelling food through the gut. One or two layers of connective tissue provide support for the epithelial lining of the gut lumen. The epithelium is richly supplied with gland cells that secrete mucus and digestive enzymes into the stomach and intestine. The general structure and function of parts of the gut and specific adaptations correlated with dietary types are described in Chapter 4.

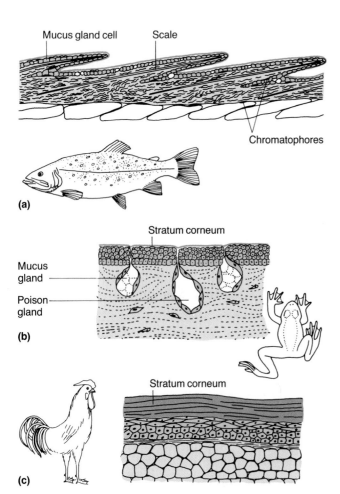

Figure 28.13 Comparison of vertebrate integuments. (a) Section through the skin of a bony fish. (b) Section of frog skin. (c) Section of bird skin (feathers not shown). Dark color: cuticle of the epidermis; intermediate color: living layers of the epidermis; light color: dermis.

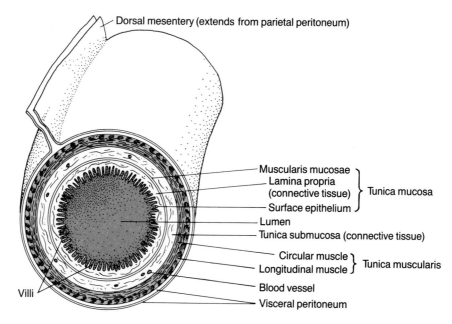

Dorsal mesentery (extends from parietal peritoneum)

Muscularis mucosae
Lamina propria
(connective tissue) } Tunica mucosa
Surface epithelium
Lumen
Tunica submucosa (connective tissue)
Circular muscle } Tunica muscularis
Longitudinal muscle
Blood vessel
Visceral peritoneum
Villi

Figure 28.14 Cross section of the vertebrate gut. Typical structures are shown. Note the three muscle layers and the outer wrapping of visceral peritoneum.

The Respiratory System (V)

Some of the most radical evolutionary changes in the vertebrate body were those that allowed vertebrates to breathe air with lungs instead of extracting oxygen from water with gills. Gills and lungs are both visceral components that develop from the pharynx, and most vertebrates have both of these organs or derivatives of them.

Agnathans and chondrichthyans have gill slits opening directly to the outside environment. They do not have any type of internal air sac. In bony fishes, gill slits open into a common chamber covered by a flap of the body wall, the **operculum** (see Figures 28.6 and 28.11). Bony fishes also have one or two internal gas-filled sacs called **swim bladders** that are homologous with tetrapod lungs (see Figure 29.5). Gas in the swim bladder makes bony fishes virtually weightless in water, allowing them to maintain their position in the water without swimming. Specialized as a buoyancy organ, the swim bladder does not function in gas exchange in most bony fishes. Fishes die on land because their gills collapse and dry, destroying the gas exchange surfaces.

One group of bony fishes, the **lungfishes**, have gills *and* air-breathing lungs and represent transitional (aquatic-terrestrial) vertebrates. (See the discussion of the Subclass Dipnoi in Chapter 29, p. 709; Figure 29.6.) Lungfishes inhale air at the water surface, and some species can survive completely dry conditions, breathing only air for months while living in dry mud cocoons. Lungfishes are distantly related to the lobe-finned fishes (lobefins), which are believed to represent the stem group that gave rise to amphibians (see Figure 28.9). In addition to lungs, lobefins also had two passageways leading from the external nostrils into the mouth cavity, allowing them to inhale air into their lungs without opening their mouth. With internal nostrils, lungs, and strongly supported paired limbs, lobefins were somewhat preadapted for terrestrial life. They may have been able to spend fairly long periods of time on land, and some of them gave rise to the first tetrapods.

Tetrapod lungs are quite varied in complexity and efficiency. Those of amphibians are small in proportion to body surface, with much less internal surface area than those of other tetrapods. As mentioned earlier, gas exchange across the skin and lining of the mouth in amphibians compensates for the rather ineffective lungs. Reptilian lungs are proportionately larger, with more gas exchange surface, and most reptiles rely entirely on their lungs for gas exchange. Birds and mammals have the greatest lung surface area, and their respiratory system is highly efficient at moving air in and out of the lungs. These systems are described in greater detail in Chapter 5 (see pp. 108, 112).

The Circulatory System (V)

Like the respiratory systems that they serve, vertebrate circulatory systems have undergone drastic evolutionary changes directly correlated with the transformation from gill to lung breathing. Pumping blood only to the gills, the typical fish heart has two main chambers, a thin-walled atrium and a highly muscular ventricle. Two accessory chambers are the **sinus venosus**, which receives blood from body tissues and passes it to the atrium, and the **conus arteriosus** (or bulbus arteriosus in bony fishes), which is valved and dispenses blood to the ventral aorta. The sinus venosus also serves as the pacemaker of the heart. The fish heart carries only oxygen-poor blood, and blood flow through the heart follows a single, linear path. (The bony fish heart and circulatory system are discussed in detail in Chapter 6, p. 128; Figure 6.11.)

With the development of lungs for air breathing, vertebrate circulatory systems became adapted to pump oxygen-poor blood to the lungs (via a pulmonary circuit) and then pump oxygen-rich blood from the lungs to other body organs (via the systemic circuit). The evolution of this dual-loop circulatory system involved major adaptations in the heart and in the large vessels carrying blood to and from the heart; these adaptations keep oxygen-poor blood separate from lung-oxygenated blood.

Lungfishes may represent primitive stages in the evolution of double (pulmonary-systemic) circulatory systems. Having gills *and* lungs, they have a circulatory system that combines features of both fishes and terrestrial tetrapods. In common with tetrapods, lungfishes have pulmonary arteries and veins that separate blood supply to and from the lungs (Figure 28.15a). The heart also provides partial separation of the pulmonary and systemic circuits. Depending on the species, there is a partial or complete septum partitioning the atrium, and some lungfishes have a partially divided ventricle. Oxygen-poor blood returning to the heart from body tissues enters the right atrium from the sinus venosus. Oxygen-rich blood from the lungs enters the left atrium via the pulmonary vein. Blood flowing from the right and left atria remains more or less separate in the heart. The partial ventricular septum helps keep the currents separate, and the ventricle has spongy internal walls that reduce swirling and mixing of the blood. Carrying blood from the ventricle to the ventral aorta, the conus arteriosus is partitioned internally by a **spiral valve**. The spiral valve separates the two blood flows. Oxygen-poor blood flows on one side of the

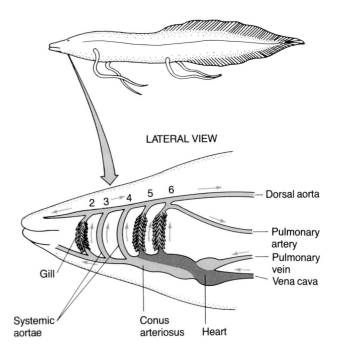

LATERAL VIEW

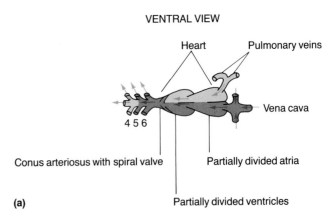

VENTRAL VIEW

(a)

Figure 28.15 Comparison of vertebrate circulatory systems. (a) Lungfishes are gill and lung breathers. In providing for oxygenation in both gills and lungs, their circulatory system combines features found in both aquatic and terrestrial vertebrates. When lungfishes are in water, their blood is oxygenated in the gills. Notice in the figure that the pulmonary artery carries blood from the posterior gill arches to the lungs. Numerals indicate aortic arches 2–6. When lungfishes are out of water, the pulmonary artery carries oxygen-poor blood, as in tetrapods. (b) (opposite) Amphibians, such as the frog shown here, have a system similar to that of the lungfish. Blood is oxygenated in the lungs, integument, and oral membranes. The frontal section of the heart and the enlargement show the spiral valve and the entrances to the left pulmocutaneous, carotid, and systemic arches. The spiral valve guides oxygen-poor blood into the pulmocutaneous arches (carrying blood to the lungs and skin for oxygenation), and oxygen-rich blood into the carotid arches (serving the head region) and into the systemic arches (carrying blood posteriorly). (c) (p. 698) Most reptiles have a nearly complete separation of pulmonary and systemic circulation. Note in this ventral view that the two systemic arches arise separately from the heart (in contrast to the amphibian plan), and both arches carry oxygen-rich blood to the dorsal aorta.

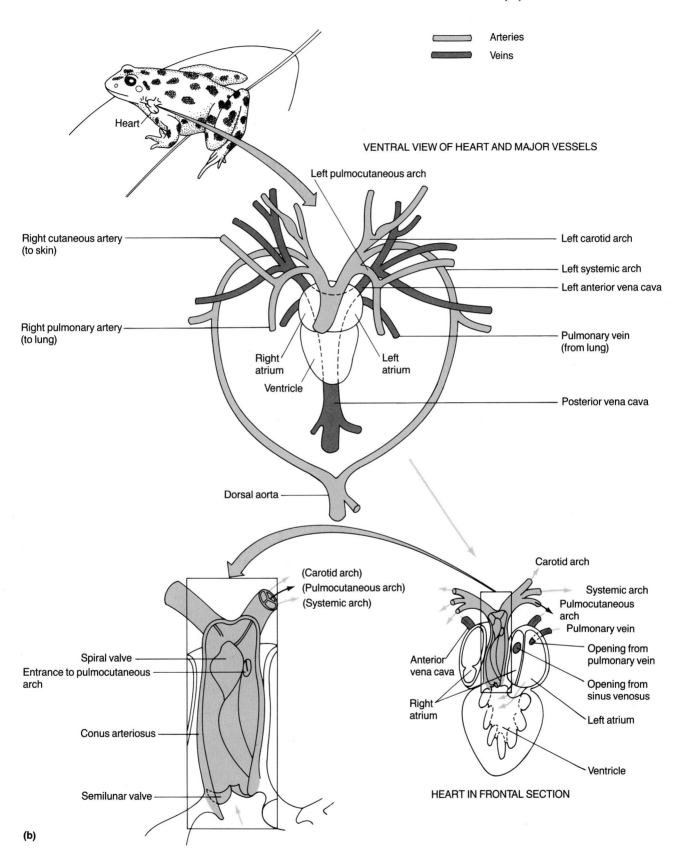

Arteries

Veins

Heart

VENTRAL VIEW OF HEART AND MAJOR VESSELS

Left pulmocutaneous arch

Right cutaneous artery (to skin)

Right pulmonary artery (to lung)

Right atrium

Ventricle

Left atrium

Left carotid arch

Left systemic arch

Left anterior vena cava

Pulmonary vein (from lung)

Posterior vena cava

Dorsal aorta

(Carotid arch)
(Pulmocutaneous arch)
(Systemic arch)

Spiral valve

Entrance to pulmocutaneous arch

Conus arteriosus

Semilunar valve

(b)

Carotid arch

Systemic arch

Pulmocutaneous arch

Pulmonary vein

Anterior vena cava

Right atrium

Opening from pulmonary vein

Opening from sinus venosus

Left atrium

Ventricle

HEART IN FRONTAL SECTION

Figure 28.15 *(Continued)*

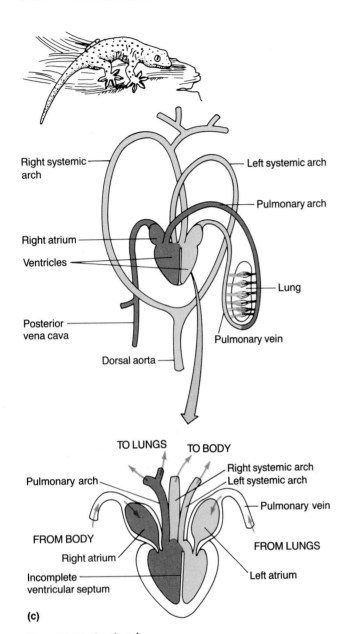

(c)

Figure 28.15 *(Continued)*

the conus arteriosus maintains separate blood flow, directing oxygen-poor blood into pulmocutaneous vessels to the lungs and skin, and oxygen-rich blood into systemic aortas (the left and right carotid and systemic arches).

Derived from an amphibian system, the circulatory systems of most reptiles provide for lung oxygenation only. The reptilian heart has two atria and typically an incomplete **interventricular septum,** which prevent virtually any mixing of oxygen-rich with oxygen-poor blood (Figure 28.15c). (Crocodiles and alligators are unique among reptiles in having two atria and two completely separate ventricles.) Other adaptations include the loss (except in turtles) of the sinus venosus and conus arteriosus as chambers. The sinus venosus has become a collection of cells called the **sinatrial node,** or **pacemaker,** located in the right atrium of the heart. The conus arteriosus and the ventral aorta have become three large vessels, a pulmonary arch and two systemic arches. Valves of the conus arteriosus have become the **semilunar valves** at the base of these three vessels. The pulmonary arch carries oxygen-poor blood from the right side of the ventricle to the pulmonary arteries. One of the systemic arches exits from the right side of the ventricle, the other from the left side of the ventricle. The arch on the right side may convey some oxygen-poor blood into the systemic circulation when the animal is not breathing air, such as when diving, but normally, both arches receive only oxygen-rich blood and convey it to the dorsal aorta.

Reflecting their origins, birds and mammals have circulatory systems similar to those of reptiles. The heart has completely separate right and left atria and ventricles. Very different from the reptilian system, however, there is only one systemic arch, or aorta, keeping pulmonary and systemic circuits completely separate, with no opportunity for mixing (see Figure 6.12). Although the systemic aorta receives blood only from the left ventricle in birds and mammals, this vessel is derived from a gill arch vessel on the left side of the body in mammals, whereas in birds it develops from a vessel on the right. This difference probably reflects the origins of birds and mammals from different groups of ancestral reptiles.

spiral valve to the gills and lungs for oxygenation. Oxygen-rich blood flows on the opposite side of the spiral valve into two pairs of **systemic aortas.** (The aortas are derived from vessels of gill arches that lack gill tissues.) The dorsal aorta receives blood from the systemic aortas and distributes it to body tissues.

Amphibian circulatory systems are similar to the lungfish system (Figure 28.15b). Blood oxygenated via the skin and mouth enters the right side of the atrium of the heart from large veins. There is some mixing of this oxygen-rich blood and oxygen-poor blood in the large veins, but as in lungfishes, little mixing occurs in the heart. The spiral valve in

The Excretory and Osmoregulatory System (V)

In aquatic vertebrates, gills are the chief excretory organs and also function in osmoregulation. Kidneys may have evolved first as osmoregulatory organs, removing excess water in freshwater vertebrates and conserving water in marine and terrestrial species. Lacking gills, terrestrial vertebrates have kidneys that dispose of most metabolic wastes. (The environmental physiology of the vertebrate kidney is discussed in Chapter 7, p. 165.)

Vertebrate kidneys have undergone several evolutionary stages that are represented in living vertebrates. Hagfishes represent a primitive stage. The kidneys are segmentally arranged with one pair of tubules per body segment. Embryonic hagfishes have a **holonephros**, which is a paired, metameric

kidney extending most of the length of the body. The tubules drain into two **archinephric ducts** that join before opening to the outside near the anus. In adult hagfishes, the anterior portion of the holonephros, called the **pronephros**, becomes a blood-cell-forming organ, the **head kidney**. The rest of the holonephros, called the **primitive opisthonephros**, remains as the functional kidney. (Figure 28.16a).

A second stage in kidney evolution is represented by most fishes and amphibians. A paired

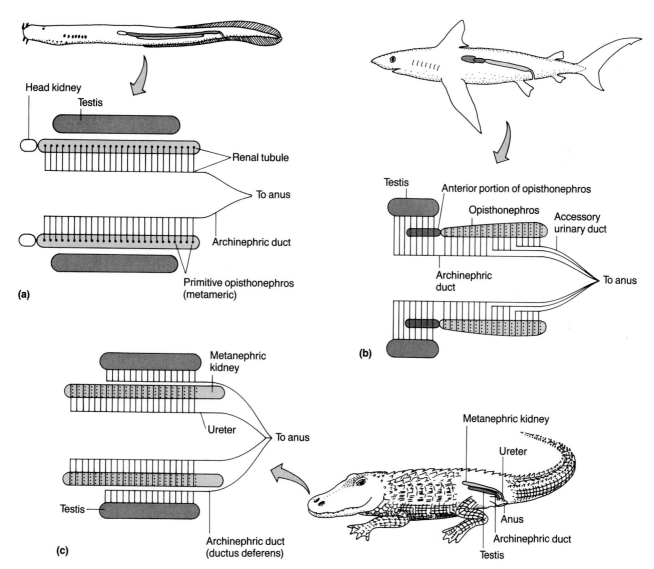

Figure 28.16 Comparison of vertebrate kidneys and their association with male reproductive systems. (a) The primitive opisthonephros seen in hagfishes. The segmental tubules are indicated by short lines that lead from the body of the opisthonephros to the archinephric duct. Note that the testes are completely separate from the kidney and its ducts. (b) The nonsegmental opisthonephros of many sharks and amphibians. The anterior part of the opisthonephros functions in blood cell formation and in secretion of secondary sex products. Note that the testis occupies the position of the anterior opisthonephros, and the archinephric duct carries both sperm and urine. (c) Amniotes (reptiles, birds, and mammals) have a pair of metanephric kidneys and ureters.

metameric pronephros functions as the kidney in embryos, but as development proceeds, the pronephros disappears. An **opisthonephros** becomes the functional kidney in adults; it is not segmentally arranged, and has many tubules per body segment (Figure 28.16b). The archinephric duct and/or accessory urinary ducts drain the kidney tubules. Fishes have a blood-cell-forming head kidney that develops from the anterior part of the opisthonephros. Bone marrow, absent in fishes, takes over the blood-cell-forming function in amphibians and amniotes.

Amniotes represent a third stage in the evolutionary derivation of the kidney. In the amniote embryo, the functional kidney, called the **mesonephros**, develops from the midportion of the holonephros and does not extend into the posterior part of the body. Although structurally similar to an opisthonephros, the mesonephros is distinguished as the embryonic kidney of amniotes. As amniotes develop, the mesonephros disappears and is replaced by a new, highly efficient kidney, the **metanephros** (metanephric kidney), which originates from the most posterior portion of the holonephros. The metanephros contains a concentrated mass of specialized tubules, and amniotes also have a new pair of urinary ducts, the **ureters**, not homologous with the kidney ducts of other vertebrates (Figure 28.16c). (The structure and function of the mammalian kidney are discussed in Chapter 7, p. 160; Figures 7.13 and 7.14.)

The Reproductive System (V)

Female and male reproductive systems evolved somewhat differently in vertebrates. Typically, the female system, including ovaries and ducts, is independent of other systems. The male system, however, is often closely associated with the kidney and its ducts (see Figure 28.16). In chondrichthyans and amphibians, the testis usurps the anterior part of the kidney, and the archinephric duct is used for sperm transport. Depending on the species, the archinephric ducts transport both sperm and urine or only sperm (in which case accessory ducts carry urine). In amniotes, the sperm duct, called the **ductus deferens** or **vas deferens**, develops from the archinephric duct; the urinary duct (ureter) is completely independent of the reproductive system. In most bony fishes, however, the archinephric duct remains an independent urinary duct, and the sperm duct is a new structure completely separate from the kidney system.

Some of the most significant evolutionary adaptations in tetrapods are the organs and sexual behaviors that ensure internal fertilization, a prerequisite for being able to reproduce out of water. (Chapter 14 discusses reproductive strategies, p. 342.) Equally important are adaptations that protect the developing embryo. Reptiles, birds, and egg-laying mammals produce a **cleidoic** (self-contained) **egg**, enclosing the embryo, food (yolk and albumin), and water in a protective shell. The calcareous or leathery shell not only protects the fertilized egg and prevents it from drying out, but also is porous, allowing gases to diffuse in and out. Placental and marsupial mammals and many snakes and lizards do not have a cleidoic egg; the female retains and nourishes developing embryos within her **uterus**. As discussed in Chapter 16, extraembryonic membranes (the **chorion**, **amnion**, and **allantois**) in the cleidoic egg or in the uterus protect the embryo and serve its metabolic needs (see p. 387 and Figures 16.12 and 16.13).

Nervous and Sensory Systems (S, V)

Visceral and somatic components of the vertebrate nervous system are highly integrated. The **central nervous system** (**CNS**), consisting of the brain and spinal cord, is entirely somatic. The **peripheral nervous system** (**PNS**) has both somatic and visceral components. Somatic components are the nerves that innervate skeletal muscles and integumentary sense organs. Visceral components include ganglia and nerves innervating the digestive tract and other internal organs. **Visceral afferent (sensory) nerves** carry impulses from sensory endings in internal organs to the CNS. **Visceral efferent (motor) nerves**, including those of the **autonomic nervous system** (see Figure 10.19), carry impulses from the CNS to glands and to smooth muscles of the viscera. The CNS dominates and controls the PNS. Coordination of visceral activities occurs in the spinal cord and in the hypothalamus of the brain. (The structure and function of the vertebrate nervous system are discussed in more detail in Chapter 10, p. 234.)

The vertebrate brain has undergone major adaptive changes concurrent with adaptations of other body systems to the challenges of life on land. Amniotes have the most complex skeletomuscular systems and the greatest diversity of body movement and behavior. Compared to fishes and amphibians, they have larger, more complex brains with a greater capacity to store and centrally integrate information about the environment.

Of the five major divisions of the adult brain (telencephalon, diencephalon, mesencephalon, metencephalon, and myelencephalon; see Figure

DORSAL VIEW　　　　　LEFT LATERAL VIEW

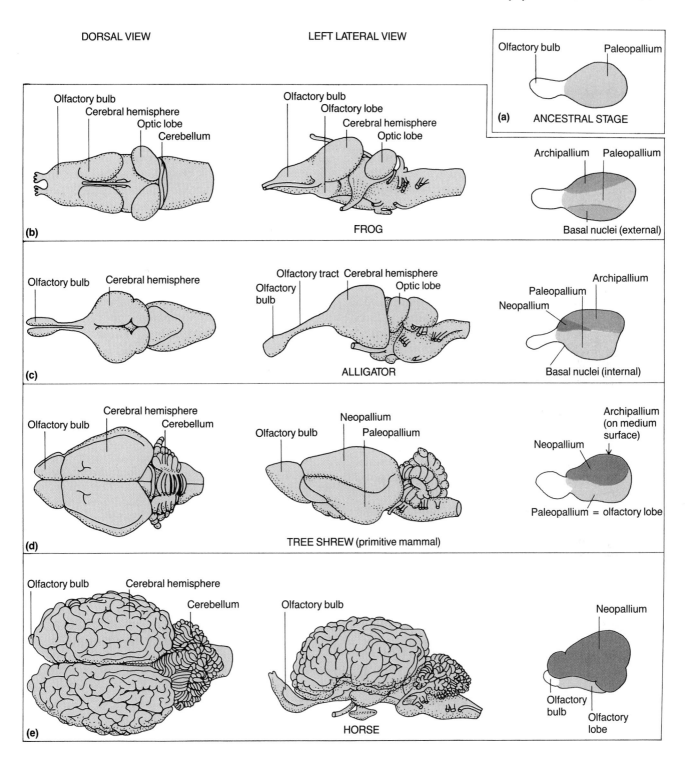

Olfactory bulb　　　　Paleopallium

(a) ANCESTRAL STAGE

(b)

Olfactory bulb
Cerebral hemisphere
Optic lobe
Cerebellum

Olfactory bulb
Olfactory lobe
Cerebral hemisphere
Optic lobe

FROG

Archipallium　Paleopallium

Basal nuclei (external)

(c)

Olfactory bulb　Cerebral hemisphere

Olfactory tract　Cerebral hemisphere
Olfactory bulb
Optic lobe

ALLIGATOR

Paleopallium　Archipallium
Neopallium

Basal nuclei (internal)

(d)

Olfactory bulb
Cerebral hemisphere
Cerebellum

Olfactory bulb
Neopallium
Paleopallium

TREE SHREW (primitive mammal)

Neopallium
Archipallium (on medium surface)

Paleopallium = olfactory lobe

(e)

Olfactory bulb
Cerebral hemisphere
Cerebellum

Olfactory bulb

HORSE

Neopallium

Olfactory bulb
Olfactory lobe

Figure 28.17　Evolution of the telencephalon. Dorsal and lateral views of the brains of selected vertebrates are accompanied by diagrammatic lateral views of the telencephalon, showing the occurrence and change in prominence of the paleopallium, archipallium, and neopallium. (a) The ancestral telencephalon probably consisted mainly of a paleopallium forming a cerebrum (cerebral hemispheres) that was chiefly olfactory in function. (b) In modern amphibians, the cerebral cortex has a distinct paleopallium (olfactory areas) and archipallium. Prominent basal nuclei (corpus striatum) occupy an external position in the cerebral hemispheres. (c) In reptiles (and also in mammals), the basal nuclei occupy an inner region of the cerebral hemispheres. Crocodiles and alligators have a distinct neopallium in addition to the paleopallium and archipallium. (d) In primitive mammals, the neopallium is greatly expanded compared to that of reptiles. The archipallium (hippocampus) is confined to the median region of the cerebral hemispheres. (e) In the horse, representing the typical condition in mammals, the cerebral cortex is composed largely of neopallium in which gray matter, or neocortex, completely surrounds white matter. The paleopallium forms the relatively small olfactory region of the telencephalon.

10.16), the most profound evolutionary changes occurred in the telencephalon, especially in the **cerebral cortex** (outer cloak) of the telencephalon. In ancestral vertebrates, the cerebral cortex probably consisted primarily of a region called the **paleopallium** (Figure 28.17a). The most primitive known telencephalon, that of agnathans, is thought to resemble the ancestral condition in consisting mainly of olfactory centers derived from the paleopallium. These include relatively large **olfactory lobes**, which make up most of the cerebral hemispheres. The main function of the olfactory bulbs and lobes is to integrate information concerning taste and smell. All of the agnathan brain, including the telencephalon, is architecturally similar to the spinal cord; white matter (nerve fibers) is external to gray matter (mainly nerve cell bodies). In addition to the paleopallium, the agnathan telencephalon also includes a region called the **archipallium**, but this is not as clearly demarcated from the paleopallium as it is in other vertebrates. (In mammals, the archipallium contributes to formation of the hippocampus; see Figure 10.17). In agnathans—in fact, in all vertebrates—the telencephalon also includes **basal nuclei (ganglia)**, which in most vertebrates form the corpus striatum. (See Figure 10.17.)

The telencephalon of sharks and amphibians, representing a later stage in brain evolution, is more complex than that of agnathans. The paleopallium and archipallium are clearly distinct from each other (Figure 28.17b). The paleopallium, generally lateral and ventral in position, forms the bulk of the olfactory lobes. The archipallium, more dorsal and medial in position, retains some olfactory function and may also serve as an integration center for complex behavior. Gray matter of the telencephalon is still largely internal to white matter.

In amniotes, the telencephalon is much larger in proportion to the rest of the brain. Compared to sharks and amphibians, reptiles have more gray matter close to the surface of the telencephalon, and in mammals, gray matter forms virtually the entire cerebral cortex. Mammals and certain reptiles have a well-developed region of cerebral cortex called the **neopallium** (Figures 28.17c–e). Gray matter of the neopallium is called the **neocortex**. This "new"cortex is most developed in mammals. Even in a primitive mammal, the neocortex forms a prominent part of the cerebral hemispheres (Figure 28.17d). In most mammals, the neocortex occupies most of the cerebral cortex and dominates all other parts of the brain (Figure 28.17e). In whales and in primates, such as monkeys, apes, and humans, the neocortex is the center of higher learning.

Birds have evolved a quite different telencephalon. Their cerebrum is large, but it develops mainly from the basal nuclei. This region of gray matter grows outward and upward to form a massively developed **corpus striatum** and an overlying **hyperstriatum**. Birds lack the neocortex, and no part of their telencephalon is homologous with the mammalian cerebral cortex. Their corpus striatum seems to be the center of programming for complex behavior patterns. Learning and memory in birds appear to be functions of the hyperstriatum.

Sensory systems in vertebrates include visceral structures that detect a variety of internal stimuli, as well as somatic sensory structures that detect external stimuli. Auditory organs underwent the most striking adaptive changes in tetrapods as these vertebrates colonized land. These changes are described in detail in Chapter 12. (See the discussion of vertebrate ears in Chapter 12, p. 284.)

❑ SUMMARY

1. Key Features of Chordates. Chordates are bilaterally symmetrical deuterostomes with four distinctive features appearing during some stage in their life cycle: a notochord; a dorsal hollow nerve cord; pharyngeal gill slits; and a postanal tail containing extensions of the notochord and dorsal nerve cord. Most chordates are metameric and have a closed circulatory system with a ventral heart that pumps blood from anterior to posterior in a dorsal aorta. Chordates may have arisen from sessile filter feeders similar to primitive echinoderms or to pterobranch hemichordates.

2. Two subphyla of marine invertebrate chordates, the Cephalochordata (lancelets) and the Urochordata (sea squirts and tunicates), have a filter-feeding gill apparatus. Larval urochordates may be similar to chordate ancestors.

3. The Subphylum Vertebrata includes agnathans, fishes, and tetrapods (amphibians, reptiles, birds, and mammals). Nearly all have cartilage or bony vertebrae surrounding the dorsal nerve cord, as well as a skeletal case called the cranium surrounding the brain. The vertebrae and muscular system are metameric.

4. Vertebrate organ systems underwent significant adaptive changes correlated with the colonization of land by tetrapods. The major adaptations sustaining vertebrates on land are: a strong internal skeleton; an integument and a kidney that limit water loss; gas exchange and circulatory systems that allow air breathing; sensory systems that receive stimuli from the air; and a reproductive system that functions in a dry environment. Most amphibians lack drought-resistant egg coverings and must return to water to reproduce. Amniotes (reptiles, birds, and mammals) have protective membranes (including the amnion) covering their developing embryos. Eggs of reptiles and birds are also shelled.

❏ FURTHER READING

Alexander, R. M. *The Chordates*. New York: Cambridge University Press, 1975. *A study of the chordate groups with emphasis on phylogeny and physiology.*

Alldredge, A. "Appendicularians." *Scientific American* 235(1976):94–102. *An illustrated account of the larvacean urochordates.*

Attenborough, D. *Life On Earth*. Boston: Little, Brown, 1979. *Contains several chapters on vertebrates, including one that describes the invasion of land; superbly illustrated.*

Barrington, E.J.W., and R.P.S. Jefferies, eds. "Protochordates." In *Symposium of the Zoological Society of London, No. 36*. New York: Academic Press, 1975. *Physiology, ecology, and phylogeny of the invertebrate chordates and hemichordates, written by experts.*

Colbert, E. H. *Evolution of the Vertebrates. A History of the Backboned Animals Through Time*. 3d ed. New York: Wiley, 1980. *A comprehensive account of all vertebrate groups, including the history of their appearance and impact on earth.*

Gould, S. J. *Ever Since Darwin*. New York: Norton, 1977. *Includes an essay on the evolution of the vertebrate brain.*

Little, C. *The Colonisation of Land. Origins and Adaptations of Terrestrial Animals*. New York: Cambridge University Press, 1983. *Contains an excellent chapter on terrestrial adaptations in vertebrates.*

Romer, A. S. and T. S. Parsons. *The Vertebrate Body*. 5th ed. Philadelphia: Saunders, 1977. *A comprehensive overview of the evolution of vertebrate structure and function.*

❏ CLASSIFICATION

Phylum Chordata (Gr. *chorde*, "string, cord"; *ata*, "state of, or having"). Chordates.

Subphylum Cephalochordata (Gr. *kephale*, "head"; *chorde*, "cord"; *ata*, "having"). Lancelets (cephalochordates); *Branchiostoma (Amphioxus)*.

Subphylum Urochordata (Gr. *oura*, "tail"; *chorde*, "cord"; *ata*, "having"). Tunicates (urochordates).

Class Ascidiacea (Gr. *askiolion*, "little bag"). Sea squirts; adults are sessile and saclike; larvae are free-swimming and planktonic; larvae show chordate features; *Ciona, Molgula*.

Class Larvacea (L. *larva*, "ghost"). Larvaceans; planktonic throughout life; no sessile stage; adult resembles ascidian larvae; *Oikopleura*.

Class Thaliacea (Gr. *thalia*, "luxuriance"). Thaliaceans (salps); barrel-shaped and planktonic, with incurrent and excurrent water openings at opposite ends of the body; *Doliolum, Salpa*.

Subphylum Vertebrata (L. *vertebratus*, "with a backbone"). Vertebrates.

Superclass Agnatha (Gr. *a*, "absence of"; *gnathos*, "jaw").

Class Agnatha (Gr. *a*, "absence of"; *gnathos*, "jaw"). Jawless vertebrates with a notochord in larval and adult stages; many lack paired appendages; many extinct groups, including heterostracans, thelodonts, osteostracans, and anaspids; two living groups: lampreys *(Petromyzon)* and hagfishes *(Myxine)*.

Superclass Gnathostomata (Gr. *gnathos*, "jaw"; *stomatos*, "mouth"). Fishes and tetrapods; jawed vertebrates with paired appendages. (For more complete classification, see Chapters 29 through 33.)

Class Chondrichthyes (Gr. *chondros*, "gristle, cartilage"; *ichthys*, "fish"). Cartilaginous fishes (sharks, skates, rays, chimaeras).

Class Osteichthyes (Gr. *osteon*, "bone"; *ichthys*, "fish"). Bony fishes (lungfishes, lobefins, paddlefishes, gars, teleosts).

Class Amphibia (Gr. *amphi*, "double"; *bios*, "life"). Amphibians (salamanders, frogs, toads, caecilians).

Class Reptilia (L. *reptile*, "a crawling animal"). Reptiles (lizards, snakes, crocodiles, turtles); ectothermic amniotes with dry, scaly skin.

Class Aves (L. *aves*, "birds"). Birds; homeothermic amniotes with feathers.

Class Mammalia (L. *mamma*, "breast"). Mammals; homeothermic amniotes with mammary glands and hair.

Fishes

Class Osteichthyes. Osteichthyans, or bony fishes, constitute the most diverse group of vertebrates. Here a school of bluestriped grunt (*Haemulon sciurus*) is swimming above a reef.

Fishes arose before any other vertebrates except agnathans, and they have been numerous and diverse for over 400 million years. Today, fishes are more diverse than ever before, and there are about as many species of fishes as all other vertebrates combined. Yet despite their numbers, we know less about fishes than about any other vertebrate group.

Fishes are so diverse that there are exceptions to nearly any generalization we can make about them. Nonetheless, we can define a fish as an ectothermic aquatic vertebrate with a head and braincase, jaws, paired fins that lack distal support elements, a heart with a single atrium and ventricle, and gills projecting outward from the gill arches. This definition excludes the agnathans, described in the previous chapter (see p. 683).

REPRESENTATIVE FISHES

Fishes are the only gnathostomes (jawed vertebrates) classified in more than one class. We have already introduced the two classes of living fishes, the Chondrichthyes and the Osteichthyes. Two extinct classes, the Acanthodii and the Placodermi, are the oldest gnathostomes in the fossil record.

Class Acanthodii

Acanthodians, also called **spiny sharks**, were probably not the first gnathostomes, but acanthodian fossils (about 450 million years old) predate fossils of other gnathostomes by 25 to 50 million years. Acanthodians were small (in the range of 10–20 cm long), and most lived in fresh water. They were more active swimmers than their contemporaries, the agnathans, and were not limited to bottom habitats. Acanthodians had a sharp spine at the leading edge of their fins; thus the name spiny sharks. They also had a bony endoskeleton, large eyes, bony scales, bony gill covers, paired pectoral fins, and a **heterocercal tail**. The heterocercal tail, a feature of primitive fishes, has a large dorsal lobe that contains the posterior end of the spinal column (Figure 29.1a). Some acanthodians were active predators, but judging from their teeth, most of them fed by straining plankton with their gills. Swimming and feeding in the open water, acanthodians were probably not in direct competition with the bottom-dwelling agnathans. The class became extinct about 300 million years ago.

Class Placodermi

The name *placoderm*, meaning "plate skin," refers to the heavy bony plates covering these early fishes. **Placoderms** were the most numerous and diverse vertebrates in the sea some 350 to 400 million years ago. They resembled spiny sharks in having a bony endoskeleton and a heterocercal tail, but they lacked fin spines, and most species were dorsoventrally compressed. The head and anterior end of the trunk were covered with shieldlike armor (Figure 29.1b and c). Most placoderms were about 20–30 cm long, but some species reached 10 m in length. Placoderms were mainly bottom dwellers, either scavengers or ambush-type predators. As bottom dwellers, they probably had more direct interaction with agnathans than did the acanthodians. Placoderms became extinct about 325 million years ago.

Class Chondrichthyes

Most of the 650 living species of **chondrichthyans** are marine predators, although a few species of

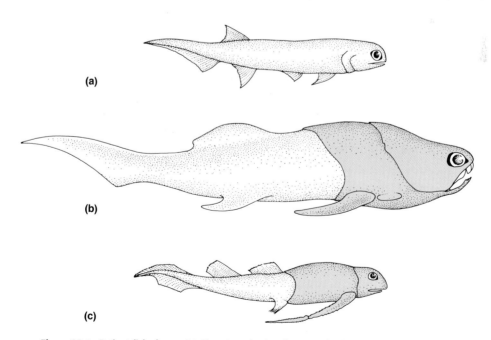

Figure 29.1 Extinct fish classes. (a) Class Acanthodii. The spiny shark *Homalcanthus* (about 15 cm long) probably ate plankton. (b) Class Placodermi. *Coccosteus*, a carnivore, was about 30 cm long. (c) *Bothriolepis*, another placoderm, was a dorsoventrally compressed bottom scavenger (about 15 cm long).

sharks and rays inhabit brackish waters and fresh-water lakes. Many extinct members of the class lived in fresh water. All chondrichthyans have a cartilaginous skeleton, although the skull and vertebrae are often reinforced with calcium deposits. Other diagnostic features are a heterocercal tail, lack of a swim bladder, the **spiral valve** (a helical coil of absorptive tissue in the intestine), and **placoid scales** and their derivatives, bony teeth (Figure 29.2). Also, unlike most other vertebrates, chondrichthyans maintain an internal osmotic pressure that is higher than that of seawater by concentrating urea in their body fluids. (See the discussion of marine sharks and rays in Chapter 7, p. 150.) They also have internal fertilization, and males have **claspers** for transferring sperm during copulation.

There are two subclasses of cartilaginous fishes. Sharks, skates, and rays are grouped in the Subclass Elasmobranchii. The probably related, but quite different, ratfishes make up the subclass Holocephali.

Subclass Elasmobranchii. **Elasmobranchs** have five to seven respiratory gills that are completely separated from one another by platelike sheets of tissue. There is no gill cover, and each gill arch has a separate opening to the outside. As discussed in Chapter 28, chondrichthyan jaws develop from the first gill arch, and in elasmobranchs, the first gill slit transforms into the **spiracle**, a small dorsal opening from the anterior part of the pharynx to the outside (see Figure 29.2). The spiracular passage bears a small gill-like structure called the **pseudobranch**. Most elasmobranchs are predators, and as an adaptation allowing them to gulp large prey, their upper jaw is not fused to the skull. They have rather poor eyesight and depend more on smell or electroreception to locate prey.

Perhaps the best-known elasmobranch—it is

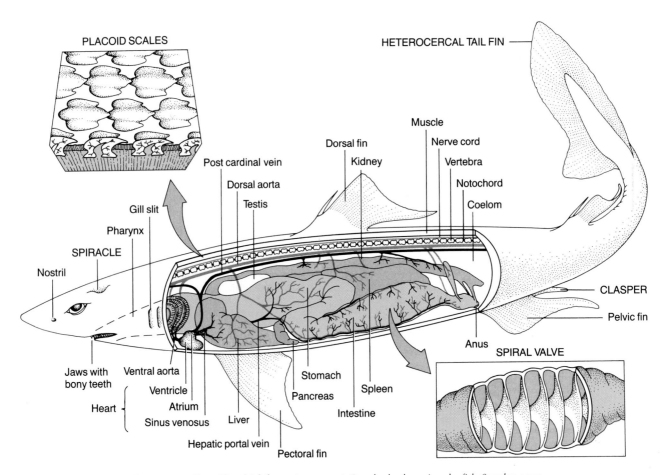

Figure 29.2 Class Chondrichthyes. A representative shark, the spiny dogfish *Squalus acanthias,* illustrating external and internal structures. Diagnostic features of the class are in capital letters. Spiny dogfish reach about 2.5 m in length. They are harvested commercially for food in the Orient and are used throughout the world for zoological teaching specimens. The inset shows the structure of the toothlike placoid scales embedded in the skin.

used widely as an anatomy teaching model—is the spiny dogfish *Squalus acanthias* (see Figure 29.2). This species is numerous along shallow coasts in the temperate Atlantic and Pacific. Tagging studies have shown that schools of dogfish migrate thousands of kilometers from one side of the Atlantic to the other and from one side of the Pacific to the other.

Sharks infamous for their tendency to attack swimmers and divers are the larger species that commonly eat seals, dolphins, cephalopod molluscs, and large fishes. The great white shark, reaching 6–7 m in length, is the most notorious (Figure 29.3a). The largest sharks—in fact, the world's largest living fishes—are the whale shark, *Rhincodon typus*, reaching 18 m in length, and the basking shark, *Cetorhinus maximus*, up to about 15 m long. Both species are filter feeders, gulping large quantities of seawater and straining plankton through a filter formed by the inner edges of their gills (gill rakers).

Skates and **rays** look quite different from sharks. They have ventral gill openings and a dorsoventrally flattened body adapted for gliding gracefully through the water or hovering just off the seafloor. The pectoral fins are greatly expanded into a flat muscular disc attached to the side of the head. The pelvic fins are much smaller. Skates and rays swim by undulating their pectoral fins.

Skates, such as *Raja*, are found in all oceans (Figure 29.3b). They have a thin tail with a caudal fin and often several small dorsal fins. Rays include species of *Torpedo* and *Diplobatis*, electric rays, which can produce strong electrical charges from electrical organs on the sides of its head (Figure 29.3c). The electricity stuns prey and wards off predators. In contrast to skates and electric rays, **stingrays**, such as *Dasyatis*, are distinguished by a thin tail bearing one to three spines with poison glands (Figure 29.3d).

Closely related to skates and rays are **sawfishes** (Figure 29.3e). The front of the skull is elongated to

(a)

(b)

(c)

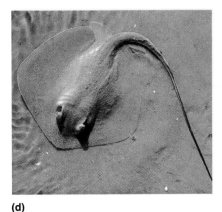

(d)

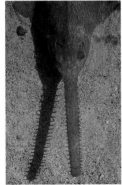

(e)

Figure 29.3 Subclass Elasmobranchii. (a) The great white shark *(Carcharodon carcharias)*, like most predaceous sharks, is attracted by blood; but recent studies indicate that it usually attacks humans in defense of its territory or because it mistakes swimmers for seals or sea lions. (b) Skates of the genus *Raja* spend much of their time groveling in shallow mud in search of bottom invertebrates. This species, *Raja binoculata* (big skate), up to about 2.5 m long, lives in the Pacific ocean. (c) Electric rays (up to about 2 m long) can produce charges of up to 200 v, enough to temporarily paralyze a diver. Shown here is *Diplobatis ommata* photographed in waters off the Galapagos Islands. (d) The southern stingray *Dasyatis americana*, an Atlantic species, is a bottom dweller, feeding on molluscs, crustaceans, and some fishes. Most stingrays live in warm oceans, but one group inhabits freshwater rivers in South America. Large rays and skates are harvested, and much of the meat is sold as "scallops." (e) Sawfishes, which may be up to 11 m long, use an elongate toothed snout to slash at prey. This species, *Pristis pectinata* (smalltooth sawfish), occurs in the western Atlantic. Sawfishes resemble saw sharks, which also have a long toothed snout. However, *Pristis* and other sawfishes have ventral gill openings; saw sharks have lateral gill openings.

Figure 29.4 Subclass Holocephali. *Hydrolagus colliei* (spotted ratfish) is common in Pacific coastal waters from southern California to Alaska. This photograph, which also shows a group of sea urchins, was taken in shallow water near Vancouver Island, Canada.

form a toothed "saw." Sawfishes inhabit salt water, brackish estuaries, Lake Nicaragua in Central America, and the Ganges River in India. They feed by slashing through schools of fish, then turning around and ingesting the dead and dying vic-

tims. A few swimmers have been seriously hurt when they have accidentally disturbed a sawfish.

Subclass Holocephali.   Ratfishes, also called chimaeras and ghostfishes, have a long, rodentlike tail (Figure 29.4). All of the 25 living species are bottom dwellers that feed on small molluscs and crustaceans, mostly in deep oceans. Their teeth have flattened surfaces adapted mainly for crushing hard shells. Unlike elasmobranchs, ratfishes lack scales, have an upper jaw that is fused to the skull, and lack a spiracle. In common with bony fishes, they have four gill openings covered by an **operculum**.

Class Osteichthyes

Bony fishes (about 21,000 species) have a bony endoskeleton, including a skull with dermal bones and an operculum covering a common gill opening (Figure 29.5). (The skeletal system is discussed in ◁ Chapter 28, p. 689.) Unlike chondrichthyans, most bony fishes also have a **swim bladder**, serving as a lung or a buoyancy organ, and most species have a **homocercal tail** (i.e., dorsal and ventral portions of the tail are equal). Bony fishes have three types of

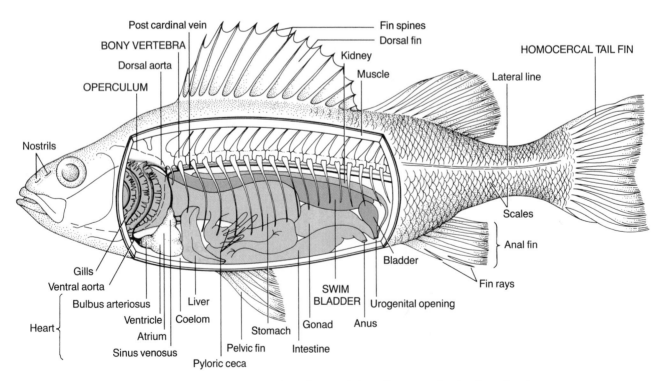

Figure 29.5 Class Osteichthyes. The yellow perch *Perca flavescens,* showing the main external and internal structures characteristic of bony fishes. Yellow perch originally occurred in North America east of the Rocky Mountains, but the species has been introduced widely throughout the continent. Diagnostic features of the class are in capital letters.

bony scales, all of which are different from the placoid scales of chondrichthyans (see the insets in Figures 29.9 and 29.10). Growth rings on the scales are a measure of bony fish growth.

There are three subclasses of osteichthyans. Two ancient groups with few species but with considerable evolutionary significance are the subclasses Dipnoi (lungfishes) and Crossopterygii (lobe-finned fishes). Most living fishes belong in the third subclass, the Actinopterygii (ray-finned fishes).

Subclass Dipnoi. **Lungfishes** (only six living species) inhabit two rivers in Australia and seasonally drying marshes in East and West Africa and South America. Unlike most bony fishes, they have **internal nostrils** (not used for breathing) and one or two air-breathing lungs. A **pneumatic duct** connects each lung to the ventral side of the esophagus.

The Australian lungfish has a single lung and bears fins with fleshy (lobed) bases (Figure 29.6a). It can live in water with little dissolved oxygen by breathing air at the surface, but it cannot survive out of water. South American and African lungfishes have two lungs, as well as long, delicate nonfleshy fins and eellike bodies (Figure 29.6b, c). Their gills are not as well developed as those of Australian lungfish, but they can live in water with little or no oxygen and can survive long dry periods. With the onset of drought, they burrow into mud and cover themselves with mucus, which dries to form a rigid cocoon. The animal becomes dormant in the cocoon, its metabolic rate becoming very low, and it slowly degrades body muscle as a source of energy. With the return of rainy weather, the cocoon floods, and the lungfish gags awake as water enters its lungs.

Subclass Crossopterygii. One living species, *Latimeria chalumnae*, the **coelacanth**, is the only survivor of a large group of lobe-finned fishes that thrived from about 100 to 400 million years ago (Figure 29.7). All other members of this subclass became extinct by about 70 million years ago. Most lobefins lived in fresh water, and until 1938, when a

(a)

(b)

(c)

Figure 29.6 Subclass Dipnoi. (a) The Australian lungfish *Neoceratodus forsteri* (up to 1.5 m long). Note its thick, fleshy fin bases. (b) The African lungfish *Protopterus dolloi* (up to about 1 m long). (c) African lungfish have been known to survive in a dried-mud cocoon for nearly four years.

Figure 29.7 Subclass Crossopterygii. *Latimeria*, the coelacanth, lives in deep water (60–600 m deep), eats mainly squid and fishes, and may reach lengths of nearly 3 m. Paired fins of this "living fossil" are supported by strong bones and muscles, enabling the fish to walk on the seafloor.

specimen of *Latimeria* was caught in the deep ocean off southeast Africa and described, all species were believed to be extinct. Actually, natives of the Comoro Islands near Madagascar had been catching coelacanths long before the scientific world learned of them in 1938.

Subclass Actinopterygii. **Ray-finned fishes** lack internal nostrils, and they have fins that are attached to the body by bony fin rays. It is convenient to consider actinopterygians in three subgroups (infraclasses): the chondrosteans and holosteans, containing only a few species of "living fossils," and a third group, the teleosteans, with over 20,000 living species.

Chondrosteans, the **sturgeons** and **paddlefishes**, have several features in common with ancestral ray-finned fishes. They have a notochord as adults, and they display certain sharklike features believed to be primitive. As the name *chondrostean* implies, their skeleton is mainly cartilaginous, although they are believed to have evolved from fishes with a bony skeleton. They also have a heterocercal tail, a sharklike intestinal spiral valve, a jaw not fused to the skull, and a pair of small spiracles.

Sturgeons inhabit fresh and marine waters of the northern hemisphere. They are bottom dwellers with a vacuum-cleaner-like mouth used to suck up small invertebrates. Five rows of platelike bony scales on the body and four sensory barbels (tentacle-like projections) around the mouth are distinctive characteristics (Figure 29.8a). Most of the 20 or more living species are large. The Russian beluga, reaching lengths of nearly 9 m, and the white sturgeon, up to 6 m long and living in the Pacific Ocean and in U.S. coastal streams, are the world's largest freshwater bony fishes.

The two living species of paddlefishes are also large. *Polyodon spathula*, of the Mississippi River system, reaches lengths of about 2 m (Figure 29.8b); its sister species, *Psephurus gladius*, of the Yangtze River in China, may be 5 m long. Mississippi paddlefish are filter feeders, straining small plankton with their fine gill rakers while swimming. The Chinese species eats small fishes. Paddlefishes have no scales except for a small patch at the base of the tail.

Holosteans, the **bowfin**, and **gars**, appear to represent a later stage in the evolution of ray-finned fishes, for some of their body features seem intermediate between those of the chondrosteans and those of the teleosteans. As their name implies, holosteans have a bony skeleton. They also lack spiracles and have a small intestinal spiral valve.

(a)

(b)

Figure 29.8 Subclass Actinopterygii: Chondrosteans. (a) Sturgeons of the genus *Acipenser* live in the Atlantic and Pacific oceans and in freshwater lakes and rivers. This photograph shows *A. sturio*, an Atlantic species. Sturgeons are commercially important, particularly in the Soviet Union, where the beluga *(Huso huso)* is prized for its tasty flesh and for its eggs, which are made into caviar. (b) The paddlefish *Polyodon spathula* is a sportfish in the Mississippi and Missouri rivers. The function of the long snout, or paddle is not known, but it is thought to be sensitive to touch and certain dissolved chemicals.

Their tail fin is heterocercal, but externally its lobes appear almost equal, like those of teleosteans. Also, the holostean skull is similar to that of teleosteans. Holosteans possess a lung and can supplement gill breathing by gulping air. They thrive in warm rivers and lakes with seasonally low dissolved oxygen levels.

There are eight living holostean species: one species of bowfin (Figure 29.9a) and seven species of gars (Figure 29.9b), all living in Gulf of Mexico drainages and rivers of the eastern United States. The bowfin, reaching about 1 m in length, has large **cycloid scales** similar to those of many teleosteans (see Figure 29.10). Gars are covered with a rigid

(b)

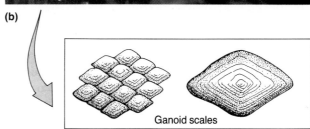

Ganoid scales

(a)

Figure 29.9 Subclass Actinopterygii: Holosteans. (a) The bowfin *Amia calva*. (b) The longnose gar *Lepisosteus osseus* is common in larger rivers and lakes throughout the Mississippi drainage. The drawing shows the ganoid scales characteristic of gars and many extinct bony fishes.

exoskeleton of **ganoid scales**, which are coated with an enamel-like substance called ganoin (see the inset in Figure 29.9b). All holosteans are aggressive predators. Gars are "ambush" predators, spending much of their time lying motionless near the surface. When approached by suitable prey, they attack suddenly with their elongate jaws and sharply pointed teeth. The alligator gar, the largest holostean, may be 3.5 m long. It inhabits the warm waters of the lower Mississippi and can also live in salt water. Gars have few natural enemies, for their eggs and newly hatched young are toxic to most predators.

Teleosteans, the most diverse group of vertebrates, exploit virtually every conceivable type of freshwater, marine, and brackish-water habitat. Such a diverse group is difficult to characterize, but in general, teleosteans have fewer vertebrae and fewer lower jaw bones than other bony fishes, and they possess a homocercal tail, lack a spiracle, and have a swim bladder that functions in maintaining neutral buoyancy. Teleostean scales are formed of flexible, nonmineralized bone and are of two types. Generally, the more primitive species (e.g. eels, minnows, and trout) have cycloid scales, whereas highly derived teleosteans (e.g, perches and their relatives) have **ctenoid scales**. The ctenoid scale resembles the cycloid type, except that its posterior edge is serrated (see insets in Figure 29.10). The serrations may be an adaptation that reduces drag in swimming. Scale size may reflect the general habits of fishes. Strong, high-speed swimmers, such as trout and tunas, typically have

smaller and more numerous scales than bottom dwellers or species that swim slowly, such as carp and other minnows.

Teleostean systematics is complex and is currently being revised as new data from cladistic analyses become available. Ichthyologists generally agree that there are at least three major teleostean subgroups (divisions), representing three lines of evolution (Figure 29.10). The largest division, the **Euteleostei**, has nearly 19,000 species and is further subdivided into primitive, intermediate, and highly derived subgroups.

HABITATS AND ENVIRONMENTAL RELATIONSHIPS

As a group, fishes inhabit just about any permanent body of water, including polar oceans, icy mountain streams, desert pools, and hot springs with temperatures near 40°C. Many species live in total darkness, either in caves or beneath the zone of light penetration in the ocean. Their vertical distribution, from mountains as high as 4.5 km above sea level to ocean abysses down to about 12 km, far exceeds that of any other group of chordates.

Fishes in Freshwater Environments

Fresh waters account for only about 0.01% of the total water on earth, yet about 40% of all living

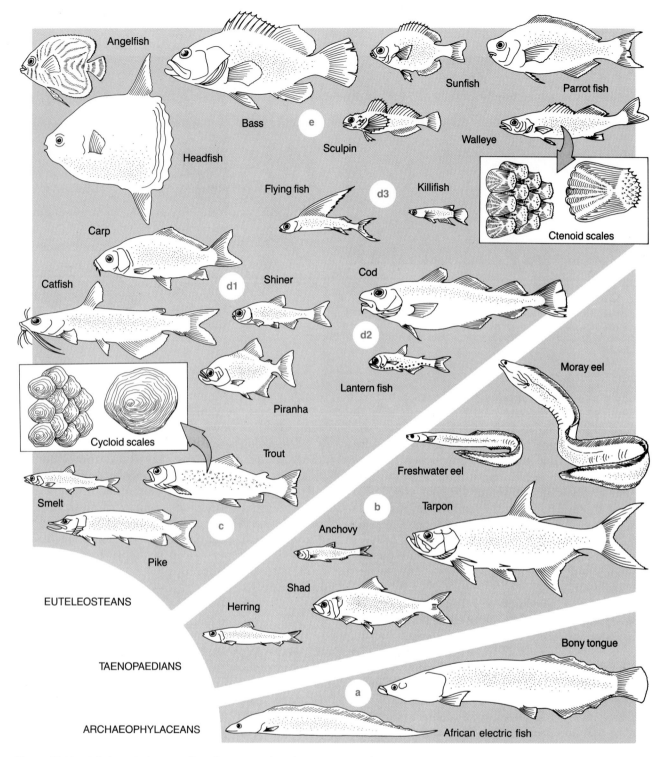

Angelfish
Sunfish
Parrot fish
Bass
Headfish
e
Sculpin
Walleye
Flying fish
d3
Killifish
Ctenoid scales
Carp
Shiner
Cod
Catfish
d1
d2
Lantern fish
Piranha
Moray eel
Cycloid scales
Freshwater eel
Trout
Smelt
c
b
Tarpon
Anchovy
Pike
Shad
EUTELEOSTEANS
Herring
Bony tongue
TAENOPAEDIANS
a
ARCHAEOPHYLACEANS
African electric fish

Figure 29.10 Subclass Actinopterygii: Teleosteans. This diverse group is composed of three large subgroups (divisions). The fishes illustrated in this diagram are present-day representatives of these subgroups. (a) The most primitive subgroup (Division Archaeophylaces) includes about 200 freshwater species, including bony tongues *(Heterotis)* and the African electric fish *(Gymnarchus)*. (b) A second subgroup (Division Taenopaedia) is made up of more than 300 species. This division includes 11 species of tarpons *(Tarpon)*; over 600 species of eels, including morays *(Echidna)* and freshwater eels *(Anguilla)*; and the herrings and their allies (sardines, anchovies, shad, alewives, and menhaden), among the most numerous fishes in the ocean. Feeding on plankton, herrings and their allies are important links in the marine food chain. (c–e) The Division Euteleostei repre-sents the mainstream of teleostean evolution. (c) Pikes, salmons, trout, and smelts are the most primitive euteleosteans. The inset shows the cycloid scales of a trout. (d) Three phylogenetically intermediate groups of euteleosteans: (1) a dominant freshwater group, the min-nows (including carps), shiners, characins (a South American and Afri-can group including piranhas and many of the fishes prized by home aquarists), and catfishes; the cycloid scales of minnows, suckers, and characins are larger than those of trouts; catfishes lack scales; (2) a marine group, the cods, deep-sea anglers, and lanternfishes; (3) the flying fishes (marine) and killifishes (mainly freshwater). (e) Perches and perchlike fishes (marine and freshwater) make up the most highly derived and by far the most diverse euteleostean group. Unlike other teleosteans, they have spiny fin rays and ctenoid scales (see the inset).

fishes (over 8000 species) live in fresh water. Three groups of teleosteans—the minnows, catfishes, and characins—contain over 75% of the freshwater species. Several factors make fresh waters conducive to fish evolution. Generally, they provide greater opportunities for genetic isolation. They are also diverse in flow rate, bottom type, temperature, pH, dissolved oxygen content, and other chemical factors; they may be drastically altered by local weather changes or geological forces; and they are usually more productive biologically than are the oceans.

Rivers and streams provide very different types of habitats at different elevations. Streams at high elevations usually have a rapid flow rate, are cool and clear, and are high in dissolved oxygen. Streamlined, highly motile fishes, such as trout, frequent high mountain streams. Less conspicuous are the sculpins, hiding on the bottom in quiet eddies around rocks. At middle elevations, streams typically have a lower gradient, temperatures may be higher, and dissolved oxygen levels may be lower. Fish populations here are quite variable and usually represent a mix of species from the upper zones with species found at lower elevations. Lower stream zones and large rivers are often murky because of suspended silt and because plentiful nutrients in the water support large populations of algae. Water currents are typically slow, and temperatures may also be seasonally warmer than at higher elevations. Fishes that thrive in the lower zones are generally intolerant of rapid currents, but are quite able to withstand wide fluctuations in temperature and amount of dissolved oxygen. Such fishes include shiners, carps and other minnows, catfishes, bass, sunfishes, perch, and pike.

Freshwater lakes are also highly varied. Deep, cold lakes are generally well oxygenated, and many of the fishes (especially trout, salmon, and whitefish) found in these lakes are the same or similar to those inhabiting high-gradient streams. Likewise, warm-water lakes, such as those common in the U.S. Midwest, have fishes similar to those of the lower stream zones.

Fishes in Marine Environments

The majority of marine fishes live in well-lighted, productive coastal waters, and fishes are most diverse and abundant on tropical coral reefs. Many fishes, especially reef inhabitants, are **benthic**, meaning they spend most or all of their time on the seafloor. Other species feed on the bottom, but spend much of their time in the overlying waters. **Pelagic** (open-water) fishes may be entirely independent of the bottom, eating other fishes or plankton in the open sea. Salmon, anchovies, herrings, and many sharks stay in shallow waters of the continental shelf, while others, such as whale sharks, tunas, flying fishes, and swordfishes, cruise continuously in the open ocean.

Deep-sea fishes are a unique group. Living in total darkness, many have light-producing luminescent organs that may help them locate mates and food. (Luminescence is discussed in more detail on p. 721.) Many species have reduced eyes and rely on detecting sound and water movement; their lateral line system is typically very sensitive, with receptors set on stalks in some species. (The acousticolateralis system is described in Chapter 12, p. 283; Figure 12.10.) Many species lack a swim bladder, and their tissues are soft and pliable with a high water content, making them less susceptible

(a)

(b)

Figure 29.11 Deep-sea fishes. (a) The striped anglerfish (*Antennarius striatus*), a predator, has a dorsal projection serving as a lure for other fishes. (b) Another predator, the viperfish (*Anoplogaster*), has long, fanglike teeth and can open its mouth very wide to engulf prey.

to high water pressures. Many deep-sea fishes are scavengers, while others are predators with exceptionally large mouths and long, daggerlike teeth (Figure 29.11).

Estuarine Fishes and Other Tolerant Species

Brackish waters in estuaries at the mouths of large rivers are rich in nutrients and support abundant plankton populations. They are also rich feeding grounds for fishes that tolerate murky waters and regular fluctuations in salinity and current. In addition to many resident species, many ocean fishes spend their larval life in estuaries and then move out into the open sea as adults.

Fishes with a wide tolerance for water conditions, especially temperature and oxygen content, may spread into new areas. If they are introduced outside their native range, such fishes may seriously disrupt the natural balance in the new habitat. The carp, *Cyprinus carpio*, has been prized and cultivated for food in Eurasia for centuries. Unfortunately, in feeding, carp uproot aquatic vegetation and stir up sediments, making the water turbid and disrupting normal feeding and reproduction of other fishes. Carp were introduced into North America from Germany in 1872; since then, the species has spread widely and has become a serious pest in the western hemisphere.

Symbiotic Relationships

Fishes have a full range of relationships with other species. Numerous examples of commensalism include the relationship between the pilotfish *Naucrates* and large sharks. The pilotfish swims along in the currents adjacent to the larger fish, eating leftovers of its host's meals. Other commensalists are the remoras, or shark suckers, which attach to large sharks, bony fishes, and even whales and sea turtles by means of a large sucking disc on the top of their head (Figure 29.12a). Remoras do not damage the surface of the host, but simply obtain a free ride. When the host slows or stops swimming, the remora detaches by swimming forward and feeds on leftover scraps. Some remoras are actually mutualistic, benefiting their hosts by eating dead skin cells and surface parasites.

Parasitism is especially common. Fishes are infected by a wide variety of microorganisms and animal parasites. Many harbor larval stages of parasitic worms that live as adults in the gut of fish-eating birds or mammals. There are also many parasitic fishes. Fishes that build nests are frequently

(a)

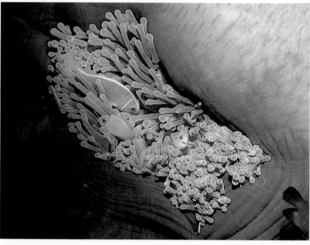
(b)

(c)

Figure 29.12 Symbiotic fishes. (a) Shark suckers *(Remora),* commensalistic or mutualistic, attach to large fishes but do no harm to the surface of their hosts. (b) The clown fish, *Amphiprion,* is unharmed by the cnidocytes of the sea anemone *Radianthus;* living among its host's tentacles, the fish is protected from predators and may also eat wastes expelled from the anemone's gut. The clown fish may benefit the anemone by cleaning its surfaces and by passively luring other fishes to the tentacles. (c) Certain wrasses, such as the small, brightly colored fish shown in this photograph, are "cleaners," obtaining their food by removing dead cells, bacteria, and small parasites from the body surfaces of larger fishes. Here, a Hawaiian cleaner wrasse, *Labroides phthirophagus,* is cleaning a sturgeonfish *Ctenochaetus strigosus.*

exploited by other fishes that lay their eggs in the host nest. The host is then obliged to expend energy caring for the "foreign" eggs and sometimes the larvae. Somewhat like a nest parasite, the bitterling, a Eurasian minnow recently introduced into eastern North America, deposits its eggs in the gills of freshwater mussels.

One of the most notorious fish parasites is the **candiru**, a small catfish (up to about 6 cm long) that inhabits the gill cavities of certain fishes in the Upper Amazon. South American Indians fear candirus, for these little fish are attracted to urine and have been known to enter the urethra, vagina, or anus of bathers. Once inside, they expand their spiny gill covers, causing intense pain that can only be relieved by surgery.

There are many examples of mutualistic relationships involving fishes. Several tropical marine species associate with cnidarians, particularly sea anemones. Clown fishes *(Amphiprion)* can live among the deadly tentacles of large fish-eating anemones (Figure 29.12b). Experimental studies show that clown fishes recognize their hosts by several species-specific chemicals in the mucus secreted by the anemones, and not by sight.

One of the most fascinating examples of mutualism involving fishes is **cleaning symbiosis**. Many small fishes and certain shrimps eat bacteria, parasites, and dead skin from the surfaces of larger fishes. The best-studied cleaners are brightly colored tropical species that establish cleaning stations around rocks, corals, or sessile invertebrates (Figure 29.12c). Cleaners advertise their stations by swimming and by displaying body colors. Other fishes learn the whereabouts of the stations and visit them regularly to be cleaned. All fishes, even large predators such as barracudas, are docile while being cleaned, letting the small cleaners swim over their body surface, including their gills, jaws, and teeth. Among temperate freshwater fishes, certain sunfishes establish cleaning stations that are used by larger predatory fishes, such as largemouth bass. Sunfishes are a favorite food of bass, but while being cleaned, bass assume a head-down position and will not harm the smaller fishes.

SOLUTIONS TO PROBLEMS OF AQUATIC LIFE

Fishes illustrate many special adaptations for aquatic life, and some of these have been described in earlier chapters. For example, fish gills are highly effective at extracting dissolved oxygen from water. (The structure and function of fish gills are discussed in Chapter 5, p. 103.) And the various osmoregulatory mechanisms that marine, freshwater, and brackish-water fishes employ to maintain internal homeostasis in internal fluids are discussed in Chapter 7 (p. 149).

Special Adaptations for Swimming

Compared to air, water is a dense medium, and fishes are adapted to move through it efficiently. Their streamlined, mucus-coated bodies slip through the water with minimal frictional drag. Forward motion results mainly from waves of muscle contraction passing along alternate sides of the body from front to rear.

Swimming efficiency generally depends on the rigidity of the fish's body. The least efficient swimmers are the highly flexible fishes, such as eels (see Figure 9.18). As an eel undulates through the water, its whole body moves sinuously, generating more drag than fishes such as trout, whose bodies are semirigid. As muscle contractions pass down the sides of a trout, most of the energy of the backwardly directed waves of muscle contraction reaches the tail fin and drives it powerfully from side to side, thrusting the fish forward. The fastest and most efficient swimmers are tunas and swordfish. Their bodies are rigid, and thrust is generated entirely by the undulating tail fin.

Myomeres (axial muscle segments) in fishes are composed of red and white muscle fibers. (See the discussion of muscle fiber type and rate of contraction in Chapter 9, p. 209.) **Red fibers** often parallel the longitudinal axis on the outer sides of the body (Figure 29.13). They are active during swimming at any speed. Species that swim continuously, such as tunas, have proportionately more red muscle, and red fibers extend deeper among the white fibers than in fishes that swim intermittently, such as trout. In certain tunas, red muscles have a countercurrent heat exchanger that warms the body (see the Essay). **White fibers** generally form the deep parts of the myomeres and provide bursts of energy for sudden movements and fast swimming. White fibers are arranged in complexly folded patterns, making the myomeres spirallike. Although the function of this arrangement is not yet fully understood, the spiraling allows the maximum number of fibers to be packed into each myomere and also allows all the fibers within each myomere to contract equally, resulting in maximum power output.

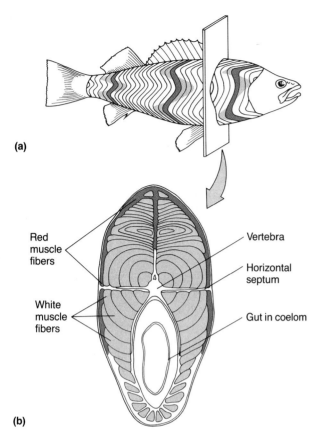

(a)

(b)

Figure 29.13   Arrangement of muscles in a bony fish. (a) Three myomeres are indicated in color. Note that they are overlapped by adjacent myomeres and that backward and forward flexures give each myomere the shape of an M lying on its side. (b) Note the relative position of the red and white fibers in this cross section. (The functional significance of this arrangement is explained in the text.)

Solutions to the Problem of Buoyancy

The specific gravity of fresh water is about 1.00, and that of seawater is about 1.03. Except for fats and oils, the tissues and organs of fishes are denser (about 1.06–1.09), meaning that without some way to lighten their bodies, fishes tend to sink. Most fishes have special adaptations that provide neutral buoyancy. Lacking a swim bladder, chondrichthyans can maintain their position in water by swimming more or less continuously. The flattened, disklike shape of the rays also helps overcome the density problem, and although most species are bottom dwellers, the mantas, or devilfishes *(Manta)*, swim actively about, feeding on plankton. Many sharks also have large, flattened pectoral fins that slow the rate of sinking and provide lift when the fish is moving forward. Moreover, certain sharks have a large liver in which they store copious amounts of a lightweight oil called **squalene**. By an unknown mechanism, they can alter the amount of squalene in the liver to adjust their body density to that of the water as they ascend or descend.

Most bony fishes maintain neutral buoyancy by storing and adjusting gas (usually oxygen or nitrogen) in their swim bladder. The bladder lies dorsally in the body; if it were ventral, gas in the bladder would make the ventral side of the fish lighter, tending to turn the body upside down. Many primitive bony fishes (especially freshwater species such as trout and minnows) have the swim bladder connected to the esophagus throughout life by the pneumatic duct (this is called the **physostomous** condition). In perches and other highly derived teleosteans, the pneumatic duct closes off in an early larval stage, and the swim bladder is gastight (**physoclistous** condition).

Gas pressure in the swim bladder is maintained by secretion or absorption of gas from the blood. A **gas gland** consisting of fine parallel capillaries (a capillary rete, or network) creates a local diffusion gradient, resulting in gas entering the bladder and lightening the body. Another cluster of parallel capillaries, called the **resorptive area**, absorbs gas back into the blood to increase the body's density. The gas gland and resorptive area can fine-tune body density as a fish swims up or down in the water. Physostomous fishes can alter gas pressures in the bladder by gulping air at the water surface or by voiding gas into the gut; but even in these fishes, the capillary retes make most of the pressure adjustments.

Feeding and Digestion

Most fishes are opportunistic carnivores, taking a variety of prey as it is available. As a fish grows, its diet usually changes. The most common foods of larval fishes are various planktonic crustaceans, a rich, readily available source of protein in most aquatic environments. Larger fishes often switch to larger prey, such as insects (in fresh waters) or various bottom-dwelling annelids, crustaceans, or other fishes. Most fishes catch and eat individual prey, whether plankton or larger animals. In feeding, a fish creates a partial vacuum in its mouth cavity and sucks in its prey. The relatively few species of fishes that filter-feed (herrings, paddlefishes, manta rays, and whale sharks) trap plankton on their long, thin gill rakers while swimming.

Fishes have several adaptations that increase digestive surface area. Chondrichthyans, sturgeons, and lungfishes have the spiral valve (see Figure 29.2), and many bony fishes have **pyloric ceca**, providing more surface area at the junction of the

Essay: Not All Fishes Are Cold-Bodied

Even though fishes as a group are ectothermic (cold-bodied), several large species that swim continuously, including the great white shark and certain tunas, keep their body temperatures relatively stable and often higher than the surrounding seawater.

Whereas in most fishes, the bulk of the blood circulates in main arteries and veins (the dorsal aorta and cardinal veins) located in the core of the body, in warm-bodied fishes, the main vessels are just under the skin (Figure E1). In ectothermic fishes, heat generated by muscle activity is carried in venous blood to the heart and gills and is lost to the environment across the gill surfaces. This means that the dorsal aorta receives and distributes cold blood to the internal organs and muscles. In a warm-bodied fish, the dorsal aorta has a small diameter and carries little blood; by contrast, the cutaneous arteries have a large diameter and carry most of the cold blood from the gills to the skin. The cutaneous arteries branch into small vessels that form a network of arterial capillaries carrying cold blood inward through the red muscles. Coursing along these fine arterial capillaries are tiny venous capillaries carrying blood that has been heated by the metabolic activity of the muscles. This network, or rete, of fine arterial and venous capillaries serves as a countercurrent heat exchanger. Most of the heat in the venous capillaries warms the cold inflowing blood. This conserves body heat and passes it to internal tissues, keeping them at a relatively constant warm temperature. Blood stripped of most of its heat then passes from the venous capillaries into the cutaneous veins and is conveyed back to the heart. Because tunas and great white sharks swim continuously, their muscles are always generating heat. Thus, there is a constant supply of warmed blood passing outward in the rete, warming the inflowing blood.

The bluefin tuna (*Thunnus thynnus*), a huge predator of the open sea, may reach a length of 4 m and weigh as much as 800 kg. As a bluefin cruises through the water, its countercurrent heat exchanger keeps its internal body temperature in the range of 22°–30°C, despite temperature changes in the surrounding water. Bluefins migrate north and south in the Atlantic, swimming in waters ranging from about 5° to 30°C. Their heat exchanger allows them to maintain a high rate of metabolism, even in icy waters. Warm muscles can generate more power and speed, for with each 10°C rise in temperature, the contraction-relaxation cycle of red muscle is about tripled. Thus, warm-bodied tunas can swim faster than most of their ectothermic prey species.

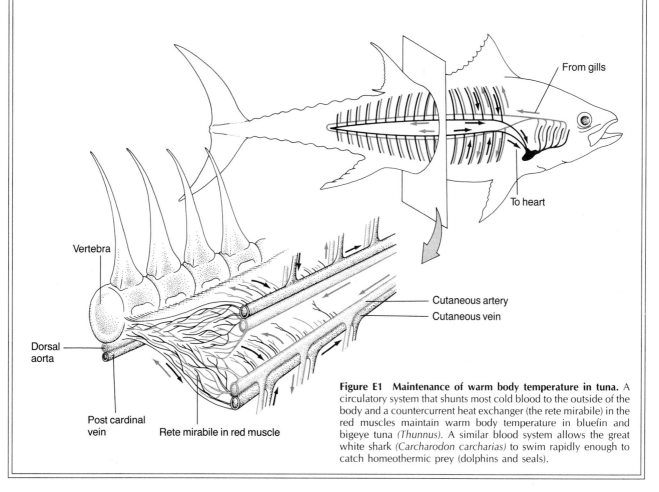

Figure E1 Maintenance of warm body temperature in tuna. A circulatory system that shunts most cold blood to the outside of the body and a countercurrent heat exchanger (the rete mirabile) in the red muscles maintain warm body temperature in bluefin and bigeye tuna (*Thunnus*). A similar blood system allows the great white shark (*Carcharodon carcharias*) to swim rapidly enough to catch homeothermic prey (dolphins and seals).

From gills

To heart

Cutaneous artery
Cutaneous vein

Vertebra

Dorsal aorta

Post cardinal vein

Rete mirabile in red muscle

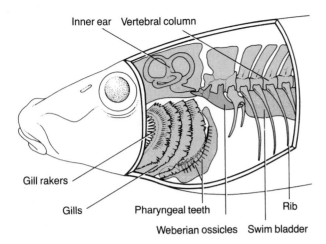

Inner ear Vertebral column

Gill rakers

Gills Pharyngeal teeth Rib

Weberian ossicles Swim bladder

Figure 29.14 Special adaptations in a freshwater teleostean. The sucker, *Catostomus commersonii*, in common with minnows and catfishes, has well-developed pharyngeal teeth used to chew food. These teeth are borne on the most posterior gill arch, which lacks respiratory filaments. Pharyngeal teeth bite against the rear surfaces of the other gill arches and against the upper surface of the pharynx. The drawing also shows the Weberian apparatus, a special "hearing aid" connecting the swim bladder to the inner ear.

stomach and intestine (see Figure 29.5). Of the relatively few strictly herbivorous fishes, most have longer intestines and/or more pyloric ceca than carnivores.

Jaws and teeth generally reflect the type of food consumed. Many herbivorous fishes have rasping teeth for removing algae from rocks. Parrot fishes (see Figure 29.10) can bite off pieces of coral with sharp teeth similar to a parrot's beak. Sharks and many bony fishes, such as piranhas and barracudas, have shearing teeth that can rip chunks from a large fish or other prey. Minnows, suckers, and catfishes lack teeth in the jaws, but have **pharyngeal teeth** on one or more gill arches (Figure 29.14). Pharyngeal teeth vary from molarlike grinders to sharp, cutting facets. Having teeth in the pharynx allows a fish's jaws to become specialized just to catch prey; the pharyngeal teeth perform the separate functions of killing and chewing.

Special Senses

Fishes have especially acute chemosensory ability. One or two openings, or nostrils, near the front of the head lead into canals whose linings are highly sensitive to chemicals dissolved in the water (see Figure 29.2). Taste buds are concentrated in the mouth region, but may occur anywhere on the body surface. The unique ability of some fishes to produce and detect electricity is discussed in Chapter 12 (p. 295).

Vision. Fish eyes are generally similar to those of land vertebrates. (Mammalian eyes are discussed in Chapter 12, p. 275; Figure 12.4.) But rather than changing the shape of the lens in accommodating to near and distant objects, the fish eye accommodates by moving the lens toward or away from the retina. Also, compared to the overall size of the eye, the pupil of many fishes is proportionately larger than that of tetrapods. This increases the eye's light-gathering capacity and is an adaptation to living in a relatively dimly lit environment. Fish eyes also lack movable lids, but because the eye is constantly exposed to water, it does not need lubrication and can be open continuously. Most sharks are color-blind, but most bony fishes can distinguish color. Several species that live exclusively in caves are blind.

Hearing. Although fishes lack outer and middle ears, their inner ear, in addition to determining body position, may supplement the lateral line system in sound reception. (The acousticolateralis system is described in Chapter 12, p. 283; Figure 12.10.) The swim bladder of bony fishes may also increase sound sensitivity. Gas in the bladder is compressible; thus, the swim bladder vibrates and causes tissues around it to vibrate in response to sounds in the water. Hearing ability is highly variable in fishes. Species with swim bladder sound receivers respond to a greater range of high-frequency sounds (up to about 6000 Hz) than other fishes. Among the most sound-sensitive species, herrings have anterior extensions of the swim bladder that project into the skull and lie against the wall of the inner ear. Suckers, minnows, and catfishes have a series of small vibratory ossicles, the **Weberian apparatus**, between the swim bladder and the inner ear (see Figure 29.14).

BEHAVIOR

Fishes have kinetic and taxic responses to a wide range of environmental stimuli. Species that feed during the day exhibit positive phototaxis, and lights of certain wavelengths are sometimes used to attract fishes to baits at night. Studies of electrotaxis are just beginning, but it is evident that many species are attracted or repelled by various types of electric fields in the water. Sharks are known to be drawn to the dipole electric fields of certain prey species.

Orientation in the water depends on locomotor responses to gravity, light, and current. Fishes orient with their dorsal surface toward light and away from the substrate. Most species orient headfirst (positive **rheotaxis**) into currents, although this may change at different periods in the life cycle. Newly hatched salmon orient upstream, but when they are ready to move out to sea, their orientation reverses, and they move downstream with the current. Many species have a so-called **optomotor response** in addition to rheotaxis. They maintain a position in a current or in still water by fixing on visible objects such as two rocks on the bottom. The optomotor response may work to a fish's advantage in avoiding nets drawn through the water. If a fish obtains a fix on the advancing sides of a net, it may be able to swim just fast enough to maintain its position relative to the front of the net.

Probably the most important stimuli to most fishes are chemicals dissolved in water. Using their finely tuned olfactory organs and numerous taste buds, fishes find food, mates, and breeding sites and avoid predators. Many can follow "scent trails" (gradients of chemicals released into the water), homing in on prey or finding breeding sites.

Homing and Migration

Many marine fishes migrate great distances in search of foods that become seasonally abundant in different areas of the oceans. Salmon and freshwater eels accomplish migrations involving enormous distances, seasonal timing, and radical changes in organ systems. (See the Chapter 12 ◁ Essay on salmon migration, p. 290.)

It is not yet known exactly how adult salmon at sea find their home streams, and less is known about how adult freshwater eels locate their deep ocean spawning grounds. Adult eels migrate through deep water in their trip from North America and Europe to the Sargasso Sea, southeast of Bermuda in the Atlantic Ocean. They may navigate by bottom landmarks. After spawning in water over 400 m deep, the adults die; larvae called **leptocephali** hatch and drift in the Gulf Stream to the coasts of North America and Europe, where they develop into adults and complete the life cycle (Figure 29.15).

What are the advantages of spawning migra-

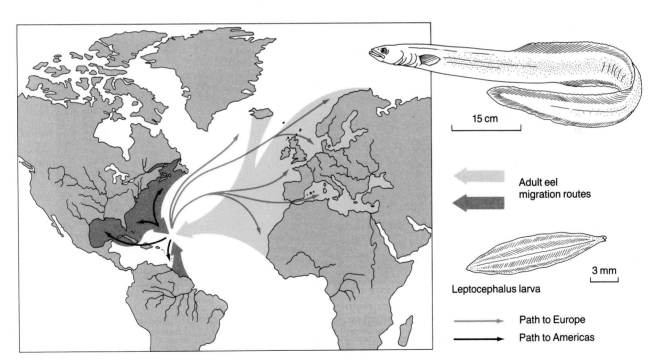

Figure 29.15 The life cycles of the American eel *(Anguilla rostrata)* **and the European eel** *(Anguilla anguilla).* The life cycles include long-distance oceanic migrations. Large gray and color arrows show the general migration routes of adults from coastal streams to the Sargasso Sea. Thin arrows show the routes of return of the planktonic larvae (leptocephali) via the Gulf Stream. Larvae reach the North American coast after drifting for about eight months; they are about 10–25 mm long when they reach the coast. Larvae of the European species reach their home streams after three to four years of drifting at sea; they are about 40–45 mm in length at that time. Females ascend freshwater streams and live there for 8 to 20 years; males remain in the estuaries.

tions of fishes? Enormous energy is expended in the process, so the benefits must be great, but so far, zoologists can only postulate what these benefits are. Salmon and eel migrations obviously center around reproduction, and they ensure that eggs are fertilized and develop in favorable habitats. Also, because adults leave the spawning area or die soon after spawning, the young fish do not have to compete with adults for food. There is also no opportunity for adults to cannibalize the young. The migration patterns also ensure that adults do most of the active swimming required in the life cycle. Young salmon are assisted in their migration to sea by the downstream current. Likewise, eel larvae move with the prevailing currents of the Gulf Stream.

Schooling

Schools are organized, temporary aggregations of fishes in which individuals, usually of about equal size, are attracted to one another (see chapter opening photograph). Schools often include individuals of more than one species, and estimates are that about half of all fish species school at some time in their lives. Precise distances are often maintained between adjacent individuals in schools. A fish seems to maintain its position relative to other members in a school by sight and by detecting water movements of adjacent fish with its lateral line system. Schools often take on different shapes, depending on whether fishes are feeding, migrating, harassing a predator, or avoiding predators (Figure 29.16).

Schooling seems to have several benefits. For one thing, fishes swim more efficiently in schools. By studying schools in the laboratory, experimenters have shown that fish maintain specific positions in the school so that each individual receives a slight forward boost from the movement of the fish in front of it. It has also been shown experimentally that small amounts of slime in the water measurably decrease drag on a swimming fish. Thus, mucus sloughed from the skin of lead fish may increase the efficiency of swimming by followers.

Schooling also seems to make fishes less susceptible to predators. A group is more likely to detect predators than is one fish, and schools can actually confuse predators. Fish scales reflect light, and when all fish in a school flash their scales at a predator, the school may look like one large animal. Also, since most predators attack by fixing on one prey individual at a time, predators may have a difficult time chasing one fish in a school.

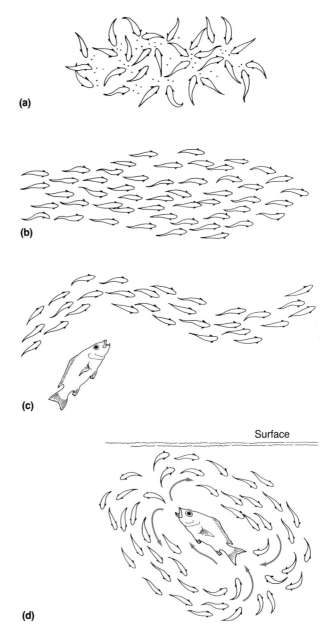

Figure 29.16 Schooling patterns. Patterns differ, depending on the activity of the fish. (a) A feeding school. (b) A traveling school. (c) A school attempting to stream by a predator. (d) Schooling fish may avoid predation by circling and then following a predator for short distances. This behavior may occur both on a vertical plane, as when the school is being driven toward the surface, and on a horizontal plane.

Other Means of Avoiding Predators

Fishes have evolved a number of adaptations that reduce their chances of being eaten. Special secretory cells in the skin of minnows and catfishes produce a chemical called **alarm substance** (chemical nature unknown). When the skin is injured, alarm substance is released into the water, warning other individuals of danger. Minnows respond immediately to alarm substance; they may dart away, stop swimming, or aggregate in protective schools.

Many fishes rely on camouflage to avoid being eaten. Most species exhibit **countershading**: Their ventral surface is light and their dorsal surface is dark. Consequently, when viewed from underneath, fishes tend to blend in with the light surface of the water, and when viewed from above, their body outline tends to disappear against the dark background of deep water or bottom objects.

Color in fishes results from pigment contained in color cells (**chromatophores**) and from light refracted and reflected from the body surface. Color changes are under hormonal and nervous control. Fish skin shows its colors most intensely when the pigment is dispersed in the color cells; as the pigment becomes concentrated at the center of the cells, the body surface becomes pale. Many species adjust their body colors to blend into or mimic objects or plants in their surroundings. Color patterns may also disrupt the outline of a fish or make the head or eyes less conspicuous, possibly confusing a predator (Figure 29.17).

Many deep-sea fishes have light-emitting organs called **photophores**. These are either clusters of light-producing cells in the skin or small external sacs full of luminescent bacteria. Cells in photophores produce cold light by chemical means. Much remains to be learned about the chemical reactions involved, and there is considerable variability among species, but the basic process appears to be similar whether light is produced by fish cells or bacteria: Luminescence results as the organic compound luciferin is oxidized by molecular oxygen in the presence of the enzyme luciferase. The function of luminescence is not well understood. Some species may foil predators by startling them with a sudden flash of light; this may be particularly effective for a large school of small fishes. Likewise, a school of small, luminous fishes may look like one large fish to a predator.

Photophores may also serve in advertisement. Poisonous fishes often have bright skin colors and are usually recognized and left alone by would-be predators. Mimicry is not common among fishes, but a few nonpoisonous species mimic the appearance of poisonous species; mimics thereby decrease their chances of being attacked by predators. There are also species that mimic the cleaner wrasses and take bites out of larger fishes instead of cleaning them. (See the discussion of Batesian mimicry in Chapter 13, p. 316, and the discussion of cleaning symbiosis on p. 715.)

(a)

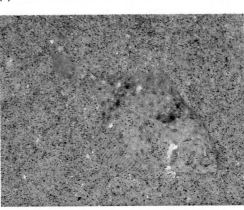

(b)

Figure 29.17 How body color helps foil predators. (a) Color patterns on this longnose butterfly fish *Chelmon rostratus,* a tropical marine species, break up the body outline. (b) Flounders, such as *Pseudopleuronectes americanus* (winter flounder) of the North American Atlantic coast, blend into the seafloor because of their flat body and because they can adjust their skin color pattern to mimic their surroundings.

Other Functions of Color and Light

Body color and light organs are often used in communication. They may be used to advertise and defend territory, act as sex attractants for mating, and communicate specific information in fish schools. Some species use color or light to attract prey. The red headlight fish, a deep-sea predator, has red-light-emitting organs under its eyes. It locates victims by shining its light on them. This strategy seems almost foolproof. Researchers have demonstrated that while the headlight fish can detect the reflected light, several of its favorite prey species cannot see the light as it is emitted by the photophores.

REPRODUCTION AND LIFE CYCLES

Sexes are separate in most fishes. Fertilization is external in most bony fishes, but is internal in sharks and other chondrichthyans. Male sharks have external claspers with grooves through which sperm is transferred to the female (see Figure 29.2). Eggs are fertilized in the oviduct.

Life cycles of fishes are extremely varied, although they generally fit into two broad strategic patterns. (See "Energy Budgeting in Reproduction" in Chapter 14, p. 347, and the discussion of our use of the word *strategy* on p. 342.) In one type of life cycle involving a parental investment or parental care strategy, relatively few eggs are produced, but the parent invests extra energy in them: Eggs and embryos are provided with yolk and protection, and there is no free-swimming larval stage. Most chondrichthyans exhibit this strategy. Many lay eggs enclosed in a leathery case that protects the embryo (Figure 29.18a). Others produce eggs that hatch in the uterus, where the embryos remain and are nourished first by yolk and then, for two to twelve months, by a type of placenta. Coelacanths (*Latimeria;* see Figure 29.7) and a few chondrichthyans lack a placenta, but retain embryos in the uterus for weeks or months (up to two years in some sharks); their embryos consume yolk, and some eat unfertilized eggs from the female's oviduct. Among bony fishes, the guppies and mosquito fish, collectively called live-bearers, produce relatively few offspring at a time and have adopted the parental investment strategy. Fertilization is internal; males have an elongate anal fin, the **gonopodium**, that transfers sperm to the female's oviduct, and embryos are retained and nourished within the female's ovary. Females may give birth to 25 to 150 young every three to four weeks.

A second, more common, type of life cycle employs a numbers strategy and is typical of bony fishes. Enormous numbers of tiny eggs are shed into and fertilized in the water. Most of the reproductive energy goes into egg production rather than into yolk or other extra investments by the parent. Eggs of many marine species are buoyant and planktonic. Freshwater species typically deposit and fertilize eggs on the bottom. Adults may cover fertilized eggs with light sand, or the eggs may stick to rocks or aquatic plants. Minute larvae with small yolk sacs usually hatch within a few days or weeks and metamorphose into adultlike young in a few weeks (Figure 29.18b). Egg and larval mortality is enormous, with the chances of any single larva surviving virtually nil; however, the massive numbers compensate for the heavy losses each generation.

The distinction between the parental investment strategy and the numbers strategy is not always clear. Several species that employ the numbers strategy also provide considerable protection for their eggs or larvae. Certain freshwater catfishes guard their eggs and escort schools of young for several months after hatching. Many tropical fishes incubate eggs and young in a special pouch in their mouth. North American sunfishes and bass construct pebble-lined nests in shallow water, and the male guards the incubating eggs until they hatch.

(a)

(b)

Figure 29.18 Reproductive adaptations. (a) An egg capsule of a skate on kelp. (b) Newly hatched larvae (alevin stage) of salmon. Notice that the yolk sacs are still attached. The larva remains on the bottom and is virtually helpless until the yolk sac is absorbed.

PHYLOGENY

Despite a fairly rich fossil record, the phylogeny of fishes is difficult to determine. Clearly, the evolution of jaws from gill arches was a key event in early fish evolution (see Table 28.1), but we do not know what types of fishes first had jaws. Representatives of the major classes (acanthodians, placoderms, chondrichthyans, and bony fishes) appear in the fossil record dating back about 400 million years, but these groups had undoubtedly diversified earlier. It is likely that jaws and the earliest fishes evolved at least 500 million years ago.

Figure 29.19 summarizes current ideas concerning the phylogenetic relationships among living and extinct classes of fishes. Based on fossil evidence, acanthodians were less numerous than their placoderm contemporaries, but as a group, acanthodians survived much longer. Acanthodians seem more closely related to chondrichthyans than to bony fishes, although this is controversial. Acanthodians had sharklike teeth, scales, and spines,

and the embryos of certain modern sharks have rows of small, acanthodian-like ventral fins. The affinities of the placoderms are more difficult to sort out, although they also seem to have resembled chondrichthyans more than they did bony fishes.

Chondrichthyans

Because of their soft cartilaginous skeleton, relatively few chondrichthyans have been fossilized, compared with the rich bony fish record. The anatomy of fossil sharks contemporary with acanthodians and placoderms indicates that all these groups arose from a common ancestor, probably a species with a bony endoskeleton like that of acanthodians and placoderms. Thus, it is generally be-

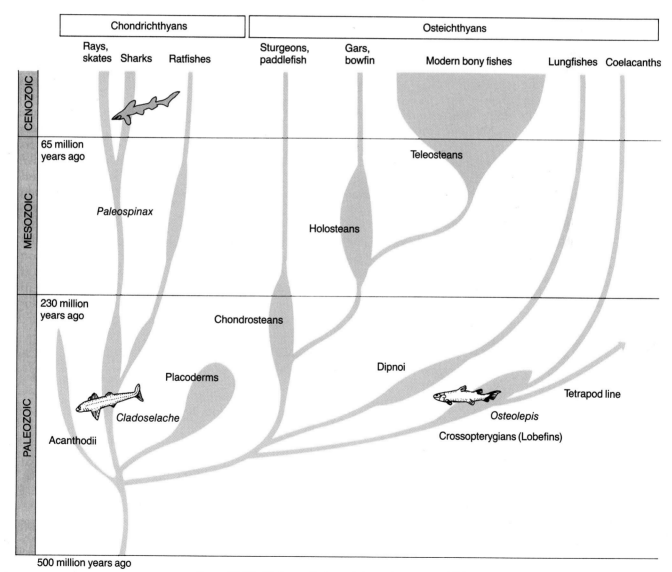

Figure 29.19 Phylogenetic tree of the major groups of fishes.

lieved that the cartilaginous skeleton of sharks and their living relatives is not primitive in the chondrichthyan lineage.

The earliest chondrichthyan fossils are those of predaceous Devonian sharks, such as *Cladoselache* (see Figure 29.19). Chondrichthyans were diverse and abundant during the Devonian and were common in both freshwater and marine environments during the Carboniferous period (280 to 350 million years ago). Following this period, many ancient groups died out, eventually to be replaced by modern chondrichthyans. Evolution of the chondrichthyans has been rather conservative. There is marked similarity between the oldest known modern-type shark *Paleospinax*, a fossil about 200 million years old, and those of the modern species such as *Carcharodon carcharias* (see Figure 29.3a).

Bony Fishes

The three subclasses of bony fishes—the Dipnoi (lungfishes), the Crossopterygii (lobefins, including coelacanths), and the Actinopterygii (ray-finned fishes)—seem to have diverged from a common ancestor with jaws and paired fins as early as 425 to 450 million years ago. Modern lungfishes and coelacanths are rather specialized "leftovers" of formerly diverse fishes adapted for spending some time on land or at least in seasonally dry rivers and ponds. The coelacanth, although specialized for life on the seafloor, is the closest living relative of extinct freshwater lobefins that were ancestral to amphibians. Species of the extinct lobefin *Osteolepis* had pectoral and pelvic bones more similar to those of amphibians than to the fin supports of any other types of fishes (see Figure 29.19). *Osteolepis* also had internal nostrils, lungs, conical teeth, and jaws similar to those of the earliest amphibians. (See the discussion of the labyrinthodonts in Chapter 30, p. 746.)

As shown in Figure 29.19, ray-finned fishes appear to have evolved on a separate line, and the fossil record shows three distinct eras when different groups of rayfins were dominant. The earliest ray-finned fishes were probably extinct relatives of the chondrosteans (paddlefishes and sturgeons). Chondrosteans originated in fresh water, diversified, and then largely died out. Their heyday was the first era of the ray-finned fishes. The second era was that of the holosteans (represented today by gars and the bowfin); note in Figure 29.19 that holosteans arose from chondrosteans. After largely replacing the chondrosteans, the holosteans be-

came diverse in freshwater and marine environments. They were the dominant fish group during much of the Mesozoic. During the early Mesozoic, holosteans gave rise to the teleosteans, beginning the third era of rayfin evolution, which continues today.

Despite their enormous diversity, teleosteans share a number of common traits, and most ichthyologists believe they arose from a single order of holostean fishes. Major trends in teleostean evolution have been the strengthening and lightening of the skeleton, a general increase in mobility, and an increase in the efficiency of the gills in gas exchange. Much teleostean diversity has resulted from the evolution of thousands of different structural and behavioral adaptations for feeding.

❑ SUMMARY

1. Key Features of Fishes. Fishes are aquatic gnathostomes with a head and skull, fins lacking distal support elements, a heart with a single atrium and ventricle, and respiratory gills with filaments projecting outward from the pharynx.

2. Two extinct classes of fishes are the Acanthodii (spiny sharks) and the Placodermi (placoderms).

3. Two living fish classes are the Chondrichthyes (about 800 species of sharks, skates, rays, and ratfishes), and the Osteichthyes, or bony fishes (about 36 orders with over 400 families and about 21,000 species).

4. Special adaptations for swimming in fishes include streamlining and complex arrangements of body muscle segments. The most efficient swimmers have semirigid bodies propelled mainly by tail thrusts. Bony fishes maintain neutral buoyancy by changing the gas pressure in their swim bladder; cartilaginous fishes lack a swim bladder, but maintain position in the water by swimming and by changing their liver content of light oil.

5. Fishes have well-developed nervous and sensory systems; many species rely heavily on chemosensory systems and on the acousticolateralis system. Complex behavior patterns include migration and homing. Many fishes exhibit parental care of young.

6. Chondrichthyans were more diverse during the Devonian and Carboniferous periods than they are today; bony fishes are more diverse and abundant today than they have ever been.

☐ FURTHER READING

Bartridge, B. L. "The Structure and Function of Fish Schools." *Scientific American* 246(1982): 114–123. *Antipredator effects of schools and how schooling behavior is studied.*

Bone, Q., and N. B. Marshall. *Biology of Fishes.* Glasgow and London: Blackie & Son, 1982. *Selected topics, with emphasis on fish physiology.*

Butler, M.J.A. "Plight of the Bluefin Tuna." *National Geographic* 162(1982): 220–230. *A pictorial essay on the decline of this large, constantly swimming species.*

Clark, E. "Flashlight Fish of the Red Sea." *National Geographic* 154(1978): 719–728. *Describes a species with a luminescent organ (symbiotic bacteria-type) below its eyes.*

Clark, E. "Sharks, Magnificent and Misunderstood." *National Geographic* 160(1981): 138–187. *A well-illustrated introduction to sharks and their behavior.*

Herald, E. S. *Living Fishes of the World.* Garden City, N.Y.: Doubleday, 1961. *A dated but excellent pictorial account of the fish groups.*

Hopkins, C. D. "Electric Communication in Fish." *American Scientist* 62(1974): 426–437. *Studies of the physical nature and behavioral significance of electricity in fishes.*

Lee, A. "Atlantic Salmon. The Leaper Struggles to Survive." *National Geographic* 160(1981): 600–615. *Describes recent increases in populations of this species, which migrates between the ocean and spawning grounds in freshwater streams.*

McCosker, J. E. "Great White Shark." *Science 81* 2(1981): 42–51. *Excellent color photographs and description of research on Carcharodon.*

Moyle, P. B., and J. J. Cech, Jr. *Fishes: An Introduction to Ichthyology.* Englewood Cliffs, N.J.: Prentice-Hall, 1982. *An excellent modern textbook in ichthyology.*

Popper, A. N., and S. Coombs. "Auditory mechanisms in Teleost Fishes." *American Scientist* 68(1980): 429–440. *An excellent description of hearing in teleosteans.*

Warner, R. R. "Mating Behavior and Hermaphroditism in Coral Reef Fishes." *American Scientist* 72(1984): 128–136. *A review of studies on sex changes in tropical fishes; includes many excellent color photographs.*

Webb, P. W. "Form and Function in Fish Swimming." *Scientific American* 251(1984): 72–82. *The mechanics of fish movement; interesting description of fishes as accelerators, cruisers, maneuvering specialists, and generalists in movement.*

Zahl, P. A. "Dragons of the Deep." *National Geographic* 153(1978): 838–845. *A pictorial essay on sea horses.*

Zahl, P. A., and T. O'Neill. "The Four-Eyed Fish Sees All." *National Geographic* 153(1978): 390–395. *Describes fish that see in water and air.*

☐ CLASSIFICATION

Class Acanthodii (Gr. *akantha*, "thorn"; *odes*, "a thing like"). Acanthodians (spiny sharks); extinct fishes with spiny fins; the earliest known gnathostomes; *Homalcanthus, Climatius; Acanthodes.*

Class Placodermi (Gr. *plax*, "a flat round plate"; *derma*, "skin"). Placoderms; extinct fishes with bony armor on the forebody; *Bothriolepis, Coccosteus; Dunkleosteus.*

Class Chondrichthyes (Gr. *chondros*, "cartilage"; *ichthys*, "fish"). Chondrichthyans (cartilaginous fishes); possess an endoskeleton of cartilage; lack a swim bladder.

Subclass Elasmobranchii (Gr. *elasmos*, "thin plate"; *branchia*, "gills"). Sharks, rays, and skates; have five to seven separate gill openings on each side; no gill cover (operculum); placoid scales (dermal denticles) or scaleless; *Squalus, Carcharodon, Raja, Torpedo, Dasyatis.*

Subclass Holocephali (Gr. *holos*, "whole"; *kephale*, "head"). Ratfishes; scaleless cartilaginous fishes with soft gill covers and a notochord in adults; *Chimaera, Hydrolagus.*

Class Osteichthyes (Gr. *osteon*, "bone"; *ichthys*, "fish"). Bony fishes; endoskeleton formed mostly of bone; a bony operculum covers a common gill chamber; most have a swim bladder or lung.

Subclass Dipnoi (Gr. *di*, "two"; *pnoe*, "breath"). Lungfishes; Australian species has lobelike fins; the swim bladder functions as a lung; three living genera: *Lepidosiren, Neoceratodus, Protopterus.*

Subclass Crossopterygii (Gr. *krossoi*, "fringe"; *pteryx*, "fin"). Lobe-finned fishes (lobefins); fins are supported by bones similar to those of tetrapod limbs; the swim bladder functions as a lung or is secondarily lost as in living coelacanths; extinct species had internal nostrils; *Osteolepis* (primitive freshwater genus, extinct, probably very similar to ancestral amphibia), *Latimeria* (living coelacanth).

Subclass Actinopterygii (Gr. *aktis*, "ray"; *pteryx*, "fin"). Ray-finned fishes (rayfins); the most diverse subclass of living fishes.

Infraclass Chondrostei (Gr. *chondros*, "cartilage"; *osteon*, "bone"). Primitive ray-finned fishes; skeleton mostly cartilaginous; notochord in adult; ganoid scales; *Polyodon* (paddlefish), *Acipenser* (sturgeon).

Infraclass Holostei (Gr. *holos*, "whole"; *ostei*, "bone"). "Intermediate" ray-finned fishes; skeleton mostly bony; swim bladder is lunglike; ganoid or cycloid scales; *Amia* (bowfin), *Lepisosteus* (gars).

Infraclass Teleostei (Gr. *teleos*, "perfect"; *osteon*, "bone"). Highly derived ray-finned fishes; bony skeleton; swim bladder functions in maintaining neutral buoyancy; cycloid or ctenoid scales, and many scaleless; about 30 living orders, 20,000 species.

  Division Archaeophylaces (Gr. *arch*, "primitive"; *phylax*, "guard"). Primitive teleosteans.

   Order Osteoglossiformes (Gr. *osteon*, "bone"; *glossa*, "tongue"; L. *forma*, "shape"). Mooneyes; bony tongues.

   Order Mormyriformes (Gr. *mormyros*, "a sea fish"; L. *forma*, "shape"). Mormyrids; electric fishes.

  Division Taenopaedia (Gr. *tainia*, "band"; *paidos*, "child"). "Intermediate teleosteans."

   Order Clupeiformes (L. *clupea*, "small river fish"; *forma*, "shape"). Herrings, sardines, anchovies.

   Order Elopiformes (Gr. *elops*, "a sea fish"; L. *forma*, "shape"). Tarpons.

   Order Anguilliformes (L. *anguilla*, "eel"; *forma*, "shape"). Freshwater eels, moray eels.

Division Euteleostei (Gr. *eu*, "true"; *teleos*, "perfect").

Order Salmoniformes (L. *salmonis*, "salmon"; *forma*, "shape"). Pikes, trout, salmon, smelts.

Order Cypriniformes (Gr. *kyprinos*, "carp"; L. *forma*, "shape"). Minnows (including carps), goldfish, shiners, suckers, North American catfishes, electric eels, sea catfishes.

Order Myctophiformes (Gr. *mykter*, "nose"; *ophis*, "serpent"; L. *forma*, "shape"). Lanternfishes.

Order Gadiformes (Gr. *gados*, "cod"; L. *forma*, "shape"). Cods, hakes.

Order Batracoidiformes (Gr. *batrachos*, "a frog"; L. *forma*, "shape"). Toadfishes.

Order Lophiiformes (Gr. *lophos*, "crest"; L. *forma*, "shape"). Anglerfishes, deep-sea anglers.

Order Atheriniformes (Gr. *atherine*, "a type of smelt"; L. *forma*, "shape"). Flying fishes, killifishes, mosquitofish, guppies, mollies, swordtails.

Order Gasterosteiformes (Gr. *gaster*, "belly"; *osteon*, "bone"). Sticklebacks, pipefishes, seahorses.

Order Scorpaeniformes (Gr. *skorpaina*, "fish with a poisonous sting"; L. *forma*, "shape"). Scorpion fishes, stonefishes, sculpins, rockfishes.

Order Perciformes (Gr. *perke*, "perch"; L. *forma*, "shape"). Perches and their relatives (the largest order of vertebrates, about 7000 species); perches, darters, sunfishes, freshwater bass, barracudas, Siamese fighting fish, tilapias, wrasses, cichlids, angelfishes, tunas, sharksuckers.

Order Pleuronectiformes (Gr. *pleuron*, "rib"; *nektos*, "swimming"; L. *forma*, "shape"). Flatfishes; flounders, halibuts, soles.

Order Tetraodontiformes (Gr. *tetra*, "four"; *odontos*, "tooth"; L. *forma*, "shape"). Headfishes, puffers.

Amphibians

Living amphibians, making up the smallest and least-studied class of tetrapod vertebrates, are classified in three orders. **Anurans**, or frogs, are the most familiar amphibians. Anyone walking along a stream in the woods or standing at the edge of a pond on a warm evening has probably startled frogs and heard their calls. **Caecilians**, constituting

a second order, are legless and somewhat wormlike. All caecilians are tropical, and the group has received comparatively little research attention. **Urodeles**, or salamanders, make up the third order and are more secretive than frogs, yet may often be seen along shady stream banks. Of some 3900 species of living amphibians, only about 160 are caecilians, some 400 are urodeles, and the rest are anurans.

Zoologists find two aspects of amphibian biology especially interesting. First, amphibians occupy a key position in vertebrate phylogeny. As the dominant land vertebrates some 300 million years

---

Class Amphibia. The Asiatic gliding frog *Rhacophorus nigropalmatus* is a member of a family of small (30–75 mm long), mostly arboreal, frogs. Note the extensive webbing between the toes; by stretching this webbing, *R. nigropalmatus* can glide from branch to branch.

ago, amphibians were the stock from which the reptiles evolved. In turn, the birds and mammals arose from reptilian stock. Second, amphibians—whose class name means "double life"—represent one of the truly momentous events in animal evolution: the conquest of land by vertebrates. Over tens of millions of years, amphibians evolved the ability to make at least limited use of land habitats.

Reflecting their phylogeny as transitional (water/land) tetrapods, most living amphibians exhibit a mixture of aquatic and terrestrial adaptations. Thus, many species have gill-breathing larvae that later transform into lung-breathing adults. (Exceptions are common, however; about a third of the anurans and half the caecilians and salamanders lack an aquatic larval stage, and many species of salamanders are lungless.) As discussed in Chapter 28, amphibian lungs are rather ineffective compared to those of other tetrapods. Consequently, in many species, the skin is soft and moist and is important in gas exchange. The development of amphibian lungs from evolutionary origins in certain fishes was accompanied by the development of a heart with separate chambers for receiving blood from the body and lungs. Also reflecting the amphibian "double life," the limbs of some amphibians are adapted for both terrestrial and aquatic locomotion; thus, the hindlimbs of most frogs are adapted for jumping on land, yet the feet of some species are webbed for swimming. (Again, there are exceptions and some amphibians lack one or both pairs of legs.) Despite such features, most amphibians are ultimately tied to water by the nature of their eggs. The majority of amphibians must lay their eggs in water or in very damp places; deposition on dry land would be fatal to the eggs because they lack water-resistant membranes or a shell.

Representative Amphibians

Order Anura

The order name **Anura** means "without a tail," and being tailless as adults distinguishes members of this group, the frogs (including toads, which zoologists consider types of frogs), from salamanders. The combination of strong, enlarged hind legs with comparatively small forelimbs is also distinctive. Anurans are relatively small animals. The largest anuran is a West African frog weighing up to 3.3 kg

and reaching about 30 cm in length, while the smallest is a tiny Cuban frog that could sit on a dime with room to spare. Other distinguishing features of the order are fused tail vertebrae, which form a rodlike **urostyle** (see Figure 30.5), and a wide head that is broadly joined to the body, making a frog appear to lack a neck. Most anurans also lack ribs, leaving the lungs and heart less protected than in most other tetrapods.

Most immature anurans are herbivorous, gilled, fishlike larvae, commonly called tadpoles, whereas typical adults are carnivorous, air-breathing tetrapods. Metamorphosis to the adult form involves many parts of the body, including the digestive tract, and is a truly dramatic event in an anuran's life cycle.

Anuran diversity is often based more on physiological and behavioral adaptations than on external features. Consequently, species adapted to live in very different habitats typically are similar in external appearance (Figure 30.1a–c). Some species are primarily aquatic, while others are terrestrial burrowers or tree dwellers.

Constituting one group of anurans, the **tree frogs** are typically small, but their large vocal sacs give them loud voices. Their toes end in adhesive discs that enable them to climb trees and cling to vegetation. Members of the **Family Hylidae** inhabit North and South America, Europe, and Australia. The spring peeper, *Pseudacris crucifer*, and the common tree frog, *Hyla versicolor*, are North American species. Another family of tree frogs, the **Rhacophoridae**, includes many species that occur in Africa, Madagascar, and Asia (see chapter opening photograph).

Members of the **Family Ranidae**, often called the **true frogs**, occur on all continents except Antarctica. Adult ranids have webbing between their toes. Of some 36 genera in the family, only one, the Genus *Rana*, is found in the United States. Familiar North American ranids include the very common species *Rana pipiens*, the leopard frog. The bullfrog, *R. catesbeiana*, is the largest North American ranid, often with a body length of nearly 20 cm (Figure 30.1a). Restricted to swampy regions, bullfrogs are almost always found in or close to water. Their tadpoles require two and sometimes three years to mature. The more tolerant leopard frog lives in a variety of habitats; its tadpoles transform to adults within one year.

Members of the **Family Pipidae**, the **tongueless frogs**, are small to moderately large (5–20 cm long) aquatic species of tropical South America and Af-

(a) (b) (c)

Figure 30.1 Order Anura. (a) The bullfrog *Rana catesbeiana* occurs naturally east of a line from Texas to southern Canada and has been introduced into western states. (b) The Surinam toad, *Pipa pipa* (about 17 cm long), lives in northern South America. The female carries her eggs in separate pockets on her back, where they hatch as tiny young. (c) The American toad *Bufo americanus* lives southward and eastward from Hudson Bay, whereas *B. boreas*, another toad common in the United States, lives in the Rocky Mountain and Pacific states north to Alaska.

rica south of the Sahara. Pipids differ from most other frogs in several ways: They lack a tongue; they have ribs that are free in juveniles and fused to transverse processes in adults; and they may have claws on several of the hind digits. The female of *Pipa pipa*, the Surinam toad, deposits and carries her eggs on her back (Figure 30.1b). *Xenopus laevis*, the African clawed frog, is cultured throughout the world and used extensively in research, especially in developmental biology, genetics, and physiology.

The **Family Bufonidae** is a large group of toads, the majority of which are in the genus *Bufo*. Bufonids occur naturally on all continents except Antarctica and Australia, and an introduced species, *Bufo marinus*, of the American tropics, has become established in Bermuda and Australia. *Bufo americanus* is common in the northeastern United States (Figure 30.1c). The bufonid body is covered by a rough, warty skin containing many poison glands. Their secretion is a poisonous **bufotoxin** that protects the toad from potential predators. Some bufotoxins may be very toxic; dogs and cats have died after catching and biting *Bufo marinus*, and humans who handle certain toads may suffer skin irritation. Unlike most adult amphibians, bufonids lack teeth, and the males of most species have a unique mass of tissue called **Bidder's organ** on the anterior part of each testis. If the testes are removed, Bidder's organs may develop into functional ovaries.

Order Gymnophiona (Apoda)

Seldom seen and the least known amphibians, **caecilians**, making up the Order Gymnophiona, are legless and tailless. They live in the tropical forests of Central and South America, Africa, and Southeast Asia (Figure 30.2). Although most caecilians are burrowers that prey on small invertebrates in the damp earth beneath the forest floor, the adults of one South American family live in ponds and streams. Caecilians range in length from about 10 cm to more than 1 m. Some species have annular folds in the skin, giving them a segmented, earthwormlike appearance, and the adults of most

Figure 30.2 Order Gymnophiona. The caecilian *Gymnopis multiplicatus* from a rain forest in Costa Rica. Little is known about these legless, cylindrical, tropical forest amphibians. Most species lay eggs, although some are viviparous. Several species exhibit parental care in guarding their eggs and young.

species have only vestigial eyes. Protrusible tentacles on the snout have replaced the eyes as the chief sensory organs, a valuable adaptation for subterranean life.

Order Urodela (Caudata)

Urodeles, or salamanders, are generally small and relatively unspecialized in form (Figure 30.3a–d). They have an elongate body, a long tail (thus the name Urodela, meaning "visible tail"), and typically four legs of similar size. In general appearance, urodeles resemble ancestral amphibians (the lepospondyls and labyrinthodonts of the late Paleozoic and early Mesozoic eras) more than do any other vertebrates (see Figure 30.18). Besides their

distinctive long tail, urodeles differ from frogs in that their hind legs are not adapted to hopping. Some salamanders have very small limbs or lack limbs altogether; the primary means of locomotion in these forms is swimming.

Urodele adults and larvae are similar in appearance, except that the larvae typically have pharyngeal gill slits and external gills, and most adults do not. Consequently, salamander metamorphosis is less extreme than the dramatic transformation typical of anurans. Adults and larvae have teeth and are carnivorous, preying on many kinds of small arthropods, worms, and molluscs.

Most urodeles live in the northern hemisphere; a single species lives above the Arctic Circle and about 130 species have evolved in the tropics. Some

Figure 30.3 Order Urodela. (a) The congo eel, or "conger," *Amphiuma tridactylum,* is a member of the aquatic Family Amphiumidae. It lives in swamps, streams, and muddy ponds from Texas north to Missouri and Virginia; it may grow to lengths of more than 100 cm. The congo eel swims by undulating its body, and it feeds on small fishes, molluscs, and crustaceans. (b) The mud puppy *Necturus maculosus* is distributed from Manitoba to Louisiana. This large (up to about 30 cm) aquatic salamander lives permanently as a neotenic (reproducing) larva with three pairs of gills as well as lungs. A related species, *N. punctatus,* the dwarf waterdog (up to about 15 cm long), lives on the coastal plain from Alabama to Virginia. Species of *Necturus* feed on small fishes and many kinds of invertebrates. (c) The tiger salamander, *Ambystoma tigrinum,* is one of nearly 40 species of mole salamanders, or ambystomatids, that occur throughout much of southern Canada, the United States, and the Mexican plateau. Adult tiger salamanders are about 35 cm in length, while the larvae, which may be neotenic, are about 20 cm long. (d) The hellbender *Cryptobranchus alleganiensis* occurs in streams and rivers in the eastern and south central United States. Adults, up to 75 cm long, live under submerged logs and among rocks.

species are totally aquatic, living in marshes, lakes, and streams. Others are terrestrial as adults, moving about in marshy wooded areas and lurking beneath logs and stones in moist, secluded places. A few salamanders are arboreal.

Members of the **Family Amphiumidae** are aquatic salamanders that live in the swamps of Florida and adjacent states. One species, the congo eel, has an eel-shaped body and four rudimentary legs (Figure 30.3a). The congo eel is highly adapted for undulatory swimming. Adults retain the larval gill slits, but lack external gills. A second family of eel-like aquatic salamanders, the **Sirenidae**, lives in the same region, burrowing in the bottom mud of ponds and ditches. Unlike the congo eel, however, mature sirenids possess gill slits and external gills, but lack lungs and hind limbs.

Another family of aquatic salamanders, the **Proteidae**, includes the mud puppies (*Necturus*; Figure 30.3b) and the European olm *(Proteus anguineus)*, which is restricted to subterranean waters in Yugoslavia. Proteids have gill slits and external gills. Five species of *Necturus*, all carnivores, live in the east central United States and are familiar to many people because they are sometimes studied in general zoology laboratories.

Ambystomatids, members of the exclusively North American **Family Ambystomatidae**, live on land as adults and in the water as neotenic larvae (larvae capable of reproducing). For example, populations of the tiger salamander, *Ambystoma tigrinum*, may include sexually mature larvae that are capable of reproduction but that continue to feed in their aquatic habitats until environmental conditions become unfavorable. Then the larvae lose their gills, develop lungs, and assume the form of a typical land-dwelling salamander (Figure 30.3c).

The largest family of salamanders is the **Plethodontidae**, with nearly 200 species. Plethodontids are lungless and rely entirely on their moist skin and oral membranes for gas exchange. Most species lack aquatic larvae and develop from eggs laid in moist sites on land. The family is largely restricted to the western hemisphere, but two species occur in Europe. *Plethodon cinereus*, the red-backed salamander of eastern North America, has been extensively studied (see Figure 5.12). Unlike most amphibians, which simply lay their eggs in water and leave them unattended, female plethodontids lay only a few eggs (as few as three) and guard them until they hatch (nearly three months); some females even stay with the young for several weeks after hatching.

The largest living salamanders are members of the **Family Cryptobranchidae**, consisting of two species in Asia and a single species, *Cryptobranchus alleganiensis*, in North America (Figure 30.3d). The Asian species reach lengths of 1.5 m, whereas *C. alleganiensis* may be up to about 75 cm long. Cryptobranchids, commonly called giant salamanders or hellbenders, are strictly aquatic; adults retain a pair of gill slits, but no external gills. Hellbenders eat invertebrates and small fishes.

HABITATS AND ENVIRONMENTAL RELATIONSHIPS

The majority of amphibians live in fresh water or damp habitats. None live in the ocean, but a few can tolerate brackish water. Salt-tolerant amphibians include the crab-eating frog *Rana cancrivora* of Southeast Asia, which develops and feeds in mangrove swamps with salinities of up to about 80% that of seawater. (Osmoregulation in this very unusual species is described in Chapter 7, p. 150.)

Oceans have been effective barriers to amphibians, and this helps us understand the distribution of present-day species in terms of earth history. Fossil records indicate that modern groups of amphibians originated some 200 to 300 million years ago, when most of the world's land masses were united as the supercontinent Pangaea. (See the discussion of the Paleozoic, or Old Animal, era in Chapter 17, p. 417.) The subsequent continental drift that subdivided Pangaea profoundly affected the distribution of amphibians. Thus, of the three modern orders of amphibians, only the anurans are worldwide, indicating that they dispersed widely before Pangaea divided. Modern urodeles are generally north temperate in distribution, and fossil records indicate that this group originated in the northern land mass derived from Pangaea. Caecilians occur in tropical Africa, South America, and Southeast Asia, indicating that they originated in the southern land mass derived from Pangaea.

The distribution of groups of amphibians is directly related to their water needs. Most species lose body water and die very quickly in dry surroundings. As we have mentioned, some amphibians, such as the mud puppies, congo eel, and hellbender, are strictly aquatic, whereas many others divide their time between land and water. Many tree frogs never enter a pond or stream; rain and dew held by vegetation meet their needs. Although urodeles and many other amphibians are plentiful in moist, temperate climates, the majority of amphibian species live in the tropics, where they thrive in the humid, relatively cool air, shaded beneath the forest canopy. A few species have managed to adapt to quite different conditions. A number of frogs and one species of salamander live above the Arctic Circle, and certain salamanders and frogs live at high altitudes in the mountains of California and South America. Special behavioral and physiological adaptations enable certain toads to live in various deserts, including those of the southwestern United States. Desert dwellers are often active at night, when the air is cooler and more humid. Many escape heat and desiccation by burrowing into the earth or by retreating into the burrows of other animals, and absorption of water from moist burrow walls may help replace lost body water.

Temperature is another key factor in the life of amphibians. Because amphibians are ectothermic, they become warm or cool in response to environmental temperature changes. In the laboratory, amphibians are readily killed by temperature extremes. However, under natural conditions, many species use behavioral adaptations for surviving seasonal temperature extremes. Toads commonly burrow below the frost line as winter approaches, and many frogs and aquatic salamanders pass the winter hibernating in the mud at the bottom of rivers or lakes. Tree frogs overwinter by burrowing into the humus on the forest floor. Their cells and tissues are protected from ice crystal formation by glycerol, an "antifreeze" that accumulates in the cells as winter approaches. At the other extreme, amphibians in southern regions may avoid the heat by aestivating, the summer equivalent of winter hibernation. During periods of heat and drought, an amphibian may simply retreat to a moist burrow or crevice. A salamander may reduce its body surface area (and therefore water loss) by folding its limbs and curling its tail and body. Going a step further, an African bullfrog *(Pyxicephalus adspersus)* envelops itself with a water-resistant sac secreted by its skin glands. Certain tree frogs of the genus *Phyllomedusa*, which inhabit dry areas in South America, are unique in that they have lipid-secreting skin glands. They are inactive during the day, sitting in a torpid state in trees and bushes, where they are exposed to drying air. Before becoming torpid, however, they spread waterproofing lipids all over their body by wiping themselves with their feet. These two adaptive features—torpidity and secretion of an oil coating—reduce the amount of water these frogs would lose from evaporation by as much as 95%.

Amphibians in Food Chains

Amphibians are generally ill equipped to fend off other predators. Their teeth and claws are small, if they exist at all, and except for the hind legs of frogs, limb musculature is lightly developed. Large frogs may kick and bite at attackers, but most amphibians defend themselves rather passively, depending mostly on poisons from skin glands, camouflaging with body colors and markings, secretive behavior, and quick jumps. Frogs are consumed in great quantities by fishes, turtles, snakes, herons, and hawks. In parts of the United States, humans are the greatest threat to frogs; millions are used each year in biology and zoology classes, and frog legs are eaten as a delicacy. Actually, most of the frogs used for these purposes are cultured commercially, and for amphibian populations in nature, a greater threat is the destruction of wetlands vital for breeding.

In addition to adult amphibians, enormous numbers of amphibian eggs and larvae are eaten by various predators, including fishes and certain insects. Predation is sometimes so intense that only a few larvae out of a large population may survive to adulthood. The largest tadpole populations occur in ponds and lakes where there are no fishes. There, tadpoles play a major role in food chains.

Adult amphibians and larval salamanders prey on a great variety of small, actively moving animals, including small molluscs, annelids, insects, and other invertebrates. Large species sometimes devour small fishes and other amphibians, even cannibalizing their own species, and bullfrogs may catch and eat small mammals and birds. The burrowing caecilians eat worms and many other kinds of small invertebrates encountered underground.

In contrast to their carnivorous parents, frog tadpoles feed mostly on algae and particles of dead plants and animals; they may also filter bacteria from the water. Frogs therefore play a dual role in food chains. In eating plant material, the tadpoles act as primary consumers, transferring some of the solar energy trapped by plants to the animal kingdom. In eating animals, the adults assume higher positions (secondary and tertiary consumers) in food chains.

COMPARATIVE STRUCTURE AND FUNCTION

This section emphasizes adaptations that distinguish the amphibians as a "transitional" (aquatic/terrestrial) class of vertebrates.

Body Support, Protection, and Locomotion

Amphibian skin is important in both gas exchange and protection. It is subdivided, as in all vertebrates, into an epidermis overlying a dermis (see Figure 28.13b). The outer layer of the epidermis in terrestrial adults is a thin layer of dead cells filled with the water-insoluble protein keratin, which helps limit water loss through the skin in terrestrial environments. The outer layer also makes a frog's thin skin relatively tough and less likely to be damaged by sand and pebbles. Amphibians molt periodically, shedding a one-cell-thick outer sheet of epidermis.

As discussed in Chapter 28, amphibian skin glands develop from the epidermis. In most species, the most prominent skin glands are the large mucous glands, whose secretions keep the skin moist for gas exchange, and the poison glands, which secrete irritating or poisonous toxins that deter predators. South American frogs of the genera *Phyllobates* and *Dendrobates* secrete alkaloid poisons that affect nerve and muscle activity (Figure 30.4). Some of these toxins, used by South American natives for their poison darts and arrows, are more deadly than the venom of the most poisonous snakes or arachnids. Bufonid toads generally have more poison glands than do other frogs, and masses of poison glands form a toad's "warts."

Many amphibians are brilliantly colored. The bright colors of *Phyllobates* and *Dendrobates* serve to warn would-be predators of the skin toxins. Some amphibians have stable color patterns, while others have colors that change in response to changing environmental conditions. Three kinds of chromatophores (color cells) are often responsible for the color changes. One kind, located just below the epidermis, contains a yellow pigment. Beneath these cells, a second kind produces a blue color by scattering certain wavelengths of light. Acting together, these two kinds of chromatophores give many frogs their greenish color. The third type, the melanophore, containing the black pigment melanin, extends pseudopodial processes between the other chromatophores. Migration of melanin into these processes causes the skin to darken. Darkening generally occurs as evening temperatures fall, whereas the skin pales at high temperature and increased light. Presumably, the darkened animal is less visible to predators at dusk, and the pale animal is less visible during daylight hours. Melanophore-stimulating hormone (intermedin) from the pituitary gland triggers dispersal of the pigment. Experimentally blinded frogs do not change color, indicating that light entering the eyes is the stimulus that causes release of intermedin.

Skeletomuscular System. Amphibians have several skeletal and muscular adaptations for moving about on land (Figure 30.5). Their vertebral column is generally more rigid than that of fishes, whose bodies are supported mainly by water. Consequently, the amphibian vertebral column supports viscera slung beneath it and at the same time transfers forces from the legs to the body during locomotion. Although capable of complex movements, an amphibian's limbs extend laterally from the body to such a degree that, when not moving, the body typically rests on the ground between them. As a result, amphibians must essentially do push-ups as they move. Adaptations for more facile locomotion on land included limbs capable of holding the body off the ground, as seen in many reptiles, birds, and mammals.

Major changes in the musculature accompanied the evolution of the amphibian skeleton. In the previous chapter, we saw that the metamerically arranged axial (trunk) musculature of fishes is well suited for side-to-side tail movements. In amphibi-

Figure 30.4 Self-defense in frogs. This brightly colored arrow-poison frog, *Dendrobates pumilio,* lives in rain forests of Central America. Its skin glands secrete highly toxic alkaloids. The deadly alkaloids produced by the skin of a single individual of the closely related species *Phyllobates terribilis* could kill about 20,000 mice!

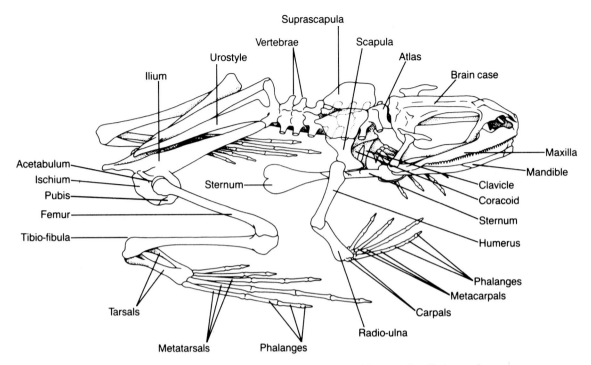

Figure 30.5 Skeleton of a frog. Elongated and closely flexed femurs, tibio-fibulae, and combined tarsals, metatarsals, and phalanges provide the great leverage required in hopping locomotion. The sternum, absent in fishes, helps support the anterior parts of the frog's body. Note the urostyle, which replaces caudal vertebrae in anurans. By contrast, salamanders typically have numerous caudal vertebrae, a more elongate body, and forelegs and hind legs of similar size.

ans, the fishlike arrangement is most obvious in salamanders, which swim by means of serpentine body movements, and is most modified in anurans, which swim and hop with powerful thrusts of the hind legs (Figure 30.6a). The ventral muscles in the abdominal region are multilayered sheets adapted for supporting the internal organs when the amphibian is on land. The dorsal muscles stiffen the vertebral column and, in urodeles, move the head, something a fish cannot do. Major changes in the musculature of the appendages accompanied changes in the axial musculature of amphibians. Whereas a fish's fin muscles generally lie within the body and produce relatively simple fin movements, various amphibians can swim, walk, hop, and climb—complex motions requiring numerous muscles located within the limb itself (Figure 30.6b).

Feeding and Digestion

The amphibian digestive tract and its functions generally follow the vertebrate pattern described in Chapter 4 (see p. 89, and compare Figures 4.11 and 30.7). Amphibians store most of their fat in **fat bodies** (see Figure 30.7), whereas sugar is stored as glycogen in the liver and muscles. In many species, fat is metabolized during hiberation or aestivation.

Most adult amphibians and larval salamanders have teeth, but the teeth are used mainly to hold food until it is swallowed whole, rather than to chew it (Figure 30.8a). Anurans swallow food by raising the floor of the mouth while pressing down with the eyeballs, a squeeze play that forces the food through the pharynx and into the esophagus. With the exception of the pipids (tongueless frogs), most amphibians have a tongue, which is important in food capture. In bufonids and ranids, the tongue is fastened at the front of the mouth and is free posteriorly (Figure 30.8b). When an insect moves into range, the tongue flips over and through the open mouth (i.e., the posterodorsal surface of the folded tongue becomes the anteroventral surface of the extended tongue), capturing the insect on the mucus-coated posterior tip. If you have ever had a chance to watch *Rana* or *Bufo* feeding, you know that the tongue action is so rapid that you cannot follow it with your eye. Equally spectacular is the tongue-feeding mechanism of some plethodontid (lungless) salamanders. Contraction of pro-

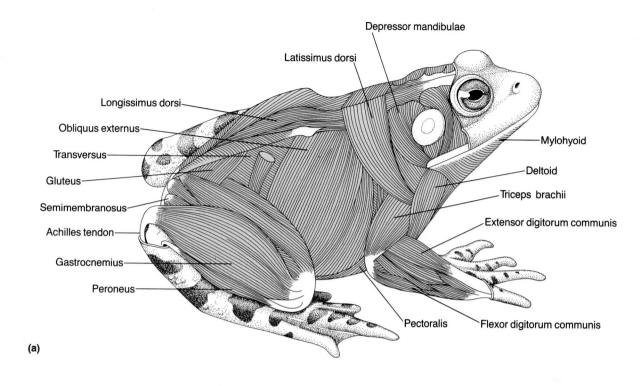

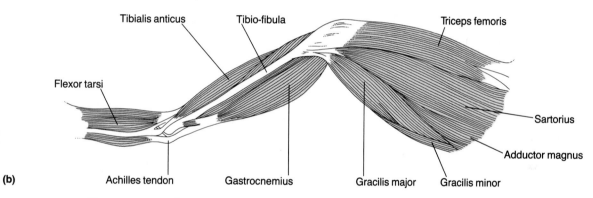

Figure 30.6 Musculature of a frog. (a) Muscles of the body. Note the layered sheets of abdominal muscles that support the internal organs in land vertebrates. (b) Musculature of the ventral surface of the left hindlimb.

tractor muscles extends the adhesive-tipped tongue distances equal to 44%–80% of the the body's length. Retractor muscles then return the tongue with its adhering prey to the mouth. In the tree-dwelling plethodontid *Bolitoglossa occidentalis*, the entire tongue sequence takes less than 50 ms, rapid enough to make insects easy targets.

Gas Exchange and Circulation

As discussed in Chapter 28, one of the more profound events accompanying the invasion of land by early amphibians was the evolution of a gas exchange system capable of extracting oxygen from air rather than from water. Corresponding evolutionary changes in the circulatory system are also outlined in Chapter 28. (The respiratory system is discussed on p. 695, the circulatory system on p. 696.)

Modern amphibians exhibit more variety in gas exchange mechanisms than any other vertebrate group. The two types of gills characteristic of many larval amphibians are illustrated in Figure 30.9a. Loss of the gills usually accompanies sexual maturity, but gills are retained in neotenic salamanders. Cutaneous (body surface and oral) gas exchange occurs in adults of most species. Oxygen diffuses into blood capillaries in the dermis, while carbon

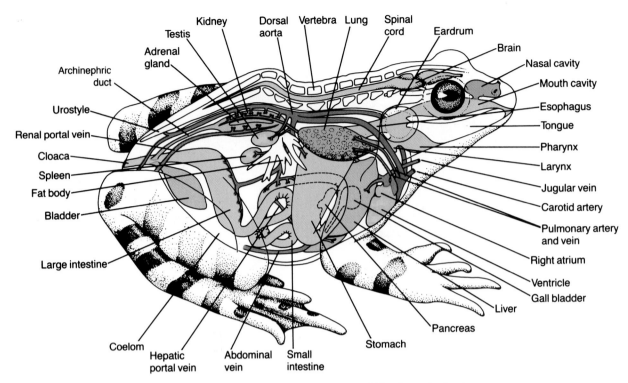

Figure 30.7 Major internal organs and blood vessels of a frog.

dioxide passes in the opposite direction. Relying exclusively on cutaneous respiration, plethodontid salamanders and a few other species of urodeles are the only vertebrates that lack both lungs and gills. Amphibian lungs are thin, vascularized sacs (see Figures 30.7 and 30.9b). As discussed in Chapter 5, however, the lung ventilation mechanism and overall system are rather ineffective compared to those of other land vertebrates (see Figure 5.11).

Excretion and Osmoregulation

Amphibians have a **urogenital system**, which carries both excretory and reproductive products. The kidneys of adult amphibians are of the opisthonephros type, usually short and compact in anurans (see Figure 30.7) and elongate in urodeles. (See the discussion of kidney evolution in Chapter 28, p. 698.) In the males of many species, both urine and sperm pass down the archinephric duct, but in females, genital and excretory products do not share ducts. The urinary bladder of amphibians is an evagination of the **cloaca**. Consequently, urine leaving the kidney tubules first enters the cloaca and then passes into the urinary bladder for storage. (In other tetrapods, urine passes from the ureters directly into the urinary bladder.)

As a group, amphibians exhibit a full range of excretory products. Larvae and most aquatic adults excrete mainly ammonia, but most terrestrial amphibians excrete urea, which can be stored temporarily in the urinary bladder. The clawed frog *Xenopus laevis* can switch from excreting ammonia when in water to excreting urea when deprived of water. A few tree frogs are well adapted for living away from water, for they can excrete virtually dry uric acid.

Aquatic amphibians have the same water control problems as other freshwater animals. (For a review of salt and water homeostasis, see Chapter 7, p. 145.) Water tends to enter an amphibian's body osmotically, primarily through its skin. Amphibians offset this diluting tendency in several ways. First, they eliminate excess water by producing dilute urine. Second, their kidney tubules return ions (particularly bicarbonate, chloride, and sodium) to the blood from the urine. And third, studies with pieces of frog skin have shown that it actively transports ions from the habitat water into the animal, helping to maintain the ionic concentration of the body fluids.

Bufonid toads are able to live in drier habitats than most other anurans can tolerate, in part because of an interesting osmoregulatory adaptation Many bufonids avoid desiccation by spending

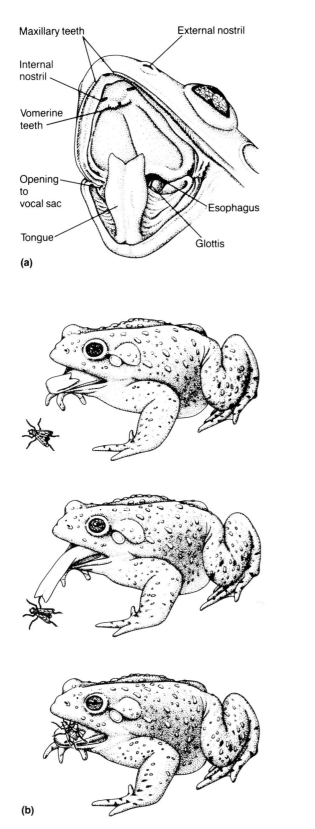

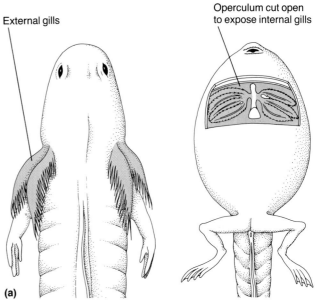

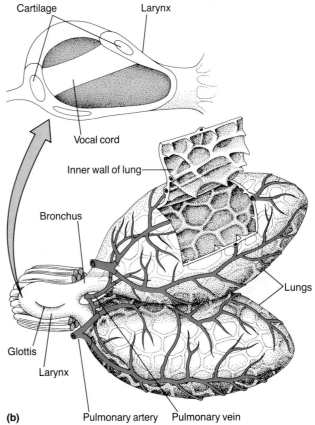

Figure 30.8 Amphibian feeding structures. (a) The mouth cavity of a frog includes maxillary and vomerine teeth and a tongue that is attached only anteriorly. (b) An anuran catching prey. Note that the tongue flips over as it extends from the mouth. The tongue also projects in many salamanders, but does not flip over.

Figure 30.9 Amphibian gas exchange structures. (a) Amphibian gills may be external, as in the neotenic tiger salamander shown here in dorsal view (left), or internal, as in the frog larva shown ventrally (right). (b) The lungs of an adult frog are vascularized and subdivided internally by delicate partitions that increase the surface area for gas exchange. Air expelled from the lungs passes through the larynx, causing the vocal cords to vibrate and giving the frog its voice.

much of the year in burrows, where they absorb soil moisture through the skin. Osmoregulation is accomplished by varying the amount of urea in the body fluids. Toads in soil having relatively little available water produce much urea and retain some of it in their blood and body tissues; increased levels of urea in the plasma raises their blood osmotic pressure enough to permit an influx of water. Conversely, toads in soil with greater water availability produce less urea.

Nervous and Chemical Coordination

The amphibian brain is fundamentally like a fish's, with only modest integrative capacity compared to the brains of birds and mammals. Most bodily activities in amphibians are directed by the medulla oblongata. (See Chapter 10, p. 237, for general brain anatomy and Figure 28.17b for frog brain anatomy.) Destruction of the medulla oblongata leads to death, but a frog with only its medulla oblongata can still catch food, swallow, hop, and swim. It may continue to breathe normally and may right itself if placed on its back. A small cerebellum, lying above the anterior end of the medulla oblongata, functions chiefly in controlling equilibrium, rather than being mainly a center of motor coordination, as in other tetrapods. Anterior to the medulla oblongata and cerebellum, a pair of relatively large, rounded optic lobes integrate visual and other sensory information and inhibit spinal cord reflexes. Each optic lobe affects the opposite side of the body. Making up the rest of the amphibian brain, the diencephalon is involved primarily in balance and vision, while the telencephalon is largely concerned with olfaction. The cerebral hemispheres, which dominate the brain of other tetrapods, appear to do very little; surgical removal of these structures makes a frog more lethargic and somewhat robotlike, but still capable of controlled movement. By contrast, as in fishes, the olfactory lobes of the telencephalon are well developed, reflecting the importance of odor detection in the moist habitats of most amphibians.

An amphibian's endocrine system is basically similar to that of other tetrapods. Releasing factors and inhibiting factors from the hypothalamus trigger release of several hormones from the pituitary gland (see Figure 11.3 and Tables 11.3 and 11.4). Pituitary hormones, in turn, stimulate other glands in the endocrine system to release their hormones. The initial triggering of the hypothalamus is generally caused by neural signals arising from shifts in hormone balance and from environmental stimuli such as day length.

In anurans, the anterior lobe of the pituitary gland secretes hormones important in growth, reproduction, and thyroid gland activity. Surgical removal of the anterior lobe from young frog tadpoles slows growth and prevents metamorphosis. In adults, the anterior lobe releases gonadotropins that stimulate egg maturation and egg laying in females and sexual maturity and sperm discharge in males. The anterior lobe also secretes melanophore-stimulating hormone (described earlier, p. 733). The posterior lobe of the pituitary stores and releases **vasotocin**, a neurohormone received from the hypothalamus. Vasotocin is important in osmoregulation. In bufonid toads, it helps conserve water by promoting uptake of water by the skin and resorption of water from the urine. (Its counterpart in mammals is a chemically similar compound called vasopressin.)

The remaining endocrine organs produce effects in amphibians similar to those in mammals and other tetrapods. The thyroid gland is of special interest because its activity is responsible for the ultimate amphibian terrestrial adaptation—the metamorphic transformation of a larva into an adult. Approximately three weeks before the emergence of front legs on a frog tadpole, the hypothalamus releases thyroid-releasing hormone (TRH). This substance triggers release of greater than usual amounts of thyroid-stimulating hormone (TSH) from the anterior pituitary gland. In response, the thyroid gland sharply increases its output of thyroxine and triiodothyronine, the hormones directly responsible for metamorphosis (Figure 30.10).

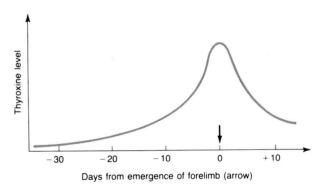

Figure 30.10 Relationship between thyroxine concentration and frog metamorphosis. Release of thyroid-releasing hormone (TRH) by the hypothalamus marks the end of premetamorphic development. TRH causes the anterior pituitary gland to release thyroid-stimulating hormone (TSH), which in turn causes release of thyroxine and triiodothyronine by the thyroid gland. The concentration of these hormones in the blood is greatest at the time that the forelimbs appear (arrow).

Sensory Structures

Except for the virtually blind caecilians, vision is an important sense in most amphibians. Evolutionary adaptation of the eye for sight in air required relatively minor modifications, including the development of eyelids and the development of tear glands, which keep the eye's surface clean and moist. Another modification was a lens that comes to rest focused on distant objects whenever the eye is not being used to view nearby objects (just the opposite of a fish eye); this is a useful feature when food may be a distantly approaching insect. The

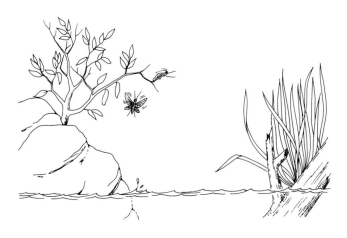

Figure 30.11 The role of retinal ganglion cells in form perception by frogs. Four kinds of ganglion cells respond to different features of a frog's visual field. (a) Edge detectors respond to sharp edges produced by stationary objects such as boulders, tree trunks, and the like. (b) Event detectors respond to moving boundaries of objects that are lighter or darker than their backgrounds, such as swaying grass and tree branches. (c) Dimming detectors respond to decreases in illumination caused, for example, by shadows. (d) Bug detectors respond to small moving objects, such as insects.

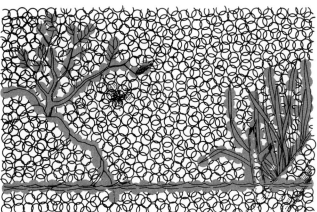

(a) Edge detectors

(b) Event detectors

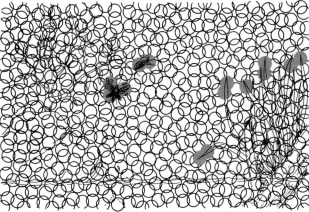

(c) Dimming detectors

(d) "Bug" detectors

amphibian eye accommodates as a fish eye does, by moving the entire lens forward toward the cornea for seeing nearby objects and away from the cornea for seeing distant objects. Amphibians also adapt to changing light conditions by dilating the pupil in dim light and constricting it in bright light. Furthermore, the presence of both cones and rods in the amphibian retina provides a basis for spectral sensitivity, and there is experimental evidence of limited ability to distinguish different wavelengths. However, there is little indication that amphibians actually perceive the color differences found in many species of anurans or the breeding colors of certain salamanders.

Many frogs and salamanders are adept at catching insects with their tongue and are usually quick to detect and evade approaching predators. The basis for such behavior is a well-developed ability to detect patterns in the visual field. Several kinds of cells in the retina provide pattern discrimination (Figure 30.11). By placing ultrathin recording electrodes within neurons in the optic nerve, investigators have found that certain ganglion cells in the retina respond to sharp edges in the visual field, such as boundaries between areas of different brightness. These cells give the frog an "outline drawing" of its surroundings and have been called **edge detectors**. Another group of cells, named **event detectors**, respond to moving edges of different brightness. In nature, these might be moving branches outlined against the sky or water. A third group of cells fires if there is a sudden dimming of light, as when shadows enter the visual field. Called **dimming detectors**, they may warn of an approaching predator, betrayed by its shadow. Lastly, some cells respond only to very small moving objects, such as flying insects or the tips of swaying cattails. Appropriately, these cells have been called **bug detectors**. Information from these four kinds of retinal cells allows a frog to construct a crude but useful picture of its surroundings (see Figure 30.11). About 90% of the visual information is processed in the retina, while the remaining 10% passes to a reflex center in the brain's optic lobes, where it is acted on very quickly. The result is a very rapid response by the frog, a matter of great importance if a fly is to be caught or a predator evaded. This research helps us understand a frog's abilities and reactions in terms of the limited information sent to the brain by the eyes. It also shows that a complex retina may compensate in some ways for a rather simple brain.

Auditory structures vary greatly among amphibians. Anurans vocalize, and their ears have been studied extensively. The anuran ear is similar to that of other tetrapods, although amphibians lack an outer ear. The most important adaptation for hearing on land is the tympanic membrane, a structure that vibrates in synchrony with arriving airborne sound waves and which, in amphibians, reptiles, and birds, transmits the vibrations to the **columella** in the middle ear (see Figure 12.12a). A frog's ears are most sensitive to relatively low sounds. For example, a female bullfrog's greatest response is to low-pitched notes of about 200 Hz, a frequency corresponding closely to the call of the adult male.

Amphibians have well-developed senses of touch, taste, and smell. Tactile receptors are plentiful under the epidermis, and taste buds occur on the tongue, roof of the mouth, and in the mucosa lining the jaws, but are not scattered on the body as in fishes. Olfactory epithelium in the nasal cavity detects airborne chemicals. Amphibians, unlike fishes, have internal nostrils that open into the nasal cavity (see Figure 30.7), permitting air movement while the mouth is closed or submerged.

BEHAVIOR

Amphibians display diverse behavioral activities, most of which are taxes (orientation behavior) coupled with complex motor programs. (Learning occurs, but is not a significant component of most amphibian behavior.) Amphibians exhibit taxes to a broad range of external stimuli. Many frogs show both positive and negative phototaxis. They may turn to face a bright area or light source, aligning their body with the light rays. This may help them to see small, dark moving objects (such as insects) against a bright background. However, when directly exposed to strong light, frogs generally move under vegetation or into other shady, moist places. Orientation to the sun (and possibly the moon and stars) appears to guide migrating anurans and some salamanders to their breeding sites. Many species congregate under artificial light at night. Besides exhibiting phototaxis, many amphibians also respond to body contact with surfaces (thigmotaxis), increasing contact by squeezing into crevices and under stones, a behavior that no doubt protects against predators and desiccation.

Amphibians generally exhibit their most complex behavioral activities during, or in preparation for, the breeding season. Mate finding and male-female interactions often involve complex social behavior, and the necessity to lay eggs in water or in moist sites requires that many species migrate to and from breeding sites. Species with aquatic adults may simply aggregate in one area of a pond or stream, but many amphibians must travel considerable distances overland to reach suitable breeding sites. Many zoologists have been intrigued by the factors that direct amphibians in their short- or long-distance migrations. Research with anurans indicates that vocalization by males at breeding ponds guides females there. The ears of a female bullfrog are "tuned" to listen at 200, 500, and 1400 Hz. Because air absorbs the high- and low-frequency calls of a male bullfrog differentially, a female can judge his distance as she moves toward him. Calls at 500 Hz stop the female; these are made by immature males.

A male's call is not the only guiding factor in migrations. Experiments with Fowler's toad *(Bufo woodhousii fowleri)*, found near the Great Lakes and in the Mississippi Valley, indicate that celestial orientation is important. Newly metamorphosed toads were collected and transported in lightproof bags to pens about 70 m from their home shore. The pens were built so that the toads could see only the sunny sky and the pen's walls. As soon as they were released and could see the sun, the toads turned and pointed directly toward their home shore. Results were identical in pens 145 km from the home shore. But when the pens were covered to obscure the sky, no orientation took place. Celestial orientation has also been demonstrated in other anurans and some salamanders. Cricket frogs *(Acris)* even appear to orient by the full moon and stars. Waterborne chemicals may also guide salamanders along streams to their breeding sites.

Many amphibians display territorial behavior during the breeding season. In research with the green frog *(Rana clamitans)*, a toe-marking system was used that allowed identification of more than 6300 individuals. When marked males arrived at their breeding site, they spread out and remained spaced 2–3 m apart during the breeding season, which lasted about two months. Even when a group moved more than 100 m to another pond during the season, individuals retained their positions relative to one another. Researchers postulate that this territorial spacing behavior may conserve energy and simplify detection of females visiting the pond. Territorial behavior also includes defensive displays. Vocalization by some frogs is threatening, and certain salamanders may nudge or bite at the snout or tail of intruders.

REPRODUCTION AND DEVELOPMENT

The emphasis in this section is on adaptations that illustrate the importance of habitat water for reproduction and development in most amphibians.

Sexual Maturity, Neoteny, and Reproductive Periodicity

Amphibians vary greatly in the length of time needed to attain sexual maturity. Variation may exist even within a species, owing in part to the effect of different environmental temperatures on developmental rate. Some amphibians reach sexual maturity and breed within one year, but many, especially larger forms, require two or more years to mature. The giant salamander *Cryptobranchus* takes five to six years to become sexually active (see Figure 30.3d).

Neoteny, which is the prolonged retention of larval body features to the point where reproduction occurs in larvalike individuals, is a common phenomenon among amphibians. Neoteny is different from **paedogenesis**, which is early (precocious) development of reproductive capacity (also in the larval state). Some salamanders are facultatively neotenic (i.e., metamorphosis may or may not take place), whereas others are obligate neotenes (i.e., they never undergo metamorphosis). The tiger salamander, *Ambystoma tigrinum*, often found in shallow temporary ponds, is facultative (see Figure 30.3c). In some populations, individuals ordinarily metamorphose, whereas members of other populations are neotenic and do not metamorphose so long as their aquatic environment remains habitable. Deteriorating environmental conditions, such as a drying pond or lack of food, may then trigger metamorphosis. By contrast, *Ambystoma mexicanum*, a lake dweller on the Mexican plateau, normally does not metamorphose in nature. Its neotenic larvae, also known as Mexican **axolotls**, are aquatic, gill-breathing forms capable of reproduction (Figure 30.12). Axolotls have been a favorite research animal of embryologists and endocrinologists ever since their discovery; yet, for many years, the adult stage of this species was unknown. In fact, no naturally occurring adult forms of *A. mexicanum* have ever been seen, but metamorphosis can be induced artificially. Axolotls may transform into adults if stressed in shipping and under certain laboratory conditions. (They will complete their metamorphosis following experimental treatment with thyroxine, the hormone that promotes

Figure 30.12 Neoteny in the Mexican axolotyl. *Ambystoma mexicanum,* also known as the Mexican axolotyl, normally retains its gills, a larval feature, after it has attained sexual maturity. Colonies of this species are reared and maintained for research purposes. Specific genetic strains of axolotyls are used extensively in modern developmental genetics laboratories throughout the world.

metamorphosis in other amphibians.) Research indicates that neoteny is maintained because the hypothalamus of the brain fails to produce the releasing factor that causes the pituitary to stimulate the thyroid gland. This hypothalamic "shutdown" may be triggered by environmental cues.

Neoteny is a developmental adaptation with considerable survival value. Prolonging the larval period allows a neotenic species to exploit its aquatic environment for as long as food and other favorable conditions last. Moreover, neotenic individuals may attain greater body size before finally transforming into adults. But even more important, they avoid exposure to potentially harsh or hazardous terrestrial conditions. It is likely that neoteny in salamanders evolved as a response to the hazards of life on land.

Reproductive activity in most vertebrates follows an annual cycle, and amphibians are no exception. Reproductive behavior and ovulation generally take place each spring, although precise timing varies with locality and species, and some salamanders develop eggs so slowly that ovulation is on a two-year cycle. The ovarian cycle may also be very prolonged, particularly in species that hibernate. Ovarian follicle growth often begins weeks before the onset of hibernation, allowing ovulation immediately after emergence from hibernation. Experiments with male salamanders indicate that increased testicular activity in the spring is a response to rising temperature and increasing day length.

Courtship and Mating

Amphibian courtship is usually a brief affair. For example, a male of the plethodontid genus *En-*

satina noses the neck and head of a female, then moves his body under the female until her throat rests over his hindlimb region (Figure 30.13). While rubbing the female's throat with his body, the male deposits a spermatophore on the ground. Moving forward, the female picks it up and inserts it into her cloaca, after which the pair separate.

As discussed earlier, male anurans use calls to attract females to breeding sites. In some species, noncalling "satellite" males lurk near the callers, intercepting approaching females and thus saving

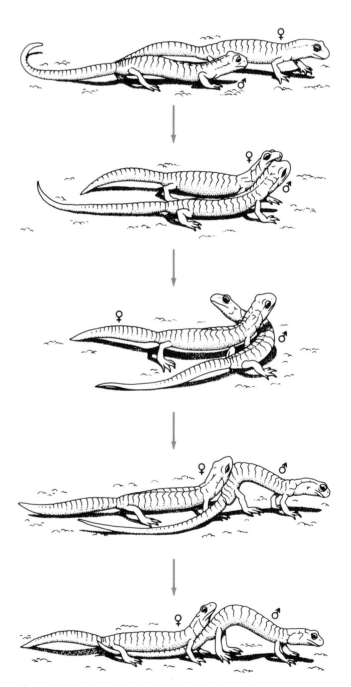

Figure 30.13 Courtship of plethodontid salamanders.

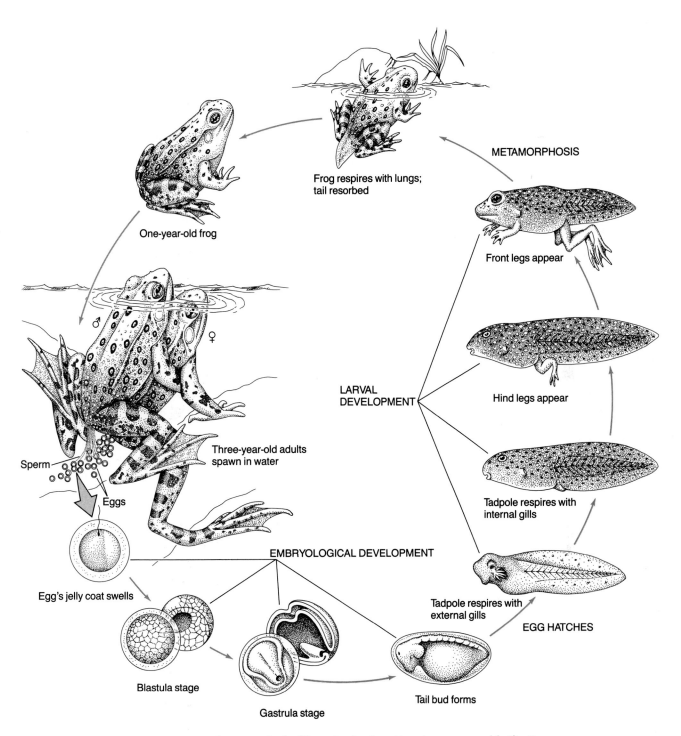

Figure 30.14 Major events in the life cycle of a frog. Note the sequence of fertilization, embryogeny, hatching, metamorphosis, and maturity.

the energy required for calling. After contact has been made, the male of most species grips the female behind the forelimbs and holds on tightly, applying pressure that helps the female expel her eggs; this type of sexual embrace is called **amplexus** (Figure 30.14).

Fertilization and Development

Amphibians have more diverse modes of fertilization than any other group of tetrapods. Nearly all anurans are oviparous, laying eggs and fertilizing them externally. During amplexus, a male frog deposits sperm on the eggs as the eggs leave the female's cloaca (see Figure 30.14). The gelatinous coat covering the eggs is protective and often has a sticky surface, making the eggs adhere to one another in masses or beadlike strings; egg masses often float freely in the water or are fastened to submerged plants, stones, or other objects.

In contrast to anurans, caecilians and most salamanders have internal fertilization. Male caecilians have a copulatory organ with which they deposit sperm in the cloaca of the female. About 75% of caecilian species are viviparous, giving birth to young. Salamanders lack copulatory organs, but most species have internal fertilization without copulation. Male salamanders deposit spermatophores in the vicinity of females, and females pick the spermatophores up through their cloacal opening.

Copulation and spermatophore transfer provide many advantages for terrestrial life: They allow fertilization to occur when the female is on land, and following fertilization, a female can brood her eggs internally or externally in a nest; she may also dispense fertilized eggs or larvae into a variety of habitats.

The number of offspring produced each generation is highly variable, and depends on the species. The Great Plains toad, *Bufo woodhousii*, lays up to 25,000 eggs, whereas the tiny Cuban frog, *Sminthillus limbatus*, deposits only one. How long the eggs must incubate before hatching also depends on the species, ranging anywhere from one to four days up to nine months for various salamanders and frogs. The eggs of spadefoot toads (*Scaphiopus*) hatch in about 36 hours, while those of the black toad (*Atelopus*) of Uruguay hatch in 24 hours. Brief incubation times allow these toads to take advantage of short-lived pools and puddles in their relatively dry habitats.

Many species have remarkable adaptations for protecting their eggs or young. The majority of salamanders and some caecilians protect their eggs simply by remaining with them until they hatch, with some salamanders attending the hatchlings as well. Males and females of the Asian tree frog *Rhacophorus* stir their egg masses with their hind feet, turning the masses into foamy nests that admit oxygen but resist water loss. In anurans, parental care may include transportation of eggs and young (Figure 30.15). The Surinam toad deposits her eggs on her own back. After the male has fertilized the eggs, a thin layer of skin grows over them, forming a fluid-filled compartment for each egg. In several species of frogs, the male incubates the eggs in his vocal pouch, but the brooding technique of

Figure 30.15 Parental care in a frog. The male of this species, *Dendrobates silverstonei,* an arrow-poison frog of South America, transports tadpoles on his back.

Australian frogs of the genus *Rheobatrachus* is truly extraordinary. The female does not feed during brooding and her digestive system becomes inactive. After fertilization, she swallows her eggs, and embryonic development occurs in her stomach. Metamorphosis also occurs in the stomach, and fully formed froglings emerge from their unique incubator in about 37 days.

Just before most amphibians hatch, glands on the larva's snout release a chemical that digests the egg capsule. Internal pressure exerted by the larva is also probably an important factor in hatching. Duration of the larval period depends on many factors, including temperature and species. Bullfrog larvae (*Rana catesbiana*) metamorphose after three summers in Nova Scotia, but after only two summers in New York and after only one summer in the heat of Louisiana. Toads that live in arid environments have the shortest larval periods, an adaptation that compensates for the brief existence of their breeding ponds. Some toads require as little as twelve days to become thick-skinned adults adapted for desert life.

PHYLOGENY

About 400 to 425 million years ago, important events taking place in fresh water and on land led to the appearance of the amphibians and later to the origin and almost explosive radiation of the reptiles.

Amphibian origins have been traced to the Devonian period (see Figure 30.18). Layers of red sandstone containing the fossilized remains of animals thought to be the forerunners of the amphibians were deposited some 350 to 400 million years ago. During that time, climates on land became increasingly dry, and it is likely that rivers and lakes were periodically reduced to stagnant pools and broad mud flats as a result of drought. Also during the Devonian, sea levels became lower, and pools of water along rocky, humid shorelines were probably inhabited by fishes. Under these rugged conditions, populations of fishes no doubt alternately thrived and shrank, and adaptations for surviving longer in drying ponds or shoreline rock pools and for crawling to more favorable environments would have had survival value.

Of the many kinds of fishes that swam in the varied aquatic environments of the Devonian, some species had lungs, and some were able to breathe air, a decided advantage for animals living in waters subject to periodic drying. The fins of one group of air-breathing Devonian fishes, the **crossopterygians**, or lobefins, were supported by a strong framework of bone and muscle. The leaflike fin of a modern lungfish (a noncrossopterygian air breather), the broader, lobelike fin of a crossopterygian, and the five-digited limb of an amphibian are compared in Figure 30.16. Although clearly not as effective for land travel as amphibian limbs, the crossopterygian fins could have provided the strength and leverage needed for a fish to struggle from a drying pond in search of another pond. Presumably, fishes with the strongest lobelike fins had the best chance to encounter ponds and plunge into them. Likewise, being able to breathe air would have allowed such fishes to tolerate oxygen-poor conditions in shallow, stagnant pools and to survive short overland excursions. Fishes lacking such adaptations would have been eliminated by natural selection if subjected to drought and dissolved oxgyen depletion. Thus, either by gradual adaptation or by a relatively rapid series of evolutionary changes, air-breathing, lobefin fishes are believed to have become better and better adapted for life on land. Ultimately, as certain crossopterygians evolved into the first amphibians, their fins evolved into the pentadactyl (five-digited) land limbs. (See the comparison of gradualism and punctuationism in Chapter 17, p. 413.)

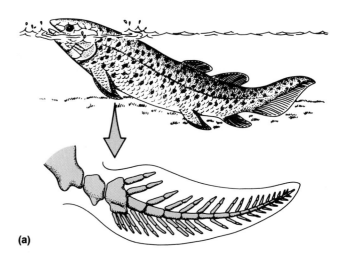

(a)

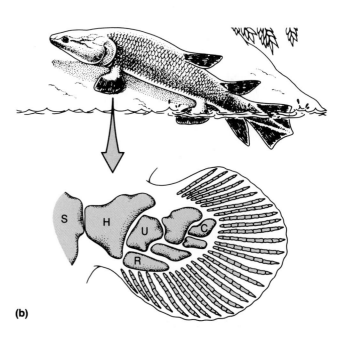

(b)

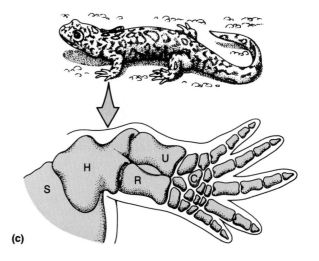

(c)

Figure 30.16 (Right) Evolution of tetrapod limbs. (a) The long, highly flexible fin of a lungfish. (b) The stout, firmly supported lobelike fin of an air-breathing crossopterygian (lobefin). (c) An amphibian (salamander) limb. The bones in the crossopterygian's fin and the amphibian limb are homologous, indicating an evolutionary relationship. (S = scapula, H = humerus, U = ulna, R = radius, and C = carpal.) Zoologists consider it likely that the pentadactyl tetrapod limb is a derivative of the lobelike crossopterygian fin.

The early amphibians were probably primarily swimmers preying on fishes. Except for worms, insects, and other amphibians, few land animals were available to serve as food for these newly evolved carnivores. Locomotion in the earliest amphibians undoubtedly resulted mainly from body undulations, as in fishes. Crossopterygians probably used undulatory movements of the body to scramble from one pond to another, and their lobed fins would have helped by pushing against the ground. *Ichthyostega*, an early amphibian that lived some 345 million years ago in what is now Greenland, had a fish-shaped body about 1.2 m in length, a fin on its tail, and sprawling legs (Figure 30.17). With its legs held laterally, *Ichthyostega* was mainly an undulator. As discussed earlier, modern salamanders have retained this primitive type of tetrapod locomotion, but nonundulatory, limb-generated locomotion evolved in the reptilian lineages, all of which were derived from amphibians, and in birds and mammals.

At their peak, about 300 million years ago, amphibians included species that were well over 2 m in length. Two major subclasses of extinct amphibians are recognized (Figure 30.18). *Ichthyostega* is classified in the **Subclass Labyrinthodontia**, the earliest amphibians. Labyrinthodonts, named for the mazelike (labyrinthine) pulp cavity of their teeth, had well-developed legs and lacked median fins, but otherwise shared many features with the crossopterygian fishes from which they were derived. The vertebral structure of labyrinthodonts resembled that of other tetrapods (the central portion of the vertebrae consisted of bony arches encasing the notochord) and is considered to represent the primitive condition from which other tetrapod vertebrae were derived. By contrast, members of a second extinct subclass, the **Lepospondyli**, were generally smaller than labyrinthodonts and had vertebrae with very different, spool-shaped central portions. Lepospondyls, probably derived from labyrinthodonts, did not last beyond the Carboniferous period, whereas the labyrinthodonts survived as a group for another 100 million years, and one (some zoologists think more than one) labyrinthodont lineage survives today as our modern amphibians. The Carboniferous period is sometimes called the Age of Amphibians because amphibians were the most diverse land vertebrates at the time. Decline of the amphibians, which oc-

Figure 30.17 Ichthyostega. The fossil record indicates that *Ichthyostega* was an early amphibian that lived almost 350 million years ago in a part of an ancient land mass that ultimately formed Greenland.

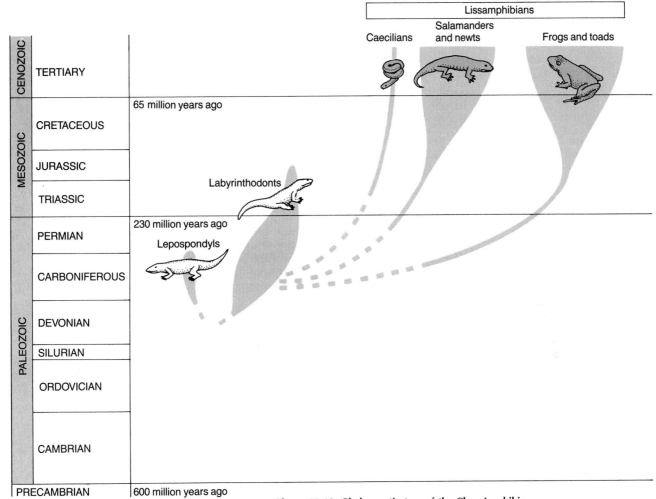

Lissamphibians

| Caecilians | Salamanders and newts | Frogs and toads |

CENOZOIC	TERTIARY	
	CRETACEOUS	65 million years ago
MESOZOIC	JURASSIC	
	TRIASSIC	Labyrinthodonts
	PERMIAN	230 million years ago
	CARBONIFEROUS	Lepospondyls
PALEOZOIC	DEVONIAN	
	SILURIAN	
	ORDOVICIAN	
	CAMBRIAN	
	PRECAMBRIAN	600 million years ago

Figure 30.18 Phylogenetic tree of the Class Amphibia.

curred throughout the ensuing Permian and Triassic periods, may have resulted in part from competition with the emerging reptiles.

Because the amphibian fossil record in the Mesozoic era is incomplete, many aspects of amphibian phylogeny and classification are controversial. The three living amphibian orders are classified in the **Subclass Lissamphibia**, but many zoologists question whether this taxon is monophyletic or polyphyletic. (The terms *monophyletic* and *polyphyletic* are defined in Chapter 18, p. 432; Figure 18.5.) There is general agreement, however, that all living species in the three orders are highly derived forms, only distantly related to ancestral amphibians. Even salamanders, which seem to exhibit the ancestral body form, are not closely related to ancestral amphibians—probably no more so than are frogs or caecilians. It is also clear that no modern amphibian represents the lineage of amphibians that gave rise to the reptiles.

☐ S U M M A R Y

1. Key Features of Amphibians. Amphibians (anurans: frogs, including toads; urodeles: salamanders; and caecilians) are ectothermic tetrapod vertebrates with a bony skeleton, an incompletely ossified skull (except in ancestral forms), and typically moist skin, with epidermal mucous and poison glands. Living forms generally lack scales. Gas exchange occurs across various combinations of the skin, gills, lungs, or mouth lining; some salamanders lack lungs and gills. Development includes a larval stage, either aquatic or within the egg membranes; larvae typically metamorphose to the adult form, which is either aquatic or terrestrial; some salamanders are neotenic. Adult amphibians and larval salamanders are carnivorous; larval anurans are herbivorous.

2. The chief nitrogenous excretory product in many adult amphibians is urea; aquatic adults and larvae excrete mainly ammonia. Behavioral adaptations are also important in maintaining body water content.

3. Vision and hearing are well developed in many species. Amphibian behavior includes taxes, complex motor programs, and simple learning.

4. Reproduction generally involves migration to breeding sites, territorial and courtship behavior by males, and external fertilization (in frogs and toads) or internal fertilization (with spermatophores in salamanders). Most amphibians lay eggs in water or in moist areas.

5. Amphibians evolved some 350 to 400 million years ago, probably from lobe-finned ancestors (crossopterygian fishes). The first amphibians were members of the extinct Subclass Labyrinthodontia. The origins and relationships of the three living orders are uncertain.

❑ FURTHER READING

Attenborough, David. *Life on Earth: A Natural History.* Boston: Little, Brown, 1979. *Includes an excellent nontechnical account of the "invasion of land" by amphibians.*

Duellman, W. E., and L. Trueb. *Biology of Amphibians.* New York: McGraw-Hill, 1986. *Definitive, current, and superbly illustrated discussion.*

Hughes, C., and D. Hughes. "Teaming Life of a Rain Forest." *National Geographic* 163(1983): 49–65. *Beautifully illustrated account, including coverage of the arrow-poison frog.*

Little, C. *The Colonisation of Land.* New York: Cambridge University Press, 1983. *Thorough, quite technical discussion of amphibian and reptilian origins, emphasizing respiratory, locomotor, and reproductive adaptations.*

Webb, J. E., J. S. Wallwork, and J. H. Elgood. *Guide to Living Amphibians.* London: Macmillan, 1981. *A well-organized, brief introduction to the class.*

❑ CLASSIFICATION

Subclass Labyrinthodontia (Gr. *labyrinthos*, "tortuous passage"; *odontos*, "tooth"). Labyrinthodonts; three orders of extinct amphibians; derived from lobe-finned fishes and possibly including the ancestors of the reptiles; notochord encased by arched vertebral bones; teeth were labyrinthine; *Ichthyostega*.

Subclass Lepospondyli (Gr. *lepos*, "a scale"; *spondylos*, "vertebra"). Lepospondyls; four orders of extinct amphibians; centers of vertebrae were spool-shaped, each consisting of a single bone pierced by a canal through which the notochord passed; limbs of some were reduced or absent; *Diplocaulus*.

Subclass Lissamphibia (Gr. *lissos*, "smooth"; *amphibios*, "leading a double life"). Modern amphibians; one extinct and three living orders of amphibians; broad, flat skull with poorly ossified brain case; larval stages common; skin often important in gas exchange; scales absent except in some caecilians.

 Order Proanura (Gr. *pro*, "before"; *an*, "without"; *oura*, "tail"). Proanurans; extinct ancestral frogs possessing ribs and a short tail; legs not adapted for jumping.

 Order Anura (Gr. *an*, "without"; *oura*, "tail"). Frogs, including toads; legs and girdles adapted for jumping; tail reduced or absent; larvae unlike adults and without teeth; not neotenic; *Rana catesbeiana* (bullfrog), *Xenopus laevis* (African clawed frog), *Hyla regilla* (Pacific tree frog), *Bufo boreas* (Western toad), *Pipa pipa* (Surinam toad).

 Order Urodela (**Caudata**) (Gr. *oura*, "tail"; *delos*, "visible"). Salamanders; legs not adapted for jumping; long tail; larvae with teeth and similar in form to adults; often neotenic; *Ambystoma tigrinum* (tiger salamander), *Necturus maculosus* (mud puppy), *Cryptobranchus alleganiensis* (hellbender).

 Order Gymnophiona (**Apoda**) (Gr. *gymnos*, "naked"; *ophioneos*, "like a serpent"). Caecilians; limbless; tail very short or absent; elongate, wormlike body; eyes reduced or vestigial; *Ichthyophis glutinosus, Gymnopis multiplicatus.*

Reptiles

Class Reptilia. The sidewinder *Crotalus cerastes,* a rattlesnake of the Mojave and Colorado deserts, makes J-shaped tracks as it crawls over desert sand.

The approximately 7000 living species of the Class Reptilia are descendants of groups of reptiles that dominated the earth during the Mesozoic era, about 65 to 230 million years ago. Even today, reptiles are found in many kinds of habitats, from fresh water and salt water to the driest of deserts.

The chief factor limiting their distribution is temperature; reptiles thrive in warm climates, and their numbers dwindle rapidly toward the higher latitudes.

Compared to amphibians, a typical reptile is exceptionally well adapted for life on land. Some features of the reptilian body, such as lungs, tetrapod limbs, and a two-layered integument, evolved before the origin of reptiles, but became highly modified in the course of reptilian evolution on land. Other features, such as the water-resistant cleidoic (self-contained) egg and the highly efficient amniote kidney, originated in the reptilian lineage.

There are four orders of living reptiles. Some 6000 species of snakes and lizards form the largest order, followed by the turtles and tortoises, and by the crocodilians (crocodiles, alligators, and their allies); the tuatara of New Zealand is the sole member of the fourth order. There are also about a dozen orders of extinct reptiles, some of which gave rise to birds and mammals. We examine the living orders and some of the extinct groups in the following survey of representative reptiles.

REPRESENTATIVE REPTILES

Order Chelonia (Testudines)

Chelonians, or turtles, are unique among reptiles in that they have an armorlike shell consisting of a domed **carapace** and a ventral **plastron** joined on each side by a bridge of bone (Figure 31.1a). The ribs and thoracic vertebrae are fused with the inner surface of the carapace, making the shell a spacious skeletal unit of exceptional strength. Turtles have a long, flexible neck, which is extended for feeding. In most species, the neck is retractible, allowing the head to be pulled safely inside the shell. The neck fits into a space in the shell, either by folding vertically into an S shape or by swinging sideways. The turtle shell has great protective value, but it imposes severe constraints on its inhabitant. Loco-motor efficiency is significantly impaired, especially in land turtles. As shown in Figure 31.1b, the limb girdles of a turtle lie within the rib cage, and the limbs project laterally through anterior and posterior openings of the shell. The lateral position of the limbs requires a turtle to expend considerable energy in lifting itself and its heavy shell off the ground. Moreover, turtle limbs are short and provide only short strides.

The largest turtles are nearly 10,000 times heavier than the smallest ones. The giant tortoise (*Geochelone elephantopus*) of the Galapagos Islands (tortoises are herbivorous land turtles) may weigh nearly 200 kilograms and live well over 100 years (Figure 31.2a). Most turtles are silent, but males of *Geochelone* often grunt loudly when mating. The largest of all living turtles is the Atlantic leather-back, *Dermochelys coriacea coriacea*, possibly weighing as much as 900 kg. Its back is covered by

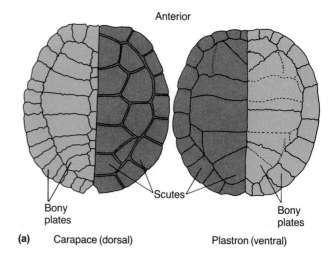

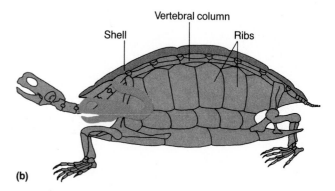

Figure 31.1 Shell of a turtle. (a) Dorsal view of the upper shell, or carapace, and ventral view of the plastron. Note that the shell has an inner layer of bony plates and an outer layer of keratinized epidermal scales (scutes) joined by strong seams. In this sketch, half of each epidermal layer has been omitted to show the underlying bone. As a turtle grows, it adds larger scales beneath the old ones. Consequently, growth rings appear as layer upon layer of scales are laid down. (b) Cutaway view of a turtle's skeleton showing the position of the limb girdles and the vertical folding of the neck, which allows it and the head to be withdrawn into the shell.

a leathery skin rather than a bony carapace. The Atlantic green turtle, *Chelonia mydas*, has flipper-like limbs and a reduced, streamlined shell and is a fast swimmer (Figure 31.2b).

Many turtles have modified shells. Snapping turtles (*Chelydra* and *Macroclemys*) have a reduced plastron that cannot fully accommodate its owner (Figure 31.2c). Instead, snappers rely for protection on their powerful jaws and aggressive nature. Snapping turtles feed largely on fishes and frogs, but they will eat almost any pond animal that their jaws can grip. Box turtles (*Terrapene*) display an-other shell specialization. They can hold their plastron tightly against their carapace, making the shell an almost impregnable closed box. This tight-fitting armor protects against predators, but it

(a)   **(b)**

(c)

Figure 31.2 Order Chelonia (turtles). (a) The giant land tortoise *Geochelone elephantopus* of the Galapagos Islands. (b) The Atlantic green turtle *Chelonia mydas* may reach lengths of more than 1.5 m and weights exceeding 300 kg. Here a female returns to the ocean after laying eggs on land; notice her trail above the surf line. All sea turtles lay eggs on land. Because of human hunting and destruction of eggs and reproductive habitat, *C. mydas* and several other species of sea turtles are in danger of becoming extinct. (c) The alligator snapping turtle *Macroclemys temminckii*, weighing up to 90 kg, is the largest freshwater turtle in North America. It is found in southern Georgia, west to Texas, and north in the Mississippi drainage to Illinois and Ohio. A voracious predator, *M. temminckii* has a tongue appendage that it uses to attract fishes into its mouth.

seems to be primarily an adaptation for terrestrial life. With the shell closed, box turtles, sometimes called dry-land turtles, are protected against water loss.

Order Rhynchocephalia

When scientists first learned of the tuatara (*Sphenodon punctatus*) in 1831, they classified it as a lizard (Figure 31.3). Nearly three decades passed before a study of its anatomy showed it to be the last of the **rhynchocephalians**, a group that otherwise became extinct about 100 million years ago. The tuatara (up to about 75 cm long) now lives on small islands off the coast of New Zealand, where it in-

habits cavities under tree roots. The New Zealand government protects the tuatara, and there is hope that its precariously small populations will survive and increase. *Sphenodon* seems to have undergone little evolutionary change during the past 200 million years.

Order Squamata

Lizards and snakes make up two suborders that constitute the largest reptilian order, the **Squamata**. The most obvious distinction between lizards and snakes is that snakes lack legs, apparently an adaptation for burrowing. Lizards and snakes also differ from each other in several other ways: The eyes of snakes are unblinking and permanently open behind a clear covering, whereas those of most lizards can be opened and closed; lizards typically have an external ear, but snakes do not; and the scales on the upper and lower surfaces of a lizard's body are quite similar, whereas the belly of

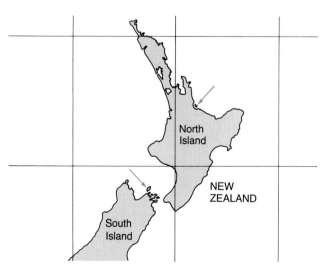

Figure 31.3 Order Rhynchocephalia. The tuatara (*Sphenodon punctatus*), the only living species in this order, and its world distribution on islands off New Zealand.

most snakes bears a single row of scales that are broader than those on the rest of the body.

Lizards. Lizards range in length from a few centimeters to several meters. The monitors, a family of terrestrial predators, are the largest of the modern lizards. The monitor *Varanus komodoensis*, often called the Komodo dragon, grows to over 3 m in length. Named for the East Indian island where it was first found, a hungry Komodo dragon uses sudden bursts of speed to run down and seize small mammals and other prey. It can swallow over 2 kg of meat in 1 minute.

At the other extreme, some members of the chameleon family of Africa and Madagascar (with a few species in Asia and southern Europe) may be only 4 cm in length. Chameleons are masters of camouflage, changing in color from gray to green or brown, and sometimes yellow, depending on the species (Figure 31.4a). Change in color may accompany changes in heat and light, daybreak, or alarm.

The Family Iguanidae is a diverse group of largely New World lizards. The common iguana *(Iguana iguana)* of Mexico and tropical America may grow to a length of 180 cm. The marine iguana *Amblyrhynchus cristatus* of the Galapagos Islands is unique among lizards in that it feeds in the ocean (see chapter opening photograph for Chapter 7, p. 145). The horned lizards *(Phrynosoma)* live in hot, arid regions in Mexico and the southwestern United States (Figure 31.4b). The "horns" that give these iguanids their peculiar appearance are actually much-enlarged scales. Another iguanid, the anole *(Anolis carolinensis)*, also called the American chameleon, changes color readily (deep brown to green) and is often kept as a pet.

Residents of the tropics throughout the world are familiar with the geckos, a fairly large family

(a)

(b)

(d)

(c)

Figure 31.4 Order Squamata (lizards). (a) A chameleon, *Chamaeleo senegalensis*, uses its projectile-like tongue to catch an insect. Except for a tiny peephole, the eyes (which move independently) are covered by eyelids fused over the eyeball. (b) The horned lizards (seven species of the genus *Phrynosoma*) are common in desert regions of the U.S. Southwest. This species, *Phrynosoma sulare* (regal horned lizard), found in Arizona, California, and the Sonoran Desert of Mexico, is about 10 cm long. (c) The blue-tailed skink *(Eumeces)* represents the large Family Scincidae. (d) The Florida worm lizard *(Rhineura floridana)* is legless.

(the Gekkonidae, 400 species) of harmless small lizards. Geckos often share people's houses, walking up walls and sometimes across ceilings by means of adhesive pads on the tips of their toes. Unique among lizards (and uncommon among reptiles as a group), some geckos have a voice. The males of certain species emit loud calls in territorial defense. Twelve species of geckos live in the United States; three of these are probably immigrants that entered the country through southern ports in cargoes of fruit, produce, or lumber.

The skinks (Family Scincidae) constitute another large group (about 700 species), but only 14 species live in the United States (Figure 31.4c). Skinks have shiny, smooth bodies covered with flat, overlapping scales. An alertness and an active nature make skinks difficult to catch and hold. When held by their long tails, they readily cast it off in a process called **autotomy**, an adaptation that may help them escape from predators. The tail may also wiggle after it is cast off, serving to distract a predator.

The least lizardlike of the lizards, in fact classified in a separate suborder by some zoologists, are the worm lizards (Family Amphisbaenidae). Cylindrical in shape and somewhat like an earthworm in appearance, they range in size from about 30 to 70 cm (Figure 31.4d). Worm lizards are adapted for subterranean life; they are blind and lack ear openings, hind legs, and usually front legs. The Florida worm lizard *(Rhineura floridana)* is sometimes uncovered by farmers cultivating their fields.

Snakes. About two-thirds of the 3000 living species of snakes belong to the Family Colubridae. Often called the common snakes, colubrids include the familiar garter snakes, kingsnakes, racers, and bull snakes.

The largest snakes in existence are the boas (Family Boidae), and the pythons (Family Pythonidae) (Figure 31.5). Boas and pythons kill by wrapping themselves around prey and squeezing until the victim suffocates. A large python can loop several coils around a goat in 2 or 3 seconds.

Cobras, coral snakes, and kraits are members of the Family Elapidae, distributed in tropical regions throughout the world. Elapids kill prey with highly toxic venom injected through immovable, erect fangs located near the front of the upper jaw. The brightly banded coral snakes (Figure 31.6a) are rather secretive burrowers in forest litter and leaf mold, but others, such as the African mamba, are fast-moving, easily provoked, and quite aggressive. Elapids range in size from tiny burrowers less than 8 cm long to the awesome, hooded king cobra of India, which reaches a length of 5.5 m (Figure 31.6b). Closely related to the elapids are the sea snakes (Family Hydrophiidae). Hydrophiids live in tropical seas, have a compressed tail used in swimming, and have special salt-excreting glands that

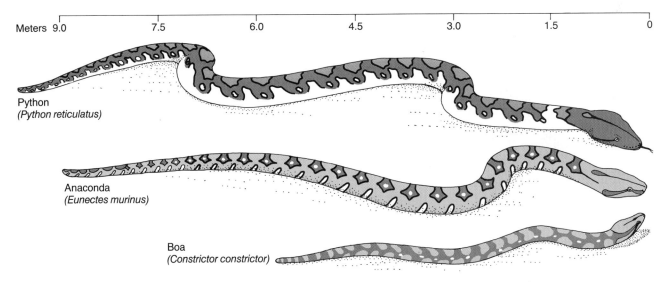

| Meters 9.0 | 7.5 | 6.0 | 4.5 | 3.0 | 1.5 | 0 |

Python
(Python reticulatus)

Anaconda
(Eunectes murinus)

Boa
(Constrictor constrictor)

Figure 31.5 Order Squamata (snakes). The largest snakes are constrictors of the Families Boidae and Pythonidae. The longest accurately measured snake was a 10 m Asian reticulate python *(Python reticulatus),* but there have been reports of a boid called the South American anaconda *(Eunectes murinus)* measuring 11.3 m long. A large python in captivity can readily swallow 100 pounds of meat and then go for more than a year before eating again.

(a)

(b)

(c)

Figure 31.6 Venomous snakes. (a) The eastern coral snake *Micrurus fulvius fulvius* ranges along the southeastern seaboard from North Carolina to Florida and west to Mississippi. The highly venomous coral snakes are widely mimicked by nonvenomous species. Coral snakes have red and yellow rings that touch, but harmless mimics have black bands separating the red and yellow rings. (b) The Indian cobra, *Naja naja,* with its hood expanded. (c) The western diamondback rattlesnake, *Crotalus atrox,* ready to strike. One of 15 species of rattlesnakes in the United States, *C. atrox* occurs from Texas to Southern California. Rattlesnakes are the largest venomous snakes in the United States; the western diamondback grows to lengths of 2.5 m.

remove excess salts from body fluids. (See the Chapter 7 Essay, "Living at Sea and Solving the Salt Problem," p. 167.)

Members of the Family Viperidae (the vipers) inhabit all continents except Australia. A viper is rather thick-bodied relative to its length, and it has a broad, wedge-shaped head that accommodates its venom glands and injection apparatus. Members of one subfamily, the **pit vipers**, have a sensory pit located between the eye and nostril. This sense organ is a heat detector that provides information about the presence and distance of homeothermic prey, such as rodents. Pit vipers *without* a rattle on the end of the tail include the bushmaster (*Lachesis*) and fer-de-lance (*Bothrops*) of tropical America, the swamp-dwelling water moccasin (*Agkistrodon piscivorus*) of the southeastern United States, the copperhead (*A. contortrix*), which lives on drier land from Massachusetts and Illinois south to the Gulf Coast, and a number of Asian species. Rattlesnakes (*Crotalus*) are pit vipers with a rattle on the end of the tail (Figure 31.6c).

Order Crocodilia

The crocodilians are the largest of the modern reptiles and the last reptilian survivors of the lineage that produced the dinosaurs. Full-grown crocodiles (*Crocodylus*) attain lengths of about 7 m; the biggest American alligator (*Alligator mississippiensis*) ever measured was about 5.8 m long (Figure 31.7a).

Crocodilians are well fitted for semiaquatic life. Although the mouth lacks lips and cannot be tightly closed, flaps at the back of the oral cavity prevent water from entering the windpipe and esophagus during dives. Likewise, ear flaps and nasal valves keep water from entering the ears and nostrils. The crocodilian armament includes a tough, leathery skin covered with keratinized scales reinforced with bone; long, powerful jaws with numerous teeth; and a muscular tail that is both a weapon and a paddle. Crocodilians are fast swimmers, but are also expert at cruising slowly on the surface, submerged except for their bulging eyes and nostrils, making scarcely a ripple as they stalk turtles or other prey. A large crocodile can sometimes drift inshore close enough to an unwary animal (or person) to sweep it into the water with a flick of its tail. Few reptiles have voices, but the crocodile's and alligator's earth-shaking bellow during the mating season is a memorable exception. In sheer volume, the roar of a crocodile ranks with that of a lion or elephant. Crocodiles and alligators prey on mammals, fishes, waterfowl, and other reptiles, including, at times, their own young. The gavial (*Gavialis gangeticus*) of India and Burma, primarily a fish eater, has a narrow snout that it uses like a sword to make slashing sideswipes at fishes.

Extinct Reptiles

One of the most striking aspects of reptilian history is the diversity that accompanied this group's exploitation of numerous terrestrial and aquatic niches following the decline of the amphibians at the end of the Paleozoic era. Reptiles so dominated

(a)

Alligator

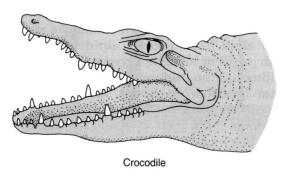

(b)   Crocodile

Figure 31.7 Order Crocodilia. (a) The American alligator, *Alligator mississippiensis,* devouring a raccoon. (b) Alligators and crocodiles are different in that the alligator's head is broader than that of a crocodile and the snout more rounded. An alligator's teeth are largely hidden when the mouth is closed, whereas the teeth of a crocodile are clearly visible, even when the mouth is shut.

the Mesozoic Era that it is often called the Age of Reptiles. More than 15 orders had been established by the middle of the Mesozoic era, but as the Mesozoic wore on, many of the great orders dwindled toward extinction.

Dominant Reptiles of the Late Paleozoic Era. Two large groups of extinct reptiles, the cotylosaurs and the pelycosaurs, were especially numerous during the Permian (final) period of the Paleozoic era. The **cotylosaurs** (literally, "stem lizards") appeared about 300 million years ago and reached their peak some 40 million years later. As some of the earliest reptiles, they were generalized in form and looked somewhat like modern-day lizards. *Seymouria* is one of the best-known cotylosaurs (Figure 31.8a). Many cotylosaurs were probably insect eaters, differing little in habits or superficial appearance from the amphibians of the period.

The **pelycosaurs**, or sail lizards, were contemporaries of the cotylosaurs. A peculiar feature of many pelycosaurs was a middorsal "fin," actually a sail-like web of skin held above the body by slender spines (Figure 31.8b). The fin may have served as a heat regulator. Turned broadside to the sun on cool mornings, perhaps the fin absorbed heat and helped warm these large ectotherms, whereas on hot afternoons, the fin might have helped cool the body by acting as a heat radiator.

Marine Reptiles of the Mesozoic Era. During their evolution, some reptilian groups secondarily returned to aquatic life. The turtles are a living example of such a group. Two much larger aquatic groups, the ichthyosaurs and the plesiosaurs, were numerous in Mesozoic seas, but became extinct toward the end of the Mesozoic era. **Ichthyosaurs**, fishlike lizards, ranged over the world's oceans during most of the Mesozoic era, feeding on fishes (Figure 31.8c). Their paddlelike limbs made it impossible for them to go ashore. One fossil of an ichthyosaur contains seven intact young individuals, indicating that the eggs of these lizards hatched within the females, with the young being born after developing to the adult form. **Plesiosaurs** were contemporaries of the ichthyosaurs. They were good swimmers, but were slower and less fishlike than ichthyosaurs. Plesiosaurs had a very long neck and large rowing paddles for limbs; they probably spent most of their time at or near the ocean surface (Figure 31.8d). Large species (up to 15 m or more in length) ranged the open oceans, while smaller forms invaded rivers and swampy estuaries, feeding on fishes and other reptiles.

Mammallike Reptiles. **Therapsids**, mammallike reptiles, were a dominant group of land vertebrates of the early Mesozoic, some 210 million years ago (Figure 31.8e). They were mammallike in several respects. Rather than the spread-eagled stance typical of many reptiles, their more or less vertically positioned limbs held the body off the ground with

(a) *Seymouria*

(b) *Dimetrodon*

(c) *Ichthyosaurus*

(d) *Elasmosaurus*

(e) *Lycaenops*

(f) Thecodont

(g) Pterosaur

(h1) *Triceratops*

(h2) Duck-billed dinosaur

(h3) *Stegosaurus*

(h) Ornithischian dinosaur

(i) Saurischian dinosaur

(i1) *Tyrannosaurus*

(i2) *Brachiosaurus*

Figure 31.8 Extinct reptiles. (a) Found fossilized near Seymour, Texas, *Seymouria,* a cotylosaur, or stem reptile, was about 50 cm in length and had a sharply tapered tail plus thick, sprawling legs suitable mainly for crawling. (b) *Dimetrodon,* a 3–4 m long carnivore, was a typical pelycosaur (sail lizard). (c) With porpoiselike bodies propelled by limbs modified as paddles, ichthyosaurs such as *Ichthyosaurus* were fast, powerful swimmers. (d) Plesiosaurs such as this *Elasmosaurus* were paddle-driven swimmers. (e) Early therapsids (mammallike reptiles), such as *Lycaenops,* of the late Paleozoic, were quite large (about 120–240 cm in length), but later forms were considerably smaller. The earliest mammals that ultimately evolved from the therapsids were smaller still, seldom exceeding the size of a mouse or rat. (f) Typical thecodonts were bipedal, with well-developed hind legs but small forelimbs generally not well suited for locomotion. Bipedal locomotion freed the forelimbs for a variety of nonlocomotor tasks, and in an evolutionary sense, made them available for modification as the wings of flying reptiles. (g) Certain pterosaurs, such as *Sordus pilosus,* were insulated with hair or hairlike feathers, evidence that these reptiles were endothermic. The elevated body temperature no doubt increased a pterosaur's metabolic rate, thus helping to meet the high energy requirements of flight. (h) Ornithischians (bird-hipped dinosaurs). The inset shows the shape of a typical pelvic girdle. (1) Great herds of *Triceratops* grazed the plains of what is now Montana and Wyoming. Members of this and related genera used their horns and tanklike bodies to ward off attacks by carnivorous dinosaurs. (2) Duck-billed dinosaurs spent much of their time wading in water, feeding on aquatic plants. (3) *Stegosaurus,* another herbivore, was a veritable walking fortress. (i) Saurischians (lizard-hipped dinosaurs). The inset illustrates a typical pelvic girdle. (1) *Tyrannosaurus* and many other saurischian dinosaurs were bipedal carnivores. (2) But some, such as the giant *Brachiosaurus* and *Brontosaurus,* were marsh-dwelling herbivores.

minimal effort. Whether or not the therapsids had hair is still debated, but there is general agreement that they used metabolic heat to elevate and maintain body temperature above that of the environment (i.e., they were endothermic). There is also general agreement that the therapsids gave rise to the mammals.

Bipedal Reptiles of the Early Mesozoic.   A large group of slender, lizard-shaped, bipedal reptiles called **thecodonts** dominated land during the early Mesozoic. Thecodonts grew to about 1.5 m in length and had large hindlimbs used in running. Their small forelegs were probably used primarily to manipulate food (Figure 31.8f). Thecodonts were an important stem group; it is generally agreed that they gave rise to the crocodilians, to the ruling reptiles (dinosaurs) of the middle and late Mesozoic, to the flying reptiles (pterosaurs), and to a lineage that led to the birds. There is some evidence that the later thecodonts, like the therapsids, were endothermic.

Dinosaurs and Pterosaurs.   From about the middle of the Mesozoic until the era ended some 70 million years ago, **dinosaurs** (ruling reptiles) dominated the land while **pterosaurs** (flying reptiles) ruled the air. Some pterosaurs were as small as a sparrow and probably ate insects; others had wingspans greater than 10 m and probably fed on fishes swimming near the surface of ponds and lagoons (Figure 31.8g). To test the principles of pterosaur flight, a free-flying, computer-controlled half-scale model of the largest known pterosaur was recently constructed and successfully flown.

Two major groups of dinosaurs, the ornithischians and the saurischians, differed in their hindlimb girdles. The **ornithischians**, sometimes called the bird-hipped dinosaurs, evolved a heavy, very strong pelvic girdle shaped somewhat like that of a bird (see the inset in Figure 31.8h). Most of the ornithischians, including *Triceratops*, were herbivores. Many kinds of duck-billed ornithischians had highly modified teeth that allowed them to grind water plants obtained from ponds and sluggish rivers. Also an ornithischian, *Stegosaurus* was slow-moving and heavily armored (Figure 31.8h). **Saurischians**, the lizard-hipped dinosaurs, had elongate pelvic bones fused to form a strong arch well suited to support bipedal locomotion (Figure 31.8i). Many, like the awesome *Tyrannosaurus*, were large bipedal carnivores. Some saurischians were plant eaters, often so heavy that they moved on all fours in spite of their arched pelvic girdles.

Brachiosaurus, the largest land animal that has ever existed (about 24 m in length and about 12 m high), and the similar (but slightly smaller) *Brontosaurus* were wetland herbivores. *Brachiosaurus* apparently spent much of its time wading, using the water's buoyancy to help support its great weight. Measurements of the depth of footprints made by *Brontosaurus* indicate that these great swamp animals could support their own weight on land.

HABITATS AND ENVIRONMENTAL RELATIONSHIPS

Compared to their Mesozoic relatives, modern reptiles exploit relatively few of the earth's varied habitats. Most living species inhabit tropical and subtropical regions, where they are important in food chains. The number of reptile species and individuals decreases sharply toward the colder latitudes and higher altitudes. Louisiana has about 70 species of reptiles, including nearly 40 kinds of snakes, but northern Alberta has only one species, a garter snake. In the United States, lizards are most plentiful in the arid Southwest, whereas snakes and turtles are most common in the humid Southeast.

Of the earth's major habitat types—water, air, and land—land supports the great majority of living reptiles. Aquatic forms include crocodilians, many turtles, and certain snakes. Sustained flight by reptiles disappeared with the pterosaurs, although some modern lizards and snakes are arboreal. Some species have broad legs and feet, flaps of skin, or expanded edges of the body wall that increase aerial buoyancy and may convert a slow fall from a tree limb into a long glide.

Many reptiles live on and in the soil. Some desert snakes and lizards are able to "submerge" out of sight by rapidly wriggling their bodies; this adaptation allows them to avoid surface heat and predators and to lie concealed as they wait for prey (Figure 31.9). Various lizards and tortoises dig permanent burrows that they inhabit when they are not feeding. A burrow serves as a hiding place and as a secure retreat during harsh weather. A few blind, virtually legless lizards remain permanently underground, foraging for worms and insects. As they burrow, they leave a trail of narrow tunnels. Soil reptiles often live in association with other animals. A reptile's burrow may be a home for many other kinds of animals. Conversely, some lizards and snakes live as commensals in ant and termite nests.

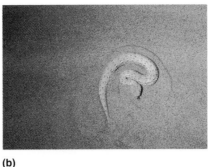
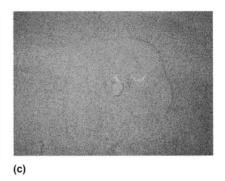

(a)   (b)   (c)

Figure 31.9 A dwarf puff adder submerging in sand. (a–c) With a quick series of shimmies and shakes, a sand viper *(Bitis peringueyi)* sinks vertically into the sand of the Namib Desert of southwestern Africa. In doing so, it may escape predators and excessive heat.

Water and Temperature

Having water-resistant skin and eggs allows reptiles to occupy a considerable range of habitats, including very dry ones. A reptile may desiccate, but its water loss is usually very slow. Consequently, many reptiles can live in deserts and dry grasslands, provided they avoid excessive heat.

Although modern reptiles are typically ectothermic, many exert considerable control over their body temperature. Reptiles rely on behavioral adaptations to govern uptake or loss of heat. They may absorb heat by basking in the sun on cool mornings or late in the day, and often they prevent overheating during hot periods by avoiding warm ground and bright sunlight, by burrowing, by submerging in sand, or by seeking shade. Compared to the relatively constant homeothermic body temperature of birds and mammals, behavioral temperature control by reptiles is less precise (i.e., heterothermic). (For a review of terms relating to body temperature, see the Chapter 3 Essay on temperature and animal life, p. 73.)

A recent hypothesis suggests that some of the great dinosaurs may have been successful in part because they were endothermic; like birds and mammals, some dinosaurs may have used metabolic heat to elevate their body temperature. The evidence is, of course, circumstantial. First, recent studies show that certain dinosaurs (and also the therapsids from which mammals evolved) had bones typical of animals with a high rate of activity. Their bones, similar to those of birds and mammals, were richly supplied with blood vessels and Haversian canals. By contrast, the bones of typical ectotherms, such as lizards, indicate a relatively low activity rate. (Bone structure is discussed in Chapter 8, p. 189.) A second piece of evidence also comes from the fossil record. Studies of predator-prey ratios in the Mesozoic indicate that dinosaur communities would have supported endothermic predators. (Endothermic predators require about ten times as much prey as the same mass of ectotherms because of the endotherms' higher energy requirements.) If we are to use this as evidence for dinosaur endothermy, we must assume that the ratio of predators and prey that were fossilized is equal to their ratio when alive; this is a risky assumption because we do not know if predators and prey were fossilized in equal proportions.

COMPARATIVE STRUCTURE AND FUNCTION

This section focuses on the adaptations that allow reptiles to be well suited for life on land.

Skin and Skeletal Adaptations

As much as any other feature, the dry, scaly skin of reptiles enables many species to occupy very dry habitats and to crawl without injury over rocky outcroppings and abrasive sand. When the skin is a rugged, almost impregnable armor, as in turtles and crocodilians, it also provides excellent defense against many predators. Many studies have shown that a dry, heavily keratinized, scaly skin is important in protecting a reptile against water loss, and as indicated in Table 31.1, reptile skin is vastly superior to typical amphibian skin in this respect.

A number of skeletal features are adaptations for terrestrial life. In contrast to most amphibians,

Table 31.1 Comparison of Evaporative Water Loss in Selected Amphibians and Reptiles

Species	Habitat	Rate of Evaporative Water Loss Through Skin (mg H_2O/g body weight/hr)
Leopard frog (*Rana pipiens*)	Moist	140
Mountain salamander (*Desmognathous ochrophaeus*)	Moist	656
Box turtle (*Terrapene carolina*)	Dry woodland	0.5
Lizard (anole, *Anolis carolinensis*)	Dry areas	1.9

Source: Adapted from C. Little, *The Colonisation of Land* (New York: Cambridge University Press, 1983). Data from J. R. Spotila and E. N. Berman, *Comparative Biochemistry and Physiology* 55A(1976): 407–411. Copyright © 1976 by Cambridge University Press. Reprinted with permission.

reptiles have well-developed **thoracic ribs**, forming a protective bony cage around the heart and lungs. The thoracic ribs and the intercostal muscles connecting them play a role in breathing (see the discussion on gas exchange, p. 763). Crocodilians (and the extinct marine reptiles) also have **abdominal (trunk) ribs**, an adaptation for supporting their heavy abdomens and protecting their vulnerable undersides. Unlike amphibians, reptiles also have an important innovation in the roof of their oral cavity: Bones from either side of the ventral surface of the skull form shelflike **palatal folds** that create a passageway between the internal and external nostrils; the palatal folds partially separate the nasal passage from the oral cavity and make the internal nasal openings open more posteriorly in the oral cavity. In the crocodilians, bones from either side of the skull fuse to form a **secondary**, or **hard**, **palate**, separating the oral cavity from the nasopharynx and allowing a crocodile to breathe even when its mouth is full of food or water.

Locomotion

Locomotion by limbed reptiles on land follows the general tetrapod pattern described in Chapter 9 (see p. 213). Aquatic locomotion by turtles is essentially an adaptation of tetrapod limb movements for paddling. Crocodilians and some lizards may also swim by means of limb movements; however, for rapid swimming, a crocodilian presses its legs against its body and surges forward with powerful, fishlike movements of its tail.

As a group lacking limbs, the snakes exhibit four modes of locomotion. **Serpentine movement**, the most familiar, has been described as "swimming on land." A snake's body undulates over the substrate, exerting forces against hard objects such as stones or bumpy soil, and these forces are resolved into forward motion (see Figures 31.10a and Figure 9.18). Using a second mode of locomotion, snakes may move along the ground by alternately curving and straightening the body in what has been called **concertina movement** (Figure 31.10b). The snake folds its body tightly (analogous to a closed accordion) and then, with its tail anchored to the ground, thrusts its head and neck forward (like an accordion opening up). Scales on the underside of the neck region then gain traction with the ground, and the body is pulled forward, compressed and ready for another forward thrust.

A third locomotor mode called **rectilinear movement** is often used by large, heavy-bodied snakes such as the boa constrictor (Figure 31.10c). The body crawls straight ahead, with no more lateral motion than a rope being pulled over the ground. This is accomplished by erecting a group of ventral scales and sliding them forward with brief movements of the skin. As the scales dig into the ground or catch on the substrate, the rest of the body is pulled forward to realign with the scales. Several points along the snake's body may alternate in this action. Rectilinear movement enables a snake to advance with a minimum of visible motion, an advantage for a constrictor in stalking a watchful monkey or other prey.

Perhaps the most remarkable of the four types of snake locomotor adaptations is **sidewinding**, a method used by the sidewinder rattlesnake (*Crotalus cerastes*) and several old-world vipers (Figure 31.10d and chapter opening photograph). Sidewinding involves a series of lateral looping movements in which no more than two parts of the body are in contact with the surface at any one time. As a sidewinder's weight presses downward, its looping body carries the points of contact posteriorly. The snake advances in a diagonal direction, producing a characteristic series of parallel

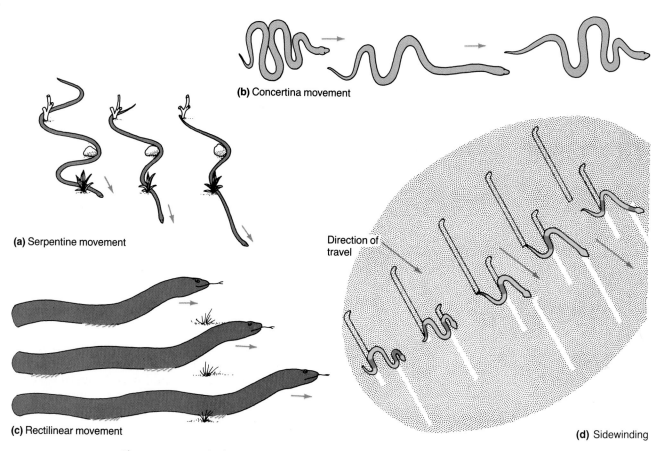

(b) Concertina movement

(a) Serpentine movement

Direction of travel

(c) Rectilinear movement

(d) Sidewinding

Figure 31.10 Snake locomotion. Snakes use one or more of four kinds of locomotion. (a) Serpentine movement. (b) Concertina movement. (c) Rectilinear movement. (d) Sidewinding. The outlined tracts indicate the snake's previous positions; nonoutlined tracts represent future positions.

J-shaped marks as it goes. As Figure 31.10d shows, the curve of the J is made as the head and neck touch the ground at the start of a new J-shaped mark. Sidewinding appears to be most advantageous in soft sand because it allows the snake to advance on two solidly established points of contact while eliminating waste motion and friction elsewhere along the body.

Feeding and Digestion

The majority of modern reptiles are carnivores with digestive systems organized like those of other carnivorous tetrapods (Figure 31.11). The most extensive adaptations of the digestive system in reptiles are associated with prey capture and involve the teeth, the use of venom, and jaw modifications that make it feasible to catch and swallow large prey.

With the exception of turtles, which lack teeth but have a horny beak, reptiles have well-developed teeth. Crocodilians and most lizards have numerous uniform (homodont) teeth. Crocodile teeth are set in deep sockets, reducing the chance that they will be pulled out during fights or while grasping prey. Typically, the teeth of lizards are simply attached to the surface of the jaw bone.

The most specialized reptilian teeth are those of snakes. Most snakes have rows of teeth on the upper and lower jaws and sometimes two longitudinal rows on the roof of the mouth. The backward curvature of the teeth makes it all but impossible for a rodent or frog to escape from a snake's mouth. A victim's struggling movements and the kneading action of the snake's jaws accelerate passage into the gut. A few venomous snakes have grooved fangs on the back of the upper jaw, but most of the dangerous venomous snakes have a pair of hollow fangs toward the anterior end of the upper jaw. Poison passes through a duct from the poison gland to the base of the fang, and then through the tubular fang into the prey (Figure 31.12). The use of venom enables snakes to subdue relatively large prey without great risk or effort. A successful attack by a

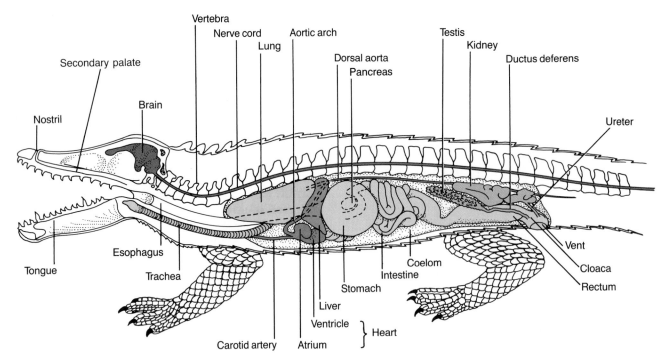

Figure 31.11   Internal anatomy of an alligator.

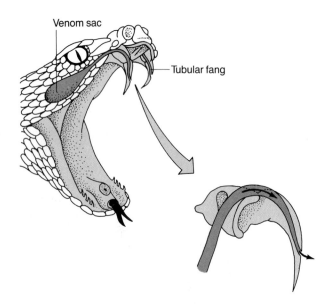

Figure 31.12   Release of venom by a viper. As a striking viper opens its mouth, muscle contractions cause its fangs to erect and swing forward, ready to penetrate the prey. As the fangs enter the prey, muscles in the snake's cheeks squeeze the venom sacs and force venom (arrows) through the tubular fangs and into the prey.

Snake venoms vary in strength and composition, and produce markedly different effects. Some are primarily **neurotoxic**, affecting the nervous system and leading to paralysis of the muscles of the respiratory system, heart, or both. Death often results from asphyxiation. Other venoms are **hemotoxic**, killing by disrupting the circulatory system. Hemotoxins destroy red blood cells and break down capillary walls, causing severe internal hemorrhaging. Cobra venom is neurotoxic, whereas that of rattlesnakes and other vipers is largely hemotoxic.

The basis of modern snakebite treatment was established a little over a century ago with the discovery that injection of small amounts of rattlesnake venom into pigeons caused production of antibodies. (Antibody production is discussed in Chapter 6, p. 139.) Now, injectable **antivenins** (blood serum from horses immunized against snake venom) are available for the venoms of many different kinds of snakes.

Many reptiles ingest relatively large food objects. Typically, a young reptile begins its predatory career by stalking and eating insects and other small invertebrates, later graduating to larger prey. For example, during its first few weeks, a young rat snake *(Elaphe)* will eat any tiny animal it can grasp, which usually means insects; but as it grows, it adds frogs to its diet and eventually is

viper requires only that the snake strike its victim. By contrast, an elapid, such as a cobra or coral snake, must grip its prey with its mouth while forcing the fangs into the flesh and injecting its venom.

large enough to ingest birds, shelled eggs, and mice.

In addition to the immobilizing effects of venom, adaptations involving jaw structure enable reptiles to exploit large prey. In reptiles, paired quadrate bones serve as hinges linking the upper and lower jaws. Quadrate bones are derived from the gill arches and are homologous with one of the middle ear ossicles in mammals. (Evolution of the gill arches is discussed in Chapter 28, p. 689.) The quadrates are firmly fused to the skull in turtles, crocodiles, and tuataras, but they are movable in snakes and lizards. Functioning like double-jointed hinges, the movable quadrates enable a snake to swallow disproportionately large prey (Figure 31.13). In addition, the right and left halves of a snake's lower jaw are loosely connected at the front by elastic ligaments, not firmly fused as in most reptiles. Stretching the ligaments allows the jaw to adjust to the size and shape of the meal. Another adaptation allows a snake to avoid suffocation while swallowing a big frog or an egg. During a meal, its glottis can be extended as a snorkellike breathing tube beyond the tip of the lower jaw.

Gas Exchange and Circulation

The adaptation of reptiles to life on dry land, which generally requires greater energy expenditure, was accompanied by improvements in the efficiency of the gas exchange and circulatory systems.

Except in turtles, gas exchange by reptiles depends entirely on lungs, not on various combinations of lungs, gills, moist skin, and mouth surfaces, as in amphibians (see Figure 31.11). Most reptiles use intercostal (rib) and abdominal musculature to change the volume (and therefore the pressure) of

the coelomic cavity around the lungs, causing air to flow in and out of the lungs. Reptiles also breathe by swallowing air, much like amphibians do. Breathing is a special problem for turtles. Because a turtle's ribs are fused with the shell, it cannot expand its chest to draw air into its lungs. Instead, contraction of a pair of limb flank muscles increases the space around the lungs, allowing air to be inhaled. Contraction of another set of muscles pushes the internal organs against the lungs, forcing air out. Supplementing lung respiration, turtles also exchange gases across moist surfaces of their cloaca.

The reptilian heart and major blood vessels are described in Chapter 28 (see p. 698 and Figure 28.15). Briefly, the heart has separate right and left atria, as in amphibians. However, the ventricle, which is a single chamber in amphibians, is partially divided in most reptiles and is completely divided into right and left ventricles in crocodilians. Oxygen-poor blood coming from the body to the right side of the crocodile's heart is kept completely separate from oxygen-rich blood returning to the left side of the heart from the lungs. Thus, crocodilians (and probably their immediate ancestors) had the first mechanism for preventing any mixing of oxygen-rich blood with oxygen-poor blood, an adaptation of critical importance to animals with high energy requirements.

Excretion and Osmoregulation

The principal nitrogenous waste product excreted by reptiles is uric acid, and this represents a significant water-saving adaptation for life in dry habitats. By excreting uric acid, reptiles can concentrate and dry their urine to the consistency of a thick paste. The special salt-excreting glands in the head of marine reptiles, such as sea snakes and the Galapagos marine iguana (*Amblyrhynchus*), are discussed in Chapter 7. After feeding on salty algae, marine iguanas often clamber onto a rocky shoreline and snort out vaporous clouds of concentrated salt solution. (See Chapter 7 opening photograph and the Essay on solving the salt problem at sea, p. 167.)

Nervous and Sensory Systems

The structure and general form of the reptilian nervous system is fundamentally similar to that of amphibians, but the forebrain (telencephalon) of reptiles is more like that of mammals than of amphibians. The most significant development involves the cerebral hemispheres. In reptiles (most

Figure 31.13 How a snake can swallow large prey. A snake's lower jaw is connected loosely to the skull by quadrate bones that act like a double hinge. When a snake, such as this timber rattlesnake (*Crotalus horridus*), opens its mouth to swallow prey (in this case a mouse), the hinge allows the lower jaw to drop at the back as well as at the front of the mouth.

notably the crocodilians), the hemispheres are much larger relative to the olfactory bulbs and optic lobes, compared to the condition in amphibians, and assume a pivotal role in integration. This condition in reptiles presages that in mammals, in which the cerebral cortex is formed mainly of large numbers of neuron cell bodies and virtually dominates the brain. (See the discussion of the telencephalon and neocortex in Chapter 28, p. 702; Figure 28.17.)

Chemical Senses. The reptilian sense of taste is served by taste buds situated mostly in the pharynx, while olfaction depends on chemosensory epithelium in the nasal passages. The ability to detect chemicals is particularly acute in snakes and lizards because they have a well-developed sensory structure called **Jacobson's organ.** Jacobson's organ consists of two internal cavities, one on each side of the snout, with ducts leading to an opening in the roof of the mouth rather than to the nasal passage. The cavities are lined with nerve endings similar to those used in olfaction. If you watch a snake for a few minutes, you will probably see its rapidly flicking tongue (Figure 31.14a). When a snake flicks its tongue in and out, it is sampling the air, picking up airborne molecules and transferring them to Jacobson's organ for analysis. The organ provides an important link between a venomous snake and its prey, particularly at night. When a rodent is struck by a rattlesnake, it will almost certainly scamper away, doomed but still mobile. Using its flicking tongue and Jacobson's organ to pick up and identify odor molecules from the dying victim, the snake is able to follow a chemical trail to the corpse. Jacobson's organ is not just a feeding adaptation. In combination with the flicking tongue, it is used during courtship by male snakes to recognize and follow the female.

Vision. Color vision and the ability to visually adapt and accommodate are well developed in most reptiles. In contrast to fishes and amphibians, which accommodate by moving the lens to change the distance between lens and retina, reptiles accommodate in the mammalian manner, by changing the shape of the lens; the lens becomes flatter for viewing distant objects and more convex for near vision.

The tuatara and many lizards have a **pineal eye** located medially on top of the skull. The tuatara's pineal eye has a retina and lens, but it cannot function in vision because it is covered by scales. In attempting to determine its function, researchers

(a)

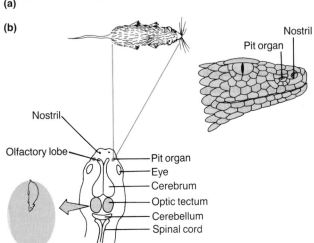
(b)

Figure 31.14 Chemical and heat detection by a snake. (a) A snake, in this case *Elaphe subocularis*, a rat snake from the Big Bend area of Texas, samples the air with its flicking tongue. Odor molecules adhere to the tongue, which transfers them to Jacobson's organ, a special sense organ associated with the sense of smell. (b) Pit vipers have a pair of pit organs, located between the eyes and nostrils; the pit organs detect heat radiating from endothermic prey such as rodents. The inset (at left) shows that sensory input from the endothermic prey becomes a neuronal projection in the optic tectum of the brain.

found that removal of the pineal eye caused lizards to stay in bright light longer than normal, indicating that the eye may help govern the duration of exposure to sunlight. This would be important in animals whose source of body heat is the sun. Similar experiments showed that the pineal eye is also involved in controlling daily (circadian) activity rhythms.

Hearing. Reptilian ear structure varies greatly. In lizards, the middle ear is well developed, and in some species, the tympanic membrane (eardrum) is set in a depression (outer ear) below the surface of the head. Snakes lack a middle ear; the quadrate bone of the jaw hinge aids in transmitting vibrations to the inner ear. Despite this relatively simple arrangement, snakes generally have good hearing. Turtles have well-developed middle and inner ears, but the tympanic membrane of terrestrial species is thick and covered with skin, and hearing is believed to be quite poor. Turtles probably depend on the shell and the skin to detect vibrations of the substrate. In crocodilians, the inner ear, where vibrational stimuli are converted into nerve impulses, is more highly developed than in any other reptile.

Heat Detection. The existence of a **pit organ** on each side of the head of a rattlesnake or other pit viper has been known for nearly 400 years, but the function of these organs as heat detectors was determined only about 50 years ago (Figure 31.14b). Rattlesnakes that have their other sense organs destroyed can still find and strike an object, provided that their pit organs are intact and the object is warmer than its surroundings. Pit organs are important to a pit viper hunting in darkness; after sunset, a rodent's high body temperature makes it a hot target against a cool background.

BEHAVIOR

Reptiles display a wide range of behavioral responses. Taxic behavior is evident in a reptile's responses to temperature and light. Many snakes and lizards warm themselves on cool mornings by orienting to the sun, often turning broadside for maximum exposure. Hikers and climbers frequently encounter snakes basking on rock ledges and outcroppings, and anyone studying a pond is likely to see turtles sunning on the bank. Some turtles extend their dark feet into direct sunlight to increase heat absorption. Reptiles must avoid overheating on hot afternoons, particularly in the desert or in the tropics. A dry skin and lack of sweat glands would seem to eliminate evaporative cooling. However, snakes and crocodilians evaporate moisture by holding their mouths wide open when hot, and the desert iguana *(Dipsosaurus dorsalis)* even pants. Another desert lizard, *Aporosaurus,* moves away from the hot sunlight into burrows in the sand when the air temperature exceeds about 40°C, a behavior that limits the animal to about 2 hours on the surface each morning and afternoon.

Complex behaviors such as reproductive activity and feeding often consist of combinations of taxes, complex motor programs, and learning. A sea turtle acts out a mainly innate program when it comes ashore to lay eggs for the first time. Without prior experience, it uses its hind legs to dig a nest of the correct size and depth, covers the eggs after depositing them in the nest, and then waddles back into the sea.

Feeding behavior in snakes often has innate and learned components. When extracts from a variety of prey animals were offered to different species of newly hatched garter snakes that had never fed, each species strongly preferred the extract of the prey it would normally eat in nature. A minnow-eating species preferred minnow extract, a frog-eating species preferred frog extract, and so on. Altering the diet of the mother snakes before they gave birth did not alter the preferences of the offspring. However, food preferences could change: A young snake with an innate preference for a certain prey would develop a preference for a different prey if it became accustomed to eating it. Thus, a snake's innate bias for the chemical traits of prey may be modified by learning experiences.

Reptiles typically exhibit courtship, territoriality, and migrations as components of their reproductive behavior. Although courtship by birds and mammals is often more elaborate, reptilian courtship may be spectacular. A bull alligator shatters swamp routine with loud bellows that attract the female, and he produces a penetrating odor (pheromone) by secreting musk from glands on his throat and near his cloaca. Prior to copulation, a male and female twist and turn, thrashing the surface and swimming wildly in circles. Courtship by turtles may also be quite "physical." Depending on the species, the male may swim backward in front of the female, waving his claws near her snout and sometimes stroking her head, or he may repeatedly nip and butt her body. Sex recognition among snakes depends mostly on odor. A male snake trails a female by her odor, using his nose, flicking tongue, and Jacobson's organ. Lizards rely primarily on vision, although secretions from preanal glands may be involved in sex recognition.

Courtship behavior by male reptiles often seems intended to intimidate rivals, and the lines between courtship behavior, territorial behavior, and displays of dominance are often indistinct (Figure 31.15a). Males in a group of lizards may arch their backs, dart and prance, erect neck crests, and display colorful throat fans, but how much of this is for other males and how much carries over to courtship is uncertain (Figure 31.15b). Behavior of North American spiny lizards *(Sceloporus)* and a Cuban lizard *(Anolis sagrei)* is more clearly territo-

(a)   (b)

Figure 31.15 Reptilian reproductive behavior. (a) Two male rattlesnakes *(Crotalus)* engage in a graceful sparring match to gain a mate; they writhe and poke, but do not bite. (b) A male anole *(Anolis polylepis* from a Costa Rican rain forest) spreads a colorful fan of skin beneath his head; in so doing, he intimidates rival males.

rial; these lizards aggressively defend territories during the breeding season. Likewise, studies of experimentally marked fence lizards *(Sceloporus occidentalis)* showed that males defend basking territory. Territories may lack sharp boundaries, but each log, stump, or rock usually has just one lizard basking in the sun.

Among the reptiles, migratory behavior is most remarkable in the sea turtles (leatherbacks, loggerheads, ridleys, and green turtles), which travel back and forth between breeding areas and feeding sites. By banding and recovering green turtles *(Chelonia mydas,* see Figure 31.2b), researchers have been able to map migration routes. These mainly herbivorous turtles (the young may eat jellyfish) spend most of their lives foraging in undersea pastures, but every two or three years, they interrupt their feeding to migrate to specific breeding beaches, where they lay eggs. Mating occurs in the sea near the beaches shortly before or after egg laying, and sperm is used to fertilize eggs that are laid at another time, perhaps two or three years later. Ascension Island, a tiny island near the center of the South Atlantic, is one of the green turtle's breeding sites. Turtles banded on Ascension Island were recovered there two and three years later, and it is now known that the turtles swim from feeding grounds along the coast of Brazil to Ascension Island, a 1400-mile journey against the south equatorial current, lay their eggs, and then return to Brazil. Although olfaction may be involved, green turtles probably use sun-compass orientation in navigating to Ascension Island.

REPRODUCTION AND DEVELOPMENT

Sexual Maturity and Mating

Because reptiles are ectotherms, the age at which they become sexually mature often depends on the effects of environmental temperature on developmental rate. Certain species of lizards in the warm tropics reproduce after just one year, but in cooler temperate zones, many species are four or five years old when they reach sexual maturity. Size is also correlated with sexual maturity: Small snakes and lizards often mature in a year or two, whereas certain turtles require six to eight years, and alligators may take more than ten years. As most temperate-zone reptiles reach sexual maturity, their reproductive cycle becomes synchronized with seasonal factors such as day length. Generally, ovulation occurs in the spring, following emergence from hibernation. However, the prairie rattlesnake, *Crotalus viridis viridis,* apparently ovulates only every second year.

All reptiles fertilize their eggs internally, a terrestrial adaptation that probably evolved in conjunction with the evolution of shelled eggs; without internal fertilization, sperm could not get to the egg before the shell was added to it. In addition, reptiles, unlike most amphibians, have a copulatory organ for sperm transfer. Only the tuatara *(Sphenodon)* lacks copulatory organs. Tuataras transfer sperm simply by pressing their cloacal openings together, probably the method used by ancestral reptiles. Other reptiles have a penis that is held outside-in within the cloaca, and is everted through the cloaca during erection. The penis of crocodilians and turtles is single, but snakes and lizards have a pair of structures called **hemipenes**. During copulation, one or the other of the hemipenes is everted and inserted into the female's cloaca.

Sperm may live for long periods in the female reproductive tract before being used, reducing the frequency (and energy expenditure) of mating. As already mentioned, female green turtles may store sperm for two or three years. Similarly, after a single mating, the diamondback terrapin, another sea turtle, continues to lay fertile eggs for four years.

Egg Laying

The majority of reptiles are oviparous, laying cleidoic (self-contained) eggs similar to those of birds (Figure 31.16a). Cleidoic eggs provide nourishment for the embryo, cushion it in amniotic fluid, and provide for gas exchange and waste disposal. (For a description of the cleidoic egg and extraembryonic membranes, see Chapter 16, p. 387; Figure 16.12.) Turtles generally bury their eggs in a hole excavated in sand or soil with the hind legs. Depending on the species, between 5 and about 100 eggs may be deposited in a single hole. The female American alligator, *Alligator mississippiensis*, combines mouthfuls of wet vegetation with mud to form a mound about 1.8 m wide at its base and nearly 1 m high. She then scoops out a hole in the top of the mound, deposits up to 80 hard-shelled eggs, and covers them with debris. The mother often stays in the vicinity to guard the eggs until, after about 2½ months, very faint squeaking sounds from the mound signal that the eggs are hatching. After helping the tiny alligators tear open the nest, the mother may stay with them for up to three years. This is one of the few known examples of post-hatching parental care among living reptiles, although some researchers consider it likely that certain duck-billed dinosaurs practiced extensive care of their young (see Figure 31.8h).

Snakes and lizards often bury their eggs in sand or soft soil, and some species place them in rotten logs, in crevices, and under stones. Parental care of the eggs is rare. However, a female python arranges her eggs into a pyramid, coils her body around it, and rests her head on the top; thus, she acts as a guard that few animals would challenge, and through her muscular contractions, provides some heat for incubation.

(a)

(b)

Figure 31.16 Modes of reproduction. (a) Like the majority of reptiles, the bull snake *(Pituophis melanoleucus),* found in the eastern United States and southwestward to the high plains of Texas, is oviparous (egg-laying). (b) The common garter snake *(Thamnophis sirtalis)* of the western hemisphere is ovoviviparous.

Toward the end of their embryonic development, most oviparous reptiles (Figure 31.16) develop a horny spine or egg tooth on the front of the mouth that aids in cutting the shell during hatching. Reptiles lack a larval stage, and the hatchling is a miniature replica of the adult, motile and ready to capture food. In some sea turtles, communication among embryos in the eggs may allow a more or less synchronous hatch. Then, by moving in a large group across an open beach from the nest to the sea, the hatchlings may overcome the effects of predation; many are killed, but the synchronous hatch may allow predators to become satiated, so that an individual turtle's chances of being eaten are lessened and some survive.

Viviparity and Ovoviviparity

Some lizards and snakes give birth to juvenile offspring. Of these, most species are ovoviviparous, retaining the eggs until they hatch in the oviduct (Figure 31.16b). Some reptiles give birth to young in one part of their range and are oviparous elsewhere. Some reptiles, such as the common European lizard *Lacerta vivipara*, are oviparous in the southern part of their range, but ovoviviparous in the northern latitudes (above the Arctic circle in Scandinavia in the case of *Lacerta vivipara*). A few reptiles are viviparous; that is, the female supplies nutrients to the developing embryos through a placenta. Viviparous lizards include the European skink *Chalcides chalcides*. Viviparity and ovoviviparity may be adaptations for protecting eggs against predators, protecting them against environmental extremes (such as low temperature), and offsetting short growing seasons. Eggs incubating within the body of a lizard basking in the sun in northern latitudes will most likely be at a higher temperature than a clutch of eggs under a rock, and they will therefore develop faster. Giving birth to young also eliminates the need to spend energy searching for suitable egg-laying sites and protecting eggs against predators. In this way, most sea snakes avoid having to return to land to lay eggs.

PHYLOGENY

The reptilian fossil record extends some 300 million years into the past, to the Carboniferous period of the Paleozoic era, a time of warmth and moisture, great inland seas, and vast expanses of low flatlands. At about this time, the amphibians were diversifying and occupying many aquatic and terrestrial habitats. Insects were widespread and abundant, providing a potential source of rich food for vertebrates colonizing land. Aside from the amphibians, the earliest reptiles must have had little competition as they populated the land and evolved adaptations that allowed them to exploit a great variety of habitats.

According to the classical view, the reptiles that appeared on this world stage evolved from labyrinthodont amphibians (see Chapter 30, p. 746). However, recent comparative studies on the development of the middle ear suggest that amphibians and reptiles may have evolved as more or less inde-

pendent lineages from crossopterygian (lobefinned fish) ancestors. Resolution of the problem of reptilian origins still eludes zoologists, and a fully satisfactory answer may never be found.

Assuming a common starting point in their ancestry, what enabled the reptilian lineage to be more successful in adapting to land than the amphibians? The most basic reason for reptilian success was probably the evolution of the cleidoic egg with its protective membranes and shell. Several ideas have been proposed to suggest why the cleidoic egg first evolved. Shelled eggs may have evolved in an aquatic ancestor as an adaptation for protecting against the dangers of larval life in the water. Certainly, the lagoons and shallow seas of the Carboniferous period teemed with predators. Another suggestion is that the cleidoic egg and internal fertilization evolved together. Zoologists have speculated that internal fertilization in the oviduct, followed by secretion of egg membranes, would have improved the reproductive success of ancestral animals living in fast-flowing streams, where external gametes would often be swept away.

Another crucial element in the reptilian conquest of land was the evolution of limbs capable of raising and holding the body off the ground without spending excessive amounts of energy doing so. Although reptilian limbs are quite variable, those of many species are positioned more under the body than to the side (as in most amphibians). Many of the great reptiles of the Mesozoic (e.g., the thecodonts and many dinosaurs; see Figure 31.8) had limbs that were positioned directly under the trunk of the body. The ability to deal vigorously and often aggressively with other predators—running to attack or escape as circumstance dictated—was in large measure due to adaptive improvements in limb mechanics. The evolution of limbs that could provide support and move the body efficiently on land was crucial in the reptilian lineages that led to birds and mammals.

The Age of Reptiles

Probably because reptilian evolution during the early Mesozoic era was not strongly channeled in any particular direction by selective pressures, many lineages developed. (Figure 31.8 illustrates the reptiles mentioned in the following account; Figure 31.17 maps their evolutionary relationships.) The cotylosaurs, or generalized stem reptiles, seem to have been the ancestral stocks from which several groups radiated. Perhaps derived

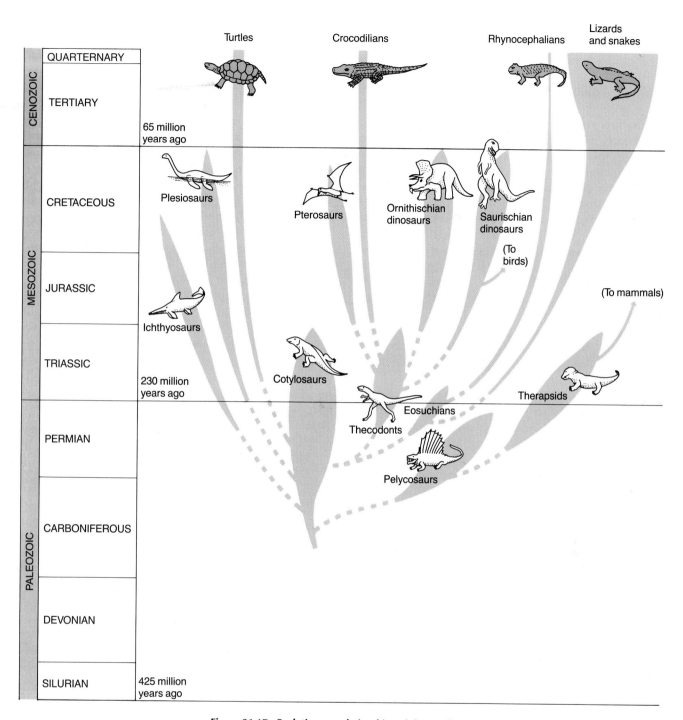

Figure 31.17 Evolutionary relationships of the reptiles.

from ancestral stocks that gave rise to the cotylosaurs, the sail lizards, or pelycosaurs, of the late Paleozoic and early Mesozoic were of great importance, because they, in turn, gave rise to the mammallike therapsids. In addition to having several skeletal features similar to those of ancestral mammals (and some that were different), there is little doubt that the later therapsids resembled mammals in being endothermic. By the middle of the Mesozic era, therapsids had given rise to tiny, shrewlike creatures, the earliest members of the Class Mammalia.

Also probably derived from immediate ancestors of the cotylosaurs were the lines of reptiles that returned to an aquatic environment. The fast-swimming, often formidably large ichthyosaurs and plesiosaurs ruled the seas throughout much of the Mesozoic and then became extinct some 100 million years ago.

Figure 31.17 shows that modern reptiles are thought to represent several lineages, all directly or indirectly derived from the cotylosaurs. Turtles, or chelonians, represent a relatively minor but ancient lineage. Another lineage gave rise to a group called the **eosuchians**, which were lizardlike reptiles, mostly of the late Paleozoic. Eosuchians never achieved much prominence, but they gave rise to two groups still in existence. One is the Squamata, the lizards and snakes. By the end of the Mesozoic, the Squamata had become the largest reptilian order, and it still holds this position today. For unknown reasons, the Squamata thrived and expanded at a time when many other groups of reptiles were becoming extinct. The other surviving eosuchian derivative, the Order Rhynchocephalia, has dwindled to the single species *Sphenodon punctatus*, the tuatara of New Zealand.

The most diverse group of early Mesozoic reptiles, the thecodonts, were ancestral to the crocodilians, pterosaurs, and dinosaurs. Apparently derived from the cotylosaurs (although some zoologists believe they arose from eosuchians), thecodonts were agile, fast-running bipeds. By freeing the forelegs for such tasks as seizing prey or fighting off predators, bipedalism was no doubt a factor in thecodont success. Bipedal locomotion may also have been a factor in the evolution of wings in pterodactyls and birds, because the forelimbs were not required for standing or walking.

As the "ruling reptiles" of the Mesozoic's Jurassic and Cretaceous periods, the dinosaurs included a vast assemblage of herbivorous and carnivorous species. Pterosaurs were also immensely successful during the middle and late Mesozoic. The end of the Cretaceous period brought with it the extinction of the dinosaurs and several other reptilian groups and the beginning of the Age of Mammals. Of the thecodonts' descendants, only the crocodilians and a lineage that eventually led to the birds survive.

The Great Extinction

The relatively sudden extinction of many plants and animals at the end of the Mesozoic, about 65 million years ago, remains one of the great mysteries of biology. Many kinds of organisms disappeared, including the giant reptiles. The pendulum of speculation about the causes of extinction has swung from catastrophic events to gradual processes and back again. Great floods, global volcanic activity, failure of the organisms to adapt to climatic change, and competition from evolving mammals have all been suggested. Recently, discovery of the mineral iridium, common in meteorites, in sediments marking the boundary between the Cretaceous and Tertiary periods leads some authorities to believe that the earth was struck by a large asteroid at the time of the great extinctions. (See "Meteors and Extinction" in Chapter 17, p. ◁ 420.) Perhaps the resulting dust in the atmosphere shaded the earth, reducing photosynthetic activity to critical levels. Insufficient energy capture by green plants could have disrupted food chains to such an extent that mass extinctions occurred, ending the Age of Reptiles. The meteor hypothesis has received considerable support in recent years, but few biologists believe that it or any other hypothesis answers all the questions about the great extinction.

▢ SUMMARY

1. Key Features of Reptiles. Reptiles are ectothermic tetrapod vertebrates, well adapted for life on land. Their major adaptations for terrestrial life include dry, scaly skin; lung respiration; a complete or partial four-chambered heart that separates pulmonary from systemic circulation; a bony skeleton with limbs that generally lift the body off the ground (snakes lack limbs, and limbs in aquatic reptiles are used as paddles); excretion of dry or pasty uric acid; internal fertilization; a cleidoic (self-contained) egg; and no larval stage.

2. Four living orders of reptiles include the Squamata (lizards and snakes), the Chelonia (turtles, including tortoises), the Crocodilia (crocodiles and alligators), and the Rhynchocephalia (the tuatara). Extinct orders include the cotylosaurs (stem reptiles), pelycosaurs (sail lizards), therapsids (mammallike reptiles), ichthyosaurs and plesiosaurs (marine reptiles), thecodonts (bipedal dinosaur ancestors), pterosaurs (flying reptiles), and ornithischian and saurischian dinosaurs. Some dinosaurs, pterosaurs, and therapsids may have been endothermic.

3. Most reptiles are carnivorous, although certain turtles and lizards are herbivorous or omnivorous. Many lizards and snakes can swallow oversized prey, and many snakes immobilize prey with venom injected by means of poison fangs.

4. Vision and the chemical senses are generally well developed, but there is great variation in hearing ability. Snakes and many lizards have an oral chemosensory

(Jacobson's) organ that analyzes molecules picked up by the tongue. Behavior includes complex feeding and migratory and reproductive activity (often including courtship displays and territoriality).

5. Most reptiles are egg layers (oviparous), but some lizards and snakes are ovoviviparous or viviparous.

6. Reptiles may have been derived from a labyrinthodont amphibian ancestor, or they may have shared an immediate ancestor (a lobe-finned fish or crossopterygian) with the amphibians.

☐ FURTHER READING

Burton, J., and D. Dixon. *Time Exposure.* New York: Beaufort Books, 1984. *A skillfully produced, simulated photographic record of the Age of Dinosaurs.*

Carr, A. *The Reptiles.* New York: Time-Life Books, 1968. *An excellent, readable text with many fine photographs.*

Colbert, E. H. *Dinosaurs, An Illustrated History.* Maplewood, N.J.: Hammond, 1983. *An authoritative, well-illustrated, readable text.*

Ferguson, M.W.J. *The Structure, Development and Evolution of Reptiles.* New York: Academic Press, 1984. *An authoritative technical account.*

Horner, J. R. "The Nesting Behavior of Dinosaurs." *Scientific American* 250(1980): 130–137. *An analysis of a group of duck-billed dinosaurs that may have cared for their eggs and young.*

Little, C. *The Colonisation of Land.* New York: Cambridge University Press, 1983, Chap. 9. *A technical account of the evolutionary relationship between amphibians and reptiles and their "invasion" of land.*

Ostrom, J. H. "A New Look at Dinosaurs." *National Geographic* 154(1978): 152–185. *A pictorial essay of dinosaurs, with a discussion of the question of temperature regulation.*

White, C. P. "Designed for Survival. Freshwater Turtles." *National Geographic* 169(1986): 40–59. *A superbly illustrated account of chelonian diversity in fresh water.*

☐ CLASSIFICATION

Groups marked with an asterisk are extinct.

Class Reptilia (L. *repere*, "to creep"). Reptiles.
Subclass Anapsida (Gr. *an*, "without"; *apsis*, "loop"). Cotylosaurs, turtles; anapsid skull (lacks temporal openings behind the eyes); quadrate bone immovable.
    Order Cotylosauria* (Gr. *kotyle*, "cup-shaped"; *sauros*, "lizard"). Cotylosaurs (stem reptiles); *Seymouria*.
    Order Chelonia (Testudines) (Gr. *chelone*, "a tortoise"). Turtles (including tortoises); *Chelydra, Macroclemys* (snapping turtles), *Chelonia* (green turtles), *Terrapene* (box turtles), *Chrysemys* (painted turtles), *Geochelone* (giant land tortoises), *Trionyx* (soft-shelled turtles.)
Subclass Synapsida* (Gr. *syn*, "together"; *apsis*, "loop"). Pelycosaurs and mammallike reptiles; synapsid skull (with a lower temporal opening behind the eye).
    Order Pelycosauria* (Gr. *pelykos*, "basin"; *sauros*, "lizard"). Pelycosaurs; *Dimetrodon, Edaphosaurus*.
    Order Therapsida* (Gr. *ther*, "wild beast"; *apsis*, "loop"). Therapsids (mammallike reptiles); *Titanosuchus, Cynognathus, Placerias, Lycaenops*.
Subclass Ichthyopterygia* (Gr. *ichthy*, "fish"; *pterygos*, "fin"). Reptiles with fishlike fins.
    Order Ichthyosauria* (Gr. *ichthys*, "fish"; *sauros*, "lizard"). Ichthyosaurs; *Ichthyosaurus*.
Subclass Sauropterygia* (Gr. *sauros*, "lizard"; *pteryx*, "fin"). Fin lizards.
    Order Protosauria* (Gr. *protos*, "first"; *sauros*, "lizard"). Protosaurs; *Protosaurus*.
    Order Sauropterygia* (Gr. *sauros*, "lizard"; *pterygos*, "wing"). Plesiosaurs; *Elasmosaurus*.
Subclass Lepidosauria (Gr. *lepidos*, "a scale"; *sauros*, "lizard"). Eosuchians, rhynchocephalians, lizards, and snakes; skull with two openings (temporal fossae) behind the eye, between which the squamosal and postorbital bones meet (diapsid skull); four-footed or limbless.
    Order Eosuchia* (Gr. *eos*, "the dawn"; *souchos*, "crocodile"). Eosuchians; *Champsosaurus*.
    Order Rhynchocephalia (Gr. *rhynchos*, "snout"; *kephale*, "head"). Rhynchocephalians; lizardlike; possess pineal eye; one living species, *Sphenodon punctatus* (the tuatara).
    Order Squamata (L. *squamatus*, "scaly"). Snakes and lizards; skull is a modified diapsid type; quadrate bone is movable; body covered with horny epidermal scales; teeth present; paired, eversible copulatory organs.
    Suborder Lacertilia (L. *lacerta*, "lizard"). Lizards; *Iguana, Dipsosaurus* (iguanas), *Chamaeleo* (chameleons), *Varanus* (monitors).
    Suborder Serpentes (L. *serpentis*, "serpent"). Snakes; *Python* (pythons), *Crotalus* (rattlesnakes), *Naja* (cobras), *Thamnophis* (garter snakes).
Subclass Archosauria (Gr. *archo*, "ruler"; *sauros*, "lizard"). Thecodonts, crocodilians, flying reptiles, dinosaurs (ruling reptiles of Mesozoic era) diapsid skull; many with tendency toward bipedal locomotion.
    Order Thecodontia* (Gr. *theke*, "case"; *odontos*, "tooth"). Thecodonts (reptiles with teeth set in sockets in the jaws); *Euparkeria, Ornithosuchus*.
    Order Crocodilia (L. *crocodilus*, "crocodile"). Crocodiles, alligators, gavials, caimans; *Crocodylus acutus* (American crocodile), *Alligator mississippiensis* (American alligator).
    Order Pterosauria* (Gr. *pteron*, "wing"; *sauros*, "lizard"). Pterosaurs; flying reptiles; *Sordus, Pteranodon, Rhamphorhynchus*.
    Order Saurischia* (Gr. *sauros*, "lizard"; *ischion*, "hip joint"). Saurischian dinosaurs; *Struthiomimus, Allosaurus, Tyrannosaurus, Diplodocus, Brontosaurus, Brachiosaurus*.
    Order Ornithischia* (Gr. *ornithos*, "bird"; *ischion*, "hip joint"). Ornithischian dinosaurs; *Camptosaurus, Iguanodon, Corythosaurus, Ankylosaurus, Stegosaurus, Triceratops*.

CHAPTER 32

Birds

Archaeopteryx lithographica. Commonly called the urvogel, this extinct bird is also called a "flying reptile" by some zoologists. This reconstruction shows several *Archaeopteryx* in a natural setting suggested by its fossil remains. Note the reptilelike teeth, free digits on the wing, scales on the head, and long, feathered tail; the inset shows an actual fossil of *Archaeopteryx*.

Charles Darwin's friend Thomas Huxley once referred to birds as "glorified reptiles," and many ornithologists (zoologists who study birds) agree with this interpretation. The general body form of birds, the presence of epidermal scales on their legs, and the nature of their beak and skin are decidedly reptilian. Feathers, homeothermy (relatively constant, warm body temperature), and a remarkable ability to communicate vocally are diagnostic features of birds, but many zoologists consider it likely that reptilian ancestors of birds had already evolved these and other birdlike features.

Living birds are diverse, ranging from the flightless penguins and ostriches to hummingbirds, woodpeckers, and hawks, but all have many features in common. Feathers insulate the body, allow flight, and are often vividly colored for courtship displays and mating rituals. All birds lay amniotic eggs, and nearly all species build nests, brood their eggs, and care for their young. Birds are also bipedal, walking or running on their hindlimbs. The forelimbs are developed into wings, which are adapted for flight in most species, but modified as paddles for swimming in penguins and are used for display in the ostrich and other flightless birds.

In birds that fly, virtually all body structures are adapted to accommodate flight, and the demands of flight have dictated a rather uniform body architecture in this group of vertebrates. Sharing a fundamentally similar body plan, bird species differ mainly in behavior, beak shape and related feeding habits, foot shape and function, flight ability, and feather coloration.

REPRESENTATIVE BIRDS

All birds belong to the Class Aves, composed of two subclasses.

Subclass Archaeornithes

The Subclass Archaeornithes is an extinct group of feathered creatures that lived some 140 to 150 million years ago, during the Jurassic period. A single species, *Archaeopteryx lithographica* (see chapter opening illustration), is known from fossils of six individuals that lived on tropical islands and may have drowned in shallow marine lagoons of the period. All the fossils have been found in limestone quarries in Bavaria, Germany.

Although generally considered a primitive bird, *Archaeopteryx* had many reptilian features and is sometimes referred to as a "feathered reptile." Like thecodont reptiles, from which it most likely evolved, *Archaeopteryx* had teeth set in sockets in the upper and lower jaws, three separate clawed fingers, and a long tail containing many caudal vertebrae. (Thecodont reptiles are described in Chapter 31, p. 758; Figure 31.8f.) The legs and head of *Archaeopteryx* were covered with reptilian scales. *Archaeopteryx* had a fairly heavy body and was probably a weak flier. It may have been mainly a glider, although its flight feathers are similar to those of modern flying birds, indicating that it supplemented gliding with flapping. *Archaeopteryx* may have used its clawed forelimbs to climb and scamper about in trees; when high enough, it could have glided from tree to tree and to the ground. In addition to feathers, *Archaeopteryx* had fused clavicles, forming the **furcula**, or wishbone, an important adaptation for flight, unique to birds. The furcula provides one site of attachment for flight muscles and helps hold the wings apart (see Figure 32.4).

Subclass Neornithes

The Subclass Neornithes includes all living birds and many extinct species. Unlike *Archaeopteryx* and reptiles, neornitheans have only a few caudal vertebrae, their visible tail is composed mainly of feathers, their wings have fused finger bones, and usually only the hindlimbs bear claws.

Toothed Neornitheans. Several extinct groups of neornitheans dating from the Cretaceous period, about 100 million years ago, retained reptilelike jaws and teeth. The extinct genera *Hesperornis* and *Baptornis* represent a group of flightless aquatic divers that were probably distributed worldwide (Figure 32.1). Fossils have been found mostly in the midwestern United States, an area that was covered by extensive seas during the Cretaceous. Another group of toothed Cretaceous birds, represented by the genus *Ichthyornis*, were gull-like fliers. Their fossils are the oldest yet discovered with a keeled sternum (breastbone) similar to that of modern birds. Like their contemporaries, the dinosaurs, toothed neornitheans became extinct near the end of the Cretaceous.

Figure 32.1 A toothed bird of the Cretaceous period. *Baptornis advenus* was about 1 m long. Note its tiny wings and toothed jaws. Its large feet with lobed toes propelled it through the water in pursuit of fishes.

(a)

(b)

Figure 32.2 Flightless (ratite) birds. (a) The ostrich *(Struthio camelus)* is omnivorous and lives in flocks on the dry African plains. It can run about 70 km/hr and can deliver formidable kicks with its hooves. Weighing up to 150 kg and being about 2.5 m tall, it is the largest living bird. Individuals may live for about 40 years. (b) Kiwis *(Apteryx)* such as the individual seen here with its rather large egg, are chicken-size nocturnal inhabitants of moist forests in New Zealand, spending the daylight hours in burrows. Unlike other birds, kiwis have nostrils at the tip of their long bill, which is used for probing worms from forest litter. The generic name *Apteryx* means ''without wings,'' and the wings of kiwis are vestigial.

Living Neornitheans. Living birds, approximately 8600 species classified in about 28 orders (see the Classification section), all lack teeth. There are two large subgroups, based on the presence or absence of the keeled sternum. The African ostrich, the South American rheas, the emus of Australia, the cassowaries of New Guinea, and the kiwis of New Zealand represent a small number of flightless, land-bound birds collectively called the **ratites** (Figure 32.2). These birds have heavy skeletons that support them on land, but they lack well-developed breast muscles, which are necessary for flight. Typical ratites have long, powerful legs and are fast runners, although the three living species of kiwis have short legs. Ratites were once believed to be closely related to flightless ancestral birds, but are now believed to be a specialized group derived from flying birds. Their breastbone is flat, providing little surface for muscle attachment.

Quite different from the ratites, other birds are **carinate**, meaning they have a **sternal keel**, or **carina**, that supports massive flight muscles. Most species can walk and hop on land, but are adapted mainly for rapid, agile flight. With wings driven by powerful flight muscles, carinate birds are the undisputed masters of aerial environments. Of the 24 orders of carinate birds, 11, representing some of the diversity of modern birds, are described briefly and illustrated in Figure 32.3. Members of two

Figure 32.3 (Opposite) Representative orders of carinate birds. (a) Order Pelecaniformes. The brown pelican *Pelecanus occidentalis* is a fish eater; note its large scooping bill and pouch. (b) Order Ciconiiformes (herons and egrets). This species, the cattle egret *(Bubulcus ibis)*, is often seen with large herbivores, such as this cape buffalo. (c) Order Anseriformes. Ducks, geese, and swans have broad, flattened beaks with serrated edges that strain water from aquatic foods. In many species, such as the wood duck *(Aix sponsa)* shown here, the male and female exhibit pronounced sexual dimorphism. (d) Order Falconiformes. Falcons, hawks, ospreys, eagles, and vultures, such as these turkey vultures *(Cathartes aura)*, have hooked beaks and strong gripping talons. (e) Order Galliformes. Turkeys, peafowl, pheasants, grouse, ptarmigan, and the jungle fowl (from which the domestic chicken was derived). Note the insulating feathers on the feet of this white-tailed ptarmigan *(Lagopus leucurus)*. (f) Order Charadriiformes. Gulls, terns, the killdeer, sandpipers, and this species, Wilson's phalarope *(Phalaropus tricolor)*, pick for small invertebrates along muddy shorelines. (g) Order Columbiformes. Pigeons and doves, such as this mourning dove *(Zenaida macroura)*, are vegetarians. (h) Order Caprimulgiformes. Whip-poor-wills and nighthawks, such as this lesser nighthawk *(Chordeiles acutipennis)*, are acrobatic flyers, twisting and turning sharply as they catch insects in midair. (i) Order Piciformes. Woodpeckers have sharp beaks used to chip into wood and an exceptionally long tongue for probing after insects under bark. Shown here is the pileated woodpecker *(Dryocopus pileatus)*, a forest dweller of North America. Unfortunately, this species has vanished over most areas of the continent. (j,k) Order Passeriformes (perching songbirds). (j) The yellow-throated warbler *(Dendroica dominica)* breeds throughout the United States and winters in the southern U.S., the West Indies, and Central America. (k) Steller's jay *(Cyanocitta stelleri)* is a year-round resident of western North American coniferous forests. (l) Order Sphenisciformes (penguins). The emperor penguin *(Aptenodytes forsteri)* of the Antarctic is the largest penguin species (up to 1.2 m tall). Penguins nest in colonies on oceanic islands or on ice in southern seas; some live as far north as the Galapagos Islands. Adult penguins have few enemies on land; at sea, swimming speed and agility are their main defenses against sharks and killer whales.

(a)

(b)

(c)

(d)

(e)

(f)

(g)

(h)

(i)

(j)

(k)

(l)

other prominent orders, the **Apodiformes** (hummingbirds and swifts) and the **Strigiformes** (owls), are illustrated in Figures 32.11 and 32.13, respectively.

By far the largest order of living birds (about 5100 species) is the **Passeriformes** (Figure 32.3j, k). Passeriforms, including sparrows, crows, warblers, finches, and many others are collectively called **songbirds** because most species have a well-developed voicebox and can produce varied, musical sounds. They are also called **perching birds** because their foot has three front toes and one hind toe, allowing them to grip a tree branch and rest in place for long periods of time.

Penguins constitute a somewhat unusual order of carinate birds in that they have lost the ability to fly and are adapted to pursue fishes in the sea (Figure 32.3l). They have strong breast muscles, a deeply keeled sternum, webbed, paddlelike forelimbs, and rudderlike hindlimbs, making them fast, agile swimmers. (Penguins have been clocked at about 9 m/s underwater.) Their feathers are small and scalelike, creating minimal drag in the water. A thick layer of fat under the skin provides insulation.

HABITATS AND ENVIRONMENTAL RELATIONSHIPS

Flight provides enormous ecological advantages. It allows birds to migrate seasonally to exploit resources in widely separated habitats. It also provides excellent protection from predators, allowing birds (in contrast to most mammals) to be active during daylight hours and to be brightly colored. Among the invertebrates, only insects have mastered flight, and flying insects provide certain birds with a vast food supply.

Predators of birds are mainly mammals, reptiles, and other birds. Birds are also afflicted by many parasites and diseases. Their feathers and nests provide homes for lice, fleas, ticks, and mites, and they are attacked by mosquitoes and other bloodsucking flies. Birds also harbor their own species of malaria and other blood parasites, and many are infected with flukes and tapeworms. Botulism, caused by a bacterial toxin, and several types of viral diseases kill large numbers of birds each year. Today, however, no natural afflictions are as potentially harmful to birds as human destruction of their habitats and pollution of the environment. Many species, some of which have not been studied at all, face extinction as large tracts of tropical rain forest are being cut down for timber and agriculture. Many birds are threatened by the heavy use of chemical poisons in industry and farming. Destruction of its breeding habitat and reckless, unregulated hunting annihilated the passenger pigeon *(Ectopistes migratorius)* in the United States; millions of these birds were shot for food and sport during the latter part of the nineteenth century, and during this same time, vast tracts of eastern hardwood forests, where the birds nested, were cut down. The last individual died in a zoo in 1914.

Not all birds are declining in numbers. A few species with a wide tolerance for climate and food types are actually increasing in numbers. The cattle egret managed to fly across the Atlantic Ocean from Africa to Brazil in the early 1900s (Figure 32.3b). Since then, it has extended its range throughout Central America and through the United States to northern Canada. The starling and the house sparrow (both songbirds) were introduced into eastern North America from Europe during the last century. Both species occupy diverse habitats, often to the detriment of native species, and thrive in cities and towns. From a few breeding pairs that were introduced, they have spread throughout North and Central America.

SPECIAL ADAPTATIONS FOR AERIAL LIFE

Most birds are highly specialized flying machines, and the uniformity of their body forms reflects the overriding demands of flight. Essentially all of their body systems are adapted in some way to accommodate four factors: lightness (high surface-area-to-weight ratio), power, balance, and mobility in air. Above all else, flight requires lightness and power, and to a large extent, we can understand birds by considering how they are adapted to accommodate these two factors.

Skeleton and Muscles

The skeleton of a bird is a remarkable product of evolutionary engineering (Figure 32.4). It is light yet amazingly strong, and in most birds it meets the very different demands of supporting the body in flight and on land. Many bones are hollow but reinforced internally with trusses similar to those used in airplane wings (see the inset in Figure 32.4). Reducing the weight even more, many long bones

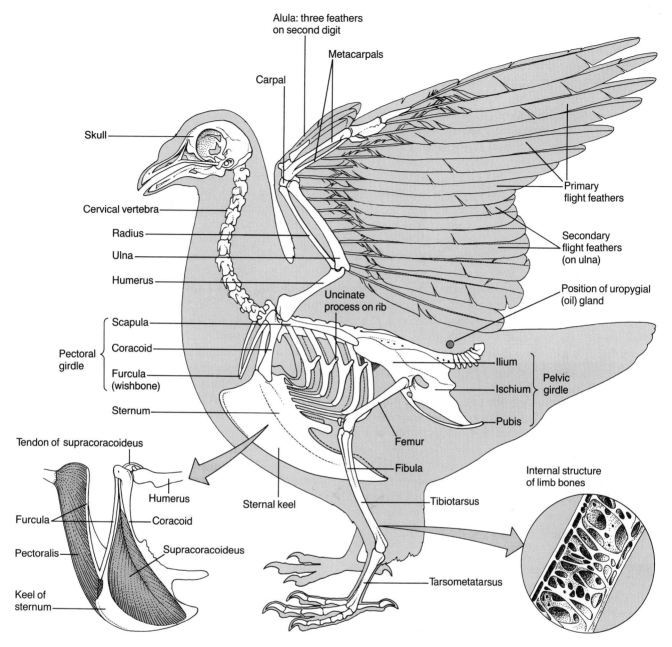

Figure 32.4 Skeleton of a pigeon. This drawing shows the main adaptations for flight. Flight feathers are shown on the left wing. The insets show the internal structure of the bones and the position of the two main flight muscles as viewed anterolaterally; the pectoralis on one side has been removed. Notice that both muscles are attached to (originate on) the sternal keel, but much of the pectoralis originates on the furcula. Note also that the pectoralis covers the supracoracoideus muscle on the sternal keel.

and certain skull bones house internal extensions of respiratory air sacs connected to the lungs. (Air sacs and their role in gas exchange are discussed in Chapter 5, p. 112.) The skull is also strong and light, and the vertebral column behind the neck shows areas of fusion that strengthen the body axis. Cervical (neck) vertebrae, on the other hand, are highly mobile, allowing the neck to turn through a wide range of motion, compensating for a general lack of movement of the bird eye.

The pectoral and pelvic girdles provide compact strength, while the long bones of the hands, legs, and feet have been reduced by fusions. The ribs are broad and strong and are braced against one another by hooklike **uncinate processes**. These and the sturdy bones of the pectoral girdle prevent the chest from collapsing when the bird is flying. Two bones, the furcula and the broad, thin sternal keel, provide sturdy surfaces for flight muscle attachment.

A bird's muscles are a good example of how highly specialized birds are for flight. Compared to other vertebrates (including flightless birds), flying birds have rather weak back muscles. The great strength of their axial skeleton makes it possible for most of the muscles to be concentrated ventrally, providing balance and power for flight. The chief flight muscles are the massive **pectoralis**, which depresses the wing, and its antagonist, the **supracoracoideus**, which elevates the wing (see the inset in Figure 32.4). These muscles may account for up to 35% of the total body weight. The pectoralis extends from the furcula and sternal keel to the underside of the humerus. The supracoracoideus is attached by a long tendon to the upper surface of the humerus. Its tendon loops over the scapula, forming a pulleylike mechanism that raises the humerus. In turkeys and chickens, these two flight muscles are the "white meat," high in glycogen, but comparatively low in fat. In active fliers, however, they are composed of red fibers with high fat content.

A bird's leg muscles are not as highly modified as the wing muscles, and most species are not especially agile on land. As a modification that brings most of the weight toward the center of the body for flight, the main leg muscles are short, with most of their mass in the upper leg around the femur. Long, thin tendons extend from the short, powerful leg muscles to operate the extremities, making bird feet light and strong.

Skin and Feathers

The integument of birds (see Figure 28.13c) serves a variety of functions, including the all-important production of feathers. A layer of fat under the dermis serves as an energy reserve and helps insulate the body against mechanical shock and temperature extremes. The legs and feet of birds are protected by a covering of cornified epidermal scales similar to those of reptiles. The single **uropygial gland** on the tail provides a lubricating and waterproofing oil for the entire body surface (see Figure 32.4). Birds rub oil from the gland onto their beaks and use it to preen their feathers. Without the oil, the feathers would become brittle and useless in flight. Water birds rely on the oil in their feathers for waterproofing; without it, they would become nonbuoyant and sink.

Birds have an enormous number of feathers. A hummingbird may have nearly 1000 feathers, a swan well over 25,000, and a penguin many thousands more. Although feathers are light, they make up a large part of a bird's total weight. Those on a bald eagle, for example, may weigh twice as much as the bird's skeleton.

Feathers resemble reptilian scales, from which they evolved. Like scales, they develop as small epidermal folds, or **papillae**, covering a core of dermal, connective tissue **pulp** (Figure 32.5). A papilla elongates and sinks into the skin, forming a pit called the **follicle**. A feather develops from epidermal cells at the base of the follicle; blood vessels and nerves in the pulp provide nourishment and regulate the developmental process. Pigment cells in the epidermis add color to the growing feathers. As the epidermal cells die and become keratinized, the feather takes shape. When it reaches full size, it unfolds and breaks out of a protective coat of dead epidermal cells. The pulp is then resorbed by the dermis, leaving a hollow feather shaft. The mature feather is entirely dead tissue, and the follicle serves only to anchor it to the body.

Birds have several kinds of feathers serving a variety of functions. Covering the outside of the body, **contour feathers** provide color and protection from mechanical injury and sunburn, and contour feathers that project beyond the body on the wings are the flight feathers. Contour feathers have a central **shaft** consisting of the **rachis**, which supports the **vane**, and a naked base, the **quill**, which is anchored in the skin follicle (Figure 32.6a). The

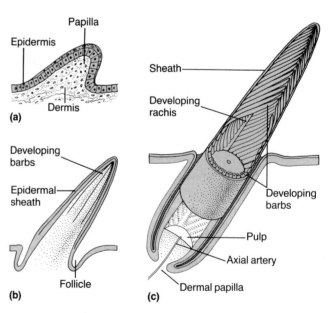

Figure 32.5 Development of a feather. (a) A papilla in the skin. (b) Formation of the feather shaft and follicle. (c) A feather takes shape and begins to break out of its epidermal sheath.

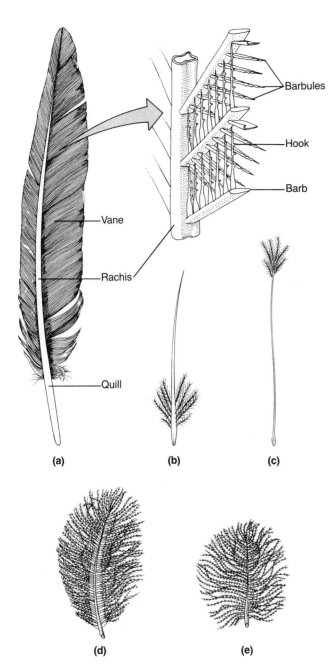

Figure 32.6 Structure and types of feathers. (a) A contour feather (a flight feather) from the wing, showing the microstructure of the vane (see enlargement). (b) A bristle. (c) A filoplume. (d) A semiplume. (e) A down feather.

feathers through its bill, hooking the barbules back together in a zipperlike manner.

As explained in Chapter 9, birds can fly because their wings are airfoils (thicker in front than in back; see Figure 9.19). The shape and arrangement of the contour feathers on the wing produce this shape. Notice the asymmetry of the flight feather's vane; the narrow side of the vane is thicker and stiffer (Figure 32.6a). In flight, this side is on the leading edge of the wing (see Figure 32.4). This feather arrangement and the forward position of the arm bones make the wing thicker in front, providing the key to flight. In contrast, the "flight" feathers of flightless birds are symmetrical.

Other kinds of feathers are shown in Figures 32.6b–e. **Bristles** are small, vaneless, hairlike feathers that form eyelashes and dust filters in the nostrils. Whip-poor-wills and nighthawks have bristles around the mouth, forming a screen that traps insects while the bird is flying (see Figure 32.3h). **Filoplumes** are long, thin, nearly naked shafts that project among, and move with, the flight feathers. As the filoplumes move, nerve endings in their follicles may detect body position in flight. **Semiplumes** and **down feathers** are vaneless insulators. They occur under contour feathers and have loose, fluffy barbs because their barbules lack hooks. Semiplumes have a central rachis, while down feathers do not. Anyone who has ever worn a down parka can attest the insulating power of down feathers and semiplumes. And in contrast to the random pile of feathers in a parka, feathers on a bird are organized into layers that more effectively conserve heat.

In addition to these types of feathers, parrots, hawks, and herons have **powder down**. These are feathers that grow continuously and are not shed. As they grow, their tips disintegrate into a fine, greasy powder that helps waterproof and adds lustre to other feathers.

Molting

Feathers, as dead structures, eventually wear out; thus, they are molted periodically. All the feathers on a bird are collectively called a **plumage**, and birds typically have several different plumages as they grow to adulthood. A newly hatched Canada goose, for example, is covered with yellowish green down that gradually molts and is replaced with grayish down (Figure 32.7). The gray plumage is replaced with adultlike feathers by about six weeks. Adult plumages are also molted and replaced, usually once a year.

vane is composed of many parallel **barbs**, each with a row of minute projections called **barbules**. Each barbule has tiny hooks that interlock with grooves on adjacent barbules, making the vane rigid. A flight feather cannot function if its barbules are detached and its vane is loose. But if the vane is not broken, separated barbules can be reconnected by preening. The bird simply draws the damaged

Figure 32.7   A family of Canada geese *(Branta canadensis)*. The juveniles (goslings) are about two weeks old.

Molting is generally regulated by photoperiod and is under endocrine control. Although the process is not fully understood, specific changes in day length seem to cause the hypothalamus and, in turn, the pituitary to stimulate production of thyroxine by the thyroid. Peak levels of thyroxine in the blood seem to stimulate growth in the epidermal papillae, and feathers begin to molt as new ones begin to grow under them.

Molting and feather replacement are energy costly and are usually accomplished during less stressful times of the year. Adult birds usually molt when food is abundant, but not when they are mating and nesting. Most birds shed their feathers in a gradual, orderly sequence, without becoming bald and without loss of flight.

Birds that need feathers in prime condition for long-distance migration, as well as those that live in dense or abrasive foliage where flight feathers are easily damaged, may molt twice a year. Ducks and geese have a complete molt and do become flightless for a short period in the summer after rearing their young. Then, before they migrate south, they molt again, shedding and replacing all but the flight feathers.

Molting in penguins is a yearly event. When temperatures warm to about −15°C, penguins spend about three weeks on land molting and growing new feathers. They do not eat during this time and may lose nearly 50% of their body weight.

Wings and Flight

Flight feathers are arranged in two distinct groups on the wing. **Primaries** are borne on the bird's hand. **Secondaries** are borne on the forearm, specifically on the posterior edge of the ulna (see Figure 32.4). When a bird is flying, movement of the hand with its large primaries describes a figure eight or an oval in the air, with the primaries providing most of the propulsive force for flight. The forearm moves little by comparison, but its secondaries provide much of the lift of the wing.

For a bird to fly, its airfoil wings must generate a lift force equal to or greater than its body weight. Tilting the wing upward (increasing its angle of attack) decreases the pressure over the wing and increases the pressure under the wing, thus increasing lift. However, at too great an angle of attack (over about 15°), air does not follow the upper wing surface, but becomes turbulent, decreasing lift and causing the wing to stall. This effect is most pronounced at low speeds. Stalling can be delayed at high angles of attack and at low speeds by **wing slots**, auxiliary airfoils that direct a layer of fast-moving air across the upper wing surface. As shown in Figures 32.4 and 32.8, many birds have a small cluster of feathers, the **alula**, on the second digit, near the middle of the wing. The alula serves as a wing slot. When it is erected, turbulence over the wing is decreased and stalling is prevented or delayed. Birds may also produce a slotting effect by spreading their flight feathers apart, and drawing air between them.

Wings have different shapes, depending on the type of flight for which they are adapted. A **cambered** (ventrally concave) wing with a high **aspect ratio** (length-to-width ratio) provides the greatest lift. Camber, however, increases drag; consequently, birds with highly cambered wings cannot fly as fast as those with flat (noncambered) ones. Wings with a high aspect ratio are less maneuverable than those with a low aspect ratio. Passeriforms, grouse, woodpeckers, and many others have wings with a low aspect ratio and high camber. Such birds fly rather slowly, but can maneuver through dense foliage and turn sharply to evade predators. Their relatively short wings are usually slotted to prevent stalling in slow flight and during sharp turns. By contrast, sandpipers, falcons, swifts, and hummingbirds are fast fliers; their wings have a high aspect ratio, low camber, and no slots. Certain species of sandpipers have been clocked at nearly 180 km/hr.

Functions of the Tail

Tail feathers can provide lift and also enable a bird to balance, brake, and maneuver in flight, depending on the angle at which the tail feathers are held. Long-tailed birds can twist and turn in flight. Although short-tailed species are generally the fastest fliers, they cannot maneuver as well as those with long tails.

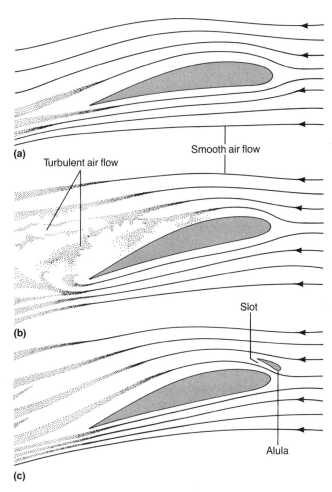

(a)

Turbulent air flow

Smooth air flow

(b)

Slot

Alula

(c)

Figure 32.8 The wing as an airfoil and the effect of wing slotting. The cross-sectional shape of a bird's wing is that of an airfoil. The wing is shown here at two different angles of attack relative to air flow. (a) Position of the wing in normal flight. The airfoil generates sufficient lift in a horizontal position by lowering the air pressure above the wing. (b) Increasing the angle of attack of the airfoil increases lift; but turbulence over the wing will eventually cause stalling. (c) Erection of a small auxiliary airfoil, the alula, creates a slot through which air is forced to flow close to the upper wing surface. Note that turbulence has decreased, even though the angle of attack is the same as in (b).

Types of Flight

The simplest, least energy costly, and probably oldest form of flying is **gliding**. A gliding bird generates no power; it simply floats slowly downward with its wings extended, using its body momentum to overcome drag from the air.

Soaring, a second type of flight, also requires minimal effort. Species that spend much of their time soaring have relatively small flight muscles, indicating the efficiency of this form of flight. To soar, a bird must be relatively large, since body

momentum is needed to carry it forward, but it can maintain or gain altitude without flapping its wings.

Eagles, vultures, condors, and large hawks that soar over land are called **static soarers**. They have broad wings with a relatively low aspect ratio. Soaring rather slowly and passively, they often gain altitude by circling on rising warm air currents, or thermals (Figure 32.9a). Their wings are usually highly cambered for maximum lift and also slotted for maneuverability, allowing them to respond to changing air currents.

Birds that soar over the sea, such as albatrosses, are called **dynamic soarers**. They take advantage of the differences in horizontal wind speed over the ocean. Because of the friction between the air and water, wind speed is lowest at the air-ocean interface, and without obstructions, wind speed regularly increases with altitude up to about 15 m above sea level. Sea birds can stay aloft for many hours, soaring for thousands of kilometers, while consuming little energy. They soar downward with the wind, then, near the water surface, often behind a wave crest, they turn into the wind. Flying against the wind, their downward momentum and the increasing vertical wind speed generate lift, and they soar rapidly upward at a sharp angle (Figure 32.9b). Dynamic soarers have long, narrow wings (an aspect ratio of 18:1 in the albatross) with low camber. The wingspan of albatrosses may reach 3.7 m—longer than that of any other living bird. Such wings are adapted for maximum lift at high speeds, but lacking slots, they are not highly maneuverable. They are also not very effective at low speeds, and for this reason, albatrosses nest on windy, hilly islands, where they can run downhill into the wind to take off.

Flapping flight is the most common, but also the most energy-consuming form of flying. It involves a complex turning motion of the wing. In its basic form, as analyzed by high-speed photography, flapping involves a downward and forward power stroke that causes air to rush over the top of the wing, lifting the bird and thrusting it forward. This is followed by a more rapid upward and backward recovery stroke (Figure 32.10). Birds often separate their primaries during the recovery stroke to reduce air resistance as the wing moves upward. Larger birds achieve added lift from the recovery stroke by twisting the wing so that the primaries push against the back-flowing air, helping to drive the animal forward. When taking off, birds usually flap vigorously and slot their wings on the downstroke (by extending the alulae or spreading the primaries) to increase lift. In flight, smaller birds

Figure 32.9   Two types of soaring flight. (a) Static soarers, such as eagles and vultures, circle upward on rising currents of warm air. (b) Seabirds are dynamic soarers, using the relatively constant wind over the ocean to remain aloft. Albatrosses and certain other seabirds spend most of their lives on the open ocean, drinking seawater and excreting excess salts via a nasal salt gland. In this drawing, broader horizontal arrows represent stronger wind.

Figure 32.10   Movements of a bird in flapping flight. This sequence was determined by high-speed photography.

Figure 32.11 The hovering flight of a hummingbird. This type of flight requires a more or less rigid, propeller-like wing. Shown here visiting a flower to obtain nectar is Allen's hummingbird *Selasphorus sasin*. Hummingbirds and their close relatives, the swifts (which catch flying insects) are masters of maneuverability in the air.

generally flap their wings more frequently than do larger ones. The tiny hummingbirds flap their wings as many as 40 to 80 times per second, while crows and pigeons may flap only 2 to 3 times per second, and large hawks only once per second.

The **hovering flight** of hummingbirds is a special type of flapping flight. It is the most energy-expensive form of flying because the bird is not moving forward and thus is not producing any lift-generating momentum. The hummingbird wing has short arm bones, and its flight feathers are almost completely supported by the hand. The wing is virtually rigid except where it joins the body at the shoulder joint. Here it can be moved in almost any direction and can rotate much like a propeller (Figure 32.11). While hovering, the bird holds itself at about 30°–45° off the vertical and achieves lift on the upstroke (recovery) as well as on the downstroke. On the upstroke, the wing is rotated so that its dorsal surface is turned over and pushes backward and downward against the air, providing almost as much power and lift as the downstroke.

Feeding, Digestion, and Excretion

Birds require high-energy foods for generating flight power; consequently, many of them eat mainly fruits, seeds, nectar, or animals. Food passes rapidly through their gut, and a high percentage of its nutrients and energy are extracted from it. Herbivorous species can digest berries and other fruits and defecate waste from them in 15–30 minutes. Meat eaters can digest small mammals, including the hard parts, in 3–4 hours.

Beaks tell a great deal about the habits and foods of birds (see Figure 32.3). The sharply hooked beak of eagles and hawks is highly specialized for cutting and tearing meat. Herons and terns have long, thin beaks effective in catching fish or aquatic invertebrates. Among the songbirds, robins and other thrushes have relatively unspecialized beaks and eat a variety of foods, including small invertebrates and fruits; cardinals and finches have short, stout beaks adapted for crushing or extracting seeds from hard shells.

Having a beak instead of teeth lightens a bird for flight and makes feather preening easier, but beaks are not as effective as teeth in chewing food. Thus, many birds swallow food more or less whole. Many species have a large **crop**, an expansion of the esophagus, where food is stored temporarily. This enables a bird to eat a large quantity of food quickly, swallow it into the crop, and then digest it later while resting or performing some other activity. In pigeons and doves, both the male and the female have a crop that secretes a watery slurry of sloughed cells called **pigeon's milk**. This substance is regurgitated and fed to nestlings. It is actually quite similar to mammalian milk and is secreted under the influence of the hormone prolactin, as in mammals.

The bird stomach consists of a glandular **proventriculus** anterior to a muscular **gizzard** (Figure 32.12). Protein digestion begins in the acidic medium of the proventriculus and continues in the gizzard, where the food is ground mechanically. To a large extent, the gizzard assumes the role of teeth and is especially strong in birds that eat large amounts of plant material. Many species swallow gravel with their food; this helps to pulverize tough fibers in the gizzard. Carnivores typically have a small, rather weak gizzard, but their stomach acid can often digest large bones. Many species also have chitin-splitting enzymes that help hydrolyze hard parts of insects and other invertebrates. Birds also have one or more pairs of **ceca** at the junction of the intestine and rectum. The ceca are large in herbivores and may be sites of bacterial fermentation.

The semisolid feces of birds consists of digestive and excretory wastes. Thick deposits of bird feces on islands frequented by large nesting colonies of seabirds are called **guano**. Deposits several meters thick on some islands are mined as a rich source of phosphate and nitrate fertilizer.

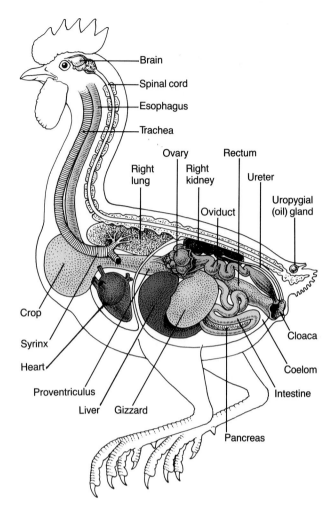

Figure 32.12 Cutaway drawing of a chicken. Birds have a cloaca into which digestive, urinary, and reproductive systems empty. Note the absence of a urinary bladder and the single ovary and oviduct; these adaptations help reduce body weight for flight.

The main excretory product of birds, uric acid, is produced primarily in the liver and precipitates out in the kidney tubules. By excreting uric acid, birds conserve water and also lighten their flight load. Uric acid passes through the ureters directly to the cloaca, where it is excreted semidry with the feces. Thus, birds do not have to carry fluid wastes, and in fact, carinate species lack a urinary bladder.

Homeothermy

Having a relatively constant warm body temperature is another important adaptation for flight. Homeothermy allows a constant high rate of metabolism and a consequent high power output. Having efficient circulatory and respiratory systems is prerequisite to maintaining a constant

warm body temperature. As described in Chapter 5 (see Figure 5.17), the respiratory system of birds extracts even more oxygen from a volume of air than does the highly effective mammalian system. Moreover, the four-chambered heart and dual (pulmonary and systemic) circulatory system of birds, similar to those of mammals (see Figure 6.12), deliver oxygen and nutrients to body tissues rapidly and under high pressure. Bird hearts are about 1½ to 2 times larger than those of equal-sized mammals, and blood pressure is generally higher in birds. Smaller, highly active birds have proportionately larger and more rapidly beating hearts than those of larger birds. The heart of an active hummingbird may beat 1000 times per minute, while that of an active domestic chicken has been recorded at about 400 beats per minute.

Body heat is generated by the metabolism of food, and the foods that birds eat generate a great deal of heat. The body temperatures of birds are generally higher and more variable (about 38°–44°C) than those of mammals (36°–39°C).

As homeotherms, birds must be able to retain heat in cold air and dissipate it when the weather is hot. Feathers minimize heat loss, and birds living in variable climates often have many more feathers in the winter than they do in the summer. Body fat is also an important insulator in some species. The emperor penguin lacks down feathers, but can molt and brood eggs in antarctic temperatures (down to −60°C) because of the thick layer of fat under its skin. During cold weather, many birds increase heat production by respiring at a greater rate. When it is very hot, they may pant and extend their wings, exposing underlying featherless skin that acts as a heat radiator.

Other means of retaining heat include **countercurrent heat exchangers** composed of closely associated arteries and veins carrying blood in opposite directions. Bird legs (except those with feathers) are sites of significant heat loss. Consequently, many species have a countercurrent heat exchanger in each leg. (See the Chapter 6 Essay on surviving extreme cold, p. 135.) Another countercurrent heat exchanger in the nostrils of many birds consists of a capillary network that cools and condenses water vapor from the respiratory system, thereby preventing heat loss that would occur if the water evaporated. In hot weather, the heat exchanger in the nostrils is bypassed, allowing heat to dissipate.

Behavior is also important in body temperature regulation. Birds puff out their feathers when it is very cold. Fluffed feathers hold air, and stable air near the body can be warmed, providing a thermal buffer. Body position is also a factor. To reduce heat loss, a bird usually sits or perches with its feathered body hunched down over its legs and with its beak tucked under its wing. Ptarmigan, living in the Arctic, sometimes bury themselves in snow for insulation against wind chill (Figure 32.3e). Moreover, many species huddle together when it is very cold; molting penguins may huddle in groups of hundreds or thousands, in this way doubling the time it takes for each bird to lose 1 kg of body fat.

A few species of birds respond to cold temperatures by becoming temporarily dormant, or torpid. Hummingbirds may use this technique to survive in high mountains. On cold nights, their metabolism may slow to a barely detectable level, and their body temperature may plunge to as low as 10°C. Without this capability, they would have to feed almost continuously at night to maintain their high daytime temperature and metabolism.

Sensory Structures

Most birds locate food by sight or sound rather than by taste or smell. This is not surprising, since chemicals are more difficult to detect in the upper, dryer air than they are in the generally moist air near the ground. The nostrils of most birds have little olfactory epithelium and function mainly in cleansing, warming and moistening inhaled air. Exceptions to this include the albatrosses and closely related seabirds, the vultures, certain pigeons, and the kiwis, all of which have a keen sense of smell and taste.

Vision. Birds have excellent eyesight, and the typical bird relies more on eyesight than on any other sensory ability. Some species—at least pigeons and hummingbirds—are sensitive to ultraviolet light. Hummingbirds, like many pollinating insects, find certain nectar-bearing flowers by detecting structures on them with the aid of ultraviolet sensors in the eye.

The bird eye is generally immovable in its socket, but otherwise is similar to the mammalian eye (see Figure 12.4). A bird turns its head to look at objects and accommodates (focuses) by changing the curvature of the lens or cornea. The bird eye can change focus very rapidly, an essential feature for birds flying through dense brush or pursuing prey on the wing. Diving birds can focus in air and underwater. Some have an accessory eyelid, the **nictitating membrane**, with a highly refractive cen-

ter that covers the cornea and functions like a swim mask.

Generally, the larger the eye, the sharper the image it can produce. The bird eye is large and has an extensive, very sensitive retina. Most birds have excellent color vision, although nocturnal species have many more rods than cones and are adapted mainly to see in dim light rather than to discern colors. The retina typically has two or three **foveae**, areas where cones are concentrated and ultrafine resolution is obtained. (By contrast, the mammalian retina has a single fovea.) Also unique to birds is a cluster of fine blood vessels called the **pecten**, which projects from the front of the retina. The pecten apparently serves to nourish and carry wastes from the retina. It is largest in daylight predators, such as eagles and hawks.

The position of the eyes on the head varies considerably among birds. Species with eyes set on the sides of their head lack binocular vision, but can see backward and forward at the same time. While feeding on the ground or flying, they can see predators approaching from virtually any angle or prey escaping in any direction. Most eagles, hawks, and other predatory birds have large eyes set toward the front of the head. This arrangement allows the bird to concentrate specifically on objects and movements directly in front of it and provides the depth perception needed when a predator dives rapidly to strike prey.

Figure 32.13 A pair of barn owls *(Tyto alba).* These and most other species of owls are specialized for nocturnal hunting. They have exceptional night vision, acute hearing, and almost soundless flight. A row of specially adapted feathers on the front of their wings muffles flight noise to below the frequency audible to most vertebrate ears. Barn owls can locate prey by sound alone.

Hearing. The typical bird ear can resolve sounds into about ten times more individual components than can the human ear. Thus, songbirds can discriminate many more specific sounds in their songs than we can. (The bird ear is described in Chapter 12, p. 285; Figure 12.12b.) Communication by sound is highly significant in the life of most birds, and several birds also hunt by sound. The barn owl can find prey in total darkness by sound detection alone (Figure 32.13). Its eardrums and cochlea are very large, and its external ears are of different sizes and are located at slightly different positions on the sides of the head. As an owl rotates its neck, its asymmetrical ears can pinpoint the position of a mouse moving on the ground.

BEHAVIOR

Most bird behavior is a mix of complex programs and learned activities, but the nature of the bird brain is such that programmed behavior is often predominant. It is sometimes argued that the adaptive advantages of wings and flight were so great that the bird brain was never challenged enough to evolve significant learning capacity. It may be tempting to adopt the attitude that the mammalian brain, with its great capacity for behavioral modification, is behaviorally and evolutionarily superior to the bird brain, but the two systems are so entirely different that value judgments about their relative merits are meaningless. To a large extent, birds are a successful group of vertebrates because their wings allow them to find food and escape predators and because their nervous system programming allows them to carry out many activities more or less automatically, without spending time or energy on learning and practice. (For a comparison of the mammalian and bird brains, see Chapter 28, p. 702.)

Complex programmed behavior in birds ranges from rituals in mating, nest building, and care of the young to various types of feeding and grooming activity. Even unhatched chicks exhibit programmed behavior. They often make peeping sounds, which may help to synchronize hatching; some are known to stop peeping when their parents make certain warning sounds. Food begging is common in young birds. (The role of innate releasing mechanisms in food begging in gulls is discussed in Chapter 13, p. 307; Figure 13.6.) Nestling songbirds gape for food when their parents return to the nest. Both the gaping and the color on the inside of the young birds' mouths act as releasers, causing the parents to open their mouths and disgorge food for the young (Figure 32.14).

Many studies show that birds can modify their behavior. Pigeons have been taught to distinguish between various colors and sounds. Many birds can learn mazes and simple puzzles, but are generally more limited than mammals in their ability to solve problems based on what they have learned. **Imprinting**, one of the simplest forms of learning, has been documented in many bird species. A newly hatched bird learns to follow the first large moving object it sees, whether the object is a parent bird, a person, or a rolling ball.

Many birds seem to have an intense fear of predators, and research indicates that fear responses are basically innate, but may be modified by experience. Quail hatched in an incubator and having no experience with adult birds or predators will scatter and attempt to hide when hawks fly over their pen. Hawklike models pulled repeatedly over a quail pen will in time elicit similar but much less intense reactions.

Communication and Social Behavior

Birds are social animals. As soon as they hatch, a bond forms between young and adult, and most

Figure 32.14 Nestling birds gaping for food. Nestling cedar waxwings *(Bombycilla cedrorum)* gape as one of their parents stands over them.

species spend at least some time in either small or large groups. Even those that are alone much of the time may roost with others at night or spend considerable time with a mate during the breeding season. Social behavior requires communication, and birds have many species-specific sounds and visual signals. Males and females are often colored very differently in what is called **sexual dimorphism**, and colored feathers, beaks, and legs serve a variety of social functions, especially during the breeding season (see Figure 32.3c).

Birds produce sounds in the **syrinx**, a voice box near the posterior end of the trachea (see Figure 32.12). They have a larynx at the top of the trachea, but it is reduced in size and does not function in sound production. Storks and herons have a long, coiled trachea that resonates sounds produced by the syrinx. Songbirds produce two types of sounds: **call notes**, which warn other individuals or advertise one's presence, and **songs**, which are full pronouncements of a bird's (usually a male's) presence (Figure 32.15). Some songbirds inherit only parts of their species' song, and young adults may have to learn to sing whole songs by listening to older birds.

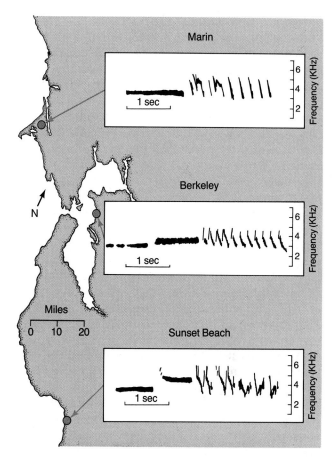

Figure 32.15 Anatomy of a bird song. Songs are species-specific. They can be reproduced as visual displays (spectrograms) that can be analyzed for number of notes, frequency, and amplitude. Here we see that representative songs of the white-crowned sparrow (*Zonotrichia leucophrys*) differ among populations in the San Francisco Bay Area. These differences indicate that the species has regional dialects.

Mobbing

Mobbing is a common form of group behavior in many species of birds. Several small birds will often attack (mob) a larger predator in flight, diving and slashing at the larger bird until it outdistances them or hides. A flock of "mobbers" may increase as other birds in the vicinity join in, attracted by the call notes of their kind. Mobbing seems to be mainly programmed behavior and may serve to harass predators out of the area or to warn other individuals of the predator's presence. In some species, it may also teach young birds how to recognize certain enemies.

Social Hierarchies

Social hierarchies are common and well documented in birds. A hierarchy of dominant and submissive individuals in a flock of domestic chickens, for example, is called a **peck order**. Each bird's position in the order is established by initial contests involving pecking, diving, or jumping at opponents. Eventually, the birds establish a social rank from the most dominant, or alpha, individual (sometimes called the despot) to the beta, then to the gamma, and finally to the most submissive, or omega, bird. Once established, social rank is maintained by nonviolent visual signals and body posturing, allowing the group to function while spending minimal energy in aggressive infighting.

Individuals with high social rank generally have first access to food and water, roosting sites, and, in many species, to mates. In chickens, alpha roosters are almost invariably the first to copulate with hens, while omega roosters may never do so. Even if an omega male is left alone with a flock of hens, he may not mate with any of them, at least not with those of higher social rank than his.

Territoriality

Most birds occupy and defend a parcel of land, at least during the breeding season. (See the discussion about the territories of the tawny owl in Chapter 13, p. 318; Figure 13.18.) As migrating birds return to their breeding grounds in the spring, males often arrive sooner than females and disperse over the suitable habitat. Each male attempts to establish himself in a specific area, advertising his presence by singing, by visual displays, or by actively chasing off other males. As the females arrive, they pair with males that have territories and may participate in territory defense. Birds frequently return to the same territory—often to the same nest site and the same mate—year after year. Yearling birds may return to the place where they hatched, but except when older breeding birds die or are not healthy, young birds are rarely successful in seizing and maintaining territory. The size of the territories varies a great deal with the species. Eagles and ospreys may occupy and defend territories of several thousand square kilometers. At the other extreme, a seabird nesting with thousands of others on a small island may have a territory defined by how far its neck can reach to peck at a neighbor.

Territoriality serves a variety of functions in different species. Most birds defend a single area containing both feeding and nesting grounds, but redwinged blackbirds and certain swans establish territories only for mating and nesting, feeding elsewhere without conflict with others of their kind. Snowy owls defend separate nesting and feeding territories. Most birds are only vaguely associated with territories during nonbreeding seasons, but a few species defend a winter feeding territory separate from that maintained while breeding.

REPRODUCTION AND LIFE CYCLES

Birds reproduce seasonally, and their gonads are inactive and small during the rest of the year, lightening their load for flying. Females of most species have only one ovary and one oviduct (usually the left ones). This lightens the flight load and also makes it possible for birds to produce large eggs that can carry enough yolk to sustain the high metabolic rate of the developing embryo. In a bird with a paired reproductive tract, two eggs could descend the adjacent oviducts simultaneously. After the eggs are covered with shell material, they may jostle against each other and break in the oviducts if the bird lands too abruptly or turns too sharply in flight. Thus, it is not surprising that in the relatively few species with paired reproductive organs, usually only one ovary produces eggs.

Males have a pair of testes. Because the high body temperature of a bird kills sperm, spermatogenesis usually occurs at night, when body temperature is lower. The testes are also cooled somewhat by the air sacs of the respiratory system, and sperm are stored in a scrotumlike seminal vesicle projecting into the cloaca, where the temperature may be nearly 5°C cooler than the rest of the body. During copulation, the male mounts the female, presses his cloaca to hers, and ejaculates sperm.

Life cycles of birds revolve around an annual reproductive period, timed to occur when weather and food supply are most favorable. The reproductive cycle, like molting, is influenced by photoperiod. Birds undergo cyclic changes correlated with reproduction even when held under constant light/dark conditions, but such changes are not automatically timed, and a bird's internal clock needs to be "set" regularly to stay on time with the seasons.

Birds, like most (if not all) animals, have **circadian** (24-hour) rhythms that are influenced by photoperiod and regulated by hormones. In many species, body temperature, metabolism, and blood glucose levels complete a full cycle about every 24 hours, decreasing while the animal is sleeping. At least in some birds (e.g., the house sparrow), these and other circadian rhythms are regulated by secretions (melatonin, serotonin, and norepinephrine) from the pineal body. Because the pineal body seems to be influenced by light that enters via the eye or through the skull, photoperiod may exert its effects daily, fine-tuning a bird's circadian rhythms to fit gradual changes in day length. As a bird's daily rhythms are adjusted, it is prepared gradually for yearly reproductive events.

For a typical bird, the annual reproductive cycle involves a number of intense activities. Many species make an exhausting migration to specific breeding areas, then establish territories, court mates, build elaborate nests, brood eggs, and care for their young, all in a matter of a few weeks.

Seasonal Migrations

Although a number of birds reside in one geographical area year-round, most species migrate seasonally between wintering grounds and summer breeding grounds. For centuries, people have wondered how birds navigate during their migrations and how homing pigeons and other species can find their way back to a winter or summer home after being displaced thousands of kilometers. Bird navigation has been extensively studied, and although some questions have been answered, we do not yet have a complete understanding of how any species of bird navigates. Many birds seem to navigate mainly by the sun and/or stars, but daytime migrants also use landmarks, particularly when near their destination. Many can also detect the earth's magnetic field and may combine this information with bearings on the sun or stars. (See the discussion of navigation in Chapter 13, p. 311.)

Seasonal migrations take an enormous amount of energy, and many birds die enroute; yet so many species migrate that we may assume that the advantages far outweigh the disadvantages. By migrating north in the spring in the northern hemisphere, a bird may gain access to mild, cooler weather, a new growing season's supply of food, and a vast breeding space that would otherwise be used only by a few resident species able to withstand the harsh winters. By migrating south in the fall, birds escape harsh weather and gain access to another supply of food.

There is great variety in the distances that birds migrate and the routes they take. The American robin *(Turdus migratorius)* migrates south in the winter only as far as necessary to find food and may even remain in its summer quarters if food is available. At the opposite extreme, the arctic tern *(Sterna paradisaea)* winters in the Antarctic and breeds in the far north. Birds of the western hemisphere generally take fairly direct routes north and south, for North and South America present no major barriers to them. Some species (including certain hummingbirds) fly over the Gulf of Mexico directly or by island hopping. By contrast, birds of Europe and Asia are confronted with deserts and east-west mountain ranges, and many take a circuitous route to wintering grounds in the Mediterranean, Africa, and Southeast Asia. Some fly nonstop over such barriers as the Sahara Desert.

Many songbirds migrate individually, covering the greatest distances at night and pausing to rest and feed during the day. Their migration routes must include suitable feeding grounds. Vultures, eagles, and other land-soaring birds may migrate along the outer edges of mountain ranges or along north-south river bluffs, where updrafts of air help keep them aloft.

Some birds fly an entire migration route without stopping. Flocks of snow geese are known to fly nonstop about 2700 km from northern Canada to Louisiana in 60 hours. Ducks have been clocked at ground speeds of up to about 80 km/hr, and several species of ducks and geese may fly at high altitudes during migrations. The mallard *(Anas platyrhynchos)* has been recorded at altitudes of about 6400 m; the bar-headed goose *(Anser indicus)* migrates over the Himalaya Mountains flying at altitudes of nearly 8900 m. Such birds are able to obtain enough oxygen from the low partial pressures in the air at such altitudes because they have highly efficient lungs, hemoglobin with a high oxygen affinity, and an unusually high number of capillaries in their flight muscles.

Courtship and Mating

When birds arrive at their breeding grounds and establish territories, they are hormonally prepared for reproduction and usually begin an elaborate series of courting and mating rituals leading to nesting and rearing of young. Intense, often prolonged courtship is vital, for it stimulates mated pairs to stay together, build nests, copulate, and perform all the other activities involved in raising young. In most species, the male postures in specific, programmed ways and exhibits his breeding plumage for the relatively passive female. In response to the male's signals, the female may begin to ovulate and become receptive to copulation. Each species has its own particular courtship activities. Some activities are simple and of short duration, while others are exceedingly complex. (See the discussion of sage grouse leks in Chapter 13, p. 321; Figure 13.20.)

Nesting and Care of the Young

Nearly all birds build nests, brood eggs, and care for their young as part of the yearly reproductive cycle. Nesting is usually stimulated by and integrated into the courtship rituals of a pair. The simplest nests are shallow, unlined depressions on the ground or on loose rocks. The most common nests are cuplike and are formed of dry grass, sticks, feathers, and a variety of loose debris that may be cemented together with mud or saliva. Songbirds generally build the most elaborate nests. The woven, domed houses of the weaverbird are among the most ornate (see Figure 13.3).

Nests have several functions. They are usually hidden and camouflaged, providing sanctuary for eggs and young. Nest materials may resemble the surroundings or the color patterns of eggs. Nesting in trees reduces the possibility of attack by ground predators. Some aquatic birds build floating nests, preventing attack from virtually all land-based predators. Many species nest in large colonies, where sheer numbers increase chances of detecting predators.

A primary role of the nest is in brooding. A nest provides a snug refuge where the parent bird can sit, and nest materials protect the incubating eggs and young from bad weather or direct sunlight when the parent bird is gone. Usually, the female incubates the eggs. Under hormonal influence, she loses feathers on her breast, developing a **brood patch**. Blood vessels in the skin of the patch swell and may provide extra heat for the eggs.

Not all birds brood their eggs themselves. The brush turkey *(Alectura lathami)*, living in the rain forests of New Guinea and Australia, incubates its eggs passively. It piles up a mound of plant debris (often 2–3 m high), then buries its eggs near the top of the mound. As the debris ferments, it generates enough heat to incubate the eggs. A few birds do not care for their eggs or young at all. The European cuckoo *(Cuculus canorus)* and the brown-headed cowbird *(Molothrus ater)* of North America do not even build nests. Both species are **brood parasites**, laying their eggs in other birds' nests and relying on the host birds to rear their young. Usually, the adult host birds will feed and care for the parasite young as though they were their own, and young cuckoos will often push hatchling host birds out of the nest.

As a significant adaptation for terrestrial life, the bird egg provides a supportive environment for the developing embryo. The typical egg shape also has survival value; it keeps an egg from rolling away from a nest, resists pressure from the outside, and, important in hatching, breaks easily with pressure from within. The eggs of most birds hatch after about two to four weeks of incubation. The newly hatched young of most birds are **altricial**, meaning they are nearly naked, have closed eyes, and are helpless. They remain in the nest under constant care of the adults until they are nearly full grown. By contrast, ducks, geese, and chickens are **precocial**, like reptiles. They hatch with a coat of down and open eyes, and they can walk, begin feeding themselves, and leave the nest soon after hatching. Unlike most reptiles, however, they remain under their parents' care for weeks or months. Family groups of the Canada goose usually remain together through an entire year. Hawks, owls, and herons are somewhat intermediate between the altricial and precocial types. As hatchlings, they are covered with down, but remain in the nest for an extended period of time and are fed by their parents.

There is an interesting correlation between brain development and length of time spent in the helpless state. Precocial birds hatch with much more nervous tissue per total body weight than do altricial species. However, by the time altricial birds reach maturity, they have acquired considerably larger brains relative to body size than the average precocial species, and altricial birds appear to be capable of more complex learning.

PHYLOGENY

Figure 32.16 indicates the probable phylogenetic position of the three main groups of birds. Note that *Archaeopteryx*, the only known member of the Subclass Archaeornithes, is placed on a side branch of the avian line. *Archaeopteryx* is not considered the ancestor of modern birds, but it was probably an offshoot of early stocks. Likewise, the toothed birds *(Hesperornis, Baptornis,* and others) of the Cretaceous period are believed to represent a later side branch of the main neornithean line.

Currently, there are two widely discussed hypotheses that attempt to reconstruct the origin of birds. There is general agreement that the bird line arose from **pseudosuchians**, a subgroup of the thecodont reptiles, which were bipedal and had many birdlike skeletal features. (The pivotal role of thecodonts in reptile evolution is discussed in Chapter 31, p. 770; Figure 31.17.) The two hypotheses differ in terms of timing—whether birds arose directly from pseudosuchians or from a group of reptiles derived from pseudosuchians. Debate centers around what reptiles could have been ancestral to a bird somewhat like *Archaeopteryx*.

The **pseudosuchian hypothesis** holds that birds arose from pseudosuchian thecodonts during the Triassic period. Extinct reptiles thought to represent bird ancestors are the pseudosuchians *Euparkeria* (about 230 million years old) or *Sphenosuchus* (about 200 million years old). The pseudosuchian hypothesis accepts that there was an immense time span (from about 50 to 80 million years) between the earliest bird ancestors and the first known bird *Archaeopteryx*.

The alternative proposition, the **dinosaur hypothesis**, postulates a much later origin of birds. In 1868, Thomas Huxley proposed that birds evolved from a group of carnivorous dinosaurs, the **coelurosaurs**, during the Jurassic period, about 180 million years ago. As shown in Figure 32.16, coelurosaurs were derived from pseudosuchian stock. Proponents of the dinosaur idea base their arguments on the similarity between *Archaeopteryx* and some of its contemporaries, especially the small (about 1 m long), birdlike coelurosaur *Comp-*

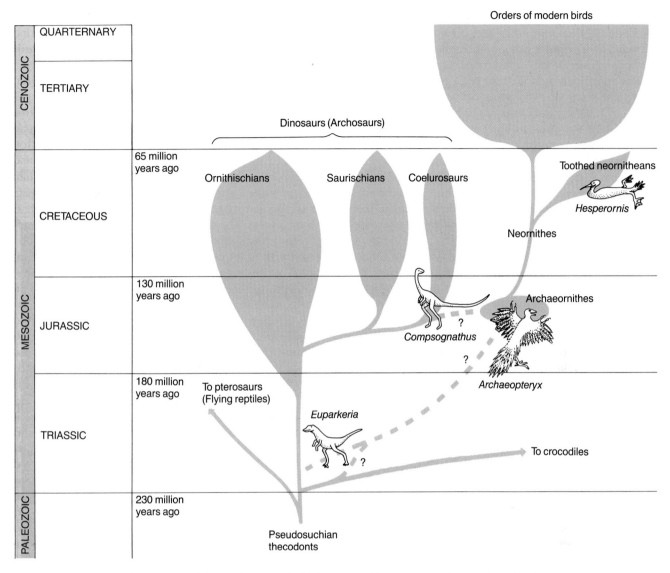

Figure 32.16 Phylogenetic tree of the birds. Two models are proposed for the origin of birds (Archaeornithes). The dinosaur model, involving the coelurosaurs, and the pseudosuchian model are indicated by broken lines.

sognathus (see Figure 32.16). The skeleton of *Compsognathus* was so much like that of *Archaeopteryx* that fossils of these two genera have sometimes been mistaken for each other. Without feather impressions and evidence of the furcula in specimens of *Archaeopteryx*, some of them are virtually indistinguishable.

There are strong arguments for and against each of these hypotheses. Adherents to the pseudosuchian hypothesis point out that the coelurosaurs were mainly a Cretaceous group, appearing later than *Archaeopteryx* (see Figure 32.16). They believe that coelurosaurs were too specialized to

have been bird ancestors. They also contend that the furcula is a significant skeletal adaptation, and its appearance in *Archaeopteryx*, but not in *Compsognathus*, means that these animals were not closely related. Thus, proponents of the pseudosuchian hypothesis believe that convergent evolution accounts for the resemblance between *Archaeopteryx* and *Compsognathus*. The pseudosuchian hypothesis is supported by recent studies showing that the inner ears of fossil birds were like those of pseudosuchians, but not like those of coelurosaurs.

Proponents of the dinosaur hypothesis argue

against the pseudosuchian view by pointing out that if *Archaeopteryx* arose directly from pseudosuchians, then intermediates between reptiles, such as *Euparkeria*, and *Archaeopteryx* would have been fossilized and discovered. Their main argument in favor of a dinosaur origin for birds is that the similarities between *Archaeopteryx* and coelurosaurs are too great to have arisen by convergence. They consider *Compsognathus* a close relative of *Archaeopteryx*, and they also think it likely that coelurosaurs had birdlike features in addition to their skeletons. They postulate that some of these dinosaurs had feathers or featherlike scales, were homeothermic, built nests, and cared for their young.

Support for one or the other of these hypotheses has waxed and waned over the past century, but the pseudosuchian hypothesis dominated thinking about bird phylogeny for about 50 years prior to the mid-1970s. Recent support for the dinosaur theory has refueled the controversy, and with the controversy has come a resurgence of interest and productive research on bird evolution.

Questions about the origin of birds are closely intertwined with questions about how flight evolved. Flight in birds depends on feathers, and the evolution of specialized flight feathers was probably coincident with the evolution of general flight ability in ancestral birds. The oldest remains of feathers are those of *Archaeopteryx*, and these were highly derived structures. *Archaeopteryx*'s wing feathers, like those of modern flying birds, had asymmetrical vanes (see Figure 32.6a). The first feathers may have been virtually indistinguishable from the elongate reptile scales from which they were derived.

Two models, first proposed in the late 1800s, attempt to explain how flight originated. Both assume that *Archaeopteryx*, and probably its ancestors (either pseudosuchians or coelurosaurs), were active bipedal carnivores, perhaps feeding largely on insects. One model, called the **cursorial hypothesis**, holds that bird ancestors were ground-dwelling dinosaurs that made a living by running and jumping after flying insects. Presumably, the faster, more agile individuals, which could leap farther off the ground and become airborne momentarily, would have had an adaptive advantage. Selection may have favored individuals with flattened, winglike forelimbs and muscles that gave added lift and control to the body. If broad, elongate scales developed on the forelimbs, these animals would have been even more effective in leaping and gliding after insects and may have eventually developed flapping ability.

An alternative model, the **arboreal hypothesis**, supposes that bird ancestors and *Archaeopteryx* were tree-dwelling gliders. Such animals may have evolved flat, feathered forelimbs as an adaptation providing more effective gliding. Flapping flight may have evolved as an improvement on the original gliding locomotion as breast muscles became adapted for moving the forelimbs up and down. Proponents believe that *Archaeopteryx* used its clawed fingers to grip and climb trees.

The arboreal model has had more support from prominent ornithologists over the past century. It is supported by the fact that other vertebrate fliers (such as bats) and gliders (such as flying squirrels, flying lemurs, and certain lizards) have all evolved from arboreal ancestors. On the other hand, proponents of the cursorial hypothesis point out some possible problems with the arboreal model. They doubt that selection pressures would have favored the development of flapping in a gliding, arboreal reptile. Their calculations indicate that if such an animal began to flap while gliding downward from a tree, it could have lost lift, rather than gained it.

As with the pseudosuchian/dinosaur debate, it is important to have these opposing views of flight evolution. They have stimulated a great deal of research in recent years, and it will be interesting to see how the controversy develops in the future.

☐ S U M M A R Y

1. Key Features of Birds. Birds are bipedal homeothermic vertebrates with feathers, wings, clavicles fused to form a furcula (wishbone), amniotic eggs, no teeth, reptilelike scales on the legs, a single oil (uropygial) gland in the skin, and a light skeleton containing air spaces. In addition, birds communicate vocally, build nests, and care for their young.

2. Avian adaptations that decrease body weight for flight include a thin skin with feathers; a light epidermal beak; a light skeleton with hollow bones often containing air sacs that branch through the body; loss of many skeletal elements, including tail vertebrae and teeth; a diet of foods that require little time to store and digest; lack of a urinary bladder (except in flightless birds) and excretion of uric acid; development of only one gonad in the female and reduction of gonadal size during nonreproductive seasons; and the laying of eggs instead of carrying and giving birth to young, as in mammals.

3. Adaptations related to providing power for flight include an insulated, homeothermic body that provides a high metabolic rate; rapid, high-pressure circulation, including separate pulmonary and systemic circuits; highly efficient ventilation of the gas exchange system; and high-energy diets and rapid digestion.

FURTHER READING

Attenborough, D. *Life on Earth. A Natural History.* Boston: Little, Brown, 1979. *Contains a thoughtful, beautifully illustrated chapter entitled "Lords of the Air."*

Birds. Readings from Scientific American. San Francisco: W. H. Freeman, 1980. *Twenty-five articles on various aspects of ornithology, written by experts.*

Dolesch, R. J., and C. Davidson. "Lord of the Shallows. The Great Blue Heron." *National Geographic* 165(1984): 540–554. *Focuses on a long-term study of a colony of herons on the Chesapeake Bay.*

Dunstan, T. C., and J. Foott. "Our Bald Eagle: Freedom's Symbol Survives." *National Geographic* 153(1978): 186–199. *A short, well-illustrated description of field studies.*

Ewald, P. W., and R. A. Tyrrell. "Hummingbirds: The Nectar Connection." *National Geographic* 161(1982): 222–227. *A brief essay with superb stop-action photographs.*

Feduccia, A. *The Age of Birds.* Cambridge, Mass.: Harvard University Press, 1980. *An authoritative treatment of bird evolution.*

Fisher, A. C., and J. Blair. "Mysteries of Bird Migration." *National Geographic* 156(1979): 154–193. *Describes some of the research on how birds navigate.*

Lofgren, L. *Ocean Birds.* London: Croom Helm, 1984. *Discussion of ecology, evolution, and behavior of seabirds, superbly illustrated.*

Nuechterlein, G. L. "Western Grebes. The Birds That Walk on Water." *National Geographic* 161(1982): 624–636. *Describes the fascinating breeding behavior of this North American species.*

Perrins, C. M., and T. R. Birkhead. *Avian Ecology.* London: Blackie and Son, 1983. *A brief, readable introduction to behavioral ecology of birds.*

Perrins, C. M., and A. L. A. Middleton, eds. *The Encyclopedia of Birds.* London: Allen & Unwin, 1985. *Birds of the world illustrated in color, with contributions from 87 authorities.*

Stacey, P. B., and W. D. Koenig. "Cooperative Breeding in the Acorn Woodpecker." *Scientific American* 251(1984): 114–121. *Describes a species that often lives in territorial groups in which members share mates and collectively store food, defend territory, and raise young.*

Welty, J. C. *The Life of Birds.* 3d ed. Philadelphia: Saunders, 1982. *A comprehensive textbook in ornithology.*

Whitson, M. A., and B. Dale. "The Roadrunner. Clown of the Desert." *National Geographic* 163(1983): 694–702. *Portrays some of the adaptations of this North American cuckoo to its desert environment.*

Zeleny, L., and M. L. Smith. "Song of Hope for the Bluebird." *National Geographic* 151(1977): 854–865. *Describes attempts to conserve the three species of bluebirds native to North and Central America.*

CLASSIFICATION

Class Aves (L. *aves,* "bird").

Subclass Archaeornithes (Gr. *archaios,* "ancient"; *ornithos,* "bird"). Ancestral, reptilelike birds with true teeth, clawed toes, and wings; *Archaeopteryx.*

Subclass Neornithes (Gr. *neos,* "new"; *ornithos,* "bird"). Modern ("true") birds; 28 extant orders; about 8600 species.

Order Struthioniformes (L. *struthio,* "ostrich"; *forma,* "shape"). Ostriches.

Order Rheiformes (Gr. *Rhea,* "mother of Zeus"; L. *forma,* "shape"). Rheas.

Order Casuariformes (L. *Casuarius,* "cassowary"; *forma,* "shape"). Emus, cassowaries.

Order Apterygiformes (Gr. *a,* "without"; *pteryx,* "wing"; L. *forma,* "shape"). Kiwis.

Order Tinamiformes (L. *Tinamus,* "tinamous"; *forma,* "shape"). Tinamous; about 60 species of weak-flying, grouselike birds of Central and South America.

Order Gaviiformes (L. *gavia,* "a type of bird"; *forma,* "shape"). Loons.

Order Podicipediformes (L. *podex,* "rump"; *pedis,* "foot"; *forma,* "shape"). Grebes.

Order Procellariiformes (L. *procella,* "tempest"; *forma,* "shape"). Albatrosses, petrels, shearwaters.

Order Sphenisciformes (Gr. *spheniskos,* "wedge"; L. *forma,* "shape"). Penguins.

Order Pelecaniformes (Gr. *pelekan,* "pelican"; L. *forma,* "shape"). Pelicans, boobies, cormorants.

Order Ciconiiformes (L. *ciconia,* "stork"; *forma,* "shape"). Herons, egrets, storks, ibises, spoonbills.

Order Phoenicopteriformes (Gr. *phoinikos,* "crimson"; *pteron,* "wing"). Flamingos.

Order Anseriformes (L. *anser,* "goose"; *forma,* "shape"). Ducks, geese, swans.

Order Falconiformes (L. *falco,* "falcon"; *forma,* "shape"). Birds of prey, including falcons, hawks, eagles, ospreys, buzzards, condors.

Order Galliformes (L. *gallus,* "cock"; *forma,* "shape"). Domestic chicken (derived from jungle fowl), turkeys, grouse, pheasants, quail, ptarmigan, peafowl.

Order Gruiformes (L. *grus,* "crane"; *forma,* "shape"). Cranes, coots, rails.

Order Charadriiformes (L. *Charadrius,* "plovers"; *forma,* "form"). Shorebirds ("waders"), gulls, terns, sandpipers, plovers, snipe, killdeer, and many others.

Order Columbiformes (L. *columba,* "dove"; *forma,* "shape"). Doves, pigeons.

Order Psittaciformes (L. *psittacus,* "parrot"; *forma,* "shape"). Parrots, parakeets, budgerigars.

Order Cuculiformes (L. *cuculus,* "cuckoo"; *forma,* "shape"). New- and old-world cuckoos, roadrunners.

Order Strigiformes (L. *strix,* "screech owl"; *forma,* "shape"). Owls.

Order Caprimulgiformes (L. *caprimulgus,* "goatsucker"; *forma,* "shape"). Nighthawks, whip-poor-will.

Order Apodiformes (Gr. *apous,* "footless"; L. *forma,* "shape"). Hummingbirds, swifts.

Order Coliiformes (Gr. *kolios,* "green woodpecker"; L. *forma,* "shape"). Colies (mousebirds).

Order Trogoniformes (Gr. *trogon,* "gnawing"; L. *forma,* "shape"). Trogons, quetzal.

Order Coraciiformes (Gr. *korax,* "raven or crow"; L. *forma,* "shape"). Hornbills, kingfishers.

Order Piciformes (L. *picus,* "woodpecker"; *forma,* "shape"). Woodpeckers, sapsuckers, honeyguides, toucans.

Order Passeriformes (L. *passer,* "sparrow"; *forma,* "shape"). Perching songbirds (sparrows, swallows, robins, crows, jays, finches, chickadees, wrens, starlings, thrushes, and over 5000 others).

Mammals

The duck-billed platypus (Ornithorhynchus anatinus). This semi-aquatic species, shown here resting on a rocky shelf, occurs in eastern Australia and Tasmania. It is one of only three living species of egg-laying mammals, the monotremes. Platypuses may live up to about 15 years.

Members of the **Class Mammalia** are endothermic (mostly homeothermic) amniotes that have hair and that nourish their young with milk secreted by the female's mammary glands. Homeothermy provides a degree of independence from environmen-

tal temperature extremes, and as a result, mammals are represented in virtually all oceans and on every continent except Antarctica. Many mammals have a dense coat of hair called the **pelage** that helps retain metabolic heat and also helps reduce absorption of heat from the environment. Species with sparse hair typically have very thick skin or a thick, insulating layer of fat under the skin. About 4500 species of mammals are known, and the class holds the distinction of containing the world's largest animal, the blue whale (*Balaenoptera musculus*), reaching a length of 30 m and weighing up to 102,000 kg.

All mammals copulate and have internal fertilization, and all mammals are viviparous except for three oviparous (egg-laying) species—the duck-billed platypus (see chapter opening photograph) and two species of spiny anteaters. Viviparous mammals are classified as either marsupials or **eutherians** (placental mammals). Marsupials give birth to tiny, embryonic offspring that complete development while attached to a nipple on the abdomen of the female. In about half of the living marsupial species, nursing young are contained in an external pouch, or **marsupium**, on the mother's abdomen. In eutherians, embryos remain in the uterus through the fetal period, and the **placenta** provides the developing fetus with nutrients from the mother's bloodstream. The eutherian placenta develops from embryonic trophoblast cells that grow out from tissue surrounding the developing embryo and make contact with the uterine wall (see Figure 16.13). Marsupials also have a type of placenta (described on p. 814), but only in one group, the bandicoots (see Figure 33.3c), is the placenta similar to that of eutherians. In all mammals, close contact and strong bonding between the mother and offspring may promote learning and may predispose the young for social interactions as adults.

Mammalian Teeth and Dental Formulas

Teeth are among the most distinctive features of mammals. They are also readily fossilized, and much of what is known about mammalian evolution has been reconstructed from fossil teeth. As a group, mammals are **heterodont**, meaning they have several (up to four) different kinds of teeth. Anterior in the jaw, **incisors** and **canine teeth** provide cutting edges that often serve to capture and kill prey; behind the canines, the **premolars** and **molars** (the so-called cheek teeth) provide cutting and grinding surfaces important in chewing food (Figure 33.1). Most mammals grow a set of **deciduous teeth**, also called milk teeth, which are shed and replaced by permanent teeth as the jaws grow. Deciduous teeth do not include molars. In many mammals, such as humans, teeth cease to grow after they reach full size; but in some species—especially herbivores, whose teeth are heavily abraded by coarse plant material—some or all teeth grow continuously. Teeth are vital to survival in most species. If they erode and are not replaced, the animal may starve.

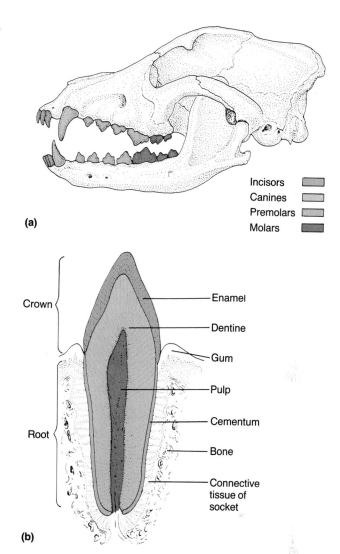

Incisors
Canines
Premolars
Molars

(a)

Crown

Enamel
Dentine
Gum
Pulp
Cementum
Bone
Connective tissue of socket

Root

(b)

Figure 33.1 Mammalian teeth. Most mammals have highly differentiated teeth. (a) The skull of a dog, showing the four types of teeth and their positions in the upper and lower jaws. (Figure 4.3 illustrates the teeth of several other mammals, including the human.) (b) The structure of an incisor.

The typical mammalian tooth consists of an inner mass of bonelike material (hydroxyapatite and organic fibers) called **dentine** (Figure 33.1b). The dentine surrounds a pulp cavity that contains blood vessels and nerves from the dermis of the surrounding gum. The tooth has a crown and one or more roots, which are set in sockets in the jawbone. A relatively soft, bonelike material called **cementum** covers the roots and helps anchor the tooth to connective tissue in the sockets. Cementum also covers part of the crown in some mammals. In most teeth, the crown is covered by **enamel**, which, being about 98% hydroxyapatite, is the hardest material in the vertebrate body.

Adaptive modification of teeth has been a central feature of mammalian evolution and diversification. Teeth in many groups of mammals are highly specialized for specific types of feeding, and many groups of mammals are distinguished by specific characteristics of their teeth. (Adaptation of teeth to different feeding habits is discussed in Chapter 4, p. 78; Figure 4.3.) Many groups are diagnosed by their **dental formula**, which is an expression of the number of incisors, canines, premolars, and molars on one side (left or right) of the upper jaw over the number of each type of tooth on one side of the lower jaw. For example, a mammal whose dental formula is 3/3, 0/1, 3/3, 2/3 would have a total of 36 teeth; each side of its upper jaw would have three incisors, no canines, three premolars, and two molars; each side of its lower jaw would have three incisors, one canine, three premolars, and three molars.

REPRESENTATIVE MAMMALS

There are 20 orders of living mammals making up two subclasses. Oviparous (egg-laying) mammals constitute the Subclass Prototheria. The viviparous marsupials and placental mammals constitute the Subclass Theria.

Subclass Prototheria

The three living species of prototherians constitute a single order, the **Monotremata**. The name Monotremata (from the Greek *monas*, "single" and *trema*, "hole") refers to the presence in these mammals of a single opening through which digestive, urinary, and reproductive products are released from a common chamber, the cloaca. The same situation prevails in birds and reptiles. Like birds, **monotremes** lay eggs and incubate them. Also in common with birds, adult monotremes are toothless, and the skull has a beaklike anterior projection called the **rostrum**. The rostrum is covered with a leathery sheath, and hard epidermal plates or spines form chewing surfaces in the mouth.

One species of monotreme, the **duck-billed platypus**, lives in eastern Australia and Tasmania (see chapter opening photograph). This species eats mainly small aquatic invertebrates caught by probing the bottom deposits of streams and lakes with its flexible bill. Female platypuses lack nipples, but secrete milk onto their abdominal fur. A female usually lays a single egg in a nest built in a burrow in a stream bank. She incubates the egg until hatching (about twelve days) and typically nurses and cares for her young for about four months.

Spiny anteaters, also called **echidnas**, occur in Australia, New Guinea, and Tasmania. These monotremes are covered with protective spines and have short, powerful limbs adapted for digging (Figure 33.2). When disturbed, they curl into a ball or burrow into the soil, thus offering only their sharp spines to predators. Spiny anteaters eat mainly ants and termites or other soil arthropods and earthworms; they catch prey with a sticky saliva that coats their tongue. In common with the platypus, females rear only one offspring at a time; a female echidna develops a temporary abdominal pouch in which she incubates her egg. Hatching usually occurs in about ten days, but the young echidna remains in the pouch for nearly two months, sucking milk from two nipplelike skin projections.

Subclass Theria

Members of this group, the marsupial and placental mammals, are classified in two infraclasses, the Metatheria and the Eutheria, respectively.

Figure 33.2 Order Monotremata. The spiny anteater, or echidna (*Tachyglossus aculeatus*), is a small (about 3–6 kg) egg-laying mammal that lives in Australia. Note its beaklike snout, which is supported internally by an elongate rostrum of the skull. In response to cold temperature, *T. aculeatus* saves energy by becoming torpid; its body temperature may decrease to about 6°C, and its heart rate may slow to about seven beats per minute. Echidnas have been known to live for 50 years in captivity.

The Metatheria contains a single order, the Marsupialia.

Infraclass Metatheria. Order Marsupialia. Some female marsupials are easily recognized by their prominent marsupium. However, many species develop only a temporary marsupium during the breeding season, and some small terrestrial species lack a pouch altogether. Marsupials have dental formulas derived from an ancestor whose formula was 5/4, 1/1, 3/3, 4/4, and most species have an unequal number of incisors in the upper and lower jaws. The rodentlike wombats (three species in Australia) are the exceptions; they have only one incisor on each side of the upper and lower jaw.

Of about 250 species of marsupials, about 175 occur only in Australia, New Guinea, Tasmania, and neighboring islands. Almost all other species occur in South or Central America. No indigenous marsupials now occur in Europe or Africa, although fossil species have been found on both of these continents. Only one family of marsupials, the Didelphidae (opossums), occurs in North America (including Central America); and only one species, *Didelphis marsupialis* (the American opossum), occurs north of Mexico (Figure 33.3a).

The island continent of Australia has been a marsupial sanctuary for much of the past 60 million years. Marsupials seem to have originated in either North or South America and apparently invaded Australia during the Cretaceous. At that time, the South American land mass was connected to the Antarctic land mass. (See the discussion of continental drift in Chapter 17, p. 417.) Australia apparently was separate at the time, but it remained close to Antarctica until it drifted north-

(a)

(b)

(c)

Figure 33.3 (Right) Order Marsupialia. (a) The American opossum *(Didelphis marsupialis)* belongs to the oldest known family of marsupials. Dating from the late Cretaceous, the Family Didelphidae arose in South America and apparently gave rise to all other living marsupials. (b, c) Australian marsupials. (b) The great gray kangaroo *Macropus giganteus* (up to about 2 m tall and about 90 kg in weight) is the largest living marsupial. This individual, a female, is carrying a young joey in her pouch. Kangaroos and wallabies make up the Family Macropodidae (about 60 species in all). Macropodids are mainly bipedal; larger species, such as *M. giganteus*, can run at speeds of nearly 90 km/h, leaping as high as 3 m and clearing about 13 m distance in a single jump! Most kangaroos are ground-dwellers, but some have strong forelimbs with large claws used for climbing trees. (c) The bandicoot *Perameles nasuta* is a rat-size omnivore, supplementing its mainly insect diet with small vertebrates and plant material. Bandicoots have a placenta like that of eutherian mammals, but the young remain in the uterus for only about two weeks. The bandicoot's pouch opens to the rear, protecting the young from dirt as the mother digs.

ward after the Cretaceous. Apparently, during the Cretaceous, some marsupials were able to cross the relatively narrow stretch of ocean separating Antarctica and Australia, but the eutherian (placental) mammals that were present in South America with marsupials were not able to do so. Why placentals failed to make the crossing is puzzling; perhaps none of them in a geographical position to disperse to Australia possessed adaptations allowing survival of the oceanic trip; or perhaps marsupials were simply more numerous and widespread at the time. Whatever the cause, Australian marsupials have been relatively isolated from placentals for over 50 million years, and they have diversified extensively, filling terrestrial niches occupied by
▷ placentals on other continents. (Convergent evolution is discussed in Chapter 17, p. 412.) Thus, Australia has kangaroos and wallabies among its dominant herbivores, occupying niches filled by cattle, deer, and antelope on other continents (Figure 33.3b). Marsupial scavengers and carnivores in Australia include the Tasmanian devil *(Sarcophilus harrisii)*, the native "cat" *(Dasyurus viverrinus)*, and the Tasmanian wolf *(Thylacinus cynocephalus)*, which is probably now extinct. Other distinctive Australian marsupials include the herbivorous wombats, the unique koala, or native "bear" (see Figure 4.2a), and the mainly insectivorous bandicoots (Figure 33.3c). Indigenous Australian placentals include several bats and rodents; a wild dog, the dingo, was apparently introduced by native peoples from New Guinea. Recent human introductions of placentals, such as cattle, sheep, horses, and rabbits, pose a serious threat to many of the native marsupials.

Infraclass Eutheria. As a group, eutherians inhabit every major type of aquatic and terrestrial environment. The dental formulas of eutherians are quite varied and usually serve to distinguish between taxa below the level of order, but all were derived from ancestors with the formula 3/3, 1/1, 4/4, 3/3. In this section, we discuss 15 of the 18 extant eutherian orders; the remaining 3 orders are described briefly in the Classification section at the end of the chapter.

Members of the **Order Insectivora** (nearly 400 species), including shrews, hedgehogs, and moles, occur in North America, Eurasia, and Africa. Insectivorans are small terrestrial or semiaquatic mammals with long, tapered snouts, minute eyes, and short, velvetlike fur. Each of their four feet bears five digits with claws. The dental formulas of insectivorans are varied, but they generally resemble the ancestral eutherian type. In many species,

the incisors are enlarged and the canines reduced. The teeth typically bear sharp cusps. Most insectivorans are nocturnal, feeding voraciously on insects, earthworms, and other small soil invertebrates. The short-tailed shrew *Blarina brevicauda* and several other species have venomous saliva (Figure 33.4a). Extinct insectivoran-like mammals are thought to have given rise to all other eutherians.

The **Order Chiroptera** (bats, about 925 species) is the second most diverse eutherian order (after rodents). Most bats are insectivorous, with crescent-shaped teeth that mascerate insects into a paste. There are also bird-, fish-, frog-, bat-, and plant-eating bats. Herbivorous bats consume fruits or nectar; fruit eaters have flattened, crushing molars. Bats range in size from small nectar-feeding species, weighing only about 3–4 g, to large fruit-eating bats, weighing up to about 1.5 kg and having wing spreads of 1.2 m. Bats occur throughout the world except in colder climates where there are no trees. At the cooler extremes of their range, they hibernate or migrate to warmer areas in the winter. As the only mammals adapted to active flight, bats have forelimbs modified as wings. Unlike the feathered wings of birds, bat wings consist of two layers of skin and connective tissue stretched between the elongated bones of fingers 2 through 5 (Figure 33.4b). Bats are mainly nocturnal, and as a result, they avoid direct competition with most birds, which sleep at night. During daytime, most bats seek shelter in caves, rock crevices, and hollow trees, where they hang head-down ready to launch into flight if disturbed. They often live in large colonies, sometimes exceeding a million individuals, although recently in some areas their numbers have been seriously depleted by humans. Large caves with resident bats may accumulate layers of bat feces thick enough to mine for fertilizer. Bats have eyes, but many species use echolocation to navigate and locate prey (see p. 811). Vampire bats *(Desmodus)* attack large mammals and birds, cutting through the skin with razor-sharp incisors and canine teeth and lapping up blood as it oozes from the wound (see Figure 4.5). They occasionally attack humans and may transmit rabies. A vampire bat can drink large amounts of blood because it has a highly distensible stomach; and it can fly while engorged with blood because it is able to lighten its load by urinating while feeding.

Members of the **Order Primates** are mainly arboreal. With few exceptions (including the human species *Homo sapiens*) primates occur only in tropical and subtropical Africa, Southeast Asia, and South America. The order is subdivided into two

(a)

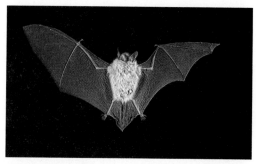

(b)

(c)

(d)

suborders, the **Strepsirhini** (about 35 species) and the **Haplorhini** (about 140 species). Strepsirhines, including the lemurs (see Figure 1.18e), lorises, and galagos, are relatively primitive primates limited to the old-world tropics. Haplorhines include the tarsiers, monkeys, marmosets, great apes, and the human (Figure 33.4c, d).

The **Order Carnivora** includes the terrestrial carnivorans (dogs, wolves, bears, raccoons, skunks, mink, otters, weasels, mongooses, hyenas, and cats, about 240 species) and a smaller group (about 35 species) of aquatic carnivorans (seals, sea lions, and the walrus) (Figure 33.5a, b). Despite the name of the order, not all carnivorans are carnivores. Most bears and raccoons are omnivores, and the giant panda, a member of the bear family, eats mainly bamboo shoots. Typical carnivorans have 3/3 incisors and 1/1 canine teeth, which are large and conical (see Figure 33.1). Terrestrial carnivorans are represented throughout the world. Their limbs are adapted for running, and they have a well-developed sense of smell. Aquatic carnivorans occur along continental coasts. In contrast to whales and dolphins, they are not completely aquatic: They must leave the water to bear young, and they use their appendages for locomotion in water and on land.

The hoofed mammals, often collectively called the **ungulates**, include the orders **Perissodactyla** (horses, rhinoceroses, and tapirs; Figure 33.5c), and the much more diverse **Artiodactyla** (camels, pigs, deer, antelope, hippopotamuses, and cattle; Figure 33.5d). Ungulates are herbivorous animals that probably originated in forests, but underwent their most extensive evolutionary radiation in grasslands. Their limbs are adapted for running, either to escape predators or to migrate. In both orders, the digits of the feet are elongated and the weight of the body is borne directly on the tips of the digits. In the perissodactyls, the main axis of the limb passes through the third digit, whereas in the artiodactyls, usually two of the digits bear the weight, giving these animals the common name cloven-hoofed mammals. As herbivores, ungulates have teeth highly modified for chewing coarse plant material. Canines are often reduced, al-

Figure 33.4 (Left) Infraclass Eutheria (placental mammals). (a) Order Insectivora. (a) The short-tail shrew *Blarina brevicauda* is among the most numerous small mammals in North America. (b) Order Chiroptera (flying mammals, or bats). This photograph of Natterers bat *(Myotis natterei)* shows how bones support the wings. (c, d) Order Primates. (c) The tarsier *Tarsius syrichta*, a haplorhine, is a nocturnal insectivore of tropical jungles in Indo-Australian islands. (d) Baboons, such as this family group of *Papio,* are ground-dwelling primates with complex social behavior.

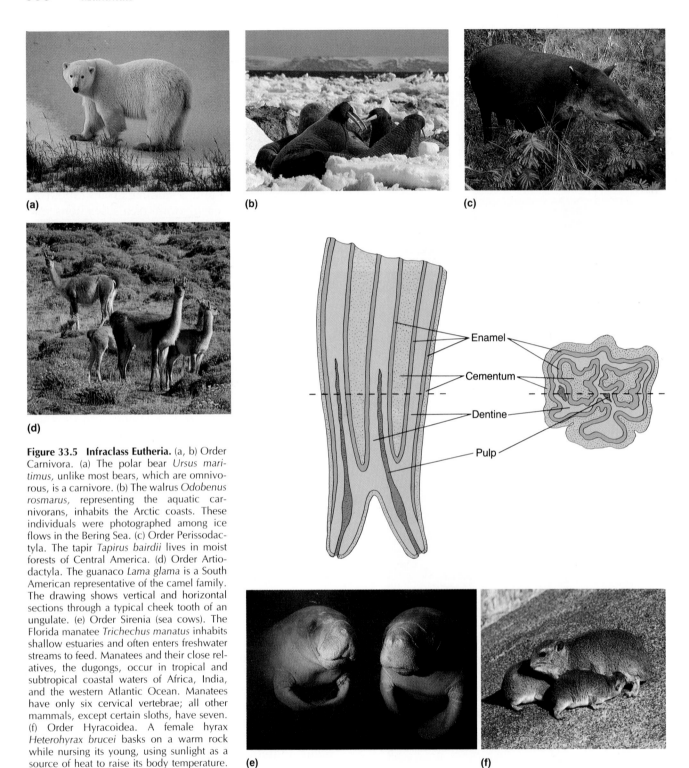

(a)

(b)

(c)

(d)

Figure 33.5 Infraclass Eutheria. (a, b) Order Carnivora. (a) The polar bear *Ursus maritimus,* unlike most bears, which are omnivorous, is a carnivore. (b) The walrus *Odobenus rosmarus,* representing the aquatic carnivorans, inhabits the Arctic coasts. These individuals were photographed among ice flows in the Bering Sea. (c) Order Perissodactyla. The tapir *Tapirus bairdii* lives in moist forests of Central America. (d) Order Artiodactyla. The guanaco *Lama glama* is a South American representative of the camel family. The drawing shows vertical and horizontal sections through a typical cheek tooth of an ungulate. (e) Order Sirenia (sea cows). The Florida manatee *Trichechus manatus* inhabits shallow estuaries and often enters freshwater streams to feed. Manatees and their close relatives, the dugongs, occur in tropical and subtropical coastal waters of Africa, India, and the western Atlantic Ocean. Manatees have only six cervical vertebrae; all other mammals, except certain sloths, have seven. (f) Order Hyracoidea. A female hyrax *Heterohyrax brucei* basks on a warm rock while nursing its young, using sunlight as a source of heat to raise its body temperature.

Enamel

Cementum

Dentine

Pulp

(e)

(f)

though they are massively developed as tusks in pigs and hippopotamuses. Premolars are generally broad and much like molars, and both types have well-developed grinding cusps (see the inset in Figure 33.5d). One group of artiodactyls, the ruminants (including cattle, sheep, goats, camels, giraffes, deer, and antelope), have a highly modified digestive system that allows them to digest cellulose. (Digestion in the cow stomach is covered in Chapter 4, p. 90; Figure 4.13.) Perissodactyls house cellulose-digesting bacteria in a well-developed cecum extending from the junction of the small and large intestines.

Elephants, constituting the **Order Proboscidea**, and a small group of aquatic mammals, the **Order Sirenia** (sea cows, or manatees and dugongs), seem to have little in common. However, the fossil record indicates that these groups and a third eutherian order, the **Hyracoidea** (hyraxes), arose from a common ancestor on the African continent. All of these eutherians are herbivorous, and because they were probably derived from hoofed mammals (ungulates) or from stocks related to ungulates, they are often called **subungulates**. In common with perissodactyls, they have a well-developed cecum containing cellulose-digesting bacteria.

There are two living species of proboscideans, the African elephant and the Indian elephant; weighing up to nearly 6000 kg, these are the largest living land animals. Elephant teeth are highly distinctive; the second upper incisors form the long tusks, and the cheek teeth, adapted for grinding plant material, are replaced as they wear down. Sirenians (four living species) lack hind limbs and are rather cumbersome looking; like whales, they have a horizontally flattened fluke used in swimming (Figure 33.5e). Manatees lack incisors, and they have numerous cheek teeth (usually five to eight are functional) that are replaced as they wear down. Hyraxes (seven species) bear superficial resemblance to rodents (see Figures 33.5f and 33.7a). Like rodents, they lack canine teeth and have long, ever-growing incisiors. The hyrax dental formula is 1/2, 0/0, 4/4, 3/3.

The eutherian **Order Xenarthra** (anteaters, tree sloths, and armadillos, about 30 species) is restricted to the new-world tropics. The name Xenarthra refers to the presence of extra articulations between vertebrae in the lumbar region of the backbone of these mammals. Xenarthrans are also distinctive in that they lack incisors and canine teeth; cheek teeth (absent in some species) lack enamel. Xenarthrans are believed to have originated in North America, and fossil species recovered from the North American continent outnumber living species by over ten to one.

Anteaters are solitary terrestrial mammals with poor eyesight and good hearing. Their feet have two large claws used to tear logs and insect nests apart. Lacking teeth, they capture ants and termites with a long, sticky tongue (Figure 33.6a). Sloths are herbivores that spend most of their time hanging upside down from tree branches. They may even sleep, mate, and give birth in this posi-

tion (Figure 33.6b). If removed from a tree, a sloth has difficulty standing and can move only by dragging itself with its forelimbs. The body temperature of sloths is low (about 24–33°C) and poorly controlled. Helping them to retain metabolic heat, sloths have long, thick fur and heat-retaining countercurrent blood flow in their legs.

In contrast to the rather specialized diets of anteaters and sloths, the armadillo diet consists of a variety of small animals and plant material. Armadillos are distinctive in having a protective double-layered body covering composed of dermal bone and horny plates. As discussed in Chapter 9, armadillos are especially strong diggers (see Figure 9.16).

The smallest order of eutherians, the **Tubulidentata**, contains a single living species, the aardvark *Orycteropus afer* (Figure 33.6c). This mammal lives in Africa south of the Sahara Desert, and the word *aardvark* is derived from an Afrikaans word meaning "earth pig." Aardvarks are nocturnal insectivores, feeding mainly on termites and ants. They have strong forelegs used mainly in digging extensive burrows. In common with xenarthrans, the aardvark lacks canine teeth and incisors; its dental formula is 0/0, 0/0, 2/2, 3/3.

Another small eutherian order, the **Pholidota**, contains one genus with eight living species of insectivores called **pangolins**, or scaly anteaters. Found in Asia and Africa, pangolins are covered dorsally with scales, which are actually fused hairs; a few single hairs project among the scales (Figure 33.6d). Pangolins lack teeth; they locate insect nests by odor and catch prey with their tongue, which is extremely long and sticky.

The totally aquatic mammals—porpoises, dolphins, and whales—are classified in the **Order Cetacea** (about 80 species), all highly intelligent with strong learning capacity. Cetaceans form social groups and communicate by means of a complex language of clicks, whistles, grunts, and groans. The cetacean body is highly adapted for life in the water. Its torpedo-like shape and general hairlessness reduce friction in the water; a thick layer of fat, or blubber, under the skin provides insulation.

Cetaceans are highly efficient swimmers; like fishes (and unlike seals and walruses), they have their locomotor muscles in the trunk of their body and are propelled by movements of their body and tail. The forelimbs of cetaceans form flippers used in steering, and the hindlimbs are vestigial (see Figure 17.17b). The tail fins, or flukes, of cetaceans contain no bony supports and are in a horizontal plane; cetaceans undulate up and down rather than from side to side as fishes do. One of the most dis-

(a)

(b)

(c)

(d)

(e)

Figure 33.6 Infraclass Eutheria. (a, b) Order Xenarthra. (a) The giant anteater *Myrmecophaga tridactyla,* weighing about 25 kg, moves slowly about on the ground in search of insect nests. (b) The 2-toed sloth *Choloepus didactylus* of the Amazon region is covered with long hair. Sloths often appear greenish in the rainy seasons because algae grows profusely in their thick fur. Sloths have from six to nine cervical vertebrae, instead of the seven typical of mammals. (c) Order Tubulidentata. The aardvark *Orycteropus afer,* a large (up to about 80 kg) insectivore, has thick skin with few hairs. (d) Order Pholidota. Pangolins *(Manis)* eat mainly termites and ants. Some species are chiefly arboreal, while others spend most or all of their time on the ground. When disturbed, a pangolin rolls into a ball and moves its tail back and forth so that its scales can cut predators. (e) Order Cetacea. The narwhal *(Monodon monoceros),* a toothed whale, inhabits open waters in arctic seas. The male narwhal has a long (up to about 2.5 m), spirally twisted tusk formed from a left upper incisor. Males sometimes fence with their tusks.

tinctive features of cetaceans is the position of the external nostrils; these are located on top of the head as one or two openings, or blowholes, with direct connections to the lungs. Cetaceans breathe air with virtually no alteration in their swimming undulations, and the position of their nostrils helps reduce drag while swimming.

The Order Cetacea is divided into two suborders based on feeding adaptations. The **Suborder Mysticeti** (the baleen whales) are toothless as adults, but have hardened epidermal plates, called **baleen**, suspended from the palate. As explained in the Chapter 4 Essay on baleen whales (see p. 82 and Figure E1), baleen is used to filter small invertebrates and fishes from seawater. In contrast to the baleen whales, members of the **Suborder Odontoceti** (toothed whales and dolphins) may have as many as 260 teeth or, as in the narwhal, as few as a single pair (Figure 33.6e). Their teeth are typically conical, and unusual among mammals, odontocetes are **homodont**, meaning all teeth have a similar shape. Odontocetes use echolocation for finding prey, such as squids and large fishes. They are mainly oceanic, but some dolphins occur in large rivers in South America, India, and China.

By far the largest order of mammals is the **Rodentia**; with nearly 1760 species, the group is distributed worldwide (Figure 33.7a). The name Rodentia is derived from the Latin word meaning "gnawing," and this refers to one of the distinctive features of the group: The upper and lower jaws of rodents each have one pair of elongate incisors. Rodent incisors grow continuously and are distinctly beveled. The beveling occurs because the posterior surfaces of these teeth lack enamel. Most rodents are herbivorous seed eaters, but many are omnivorous, and some eat mainly insects.

Superficially resembling rodents, and apparently related to them through an immediate common ancestor, the **Order Lagomorpha** includes about 65 species of rabbits and pikas (Figure 33.7b). Like rodents, lagomorphs have two pairs of long, constantly growing incisors; but unlike rodents, they have a second pair of small incisors in the upper jaw.

(a)

(b)

Figure 33.7 Infraclass Eutheria. (a) Order Rodentia. The woodchuck *Marmota monax* (also known as the groundhog), a North American species, is a ground-dwelling member of the squirrel family. (b) Order Lagomorpha. The pika *Ochotona princeps* inhabits rock slides in mountains in western North America. It stores plant material in "haystacks" for winter food.

MAJOR ADAPTATIONS OF MAMMALS

In this section, we highlight the adaptive features that distinguish mammals from their reptilian ancestors and that allow mammals to inhabit virtu-

ally all types of land, aerial, and aquatic environments.

Endothermy and Homeothermy

Much of the uniqueness of mammals centers around endothermy, and virtually all aspects of the mammalian body represent evolutionary adaptations related in some way to temperature regulation. Most mammals are both endothermic and homeothermic. (For a review of temperature relationships of animals, see the Chapter 3 Essay, p. 73.)

To regulate body temperature, an animal must be able to generate and retain considerable metabolic heat. A typical resting mammal exhibits a metabolic rate that is ten to twelve times higher than that of reptiles. Some of the organ systems that support this high metabolic rate were discussed in earlier chapters. The efficient gas exchange system of mammals supplies oxygen to tissues at a rate that supports a high metabolic rate. (Gas exchange in mammals is described in Chapter 5, p. 108; Figure 5.13.) Likewise, mammals have a high-pressure circulatory system that distributes oxygen and nutrients to body tissues quickly. (The mammalian heart and circulatory system are discussed in Chapter 6, p. 130; Figure 6.12.) Mammalian gas exchange and circulatory systems also work in concert with a highly effective excretory system in maintaining body fluid homeostasis and in processing and disposing of metabolic wastes as rapidly as they are generated. (The function of the mammalian kidney is described in Chapter 7, p. 162; Figures 7.13 and 7.14.) Maintaining a high metabolic rate also requires that a mammal provide its energy-generating systems with adequate fuel. Hence, mammals generally eat more food than reptiles do and process it more efficiently than reptiles. The heterodont teeth of most mammals are usually specialized for chewing particular types of food. Many mammals, especially herbivores, masticate their food thoroughly, increasing surface area for more efficient digestion before swallowing. In contrast, reptiles do not chew, but swallow their food whole, and their digestion is much less efficient.

Sources of Heat. Mammals derive much of their body heat from the metabolic activity in contracting muscles, and exercise provides great quantities of heat. Shivering is an adaptation that makes muscles contract and generate heat when a mammal is cold. Other heat sources include deposits of so-called **brown fat**, located between the shoulder blades or in the neck region of many mammals. Brown fat cells are rich in mitochondria, and brown fat derives its color from the brownish cytochrome pigments in these organelles. The sole function of brown fat is heat production, and deposits of brown fat are especially pronounced in newborn mammals and in species that hibernate. When a mammal's body temperature drops, sympathetic nerve endings in the brown fat release norepinephrine, which activates the enzyme lipase, which in turn breaks down fats into triglycerides. The triglycerides are oxidized in the mitochondria, which, instead of making ATP, release the energy as heat.

Regulation of Heat Exchange. Heat exchange between an animal and its environment occurs by four means: conduction, convection, evaporation, and radiation. Mammals are able to regulate each of these exchange mechanisms and to balance them against heat production to maintain stable body temperatures.

The insulating properties of the coat of a terrestrial mammal typically change with the seasons in temperate zones. Winter coats are usually thicker and are shed with the advent of warm weather. Dark hair on the nose, ears, and feet of many species is also a thermal adaptation, promoting heat absorption in the extremities. Many mammals also have the ability to change the insulating properties of their coat almost instantly by raising or lowering their hairs. **Arrector pili muscles** originate at the base of the dermis, with each muscle inserting on a hair follicle (see Figure 8.21). Contraction of these muscles in response to cold (and sometimes in response to anger and fear) erects the hairs, trapping a layer of air in the fur. The trapped air is warmed by the body, providing excellent insulation. (The "goose bumps" that humans experience when they are cold result from the contraction of arrector pili muscles in the skin, but humans have too little hair to prevent heat loss in this way.)

Despite mechanisms that conserve heat, different regions of the mammalian body may be cooler than the body core temperature. Measurements of temperatures in the forelegs of arctic sled dogs show that the temperature of the foot pad may be 0°C, the top of the foot 8°C, and the lower leg 14°C when the body core temperature is 38°C. The muscle and nerve cells in the legs function at lowered temperatures because their cells contain cold-adapted phospholipids and enzymes. Having cold extremities is an effective way to conserve body heat; if an arctic sled dog's extremities were maintained at 38°C, the rate of heat loss from the extremities would be much greater.

Heat loss is a special problem for aquatic mammals because water conducts heat 10 to 100 times faster than does air. Beavers, seals, and the muskrat are active in water near 0°C, yet do not suffer **hypothermia**, a critical lowering of the body temperature. In these and many other species, fine undercoat hairs form a layer so dense that it does not wet; and the dry undercoat traps a layer of air that insulates the skin. Insulating blubber may account for about 45% of the body weight of seals and whales, and many aquatic and semiaquatic mammals also have circulatory adaptations that regulate heat loss. In many species, relatively little blood circulates to the skin in cold water; but when the animal is in warm water, dermal arterioles open, allowing blood to lose heat as it circulates close to the base of the epidermis. Muskrat legs and the flippers and flukes of many marine mammals have countercurrent heat exchangers consisting of arteries carrying blood outward, from the body core, lying next to veins returning blood from the extremities. In cold water, the extremities may be near environmental temperature, and the returning venous blood is warmed by the arterial blood. This lessens the heat drain on the body core. In warm water, to prevent overheating, blood returning from the extremities can be shunted through veins that bypass the countercurrent system.

Terrestrial animals in environments with temperatures higher than their body temperature can only lose heat by evaporation. Many mammals have numerous sweat glands in the skin, and evaporative cooling may occur over the whole body surface. Others, such as dogs, cats, and sheep have few or no sweat glands, but evaporative cooling occurs from moist oral and nasal surfaces as a result of panting. In some mammals, blood that is cooled in the nose flows back to the heart in veins situated next to arteries circulating blood to the head—another countercurrent heat exchanger. The venous blood cools the arterial blood, preventing the brain from overheating.

Behavioral adaptations may also be involved in heat exchanges. Many rodents are active beneath the snow in winter and only rarely venture to the surface, where temperatures are much lower. Some species become inactive when it is cold, minimizing heat loss by curling up to reduce their exposed surface area. When temperatures are high, a mammal may remain inactive, often in a burrow; it may also increase its heat loss by stretching out and by exposing skin areas where hair is sparse.

Adaptive Hypothermia. Many mammals cope with various types of environmental stress by reducing their metabolic rate and becoming dormant, or torpid. Reduced metabolism is accompanied by reduced body temperature, or hypothermia. In such cases, hypothermia is an adaptation that involves a resetting of the body's metabolic rate and temperature so that less energy is consumed when it is difficult to obtain or is in short supply. Adaptive hypothermia in response to seasonal cold temperatures is called **hibernation**, or winter dormancy; similarly, some animals undergo **aestivation**, or summer dormancy, as a way to cope with temperature extremes or seasonal shortages of food or water. Daily dormancy, or **diurnation**, is also common in animals exposed to marked diurnal temperature extremes. The physiological state of mammals during these various forms of dormancy varies with the species. Certain ground squirrels and bats enter a deep sleep during hibernation; their heart rate and oxygen consumption rate decrease markedly, and their body temperature may drop to within a few degrees of environmental temperatures. Such animals remain in this state for weeks or months, until environmental conditions become favorable. By contrast, hibernation in the black bear (*Ursus americanus*) typically involves various levels of drowsiness, during which the heart rate and breathing rate may slow to as little as 10% of the active rate, but body temperature, maintained by brown fat consumption, may be only a few degrees lower than when the animal is active. Also in contrast to "deep sleepers," which may arouse only after marked temperature increase, black bears generally exhibit visible responses to stimuli such as loud noises while hibernating.

The Mammalian Integument

The skin of mammals plays a number of significant roles in helping to maintain internal homeostasis. (The structure of mammalian skin is discussed in ◁ Chapter 8, p. 191; Figure 8.21.) Mammalian skin is important in reducing water loss, in helping to prevent infection, and in facilitating locomotion in bats, gliding squirrels, and aquatic mammals. The skin is also important in communication: Scent glands derived from the epidermis produce various pheromones (chemical signals), and skin muscles allow some mammals to change facial expressions and raise hairs in threat postures.

Hair, an epidermal derivative, is composed of fine scales of the protein keratin (Figure 33.8).

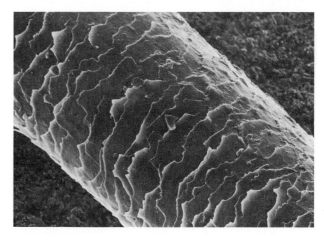

Figure 33.8 The structure of mammalian hair. This scanning electron micrograph shows the scaly outer layer (1500×). Different species have different scale patterns.

Mammals have several kinds of hair, and hair serves a variety of functions in addition to its vital role in thermoregulation. Many species have an undercoat of soft, fluffy hair close to the skin; longer, coarse **guard hairs** protect the undercoat. Certain ungulates have hollow guard hairs, providing a layer of minute dead air spaces with superb insulating properties. The pronghorn antelope *(Antilocapra americana)* of western North America, actually not an antelope, but the sole member of a unique ungulate family, has a thick undercoat and hollow, bristlelike guard hairs. Pronghorn are often seen feeding on open prairies in steady wind, when temperatures are well below freezing (Figure 33.9).

Other types of hair include stiff bristles, or **vibrissae**, on the faces and legs of many mammals. The bases of vibrissae are highly innervated and serve as tactile receptors. In porcupines and the spiny anteaters, some hairs are thickened into spines or quills that function in defense.

Hair derives its color from pigments in hair shafts. Hairs are lubricated and made water-repellent by oily secretions from one or several **sebaceous glands** associated with each hair follicle. In most mammals, hair growth is determinate, and hairs are shed after they reach a certain length. However, the hair of sheep and humans, as well as the hair composing the tails and manes of horses, grows continuously.

Claws, hooves, horns, and antlers are also epi-

Figure 33.9 The pronghorn *(Antilocapra americana).* This ungulate lives on wind-swept grasslands of the western United States. Both sexes have horns, but the male has larger ones. The pronghorn is the only horned mammal that sheds its outer horn sheath annually. In this photograph, a large buck is guarding a doe from contact with a smaller buck on the right.

dermal derivatives. Rhinoceros horns are composed entirely of keratin fibers, whereas the horns of cows, sheep, and goats consist of hollow cones of epidermal keratin covering dermal bone. The antlers of deer, elk, and caribou are formed of dermal bone. Antlers, unlike horns, are shed and grow anew each year under the control of hormones in the hypothalamic-pituitary-gonadal system. (This hormonal control system is described in Chapter 11, p. 251.) While regrowing, antlers are living bone and are covered with epidermal tissue called velvet. As testosterone (male hormone) levels in the blood reach a peak during the breeding season, antler growth ceases and the velvet is shed. Loss of antlers following the breeding season correlates with decreased testosterone levels (Figure 33.10).

Skeletal and Locomotor Adaptations

The mammalian skeleton contains over 200 bones, providing protection for internal organs and supporting a complex locomotor system. The mammalian skull, composed mainly of dermal and endochondral bones, forms a strong, resilient braincase (see Figure 28.10c). The lower jaw and its articulation with the upper jaw distinguish mammals from reptiles. In mammals, the lower jaw is composed of a single dermal bone, the **dentary**, whereas in reptiles, the lower jaw is composed of several bones. The mammalian jaw articulation is formed by the dentary bone and the squamosal bone of the upper jaw. In a typical reptile, the articular bone of the lower jaw and the quadrate bone of the upper jaw form the jaw articulation. The articular and quadrate bones are both gill arches derivatives; homologues of these bones in mammals are two of the ossicles in the inner ear. (Evolution of the gill arches is discussed in Chapter 28, p. 689.)

Compared to other vertebrates, terrestrial mammals have an enormous range of movement in the forelimbs, and the distal elements of both fore- and hindlimbs show a variety of locomotor modifications. Bears, primates, and many insectivorans are **plantigrade**, meaning they walk on the plantar surface of their feet (see Figure 28.12). This is thought to be the ancestral condition; the bones of the hands and feet are not fused, and the wrist and ankle joints allow rotation in three dimensions. The bones of the foot of **cursorial** (running) mammals are usually elongated. Many carnivorans, such as cats and dogs, are **digitigrade**, meaning that they walk with the toes on the ground but with the heel raised. **Unguligrade** mammals, including

Figure 33.10 Horns and antlers of mammals. (a) Horns of the ibex *Capra ibex* have a core of bone and are not shed. (b) Antlers of the moose *Alces alces* are shed at the end of the breeding season and regrow completely each year. (c) The sequence of antler regrowth. Starting as a small protuberance on the frontal bone of the skull, the bony core of the antler enlarges. Velvet, the soft skin covering the antlers, is rubbed off prior to the breeding season.

(a)

(b)

(c)

the ungulates, walk on the tips of their toes, and the metacarpal and metatarsal bones of the feet are elongated and reduced in number. The middle digit of the foot of perissodactyls is enlarged and bears the body weight (see Figure 28.12), while in artiodactyls, the body weight is on the enlarged third and fourth digits.

Mammals with **saltatory** (jumping) locomotion tend to have stocky bodies with the center of gravity shifted to the posterior, as in kangaroos and rabbits. The neck is also generally short, and the hind feet are elongated for propulsive force. Digging, or **fossorial**, mammals, such as moles, badgers, and pocket gophers, generally have short legs with heavy bones and large areas for muscle attachment. (See the discussion of mechanical advantage in the armadillo leg in Chapter 9, p. 211; Figure 9.16.) **Arboreal** mammals have skeletons adapted to various types of locomotion in trees. Certain monkeys and apes that **brachiate** (i.e., that swing through tree branches using mainly their forelimbs) have an especially strong shoulder girdle. Moreover, their forelimbs are typically longer than their hindlimbs, and their digits are elongated. Also, the hand is often hooklike, and some species have hands and feet with opposable digits that enhance grasping.

The locomotor systems of semiaquatic mammals, such as beavers, otters, and the muskrat, and those of the aquatic carnivorans (e.g., seals and sea lions) and cetaceans clearly reflect the relative amounts of time these animals spend on land and in the water. Semiaquatic species spend considerable time in water, but also travel extensively on land. Correlated with this mode of life, their limbs have relatively minor adaptations for swimming, such as webbing between the digits, but are adapted primarily for transporting the body on land. In seals and sea lions, which spend much more time in the water, the torpedo-like body shape and flippers allow limited movement on land, but are mainly adapted for swimming. Finally, the fully aquatic cetaceans (whales, dolphins, and porpoises), powered entirely by body and tail movements rather than by limbs, are completely helpless if stranded on land.

Special Adaptations of the Circulatory and Respiratory Systems

The mammalian circulatory system, with its separation of pulmonary and systemic circuits, and the respiratory system, with its effective mechanism for ventilating the lungs and exchanging gases in moist alveoli, are significant adaptations for life on land. Mammals are unique in having a muscular diaphragm that assists lung ventilation (see Figures 5.14 and 6.12). Aquatic mammals have special adaptations that allow them to remain underwater for extended periods of time and to return to the surface from considerable depths without ill effects. (See the Chapter 5 Essay on diving adaptations in mammals, p. 115.) Semiaquatic mammals, such as beavers, hippopotamuses, and the muskrat, exhibit similar adaptations.

The circulatory system of tall mammals, such as giraffes, illustrates other kinds of special adaptations. The long neck of the giraffe is useful for feeding on trees, but it presents the animal's heart and vascular system with special problems. When an adult giraffe raises its head, the heart must be able to pump blood a vertical distance of 4 m up to the brain. Then, when the giraffe lowers its head, the blood pressure must be controlled so that it

does not rupture blood vessels around the brain (Figure 33.11).

Gas Exchange and Circulation in the Eutherian Fetus. Gas exchange in the fetus (the late developmental stage of placental mammals) occurs by diffusion between maternal and fetal blood vessels across the placenta. Fetuses have lungs, but these organs do not become functional until birth. Because of several adaptations in the fetal circulatory system, the fetal lungs receive only enough blood to sustain their cells, while oxygenated blood from the placenta passes more or less directly to the general fetal circulation.

Fetal blood circulates to the placenta via the **umbilical arteries**, which branch from the internal iliac arteries in the legs of the fetus (Figure 33.12). Blood is oxygenated and picks up nutrients in the placenta and then returns to the fetus via the **umbilical veins**; these veins pass into the fetal liver, but most of the blood they carry bypasses the liver and enters a special vessel called the **ductus venosus**. The ductus venosus conveys blood to the inferior vena cava, which enters the right atrium of the heart. Because the inferior vena cava also receives blood from the fetal body, oxygen-poor blood mixes with oxygen-rich blood in this vessel and in the heart. From the right atrium of the fetus, blood flows into the right ventricle; from here some of it passes into the pulmonary arteries. (In adults, all blood from the right ventricle takes the pulmonary route to the lungs.) However, because the lungs are collapsed in the fetus, they offer high resistance to blood flow. Consequently, most blood in the pulmonary artery is shunted into the systemic aorta through a fetal vessel called the **ductus arteriosus**. As in the adult, the systemic aorta conveys blood throughout the body.

The fetal heart has another feature not present in the adult. An opening called the **foramen ovale** connects the right atrium with the left atrium, and this opening shunts most of the blood returning from the placenta via the inferior vena cava into the left atrium. A flap of tissue covering the foramen ovale allows blood to flow only from right to left.

At birth, the fetal circulatory pattern undergoes immediate reorganization. Separation of the placenta from the uterine wall or severing of the umbilical cord deprives the fetus of an oxygen supply and of a means of disposing of carbon dioxide. Consequent changes in blood gas composition trigger the respiratory center in the brain, and the new-

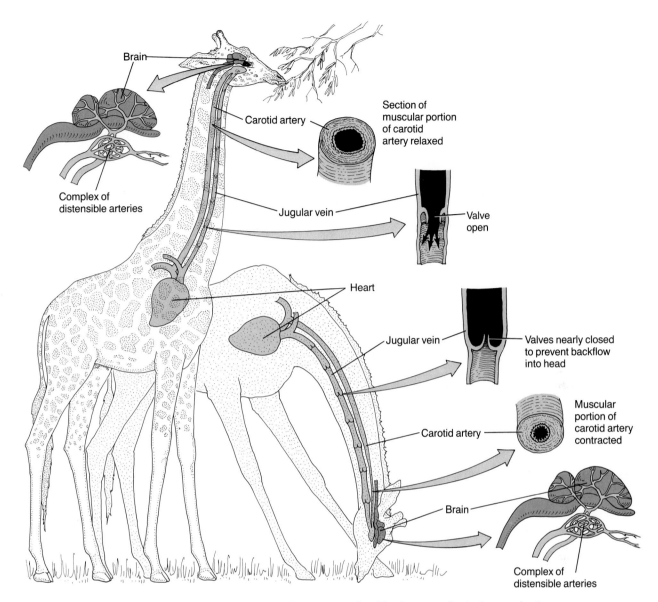

Figure 33.11 Special adaptations that accommodate blood pressure in the long neck of the giraffe. The heart is large, and its heavy musculature develops adequate pressure to drive blood up to the brain when the animal is feeding on tree leaves. When the head is down, blood collects in the large-diameter jugular veins, which contain sets of valves. With less blood returning to the heart, cardiac output decreases. Also, when the head is down, blood pressure is controlled by the heavy musculature of the carotid artery. A complex set of arteries at the base of the brain was once thought to aid in controlling blood pressure, but more recent studies show it is involved in temperature regulation.

born infant takes its first breath. Then, a few minutes after birth, as a result of changing blood pressure dynamics, the young mammal's circulatory system is transformed. As the lungs expand, their resistance to blood flow drops markedly; consequently, blood flows into the lung capillaries and returns through the pulmonary vein to the left atrium. This, in turn, raises blood pressure in the left atrium and closes the flap of tissue covering the

foramen ovale, blocking this shunt. (Eventually, the foramen ovale fills with connective tissue and is blocked permanently.) Lowered lung resistance also means that blood in the pulmonary artery is no longer forced through the ductus arteriosus into the systemic aorta. Shortly after birth, the ductus arteriosus constricts and closes off. (Eventually, the ductus arteriosus fills with connective tissue and becomes the **ligamentum arteriosum** in adults.)

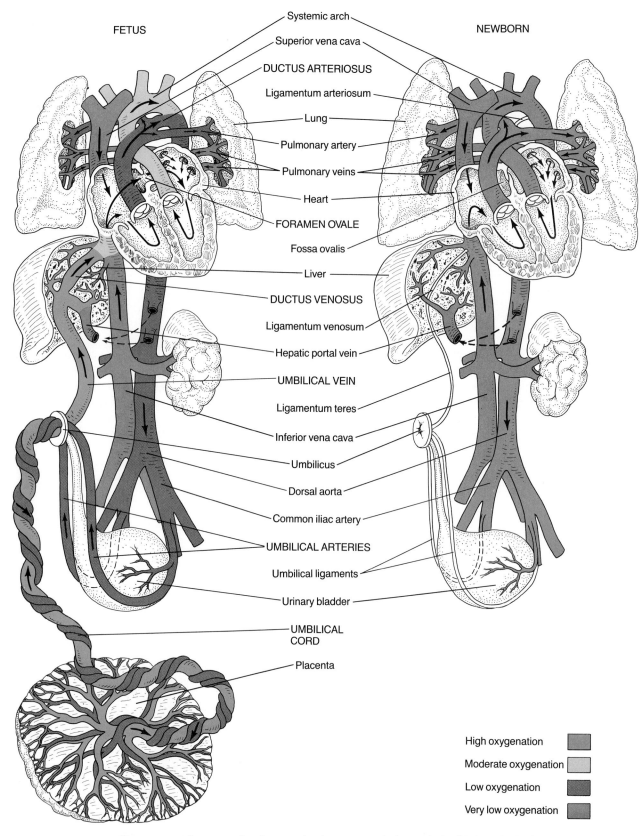

FETUS   NEWBORN

Systemic arch

Superior vena cava

DUCTUS ARTERIOSUS

Ligamentum arteriosum

Lung

Pulmonary artery

Pulmonary veins

Heart

FORAMEN OVALE

Fossa ovalis

Liver

DUCTUS VENOSUS

Ligamentum venosum

Hepatic portal vein

UMBILICAL VEIN

Ligamentum teres

Inferior vena cava

Umbilicus

Dorsal aorta

Common iliac artery

UMBILICAL ARTERIES

Umbilical ligaments

Urinary bladder

UMBILICAL CORD

Placenta

High oxygenation

Moderate oxygenation

Low oxygenation

Very low oxygenation

Figure 33.12 The mammalian (human) circulatory system before and after birth. In the fetal system, note the locations of the foramen ovale, ductus arteriosus, and ductus venosus in this ventral view of the fetal heart and major vessels. Arrows indicate the direction of blood flow. The drawing to the right illustrates the major changes that occur at birth.

Finally, because blood no longer flows in the umbilical arteries and veins, the ductus venosus (over several weeks) constricts, forcing more and more blood through the liver, which is the adult mammal's major blood filter.

Special Senses

Except for the cetaceans and some primates, mammals have a highly developed sense of smell that they use in tracking and locating prey and in communication.

Vision is also highly developed in most mammals. Most species are nocturnal or crepuscular (active mainly in the twilight hours around dawn and dusk) and have eyes adapted for seeing in dim light, but most are also color-blind. Primates are an important exception. (See the description of the human eye in Chapter 12, p. 275; Figure 12.4.) Squirrels are another exception; active in daylight, they have excellent color vision; in fact, their retinas are composed entirely of color-sensitive cones. Strictly nocturnal mammals generally have larger eyes, relative to their body size, than species that are active during daylight hours. Species with eyes on the sides of the head can see backward and forward without turning their heads. Mammals with forward-facing eyes, such as humans and apes, have excellent depth perception.

Most mammals also have very acute hearing. Many species have large, movable **pinnae** (external ear flaps, or auricles) that capture sound and help locate its source. (The mammalian ear is described in detail in Chapter 12, p. 284; Figures 12.12, 12.13, and 12.14.)

Echolocation. Insect-eating bats and the toothed whales (odontocetes) are unusual in that they use reflected sound to locate and identify objects. (See Figure 33.13.) Such a sense is useful to animals seeking prey in the dark depths of the sea or in the air at night. Odontocetes emit sounds (variously described as hums, grunts, growls, and whistles), ranging from about 500 to 15,000 Hz, depending on the species. They use these relatively low frequency sounds mainly in communicating with one another. Toothed whales use other (supersonic) sounds (in the range of 20,000–200,000 Hz) emitted in short pulses to echolocate prey or objects in the water.

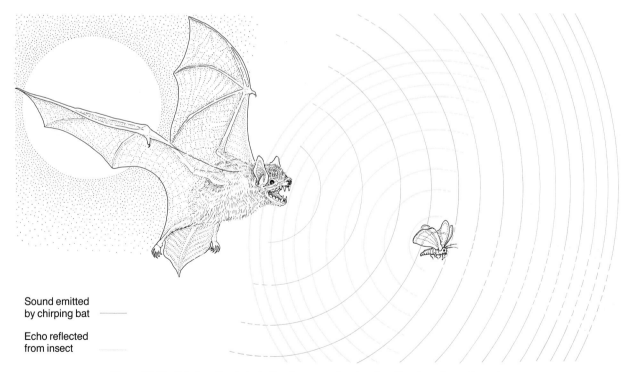

Sound emitted
by chirping bat ————

Echo reflected
from insect ————

Figure 33.13 Echolocation in bats. Bats have excellent hearing. They have large pinnae (external ear auricles) that enhance sound reception. Insectivorous bats locate flying prey by emitting ultrasonic chirps and detecting returning sounds. Here, high-frequency chirps are being reflected from a moth in a bat's flight path (color). A muscle in the bat's ear damps the sound of the chirp as it is produced so that the bat receives the echo with less interference.

Echolocation has been investigated most thoroughly in bats because their small size makes them amenable to laboratory study. A bat's larynx, or voice box, produces a high-frequency (30,000–140,000 Hz) chirp of ultrasound. Chirps are emitted from the animal's mouth or nostrils and may be directed by shields and flaps on the face. Some bats produce a chirp that consists of a frequency-modulated (FM) pulse; others produce a constant-frequency (CF) pulse with a slight frequency modulation at the end. Still others can switch back and forth between FM and mainly CF pulses. If an object is in the path of the radiating sound, an echo returns to the bat. Bats have excellent hearing, and their auditory system damps reception of the chirps as they are produced, so that their ears receive returning sounds with less interference. Returning sounds contain a considerable amount of information. FM pulses provide a change in pitch so that the returning sounds can be timed. The time from sound emission to receipt of an echo is proportional to the distance between the target and the bat. If the target is moving toward or away from the bat, a Doppler shift occurs in the returning sound. An object moving toward a bat shifts the echo to a higher frequency than the emitted chirp, but an object moving away produces a lower-frequency echo. The FM portion of a chirp also yields information about the size and shape of a target. Small objects reflect higher-frequency sound better than low-frequency sound, and objects of different shapes and textures reflect sound in different ways. By interpreting the information in the echos, a bat can decide whether an object should be avoided or pursued.

BEHAVIOR

A typical mammal has a greater ability to learn, remember, and solve problems than any other kind of animal. This ability is correlated with the large size of the mammalian cerebral cortex relative to body size. (The mammalian neocortex is discussed in Chapter 28, p. 702.) Social behavior is highly developed in some mammals and typically involves both innate behavior patterns and learning.

Mammals use various forms of communication in their social groupings. Visual cues are important in many species. Well-developed skin muscles allow some mammals to communicate by manipulating facial expression, changing position of the ears, and erecting specific patches of hair. A dog with its ears down, the hair on its back erect, its tail arched over its back, and its fangs bared is exhibiting an aggressive posture. Signals of appeasement are often more or less the opposite of aggressive signs. Thus, a dog may express appeasement by wagging its tail and opening its mouth and letting its tongue hang out. Other mammals may communicate in similar ways, although the postures and positions differ somewhat from species to species. Figure 33.14 illustrates some of the diverse facial expressions typical of primates.

Mammals also use olfactory signals in communication. Dogs, cats, rodents, deer, and small African antelope commonly use pheromones produced by glands in the skin, as well as urine or feces, to mark the boundaries of their territories. Other members of the species recognize the boundaries and usually do not cross over them, thus reducing fighting and energy expenditure in territorial defense. When rutting and maintaining a harem of females, a bull elk advertises his status and sexual condition by urinating on his chest and abdomen. Rival bulls seem to determine the physical state of a bull with a harem by the scent of his urine. The rivals usually do not offer challenge while the urine indicates that the bull is in prime breeding condition; however, when a bull with a harem becomes exhausted, his urine may signal rivals to challenge him. Thus, so long as a breeding bull is in prime condition, his harem will most likely not be disturbed, and copulation can occur without interruption. Once out of condition, however, he may soon be replaced by a fresh rival.

Auditory communication is also common among mammals, and many species have diverse vocal repertoires. Whales and dolphins make almost continuous clicks and groans that keep members of a herd informed of one another's whereabouts. Some rodents communicate through ultrasound. Upon ejaculation, the males of some species of rodents emit a sound at about 22,000 Hz, and then the male and female disengage. When a male tries to mount a female that is not receptive, she will emit a sound at a similar frequency.

Social behavior is involved in the capture of food by some carnivorous mammals. Whales may swim beside one another in a line, presumably sweeping an area for prey. Wolves typically hunt in packs, allowing them to kill larger prey than any single wolf could handle. Group hunting by canines contrasts with the method employed by most cats, where individuals stalk and kill prey.

Many mammals are territorial, and many also

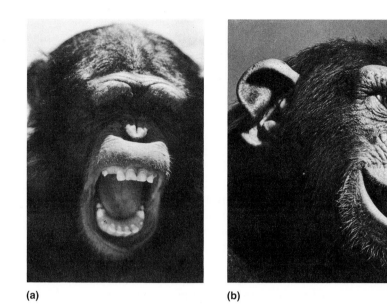

(a)

(c)

(b)

Figure 33.14 The chimpanzee *(Pan troglodytes)* communicating by facial expressions. Although it is generally not appropriate to attribute human emotions to other animals, studies of the chimpanzee and other species of great apes show that these animals use a wide variety of facial expressions and body postures to communicate mental sensations (emotions?) to one another. The three expressions shown here have been interpreted as (a) anger, (b) pleasure, and (c) uneasiness.

exhibit dominance hierarchies. (These social phenomena are described in Chapter 13, p. 317.) Territories are usually breeding areas, but they may also include den sites. Males of the northern fur seal (*Callorhinus ursinus*) defend specific areas of arctic beach against other males. Some males accumulate harems when females come ashore and enter the territories. Other species of mammals have mobile territories. A male bison (*Bison bison*), for example, defends an area around a female that is in mating condition.

Social behavior is an important component of reproduction in most mammals, and numerous studies have documented the complex breeding systems of many large mammals. Unfortunately, there have been relatively few studies on the more numerous small mammals, such as rodents and bats. Many mammals, such as wolves, some other members of the dog family, beavers, and gibbons are monogamous; a female-male pair bond forms and lasts for at least one mating season. Many rodents and some bats are promiscuous, with no pair bond forming between the sexes before or after mating. By contrast, larger ungulates, marine carnivorans, some bats, and even some rodents (e.g., marmots) are polygamous, with a male accumulating a harem.

REPRODUCTION AND LIFE CYCLES

Mammals are dioecious, and providing for internal fertilization, all species have well-developed copulatory organs. Males have an erectile penis, which either lies in a cloaca and conveys only sperm, as in monotremes, or has its own sheath and conveys both sperm and urine, as in marsupials and eutherians. In many species, the penis contains a bone called the **baculum**, or **os penis**. The tip of the penis in some marsupials is bifurcated, accommodated by a double vagina in the females. In monotremes and some eutherians (e.g., most aquatic mammals, elephants, and some insectivorans), the testes are permanently housed in the abdominal cavity. In marsupials and in most eutherians, however, the testes are periodically or permanently housed outside the body in a saclike extension of the body cavity called the scrotum. The sperm of most species cannot develop or remain viable except at the somewhat cooler temperatures in the scrotum. (The structure and function of the human male reproductive system are described in Chapter 14, p. 336; Figure 14.9.)

The female reproductive system consists of two

ovaries, two oviducts (fallopian tubes), where fertilization occurs, one or two uteri, where embryos develop, and one or two vaginae. (The structure of the human female reproductive system is described in Chapter 14, p. 338; Figure 14.11.) Only monotremes and one eutherian, the pika *Ochotona princeps*, have a cloaca instead of a vagina (see Figure 33.7b). Most mammals are **spontaneous ovulators**, meaning that their eggs are released from the ovary as a result of cyclic changes in hormone levels. (See the discussion of the mammalian estrous cycle in Chapter 11, p. 265.) By contrast, some species (e.g., certain rodents and rabbits) are **induced ovulators**: Copulation stimulates their ovaries to release eggs.

Intrauterine Development and Maternal Care

All mammals, including monotremes, have some amount of intrauterine development and nurse their young after hatching or birth. The eggs of monotremes are like those of birds in being shelled and in having much yolk in the vegetal hemisphere. Also as in birds, early cleavage occurs in a small disc at the animal pole of the egg. The monotreme eggshell is elastic, and absorption of uterine fluid through the shell increases the egg's size from a diameter of about 4 mm to about 15 mm by 17 mm by laying time. The uterine fluid may also nourish and protect the embryo. Newly hatched young (incubation is about ten to twelve days) are tiny (about 10–12 mm long), and maternal care lasts from about four to six months.

Intrauterine development in marsupials involves some type of connection between the membranes surrounding the blastocyst and the uterine wall. Bandicoots (see Figure 33.3c) have a placenta similar to that of eutherian mammals (described in the next paragraph). In most marsupials, however, the yolk sac of the blastocyst loosely attaches to the uterine wall, forming a primitive type of placenta called a **choriovitelline placenta**. This is markedly different from the eutherian placenta and allows little exchange of materials between the embryo and the maternal blood. While in the uterus, a marsupial embryo obtains most of its nutrients by absorbing a substance called uterine "milk" secreted by glands in the uterine wall. At birth, an offspring, actually still a young embryo, is expelled from the

An older joey that returns periodically to nurse

A young joey confined inside mother's pouch

Figure 33.15   A female red kangaroo *(Megaleia rufa)* with young of two different ages. An older, relatively independent joey returns periodically to the mother to drink high-fat, low-protein milk, while a younger, relatively helpless joey is confined to the mother's pouch (marsupium), where it drinks from a different nipple, receiving high-protein, low-fat milk.

uterus and in pouched species makes its way into the pouch and attaches to a nipple. Female kangaroos give birth to only one offspring (about 2–6 cm long) per reproductive cycle; however, because the young spend much less time in the uterus than as sucklings, kangaroos often have a very young offspring, called a young joey, attached to a nipple in the pouch while an older joey returns periodically to suckle a different nipple (Figure 33.15).

Eutherian mammals carry their young in the uterus for the entire fetal period. Their placenta, called the **chorioallantoic placenta**, is firmly attached to the uterine wall by the allantois, and fingerlike villi that increase surface area for exchange develop between the fetal and maternal circulatory systems. (Gastrulation and later development of the human embryo are dealt with in Chapter 16, p. 387; Figure 16.13.) Pregnancy and birth of the eutherian fetus involve several hormone-coordinated events. (Hormonal control of pregnancy is discussed in Chapter 11, p. 264; Figures 11.14 and 11.15.) During pregnancy, in preparation for birth, a hormone called **relaxin**, produced by the corpus luteum of the ovary, softens the cartilages in the pelvic region so that the relatively large young can pass through the birth canal. As birth nears, another hormone, **oxytocin**, released from the posterior pituitary, stimulates uterine contractions, which eventually expel the young from the birth canal.

Mammary Glands and Milk

Active mammary glands consist of grapelike clusters of milk-secreting tissue and ducts leading from the secretory tissue to the apex of a papilla, or nipple (Figure 33.16). Prolactin, a hormone from the anterior pituitary gland, stimulates milk production during pregnancy. After birth, suckling by the young stimulates release of oxytocin from the posterior pituitary; this hormone causes contraction of smooth muscle surrounding the ducts, forcing milk toward the nipple.

Milk contains the sugar lactose, various proteins called caseins, fat, salts, vitamins, and certain antibodies. Just after birth, a mammary gland produces a type of milk called **colostrum**, which contains significant amounts of antibodies and white blood cells. Thus, the suckling newborn receives some of its mother's immunity and is protected against certain diseases until its own immune system develops.

Reproductive Timing

Many mammals are **polyestrous**, mating and rearing young several times a year, whereas others are **monoestrous**. Large mammals, such as the elephant, have a cycle that takes more than a year. Most temperate-zone mammals are seasonal

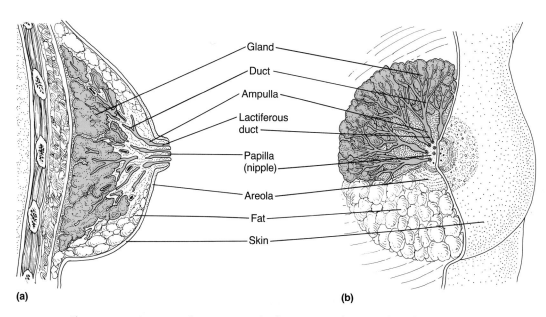

(a)

(b)

Gland
Duct
Ampulla
Lactiferous duct
Papilla (nipple)
Areola
Fat
Skin

Figure 33.16 Anatomy of a mammary gland. (a) A vertical section through the mammary gland of a human. (b) Structure at increasing depths (clockwise) of a lactating mammary gland.

breeders, with the young born in the most favorable seasons. In species with long gestation periods, breeding often occurs in the summer or fall, and the young are born the following spring. Other mammals breed in the spring and produce their young during that spring or during the summer. Certain species copulate in the fall or winter, but fertilization may be delayed, or if it occurs, development beyond the blastocyst stage may be delayed. Delayed fertilization occurs in certain bats. A variety of other mammals (e.g., the black bear, northern fur seal, and western spotted skunk) exhibit **delayed implantation**, in which blastocysts lie dormant in the uterus for several months, usually over the winter; implantation and further development are timed so that the young are born when climatic conditions favor their survival. Delayed implantation is also part of the life cycle of many marsupials. In some kangaroos, as soon as one joey is in a female's pouch, she becomes sexually receptive and mates. The fertilized egg from this mating enters her uterus and undergoes cleavage, but the resultant blastocyst does not implant and develop so long as she has a joey in her pouch. After about six months, the joey leaves the pouch and forages on its own, although, as mentioned earlier, it may return to its mother periodically for milk (see Figure 33.15). After the joey leaves the pouch, the blastocyst implants and develops in the uterus.

PHYLOGENY

As discussed in Chapter 31, mammals are believed to have originated from mammallike reptiles called therapsids (Subclass Synapsida), about 225 million years ago during the Triassic period of the Mesozoic era. (The phylogeny of reptiles is discussed in Chapter 31, p. 768; Figure 31.17.)

Fossils of early mammals are distinguished from those of mammallike reptiles by several features. Some of the therapsids had pronounced mammallike features, such as heterodont teeth, a secondary palate, and a lower jaw and jaw articulations that were very similar to those of mammals. However, several other features distinguished the early mammals: Their skulls generally accommodated larger brains relative to body size than did those of reptiles; their cheek teeth were differentiated as premolars and molars; their vertebrae were more highly differentiated than those of the therapsids; and the middle region of the backbone was markedly arched, providing more effective movement on land.

The first mammals were a rather insignificant group of mouse-size species. Small size may have been a highly adaptive feature during the early Mesozoic because it would have allowed early mammals to occupy habitats where they were less

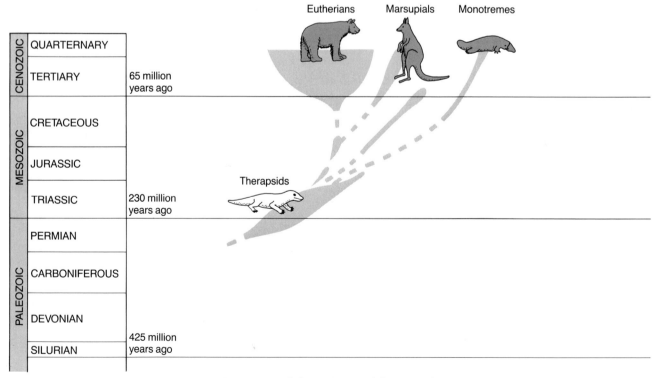

Figure 33.17 Phylogenetic tree of the mammals.

conspicuous to carnivorous dinosaurs. With the demise of most of the ruling reptiles at the end of the Mesozoic, mammals radiated extensively in the early Cenozoic era (Figure 33.17). (See the discussion of the Cretaceous extinctions in Chapter 31, p. 770.)

The mammalian lineage underwent two periods of significant divergence, which resulted in the three distinct groups of living mammals (monotremes, marsupials, and eutherians). The monotremes apparently separated from the ancestral mammalian stock sometime in the early Mesozoic era. The three living species of monotremes exhibit decidedly reptilelike features (e.g., having a cloaca and laying eggs). The marsupials and the eutherians probably arose as a result of a second major split in the mammal lineage about 135 million years ago, during the Cretaceous period. Marsupials evolved in North or South America, most likely from a stem group similar to the opossum family (Didelphidae) of today. Eutherians probably evolved from insectivorous or omnivorous shrewlike stocks, and as evidenced by the fossil record, there were several different types of placental mammals by the beginning of the Cenozoic. From this radiation, several general trends evolved among eutherians: Species tended to be larger, to have fewer teeth, and to have larger brains relative to body size. Most of the orders of living placentals were established by the Eocene period of the Cenozoic.

Human Evolution

It seems appropriate to end this chapter with a discussion of human evolution because the human species *Homo sapiens* is, in terms of its effect on world ecosystems, the single most successful and also the most ecologically harmful species that has ever lived. No other species has approached the human level of controlling and manipulating its environment or of overpopulating the earth. The abilities to manipulate and overpopulate go hand in hand with the capacity to alter the natural balance of environments at a rate that is detrimental to most other living species and to the life-sustaining capacity of the earth for virtually all organisms, *Homo sapiens* included. The last two chapters in this book examine population growth and environmental biology and how they relate to some of the modern problems caused by and facing our species. Here we look briefly at the phylogeny of *Homo sapiens*.

The ancestral lineage of *H. sapiens* probably diverged from ancestral primate stocks in the Oligocene epoch of the Cenozoic era (Figure 33.18). The oldest fossils of the **Superfamily Hominoidea**, to which humans, great apes (the chimpanzee, gorilla, and orangutan), and gibbons belong, are about 25 to 30 million years old. Following the Oligocene, the Miocene (about 5 to 24 million years ago) is sometimes referred to as the "heyday of the apes," because the primate order underwent marked radiation, especially in the old world. Unfortunately, little is known about hominoid evolution during the Miocene, but some of the features that humans share with other primates (e.g., the hand with its opposable, gripping thumb; the large brain relative to body size; the short snout with forward-facing eyes that provide depth perception; and a skeletomuscular system adapted in part for climbing and brachiating) presumably were developed in ancestral hominoids during this time. One of the best-known Miocene hominoids, a fossil species called *Proconsul africanus*, was a baboon-size ape that probably was similar to the common ancestor of all hominoids.

Recent molecular and biochemical studies have clarified some of the relationships among the hominoids. (Phylogenic data from cell and molecular biology are discussed in Chapter 18, p. 431; Table 18.2.) Asian hominoids, represented by the orangutan, probably diverged from the African hominoids about 15 million years ago. African hominoids (the gorilla and chimpanzee) are biochemically very similar to humans. (Some biochemical data even indicate that the chimpanzee is more closely related to the human than to the gorilla.) It is likely that the human and African ape lineages diverged only about 5 to 8 million years ago.

The earliest fossils of the **Family Hominidae** (humans and extinct humanlike species) date from about 4 million years ago in Africa (Figure 33.18). The primary diagnostic feature of hominids (note that this group is a family within the homin*oid* superfamily) is more or less constant bipedalism. The oldest known genus of hominid is *Australopithecus*, represented by many fossil specimens, including a famous one called "Lucy," of the species *A. afarensis*. Judging from the fossil record, *A. afarensis* lived mainly in East Africa from about 3 to 4 million years ago, was bipedal, may have been partially arboreal, and had short legs and long arms; its canine teeth were larger than those of *H. sapiens*, but smaller than those of apes. The cranial capacity of *A. afarensis* was about 450 cubic centimeters (cm^3), greater than that of apes, but less than that of the genus *Homo*, which varies from about 900 to 2000 cm^3. Other fossil hominids of Africa indicate that the hominid lineage split some

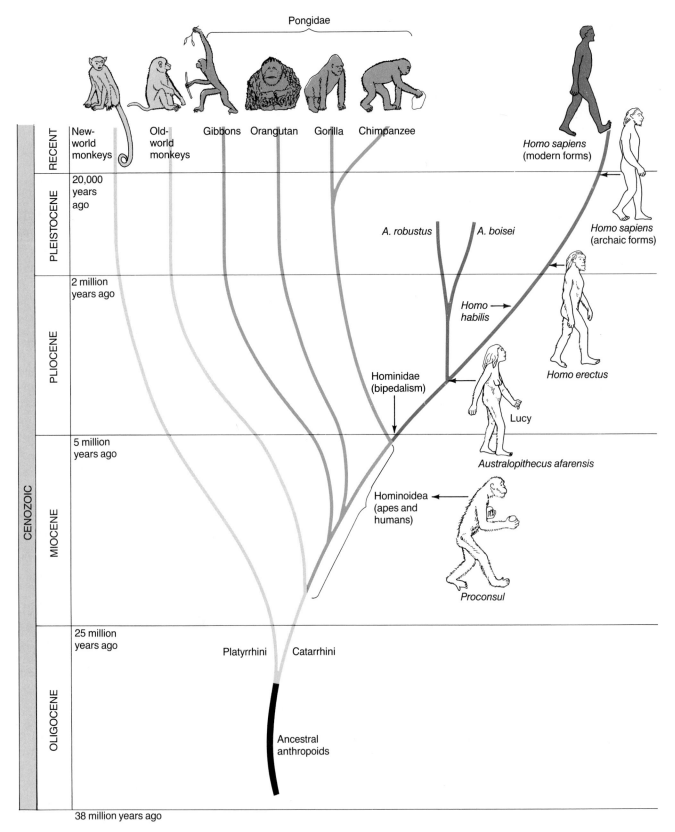

Figure 33.18 Phylogeny of hominoids (great apes and humans) and hominids (humans and humanlike extinct species). Figures are not drawn to the same scale: *Homo erectus* was about a head shorter than a modern human, Lucy was about half as tall as a human, and *Proconsul* was about two-fifths as tall.

2.5 million years ago. One lineage, the australo-pithecines (represented by the fossil species *A. africanus*, *A. robustus*, and *A. boisei*), became extinct about 1.5 million years ago in Africa. The other hominid lineage, that of the genus *Homo*, was represented by the extinct species *H. habilis*, known from fossils in East and South Africa. *Homo habilis* was probably the first species to make wooden and stone tools; its cranial capacity (600–800 cm^3) approached that of modern humans. As indicated in Figure 33.18, *H. habilis* was eventually replaced in Africa by *Homo erectus*, a more culturally advanced species that was probably the first hominid to disperse throughout the old world; fossils of *H. erectus* have been found throughout Africa and Eurasia. *Homo erectus* had a cranial capacity of about 800–1200 cm^3, its tools were more sophisticated than those of its predecessors, and this species may have used fire.

Homo sapiens, the only living representative of the Family Hominidae, probably evolved from *H. erectus* stocks some 500,000 years ago in Africa or Eurasia. The oldest known fossils of *H. sapiens* are those of robust, archaic forms called the neanderthals, dating from 100,000 to 300,000 years ago in Europe. The neanderthals used stone tools and fire and buried their dead with ceremony. The oldest fossils of truly modern humans, by far the most culturally advanced of any hominid and known as the Cro-Magnon peoples, date from about 45,000 years ago in Europe. There is also some evidence that modern humans appeared as early as about 125,000 years ago in South Africa. As indicated in Figure 33.18, the phylogeny of *H. sapiens* from *Homo habilis* probably involved a single phyletic lineage. (See the discussion of patterns of evolution in Chapter 17, p. 411.)

SUMMARY

1. Key Features of Mammals. Two characteristics separate living members of the Class Mammalia from all other animals: the presence of hair and of mammary glands. Mammals are also endothermic (most are homeothermic) and have highly glandular skin, a secondary palate, three middle ear ossicles, ribs only on thoracic vertebrae, and a lower jaw formed of a single bone, the dentary, which articulates with the squamosal bone of the upper jaw. Most mammals also have seven cervical vertebrae and are heterodont (typically with incisors, canine teeth, premolars, and molars). Mammals inhabit all types of aquatic, terrestrial, and aerial environments.

2. Two subclasses of mammals are the Prototheria (oviparous mammals) and the Theria, which are viviparous. Therians are subdivided into two infraclasses, the Metatheria (marsupials) and the Eutheria (placentals).

3. Mammals have separate systemic and pulmonary circulatory pathways and a four-chambered heart. Blood supply to the body is through the left aortic arch. A unique muscular diaphragm aids in ventilating the lungs.

4. The ratio of brain mass to body mass is greater in mammals than in other animals; the highly developed mammalian cerebral cortex allows complex learning. The senses of vision, olfaction, and hearing are highly developed. Toothed whales and insectivorous bats use echolocation to communicate and find prey.

5. Reproduction involves internal fertilization and extensive maternal care of the young. Eutherian mammals have a placenta that nourishes the young through the fetal period.

FURTHER READING

Harrison, R. J., and J. E. King. *Marine Mammals*. London: Hutchinson, 1980. *A brief, but authoritative introduction to marine mammals.*

Macdonald, D. W., ed. *The Encyclopedia of Mammals*. New York: Facts on File Publications, 1984. *A large volume containing descriptions and photographs of all mammalian orders.*

Napier, J. R., and P. H. Napier. *The Natural History of the Primates*. London: British Museum (Natural History), 1985. *A brief, yet authoritative review of primate systematics, behavior, and evolution.*

National Geographic Society. *Book of Mammals*. Vols. 1 and 2. Washington, D. C.: The National Geographic Society, 1981. *A superbly illustrated introduction to the mammals of the world.*

Nelson, C. H., and K. R. Johnson. "Whales and Walruses as Tillers of the Sea Floor." *Scientific American* 256(1987): 112–117. *Describes how these bottom-feeding mammals move more sediment than do rivers flowing into the Bering Sea.*

Pilbeam, D. "The Descent of Hominoids and Hominids." *Scientific American* 250(1984): 84–96. *A brief introduction to the study of fossilized ancestors of the modern apes and humans.*

Sharman, G. B. "They're a Marvelous Mob, Those Kangaroos." *National Geographic* 155(1979): 192–209. *A beautifully illustrated survey of these Australian herbivores.*

Vaughan, T. A. *Mammalogy*. 3d ed. Philadelphia: Saunders, 1986. *An up-to-date textbook on the taxonomy and distinguishing features of mammals.*

Young, J. Z. *The Life of Mammals*. Oxford: Clarendon Press, 1983. *A standard textbook on the comparative anatomy of mammals.*

☐ CLASSIFICATION

Class Mammalia (L. *mamma*, "breast"). Mammals.

Subclass Prototheria (Gr. *protos*, "first"; *ther*, "wild animal"). Egg-laying mammals.

Order Monotremata (Gr. *monos*, "single"; *trema*, "hole"). Monotremes (egg-laying mammals); duck-billed platypus (*Ornithorynchus anatinus*) and two spiny anteaters (*Tachyglossus* and *Zaglossus*).

Subclass Theria (Gr. *ther*, "wild animal"). Viviparous mammals.

Infraclass Metatheria (Gr. *meta*, "later"; *ther*, "wild animal"). Marsupial mammals.

Order Marsupialia (Gr. *marsypos*, "bag"). Marsupial mammals; *Didelphis* (opossums), *Thylacinus* (Tasmanian "wolf"), *Sarcophilus* (Tasmanian devil), bandicoots, koala, kangaroos, and wallabies.

Infraclass Eutheria (Gr. *eu*, "true"; *ther*, "wild animal"). Placental mammals.

Order Insectivora (L. *insectum*, "insect"; *voro*, "to devour"). Insectivorans; moles, shrews, hedgehogs.

Order Macroscelida (Gr. *makros*, "large"; *skelis*, "leg"). Elephant shrews; small insectivores with a trunklike snout and hind legs longer than the forelimbs; in dry African grasslands and savannahs; about 20 species.

Order Scandentia (L. *scandentis*, "climbing"). Tree shrews; arboreal insectivores of Southeast Asia; about 16 species.

Order Dermoptera (Gr. *derma*, "skin"; *pteron*, "wing"). Flying lemurs; two species of arboreal herbivores of Southeast Asia and the Philippines; can extend limbs and connecting skin to glide.

Order Chiroptera (Gr. *cheir*, "hand"; *pteron*, "wing"). Bats.

Order Primates (L. *prima*, "first"). Primates; lemurs, marmosets, monkeys, tarsiers, baboons, apes, humans.

Order Carnivora (L. *caro*, "flesh"; *voro*, "to devour"). Carnivorans; wolves, foxes, jackals, bears, raccoons, skunks, badgers, weasels, otters, mongooses, hyenas, civets, cats, seals, walruses.

Order Perissodactyla (Gr. *perissos*, "odd"; *daktylos*, "toe"). Odd-toed hoofed mammals; asses, horses, rhinoceroses, tapirs, zebras.

Order Artiodactyla (Gr. *artios*, "even"; *daktylos*, "toe"). Even-toed mammals, antelope, bison, camels, cattle, deer, giraffes, elk, goats, hippopotamuses, llamas, guanacos, vicunas, moose, swine.

Order Proboscidea (Gr. *proboskis*, "elephant's trunk"). Elephants; two living species.

Order Sirenia (Gr. *seiren*, "sea nymph"). Manatees ("sea cows"), dugongs.

Order Hyracoidea (Gr. *hyrax*, "shrew"). Hyraxes.

Order Xenarthra (Edentata) (Gr. *xenos*, "foreign"; *arthron*, "joint"). Anteaters, tree sloths, armadillos.

Order Tubulidentata (L. *tubulus*, "tube"; *dentatus*, "toothed"). Aardvark; one living species.

Order Pholidota (Gr. *pholis*, "horny scale"). Pangolins (scaly anteaters).

Order Cetacea (L. *cetus*, "whale"). Cetaceans; whales, dolphins, porpoises.

Order Rodentia (L. *rodere*, "to gnaw"). Rodents; rats, mice, squirrels, beaver, marmots, lemmings, prairie dogs, chipmunks, and many others.

Order Lagomorpha (Gr. *lagos*, "hare"; *morphe*, "form"). Rabbits, hares, pikas.

ECOLOGY

Environmental Relationships
of Animals

In this final unit, we look at the subfield of biology called **ecology**, the study of relationships between organisms and their environment. Actually, much of this book is about ecology. The chapters in Unit III contain sections on the habitats and environmental relationships of individual animal phyla. Physiological and behavioral adaptations of animals, as well as evolution and phylogeny, are all "ecological" subjects in the sense that they are inseparable from how animals interact with their environment. In contrast to what we have emphasized so far, however, this unit focuses on ecology at the level of populations, communities, and ecosystems.

We have already defined the biological population and discussed its role in the evolutionary process (see Chapter 17, p. 398). A **community** is a group of interacting populations. The living organisms in a forest or lake are examples of communities. An **ecosystem**—the highest level of biological organization—is a more or less complete, functional environmental unit, consisting of a biotic community and the abiotic (nonliving) factors on which the organisms in that community depend. Lakes, forests, and deserts—that is, the living organisms and their nonliving surroundings—are examples of ecosystems. On a global scale, the total of all ecosystems, or all the inhabited area on earth, is called the **biosphere**.

In concentrating on populations, communities, and ecosystems, this unit also focuses on three concepts that are central to biology and that have been recurrent themes in this book: evolution, homeostasis, and energy. Evolution and ecology are inseparable, for evolution through natural selection results from the interaction between organisms and their environment. We have already considered certain aspects of evolution and ecology, especially at the level of the individual animal species, in studying physiological systems and animal behavior as results of evolutionary adaptations to environmental challenges. On a larger scale, the evolution of each animal phylum, as discussed in the phylogeny sections in Unit III, is the result of interactions between the organism and the environment across eons of geological time.

The concept of homeostasis (a dynamic, steady state in living organisms) is important at all levels of organization in biology. We have seen that maintenance of homeostatic balance is vital at the subcellular, cel-

823

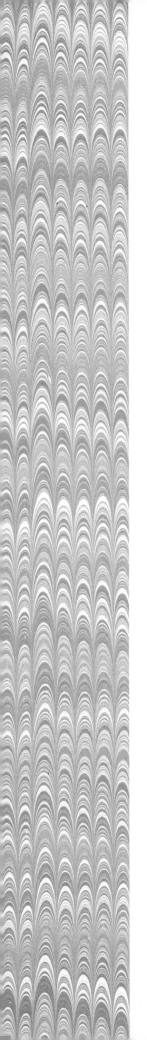

lular, tissue, organ, organ system, and organismal levels; homeostatic balance must also be maintained in populations, communities, and ecosystems. In Chapter 34, we examine some of the checks and balances that regulate population growth among animals, and in Chapter 35, we look at the homeostatic cycling of abiotic nutrients. As the largest homeostatic units in nature, ecosystems represent hundreds of millions of years of changes in environments and the consequent evolutionary adjustments in organisms.

The role of energy, the third theme of this unit, has been prominent throughout this book. We have described how animals obtain and use energy, how the availability of energy affects animal activities, and how adaptations in animals often represent ways to obtain and efficiently use the limited energy available to them. As we will see in Chapter 35, the limitations on energy availability implied by the second law of thermodynamics apply just as much to energy relationships in ecosystems as they do to the chemical processes of aerobic and anaerobic respiration in cells. In fact, energy relationships in ecosystems depend directly on energy use in cellular metabolism, and just as in Chapter 3 we analyzed the efficiency of cells in obtaining energy from organic compounds, in this unit we will examine the relative efficiency of energy use in food chains in ecosystems.

As a final note, this unit ends with a short analysis of the impact that our own species has had on world ecosystems and points out some of the implications of the rapid environmental changes that are occurring as a result of human activity all over the planet.

Population Ecology

A herd of zebras in East Africa. A zebra population may consist of several such herds. Zebras are members of the same genus *(Equus)* as the domestic horse. They are among the most numerous large herbivores on African grasslands.

Population ecology is the study of how natural processes (including human activities) affect the size, composition, and dynamics of plant and animal populations. Population ecologists attempt to predict such things as how many animals in a population can die, through predation or disease, without leading to long-term depletion of a population. For economic reasons, they would also like to be able to predict pest outbreaks so that such outbreaks could be controlled before crop damage became too extensive. Today there is an urgent need to monitor the status of rare and endangered species and to protect them from extinction.

CHARACTERISTICS OF POPULATIONS

As discussed in Chapter 17, a biological population is an interbreeding group of individuals of a particular species. The term *population*, however, is often used rather loosely. For instance, one might refer to the elephant population of Kenya, the human population of New York City, or the meadow vole population of an Illinois grassland. But these so-called populations are defined by boundaries established by humans, not by biological limits, as in the case of species populations. As used in this unit, the term *population* refers only to biological populations.

Each population has certain definable characteristics that individuals do not have. Structural features of a population, including its size, density, distribution, and age structure, describe aspects of the population at some point in time. Functional features, such as natality (birthrate) and mortality (death rate), are dynamic processes and describe rates of change over time.

Size and Density

In general, before any other features are studied, population size, or the number of individuals in the population, is determined or estimated. Because many populations are large and widely distributed, all individuals composing them cannot be counted directly. In fact, if all individuals *can* be counted, the species may be so small as to be endangered. It is possible, for example, to determine the total number of living individuals of the whooping crane *(Grus americana);* likewise, biologists know how many individuals—fewer than 30—of the California condor *(Gymnogyps californianus)* remain. Both of these species are endangered, and individuals are large enough to be easily seen; furthermore, nearly all California condors are in captivity. In contrast, studies of large populations rely on sampling methods that allow researchers to estimate statistically the size of a small portion of the population in a defined sampling area. Thus, population ecologists usually work with only samples of species populations and must clearly define the boundaries arbitrarily placed on a population under study.

Population size alone does not convey much information. It means little, for example, to say that there are 100 deer in a population. This says nothing about whether the deer are in a very small or very large area. The critical issue—the one that allows a researcher to begin to understand the status of a deer (or other) population—is **density**. Rather than simply knowing that there are 100 deer in a population, it is much more meaningful to know how many deer there are per hectare (10,000 m^2) in a study area. Before a population's density can be determined, the area being sampled must be scaled to the animal being studied. For deer and many other large animals, the number of individuals per hectare is appropriate, but for small terrestrial invertebrates, the number per square meter may be a more realistic scale.

Generally, expressing population density as the number of individuals per unit area does not give an adequate picture of the density experienced by animals. Individuals usually occupy only certain habitats or portions of habitats within a study area, and it is often in the specific habitats where individuals are concentrated that density has its greatest meaning. Therefore, in addition to determining population density, population ecologists usually must determine *how* individuals in the population are distributed in a study area.

Distribution Patterns

Hypothetically, three distribution patterns are possible: random, uniform, and clumped (Figure 34.1). If individuals in a population exhibit *random* distribution, each individual has an equal chance of being found at any place in a given area. Random distribution is a hypothetical condition that rarely (if ever) occurs among animals in nature, because resources are not randomly distributed. In *uniform* distribution, animals are evenly, or regularly, spaced throughout the habitat, and this may occur, or at least be approached, in nature. Certain territorial animals may show relatively uniform spacing, but absolute uniformity is probably rare (Figure 34.1b). By far the most common pattern of distribution is the *clumped* pattern, and this is not surprising, because resources required by organisms generally have a patchy or clumped distribution, and animals using these resources are typically distributed in a similar fashion. Reproductive patterns, especially in animals displaying some degree of parental care, can also lead to clumping. The resulting family groups are small aggregates of individuals. Many social interactions of individuals also result in aggregations, or clumping, as seen in herds, flocks, and schools. (Schooling behavior in fishes is discussed in Chapter 29, p. 720; Figure 29.16.) At times, however, individuals *within* flocks or schools exhibit more or less uniform spacing because of social interactions among them.

Many animal populations maintain different

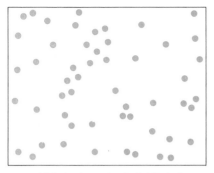

(a) Animals randomly distributed

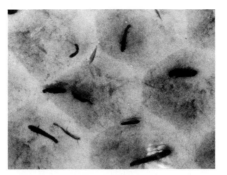

(b) Animals uniformly distributed

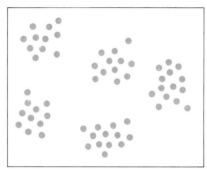

(c) Animals in clumped distribution

Figure 34.1   Random, uniform, and clumped population distributions. (a) In this hypothetical case of random distribution, the spacing of individuals is entirely due to chance, with each individual having an equal chance of occurring anywhere in the population's range. The dots represent individuals of the same species. (b) Territorial males of the bony fish *Oreochromis mossambica* are more or less uniformly distributed. Each male's territory is a pit dug in the sand by its occupant. The breeding males are the black fish; the gray fish are females, juveniles, and nonbreeding males. (c) Many species show two distinct patterns of distribution. As shown in this photograph, the snow goose *Chen hyperborea* forms large flocks (clumps) when migrating to and from its winter feeding grounds. By contrast, the snow goose becomes territorial and exhibits a more uniform distribution on its breeding grounds (not shown). This species breeds in the Arctic tundra and winters in Japan, California, and in the southern U.S., especially in coastal areas.

patterns of distribution at different times. This is not surprising when we consider that their habitats often vary both spatially and temporally. Most bats congregate in caves or hollow trees during the day and disperse at night. During the dry season on the central African plains, animals congregate near water sources, whereas at other times of the year, when water is relatively plentiful, they are distributed more widely over the grasslands. Many species of ducks and geese form dense flocks during migration in the spring and fall but disperse during the summer breeding season (Figure 34.1c).

Age Structure

The age structure of a population may tell us more about population dynamics than does either density or distribution. Knowledge of the age structure of a population helps us to predict its growth trends. Populations are typically divided into three

functional age classes: prereproductive, reproductive, and postreproductive. The percentage of an individual's life span spent in each age class is quite variable, but in general, the shortest phase is the postreproductive stage (Figure 34.2).

Population age structure can be represented by **age profiles**, which are constructed by estimating the number of individuals in each age interval (Figure 34.3). When the number of individuals in the prereproductive ages form a large percentage of the population, the profile is broad-based, and the population is likely to be increasing in size. In contrast, relatively stable populations have a more even distribution of individuals in the three age classes and have more triangular profiles. When the percentage of prereproductive animals is relatively small and the number of postreproductive animals is relatively large, the population is declining, as proportionately few individuals are maturing to provide population replacement.

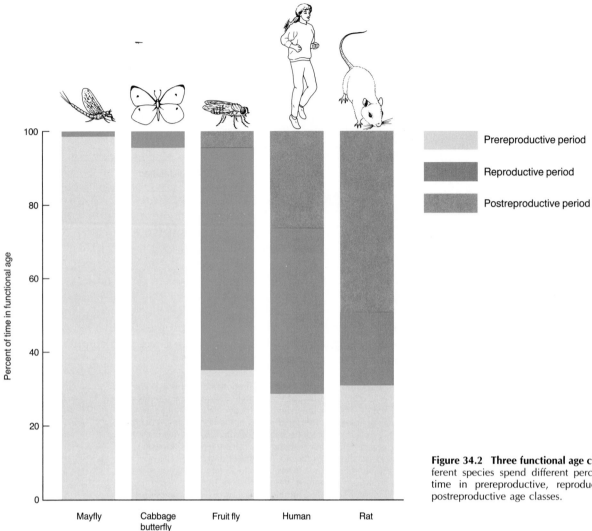

Figure 34.2   Three functional age classes. Different species spend different percentages of time in prereproductive, reproductive, and postreproductive age classes.

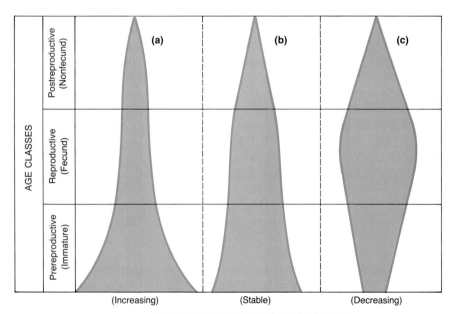

Figure 34.3   Age profiles of three general types of populations. (a) A population that is increasing in size has most individuals in the prereproductive age class. (b) A stable population has a more nearly equal number of individuals in the three age classes. (c) A declining population often has most individuals in the reproductive age class.

It is important to realize that species in which individuals have short life spans may show all three types of age profiles during a seasonal cycle. In honeybees, for example, young are not produced during the fall, but varying numbers of offspring are produced during the rest of the year. Thus, an age profile would show that the population is stable during the summer but that it declines in the fall and early winter. By mid-winter, most old individuals have died, but young are again being produced, and the population is expanding; at this time the age profile has a broad base and a narrow top, typical of an increasing population.

Natality and Mortality

Considering the structural characteristics of populations just discussed, it is apparent that most populations are dynamic, changing with time. Although structural population characteristics indicate that change occurs, they tell us little about how changes are brought about. An investigation into the functional characteristics of populations is required to provide this information.

The balance between births and deaths is often the primary determinant of population change, with the balance between immigration and emigration usually being secondary in importance. **Natality** and **mortality** are the ratios of the number of births and deaths, respectively, in a given time to the total population. These ratios are often expressed as a number per 1000 or 10,000 individuals in a population.

A statistical device called a **life table** provides a

picture of mortality in a population. To construct a life table, the number of individuals in various age classes is determined, and the mortality is calculated for each age class. Note in Table 34.1 that cottontail rabbits tend to die young; relatively few survive to two years of age. If we express the data from the second column of the table as percent survival and plot the resulting values against age intervals (months, in this case), we get a **survivorship curve** (Figure 34.4). The vertical scale of a survivorship curve may be arithmetic or logarithmic; if it is logarithmic, as in Figure 34.4, the curve allows us to classify populations into one of three general types (Figure 34.5). In comparing Figures 34.4 and 34.5, note that the survivorship curve for rabbits most closely resembles that of birds. This is a nearly straight-line relationship (the type II curve in Figure 34.5) characteristic of many animals for which the probability of survival is relatively constant throughout a typical lifetime. In contrast, human populations (although they are not true biological populations) in developed countries such as

Table 34.1   Life Table for a Population of Cottontail Rabbits (*Sylvilagus*)

Age Interval (Months)	Number Living at Start of Age Interval*	Number Dying During Age Interval	Mortality (Death Rate)
0–5	10,000	7,722	0.77
5–10	2,278	1,106	0.49
10–15	1,172	672	0.57
15–20	500	205	0.41
20–25	295	151	0.51
25–30	144	94	0.65
30–35	50	46	0.92
35–40	4	4	1.00

*For convenience, data have been converted to proportions of a standard-size group of 10,000 individuals.

Source: Modified from R. D. Lord, Jr., "Mortality Rates of Cottontail Rabbits," *Journal of Wildlife Management* 25(1961): 33–40. Copyright © The Wildlife Society. Reproduced by permission.

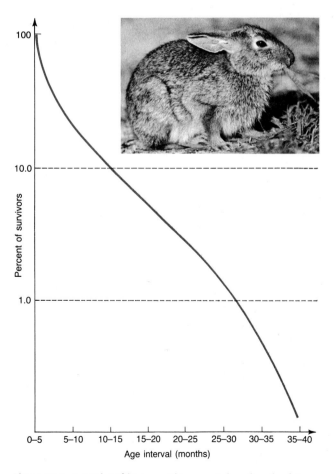

Figure 34.4   A survivorship curve. This curve is based on the data on cottontail rabbits presented in Table 34.1.

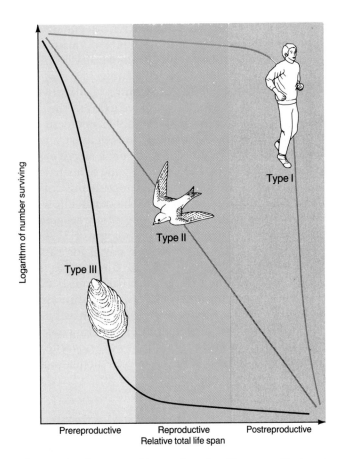

Logarithm of number surviving

Type I

Type II

Type III

Prereproductive | Reproductive | Postreproductive
Relative total life span

Figure 34.5 Three general types of survivorship curves. The type I survivorship curve is typical of animals that tend to live out the full physiological life span of their species. The type II curve indicates that mortality is fairly constant at all age levels and that there is a more or less uniform percentage decrease in the number of individuals that survive. The type III curve is typical of animals that show high mortality early in life.

the United States may have a type I survivorship curve; following a short period of relatively high mortality in infancy, there is a relatively low probability of dying until late in life. In contrast to humans and birds, many fishes and invertebrates (e.g., oysters) produce enormous numbers of juveniles or larvae, most of which are consumed by predators or fail to find a habitat suitable for further development; such organisms have a type III survivorship curve.

Population Growth Models

The overall roles of natality and mortality in population growth are best seen by examining two idealized models for population growth.

Intrinsic Rate of Increase and the Exponential Growth Model. Theoretically, if environmental

conditions are ideal, placing no restrictions on reproduction and keeping mortality extremely low, a population could display its **biotic potential**, or **maximum intrinsic rate of population growth**. Essentially, this is the maximum growth rate of which a population is physiologically capable. A population rarely, if ever, achieves its biotic potential in nature because environmental conditions are rarely ideal. Instead, a population may exhibit what is called a **realized intrinsic rate of growth**. But even this rate can provide for rapid growth in populations. Rapid growth is most pronounced in species with high birthrates and short life spans, but even species with long life spans have an enormous potential for population growth.

The realized intrinsic rate of population growth, designated r, is measured by the difference between natality n and mortality m. Thus,

$$r = n - m$$

When $r = 0$, natality equals mortality, and population size remains constant, even though individuals are being born and are dying. This results in **zero population growth**. When r is negative, the population is decreasing, and when r is positive, the population is increasing.

In addition to r, population growth is also directly related to the number of individuals N already in the population. Table 34.2 illustrates the effect of N for a population of houseflies. Note that the production of houseflies by the seventh generation is much greater than at the end of the first generation. In other words, a large population produces a larger number of offspring in a given time

Table 34.2 Hypothetical Reproductive Potential of the Housefly (*Musca domestica*)

Generation	Numbers if Each Female Lays 120 Eggs Per Generation and All Offspring Survive*
1	120
2	7,200
3	432,000
4	25,920,000
5	1,555,200,000
6	93,312,000,000
7	5,598,720,000,000

*The model assumes that each fly survives only one generation and that half of the offspring each generation are females. This species can produce about seven generations per year.

Source: Adapted from E. J. Kormondy, *Concepts of Ecology.* 3d ed. (Englewood Cliffs, N.J.: Prentice-Hall, 1984), by permission of Prentice-Hall, Inc. © 1986.

period than a small population, even though the two populations have the same *r* value. If we use the term $\Delta N/\Delta t$ to represent the change (Δ) in population numbers *N* per unit of time *t*, $\Delta N/\Delta t$ is equal to *r* times *N*. Time in this case might be measured in hours, days, or years. By plotting *N* against *t* on graph paper, we obtain a curve that describes how a population may grow if its growth depends only on *r* and *N* (Figure 34.6). The shape of this curve is described by the equation

$$dN/dt = rN$$

and population biologists use this equation as a model to describe unrestricted or **exponential population growth**. The term *dN*, pronounced "dee N," is the amount of population growth that occurs in an infinitely small amount of time *dt* ("dee t"). (The letter *d* in *dN/dt* is a mathematical expression for instantaneous change in *N* and *t*.)

Environmental Resistance and the Logistic Growth Model. From many projections of population increase, such as the housefly example in Table 34.2, it is clear that if exponential growth occurs, it cannot continue for many generations before the population becomes too large for any habitat to support it. In most, if not all, populations, a variety of factors tend to prevent exponential growth. These factors are discussed individually later, but for now we can refer to them collectively as **environmental resistance**. As a result of environmental resistance,

exponential growth slows down because of a combination of declining birthrate and increasing death rate. Eventually, the population stops growing as it reaches or fluctuates around zero population growth at the habitat's **carrying capacity**, the maximum population density that the environment can support for an extended time.

It is possible to obtain a mathematical representation of restricted growth by adding a factor called *K* to the equation for exponential growth. The factor *K*, which represents the carrying capacity, is introduced into the equation as part of the expression $(K - N)/K$. Thus, the equation becomes

$$dN/dt = rN \times (K - N)/K,$$

which describes restricted or **logistic population growth**. The effect of the expression $(K - N)/K$ can be seen by considering some possible conditions in a growing population. For example, when *N* (the number of individuals) is very small compared to the carrying capacity, $(K - N)/K$ nearly equals 1, or 100%, and the population approximates exponential growth (described by $dN/dt = rN$). As population size increases, the carrying capacity expression exerts a greater effect. When *N* equals *K*, 0% growth occurs.

When the logistic growth equation is plotted, it yields an S-shaped, or sigmoidal, growth curve. Figure 34.7 shows how a plot of logistic growth compares to that of exponential growth. Note that the curve shows the increasing effect of the term $(K - N)/K$ on population growth, making it increasingly less like the exponential curve as the population gets larger. As Figure 34.7 indicates, the term $(K - N)/K$ represents the effect of environmental resistance, that is, of slowing the birthrate and/or increasing the death rate, and keeping the growth rate from being exponential. The logistic growth curve shows a period of accelerating growth followed by a pronounced deceleration phase as population numbers approach the carrying capacity *K*. Finally, population size described by the logistic model stabilizes when zero population growth rate is reached at the environment's carrying capacity.

Critique of the Models. It is important to realize that both the exponential and logistic growth equations are mathematical models. They serve a useful function in helping us to understand—but not necessarily describe precisely—how populations actually grow. Population ecologists may use these equations in studying and comparing growth patterns of natural populations. In nature, some populations seem to approximate the exponential model

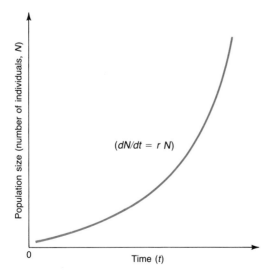

Figure 34.6 An exponential population growth curve. This model assumes that nothing (including such factors as food and space) limits population growth. The curve is described by the relationship *dN/dt = rN*.

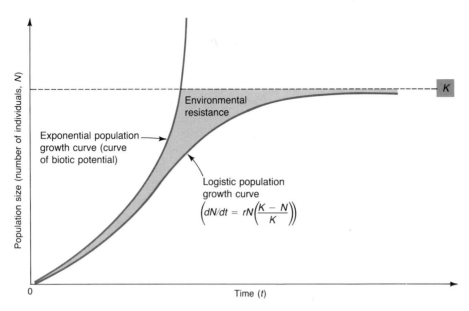

Figure 34.7   A logistic population growth curve. This graph compares the sigmoidal growth curve of the logistic growth model with the exponential curve of unrestricted population growth, shown here and in Figure 34.6. The logistic model illustrates the relationships between biotic potential, environmental resistance, and carrying capacity, *K*.

more closely than the logistic model, and vice versa.

Both of these models include important simplifying assumptions. The exponential model assumes that populations grow without being influenced by environmental resistance. The logistic model may seem more realistic, but it makes several assumptions that may not represent natural conditions. First, it assumes that carrying capacity *K* is constant, an assumption that is seldom true. Seasonal and year-to-year variability in carrying capacity is common; consider, for example, the effects of droughts and floods. Second, the logistic model assumes that each individual affects the growth of the entire population by increasing environmental resistance as soon as it is born; often, the effects of population increase are not seen until new individuals mature or until some particular density is reached. A third assumption is that the number of surviving offspring produced by an individual relates directly to the resources available at the time of birth. The logistic equation does not take into account that conditions can change during the long time periods between conception, birth, and maturity in many species.

Carrying Capacity and Time Lag. Populations do not usually increase to a specific carrying capacity and remain there indefinitely. Not only may the carrying capacity vary because of environmental

conditions, but populations may actually increase beyond the carrying capacity. The time between the origin of a cause (e.g., the fertilization of an egg) and the resultant effect (overpopulation) is called **time lag.** In species with extended time lags, populations may grow well beyond the carrying capacity. In such populations, overpopulation may even reduce the carrying capacity. This may occur because large numbers of individuals seriously reduce vital resources. In turn, mortality increases, and the population size may decrease to a level *below* the original carrying capacity.

Figure 34.8 shows three hypothetical populations with varying time lag. When the lag is short, as is common with many small animals with brief generation times, but also with some larger species, such as the wolf *(Canis lupus),* a logistic curve may actually be obtained and may show only slight oscillation around the carrying capacity. The most common populations in nature probably show a medium lag, wherein the population oscillates around, but near, carrying capacity. In populations with a long lag, such as white-tail deer *(Odocoileus virginianus),* population growth may more closely approximate exponential increase, and the population may greatly exceed the carrying capacity; the resultant overpopulation may then be followed by increased mortality and reduced natality, and the population may eventually "crash." One of the main causes of population crashes is habitat de-

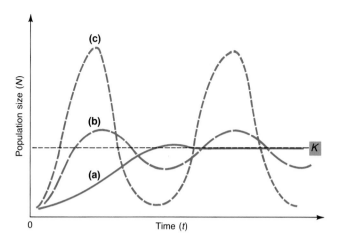

Figure 34.8 Effects of time lags on the extent of population oscillation. Carrying capacity K is the maximum population size if population growth does not oscillate. When lags are small (curve a), near logistic growth is obtained. When lags are extensive (curve c, and to a lesser degree, curve b), population growth more closely approximates the exponential model, and great fluctuations can occur.

struction due to overpopulation; in this case, the carrying capacity may decline prior to a population decline. Following a crash, surviving individuals may reproduce rapidly because of reduced environmental resistance; the population may again grow rapidly and may be followed by yet another crash. If critical environmental fluctuations happen to coincide with the lows of an overpopulation/crash cycle, populations may die out.

Problems in Controlling Populations. An understanding of population growth helps make clear how some populations are able to withstand large sustained mortality and why controlling pests, such as rats and mosquitoes, by trapping and poisoning often yields only temporary success. Many relatively stable populations remain near carrying capacity. Natality is "adjusted" to limited resources. When populations that approximate logistic growth sustain heavy mortality, numbers may be decreased to levels where the growth rate approaches exponential growth. Pest control campaigns seldom kill all of the target animals, and following such control programs, pest populations often grow back rather rapidly toward the carrying capacity. To control certain pests, it is more effective to decrease the carrying capacity than to kill large numbers of individuals. For example, improving sanitation and thus reducing food sources may be a more effective way to control rats than poisoning them.

Humans, of course, affect the population growth rate of many species. We increase mortality directly by harvesting animals for food, profit, or sport, and indirectly by altering habitats and lowering or raising the environment's carrying capacity for certain species. As human populations grow, we increase our demands on the environment to meet our needs for food and other resources, and in the process, we intensify our impacts on other species. Currently, habitat destruction is the most important effect humans have on other species populations. The draining of wetlands for agriculture, for example, has reduced waterfowl populations in North America far more than hunting has. For most species now near extinction, habitat loss is the critical threat. On the other hand, if animals are sparse but conspicuous in their environment, and if they reproduce only at long intervals, then hunting can have a devastating effect. Serious declines in whale populations are a classic example of hunting pressure (not habitat reduction) that has pushed some species toward extinction.

K Selection and *r* Selection

Each species has a particular life history and population growth pattern adapted to the environmental conditions under which it evolved. Some species are capable of very rapid population growth and may exhibit, or approximate, an exponential growth pattern followed by population crashes. Such species are referred to as ***r*-selected species** because their population growth may be nearly exponential and thus affected mainly by the factor rN rather than by environmental resistance. Populations of r-selected species seldom approach carrying capacity, and crashes in the adult population are common and usually due to abiotic variables such as drought, cold, or extremely variable resources. Hardy life forms, such as cysts or resistant zygotes, carry the population through the stressful times. Populations of r-selected species are usually adapted to environments that change rapidly, although some occur in more stable environments. Most of what we call pest or weed species are r-selected. Selective pressures on their populations favor individuals that produce large numbers of small, rapidly maturing individuals. Because the population is not at carrying capacity, an ability to outcompete others for scarce resources is not a necessary qualification for success. Fitness depends instead on producing the maximum number of offspring before environmental change brings about stressful conditions and perhaps mass mortality. Mosquitoes are good examples of r-selected species. During warm, moist months, when small, shallow areas of standing water abound, their populations often explode. Populations typically remain high until the waters dry up or cold weather kills the adults and limits reproduction. Then the populations crash.

K-selected species are at the other extreme; they generally produce few offspring, but spend enormous amounts of energy ensuring that the offspring survive. In contrast to *r*-selected species, *K*-selected species have more or less stable populations adapted to exist at or near carrying capacity, often in relatively stable habitats. Fitness in *K*-selected species usually depends on the ability to compete with others for limited resources. *K*-selected species generally produce small numbers of large, relatively long-lived, slowly maturing offspring, and adults usually provide a great deal of parental care. Most large mammals are considered *K*-selected species.

K selection and *r* selection represent two very different, but successful, approaches or strategies for survival. For *r*-selected species, regulating forces are often abiotic and independent of population size. For *K*-selected species, regulating forces are often biotic and are effective mainly when the population's density is near carrying capacity and resources such as food or nest sites become limiting. Relating this discussion back to Figure 34.5, *K*-selected species tend toward a type I survivorship curve, and *r*-selected species tend toward a type III curve. In either case, what actually determines success is whether sufficient offspring are produced to continue the population into the future.

As with the exponential and logistic growth models, the concepts of *r* and *K* selection are idealized categories. In nature, species form a general continuum between these two categories, and many species are not easily placed in either one. Nonetheless, in looking at life cycles and reproductive patterns in several groups of animals, we have seen how a general "numbers strategy" seems to characterize many species and a "care strategy" characterize many others. (This topic is discussed in more detail in Chapter 14; see "Energy Budgeting in Reproduction," p. 347; and in Chapter 29, see "Reproduction and Life Cycles," p. 722.)

ENVIRONMENTAL RESISTANCE AND POPULATION CONTROL

We can make two fundamental observations about populations: Population density varies from habitat to habitat, and no population increases indefinitely. Many different environmental resistance factors may control population growth, and most populations are subject to simultaneous control by more than a single factor. In this section, we examine some of the major types of environmental resistance that may curb population growth.

Tolerance Limits

Populations are limited in their distribution and density by the effects of the physical environment on individuals. Some habitats are optimal for a population, and some are not. A wide variety of environmental variables, such as temperature, moisture, light, and chemical concentrations, determine whether the environment is optimal, marginal, or uninhabitable for a species population. In 1840, the German chemist Justus von Liebig proposed the **law of the minimum** to explain the growth of plants. Liebig's "law" states that the essential material available in amounts most closely approaching the critical minimum needed by an organism will tend to limit the organism's growth and development. As applied to animals, a short supply of vitamins and trace minerals, for example, may determine that certain species cannot survive in some habitats. In 1913, American ecologist Victor Shelford expanded this concept by stating that too little *or* too much of an environmental factor can limit population density by approaching or exceeding the **limits of tolerance** of individuals. In other words, an animal has limits of tolerance to *extremes* of temperature, moisture, pH, light, oxygen, copper, and so forth. Also, as shown in Figure 34.9, each animal has optimal levels for each environmental factor, and animal populations grow best under optimal conditions. Populations cannot exist outside the tolerance limits of its individuals.

Concepts of tolerance help us understand patterns of distribution of species and populations. In general, species that are widely distributed have wide tolerance limits for many environmental factors. Such species are said to be **euryecious** (from the Greek *eury*, "wide," and *oikos*, "house"). By contrast, species that have narrow distribution patterns are called **stenoecious** (from the Greek *stenos*, "narrow"). Trout are stenothermal fishes; most species require relatively cold waters. In contrast, carp are eurythermal, thriving in a wide range of water temperatures.

Density-Independent Control

Density-independent controls are forces whose effect on a population is not influenced by the density

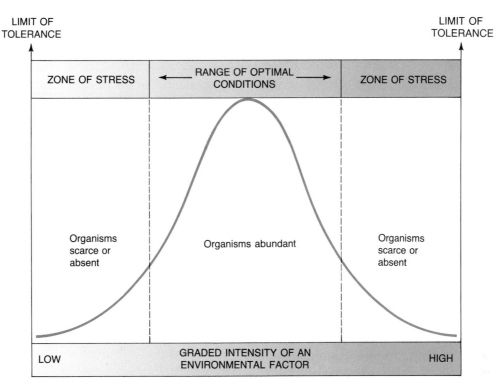

Figure 34.9 A generalized tolerance curve. The largest population is usually found within a range of optimal conditions along an environmental gradient; numbers typically decline outside that range in the zone of stress.

of the population. Human destruction of habitats is one example. Others include abiotic forces, such as severe storms, droughts, and volcanic eruptions. Imagine the effect that a volcanic eruption has on animal populations in the region of impact. Volcanic dust, lava, and mud slides cover an entire area, killing large numbers of plants and animals.

The increased mortality is not related to population density (Figure 34.10a), and populations well below carrying capacity are usually affected just as much as populations near carrying capacity.

Many populations crash at the onset of a dry season or winter, leaving behind a few survivors or resistant stages to repopulate when conditions for

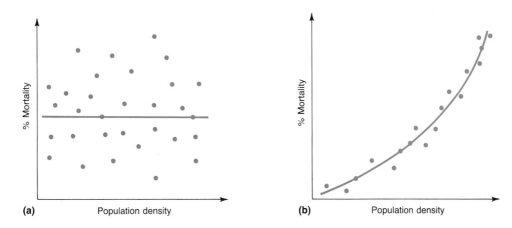

Figure 34.10 Comparison between density-dependent and density-independent population control mechanisms. (a) In density-independent mechanisms, there is no relationship between mortality and population density. (b) In density-dependent mechanisms, the percent mortality is directly related to the population density.

population growth become more favorable. Density-independent controls, such as drought and freezing temperatures, commonly affect *r*-selected species, such as insects and other small invertebrates that produce resistant zygotes or cysts able to survive harsh environmental conditions.

Abiotic controls are not always entirely density-independent. In cases where protective microhabitats are limited, as when an abiotic factor such as freezing temperature exerts its effects, only those animals having access to protective cover are likely to survive.

Density-Dependent Control

Density-dependent controls decrease population growth most effectively when the population density is high, and have less controlling influence at low density (Figure 34.10b). Examples include intraspecific competition, emigration, and certain types of stress. Because they depend on the density of the population itself and not entirely on some outside influence, density-dependent controls may provide a finely tuned regulatory mechanism that maintains populations at fairly stable levels near carrying capacity.

Intraspecific Competition. **Intraspecific competition** is competition among members of a species population. It occurs when population density increases to the point where resources, such as food, shelter, or nesting sites, become limiting. Such competition usually does not occur at low density, where resources are not limiting. Intraspecific competition is often an important feature of *K*-selected species in populations that are near the carrying capacity; limited resources keep these populations in check.

Intraspecific competition can be categorized as scramble or contest competition. **Scramble competition** occurs when all members of a population have about equal access to a limiting resource, usually food. When the resource is scarce, few individuals may obtain adequate amounts, and most individuals suffer ill effects. Scramble competition is common where the resource is more or less evenly distributed and cannot readily be defended by competitors. Browsing herbivores, such as whitetail deer, often show this type of interaction. In dense deer populations, scramble competition for choice green twigs and buds is often intense, and during the winter or other times when these resources are scarce, many individuals may starve.

In **contest competition**, successful individuals obtain adequate amounts of limited resources and prevent less successful individuals from obtaining any (or at least large quantities) of the resource. In this case, only the losers of the contest suffer the consequences of the resource limitation. They may not obtain sufficient quantity or quality of food to survive, or they may not gain access to mates for reproduction. Contest competition is seen most often in *K*-selected species with highly developed social behavior. Territoriality and dominance hierarchies are two social mechanisms that result in contest competition and produce unequal allocation of resources. (These types of social behavior are discussed in Chapter 13, p. 317.)

Emigration and Stress. **Emigration** occurs in many populations and may have a significant effect on population growth. Individuals often respond to crowded conditions by moving to new habitats. Emigrants often do not return to the parent population, and many fail to survive. **Stress** may result from a variety of forces, but when it results from crowding and the consequent increased competitive and aggressive interactions among individuals within a population, it may serve as a density-dependent population control. Actually, the role of stress in population regulation is not well understood and is difficult to study, especially under natural conditions. Crowding undoubtedly stresses animals, perhaps making them less reproductively active and more susceptible to harsh weather and disease; but studies on the effect of stress on population regulation are inconclusive. (For a review of a hormonal mechanism whereby stress may alter normal body functions in vertebrates, see the discussion of the vertebrate stress response in Chapter 11, p. 268; Figure 11.17.)

Interspecific Interactions

Many types of **interspecific interactions**, or interactions between populations of two or more species, may affect population growth. Examples include the various forms of symbiosis, such as parasitism and commensalism. In parasitism, both the host and the parasite populations are affected by the relationship; in commensalism, the host population may be unaffected. (Symbiotic relationships are discussed in the Trends and Strategies section beginning on p. 509.) In this section, we examine the role of two other types of interspecific interaction—interspecific competition and predation—in controlling population growth.

Interspecific Competition. **Interspecific competition** involves the use of a limited resource by two or more species and results in some negative effects on each competing population. Depending on the intensity of competition, as determined by the degree of overlap in resource use and by the relative scarcity of the resource, effects can range from reduction in growth of individuals to extinction of populations. Interspecific competition for the same finite resources may also intensify the effects of intraspecific competition.

The Russian biologist G. F. Gause carried out early experiments on interspecific competition. Gause used two species of the protist *Paramecium (P. caudatum* and *P. aurelia)* to study how interspecific competition affects population growth. His results led him to postulate that two or more species could not coexist in the same culture, although populations of either species grew well in the same environment when alone (Figure 34.11). These and similar experiments with other species eventually led to the formulation of the **competitive exclusion principle**, which states that competition between populations of two species for the same limiting resource eventually leads to the elimination of one of the species populations.

The concept of **ecological niche** is critical to an understanding of how competition affects the densities and the patterns of resource use of interacting populations. An animal's niche is determined by all the ways the animal interacts with its environment, including what it eats, how it obtains its food, what physical and chemical conditions it will tolerate, what conditions are optimal for its well-being, and what its predators and parasites are and how it interacts with them. Thus, to understand an organism's niche, we must understand a great deal about the organism.

What is conveyed by the competitive exclusion principle is that no two species can coexist for long if they require and use the same resources in the same ways, assuming that the resources are in limited supply. Another of Gause's experiments illustrates this point. Whereas the protists *P. caudatum* and *P. aurelia* exhibit competitive exclusion, when Gause used *P. caudatum* and another species, *P. bursaria,* in the same experiment depicted in Figure 34.11, populations of both species grew as well together as they did alone. Both species may have eaten the same food, but *P. caudatum* fed in the water column, while *P. bursaria* fed on food that had settled to the bottom of the test tube. *Paramecium bursaria* also contains symbiotic green algal cells that may supply it with some nutrients.

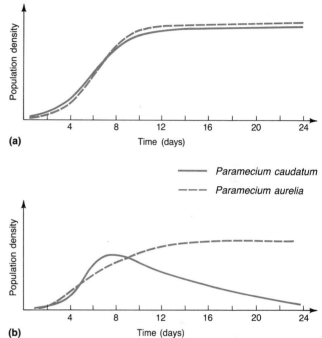

Figure 34.11 Population growth curves of two species of *Paramecium.* (a) These curves represent the growth patterns when each species was grown alone. (b) Different growth patterns generally result when the species are cultured together, competing for the same food.

Therefore, these two species do not compete directly for food, and they do not have significant overlap in resource use. In a sense, their niches partition the habitat, and each species lives in a different microhabitat.

Predation. **Predation**, the feeding of free-living organisms on other organisms, can be a very important interaction contributing to population control. Rarely, however, is a predator's role in regulating its prey clear-cut. One of the most commonly cited examples of a predator-prey relationship is that of the lynx and one of its principal prey species, the snowshoe hare. Populations of these two species undergo oscillations that seem to be linked, but this does not necessarily imply that the causal

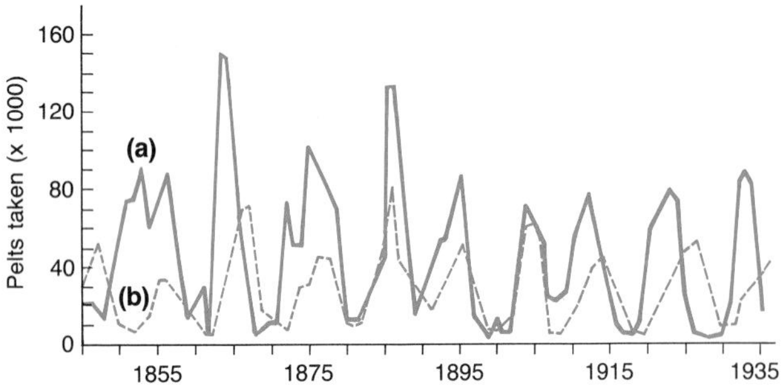

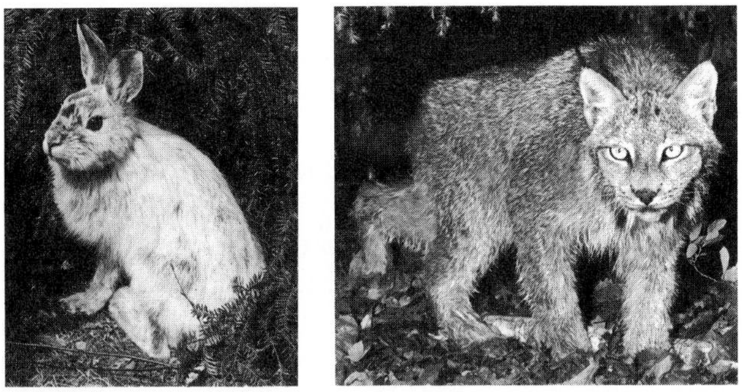

(a) (b)

Figure 34.12 Predator-prey cycles. This classical model shows an apparent relationship between populations of two species, (a) the snowshoe hare *(Lepus americanus)* and (b) the Canada lynx *(Lynx canadensis)*. This may not be a cause-and-effect relationship.

mechanism is the predator-prey relationship (Figure 34.12). In this case, the reliability of data (the number of pelts of the two species obtained from trappers by the Hudson Bay Company) is questionable. There is also evidence that even without predation pressure, snowshoe hare populations fluctuate in response to changes in their food resources. This is also true of many rodent populations that fluctuate markedly (Figure 34.13). The question often becomes, Do predators control prey populations, or do prey populations control predators? In any type of environment, there may be several species of prey and predators interacting; consequently, these systems are complex and not easily analyzed.

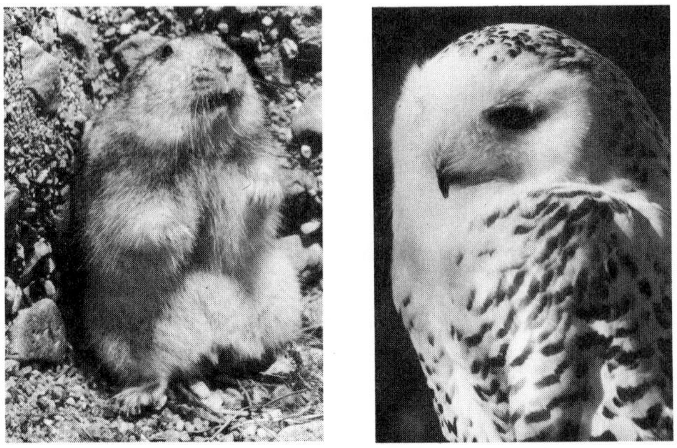

(a) (b)

Figure 34.13 A predator-prey relationship. (a) A collared lemming *(Dicrostonyx torquatus)*, a common herbivorous rodent in Arctic North America and Eurasia, exhibits marked population expansions often followed by drastic declines. (b) The snowy owl *(Nyctea scandiaca)* preys heavily on lemming populations. Predation affects the lemming's population growth, but does not cause major declines in numbers. The marked population declines, a characteristic of lemmings and many other rodent populations, may result more from the stresses of crowding and the consequent behaviorial changes and, in some cases, from a decline in the quality of plant food in an area than from predation.

Further complicating predator-prey effects, predators tend to switch to other prey when populations of one species become scarce. This usually occurs long before complete depletion of a prey population, so predation is seldom a cause for extinction of a prey population. Predators are generally adapted to balance the energy costs of pursuing, capturing, and handling the prey with the energy value of the prey. They do not usually spend great effort pursuing rare, energy-poor, or hard-to-handle prey. **Optimal foraging** brings a predator the maximum net food energy gain. The foraging behavior of the bluegill *(Lepomis macrochirus)*, a common bony fish inhabiting freshwater lakes and ponds, illustrates this principle. Bluegills can select from a wide variety of prey, ranging in size from relatively large invertebrates, such as immature dragonflies congregating on vegetation near shore, to small water fleas *(Daphnia)* living primarily in the open water beyond the beds of vegetation. The dragonflies contain more energy per individual than the water fleas, but the energy costs of capturing and swallowing the larger prey are great. When the larger invertebrates are abundant and easy to find, as they often are in early summer, bluegills tend to forage in the vegetation. However, later in the summer, dragonflies become scarce, and the bluegills usually switch to eating smaller prey in the open water. Thus, the fish appear to select the prey that provides the greatest available net energy gain at a particular time.

Prey availability is a key factor in the foraging behavior of predators. Behavioral studies indicate that increasing numbers of prey elicit two types of responses from predators (Figure 34.14). In one type, called the **functional response**, there is a relationship between prey density and the number of prey consumed per predator per unit time. In one kind of functional response, a predator seems to

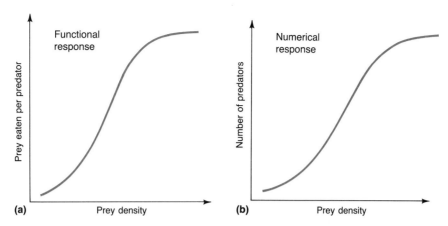

Figure 34.14 The responses of predators to increased densities of a prey population. (a) A functional response occurs when the predator switches its diet to include an increasing amount of the prey. (b) In a typical numerical response, the number of predators increases, by immigration or reproduction, in response to the increasing prey.

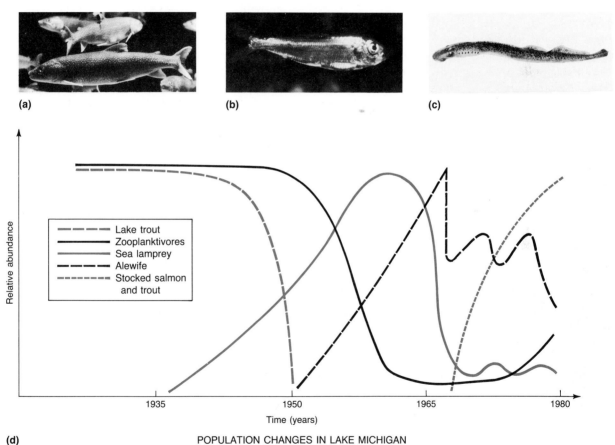

Figure 34.15 The effects of predation on prey populations. Dramatic changes in (a) lake trout *(Salvelinus namaycush)* and (b) alewife *(Alosa pseudoharengus)* populations occurred as the (c) sea lamprey *(Petromyzon marinus)* became established in the Great Lakes. As the graph (d) illustrates for Lake Michigan, increased abundance of the lamprey was accompanied by a decline of the lake trout through predation. With the drop in numbers of lake trout, the alewife was released from trout predation pressure, and alewife numbers increased. The alewife apparently out-competed other species of fishes that prey on zooplankton, causing a decline in native zooplanktivorous fishes and a change in the species composition of the zooplankton. Lamprey populations have now declined through management with poisons, and alewife populations are being controlled by predation by stocked salmon and trout. Clearly, predator-prey interactions have been and continue to be important.

develop a **searching image** for an increasingly abundant prey. In forming a searching image, the predator seems to focus its attention on the shape and general appearance of the abundant prey, filtering out other potentially distracting stimuli. Consequently, the abundant prey makes up an increasing percentage of the predator's diet. In a second type of response, called the **numerical response**, there is a relationship between prey density and predator density, and in a typical case, predator numbers increase as a prey population increases. A predator population may increase because of predator immigration into the area or, in *r*-selected species, because of rapid population expansion. Sometimes, predators show both functional and numerical responses, but in all of these situations, the prey control the predator populations. It is not clear whether increased predation pressure controls the prey.

The effects of predation on prey populations have become most evident when predators have been markedly reduced or removed from a system. For example, predation by the sea lamprey *(Petromyzon marinus)* and commercial fishing (a form of predation) greatly reduced the population of lake trout in Lake Michigan and other Great Lakes during the 1950s (Figure 34.15). (See the Chapter 28 Essay on the destruction of the Great Lakes fishery, p. 685.) The reduction of lake trout seemed to allow a population explosion of the alewife, a small fish in the herring family that eats small, free-swimming animals (zooplankton). The lake trout was the main predator of the alewife. Without the controlling influence of lake trout, alewife populations reached such high densities in the late 1960s that alewives seriously depleted zooplankton populations and outcompeted and caused the decline of many other fish species that depend on zooplankton.

As discussed in the Essay in Chapter 28, sea lamprey populations in the Great Lakes are now controlled by the poisoning of lamprey larvae in the streams they inhabit. Salmon and trout have also been stocked in Lake Michigan. As a result of predation by these species, alewife populations are now relatively low, and populations of other zooplanktivorous fishes are increasing.

◻ SUMMARY

1. Biological populations have structural features (size, density, distributional patterns, and age structure) and functional features (natality, mortality, and growth characteristics).

2. Individuals within most populations show a clumped distribution. Age profiles describe population age structure. Life tables and survivorship curves depict the role of mortality in population growth and change.

3. Exponential (unrestricted) population growth is described by the relationship $dN/dt = rN$, where dN/dt is the amount of population growth (dN) in an infinitely small amount of time (dt); r is a realized intrinsic rate of population growth and is equal to n (natality, or birthrate) minus m (mortality, or death rate); N is the number of individuals in a population.

4. Logistic (restricted) population growth is described by the relationship $dN/dt = rN \times (K - N)/K$, where K is the carrying capacity of the environment and the factor $(K - N)/K$ represents environmental resistance, or the various environmental factors that limit population growth.

5. Species may be classified as relatively *r*-selected or *K*-selected. The *r*-selected species may exhibit exponential growth and often experience significant population declines, or crashes; *K*-selected species generally have relatively low reproductive rates and stable populations.

6. Population distribution, growth, and size are largely determined by the tolerance levels of species.

7. Population growth may be controlled by density-independent physical factors of the environment, density-dependent regulating mechanisms (including intraspecific competition, emigration, and stress), and interspecific interactions, including symbiosis, interspecific competition, and predation. Competition is most intense between species that inhabit or compete for the same finite resources. Species with highly intense competition usually cannot coexist.

◻ FURTHER READING

Begon, M., J. L. Harper, and C. R. Townsend. *Ecology: Individuals, Populations, and Communities.* Sunderland, Mass.: Sinauer Associates, 1986. *A modern textbook with a quantitative approach.*

Colinvaux, P. *Ecology.* New York: Wiley, 1986. *A clear presentation of all aspects of ecology.*

Ehrlich, A. H. "The Human Population: Size and Dynamics." *American Zoologist* 25(1985): 395–406. *Discusses the current status of the human population and the implications of overpopulation.*

Hardin, G. "Cultural Carrying Capacity: A Biological Approach to Human Problems." *Bioscience* 36(1986): 599–606. *An essay on carrying capacity and its application to human population problems.*

Ricklefs, R. E. *Ecology.* 3d ed. New York: Chiron Press, 1986. *Includes strong coverage of population genetics and adaptation.*

Smith, R. L. *Ecology and Field Biology.* 3d ed. New York: Harper & Row, 1980. *Includes two chapters on population ecology.*

Communities
and Ecosystems

The tropical rain forest, one of the most important ecosystems on earth. Also one of the most fragile, it is currently being destroyed at an alarming rate. Biologists estimate that hundreds of thousands of undescribed species of animals (mostly insects) live in these unique forests. Destruction of vast tracts of tropical rain forest is likely to alter world weather patterns and cause significant changes in the balance of gases in the atmosphere. This photograph shows the canopy of a rain forest in Southeast Mexico.

As discussed in Chapter 34, animal populations are not isolated from one another, and ecologists studying them cannot focus all their attention on one population alone. The assemblage of species populations inhabiting the same area forms a **community**. Communities interact with their physical and chemical environments to form **ecosystems**, which are the largest units of biological organization that biologists attempt to study. A natural ecosystem is very complex and difficult to study, requiring the simultaneous analysis of many uncontrollable environmental variables and interactions. Yet, it is almost impossible, using data only from experiments run in the laboratory under controlled conditions, to accurately measure the behavior of ecosystems. The study of ecosystems requires a broad knowledge of many scientific disciplines, including agronomy, chemistry, geology, meteorology, botany, and zoology.

PROPERTIES OF ECOSYSTEMS

In viewing a forest as a community, we can consider it a structural entity consisting of diverse plants and animals. However, if we view a forest as an ecosystem, we must also consider abiotic factors that sustain the living community. All biological systems, from cells to ecosystems, require energy and nutrients. Ecosystem biologists are concerned with such questions as, Where do the energy and nutrients come from that support community structure, and how are these resources incorporated into organisms and passed from one organism to another in food chains?

Energy Flow

Ecosystems are open energy systems, meaning that they can continue to function only if they have sufficient energy flowing through them continuously. Solar energy, captured in the process of photosynthesis by plants, is the driving force for all but a few ecosystems. (Deep-sea vent communities, described in the Chapter 25 Essay, p. 606, are an exception; they derive energy chemosynthetically from bacterial oxidation of inorganic sulfur compounds.) Energy flows *through* ecosystems, rather than cycling within them, because as energy is used by living organisms, it is constantly being lost as heat. This process is described by the second law of thermodynamics. As a result of this universal tendency toward heat loss, and because living organisms do not have the metabolic machinery to capture and use heat energy, all energy that plants and chemosynthetic microorganisms make available to the biosphere is eventually lost. (For a review of the general principles of bioenergetics, including the concept of free energy, see Chapter 3, p. 56.)

The four major forms of energy important in ecosystems are solar energy, chemical energy (available as the potential energy of chemical bonds in organic molecules), kinetic energy, and heat energy, all of which can be expressed as calories or kilocalories (for definitions, see Chapter 3, p. 55). Only a fraction (usually about 1%) of the solar energy that enters the earth's atmosphere is converted to chemical energy by plants via photosynthesis. The amount of solar energy captured in photosynthesis also varies with different plant species and with the density of plant cover in an area. As a result, different ecosystems vary in the percentage of the incoming solar energy they capture.

Plants use some of the chemical energy they capture for their own cellular respiration, maintenance, and growth; some of the energy not used by plants is transferred through feeding to animals, where it is used for growth and maintenance and converted to kinetic energy of motion.

One of the keys to understanding energy flow in ecosystems is understanding ecosystem **productivity**. Of course, plants do not really produce, or make, energy; energy productivity, in this context, means the rate of assimilation (accumulation) of chemical energy as organic matter. Assimilated organic matter can be measured as **biomass**, the weight of living matter in a particular area. The rate at which energy is fixed in photosynthesis by autotrophs is called **primary productivity**, and this has two components: gross primary productivity and net primary productivity. **Gross primary productivity (GPP)** is the total rate of photosynthesis. **Net primary productivity (NPP)** is the storage rate of organic matter in plant tissues in excess of plant respiration (R_A). Only the fraction of the GPP represented by the NPP is available for use by heterotrophs, including animals. The following equation gives the relationship between GPP and NPP:

$$GPP - R_A = NPP$$

Levels of primary productivity vary from ecosystem to ecosystem. When GPP for an ecosystem is combined with the area of the earth's surface occupied by that ecosystem, we see the importance of such highly productive systems as tropical forests (Figure 35.1, and see chapter opening photograph). In contrast, the open ocean is not highly productive, but marine algae contribute a high percentage of the earth's total gross productivity because of the vastness of the open sea.

Figure 35.2 illustrates energy flow at the heterotroph level. Because heterotrophs do not produce organic food from sunlight, we do not use the terms *gross* and *net* in reference to their productivity. Instead, the rate of conversion of food energy into metabolizable energy is called the **assimilation rate**; and the rate of energy storage at the heterotroph level is called **secondary productivity**.

Energy relationships can also be considered at

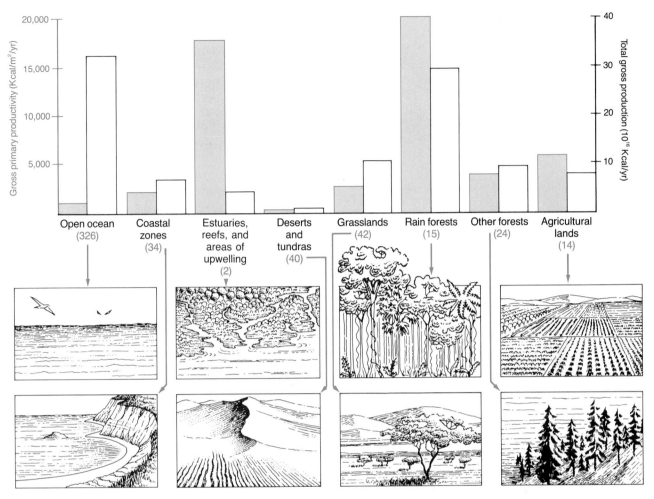

Figure 35.1   A comparison of primary productivity in different types of ecosystems. The vertical scale and bars on the left represent the rate of gross primary productivity (GPP) per square meter. The vertical scale and bars on the right represent the total organic material (biomass) produced by the ecosystems in a year. The numerical values under the names of the ecosystems are total areas ($\times 10^6$ km^2). Note that while the productivity rate of the open ocean is low, because of its large area (about 326×10^6 km^2), the open ocean has the highest total gross production.

the community level. We can estimate **net community productivity** (**NCP**) by subtracting community respiration (both autotrophic, R_A, and heterotrophic, R_H) from GPP:

$$NCP = GPP - (R_A + R_H)$$

Net community productivity is the rate at which biomass is accumulated and stored in an ecosystem. Young, rapidly growing ecosystems typically have a relatively high NCP; young forests, for example, grow rapidly and accumulate biomass rapidly. In contrast, old ecosystems, such as mature forests, typically grow slowly and accumulate little biomass; old ecosystems usually consume (in respiration) most of the energy that they produce, returning most of it to the atmosphere as heat and leaving little for ecosystem growth.

Trophic Relationships.   We introduced the concept of trophic (feeding) relationships in Chapter 4 (see p. 77). Viewing a community or an ecosystem as a trophic system allows us to visualize energy flow. Because they produce organic energy, autotrophs are called **primary producers**. In consuming autotrophs, herbivorous animals are **primary consumers**. In eating herbivores, carnivorous animals are **secondary consumers**. Higher-level carnivores (tertiary consumers and quaternary consumers) eat other carnivores. Omnivores are multilevel consumers. Organic matter that is not utilized by these consumer trophic levels enters the **decomposer** trophic level. Decomposers break dead organic matter down into progressively smaller organic materials and ultimately into inorganic matter. Bacteria and fungi are the main decomposer orga-

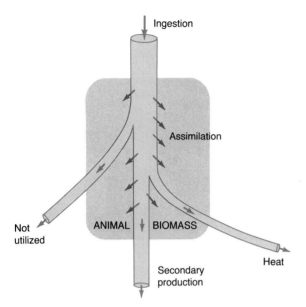

Figure 35.2 A model of secondary productivity. Animal biomass may be an individual, a population, or several populations in a community. Energy enters the system as organic food molecules. Some passes through the system and is released as undigested waste. Of the amount that is assimilated, a portion is used in respiration and is eventually lost as heat. The remaining energy, called secondary productivity, contributes to the growth of the system. The width of the pathways indicates the relative amount of energy flowing through each component of the system.

nisms, but certain animals, including earthworms, certain soil nematodes, millipedes, mites, and many aquatic insects, are also important. As a group, these animals are called **detritivores** because they eat mainly dead, fragmented plant and animal material, collectively called **detritus**. Detritivores also consume bacteria and fungi growing on detritus, and as other animals eat the detritivores, much of the energy in the organic matter that enters the decomposer level may be returned to higher trophic levels.

As we have seen many times in this book, trophic relationships among animals are usually quite complex, with many animals assuming more than a single trophic level. Most animals, whether herbivores, carnivores, detritivores, or omnivores, do not consume only one species of plant or animal, and most species can switch between alternate food sources, depending on food availability; moreover, larval forms often differ from adults in food habits. Food webs, such as that illustrated in Figure 26.11, show some of the complex feeding relationships typical of communities, but even complex diagrams of food webs invariably understate all of the feeding relationships found in most communities.

Rather than trying to study energy flow in complex food webs, ecologists usually lump species into trophic levels and analyze trophic systems as simplified food chains. In this way, the significance of the second law of thermodynamics to ecosystems becomes clear. As shown in Figure 35.3, because of the constant loss of energy as heat as organic matter is transferred from organism to organism and as energy is converted from one chemical form to another, there is always less energy available to

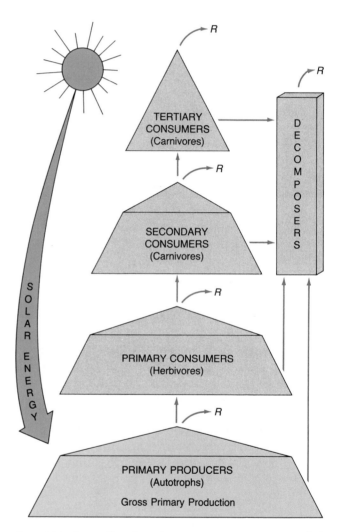

Figure 35.3 Energy flow through an ecosystem. By a process called gross primary productivity (GPP), primary producers (autotrophs) convert solar energy from the sun to chemical energy in organic molecules. Some of this chemical energy is used in respiration (R), and some is lost as heat in the process. Energy remaining after the autotrophs' respiratory needs are met may enter other trophic levels. By subtracting all the values for respiration from gross primary productivity, we obtain the net community productivity (NCP). All trophic levels contribute energy to the decomposer level. As stated by the second law of thermodynamics, with each transformation of energy, the system loses some energy as heat; therefore, energy models are often depicted as pyramids.

Figure 35.4 The osprey (Pandion haliaetus). This bird feeds exclusively on fishes and is a top carnivore in its food chain. As such, it has much less energy available to it than does a species of small fish that eats zooplankton.

successively higher trophic levels. Thus, in all ecosystems, energy flow through the system describes a pyramid. This is why there is a limit to the number of trophic levels that any ecosystem can have (Figure 35.4).

Biogeochemical Cycles

In contrast to the linear flow of energy through ecosystems, chemical materials, generally called nutrients, pass cyclically *within* ecosystems. The cyclic passage of nutrients is described by what ecologists call **biogeochemical cycles**. The processes of energy flow and the cycling of chemicals are intimately related in ecosystems. Energy storage cannot occur unless raw materials (i.e., chemicals) are available from which to synthesize energy-containing organic molecules; and materials will not pass within and between organisms unless energy is available to power the system (Figure 35.5). Both energy flow and nutrient cycling are vital factors determining where organisms live, how rapidly they can multiply, and the density of their populations.

The most prevalent types of biogeochemical cycles are gaseous cycles and sedimentary cycles. **Gaseous cycles** involve those elements, such as carbon, hydrogen, oxygen, and nitrogen, that have their major reservoir in the atmosphere. **Sedimentary cycles** involve elements that have their major reservoir in the **lithosphere**, that is, in the earth's crust and soil.

Gaseous Cycles. Elements that undergo gaseous cycles typically have large inorganic pools that are more accessible to the living community than are sedimentary elements.

The nitrogen cycle serves as an example of a gaseous cycle. It is one of the most complex of all biogeochemical cycles (Figure 35.6). As a component of all proteins and nucleic acids, nitrogen is an essential element. The primary reservoir form of nitrogen is the nitrogen gas (N_2) that forms about 79% of the atmosphere. Few species can utilize N_2 directly; it must first be converted to ammonia (NH_3), nitrite (NO_2^-), or nitrate (NO_3^-). The process by which nitrogen gas is converted to ammonia or nitrate is called **nitrogen fixation**. Some nitrogen is fixed abiotically. Volcanic activity accounts for some of this fixation, and during electrical storms, lightning converts N_2 to a fixed form that is dissolved and carried to the ground by rain. Certain bacteria, including cyanobacteria (blue-green algae), are important in biological nitrogen fixa-

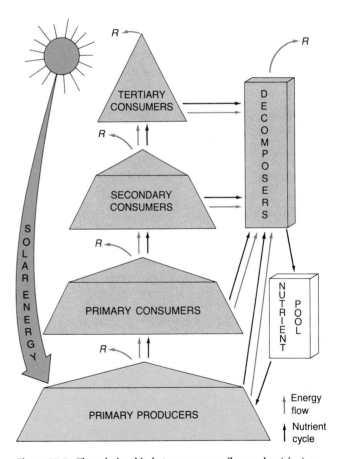

Figure 35.5 The relationship between energy flow and nutrient cycling. Energy takes a linear (noncyclic) path through ecosystems, whereas nutrients cycle; however, the two processes are intimately associated.

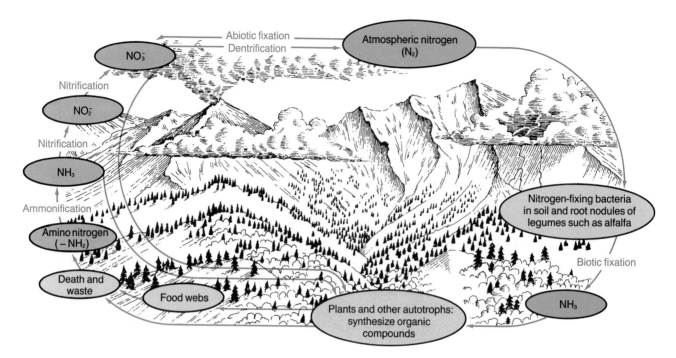

Figure 35.6   The nitrogen cycle. Many of the most important processes in this gaseous cycle involve microorganisms (bacteria and fungi) and take place in the soil or in aquatic ecosystems. Microorganisms play a major role in converting nitrogen from one form to another. Biotic fixation by nitrogen-fixing bacteria, ammonification, nitrification, and denitrification are all primarily microorganism-mediated processes, whereas abiotic fixation is a physicochemical process that involves volcanism and lightning.

tion. Cyanobacteria are major nitrogen fixers in aquatic ecosystems. On land, a relatively small number of species of nitrogen-fixing bacteria of the genus *Rhizobium* live symbiotically in the root nodules of some 12,000 species of legume plants. Other nitrogen-fixing bacteria inhabit leaf nodules in certain other plants, such as alders. As the bacteria fix nitrogen in the root or leaf nodules, the host plant and the bacteria use the products of fixation, and the bacteria survive by obtaining other nutrients from the plant cells.

Animals obtain nitrogen-containing organic compounds from plants through food chains. Nitrogen is released as ammonia in wastes or when plants and animals die and are decomposed by bacterial and fungal action. This decomposition step in the nitrogen cycle is called **ammonification** and is important in returning ammonia from biomass back to the soil, where it may contribute to future plant growth. A portion of this ammonia is utilized as an energy source by certain chemosynthetic bacteria. By a process called **nitrification**, ammonia is converted first to nitrite by one group of microorganisms and then to nitrate by another group of microorganisms; the two groups of microbes are called **nitrifiers**. Nitrate is also a valua-

ble plant nutrient and is readily absorbed from soil water by plant roots. Under certain conditions, especially anaerobic ones, the opposite of nitrification, called **denitrification**, occurs. **Denitrifiers** (certain bacteria and fungi) use nitrate as an alternative to oxygen as the hydrogen acceptor in the respiratory chain; nitrogen gas is released as a byproduct, thus returning nitrogen to the atmosphere and completing the cycle.

Sedimentary Cycles.   In contrast to materials that have gaseous cycles, elements with sedimentary cycles form large lithospheric reservoirs that are available to living communities mainly through the slow processes of weathering. Sedimentary nutrients are usually weathered from the lithosphere by being dissolved in water.

Phosphorus is a good example of an element with a sedimentary cycle (Figure 35.7). In the form of phosphate (PO_4^{2-}), phosphorus is an essential element in biological systems because of its role in energetics (in generating ATP), nucleic acids, and cell membranes (phospholipids). Under natural conditions in most ecosystems, phosphorus is in short supply and may therefore limit primary productivity. Small amounts of phosphate, released from rocks by weathering, enter the pool of nutrients available to living communities by becoming dissolved in soil water or aquatic ecosystems. From here, phosphates are taken up by plants, and animals gain phosphate from food and water. An important group of decomposers, called **phosphatizing bacteria**, convert some of the phosphorus-containing remains of dead plants and animals back into dissolved phosphates. Note that the phosphorus cycle includes an important pathway leading to deep marine sediments. This constitutes a net loss from the biotic cycling of phosphate because the recycling of marine sediments is extremely slow. To make such sediments available for cycling within living communities requires an upwelling of deep ocean currents or significant geological activity, such as mountain building. Thus, in the phosphorus cycle, which is a typical sedimentary cycle, most of the actual cycling occurs between living organisms and the pool of nutrients (in this case PO_4^{2-}) dissolved in water in the soil or in aquatic ecosystems.

The Hydrologic Cycle. The **hydrologic (water) cycle** does not fit neatly into either the gaseous or the sedimentary category. The cycle is unique in that it involves a liquid reservoir and mainly liquid cycling. The hydrologic cycle also involves part of the hydrogen and oxygen gaseous cycles. Most of the water available for cycling (i.e., not firmly bound in sedimentary rocks) is in the oceans, subsurface stores, and glaciers. Surface fresh water accounts for only a small fraction of the total.

The hydrologic cycle is driven by solar energy, which causes evaporation followed by transport in the atmosphere and return to the earth by precipitation. The atmosphere holds relatively little water, and yet large amounts of water are moved rapidly into and out of the atmosphere. Figure 35.8 shows that there is a net loss to the atmosphere from the ocean of approximately 36×10^{18} g of water per year (319×10^{18} g/yr minus 283×10^{18} g/yr). This is most of the water that is transported in the atmosphere and is seen as a net increase in water from precipitation to terrestrial ecosystems. The global hydrologic cycle is completed with water running off the land, collecting in rivers and returning to the oceans. This flowing water draining from the land weathers rocks and provides the energy to erode and leach materials from the land into the oceans.

Human Impact on Materials Cycling. Biogeochemical cycles are homeostatic systems, and as such are self-regulating. Under natural conditions, if one point in a biogeochemical cycle becomes overloaded, another point in the cycle can become more or less active, and as a result, the overload will usually be dissipated. For example, if the

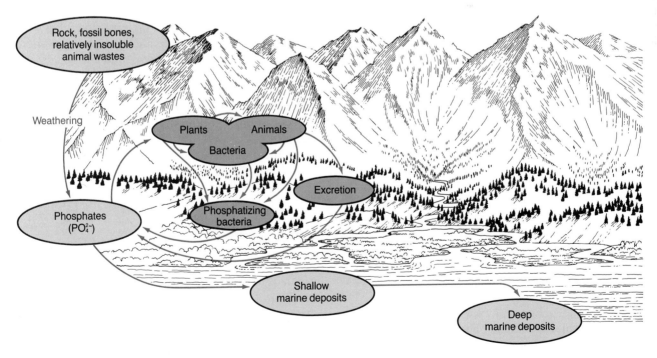

Figure 35.7 The phosphorus cycle. This biogeochemical cycle does not have a gaseous phase. Note that the large-scale movement of phosphorus is from mineral deposits to the ocean floor.

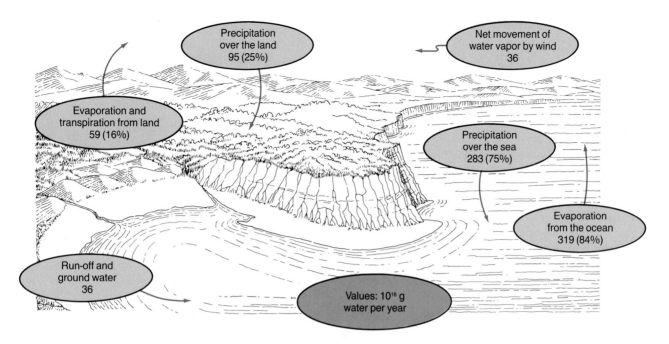

Figure 35.8 The global water cycle. The 378×10^{18} g of water that passes from the land and oceans to the atmosphere each year is balanced by an equal amount falling as precipitation. Note the relative importance of the oceans: 75% of the precipitation falls on the oceans, and 84% of the evaporation arises from them. Runoff and groundwater total about 36×10^{18} g per year, while an equivalent amount of water in the atmosphere is moved by winds.

amount of ammonia in the soil becomes higher than normal as a result of decomposition, bacterial populations may increase and convert some of the ammonia to nitrate, which is readily taken up by plants. In general, gaseous cycles have more precise feedback control than sedimentary cycles because the latter depend so heavily on slow geological processes.

Human activities often overload biogeochemical cycles to the point where natural controls cannot restore balance, and this leads to a variety of environmental problems. The addition of massive amounts of phosphorus and nitrogen fertilizers onto farmlands disrupts these two cycles. In both cases, the cycles become overloaded at the points where soluble nutrients are made available to plants. An overload of nutrients in the soil soon becomes an overload in aquatic ecosystems and groundwater as the materials dissolve in water and wash out of the soil. An excess of nitrates and phosphates in aquatic ecosystems causes overgrowth of algae, and large amounts of nitrates in groundwater pose a serious human health problem.

Other biogeochemical cycles may also be unbalanced by human activities. The burning of coal and other fossil fuels seems to be upsetting the homeostatic balance in the global cycling of carbon and oxygen. Coupled with the destruction of tropical rain forests, which use large amounts of carbon dioxide and give off much oxygen, fossil fuel burning may be increasing atmospheric amounts of carbon dioxide to levels that could increase global temperatures to levels higher than those of millions of years ago. Fossil fuel consumption also adds large amounts of sulfur oxides to the atmosphere, leading to acid precipitation, which damages forest and aquatic ecosystems. There are limits to how far biogeochemical cycles can be modified without serious consequences. When the rate of change in these systems exceeds that which the systems can accommodate, we may pay a large price in environmental damage.

PROPERTIES OF COMMUNITIES

As an association of species populations living in a particular area, a community may be any natural biotic unit, regardless of size, from the **biota** (living organisms) of a small field to that of a large forest. Each community functions as a more or less discrete unit in processing energy and nutrients. Aquatic communities are named according to the features of the water itself (e.g., freshwater, marine, and brackish-water communities). Specific types of aquatic communities have more explicit names (e.g., mussel bed communities, kelp forest communities, and coral reef communities). Terrestrial communities are usually named according to the dominant plant species they support. The dominant plants are those that contribute the most biomass to the system. Each assemblage of plants has its own assemblage of animals.

Many factors, including energy and nutrient availability, climate, soil type, interactions with other organisms, and geological and biological history, may determine the kinds of plants, animals,

and microorganisms composing a community. Two physical factors—precipitation and latitude (which determines solar energy availability and temperature)—are key determinants of community composition. Communities of large biomass, such as tropical rain forests, are generally located at low latitudes (characterized by warm temperatures and abundant light throughout the year) and high moisture levels. Community biomass generally declines as latitude increases or moisture decreases. As shown in Figure 35.9, the effect of altitude on the distribution of communities is similar to the effect of latitude. Each 100 m of elevation is

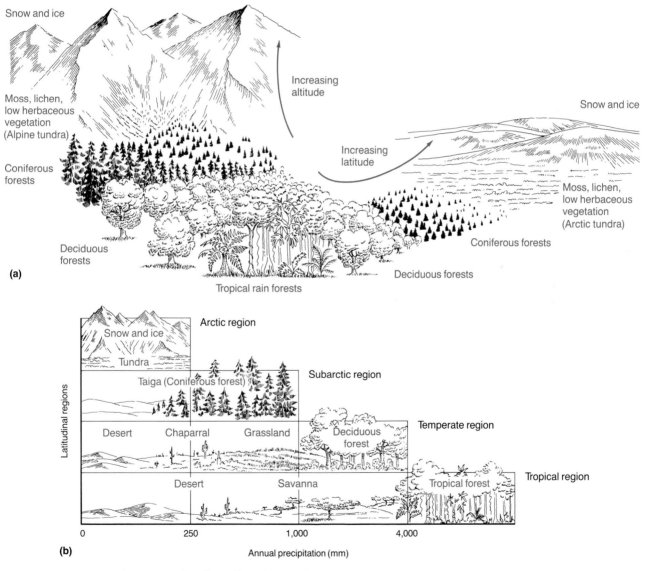

Figure 35.9 The effect of latitude, altitude, and precipitation on the structure of plant communities. (a) Latitude directly affects solar input and temperature, both of which directly influence plant communities. The effect of altitude is similar to that of latitude. (b) The amount of annual precipitation directly influences plant communities. In the temperate region, for example, annual precipitation ranges from near zero to about 400 cm, supporting plant communities ranging from deserts and dry shrub (called chaparral) to grasslands and deciduous forests.

correlated with a reduction of air temperature equivalent to about 1° increase in latitude. The types of communities encountered while climbing Africa's Mt. Kilimanjaro (5965 m above sea level) are similar to those seen on a trip from the equator to the poles.

The interactions of mountains and prevailing winds also have a profound effect on community distribution. An air mass cools as it rises over mountain ranges, causing precipitation on the windward side (Figure 35.10). On the lee side of the mountain, the descending air warms and picks up moisture, forming a **rain shadow**. This is why in western North America, with westerly prevailing winds, dry conditions are found east of the Sierra Nevada, Cascade, and Rocky mountains.

Community Change

No community is completely stable, and some may exhibit rather rapid change. If a community is drastically altered, as by hot volcanic ash or a fire, a progression of communities, each characterized by its own dominant plant species and the animals that depend on them, may appear on the site over time. This type of community development pattern is called **ecological succession**. Young communities change rapidly (they accumulate biomass rapidly), but mature communities are relatively stable and if not disturbed may remain with little significant change for many hundreds of years. Each community in a successional series is called a **sere**, and the mature stage is the **climax community**. Pioneering communities are characterized by *r*-selected species, whereas climax communities are dominated by *k*-selected species. (See the discussion of *r* and *K* selection in Chapter 34, p. 833.) Plant and animal

diversity is often relatively low in early seres, but species that are able to use these areas are usually abundant. Pioneering plants, especially, are characterized by rapid growth and maturation. They may produce large numbers of seeds that readily disperse, rapidly spreading the plants over the entire disturbed area. Pioneering plants are usually short and not shade-tolerant; consequently, they are eventually overgrown by larger plants. The dominant plants in later stages of succession are generally large, shade-tolerant species, many of which tend to produce relatively few, large seeds (such as acorns), which disperse relatively short distances.

Succession in disturbed areas that were formerly occupied by communities is called **secondary succession** (Figure 35.11). In this case, the soil is already established and succession is relatively rapid. In contrast, **primary succession** occurs on sites where no community has previously existed (e.g., new volcanic islands or recently formed sand dunes). Primary succession is a slow process because soil must be formed as one community grades into the next. Here, succession results from the modification of the physical environment by the community, and specific limiting factors, such as shade, soil depth, and nutrient richness, change with time. In primary succession on sand dunes, for example, soil building is an important process that must take place before later communities can occupy the site. Organic matter from dead organisms increases with time, and the developing soil stores increasing amounts of moisture and nutrients. Soil building is largely a biological process. As it occurs, conditions become favorable for larger plants, which can shade early-sere plants and outcompete them for soil nutrients.

Figure 35.10 The effect of mountain ranges on patterns of precipitation. Air rising over a mountain cools by compression and drops precipitation on the windward side. On the lee side, descending air warms as it expands, absorbing moisture and creating dry areas in the rain shadow.

(a)

(b)

Figure 35.11 Secondary succession. Secondary succession replaces communities destroyed by natural catastrophes. Hot volcanic ash from the eruption of Mt. St. Helens in Washington State destroyed the coniferous forest on the mountain slopes. Eventually, following a series of different communities, this area will be repopulated by a forest similar to the one that was destroyed. (a) Mt. St. Helens before eruption. (b) Mt. St. Helens after eruption; notice the vast track of felled trees.

Succession and the development of climax communities are actually more complex and often less predictable than we have indicated; in fact, the nature of these processes and how they occur are still debated by ecologists. In some areas, climate may be the major factor that regulates community development. In such cases, climate may induce the eventual establishment of a single **climatic climax community**. In many areas, however, factors such as recurring fires or moisture and soil limitations result in more localized, persisting communities called **polyclimax communities** (also called edaphic climaxes or pattern climaxes). Moreover, fire, pest outbreaks, and storms may tend to return local areas to earlier seres. The result is often a patchwork of many smaller communities representing many seres interspersed in a larger area occupied mostly by the climax community.

MAJOR TERRESTRIAL ECOSYSTEMS: BIOMES

As discussed in Chapter 17, the earth's surface has undergone many significant changes throughout geological time (see Figure 17.18). As continents have drifted through different latitudes, their climates, and consequently their plant and animal life, have changed drastically. The world map shown in Figure 35.12 illustrates the present-day distribution of major terrestrial ecosystems, also called **biomes**. As has been true throughout earth history, these ecosystems have resulted from regional climates influencing regional biota and soils (see Figure 35.9). The biota of each major biome is generally viewed as a climax community, and biomes are often named for the dominant plants of their communities.

Tundra

The **arctic tundra** is a treeless plain of North America, Europe, and Asia. It extends northward from the limits of forest growth to the region that is covered permanently with ice and snow. Dominant plants of the arctic tundra are small woody shrubs, grasses, sedges, and lichens (Figure 35.13). The climate is cold, and the growing season lasts for only about two months. In summer, the sun remains above the horizon for up to 24 hours a day, and most organisms display an intense burst of activity at this time. Numerous shallow ponds provide larval habitat for many insects, and hordes of mosquitoes and other biting flies emerge during the short

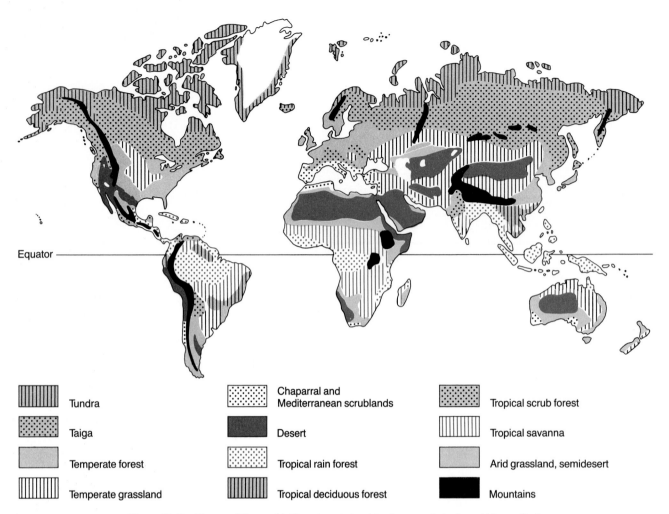

Figure 35.12 Biomes of the world. Note the relationships between latitude and biome distribution across the northern hemisphere, as well as the similarities and differences between the northern and southern hemispheres. Biome distribution is much more limited in the southern hemisphere. Also note that there are many small biomes in addition to those described in the text. For example, the region of temperate forest on the Pacific Northwest coast of the United States is actually a type of rain forest.

growing season. Most organisms spend the winter in dormancy or at lowered activity levels. Animals that are active throughout the year include several homeotherms, including such herbivores as caribou, musk-ox, arctic hare, lemmings, and ptarmigan and such carnivores as the arctic fox and snowy owl. Other species, especially most of the birds that inhabit this region during reproductive periods, migrate south to warmer climates for the winter.

The word *tundra* means "marshy plain," and yet precipitation is relatively low in this biome. The numerous small ponds exist because of **permafrost**, where the soil is permanently frozen beneath the surface. The permafrost provides poor drainage, and precipitation results in standing water. Cold

temperatures and low rates of evaporation contribute to the persistence of the wet areas.

As shown in Figure 35.9, **alpine tundra**, a biome with dominant vegetation similar to that of arctic tundra, exists in mountainous regions at lower latitudes but at altitudes above the tree line. Life forms here are adapted to low temperatures and to freezing conditions most nights of the year.

Taiga

Taiga refers to the northern coniferous (cone-bearing) forest, dominated by trees that have needlelike leaves. The taiga extends southward from the arctic tundra and across North America, Europe, and Asia (Figure 35.14). As shown in Figure 35.12, taiga

Figure 35.13 The arctic tundra. The tundra is shown here during the brief summer, when the sun shines 24 hours a day. The area is under snow for two-thirds of the year, and the sun never shines for several winter months. This photograph shows a bull caribou *(Rangifer tarandus)*, a common tundra species, framed by Mt. McKinley in Alaska.

Figure 35.14 The taiga. Long, cold winters and coniferous trees are distinguishing characteristics of this biome, as are the moose cow and calf *(Alces alces)* seen here.

also extends southward in mountain ranges, such as those of western North America. Evergreen trees, including various species of pine, fir, spruce, and hemlock, often dominate the taiga, but in large expanses of Siberia, the dominant trees are deciduous conifers called larches. There is little undergrowth in coniferous forests as a result of shading from the dense tree growth. Needles resist decomposition and build up on the surface of the soil, resulting in relatively poor soil development and few soil organisms.

The taiga growing season lasts three to five months, with very pronounced seasonality. Precipitation varies from 25–100 cm per year and often results in a very deep snow cover throughout the winter. Representative vertebrates in this biome include herbivores such as elk, moose, beaver, porcupines, snowshoe hare, mice, and grouse. Carnivores include owls, foxes, wolves, wolverines, lynx, and shrews. Many tundra animals move into the taiga during the winter months.

Temperate Deciduous Forests

Temperate deciduous forests occupy areas with warm growing seasons of up to nine months, moderate precipitation (75–150 cm annually), and cold winters that are not suited to plant growth. Loss of leaves in the autumn contributes organic matter to the soil, making it relatively rich. Communities are generally much more diverse than those of the more northern biomes, and communities vary geographically. For example, in northern and upland areas of North America, beech, maple, oak, and birch trees predominate. More southern and lowland areas are often dominated by oak-hickory forests. Many of the vast deciduous forests that covered eastern North America in the past have been cleared and replaced by human population centers or converted to agricultural lands.

Temperate deciduous forests have well-developed **vertical stratification,** or layering. Vertical strata include a ground layer consisting of mosses, liverworts, and leaf litter; a layer of grasses and/or herbaceous (nonwoody) plants; a layer of woody shrubs and young trees; and mature trees forming a canopy. This layering increases overall primary productivity by intercepting much of the light before it reaches the ground.

The many distinct habitats, provided in part by the stratification, provide for a diverse assemblage of animals in temperate deciduous forests. A variety of small mammals forage for fruits, nuts, and insects, especially on the forest floor. These include

squirrels, chipmunks, raccoons, opossums, voles, and mice (Figure 35.15). Whitetail deer browse among the shrubs and seedlings at the forest edge. Populations of large carnivores are much reduced from former times. These include (or once included) the bobcat, mountain lion, foxes, and wolves. Much of the energy that flows through these ecosystems is derived from detritus on the forest floor; much less is derived directly from living parts of plants by herbivores.

Tropical Rain Forests

Tropical rain forests are found in Central and South America, Africa, and Asia in areas where the climate is characterized by warm, stable temperatures, high humidity, and annual rainfall above 200 cm (see chapter opening photograph). The abundant rainfall does not fall evenly throughout the year, but even during the driest periods, the forests rarely receive less than 10 cm per month.

Most rain forest does not fit the image we often have of impenetrable jungle. In undisturbed areas away from waterways, the forest floor is often very open, but vertical stratification is highly developed. The variety of plant and animal species is truly astounding; more species exist in this biome than in all other biomes of the world combined. Some biologists estimate that 2 to 20 million tropical species have yet to be discovered and classified.

With so much biological diversity found in the tropical rain forest, it is perhaps surprising to find that the forest floor has only a thin layer of litter,

Figure 35.15 A temperate deciduous forest. Winter dormancy of the dominant vegetation is a prominent feature of this type of biome. The raccoon *(Procyon lotor)* is a common omnivore in temperate deciduous forests.

and the soil is very poor. Dead vegetation and other detritus are eaten or decomposed rapidly, and the nutrients released are quickly absorbed by plants. Nutrients are stored in the living biomass, not in the soil as with temperate biomes. When tropical forests are cleared, the nutrients are removed with the lumber or are quickly leached from the soil and lost from the system. Typically, these soils produce crops for only a year or two before they must be abandoned.

In the past, tropical rain forests have been little studied, in part because of their remoteness, but also because it is difficult to work at the great tree heights found in these highly stratified ecosystems. Now, a great deal of research is being focused on tropical forests in an attempt to understand them and protect them from destruction before it is too late. Innovative methods are being employed in these investigations. Biologists are able to observe animals 100 m above the ground by using mountain-climbing techniques to scale trees and by making webs of cables from one tall tree to another (Figure 35.16). Monkeys, sloths, anteaters, lizards, many species of birds, and enormous numbers of insects may spend their entire lives in this arboreal environment. Destruction of the rain forests is leading to the extinction of species at an alarming rate, and global climate may also be affected.

Grasslands

Grasslands usually exist in the centers of continents. The dominant plant species in these ecosystems are various species of grasses and forbs (broad-leaved, nonwoody plants). Grasslands include the North American prairies, the African velds, the Asian steppes, and the South American pampas (see Figure 35.12). They are characterized by uneven but often substantial amounts of rainfall. Actually, climate is not always as much a force in maintaining grasslands as it is in most other biomes. Many grasslands would become shrub or woodland communities except for the effects of grazing and fire. In North America, the prairies of the Midwest are characterized by abundant precipitation, deep, rich soils, and species of tall, sod-forming grasses. These are now the richest agricultural areas in the world. Prior to being converted to farmland, these areas existed as grasslands interspersed with trees. (This type of biome is actually called a **savannah**.) Areas of midwestern grassland were largely maintained by herds of grazing animals, such as bison, and by fire, which kills trees but not grass roots. Farther west in North America, precipitation becomes limiting, short-grass species

Figure 35.16 Study of a tropical rain forest. Here a biologist works about 30 m above the forest floor, in the canopy of a monkey pod tree (*Leucithus*).

prevail, and cultivation (except with irrigation) gives way to the grazing of livestock.

Animals adapted to grasslands are mostly burrowers and runners. These ecosystems support numerous species of small, burrowing, herbivorous rodents and an abundance of invertebrates (Figure 35.17). Carnivores include badgers, skunks, foxes, and coyotes. The most prominent grassland species are the large herbivores; bison and pronghorn in North America and gazelles, zebras, and other large ungulates in Africa are examples. With human exploitation of grasslands for agriculture in developed countries, the large herds are almost a thing of the past. In short-grass ecosystems, herds of domestic herbivores largely replace the native species.

Chaparral and Other Dry Shrub Biomes

Ecosystems characterized by short growing seasons, mild winters, and long, hot, dry summers are often dominated by tall shrubs with thick, broad

Figure 35.17 A temperate grassland. Dominant plant species are forbs and grasses. Burrowing rodents, such as these blacktail prarie dogs *(Cynomys ludovicianus),* and large ungulates are abundant here.

tems occur in areas of the Mediterranean, Chile, southern Africa, and Australia.

Deserts

Deserts generally receive less than 25 cm of precipitation per year (Figure 35.19). Precipitation is distributed over the earth in such a way that most wet regions are near the equator and regions of hot, dry climate are found in bands around the earth at about 30° latitude north and south of the equator. Because of extremely low humidity, large areas of bare ground, and very little standing water, there is little moisture to moderate temperatures; therefore, temperatures fluctuate markedly between daytime highs and nighttime lows. Taken together, these conditions present a very harsh environment.

Many plants and animals have adapted to desert conditions. Plants fall into three broad categories of adaptation: annuals, which grow and reproduce explosively after rain events, then remain as seeds until the next rain; succulents, such as cacti, which store water in their tissues; and desert shrubs, with short trunks, many branches, and thick, waxy leaves that fall off during long dry periods. Most desert animals are nocturnal, thus avoiding exposure to daytime heat and water loss. Desert animals generally spend the daytime in burrows, where temperatures are relatively moderate and humidity is much above that of the desert air. Some animals depend on metabolic water, seldom

evergreen leaves (Figure 35.18). The climate of these areas is intermediate between deserts and dry grasslands. In North America, dry shrub areas are known as **chaparrals**. Similar kinds of ecosys-

(a)

(b)

Figure 35.18 A chaparral. (a) This type of shrub-dominated ecosystem occurs in North America on the Pacific coast of central and southern California. (b) The scrub jay *(Aphelcoma coerulescens)* is a resident of the Pacific coast chaparral and other shrublands of the western U.S. and Florida.

Figure 35.19 A desert biome. This area in Arizona is dominated by large saguaro cacti adapted for storing water.

or never needing to drink water. (See the discussion of the kangaroo rat, *Dipodomys*, in Chapter 7, p. 151.) Others have integuments that are impervious to water, thus preventing dehydration from water loss through the skin. Most desert animals also conserve water by producing dry feces and concentrated urine. Reproduction of most desert animals is synchronized with the wet season, and many are *r* strategists, multiplying rapidly when conditions are favorable.

AQUATIC ECOSYSTEMS

Water is a truly remarkable substance. Its thermal properties are critical to all types of ecosystems. Water tends to absorb solar energy, thus moderating high temperatures; and heat released from water when the atmosphere is cool moderates low temperatures. Water density is also affected by temperature, and water is most dense at 4°C. Below this temperature, water begins to crystallize, and an open porous structure gives the crystals a low density, allowing ice to float on liquid water. Water temperature and density greatly affect many aspects of aquatic ecosystems. As we will see, the thermal properties of water are very important in

the seasonal changes that occur in lakes in temperate regions of the world.

Freshwater Ecosystems

There are many kinds of freshwater ecosystems. Standing water, or lentic ecosystems, such as lakes, ponds, marshes, bogs, and swamps, are markedly different from one another and from flowing waters, or lotic ecosystems, such as streams and rivers.

Lentic Ecosystems. Lentic ecosystems are very diverse, but most have similar habitats or zones (Figure 35.20a). The **littoral zone** is the shallow water, or shore zone, penetrated by light and often characterized by rooted vegetation. The **limnetic zone** is the zone of open water, extending from the littoral zone to the limits of light penetration. Together, the littoral and limnetic zones form the **euphotic zone**, the upper waters that support photosynthesis. Below the euphotic zone is the **profundal (aphotic) zone**, where light does not penetrate in sufficient quantity or quality to support plant growth.

The inhabitants of lentic systems can be broadly categorized. Flowering plants contribute significantly to primary productivity in shallow lakes and marshes. Rooted plants are emergent (such as cattails), floating (such as pond lilies), or submergent (such as pond weeds, with most leaves below the water surface). Phytoplankton (floating and suspended algae) replace rooted plants as primary producers in the limnetic zone, and algal photosynthesis often accounts for much of the primary productivity in lakes with extensive limnetic zones. Attached to substrates such as rocks and rooted plants are microscopic plants and animals, collectively called **periphyton**. Various species of zooplankton (floating and suspended animals), including water fleas *(Daphnia)*, copepods *(Cyclops)*, and rotifers (see Figures 23.9 and 26.6), eat phytoplankton, while other species are carnivorous. Larger animals, collectively called **nekton**, are generally more motile and not at the mercy of currents, as are zooplankton; nekton include many insects (mostly associated with the vegetation) and fishes. **Benthos** are bottom-dwelling organisms, such as sponges, insect larvae, molluscs, and annelid worms. **Neuston** live at or on the water surface; water striders are a familiar example.

In temperate regions of the world, a pronounced seasonal cycle occurs in many lentic ecosystems. In the spring, the temperature (and thus the water

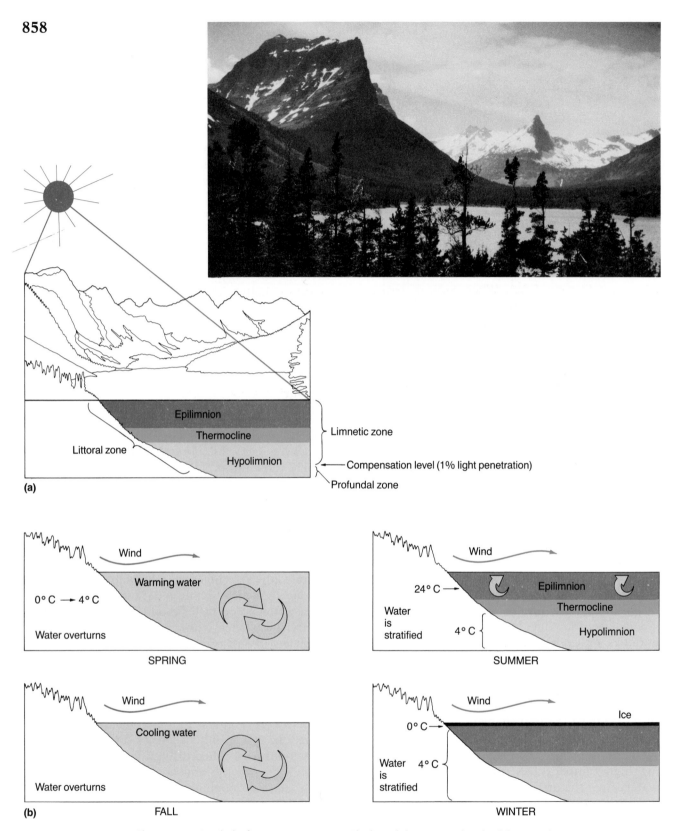

(a)

Epilimnion

Thermocline

Littoral zone

Hypolimnion

Limnetic zone

Compensation level (1% light penetration)

Profundal zone

Wind

Warming water

0° C → 4° C

Water overturns

SPRING

Wind

24° C →

Water is stratified

4° C

Epilimnion

Thermocline

Hypolimnion

SUMMER

Wind

Cooling water

Water overturns

(b)   FALL

Wind

Ice

0° C →

Water is stratified

4° C

WINTER

Figure 35.20 Lentic freshwater ecosystems. (a) The littoral, limnetic, and profundal zones of freshwater lakes are determined by characteristics of solar energy penetration. At and below the compensation level, insufficient light exists to support photosynthesis. (b) During spring and fall in temperate zones, water temperatures are the same throughout the lake and complete mixing occurs. During the summer, the lake is thermally stratified, with the epilimnion the only area of water circulation. The thermocline is a zone of rapid temperature change, and the hypolimnion is the deep, cold, generally unlighted zone. In winter, stratification also occurs, with ice and a narrow layer of cold, lighter water over the denser 4°C main body of water. The photograph is of St. Mary's Lake in Glacier National Park, Montana.

density) is virtually the same at all depths (Figure 35.20b). Because the water at all depths is of equal density, there is little resistance to water movement, and wind and wave action mix the water throughout the lake. As air temperatures rise in early summer, the surface waters warm, while deeper, denser waters remain cold. This density differential creates a resistance to mixing and allows **thermal stratification** to occur. The upper, circulating zone is called the **epilimnion**, and the deep, noncirculating layer is the **hypolimnion**. Between the epilimnion and the hypolimnion is a zone where temperatures decrease rapidly with depth. This transition zone is called the **thermocline** (see Figure 35.20). When air temperatures drop in the fall, the surface waters cool and sink, obliterating the thermocline. Eventually, temperatures become uniform throughout, and all areas of the lake mix together. In the winter, the surface water becomes cooler and less dense than water at 4°C. Winter stratification occurs as very cold, less dense water floats on water that is mostly at 4°C. When ice forms, the lake is sealed from the wind, and circulation ceases until the ice melts.

Temperature is the primary factor causing this seasonal cycle, but the effects of the cycle depend greatly on the biological productivity of the lake. Deep, cold lakes with very limited nutrient levels, called **oligotrophic lakes**, support relatively little plant growth or algae. With limited primary productivity, there is also limited community respiration and limited consumption of the oxygen dissolved in the water. As a result, dissolved oxygen concentrations usually remain high from the surface to the bottom throughout the year. The opposite is generally true of shallow, warmer, nutrient-enriched **eutrophic lakes**. Algal productivity and community respiration are often high, and oxygen depletion may occur as a result of bacterial decomposition of algae that die and sink to the bottom, and at night when algae are respiring but not photosynthesizing. Oxygen depletion and consequent die-offs of aquatic organisms are especially common in the deeper waters of the hypolimnion in the summer and may occur throughout the lake in the winter.

Lotic Ecosystems. Precipitation is important to all aquatic systems, but short-term changes in precipitation affects lotic systems more than lentic ones. The water flow rate (called discharge), the size and gradient of the channel, and the rate of stream bed erosion all depend on and vary with precipitation. Lotic systems typically have a more unstable shoreline and bottom than do lentic ecosystems.

Streams often have a great variety of habitats created by the differences in gradient, substrate, and channel meanders. Quiet, deep pools often alternate with riffles, where fast-flowing water moves over shallow, gravel-covered areas. The effect of current and the transport of suspended sediment create special problems for organisms, but species diversity, especially among benthic invertebrates, is often high in lotic ecosystems. Because of prevailing currents, planktonic organisms are not common in lotic systems.

Many lotic ecosystems derive much of their energy from outside sources, such as detritus in the form of leaves that enter the system seasonally. As a result, many lotic animals are detritivores. In streams with little shading from terrestrial vegetation, photosynthesis by attached algae and rooted aquatic plants may supply much of the system's energy.

Estuarine Ecosystems

Estuaries are zones of transition where freshwater rivers enter the ocean. The Mississippi Delta, Chesapeake Bay, and Puget Sound are large estuaries of major North American rivers. Estuarine environments are highly variable, changing with the tides and river discharge. As discussed in Chapter 7 (see p. 148), animals inhabiting these areas must be able to tolerate rapidly changing salinity or migrate in and out daily. Because estuaries are relatively shallow and are more or less constantly enriched by organic matter brought in by the rivers, most of these ecosystems are extremely productive. They are breeding and nursery areas for many fishes and invertebrates.

As productive and important as estuaries are in providing food for humans, we also use these waters as major waste disposal sites. Most major ports are built on estuaries, and many major population centers are within 100 km of them. Major river systems are collectors of pollutants, which they then deliver to estuaries. These incompatible uses of estuaries form one of the major environmental conflicts now facing many nations of the world.

Marine Ecosystems

The major difference between ocean water and fresh water is salinity. (Salinity is discussed in Chapter 7, p. 147.) Estuaries have a broad salinity range between these two extremes. Oceans occupy about 70% of the earth's surface, but much of this area is relatively unproductive biologically. Exceptions include coastal areas, coral reefs, and regions where nutrients from the ocean depths are brought to the surface periodically by upwelling. (Coral reef ecosystems are described in some detail in the Chapter 21 Essay, p. 496.)

Ocean zonation is marked by differences in depth and light penetration. The littoral, or intertidal, area is a highly variable environment. Organisms alternately exposed to air and water as the tides change have special adaptations that prevent water loss. As we have seen in earlier chapters, many intertidal invertebrates burrow into soft mud at low tides. Others have thick outer body coverings or impervious shells. Those animals that can best withstand desiccation are found highest in the intertidal zone. (Intertidal zonation is discussed in Chapter 24, p. 565; Figure 24.11.)

Two major life zones in the ocean are the open water, or **pelagic zone**, and the ocean floor, or **benthic zone**. Waters of the pelagic zone are subdivided into highly productive **neritic zones**, lying over continental shelves, and less productive **oceanic zones**, lying over deep areas (Figure 35.21). The oceanic zone is subdivided vertically according to light penetration. As in freshwater lakes, the uppermost **euphotic zone** is well lighted and supports photosynthetic algae. A zone of twilight water below the euphotic zone, called the **dysphotic zone**, does not support algal growth. Below the dysphotic zone is a zone of permanently dark water, called the **aphotic zone**. Subdivisions of the benthic zone according to depth are the **sublittoral zone** of the continental shelf; the **bathyal zone** from the continental shelf down to about 4000 m; the **abyssal zone**, often an extensive deep-sea plain, from about

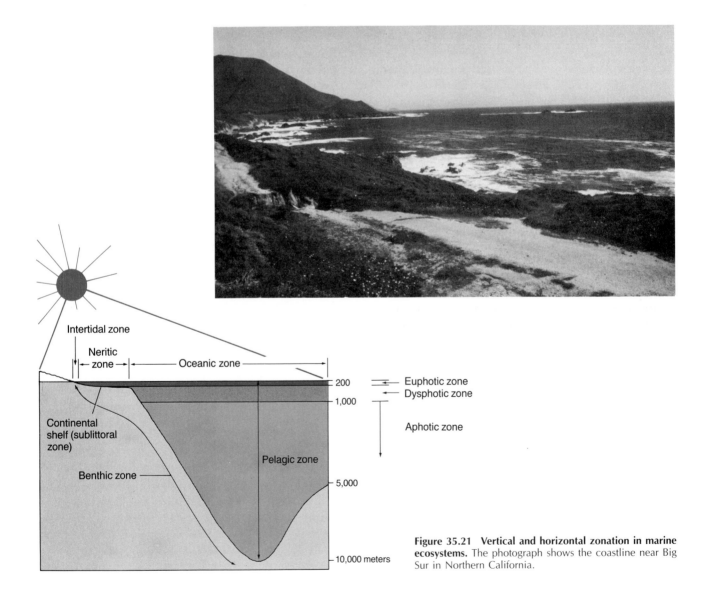

Figure 35.21 Vertical and horizontal zonation in marine ecosystems. The photograph shows the coastline near Big Sur in Northern California.

4000 m down to about 6000 m; and the **hadal zone** of the deepest ocean trenches, from about 6000 m to 10,000 m. The terms *bathyal, abyssal,* and *hadal* also refer to subdivisions of aphotic waters of the oceanic zone.

Marine zooplankton consist mainly of small adult invertebrates and an enormous number of larval fishes and invertebrates that drift, feed, and eventually settle to the bottom to complete their life cycle. Nekton in the marine environment include larger animals, such as fishes, squids, and aquatic mammals. Benthic communities of deep oceans are often diverse, but populations are often sparse because of a poor energy base. Deep-ocean organisms have fascinating adaptations to the extremes in pressure, darkness, and limiting nutrients and energy. (Deep-sea fishes are discussed in Chapter 29, p. 713; Figure 29.11.)

HUMAN IMPACT ON ECOSYSTEMS

Ecosystems are not static; environmental disturbance is a frequent, natural occurrence to which ecosystems respond in a variety of ways. Species in communities adapt to environmental change over long time periods through evolutionary processes. Communities may adapt in shorter time periods through successional processes. Ecologists studying the effects of disturbance on ecosystems are concerned with ecosystem *resistance,* that is, the capacity of an ecosystem to remain relatively unchanged and homeostatic when confronted by a disturbance. Also of importance in analyzing the effect of disturbance is ecosystem *resilience,* which is the speed at which the system returns to its former state following disturbance. Ecosystems vary markedly in their resilience and resistance, and their responses vary with the nature of the disturbance.

Ecosystem disturbances due to human activities generally go well beyond natural disturbances in scale and kind, and the extremely rapid rate of human-caused changes typically exceeds ecosystem resistance and resilience. Throughout most of the world, it is more difficult to find undisturbed areas than to find areas that have been significantly altered by humans. Major environmental problems result mainly from human population growth and from the increasingly widespread and more intensive use of modern technology. The demand of humans for environmental resources depends on population size, the state of technology, and the time scale involved. The pressures exerted by each of these factors differ from nation to nation, but all species, directly or indirectly, experience the effects.

Today, more than 5 billion people are seeking an existence from the earth, and because the human population growth curve is virtually exponential, nearly 80 million more are added each year (Figure 35.22). This puts an enormous strain on world ecosystems, and many people believe that human population growth is the most important factor affecting the future of humanity and all life on earth. The central question facing humankind is, What is our carrying capacity? At what population density can the earth continue to meet the basic needs of the human species for food, water, shelter, and space? Many ecologists feel that we are already well beyond our carrying capacity and that our modern technology is allowing us to develop a large lag time between population size and carrying capacity. (The concepts of carrying capacity and lag time are discussed in Chapter 34, p. 832.) If this is true, then the central question facing us is, What form will environmental resistance take in modifying the human population growth curve? Clearly, population growth cannot continue for many more generations at its present rate.

Some people look to our modern technology, both present and future, to continually raise the world's carrying capacity for humans. But what price are we already paying for the disturbances in the biosphere caused by our present numbers and technology? Growing food and providing housing

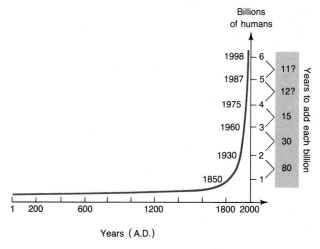

Figure 35.22 The human population growth curve. After an extensive period of relatively slow growth, the current growth pattern is essentially exponential.

for billions of people require that natural habitats be replaced by human-support systems. In addition to direct habitat destruction, the products and by-products of technological development severely stress ecosystems and may change them at rates that overwhelm the ability of most living species to adapt. Species are becoming extinct some 50 times faster today than they were even a century ago, and we are probably just beginning to see the effects of pesticides, numerous other toxic substances, acid precipitation, radioactive wastes, and synthetic fertilizers on the world's ecosystems.

In an ecological sense, the human species fills the broadest niche ever occupied by a single species. As a result, our species competes with many other organisms for natural resources. It is incumbent on us, as the dominant species, to work toward solving, rather than trying to escape the realities of, the problems created by our own success. Our survival and that of most other species depends on our acting to avoid ecological collapse.

Fifty years ago, zoologists might have been able to end a textbook such as this on an optimistic note. To do so today, we would have to ignore the sad fact that some (in fact, perhaps much) of the animal diversity we have described in this book, and much more that may never be described, is already doomed to extinction, largely because of human-caused changes in ecosystems. It is imperative that we find ways to sustain ourselves without jeopardizing the existence of most other species and ultimately ourselves.

☐ SUMMARY

1. A biological community is an interacting assemblage of species populations. An ecosystem, the largest unit of biological organization, consists of one or more communities and the abiotic environmental factors on which the communities depend.

2. Energy flow in ecosystems is linear; photosynthetic plants and chemosynthetic microbes synthesize the organic chemical energy that powers ecosystems. Gross primary productivity (GPP) is the total rate of photosynthesis. Net primary productivity (NPP) is the rate of storage of organic matter (biomass) by plants in excess of plant respiration (R_A). $GPP - R_A = NPP$. Secondary productivity is the rate of energy storage by heterotrophs. Net community productivity is found by the equation $NCP = GPP - (R_A + R_H)$, where R_H represents heterotroph respiration.

3. Energy flows through trophic systems, which consist of primary producers (plants, and chemosynthesizers),

primary consumers (herbivores), decomposers, including detritivores (animals that eat dead organic matter), and secondary, tertiary, and quaternary consumers (carnivores).

4. Nutrients cycle in ecosystems; some biogeochemical cycles (e.g., the nitrogen cycle) have gaseous reservoirs; others (e.g., the phosphorus cycle) have sedimentary reservoirs.

5. If disturbed, communities may undergo a regular series of changes, called secondary succession. Primary succession leads to the establishment of new communities where none existed previously.

6. Major terrestrial ecosystems, called biomes, include tundra, taiga, temperate deciduous forests, tropical rain forests, grasslands, dry shrub biomes, and deserts. Aquatic ecosystems are described as being lentic (standing fresh waters), lotic (flowing fresh waters), estuarine, and marine. Freshwater lakes may undergo thermal stratification in summer and winter. Oligotrophic lakes have relatively low productivity, whereas eutrophic lakes are highly productive.

7. The human population is currently growing at a virtually exponential rate and may have already exceeded the earth's carrying capacity. The demands of overpopulation and human cultural changes are causing rapid, often seriously detrimental changes in ecosystems.

☐ FURTHER READING

Brown, L. R., W. U. Chandler, C. Flavin, J. Jacobson, C. Pollock, S. Postel, L. Starke, and E. C. Wolf. *State of the World. 1987.* A Worldwatch Institute Report on Progress Toward a Sustainable Society. New York: Norton, 1987. *Fourth in a continuing series of yearly reports on environmental deterioration and its political, social, and economic consequences.*

Colinvaux, R. L. *Ecology.* New York: Wiley, 1986. *An excellent modern textbook.*

Kormondy, E. J. *Concepts of Ecology.* 3d ed. Englewood Cliffs, N.J.: Prentice-Hall, 1984. *A brief, but excellent introduction to populations, communities, and ecosystems.*

Miller, G. T., Jr. *Environmental Science: An Introduction.* Belmont, Calif.: Wadsworth, 1986. *A clearly written introduction to basic ecological principles and their application to modern environmental problems.*

Myers, N. *The Primary Source. Tropical Forests and Our Future.* New York: Norton, 1984. *Examines tropical forest ecosystems and the ongoing process of tropical deforestation.*

Smith, R. L. *Ecology and Field Biology.* 3d ed. New York: Harper & Row, 1980. *Includes several chapters on communities and ecosystems.*

Vitousek, P. M., P. R. Ehrlich, A. H. Ehrlich, and P. A. Matson. "Human Appropriation of the Products of Photosynthesis." *Bioscience* 36(1986): 368–373. *Discusses implications of estimates that the human species directly or indirectly consumes about 40% of the net primary production of land ecosystems.*

Appendix

Major Characteristics
of Protozoa and Animals

		Major Characteristics				
Taxonomic Subdivision	*Habitat*	*Body Cavity and Germ Layers*	*Skin/Skeleton Body Form*	*Movement*	*Digestive*	*Gas Exchange*
Kingdom Protista						
Phylum Sarcomastigophora	Free-living, endoparasitic, marine, fresh water, moist soil, depending on species.	None.	Tests, ectoplasm, pellicle, depending on the species.	Flagella and/or pseudopodia.	Heterotrophic, autotrophic. Phagocytosis, pinocytosis with digestion in food vacuoles. Diffusion across cell surface.	Diffusion across cell surface.
Phylum Apicomplexa	Endoparasitic.	None.	Hydrostatic pressure, ectoplasm, and pellicle in some.	Some move by means of contractile filaments in the cytoplasm.	Absorption across cell surface.	Diffusion across cell surface.
Phylum Microspora	Endoparasitic.	None.	Hydrostatic pressure, ectoplasm.	Little known.	Absorption across cell surface.	Diffusion across cell surface.
Phylum Ciliophora	Free-living, endoparasitic, marine, fresh water, depending on species.	None.	Pellicle, ectoplasm, skeletal plates in some.	Cilia. Myonemes in some. Tentacles in some.	Cytopharynx and food vacuoles in some. Channels in tentacles in suctorians.	Diffusion across cell surface.
Kingdom Animalia						
Phylum Myxozoa	Endoparasitic.	None.	Hydrostatic pressure.	Little known.	Plasmodium erodes host tissue (e.g., fish's cartilage).	Diffusion across body surface.
Phylum Placozoa	Marine.	None.	Hydrostatic pressure.	Flagellated outer cell layer.	Body surface cells ingest food particles by phagocytosis.	Diffusion across body surface.
Phylum Mesozoa	Endoparasitic in marine invertebrates.	None.	Hydrostatic pressure.	Cilia on adults and larvae.	Absorb nutrients, and phagocytic.	Diffusion across body surface.
Phylum Porifera	Marine or fresh water, depending on species.	None (the spongocoel is not a body cavity). No germ layers.	Pinacoderm of pinacocytes; with pores, canals that penetrate body wall. Endoskeleton of spicules and/ or spongin.	Adults immobile; free-swimming larvae propelled by flagellated cells.	No alimentary tract; choanocytes ingest food particles that are passed to amoebocytes where digestion is intracellular.	Diffusion across cell surfaces.

Taxonomic Subdivision	Major Characteristics					
	Circulatory	**Excretory**	**Nervous**	**Sensory**	**Reproductive**	**Developmental**
Kingdom Protista						
Phylum Sarcomastigophora	Movement of cytoplasm.	Diffusion across cell surface.	No specific structures.	Sensitive surface membranes and flagella. Photosensitive cytoplasm near light-shading pigment (stigma).	No specific reproductive organelles. Asexual reproduction by binary fission. Sexual in some.	Development may include amoeboid, cyst, and flagellated stages.
Phylum Apicomplexa	Movement of cytoplasm.	Diffusion across cell surface.	No specific structures.	Sensitive surface membranes.	No specific reproductive organelles. Asexual and sexual reproduction in complex life cycles.	Complex life cycles may include multiple fission, fusion of gametes, and development of spores.
Phylum Microspora	Movement of cytoplasm.	Diffusion across cell surface.	No specific structures.	Sensitive surface membranes.	No specific reproductive organelles.	Life cycles in which complex spore injects sporoplasm into host cell.
Phylum Ciliophora	Movement of cytoplasm.	Diffusion across cell surface.	No specific structures.	Sensitive surface membranes and cilia.	No specific reproductive organelles. Asexual binary fission; sexual reproduction with conjugation, syngamy, autogamy.	Life cycle includes meiosis of micronuclei, and exchange of haploid micronuclei between cells.
Kingdom Animalia						
Phylum Myxozoa	No specific structures.	Diffusion across body surface.	No specific structures.	Little known.	Generative cells.	Spore stage infective to hosts; plasmodium feeding stage.
Phylum Placozoa	No specific structures.	Diffusion across body surface.	No specific structures.	Sensitive surface membranes.	Asexual reproduction by budding and fission.	Eggs develop in inner cell mass. Body has somatic and reproductive cell lines.
Phylum Mesozoa	No specific structures.	Diffusion across body surface.	No specific structures.	Sensitive surface membranes.	Asexual production of vermiforms; sexual production of dispersal larvae.	Vermiforms from axoblasts within axial cell.
Phylum Porifera	No specific structures.	Diffusion from body surface; transport of wastes by amoebocytes.	No nerve cells; impulses conducted by body cells may cause changes in pore size.	No sensory structures; general reaction of ectoplasm to stimulation.	Monoecious or dioecious; no gonads, ducts; asexual reproduction by budding, fragmentation, gemmules, and reduction bodies.	Eggs usually incubated in parent body; free-swimming flagellated larvae attach to bottom and develop into parent form.

APPENDIX

			Major Characteristics			
Taxonomic Subdivision	Habitat	Body Cavity and Germ Layers	Skin/Skeleton Body Form	Movement	Digestive	Gas Exchange
Class Calcarea (Calcispongiae)	Marine.	None. No germ layers.	Calcareous spicules. Asconoid, syconoid, leuconoid, depending on species.	Flagellated collar cells and larvae. Myocytes.	Digestion intracellular.	Diffusion across cell surfaces.
Class Hyalospongiae (Hexactinellida)	Marine.	None. No germ layers.	Six-rayed siliceous spicules. Syconoid, leuconoid, depending on species.	Flagellated collar cells and larvae.	Digestion intracellular.	Diffusion across cell surfaces.
Class Demospongiae	Marine, fresh water (one family).	None. No germ layers.	Two- or four-rayed siliceous spicules and/or spongin. Leuconoid.	Flagellated collar cells and larvae.	Digestion intracellular.	Diffusion across cell surfaces.
Phylum Cnidaria	Marine or fresh water, depending on species.	None (gastro-vascular cavity is not a body cavity). Have ectoderm and endoderm; Middle layer (mesenchyme) cellular in some.	Epidermis. Hydrostatic skeleton; calcareous or coral exoskeleton in some; perisarc in some. Radial or biradial symmetry.	Contractile fibers within epithelio-muscular cells in epidermis and in nutritive-muscular cells in gastrodermis.	Incomplete gut; digestion both extracellular and intracellular. Tentacles and nematocysts for capture of food.	Diffusion across body surface.
Class Hydrozoa	Majority are marine.	None. Mesenchyme generally acellular.	Epidermis, and perisarc in some. Polyp and medusa forms. Radial symmetry.	Epithelio- and nutritive-muscular cells.	Extracellular and intracellular digestion. Incomplete gut.	Diffusion across body surface.
Class Scyphozoa	Marine. Most are free-floating jellyfishes.	None. Thick mesenchyme variously cellular.	Cellular or syncytial epidermis. Medusa; polyp form reduced or absent. Radial symmetry.	Epithelio-muscular cells.	Extracellular and intracellular digestion. Incomplete gut.	Diffusion across body surface.
Class Cubozoa	Marine.	None. Thick mesenchyme contains cells.	Cellular or syncytial epidermis. Medusa; polyp form reduced. Radial symmetry.	Epithelio-muscular cells.	Extracellular and intracellular digestion. Incomplete gut.	Diffusion across body surface.
Class Anthozoa	Marine. Sessile.	None. Thick mesenchyme contains cells.	Epidermis. Calcareous exoskeleton, coral in some. Polyp; medusa absent. Biradial symmetry.	Epithelio- and nutritive-muscular cells.	Extracellular and intracellular digestion. Incomplete gut.	Diffusion across body surface.
Phylum Ctenophora	Marine. Most free-floating; some creep or are sessile.	None. Have ectoderm, endoderm, and cellular mesenchyme.	Hydrostatic. Biradial symmetry.	Ciliated comb plates in 8 meridional rows.	Complete gut and gastro-vascular canal system. Adhesive collocytes for food capture.	Diffusion across body surface.
Phylum Platyhelminthes	Free-living (one class), ectoparasitic, endoparasitic, fresh water, marine, moist situations on land, depending on species.	Acoelomate. Triploblastic.	Hydrostatic. Bilateral symmetry.	Layers of muscle; cilia in some, suckers in others.	Gut incomplete, absent in some.	Diffusion across body surface.

			Major Characteristics			
Taxonomic Subdivision	**Circulatory**	**Excretory**	**Nervous**	**Sensory**	**Reproductive**	**Developmental**
Class Calcarea (Calcispongiae)	No specific structures.	Diffusion from body surface; amoebocytes.	No specific structures.	No sensory cells.	Asexual reproduction by budding; sexual by gametes.	Free-swimming amphiblastula larva; some have a parenchymella larva.
Class Hyalospongiae (Hexactinellida)	No specific structures.	Diffusion from body surface; amoebocytes.	No specific structures.	No sensory cells.	Asexual reproduction by budding; sexual by gametes.	Free-swimming parenchymella larva.
Class Demospongiae	No specific structures.	Diffusion from body surface; amoebocytes.	No specific structures.	No sensory cells.	Asexual reproduction by budding, reduction bodies, and gemmules in freshwater forms; sexual by gametes.	Free-swimming parenchymella larva in most. Asexual reduction bodies and gemmules in freshwater forms.
Phylum Cnidaria	Gastrovascular cavity and associated canals.	Diffusion across body surface.	One or more nerve nets in body wall.	Sensory cells in epidermis; statocysts in some.	Dioecious or monoecious; gonads, without ducts; fertilization external. Asexual reproduction is common.	Typically, the zygote develops into a free-swimming planula larva that becomes a polyp, which produces medusae by budding.
Class Hydrozoa	Gastrovascular cavity and associated canals.	Diffusion across body surface.	Nerve net(s).	Sensory cells in epidermis.	Dioecious or monoecious.	Asexual polyps, sexual medusae, one of which may be suppressed.
Class Scyphozoa	Gastrovascular cavity and associated canals.	Diffusion across body surface.	Nerve net(s).	Epidermal sensory cells. Rhopalia with statocysts and ocelli in some.	Most are dioecious.	Prominent medusa; polyp absent or reduced.
Class Cubozoa	Gastrovascular cavity and associated canals.	Diffusion across body surface.	Nerve net(s).	Epidermal sensory cells. Rhopalia.	Dioecious.	Prominent medusa; polyp reduced. No ephyra larva.
Class Anthozoa	Gastrovascular cavity.	Diffusion across body surface.	Nerve net(s).	Epithelial sensory cells.	Dioecious or monoecious.	Polyp; no medusa stage.
Phylum Ctenophora	Gastrovascular canals.	Diffusion across body surface.	Nerve net.	Single aboral statocyst near origin of comb plates.	Monoecious; gonads, some species with ducts.	Cydippid larva; no polyp.
Phylum Platyhelminthes	Gut, highly branched in some.	Diffusion across body surface; protonephridia mainly osmoregulatory.	Ladderlike, with cerebral ganglia and longitudinal nerve cords linked by transverse nerves; nerve net in some.	Pigment-cup eyespots, chemo- and mechanoreceptors; statocyst in some.	Dioecious or monoecious; gonads with ducts; copulatory organs; asexual reproduction by fragmentation and/or fission in some.	Development direct, or life cycles with one or more larval stages in more than one host species.

APPENDIX

			Major Characteristics			
Taxonomic Subdivision	*Habitat*	***Body Cavity and Germ Layers***	***Skin/Skeleton Body Form***	***Movement***	***Digestive***	***Gas Exchange***
Class Turbellaria	Majority free-living. In marine, fresh water, or wet places on land.	Acoelomate. Triploblastic.	Hydrostatic skeleton. Ciliated epidermis. Bilateral symmetry.	Circular, longitudinal, oblique muscle layers; cilia.	Incomplete gut, absent in some.	Diffusion across body surface.
Class Trematoda	Flukes. Endoparasites of vertebrates.	Acoelomate. Triploblastic.	Hydrostatic skeleton. Tegument. Bilateral symmetry.	Layers of muscle. Suckers.	Incomplete gut.	Diffusion across body surface.
Class Monogenea	Flukes. Ectoparasites.	Acoelomate. Triploblastic.	Hydrostatic skeleton. Tegument. Bilateral symmetry.	Layers of muscle. Suckers.	Incomplete gut.	Diffusion across body surface.
Class Cestoidea	Tapeworms. Endoparasites.	Acoelomate. Triploblastic.	Hydrostatic skeleton. Tegument. Bilateral symmetry; divided into proglottids.	Layers of muscle. Suckers.	No gut; nutrients absorbed across body wall.	Diffusion across body surface.
Phylum Gnathostomulida	Marine. Live among the sand grains of coastal sediments.	Acoelomate. Triploblastic.	Hydrostatic skeleton. Ciliated epidermis. Bilateral symmetry.	Longitudinal muscles. Cilia.	Jaws. Incomplete gut.	Diffusion across body surface.
Phylum Nemertea (Rhynchocoela)	Ribbon worms. Mostly marine, a few in damp soil, fresh water.	Acoelomate, except for rhynchocoel. Triploblastic.	Hydrostatic skeleton. Ciliated epidermis. Bilateral symmetry.	Longitudinal and circular musculature. Cilia.	Complete gut; digestion is largely extracellular.	Diffusion across body surface.
Phylum Nematoda	Roundworms. Marine, fresh water, moist soil, free-living, parasitic, depending on species.	Pseudocoelomate. Triploblastic.	Hydrostatic skeleton. Epidermis and multilayered cuticle. Bilateral symmetry.	Longitudinal muscle fibers.	Mouth with or without jaws. Complete gut.	Diffusion across body surface.
Phylum Nematomorpha	Horsehair worms. Marine or fresh water, depending on species.	Pseudocoelomate. Triploblastic.	Hydrostatic skeleton. Epidermis and multilayered cuticle. Bilateral symmetry.	Longitudinal muscle fibers.	Gut degenerate (probably not functional). Food absorbed across body wall.	Diffusion across body surface.
Phylum Acanthocephala	Spiny-headed worms. Endoparasites.	Pseudocoelomate. Triploblastic.	Hydrostatic skeleton. Tegument with crypts and canals. Bilateral symmetry.	Circular and longitudinal muscles. Retractable, spiny proboscis.	No gut; food absorbed across body wall.	Diffusion across body surface.

	Major Characteristics					
Taxonomic Subdivision	**Circulatory**	**Excretory**	**Nervous**	**Sensory**	**Reproductive**	**Developmental**
Class Turbellaria	Gut, highly branched in some.	Diffusion across body surface.	Ladderlike system; nerve net in some.	Pigment-cup eyespots. Statocyst in some; other mechano-receptors.	Typically hermaphroditic with internal fertilization. Fragmentation, fission may occur.	With some exceptions, life cycles simple, with no larval stages.
Class Trematoda	Gut, highly branched in some.	Diffusion across body surface.	Ladderlike system.	Eyespots in some. Mechano-receptors.	Monoecious; some dioecious.	Life cycles with one or more larval stages in one or more intermediate hosts.
Class Monogenea	Gut, highly branched in some.	Diffusion across body surface.	Ladderlike system.	Eyespots, mechano-receptors.	Monoecious.	Egg hatches into free-swimming larva that attaches to amphibian or fish and becomes adult.
Class Cestoidea	None.	Diffusion across body surface.	Ladderlike system.	No primary sense organs. Sensory endings around scolex.	Monoecious.	Life cycles with several larval stages, in one or more intermediate hosts.
Phylum Gnathostomulida	None.	Protonephridia.	Epidermal nerve plexus; cerebral ganglia.	Sensory pits and cilia.	Monoecious. With gonads, copulatory structures.	Little known.
Phylum Nemertea (Rhynchocoela)	Closed system; a dorsal and two lateral vessels; no heart.	Protonephridia.	Lateral and dorsal longitudinal nerve cords, and cerebral ganglia.	Tactile receptors, sensory pits, ocelli.	Most are dioecious.	Some with a helmet-shaped pilidium larva, direct development in others.
Phylum Nematoda	Pseudocoelomic fluid.	Renette cells or canals.	Longitudinal nerves from nerve ring encircling gut. Anterior ganglia.	Chemosensory phasmids and amphids. Mechanoreceptive setae and papillae.	Some partheno-genetic, but most dioecious. Gonads, copulatory organs and spicules.	Four juvenile stages; no larval stages.
Phylum Nematomorpha	Pseudocoelomic fluid.	Diffusion across body surface.	Midventral nerve cord from anterior nerve ring.	Mechanoreceptive setae and papillae. Some with ocelli.	Dioecious.	Larval and juvenile stages endo-parasites of arthropods.
Phylum Acanthocephala	Canals in tegument may have circulatory function.	Diffusion across body surface. Protonephridia in some.	Anterior ganglion and several nerve cords.	Lateral and apical sense organs on proboscis.	Dioecious.	Larval stages in arthropods, adults in vertebrate intestine.

			Major Characteristics			
Taxonomic Subdivision	Habitat	Body Cavity and Germ Layers	Skin/Skeleton Body Form	Movement	Digestive	Gas Exchange
Phylum Rotifera	Most live in fresh water; a few are marine, some in wet soil and sand, or parasitic.	Pseudocoelomate. Triploblastic.	Hydrostatic skeleton. Cuticle within epidermis, sometimes thickened into a lorica. Bilateral symmetry.	Ciliated corona. Paired muscle bands in body wall.	Complete gut; mastax adapted to food type.	Diffusion across body surface.
Phylum Gastrotricha	Marine or fresh water.	Pseudocoelomate (though functionally acoelomate). Triploblastic.	Hydrostatic skeleton. Cuticle over syncytial epidermis. Bilateral symmetry.	Ciliated underside. Paired muscle bands in body wall.	Complete gut. Muscular pharynx and head cilia used in feeding.	Diffusion across body surface.
Phylum Kinorhyncha	Marine, in bottom mud.	Pseudocoelomate. Triploblastic.	Hydrostatic skeleton. Cuticle over syncytial epidermis. Bilateral symmetry.	Bands of circular, longitudinal, diagonal muscles.	Complete gut with retractable snout and muscular pharynx.	Diffusion across body surface.
Phylum Loricifera	Marine, in interstices of sand or gravel.	Pseudocoelomate. Triploblastic.	Hydrostatic skeleton. Adults with armored case. Bilateral symmetry.	Musculature in body wall; larvae have locomotory spines and posterior "toes."	Complete gut; flexible, tubular mouth.	Diffusion across body surface.
Phylum Priapulida	Marine, in bottom mud or sand.	Pseudocoelomate. Triploblastic.	Hydrostatic skeleton. Epidermis and cuticle. Bilateral symmetry.	Body wall musculature.	Complete gut. Retractable pharynx; spiny proboscis.	Posterior appendages possibly respiratory.
Phylum Entoprocta	Marine, except for *Urnatella*, a freshwater genus.	Pseudocoelomate. Triploblastic.	Hydrostatic skeleton. Epidermis and cuticle. Bilateral or asymmetrical.	Longitudinal muscles and muscular tentacles.	Complete, U-shaped gut.	Diffusion across body surface.
Phylum Mollusca	Majority are marine, but many live in fresh water and on land.	Eucoelomate; coelom reduced and main body cavity a hemocoel. Triploblastic.	Hydrostatic skeleton and, typically, an external calcareous shell of one or more parts. Bilateral or asymmetrical.	Typically, body musculature and muscular foot.	Complete gut, with radula in many species.	Ciliated gills or lungs in mantle cavity in many.
Class Monoplacophora	Marine, in deep water.	Eucoelomate. Triploblastic.	Single shell. Epidermis. Bilateral symmetry.	Body and foot musculature.	Complete gut; crystalline style; radula.	Five or six paired gills.
Class Aplacophora	Marine, mostly in deep water.	Eucoelomate. Triploblastic.	Shell-like granules in epidermis. Bilateral symmetry.	Body wall musculature (foot reduced or absent).	Complete gut; usually a radula.	Small gills in some; reduced or absent in others.
Class Polyplacophora	Chitons. Marine, often intertidal or subtidal.	Eucoelomate. Triploblastic.	Shell of 8 dorsal plates. Epidermis. Bilateral symmetry.	Body and foot musculature.	Complete gut; radula.	Numerous paired gills.

	Major Characteristics					
Taxonomic Subdivision	*Circulatory*	*Excretory*	*Nervous*	*Sensory*	*Reproductive*	*Developmental*
Phylum Rotifera	Movement of pseudocoelomic fluid.	Protonephridia and posterior bladder.	Cerebral ganglion; nerves to corona and posterior region.	Eyespots. Clusters of cilia and sensory bristles on antennae.	Dioecious. Copulatory organs.	Development is direct. Parthenogenesis common; females dominant in all species.
Phylum Gastrotricha	None.	Protonephridia.	Cerebral ganglia; lateral nerves.	Eyespots. Sensory cilia and bristles.	Monoecious.	Parthenogenesis common.
Phylum Kinorhyncha	Movement of pseudocoelomic fluid.	Protonephridia.	Circumpharyngeal nerve ring and cerebral ganglion. Ventral nerve cord.	Eyespots. Sensory bristles.	Dioecious.	Little known.
Phylum Loricifera	Little known.	Little known.	Two or more ventral ganglia, ganglia around mouth, and cerebral ganglion.	Little known.	Dioecious.	Little known.
Phylum Priapulida	Pseudocoelomic fluid; hemerythrin in coelomocytes.	Protonephridia.	Anterior nerve ring and ventral nerve cord.	Sensory spines and papillae.	Dioecious.	With radial cleavage, atypical schizocoels.
Phylum Entoprocta	Pseudocoelomic fluid.	Protonephridia.	Ganglion in lophophore with radiating nerves.	Bristled sensory cells on tentacles.	Majority are dioecious. Budding occurs in some.	Spiral cleavage and trochophore-like larva.
Phylum Mollusca	Typically, an open system. A 2- or 3-chambered heart and spacious hemocoel.	Tubular nephridia associated with pericardial cavity.	Basically, a circumesophageal nerve ring with nerve cords to foot, mantle, viscera.	Typically, statocysts in foot, chemosensors (osphradia) on gills, simple or complex eyes.	Majority are dioecious. Reproduction solely sexual. Fertilization internal or external.	With or without trochophore and veliger larval stages. Spiral cleavage. Protostomate.
Class Monoplacophora	Open system with heart and arteries.	Nephridia.	Ganglia and nerve cords.	Receptors scattered on body.	Dioecious.	Protostomate. Trochophore larva.
Class Aplacophora	Open system; little known.	Nephridia.	Two pairs of longitudinal nerve cords.	Little known.	Monoecious.	Protostomate. Trochophore larva.
Class Polyplacophora	Open system with heart and arteries.	Nephridia.	Nerve ring and longitudinal nerve cords.	Osphradia, eyes, and other sense organs.	Monoecious, dioecious.	Protostomate. Trochophore larva.

Major Characteristics

Taxonomic Subdivision	Habitat	Body Cavity and Germ Layers	Skin/Skeleton Body Form	Movement	Digestive	Gas Exchange
Class Gastropoda	Marine, fresh water, terrestrial, depending on species.	Eucoelomate. Triploblastic.	Typically, a single coiled shell. Shell absent or uncoiled in some. Epidermis. Bilateral or asymmetrical.	Body and foot musculature.	Complete gut; crystalline style in some; radula.	One or two gills, or secondary gills or lungs formed from mantle.
Class Bivalvia	Marine, fresh water, depending on species.	Eucoelomate. Triploblastic.	Two-part shell forms lateral valves. Epidermis. Bilateral symmetry.	Body and foot musculature.	Complete gut; no radula; crystalline style.	Mantle and paired flap-like gills.
Class Scaphopoda	Marine. Most live in deep water to about 3000 m.	Eucoelomate. Triploblastic.	Single tubular shell. Epidermis. Bilateral symmetry.	Body and foot musculature.	Complete gut; radula.	Mantle; no gills.
Class Cephalopoda	Marine.	Eucoelomate. Triploblastic.	Shell reduced or absent in many. Epidermis. Bilateral symmetry.	Body muscles; muscular arms with suckers.	Complete gut; radula; jawed pharynx.	Gills.
Phylum Annelida	Marine, fresh water, terrestrial, depending on species.	Eucoelomate. Triploblastic.	Hydrostatic skeleton. Bilateral symmetry.	Longitudinal and circular muscles. Setae in most; parapodia, suckers in some.	Complete gut. Jaws in some.	Diffusion across body surface. Some with gills, parapodia, tentacles.
Class Polychaeta	Typically marine, some in brackish water, a few in fresh water.	Eucoelomate. Triploblastic.	Hydrostatic skeleton. Bilateral symmetry.	Longitudinal and circular muscles, parapodia, setae.	Complete gut. Jawed pharynx in some. Feeding tentacles in some.	Diffusion across body surface. Some with gills, parapodia, tentacles.
Class Oligochaeta	Majority terrestrial or in fresh water, a few in brackish water. Some parasitic.	Eucoelomate. Triploblastic.	Hydrostatic skeleton. Bilateral symmetry.	Longitudinal and circular muscles, setae.	Complete gut, including grinding gizzard in some.	Diffusion across body surface.
Class Hirudinea	Mostly in fresh water, some terrestrial or marine, depending on species.	Eucoelomate. Triploblastic.	Hydrostatic skeleton. Bilateral symmetry.	Longitudinal and circular muscles, suckers.	Complete gut. Jawed mouth. Crop with ceca.	Diffusion across body surface.
Class Branchiobdellida	Ectoparasitic on freshwater crustacea.	Eucoelomate. Triploblastic.	Hydrostatic skeleton. Bilateral symmetry.	Longitudinal and circular muscles, suckers.	Complete gut.	Diffusion across body surface.
Phylum Sipuncula	Peanut worms. Marine; bottom dwellers in shallow water.	Eucoelomate. Triploblastic.	Hydrostatic skeleton. Cuticle, epidermis and dermis. Bilateral symmetry.	Longitudinal and circular muscles. Muscular introvert.	Complete gut. Retractable introvert and feeding tentacles; gut U-shaped.	Diffusion across body surface. Respiratory pigment in coelomocytes.
Phylum Echiura	Marine; most in shallow water.	Eucoelomate. Triploblastic.	Hydrostatic skeleton. Cuticle, epidermis, dermis. Bilateral symmetry.	Longitudinal, circular, and diagonal muscles.	Complete gut.	Diffusion across body and cloacal surfaces. Respiratory pigment in coelomocytes.

APPENDIX

Taxonomic Subdivision	Major Characteristics					
	Circulatory	**Excretory**	**Nervous**	**Sensory**	**Reproductive**	**Developmental**
Class Gastropoda	Open system with heart and arteries.	Nephridia.	Ganglia (cerebral, buccal, pedal) and nerve cords.	Osphradia, eyes, statocysts, tentacles, other sense organs.	Monoecious, dioecious.	Protostomate. Trochophore larva in some; typically a veliger larva. Torsion in many.
Class Bivalvia	Open system with heart and arteries.	Nephridia.	Ganglia (cerebral, pedal, visceral) and nerve cords.	Osphradia, eyes, statocysts, other sense organs.	Dioecious.	Protostomate. Typically trochophore and veliger larvae.
Class Scaphopoda	Open system, no heart.	Nephridia.	Ganglia (cerebral, pleural, pedal), and nerve cords.	Statocysts, tentacles, and other sense organs.	Dioecious.	Protostomate. Trochophore and veliger larvae.
Class Cephalopoda	Closed system with systemic and branchial hearts and capillaries.	Nephridia.	Ring of ganglia (cerebral visceral, pedal, pleural) and nerve cords.	Tactile and chemosensory receptors, statocysts, and complex eyes.	Dioecious.	Protostomate. No free larval stages.
Phylum Annelida	Closed or open system with arteries, veins, heart(s), respiratory pigments; many lack capillaries.	Segmentally arranged metanephridia.	Cerebral ganglia and ventral nerve cord with segmental ganglia.	Simple and complex eyes, statocysts, chemoreceptors, other sensory structures.	Monoecious, dioecious.	Protostomate. Direct or indirect development (via trochophore larva); budding in some.
Class Polychaeta	Closed or open system, depending on species.	Segmentally arranged metanephridia.	Cerebral ganglia and ventral nerve cord with segmental ganglia.	Simple and complex eyes, statocysts, tactile receptors, chemoreceptors, other sense organs.	Dioecious. Gonads temporary. Fertilization external.	Protostomate. Typically, a trochophore; asexual budding in some.
Class Oligochaeta	Closed or open system, depending on species.	Segmentally arranged metanephridia.	Cerebral ganglia and ventral nerve cord with segmental ganglia.	Epidermal photoreceptors, tactile receptors, chemoreceptors, other sensory structures.	Monoecious. Gonads permanent. Fertilization external in cocoon (earthworms).	Protostomate. Direct development; regeneration in some.
Class Hirudinea	Typically open system with hemocoelic spaces.	Segmentally arranged metanephridia.	Cerebral ganglia and ventral nerve cord with segmental ganglia.	Simple eyes and other photo-, chemo-, and mechanoreceptors.	Monoecious. Gonads permanent. Fertilization internal.	Protostomate. Direct development.
Class Branchiobdellida	Little known.	Segmentally arranged metanephridia.	Cerebral ganglia and ventral nerve cord with segmental ganglia.	Little known.	Monoecious.	Protostomate. Direct development.
Phylum Sipuncula	Coelomocytes containing respiratory pigment.	Metanephridia remove waste-collecting amoebocytes.	Cerebral ganglion and ventral nerve cord.	Simple eyes in some. Sensory endings on skin and on introvert.	Dioecious. Gonads temporary.	Protostomate. Development direct or with trochophore larva.
Phylum Echiura	Closed system. Coelomocytes with respiratory pigment.	Several pairs of metanephridia.	Anterior nerve ring and ventral nerve cord.	Sensory endings in skin.	Dioecious.	Protostomate. Trochophore larva.

			Major Characteristics			
Taxonomic Subdivision	*Habitat*	***Body Cavity and Germ Layers***	***Skin/Skeleton Body Form***	*Movement*	*Digestive*	*Gas Exchange*
Phylum Pogonophora	Marine, along continental slopes below 100 m.	Eucoelomate. Triploblastic.	Hydrostatic skeleton. Cuticle and epidermis. Bilateral symmetry.	Longitudinal and circular muscles.	Gut absent. Extracellular digestion, and symbiotic bacteria in vestimentiferans.	Diffusion across body surface. Hemoglobin in plasma.
Phylum Arthropoda	Marine, terrestrial, fresh water. Symbiotic.	Eucoelomate; coelom reduced and main body cavity a hemocoel. Triploblastic.	Exoskeleton of chitinous cuticle. Epidermis. Bilateral symmetry.	Jointed appendages and striated (skeletal) muscles.	Complete gut.	Book gills, book lungs, gills, tracheae, body surface.
Subphylum Trilobita (Extinct)	Marine.	Eucoelomate. Triploblastic.	Exoskeleton of chitinous cuticle. Epidermis. Bilateral symmetry.	Jointed appendages and skeletal muscles.	Complete gut.	Perhaps gills on legs.
Subphylum Chelicerata Class Merostomata	Marine, on soft bottoms in shallow water.	Eucoelomate. Triploblastic.	Exoskeleton of chitinous cuticle. Epidermis. Bilateral symmetry.	Jointed appendages and striated muscles.	Complete gut.	Book gills.
Class Arachnida	Majority terrestrial and free-living, some aquatic, some parasitic.	Eucoelomate. Triploblastic.	Exoskeleton of chitinous cuticle. Epidermis. Bilateral symmetry.	Jointed appendages and striated muscles.	Complete gut.	Book lungs, tracheae, gills.
Class Pycnogonida	Sea spiders. Marine.	Eucoelomate. Triploblastic.	Exoskeleton of chitinous cuticle. Epidermis. Bilateral symmetry.	Jointed appendages and striated muscles.	Complete gut.	Diffusion across body surface.
Subphylum Crustacea Class Branchiopoda	With a few exceptions (e.g. brine shrimps), restricted to fresh water.	Eucoelomate. Triploblastic.	Exoskeleton of chitinous cuticle. Epidermis. Bilateral symmetry.	Jointed appendages and striated muscles.	Complete gut.	Gills on trunk appendages.
Class Copepoda	Most marine, many in fresh water, a few in moist soil, a few parasitic.	Eucoelomate. Triploblastic.	Exoskeleton of chitinous cuticle. Epidermis. Bilateral symmetry.	Jointed appendages and striated muscles.	Complete gut.	Diffusion across body surface.
Class Ostracoda	Seed shrimps. In marine and fresh water.	Eucoelomate. Triploblastic.	Exoskeleton of chitinous cuticle. Epidermis. Bilateral symmetry.	Jointed appendages and striated muscles.	Complete gut.	Diffusion across body surface.
Class Cirripedia	Barnacles. Marine.	Eucoelomate. Triploblastic.	Exoskeleton of chitinous cuticle. Epidermis. Bilateral symmetry.	Jointed appendages and striated muscles.	Complete gut.	Diffusion across mantle and cirri.
Class Malacostraca	Marine, fresh water, terrestrial, depending on species.	Eucoelomate. Triploblastic.	Exoskeleton of chitinous cuticle. Epidermis. Bilateral symmetry.	Jointed appendages and striated muscles.	Complete gut.	Thoracic epipodites form gills. Some have abdominal gill books.

Taxonomic Subdivision	Major Characteristics					
	Circulatory	**Excretory**	**Nervous**	**Sensory**	**Reproductive**	**Developmental**
Phylum Pogonophora	Closed system, with hemoglobin dissolved in plasma.	Two metanephridia.	Cerebral ganglion, dorsal nerve cord, tentacle nerves.	Photoreceptor and other sensory cells on body surface.	Dioecious.	Probably protostomate.
Phylum Arthropoda	Open system with hemocoel, heart, vessels. Respiratory pigment in some.	Maxillary, antennal, or coxal glands, or Malpighian tubules.	Cerebral ganglia (brain) and paired ventral nerve cords with segmental ganglia.	Various well-developed photo-, chemo-, and mechano-receptors.	Dioecious.	Protostomate. Development is direct or indirect. Metamorphosis common. Some species partheno-genetic.
Subphylum Trilobita (Extinct)	Open system.	Little known.	Cerebral ganglia and ventral nerve cords.	Compound eyes, antennae.	Dioecious.	Little known.
Subphylum Chelicerata Class Merostomata	Open system with hemocoel, heart, vessels.	Four pairs of coxal glands.	Cerebral ganglia, paired ventral nerve cords.	Two simple and two compound eyes. Chemo-receptive frontal organ.	Dioecious.	Protostomate. Horseshoe crab larva resembles trilobite.
Class Arachnida	Open system, typically with hemocoel, heart, aorta.	Coxal glands, Malpighian tubules.	Cerebral ganglia, paired ventral nerve cords.	Groups of simple eyes; pedipalps; sensory hairs on body.	Dioecious.	Protostomate. Development direct; no true metamorphosis.
Class Pycnogonida	Open system with dorsal heart.	Absent.	Cerebral ganglia, paired ventral nerve cords.	Four simple eyes; pedipalps; sensory hairs.	Dioecious.	Protostomate. Development via proto-nymphon larva.
Subphylum Crustacea Class Branchiopoda	Hemocoel, glob-ose or tubular heart, and anterior aorta.	Maxillary glands.	Cerebral ganglia, paired ventral nerve cord.	Compound eyes, nauplius eye; antennae.	Dioecious.	Protostomate. Many with par-thenogenesis. De-velopment direct or indirect with nauplius larva.
Class Copepoda	Hemocoel but no heart or vessels in most.	Maxillary glands.	Cerebral ganglia; paired, reduced ventral nerve cord.	Nauplius eye; antennae; setae on telson.	Dioecious.	Protostomate. Free-living forms with no or slight meta-morphosis; parasitic forms strongly meta-morphic.
Class Ostracoda	Hemocoel. Heart and vessels absent in most.	Maxillary glands.	Cerebral ganglia; ventral nerve cord ganglia vari-ously concen-trated.	Nauplius eye; compound eyes in some; an-tennae.	Dioecious.	Protostomate. Development includes nauplius larva.
Class Cirripedia	Hemocoel. No heart, arteries; sinus pump circulates hemolymph.	Maxillary glands.	Cerebral ganglia; ventral nerve cord ganglia a sin-gle mass in sessile barn-acles.	Sensory hairs on appendages; antennae in larvae.	Most are monoecious; self-fertil-ization pos-sible in some.	Protostomate. Strongly meta-morphic; naup-lius larva followed by cyprid larva.
Class Malacostraca	Open system with hemocoel, heart, vessels.	Antennal glands (deca-pods, amphi-pods); maxil-lary glands (isopods).	Cerebral gan-glia; ventral nerve cord ganglia vari-ously fused.	Compound eyes; statocysts; antennae; sensory hairs.	Dioecious.	Protostomate. Strongly or slightly meta-morphic with nauplius larva or nonmeta-morphic, depending on order.

		Major Characteristics				
Taxonomic Subdivision	*Habitat*	*Body Cavity and Germ Layers*	*Skin/Skeleton Body Form*	*Movement*	*Digestive*	*Gas Exchange*
Subphylum Uniramia Class Diplopoda	Millipedes. Terrestrial; typically burrowing.	Eucoelomate. Triploblastic.	Exoskeleton of chitinous cuticle. Epidermis. Bilateral symmetry.	Jointed appendages and striated muscles.	Complete gut.	Tracheae.
Class Pauropoda	Terrestrial; in soil, leaf litter, and other moist places.	Eucoelomate. Triploblastic.	Exoskeleton of chitinous cuticle. Epidermis. Bilateral symmetry.	Jointed appendages and striated muscles.	Complete gut.	Diffusion across body surface.
Class Chilopoda	Centipedes. Terrestrial; under stones, logs, and other moist situations.	Eucoelomate. Triploblastic.	Exoskeleton of chitinous cuticle. Epidermis. Bilateral symmetry.	Jointed appendages and striated muscles.	Complete gut.	Tracheae.
Class Symphyla	Terrestrial; in soil, leaf litter, and other moist places.	Eucoelomate. Triploblastic.	Exoskeleton of chitinous cuticle. Epidermis. Bilateral symmetry.	Jointed appendages and striated muscles.	Complete gut.	Tracheae.
Class Insecta	Majority terrestrial many in fresh and brackish water, few marine. Free-living, symbiotic, social, depending on species.	Eucoelomate. Triploblastic.	Exoskeleton of chitinous cuticle. Epidermis. Bilateral symmetry.	Jointed appendages and striated muscles.	Complete gut.	Tracheae.
Phylum Tardigrada	Water-bears. Fresh water (including droplets on lichens and mosses).	Eucoelomate. Triploblastic.	Hydrostatic skeleton, and exoskeleton of chitinous cuticle. Epidermis. Bilateral symmetry.	Bands of longitudinal muscle fibers along body wall.	Complete gut. Sucking mouthparts.	Diffusion across body surface.
Phylum Pentastomida	Endoparasites in respiratory system of reptiles and a few birds and mammals.	Eucoelomate. Triploblastic.	Hydrostatic skeleton, and exoskeleton of chitinous cuticle. Epidermis. Bilateral symmetry.	Muscular body wall.	Complete gut adapted for sucking.	Diffusion across body surface.
Phylum Onychophora	Walking-worms. Terrestrial; under leaf litter, logs, other moist situations. Tropical and south temperate.	Eucoelomate. Triploblastic.	Hydrostatic skeleton, and exoskeleton of cuticle. Epidermis. Bilateral symmetry.	Muscular body wall.	Complete gut. Jawed mouth.	Tracheae.
Phylum Bryozoa	Moss animals. Mostly marine, some in fresh water.	Eucoelomate. Triploblastic.	Hydrostatic skeleton and calcareous or chitinous case. Epidermis. Bilateral or asymmetrical.	Muscular movement of lophophore.	Complete gut. Ciliary feeding with tentacles on lophophore; gut U-shaped.	Diffusion across body and tentacle surfaces.
Phylum Phoronida	Marine.	Eucoelomate. Triploblastic.	Hydrostatic skeleton. Cuticle over epidermis. Bilateral or asymmetrical.	Longitudinal and circular muscles.	Complete gut. Ciliary feeding with tentacles on lophophore; gut U-shaped.	Diffusion across body and tentacle surfaces.

Taxonomic Subdivision	Major Characteristics					
	Circulatory	**Excretory**	**Nervous**	**Sensory**	**Reproductive**	**Developmental**
Subphylum Uniramia Class Diplopoda	Hemocoel, and dorsal heart and aorta.	Malpighian tubules.	Cerebral ganglia, paired ventral nerve cords.	Clustered ocelli; antennae; sensory hairs.	Dioecious.	Protostomate. Slightly metamorphic, with segments and legs added at each molt.
Class Pauropoda	Hemocoel. No heart or vessels.	Little known.	Cerebral ganglia, paired ventral nerve cord.	Eyes; two-branched antennae; sensory hairs.	Dioecious.	Protostomate. Metamorphosis, as in millipedes.
Class Chilopoda	Hemocoel. Dorsal tubular heart with pair of arteries to each segment.	Malpighian tubules.	Cerebral ganglia, paired ventral nerve cords.	Clusters of few to many ocelli, absent in some; antennae; sensory hairs.	Dioecious.	Protostomate. Nonmetamorphic, or with legs and segments added at each molt.
Class Symphyla	Hemocoel and dorsal tubular heart.	Little known.	Cerebral ganglia, paired ventral nerve cords.	Antennae; sensory hairs; no eyes.	Dioecious.	Protostomate. Slightly metamorphic.
Class Insecta	Hemocoel, dorsal tubular heart and aorta, diaphragms, and accessory pumps.	Malpighian tubules.	Cerebral ganglia; ventral nerve cord ganglia variously fused.	Compound and simple eyes; antennae and palps; various sensilla.	Dioecious.	Protostomate. Without metamorphosis or with gradual or complete metamorphosis.
Phylum Tardigrada	Open system; hemocoel but no heart or vessels.	Glands (Malpighian tubules?) at anterior end of rectum.	Cerebral ganglion and ventral nerve cord.	Simple eyespots and sensory spines and bristles.	Dioecious.	Protostomate? Development is direct.
Phylum Pentastomida	Open system; hemocoel but no heart or vessels.	No specific structures.	Three pairs of ganglia along ventral nerve cord.	Little known.	Dioecious.	Protostomate. Mite-like larval stages in intermediate host.
Phylum Onychophora	Open system. Hemocoel and tubular heart.	Pairs of segmental metanephridia.	Cerebral ganglia, paired ventral nerve cords.	Simple eyes; antennae; and sensory tubercles on body.	Dioecious.	Protostomate. Cleavage is superficial or holoblastic.
Phylum Bryozoa	No specific structures.	No specific structures.	Cerebral ganglia, nerves.	Chemo- and mechano-receptors.	Monoecious, dioecious.	Deuterostomate. Radial cleavage.
Phylum Phoronida	Closed system; vessels but no heart.	Metanephridia.	Nerve ring below lophophore, and nerve net.	Sensory cells on tentacles and body.	Monoecious, dioecious.	Deuterostomate? Radial cleavage produces modified trochophore (an actinotroch).

		Major Characteristics				
Taxonomic Subdivision	Habitat	Body Cavity and Germ Layers	Skin/Skeleton Body Form	Movement	Digestive	Gas Exchange
Phylum Brachiopoda	Marine.	Eucoelomate. Triploblastic.	Hydrostatic skeleton, and dorsal and ventral calcareous valves. Epidermis. Bilateral or asymmetrical.	Muscular movement of valves and contraction of stalk.	Complete or incomplete gut. Ciliary feeding with tentacles on lophophore; gut U-shaped.	Diffusion across body and mantle surfaces.
Phylum Echinodermata	Marine.	Eucoelomate. Triploblastic.	Typically, a dermal endoskeleton of unique calcareous plates. Epidermis and thin cuticle. Radial, bilateral, or asymmetrical.	Water vascular system and tube feet.	Complete gut, some with chewing structure.	Dermal branchiae in some; diffusion across surface of body and tube feet.
Class Crinoidea	Feather stars, sea lilies, Marine.	Eucoelomate. Triploblastic.	Endoskeleton of calcareous plates. Syncytial epidermis. Radial symmetry.	Stalkless forms move arms to swim. Water vascular system and tube feet.	Complete gut.	Diffusion across surface of body and tube feet.
Class Asteroidea	Sea stars. Marine.	Eucoelomate. Triploblastic.	Endoskeleton of calcareous plates. Ciliated epidermis Radial symmetry.	Water vascular system and tube feet, and limited arm motion.	Complete gut.	Dermal branchiae and tube feet.
Class Ophiuroidea	Brittle stars. Marine.	Eucoelomate. Triploblastic.	Endoskeleton of calcareous plates. Epidermis. Radial symmetry.	Highly mobile; rowing or coiling movements of arms. Water vascular system and tube feet.	Gut lacks anus.	Passage of water in and out of internal sacs (bursae).
Class Concentricycloidea	Marine; on submerged wood at depths of 1200 m.	Eucoelomate. Triploblastic.	Endoskeleton of calcareous plates. Epidermis. Radial symmetry.	Circular water vascular system and tube feet.	No gut. External velum may absorb nutrients.	Diffusion across body surface.
Class Echinoidea	Sea urchins. Marine.	Eucoelomate. Triploblastic.	Endoskeleton of calcareous plates. Ciliated epidermis. Radial, bilateral, or asymmetrical.	Water vascular system and tube feet.	Typically, a complete gut with Aristotle's lantern for chewing.	Diffusion across surface of body and tube feet.
Class Holothuroidea	Sea cucumbers. Marine.	Eucoelomate. Triploblastic.	Endoskeletal calcareous plates much reduced. Epidermis. Radial, bilateral, or asymmetrical.	Water vascular system with podia in some. Longitudinal and circular muscles in body wall.	Complete gut.	Respiratory tree; tube feet in some; body surface.
Phylum Chaetognatha	Arrow worms. Marine.	Eucoelomate. Triploblastic.	Hydrostatic skeleton. Stratified epidermis. Bilateral symmetry.	Body wall musculature.	Complete gut.	Diffusion across body surface.
Phylum Hemichordata	Acorn worms, pterobranchs. Marine.	Eucoelomate. Triploblastic.	Hydrostatic skeleton. Glandular, ciliated epidermis. Bilateral or asymmetrical.	Body wall musculature.	Complete gut.	Diffusion across body surface. Gill slits in some.

Major Characteristics

Taxonomic Subdivision	Circulatory	Excretory	Nervous	Sensory	Reproductive	Developmental
Phylum Brachiopoda	Open system, with heart and blood sinuses.	Metanephridia.	Nerve ring below lophophore, and nerves.	Some with pair of statocysts.	Dioecious.	Deuterostomate. Radial cleavage leads to free-swimming juvenile or larva.
Phylum Echinodermata	Movement of coelomic fluid and coelomocytes. Open hemal system.	Diffusion across thin areas of body surface. Coelomocytes also collect waste.	Nerve rings, radial nerves, nerve net below epidermis.	Eyespots, epidermal and epithelial sensory cells. Statocysts in some.	Majority are dioecious.	Deuterostomate. Bilaterally symmetrical larvae metamorphose to radially symmetrical adults; some adults revert to bilateral condition.
Class Crinoidea	Coelomic fluid, coelomocytes. Open hemal system.	Coelomocytes gather wastes.	Nerve rings, radial nerves.	Epidermal sensory cells.	Dioecious.	Deuterostomate. Doliolaria larva metamorphoses to adult.
Class Asteroidea	Coelomic fluid, coelomocytes. Open hemal system.	Diffusion across body surface. Coelomocytes.	Nerve ring, radial nerves, nerve net.	Eyespots, epidermal sensory cells.	Dioecious; some monoecious.	Deuterostomate. Bipinnaria larva becomes a brachiolaria that metamorphoses to adult.
Class Ophiuroidea	Coelomocytes. Open hemal system.	The respiratory bursae may also remove wastes.	Nerve ring, radial nerves.	Epithelial sensory cells.	Dioecious; some monoecious.	Deuterostomate. Ophiopluteus larva metamorphoses to adult.
Class Concentricycloidea	Little known.	Little known.	Nerve ring.	Not known.	Little known.	Deuterostomate. No free-swimming larvae; adults brood embryos.
Class Echinoidea	Coelomic fluid, coelomocytes. Open hemal system.	Diffusion across body surface. Coelomocytes.	Nerve ring, radial nerves, nerve net.	Epithelial sensory cells.	Dioecious.	Deuterostomate. Echinopluteus larva metamorphoses to adult.
Class Holothuroidea	Coelomic fluid, coelomocytes. Open hemal system.	Coelomocytes. Lumina of gut, respiratory trees, gonads.	Nerve ring, radial nerves, nerve net.	Epidermal sensory cells. Statocysts.	Dioecious; some monoecious.	Deuterostomate. Auricularia larva becomes a doliolaria that metamorphoses to adult.
Phylum Chaetognatha	Coelomic fluid movement.	Diffusion across body surface.	Cerebral, ventral, lateral ganglia; circumpharyngeal nerve ring.	Pigment-cup eyes. Ciliary clusters over body. Ciliary loop at back of head.	Monoecious.	Deuterostomate. Development direct.
Phylum Hemichordata	Sinuses, and dorsal and ventral vessels.	Glomerulus (acorn worms) may be excretory.	Dorsal tubular nerve cord in collar (acorn worms); nerve cords in trunk; nerve net.	Scattered epithelial sensory cells.	Dioecious.	Deuterostomate. Development indirect; many have a tornaria larva.

APPENDIX

			Major Characteristics			
Taxonomic Subdivision	*Habitat*	*Body Cavity and Germ Layers*	*Skin/Skeleton Body Form*	*Movement*	*Digestive*	*Gas Exchange*
Phylum Chordata	Terrestrial, marine, fresh or brackish water, depending on species.	Eucoelomate. Triploblastic.	Hydrostatic, notochord, bony and/or cartilaginous endoskeleton. Dermis, stratified epidermis; simple epidermis in some. Bilateral symmetry.	Cilia. Skeletal muscles	Complete gut, including mouth, oral cavity, pharynx, esophagus, stomach, small and large intestines, cecum, and anus.	Gill slits, gills, lungs, swim bladder, and/or diffusion across body surface.
Subphylum Cephalochordata	Lancelets. Marine; bottom dwellers in warm, shallow waters.	Eucoelomate. Triploblastic.	Notochord. Cuticle and simple epidermis, dermis. Bilateral symmetry.	Cilia, and skeletal muscles arranged as myomeres.	Complete gut.	Gill slits.
Subphylum Urochordata	Tunicates. Marine; benthic or pelagic, depending on species.	Eucoelomate. Triploblastic.	Hydrostatic skeleton. Notochord in free-swimming stages. Dermis and epidermis; some with cellulose-like tunic. Bilateral or asymmetrical.	Cilia and muscles.	Complete gut.	Gill slits.
Subphylum Vertebrata Class Agnatha	Lampreys, hagfishes. Marine.	Eucoelomate. Triploblastic.	Notochord. Cartilaginous skeleton. Dermis and stratified epidermis. Bilateral symmetry.	Skeletal muscles.	Complete gut.	Gill slits, gills.
Class Chondrichthyes	Sharks, rays, skates. Marine, a few in fresh water.	Eucoelomate. Triploblastic.	Cartilaginous skeleton with vertebral column. Dermis and stratified epidermis. Placoid scales. Bilateral symmetry.	Skeletal muscles.	Complete gut.	Gill slits, gills.
Class Osteichthyes	Bony fishes. Marine, fresh or brackish water, depending on species.	Eucoelomate. Triploblastic.	Skeleton of cartilage and bone, with vertebral column. Dermis and stratified epidermis. Cycloid, ctenoid, or ganoid scales. Bilateral symmetry.	Skeletal muscles.	Complete gut.	Gill slits, gills. Lungs in some.
Class Amphibia	To varying degrees, terrestrial or aquatic (i.e. quasiterrestrial).	Eucoelomate. Triploblastic.	Skeleton of cartilage and bone, with vertebral column. Dermis and lightly keratinized, stratified epidermis. Bilateral symmetry.	Skeletal muscles.	Complete gut.	Gill slits, gills, lungs, diffusion across body surface.
Class Reptilia	Terrestrial; some with adaptations for life in aquatic environments.	Eucoelomate. Triploblastic.	Skeleton of cartilage and bone, with vertebral column. Dermis and stratified, keratinized epidermis with scales. Bilateral symmetry.	Skeletal muscles.	Complete gut.	Lungs.

Taxonomic Subdivision	Major Characteristics					
	Circulatory	*Excretory*	*Nervous*	*Sensory*	*Reproductive*	*Developmental*
Phylum Chordata	Typically, closed system with a chambered heart. Blood vessels. Respiratory pigment.	Pro-, opistho-, or metanephros. Diffusion across body surface, Nephrocytes. Protonephridia.	Dorsal tubular nerve cord, nerves, and usually a brain.	Typically, sense organs of touch, taste, smell, and sight. Various sensory corpuscles and endings.	Dioecious; some monoecious.	Deuterostomate. Enterocoelous developmental pattern ancestral in the phylum.
Subphylum Cephalochordata	Closed system. No heart. Dorsal, ventral aortae.	Numerous protonephridia.	Dorsal tubular nerve cord, nerves.	Eyespots. Sensory cells in epidermis.	Dioecious.	Deuterostomate. Holoblastic radial cleavage. Ciliated planktonic larva.
Subphylum Urochordata	Open system. Tubular heart.	Waste collected by nephrocytes. Diffusion across body surface.	Cylindrical to spherical cerebral ganglion and nerves.	Eyespots. Epidermal sensory endings. Statocysts.	Mostly monoecious.	Deuterostomate. Tadpole-shaped ascidian larva transforms into a sessile adult. Non-ascidian urochordates permanently free-swimming plankton.
Subphylum Vertebrata Class Agnatha	Closed system. Four-chambered heart; accessory hearts in hagfishes.	Primitive opisthonephros (hagfish); opisthonephros (lamprey).	Dorsal tubular nerve cord; brain; spinal and cranial nerves.	Sense organs of touch, taste, smell, and usually sight. Lateral line system.	Dioecious (lamprey); Monoecious (hagfish).	Deuterostomate. Lamprey with long larval period; hagfish development direct.
Class Chondrichthyes	Closed system. Four-chambered heart.	Opisthonephros.	Dorsal tubular nerve cord; brain; spinal and cranial nerves.	Sense organs of touch, taste, smell, and sight. Other sensory structures, including lateral line system.	Dioecious.	Deuterostomate. Development direct. Ovoviviparous, oviparous, or viviparous.
Class Osteichthyes	Closed system. Four-chambered heart.	Opisthonephros.	Dorsal tubular nerve cord; brain; spinal and cranial nerves.	Sense organs of touch, taste, smell, and sight. Other sensory structures, including lateral line system.	Typically, dioecious.	Deuterostomate. Majority are oviparous. Metamorphosis of larva to adult.
Class Amphibia	Closed system. Three-chambered heart.	Opisthonephros.	Dorsal tubular nerve cord; brain; spinal and cranial nerves.	Sense organs of touch, taste, smell, hearing, and sight. Other sensory structures, including lateral line system in larval forms.	Dioecious.	Deuterostomate. Mostly oviparous; some ovoviviparous, or viviparous. Usually, a larval stage and metamorphosis.
Class Reptilia	Closed system. Three- or four-chambered heart.	Metanephros.	Dorsal tubular nerve cord; brain; spinal and cranial nerves.	Sense organs of touch, taste, smell, hearing, and sight. Other sensory structures.	Dioecious.	Deuterostomate. Mostly oviparous; some ovoviviparous. Shelled egg with extraembryonic membranes; amniotic.

APPENDIX

APPENDIX

		Major Characteristics				
Taxonomic Subdivision	*Habitat*	***Body Cavity and Germ Layers***	***Skin/Skeleton Body Form***	*Movement*	*Digestive*	*Gas Exchange*
Class Aves	Terrestrial; some with adaptations for life in aquatic environments.	Eucoelomate. Triploblastic.	Skeleton of cartilage and bone, with vertebral column. Dermis and stratified, keratinized epidermis; feathers. Bilateral symmetry.	Skeletal muscles.	Complete gut.	Lungs.
Class Mammalia	Terrestrial; some with adaptations for life in aquatic environments.	Eucoelomate. Triploblastic.	Skeleton of cartilage and bone, with vertebral column. Dermis and stratified, keratinized epidermis; hair. Bilateral symmetry.	Skeletal muscles.	Complete gut.	Lungs.

Taxonomic Subdivision	Major Characteristics					
	Circulatory	*Excretory*	*Nervous*	*Sensory*	*Reproductive*	*Developmental*
Class Aves	Closed system. Four-chambered heart.	Metanephros.	Dorsal tubular nerve cord; brain; spinal and cranial nerves.	Sense organs of touch, taste, smell, hearing, and sight. Other sensory structures.	Dioecious.	Deuterostomate. Oviparous. Shelled egg with extraembryonic membranes; amniotic.
Class Mammalia	Closed system. Four-chambered heart.	Metanephros.	Dorsal tubular nerve cord; brain; spinal and cranial nerves.	Sense organs of touch, taste, smell, hearing, and sight. Other sensory structures.	Dioecious.	Deuterostomate. Some oviparous. Most viviparous; uterine development with placenta. Egg with extraembryonic membranes; amniotic. Milk from mammary glands.

APPENDIX

Glossary

A band The band that corresponds to the fixed length of the thick (myosin) filaments in a striated muscle fiber.

Aboral Pertaining to a region of the body opposite or away from the mouth, or oral region.

Absorption Taking up of substances by osmotic, capillary, solvent, or chemical action, as by the lining of the digestive tract or integument.

Accommodation Adjustment of the focal length of the eye to focus sharply on an object in the visual field.

Acetylcholine An acetic acid ester of choline; a neurotransmitter released and hydrolyzed in the course of neural transmission at many synapses and junctions with effector organs.

Acoelomate (1) Body plan in which a body cavity is lacking, as in flatworms. (2) Pertaining to the condition of not having a body cavity.

Acousticolateralis system A system that mediates sound reception in fishes and some amphibians; consists of the inner ear and lateral line system.

Acrosome A caplike organelle in the tip of a sperm; its rupture causes formation of an acrosomal filament (in echinoderms) and release of enzymes that play a role in the penetration of the egg cell membrane during fertilization.

Actin A cytoplasmic protein associated with cellular movement; in striated muscle, actin forms the thin myofilaments that interact with the thick myosin myofilaments during muscle contraction.

Action potential (spike) The altered, self-propagating electrical potential (depolarization) that develops across the membrane of a stimulated nerve or muscle fiber.

Activation energy The minimum energy required to start a chemical reaction.

Active transport Transport of a substance across a cell membrane, against a concentration gradient, by a mechanism that requires expenditure of energy.

Adaptation (1) A change in the structure or function of an organism or any of its parts which makes the organism better able to survive and reproduce. (2) Reduction in excitability of sense cells or organs in response to repeated stimulation.

Adaptive radiation Evolution from a common ancestor of two or more species by adaptation to different habitats.

Adenohypophysis The anterior lobe of the pituitary gland.

Adenosine monophosphate (AMP) A compound consisting of one phosphate group, ribose, and adenine.

Adenosine triphosphate (ATP) A compound consisting of three phosphate groups, ribose, and adenine; ATP is ubiquitous in cells, where it takes part in many energy exchange reactions.

Adrenaline *See* Epinephrine.

Adrenergic Pertaining to a neuron that releases epinephrine or norepinephrine.

Aerobic Pertaining to the utilization of oxygen in respiration.

Aestivation Dormancy or torpor during hot or dry periods.

Afferent Referring to a neuron or vessel that conveys nerve impulses or fluid toward a specific point of reference; for example, afferent (sensory) neurons carry impulses toward the central nervous system, and afferent arteries carry blood into a fish's gill.

Agglutination The clumping of cells in a fluid; e.g., agglutination of erythrocytes or other cells by an antibody.

Agonistic behavior A threat or offensive action directed by one animal toward another; an offensive or defensive behavior involved in a confrontation between two or more animals.

Airfoil A surface, such as a bird or airplane wing, that responds to air movement by producing lift.

Aldosterone A hormone of the adrenal cortex that increases reabsorption of sodium from the kidney tubules.

Allantois (1) One of the extraembryonic membranes of amniotes (reptiles, birds, and mammals). (2) A pouch, derived from the posterior region of the amniote embryo's digestive tract, that provides blood vessels serving the placenta or chorion and that functions in waste storage in embryonic birds and reptiles.

Allele One of two or more forms of a gene that may occur at a particular locus on a chromosome.

Allopatric Pertaining to species or to populations of a species living in different geographical regions or separated by geographical barriers.

Altricial Pertaining to the young of many birds (e.g., songbirds) and mammals (e.g., marsupials, carnivorans, and rodents) that are small and helpless when hatched or born and that must receive parental care or they will die. *Compare* Precocial.

Altruism Behavior that benefits others at some risk or cost to the altruistic animal.

Alveolus Microscopic air sac of the vertebrate lung, formed by the terminal dilation of minute air passages.

Ambulacral Pertaining to the radially arranged grooves from which the tube feet of the water vascular system of echinoderms project.

Amictic Referring to diploid (unfertilizable) rotifer eggs or to the female rotifers that produce only diploid eggs.

Amino acid Any one of the 20 monomers that may be linked by peptide bonds to form polypeptides; an amino acid contains one or more carboxyl (–COOH) and amino (–NH_2) groups.

Ammocoete The free-swimming, filter-feeding larval stage of lampreys.

Ammonotelic Pertaining to an animal whose main nitrogenous excretory product is ammonia.

Amnion In mammals, birds, and reptiles, the extraembryonic membrane that forms the fluid-filled sac that encloses and suspends the embryo.

Amniote A vertebrate that has an amnion; reptiles, birds, and mammals.

Amoebocyte Any unattached somatic cell capable of locomotion by means of amoeboid movement.

Amoeboid (pseudopodial) movement Cell movement by formation of cytoplasmic extensions, or pseudopodia.

Amphiblastula A blastula-like, ciliated, free-swimming larval stage of certain marine sponges.

Amplexus Embrace of male and female during copulation, as in amphibians.

Ampulla A vesicle formed as a dilatation in a duct or canal; includes the ampulla associated with the semicircular canals of the vertebrate inner ear and with the tube feet of sea stars.

Anabolism Energy-requiring synthetic reactions whereby chemical compounds are combined to form new, more complex cell substances.

Anadromous Pertaining to fishes that migrate from the ocean into streams to spawn.

Anaerobic Pertaining to respiration without the utilization of oxygen.

Anagenic evolution Anagenesis. *See* Phyletic evolution.

Analogous Referring to structures that are alike in function and which may or may not be similar in appearance, but whose embryonic precursors differ.

Anaphase The mitotic or meiotic stage that follows metaphase; the centromeres divide, or the homologous chromosomes separate, and the division products move toward opposite poles of the spindle.

Ancestral *See* Primitive.

Androgen A substance having male sex hormone activity, such as testosterone in vertebrates and androgenic gland hormone in crustaceans.

Antennal gland Crustacean excretory organ located in the antennal body segment.

Anterior (1) Toward the head end of an animal. (2) The head end of an animal.

Antibody A protein produced in response to an antigen (usually a foreign substance in the tissues or circulating in the blood) and capable of binding specifically with that antigen.

Anticodon In transfer RNA, a triplet of nucleotides that is complementary to, and which bonds with, a triplet of nucleotides (a codon) on messenger RNA.

Antigen Any substance, but usually a foreign protein, that stimulates production of specific antibodies or a specific white blood cell response by an animal's immune system.

Antiserum Blood serum that contains antibodies specific for an antigen.

Aortic arch In a vertebrate embryo, one of a series of paired blood vessels that branch from the ventral aorta, pass dorsally, and unite to form the dorsal aorta.

Apical complex A group of organelles that characterize protozoa of the Phylum Apicomplexa.

Apodeme An invagination of the arthropod integument to which skeletal muscle is attached.

Appendicular skeleton In vertebrates, the pectoral and pelvic girdles and the bones of the appendages.

Archenteron The endoderm-lined cavity in the gastrula, which represents the future digestive tract.

Arteriole A small artery; in animals with closed circulatory systems, blood enters capillary beds through arterioles.

Artery A vessel carrying blood or hemolymph away from the heart toward other parts of the body.

Articulation (1) Union of bones (or other skeletal elements such as cartilage in vertebrates or cuticle in arthropods) forming a joint. (2) Being jointed together, or a jointing.

Artificial selection The systematic selection by humans of breeding stock of plants and animals on the basis of desired traits.

Asconoid The simplest type of sponge, as compared to leuconoid and syconoid sponges; water canals in an asconoid sponge follow a direct course from the exterior to the spongocoel.

Aspect ratio The ratio of length to width in an airfoil.

Assimilation   The absorption of molecules (the products of digestion in most animals) and their incorporation into an organism's biochemical pathways.

Association neuron   *See* Interneuron.

Atrium   A chamber or cavity; e.g., a receiving chamber of a heart.

Autogamy   The fusion of two haploid gametic nuclei within the organism that produced them, thus restoring the diploid number.

Autosome   Any chromosome other than a sex chromosome.

Autotomy   Self-amputation, usually of appendages that can be regenerated.

Autotroph   An organism that synthesizes organic nutrients from inorganic precursors.

Axial   Pertaining to the axis, or central plane, bisecting a body; the plane of reference of a symmetrical body.

Axial skeleton   The cranium, vertebral column, and rib cage of vertebrates.

Axon   A nerve cell fiber that transmits impulses away from the cell body toward a synapse or effector; a neurite.

Basal body   A centriole, or circle of nine triplets of short microtubules at the base of a cilium or flagellum in eukaryotic cells; also called a basal granule, blepharoplast, or kinetosome.

Basal lamina   A thin sheet of collagenous fibers underlying and giving support to the cells of a tissue; often called the basement membrane.

Basal metabolic rate   In an animal at rest, the amount of oxygen consumed per minute per kilogram of body weight.

Batesian mimicry   A form of mimicry in which a palatable species gains protection from predators by mimicking an unpalatable species.

B cell   A lymphocyte that matures in the bone marrow of mammals or in the bursa of Fabricius of birds, circulates in the blood, and is important in the production of circulating antibodies during humoral immune responses.

Benthic   Pertaining to the bottom zone of an aquatic ecosystem or to organisms (benthos) living in and on the bottom.

Bilateral symmetry   Having only one plane of symmetry, with right and left sides that are approximately mirror images of each other.

Binary fission   A method of reproduction in which the parent organism divides into two parts of approximately equal size, each of which then develops into an individual similar to the parent.

Binominal   Consisting of two names; refers to the two names (genus and species epithet) used for each species; also called binomial.

Biogeochemical cycle   The cyclic passage of materials (nutrients) in an ecosystem.

Biological clock   An innate timing mechanism that enables animals to monitor the passage of time; important in regulating many cyclic activities, including circadian rhythms.

Biomass   The mass of living matter measured in weight or calories per unit area.

Biome   A terrestrial ecosystem, such as a grassland, tundra, or coniferous forest.

Biosphere   The entire inhabited area on earth.

Biotic potential   The theoretical maximum growth rate of which a species population is physiologically capable; maximum (never realized) intrinsic rate of population increase. *See* Intrinsic rate of increase.

Bipedal locomotion   Walking, running, or hopping on two feet, as by humans, kangaroos, birds, and some dinosaurs.

Bipinnaria   In sea stars, the ciliated, bilaterally symmetrical larval stage that precedes the brachiolaria stage.

Biradial symmetry   Having two planes of symmetry that are at right angles to each other, as in sea anemones and ctenophores.

Bisexual   (1) Having both male and female sex organs; hermaphroditic. (2) An animal sexually attracted to both sexes.

Blastocoel   The cavity within a blastula.

Blastocyst   A mammalian blastula; the blastocyst consists of a trophoblast that forms part of the placenta and an inner cell mass that develops into the embryo.

Blastoderm   The single-celled layer that encloses the blastocoel in a blastula.

Blastodisc   A small disc of cytoplasm that lies on the yolk of the egg of a reptile or bird, and which contains the egg nucleus.

Blastomere   Any cell produced during development of the blastula.

Blastopore   The opening of the archenteron formed by the process of gastrulation.

Blastula   The early embryonic stage that results from cleavage; consists of a sphere of cells surrounding a central cavity called the blastocoel (in some species, the blastocoel is filled with cells). *See also* Coeloblastula and Stereoblastula.

Book lung   In terrestrial arachnids, a respiratory organ in which thin membranes are arranged like the leaves of a book; similar to the book gill of horseshoe crabs.

Bowman's capsule   A double-walled, saclike membranous capsule enclosing a glomerulus and located at the proximal end of each nephron in the kidneys of vertebrates.

Brachial   Pertaining to an arm or armlike structure.

Brachiation Arboreal locomotion, characterized by swinging by the arms from one handhold to another.

Brain In vertebrates, the enlarged anterior end of the central nervous system enclosed in the cranium; in invertebrates, an anterior ganglionic mass of neurons that more or less functions like the brain of vertebrates.

Branchial Pertaining to gills or to the region where gills are located.

Budding A form of asexual reproduction in which a relatively small mass of cells separates from the parent's body as a bud that develops into a new individual.

Buffer A mixture of substances which, when in solution, can neutralize both acids and bases, thereby maintaining an equilibrium pH.

Calorie The amount of energy (heat or the equivalent chemical energy) required to raise the temperature of 1 g of water from 14.5°C to 15.5°C.

Canaliculus In bone, a tiny canal that communicates between osteocytes.

Capillary A microscopic blood vessel which, as part of a capillary bed made up of many capillaries, helps to channel blood between arterioles and venules.

Carbonic anhydrase An enzyme that catalyzes the formation of carbonic acid from carbon dioxide and water, as well as the reverse reaction, the conversion of carbonic acid to carbon dioxide and water.

Cardiac Pertaining to the heart.

Carinate Having a carina, or keel.

Carnivore An organism that consumes animals.

Carotid artery In vertebrates, either of a pair of arteries located on each side of the head and neck; the site of carotid bodies, which contain chemoreceptors responsive to amounts of carbon dioxide and oxygen in the blood.

Carrying capacity The greatest population density that an environment can support for an extended period of time.

Caste One of several types of individuals in a colony of social insects, such as the worker, soldier, king (male), and queen castes in a termite colony, wherein each performs specific tasks.

Castle-Hardy-Weinberg law A statement of the probabilities of genotype frequencies, based on known frequencies of alleles in a population of diploid, randomly mating individuals.

Catabolism The breakdown phase of metabolism, during which energy is released from organic molecules.

Catalyst An agent that increases the rate of a chemical reaction without undergoing permanent change itself; enzymes are biological catalysts.

Cecum A saclike diverticulum of the intestine; in vertebrates, it is located at the junction of the small and large intestines.

Central pattern generator Interneurons that provide rhythmic outputs to motor neurons in the absence of sensory or other CNS inputs; central pattern generators produce patterns of motor activity (motor programs) that result in the basic movements of repetitive and stereotyped behaviors, such as peristaltic locomotion of earthworms and walking by arthropods and vertebrates.

Centriole One of a pair of cylindrically arranged groups of nine triplets of short microtubules located near the nucleus of eukaryotic cells; apparently involved in forming the spindle and asters during meiosis and mitosis.

Centrolecithal Pertaining to a type of arthropod egg in which the abundant yolk forms a central mass surrounding the nucleus.

Centromere The region of a chromosome where sister chromatids are attached to each other and to which spindle fibers are attached during meiosis and mitosis.

Cephalization The evolutionary development of the head as a large, dominant body part; the concentration of nervous and sensory elements in a head.

Cercaria A motile larval stage of parasitic flukes (trematodes); a cercaria usually infects a definitive host and develops into an adult worm.

Cerebrum The cerebral hemispheres of the vertebrate brain.

Chaparral A dry brush biome.

Character displacement An increase in the heritable differences between two populations or closely related species living in the same geographical area.

Chelicera One of the anteriormost pair of appendages of chelicerates (e.g., horseshoe crabs, spiders, ticks, and mites).

Chemosynthesis Production of energy-containing organic molecules by oxidation and reduction of inorganic molecules and elements.

Chiasma A place of contact between pairs of homologous chromatids during early stages of meiosis; a place where crossing over and exchange of homologous parts has occurred between nonsister chromatids.

Chitin A class of nitrogen-containing polysaccharides that are important exoskeletal elements in arthropods and many other invertebrates; ($C_8H_{13}O_5N$).

Choanocyte A filter-feeding cell that is unique to sponges, and which has a flagellum that creates water currents and a sievelike funnel, or collar, that serves as a particle trap; also called a collar cell.

Choanoderm A cell layer in sponges, consisting of choanocytes.

Cholinergic Pertaining to, or stimulated by, acetylcholine or chemicals with similar activity.

Chondrocyte A cartilage-producing cell; a cell housed in a lacuna in cartilage.

Chorion The outermost of three extraembryonic membranes of amniotes.

Chromatid One of the two subunits of a duplicated chromosome, visible during mitosis and meiosis; one of a pair of sister chromatids that together constitute a doubled chromosome.

Chromatin DNA and associated RNA and proteins in a nondividing nucleus.

Chromatophore A pigment-containing cell; chromatocyte.

Chromosome A distinct subunit of the genetic material of a cell's nucleus; becomes distinct during meiosis and mitosis and consists of DNA and associated proteins.

Cilium A hairlike, typically short, locomotor extension of a cell, formed of a cell's plasmalemma surrounding a ring of nine doubled microtubules encircling a single central pair of unfused microtubules.

Circadian rhythm A repeated body activity, or cyclic body function, that occurs at intervals of approximately 24 hours.

Cirrus (1) A coarse, hairlike projection, often formed of fused cilia or hairlike cellular processes. (2) The male copulatory organ in many flatworms and other invertebrates. (3) A bristlelike appendage or part of an appendage formed of setae.

Cisterna A space, or cavity, in an organelle or organ.

Cladistics Phylogenetic classification based on the identification of points of divergence of evolutionary lineages; classification based entirely on monophyletic groups.

Cladogenesis Branching (divergent) evolution; evolution of lineages by divergence from common ancestors.

Cleavage Early cytoplasmic and mitotic nuclear division in an embryo, generally involving little or no increase in overall size of the embryo; successive divisions producing a multicellular blastula from a zygote.

Cline A character gradient, or a gradual variation in a phenotypic character over a geographical range of a species.

Clitellum A series of enlarged, fused, midbody segments in many annelids (oligochaetes and leeches); secretes mucus during copulation and forms cocoons in which eggs are deposited.

Cloaca Posterior organ of the digestive tract of many animals, serving as a common terminal chamber for digestive and excretory wastes, as well as for sex cells.

Clone Genetically identical offspring of asexual reproduction by a single individual.

Cnida An eversible organelle produced by, and contained in, a cnidocyte. *See also* Nematocyst.

Cnidocyst A cellular component of a myxozoan spore; consists of a nematocyst-like eversible organelle surrounded by a protein capsule.

Cnidocyte A cell containing a cnida.

Cochlea The elongate tube housing the hearing organ (organ of Corti) in the inner ear of crocodilians, birds, and mammals; formed as an outgrowth (the lagena) of the sacculus.

Codon Three adjacent nucleotides on a messenger RNA molecule; each codon carries the genetic code, transcribed from DNA, for one amino acid or for terminating a polypeptide chain during protein synthesis.

Coeloblastula A blastula consisting of a sphere of cells surrounding a cavity, the blastocoel.

Coelogastrula A gastrula containing a central cavity.

Coelom The body cavity (between the digestive tract and the body wall) formed by mesoderm and lined by peritoneum derived from mesoderm.

Coelomate (1) An animal with a mesoderm-lined body cavity (coelom). (2) Pertaining to the condition of having a coelom.

Coenzyme An organic molecule whose association with an enzyme is necessary in many enzyme-catalyzed reactions.

Coevolution The evolution of two or more lineages whose adaptations strongly affect each other.

Cofactor A chemical substance that interacts with, and activates, an enzyme.

Collar cell *See* Choanocyte.

Colony A group of closely associated individuals of one species, typically formed by asexual budding.

Commensalism A type of symbiosis in which individuals of one species (the commensal, or commensalist) benefit, while individuals of the other species (the host) are unaffected.

Community All living species in an ecosystem; all living organisms that interact in a particular environment.

Conjugation A temporary union (mating) between two individuals (two cells of a species of ciliate) during which micronuclei are exchanged.

Connective tissue An integrated aggregate of specialized cells in a noncellular matrix secreted by the cells.

Convergent evolution The evolution of similarity (e.g., analogous structures) in organisms in distinct lineages.

Coprophagy Feeding on feces.

Corona A crownlike body part; e.g., the ciliated anterior disc of a rotifer.

Countercurrent flow The opposite flow of fluids or gases in adjacent tubes, providing for the maintenance of diffusion gradients and, consequently, the efficient exchange of dissolved gases (in fish gills), heat (in heat exchangers in bird legs), and solutes (in the vertebrate kidney) between the opposite-flowing fluids.

Courtship Behavior (typically ritualized) that serves to stimulate mating in a member of the opposite sex.

Covalent bond A chemical bond involving the sharing of electrons between two or more atoms.

Cranium Cartilages and/or bones enclosing the vertebrate brain.

Crop An anterior enlargement of the digestive tract, serving as a temporary storage organ where newly ingested food is moistened and where some chemical digestion may occur.

Crosscurrent flow The flow of fluids or gases in adjacent tubes such that the flow of one crosses the direction of flow of the other; e.g., the flow of gases versus blood in a bird lung.

Crossing over The intertwining of adjacent nonsister chromatids, resulting in an exchange of chromatid segments between homologous chromosomes during meiosis.

Cryptobiosis A state of suspended animation as seen in many nematodes, rotifers, and tardigrades; it involves virtual cessation of metabolism and desiccation of the body during unfavorable periods, and the ability to become active when favorable conditions return.

Ctenidium A gill; usually used in reference to a gill of a mollusc.

Cutaneous Pertaining to the integument, the covering of the general body surface and oral membranes.

Cuticle A nonliving, noncellular outer layer of the integument, secreted by underlying cells.

Cyclic adenosine monophosphate (cAMP) A nucleotide that acts as a regulator of cellular activity; a second messenger, mediating the effect of many hormones and transmitter substances; 3′,5′-cyclic adenosine monophosphate.

Cytochrome Any of a group of proteins with an iron-containing (heme) group, and which function as hydrogen (electron) carriers in the electron transport system of aerobic respiration.

Cytokinesis Division of a cell's cytoplasm.

Cyton The cell body, or portion of a cell containing the nucleus.

Cytoplasm All of the constituents of a cell except the nucleus and the cell membrane.

Cytoskeleton A meshwork of interconnected, fine protein filaments and microtubules that collectively provide structure in the cytoplasm of a cell.

Cytosol The fluid, or aqueous, portion of a cell's cytoplasm, situated outside the organelles.

Cytostome A region on the surface of many protozoa where nutrients are ingested; the cell "mouth."

Daughter chromosome A single chromosome derived from a double chromosome by mitosis or by the second meiotic division; a chromatid after the doubled chromosomes have divided during anaphase of mitosis or meiosis II; an offspring chromosome present at one pole of a cell in telophase of mitosis or meiosis II.

Decomposer An organism whose metabolism converts dead organic matter into inorganic material.

Definitive host In a parasitic relationship, the host organism infected by the sexually mature stage of the parasite.

Delamination Formation of two or more sheets of cells from a single layer of cells during embryonic development.

Deme A population whose members inhabit a limited geographical area; a local population.

Dendrite A neuron process (neurite), usually highly branched, that transmits nerve impulses toward a nerve cell body.

Deposit feeding Ingestion of materials composing the substrate.

Derived condition Body features or adaptations that are evolved late in a lineage (compared to primitive ones, which were evolved by, and characterize, the ancestor); may also refer to a more recently evolved taxon in a lineage; contrasts with primitive, or ancestral features or adaptations.

Dermal Pertaining to, or derived from, the dermis.

Dermis The connective tissue layer underlying the epidermis of the integument.

Detritus Organic and inorganic debris.

Deuterostome A coelomate animal that is a member of a phylum containing species characterized by enterocoelous mesoderm and coelom formation, a mouth not derived from the blastopore, and radial cleavage; e.g., lophophorates, echinoderms, chaetognaths, hemichordates, and chordates.

Diapause A period of developmental arrest that may occur in any life cycle stage; depending on the species, the arrest may be obligatory and occur each generation, or it may occur in response to environmental signals.

Differentiation The change from an unspecialized to a specialized condition; the process of cell specialization and the development of cell diversity.

Diffusion The net movement of molecules from a region of greater concentration to a region of lesser concentration.

Dioecious Pertaining to species that have separate male and female individuals.

Diploblastic Pertaining to an animal body composed of only two cell layers; often said of cnidarians, although the cnidarian body has three cell layers, or regions; best used only in reference to embryonic stages consisting only of ectodermal and endodermal cells.

Diploid (2n) Having a double set of chromosomes.

Divergent evolution *See* Cladogenesis.

Dominance hierarchy A type of social organization within a species group involving the sequential ranking of individuals and usually established and maintained by ritualized agonistic behavior.

Dorsal Pertaining to the upper surface, or back, of a bilateral animal; opposite of ventral.

Duodenum The anteriormost region of the vertebrate small intestine, into which bile and digestive enzymes from the pancreas are secreted.

Ecdysis *See* Molting.

Ecdysone A steroid hormone that promotes growth and molting in arthropods and is secreted by the prothoracic glands of insects and by the Y organs of crustaceans.

Ecological niche *See* Niche.

Ecosystem A community or communities of living organisms and the nonliving (abiotic) environmental factors with which the organisms associate.

Ectoderm The outer cell layer of the two primary germ layers.

Ectotherm An organism whose body temperature depends on the temperature of the outside environment.

Effector A cell, tissue, or organ that can react to a stimulus.

Efferent Pertaining to a circulatory vessel that conveys fluid away from a particular organ or tissue, or to a neuron that conducts nerve impulses outward from the central nervous system to an effector.

Egg (1) A haploid female gamete (ovum), including its membranes and shell, if present. (2) A zygote and its surrounding membranes and shell. (3) Also used loosely in reference to a secondary oocyte released from an ovary.

Endergonic Pertaining to an energy-absorbing chemical reaction, in which the quantity of potential energy is increased.

Endocrine gland A ductless gland that secretes a hormone into blood or hemolymph; the hormone has an effect elsewhere in the body.

Endocytosis Ingestion of fluid or particulate material by a cell. *See also* Phagocytosis and Pinocytosis.

Endoderm The inner cell layer of the two primary germ layers, which develops into the lining of the gut and gut-derived organs.

Endoskeleton A system of support consisting of rigid spicules, fibrous connective tissue, cartilage, bone, or bonelike tissue contained within an animal's body; characteristic of sponges, echinoderms, and chordates.

Endotherm An animal whose body temperature is maintained by its own internal homeostatic mechanisms; body heat is derived from metabolism and maintained as needed by body insulation (fur, feathers, or fat deposits).

Enterocoel (1) A body cavity formed by an outpocketing from the archenteron of the gastrula. (2) An animal with enterocoelous body cavity formation; enterocoelomate.

Environmental resistance Limitation of population growth by environmental forces (e.g., harsh weather, limited food, predation).

Enzyme Any of a class of proteins that catalyze chemical reactions; an organic catalyst that temporarily combines with a substrate and increases the rate of a reaction involving the substrate by decreasing the energy required to initiate the reaction.

Epiboly Gastrulation by the spreading of an outer sheet of cells around other cells.

Epidermis The outermost layer(s) of cells of an integument.

Epinephrine A hormone secreted by the adrenal medulla, and by neurons of the sympathetic nervous system; prepares the body for emergency "fight or flight" activity; $C_9H_{13}O_3N$; also called adrenaline.

Epithelium Tissue consisting of one or more layers of tightly associated cells forming a covering or lining; epithelial tissue.

Estivation *See* Aestivation.

Estrogen Any of a class of female steroid hormones that produce female secondary sex characteristics, elicit estrus, and, in mammals, prepare the female reproductive tract for implantation of the zygote; synthesized by the ovary, adrenal cortex, and testis (in males).

Estrus A period of sexual excitement and receptivity ("heat") of female mammals (except primates); estrus is part of the estrous cycle, which includes ovulation, growth and disintegration of a corpus luteum, and correlated changes in the uterine wall.

Eucoelomate An animal with a mesoderm-lined body cavity (coelom); coelomate.

Eukaryote A cell with a membrane-bounded nucleus and diverse membrane-bounded cytoplasmic organelles.

Euryhaline Able to tolerate a wide variation of dissolved salt concentration (salinity).

Eutely The condition whereby an adult animal body has a more or less constant number of cells or nuclei; results from cessation of mitosis in somatic organs during embryonic development.

Evolution Nonreversible changes in gene frequency in populations; the development of genetically determined adaptations in lineages of organisms; the derivation of new organisms from ancestral species; includes anagenesis and cladogenesis.

Excretion Disposal of metabolites (metabolic wastes).

Exergonic Pertaining to chemical reactions that yield free energy and heat; in exergonic reactions, there is a change from a condition of higher potential energy to one of lower potential energy.

Exoskeleton A system of support that forms the outside covering of an animal.

Eyespot A type of eye or photoreceptive cell organelle that lacks a lens and does not form images. *See also* Stigma.

Feedback mechanism A system in which control is mediated by the products, or output, of the system itself; an increasing amount of the system's product speeds (positive feedback) or slows (negative feedback) the activity rate of the system.

Fertilization Fusion of egg and sperm nuclei, forming a fusion nucleus within a new cell called a zygote, or fertilized egg.

Filter feeding Obtaining nutrients by filtering (straining or actively capturing) suspended particles or small planktonic organisms from water.

First law of thermodynamics A natural law concerning energy transformations; in any system or process, energy is neither created nor destroyed; although energy and matter can be interconverted, the total energy of a system and its environment remains constant.

Fission Asexual reproduction by division of an individual into equal or nearly equal halves (binary fission) or into many offspring (multiple fission).

Fitness A measure of the evolutionary success of an individual or phenotype, measured as the number of genes contributed to the gene pool of future generations relative to the number contributed by other individuals.

Flagellum A relatively long, whiplike locomotor cell organelle having the same internal structure as a cilium.

Food chain An idealized trophic (feeding) system composed of a linear sequence of predator and prey species; involves producers (chemosynthetic or photosynthetic organisms), herbivores, carnivores, omnivores, and decomposers.

Food web A complex trophic (feeding) system in a living community; all of the feeding relationships (combined food chains) involving passage of materials and energy in a living community.

Fossil The remains or impression of an organism that lived in a former geological age.

Fragmentation The breaking of an animal into two or more fragments; may result in asexual reproduction if followed by regeneration of the fragments. *See also* Regeneration.

Free energy Any form of energy that is available to do work.

Fusion nucleus The diploid nucleus resulting from fertilization, or fusion of an egg nucleus and a sperm nucleus.

Gamete A mature sex cell; a reproductive cell (sperm or egg) that can participate in fertilization to form a zygote.

Gamont A cell that fuses (conjugates) and exchanges nuclei with another cell of the same species, prior to reproduction.

Ganglion A distinct aggregate of nervous tissue consisting mainly of nerve cell bodies.

Gastrodermis Epithelial tissue lining the gastrovascular cavity of cnidarians and flatworms; derived from endoderm.

Gastrovascular cavity The cavity of an incomplete gut, as in cnidarians and flatworms; functions in digestion and circulation.

Gastrula An early embryo consisting of an outer layer of ectoderm and an inner layer of endoderm; usually has mesoderm proliferated from one or both of the other layers.

Gastrulation Formation of a gastrula from a blastula; involves proliferation of endoderm cells surrounded by ectoderm; a blastopore often marks the region from which endoderm proliferates.

Gene (1) A nucleotide sequence on DNA that carries the genetic code for a polypeptide, for an RNA molecule, or for allowing another gene to be transcribed. (2) A DNA nucleotide sequence that has a specific function. (3) A hereditary unit contained in DNA in a chromosome.

Gene flow Exchange of genes between two or more populations.

Gene pool All the genes in a biological population.

Genetic drift The change in gene frequency from one generation to another in a gene pool due to chance.

Genome (1) The haploid number of chromosomes of a diploid species. (2) All of the genes contained in the haploid number of chromosomes.

Genotype (1) The genetic makeup of an individual (total genotype). (2) The genetic makeup of an individual with respect to a specific trait or traits under consideration.

Genus A taxonomic category consisting of a group of related species; a taxonomic category above species and below family in the classification hierarchy.

Germinal epithelium The outermost layer of cells in an ovary and testis; proliferates mitotically, producing cells that may give rise to gametes.

Gill A gas exchange organ that extracts dissolved oxygen from water; an extension of the body surface that functions in aquatic gas exchange.

Gizzard A muscular digestive organ that functions in mechanical grinding and softening of food material; usually contains abrasive plates or inorganic abrasives from the outside environment.

Gluconeogenesis The synthesis of carbohydrates from amino acids, fatty acids, or other noncarbohydrates.

Glycocalyx A layer of glycoproteins covering the outside surface of the plasmalemma of a cell.

Glycolysis The enzymatic, anaerobic, energy-releasing breakdown of glucose and other simple sugars to pyruvate.

Glycoprotein Any of a class of large proteins covalently bonded to carbohydrates; glycoproteins include many cell surface molecules, antibodies, mucus, albumins, and proteoglycans.

GLOSSARY F–G

Gonad A sex organ (ovary or testis) that produces gametes by meiosis and synthesizes and may secrete female and male sex hormones.

Gradualism A classical hypothesis stating that evolution proceeds in slow, gradual stages, rather than abruptly. *Compare* Punctuated equilibrium.

Guanine One of the purine bases in DNA and RNA; $C_5H_5N_5O$; a nitrogenous waste product of many terrestrial animals, especially arachnids.

Habitat The place in the environment in which an organism lives.

Habituation A simple type of learning involving the loss of response(s) to irrelevant stimuli.

Haploid number (1n) (1) A single set of chromosomes and genes; *see also* Genome and Diploid. (2) The number of chromosomes in a gamete in a species whose somatic cells are diploid.

Hemocoel The collective body spaces (sinuses and lacunae) in an animal with an open circulatory system; the hemocoel is bathed with hemolymph.

Hemolymph Circulatory fluid in an animal with an open circulatory system.

Hepatic portal system A complex of veins conveying blood from capillaries in the digestive tract and spleen to capillaries in the liver; found in all vertebrates.

Herbivore An animal that eats plants primarily or exclusively.

Hermaphroditism Having both female and male reproductive systems in a single individual.

Heterotherm An animal whose body temperature fluctuates markedly; in some heterotherms, body temperature may at times be self-regulated and relatively stable, while at other times it fluctuates with the environmental temperature.

Heterotroph An organism that is unable to synthesize energy-rich organic molecules from inorganic substances and that must obtain them by eating other organisms or products of other organisms.

Heterozygous Pertaining to an organism or cell that has two or more different alleles at the same locus on homologous chromosomes.

Hibernation Dormancy, or torpor, during cold seasons; usually involves a lowering of the metabolic rate and, in mammals and birds, a marked decrease in body temperature; winter dormancy.

Homeostasis The maintenance of dynamic, steady-state balance in living systems.

Homeotherm An animal (bird or mammal) that maintains its body temperature at a constant, high level and that may remain equally active in cold and warm surroundings.

Home range The total area over which an individual or social group of animals moves in all its activities; in territorial species, the home range contains the territory, or that part of the home range that is defended.

Homology (1) Evolutionary homology: Resemblance between two or more body structures due to derivation from a common structure in an ancestor. (2) Developmental homology: Resemblance due to derivation from a common, undifferentiated embryonic source. (3) *See* Serial homology.

Homozygous Pertaining to an organism or cell that has identical alleles at a particular locus on homologous chromosomes.

Hormone A regulatory chemical synthesized and secreted by an endocrine (ductless) gland, and which is conveyed by circulatory fluid to target cells, tissues, or organs, whose function it regulates. *See* Neurohormone.

Humoral immunity A component of an animal's defense mechanisms involving protective cell secretions (e.g., antibodies, or immunoglobulins, in vertebrates) dissolved in circulatory or other body fluids.

Hybrid (1) An offspring of parents of two different species, varieties, races, or genera. (2) An offspring of parents differing in one or more heritable traits.

Hydraulic movement Movement resulting from manipulation of fluid in a hydrostatic skeleton.

Hydrogen bond A weak chemical bond resulting from electrostatic attraction between two polar molecules; the attraction between a negatively charged portion of one molecule and a positively charged portion of another molecule.

Hydrolysis The breakdown, or catabolism, of a chemical substance by the enzymatic addition of water.

Hydrostatic skeleton A supportive system consisting of fluid contained in a rigid body wall or organ; muscles in the rigid body part may manipulate the fluid, providing movement and locomotion.

Hyperosmotic Pertaining to animals whose body fluids have a higher solute concentration, and consequently a higher osmotic pressure, than the outside environment.

Hypoosmotic Pertaining to animals whose body fluids have a lower solute concentration, and consequently a lower osmotic pressure, than the outside environment.

Hypothesis A tentative, educated guess, or testable supposition, to explain a phenomenon.

I band The isotropic, or light, band in striated muscle, consisting mainly of actin filaments.

Ileum Posteriormost region of the small intestine in mammals.

Immunity Resistance or nonsusceptibility to infection.

Immunoglobulin Any of a class of vertebrate plasma proteins synthesized and secreted by lymphocytes (B cells) in response to specific antigens; an antibody.

Imprinting A simple form of learning; formation of a strong bond by a hatchling or newborn animal with the first large body (usually a parent) that it sees or contacts.

Incomplete gut A digestive tract that lacks an anus.

Indirect development Animal development involving a larval stage that is different in structure and often habitat from that of the adult.

Induction Influence of one tissue, the organizer, on the embryonic differentiation of another tissue; may be caused by inducer chemicals produced by cells of organizer tissues.

Inflammation The nonspecific reaction of tissue to infection or injury; usually involves swelling, reddening, increase in temperature, and local pain, all caused by histamines and other tissue irritants released by the injured cells.

Ingression The movement of outer cells into the blastocoel during embryonic development.

Innate behavior Behavior that is genetically determined (not learned).

Insertion The more movable attachment site of a muscle. *See* Origin.

Instar A stage (especially in arthropods and nematodes) between molts.

Insulin A hormone (polypeptide) secreted by the pancreatic islets; has many physiological effects, including stimulating the conversion of glucose to glycogen.

Integration Incorporation of parts into a whole; in nervous systems, interneurons integrate inputs from sensory neurons, producing an adaptive response to environmental stimuli.

Integument An external covering or skin.

Intercellular fluid The fluid situated among the cells of an animal's body; interstitial fluid.

Interneuron A neuron (nerve cell) that links sensory and motor neurons; an association neuron.

Interphase All nonmitotic stages of the cell cycle.

Interstitial Pertaining to inorganic or organic material, or cells situated among the main structural elements of a tissue, organ, or inorganic substance; e.g., interstitial cells of the testis, interstitial water in soil or sand.

Intrinsic rate of increase (1) The growth rate (realized intrinsic rate of increase), symbolized by r, of a population; r equals the birthrate, or natality (symbolized by n), minus the death rate, or mortality (symbolized by m). (2) The maximum (theoretical) rate of population growth. *See* Biotic potential.

Invagination A folding or pushing inward of a cell layer or tissue; formation of a cavity by infolding.

Inversion Reversal of a chromosome segment, causing a chromosome mutation (change in gene order).

Involution The inward movement and expansion of outer cells, or an outer cell layer, under other cells.

Islet of Langerhans A cluster of endocrine cells within the pancreatic tissues that secrete digestive enzymes; α cells of the islets secrete glucagon; β cells secrete insulin.

Isolecithal Pertaining to an egg or zygote that has a small amount of yolk evenly distributed throughout the cytoplasm.

Isomer One of two or more compounds or molecules composed of the same number and kind of atoms, but with different structural formulas.

Isosmotic Having equal osmotic pressure; pertains to an animal whose body fluids have virtually the same osmotic pressure as the aquatic environment.

Jejunum The middle portion of the small intestine of mammals, located between the duodenum and the ileum.

Juvenile An adultlike, sexually immature individual.

Keratin Tough, sclerotized protein common in hardened regions (e.g., hooves, claws, hair, feathers, scales) of the tetrapod integument.

Kinesis A random, nondirected movement in response to a stimulus.

Kinetic energy Energy of motion.

Kinetosome *See* Basal body.

Kin selection Natural selection favoring closely related individuals.

Krill The common name of certain planktonic crustaceans, especially those of the genus *Euphausia;* planktonic crustaceans eaten by baleen (toothless) whales.

***K* selection** Natural selection favoring species that tend to have relatively stable populations and whose population growth rate follows a logistic (restricted) form. *Compare r* selection.

Labyrinthodont Literally, "maze-teeth," a group of extinct Paleozoic amphibians that probably gave rise to most other amphibians.

Lacteal (1) A lymphatic vessel in an intestinal villus. (2) Pertaining to milk.

Lacuna A small cavity in bone or cartilage containing an osteocyte or a chondrocyte; a sinus.

Lagena An evagination of the sacculus of the vertebrate inner ear; becomes the cochlea of birds and mammals.

Lamella A thin, platelike structure, often one of many layers of thin tissue forming gas exchange surfaces in gills or concentric rings in bone.

Lancelet An invertebrate chordate of the genus *Branchiostoma* (formerly *Amphioxus*).

Larva An immature stage that is markedly different from the adult.

Larynx An expansion at the junction between the pharynx and the trachea in tetrapods; the voice box of amphibians, reptiles, and mammals.

Lateral   Pertaining to the right or left side of an animal.

Lateral line system   *See* Acousticolateralis system.

Lek   An area where a group of males and females of a species assembles for courtship and mating.

Lentic ecosystem   A standing body of fresh water; e.g., a lake, pond, or marsh.

Leuconoid   Pertaining to the multibranched canal system of most sponges.

Leukocyte   A white blood cell, usually amoeboid.

Life cycle   The complete sequence of life stages undergone by a species; e.g., from adult to adult, or from egg to egg.

Ligament   A band of dense connective tissue that connects bones.

Linkage   The location of gene loci on the same chromosome; linked genes (linkage groups) do not assort independently.

Lipid   A general term for fats and fatlike compounds (e.g., oils and waxes).

Littoral zone   The shallow water zone of a lake or ocean.

Lobe-finned fishes   A group of bony fishes whose fin bases are thick and contain muscles and bones similar to those of tetrapods; lobefins.

Locus   In genetics, a specific site of a gene on a chromosome.

Logistic (restricted) growth equation   A mathematical model of population growth that is restricted by environmental resistance; sigmoidal growth.

Lophophore   A hollow crown or arm, bearing ciliated tentacles; a feeding structure of lophophorate animals and certain invertebrate chordates.

Lorica   A loose-fitting protective case formed of secreted materials or environmental detritus.

Lotic ecosystem   A flowing water system.

Lumen   The cavity or space within an organ.

Lung   An internal sac containing moist surfaces for gas exchange.

Lymph   Fluid in the vertebrate lymphatic system; similar to tissue fluid, but usually with fewer nutrients and less dissolved oxygen.

Lymphatic system   An accessory drainage system of finely branched vessels that collect lymph from tissue fluid and convey it to large veins of the blood vascular system in vertebrates.

Lymphocyte   A white blood cell that functions in the vertebrate immune response; B cell or T cell.

Lysosome   A membrane-bound cell organelle containing hydrolytic enzymes.

Macroevolution   The evolution of taxa (e.g., genera, classes, phyla) above the species level; the evolution of phylogenetic lineages.

Macrophage   A phagocytic cell in body tissues other than the blood.

Mantle   A thin, sheetlike extension of the body wall in molluscs and brachiopods; usually produces, and is covered by, a shell; gills or a lung develop in the mantle cavity between the mantle and the main body surface in many molluscs.

Mantle cavity   The space between the mantle and the body wall of molluscs and brachiopods.

Marsupium   A pouch or sac in which young are nurtured; an external pouch in many marsupial mammals; an expanded gill sac in certain freshwater mussels.

Matrix   Noncellular organic or inorganic substance in which living cells are embedded.

Mechanoreceptor   A sensory cell, tissue, or organ that detects mechanical stimuli (e.g., sound, touch, gravity, or pressure).

Meiosis   In sexually reproducing species, nuclear division that reduces the chromosome number by half; also called reduction division.

Melanin   A class of dark brown and blackish pigments in animals; polymers of dopa or tyrosine.

Membrane potential   The difference in electrical potential between two sides of a membrane.

Menstrual cycle   Hormonally regulated cyclic changes in the female reproductive system in primates.

Merozoite   A uninucleate life cycle stage in certain protozoa (apicomplexans) produced by multiple fission, or schizogony.

Mesenchyme   Connective tissue composed of undifferentiated cells (often many amoebocytes) in a fibrous or gelatinous matrix; embryonic connective tissue; the middle layer (between the epidermis and gastrodermis) of cnidarians.

Mesentery   The thin connective tissue "wrapping" that surrounds organs and may suspend them in the body cavity from the body wall; consists of two layers of peritoneum in many eucoelomates.

Mesoderm   The middle germ layer, which lies between the ectoderm and the endoderm; considered a secondary germ layer because it develops from either ectoderm or endoderm or both.

Mesoglea   The fibrous or gelatinous matrix of mesenchyme; the middle (noncellular) body layer (between the epidermis and gastrodermis) of some cnidarians.

Mesohyl   The middle body region of sponges, containing skeletal elements and many amoeboid cells.

Mesolecithal   Pertaining to an egg or zygote with a moderate amount of yolk concentrated in the vegetal hemisphere.

Mesonephros   A tubular kidney that develops in the middle portion of the coelomic cavity; the functional kidney of embryonic amniotes.

Metabolism All of the chemical and physical processes involved in catabolism and anabolism in an organism.

Metachronal rhythm Wavelike, coordinated movement; usually used in reference to the coordinated beating of a large number of cilia or flagella.

Metamerism Repetition of body parts as segments, or metameres, along the longitudinal axis; also called segmentation; seen in annelids, arthropods, and chordates.

Metamorphosis Marked alteration in body form from one life cycle stage to another; transformation of a larval stage to an adult; e.g., the process of change from a tadpole to a frog.

Metanephridium A tubular excretory/osmoregulatory organ that filters fluid from the coelom via a ciliated nephrostome and passes urine to the environment through a nephridiopore; found in molluscs, annelids, and many annelid-like worms.

Metaphase The stage in mitosis or meiosis when the chromosomes are situated at the equatorial plane of the cell; precedes separation of chromatids or homologous chromosomes.

Metazoan An animal; a multicellular eukaryotic heterotroph.

Microevolution Evolution at the species level or below; changes in gene frequencies in biological populations.

Microfilament A minute (diameter less than 100 Å) solid rod forming part of the cytoskeleton in the cytoplasm of cells.

Microtubule A minute hollow tube forming part of the cytoskeleton of cells; formed of the protein tubulin.

Microvillus A minute, fingerlike projection that increases the surface area of a cell.

Mictic Pertaining to a haploid egg that can be fertilized, or to a female that produces such an egg; used in reference to parthenogenetic rotifers.

Mimicry Resemblance of one organism to another organism or to a nonliving object. *See* Batesian mimicry and Müllerian mimicry.

Mitochondrion An organelle in eukaryote cells and the site of most oxidative phosphorylation (Krebs cycle and electron transport system) in cellular respiration.

Mitosis Nuclear division involving chromosome division but no change in chromosome number.

Mobility Movement from place to place.

Molecule A covalently bonded chemical entity, either a compound or an entity formed of two or more atoms of a single element.

Molting Shedding of the outer dead layer (cuticle) of the integument; shedding of feathers by a bird; ecdysis.

Molt-inhibiting hormone A hormone secreted by the X organ in the eyestalk of crustaceans, and which prevents molting (ecdysis).

Monoecious Pertaining to species that have both male and female sex organs in the same individual.

Monophyletic Pertaining to a taxon that has been derived from a single ancestral species.

Morphogenesis Development, or origin, of form in an animal, resulting from differentiation of cells and tissues.

Motility General body movement; movement of an animal in place or from place to place. *See* Mobility.

Motor neuron A nerve cell (neuron) that innervates effectors (muscles, chromatophores, glands, or cnidocytes).

Mucus A class of glycoproteins that serve as lubricants on cells, tissues, organs, and body surfaces; synthesized and secreted by unicellular or multicellular mucous glands.

Müllerian mimicry Resemblance between two or more species that are unpalatable to predators.

Multiple fission *See* Schizogony.

Mutation Stable inheritable change in genetic material (genes or chromosomes).

Mutualism A type of symbiosis involving the mutual exploitation and mutual benefit of two or more species; may be facultative or obligatory for either or both species.

Myelin A fatty substance formed of layers of membranes; e.g., the myelin sheath, formed of layers of Schwann cell membranes, surrounding certain axons.

Myofibril The contractile fibrils in a muscle cell, composed of contractile units called sarcomeres.

Myofilament A thick (myosin) or thin (actin) filament that is part of the contractile units (sarcomeres) of myofibrils in a muscle cell.

Myogenic Generated by muscle contractions; heart and wing muscles, whose contractions are generated or sustained by muscle contractions, are myogenic.

Myomere A muscle block, or segment, in an adult animal.

Myoneme A contractile thread, or fibril, in protozoa (e.g., *Stentor* and *Vorticella*).

Myotome A block, or segment, of embryonic tissue that develops into a myomere.

Natality The birthrate, usually expressed as the number of births per 1000 individuals in a population.

Natural selection Differential (nonrandom) reproduction; differential selection of the most fit individuals (those that pass the greatest number of genes to ensuing generations).

Nauplius The early larval stage of many crustaceans, characterized by three pairs of appendages.

Nekton Strong-swimming aquatic organisms that are more or less independent of water currents and wave action.

Nematocyst In cnidarians, a microscopic stinging organelle that can be discharged in defense or for capturing food; a type of eversible organelle, or cnida.

Neoteny Reproduction by a larvalike adult; retention of larvalike features by adults.

Nephridium A tubular excretory-osmoregulatory organ; e.g., the protonephridia of flatworms and rotifers and the metanephridia of molluscs, annelids, and annelid-like worms.

Nephron A structural and functional unit of the vertebrate kidney, consisting of a renal corpuscle and renal tubule.

Nephrostome A ciliated opening into the body cavity from the tubule of a nephridium.

Nerve A group, or bundle, of nerve fibers held together by connective tissue.

Nerve impulse A rapid, self-propagating wave of depolarization (change in membrane potential) in a nerve cell.

Nerve net A diffuse, noncentralized network of nerve cells.

Nested set A series of groups within groups, as in the hierarchical arrangement of taxa in Linnaean classification.

Neurilemma The outer covering of a nerve cell axon, consisting of the outer membrane of Schwann cells and their glycoprotein coat.

Neurite The general name for a nerve fiber; an axon or dendrite.

Neurogenic Generated by nerve impulses.

Neuroglia The supportive elements of the nervous system; include Schwann cells, astrocytes, microglia, and ependymal cells; also called glial cells.

Neurohormone A regulatory chemical, or hormone, secreted by nerve cells.

Neuromast A nipplelike sensory projection, or papilla, often consisting of a cluster of sensory hair cells covered by a gelatinous cupula.

Neuron A nerve cell.

Neurosecretion Secretion of neurohormones.

Neurosecretory hormone *See* Neurohormone.

Neurotransmitter A chemical secreted by a presynaptic neuron that stimulates a postsynaptic neuron at a synapse.

Niche The total role of a species in relation to its environment; includes the species habitat.

Nitrogen fixation Conversion of atmospheric nitrogen gas (N_2) to nitrites or nitrates in the soil.

Node of Ranvier A gap on a myelinated nerve fiber between Schwann cells, and thus a point that lacks myelin; saltatory conduction of nerve impulses occurs from one such node to another.

Noradrenaline *See* Norepinephrine.

Norepinephrine A neurotransmitter and neurohormone, secreted by certain cells in the vertebrate central nervous system, by the adrenal medulla, and by neurons of the sympathetic nervous system.

Notochord A cartilage-like, dorsal stiffening rod formed of fluid-filled cells in chordates; a chordate diagnostic feature.

Nucleic acid DNA or RNA; a polymer of deoxyribose or ribose nucleotides; DNA occurs in cell nuclei and mitochondria; RNA occurs in nuclei and cytoplasm.

Nucleolus An intranuclear organelle consisting of RNA and protein; functions in synthesis and storage of ribosomes.

Ocellus A type of eye that has a lens, but little focusing ability; may also refer to an eye without a lens.

Olfaction The sense of smell; detection of chemical odors.

Olfactory (nasal) epithelium A layer of chemosensory (odor-sensitive) cells in the nasal region.

Ommatidium A functional unit of a compound eye, characterized by lenses and retinal cells.

Omnivore An organism that eats both plant and animal matter.

Oncotic pressure The osmotic pressure of blood plasma; results mainly from albumen, a protein dissolved in plasma.

Ontogeny The development (life history) of an individual.

Oocyte A female germ cell that may divide meiotically to yield an egg; a developing egg.

Oogenesis The development of an egg, or ovum, from an undifferentiated germ cell; usually includes meiotic division.

Oogonium An ovarian cell that develops by mitosis into oocytes.

Open (lacunar) circulatory system A vascular, or internal transport, system that lacks capillaries.

Operculum (1) A covering or lid. (2) The gill covering in bony fishes. (3) A hardened plate that seals the shell aperture of some snails.

Opisthonephros The tubular kidney of most adult fishes and amphibians; develops from mesonephric kidney tissue in the middle and posterior regions of the coelomic cavity.

Opisthosoma The posterior, or abdominal, region of the body of pogonophorans and most arachnids.

Organ An integrated aggregate of several types of tissues specialized to perform a specific function or several related functions.

Organ system An integrated group of organs operating as a unit in performing specific functions.

Organelle A structurally and functionally specialized structure within a cell.

Origin (1) Evolutionary (phylogenetic) ancestry. (2) A muscle attachment that is relatively immobile and usually relatively close to the trunk of the body. *Compare* Insertion.

Osmoconformer An organism whose body (interstitial) fluids have the same or very similar osmotic pressure as that of its aquatic environment; a marine organism that does not utilize energy in osmoregulation.

Osmoregulation Active (energy-consuming) maintenance of internal fluid homeostasis; active maintenance of the osmotic pressure of body (interstitial) fluids.

Osmosis Diffusion of solvent molecules through a membrane that is differentially permeable (permeable to the solvent, but impermeable to most solutes); in biological systems, water is the universal solvent.

Osmotic pressure (1) Force (pressure) that a solute exerts on a membrane through which the solute cannot pass. (2) Pressure that must be applied to prevent osmosis between two solutions separated by a differentially permeable membrane. (3) Pressure due to osmotic flow between solutions separated by a differentially permeable membrane.

Ossicle A small bone, or bonelike structure, such as those that transmit sound vibrations in the middle ear of vertebrates or those composing the endoskeleton of many echinoderms.

Osteon A structural and functional unit of vertebrate bone; a Haversian system.

Ostium An opening from or to an internal tube or chamber.

Out-group In cladistic analyses of phylogeny, a taxon that is not closely related to the taxa actually being studied; an out-group is used to determine which features of taxa being studied are unique to these taxa and which ones are shared by more distantly related species.

Ovary A female gonad, which produces female gametes and, in some animals, female hormones.

Oviduct A tube or duct through which eggs pass from the ovary to the outside environment or to a uterus.

Oviparity Production of eggs that develop and hatch outside the female's body; the eggs may be fertilized inside or outside the female.

Ovipositor An egg-laying organ at the posterior end of the abdomen in many female insects.

Ovoviviparity Production of yolk-laden eggs that are fertilized within the female body and that develop and hatch within the female.

Ovulation Release of an egg, several eggs, or oocytes from an ovary.

Ovum A mature, unfertilized female gamete; a haploid female sex cell.

Oxidation-reduction Loss of electrons (or hydrogen atoms) from one molecule, and the simultaneous gain of electrons (or hydrogen atoms) by another molecule; a molecule that loses an electron is oxidized; one that gains an electron is reduced; also called redox.

Paedogenesis Reproduction by larval stages, often by parthenogenesis.

Pallium A mantle, such as that of molluscs.

Pangaea The world continent, or supercontinent, consisting of all of the earth's major land masses fused together.

Papilla A small, nipplelike projection.

Parallel evolution Evolution of similar adaptations in two closely related lineages; also called parallelism.

Paraphyletic Pertaining to a taxonomic group consisting of an ancestral group and some of its descendants.

Parapodium One of the paired, segmentally arranged, lateral appendages in polychaete annelids.

Parasitism A type of symbiosis in which one species, the parasite, lives at the expense of another species, the host; the parasite is physiologically dependent on the host.

Parathormone A polypeptide hormone secreted by the parathyroid glands, and which regulates absorption, excretion, and deposition of calcium; also called parathyroid hormone.

Parthenogenesis Reproduction without fertilization; development and maturation of an unfertilized (either haploid or diploid) egg; unisexual reproduction.

Partial pressure Pressure exerted by each gas in a mixture of gases; the part of a total gas pressure produced by a specific gas in a mixture.

Peck order A social hierarchy of dominance and recessiveness; usually refers to social groups of chickens or other birds.

Pectoral Pertaining to the chest or breast; in vertebrates, pertains to the anterior (shoulder) limbs and their supportive girdle.

Pedicel (1) A stalk that supports an animal body or its parts. (2) A small, footlike structure. (3) The narrow waist connecting the thorax and abdomen of ants, wasps, and bees.

Pedipalp One of the second appendages variously specialized for grasping, stimulus reception, chewing, etc., situated posterior to the mouth in chelicerates (e.g., horseshoe crabs, spiders, ticks, and mites).

Pelagic In marine environments, pertaining to the zone of open water from the surface to the bottom; pertaining to waters of the open sea, excluding waters of the littoral (or intertidal) zone.

Pellicle The semirigid or rigid surface covering of many protozoa.

Pelvic   Pertaining to the hip, or pelvic, region of the vertebrate body; the pelvic girdle supports the hindlimbs in vertebrates.

Penis   The male copulatory, or intromittent, organ.

Pentadactyl   Pertaining to a limb with five digits.

Peptide bond   A covalent bond linking two amino acid residues; chemical linkage between an amino group and a carboxyl group.

Pericardial cavity   A coelomic chamber that surrounds the heart.

Pericardial sinus   A hemocoelic space surrounding the heart of many animals with open circulatory systems.

Peristalsis   Rhythmic waves of muscle contraction and relaxation passing from one end to the other of a tubular organ (e.g., the gut) or elongate animal (e.g., an earthworm); functions in moving materials within the tubular organ, or in locomotion of the whole animal.

Peritoneum   The mesodermally derived connective tissue membrane that lines the coelomic cavity in eucoelomate animals.

pH   A measure of the acidity and basicity of solutions; the negative logarithm of the hydrogen ion concentration, ranging from 1 (most acidic) to 14 (most basic); pH 7 is neutral.

Phagocytosis   Ingestion (engulfment) of particulate matter (nutrients or foreign bodies) by cells; a type of endocytosis.

Pharynx   The anterior, usually muscular, region of the gut between the oral cavity and the esophagus; contains the gill structures in vertebrates.

Phenotype   The actual appearance of an individual; an individual's observable features (i.e., body form and behavior), resulting from the interaction between the genotype and the external environment.

Pheromone   A chemical signal that is released into the environment by one animal and that influences the behavior and/or physiology of another individual of the same species.

Phoresis   A type of symbiosis in which a member of one species, the phoront, obtains transport on an individual of another species, the host; the host is unaffected; the phoront may or may not benefit.

Phototrophic   Producing energy-containing organic food molecules by photosynthesis.

Phyletic evolution   Direct speciation, that is, the evolution of one species from another without branching, or divergence; macroevolution along linear paths; also called anagenesis.

Phylogeny   Evolutionary history; macroevolution of major taxonomic lineages.

Phytoplankton   Floating and suspended algae in aquatic environments.

Pinacoderm   The outer cell layer of many sponges; the pinacoderm is synctial or cellular (composed of pinacocytes).

Pineal body   A small outgrowth on the dorsal surface of the diencephalon of the vertebrate brain; secretes the hormone melatonin, which influences control of photoperiodic sexual cycles by the hypothalamus.

Pinocytosis   Ingestion of fluid by cells; endocytosis of fluids.

Placenta   A composite organ formed by a developing embryo and the uterine tissues of its mother; provides oxygen and nutrients to the embryo, and removes wastes from the embryo; also secretes hormones that regulate pregnancy.

Placoid scale   A type of scale, derived from the dermis of the skin and homologous with vertebrate teeth, found in many cartilaginous fishes; composed of a dentine base and toothlike spine covered with enamel.

Plankton   Organisms that float or are suspended in aquatic environments and that are moved about by waves and currents. *See* Phytoplankton and Zooplankton.

Planula   A free-swimming, ciliated larval stage that develops following gastrulation in many cnidarians; composed of two body regions (an outer region formed from ectoderm and an inner endodermal region).

Plasma   The liquid portion of circulatory fluid (blood, lymph, or hemolymph) in which formed elements (cells and cell parts) are suspended; vertebrate plasma consists of serum plus dissolved fibrinogen.

Plasmalemma   The plasma membrane.

Plasmodium   (1) A growing (usually multinucleate) cell in the life cycle of many protozoa. (2) The genus of apicomplexan parasites that causes malaria in vertebrates.

Pleopod   One of the paired, biramous swimming appendages on the abdomen of many crustaceans; also called a swimmeret.

Pleura   A membrane that lines one of the coelomic pleural cavities surrounding a lung in vertebrates, and which also covers the lung surface. (Also, the pleural of *pleuron*.)

Pleuron   Lateral plate of a segment of the exoskeleton of many arthropods.

Polyembryony   Production of two or more embryos from a single zygote.

Polyp   A single, sessile individual of many cnidarians; usually reproduces asexually.

Polyphyletic   Pertaining to a taxon derived from more than one ancestral group; pertaining to a taxon that does not include a species ancestral to all members of the taxon.

Polyploid   Having three or more sets, or a multiple greater than two, of the haploid number of chromosomes.

Population A group of interacting (among sexually reproducing species, interbreeding) individuals of a species; a biological population; members of a biological population have a common gene pool and interact because they inhabit the same geographical range.

Portal system A system of vessels that conveys blood from one capillary bed to another.

Posterior Pertaining to the rear, or caudal, region of a bilaterally symmetrical animal.

Potential energy Stored energy with the capacity (that can be released) to do work; e.g., the energy in chemical bonds, or the energy resulting from the position of an object above a surface.

Precocial Covered with feathers (down) or hair and able to move actively about soon after birth (e.g., in horses and sheep) or hatching (e.g., in ducks and geese); opposite of altricial.

Predation The capture and killing of living organisms (animals or plants) for food.

Primary consumer *See* Herbivore.

Primary germ layer Either of the two germ layers (ectoderm or endoderm) that develop first during gastrulation; primary germ layers are the source of all cell layers and tissues of an adult animal. *See also* Mesoderm.

Primitive Pertaining to features or organisms that resemble the ancestral condition; ancestral; contrasts with derived.

Productivity The rate of accumulation (synthesis or assimilation) of energy-containing organic matter; primary productivity is the rate of sunlight energy fixed by plants in photosynthesis; secondary productivity is the rate of energy storage by heterotrophs.

Prokaryote A cell that lacks a nucleus; the genetic material is not bounded by a nuclear membrane; bacteria (including blue-green algae) are prokaryotes.

Pronephros A tubular kidney that forms first and occupies the anterior part of the coelomic cavity in vertebrate embryos; the functional kidney in embryonic fishes and amphibians; forms the head kidney, a blood-forming organ, in adult hagfishes; a pronephric tubule has a nephrostome opening into the coelom.

Prophase The first stage of mitosis and meiosis, during which the doubled chromosomes become visible and the nuclear membrane and nucleolus disappear.

Proprioceptor A sensory receptor that detects body motion and position; found on body surfaces, in muscles, tendons, and in the vertebrate inner ear.

Prosoma Anterior region of the body of an invertebrate, especially an arthropod.

Prostaglandin Any of a group of fatty acid-derived regulatory chemicals that effect changes in the tissues that produce them.

Protandry (1) Maturation of the testes and production of sperm before ovarian maturation and egg production in a hermaphroditic animal. (2) Development of sexually mature males before the appearance of mature females.

Proteoglycans Glycoproteins that consist of a protein core from which sugar molecules project.

Protogyny (1) In a hermaphroditic animal, the maturation of ovaries and the production of eggs prior to the production of sperm. (2) Development of sexually mature females prior to the appearance of mature males.

Protonephridium The flame cell and tubular osmoregulatory-excretory structure of certain invertebrates (e.g., flatworms, rotifers, and lancelets).

Protostome A coelomate animal that is a member of a phylum containing species with schizocoelous body cavity formation, a mouth derived from the blastopore or region of the blastopore, spiral cleavage, and a trochophore larva; e.g., molluscs, annelids, arthropods, and several related groups.

Pseudocoel A body cavity that develops from the blastocoel and that is not lined by mesodermal peritoneum; a pseudocoelom.

Pseudocoelomate (1) An invertebrate with a pseudocoel; e.g., nematodes, rotifers, and acanthocephalans. (2) Pertaining to the condition of having a pseudocoel.

Pseudopodium A temporary cytoplasmic extension, functioning in locomotion or phagocytosis in an amoeboid cell or amoebocyte; types include lobopodia, filopodia, reticulopodia, and axopodia.

Pulmonary Pertaining to the lungs (as in the pulmonary artery and vein).

Punctuated equilibrium The hypothesis that periods of speciation are relatively short and are followed by long periods of stasis (little or no evolutionary change); punctuationism. *Compare* Gradualism.

Pupa The immature stage between the larva stage and the adult in insects with complete metamorphosis; nonfeeding and immobile in most species.

Purine Any of a class of nitrogen-containing chemicals, $C_5H_4N_4$ (including two rings of nitrogen and carbon atoms), from which purine bases (e.g., adenine, guanine, uric acid) are derived.

Pyrimidine Any of a class of nitrogen-containing chemicals, $C_4H_4N_2$ (including a single ring of nitrogen and carbon atoms), from which pyrimidine bases (e.g., cytosine, thymine, uracil) are derived.

Q_{10} Pertaining to the increase in the rate of oxygen consumption (and other metabolic rates affected by temperature) by an animal, correlated with an increase in temperature; in general, an increase of 10°C causes the oxygen consumption rate to increase two- to threefold, indicated by a Q_{10} value of 2 or 3.

GLOSSARY P–Q

Radial cleavage Cleavage that produces tiers of blastomeres, each lying directly above the one beneath it; cleavage planes are at right angles to one another; characteristic of deuterostomes.

Radial symmetry The arrangement of body parts such that planes passing through the central longitudinal axis (oral-aboral axis) divide the body into equal, mirror-image halves.

Radula A chitinous, toothed, tonguelike structure in the mouth cavity of many molluscs (e.g., many snails, slugs, chitons, and squids) and used for rasping food and conveying it into the gut.

Receptor A sensory cell, tissue, or organ that detects internal or external stimuli.

Recombinant DNA A DNA molecule derived from the DNA of two or more species.

Recombination The production of new combinations of genes in offspring (different from those in the parents); occurs as a result of crossing over, which produces different chromosomes, and also as a result of new associations of chromosomes, which in turn result from the union (fertilization) of gametes from two parents.

Reflex An automatic motor response to a stimulus, mediated by a reflex arc, consisting of a receptor, sensory (afferent) neuron, usually one or more interneurons, a motor neuron, and an effector (usually one or more muscles).

Refractory period The short period following a response to a stimulus by a neuron or muscle cell when the neuron or cell will not respond to additional stimuli.

Regeneration Replacement of lost or injured body parts; regrowth of an individual following disruption.

Releaser A stimulus, or sensory cue, that initiates an innate behavior pattern.

Renal Pertaining to the kidney (as in renal artery and renal portal system).

Rennin An enzyme that coagulates and helps digest milk.

Resolving power The capacity of an eye, microscope, or other optical device to separate images of adjacent objects.

Respiration (1) The catabolic (breakdown) processes involved in energy release from organic molecules. (2) Gas exchange between an organism and its environment.

Resting potential The electrical potential of a cell membrane at rest (unstimulated).

Rete A network of nerve fibers, or a network of fine blood vessels in which venous blood usually flows in a direction opposite that of arterial blood.

Reticuloendothelial system (RES) In vertebrates, the system of phagocytic cells in blood and fixed in tissues.

Retina The part of an eye containing light-sensitive cells and fibers of neurons carrying photosensory stimuli to the central nervous system.

Rhabdite A rodlike organelle in certain cells in the epidermis of turbellarian flatworms; may form mucus and have a protective function.

Ribosome An organelle active in protein synthesis in eukaryotic cells; consists of ribosomal RNA and protein; occurs attached to membranes of the endoplasmic reticulum or free in the cytoplasm.

***r* selection** Natural selection favoring species that colonize and reproduce rapidly. *Compare K* selection.

Salinity (‰) The total amount of dissolved solids in a solution, expressed in parts per thousand.

Salt A chemical compound formed from an acid by the replacement of one or more hydrogen atoms with a chemical group that ionizes in solution.

Saltatory Pertaining to a leaping or jumping movement or to the conduction of nerve impulses along myelinated nerve fibers.

Saprozoic Pertaining to an organism that absorbs nutrients from its surroundings (e.g., a tapeworm).

Sarcolemma The plasmalemma (cell membrane) of a muscle fiber.

Sarcomere A structural and functional (contractile) unit of a muscle cell; consists of the region between two Z lines on a myofibril.

Savannah A biome (terrestrial ecosystem) consisting of grassland interspersed with trees.

Schizocoel (1) A eucoelomate animal, also called a schizocoelomate, whose mesoderm forms from cell bands that proliferate from the region of the blastopore, and whose coelom forms as the mesodermal bands split internally; e.g., molluscs and annelids. (2) A body cavity (coelom) formed by splits in the mesodermal bands.

Schizogony A type of asexual reproduction involving division of a multinucleate cell into many uninucleate cells called merozoites; also called multiple fission.

Schwann cell A type of neuroglial cell that enwraps nerve fibers with its membrane, forming the myelin sheath.

Sclerotization The hardening, and usually, tanning, of a cuticle (e.g., the outer cuticular layer of arthropods) by the chemical cross-linking of protein molecules.

Second law of thermodynamics A universal principle of energy transformation that states that all natural processes involve an increase in disorder, or randomness, and that changes in energy always involve dissipation of concentrated energy.

Second messenger Any of a class of intracellular regulatory substances (e.g., cAMP) that are activated by a first messenger (e.g., a hormone) to produce a specific effect on a target cell.

Secretion Release of a chemical by a cell or organ into an animal's body or onto its surface.

Sedentary Living in one place; moving little, if at all, from a specific site.

Segment A linearly repeated body part, or subdivision; a metamere.

Segmentation *See* Metamerism.

Seminal receptacle An organ in a female in which sperm are stored following copulation; also called a spermatheca.

Seminal vesicle An organ in a male in which sperm are stored prior to copulation; secretes part of the seminal fluid in some animals (e.g., mammals).

Sensory (afferent) neuron A nerve cell that carries impulses away from a receptor and, in most animals, into the central nervous system.

Septum A partition (e.g., connective tissue or a membrane) that separates two tissues or cavities.

Serial homology The resemblance between repeated body parts derived from a common source; e.g., the appendages of polychaete annelids and arthropods, and the vertebrae of vertebrates.

Serotonin A neurotransmitter; 5-hydroxytryptamine.

Serum Vertebrate plasma without fibrinogen; the clear fluid that is expelled when fibrin contracts while forming a clot.

Sessile Nonmobile (not moving from place to place) and usually attached to a substrate.

Seta A hairlike, or bristlelike, projection of the body surface; usually chitinous.

Sexual dimorphism Marked differences between females and males of a species.

Sinus A body space, or cavity.

Sister chromatid Half of a doubled chromosome, visible during mitosis and meiosis; the two sister chromatids in a doubled chromosome are held together by a centromere.

Social behavior Behavioral interactions between two or more individuals, involving one or more forms of communication.

Social hierarchy A social structure in a group of animals based on the ranking of individuals in the group; usually maintained by some form of aggressive interaction among group members.

Somatic Pertaining to cells of a living organism other than the reproductive (germ) cells; nonreproductive cells.

Species In sexual organisms, a group of individuals that is capable of interbreeding and producing fertile offspring and that is reproductively isolated from other interbreeding groups (reproductive isolation implies an inability to produce fertile offspring); an evolutionary unit consisting of biological populations whose gene pools can intermix.

Speciation The formation of new species from preexisting ones by evolutionary mechanisms.

Spermatheca *See* Seminal receptacle.

Spermatogenesis The formation of sperm from undifferentiated germ cells; in species in which meiosis is part of spermatogenesis, four mature sperm result from a single undifferentiated germ cell in the testis.

Spermatophore A packet, or case, containing a number of sperm; produced by the male reproductive system as a means to protect sperm in their transfer to a female.

Sphincter A circular band of muscle surrounding the opening of a tubular organ; contraction of the muscle constricts or closes the opening.

Spicule A microscopic, rodlike or needle-shaped mineralized (calcareous or siliceous) skeletal element.

Spiracle An opening on an animal's body surface that admits water or air.

Spiral cleavage Cleavage that produces alternating tiers of blastomeres, each lying in the groove between the two blastomeres beneath it; cleavage planes are diagonal to the polar axis of the egg; characteristic of protostomes.

Spongin Fibrous proteins forming part or all of the endoskeleton of many sponges.

Spongocoel The main, central water cavity in asconoid and syconoid sponges; the spongocoel is subdivided into many small water channels in leuconoid sponges.

Spore A reproductive body surrounded by a resistant case; in apicomplexan protozoa, a case containing single-celled sporozoites, each capable of infecting a host and developing into a multinucleate stage.

Sporogony Asexual reproduction (multiple fission) resulting in the formation of spores or sporozoites; occurs in the apicomplexan protozoa.

Stasis Evolutionary stability; little or no evolutionary change in a phylogenetic lineage over a long time period.

Statocyst A mechanoreceptor that detects the position of an animal's body in reference to the direction of gravity; an organ of equilibrium; contains one or more hard granules (statoliths) that stimulate hairlike projections on receptor cells.

Stenohaline Having little tolerance for changes in environmental salinity.

Stereoblastula A blastula in which the blastocoel is filled with cells.

Stereogastrula A gastrula consisting of a solid mass of cells; the central cavity becomes filled by cells.

Stigma (1) An eyespot in certain protozoa; usually consists of pigmented granules adjacent to light-sensitive cytoplasm. (2) Another name for a spiracle.

Stridulation Sound production (e.g., by grasshoppers and crickets) by rubbing body parts (e.g., legs) together.

Strobila (1) An animal body consisting of a chain, or series, of individuals (zooids) produced by budding (e.g., the body, exclusive of the scolex, of an adult tapeworm). (2) An asexual stage that produces medusae in the life cycle of scyphozoan cnidarians.

Style A pencillike pointed process or body part; e.g., the crystalline style in the stomach of a clam.

Substrate (1) A surface, or material base, on which an organism lives; substratum. (2) The chemical substance acted on by an enzyme.

Suture An immovable junction between two bones or between two hard skeletal plates.

Syconoid Pertaining to a type of water canal system in certain sponges; the body wall is folded and contains choanocyte chambers.

Symbiosis The close association between individuals of two or more species; includes phoresy, commensalism, parasitism, and mutualism.

Symbiote A symbiotic organism that lives in or on another organism, the host; also called the symbiont.

Sympatric speciation The formation of new species without the geographical isolation of populations.

Synapomorphy A shared, derived characteristic; a homologous feature (i.e., one evolved from the same or similar source in a common ancestor) possessed by two or more taxa.

Synapse The junction between two interacting neurons (a presynaptic neuron and a postsynaptic neuron); chemicals or electrical stimuli from the presynaptic neuron may excite or inhibit the postsynaptic neuron.

Synapsis Close, side-by-side association (pairing) of homologous chromosomes during meiosis.

Syncytium A multinucleate body or cell.

Syrinx The voicebox, or sound-producing organ, of birds; located where the trachea divides to form the bronchi.

Systematics The study of relationships among taxa of organisms, and the classification of organisms according to evolutionary relationships.

Systemic Pertaining to the main body, as opposed to its specific parts.

Tagma In arthropods, a functional group of fused body segments; the head, thorax, and abdomen of an insect are three tagmata.

Taste The ability to detect chemicals in solution; gustation.

Taxis Orientation toward or away from a stimulus or stimulus gradient.

Taxon A group of actual organisms (e.g., Class Aves; the species *Homo sapiens*); a unit of classification. *See also* Taxonomic category.

Taxonomic category A level, or group, in the classification hierarchy; e.g., phylum, class, or order. *See also* Taxon.

Taxonomy The scientific study of classification.

Tegument The outer syncytial covering of parasitic flatworms (flukes and tapeworms) and acanthocephalans.

Telolecithal egg An egg with a moderate to large amount of yolk concentrated in the vegetal hemisphere.

Telophase The fourth and final phase of mitosis and meiosis; the chromosomes or chromatids are at opposite poles of the dividing cell, and the chromosomes transform into chromatin.

Tendon Fibrous connective tissue attaching muscle to bone.

Tentacle An elongate, unsegmented extension of the body wall of a protozoan or animal.

Territory A specific region that is actively defended by one or more individuals against trespass by other members of the species or by members of other species.

Test The case, or outer protective covering of many protozoa; the fused endoskeletal plates of a sea urchin.

Testosterone A steroid hormone that mediates development and maintenance of male secondary sex characteristics in vertebrates.

Tetrapods Four-footed vertebrates, including limbless and bipedal species derived from them; amphibians, reptiles, birds, and mammals.

Theory A scientific generalization that has withstood repeated testing and has a high probability of being a correct interpretation of a natural phenomenon.

Thermodynamics The study of energy flow in physical or biological systems.

Thigmotaxis Orientation toward or away from an object that provides touch stimuli.

Thorax Generally, the middle region of an animal's body (i.e., between the head, or neck, if present, and the abdomen); the chest region of tetrapod vertebrates; the leg-bearing region of arthropods.

Thrombocyte A colorless blood cell that functions in clotting in vertebrates (except mammals, whose clotting elements are cell fragments called platelets).

Tissue An integrated group of cells differentiated structurally and functionally to perform a specific function.

Tornaria A free-swimming, ciliated larva of many hemichordates; similar to the bipinnaria of sea stars.

Torpor Inactivity, or dormancy, accompanied by decreased metabolic rate and, in homeotherms, decreased body temperature.

Trachea (1) In a tetrapod vertebrate, the tube, or windpipe, that conveys respiratory gas between the larynx and the bronchi. (2) One of the respiratory tubes in an arthropod tracheal system; located between the spiracles on the body surface and the tracheoles that branch among tissue cells.

Triploblastic Having three fundamental body layers (ectoderm, endoderm, and mesoderm) that develop into adult body parts.

Trochophore larva The free-swimming, ciliated, top-shaped larva of many marine molluscs and polychaete annelids; a hallmark of the protostomes.

Tubule A small tube, or channel.

Tympanum A vibratory membrane, or eardrum; covers a cluster of sound receptors on the surface of many insects; in vertebrates, receives sound waves in the air and transmits them to the middle ear; also called the tympanic membrane.

Typhlosole An inpocketing, or infolding, of the intestine, which increases surface area for digestion and absorption in many annelids and bivalve molluscs.

Uniramous Pertaining to an unbranched appendage.

Urea A nitrogen-containing waste product of many aquatic and terrestrial animals; $CO(NH_2)_2$.

Ureotelic Pertaining to an animal whose main nitrogenous excretory product is urea.

Ureter The duct that conveys urine from the metanephric kidney to the urinary bladder in amniote vertebrates.

Urethra In mammals, the tube, or duct, that conveys urine from the urinary bladder to the outside.

Uric acid The chief nitrogen-containing waste product of many terrestrial animals; uric acid is crystalline and of very low solubility in water; $C_5H_4O_3N_4$.

Uricotelic Pertaining to an animal whose main nitrogenous excretory product is uric acid.

Urine The final excretory product containing waste metabolites voided by an animal.

Uterus An expanded portion of the female reproductive tract through which eggs pass to the outside environment, or in which young are nurtured.

Vacuole A membrane-bounded vesicle in the cytoplasm of a cell.

Vagina The distal portion of the female reproductive tract; receives the penis during copulation.

Vascular Pertaining to a system of vessels or ducts that transport fluid.

Vas (ductus) deferens A duct that conveys sperm from the epididymis of the testis to the urethra in mammals (and to the cloaca in amphibians, reptiles, and birds); sperm duct.

Vein A vessel that conveys circulatory fluid toward the heart from body tissues.

Veliger A free-swimming larva that develops from the trochophore of many marine molluscs; has a ciliated velum that functions in locomotion and filter feeding, as well as a developing mantle, shell, and foot.

Velum (1) A membranous band or veillike, ciliated projection serving in locomotion and filter feeding in molluscan veliger larvae. (2) A shelflike, subumbrellar projection of many hydrozoan jellyfish.

Ventral Pertaining to the lower or undersurface of an animal.

Venule A small vein that carries blood from a capillary bed to a vein.

Vestigial Pertaining to an anatomically degenerate or functionless structure; in an ancestral species, the structure was well developed and functionally important.

Villus A minute, fingerlike projection containing vascular capillaries; villi lining the vertebrate intestine increase the surface area for digestion and absorption; chorionic villi project from the chorion surrounding the amniote embryo, embed in the uterine wall, and increase surface area for exchange across the placenta.

Vitellarium A gland that synthesizes and secretes yolk.

Vitelline Pertaining to yolk.

Viviparity Birth of young following some amount of embryonic development within the mother's body.

Water expulsion vesicle An osmoregulatory organelle in freshwater protozoa and sponges; functions in collecting and pumping excess water out of the cell; may remove some excretory wastes with the voided water; a contractile vacuole.

Water vascular system A hydraulic system unique to echinoderms; consists of an internal system of branching, fluid-filled, closed tubes; various functions include locomotion, feeding, and gas exchange.

X chromosome A sex chromosome; in many species, including fruit flies and humans, *XX* individuals are females, and *XY* individuals are males.

Y chromosome A sex chromosome; *see X* chromosome.

Yolk Fat and protein food reserves in an egg, which serve to nourish the developing embryo.

Zoea An early larval stage of many marine crustaceans.

Zooid One individual in a colony of animals.

Zoology The scientific study of animals; animal biology.

Zooplankton Minute animals suspended or floating in aquatic environments; not independent of wave and current motion.

Zygote A diploid cell resulting from fertilization (union of a sperm and egg); will undergo the first mitotic division of cleavage.

Zymogen An inactive form of an enzyme.

Credits for Photographs and Illustrations

Photograph Credits

Chapter 10 Opener: © Manfred Kage/Peter Arnold, Inc. 10.1b: Courtesy G.L. Scott, J.A. Feilbach, and T.A. Duff. 10.3: E.R. Lewis, University of California, Berkeley/BPS.

Chapter 11 Opener: © Science Source/Photo Researchers, Inc. 11.5b: © Manfred Kage/Peter Arnold, Inc. 11.8a, b: © Leonard Lee Rue III/Animals Animals. 11.14b: © Biophoto Associates/Photo Researchers, Inc.

Chapter 12 Opener: © Tom McHugh/Photo Researchers, Inc. 12.6a: From *Tissues and Organs: A Text-Atlas of Scanning Electron Microscopy* by Richard G. Kessel and Randy H. Kardon. W.H. Freeman and Company. Copyright © 1979. 12.6b: Dr. E.R. Lewis, University of California, Berkeley. 12.11: From *Fundamentals of Entomology, Third Edition* by Richard J. Elzinga (1987). 12.17a, b: Dr. R.A. Steinbrecht. 12.18: © Omikron/Photo Researchers, Inc. Essay E1: © Joe Monroe/Photo Researchers, Inc.

Chapter 13 Opener: © Frans Lanting. 13.9: © Toni Angermayer/Photo Researchers, Inc. 13.12: © Ed Porowski. 13.13: © Klaus Paysan/Peter Arnold, Inc. 13.14: © Hans and Judy Beste/Animals Animals. 13.15a: James L. Nation, Jr. 13.15b: Thomas C. Emmel. 13.16a: © Breck P. Kent/Animals Animals. 13.16b(1): © Michael Fogden-Oxford Scientific Films/Animals Animals. 13.16b(2): © J.H. Robinson/Photo Researchers, Inc. 13.17 © F. Gohier/Photo Researchers, Inc. 13.18a: © Hans Reinhard/Bruce Coleman, Inc. 13.19: © Art Wolfe. 13.20: © Leonard Lee Rue III.

Chapter 14 Opener: © Hans Pfletschinger/Peter Arnold, Inc. 14.1a: R.D. Campbell, University of California, Irvine/BPS. 14.1b: © Fred Bavendam, Peter Arnold, Inc. 14.3a: Wilmar Jansma, University of Southern Illinois; Harry Schafer, NIH; Bruce Wetzel, NIH. 14.3b: Ray Richardson/Animals Animals. 14.8: J. G. Gall, Carnegie Institution of Washington. 14.14: © Tom McHugh/Photo Researchers, Inc. 14.15: © C. Allan Morgan/Peter Arnold, Inc. 14.16: © Dale Stierman. 14.17: © Jen and Des Bartlett/Photo Researchers, Inc.

Chapter 15 Opener: © M. Calder/Photo Researchers, Inc. 15.6(1), 15.6(2): © Martin Rotker/Taurus Photos, Inc. 15.8: © Oxford Scientific Films/Animals Animals.

Chapter 16 Opener: © Zig Leszczynski/Animals Animals. 16.1a: G. Schatten and D. Mazia, *J. Supramolecular Structure* 5:343 (1976).

Chapter 17 Opener: Larry Burrows, Life Magazine © Time Inc. 17.7a, b: © M.W.F. Tweedie/Photo Researchers, Inc. 17.9a, b: © Pat and Tom Leeson/Photo Researchers, Inc. 17.10a: © Leonard Lee Rue III/Animals Animals. 17.10b: © Ian C. Tait.

Chapter 18 Opener: Charles H. Phillips/Smithsonian Books. 18.7a: © Breck P. Kent/Animals Animals. 18.7b: © Leonard Lee Rue III/Animals Animals. 18.7c: © Arthur Panzer/Photo Researchers, Inc. 18.7d: © Lawrence E. Naylor/Photo Researchers, Inc.

Chapter 19 Opener: © Roland Burke/Peter Arnold, Inc. 19.1b: © Phillip A. Harrington/Peter Arnold, Inc. 19.1c(1), (2): © Manfred Kage/Peter Arnold, Inc. 19.1d(1): © Eric V. Grave/Photo Researchers, Inc. 19.1d(2): © Manfred Kage/Peter Arnold Inc. 19.2a: © M.I. Walker/Science Source. 19.2b(1): © Manfred Kage/Peter Arnold, Inc. 19.2b(2): Courtesy of John Donelson and Steven Brentano, University of Iowa. 19.2c: © Eric V. Grave/Photo Researchers, Inc. 19.3a: © Manfred Kage/Peter Arnold, Inc. 19.8b: © Biophoto Associates/Photo Researchers, Inc. 19.8c: © Tom Branch/Photo Researchers, Inc. 19.9a: J.N.A. Lott, McMaster University/BPS. 19.10b, c: Courtesy of Charles Ehret. 19.14c: Courtesy Karl G. Grell, Tubingen. 19.16: Photograph by Jack Salmon, courtesy of Louis Roth.

Chapter 20 Opener: © Breck P. Kent. 20.2: Courtesy of Karl G. Grell, Tubingen. 20.5: © G.I. Bernard-Oxford Scientific Films/Animals Animals. 20.7a © Fred Bavendam/Peter Arnold, Inc. 20.7b: © Runk/Schoenberger/Grant Heilman Photography.

Chapter 21 Opener: © Fred Bavendam/Peter Arnold, Inc. 21.2a: © D.P. Wilson/Science Source. 21.2b: © Oxford Scientific Films/Animals Animals. 21.2c: Carolina Biological Supply. 21.3a: © Runk/Schoenberger/Grant Heilman Photography. 21.4a: © Breck P. Kent. 21.4b: © Carl Roessler. 21.6a: © Chuck Davis/Tidal Flats, Ltd. 21.7a: H.W. Pratt/BPS. 21.7b: © Jim Doran/Animals Animals. 21.7c: © S. Webster. 21.8a, b: © Fred Bavendam/Peter Arnold, Inc. 21.8c: © Neil G. McDaniel/Tom Stack and Associates. 21.14a-c: © F. Stuart Westmoreland/Tom Stack and Associates. 21.18a: © Chris Newbert. Essay E1: © Matthew D. Ross.

Chapter 22 Opener: © Cath Ellis, Department of Zoology, University of Hull/Science Photo Library. 22.1b © Alex Kerstitch. 22.1c: Larry Mitchell. 22.1d: © Dwight Kuhn. 22.2: Photograph courtesy of the Harold W. Manter Laboratory, University of Nebraska State Museum, Lincoln. 22.3: © Ed Reschke, Muskegon, MI. 22.4: © Dwight Kuhn. 22.8a: Courtesy Robert D. Specian. 22.13c(1): © M.I. Walker/Photo Researchers, Inc. 22.13c(2): © Eric V. Grave/Photo Researchers, Inc. 22.17: Courtesy of Richard Farris. 22.18a: © Fred Bavendam/Peter Arnold, Inc.

Chapter 23 Opener: N. Allin and G.L. Barron, University of Guelph/BPS. 23.2a: Photograph by R.E. Kuntz, courtesy of the Harold W. Manter Laboratory, University of Nebraska State Museum, Lincoln. 23.2b: © Science Source/Photo Researchers, Inc. 23.2c: Courtesy of Mayo Foundation. 23.3: © Ward's Natural Science Establishment/Photo Researchers, Inc. 23.7a: © Hugh Spencer/Photo Researchers, Inc. 23.8a(1): Courtesy of R.A. Hammond. 23.8b: Courtesy of J. Mahrt. 23.9b: Courtesy of John J. Gilbert. 23.11a: © Dennis P. Levy, Yardley, PA (unpublished). 23.12a, 23.13: Courtesy of Robert Higgins. 23.14a: C.B. Calloway, *Marine Biology* 31(1975), 161–74. 23.15a: Sea Studios, Inc. Essay E1: Courtesy of Einhard Schierenberg.

Chapter 24 Opener: © Kevin Schafer. 24.2b: © Jeff Foott/Tom Stack and Associates. 24.2c: © Carolina Biological Supply. 24.4a: © Sea Studios, Inc. 24.4b: © Oxford Scientific Films/Animals Animals. 24.4c: © Scott Johnson/Animals Animals. 24.4d, 24.5a, b: © Fred Bavendam/Peter Arnold, Inc. 24.6a © Adrian Davies/Bruce Coleman, Inc. 24.6b: © Milton Rand/Tom Stack and Associates. 24.7a: Courtesy of Diane Waller, U.S. Fish and Wildlife Service. 24.7b: C.R. Wyttenbach, University of Kansas/BPS. 24.8a: © Alex Kerstitch. 24.8b: © Tim Rock/Animals Animals. 24.10a(1): © Douglas Faulkner/Photo Researchers, Inc. 24.10a(2): © Alex Kerstitch. 24.10b: © Tom McHugh, Steinhart Aquarium/Photo Researchers, Inc. 24.10c: © Carl Roessler/Animals Animals. 24.10d: J.W. Porter, University of Georgia/BPS. 24.13: H.W. Pratt/BPS. 24.14a,b: Courtesy of Carole Hickman. 24.16a: © Scott Johnson/Animals Animals. 24.16b(1): C.R. Wyttenbach, University of Kansas/BPS. 24.16b(2): Larry Mitchell. 24.21a: B.J. Miller/BPS. 24.21b: Courtesy of John Arnold. 24.22a, b: C.B. Calloway.

Chapter 25   Opener: © Eugene Kozloff. 25.1a: © Carolina Biological Supply. 25.2a: © Milton Rand/Tom Stack and Associates. 25.2b: Courtesy of Mark Zoran. 25.13b: © Hans Pfletschinger/Peter Arnold, Inc. 25.15a: Courtesy Mary E. Rice. Essay E1b: © Jack Donnelly/Woods Hole Oceanographic Institution.

Chapter 26   Opener: © Raymond Coleman/APSA. 26.1a: © Kevin Schafer. 26.3a © John Shaw/Bruce Coleman, Inc. 26.3b: © Z. Leszczynski/Animals Animals. 26.3c: © Rod Planck/Photo Researchers, Inc. 26.3e: © Robert and Linda Mitchell. 26.3f: © E.R. Degginger. 26.3g-i: S. Murphree, Auburn University/BPS. 26.4: © Dana O'Connor. 26.6a: P. Gates, University of Durham/BPS. 26.6b: P.J. Bryant, University of California, Irvine/BPS. 26.6c: Courtesy Louis Kornicker. 26.6d: © Frans Lanting. 26.6f: C.R. Wyttenbach, University of Kansas/BPS. 26.7a © Milton Rand/Tom Stack and Associates. 26.7b: © Eugene Kozloff. 26.7c: © Tom Cawley/Tom Stack and Associates. 26.7d, e: © Alex Kerstitch. 26.7f: D.J. Wrobel, Monterey Bay Aquarium/BPS. 26.8a: © Zig Leszczynski/Animals Animals. 26.8b: © Raymond A. Mendez/Animals Animals. 26.12: © Tony Florio/Photo Researchers, Inc. 26.20c: P.J. Bryant, University of California, Irvine/BPS. 26.21a: © Carolina Biological Supply. 26.21b, c: © Dwight Kuhn. 26.21d: © William H. Amos/Bruce Coleman, Inc. 26.24b: © John Gerlach/Tom Stack and Associates. 26.26e: Patti Murray/Animals Animals. 26.28a: Carolina Biological Supply Co. 26.29: © Omikron/Photo Researchers, Inc. 26.30a: © C.B. and D.W. Frith/Bruce Coleman, Inc. 26.30b: © Raymond A. Mendez/Animals Animals.

Chapter 27   Opener: J.N.A. Lott, McMaster University/BPS. 27.3a: © Eugene Kozloff. 27.4a: © Fred Bavendam/Peter Arnold, Inc. 27.5a: © Dr. David Schwimmer/Bruce Coleman, Inc. 27.6a: Field Museum of Natural History (85l07c), Chicago. 27.6b: © Fred Bavendam/Peter Arnold, Inc. 27.8a, b: © Alex Kerstitch. 27.9: Courtesy Alan N. Baker, National Museum of New Zealand. 27.10a: © Brian Parker/Tom Stack and Associates. 27.10b: © Fred Bavendam/Peter Arnold, Inc. 27.11a: © Eugene Kozloff. 27.11b: © Gary Milburn/Tom Stack and Associates. 27.13b: C.B. Calloway. 27.14: © Gary Milburn/Tom Stack and Associates. 27.17a: Field Museum of Natural History (80872c), Chicago. 27.21: Courtesy Russell Zimmer.

Chapter 28   Opener: Ralph Buchsbaum from *Living Invertebrates* (Blackwell/Boxwood). 28.2a: H.W. Pratt/BPS. 28.4a: Sea Studios, Inc. 28.4b: Courtesy of L.P. Madin. 28.7a: © David M. Dennis/Tom Stack and Associates. 28.7c: © Tom Stack.

Chapter 29   Opener: © C.C. Lockwood/DRK Photo. 29.3a: © Ron and Valerie Taylor/Bruce Coleman, Inc. 29.3b: © Alexandra Edwards/Sea Studios, Inc. 29.3c: © Alex Kerstitch. 29.3d: © John L. Tveten. 29.3e: © Fred Whitehead/Animals Animals. 29.4: © Fred Bavendam/Peter Arnold, Inc. 29.6a: © Tom McHugh, Steinhart Aquarium/Photo Researchers, Inc. 29.6b: © Zig Leszczynski/Animals Animals. 29.6c: © Fran Allen/Animals Animals. 29.7 © Tom McHugh/Photo Researchers, Inc. 29.8a: © Hans Reinhart/Bruce Coleman, Inc. 29.8b: © Zig Leszczynski/Animals Animals. 29.9a: © Carlton Ray/Photo Researchers, Inc. 29.9b: © Gary Milburn, Tom Stack and Associates. 29.11a: © Zig Leszczynski/Animals Animals. 29.11b: Courtesy Bruce H. Robison. 29.12a: © Al Grotell. 29.12b: © Tim Rock/Animals Animals. 29.12c, 29.17a: © Zig Leszczynski/Animals Animals. 29.17b: © Breck P. Kent. 29.18a: © Jeff Foott/Tom Stack and Associates. 29.18b: © Jeff Foott.

Chapter 30   Opener: © Zig Leszczynski/Animals Animals. 30.1a: © Breck P. Kent. 30.1b: © Tom McHugh, Steinhart Aquarium/Photo Researchers, Inc. 30.1c: © R.C. Simpson/Tom Stack and Associates. 30.2: © Michael Fogden/Animals Animals. 30.3a: © R.W. VanDevender. 30.3b: © R.J. Erwin/DRK Photo. 30.3c: © David M. Dennis/Tom Stack and Associates. 30.3d: © Zig Leszczynski/Animals Animals. 30.4 © M.P.L. Fogden/Bruce Coleman, Inc. 30.12: © Jane Burton/Bruce Coleman, Inc. 30.15: © E.S. Ross.

Chapter 31   Opener: © Bob McKeever/Tom Stack and Associates. 31.2a: © Mark I. Jones/Bruce Coleman, Inc. 31.2b: © James Hanken/Bruce Coleman, Inc. 31.2c: © C.C. Lockwood/DRK Photo. 31.3a: © Zig Leszczynski/Breck Kent. 31.4a: © Dwight R. Kuhn. 31.4b: © C. Allan Morgan. 31.4c: © Breck P. Kent. 31.4d: © R.W. VanDevender. 31.6a, b: © Breck P. Kent. 31.6c: © John Cancalosi/Peter Arnold, Inc. 31.7a: © Jeff Foott. 31.9a-c: © E.S. Ross. 31.13: © Breck P. Kent. 31.14a: © Robert and Linda Mitchell. 31.15a: © Gordon Wiltsie (1983). 31.15b: © Doug Wechsler/Animals Animals. 31.16a: © Robert and Linda Mitchell. 31.16b: © Breck P. Kent/Animals Animals.

Chapter 32   Opener: Neg. No. 5509 (Photo by Thomson) Courtesy Department of Library Services, American Museum of Natural History. 32.2a: © Nicholas DeVore III/Bruce Coleman, Inc. 32.2b: © Buff and Gerald Corsi. 32.3a: © Larry Minden. 32.3b: © Frans Lanting. 32.3c: © Breck P. Kent/Animals Animals. 32.3d: © Gary R. Zahm/DRK Photo. 32.3e: © Tom and Pat Leeson. 32.3f: © David C. Fritts/Animals Animals. 32.3g: © A. Nelson/Tom Stack and Associates. 32.3h: © C. Allan Morgan. 32.3i: © Wayne Lankinen/DRK Photo. 32.3j: © Ron Austing/Bruce Coleman, Inc. 32.3k: © Mary Clay/Tom Stack and Associates. 32.3l: © Kevin Schafer. 32.7: © Lewis Kemper/DRK Photo. 32.11: © Bob and Clara Calhoun/Bruce Coleman, Inc. 32.13: © Hans Reinhard/Bruce Coleman, Inc. 32.14: © John Gerlach/Tom Stack and Associates.

Chapter 33   Opener: © Hans and Judy Beste/Animals Animals. 33.2: © Oxford Scientific Films/Animals Animals. 33.3a: © E.R. Degginger/Animals Animals. 33.3b, c: © Fritz Prenzel/Animals Animals. 33.4a: © Breck P. Kent. 33.4b: © Press-Tige Pictures-Oxford Scientific Films/Animals Animals. 33.4c: © Gary Milburn/Tom Stack and Associates. 33.4d: © Frans Lanting. 33.5a: © Art Wolfe. 33.5b: © Steven C. Kaufman/Peter Arnold, Inc. 33.5c: © Kevin Schafer. 33.5d: © Gary Milburn/Tom Stack and Associates. 33.5e: © Jeff Foott. 33.5f: © Stephen J. Kraseman/DRK Photo. 33.6a, b: © Luiz Claudio Marigo/Peter Arnold, Inc. 33.6c: © Gary Milburn/Tom Stack and Associates. 33.6d: © M.P.L. Fogden/Bruce Coleman, Inc. 33.6e: © Richard Ellis/Photo Researchers, Inc. 33.7a: © Patti Murray/Animals Animals. 33.7b: Jeff Foott. 33.8: © Biophoto Associates/Photo Researchers, Inc. 33.9: © Tom and Pat Leeson. 33.10a: © Brian Parker/Tom Stack and Associates. 33.10b: © Johnny Johnson/DRK Photo. 33.14a: © Tom McHugh/Photo Researchers, Inc. 33.14b: © Rapho by Ylla/Photo Researchers, Inc. 33.14c: © Tom McHugh/Photo Researchers, Inc.

Chapter 34   Opener: © Frans Lanting. 34.1b: G.W. Barlow, *Animal Behaviour*, 22:876, 1974. 34.1c: © Stephen J. Kraseman/DRK Photo. 34.4b: © Leonard Lee Rue III/Animals Animals. 34.11a: © Patrick Lynch/Photo Researchers, Inc. 34.12a, b: © Leonard Lee Rue III/Animals Animals. 34.13a: © Tom McHugh/Photo Researchers, Inc. 34.13b: © Wayne Lynch/DRK Photo. 34.15a: © Tom McHugh, Steinhart Aquarium/Photo Researchers, Inc. 34.15b: © Tom McHugh/Photo Researchers, Inc. 34.15c: New York State Department of Environmental Conservation.

Chapter 35   Opener: © M.P.L. Fogden/Bruce Coleman, Inc. 35.4: © Laura Riley/Bruce Coleman, Inc. 35.11a: © Gary Rosenquist/Earth Images. 35.11b: © Annie Griffiths/Bruce Coleman, Inc. 35.13: © Mark Newman/Tom Stack and Associates. 35.14: © Erwin and Peggy Bauer/Bruce Coleman, Inc. 35.15: © Marlene and Bruce Ehresman. 35.16: C.M. Pringle, University of California, Berkeley/BPS. 35.17: © Jeff Foott. 35.18a: © Leonard Lee Rue III/Earth Scenes. 35.18b: © Frans Lanting. 35.19: © Leonard Lee Rue III/Earth Scenes. 35.20a: Larry Mitchell. 35.21a: Larry Mitchell.

ILLUSTRATION CREDITS

Chapter 1 1.7, 1.9, 1.11: Linda McVay.

Chapter 2 2.1-2.6a: Linda McVay. 2.6b: David Freifelder. 2.7: Linda McVay. 2.8, 2.9: David Freifelder. 2.10: Linda McVay; c, d: © Irving Geis. 2.11, 2.12: David Freifelder. 2.13: Linda McVay, David Freifelder. 2.14: Linda McVay. 2.15: Carla Simmons. 2.16: Linda McVay. 2.17 Cecile Duray-Bito. 2.18. Fran Milner, Linda McVay. 2.19: Linda McVay. 2.20b: Carol Verbeeck, adapted from "The Ground Substance of the Living Cell," by Keith R. Porter and Jonathan B. Tucker. Copyright © 1981 by Scientific American, Inc. All rights reserved. 2.23: Fran Milner, Linda McVay. 2.24, 2.25: Linda McVay. 2.26, 2.27: Fran Milner, Linda McVay. Essay E1-E3: Linda McVay.

Chapter 3 3.1: Carl Brown. 3.2: David Freifelder. 3.3: Carl Brown. 3.4: Carla Simmons. 3.5, 3.6: David Freifelder. 3.7: Carl Brown, David Freifelder. 3.8: Carl Brown. 3.9: Carl Brown, David Freifelder. 3.10, 3.11: Carl Brown. 3.12: Carla Simmons. 3.13: Carl Brown, David Freifelder. 3.14: Carl Brown. 3.16: Carl Brown, David Freifelder. 3.17: Carl Brown. 3.18: Carla Simmons. 3.19: John and Judy Waller.

Chapter 4 4.1: Carla Simmons. 4.3: Carla Simmons. 4.6b, 4.7: Carla Simmons. 4.8: Fran Milner, Carla Simmons. 4.9: Carla Simmons. 4.10b-4.12: Fran Milner, Carla Simmons. 4.13: Darwen and Vally Hennings, Elizabeth Morales. 4.14: Darwen and Vally Hennings. 4.15: Darwen and Vally Hennings, Elizabeth Morales. 4.16, 4.17: Fran Milner, Carla Simmons. Essay E1a: Carla Simmons.

Chapter 5 5.1 Linda McVay. 5.2-5.4: Fran Milner, Linda McVay. 5.5: Linda McVay. 5.6: Fran Milner, Elizabeth Morales. 5.7a: Linda McVay. 5.7b: Fran Milner, Linda McVay. 5.8, 5.9: Linda McVay. 5.10, 5.11: Fran Milner, Linda McVay. 5.13a: Fran Milner, Linda McVay. 5.14 Linda McVay. 5.15: Carl Brown. 5.16: Fran Milner, Linda McVay. 5.17: Linda McVay. 5.18, 5.19: Carl Brown. 5.20: Linda McVay.

Chapter 6 6.1: Fran Milner, Elizabeth Morales. 6.2, 6.3: Elizabeth Morales. 6.4-6.7: Fran Milner, Elizabeth Morales. 6.8, 6.9: Elizabeth Morales. 6.10, 6.11: Fran Milner, Elizabeth Morales. 6.12a: Elizabeth Morales. 6.12b, 6.12c: Darwen and Vally Hennings, Elizabeth Morales. 6.13: Fran Milner, Elizabeth Morales. 6.14b: Darwen and Vally Hennings, Elizabeth Morales. 6.15: Carl Brown. 6.16: Elizabeth Morales. 6.17a: Darwen and Vally Hennings, Elizabeth Morales. 6.18 Carl Brown. 6.19-6.21: Elizabeth Morales.

Chapter 7 7.1, 7.2: Carl Brown. 7.3, 7.4: Fran Milner, Martha Blake. 7.5: Martha Blake. 7.6: Martha Blake, David Freifelder. 7.7-7.10: Fran Milner, Martha Blake. 7.11-7.15: Martha Blake. Essay E1: Martha Blake.

Chapter 8 8.1-8.3: Carla Simmons. 8.5: Carla Simmons. 8.6, 8.7: Fran Milner, Carla Simmons. 8.8: David Freifelder. 8.9, 8.10: Carla Simmons. 8.12-8.20: Carla Simmons. 8.21: Fran Milner, Carla Simmons. Essay E1a: Carla Simmons.

Chapter 9 9.1-9.2b: Fran Milner, Carla Simmons. 9.2c, 9.3: Carla Simmons. 9.4: Carl Brown. 9.5, 9.6: Carla Simmons. 9.7-9.9: Fran Milner, Carla Simmons. 9.10-9.13: Carla Simmons. 9.14: Carl Brown. 9.15: Carla Simmons. 9.16: Fran Milner, Carla Simmons. 9.17: Carla Simmons. 9.18, 9.19: Carl Brown. Essay E1: Carl Brown.

Chapter 10 10.1, 10.2: Fran Milner, Cecile Duray-Bito. 10.4: Fran Milner, Cecile Duray-Bito. 10.5: Carl Brown. 10.6-10.8: Fran Milner, Cecile Duray-Bito. 10.9: Cecile Duray-Bito. 10.10: Fran Milner, Cecile Duray-Bito. 10.11: Cecile Duray-Bito. 10.12: Fran Milner, Cecile Duray-Bito. 10.13: Cecile Duray-Bito. 10.14a: Fran Milner, Cecile Duray-Bito. 10.14b: Cecile Duray-Bito. 10.15: Fran Milner, Cecile Duray-Bito. 10.16-10.20: Cecile Duray-Bito.

Chapter 11 11.1, 11.2: Cyndie Clark-Huegel. 11.3: Carl Brown. 11.4: Cyndie Clark-Huegel. 11.5a: Carl Brown. 11.5b: Cyndie Clark-Huegel. 11.6, 11.7: Fran Milner, Cyndie Clark-Huegel. 11.9: Fran Milner, Cyndie Clark-Huegel. 11.10-11.13: Cyndie Clark-Huegel. 11.14-11.16: Fran Milner, Cyndie Clark-Huegel. 11.17: Cyndie Clark-Huegel. Essay E1: David Freifelder.

Chapter 12 12.1-12.7: John and Judy Waller. 12.8: Carl Brown. 12.9-12.16: John and Judy Waller. 12.18-12.20: John and Judy Waller.

Chapter 13 13.1-13.6: Martha Blake. 13.7a-d: Martha Blake, b-d after V.B. Wigglesworth, *The Principles of Insect Physiology*, 7th ed. London: Chapman and Hall, 1972, Fig. 210. 13.8: Martha Blake. 13.10, 13.11: Martha Blake. 13.18b: Carl Brown after H.M. Southern, "Natural Control of a Population of Tawny Owls (*Strix aluco*)." *J. Zool., Lond.* 162:197-285, 1980. Adapted by permission of the Zoological Society of London. Essay E1: Martha Blake.

Chapter 14 14.2: Darwen and Vally Hennings. 14.4-14.7: Martha Blake. 14.9-14.11: Darwen and Vally Hennings. 14.12: Martha Blake. 14.13: Darwen and Vally Hennings.

Chapter 15 15.1a: Fran Milner, Darwen and Vally Hennings. 15.1b: John and Judy Waller. 15.1c, 15.2: Carl Brown. 15.3: Fran Milner, John and Judy Waller. 15.4, 15.5: John and Judy Waller. 15.7: Carl Brown. 15.9a,b: David Freifelder. 15.9c: Fran Milner, John and Judy Waller. 15.10-15.13: John and Judy Waller.

Chapter 16 16.1-16.15: Cecile Duray-Bito.

Chapter 17 Opener : John and Judy Waller. 17.1: Carl Brown. 17.2, 17.3: Darwen and Vally Hennings. 17.4: Carl Brown. 17.5: Darwen and Vally Hennings. 17.6: Carl Brown. 17.8: John and Judy Waller. 17.11: Darwen and Vally Hennings. 17.12: John and Judy Waller. 17.13: Darwen and Vally Hennings. 17.14: Carl Brown. 17.15: Darwen and Vally Hennings. 17.16: Carl Brown. 17.17: Darwen and Vally Hennings. 17.18: Carl Brown. Essay E1: Carl Brown.

Chapter 18 18.1, 18.2: Elizabeth Morales. 18.4: Carl Brown. 18.5, 18.6: Elizabeth Morales. 18.8-18.10: Elizabeth Morales.

Chapter 19 19.1-19.17: Elizabeth Morales. 19.10, center: Darwen and Vally Hennings.

Chapter 20 20.1: Fran Milner, Linda McVay. 20.3: Linda McVay. 20.4: Fran Milner, Linda McVay. 20.5b: Linda McVay. 20.6: Fran Milner, Linda McVay. 20.8: Linda McVay. 20.9: Fran Milner, Linda McVay. 20.10, 20.11: Linda McVay. 20.12: Fran Milner, Linda McVay.

Chapter 21 Trends and Strategies A,B: Darwen and Vally Hennings. 21.1: Fran Milner, Darwen and Vally Hennings. 21.3: Darwen and Vally Hennings. 21.5: Darwen and Vally Hennings. 21.6b: Darwen and Vally Hennings. 21.9-21.13: Darwen and Vally Hennings. 21.15, 21.16: Fran Milner, Darwen and Vally Hennings. 21.17: Darwen and Vally Hennings. 21.18b: Darwen and Vally Hennings. Essay E2, E3: Darwen and Vally Hennings.

Chapter 22 22.1: Linda McVay. 22.5-22.7: Fran Milner, Linda McVay. 22.8b: Linda McVay. 22.9: Fran Milner, Linda McVay. 22.10: Carl Brown. 22.11-22.16: Linda McVay. 22.18b, c: Fran Milner, Linda McVay. Essay E1: Linda McVay.

Chapter 23 Trends and Strategies A: Darwen and Vally Hennings. 23.1: Fran Milner, Darwen and Vally Hennings. 23.4, 23.5: Fran Milner, Darwen and Vally Hennings. 23.6-23.9: Darwen and Vally Hennings. 23.10: Fran Milner, Darwen and Vally Hennings. 23.11, 23.12: Darwen and Vally Hennings. 23.14b; John and Judy Waller. 23.15b: Darwen and Vally Hennings.

Chapter 24 24.1: Fran Milner, Martha Blake. 24.2a: Martha Blake. 24.3: Fran Milner, Martha Blake. 24.9: Fran Milner, Martha Blake. 24.11: Martha Blake. 24.12: Fran Milner, Martha Blake. 24.14c: Martha Blake. 24.15: Fran Milner, Martha Blake. 24.17: Fran Milner, Martha Blake. 24.18-24.21: Martha Blake. 24.23: Fran Milner, Martha Blake.

Chapter 25 Trends and Strategies A,B: John and Judy Waller. 25.1b: John and Judy Waller. 25.3-25.9: John and Judy Waller. 25.10: Carl Brown. 25.11-25.17: John and Judy Waller. Essay E1a: John and Judy Waller.

Chapter 26 26.1-26.3: Martha Blake. 26.5, 26.6: Martha Blake. 26.9-26.11: Martha Blake. 26.13-26.20: Martha Blake. 26.22: Martha Blake. 26.23: Carl Brown. 26.24-26.27: Martha Blake. 26.28b: Fran Milner, Martha Blake. 26.30c: Martha Blake.

Chapter 27 27.1-27.5: Linda McVay. 27.7: Fran Milner, Linda McVay. 27.12, 27.13a: Fran Milner, Linda McVay. 27.15, 27.16: Fran Milner, Linda McVay. 27.17b,c: Linda McVay. 27.18: Fran Milner, Linda McVay. 27.19: Linda McVay. 27.20: Fran Milner, Linda McVay. 27.22: Fran Milner, Linda McVay.

Chapter 28 28.1-28.17: Cecile Duray-Bito. Essay E1: Cecile Duray-Bito.

Chapter 29 29.1, 29.2: Elizabeth Morales. 29.5: Elizabeth Morales. 29.9b, 29.10: Elizabeth Morales. 29.13-29.16: Elizabeth Morales. 29.19: Cecile Duray-Bito. Essay E1: Elizabeth Morales.

Chapter 30 30.5-30.9: John and Judy Waller. 30.10: Carl Brown. 30.11: John and Judy Waller. 30.13, 30.14: John and Judy Waller. 30.16, 30.17: John and Judy Waller. 30.18: Cecile Duray-Bito.

Chapter 31 31.1: Cecile Duray-Bito. 31.3b: Cecile Duray-Bito. 31.5: Cecile Duray-Bito. 31.7, 31.8: Cecile Duray-Bito. 31.10-31.12: Cecile Duray-Bito. 31.14b: Cecile Duray-Bito. 31.17: Cecile Duray-Bito.

Chapter 32 Opener: Darwen and Vally Hennings. 32.1: Darwen and Vally Hennings. 32.4-32.6: Darwen and Vally Hennings. 32.8-32.10: Darwen and Vally Hennings. 32.12: Darwen and Vally Hennings. 32.15: Darwen and Vally Hennings. 32.16: Cecile Duray-Bito.

Chapter 33 33.1: Martha Blake. 33.5d: Martha Blake. 33.10c-33.13: Martha Blake. 33.15, 33.16: Martha Blake. 33.17, 33.18: Cecile Duray-Bito.

Chapter 34 34.1, 34.2: John and Judy Waller. 34.3, 34.4a: Carl Brown. 34.5: John and Judy Waller. 34.6-34.12: Carl Brown. 34.14, 34.15: Carl Brown.

Chapter 35 35.1: Elizabeth Morales. 35.2, 35.3: Carl Brown. 35.5: Carl Brown. 35.6-35.10: Elizabeth Morales. 35.12: Elizabeth Morales. 35.20, 35.21: Elizabeth Morales. 35.22: Carl Brown.

The art for the marbling panels in the Trends and Strategies sections was created by Peggy Skycraft. The endpaper map was prepared by John Parsons.

Index

Note: Italicized page numbers refer to figures and tables.